Chemical Reaction and Reactor Engineering

CHEMICAL INDUSTRIES

A Series of Reference Books and Textbooks

Consulting Editor
HEINZ HEINEMANN
Heinz Heinemann, Inc.,
Berkeley, California

Volume 1: Fluid Catalytic Cracking with Zeolite Catalysts, *Paul B. Venuto and E. Thomas Habib, Jr.*

Volume 2: Ethylene: Keystone to the Petrochemical Industry, *Ludwig Kniel, Olaf Winter, and Karl Stork*

Volume 3: The Chemistry and Technology of Petroleum, *James G. Speight*

Volume 4: The Desulfurization of Heavy Oils and Residua, *James G. Speight*

Volume 5: Catalysis of Organic Reactions, *edited by William R. Moser*

Volume 6: Acetylene-Based Chemicals from Coal and Other Natural Resources, *Robert J. Tedeschi*

Volume 7: Chemically Resistant Masonry, *Walter Lee Sheppard, Jr.*

Volume 8: Compressors and Expanders: Selection and Application for the Process Industry, *Heinz P. Bloch, Joseph A. Cameron, Frank M. Danowski, Jr., Ralph James, Jr., Judson S. Swearingen, and Marilyn E. Weightman*

Volume 9: Metering Pumps: Selection and Application, *James P. Poynton*

Volume 10: Hydrocarbons from Methanol, *Clarence D. Chang*

Volume 11: Foam Flotation: Theory and Applications, *Ann N. Clarke and David J. Wilson*

Volume 12: The Chemistry and Technology of Coal, *James G. Speight*

Volume 13: Pneumatic and Hydraulic Conveying of Solids, *O. A. Williams*

Volume 14: Catalyst Manufacture: Laboratory and Commercial Preparations, *Alvin B. Stiles*

Volume 15: Characterization of Heterogeneous Catalysts, *edited by Francis Delannay*

Volume 16: BASIC Programs for Chemical Engineering Design, *James H. Weber*

Volume 17: Catalyst Poisoning, *L. Louis Hegedus and Robert W. McCabe*

Volume 18: Catalysis of Organic Reactions, *edited by John R. Kosak*

Volume 19: Adsorption Technology: A Step-by-Step Approach to Process Evaluation and Application, *edited by Frank L. Slejko*

Volume 20: Deactivation and Poisoning of Catalysts, *edited by Jacques Oudar and Henry Wise*

Volume 21: Catalysis and Surface Science: Developments in Chemicals from Methanol, Hydrotreating of Hydrocarbons, Catalyst Preparation, Monomers and Polymers, Photocatalysis and Photovoltaics *edited by Heinz Heinemann and Gabor A. Somorjai*

Volume 22: Catalysis of Organic Reactions, *edited by Robert L. Augustine*

Volume 23: Modern Control Techniques for the Processing Industries, *T. H. Tsai, J. W. Lane, and C. S. Lin*

Volume 24: Temperature-Programmed Reduction for Solid Materials Characterization, *Alan Jones and Brian McNicol*

Volume 25: Catalytic Cracking: Catalysts, Chemistry, and Kinetics, *Bohdan W. Wojciechowski and Avelino Corma*

Volume 26: Chemical Reaction and Reactor Engineering, *edited by James J. Carberry and Arvind Varma*

Volume 27: Filtration: Principles and Practices, second edition, *edited by Michael J. Matteson and Clyde Orr*

Additional Volumes in Preparation

Catalysis and Surface Properties of Liquid Metals and Alloys, *Yoshisada Ogino*

Catalyst Deactivation, *edited by Eugene E. Petersen and Alexis T. Bell*

Chemical Reaction and Reactor Engineering

edited by

James J. Carberry / Arvind Varma

Department of Chemical Engineering, University of Notre Dame
Notre Dame, Indiana

Marcel Dekker, Inc. **New York and Basel**

Library of Congress Cataloging in Publication Data

Chemical reaction and reactor engineering.

(Chemical industries ; 26)
Includes index.
1. Chemical reactors. I. Carberry, James J.
II. Varma, Arvind
TP157.C416 1986 660.2'83 86-19673
ISBN 0-8247-7543-0

MARCEL DEKKER, INC.
270 Madison Avenue, New York, New York 10016

Current printing (last digit):
10 9 8 7 6 5 4 3 2 1

PRINTED IN THE UNITED STATES OF AMERICA

Preface

There exist few, if any, Solomons in our midst capable of single-handedly fashioning a comprehensive treatise embracing the diverse aspects of chemical reaction and reactor engineering. This topic, by its very scope and complexity, inevitably demands that each subdiscipline be treated individually by recognized authorities. Thus the genesis of this volume. Herein leading authorities set forth their knowing analyses of the many facets of the general topic. It is hoped that this compilation, while hardly a textbook or a handbook, will prove to be of value to academic and industrial practitioners of the art and science of chemical reaction engineering, a field which, by definition, must embrace the signal complexities of chemical reaction and reactor systems.

The scope of this treatise is general, hardly universal. New insights will, in time, emerge. *Chemical Reaction and Reactor Engineering* presents an authoritative progress report that will remain germane to the topic and, we trust, prove to be a substantial inspiration to further progress.

James J. Carberry
Arvind Varma

Contributors

RUTHERFORD ARIS Department of Chemical Engineering and Materials Science, University of Minnesota, Minneapolis, Minnesota

SERGIO CARRÀ Dipartimento di Chimica Fisica Applicata, Politecnico di Milano, Milan, Italy

ALBERTO E. CASSANO Instituto de Desarrollo Tecnológico para la Industria Quimíca–INTEC, Universidad Nacional del Litoral–Consejo Nacional de Investigaciones Cientifícas y Técnicas, Santa Fe, Argentina

MARÍA A. CLARIÁ Instituto de Desarrollo Technólogico para la Industria Química–INTEC, Universidad Nacional del Litoral–Consejo Nacional de Investigaciones Cientifícas y Técnicas, Santa Fe, Argentina

ELIANA R. DE BERNARDEZ Instituto de Desarrollo Tecnológico para la Industria Química–INTEC, Universidad Nacional del Litoral–Consejo Nacional de Investigaciones Científicas y Técnicas, Santa Fe, Argentina

W. NICHOLAS DELGASS School of Chemical Engineering, Purdue University, West Lafayette, Indiana

MORTON M. DENN Department of Chemical Engineering, University of California, Berkeley, California

L. K. DORAISWAMY National Chemical Laboratory, Pune, India

VICTORIA EDWARDS Department of Chemical Engineering, University of California, and Materials and Molecular Research Division, Lawrence Berkeley Laboratory, Berkeley, California

LARRY E. ERICKSON Department of Chemical Engineering, Kansas State University, Manhattan, Kansas

GILBERT F. FROMENT Laboratorium Voor Petrochemische Techniek, Rijksuniversiteit, Gent, Belgium

RAFAEL GALVAN Department of Chemical Engineering and Materials Science, University of Minnesota, Minneapolis, Minnesota

HANNS P. K. HOFMANN Institut für Technische Chemie I (Reaktionstechnik), Universität Erlangen—Nürnberg, Erlangen, Federal Republic of Germany

B. D. KULKARNI Chemical Engineering Division, National Chemical Laboratory, Pune, India

ROBERT L. LAURENCE Department of Chemical Engineering, University of Massachusetts, Amherst, Massachusetts

DAN LUSS Department of Chemical Engineering, University of Houston, Houston, Texas

MICHAEL L. MICHELSEN Instituttet for Kemiteknik, Danmarks Tekniske Højskole, Lyngby, Denmark

MASSIMO MORBIDELLI Dipartimento di Chimica Fisica Applicata, Politecnico di Milano, Milan, Italy

JOHN S. NEWMAN Department of Chemical Engineering, University of California, and Materials and Molecular Research Division, Lawerence Berkeley Laboratory, Berkeley, California

PETER N. ROWE Department of Chemical and Biochemical Engineering, University College London, London, England

YATISH T. SHAH Department of Chemical and Petroleum Engineering, University of Pittsburgh, Pittsburgh, Pennsylvania

MAN MOHAN SHARMA Department of Chemical Technology, University of Bombay, Bombay, India

REUEL SHINNAR Department of Chemical Engineering, The City College of the City University of New York, New York, New York

GREGORY STEPHANOPOULOS* Department of Chemical Engineering, California Institute of Technology, Pasadena, California

MATTHEW TIRRELL Department of Chemical Engineering and Materials Science, University of Minnesota, Minneapolis, Minnesota

---

**Current affiliation*: Department of Chemical Engineering, Massachusetts Institute of Technology, Cambridge, Massachusetts

GARY G. TROST* Department of Chemical Engineering, University of California, and Lawrence Berkeley Laboratory, Berkeley, California

ARVIND VARMA Department of Chemical Engineering, University of Notre Dame, Notre Dame, Indiana

JOHN VILLADSEN Instituttet for Kemiteknik, Danmarks Tekniske Højskole, Lyngby, Denmark

EDUARDO E. WOLF Department of Chemical Engineering, University of Notre Dame, Notre Dame, Indiana

JOHN G. YATES Department of Chemical and Biochemical Engineering, University College London, London, England

**Current affiliation*: Raychem Corporation, Menlo Park, California

Contents

Chemical Reaction and Reactor Engineering

1

Numerical Methods in Reaction Engineering

JOHN VILLADSEN and MICHAEL L. MICHELSEN *Instituttet for Kemiteknik, Danmarks Tekniske Højskole, Lyngby, Denmark*

INTRODUCTION

Mathematical modeling is an integral part of reaction engineering and a working knowledge of a range of numerical methods is as essential to the engineer as is a set of keys to a locksmith.

Numerical problems crop up in many forms. Trivial numerical methods are used almost unconsciously by the professional engineer to solve routine problems, while much more elaborate methods—some only recently described in the literature—are finding applications in the solution of complex research problems or in industrial design problems.

Much of this chapter deals with numerical methods that were developed to solve rather difficult problems—most of them related to research in reaction engineering in our group at Lyngby, but hopefully reminiscent of problems faced by others. With this in mind, it is befitting to start with some philosophical remarks that may perhaps appear to contradict the main body of the text, but which are still important:

1. The large majority of numerical problems in reaction engineering can be solved by primitive, ad hoc methods if these are used in a sensible way. Model simplifications frequently point to limiting solutions that can be computed on a pocket calculator. These limiting solutions may indicate what further insight one may gain into the problem by a full-scale numerical attack. Also, the limiting solutions are extremely valuable when it comes to checking the results of more elaborate calculations.
2. It is a regrettable observation that many numerical techniques are stretched to the limit of their applicability in order to solve problems which are physically unrealistic. It is a barren victory to file a stack of computer output, obtained after endless modifications of a basic algorithm, when it is discovered that the phenomenon which was finally tracked down numerically could be neglected altogether because it exists only for unreasonable parameter values.

To stress the importance of ad hoc methods, we shall give three examples which together contain embryonic forms of all the numerical methods to be treated in the following.

Example 1 The rate of deacylation of a 0.250 M solution of penicillin V on a commercial immobilized enzyme is given by

$$r = 3.20m_c[(1 - x) + 1.065(1 - x)^2] \qquad \text{mol/h} \tag{1}$$

where x is conversion and m_c is the amount of immobilized enzyme in kilograms. It is desired to obtain 90% conversion of v = 670 L/h penicillin solution in two continuous tank reactors, each with $m_c/2$ kg of enzyme. The intermediate conversion x_1 and the required catalyst weight are calculated from

$$vc_A^\circ x_1 = 167.5x_1 = 3.20m_c/2[(1 - x) + 1.065(1 - x_1)^2]$$

$$vc_A^\circ(0.9 - x) = 167.5(0.9 - x_1) = 3.20m_c/2[0.1 + 1.065(0.1)^2]$$

or

$$\frac{x_1}{0.9 - x_1} = \frac{(1 - x_1) + 1.065(1 - x_1)^2}{0.11065} \tag{2}$$

There are many ways in which Eq. (2) can be solved for x_1, and some are definitely better than others. Following are three possibilities:

1. $x_1 = f(x_1) = 9.037(0.9 - x_1)[(1 - x) + 1.065(1 - x_1)]^2$. A value of x_1 is inserted on the right-hand side. x_1 is calculated and reinserted on the right-hand side to give a successive substitution algorithm.
2. $F(x_1) = 0 = x_1^3 - 3.839x_1^2 + 4.688x_1 - 1.745$. After reformulation of Eq. (2) into a polynomial expression for x_1, one applies a general routine to find the zeros of $F(x_1)$.
3. The left-hand side of Eq. (2) is plotted against the right-hand side.

For a quick estimate of x_1, one would choose method 3. For $0 < x_1 < 0.9$, the left-hand side of Eq. (2) increases with increasing x_1, while the right-hand side decreases with increasing x_1. A nice intersection ($x_1 = 0.7028$, $m_c = 187$ kg) is easily obtained.

Algorithm 2 is unwieldy compared to algorithm 3, while algorithm 1 does not work at all since $(df/dx_1)_{x_1=0.702} = -6.48$, and a successive approximation algorithm diverges at any point where $|df/dx| > 1$.

The moral is that even a simple problem may be confounded by the choice of a poor numerical scheme. It is particularly important to choose the right numerical formulation when the dimensionality of the problem increases—as will be shown later in many of our examples.

Example 2 Ketene (A) is absorbed in acetone (B), where it reacts—with H_2SO_4 as catalyst—to form isopropenyl acetate by an elementary bimolecular reaction. At 37°C the rate constant is k = 0.0128 mol/L·min.

The reaction is carried out in a flask which initially contains 150 L of pure acetone ($c_B^o = 13.62$ mol/L). The reactor is equipped with a reflux condenser. For a constant flow of F = (7.5 mol of ketene/min + 7.5 mol of methane/min) to the reactor, no ketene can be traced in the outlet stream until the liquid is saturated with ketene—which indicates that the reflux of acetone is sufficient to reabsorb all ketene from the outlet stream. The solubility of ketene in acetone (and in mixtures of acetone, isopropenyl acetate, and H_2SO_4) is taken to be 0.834 mol/L at $p_A = 0.5$ atm and T = 37°C.

It is desired to calculate the time t* until the liquid is saturated with ketene:

$$\frac{dc_A}{dt} = \frac{F}{V} - kc_A c_B \qquad \frac{dc_B}{dt} = -kc_A c_B \tag{3}$$

where $c_A^o = 0$, $c_B^o = 13.62$ mol/L; and F/V = 7.5/150 = 0.05 mol/L·min.

Elimination of c_B yields

$$\frac{dc_A}{dt} = f(c_A) = \frac{F}{V} - kc_A\left(c_B^o + c_A - \frac{F}{V}t\right) \tag{4}$$

If no chemical reaction takes place, the liquid is saturated when $t = t_1 = t_1 = 0.834/(F/V) = 17$ min. If the reaction is very fast, saturation is reached when all acetone is consumed and the isopropenyl acetate is saturated with ketene [e.g., at $t_2 = (0.834 + 13.82)/(F/V) = 289$ min]. For a finite k we obtain $t_1 < t^* < t_2$.

For t = 0 the ketene is consumed with a time constant $t_r = 1/kc_B^o = 5.7$ min; hence the reaction is fairly rapid, and an explicit integration method (e.g., Euler's method) is numerically stable only when the time step is comparable with t_r. This could lead to a substantial number of integration steps. With an implicit integration method a single differential equation can be integrated in large steps. Consequently, $c_A(t)$ is computed such that $f(c_A)$ is small. A much simpler interpretation of the limit $\Delta t \to \infty$ is the quasi-stationarity principle, by which

$$0 \sim \frac{F}{V} - kc_A\left(c_B^o + c_A - \frac{F}{V}t\right) \tag{5}$$

Inserting $c_A = 0.834$ yields t = t* = 195.4 min (and $c_B = 4.68$ mol/L).

An improvement in this result is obtained by setting $d^2c_A/dt^2 = 0$ for t = t*:

$$\frac{d^2c_A}{dt^2} = -k\left\{\left(c_B^o - \frac{F}{V}t + 2c_A\right)\frac{dc_A}{dt} - \frac{F}{V}c_A\right\} = 0 \tag{6}$$

Inserting $c_A = 0.834$ at t = 195.4 yields an estimate for $dc_A/dt|_{t=t^*}$.

$$\left.\frac{dc_A}{dt}\right|_{t=t^*} \sim 0.0076 = \frac{F}{V} \quad kc_A\left(c_B^o \mathrel{\text{ı}} c_A - \frac{F}{V}t\right)$$

by Eq. (4). Hence t* = 209.5 min, which is almost the same as the result t* = 208.5 min obtained with a fourth-order Runge-Kutta method using Δt = 5 min.

Example 3 A batch reactor contains M_0 kilograms of solvent + catalyst. Reactant A is poured into the reactor at constant speed n_A°/t_f (kg/min) during the time interval $[0,t_f]$. At time = 0, reactant and solvent are at the temperature T_1 of the surroundings, and natural convection carries the heat of reaction ($-\Delta H$ kcal/kg) away. Heat capacity c_p (kcal/kg K) and density ρ(kg/m^3) are independent of T and of conversion $1 - n_A/n_A^\circ = 1 - y$. The reaction is first order in A.

It is desired to calculate the reaction temperature T(t) for given values of the operating parameters M_0, n_A°/M_0, T_1, and t_f.

Mass balance:

$$\frac{dy}{d\tau} = \begin{cases} -kt_f y + 1 & \text{for } \tau < 1 \\ -kt_f y & \text{for } \tau \geqslant 1 \end{cases} \qquad (y = 0,\ \tau = 0) \tag{7}$$

Energy balance:

$$\frac{dT}{d\tau} = \frac{-\Delta H}{c_p} k(T_1)t_f \frac{n_A^\circ/M_0}{1 + (n_A^\circ/M_0)\tau a} y \exp\left[\gamma\left(1 - \frac{T_1}{T}\right)\right]$$
$$- \frac{3ht_f(T - T_1)^{1.25}}{\rho^{2/3} c_p M_0[1 + (n_A^\circ/M_0)\tau a]^{1/3}} - \frac{(n_A^\circ/M_0)b}{1 + (n_A^\circ/M_0)\tau a}(T - T_1) \tag{8}$$

where

$$T = T_1 \quad \text{at } \tau = 0; \quad a = \begin{cases} 1 & \tau < 1 \\ 1/\tau & \tau \geqslant 1 \end{cases} \qquad b = \begin{cases} 1 & \tau < 1 \\ 0 & \tau \geqslant 1 \end{cases}$$

The heat transfer coefficient $h \sim 5$ kcal/m^2·h·K$^{1/4}$ and the heat transfer area is taken to be $3(M/\rho)^{2/3}$; $M = M_0(1 + n_A^\circ/M_0 \tau a)$. The kinetic parameters $k(T_1)$ and $\gamma = E/RT_1$ and the thermodynamic parameters ($-\Delta H$) and c_p have given values.

A crude numerical method (e.g., an explicit Euler method) may be used to solve the important design problem of finding the maximum permissible M_0 (to avoid temperature runaway) for given n_A°/M_0, t_f, and T_1. A higher-order method (e.g., an explicit fourth-order Runge-Kutta method) may give a shorter computation time for the same accuracy, but it is unnecessary to use an implicit method. The reason is that the structure of Eqs. (7) and (8) is such that numerical instability of an explicit integration method is not a problem. When the numerical method becomes unstable, it is a reflection of the inherent physical instability of the problem: for M_0 larger than a certain critical value, the whole batch will eventually ignite and T increases to a value close to $(-\Delta H/c_p)n_A^\circ/(n_A^\circ + M_0)$.

Many problems in reaction engineering can be handled with such crude tools and it becomes a matter of experience to recognize those situations where the more elaborate methods to be discussed in the following sections have to be utilized.

ELEMENTS OF NUMERICAL WORK

The three introductory examples illustrated ad hoc solutions of two standard numerical problems:

1. Solution of algebraic equations
2. Solution of coupled first-order differential equations

It is difficult to imagine a simulation job in reaction engineering which cannot be analyzed in terms of these fundamental numerical problems. As a preliminary to a more systematic discussion of typical reaction engineering simulation problems, we shall briefly comment on the techniques that we recommend for solution of the two problem types.

Solution of Algebraic Equations

Systems of Linear Algebraic Equations

The most efficient technique for solving

$$\underline{\underline{A}}\,\underline{x} = \underline{b} \tag{9}$$

is probably Gaussian elimination (with pivotation of rows and perhaps equilibration before each elimination step) in the particular version where $\underline{\underline{A}}$ is factorized into a lower triangular matrix $\underline{\underline{L}}$ and an upper triangular matrix $\underline{\underline{U}}$.

$$\underline{\underline{A}} = \underline{\underline{L}}\,\underline{\underline{U}} \tag{10}$$

$$\underline{\underline{L}} = \begin{pmatrix} 1 & 0 & 0 & \cdots & 0 \\ c_2^{(1)} & 1 & 0 & \cdots & 0 \\ c_3^{(1)} & c_3^{(2)} & 1 & \cdots & 0 \\ \cdot & \cdot & \cdot & \cdots & \cdot \\ c_N^{(1)} & \cdots & \cdots & c_N^{(N-1)} & 1 \end{pmatrix} \quad U = \begin{pmatrix} a_{11} & a_{12} & \cdots & & a_{1N} \\ 0 & a_{22}^{(1)} & \cdots & & a_{2N}^{(1)} \\ 0 & 0 & a_{33}^{(22)} & \cdots & a_{3N}^{(2)} \\ 0 & 0 & \cdots & \cdots & a_{NN}^{(N-1)} \end{pmatrix}$$

where $\underline{\underline{U}}$ and the subdiagonal elements of $\underline{\underline{L}}$ are stored in $\underline{\underline{A}}$. $c_i^{(k)}$ are the elimination constants for the kth elimination step:

$$c_i^{(k)} = \frac{a_{ik}^{(k-1)}}{a_{kk}^{(k-1)}} \quad \begin{cases} k = 1, 2, \ldots N - 1 \\ i = k + 1, \ldots N \end{cases}$$

and $\underline{\underline{U}}$ is the triangular matrix that appears after the N − 1 elimination steps on matrix $\underline{\underline{A}}$.

With $\underline{\underline{A}} = \underline{\underline{L}}\,\underline{\underline{U}}$, the solution of Eq. (10) follows a two-step procedure:

$$\underline{\underline{L}}\,\underline{z} = \underline{b} \quad \text{and subsequently} \quad \underline{\underline{U}}\,\underline{x} = \underline{z} \tag{11}$$

In an iterative or multistep process,

$$\underline{\underline{A}}\underline{x}_k = \underline{b}(\underline{x}_{k-1}) \qquad k = 1, 2, \ldots, M \tag{12}$$

with a constant matrix $\underline{\underline{A}}$, the LU decomposition followed by M applications of Eq. (11) is preferable to any alternative method [e.g., $\underline{x}_k = \underline{\underline{A}}^{-1}\underline{b}(\underline{x}_{k-1})$].

The specific problem

$$\underline{\underline{A}}\underline{x} = \lambda\underline{x} \tag{13}$$

has a nontrivial solution only when λ is one of the N eigenvalues of $\underline{\underline{A}}$. The solution $(\underline{x}_i, \lambda_i)$, i = 1, 2, . . . N, is found by diagonalization of $\underline{\underline{A}}$ using (most efficiently) a QR alogorithm for a general nonsymmetric $\underline{\underline{A}}$. Using only real arithmetic, one arrives at

$$\underline{\underline{A}} = \underline{\underline{U}}\,\underline{\underline{\Lambda}}\,\underline{\underline{U}}^{-1} \tag{14}$$

If all eigenvalues are real and distinct, $\underline{\underline{\Lambda}} = \text{diag}\,\{\lambda_i\}$ and $\underline{u}_i$ is the ith eigenvector $\underline{x}_i$, while the eigenrows $\underline{y}_i$ rows in $\underline{\underline{U}}^{-1}$. If $(\lambda_j, \lambda_{j+1})$ forms a complex pair of eigenvalues, $\underline{\underline{\Lambda}}$ has a (2×2) block

$$\underline{\underline{D}} = \begin{pmatrix} a & b \\ -b & a \end{pmatrix} \tag{15}$$

on the main diagonal in rows j and j + 1 and $(\lambda_j, \lambda_{j+1}) = a \pm ib$. A typical computer code based on the QR algorithm is discussed in (Villadsen and Michelsen, 1978).

Nonlinear Systems of Algebraic Equations

In a majority of problems, the multidimensional Newton's method (Newton-Raphson's method) has been our preferred choice. To solve $\underline{F}(\underline{x}) = \underline{O}$, the iteration

$$\underline{x}_{k+1} = \underline{x}_k - \underline{q} \quad \text{with} \quad \underline{\underline{J}}\,\underline{q} = \underline{F}(\underline{x}_k) \qquad J_{ij} = \left.\frac{\partial F_i}{\partial x_j}\right|_{\underline{x}=\underline{x}_k} \tag{16}$$

is carried out until $\|\underline{x}_{k+1} - \underline{x}_k\| < \varepsilon$.

The elements of the Jacobian matrix $\underline{\underline{J}}$ depend on the current value of $\underline{x}$, and in principle $\underline{\underline{J}}$ must be reevaluated in each iteration step, although one may possibly reduce the computational work by taking a few successive substitution steps (12) before updating and refactorizing $\underline{\underline{J}}$. Frequently, the elements of $\underline{\underline{J}}$ can be calculated quite simply from F_i—in particular, this is true when the algebraic equations result from discretization of boundary value problems with a nonlinear kinetic expression. Here the nonlinearities are confined to the main diagonal of $\underline{\underline{J}}$ and reevaluation of $\underline{\underline{J}}$ is trivial (but not, of course, the refactorization of $\underline{\underline{J}}$).

As with any other algorithm for solution of nonlinear algebraic equations, Newton's method does not guarantee convergence for all initial guesses $\underline{x}_o$, but in the vicinity of the solution $\underline{x}$ convergence is very rapid except in pathological cases where $\underline{\underline{J}}$ has one or more eigenvalues equal to zero at $\underline{x}$.

Calculations on a simplified model are often useful in selecting a proper $\underline{x}_0$. When the solution is known for one set of parameters $\underline{p}_0$ close to the set that is currently being investigated, an excellent $\underline{x}_0$ is found by means of the sensitivity of $\underline{x}$ to a variation in $\underline{p}$ at $\underline{p} = \underline{p}_0$. This important numerical approach is discussed later in this section.

In large problems it may be advantageous to expend considerable effort in selecting an optimal structuring of the equations before a solution is attempted. Considerations concerning partitioning of the equations are given below.

Solution of Coupled Ordinary Differential Equations

First-Order Differential Equations with Constant Coefficients

Together with Eqs. (9) and (13), the following problem is a key element in many numerical calculations.

$$\frac{d\underline{y}}{dt} = \underline{\underline{A}}\underline{y} + \underline{b} \qquad \underline{y}(0) = \underline{y}_0 \tag{17}$$

Factorizing A as in Eq. (14) leads to

$$\frac{d\underline{y}}{dt} = \underline{\underline{U}}\,\underline{\underline{\Lambda}}\,\underline{\underline{U}}^{-1}y + \underline{b}$$

or

$$\frac{d\underline{z}}{dt} = \underline{\underline{\Lambda}}\underline{z} + \underline{b}_1 \quad \text{with } \underline{z}(t = 0) = \underline{\underline{U}}^{-1}\underline{y}_0 \tag{18}$$

where $\underline{z} = \underline{\underline{U}}^{-1}\underline{y}$ and $\underline{b}_1 = \underline{\underline{U}}^{-1}\underline{b}$. Hence the components of $\underline{z}$ are found by solution of either

$$\frac{dz_i}{dt} = \lambda_i z_i + b_{1i} \tag{19}$$

or

$$\frac{d}{dt}\begin{pmatrix} z_i \\ z_{i+1} \end{pmatrix} = \begin{pmatrix} a & b \\ -b & a \end{pmatrix}\begin{pmatrix} z_i \\ z_{i+1} \end{pmatrix} + \begin{pmatrix} b_{1i} \\ b_{1i+1} \end{pmatrix} \tag{20}$$

The solution of the individual problems (19) and (20) is trivial and the final solution of Eq. (17) is

$$\underline{y} = \underline{\underline{U}}\underline{z} \tag{21}$$

For a nonsingular $\underline{\underline{A}}$, one obtains the explicit form

$$\underline{y} = \underline{q} + \underline{\underline{U}}\,\underline{\underline{\Omega}}(t)\underline{\underline{U}}^{-1}(\underline{y}_0 - \underline{q})$$

where $\underline{\underline{\Omega}}$ is diagonal except for the (2×2) blocks (15):

$$\Omega_{ii} = \exp(\lambda_i t) \tag{22a}$$

or

$$(\Omega_{ii}) = \begin{pmatrix} \cos bt & \sin bt \\ -\sin bt & \cos bt \end{pmatrix} \exp(at) \tag{22b}$$

and $\underline{q}$ is the solution of $\underline{\underline{A}}\underline{q} = \underline{b}$.

Runge-Kutta (or "Single-Step") Methods

To integrate the set of N autonomous nonlinear first-order equations

$$\frac{d\underline{y}}{dt} = \underline{f}(\underline{y}) \tag{23}$$

one step forward from $\underline{y}_n$ to $\underline{y}_{n+1}$ with $\Delta t = h$, the algorithm (24)–(25) is used:

$$\underline{y}_{n+1} = \underline{y}_n + \sum_{i=1}^{M} R_i \underline{k}_i \tag{24}$$

$$\underline{k}_i = h\underline{f}\left(\underline{y}_n + \sum_{j=1}^{M} b_{ij}\underline{k}_j\right) \tag{25}$$

When $b_{ij} = 0$ for $j > i - 1$, the method is explicit and the M Runge-Kutta constants $\underline{k}_i$ are found one by one, starting with $i = 1$. In the explicit Euler method ($M = 1$), $R_1 = 1$ and $b_{11} = 0$. The fourth-order Runge-Kutta-Gill method,

$$\underline{\underline{b}} = \begin{matrix} 0 & 0 & 0 & 0 \\ \frac{1}{2} & 0 & 0 & 0 \\ \frac{1}{2}(\sqrt{2} - 1) & \frac{1}{2}(2 - \sqrt{2}) & 0 & 0 \\ 0 & -\frac{1}{2}\sqrt{2} & \frac{1}{2}(2 + \sqrt{2}) & 0 \end{matrix}$$

$$\underline{R} = \frac{1}{6}(1,\ 2 - \sqrt{2},\ 2 + \sqrt{2},\ 1) \tag{26}$$

has better round-off properties than the slightly simpler "classical" Runge-Kutta method, where the elements of $\underline{\underline{b}}$ are zero except for

$$b_{i+1,i} = (\tfrac{1}{2},\ \tfrac{1}{2},\ 1) \qquad i = 1, 2, 3$$

$$\underline{R} = \frac{1}{6}(1,\ 2,\ 2,\ 1) \tag{27}$$

If the N right-hand sides depend on t, one extra equation,

$$\frac{dy_{N+1}}{dt} = 1 \qquad y_{N+1,n} = t_n \tag{28}$$

is added to the N original equations to give a set of N + 1 autonomous equations of the form (23).

In a truly implicit method all elements in $\underline{\underline{b}}$ can in principle be nonzero. Here the determination of $\underline{k}_i$, i = 1, 2, . . . , M, requires simultaneous solution of (N × M) nonlinear algebraic equations, an exasperating computation within each integration step which should be avoided except when required by the structure of the problem (even though the order of the implicit method may be as high as 2M).

An implicit integration is necessary when at $\underline{y}_n$ the Jacobian $\underline{\underline{J}}$ of $\underline{f}$ has a set of widely separated, negative eigenvalues. The existence of positive eigenvalues (or eigenvalues with positive real part) indicates that the system is physically unstable, and the numerical solution diverges in response to this inherent instability. When all λ_i have real negative parts the numerical solution should converge to some finite value, but this does not happen for explicit methods unless h is smaller than some constant multiplied by $|\lambda_N|^{-1}$, where λ_N is the eigenvalue of largest modulus. The natural time scale of the problem is, however, given by $|\lambda_1|^{-1}$, where λ_1 is the eigenvalue of the smallest modulus. Thus

$$\begin{aligned} &h < c_1 |\lambda_N|^{-1} && \text{to avoid numerical instability} \\ &nh > c_2 |\lambda_1|^{-1} && \text{to reach the required integration time} \end{aligned} \tag{29}$$

c_1 is 2 for the explicit Euler method and 2.83 . . . for the explicit fourth-order Runge-Kutta method.

With eigenvalues spanning many orders of magnitude, the computation time of an explicit method becomes intolerably large, and this is where implicit methods should be used since these can be made numerically stable for any h.

Typical applications of implicit methods are:

1. Integration of kinetic equations for radical reactions
2. Integration of a discretized version of a parabolic partial differential equation (e.g., for transient diffusion/reaction problems with nonlinear kinetics).
3. Differential equations coupled with algebraic equations.

The following approximation does, however, give an implicit method that works almost as an explicit method.

Consider a method where the b_{ij} in Eq. (25) are zero for j > i (i.e., where the equation for $\underline{k}_i$ involves $\underline{k}_j$, j = 1, 2, . . . , i). Rather than solving Eq. (25) by Newton iteration, one expands

$$\underline{f}(\underline{z}) = \underline{f}\left(\underline{y}_n + \sum_{j=1}^{i} b_{ij}\underline{k}_j\right) \quad \text{from} \quad \underline{z}_0 = \underline{y}_n + \sum_{j=1}^{i} b_{ij}\underline{k}_j$$

which results in

$$\underline{f}(\underline{z}) = \underline{f}(\underline{z}_0) + \left(\frac{\partial \underline{f}}{\partial \underline{z}}\right)_{\underline{z}_0} b_{ii}\underline{k}_i \equiv \underline{f}(\underline{z}_0) + a\left(\frac{\partial \underline{f}}{\partial \underline{y}}\right)_{\underline{y}_n} \underline{k}_i \tag{30}$$

or

$$(\underline{\underline{I}} - ah\underline{\underline{J}})\underline{k}_i = h\underline{f}\left(\underline{y}_n + \sum_{j=1}^{i-1} b_{ij}\underline{k}_j\right) \tag{31}$$

The constant a is determined such that the method integrates $f(y) = \lambda y$ correctly for $\lambda \to -\infty$, while the remaining constants in Eqs. (31) and (24) are fixed by the requirement that a certain order of truncation error should be obtained. Details on the fitting of Runge-Kutta constants in general are given in, for example, Gear (1971).

The point is that Eqs. (31) and (24) work like an implicit method, but compared to an explicit method, the only additional work is a factorization of the matrix $(\underline{\underline{I}} - ah\underline{\underline{J}})$ at the beginning of each integration step followed by a number of back-substitution steps (12) (this is, of course, not a trivial complication, but it is still much simpler than solution of $N \times M$ nonlinear equations in each step).

Our preferred choice of a semi-implicit method [the nomenclature used for an algorithm of the type (31) and (24)] was first derived by Caillaud and Padmanabhan (1971). It is a third-order method [$M = 3$, integration error from $t = 0$ to $t_{final} \sim O(h^3)$], and instead of the last function evaluation in Eq. (31), one takes $\underline{k}_3$ to be a linear combination of $\underline{k}_1$ and $\underline{k}_2$. Thus the method is defined as follows:

$$(\underline{\underline{I}} - ah\underline{\underline{J}})\underline{k}_1 = h\underline{f}(\underline{y}_n) \tag{32}$$

$$(\underline{\underline{I}} - ah\underline{\underline{J}})\underline{k}_2 = h\underline{f}(\underline{y}_n + b_2\underline{k}_1) \tag{33}$$

$$(\underline{\underline{I}} - ah\underline{\underline{J}})\underline{k}_3 = b_{31}\underline{k}_1 + b_{32}\underline{k}_2 \tag{34}$$

$$\underline{y}_{n+1} = \underline{y}_n + \sum_1^3 R_i\underline{k}_i + O(h^4) \tag{35}$$

Values for b_2, b_{31}, b_{32}, R_1, R_2, R_3, and a are given in (Villadsen and Michelsen, 1978, p. 320 and the computer code STIFF3 on p. 322).

A semi-implicit method is equally suitable for solving N_1 differential equations coupled to N_2 algebraic equations:

$$\begin{aligned} \frac{d\underline{y}_1}{dt} &= \underline{f}(\underline{y}_1, \underline{y}_2) \\ \underline{0} &= \underline{g}(\underline{y}_1, \underline{y}_2) \end{aligned} \tag{36}$$

$\underline{y} = [\underline{y}_1$ (with dimension N_1), $\underline{y}_2$ (with dimension N_2)]. Equation (36) can be interpreted as a limiting case of

$$\frac{d}{dt}\begin{pmatrix} \underline{y}_1 \\ \underline{y}_2 \end{pmatrix} = \begin{pmatrix} \underline{f} \\ \frac{1}{\varepsilon}\underline{g} \end{pmatrix} = \underline{F}(\underline{y}) \quad \text{for } \varepsilon \to 0$$

Let $\underline{\underline{J}}$ be the Jacobian of $\underline{F}$ at $\underline{y}_n$, let $\underline{\underline{B}} = (\underline{\underline{I}} - ah\underline{\underline{J}})$, and define $\underline{\underline{B}}^*$ as $\underline{B}$ multiplied by a diagonal matrix:

$$\underline{\underline{B}}^* = \underline{\underline{B}} \text{ diag } (\underbrace{1,1,\ldots,1}_{N_1}, \underbrace{\varepsilon,\varepsilon,\ldots,\varepsilon}_{N_2}) \xrightarrow{\varepsilon\to 0} \begin{matrix} \underline{\underline{I}} - ah\underline{\underline{f}}_{y_1} & -ah\underline{\underline{f}}_{y_2} \\ -ah\underline{\underline{g}}_{y_1} & -ah\underline{\underline{g}}_{y_2} \end{matrix}$$

where $\underline{\underline{f}}_{y_1}$ is $\partial \underline{f}/\partial \underline{y}_1$, $\underline{\underline{g}}_{y_1}$ is $\partial \underline{g}/\partial \underline{y}_1$, and so on. When the device is applied to the semi-implicit Runge-Kutta method [Eqs. (32) to (34)], it is seen that multiplication by ε from both sides of Eqs. (32) and (33):

$$\underline{\underline{B}}^*\underline{k}_1 = h \begin{pmatrix} \underline{f}(\underline{y}_n) \\ \underline{g}(\underline{y}_n) \end{pmatrix}$$

$$\underline{\underline{B}}^*\underline{k}_2 = h \begin{pmatrix} \underline{f}(\underline{y}_n + b_2\underline{k}_1) \\ \underline{g}(\underline{y}_n + b_2\underline{k}_1) \end{pmatrix} \tag{37}$$

while

$$\underline{\underline{B}}^*\underline{k}_3 = \begin{pmatrix} b_{31}\underline{k}_{1,\underline{y}_1} & + & b_{32}\underline{k}_{2,\underline{y}_1} \\ & \underline{O} & \end{pmatrix}$$

In Eq. (37) $\underline{k}_{1,\underline{y}_1}$ and $\underline{k}_{2,\underline{y}_1}$ are the first N_1 components of $\underline{k}_1$ and $\underline{k}_2$, respectively.

Any semi-implicit (or implicit) Runge-Kutta method, and the implicit versions of multistep methods to be discussed below (Gear, 1971, p. 226), can be extended to solve a set of N_1 differential equations and N_2 algebraic equations. It is advantageous to treat the algebraic equations and the differential equations in the same manner, since all components of the solution vector $\underline{y}$ are found with the same, specified accuracy.

The STIFF3 algorithm based on Eqs. (32) to (35) (Villadsen and Michelsen, 1978, pp. 316–322), and the STIFFALG algorithm (Michelsen and Villadsen, 1981) with a slight modification [Eqs. (36)–(37)], control the step length h by integrating the differential equations twice in every integration interval. First the step length h is used to get $y_{n+1}(h)$ and the next two steps h/2 are taken to obtain $\underline{y}_{n+1}(h/2)$. The difference between the two estimates $\underline{e} = \underline{y}_{n+1}(h/2) - \underline{y}_{n+1}(h)$ is first used to check the accuracy by comparing e_i with ε_i, the tolerance for the ith component of $\underline{y}$. If the step length h is accepted, a suitable value for h in the next step is predicted, and by extrapolation of $\underline{y}_{n+1}(h/2)$ to $\underline{y}^*_{n+1} = \underline{y}_{n+1}(h/2) + \underline{e}/7$, one obtains a result $\underline{y}^*_{n+1}$ with single-step accuracy $\sim O(h^5)$. A step-size selection strategy can also be based on integration from t_n to t_{n+1}, first with a formula of order M and next with a formula of order M − 1. This is the method used in the IMSL subroutine DVERK and in the versions of Gear's method that are discussed briefly below.

Multivalue (or Multistep) Methods

For k = 1, 2, . . ., N, let the kth component of $\underline{y}$ and of $\underline{f}(\underline{y})$ be given at M previous values of t: $t_n, t_{n-1}, \ldots, t_{n-i}, \ldots, t_{n-M+1}$. These values are ordered in vector $\underline{v}_{n,k}$ with 2M components:

$$\underline{v}_{n,k} = (y_n, y_{n-1} \cdots y_{n-i} \cdots y_{n-M+1}, f(y_n) \cdots f(y_{n-M+1}))_k$$

Using Eq. (38), where the elements of the (2M × 2M) matrix $\underline{\underline{B}}$ are known for any particular multistep method, one may update $\underline{v}_{n,k}$ to $\underline{v}_{n+1,k}$:

$$\underline{v}_{n+1,k} = \underline{\underline{B}}\underline{v}_{n,k} \tag{38}$$

The N vectors $v_{n+1,k}$ are arranged in an (N × 2M) matrix $\underline{\underline{V}}_{n+1}$. Take the first and the (M + 1) column $\underline{V}_1$ and $\underline{V}_{M+1}$ of $\underline{\underline{V}}_{n+1}$ and compute the residual $\underline{G}$ of the N differential equations:

$$\underline{G} = h\underline{V}_{M+1} - h\underline{f}(\underline{V}_1) \tag{39}$$

Correct $\underline{\underline{V}}_{n+1}$ by means of the vector product of $\underline{G}$ and the constant vector $\underline{c}$:

$$\underline{\underline{V}}_{n+1} := \underline{\underline{V}}_{n+1} + \underline{G}\underline{c} \tag{40}$$

Reevaluate Eq. (39), and iterate between Eqs. (39) and (40) until $||G|| < \varepsilon$, or solve $\underline{G} = \underline{0}$ by a Newton method (which is the safer approach).

This is in principle the multistep method described by Gear (1971, p. 102, 215). There are explicit and implicit versions, and they are incorporated into very popular integration packages by Gear and by Hindmarsh [GEAR, GEARB for a band structure of the Jacobian of $\underline{f}$, and GEARIB, a strictly implicit method that can also be used to solve $\underline{\underline{A}}(\underline{y})\underline{y}^{(1)} = \underline{f}(\underline{y})$, at least when $\underline{\underline{A}}$ is a slowly varying function of $\underline{y}$]. Byrne (1981) reviews several commercial packages based on multistep integration of the differential equations. It is our experience that STIFF3 and the implicit GEAR packages work equally well on many different problems. This is also the conclusion of Finlayson (1980, p. 54).

Sensitivity Functions

We shall introduce the sensitivity functions and point to their application in parametric studies to solve three major problems: algebraic equations, initial value problems, and boundary value problems.

First consider a set of N algebraic equations that contain M parameters $\underline{p}$:

$$\underline{F}(\underline{x},\underline{p}) = \underline{0} \tag{41}$$

The solution $\underline{x}_0$ of Eq. (41) is known for $\underline{p} = \underline{p}_0$, and we desire to compute the variation of the solution $\underline{x} - \underline{x}_0$ as a function of $\underline{p} - \underline{p}_0$. Now, with the constraint (41),

$$\underline{dF} = \underline{0} \cong \left.\frac{\partial \underline{F}}{\partial \underline{x}}\right|_{\underline{x}=\underline{x}_0} \underline{dx} + \left.\frac{\partial \underline{F}}{\partial \underline{p}}\right|_{\underline{p}=\underline{p}_0} \underline{dp} \tag{42}$$

$$\underline{\Delta x} = \underline{x} - \underline{x}_0 \cong (-\underline{\underline{J}})^{-1}_{\underline{x}_0} \left.\frac{\partial \underline{F}}{\partial \underline{p}}\right|_{\underline{p}_0} (\underline{p} - \underline{p}_0) \tag{43}$$

where $\underline{\underline{J}}$ is the Jacobian of $\underline{F}$ evaluated at $(\underline{x}_0, \underline{p}_0)$ and is available when a Newton method has been used to solve Eq. (41) with $\underline{p} = \underline{p}_0$. Consequently, as long as $\underline{\underline{J}}$ is reasonably well conditioned, one obtains an excellent starting value $\underline{x}$ for solution of Eq. (41) at $\underline{p}$ by

$$\underline{x} = \underline{x}_0 + \underline{\Delta x} \quad \text{where} \quad \underline{\underline{J}}\ \underline{\Delta x} = -\frac{\partial \underline{F}}{\partial \underline{p}} (\underline{p} - \underline{p}_0) \tag{44}$$

Next, let $\underline{y}$ be the N-dimensional solution of the initial value problem

$$\frac{d\underline{y}}{dt} = \underline{f}(\underline{y}, \underline{p}) \qquad \underline{y}(t = 0) = \underline{y}_0(\underline{p}), \quad \underline{p} = \underline{p}_0 \tag{45}$$

Assume that the variation of $\underline{y}$ with respect to the parameters $\underline{p}$ is desired—the essential problem in parameter estimation. The sensitivity functions $\underline{z}_k(t)$, k = 1, 2, . . . , M, are calculated as the solution of M sets of N coupled differential equations.

$$\underline{z}_k \equiv \frac{\partial \underline{y}}{\partial p_k} \qquad \frac{d\underline{z}_k}{dt} = \left.\frac{\partial \underline{f}}{\partial \underline{y}}\right|_{\underline{p}_0} \underline{z}_k + \frac{\partial \underline{f}}{\partial p_k} \tag{46}$$

with $\underline{z}_k(t = 0) = d\underline{y}_0/dp_k$.

The differential equations for the sensitivities are all linear in the dependent variable $\underline{z}_k$. If Eq. (45) is solved for $\underline{p} = \underline{p}_0$ from $\underline{y}_n$ to $\underline{y}_{n+1}$ by an *implicit* method, the Jacobian of $\underline{f}$ is available at t_{n+1} when $\underline{y}_{n+1}$ has been determined with the required accuracy. Subsequent solution of the M sets of sensitivity equations (46) now follows from Eqs. (19) and (20) by relatively trivial operations. Only one diagonalization (14) is necessary since the coefficient matrix is the same for all M sets of equations, while the constant vector $\underline{b}$ in Eq. (17) depends on p_k. The implicit version of Gear's method has proved to be well suited for this type of calculation if the Jacobian is computed with sufficient accuracy. Other implicit methods (e.g., collocation) are also suitable. A semi-implicit Runge-Kutta method is less satisfactory, for the following reasons. One may integrate Eq. (45) from t = 0 to t_{final} with a fixed $\underline{p}_0$ and thereafter solve the M sets of equations in (46) from t = 0 to t_{final}. This would require storage of a vast amount of data, and no step-length control is possible since the sequence of t values is fixed by the solution of Eq. (45). On the other hand, one may simultaneously integrate Eq. (45) and

$$\frac{dz}{dt} = f_p + f_y z = g(y, p, z) \tag{47}$$

where f_p is a short notation for $\partial \underline{f} / \partial \underline{p}$ and f_y a short notation for $\underline{\underline{J}}$. This would, however, require that f_{py} and f_{yy} be calculated to form the

Jacobian for the N(M + 1) systems of equations, and this is hardly attractive compared to an implicit solution of Eq. (45), followed by solution of Eq. (46) with the correct f_y and f_p at $\underline{y}_{n+1}$.

The application of sensitivity functions in the solution of boundary value problems is also very important. One may consider a typical situation where it is desired to solve Eq. (48) by an initial value technique:

$$\frac{d^2y}{dx^2} = f(x,y,y^{(1)}) \tag{48}$$

$$\alpha_1 \frac{dy}{dx} + y = a \quad \text{at } x = 0 \quad \text{and} \quad \alpha_2 \frac{dy}{dx} + y = b \text{ at } x = 1$$

To integrate Eq. (48) as an initial value problem, the initial element (y, dy/dx) at x = 0 is required. Let the estimated value of $y_{x=0}$ be y_0. Then the boundary condition at x = 0 yields $(dy/dx)_{x=0} = (1/\alpha_1)(a - y_0)$, and the integration can be performed resulting in (y,dy/dx) at x = 1.

The value of y_0 must be chosen such that the second boundary condition is satisfied:

$$b^*(y_0) \equiv \left(\alpha_2 \frac{dy}{dx} + y\right)_{x=1} = b$$

The correct value of y_0 is determined iteratively (e.g., by Newton iteration). Differentiation of Eq. (48) with respect to y_0 yields the following differential equation for $z = \partial y/\partial y_0$:

$$\frac{d^2z}{dx^2} = \frac{\partial f}{\partial y} z + \frac{\partial f}{\partial y^{(1)}} \frac{dz}{dx} \tag{49}$$

with $(z = 1,\ dz/dx = -1/\alpha_1)$ at x = 0.

Equation (49) is integrated simultaneously with Eq. (48). The desired derivative is

$$\frac{db^*}{dy_0} = \left(\alpha_2 \frac{dz}{dx} + z\right)_{x=1}$$

and a corrected estimate of y_0 is found from

$$y_0 := y_0 - \frac{(b^* - b)}{db^*/dy_0} \tag{50}$$

The method is easily extended to a case with an N-component solution vector $\underline{y}$ and N guessed initial values at x = 0. Also, one might find the sensitivity of $\underline{y}$ to a parameter p which appears in the differential equation and determine p to satisfy given side conditions at x = 0 and x = 1. Several applications will be shown in the following.

Partitioning of Large Problems

A large problem should very rarely be attacked numerically before a careful analysis of its structure has shown the following:

1. That the problem is well posed
2. That the proposed numerical method is applied to a reasonable formulation of the model

A number of algorithms exist for testing the consistency of the model—no missing equations or no duplication of equations—and these automatic procedures are definitely helpful, since the models can rapidly become too large and obscure to make a manual check safe enough. An algorithm by Sørensen (1982) specifically treats reacting systems. He divides the key reactant species (those that cannot be obtained from stoichiometric balances) into four groups:

1. *Mobile species*: When set up in a stationary coordinate system the mass balances must include convective or diffusive outflow of these species from the volume element.
2. *Immobile species*: The net outflow of these species from the stationary volume element is neglected.

3,4. In either group one may find species given by either equilibria or by kinetic equations with finite time constants.

Sørensen's algorithm makes a systematic search of any reaction model posed in these very general terms and checks it for consistency, ending up with some minimum set of key species, which is, however, not necessarily the set that is best suited for the numerical solution of the model.

Concerning optimal formulation of the numerical solution scheme, it may sometimes be advantageous to divide the model into blocks which are solved in a nested fashion. Let $\underline{F}(\underline{x}) = \underline{0}$ be partitioned into two sets of equations:

$$\begin{aligned} \underline{f}(\underline{x}_1, \underline{x}_2) &= 0 \\ \underline{g}(\underline{x}_1, \underline{x}_2) &= 0 \end{aligned} \tag{51}$$

where the dimension of $\underline{f}$ and of $\underline{x}_1$ is N_1 and the dimension of $\underline{x}_2$ and of $\underline{g}$ is N_2. For a set of linear equations

$$\begin{aligned} \underline{\underline{A}}_1\underline{x}_1 + \underline{\underline{A}}_2\underline{x}_2 + \underline{a} &= \underline{0} \\ \underline{\underline{B}}_1\underline{x}_1 + \underline{\underline{B}}_2\underline{x}_2 + \underline{b} &= \underline{0} \end{aligned} \tag{52}$$

partitioning leads to the explicit solution

$$\begin{aligned} \{\underline{\underline{A}}_1 - \underline{\underline{A}}_2\underline{\underline{B}}_2^{-1}\underline{\underline{B}}_1\}\underline{x}_1 &= \underline{\underline{A}}_2\underline{\underline{B}}_2^{-1}\underline{b} - \underline{a} \\ \underline{x}_2 &= \underline{\underline{B}}_2^{-1}(-\underline{\underline{B}}_1\underline{x}_1 - \underline{b}) \end{aligned} \tag{53}$$

which compared to the straightforward procedure [Eqs. (10) and (11)], is advantageous only when the matrix $\begin{pmatrix} A_1 & A_2 \\ B_1 & B_2 \end{pmatrix}$ has a pronounced banded structure in $\underline{\underline{A}}_1$ or $\underline{\underline{B}}_2$.

An algorithm for nested iterative solution of a set of nonlinear equations (51) may work in the following way. First $\underline{g}(\underline{x}_1,\underline{x}_2) = \underline{0}$ is solved for a fixed $\underline{x}_1$:

$$\underline{x}_2 := \underline{x}_2 - \underline{q} \qquad \left(\frac{\partial \underline{g}}{\partial \underline{x}_2}\right)_{\underline{x}_1} \underline{q} = \underline{\underline{g}}_{\underline{x}_2} \underline{q} = \underline{g} \tag{54}$$

Next, the sensitivity of the solution $\underline{x}_2$ is calculated:

$$\underline{\underline{g}}_{\underline{x}_2} \left(\frac{\partial \underline{x}_2}{\partial \underline{x}_1}\right)_{\underline{g}=\underline{0}} = -\left(\frac{\partial \underline{g}}{\partial \underline{x}_1}\right)_{\underline{x}_2} \tag{55}$$

The matrix $\underline{\underline{g}}_{\underline{x}_2}$ is available when the iteration (54) has converged, and the $(N_2 \times N_1)$ matrix $(\partial \underline{x}_2/\partial \underline{x}_1)_{\underline{g}=\underline{0}}$ is found by N_1 applications of the back substitution procedure (12).

In the third step the outer iteration $\underline{f} = \underline{0}$ is performed once:

$$\left(\frac{\partial \underline{f}}{\partial \underline{x}_1}\right)_{\underline{g}=\underline{0}} \Delta \underline{x}_1 = -\underline{f} \tag{56}$$

or by the chain rule for differentiation,

$$\left\{\left(\frac{\partial \underline{f}}{\partial \underline{x}_1}\right)_{\underline{x}_2} + \left(\frac{\partial \underline{f}}{\partial \underline{x}_2}\right)_{\underline{x}_1} \left(\frac{\partial \underline{x}_2}{\partial \underline{x}_1}\right)_{\underline{g}=\underline{0}}\right\} \Delta \underline{x} = -\underline{f} \tag{57}$$

Now Eq. (54) is repeated with the new $\underline{x}_1$ and the nested iteration ends by a final correction of $\underline{x}_2$:

$$\underline{x}_2 := \underline{x}_2 + \left(\frac{\partial \underline{x}_2}{\partial \underline{x}_1}\right)_{\underline{g}=\underline{0}} \Delta \underline{x}_1 \tag{58}$$

to get the same accuracy of $\underline{x}_1$ and $\underline{x}_2$.

Having obtained the solution $(\underline{x}_1,\underline{x}_2)$, one may determine its sensitivity with respect to a set of parameters $\underline{p}$ which appear in $\underline{f}$ and $\underline{g}$. First, the sensitivity is found by solution of Eq. (59) with $(\partial \underline{f}/\partial \underline{x}_1)_{\underline{g}=\underline{0}}$ given in Eqs. (56) and (57):

$$\left(\frac{\partial \underline{f}}{\partial \underline{x}_1}\right)_{\underline{g}=0} \left(\frac{\partial \underline{x}_1}{\partial \underline{p}}\right)_{\underline{g}=\underline{f}=\underline{0}} = -\left(\frac{\partial \underline{f}}{\partial \underline{p}}\right)_{\underline{g}=0} =$$

$$-\left\{\left(\frac{\partial \underline{f}}{\partial \underline{p}}\right)_{\underline{x}_1,\underline{x}_2} + \left(\frac{\partial \underline{f}}{\partial \underline{x}_2}\right)_{\underline{x}_1} \left(\frac{\partial \underline{x}_2}{\partial \underline{p}}\right)_{\underline{g}=\underline{0}}\right\} \tag{59}$$

where $(\partial \underline{x}_2/\partial \underline{p})_{\underline{g}=\underline{0}}$ is the solution of

$$\underline{\underline{g}}_{\underline{x}_2}\left(\frac{\partial \underline{x}_2}{\partial \underline{p}}\right)_{\underline{g}=\underline{0}} = -\left(\frac{\partial \underline{g}}{\partial \underline{p}}\right)_{\underline{x}_1,\underline{x}_2} \tag{60}$$

Finally, the sensitivity of $\underline{x}_2$ is calculated from

$$\left(\frac{\partial \underline{x}_2}{\partial \underline{p}}\right)_{\underline{f}=\underline{g}=\underline{0}} = \left(\frac{\partial \underline{x}_2}{\partial \underline{p}}\right)_{\underline{g}=\underline{0}} + \left(\frac{\partial \underline{x}_2}{\partial \underline{x}_1}\right)_{\underline{g}=\underline{0}}\left(\frac{\partial \underline{x}_1}{\partial \underline{p}}\right)_{\underline{f}=\underline{g}=\underline{0}} \tag{61}$$

The treatment of a set of N equations by nested iteration is in principle not different from what has already been shown [see Eqs. (41) to (44)]. The method is, however, less stringently defined than a straightforward approach [should Eq. (54) be solved until convergence has been reached, as suggested by Sørensen (1982, Chap. 4), or should one iteration in the inner loop be followed by one iteration in the outer loop?], and it may be questioned whether an individual treatment of each problem is advantageous on the most modern computing machinery (e.g., a vector processor).

Certain problems in reactor simulation do, however, appear to lend themselves quite naturally to solution in a nested fashion:

1. Some variables appear locally and the structure of the problem is somewhat like

$$\underline{f}(\underline{x}_1,\underline{x}_2,\ldots,\underline{x}_i,\ldots,\underline{x}_M) = \underline{0} \tag{62}$$

$$\underline{g}_k(\underline{x}_1,\underline{x}_k) = \underline{0} \qquad k = 2, 3, \ldots, M \tag{63}$$

   This could typically be a combination of fluid balances ($\underline{f}$) and pellet balances ($\underline{g}$) which may be solved individually as $\overline{M} - 1$ inner-loop problems for given values of the outer-loop variables $\underline{x}_1$. A sequence of reactors where the inlet conditions to reactor k + 1 form the only tie between reactor k and reactor k + 1 is another example that greatly resembles the catalyst pellet-fluid phase problem. Here the concept of "tearing" is, however, commonly recognized as a useful tool.
2. Frequently, modifications of a well-established program system can be made via nested iterations. A reactor simulation package can easily be modified to include diffusion restriction in catalyst pellets through addition of another well-tested program for effectiveness factor calculation.

The choice of $\underline{f}$ and $\underline{g}$ can only be made for the individual problem, and considerable effort may have to be spent in selecting the best partitioning of the system. If numerical difficulties can be associated with the solution of a small set of equations, these might be selected as $\underline{g}$, especially if the outer-loop variables $\underline{x}_1$ are relatively insensitive to variations in $\underline{x}_2$; but a general rule of thumb cannot be given.

SIMULATION OF REACTION ENGINEERING MODELS

A number of fundamental numerical techniques were presented in the preceding section. We consider next some standard numerical problems which occur specifically in reaction engineering: boundary value problems (exemplified by "catalyst pellet" problems), parabolic partial differential equations ("the steady-state fixed-bed reactor"), and hyperbolic equations with a dispersive element ("moving-front problems," exemplified by "transient fixed-bed reactor models").

Having presented our views on the proper numerical treatment of these simulation problems, in the final section of the chapter we discuss how parameters can be fitted to standard reaction engineering models.

Boundary Value Problems

The archetypal boundary value problem in reaction engineering is given by the simultaneous mass and heat balances for reaction and diffusion on catalyst pellets. With a very simplified model for the diffusion fluxes [see Jackson (1977) for a detailed discussion of realistic physical models], one obtains.

$$\nabla^2 y_i - \phi_i^2 R_i = 0 \qquad i = 1, 2, \ldots, M \tag{64}$$

$$\nabla^2 \theta + \Sigma\, \beta_i \phi_i^2 R_i = 0 \tag{65}$$

where

$$\nabla^2 = x^{-s} \frac{d}{dx}\left(x^s \frac{d}{dx}\right) \qquad s = 0, 1, \text{ or } 2 \text{ in three standard pellet geometries}$$

$$\frac{dy_i}{dx} + Bi_{M,i}(y_i - y_{b,i}) = 0 \qquad \text{at } x = 1 \tag{66}$$

$$\frac{d\theta}{dx} + Bi_H(\theta - \theta_b) = 0 \qquad \text{at } x = 1 \tag{67}$$

$$\theta \text{ and } y_i \text{ are finite at } x = 0 \tag{68}$$

where

$y_{b,i}$ = fluid-phase concentration relative to a reference concentration c_r, which may be the inlet concentration of a specific reacting species

y_i = pellet-phase concentration of species i relative to c_r

(θ_b, θ) = temperatures relative to a reference temperature T_r

The M mass balances (64) correspond to the M independent reactions in M key species. The rate terms are also functions of non-key species ($k = M + 1, \ldots, M_{max}$). The concentrations of these species (c_k, $k = M + 1, \ldots, M_{max}$) can be expressed in terms of their fluid-phase values and the pellet concentrations of key species. Similarly, the energy balance can be eliminated (i.e., Θ is expressed in terms of $\ddot{\Theta}_b$ and y_k, k = 1, 2,. . ., M).

Consequently, the fully reduced pellet model contains only the M mass balances (64) expressed as M nonlinear coupled boundary value problems in y_i (i = 1, 2,. . ., M), $y_{b,i}$ (i = 1, 2,. . ., M_{max}), and Θ_b. They are solved with the linear boundary conditions (66), i = 1, 2,. . ., M, and the condition that the y_i are finite at x = 0.

The pellet model is thoroughly discussed in Aris (1975) and elsewhere in this book and hardly needs further explanation.

ϕ_i = Thiele modulus for ith mass balance

β_i = adiabatic temperature rise for reaction i

$$Bi_{M,i} = \frac{D_i L_p}{h_{M,i}} \quad \text{and} \quad Bi_H = \frac{kL_p}{\rho c_p h_H}$$

Collocation was originally introduced to handle simple versions of Eqs. (64)–(68) (Villadsen and Stewart, 1967). Since then orthogonal collocation has been proved to handle many boundary value problems—especially those occurring in reaction engineering—in a very efficient way. The reader is referred to Villadsen (1970), Finlayson (1972, 1974, 1980), Villadsen and Michelsen (1978), Michelsen and Villadsen (1981), and Sørensen (1982) for detailed discussions of the method. The paper by Michelsen and Villadsen (1981) firmly establishes polynominal orthogonal collocation as a close relative of Galerkin's method.

Basically, what (polynomial) collocation does is to interpolate each concentration profile $y_i(x)$ by a polynomial with N interior interpolation points [defined as the collocation points, and chosen as zeros of certain Jacobi polynomials $P_N^{(\alpha,\beta)}(x)$ which are given through a choice of α, β, and N] and one or two extra interpolation points, at x = 1 [to include the boundary condition (66)] and possibly at x = 0 (if the symmetry of the problem is not used to eliminate the boundary condition at x = 0).

When $y_i(x)$ has been approximated by an Nth-degree polynomial (in x or in $u = x^2$), the derivatives can be written exactly as linear combinations of the N unknown interpolation ordinates and the boundary ordinate(s). After discretizing Eq. (64) by inserting these linear combinations of ordinates for the derivatives at each interior interpolation point and equating Eq. (64) to zero at x_i (or u_i), i = 1, 2, . . ., N, the original set of M boundary value problems is converted into N × M algebraic equations in $y_{i,j}$ (i = 1, 2, . . . , M; j = 1, 2, . . . , N). The structure of the equations is dense along the main diagonal (blocks of N × N full matrices) and sparse elsewhere (see Villadsen and Michelsen, 1978, Chap. 8) and may be suitable for solution by partitioning.

For a problem in one of the three standard pellet geometries ($u = x^2$) and a single mass balance, one obtains

$$\nabla^2 y = 4\left\{u\frac{d^2y}{du^2} + \frac{s+1}{2}\frac{dy}{du}\right\} \tag{69}$$

where s = 0 (planar), s = 1 (cylinder), and s = 2 (spherical) geometry. At collocation point j, j = 1, 2, . . . , N:

$$\underline{y} = \left(y_1, y_2, \ldots, y_{N+1} = y(u = 1) \right)$$

$$\nabla^2 y \Big|_{u=u_j} = 4 \left\{ u_j \underline{B}_j^T + \frac{s+1}{2} \underline{A}_j^T \right\} \underline{y} = \underline{C}_j^T \underline{y} \qquad (70)$$

where B_{ji} and A_{ji} are coefficients to y_i in the linear combinations of ordinates that are used to express the second and first derivative of the interpolation polynomial for y(u) at collocation point j.

The final result is the set of N algebraic equations

$$\sum_{1}^{N} C^*_{ji} y_i - \phi^2 R(y_j) + b_j = 0 \qquad j = 1, 2, \ldots, N$$

$$C^*_{ji} = C_{ji} - \frac{C_{j,N+1} A_{N+1,i}}{d} \qquad b_j = \frac{C_{j,N+1} Bi_M y_b}{2d} \qquad (71)$$

$$d = A_{N+1,N+1} + \frac{Bi_M}{2}$$

Equation (71) is particularly well suited for solution by Newton's method since the nonlinearities appear only in the main diagonal of the Jacobian, while the remaining elements are constants = C^*_{ji}.

When the catalyst pellet model is solved as part of a fixed-bed reactor simulation, all that is needed is the flux of reactant at the pellet surface or equivalently the average rate of reaction on the pellet:

$$\eta = \text{pellet effectiveness factor} = \int_0^1 \frac{R(y)}{R(y_b)} \, dx^{s+1} \qquad (72)$$

where $R(y_b)$ is the rate taken at bulk-phase conditions.

The appendixes of Villadsen and Michelsen (1978) contain a well-tested (nonproprietary) package of subroutines that can be arranged to solve pellet simulation problems as exemplified by Eqs. (69) to (72). Other programs are published in Finlayson (1980). In general, these programs can be expected to work well as long as the pellet profile y(u) can be reasonably well approximated by a polynomial. A severe diffusion resistance (i.e., the case where most of the pellet is inactive) is perhaps better solved using other methods (e.g., asymptotic methods), but collocation variants that treat these situations have also been developed. Some comments on these methods are given below.

Spline collocation [or "collocation on finite elements" in the nomenclature of Finlayson (1980)] combines the property of finite-difference methods to represent a rapidly varying function by a series of connected curve segments with the accurate approximation of the function within each interval by interpolation based on orthogonal polynomials.

The interval $0 \leqslant x \leqslant 1$ is divided into subintervals. Collocation is applied to each subinterval and the $M - 1$ spline point ordinates are determined by extra equations that impose continuity of the first derivative of $y(x)$ at each spline point. With N_k collocation points in the interval of length δ_k to the left of spline point k and N_{k+1} collocation points in the interval of length δ_{k+1} to the right of spline point k, one obtains

$$\frac{1}{\delta_k}\left(\sum_{1}^{N_k} A^{(k)}_{N_k+1,i}y_i + A^{(k)}_{N_k+1,N_k+1}y_s\right) = \frac{1}{\delta_{k+1}}\left(\sum_{2}^{N_{k+1}} A^{(k+1)}_{1,i}y_i + A^{(k+1)}_{1,1}y_s\right) \quad (73)$$

where y_s is the spline point ordinate and $\underline{\underline{A}}^{(k)}$ and $\underline{\underline{A}}^{(k+1)}$ are discretization matrices for the first derivative to the left and to the right of the spline point respectively.

The collocation equations are written up for each subinterval and the final collocation procedure with $N = N_1 + N_2 + \cdots + N_{M-1} + N_M$ collocation points becomes a set of $N + M - 1$ algebraic equations with a pronounced diagonal block structure. The diagonal elements of $\underline{\underline{C}}^*$ and of $\underline{\underline{A}}$ are usually larger than the other elements, which means that the spline point equations may retain their position, and factorization of the coefficient matrix can be done block by block, with a resulting reduction of computational effort.

Quite often, one spline point may be enough to represent an outer pellet region with rapid reaction followed by an inner, almost inactive region. In recent publications these "penetration front" problems have frequently been solved by one-point spline collocation (typical examples are transient gas-solid reactions where the gaseous reactant is used up in an outer core of the pellet until the solid reactant concentration has decreased appreciably).

The original idea derives from Patterson and Cresswell (1971), who proposed to equate y and the derivative of y to zero at the spline point u_s (or x_s) and determine the spline point position and the effectiveness factor by collocation in the outer interval. An improvement was suggested in Villadsen and Michelsen (1978, Chap. 7). Here only the derivative of y was equated to zero at x_s, while the position of x_s was chosen somewhat arbitrarily (e.g., by solution of a linearized problem). For a first-order irreversible, exothermic reaction on spheres, one obtains

$$y_1(x) = \frac{\sinh \lambda_r x}{x \sinh \lambda_r x} \qquad \text{where } \lambda_r = \phi \exp\left[\frac{\gamma\beta}{2(1+\beta)}\right] \quad (74)$$

if the linearization of $R(y)$ is made at $y = 0$, which is quite reasonable when the major part of the catalyst pellet is inactive. The spline point may be chosen as the x value where $y_1(x) = 0.05y_b$, and collocation with $N = 5 - 6$ in the interval $(x_s, 1)$ usually gives an excellent value for the true flux at the pellet surface, at least if the solution $y(x)$ in the truncated interval satisfies $y(x_s) < 0.05y_b$.

If the original interval is split up into only two parts with N_a collocation points to the left of x_s and N_b collocation points to the right of x_s, it is not difficult to express y_s is terms of the $N = N_a + N_b$ collocation ordinates and y $(x = 1)$. Explicit formulas are given in Holk-Nielsen and Villadsen

(1983) and their final method is indistinguishable from a global collocation solution of order $N_a + N_b$.

Hyperbolic collocation is another variant of collocation, which is specifically developed to treat the wedge-shaped profiles associated with near-asymptotic behavior of the pellet model. This method was originally proposed by Guertin et al. (1977) for one-dimensional homogeneous reactor problems when these are solved by collocation, applied to the total reactor length. The method is fully developed and explained in Sørensen and Stewart (1982) and Sørensen (1982).

The rate expression for M key species may be expanded from a reference state $\underline{y}_R$ [as exemplified in Eq. (74)]:

$$\underline{R}(\underline{y}) = \underline{R}(\underline{y}_R) + \left.\frac{\partial \underline{R}}{\partial \underline{y}}\right|_{\underline{y}_R} (\underline{y} - \underline{y}_R) \tag{75}$$

This gives a linearized version of Eq. (64):

$$\nabla^2 \underline{y} - \underline{\underline{J}}\,\underline{y} = \underline{b}$$

$$J_{ij} = \phi_i^2 \left.\frac{\partial R_i}{\partial y_j}\right|_{\underline{y}_R} \qquad b_i = \phi_i^2 \left[R_i(y_R) - \left(\frac{\partial R_i}{\partial y}\right)^T \underline{y}_R\right] \tag{76}$$

which may be solved by standard methods, analogously to Eqs. (17) to (22), as a linear combination of $\sinh(\lambda_r x)/x$ (for spherical geometry and λ_r distinct and nonzero). The eigenvalues of $\underline{\underline{J}}$ are Λ_r^2 and $\lambda_r = |\Lambda_r|$ are used to construct the approximation functions.

If there are many independent reactions, the λ_r may give a sufficient number of approximation functions $\sinh(\lambda_r x)/x$ to provide an accurate approximation for $\underline{y}$. If the number of independent reactions is small (there may be only one reaction), the hyperbolic approximation functions are augmented by new functions which appear by differentiation of the original set with respect to λ_r. For typical pellet problems, Sørensen (1982) suggests $(1 - x^2)^j$, j = 1, 2, . . . , as multipliers in the higher approximation functions. When $\lambda_r \lesssim 0.5$, the hyperbolic functions are easily represented by low-order polynomials, and hyperbolic collocation merges into the better known polynomial collocation method; but clearly the method is a valuable generalization of the original collocation method, since the approximate solution is based on the structure of the individual problem, rather than on a fixed set of approximation functions.

Sørensen and Stewart (1982) and Sørensen (1982) illustrate the application of hyperbolic collocation on some very complicated problems, and the literature contains numerous examples of the application of high-order polynomial collocation. Consequently, we shall be content with two illustrations. In the first illustration the one-point collocation method is discussed as a valuable link between heterogeneous models and homogeneous models for reactors (e.g., a stirred-tank approximation for a pellet). In the second illustration collocation is applied to solve the pellet equations in a fixed-bed reactor model.

Example 4 One-Point collocation Consider a first-order exothermic reaction on catalyst pellets, and let the pellet model (64) to (68) be approximated by one interior interpolation point at (u_1, y_1):

$$-C_1[y(1) - y_1] = y_1\phi^2 \exp\left[\gamma\left(1 - \frac{1}{\Theta_1}\right)\right] \tag{77}$$

where

$$\Theta_1 + \beta y_1 = \text{constant} = \Theta(1) + \beta y(1) \tag{78}$$

$$\frac{d}{dx}\begin{pmatrix} y \\ \Theta \end{pmatrix}_{x=1} = 2\frac{d}{du}\begin{pmatrix} y \\ \Theta \end{pmatrix}_{u=1} = \begin{pmatrix} -qy_1 + qy(1) \\ -q\Theta_1 + q\Theta(1) \end{pmatrix} = \begin{pmatrix} Bi_M(1 - y(1)) \\ Bi_H(1 - \Theta(1)) \end{pmatrix} \tag{79}$$

where $q = 2/(1 - u_1)$

Equations (78) and (79) with $y_b = \Theta_b = 1$ are used to express Θ_1 and $\Theta(1)$, $y(1)$ in terms of y_1:

$$y(1) = \frac{Bi_M + qy_1}{Bi_M + q} \qquad \Theta(1) = \frac{Bi_H + q\Theta_1}{Bi_H + q} \tag{80}$$

$$\Theta_1 = 1 + \frac{1 + q/Bi_H}{1 + q/Bi_M}(1 - y_1) \tag{81}$$

Now the collocation equation (77) is simplified to

$$-C_1(1 - y_1) = (\phi^*)^2 y_1 \exp\left[\gamma\beta^* \frac{1 - y_1}{1 + \beta^*(1 - y_1)}\right] = (\phi^*)^2 R^*(y_1) \tag{82}$$

$$\beta^* = \beta\frac{1 + q/Bi_H}{1 + q/Bi_M} \quad \text{and} \quad \phi^* = \left(1 + \frac{q}{Bi_M}\right)^{1/2}\phi \tag{83}$$

Finally,

$$\eta = \int_0^1 \frac{R(y)}{R(y_b)}\, dx^{s+1} = \int_0^1 y \exp\left[\gamma\left(1 - \frac{1}{\Theta}\right)\right] dx^{s+1} \sim y_1 \exp\left[\gamma\left(1 - \frac{1}{\Theta_1}\right)\right] \tag{84}$$

The left half of Table 1 shows the values of the collocation constants $(u_1, -C_1)$ which are needed to construct a one point collocation solution of Eq. (77). u_1 is the zero of $P_1^{(0_1(s-1)/2)}(u)$, where s is the geometry factor.

TABLE 1 Parameters for One-Point Collocation

	Bi_H, Bi_M finite			Bi_H, $Bi_M \to \infty$			
s	u_1	$-C_1$	q	u_1	$-C_1$	w_1	w_2
0	1/3	3	3	1/5	5/2	5/6	1/6
1	1/2	8	4	1/3	6	3/4	1/4
2	3/5	15	5	3/7	21/2	7/10	3/10

In the specific case $Bi_H = Bi_M \to \infty$, the zero of $P_1^{(1,(s-1)/2)}(u)$ is preferred and η is approximated by a more accurate two-point quadrature formula:

$$\eta = w_1 \frac{R(y_1)}{R(y_b)} + w_2 \tag{85}$$

Constants for this specific situation (from Villadsen and Michelsen, 1978, Chap. 5) are shown on the right-hand side of Table 1.

Equation (82) is an approximate solution of the general catalyst pellet problem. There is only one unknown (y_1), and the form of the equation is the same as one would obtain from a stirred-tank reactor model. A graphical or algebraic solution can be found for any value of the parameters, but the solution is reasonable only when y(u) can be represented by a fairly low order polynomial. Note in Eq. (83) that the original β [$= c_r(-\Delta H)D/kT_r$] is modified by a factor that can be very large when Bi_H is small. Typical values of Bi_H and Bi_M in a fixed-bed reactor are 2 and 40, respectively (an industrial SO_2 converter) and with spherical catalyst particles, Eq. (83) predicts amplification by a factor 3.11.

The application of Eq. (82) will be illustrated by two numerical examples. First consider the trivial case of a first-order isothermal reaction with $Bi_H = Bi_M \to \infty$ on spherical pellets. The collocation equation has the explicit solution

$$y\left(u_1 = \frac{3}{7}\right) = \frac{21}{21 + 2\phi^2} \quad \text{and} \quad \eta = \frac{7}{10}\,\frac{21}{21 + 2\phi^2} + \frac{3}{10} \tag{86}$$

For $\phi = 1$, one obtains $\eta = 0.93913$, which to the first four figures agrees with the exact solution $\eta = 3/\phi^2\,(\phi \coth \phi - 1) = 0.93911$.

Also, at the much higher value $\phi = 4$, one obtains a good approximation to η:

$$\phi = 4: \quad y\left(\frac{3}{7}\right) = 0.3962 \quad \text{and} \quad \eta = 0.577 \quad (\eta_{exact} = 0.563)$$

When ϕ is as large as 4, one may safely use an asymptotic formula [e.g., the general formula of Wedel and Luss (1980) for an nth order reaction on spheres]:

$$\eta = \sqrt{\frac{2}{n+1}\ \phi^{-1} - \frac{12}{n+3}\ \phi^{-2}} \tag{87}$$

which gives $\eta = 0.5625$ for $n = 1$ and $\phi = 4$. Thus the whole region $0 < \phi < \infty$ is nicely covered with the two simple formulas (86) and (87).

We shall next consider the solution of Eq. (64) for a first-order irreversible reaction and the following set of parameters:

$$\beta = 0.05,\ \gamma = 30,\ Bi_H = 2,\text{ and } Bi_M = 40 \tag{88}$$

The curve marked "e" on Fig. 1 is the "exact" solution η versus ϕ obtained by high-order collocation. The surface temperature $\theta(1)$ increases along the curve from 1 at $(\eta,\phi) = (1,0)$ to $1 + \beta(Bi_M/Bi_H) = 2$ on the upper, ignited branch of the curve where the effectiveness factor reaches extraordinary high values [$\max(\eta) = 2366$] before its final decline, proportional to ϕ^{-2} when $\phi \to \infty$.

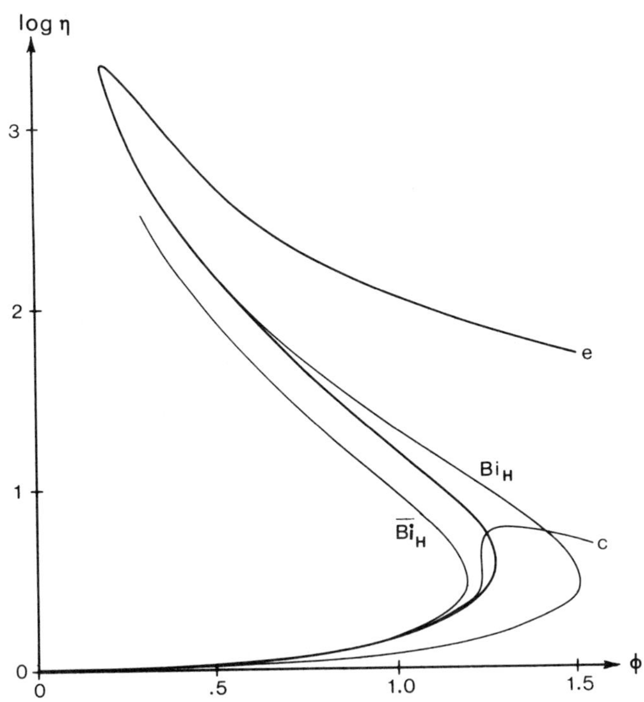

FIGURE 1 Effectiveness factor η versus Thiele modulus ϕ for a first-order nonisothermal reaction. $\gamma = 30$, $\beta = 0.05$, $Bi_M = 40$, $Bi_H = 2$. Curve e: exact solution by high-order collocation; merges with curve marked Bi_H for large Θ. Curve c: one point collocation. Curve Bi_H: isothermal pellet; pellet temperature Θ_p = surface temperature $\Theta(1)$. Curve $\overline{Bi}_H$: $\Theta_p > \Theta(1)$; $\overline{Bi}_H$ calculated from Eqs. (91) and (92).

The lower branch of the η versus ϕ curve is well approximated by one-point collocation (curve marked "c" on Fig. 1). For example, at point A, where $\phi^2 = 1.2$, one obtains $(\phi^*)^2 = 1.35$, and inserting the modified value of β [$\beta^* = 0.05 \cdot 28/9 = 0.1555$] in Eq. (82), one obtains $y_1 = 0.8468$ and $\eta \sim R(y_1) = 1.702$, which may be compared with the exact result, 1.683. The value of $\Theta_1 = 1.0238$, while $\Theta(1)$ is 1.0170. One-point collocation breaks down near the first bifurcation point, where the parabolic approximation for the concentration profile inside the pellet becomes very poor.

There is another model approximation that appears to give remarkably good results. The energy balance (65) is integrated over the pellet volume, and if β is small, one may assume that the pellet is nearly isothermal, in which case the first-order rate expression can be integrated analytically to give the right-hand side of Eq. (89):

$$\mathrm{Bi}_H[\Theta(1) - 1] = \beta \mathrm{Bi}_M \frac{\phi_1 \coth \phi_1 - 1}{\phi_1 \coth \phi_1 + \mathrm{Bi}_M - 1} \tag{89}$$

$$\phi_1 = \phi \exp\left[\frac{\gamma}{2}\left(1 - \frac{1}{\Theta_p}\right)\right] \tag{90}$$

and Θ_p is a representative pellet temperature.

If we are on the upper branch of the η versus ϕ curve of Fig. 1, the pellet temperature is so far above the bulk temperature 1 that the reaction is complete in a thin boundary layer. Here the representative pellet temperature Θ_p should be equated to $\Theta(1)$ and (89) is an equation to determine $\Theta(1) = \Theta_p$.

On the lower branch of the curve, most of the reaction occurs in the bulk of the pellet. The case $\phi^2 = 1.2$ illustrates the difference between $\Theta(1) = 1.017$ and Θ_p, which in the one-point collocation approximation is taken to be $\Theta_1 = 1.0238$. In Villadsen and Michelsen (1978, p. 28 and Chap. 6) it is shown that the factor $\Theta(1) - 1$ on the left-hand side of Eq. (89) can be replaced by the larger factor $\Theta_p - 1$ if at the same time Bi_M is replaced by a smaller, modified value $\overline{\mathrm{Bi}_H}$ defined by

$$\mathrm{Bi}_H[\Theta(1) - 1] \equiv \overline{\mathrm{Bi}_H}(\Theta - 1) \tag{91}$$

where $\overline{\mathrm{Bi}_H}$ is found by perturbation analysis to be

$$\overline{\mathrm{Bi}_H} = \frac{\mathrm{Bi}_H}{1 + \alpha \mathrm{Bi}_H} \qquad \alpha = \frac{1}{s + 3} \qquad s = 0,\ 1,\ \text{or } 2 \tag{92}$$

Thus, also for a modest reaction rate, one obtains an algebraic equation for the representative pellet temperature Θ_p. Whether Θ_p is taken to be $\Theta(1)$ or the left-hand side of Eq. (89) is replaced by Eq. (91) to determine a $\Theta_p > \Theta(1)$, the effectiveness factor is calculated by

$$\eta = \frac{3\mathrm{Bi}_M}{\phi^2} \frac{\phi_1 \coth \phi_1 - 1}{\phi_1 \coth \phi_1 + \mathrm{Bi}_M - 1} \tag{93}$$

For Θ_p larger than ~ 1.2, one obtains values of ϕ_1 in excess of 10 and $\coth \phi_1 \approx 1$. Now Eq. (89) can be solved explicitly for ϕ as a function of Θ_p:

$$\phi = \exp\left[-\frac{\gamma}{2}\left(1 - \frac{1}{\Theta_p}\right)\right]\left\{1 + \mathrm{Bi}_M \frac{\Theta_p - 1}{q + 1 - \Theta_p}\right\} \qquad q = \frac{\mathrm{Bi}_M \beta}{\mathrm{Bi}_H} \tag{94}$$

which is easily differentiated to obtain (Θ_p, ϕ) at the point where $d\phi/d\Theta_p = 0$ (i.e., at the upper "nose" of the η versus ϕ curve). For $q = 1$, one obtains

$$\frac{d\phi}{d\Theta_p} = 0 \text{ at } \Theta_p = 1.7291 \qquad \phi = 0.1946 \text{ and } \eta = 2310 \tag{95}$$

Also,

$$\frac{d\eta}{d\Theta_p} = 0 \text{ at } \Theta_p = 1.765 \qquad \phi = 0.1970 \text{ and } \eta = 2366$$

For practical applications of the present model all that is really needed is calculated from very simple formulas; by one-point collocation, or the approximation Eqs. (89) and (91) to (93), the lower branch of the η versus ϕ curve can be calculated.

The upper branch of the curve is found by Eqs. (89) and (93). The lower limit of the ignited branch is found with high accuracy by Eq. (95), while the upper limit of the quenched branch is found with reasonable accuracy by one-point collocation.

Example 5 The catalytic reaction

$$A \xrightarrow{k_1 c_A} B \xrightarrow{k_2 c_B^2} C$$

is carried out in a reactor with a given space time $\tau_0 = Vc_{A0}/F_{A0}$. The conversion A is $x_A = 1 - c_{As}/c_{A0}$, while the conversion to B is $x_B = c_{Bs}/c_{A0}$. c_{As} and c_{Bs} are fluid-phase concentrations of A and B at a given position in the reactor. Inside the catalyst particle the reactant concentrations are c_A and c_B, respectively. The effective diffusivity D is the same for A and B.

It is desired to design a reactor with the proper τ_0 to achieve the largest possible yield of component B for given values of $\phi_1^2 = L_p^2(k_1/D)$ and $\phi_2^2 = L_p^2(k_2 c_{A0}/D)$. $L_p = V_p/S_p$ and the particles are described by slab geometry.

One obtains the following mass balances for the particle phase:

$$\frac{d^2 c_A}{d\zeta^2} = \phi_1^2 c_A \qquad \frac{c_A}{c_{As}} = \frac{\cosh \phi_1 \zeta}{\cosh \phi_1} \tag{96}$$

$$\frac{d^2(c_B/c_{As})}{d\zeta^2} = \frac{d^2y}{d\zeta^2} = -\phi_1^2\left[\frac{\cosh \phi_1\zeta}{\cosh \phi_1} - \alpha^2(1 - x_A)y^2\right] \tag{97}$$

$$y = \frac{c_B}{c_{As}} = \frac{x_B}{1 - x_A}\frac{c_B}{c_{Bs}} \quad \text{and} \quad y(1) = \frac{x_B}{1 - x_A} \qquad y^{(1)}(0) = 0$$

$$\alpha = \frac{\phi_2}{\phi_1} = \left(\frac{c_{A0}k_2}{k_1}\right)^{1/2}$$

The fluid balances are

$$\frac{dx_A}{d\tau} = \tau_0 k_1 \frac{\overline{c_A}}{c_{A0}} = \frac{\tau_0 D}{L_p^2}(1 - x_A)\phi_1 \tanh \phi_1 \tag{98}$$

$$\frac{dx_B}{d\tau} = \frac{dx_A}{d\tau}\left[1 - \frac{\phi_2^2(1 - x_A)}{\phi_1 \tanh \phi_1}\overline{y^2}\right] \quad \text{where } \overline{y^2} = \int_0^1 y^2\, d\zeta \tag{99}$$

The value of τ_0 that is required to give a certain x_A value is

$$\tau_0 = -\frac{L_p^2}{D\phi_1 \tanh \phi_1}\ln(1 - x_A) \tag{100}$$

while the x_B value is obtained by integration of

$$\frac{dx_B}{dx_A} = \left[1 - \frac{\phi_2^2(1 - x_A)}{\phi_1 \tanh \phi_1}\overline{y^2}\right] = -\frac{dy/d\zeta\big|_{\zeta=1}}{\phi_1 \tanh \phi_1} \tag{101}$$

from $(x_A, x_B) = (0,0)$.

For moderate values of ϕ_1 and ϕ_2 (e.g., <5) solution of Eq. (97) by a straightforward collocation method causes no problems. Every evaluation of the right-hand side of (101) is, however, expensive since it requires solution of the nonlinear boundary value problem (97) transformed into N coupled nonlinear algebraic equations. After having solved the collocation equations, it is advantageous to find the sensitivity of the collocation ordinates with respect to x_A in order to obtain good starting values for the ordinates at the next value of x_A. Hence the linear boundary value problem (102) is solved by collocation before x_A is changed:

$$\frac{d^2z}{d\zeta^2} = \phi_2^2[2y(1 - x_A)z - y^2]$$

$$z \equiv \frac{\partial y}{\partial x_A} \qquad z^{(1)} = 0 \text{ at } \zeta = 0 \text{ and } z = \frac{x_B}{(1 - x_A)^2} \text{ at } \zeta = 1 \tag{102}$$

Throughout the calculations values of $(\cosh \phi_1\zeta/\cosh \phi_1)$ at the collocation abscissas are, of course, stored in a table.

Numerical difficulties arise for fast reactions. Figure 2 shows profiles $y(\zeta)$ for $\phi_1 = \phi_2 = 10$, six collocation points, and different x_A values. For $x_A = 0$ the profile increases to a sharp maximum at $\zeta = \zeta_{max}$ (the overall rate of production of B is zero in the inner part of the pellet), then decreases through an inflection point (where the local rate of production of B is zero) to a long, flat part in the interior of the pellet. To obtain an accurate value of $\overline{y^2}$, six collocation points may not be enough and spline collocation or some other method may have to be used.

The time-consuming iterations on high-order collocation equations are avoided if the problem is treated as two differential equations (98) and (99) coupled to N + 1 algebraic equations:

$$\sum_{i=1}^{N+1} C_{ji}y_i + [p_j - \alpha^2 y_j^2(1 - y_{N+2})]\phi_1^2 = 0 \tag{103}$$

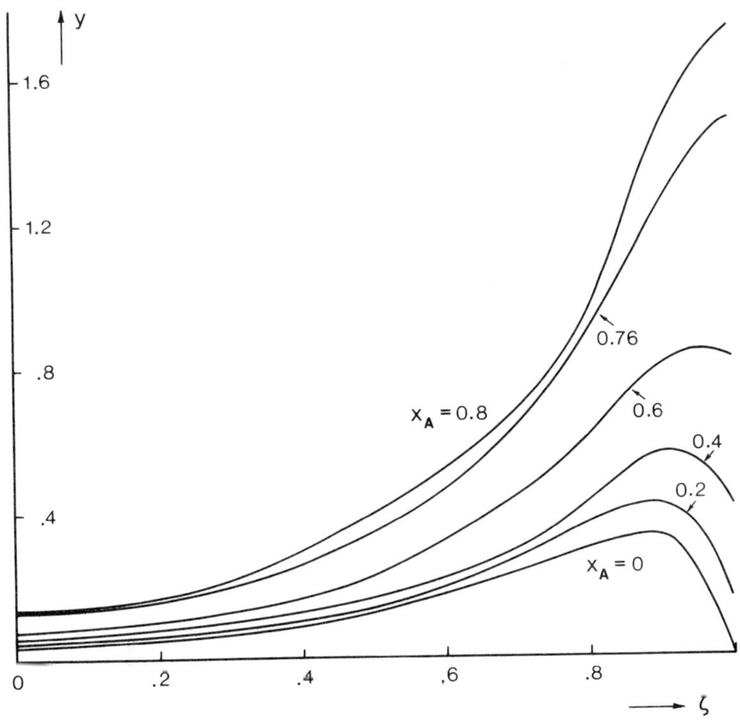

FIGURE 2 Pellet concentration profiles for component B in $A \xrightarrow{k_1 c_A} B \xrightarrow{k_2 c_B^2} C$. Parameter on curves: x_A = fluid phase conversion of A. Maximum yield of B is obtained for $x_A \sim 0.75$ where the pellet profile has zero slope at $\zeta = 1$.

$$y_{N+1} - \frac{y_{N+3}}{1 - y_{N+2}} = 0 \tag{104}$$

$$\frac{dy_{N+2}}{dt} = 1 \tag{105}$$

$$\frac{dy_{N+3}}{dt} = -\frac{\Sigma_1^{N+1} A_{N+1,i} y_i}{\phi_1 \tanh \phi_1} \tag{106}$$

where

$y_{N+2} = x_A$, $y_{N+3} = x_B$
t = dummy integration variable
p_j = pretabulated values of $\cosh \phi_1 \zeta_j / \cosh \phi_1$

The boundary ordinate y_{N+1} is eliminated and the resulting set of N + 2 differential-algebraic equations is solved [e.g., by the STIFFALG package based on Eqs. (36) and (37)]. Initial values of y_i, i = 1, 2,. . ., N, are found by solution of Eq. (103) with $y_{N+1} = y_{N+2} = y_{N+3} = 0$. When 10 or 12 collocation points are used, the optimum $(x_A, x_B) = (0.7514, 0.3232)$ for $\phi_1 = 10$ and $\alpha = 1$ is found with four- to five-digit accuracy.

The advantages of the present method are easily summed up: no unnecessary computation is spent in solving Eq. (97) "exactly" at every integration step, and all components $y_1, y_2, \ldots, y_{N+2}, y_{N+3}$ are calculated with a balanced accuracy.

For $\Phi_1 \to \infty$ at a fixed value of α, the pellet equation cannot be solved by collocation as it stands in Eq. (97). An asymptotic solution is, however, easily derived using the following transformation. Let $u = \phi_1(1 - \zeta)$ and rewrite (97) as

$$\frac{d^2y}{du^2} = -\exp(-u) + \alpha^2(1 - x_A)y^2 \tag{107}$$

with $y(u = 0) = x_B/(1 - x_A)$ and $y(u \to \infty) = 0$,

$$\frac{dx_B}{dx_A} = \left.\frac{dy}{du}\right|_{u=0} \tag{108}$$

The boundary condition at "$u \to \infty$" can be applied at two different u values, say u = 5 and 8, and Eq. (107) is solved by collocation for each of these values. If there is no appreciable change in $dy/du|_{u=0}$, the solution is accepted; otherwise, u > 8 is tried. Alternatively, one may make a second transformation to obtain a finite interval:

$$v = (1 + u)^{-2} \quad \text{where } v = 1 \text{ at } u = 0 \text{ and } v = 0 \text{ at } u \to \infty \tag{109}$$

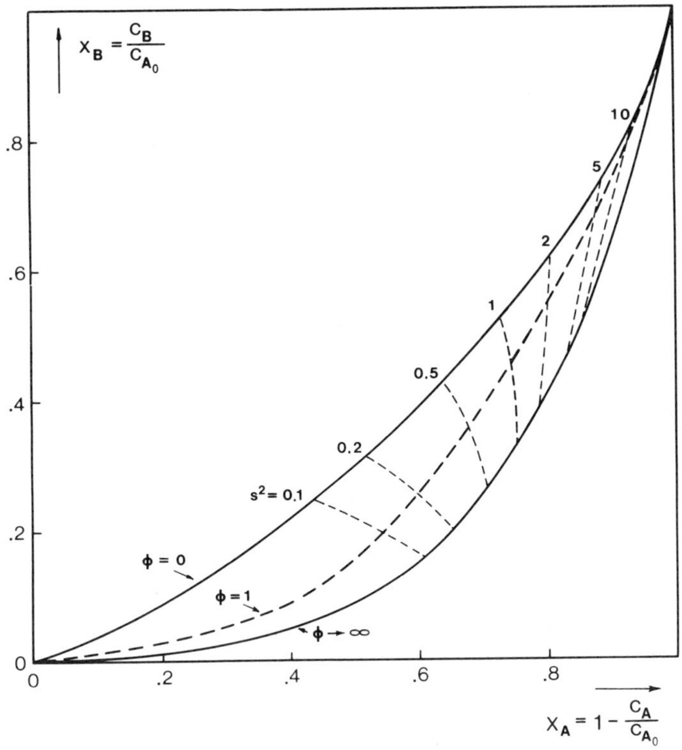

FIGURE 3 Maximum yield of B as a function of $s^2 = k_1/c_{A0}k_2$ and of $\phi = L_p(k_1/D)^{\frac{1}{2}}$. Curves for ϕ = 0, 1, and ∞ are shown, and points with s^2 = 0, 1, . . . , 10 are connected. Compare with Fig. 7.7 of Levenspiel (1972) for consecutive first-order reactions, $\phi = 0$.

The transformation (109) is well chosen in the present example since from Eq. (107) it is easily seen that $y \sim u^{-2}$ for large u.

Figure 3 shows the boundary curves $\phi_1 = 0$ and $\phi_1 \to \infty$ for the optimum (x_A, x_B) as a function of $1/\alpha^2$ [which is called s^2 in the classical paper of Wheeler (1951) on first-order consecutive reactions]. A dashed line indicates the locus of the $\phi_1 = 1$ optimum solutions, and at seven different values of $1/\alpha^2$ (10, 5, 2, 1, 0.5, 0.2, and 0.1) other dashed lines connect solutions with the same value of $1/\alpha^2$. The solution was found by the technique (103) to (106), using Eq. (97) to obtain the collocation equations for $\phi_1 < 20$ and Eq. (107) when $\phi_1 \to \infty$. The asymptotic solution $\phi_1 \to \infty$ is a good approximation when $\phi_1 > 10$.

Even though in the present example a collocation solution could be found for all ϕ_1 values, it is worthwhile to remember than an accurate solution of the boundary value problem can also be obtained by forward integration from $\zeta = 0$ or from $\zeta = 1$. There are positive eigenvalues in the Jacobian of f, and the solution may easily diverge, but an explicit Runge-Kutta method is just as good as an implicit method.

To find the solution that satisfies $y(1) = x_B/(1 - x_A)$, one proceeds as in Eqs. (48) to (50), integrating Eq. (97) together with the sensitivity equation for $z_{y_o} = \partial y/\partial y_o$:

$$\frac{d^2 z_{y_0}}{d\zeta^2} = 2\phi_1^2 \alpha^2 (1 - x_A) y z_{y_0} \tag{110}$$

with

$$z_{y_0}(\zeta=0) = 1 \quad \text{and} \quad \left.\frac{dz_{y_0}}{d\zeta}\right|_0 = 0$$

Having obtained the correct value of y_0 for a given (x_A, x_B), a single integration of Eqs. (97) and (111) from $\zeta = 0$ to $\zeta = 1$ gives the sensitivity of y with respect to x_A:

$$\frac{d^2 z_{x_A}}{d\zeta^2} = \alpha^2 \phi_1^2 [-y^2 + 2(1 - x_A) y z_{x_A}] \qquad z_{x_A}(0) = \left.\frac{dz_{x_A}}{d\zeta}\right|_0 = 0 \tag{111}$$

From the computed values of z_{x_A} ($\zeta = 1$) and $dz_{x_A}/d\zeta|_{\zeta=1}$, one may obtain the second derivative of x_B with respect to x_A. First, calculate dy_0/dx_A, using the following equivalent expressions for dy(1):

$$dy(1) = \left.\frac{\partial y}{\partial x_A}\right|_1 dx_A + \left.\frac{\partial y}{\partial x_B}\right|_1 dx_B = \left[\frac{x_B}{(1 - x_A)^2} + \frac{1}{1 - x_A}\frac{dx_B}{dx_A}\right] dx_A \tag{112a}$$

or

$$dy(1) = \left[z_{x_A}(1) + z_{y_0}(1)\frac{dy_0}{dx_A}\right] dx_A \tag{112b}$$

from which Eq. (113) is obtained:

$$\frac{dy_0}{dx_A} = \left[-z_{x_A}(1) + \frac{y(1) + dx_B/dx_A}{1 - x_A}\right] z_{y_0(1)}^{-1} \tag{113}$$

Next,

$$d\left(\left.\frac{dy}{d\zeta}\right|_{\zeta=1}\right) = \left(\left.\frac{dz_{x_A}}{d\zeta}\right|_{\zeta=1}\right) + \left(\left.\frac{dz_{y_0}}{d\zeta}\right|_{\zeta=1}\frac{dy_0}{dx_A}\right) dx_A \tag{114}$$

and finally,

$$x_B = x_B(x_A°) + f_1(x_A°)\, dx_A + \frac{1}{2} f_1^{(1)}(x_A°)\, dx_A^2 \tag{115}$$

where $f_1(x_A) = -(\phi_1 \tanh \phi_1)^{-1} \, dy/d\zeta|_{\zeta=1}$, and $f_1^{(1)}(x_1)$ is found from (114).

As a comment on the procedure of Eqs. (97) and (110) to (115), one might add that for explicit integration of the differential equation (97), it is computationally simpler to find y_0 by the secant method. Also, for an explicit integration it may be preferable to calculate z_{x_A} by numerical perturbation of Eq. (97) rather than by Eq. (111). The computational cost is the same and the computer program is simpler.

Very difficult catalyst pellet problems have recently been solved by ingenious forward integration techniques: Kaza et al. (1980; methanation on nonisothermal pellets), Sundaresan and Amundson (1980; diffusion and reaction in a boundary layer surrounding a carbon particle), and Holk and Villadsen (1983; absorption followed by exothermic reaction in a liquid film). Other examples are given in Villadsen and Michelsen (1978, Chaps. 5 and 9). We may conclude that the model for a single reaction on a catalyst pellet can be treated by a standard numerical approach: finite-difference methods, collocation, or forward integration from the center of the particle, as the case may be. Examples with several independent reactions will frequently require a numerical technique which is tuned to the specific problem. Example 5 has shown some of the methods that can be used to construct an efficient numerical solution, and the references given above illustrate other techniques.

As a final comment it should be acknowledged that the catalyst pellet model is virtually the same as the model used to calculate absorption with chemical reaction. In particular, the catalyst effectiveness factor is closely related to the enhancement factor which is used in the film model for chemical absorption.

The design of multicomponent countercurrent gas-liquid absorbers [or moving-bed reactors, as in Peytz et al. (1982)] presents some nasty numerical problems—far more difficult than those encountered in the boundary value problems associated with catalyst pellets. A set of exit concentrations in the gas phase must be guessed, and after integration of the gas-phase and liquid-phase mass balances through the column, one makes a comparison with the inlet gas composition to obtain an iterative calculation procedure. The differential equations are often very stiff, and it requires careful programming to achieve a stable iteration.

Steady-State Tubular or Fixed-Bed Reactor Models

Detailed discussion of reactor models appears in other chapters of this book; our task is to bring up some of the common techniques which are applicable in a numerical study of the models. For this purpose we need only one mass balance (116) and the energy balance (117). Axial diffusion terms can be neglected in the steady-state model since they are small compared to the convective terms.

$$\frac{v_z}{v_{av}} \frac{\partial y}{\partial z} = \frac{LD}{r_t^2 v_{av}} \frac{1}{x} \frac{\partial}{\partial x}\left(x \frac{\partial y}{\partial x}\right) - \frac{L}{v_{av}} R \tag{116}$$

$$\frac{v_z}{v_{av}} \frac{\partial \theta}{\partial z} = \frac{Lk}{r_t^2 v_{ac} \rho c_p} \frac{1}{x} \frac{\partial}{\partial x}\left(x \frac{\partial \theta}{\partial x}\right) + \frac{L(-\Delta H)c_r}{v_{av} \rho c_p T_r} R \tag{117}$$

where

z = axial distance in reactor relative to the total reactor length L

x = radial coordinate relative to the tube radius r_t

y, θ = fluid-phase reactant concentration and temperature relative to a reference state with concentration c_r and temperature T_r

R = reaction rate divided by the reference concentration c_r

v_z = radial velocity distribution in the reactor tube

$\frac{k}{\rho c_p}, D$ = radial diffusivity of heat and mass

In a "homogeneous" reactor model R is a function of fluid-phase properties only; the tubular reactor with or without an inert solid packing is described by this type of model. The catalytic fixed-bed reactor is described by a "heterogeneous" model in which R is a function of particle-phase temperature and concentration, but these extra variables do not appear explicitly, and in the numerical treatment of a catalytic fixed-bed reactor model, the pellet problem is treated separately from the fluid-phase model [Eqs. (116) and (117)]—an ideal case for solution by partitioning. The influence of pellet-phase variables on the total model appears only in terms of the effectiveness factor η, and R is an implicit function of the fluid-phase variables y and θ. Calculation of η may be more or less complicated, ranging from the solution of simple algebraic equations for a surface reaction on impervious pellets to the solution of coupled nonlinear boundary value problems, as in examples 4 and 5 earlier in this section. In all circumstances one should try to reduce the complexity of the pellet model as far as possible (Example 4 showed us how far one can get in this respect without sacrificing any important feature of the pellet model). The fluid-phase model is itself quite complicated, and the influence of reactor macrovariables (e.g., the large temperature gradients which occur radially in the tubes of a reforming reactor) can be studied with sufficient accuracy without too many details in the pellet-phase description.

If the rate of reaction is moderate, the radial diffusion terms are small compared to the axial gradients, and one may reduce Eqs. (116) and (117) to a one-dimensional model. There are certainly situations where this approximation may lead to loss of major features of the model, but the potential reduction in computer expenditure is so significant that we feel it necessary to comment on proper averaging techniques for the radial gradients before we discuss suitable numerical techniques for solution of the full model.

Averaging of the Steady-State Model over the Cross Section of the Tube

Assume that the physical properties $k/\rho c_p$ and D are independent of x and integrate over the cross section of the tube to obtain two coupled ordinary differential equations (118) and (119) for the averages of y and θ at reactor position z:

$$\frac{d\bar{y}}{dz} = -\frac{L}{v_{av}}\bar{R} \tag{118}$$

where $\bar{R} = \int_0^1 R\, dx^2$, which is approximated by $R(\bar{y},\bar{\theta})$:

$$\frac{d\bar{\theta}}{dz} = \frac{2LU}{r_t v_{av} \rho c_p}(\theta_w - \theta_{x=1}) + \frac{L(-\Delta H)c_r}{v_{av}\rho c_p T_r}\bar{R} \tag{119}$$

where θ_w is the value of θ at the reactor wall and $\theta_{x=1}$ is the value of θ just inside the wall. In the same manner in which we introduced a modified pellet heat transfer coefficient in Eq. (91) we shall define a modified wall heat transfer coefficient $\bar{U}$:

$$\frac{k}{r_t}\left.\frac{\partial \theta}{\partial x}\right|_{x=1} = U(\theta_w - \theta_{x=1}) \equiv \bar{U}(\theta_w - \bar{\theta}) \tag{120}$$

Expression (120) is inserted in Eq. (119) and now only $\bar{\theta}$ and $\bar{y}$ appear in the approximate reactor model, which can be solved by integration from z = 0 to z = 1.

It remains to calculate $\bar{U}$ in terms of U and the radial heat conductivity k. In general, $\bar{U}$ is obtained from

$$\frac{1}{\bar{U}} = \frac{1}{U} + \alpha\frac{r_t}{k} \tag{121}$$

where the constant α depends on the radial velocity distribution. Villadsen and Michelsen (1978, Chap. 6) discuss a perturbation technique by which α may be calculated. The value of α will fall between 1/4 for a flat velocity profile $v_z = v_{av}$ and 11/24 = 0.4583 for the parabolic velocity profile of the laminar flow tubular reactor. Typical results between these values are

$$\alpha = \frac{19}{48} = 0.3958 \qquad \text{for } \frac{v_z}{v_{av}} = \frac{3}{2}(1 - x^4)$$

an almost flat profile, and

$$\alpha = \frac{495b^2 + 234b + 31}{120(3b + 1)^2} \qquad \text{for } \frac{v_z}{v_{av}} = \frac{6(1 - x^2)(x^2 + b)}{3b + 1}$$

For b = 1/4 this profile has a minimum at x = 0 and a maximum at x ∿ 0.7, a feature that is observed in many experimental studies on packed-bed velocity distributions. α = 0.3277 for b = 1/4.

The result (118) obtained by straightforward averaging of the mass balance can be improved by including one more term in perturbation analysis from R = 0. For a first-order, isothermal reaction ($R = k_R\, c/c_r$)

in a tubular reactor with parabolic velocity distribution, one obtains, instead of Eq. (118),

$$\frac{d\bar{y}}{dz} = -\, Da\left(1 - \frac{1}{48}\, Da\, \frac{r_t^2 v_{av}}{DL}\right)\bar{y} \tag{122}$$

where $Da = k_R L/v_{av}$ and the numerical constant 1/48 is calculated by perturbation analysis as discussed in Villadsen and Michelsen (1978, p. 271). This result was first given in a famous paper by Sir Geoffrey Taylor (1953). Qualitatively speaking, he interpreted the radial diffusion term in Eq. (116) where D is now a true molecular diffusivity—in terms of fictitious "axial dispersion term" in a one-dimensional model:

$$\frac{d\bar{y}}{dz} = -\frac{k_r L}{v_{av}}\bar{y} + \frac{1}{Pe_{ef}}\frac{d^2\bar{y}}{dz^2} \tag{123}$$

where

$$Pe_{ef} = 48\,\frac{DL}{v_{av} r_t^2} = 192\,\frac{DL}{v_{av} d_t^2}$$

Equation (123) may be solved using the simple "semi-infinite" boundary condition (e.g., $\bar{y} = 1$ at $z = 0$ and finite for $z \to \infty$) or with the more complicated boundary condition

$$\bar{y}(0) + \frac{1}{Pe_{ef}}\left.\frac{d\bar{y}}{dz}\right|_{z=0} = 1 \quad \text{and} \quad \left.\frac{d\bar{y}}{dz}\right|_{z=1} = 0$$

In both cases, one obtains

$$\bar{y}(z = 1) = \exp\left[-Da\left(1 - \frac{Da}{Pe_{ef}}\right)\right] + O\left(\frac{Da}{Pe_{ef}}\right)^2 \tag{124}$$

which is the same result as that obtained by perturbation analysis (122).

The final result (124) is well known from elementary textbooks in reaction engineering [e.g., Levenspiel (1972, pp. 283–287)], where a "dispersion number"

$$\frac{D_{ef}}{v_{av} L} = \frac{1}{192}\,\frac{v_{av} d_t^2}{DL} = \frac{1}{Pe_{ef}}$$

is used to correct "near plug flow" data obtained in a tubular reactor. The improvement is substantial, at least for small values of the dispersion number. Thus at Da = 1, where Eq. (118) yields the result $\bar{y}(z = 1) = \exp(-1) = 0.3679$ regardless of the contribution from the radial dispersion, one obtains the following results from Eq. (124):

$192 \frac{D_{ef}}{v_{av}L}$	1	2	5	10	∞
$\bar{y}$ by Eq. (124)	0.3752	0.3818	0.3892	0.4179	—
$\bar{y}$ by a "true model"	0.3750	0.3809	0.3919	0.4038	(0.443)

The bottom line of the table is calculated by high-order collocation—as described below—applied to Eqs. (116) for a first-order isothermal reaction

$$2(1-x)^2 \frac{\partial y}{\partial z} = \frac{v_{av}L}{192D_{ef}} \frac{1}{x} \frac{\partial}{\partial x}\left(x \frac{\partial y}{\partial x}\right) - \text{Da } y \tag{125}$$

$$\bar{y}(z=1) = \int_0^1 4(1-x^2)y(z=1,x)x\,dx$$

The case $D_{ef} \to \infty$ (which may be interpreted as $D \sim 0$) corresponds to a completely segregated flow with no cross-sectional mixing caused by radial gradients.

Numerical Solution of Steady-State Reactor Model

When Eqs. (116) and (117) are discretized in the x direction, there appears a set of coupled ordinary differential equations (j = 1, 2, . . ., N):

$$\frac{dy_j}{dz} = \alpha_M \underline{C}_j^T \underline{y} - \frac{L}{v_{av}} R(y_j, \theta_j) \tag{126}$$

$$\frac{d\theta_j}{dz} = \alpha_H \underline{C}_j^T \underline{\theta} + \frac{L}{v_{av}} \frac{(\Delta T)_{max}}{T_r} R(y_j, \theta_j) \tag{127}$$

The coefficients C_{ji} are defined in Eq. (70). They have different values depending on the method of discretization. Ordinary finite-difference methods, global collocation, Galerkin's method, or spline collocation have been used by various authors in numerous computer studies over the last 20 to 25 years.

Equations (126) and (127) can be solved from z = 0 by any of the standard packages for coupled initial value problems. An implicit or a semi-implicit method is preferable to an explicit method because of the large spread of the eigenvalues of $\underline{\underline{C}}$. This is true in particular when orthogonal collocation is used to discretize the radial derivative; even for a relatively low order method (N = 4 or 5) the ratio between the largest and the smallest eigenvalue can be of the order of 1000. The structure on the right-hand sides of Eqs. (126) and (127) is, however, so simple that the Jacobian can be computed very easily.

It is also possible to discretize Eqs. (126) and (127) in the z direction to obtain a so-called double-collocation method. This method was originally proposed by Villadsen and Sørensen (1969), but it was only recently elaborated into efficient computer codes by Sørensen (1982). These codes appear to be particularly suitable when applied in a parameter estimation problem.

A situation that lends itself quite naturally to solution by double collocation is that of a wall-catalyzed chemical reaction: the rate term appears in the boundary condition of an otherwise linear partial differential equation [Eqs. (116) and (117) without the rate terms]. Michelsen and Villadsen (1981) give a detailed discussion of this particular type of chemical reactor model, which may be solved either by double collocation or by the STIFFALG routine [Eqs. (36) and (37)]; there is one nonlinear algebraic equation (the boundary condition) and N linear differential equations which can be transformed into the diagonal form (18) and used in this form throughout the calculation.

The solution of Eqs. (116) and (117) for a rate expression which is linear in y and θ (or has been linearized from a given reference state) deserves particular attention. Crank (1957) has discussed the multitude of practical problems in reaction engineering which are generated from the same basic equation, the linear "diffusion equation." Numerically, all these different problems are handled by the same general technique, but with slight modifications for different boundary conditions.

Let us use a linear mass balance as our "standard problem":

$$v(u)\,\frac{\partial y}{\partial z} = 4u^{(1-s)/2}\,\frac{\partial}{\partial u}\left(u^{(s+1)/2}\,\frac{\partial y}{\partial u}\right) - \mathrm{Da}\;y \tag{128}$$

where $u = x^2$ and s = 0, 1, and 2. When s = 1 and v(u) is the radial velocity distribution, we have a steady-state tubular reactor problem. When v(u) = 1 and z is interpreted as contact time, we have a linear transient pellet problem. Here we use $\phi^2 = k_R r_p{}^2/D$ instead of $\mathrm{Da} = k_R L/v_{av}$ as the dimensionless group of parameters.

Equation (128) is discretized by Nth-order collocation:

$$\underline{\underline{V}}\,\frac{d\underline{y}}{dz} = (\underline{\underline{C}}^* - \mathrm{Da}\underline{\underline{I}})\underline{y} + \underline{b} \tag{129}$$

$\underline{\underline{V}}$ is diagonal with $V_{jj} = v(u_j)$, and the (C^*_{ji}, b_j) are given in Eq. (71). Different side conditions at $u = u_{N+1} = 1$ lead to different problem modifications.

Diagonalization of $\underline{\underline{M}} = \underline{\underline{V}}^{-1}(\underline{\underline{C}}^* - \mathrm{Da}\underline{\underline{I}})$ yields the standard form (18). Michelsen and Villadsen (1981) prove that all eigenvalues of $\underline{\underline{M}}$ are real and distinct if collocation is made at the zeros of $P_N^{(0,0)}(u)$, and the boundary condition at u = 1 is any of the linear expressions discussed in Crank (1957).

The solution appears as an N-term expansion in eigenfunctions $\exp(\lambda_j z)$, and the modified Fourier coefficients are found by simple matrix-vector manipulations [see Villadsen and Michelsen (1978, Chap. 4) for the solution of a particular example].

For the case Da = 0, Michelsen (1979) proves that the eigenvalues and eigenfunctions of Eq. (128) with boundary condition

$$\frac{dy}{du} + \frac{\mathrm{Bi}_M}{2}\,y = \text{constant at } u = 1 \tag{130}$$

can be derived for any value of Bi_M by a simple algorithm that utilizes the eigenvalues and eigenfunctions obtained at any specific reference value $(Bi_M)_r$, which, for example, may be either zero or infinite. This leads to substantial savings in computer time when estimating the value of Bi_M (and a diffusivity) from a series of measurements of y at different z and x. When Da ≠ 0, the diagonalization of $\underline{\underline{M}}$ must, however, be performed for each value of Da.

Another important special case, that of a finite exterior medium, is treated in Sotirchos and Villadsen (1981). The boundary condition at u = 1 is given by an overall mass balance

$$q\ y_b(z) + \int_0^1 y(x,z)\ dx^{s+1} = \alpha - \int_0^z \left(\int_0^1 R\ dx^{s+1} \right) dz' = \alpha - f \tag{131}$$

Outside the reaction medium of volume 1 (or the catalyst pellets as the case may be) there is a radially well mixed volume q into which the reactant may diffuse. The reactant concentration is y_b in the finite exterior medium at position z (or at contact time t). In Eq. (131) the total "mass" in the reactor and the outer medium is equated to the initial "mass" α, less the total amount of reacted material. Concentrations in the two media are connected by a boundary condition of type (130) where the right-hand-side "constant" = $(Bi_M/2)y_b(z)$.

Let v(u) = 1 and write the jth collocation equation,

$$\frac{dy_j}{dz} = \sum_{i=1}^{N} C_{ji}y_i - \frac{b_j}{q} f - R(y_j) + b_j \frac{\alpha}{q} \tag{132}$$

where f is given in Eq. (131).

$$C_{ji} = 4\left(u_jB_{ji} + \frac{s+1}{2} A_{ji}\right) - b_j\left(\frac{w_i}{q} + \frac{2A_{N+1,i}}{Bi_M}\right)$$

$$b_j = 4\left(u_jB_{j,N+1} + \frac{s+1}{2} A_{j,N+1}\right) \frac{Bi_M}{2A_{N+1,N+1} + Bi_M}$$

$$\int_0^1 y\ dx^{s+1} \cong \sum_1^N w_iy_i$$

For a nonlinear reaction rate the N equations (132) are solved together with an equation for the average rate of reaction:

$$\frac{df}{dz} = \int_0^1 R\ dx^{s+1} \cong \sum_1^N w_iR(y_i) \tag{133}$$

but for a first-order reaction the numerical approach is just like that used in Eq. (129):

$$\frac{d\underline{Y}}{dz} = \underline{\underline{M}}\underline{Y} + \underline{b}^*$$

$$\underline{Y} = (y_1, y_2, \ldots, y_N, f)^T \qquad \underline{\underline{M}} = \begin{pmatrix} \underline{\underline{C}} - Da\underline{\underline{I}} & -\frac{b}{q} \\ Da\underline{w}^T & 0 \end{pmatrix} \tag{134}$$

$$\underline{b}^* = \frac{\alpha}{q}\{b_1, b_2, \ldots, b_N, 0\}^T$$

In Eqs. (132) to (134) the unfortunate misprints in Sotirchos and Villadsen (1981) have been corrected. A number of important applications of Eq. (132), (133), or (134) can be easily listed:

1. Absorption of a reactant gas from a gas phase and reaction in the liquid phase [gas flow rate $v_G(m^3/h)$, liquid flow rate $v_L(m^3/h)$, and $q = v_L/v_G$].
2. Determination of reaction rate constant and pellet diffusivity by concentration measurements in a finite volume of well stirred liquid in which the catalyst particles are suspended. All reactant is originally in the fluid outside the particles.
3. Analysis of membrane reactors: a certain amount of reactant $v(m^3/h)$ flows in a tube coated with a catalyst layer of thickness δ. Reactant diffuses into the catalyst layer and reacts there. It is desired to calculate the length of tube required to obtain a certain average concentration in the flowing liquid.

As a final illustration of the general numerical approach to linear steady-state reactor models, one may consider Eqs. (128) and (130) with the addition of a small axial diffusion term $(1/Pe_M)(\partial^2 y/\partial z^2)$ on the right-hand side of Eq. (128). The problem has only marginal interest in industrial reactor design (radial diffusion may be accounted for by a fictitious axial dispersion term as discussed earlier, but an axial diffusion term per se is almost always insignificant). Addition of the second derivative with respect to z in the partial differential equation (128) does, however, lead to formidable numerical complications which have intrigued many authors [see, e.g., Papoutsakis et al. (1980) or Michelsen and Villadsen (1981)].

It is interesting that with the exception of the inlet zone of the reactor, one may study the influence of an axial diffusion term $(1/Pe_M)(\partial^2 y/\partial z^2)$ on the solution of Eq. (128), using only the eigenfunctions and eigenvalues of the radial diffusion operator. The technique and its application to the "extended Graetz problem" discussed here and to the much more interesting asymptotic stability analysis of catalyst pellet models are developed in Villadsen and Michelsen (1978, Chap. 9).

Every one of the numerical methods that have been described in this section have been based on an approximation of the x profile by a polynomial. High-order methods such as collocation will not be able to handle the near discontinuities that appear in the profiles close to the reactor entrance (the inlet concentration may be 1 at all interior points and zero at u = 1 for z = 0+). Spline collocation with a spline point that gradually moves away from u = 1 as z increases has been successful in the computation of "penetration fronts" in the inlet zone of the reactor. One example is given in Holk-Nielsen and Villadsen (1983), and other examples are shown in Villadsen and Michelsen (1978, Chap. 7).

Another, quite different but apparently versatile approach is to introduce a variable transformation $\eta = f(z,x)$ which as far as possible describes the combined influence of z and x on the solution near z = 0. This is called "similarity transformation" and is treated in most textbooks on partial differential equations. The reduction of the model into a problem with only one independent variable succeeds only in trivial cases, and one will usually end up with two independent variables η and z in the transformed equation as well as in the boundary conditions. If, however, η has been judiciously chosen, a perturbation analysis of the transformed equation from z = 0 will give an accurate solution of the original problem in the inlet zone. There are several examples in Villadsen and Michelsen (1978, Chap. 4) where this technique has been used to solve linear reactor problems, and a pellet model with a concentration dependent diffusivity is treated in the same reference p. 339).

Unsteady-State Fixed-Bed Reactor Models

Except for regions of hot spots, the steady-state axial temperature and concentration profiles in a fixed bed are smooth functions of z, and the radial profiles can almost always be represented by low-order polynomials—if they are not averaged away as discussed in the preceeding section. Thus steady-state reactor simulation is a rather modest affair on a high-speed computer, requiring at most a couple of seconds computing time to obtain $\theta(x,z)$ and $y(x,z)$.

When a time derivative is included in Eqs. (116) and (117) there is a fundamental change in the nature of the solution. Sharp concentration and temperature fronts may be formed, and these fronts move slowly through the reactor. To simulate the dynamic response of the reactor to a control action in the inlet (e.g., a jump in reactant concentration), one has to follow the reactor profiles of the dependent variables through many thousands of fluid-phase time constants. This, of course, means that the amount of computation increases drastically, and 25 to 40 sec of computer time per simulation is not at all extravagant. Quite apart from the cost of the computation, the time it takes to make a computer simulation of the unsteady-state reactor makes it very difficult to use the result, for instance, in a computer control of the reactor.

A compromise must be made between the demand for accurate modeling of the reactor and a reasonable computational effort. The literature on unsteady-state reactor simulation offers a bewildering array of model simplifications and numerical shortcuts. Some of the simplifications are justified and should certainly be generally accepted, but others are downright silly and will lead to order-of-magnitude errors in, for example, the breakthrough time of a catalyst poison.

Thus for gaseous reactants it is reasonable to neglect accumulation terms in the reactor fluid-phase mass and energy balances and in the pellet mass balances. The time constant for convective transport of mass through the reactor and for diffusion into the pellet are orders-of-magnitude smaller than the velocity of the reaction zone. The pellet temperature is taken to be independent of position in the pellet (as in Example 4), and while the pellet time constant for energy transport is certainly larger than the time constant for mass transport, it is still much smaller than the thermal residence time for the bed, an observation that may justify a further assumption that $\theta_p = \theta_g$ at any time. To reduce the dimensionality of the problem, it is common practice to neglect the cross-sectional conductive term in Eq. (117) when solving unsteady-state reactor models. This may be justifiable when the diameter-to-length ratio of the bed is larger than 20, but a radial gradient can also be incorporated as a fictitious axial dispersion term and there is experimental evidence that true axial conductive terms may have to be taken into account in modeling of transient fixed-bed reactor behavior.

It is usually dangerous to tamper with the rate expression to obtain computational advantages. If it is reasonable to suspect a substantial resistance to mass transfer in the pellets, it will not do to neglect this dispersing effect, since breakthrough times that are much different from those observed experimentally are likely to be calculated. Our discussion of the pellet problem for a single reaction has also shown that a complicated rate expression is treated with almost the same effort as a simpler rate expression. Therefore, it is usually not advisable to jeopardize the trustworthiness of the simulation by neglecting the influence of one or more reactants on the rate of reaction.

Unsteady-state reactor models display such a variation in complexity that it becomes impossible to discuss suitable solution techniques in general. At one end of the spectrum one finds the one-component isothermal gas adsorption unit which can be treated analytically (Aris and Amundson, 1973). At the other end of the spectrum there are studies of reactor control with several independent variables, steep temperature fronts, and with time-changing kinetic and transport parameters. Simulation of reactors for reduction of mineral ore and simulation of explosion fronts are other complex problems.

An example that illustrates major features of unsteady-state reactor behavior without being very complicated is discussed in Michelsen et al. (1973) and Villadsen and Michelsen (1978, Chap. 9): The reaction takes place in the fluid phase, but the inert packing material of the reactor introduces a large thermal residence time. After linearization the model is analyzed in terms of transfer functions, using collocation to account for the axial variation of the variables y and θ.

In the present chapter we use an example of moderate complexity to illustrate certain numerical techniques which we believe can also be applied in many other situations.

Example 6: Simulation of coke burning in a fixed-bed reactor A reforming catalyst is slowly deactivated due to deposition of tarry material in the pores. This so-called "coke" has to be burned off periodically, and a reliable simulation is important to minimize the length of the burn-off period without sintering the metal crystallites by excessive overheating of the pellets. Originally, the coke is homogeneously distributed on the particles with concentration c_{co}. Nitrogen containing a small proportion of oxygen (concentration c_{oi}) is fed to the reactor inlet at temperature T_{gi}.

A reaction zone develops and moves slowly through the reactor. When the reaction zone passes out of the reactor the burn is complete, and it is successful if only a small residual coke concentration $c_c(z)$ is left behind and if the temperature at no time has exceeded T_{max} anywhere in the reactor. During the operation of the reactor a temperature wave passes through the bed. It may move faster or slower than the reaction zone—the inlet oxygen concentration c_{oi} can be used to control the relative movement of the two fronts. Both fronts move exceedingly slowly compared to the gas residence time in the reactor. Typical velocities are 0.1 m/h for the reaction zone and 0.9 m/h for the temperature front.

A model with three equations [(135) to (137)] for the pellet phase and two [(138) and (139)] for the fluid phase was proposed by Liaw et al. (1982).

$$t_R \frac{\partial y_c}{\partial t} = -y_c y_p y_g \exp\left[\gamma\left(1 - \frac{1}{\theta_p}\right)\right] \qquad y_c = \frac{c_c}{c_{co}} = 1 \text{ at } t = 0 \tag{135}$$

$$\frac{1}{x^2}\frac{\partial}{\partial x}\left(x^2 \frac{\partial y_p}{\partial x}\right) - \alpha_c \phi^2 y_c y_p \exp\left[\gamma\left(1 - \frac{1}{\theta_p}\right)\right] = 0 \tag{136}$$

$$y_p = \frac{c_{op}}{c_{og}}$$

$$\frac{\partial y_p}{\partial x} = Bi_M[1 - y_p(1)] \text{ at } x = 1$$

$$t_H \frac{\partial \theta_p}{\partial x} = H_p(\theta_g - \theta_p) + \beta \, Da \, \eta y_g \qquad \theta_p = \frac{T_p}{T_{gi}} \qquad \theta_g = \frac{T_g}{T_{gi}} \tag{137}$$

$$\frac{\partial y_g}{\partial z} = -Da \, \eta y_g \qquad y_g = \frac{c_{og}}{c_{oi}} \tag{138}$$

$$\frac{\partial \theta_g}{\partial z} = H_p(\theta_p - \theta_g) \tag{139}$$

The many constants in Eqs. (135) to (139) (t_R = characteristic reaction time, t_H = thermal residence time, H_p = fluid-to-pellet heat transfer units, α_c = moles of O_2 consumed per mole coke burned) are all defined in Liaw et al. (1982).

There are three reference concentrations: c_{co} is used to scale the coke concentration c_c, and c_{oi} is used to scale the fluid-phase oxygen concentration c_{og} at position z and time t. Finally, in the local oxygen balance (136) c_{og} is used as a reference concentration. It had been easier to read the model if c_{oi} had also been used to scale the pellet oxygen concentration c_{op}, but we have used the same scaling procedure as in Liaw et al. (1982) to facilitate cross-reference to the published paper.

The inlet gas temperature T_{gi} is used to scale T_g as well as T_p. There is no temperature variation in the pellet. The previously mentioned model simplifications (no energy or mass accumulation in fluid phase, no mass accumulation in the pellets and no radial or axial conductive terms) have all been used, and the model is still quite formidable.

There is a large oxygen gradient in those pellets which are close to the reaction zone, and a much too sharp burn front will result if the burning rate is expressed as $kc_c c_{og}$ rather than $kc_c c_{op}$. This is one of the less justified assumptions that have been made in previous studies of the problem. The present model does not assume "a constant burning rate," low-temperature burning" (where $c_p \approx c_g$), or "diffusion-controlled burning" (where the rate is almost independent of temperature), and consequently the predicted burnout history is close to that obtained in laboratory experiments.

The model has recently been critically reviewed from a numerical point of view by Funder (1982), and as a result the computation time per burnout simulation was reduced from about 5 min in Liaw et al. (1982) to about 35 to 40 s. We shall review some of his conclusions.

1. It is meaningless to simplify the model by the assumption $\theta_g = \theta_s = \theta$. This is frequently done in numerical studies of unsteady-state reactor models, and the resulting "homogeneous" model with

$$\frac{\partial \theta}{\partial t} + \frac{1}{t_H}\frac{\partial \theta}{\partial z} = \frac{1}{t_H}\beta \mathrm{Da}\eta y_g \tag{140}$$

instead of Eqs. (137) and (139) is amenable to solution by the method of characteristics using increments related by

$$dt = t_H\, dz = \frac{d\theta}{\beta \mathrm{Da}\eta y_g} \tag{141}$$

There may be about 10% saving in computer time when this approximation is made, but significant errors in the temperature peak may also occur.

2. The implicit three-point integration scheme suggested by Christiansen and Andersen (1980) is very suitable for integration of the fluid balances in a rectangular grid. The pellet mass balance is approximated by collocation [using $P_N^{(0,0)}(x^2)$ with N = 3 or 4], and the resulting equations for y_g, θ_g, θ_p, and $\{(y_p, y_c);\ j = 1, 2, \ldots, N\}$ are solved by a Newton method for each grid point. All z values at a given t are treated before t is increased.

3. Funder (1982) points to an interesting model approximation: whereas it is incorrect to assume a constant oxygen level in the pellets (except when the pellet is far from the reaction zone), it is quite reasonable to assume a constant coke level or a coke profile that changes in a few steps. When y_c is assumed to be piecewise constant in the pellet, it is possible to solve the pellet model analytically, and since η is the only "output" of the pellet mass balance, the problem is reduced from one in 2N + 3 variables to (ideally) one in four or five variables. The resulting reduction in computer time is very significant (from 35 to 4 to 6 s per burnout simulation) and the results are only a few percentage points different from those obtained by the full model, probably because the sharply decreasing oxygen profile in the pellet is very little influenced by the coke profile as long as the correct average coke concentration is used.

Characteristically, the solution of the unsteady-state reactor model is smooth except in the reaction zone and in the temperature front. In Example 6 the oxygen concentration is zero downstream from the reaction zone, and it increases rapidly to the inlet value c_{oi} at positions upstream from the reaction zone.

Global collocation does not give a satisfactory representation of the profile when two, more or less flat parts are joined by a steep front.

Spline collocation is a possibility, but a suitable algorithm must be constructed to move the zone with many collocation points at the same speed as the reaction zone. With several reacting species and an energy balance, the fronts will move with different speeds, and unless the program automatically follows each of the fronts, spline collocation will also be impractical.

An ideal code should:

1. Ensure that each front is followed by a sufficient number of nodal points.
2. Move nodal points in and out of the fronts—not too fast (since otherwise all nodes would be sucked into the fronts) and not too slow (the fronts will be inaccurately represented)

When the profiles are given in terms of $y_j(t)$ at positions $z_j(t)$ [e.g., as a set of piecewise linear functions between (y_j, z_j) and (y_{j+1}, z_{j+1})], one must require that (y_j, z_j) quickly settle down to become smooth functions of t. This certainly requires that condition 1 is satisified. If a number of nodes pass quickly through the front, the profiles (y_j, z_j) will oscillate wildly with t.

Since the equations that determine the (y_j, z_j) are surely going to be "stiff," an implicit or semi-implicit integration method in t must be used.

The most promising recent development toward a general algorithm for solving problems with steep fronts (or schocks) is due to Miller and coworkers (Miller and Miller, 1981). Their method employs a complicated system of "spring forces" and "viscosity forces" to move the nodes—quickly but not too quickly—in and out of the fronts. Computing times (with an implicit Gear method to solve the differential equations for the time development of the profiles) appear to be remarkably small. We have used the information given in the published articles to test their method, both on Burgers' equation (their test example) and on a chromatographic separation example. Our computation times are, however, three to four times as high as stated by Miller, but undoubtedly the method deserves further study, since the basic ideas seem to be just right for solution of some extremely complicated problems of great practical interest.

PARAMETER FITTING IN REACTION ENGINEERING MODELS

Up to this point we have presented various numerical techniques which are suitable for reactor simulation (i.e., computation of reactant concentration, temperature, etc.) as functions of the independent variables in the reactor model.

We shall now apply these techniques to determine a set of parameters in a postulated rate expression, heat transfer model, or reactor model. At n specified values of the (possibly multidimensional) independent variable $\underline{x}$, there are measurements of some or all the components of the M-dimensional state variables in the postulated model. The experimental values are $(y_1, y_2, \ldots, y_M)_k$ for k = 1, 2, . . . , n. They are collected into an experimental state vector $\underline{y}_e$. The experimental values of $\underline{x}$ are supposed to be free of experimental error, while the experimental error of $\underline{y}_e$ is supposed to be normally distributed with a mean of zero (this is possibly not true, but it is often as good as any other guess, and the bias in the estimated $\underline{p}$ that results if the assumption is incorrect will be difficult to assess).

The postulated model contains m parameters $(p_1, p_2, \ldots, p_m) = \underline{p}$, and for a given value of $\underline{p}$ the model can be used to calculate values $\underline{y}_s$ of the state vector at the experimental positions $\underline{x}$.

The parameters will be determined by minimization of an object function $\varphi(\underline{y}_e, \underline{y}_s, (\underline{p}))$ with respect to $\underline{p}$. Weighted least squares, maximum likelihood fitting, or any of the statistical principles advanced in textbooks on statistics may be used.

Our choice of object function will be

$$\varphi = (\underline{y}_s - \underline{y}_e)^T \underline{\underline{\Sigma}}^{-1} (\underline{y}_s - \underline{y}_e) \tag{142}$$

The $(n \times n)$ covariance matrix $\underline{\underline{\Sigma}}$ for the experimental data is assumed to be diagonal (uncorrelated data) and the diagonal elements are $\Sigma_{ii} = V(y_i) = \sigma_i^2$, the variance of the ith experimental point. $\underline{\underline{\Sigma}}$ is assumed to be known except for a scalar σ^2.

Differentiation of φ with respect to $\underline{p}$ yields the following algorithm for improvement of an initial estimate $\underline{p}_0$ for $\underline{p}$:

$$\underline{p} = \underline{p}_0 + \Delta\underline{p}$$

where

$$\left[\left(\frac{\partial \underline{y}_s}{\partial \underline{p}}\right)^T \underline{\underline{\Sigma}}^{-1} \frac{\partial \underline{y}_s}{\partial \underline{p}}\right]_{\underline{p}=\underline{p}_0} \Delta\underline{p} = \underline{\underline{G}}\,\Delta\underline{p} = -\left(\frac{\partial \underline{y}_s}{\partial \underline{p}}\right)^T_{\underline{p}=\underline{p}_0} [\underline{y}_s(\underline{p}_0) - \underline{y}_e] \tag{143}$$

This is the Gauss-Newton algorithm, which can be applied iteratively until $|\Delta\underline{p}| < \varepsilon$ and $\underline{p} = \hat{\underline{p}}$, the maximum likelihood estimate for $\underline{p}$.

The variance-covariance matrix $\underline{\underline{V}}$ is obtained from the estimate $\underline{\underline{G}}(\hat{\underline{p}})$ for the Hessian matrix that results after convergence has been reached in Eq. (143):

$$\underline{\underline{V}} = E[(\hat{\underline{p}} - \underline{p})^T(\hat{\underline{p}} - \underline{p})] = s^2 \underline{\underline{G}}^{-1}(\hat{\underline{p}}) \tag{144}$$

where

$$\sigma^2 \sim \hat{\sigma}^2 = s^2 = \frac{1}{n - m}\,\varphi\,(\hat{\underline{p}}) \tag{145}$$

Gauss-Newton's algorithm is at least linearly convergent, but it may fail to converge even when excellent starting values $\underline{p}_0$ are provided. Most standard codes for parameter estimation with an explicit object function employ one of the several variants of Levenberg-Marquardt's restricted step algorithms. The simplest version is that in which $\underline{\underline{G}} + \nu\underline{\underline{I}}$ is used instead of $\underline{\underline{G}}$ in Eq. (143), but an arbitrary positive-definite matrix may be used instead of the identity matrix $\underline{\underline{I}}$. At the start of the iteration a rather large value is chosen for the Levenberg parameter ν, and ν is gradually reduced when $\Delta\underline{p} \to \underline{0}$. Thus the method starts as a steepest-descent method and it may end as a Gauss-Newton method in the last iterations. It is robust and often works well, although it is not spectacularly fast. A good, recent description of Levenberg-Marquardt methods is Fletcher

(1980, Chaps. 5 and 6). In the following we assume that one of the Levenberg-Marquardt library codes (e.g., IMSL or Harwell) with externally calculated values of $\underline{y}_s$ and of its gradient $\partial \underline{y}_s / \partial \underline{p}$ is available.

The parameters to be determined fall in two groups:

1. Kinetic parameters (rate constants and equilibrium constants)
2. Reactor parameters [residence-time distribution (RTD), cross-sectional velocity—and heat conductivity distribution, wall heat transfer coefficients, pellet-to-fluid transfer parameters, axial dispersion, etc.]

The measurements can be of wildly different complexity. They may include:

1. Direct measurement of the reaction rate at given conversion and space time (recirculation reactor)
2. Measurement of the average conversion and temperature profiles in the axial direction of a fixed-bed reactor, or profiles of conversion versus time in a batch reactor
3. Radial and axial temperature and (possibly) concentration profiles in a fixed-bed reactor at different values of time

The easiest experiments to analyze are those obtained in a recirculation reactor. The experimental procedure is slow compared to batch reactor experiments—only one set of outlet concentration and temperature is found for each steady-state experiment—but in principle it is possible to obtain a complete picture of the rate expression without worrying about interference from the reactor parameters. Experiments over the whole range of conversion should be used to distinguish as far as possible between different kinetic parameters. Thus, for oxidation of CO, the pure first-order behavior is seen for measurements taken at conversion close to 1, while the "negative order kinetics" (i.e., the influence of CO inhibition) is observed for small conversion and large inlet CO concentration.

Intraparticle mass transfer resistance is measured on pellets of different size. Jackson (1977, Chap. 10) reviews the subject of diffusivity measurements by steady-state methods as well as by transient methods. It may be safer to measure the transport parameters under reacting conditions. One or more standard reactions with known (and simple) kinetics may be used, but in many studies it has also been proved that kinetic parameters can be found simultaneously with the pellet mass transfer resistance.

Similarly, the pellet heat conductivity can be measured independently, followed by measurements of the pellet-to-fluid phase transfer properties, radial heat conductivity, wall heat transfer coefficient, and so on, in the packed bed operated as a heat recuperator. Unfortunately, different experimental methods may give very different values of these parameters (see, e.g., Hoffmann et al., 1978), and many authors doubt that the parameters can also be applied for simulation of the reactor under reacting conditions.

A word of warning should finally be given concerning recirculation reactor experiments: when the conversion is very high, the recirculation ratio is likely to be too small, and for measurements at small conversion the amount of catalyst may be too small to give a homogeneous bed—or there may be severe short-circuiting in the reactor itself. Wedel and Villadsen (1983) show some nasty systematic errors that may arise in the interpretation of recirculation reactor data errors which could not be avoided by including RTD measurements in the data analysis.

Still, as far as possible, one should try to proceed as indicated above, using different equipment to extract information regarding different groups of parameters. It is far more difficult, from both an experimental and from a simulation point of view, to determine kinetic and transport parameters simultaneously. The most probable outcome is that the physical parameters (which frequently have a much smaller influence on the conversion than the kinetic parameters) will be determined very poorly. Many authors (see, e.g., Clement and Jørgensen (1983)) have observed that the parameters obtained by the "building-block" approach fail to give satisfactory prediction of the reactor performance under reacting conditions. Hence they feel that it is necessary to determine all (or most) of the parameters simultaneously in order to give a reactor model that can be used for interpolation and for control purposes. One may argue that the failure of the building-block method must indicate that there is some misconception of the kinetic model or of the flow pattern in the reactor, and that these uncertainties had better be resolved. If there are errors in the model, the predicted parameters are of dubious value for large-scale extrapolation, and for control purposes a very simple but self-adapting algorithm is satisfactory if the disturbances are small in magnitude.

The final goal of the model work is to understand the underlying physics and chemistry of the system, and this goal will not be reached unless the complete reactor performance can be patched together by the building block method.

Numerical Techniques

The library routines for parameter estimation require that φ and $d\varphi/d\underline{p}$ be generated externally, and consequently the state vector $\underline{y}_c$ and its gradient $d\underline{y}_s/d\underline{p}$ must be calculated for given $\underline{p}$. It is often claimed that it is more difficult to extract $\underline{y}_s$ from a plug-flow or batch reactor model than from a recirculation reactor model, but this is not the case when there is only one reaction. In the schematic representation shown in Table 2, the steps that lead to $\hat{\underline{p}}$ are shown for both recirculation reactor experiments anf for plug-flow reactor experiments. The state vector is a set of n conversions $\underline{\xi} = \underline{y}_s$ measured at given values of space time, τ_i (i = 1, 2, . . ., n). In both cases n algebraic equations must be solved to find ξ_i at the known values of $\tau = \tau_i$. The only difference is that evaluation of the integral in the plug-flow reactor model requires a subroutine for numerical quadrature, and excellent codes based on Gauss quadrature are available for this purpose.

When there is more than one component in the state vector, the analogy between the backmix reactor model and the plug-flow (batch) reactor model becomes less apparent. The procedure for the backmix reactor is the same; there are now M coupled algebraic equations which have to be solved for $\underline{y}_s$ at each i (i = 1, 2, . . ., n), but this is in principle not different from the case of one reaction. Solution of M coupled differential equations in the plug-flow reactor model to find $\underline{y}_s$ at τ_i (i = 1, 2, . . ., n) is much more difficult than evaluating n quadratures.

TABLE 2 Steps leading to $\hat{\underline{p}}$

	Recirculation (or backmix) reactor model	Plug-flow (or batch) reactor model
Model:	$\xi = \tau R(\xi,\underline{p})$	$\tau = \int_0^{\xi} \frac{d\zeta}{R(\zeta,\underline{p})}$
Algebraic equation:	$F_i = \frac{\xi_i}{\tau_i} - R(\xi_i,\underline{p}) = 0$ $i = 1, 2, \ldots, n$	$F_i = \tau_i - \int_0^{\xi_i} \frac{d\zeta}{R(\zeta,\underline{p})} = 0$ $i = 1, 2, \ldots, n$
Derivative of F_i:	$\frac{dF_i}{d\xi_i} = \frac{1}{\tau_i} - \left.\frac{dR}{d\xi}\right\|_{\xi_i}$	$\frac{dF_i}{d\xi_i} = -\frac{1}{R(\xi_i,\underline{p})} = -\frac{1}{R_i}$
Iteration:	$\xi_i := \xi_i - \left(\frac{dF_i}{d\xi_i}\right)^{-1} F_i$	$\xi_i := \xi_i + R_i F_i$

For each i repeat the calculation of F_i, $dF_i/d\xi_i$, and $\Delta\xi_i$ until convergence. This leads to $\underline{\xi} = \underline{y}_s$ at the given $\underline{p} = \underline{p}_0$. Next, update $\underline{p}$ using the following procedure with a diagonal matrix $(\partial \underline{F}/\partial \underline{\xi})$:

Gradient of $\underline{y}$:

$$\left(\frac{\partial \underline{F}}{\partial \underline{\xi}}\right)_{\underline{y}_s,\underline{p}_0} \frac{d\underline{y}}{d\underline{p}} = -\left(\frac{\partial \underline{F}}{\partial \underline{p}}\right)_{\underline{y}_s,\underline{p}_0} \quad \text{for both reactor models}$$

Recirculation (or backmix) reactor model:

$$\left(\frac{\partial \underline{F}}{\partial \underline{p}}\right)_{\underline{y}_s,\underline{p}_0} = \left(\frac{\partial \underline{R}}{\partial \underline{p}}\right)_{\underline{y}_s,\underline{p}_0}$$

Plug-flow (or batch) reactor model:

$$\left(\frac{\partial \underline{F}}{\partial \underline{p}}\right)_{\underline{y}_s,\underline{p}_0} = \int_0^{\underline{y}_s} \frac{1}{R^2} \left(\frac{\partial R}{\partial \underline{p}}\right) d\zeta$$

Finally, use Eq. (143) or a similar formula to update $\underline{p}$.

Still, a close analogy exists if the differential equation model is approximated by a discrete model from $\tau = 0$ to τ_{max}. This can be done, for example, by global collocation. If the collocation order N is high enough, the discrete model is an adequate representation of the original model, and the solution $\underline{y}_{sc}$ is close to $\underline{y}_s$. Hence, when there is the usual noise in the data, the estimate $\hat{\underline{p}}$ is not statistically inferior to the estimate $\hat{\underline{p}}$ which would be obtained by an "exact" solution of the model.

When the Nth-order collocation model is solved to machine accuracy, the sensitivities are calculated with only a small extra effort, and they are based on the exact solution of the (approximate) model at τ_i, just as is the case for solution of the (exact) algebraic equation model which describes recirculation reactor data. This is important: if there is an inconsistency between $\underline{y}_s$ and $\partial \underline{y}_s / \partial \underline{p}$, the parameter-fitting algorithm converges slowly or not at all. A remark on this point was made earlier in the section on sensitivity functions. Most automatic integration routines, whether they are based on explicit or semi-implicit Runge-Kutta procedures or on Gear's method, operate with a fixed accuracy in each step or with some other device that is outside the user's control, and it is difficult to obtain corresponding values of $\underline{y}_s$ and $\partial \underline{y}_s / \partial \underline{p}$.

A truly implicit, stepwise integration routine will give consistent sensitivities when the sensitivity equations are solved one by one after convergence of $\underline{y}_s$ has been reached in the step [see Eq. (46)].

When using an explicit or a semi-implicit integration method, one should solve the state equations together with m numerically perturbed equations ($\underline{p} = p_1, p_2, \ldots, p_m$). This is no more difficult than the procedure (46) when the method is explicit. For a semi-implicit method where all $(m + 1)M$ equations have to be solved simultaneously, it is highly profitable to modify the LU decomposition routine to utilize the $(M \times M)$ block diagonal structure of the $[M(m + 1)]$-dimensional Jacobian matrix.

Collocation applied to parameter estimation in differential equations was first proposed by Van den Bosch and Hellinckx (1974). As explained above, it may be an excellent approach to transform each differential equation into N collocation equations, but unfortunately a number of misconceptions about the proper use of the method are apparent from the literature, the most misleading being that "the experimental points should be chosen at values of the independent variable (axial position, etc.) which are zeros of a specific Jacobi polynomial." Obviously, the N collocation points should be chosen to make the approximate solution $\underline{y}_{sc}$ as close to $\underline{y}_s$ as possible. The collocation order N in one particular coordinate may have to be much higher than the available number of data points n in this direction—a situation that occurs if $\underline{y}_s$ is not a smooth function of this particular independent variable. After having solved the model by a sufficiently high order collocation method, one may (without loss of accuracy) interpolate the solution to the values of the independent variable where the experimental values $\underline{y}_e$ are available.

If, on the other hand, there are many more experiments than required for comparison with a collocation solution of sufficiently high order N (this may be the case if the experimental results are recorded automatically), one applies Nth-order polynomial least-squares fitting of the original data to obtain a synthetic data base $\underline{y}_{ec}$ at the same values of the independent variable where the collocation solution is constructed.

Baden and Villadsen (1982) give a critical review of collocation-based parameter estimation methods. They emphasize the importance of choosing a collocation method that gives the best possible fit between the solution $\underline{y}_{sc}$ to the approximate model and the (unknown) solution $\underline{y}_s$ of the exact model. Their model B is the same as that proposed above: solve the collocation equations, find the gradient $d\underline{y}_{sc}/d\underline{p}$, and update $\underline{p}$. One may either choose to solve the collocation equations to machine accuracy before calculating $\Delta\underline{p}$ or take just one step in the iteration toward the solution of the collocation equations before updating $\underline{p}$. A particular variant (their method A) may be valuable to find good starting values $\hat{\underline{p}}$ for the parameters. If the differential equation model is linear in the parameters (or only slightly nonlinear), a situation that frequently occurs, one obtains a very good estimate for $\underline{p}$ by

$$\left[\left(\frac{\partial O_N}{\partial \underline{p}}\right)^T \underline{\underline{W}} \left(\frac{\partial O_N}{\partial \underline{p}}\right)\right] (\hat{\underline{p}} - \underline{p}_0) = -\left(\frac{\partial O_N}{\partial \underline{p}}\right)^T \underline{\underline{W}} O_N(\underline{y}_{ec}, \underline{p}_0) \tag{146}$$

O_N is the residual of the differential equations when the dependent variable $\underline{y}$ is replaced by $\underline{y}_{ec}$, the "synthetic" data base referred to the collocation points as described above, $\underline{\underline{W}}^{-1}$ is an estimate for the dispersion matrix for the residual O_N. It is only a weak function of $\underline{p}$ and may be evaluated at any reasonable point $\underline{p}_0$.

$$\underline{\underline{W}} = \left[\frac{\partial O_N}{\partial \underline{y}} \underline{\underline{\Sigma}} \left(\frac{\partial O_N}{\partial \underline{y}}\right)^T\right]^{-1}_{\underline{y}=\underline{y}_{ec}} \tag{147}$$

The general collocation method may, however, be applied for models that are nonlinear in the parameters as well as in the state variable $\underline{y}$. It is equally applicable for parameter estimation in nonlinear coupled ordinary-differential-equation models (where it competes with the explicit or semi-implicit initial value techniques) and for parameter estimation in boundary value problems.

The best documented general code for parameter estimation using collocation is developed by Sørensen and Stewart. The software is shown in appendixes to Sørensen (1982).

Special attention should be given to the linear boundary value problems which are the basis for Fourier expansion of linear PDE [e.g., variants of Eq. (128)], since these models occur so frequently.

Michelsen (1979) convincingly demonstrated that the transport parameters (radial dispersion and a wall transfer number) can be estimated with only one diagonalization of the collocation matrix (for $Bi_M \to \infty$) and sensitivities calculated analytically from the eigenvalues and eigenvectors of $\underline{\underline{C}}$. The resulting reduction of computer time compared to previous calculations on the same data base was at least by a factor of 20.

When $v_z(x) = 1$, a situation that certainly occurs in pellet problems, one may find the radial transport parameter, the wall transport parameter, and Da simultaneously. There are only trival modifications in the procedure of Michelsen (1979): The solution with Da $\neq 0$ is simply

$$\underline{y}(\tau, Da) = \exp(-Da\tau)\underline{y}(\tau, Da = 0) \tag{148}$$

where $d\underline{y}/d\tau = (\underline{\underline{C}} - Da\underline{\underline{I}})\underline{y}$ and $\underline{y}(\tau, Da = 0)$ is found by the procedure (128) and (129) with Da = 0.

When $v_z(x)$ depends on x, a new diagonalization is required for each value of Da, but the sensitivity with respect to Da can still be derived quite inexpensively. Consider the system

$$\frac{d}{d\tau}(\underline{y}) = \underline{\underline{M}}\underline{y} \qquad \text{where } \underline{\underline{M}} = \underline{\underline{V}}^{-1}(\underline{\underline{C}} - Da\underline{\underline{I}})$$

with solution

$$\underline{y} = \exp(\underline{\underline{M}}\tau)\underline{y}_0 = \underline{\underline{Q}} \exp(\underline{\underline{\Lambda}}\tau)\underline{\underline{Q}}^{-1}\underline{y}_0 \tag{149}$$

The sensitivity z_{Da} with respect to Da is found by solution of the inhomogeneous linear differential equation

$$\frac{d}{d\tau}(\underline{z}_{Da}) = \underline{\underline{M}}\underline{z}_{Da} - \underline{\underline{V}}^{-1}\underline{y} \qquad \text{where } \underline{z}_{Da}(\tau = 0) = \underline{0} \tag{150}$$

with solution

$$\underline{z}_{Da} = -\exp(\underline{\underline{M}}\tau)\int_0^{\tau} \exp(-\underline{\underline{M}}t)\underline{\underline{V}}^{-1}\underline{y}(t)\, dt = -\underline{\underline{Q}}\underline{\underline{S}}(\tau)\underline{\underline{Q}}^{-1}\underline{y}_0$$

where

$$S_{ij} = \begin{cases} (\lambda_j - \lambda_i)^{-1}[\exp(\lambda_j\tau) - \exp(\lambda_i\tau)]r_{ij} & \text{for } i \neq j \\ \tau r_{ii} \exp(\lambda_i\tau) & \text{for } i = j \end{cases} \tag{151}$$

and

r_{ij} is an element of $\underline{\underline{r}} = \underline{\underline{Q}}^{-1}\underline{\underline{V}}^{-1}\underline{\underline{Q}}$

When the rate term is nonlinear in $\underline{y}_s$, one proceeds more or less as described above for the linear partial differential equation. Collocation is used to discretize the PDE in the x direction, but now the collocation equations are integrated in the τ direction, either by collocation as proposed by Sørensen (1982) or by a stepwise procedure.

Sensitivities can be calculated by numerical perturbation, but at least for the case of a single reaction it is also possible to use a semi-implicit Runge-Kutta routine to integrate the N state equations and the N × m analytically derived sensitivity equations. The calculation of the two higher derivatives f_{yy} and f_{py} in Eq. (47) is not difficult since, in contrast to the case of M coupled independent reactions, the collocation differential equations have a very simple structure. The nonlinearities are found only in the main diagonal of the Jacobian $\underline{\underline{J}}$, whereas all the other elements in $\underline{\underline{J}}$ are constants.

It is nearly impossible to give a balanced picture of the subject of parameter estimation within the framework of one section of one chapter in a book. So many things are involved, and most of them have more to do with common sense engineering that with mathematics: the proper choice of experimental equipment and analytical method, planning of experiments, and a general pessimism concerning the influence of all the variables which for one reason or another were not included in the investigation.

The methods that we have chosen to discuss have been used by us and by others in many different situations, but still many topics have been left out. It is hoped that Examples 7 and 8, which mark the end of our review, will also emphasize some of the facets of parameter estimation that could not be adequately covered in the preceding text.

Example 7: Determination of a first-order rate constant and a dispersion coefficient from experiments in a laminar flow tubular reactor Consider the mass balance (125) with zero mass flux at the wall, and consider a series of experiments in which the velocity-averaged outlet concentration $\bar{y}$ in Eq. (125) has been measured at different axial positions z in a reactor tube of given diameter. It is desired to estimate the rate constant k_R and the radial diffusivity D. Experimental results of this kind were obtained by Cleland and Wilhelm (1956), who determined D assuming that k_R was known. The experimental setup is also described by Seinfeld (1969), Seinfeld and Lapidus (1974), and Van den Bosch and Hellinckx (1974), where k_R was to be determined while the dispersion number was assumed to be known from independent measurements.

We shall consider the possibility of simultaneous estimation of the two parameters. Write Eq. (125) as

$$2(1 - x^2)\frac{\partial y}{\partial z} = \text{Da}\left(\frac{D}{k_R r_t^2}\nabla^2 y - y\right) = \text{Da}(\psi\nabla^2 y - y) \tag{152}$$

where $\psi = D/k_R r_t^2$. The solution of Eq. (152) by collocation is averaged to give

$$\bar{y} = \sum_1^N c_i \exp(\lambda_i \text{Da}\ z) \tag{153}$$

where the constants c_i and λ_i are functions of ψ only.

A set of equivalent experiments would comprise variation of v_{av} at fixed total reactor length and diameter. Let $\text{Da}_r = k_R L/v_{av,r}$ and

$$\text{Da}\ z = \text{Da}_r z' \qquad \text{where } z' = z\frac{v_{av,r}}{v_{av}} \tag{154}$$

and the outlet, z = 1, the value of $z' = v_{av,r}/v_{av}$.

Numerical values of c_i and of λ_i are calculated for $\psi = 0.1$, the value used by Seinfeld (1969) and Lapidus and Seinfeld (1974):

i	λ_i	c_i
1	−0.8429	0.9211
2	−2.30	0.069
3	−5.26	0.007

A perturbation analysis for large ψ [see Eq. (122) for the first term] yields

$$\lambda_1 = -1 + \frac{1}{48\psi} - \frac{1}{1920\psi^2} + O(\psi^{-3})$$
$$c_1 = 1 - \frac{1}{768\psi^2} + O(\psi^{-3}) \tag{155}$$

while $\lambda_i \sim O(\psi^{-1})$ and $c_i \sim O(\psi^{-2})$ for $i > 1$.

When ψ is large enough to allow terms in ψ^{-2} to be neglected ($\psi = 0.1$ is barely large enough to permit this approximation), the expansion (153) can be truncated after the first term:

$$\bar{y} \sim \exp\left[Da_r\left(-1 + \frac{1}{48\psi}\right) z'\right] \tag{156}$$

and measurements of $\bar{y}$ alone are not sufficient to determine separate values of Da_r and ψ. Only the product $Da_r\,(-1 + 1/48\psi)$ will be found by plotting $\ln \bar{y}$ versus z'.

We have constructed the following "measurements" of $\bar{y}$ for $Da_r = 2$ and $\psi = 0.1$ by high-order collocation of Eq. (152) and truncating the results after the third digit:

z'	$\bar{y}$	$-\ln \bar{y}$	$-\ln \bar{y}/z'$
0.1	0.826	0.1912	1.912
0.2	0.687	0.3752	1.877
0.3	0.555	0.5551	1.850
0.4	0.481	0.7319	1.830
0.5	0.404	0.9063	1.813
0.6	0.340	1.079	1.798
0.7	0.286	1.252	1.788
0.8	0.241	1.423	1.779
0.9	0.203	1.595	1.772
1	0.172	1.760	1.760

According to Eq. (156), the ratio $\ln \bar{y}/z'$ should be constant, and since this is not quite so, we may be able to determine both parameters Da_r and ψ.

Linear regression of the data with $z' > 0.5$ yields $\ln \bar{y} = -0.053 - 1.71z'$ and now, from Eq. (153), truncated after the first term (which is certainly adequate for large z', say $z' > 0.5$), one obtains initial estimates for Da_r and ψ:

$$\bar{y} \sim \left(1 - \frac{1}{768\psi^2}\right) \exp\left[-Da_r z'\left(1 - \frac{1}{48\psi} + \frac{1}{1920\psi^2}\right)\right] \tag{157}$$

The result is $\underline{p}_0 = (Da_r, \psi)_0 = (1.93, 0.157)$.

These values are used as initial values in a complete least-squares procedure based on Eq. (152), which is solved by three-point collocation, and on all the "experimental" points $(z', \bar{y})$.

The sensitivity with respect to Da_r is found by differentiation of Eq. (152):

$$\frac{d}{dDa_r} \sum_1^N \lambda_i z' c_i \exp(\lambda_i Da_r z') \tag{158}$$

while the sensitivity with respect to ψ is found by numerical perturbation (although analytical differentiation could also be used).

Convergence is obtained in a few iterations, and total computing time (WATFIV compiler, Harwell VBO1AD parameter estimation program) is 0.2 s. The results are

$$\begin{aligned} Da_r &= 1.994 \pm 0.1\% \\ \psi &= 0.104 \pm 2\% \end{aligned} \tag{159}$$

The error of ψ is substantial, considering that the data are accurate to three digits, an accuracy that is not likely to be found in practice. The rate constant is, however, determined with the expected small standard deviation. The two parameters are strongly correlated (correlation coefficient 0.95) and this, combined with the large standard deviation of ψ, shows that the use of laminar flow reactor data for $\bar{y}$ are unsuitable for determination of ψ, based on a "known" value of Da. A small error in the rate constant will be strongly magnified when ψ is estimated. Conversely, an approximate value for ψ is all we need to determine a very good estimate for the rate constant, using exactly the same data.

Finally, it should be mentioned that concentration measurements at the exit of the tube, but at the centerline (i.e., measurements on samples taken through a small-bore tube which is centrally placed in the reactor tube) will be much better suited for determination of the two parameters. We have repeated the estimation, but based on 10 "measurements" taken at $x = 0$ and with the same "experimental error." Now ψ is found to 0.1004 with a standard deviation of only 0.4%, while the value of Da is 1.997 and has the same standard deviation (0.4%) as before.

The result, which is obtained by a simple numerical analysis of the model from which the parameters are to be extracted, illustrates how an experimental program can be guided by simulation studies.

Example 8: Kinetics of a homogeneous liquid-phase substitution reaction ("the Dow Chemical Company test problem") In 1981, Blau et al. from the Dow Chemical Company published the formulation of a multi-response-parameter estimation problem. The problem is concerned with an industrial reaction, "hidden" behind the following symbolic nomenclature:

$$HA + 2BM = AB + HMBM \tag{160}$$

It is a liquid-phase reaction catalyzed by QM, which is fully dissociated to M^- and Q^+.

The problem was given to a number of research groups in the United States and in Europe who were supposed to contribute with a solution. A comparison of the solutions would, it was hoped, give a review of the state of the art in parameter fitting for reaction engineering models. Here we summarize the more general conclusions from the solution that was submitted by our group (Sarup, 1982).

The postulated mechanism is (Blau et al., 1981):

$$HA \overset{K_2}{\longleftrightarrow} A^- + H^+$$

$$A^- + BM \overset{K_3}{\longrightarrow} ABM^- \overset{K_3}{\longleftrightarrow} HABM(-H)^+$$

$$M^- + BM \underset{k_{-1}}{\overset{k_1}{\rightleftharpoons}} MBM^-$$

$$ABM^- \underset{k_{-3}}{\overset{k_3}{\rightleftharpoons}} AB + M^- \qquad (k_3 = \tfrac{1}{2}k_{-1},\ k_{-3} = k_1)$$

$$MBM^- + H^+ \overset{K_1}{\longleftrightarrow} HMBM$$

The reaction is carried out in a batch reactor. Table 3 summarizes the concentration-time profiles for one of the three temperatures used in the experimental work.

There are three rate constants, k_1, k_{-1}, and k_2, and there are three equilibrium constants, K_1, K_2, and K_3. Activation energy and frequency factor for the three rate constants and values of the three temperature independent equilibrium constants are desired.

The 10 species HA, BM, HABM, AB, HMBM, M^-, A^-, H^+, ABM^-, and MBM^- are given by mass balances—six ordinary differential equations coupled with three equilibrium relations and one condition that expresses electroneutrality. At $t = 0$ only HA and BM are present with a small amount of catalyst QM.

Initial estimates for all parameters are given by Blau et al. (1981). From the set of initial estimates, we shall mention only that $K_1 = K_3 = 10^{-17}$ and $K_2 = 10^{-11}$.

Our solution of the problem proceeds in the following three steps:

1. *Analysis of the original data.* A closer study of the data in Table 2 reveals an alarming fit of the mass balance for B. Except for obvious misprints it is seen that

$$c_{BM} = c^{\circ}_{BM} - c_{HABM} - 2c_{AB} \tag{161}$$

The data were analyzed by the method of Box (1973) and it was found that there is not only one, but two linear relations between the "original" data.

Correspondence with Blau gave the following supplementary information:

a. c_{BM} was not measured, but back-calculated from the mass balance (161).
b. The concentrations of HA, HABM, and AB were all measured, but their sum was normalized to agree with a "known" initial concentration of HA:

TABLE 3 Concentration Versus Time Profiles for The Dow Chemical Company Test Example (T = 40°C)

Time (hrs)	Concentrations (g mol/kg)			
	BM	HA	HABM	AB
0.00	8.3200	1.7066	0.0000	0.0000
0.08	8.3065	1.6960	0.0077	0.0029
0.58	8.2954	1.6826	0.0234	0.0006
1.58	8.2730	1.6596	0.0470	—
2.75	8.2437	1.6305	0.0763	—
3.75	8.2277	1.6143	0.0923	—
4.75	8.2026	1.5892	0.1174	—
5.75	8.1781	1.5673	0.1371	0.0024
8.75	8.1265	1.5133	0.1935	—
23.05	8.0167	1.4075	0.2949	0.0042
21.75	7.8440	1.2308	0.4760	—
28.75	7.6977	1.0931	0.6047	0.0088
46.25	7.3234	0.7268	0.9530	0.0268
52.25	7.1495	0.5773	0.0881	0.0412
76.25	6.6123	0.2065	1.2929	0.2074
106.25	6.2309	0.0650	1.1941	0.4475
124.25	6.1220	0.0391	1.1370	0.5305
147.75	6.0084	0.0244	1.0528	0.6294
172.25	5.9193	0.0145	0.9835	0.7086
196.25	5.8556	0.0083	0.9326	0.7659
219.75	5.8037	0.0074	0.8821	0.8171
240.25	5.7680	0.0050	0.8492	0.8514
274.25	5.7222	0.0047	0.8064	0.8957
292.25	5.7021	0.0042	0.7869	0.9155
316.25	5.6722	0.0015	0.7628	0.9425
340.75	5.6593	0.0017	0.7495	0.9556
364.25	5.6351	—	0.7263	0.9793
386.75	5.6176	—	0.7112	0.9956
412.25	5.6131	—	0.7063	1.0003
442.75	5.5991	—	0.6927	1.0141
460.75	5.5959	—	0.6871	1.0195

TABLE 2 (Continued)

Time (hrs)	Concentrations (g mol/kg)			
	BM	HA	HABM	AB
483.75	5.5905	—	0.6837	1.0229
507.25	5.57.36	—	0.6672	1.0396
553.75	5.5568	—	0.6494	1.0574
580.75	5.5631	0.0046	0.6467	1.0551
651.25	5.5472	—	0.6408	1.0660
673.25	5.5516	—	0.6452	1.0616
842.75	5.5465	—	0.6397	1.0669

Source: Blau (1981).

$$c^{\circ}_{HA} = c_{HA} + c_{HABM} + c_{AB} \tag{162}$$

This normalization is common practice in analysis of chromatographic data and the error is not as "crude" as that involved in (a). Still, when normalized measurements are used, the error structure of the true measurements is completely distorted and one may use a standard least-squares approach rather than one of the more sophisticated statistical criteria.

The comments regarding the data base should not be construed as a criticism of the author of the problem. A preliminary analysis of the raw data is always valuable, and the flaws in the original data material are probably typical for what can be expected in practice.

Our conclusion has been to delete the concentration measurements of c_{BM} from the data base since they contain no independent information.

2. *Preliminary analysis of the model.* Total mass balances for A, B, and M and the three equilibrium relations can be used to eliminate 6 out of 10 dependent variables to give a set of three coupled differential equations and one algebraic equation.

The postulated values of the dissociation constants K_1, K_2, and K_3 are so small that the acids are almost totally undissociated. One might assume that the concentration of H^+ is nearly zero, and this reduces the model to three differential equations in three components, chosen as HA, BM, and M^-. c_{A^-}, c_{ABM^-}, and c_{MBM^-} are given by

$$c_{A^-} = \left(1 + \frac{K_3}{K_2}\frac{c_{HABM}}{c_{HA}} + \frac{K_1}{K_2}\frac{c_{HMBM}}{c_{HA}}\right)^{-1}(c_{Q^+} - c_{M^-})$$

$$c_{ABM^-} = \frac{K_3}{K_2}\frac{c_{HABM}c_{A^-}}{c_{HA}} \tag{163}$$

$$c_{HBM^-} = \frac{K_1}{K_2}\frac{c_{HMBM}c_{A^-}}{c_{HA}}$$

while c_{HABM}, c_{AB}, and c_{HMBM} are calculated from the total mass balances for A, B, and M. Except for the assumption $c_{H+} \sim 0$, there are no approximations involved relative to the original model, and we have checked that our results for the parameters are virtually unaffected by the assumption $c_{H+} \sim 0$.

With this assumption—which appears to be quite realistic—it is impossible to find absolute values for the equilibrium constants K_1, K_2, and K_3. Only their ratios K_3/K_2 and K_1/K_2 can be found.

If these ratios are as small (ca. 10^{-6}) as indicated in the problem description, the system is extremely stiff: c_{HA} will decrease almost to zero, following pseudo-zero-order kinetics. When c_{HA} has become of the order of K_1/K_2, the rate expression changes to pseudo first order in c_{HA}. All HA would essentially have to react before the "dead-end" reaction to HABM reverses direction and starts to send ABM^- back into the main reaction sequence.

3. *Simulation and parameter estimation.* To simulate the solution $\underline{y}$ of the model, we used the IMSL routine DGEAR, which was modified to include algebraic ("exact") evaluation of the sensitivities with respect to the parameters.

There are in all eight parameters in k_1, k_{-1}, k_2, K_1/K_2, and K_3/K_2, but sensitivities with respect to activation energy E_i and frequency factor α_i of the rate constants k_i can easily be computed from the sensitivities with respect to k_i:

$$\frac{\partial \underline{y}}{\partial \alpha} = \frac{\partial \underline{y}}{\partial k}\frac{\partial k}{\partial \alpha} = \exp\left(-\frac{E}{T}\right)\frac{\partial \underline{y}}{\partial k}$$

$$\frac{\partial \underline{y}}{\partial E} = \frac{\partial \underline{y}}{\partial k}\frac{\partial k}{\partial E} = -\frac{\alpha}{T}\exp\left(-\frac{E}{T}\right)\frac{\partial \underline{y}}{\partial k} \qquad (164)$$

With this simple device the original $3(2 \cdot 3 + 2) = 24$ sensitivity equations are reduced to 15 equations.

Also, it is important to rescale the parameters since the magnitude of α and E are so different that it becomes difficult to harmonize the sensitivities. Instead of α and E, we use $k(T_0)$ and E/T_0 defined by

$$k = \alpha \exp\left(-\frac{E}{T}\right) = k(T_0)\exp\left[\frac{E}{T_0}\left(1 - \frac{T_0}{T}\right)\right] \qquad (165)$$

where T_0 is a suitable reference temperature (chosen as the middle of the three temperatures used in the experimental investigation). With the rescaled parameters a large part of the inherent correlation between the two parameters involved in k is removed.

In general, if a new parameter $p_1 = f(p_0)$ is introduced instead of p_0, one obtains the sensitivity with respect to p_1 by the simple algorithm

$$\frac{\partial y}{\partial p_1} = \left[\frac{\partial f(p_0)}{\partial p_0}\right]^{-1} \frac{\partial y}{\partial p_0} \tag{166}$$

The Harwell code VB01AD was used for the least-squares minimization. The routine is based on the Levenberg-Marquardt method, and it requires analytic derivatives of the error vector with respect to the parameters. There was no difficulty in obtaining the minimum of the sum of least squares—about 10 s of CPU time to determine all eight parameters (which corresponds to ca. 0.5 s per integration of the model and the 15 sensitivity differential equations).

It is remarkable that the integration of the three equations for c_{HA}, c_{BM}, and c_M—together with all the sensitivity equations—requires only twice the time used for the three quations of the state vector. Here it is much better to use exactly derived sensitivity equations than to use numerical perturbation of the parameters, which is considerably slower (by a factor of 6) and requires a much higher integration accuracy to give meaningful results.

Also, a comment on the initial values of K_3/K_1 and K_1/K_2 is necessary. To obtain the best estimates for these parameters (1.43 and 0.0672, respectively) it was necessary to start with an initial estimate of K_1/K_2 which was decidedly larger than the small value 10^{-6} suggested by Blau (1981).

It is not difficult to see from Table 3 that c_{HABM} starts to decrease much before HA has been consumed, and consequently that K_1/K_2 has to be larger than 10^{-6}. It is, however, not at all obvious that the routine fails to predict the correct values of $q_1 = K_1/K_2$ and $q_3 = K_3/K_2$ if the original estimate of the parameters is used. The reason is that the object function is affected only by the ratio q_3/q_1, not by their absolute values when they are as small as 10^{-6}, and it becomes impossible for the routine to drag the parameter values all the way from 10^{-6} to their correct values of about 10^{-1}.

The integration routine is much faster when $(K_1/K_2, K_3/K_2)$ are of the order of 1 than when they are ca. 10^{-6}. The reason is the sudden increase of stiffness that occurs when $c_{HA} \sim 0$, and the apparent reaction order of the reaction $A^- + BM \rightarrow ABM^-$ changes from zero to 1. The problem may occur quite frequently in the application of standard "kinetic codes," and it may be illustrated by the simple rate expression (167):

$$R = \frac{k y_A}{K + y_A} \quad \text{with } k \sim 1 \text{ and } K \sim 10^{-6} \tag{167}$$

Even a good numerical integration routine (e.g., DGEAR) will fail to notice the rapid change in the structure of the solution when $y_A \sim 10^{-6}$ and will happily integrate to negative concentrations of A. This may have a disasterous effect on the computation of other reaction species, which may be heavily influenced by the concentration level of component A.

REFERENCES

Aris, R. *The Mathematical Theory of Diffusion and Reaction in Permeable Catalysts*, Clarendon Press, Oxford (1975).

Aris, R., and N. R. Amundson, *First Order Partial Differential Equations with Applications*, Prentice-Hall, Englewood Cliffs, N.J. (1973).

Baden, N., and J. Villadsen, A Family of collocation based methods for parameter estimation in differential equations, Chem. Eng. J., *23*, 1 (1982).

Blau, G., L. Kirkby, and M. Marks, An industrial kinetics problem for testing nonlinear parameter estimation algorithms, The Dow Chemical Company, Midland, Mich. (1981).

Box, G. E. P., W. G. Hunter, J. McGregor, and J. Erjavec, Some problems associated with the analysis of multiresponse data, Technometrics, *15*, 33 (1973).

Byrne, C. D., Some software for solving ordinary differential equations, in *Foundations of Computer-Aided Chemical Process Design*, Vol. 1, S. H. Mah and W. D. Seider, eds., Engineering Foundation, New York, 1981, p. 403.

Caillaud, J. B. and L. Padmanhaban, An improved semi-implicit Runge-Kutta method for stiff systems, Chem. Eng. J., *2*, 227 (1971).

Christiansen, L. J. and S. L. Andersen, Transient profiles in sulphur poisoning of steam reformers, Chem. Eng. Sci., *35*, 314 (1980).

Cleland, F. A. and R. H. Wilhelm, Diffusion and reaction in viscous flow tubular reactor, AIChE J., *2*, 489 (1956).

Clement, K. and S. B. Jørgensen, Experimental investigation of axial and radial thermal dispersion in a packed bed, Chem. Eng. Sci., *38*, 835 (1983).

Crank, J., *Mathematics of Diffusion*, Clarendon Press, Oxford (1957).

Finlayson, B. A., *The Method of Weighted Residuals and Variational Principles*, Academic Press, New York (1972).

Finlayson, B. A., Orthogonal collocation in chemical reaction engineering, Catal. Rev. Sci. Eng., *10*, 69 (1974).

Finlayson, B.A., *Nonlinear Analysis in Chemical Engineering*, McGraw-Hill, New York (1980).

Fletcher, R., *Unconstrained Optimization*, Wiley, New York, (1980).

Funder, C. R., M.Sc. thesis (in Danish), Instituttet for Kemiteknik (1982).

Gear, C. W., *Numerical Initial Value Problems in Ordinary Differential Equations*, Prentice-Hall, Englewood Cliffs, N.J. (1971).

Guertin, E. W., J. P. Sørensen, and W. E. Stewart, Exponential collocation of stiff reactor models, Comp. Chem. Eng., *1*, 197 (1977).

Hoffman, U., G. Emig, and H. Hofmann, Comparison of different methods for determination of effective thermal conductivity of porous catalysts, ACS Symp. Ser., *65*, 189 (1978).

Holk-Nielsen, P., and J. Villadsen, Absorption with exothermic reaction in a falling film column, Chem. Eng. Sci., *38*, 1439 (1983).

Jackson, R., *Transport in Porous Catalysts*, Elsevier, Amsterdam (1977).

Kaza, K. R., Villadsen, J., and R. Jackson, Intraparticle diffusion effects in the methanation reaction, Chem. Eng. Sci., *35*, 17 (1980).

Levenspiel, O., *Chemical Reaction Engineering*, 2nd ed., Wiley, New York, (1972).

Liaw, W. K., J. Villadsen, and R. Jackson, A simulation of coke burning in a fixed bed reactor, ACS Symp. Ser., *196*, 39 (1982).

Michelsen, M. L., A fast solution technique for a class of linear PDE, and Estimation of heat transfer parameters in packed beds from radial temperature measurements, Chem. Eng. J., *18*, 59, 67 (1979).

Michelsen, M. L. and J. V. Villadsen, Polynomial solution of differential equations, in *Foundations of Computer-Aided Chemical Process Design*,

Vol. 1, S. H. Mah and W. D. Seider, eds., Engineering Foundation, New York, 1981, p. 341.
Michelsen, M. L., H. B. Vakil, and A. S. Foss, State space formulation of fixed bed reactor dynamics, Ind. Eng. Chem. Fundam., *12*, 323 (1973).
Miller, K. and R. Miller, Moving finite elements, Siam J. Numer. Anal., *18*, 1010, 1033 (1981).
Papoutsakis, E., D. Ramkrishna, and H. C. Lim, The extended Graetz problem with prescribed wall flux, AIChE J., *26*, 779 (1980).
Patterson, W. R., and D. L. Cresswell, A Simple method for the calculation of effectiveness factors, Chem. Eng. Sci., *26*, 605 (1971).
Peytz, M. J., H. Livbjerg, and J. Villadsen, Steady state operation of a dual reactor system with liquid catalyst, Chem. Eng. Sci., *37*, 1095 (1982).
Sarup, B., M.Sc. thesis (in Danish), Instituttet for Kemiteknik (1982).
Seinfeld, J. H., Identification of parameters in PDE, Chem. Engr. Sci., *24*, 65 (1969).
Seinfeld, J. H. and L. Lapidus, *Process Modeling, Estimation, and Identification*, Prentice-Hall, Englewood Cliffs, N.J. (1974).
Sotirchos, S. V. and J. Villadsen, Diffusion and reaction with a limited amount of reactant, Chem. Eng. Commun., *13*, 145 (1981).
Sundaresan, S. and N. R. Amundson, Diffusion and reaction in a stagnant boundary layer about a carbon particle, Ind. Eng. Chem. Fundam., *19*, 344, 351 (1981).
Sørensen, J. P., *Simulation, Regression, and Control of Chemical Reactors by Collocation Techniques*, Polyteknisk Forlag, Lyngby, Denmark (1982).
Sørensen, J. P. and W. E. Stewart, Collocation analysis of multicomponent diffusion and reactions in porous catalysts, Chem. Eng. Sci., *37*, 1103 (1982).
Taylor, G., Dispersion of soluble matter in solvent flowing slowly through a tube, Proc. R. Soc., A*219*, 186 (1953).
Van den Bosch, B., and L. Hellinckx, A new method for the estimation of parameters in differential equations, AIChE J., *20*, 250 (1974).
Villadsen, J., *Selected Approximation Methods for Chemical Engineering Problems*, Instituttet for Kemiteknik, Lyngby, Denmark (1970).
Villadsen, J. and M. L. Michelsen, *Solution of Differential Equation Models by Polynomial Approximation*, Prentice-Hall, Englewood Cliffs, N.J. (1978).
Villadsen, J. and W. E. Stewart, Solution of boundary-value problems by orthogonal collocation, Chem. Eng. Sci., *22*, 1483 (1967).
Villadsen, J. and J. P. Sørensen, Solution of parabolic PDE by a double collocation method, Chem. Eng. Sci., *24*, 1337 (1969).
Wedel, S. and D. Luss, A rational approximation of the effectiveness factor, Chem. Eng. Commun., *7*, 245 (1980).
Wedel, S. and J. Villadsen, Falsification of kinetic parameters by incorrect treatment of recirculation reactor data, Chem. Eng. Sci., *24*, 1346 (1983).
Wheeler, A., Reaction rates and selectivity in catalyst pores, Adv. Catal., 3, 316 (1951).

2
Use of Residence- and Contact-Time Distributions in Reactor Design

REUEL SHINNAR *The City College of the City University of New York, New York, New York*

INTRODUCTION

Tracer experiments and residence-time distributions measurements have been used by reaction engineers for 50 years. They are a powerful tool with an increasing number of applications. The first rigorous presentation was given by Danckwerts (1953). Since then a large body of literature has accumulated and the topics are discussed in several textbooks (Froment and Bischoff, 1979; Himmelblau and Bischoff, 1968; Seinfeld and Lapidus, 1974; Nauman, 1983; Levenspiel and Bischoff, 1963) and review articles (Shinnar, 1977; Petho and Noble, 1982; Nauman, 1981; Weinstein and Adler, 1967).

Tracer experiments and residence-time distributions are useful to reaction engineers in several ways. They provide a diagnostic tool for the detection of maldistributions and the flow pattern inside a reactor. They are useful in experimentally measuring the parameters of simplified flow models and provide ideas and guidelines for creating and testing such models. Finally, understanding the concept of residence-time distribution greatly simplifies the solution of many problems in reactor design, including the choice of reactor configuration, optimization, and scaleup, and allows one to solve them without complex and time-consuming computations.

One must, however, be careful not to expect more from the techniques than they can deliver. It is extremely rare that one can either construct a reactor model or predict a reactor performance based solely on a residence-time distribution (Shinnar, 1978) in a situation involving complex flows. Most reactors are heterogeneous, and the theory as presented in most textbooks is rigorously applicable only to homogeneous reactors. It is therefore important to understand the advantages as well as the limitations of the method.

This writer has been involved both in developing some of the theoretical concepts related to their use and in their practical applications (Naor and Shinnar, 1963; Shinnar and Naor, 1967; Krambeck et al., 1969; Shinnar et al., 1972; Naor et al., 1972; Zvirin and Shinnar, 1975; Glasser et al., 1973; Krambeck et al., 1967; Silverstein and Shinnar, 1975; Shinnar et al., 1973). In the following sections I try to present the subject in a way that stresses practical applications, discussing both advantages and pitfalls. Thus, by necessity, I must exclude a large amount of theoretical work which may become useful in the future but has yet to prove its applicability in practice, and in that sense this chapter presents a personal viewpoint based on my own experience. Although I try to provide the reader with references to the literature, no attempt will be made to give full proper credits to original papers or to historical development.

The theoretical concepts for homogeneous systems are derived in the following section. Then we discuss experimental techniques for their measurements and their use and applications in reactor modeling and in the testing of industrial reactors. Next, we deal with applications to the design of homogeneous reactions and with multiphase systems and the use of multiple tracers. Following that, we introduce contact-time distributions, a concept similar to residence-time distribution that is useful in catalytic reactors. Finally we summarize the use of residence-time and contact-time distributions in catalytic reactor design.

DEFINITIONS OF THE RESIDENCE-TIME DISTRIBUTION, AGE, AND FUTURE LIFETIME DISTRIBUTION

Residence-Time Distribution

Consider a homogeneous reactor as shown in Figure 1. Assume that the flow is stable and steady and that the inlet and outlet streams are completely mixed. Let us now assume that each fluid particle is equipped with a timer that is activated when it enters the reactor and stops when it exits. It is important to define a single exit from which the particle cannot return to the reactor. Each particle at the exit has a time t associated with it, which measures the time it spent in the reactor. This quantity is defined as the residence time for that particle. As with any scalar property, one can define the distribution of this property by the fraction of the particles for which the residence time is less than a given value t (see Fig. 2). This fraction F(t) is defined as the residence-time distribution. [Alternatively, one can define the residence-time distribution as the fraction of particles or molecules which has a residence time larger than t, simply $1 - F(t)$.] Clearly, $F(O) = 0$ and F(t) must be a monotonically increasing function with $F(\infty) = 1$.

Density Function

For any probability distribution, F(t), the derivative that is defined by

$$f(t) = \frac{dF(t)}{dt} \tag{1}$$

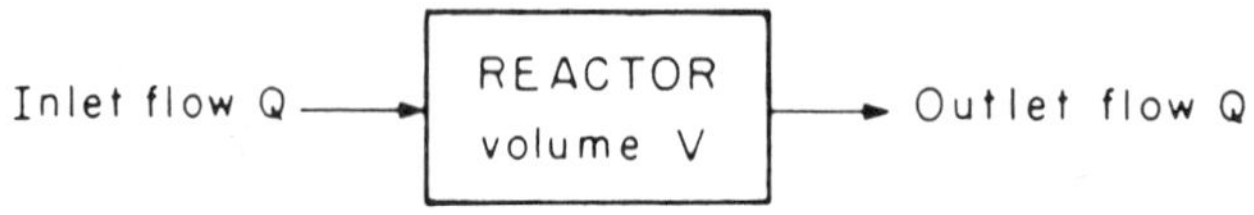

FIGURE 1 Steady-state homogeneous reactor.

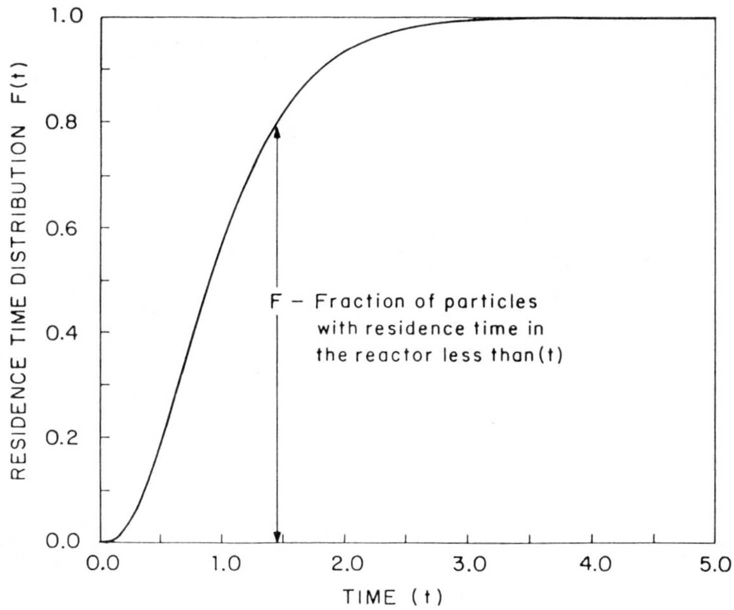

(a)

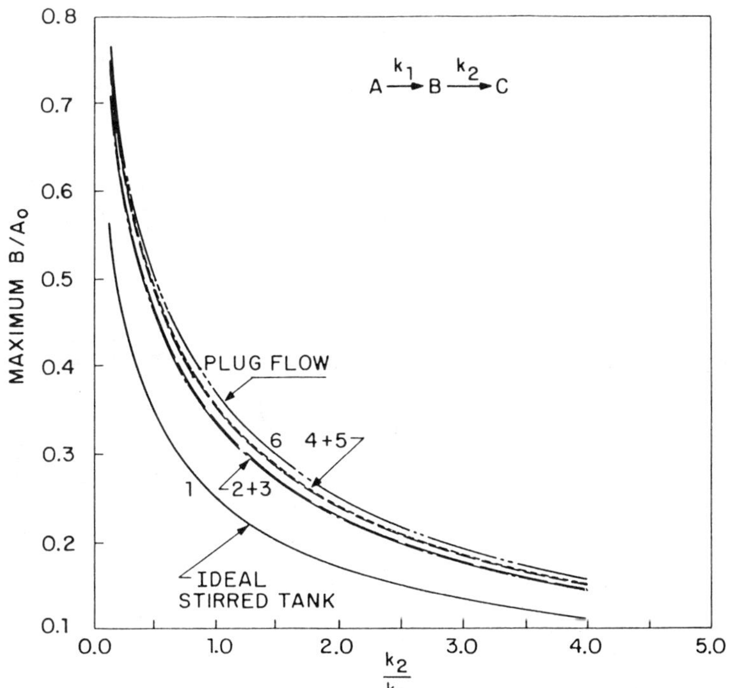

(b)

FIGURE 2 Definition of residence time distribution.

is called the *density function* and is often of great interest. In the case where F(t) is a residence-time distribution, f(t) dt corresponds to the fraction of fluid particles whose residence time is between t and t + dt. Some typical distributions of F(t) and f(t) are given in Figure 3. The way such distributions are measured and computed will be shown in the next section. Some properties of the function will be discussed next.

One should note that in the chemical engineering literature the density function f(t) is often designated E(t) and is sometimes called the *residence-time distribution*. [1 − F(t)] is often designated as I(t). As there is a large literature on the mathematical properties of residence-time distribution outside the chemical engineering literature, the standard notation F(t) and f(t) will be used in this chapter.

Mean, Variance, and Moments of f(t)

The expected value or mean of the density function f(t) is defined by

$$\tau = \mu_1 = E_f(t) = \int_0^\infty tf(t)\,dt = \int_0^\infty [1 - F(t)]\,dt \tag{2}$$

This is also equal to the first moment of f(t) and is a widely accepted measure of the central value of a distribution. In reactor systems with constant density, it is equal to the volume or liquid holdup of the reaction, divided by the flow rate:

$$\tau = \frac{V}{Q} \tag{2a}$$

and is also known as the *time constant* of the system. For purposes of comparison, it is often useful to normalize the residence-time distribution such that the time constant or mean value is unity. Therefore, we define

$$\theta = \frac{t}{\tau} \tag{2b}$$

where θ is a dimensionless normalized variable. Most of the graphs in this chapter use θ.

Another important measure is the variance of a distribution, $V_f(t)$, which is defined by

$$V_f(t) = \int_0^\infty [t - E_f(t)]^2 f(t)\,dt = E_f(t^2) - E_f^2(t) \tag{3}$$

$V_f(t)$ can also be computed directly from F(t) by

$$V_f(t) = 2\int_0^\infty t[1 - F(t)]\,dt - \left[\int_0^\infty (1 - F(t))\,dt\right]^2 \tag{3a}$$

The variance (or the second moment with respect to the average) measures the dispersion of a distribution. Relative disperison is also frequently characterized by the coefficient of variation,

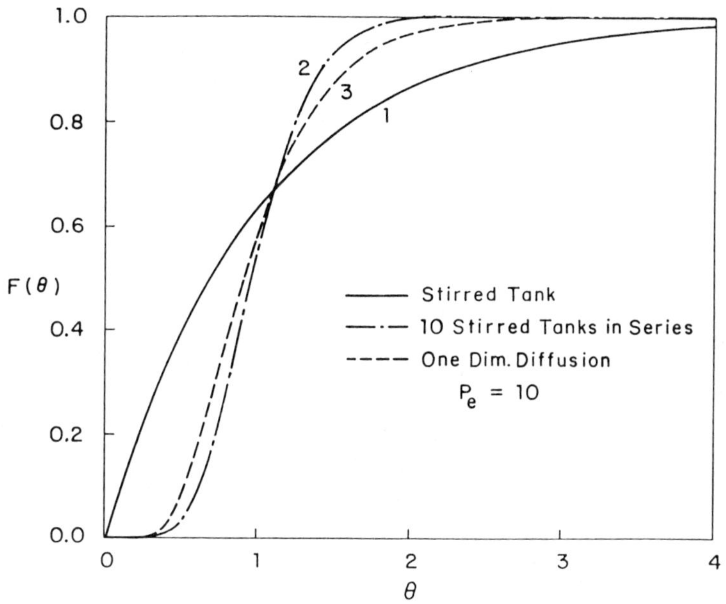

(a)

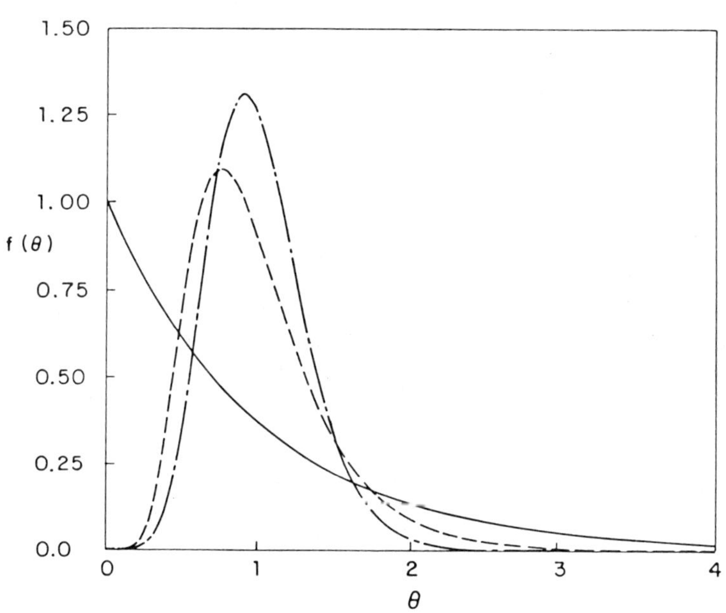

(b)

FIGURE 3 (a) Residence-time distribution, (b) residence-time density function for some simple theoretical flow models.

$$\gamma(t) = \frac{\sqrt{V_f(t)}}{E_f(t)} \tag{4}$$

When needed, higher moments can be computed using standard methods. The rth moment of the density function f(t) is defined as

$$\mu_r(f) = E_f(t^r) = \int_0^\infty t^r f(t)\, dt = \gamma \int_0^\infty t^{\gamma-1} [1 - F(t)]\, dt \tag{5}$$

It should be stressed that the functions f, F, and (1 − F) refer to the distribution of the residence times of particles *as they appear at the exit of the flow system.*

Linear Properties of Residence-Time Distributions

Residence-time distributions have the properties of linear functions. Thus if two systems are put in series (see Fig. 4a), the density function of the total system is the convolution of the two systems, and if they are parallel, it is the weighted average (Fig. 4b). Thus for mathematical operations it is useful to apply the Laplace transform of f(t),

$$L[f(t)] = \hat{f}(s) = \int_0^\infty e^{-st} f(t)\, dt \tag{6}$$

The Laplace transform can also be used as a moment-generating function. The rth moment of f(t), μ_r, is then given by

$$\mu_r = E_f(t^r) = \lim_{s\to 0} (-1)^r \frac{d^r \hat{f}(s)}{ds^r} \tag{6a}$$

Flow systems are nonlinear systems insofar as the Navier-Stokes equation is nonlinear. The fact that residence-time distributions have the properties of a linear system indicates that they provide an incomplete description of the flow process in the reactor. Therefore, they cannot completely describe the flow, but represent only some important properties of it. If one designates the complete hydrodynamic description of all the flows and mass transfer processes in the reactor as M, then F is a linear property associated with M. While the transformation of M into F is unique, the opposite is not true. As F is an incomplete description of M, a single F can have an infinite number of functions M associated with it. In other words, *there can be a large number of completely different hydrodynamic flow systems with the same residence-time distribution.*

Intensity Function or Escape Probability

A residence-time distribution is a deterministic property of the flow system. It is sometimes advantageous to interpret it probabilistically. One can also look at F(t) as the probability that a single molecule will have a residence

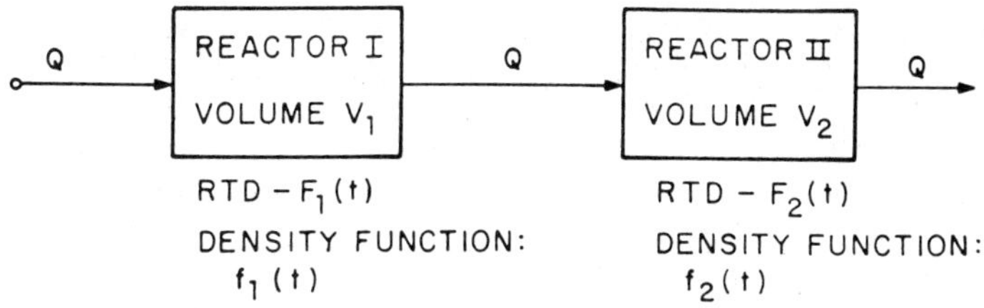

RTD OF TOTAL SYSTEM:

$$f_{(1+2)}(t) = \int_0^t f_1(t-\tau)\ f_2(\tau)\, d\tau$$

OR IN LAPLACE DOMAIN:

$$\hat{f}_{(1+2)}(s) = \hat{f}_1(s) * \hat{f}_2(s)$$

AVERAGE RESIDENCE TIME: $\tau_{(1+2)} = \dfrac{V_1 + V_2}{Q}$

(a)

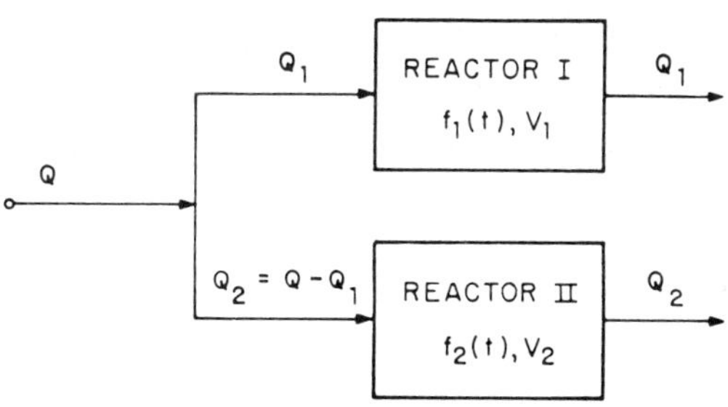

$$F_{(1+2)}(t) = \frac{Q_1}{Q} F_1(t) + \frac{Q_2}{Q} F_2(t)$$

$$f_{(1+2)}(t) = \frac{Q_1}{Q} f_1(t) + \frac{Q_2}{Q} f_2(t)$$

AVERAGE RESIDENCE TIME: $\tau = \dfrac{V_1 + V_2}{Q}$

(b)

FIGURE 4 Residence-time distribution of two reactors: (a) in series; (b) in parallel.

time less than t in the reactor. This presents the information contained in F(t) in a more accessible way. Consider a system characterized by the residence-time density f(t). In probability terms, f(t) can be interpreted as follows. If a particle enters the system at time t = 0, the probability of its leaving the system within the time interval (t to t + dt) is equal to f(t) dt. However, a slightly different problem may be posed. A particle has already stayed in the system for a time t; one wishes to know the probability of its leaving the system within the next time element dt. Let this probability be denoted by $\Lambda(t)$ dt. The function $\Lambda(t)$ may be evaluated as follows. The probability of a particle leaving the system between time t and t + dt equals f(t) dt. This probability is the product of two terms: The probability of the particle not leaving before t, which is 1 − F(t), and the probability of its leaving between t and t + dt, assuming that it is still in the system and has not left before—this was defined as $\Lambda(t)$ dt.

Thus we obtain

$$f(t)\,dt = [1 - F(t)]\Lambda(t)\,dt \tag{7}$$

and by simple rearrangement $\Lambda(t)$ is derived as

$$\Lambda(t) = \frac{F(t)}{1 - F(t)} = \frac{-d\ \ln[1 - F(t)]}{dt} \tag{8}$$

$\Lambda(t)$ is the conditional probability distribution and is also known as the intensity function or the escape probability. It must be stressed that there exists a one-to-one correspondence between $\Lambda(t)$ and f(t). Indeed, the transformation of $\Lambda(t)$ into f(t) is governed by

$$f(t) = \Lambda(t)\exp\left[-\int_0^t \Lambda(t')\,dt'\right] \tag{8a}$$

Physically, $\Lambda(t)$ dt is a measure of the probability of escape for a particle that has stayed in the system for a period t during the time interval t and t + dt. Thus, in the case of an ideally (exponentially) mixed vessel, one expects the intensity function Λ to be a constant, as the chance of escape during any time interval dt is independent of previous history and is equal to dt/τ. This is indeed the case. Although there is a one-to-one correspondence between f(t), F(t), and $\Lambda(t)$, information about the properties of the residence-time distribution and the nature of the flow system is more easily accessible from $\Lambda(t)$.

Utilizing the intensity function, it is easy to demonstrate a previously mentioned problem with residence-time distributions, namely, that the relation between F and M is not unique. This is illustrated in Figure 5, which describes the properties of a number of theoretical flow models. Thus, for an ideal uniformly stirred tank, $\Lambda(t)$ must be a constant (see Fig. 5a, curve A). However, the converse is not true. A $\Lambda(t)$ that is constant implies that the escape probability of a particle is independent of its previous history. In other words, the information as to how long a particle has been in the system tells us nothing about its future behavior. The fact that a system has this property does not prove or indicate that it is an ideally stirred reactor. The latter also implies that the concentration of

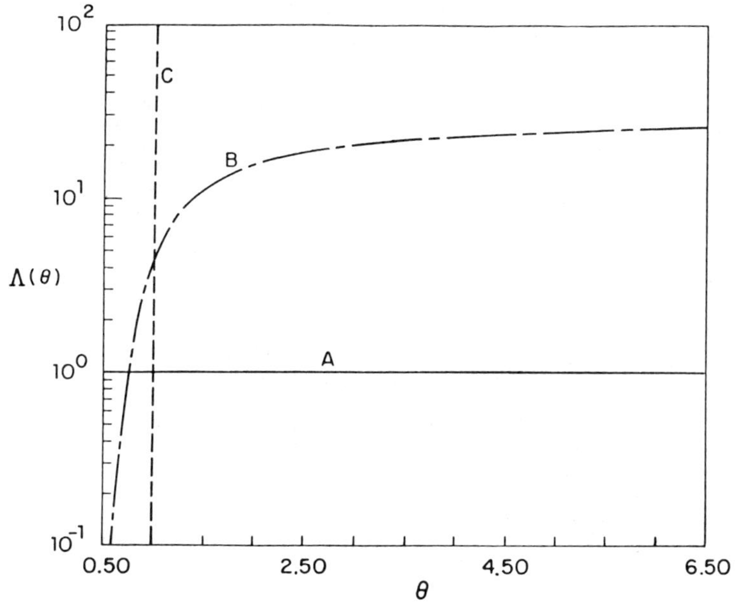

(a)

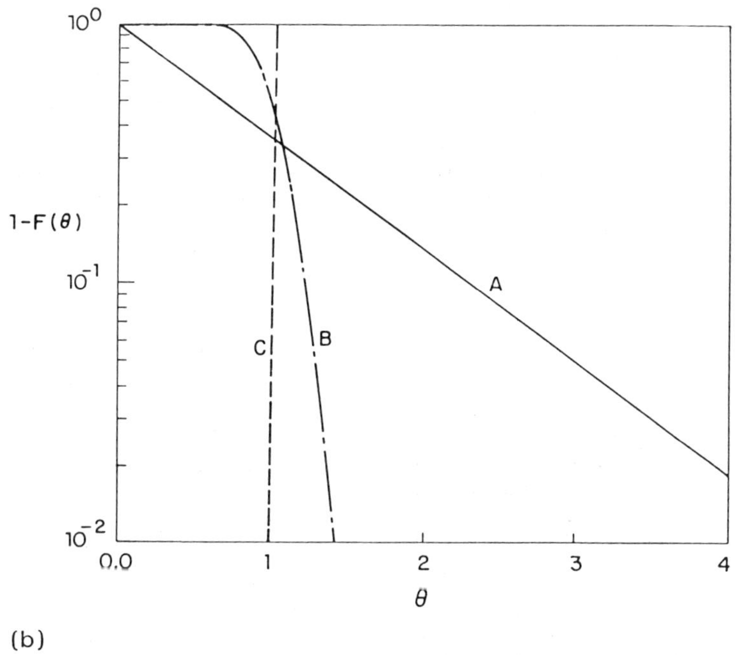

(b)

FIGURE 5 (a) Intensity function for some simple flow models: A, stirred tank; B, approach to plug flow (estimated by 30 tanks in series); C, ideal plug flow (δ function at $\theta = 1$). (b) Cumulative residence-time distribution for the models in part (a).

all different species in the reactor is completely uniform at all points, which is not a necessary condition for $\Lambda(t)$ to be constant.

In many reactors, such as packed beds or baffled fluid-bed reactors, one wishes to approach plug flow as closely as possible. In reality, one cannot achieve ideal plug flow and the escape probability will resemble the one shown in Fig. 5a, curve B. This condition will be called near plug flow. Such a flow can be modeled by either a series of stirred tanks or by a one-dimensional diffusion model. Both of these share one property, which is physically very well represented by the escape probability. For values of t that are small compared to the average residence time ($t \ll \tau$), the escape probability is close to zero. As t approaches τ the escape probability rises until it reaches a final constant value. Physically, that means that for $t \gg \tau$ the probability of the molecule being near the exit if it is still in the system is high. In a series of stirred tanks one can therefore assume that for $t \gg \tau$ the particle is probably in the last tank and its escape probability is the probability of escaping from the last tank (see Fig. 6a, curve A). It can be shown (see Shinnar and Naor, 1967), that in a series of stirred tanks with forward and backward flow between the tanks (Fig. 7), the escape probability has a similar form. The same applies to the one-dimensional diffusion model (Fig. 6a, curve B).

In a real packed bed the flow is not necessarily one-dimensional. There might be a flow maldistribution due to a bad distributor. If there are regions in which the fluid flows faster, then to maintain the same average residence time, there must also be regions of slower flow. Three such models are given in Figure 8. In such a case, if $t \gg \tau$, the particle is probably in the slow region and has a lower escape probability. The same is true if we consider a system in which there are stagnant regions which particles can diffuse in and out of, or if there are adsorption phenomena (Fig. 8a, curve B). Whenever any $\Lambda(t)$ has a maximum and then decreases, it implies that the system either has regions of fast and slow flows or stagnant regions in which a particle can get trapped. Curve C in Fig. 8a shows that adsorption phenomena do not necessarily lead to decreasing values of Λ. If the mass transfer processes are rapid compared to the overall flow velocity, no stagnation will be occur and the escape probabilities will not be significantly affected.

The exact value of the intensity function is often less important than its form, which provides physical insight into the nature of the flow. The critical features of $\Lambda(t)$ are recognizable from the form of $\ln(1 - F)$ versus t as $\Lambda(t)$ is the derivative of $-\ln(1 - F)$. This can be seen from Figs. 6b and 8b, where $\ln(1 - F)$ plotted versus t for the flow models given in Figs. 6a and 8a. Figure 8c illustrates f(t) for cases with stagnancy. Note that it is much harder to interpret f(t) then $\ln(1 - F)$ or $\Lambda(t)$.

Age Distribution

Consider the thought experiment where a timer is attached to all the particles entering the system. Instead of collecting them at the outlet, the whole system is stopped at some time t_0 to collect all the particles inside the reactor. One can then construct a distribution G(t) such that G(t) measures the fraction of particles inside the reactor that entered after $t_0 - t$. G(t) expresses the probability that a particle chosen at random inside the reactor has an age less than t. It is also called the *age distribution*. G(t) is related to F(t) by the relation

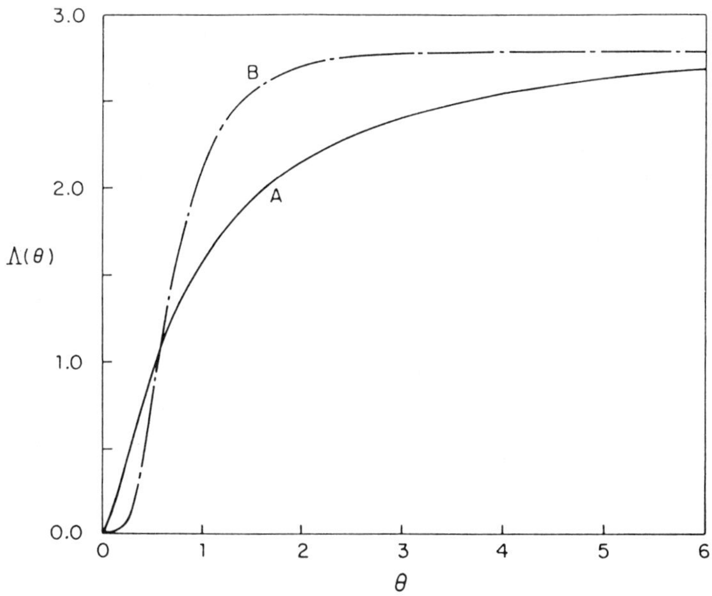

(a)

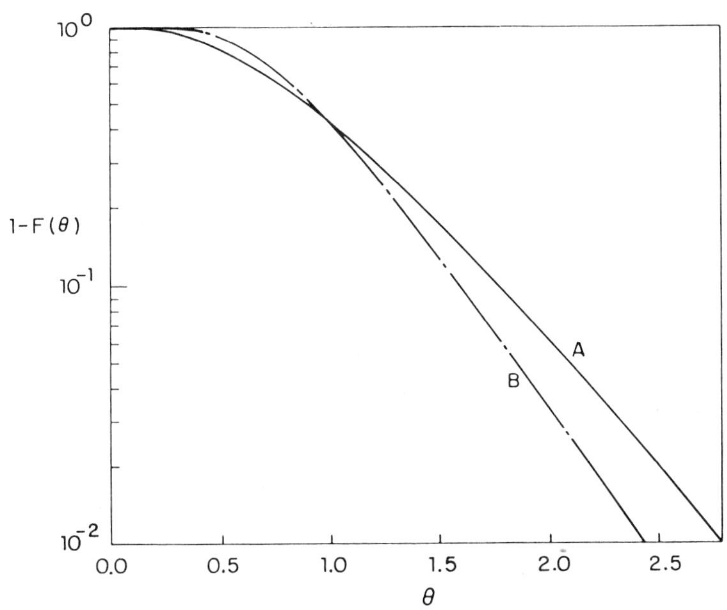

(b)

FIGURE 6 (a) Intensity function for some simple flow models: A, three stirred tanks in series; B, one-dimensional diffusion model. (b) Cumulative residence-time distribution for the models in part (a).

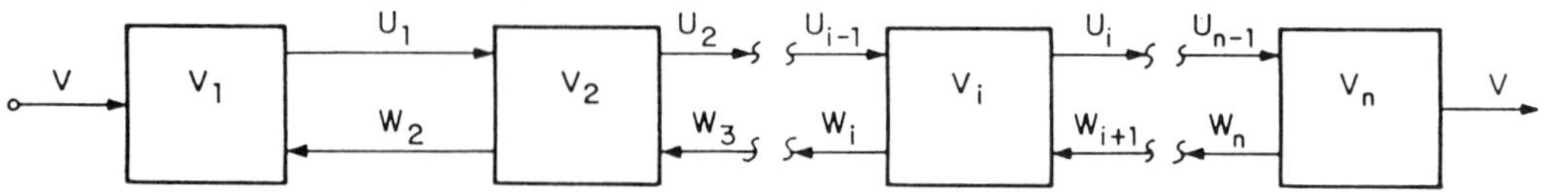

FIGURE 7 Schematic diagram for a series of stirred tanks in series with forward and backward flow between the tanks.

$$g(t) = \frac{1 - F(t)}{E_f(t)} \tag{9}$$

where g(t) is the density function of G(t).

In a multistage system, one can examine a single stage Z and define a local age distribution G_Z with respect to the inlet in the same way as one defines a local residence-time distribution F_Z (see Shinnar and Naor, 1967; Zvirin and Shinnar, 1975). The same can be done for a local zone in a reactor if the flow is not turbulent and the properties of the reactor are steady in this zone.

A related function is the future lifetime distribution A(t). Consider again the thought experiment where all molecules inside the reactor are

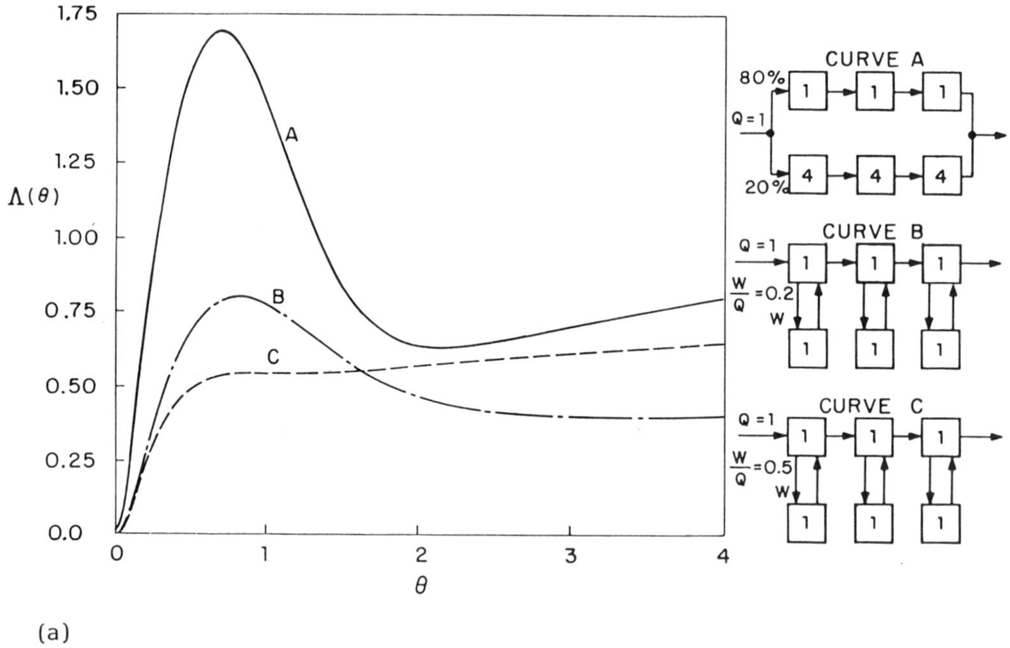

(a)

FIGURE 8 (a) Intensity function for three simple flow models described above. (b) Cumulative residence-time distribution for the models in part (a). (c) Residence-time density function for the models in part (a).

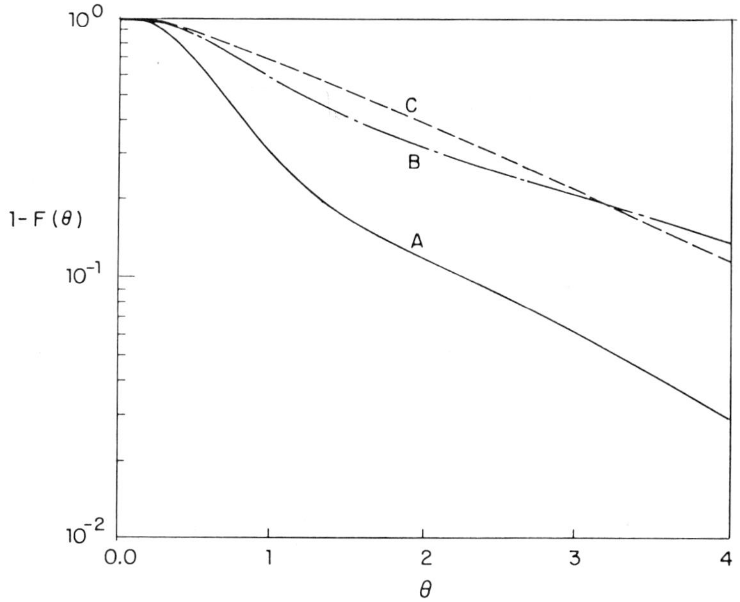

(b)

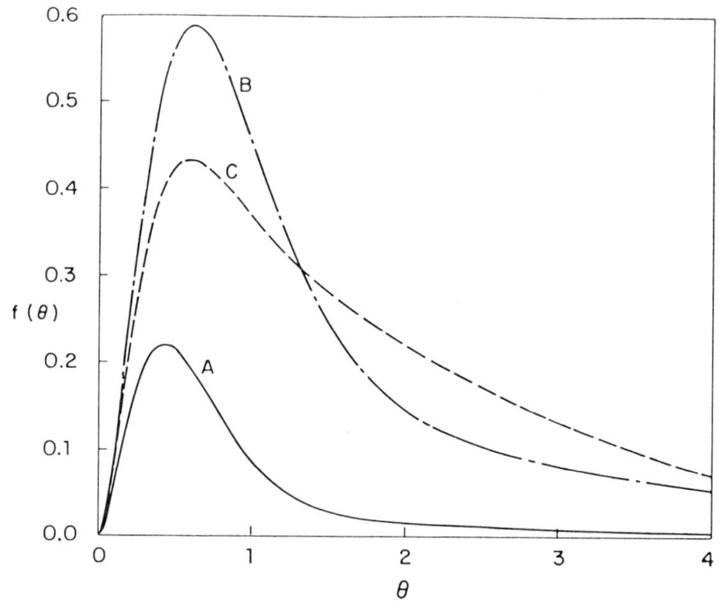

(c)

FIGURE 8 (Continued)

provided with a timer. This time, at $t = t_0$ set all the timers to zero. Then record the time t for each molecule as it exits and construct the distribution A(t). Thus A(t) is the probability that a particle which is in the reactor at time t_0 has exited before $t_0 + t$.

One can also define another related distribution inside the reactor. Consider the thought experiment where all the particles inside the reactor are again provided with a timer. At a set time t = 0 all the clocks are set to 0. One can then record the clock times on the particles at the exit and construct a distribution A(t). A(t) is called the future lifetime distribution. It can be shown that A(t) is equal to G(t) (see Noar and Shinnar, 1963). Once again, one can define a local future lifetime distribution $A_Z(t)$. However, $A_Z(t) \neq G_Z(t)$. Both $A_Z(t)$ and $G_Z(t)$ can be measured and are sometimes useful in modeling (Zvirin and Shinnar, 1975), but this is outside the scope of this chapter.

Multiphase Systems

All the definitions given until now refer to a single fluid. If a fluid is made up of different components, each of them may have a different residence-time distribution. In a homogeneous reactor, the differences between the residence-time distributions are in general negligible. However, in heterogeneous reactors, the differences may be quite large, as different compounds may adsorb differently or distribute themselves differently between the phases in multiphase reactors. Here the initial hypothetical definition can be very helpful in keeping things straight. The system has to be steady state and one must be able to tag a particle with a clock such that it retains its identity in passing through the reactor. If the compounds react, one may have to tag an element that does not change and obtain a set of residence-time distributions $F_i(t)$ for each such element. For each $F_i(t)$ the definitions of $f_i(t)$ and $\Lambda_i(t)$ do not change. The only relation that does not hold is that τ_i, the first moment of f(t), is no longer equal to V/Q. The holdup or average residence time τ_i is now different for each element and not known a priori. A detailed discussion of this problem is given in a later section.

TRACER EXPERIMENTS: MEASUREMENT OF RESIDENCE-TIME DISTRIBUTIONS

Methods of Measurement

Step Changes in Tracer Concentration

The residence-time distribution discussed in the preceding section can be measured in any steady-state system by use of a suitable tracer. Assume that at time t = 1 one starts labeling all the particles in the feed. One then measures the fraction of labeled particles at the exit from the reactor at time t. Any labeled particle that appears at the exit at time t must have entered the reactor after time t = 0 since before that time, no particles were labeled. Similarly, any unlabeled particle must have entered the reactor prior to time t = 0 since after that time all particles were labeled. Therefore, the fraction of labeled particles is, by definition, equal to F(t), which is the probability of staying in the reactor less than time t and the fraction of unlabeled particles is equal to (1 − F). For small time intervals it is easier to measure F, whereas for long time spans it is more convenient to measure 1 − F.

It is not necessary to label all the particles. Assume that at some instant in time t = 0, instead of labeling all the particles, one starts labeling a constant fraction C_f of the particles in the feed and then measures the fraction of labeled particles C(t) at the exit. Clearly,

$$C(t) = F(t)C_f$$

Since one is usually more interested in estimating F(t), this is usually expressed as

$$F(t) = \frac{C(t)}{C_f} \tag{10}$$

In this case, the fraction of unlabeled particles is no longer directly related to (1 − F), as some unlabeled particles have entered the reactor after time t = 0.

If one waits long enough, then as $t \to \infty$, $F(t) \to 1$ and $C(t) \to C_f$. In practice, equilibrium is reached when $t \gg \tau$, the time constant of the system. At that point the fraction of labeled particles at the exit will equal C_f, the fraction of labeled particles at the feed. Once the system reaches equilibrium it is useful to stop suddenly the flow of labeled particles (tracer). From this point on, the fraction of labeled particles at the exit will decrease monotonically. The fraction of labeled particles, C(t) at the exit, measured from the new t = 0 is then given by

$$C(t) = [1 - F(t)]C_f$$

or

$$1 - F(t) = \frac{C(t)}{C_f}$$

This is identical to the case in the preceding paragraph except that now it is the labeled particles that have entered the system prior to time t = 0.

A step change in tracer is the most accurate way of measuring a residence-time distribution. $C(t)/C_f$ should be measured after the tracer is introduced until equilibrium is achieved and then after the tracer is stopped. If one could measure the average concentration $\langle C/C_f \rangle$ inside the reactor after a step input at time zero (averaged over the total volume at each instant of time), this yields an estimate of G(t), the age distribution of the system. This can be achieved experimentally by using a radioactive tracer and measuring the total radiation emanating from the reactor.

Pulse Experiments

The step input experiments described in the preceding section directly measure F(t) and 1 − F(t). Another more commonly used tool is the injection of a fixed amount m of tracer at t = 0 and then measuring the fraction of labeled particles C(t) at the exit. This leads to a direct measurement of the density function f(t) given by

$$mf(t) = QC(t)$$

or

$$f(t) = \frac{QC(t)}{m} \tag{11}$$

where m is the amount of tracer injected at t = 0 and Q is the volumetric flow rate. To have reliable estimates of f(t), one needs independent measurements of Q and m. The relation

$$m = Q \int_0^\infty C(t)\, dt \tag{12}$$

allows an estimate of the consistency of available estimates. In the absence of accurate knowledge of either Q or m, one can use Eq. (12) to reformulate Eq. (11) as

$$f(t) = \frac{C(t)}{\int_0^\infty C(t)\, dt} \tag{13}$$

This allows an estimate of f(t) from a pulse injection in any steady-state system directly from the measured fraction C(t). It is the most common way in which tracer pulse experiments are utilized.

Sinusoidal Inputs

Consider the case where the fraction of labeled particles C_f at the inlet is varied as a function of time. It can be shown that f(t) is equal to the linear response function for the tracer, or

$$C(t) = f(t) * C_f(t) \tag{14}$$

where C(t) is the outlet concentration, $C_f(t)$ is the inlet concentration, and * denotes the convolution integral. If one writes Eq. (14) in the Fourier transform, then

$$\hat{C}(j\omega) = \hat{f}(j\omega) \times \hat{C}_f(j\omega) \tag{15}$$

Equation (15) is well known from control theory. One can therefore evaluate f(t) in the same way that one evaluates the process transfer function in controller design.

In this case $C_f(t)$ is varied in a sinusoidal form. After waiting for all transients to die out and for steady state oscillation in the output to be established, the amplitude ratio $C(t)/C_f(t)$ and the phase angle are found by recording input and output data (see Fig. 9a). The frequency of the input signal is changed and a new amplitude ratio and phase angle are determined. Thus the complete frequency-response curve is determined experimentally by varying the input frequency over the range of interest. These measurements are direct measurements of the Fourier transform of

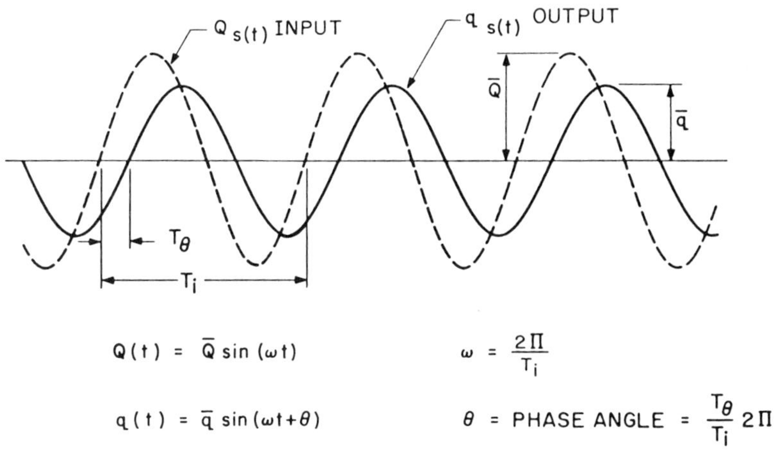

(a)

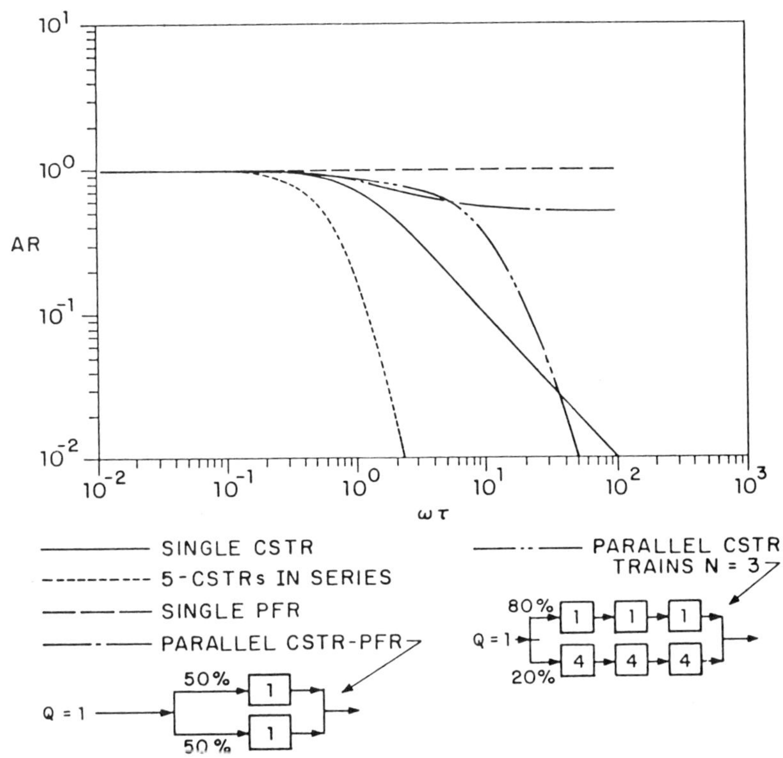

(b)

FIGURE 9 (a) Sinusoidal input-output curves, ω is frequency of input signal. (b) Frequency response for some simple flow models—amplitude ratio. (c) Frequency response for the flow models in part (b)—phase lag. (d) Laplace transform of the density function for the flow models in part (b).

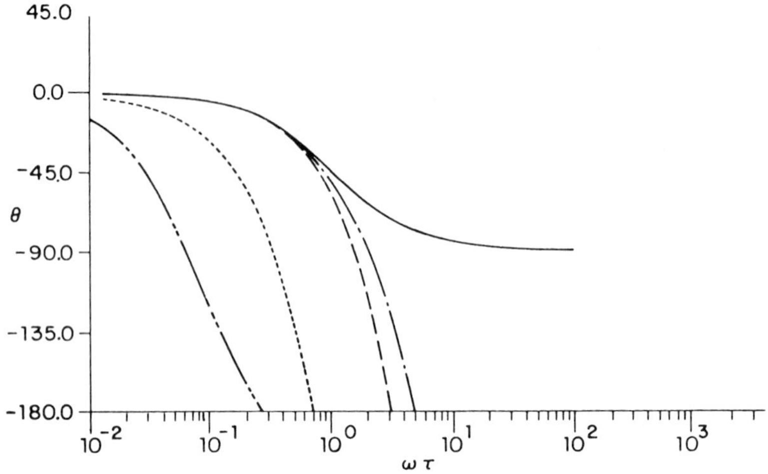

(c)

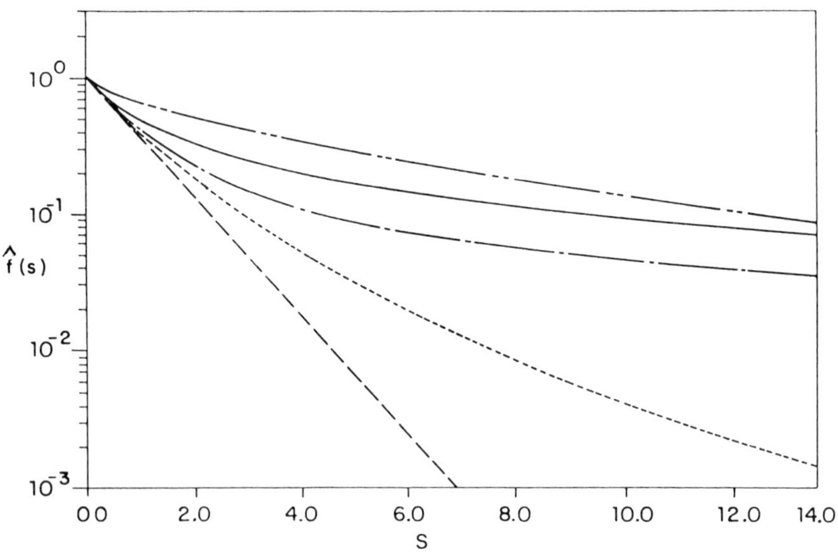

(d)

FIGURE 9 (Continued)

the density function $\hat{f}(j\omega)$. One can look at $\hat{f}(j\omega)$ as a vector in polar coordinates (r,ϕ). The amplitude ratio defines the magnitude of the vector r for each w, and the phase lag defines ϕ. If one can measure both the amplitude ratio and the phase lag over a wide range of frequencies, one has a good estimate of the Fourier transform of f(t). Some typical forms of these functions are given in Fig. 9b and c.

Measurement of the Laplace Transform from Arbitrary Inputs

In some cases it is hard to perform an accurate step or pulse experiment. One might not be able to feed the tracer directly to the reactor, or one might have a two-stage system where the inlet concentration to the second stage can be measured, but the tracer cannot be fed directly to it (Østergaard and Michelsen, 1969). For all such cases, one can utilize Eq. (14) and obtain the Laplace transform $\hat{f}(s)$ of the density function f(t) by

$$\hat{f}(s) = \frac{\hat{C}(s)_{\text{output}}}{\hat{C}(s)_{\text{input}}} \tag{16}$$

As C(t) is a measured function, $\hat{C}(s)$ must be obtained by numerical integration. This is illustrated in Fig. 10b for several values of s.

Measurement of $\hat{f}(j\omega)$ and $\hat{f}(s)$ poses a problem. Although in theory it is possible to obtain f(t) by inversion,

$$f(t) = \frac{1}{2\pi j} \int_{a-\delta}^{a+\delta} e^{st} \hat{f}(s)\, dt$$

$$f(t) = \frac{1}{2\pi j} \int_{-\delta}^{\delta} e^{j\omega t} \hat{f}(j\omega)\, d(j\omega) \tag{17}$$

the results depend strongly on the values of $\hat{f}(s)$ as $s \to 0$ or $\hat{f}(j\omega)$ as $w \to 0$, which are hard to measure. As $\hat{f}(s)$ is hard to invert, either one can try and fit $\hat{f}(s)$ with a reasonable model and then invert it, or one can use the measurement of $\hat{f}(s)$ to evaluate directly the properties of the residence-time distribution. Some typical forms of $\hat{f}(s)$ are shown in Fig. 9d.

Indirect Estimates of f(t) and F(t)

Indirect ways of measuring residence-time distribution are discussed later. These are based on the effect of residence-time distribution on conversion, especially on product distribution in complex reactions.

Measurement of Moments of Residence-Time Distribution

The moments of the residence-time distribution can be computed from either f(t) or F(t) by using Eq. (2) or

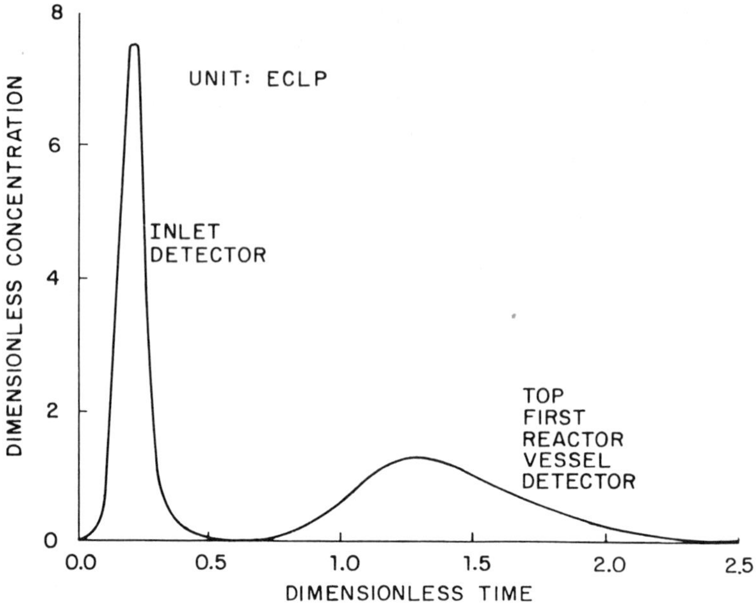

(a)

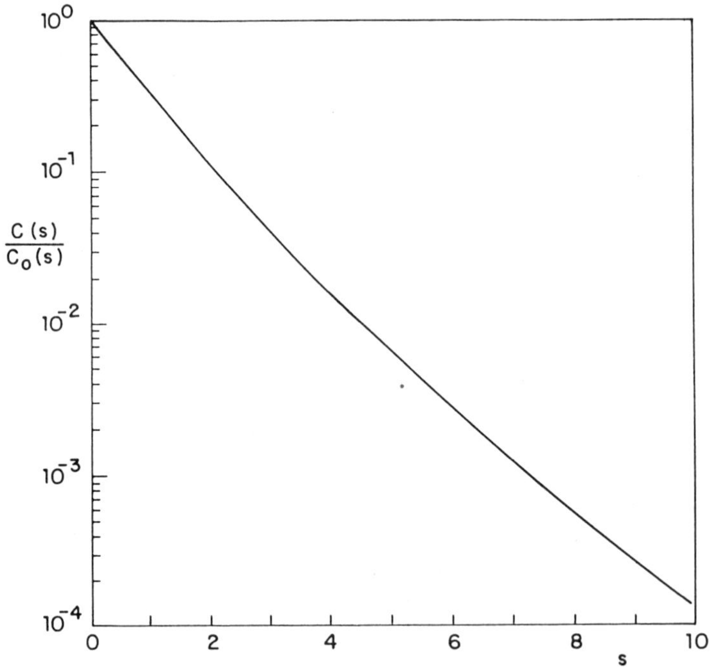

(b)

FIGURE 10 (a) Tracer response of a radioactive gas tracer in an Exxon Donor Solvent coal liquefication reactor. (b) Laplace transform of the residence-time distribution from part (a) (top of reactor) obtained by

$$\mu_r(f) = E_f(t^r) = r \int_0^\infty t^{r-1}(1 - F)\, dt$$

The comments on accuracy made before apply also here and the step response has a significant advantage. In homogeneous systems, the first moment can also often be estimated independently from

$$\mu_1 = \tau = \frac{V}{Q} \tag{18}$$

where V is the volume of the fluid in the reactor. This is an important check on the accuracy of the measurement of f(t0. Relation (18) can only be used in homogeneous systems, as in a system where there is adsorption (or solution in another phase) independent knowledge or estimate of the holdup is difficult.

In the case where, instead of directly measuring f(t) or F(t), the Laplace transform $\hat{f}(s)$ is measured, there are two ways to obtain the moments. When the Laplace transform is obtained from measuring the response $C_2(t)$ to an arbitrary input signal $C_1(t)$ (see Fig. 10), one can get the moments from the fact that the moments for two systems in series are additive. Recall that

$$f(t) = \frac{Q}{m} C(t) \tag{11}$$

Thus if one now writes

$$\mu_1 = \tau_1 = \frac{Q}{m} \int_0^\infty tC(t)\, dt \tag{19a}$$

and

$$V_1 = \frac{Q}{m} \int_0^\infty (t - u)^2 C(t)\, dt$$

where μ_1 and V_1 are the mean and variance, then

$$E_f(t) = \tau = \tau_2 - \tau_1 \tag{19b}$$

$$\hat{f}(s) = \frac{\int_0^\infty 1^{-st}\, C(t)\, dt}{\int_0^\infty 1^{-st}\, C_0(t)\, dt}$$

where C is the concentration at the top and C_0 the concentration at the inlet. The curve can be fitted by $[1/(1 + \tau s/N)]$, where $\tau = 1.18$ and N = 17. (From Tarmy, 1983.)

$$V_f(t) = V_2 - V_1$$

where V_2 and τ_2 are defined analogously to Eq. (19a). There is another way to extract the moments from $\hat{f}(s)$, which is especially important for cases where $\hat{f}(s)$ is obtained directly, as will be discussed later.

One can obtain the moments of f(t) directly from $\hat{f}(s)$ as shown in Eq. (17). Unfortunately, it is hard to estimate accurately the higher derivatives near $s \to 0$, but one can usually get a good estimate for the first derivative by using a Taylor expansion, get a reasonable estimate for V and γ, the coefficient of variation. If γ is small (the flow is close to plug flow), V and γ can be estimated by a filtering model such as

$$\hat{f}(s) = \frac{1}{(1 + \tau s/n)^n} \tag{19c}$$

which is a good approximation for flows that are close to plug flow. Equation (19c) has the form of transfer function for stirred tanks in series, but here n is used as a fitting parameter. If one can fit $\hat{f}(s)$ with Eq. (19c), then

$$\gamma^2 = \frac{1}{n} \tag{19d}$$

Another way to get the variance from the Laplace transfer for cases close to plug flow can be derived as follows. The measured quantities in Fig. 10 are $C_1(t)$ and $C_2(t)$. From this one can derive not only $\hat{f}(s)$ but also $d\hat{f}/ds$:

$$\frac{1}{f(s)} \frac{d\hat{f}(s)}{ds} = \frac{1}{\hat{C}_2(s)} \frac{d\hat{C}_2(s)}{ds} \quad \frac{1}{\hat{C}_1(s)} \frac{d\hat{C}_1(s)}{ds} \tag{20}$$

where

$$\frac{d\hat{C}_2(s)}{ds} \int_0^{\infty} t\hat{C}_2(t)e^{-st}\, dt \tag{20a}$$

Evaluation of the derivative of $\hat{f}(s)$ therefore does not require any numerical differentiation, but can be evaluated from the concentration curves by integration. If $\hat{f}(s)$ can be approximated by Eq. (19c),

$$\frac{1}{\hat{f}(s)} \frac{d\hat{f}(s)}{ds} = -\tau \left(1 + \frac{\tau s}{n}\right)^{-1} \tag{20b}$$

and

$$\frac{\hat{f}(s)}{d\hat{f}(s)/ds} = \frac{s}{n} - \frac{1}{\tau} \tag{20c}$$

Plotting the left side of Eq. (20c) versus s gives a straight line, the slope of which is equal to the coefficient of variation.

Required Properties of Reactors and Tracers for Measurements of Residence-Time Distributions

The most frequent problem the author has encountered in the practical measurement of residence-time distribution is that not enough though is given to verify that $C(t)/C_f$ is really an estimate of the residence-time distribution desired. There are several important conditions that need to be checked.

The choice of the tracer is very important. It should be accurately measurable over a wide range of concentrations. There are several other properties that also affect its choice. Furthermore, there are a number of other conditions that have to be fulfilled in order that the tracer experiments mentioned earlier will lead to a proper estimate of the residence-time distribution.

1. *The system must be a steady state and stationary in the sense that $C(t)/C_f$ is independent of the time at which the tracer is introduced into the feed.* As most flows are turbulent, this is not necessarily true for all reactors. However, if the length-to-diameter ratio is large, the turbulence effects are averaged out and at the outlet, this condition is satisfied at least to a first approximation. The same applies to systems equipped with mechanical agitators. In that case the size of the large eddies is much smaller than the diameter of the reactor and therefore are often averaged out independent of the length-to-diameter ratio. This can be tested in two ways. First, the step response must be monotonous. A response such as the one shown in Fig. 11 cannot be considered as a residence-time distribution. The ability to apply this criterion directly is a significant advantage in using the step resonse. Furthermore, two successive tracer experiments should give the same result within the expected experimental error. This condition is not just a question of measurement. The hypothetical experiment of equipping each molecule with a clock had exactly the same requirement. It was assumed that it did not matter when the particle entered.

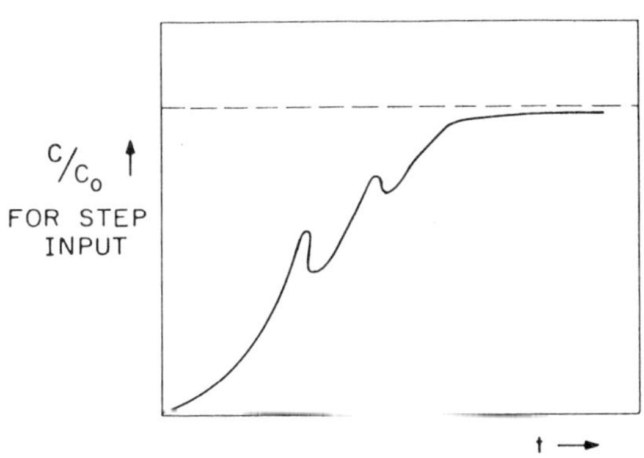

FIGURE 11 Response to step input for a system in which the response to a tracer is a function of the time of injection. In this case, $C(t)/C_f$ does not give the residence-time distribution of the system. This may be due to turbulent mixing, fluctuation in the flow rate, or large bubbles, or other factors.

In Krambeck et al. (1969) the concept of residence-time distribution is extended to systems that lack this property, but this is outside the scope of this chapter.

2. *The tracer response must be linear in the range of the step change.* If the tracer particles behave exactly as one of the molecular compounds in the steady-state system, this condition will be fulfilled automatically. If one can really tag particles in a steady-state flow, it cannot make any difference how many particles are tagged. As long as the system is in steady state, it also does not matter if the transport processes are nonlinear. This also applies to multiphase systems in which one of the compounds is either adsorbed on a solid surface or dissolved in the liquid phase. As long as the system is at steady state in the sence defined in the first criterion, each compound has a residence-time distribution, and if one finds a real tracer for this compound, its response will be linear, as it cannot matter how many particles are tagged. However, the only way to obtain tracers with such properties (at least approximately) is to take the real reactant feed and mark it by exchanging one of its atoms with an isotope. Such tracers are seldom used and in most cases one uses tracers that are inherently different from the fluid at steady state. There is nothing wrong with this method as long as one recognizes its shortcomings and takes them into account.

If the tracer is different from the steady state feed, it may upset the steady-state conditions, and there is no a priori reason for a linear response. If one wants to obtain a residence-time distribution, one must verify experimentally that the response is linear. This is accomplished by two successive tracer experiments with two different step magnitudes which should give the same response. What one obtains in this case is a residence-time distribution of the tracer compound, which is not necessarily identical with the residence-time distribution of the fluid in the reactor. Therefore, one must add yet another criterion.

3. *The tracer should be similar in behavior to the fluid in the reactor.* It is hard to specify this condition in a general way, as it depends strongly on the specific situation. In a homogeneous stirred tank, almost any tracer will do, as diffusion processes have negligible effects on F and there is no potential for adsorption. The tracer is carried along with the liquid. This applies to most homogeneous reactors. In such cases one seldom has to worry about linearity. On the other hand, in a fluidized-bed reactor, diffusion and adsorption are very important factors, and the results depend strongly on the tracer chosen. Which tracer to choose is therefore a question not only of availability but also of what information is sought from the experiment. In multiphase systems it is often appropriate to use several tracers, each simulating a different compound, or fluid property. This is discussed in detail later.

Relative Accuracy of Tracer Experiments Utilizing Step and Pulse Inputs

In principle, f(t) and F(t) contain the same information and each one can be derived from the other. In practice, the experimental errors are different for pulse inputs and step inputs. Step inputs tracer experiments have significant advantages over pulse experiments, provided that they are carried out in both directions. These advantages include:

1. Immediate recognition of some unsteady phenomena. These are illustrated in Fig. 11.
2. In a step experiment all one needs is a knowledge of C_f and C(t). As they are measured by the same instruments, any bias in measuring C_f will also appear in C(t) and will tend to cancel. However, it is important that the step be carried out in both directions, as it is difficult to measure $C(t)/C_f$ accurately if it is close to unity (large t for F and small t for 1 − F).

If one can do this, a direct estimate of F and 1 − F is obtained. On the other hand, in a pulse experiment C(t) does not give a direct estimate of f(t). One must derive it utilizing either Eq. (11) or (13). If one uses Eq. (13), the problem of bias in measuring C(t) is eliminated, but one faces the problem of estimating the integral. If the residence-time distribution is simple and well behaved and has an exponential tail, a fairly accurate estimate of f(t) is easily obtained. Otherwise, the overall form of f(t) can usually be derived, but the relative magnitude may not be accurate. One way to verify the accuracy of the measurements utilizes Eq. (12) and compares $\int_0^\infty C(t)\,dt$ to m/Q. If possible, this should always be done, as it greatly improves the reliability of the measurement. However, in industrial reactors it is often impossible or very difficult to measure either Q or m accurately. A step experiment measures F and 1 − F directly and therefore also allows a direct measurement of the escape probability. If one plots ln(1 − F) versus time, all the interesting features of the escape probability (or intensity function) will be directly apparent from the plot. On the other hand, to derive the intensity function $\Lambda(t)$ from f(t) using Eq. (7), one must first compute [1 − F(t)] by integration of f(t), which introduces inaccuracies. In order To minimize the error in estimating $\Lambda(t)$, one should use the relation

$$\Lambda(t) = \frac{C(t)}{\int_t^\infty C(t)\,dt} \tag{21}$$

The value of $\int_0^\infty C(t)\,dt$ is estimated by plotting C(t) on a logarithmic plot and approximating the tail by an exponential function. The time constant of the exponent is then estimated from the tangent of log C(t) for the largest values of t for which the measurements are still reliable.

In general, the author does not recommend the use of frequency-response techniques or sinusoidal inputs for measuring residence-time distributions. The only reason they are discussed is that they might be available from control studies and can therefore serve as a starting point. The reason for this broad statement is that whenever such experiments are feasible, a proper step response in both directions is also feasible, is less time consuming, and provides far better information. In control design the opposite is true. It is always best to get a direct estimate of the property of the function required for design. In chemical controller design one needs to know $\hat{f}(j\omega)$ and it is preferable to measure it directly, compared to estimating it from a step response. On the other hand, to evaluate the flow pattern, one needs to know the escape probability and therefore a direct estimate of $[1 - F(t)]$ is preferable. Furthermore, the accuracy and the range of ω at which $\hat{f}(j\omega)$ is needed is different in the two cases. In controller design, one needs the overall form of $\hat{f}(j\omega)$ and accurate values near the phase lag of $-180°$. Accurate estimates of $\hat{f}(j\omega)$ for very high and very small ω are of little interest in controller design, but are important in evaluating residence-time distribution.

In cases where an experiment with an input that reasonably approximates a step or a pulse cannot be performed, and one must often compromize and utilize other inputs. A practical example is shown in Fig. 10. In this setting one cannot directly measure F(t) of f(t) but can still measure an estimate of the Laplace transform. If one has an approximate flow model, one can fit it directly to $\hat{f}(s)$ as shown in Fig. 9d. This is less accurate than estimating $[1 - F(t)]$ by a step or pulse experiment. With the latter one can often get a good estimate of the tail of the function which determines the moments. On the other hand, to obtain reliable estimates of the tails from $\hat{f}(s)$, one needs to have accurate measurements of $\hat{f}(s)$ for values of s where $s \to 0$. Estimating the second derivative near $s \to 0$ from experimentally determined $\hat{f}(s)$ is much more difficult, and therefore estimates of variance are more difficult and have to be obtained by first fitting a two-parameter model of $\hat{f}(s)$ and then estimating the variance from this model. This procedure involves some mistakes, but if the variance is small, the results are often reasonably accurate [Østergaard and Michelsen (1969)]. An example is shown in Fig. 10. If one could plot $\hat{f}(s)$ in a way equivalent to an escape probability, it would allow better insight into the meaning of any deviations of $\hat{f}(s)$ from the simplified flow models used to fit the experimental results. However, the author is not aware of any method that would make this possible.

It is sometimes possible to use Eq. (16),

$$\hat{f}(s) = \frac{\hat{C}(s)_{\text{output}}}{\hat{C}(s)_{\text{input}}} \tag{16}$$

to filter out the effects of measurement noise, especially on the tail. It should be stressed that this method is always one of last resort and that one always loses significant information. In many cases the most important information of the tracer experiment is contained in the tail ($t \to \infty$) and there is no way to replace getting an accurate measurement of it. Using Eq. (16) merely obscures the fact that the measurement noise prevented an accurate measurement of the tail of f(t). The method is sometimes useful when one knows that the contribution of the tail is small or when one is interested in bypass phenomena [which depend on the properties of f(t) near zero]. In all such cases it is important to use an adequate number of values of s in the range $0 < s < \tau$, where τ is the expected average residence time. But one should always be aware of the shortcomings of this method.

USE OF RESIDENCE-TIME DISTRIBUTIONS IN REACTOR MODELING AND TESTING OF INDUSTRIAL REACTORS

Testing for Maldistribution in Industrial Reactors

Residence-time distributions are an important diagnostic tool inevaluating the performance of industrial reactors as well as in setting up simplified flow models. They can be utilized for modeling both homogeneous and heterogeneous reactors. The most important application is probably checking for maldistribution in flows. A large number of industrial reactors are designed to approximate either a stirred tank or a plug-flow reactor. In the first case good mixing is essential. In the second case, such as a packed-bed reactor, it is very important that the feed is well distributed over the cross section and that there be no bypassing due to faulty internals or defective catalyst loading. If a reactor does not perform as predicted, it is important to find out whether the poor performance is due to an inherent scaleup problem, a faulty catalyst, or a wrong mechanical design. A tracer experiment can be a valuable tool in distinguishing between these possibilities.

If there are similar reactors in operation which are accessible, one can conduct a tracer experiment on both reactors and compare the results. Otherwise, one can compare the results of the tracer experiment to those predicted by an idealized model. They key question is how one interprets any deviations between the results of two reactors or between the actual performance and the model. Here the escape probability or intensity function, $\Lambda(t)$, is a useful tool, as it provides one with immediate insight as to what the deviations mean. One does not even always have to actually plot or compute $\Lambda(t)$. With some experience one can recognize the important features of the escape probability from a plot of $\ln(1 - F)$ versus t. Therefore, what is really needed is a reliable estimator of $(1 - F)$ which can usually be obtained from tracer experiments. The following two examples will illustrate the method.

Imperfect Mixing and Bypassing in Stirred-Tank Reactors

There are several problems that can occur in the scaleup of stirred-tank reactors. One of them is that the mixing is insufficient to ensure that the entering fluid mixes sufficiently rapidly with the contents of the reactor [(Evangelista et al., 1969b; Shinnar, 1961). In heterogeneous stirred-tank reactors there are the additional problems of mass transfer between the phases, uniformity of turbulence, and the other complicating factors which are discussed in Shinnar (1961) and Tarmy (1983). Although these problems can be studied using tracer experiments, they involve methods that are outside the scope of this chapter. There is one problem that commonly occurs in the scaleup of a stirred tank reactor and can be effectively dealt with using the residence-time distribution. This is the case of a design where the escape probability is to remain constant and is discussed below.

When trying to analyze whether the escape probability is really constant, consider the type of deviations to be expected. One would always expect some deviation as $t \to 0$, as there is a minimum finite time for the particle

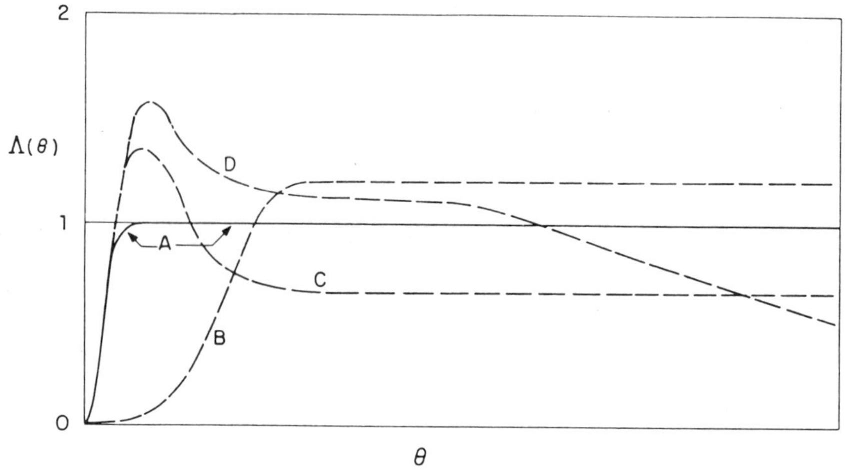

FIGURE 12 Intensity function for imperfectly mixed stirred-tank reactors: A, short delay between inlet and outlet (normal case); B, delay between inlet and outlet due to insufficient stirring; C, bypass between inlet and outlet; D, bypass between inlet and outlet and stagnant regions due to insufficient stirring.

to reach the exit (see Fig. 12, curve A). However, this time is usually small compared with the average residence time τ, and if τ is large, one would not notice it. If the mixing intensity is too low, this initial deviation might be quite large, especially in tall reactors with viscous fluids. In that case one would expect $\Lambda(t)$ to look like curve B in Fig. 12. This could be remedied by increasing the agitator speed or the agitator design. Another common type of faulty performance is shown by curve C in Fig. 12. It occurs if the inlet and outlet pipes are close to each other and mixing intensity is insufficient to distribute the incoming fluid sufficiently fast over the total reactor. A fraction of the liquid will reach the outlet directly creating a bypass. For small t one could then expect $\Lambda(t)$ to be significantly higher than the average expected value, as entering particles have a chance to be in that bypass. If a particle has stayed in the reactor longer than it takes to reach the exit via the bypass, its chance to exit is now lower than in the beginning and $\Lambda(t)$ will exhibit a maximum at low values of t. If the vessel is really badly mixed, $\Lambda(t)$ will even have a decreasing tail (Fig. 12, curve D).

A significant advantage of representing the information contained in the residence-time distribution by plotting $\Lambda(t)$ is that one can plot the expected form of $\Lambda(t)$ without any computations, as $\Lambda(t)$ corresponds to the physical concept of the system. An actual example taken from an industrial reactor is shown in Fig. 13 (Murphree et al., 1964). The conversion in the reactor was much lower than expected from the pilot plant results. A step input tracer experiment was performed and the result is replotted in Fig. 13a, curve E (one impeller). Figure 13b gives $\ln(1 - F)$ and Fig. 13c the corresponding curve. The escape probability is similar to curve B, in Fig. 12. It indicates imperfect mixing. The mixer was modified and equipped with two impellers and again tested by a tracer step input. The results are also given in Fig. 13, curve G, and the escape probability is similar to curve A, indicating good mixing. The new mixer also increased the conversion to that expected from the pilot plant.

The tracer results in Fig. 13, illustrate another aspect of the use of residence-time distributions that is not evident from the escape probability as plotted. Both curves refer to a reactor with the same volume and volumetric feed rate, and therefore the same average residence time τ. One can in both cases estimate τ from $\int_0^\infty (1 - F)\,dt$ using an exponential tail and compare it to its known value. For curve G the estimate is correct, whereas curve E underestimates τ by 30%. The fact that curve E underestimates τ shows that the reactor is not well perfused and has a strongly stagnant region which does not appear in the plot. If one could measure F(t) accurately for long times, curve E must cross curve G for the two integrals to be equal. The true escape probability would look like curve D in Fig. 12. If τ is equal for the two cases, then curves of F(t) must cross each other. Therefore, independent knowledge of τ provides some important checks on the accuracy of a tracer experiment.

A fairly constant escape probability, independent of history, which is characteristic of a well-stirred tank, occurs in quite a number of other situations, such as solids reacted in fluid beds, and the test described here applies to all these situations.

Packed-Bed and Trickle-Bed Reactors

In the design of packed-bed and trickle-bed reactors, it is important to ensure that the feed is well distributed and that the catalyst is arranged in such a way that there are no preferred pathways with lower pressure drops

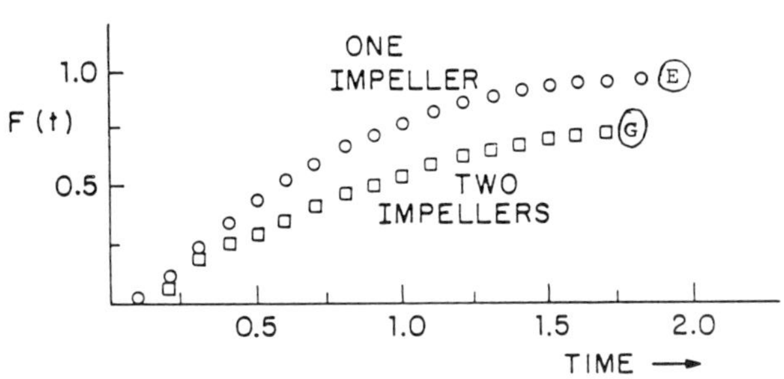

(a)

FIGURE 13 Residence-time distribution for reactors with imperfect mixing. (a) Experimental residence-time distribution for a stirred-tank reactor (from) Woodrow, 1978). The reactor did not match the pilot plant and gave low conversion. A tracer experiment showed a severe maldistribution of the flow. A second impeller was installed, which improved the flow pattern as measured by a residence-time distribution approaching an ideally stirred tank. The reactor then matched the pilot plant. Here the imperfect mixing is not directly recognizable from curve E. It can only be recognized by comparing the estimate of $\int (1 - F)$ with the known value of τ estimated from volume divided by flow rate (see the text). This indicates that in case E the reactor is not well perfused. (b) Cumulative residence-time distribution. (c) Escape probability (intensity function).

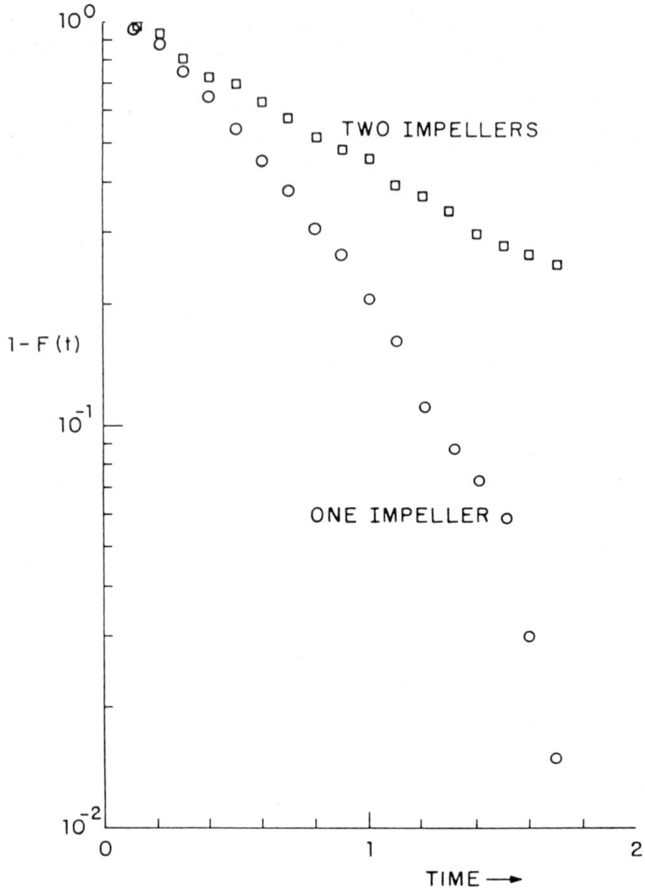

(b)

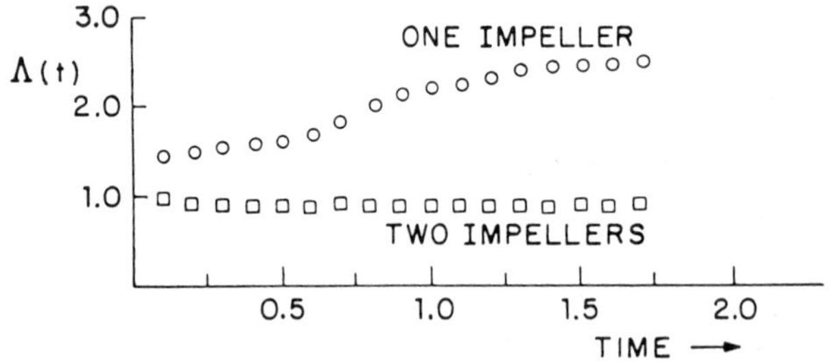

(c)

FIGURE 13 (Continued)

which create a bypass. Before proceeding with analytical examples, one must first define the concept of bypass and stagnant regions.

A well-designed packed bed approaches plug flow and will lead to a $\Lambda(t)$ function such as curve B in Fig. 5a. Most of this is due to the fact that in a randomly packed bed, the path length for different molecules is unequal and part of it is due to mixing in the interstices between the particles. Whenever f(t) is not a delta function, one can in theory represent the pathways of the individual molecules by an organ pipe model such as the one shown in Fig. 14. As the pathways are of unequal length, one can conceive of the shorter one as a bypass. In general, this is not the prime concern of the diagnostic test. One is looking for much stronger differences caused by maldistribution of flow. One can utilize the residence-time distribution in this setting because the typical flow distribution in a packed bed has a Λ function similar to that of curve B in Fig. 5a with no peaks. If the maldistribution is strong enough to cause a peak in the Λ function, it indicates that one is dealing with maldistribution in flow and not simply a backmixing process.

There are some exceptions to the foregoing statement. If a tracer is adsorbed, it could lead to a peak in Λ. The same applies if there is a strong diffusional resistance inside the catalyst particle or a mass transfer resistance to the particle. Theoretical examples of such cases are illustrated in Fig. 15, which shows two examples of a packed bed with a significant chromatographic effect. The residence-time distributions in Fig. 15 are based on the standard equation used in chromatography or ion exchange (Shinnar et al., 1972; Hiester and Vermeulen, 1952). These cases demonstrate that to interpret a residence-time distribution properly, one needs either a good understanding of the system or a good data base, such as a tracer experiment from a pilot plant or a similar industrial reactor. Examples are illustrated in Fig. 16, which shows tracer studies from an industrial trickle bed used for hydrotreating fuel oil. Curve A is from a pilot plant and is similar to curve B in Fig. 5a. Curve B is from an industrial reactor that performs well and shows a small peak in $\Lambda(t)$. Curve C is from an industrial reactor that did not perform well. It shows a strong peak in Λ and indicates a strong bypass. Shinnar (1978) reports that this bypass was correctable by a change in the design which led to better reactor performance. If the bypass is really strong, one would expect f(t) to be double peaked. This occurs rather infrequently and a strong peak in Λ is a much more sensitive test.

For a packed bed one can also check the flow distribution from the Laplace transform. For near-plug-flow conditions the Laplace transfer function should be of the form $\exp[-\tau\hat{P}(s)]$. One can check the form of $\hat{P}(s)$ by plotting $\ln[\hat{f}(s)$ versus s. For one-dimensional diffusion with very large Péclet numbers,

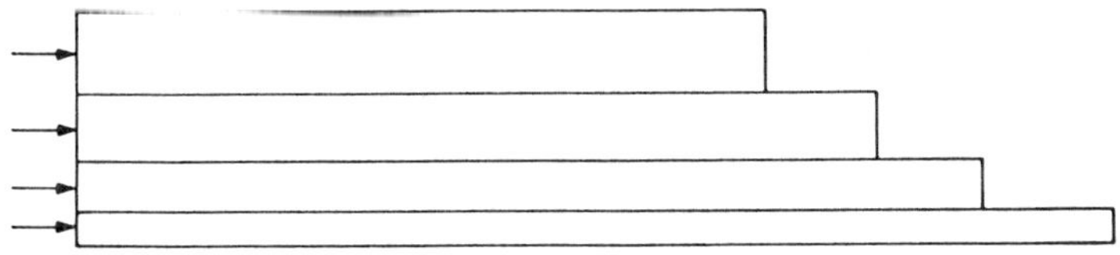

FIGURE 14 Alternative flow model for a packed bed.

$$\hat{f}(s) \approx \exp(\text{Pe}/2)[1 - (1 + 4\tau s/\text{Pe})^{1/2}] \quad (22)$$

Equation (22) is approximate only for high values of Pe (Pe > 20). One can check the fit to Eq. (22) by plotting $\ln[1/f(s)]^{-1}$ versus $\tau s\{\ln[1/f(s)]\}^{-2} - 1/\text{Pe}]$, which should give a straight line with slope τs and intercept 1/Pe (Østergaard and Michelsen, 1969). However, plotting $\ln[\hat{f}(s)]$ versus s might be sufficient and should also be close to a straight line with slope $-\tau$ near $s \to 0$. If in this plot the slope near zero is less than τ or there are strong changes in the slope (see Fig. 9), it indicates a strong deviation from a well-distributed plug-flow reactor. It is extremely difficult to distinguish solely by means of a tracer experiment between maldistributions caused by a bad distributor and those caused either by diffusion processes inside the particle or by adsorption processes (see Figs. 15 and 26). Fortunately, in an industrial case, one often has a reference case from another well-performing reactor which can be used for comparison. One should,

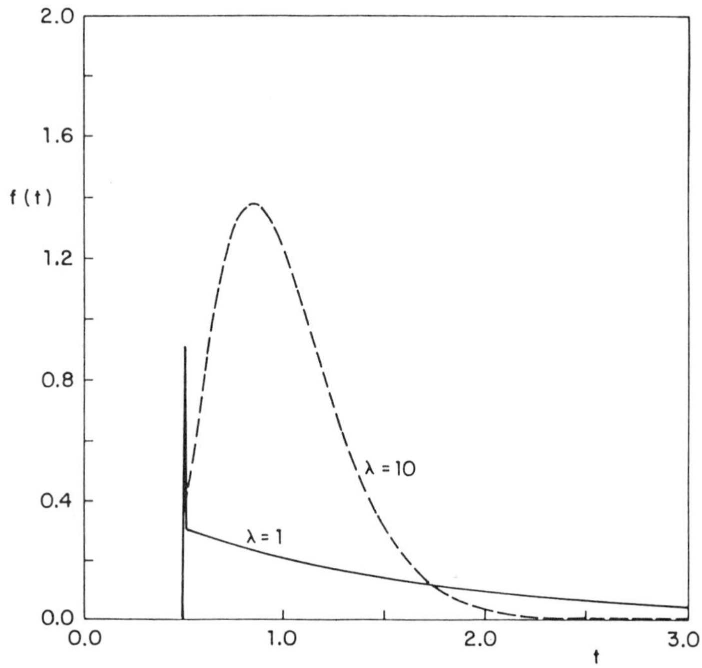

(a)

FIGURE 15 (a) Residence-time distribution of a chromatographic column: (From Shinnar et al., 1972; Hiester and Vermeulen, 1952.)

$$f = e^{-\lambda\tau}\left[\delta(t - \tau) + \frac{\lambda}{r} e^{-\lambda(t-\tau)/r}\sqrt{\frac{\tau\lambda}{t - \tau}}\, I\left(2\lambda\sqrt{\frac{\tau(t - \tau)}{r}}\right)\right] \tau = 0.5 \quad r = 1$$

(b) Cumulative residence-time distribution for the column in part (a).
(c) Escape probability (intensity function) for the column in part (a).

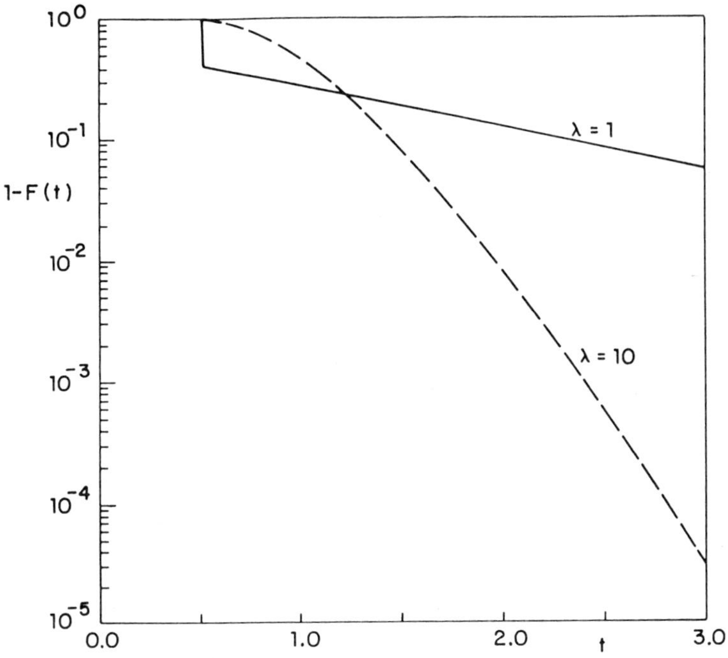

(b)

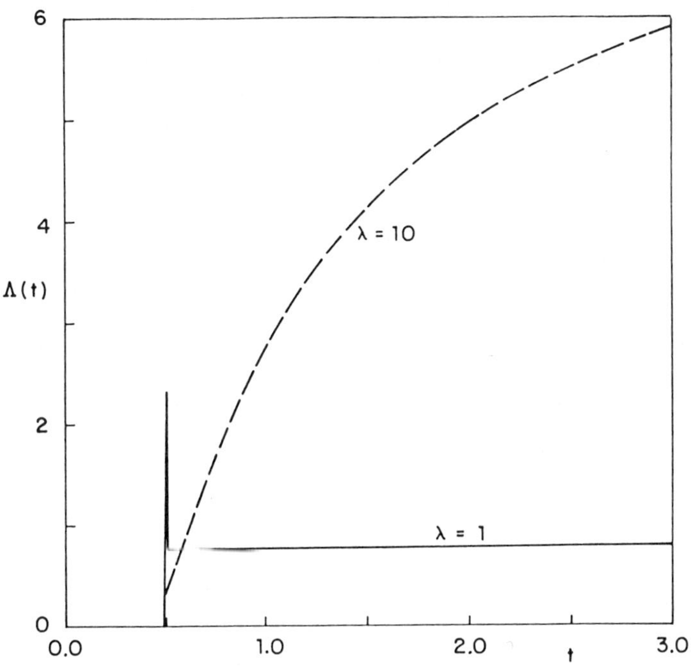

(c)

FIGURE 15 (Continued)

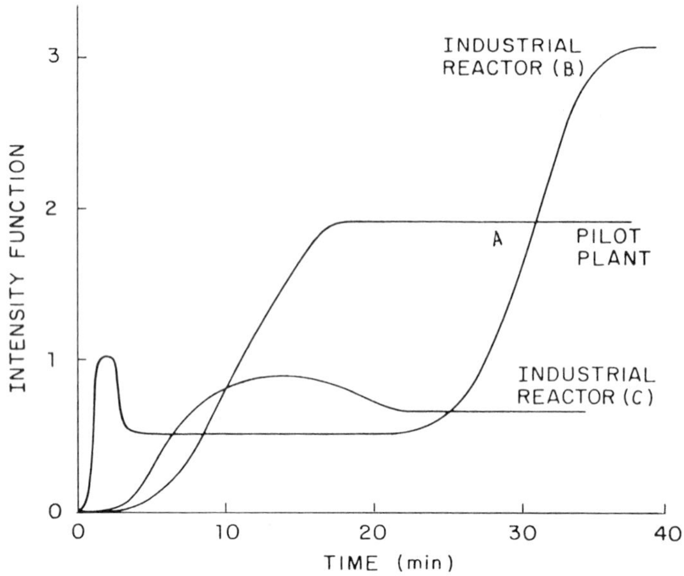

FIGURE 16 Tracer experiments in a trickle-bed reactor. (From Shinnar et al., 1978.)

however, be careful when using pilot plant results as a comparison unless the pilot plant has the same linear velocity. A change in linear velocity changes the mass transfer coefficients and therefore also the residence-time distribution in a packed bed. It also changes the flow regime in a trickle bed. Often this can be predicted by fluid dynamic conditions.

Use of RTD in Reactor Modeling

Tracer experiments and residence-time distributions are just one of many tools in studying flow systems in reactors. One seldom starts in a vacuum and usually has other information about the reactor. In principle, one wants to understand its fluid dynamics. As F(t) measures a linear property of a nonlinear system, the information obtainable from it is always incomplete, although it can be quite important.

If a reactor model is available, one can compute the expected results of a tracer experiment and compare it with the actual experimental results. If one wants only to confirm the model or measure its parameters, doint it via a residence-time distribution is an unnecessarily complicated technique. However, the use of residence-time distribution can be of benefit in two important ways.

Firstly, one never expects models to be completely accurate, so that results which deviate slightly from the predictions of the model are not surprising. Using $\Lambda(t)$ gives a physical insight as to the meaning of the deviations. Second, $\Lambda(t)$ gives information in a form that is at least partially model free. One still has a physical model of steady-state and tracer behavior but does not need any a priori assumptions as the the exact flow model. Interpreting the results in such a way forces one to consider what other explanations or alternative flow models may be consistent with the observed data. This type of approach is very important in any reactor design problem (Overcashier et al., 1959).

As an example, one might cite some of the early tracer experiments in fluid beds, which are represented in Fig. 17. A low fluid bed has a tracer response quite similar to a stirred tank, but representing it in the $\Lambda(t)$ domain immediately indicates that one is dealing with a bypass problem, which was subsequently confirmed in other ways.

It is often useful to have a knowledge of the properties of residence-time distributions of different flow models. These are shown in Table 1.

One must always remember that transport processes are nonlinear and cannot be described adequately by simplified models. F(t) is a linear property of such a model and the only time one can use it to predict reactor performance is in first-order reactions in homogeneous systems, a problem that is not often encountered by the reaction engineer. However, simplified models are a very valuable tool. In Shinnar (1978) the term *learning models* was introduced to distinguish them from models used in actual design predictions. They provide an understanding of how the transport processes might affect the chemical reactions and give some guidance for designing the scaleup. Fortunately, many reactions are not very sensitive to scaleup. Nevertheless, it is important to be able to recognize those cases where significant scaleup problems may be expected. There is one important case where models derived from tracer experiments are directly useful in reactor design. In many reactor problems, a plug-flow reactor is the optimal configuration or if not optimal, is the only design that is safe for scaleup. As real reactors are seldom true plug-flow reactors, one wants to know how closely the design approaches plug-flow and how the deviations could affect reactor performance. Here one utilizes the fact that if deviations from plug flow are small, one should get a reasonable estimate about their impact from any model that has a similar residence-time distribution. This is equivalent to an asymptotic expansion around a solution retaining only the first-order terms. In such a case, one could use either a model based on one-dimensional diffusion or a stirred tank followed by plug flow or a series of stirred tanks. The latter is preferred as it is easier to compute, and the additional complexity of a diffusion model is not justified for cases where the real physical transport processes are not molecular diffusion. It is also more similar in its form to actual measure tracer responses, as compared to a single stirred tank followed by a plug-flow reactor.

It is common to derive these simplified asymptotic models by demanding that the variance of the residence-time distribution of the model is equal to that of the actual measured residence-time distribution. The variance is then expressed as an equivalent Péclet number by the relation $Pe = 2/\gamma^2$, where γ^2 is the coefficient of variation as defined earlier. For a series of stirred tanks the equivalent Péclet number is approximately 2n. It will be shown later that such models give similar kinetic performance as long as the deviation from plug flow is small.

These relations make sense for Péclet numbers larger than 10 (preferably 20). For reactors in which the Péclet number is smaller, simplified models based on one-dimensional diffusion or series of stirred tanks make no physical sense (unless, of course, one is really dealing with three stirred tanks hooked up in series).

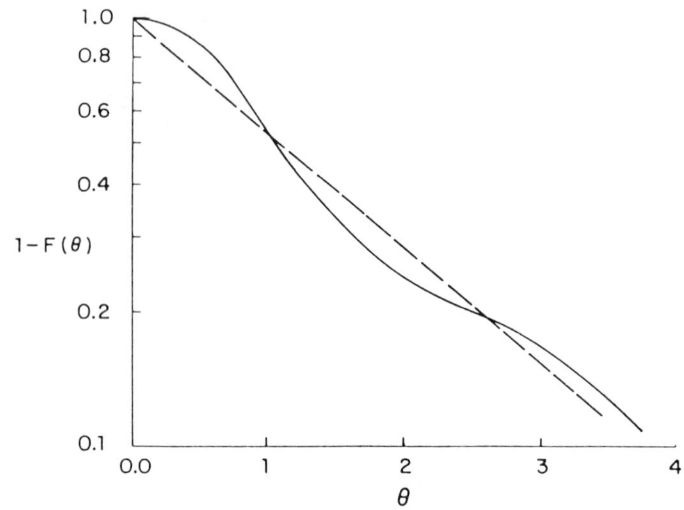

(a)

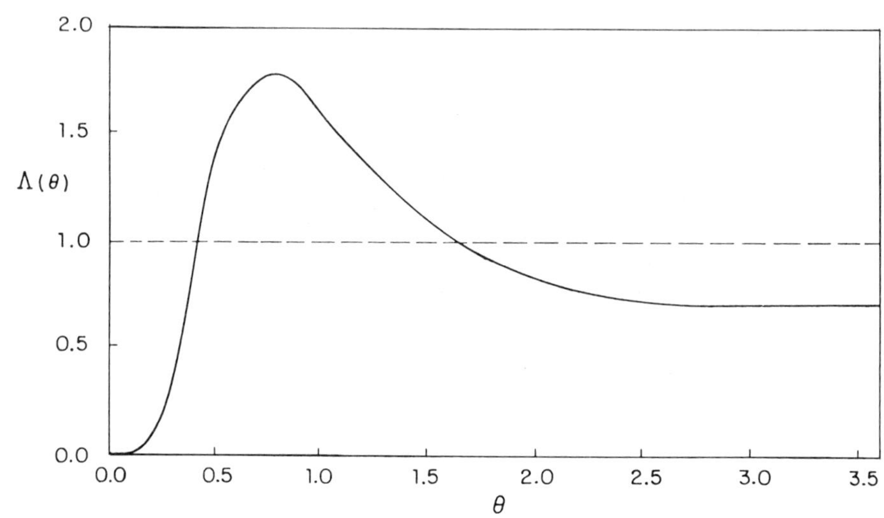

(b)

FIGURE 17 (a) Experimental cumulative residence-time distribution for a fluidized bed. (b) Experimental intensity function for a fluidized bed. Solid line: flow of gas in fluidized-bed reactor; dashed line: completely mixed reactor (for comparison).

TABLE 1 Properties of Some Theoretical Residence-Time Distributions

Ideally mixed vessel

Density: $f(t) = \frac{1}{\tau} e^{-t/\tau}$

Cumulative distribution: $1 - F(t) = e^{-t/\tau}$

Expected value: $E_f(t) = \tau$

Coefficient of variation: $\gamma_f = 1$

Laplace transform: $L(f,s) = E(e^{-st}) = \frac{1}{1 + \tau s}$

Intensity function: $\Lambda(t) = \frac{1}{\tau}$

Ideally mixed vessels in series (τ is the parameter of each single vessel)

Density: $f(t) = \frac{(t/\tau)^{n-1} e^{-t/\tau}}{\tau(n - 1)!}$

Cumulative distribution: $1 - F(t) = e^{-t/\tau} \sum_{i=0}^{n-1} \frac{(t/\tau)^i}{i!}$

Expected value: $E_f(t) = n\tau$

Coefficient of variation: $\gamma_f = \frac{1}{\sqrt{n}}$

Laplace transform: $L(f,s) - E(e^{-st}) = \frac{1}{1 + \tau s}$

Intensity function: $\Lambda(t) = \frac{(t/\tau)^{n-1}/(n - 1)!}{\tau \Sigma_{i=0}^{n-1} (t/\tau)^i/i!}$

Special values: $\left.\begin{aligned} \Lambda(0) &= 0 \\ \Lambda(\infty) &= \frac{1}{\tau} \end{aligned}\right\} (n > 1)$

Two nonidentical ideally mixed vessels in series (τ_1 and τ_2 are the parameters of the two vessels)

Density: $f(t) = \frac{1}{\tau_1 - \tau_2} (e^{-t/\tau_1} - e^{-t/\tau_2})$

TABLE 1 (Continued)

Cumulative distribution:	$1 - F(t) = \frac{1}{\tau_1 - \tau_2}(\tau_1 e^{-t/\tau_1} - \tau_2 e^{-t/\tau_2})$
Expected value:	$E_f(t) = \tau_1 + \tau_2$
Coefficient of variation:	$\gamma_f = \sqrt{\frac{\tau_1^2 + \tau_2^2}{\tau_1^2 + 2\tau_1\tau_2 + \tau_2^2}}$
Laplace transform:	$L(f,s) = E(e^{-st}) = \frac{1}{(1 + \tau_1 s)(1 + \tau_2 s)}$
Intensity function:	$\Lambda(t) = \frac{e^{-t/\tau_1} - e^{-t/\tau_2}}{\tau_1 e^{-t/\tau_1} - \tau_2 e^{-t/\tau_2}}$
Special values:	$\Lambda(0) = 0$
	$\Lambda(\infty) = \min\left(\frac{1}{\tau_1}, \frac{1}{\tau_2}\right)$
Plug-flow vessel	
Density:	$f(t)$ = unit Dirac function
Cumulative distributuon	$1 - F(t) = \begin{cases} 1 & t < \tau \\ 0 & t \geq \tau \end{cases}$
Expected value	$E_f(t) = \tau$
Coefficient of variation:	$\gamma_f = 0$
Laplace transform:	$L(f,s) = E(e^{-st}) = e^{-s\tau}$
Intensity function:	$\Lambda(t) = \begin{cases} 0 & t \neq \tau \\ \infty & t = \tau \end{cases}$
Two parallel, nonidentical, ideally mixed vessels [fraction Π passes through one vessel with parameter τ_1; fraction $(1 - \Pi)$ passes through second vessel with parameter τ_2]	
Density:	$f(t) = \frac{\Pi}{\tau_1} e^{-t/\tau_1} + \frac{1 - \Pi}{\tau_2} e^{-t/\tau_2}$

TABLE 1 (Continued)

Cumulative distribution: $1 - F(t) = \Pi e^{-t/\tau_1} + (1 - \Pi)e^{-t/\tau_2}$

Expected value: $E_f(t) = \Pi\tau_1 + (1 - \Pi)\tau_2$

Coefficient of variation: $$\gamma_f = \sqrt{\frac{\tau_2^2(1 - \Pi) + \tau_1^2\Pi(2 - \Pi) - 2\tau_1\tau_2\Pi(1 - \Pi)}{\tau_2^2(1 - \Pi)^2 + \tau_1^2\Pi^2 + 2\tau_1\tau_2\Pi(1 - \Pi)}}$$

Laplace transform: $$L(f,s) = E(e^{-st}) = \frac{\Pi}{1 + \tau_1 s} + \frac{1 - \Pi}{1 + \tau_2 s}$$

Intensity function: $$\Lambda(t) = \frac{(\Pi/\tau_1)e^{-t/\tau_1} + (1 - \Pi/\tau_2)e^{-t/\tau_2}}{\Pi e^{-t/\tau_1} + (1 - \Pi)e^{-t/\tau_2}}$$

Special values: $$\Lambda(0) = \frac{\Pi}{\tau_1} + \frac{1 - \Pi}{\tau_2}$$

$$\Lambda(\infty) = \min\left(\frac{1}{\tau_1}, \frac{1}{\tau_2}\right)$$

Two parallel vessels, one ideally mixed, one plug flow [fraction Π passes through plug flow vessel with parameter τ_1; fraction $(1 - \Pi)$ passes through ideally mixed vessel with parameter τ_2]

Density: $$f(t) = \frac{1 - \Pi}{\tau_2} e^{-t/\tau_2} \qquad (t \neq \tau_1)$$

$$f(t) = \Pi \times \text{unit Dirac function } (t = \tau_1)$$

Cumulative distribution: $$1 - F(t) = \begin{cases} \Pi + e^{-t/\tau_2}(1 - \Pi) & t < \tau_1 \\ e^{-t/\tau_2}(1 - \Pi) & t \geq \tau_1 \end{cases}$$

Expected value: $E_f(t) - \Pi\tau_1 + (1 - \Pi)\tau_2$

Coefficient of variation: $$\gamma = \sqrt{\frac{\tau_2^2(1 - \Pi^2) - 2\tau_1\tau_2\Pi(1 - \Pi) + \tau_1^2\Pi(1 - \Pi)}{\tau_2^2(1 - \Pi)^2 + 2\tau_1\tau_2\Pi(1 - \Pi) + \tau_1^2\Pi^2}}$$

TABLE 1 (Continued)

Laplace transform:	$L(f,s) = E(e^{-st}) = \Pi e^{-s\tau_1} + (1 - \Pi)\dfrac{1}{1 + \tau_2 s}$
Intensity function:	$\Lambda(t) = \dfrac{[(1 - \Pi/\tau_2]e^{-t/\tau_2}}{(1 - \Pi)e^{-t/\tau_2} + \Pi} \quad t < \tau_1$
	$\Lambda(t) = \dfrac{1}{\tau_2} \quad t > \tau_1$
Special value:	$\Lambda(0) = \dfrac{1 - \Pi}{\tau_2}$

Estimation of Model Parameters from Residence-Time Distribution

As discussed in the preceding section, to estimate the kinetic impact of small deviations from plug flow, one estimates an equivalent Péclet number from the coefficient of variation of the residence-time distribution. One can directly use either f(t) or F(t) [see Eqs. (1) to (3)]. In the literature, residence-time distributions are often used to estimate simultaneously several parameters of a complex mocel. The author does not recommend this.

Residence-time distributions are hard to measure accurately and are quite constrained. In a homogeneous flow, the first moment is fixed by V/Q. The limits of experimental accuracy seldom allow one to measure more than two additional moments. Often only one additional moment, namely the variance, is all one can reasonably expect to obtain. Therefore, one cannot estimate more than one or two parameters from such experiments. Furthermore, the models themselves are not accurate, and estimating multiple parameters for approximate models from a single experiment is a doubtful procedure at best (Overcashier et al., 1959).

Use of Tracer Experiments in Cold-Flow Models and Design of Reactor Internals

Tracer experiments and residence-time distributions are becoming a widely used tool in this area. In principle, there is very little difference in the approach from that which was discussed in the previous sections. One wants to make sure that one of the limiting models, such as a stirred tank or plug flow, is approximated as closely as possible. Quite often one uses multiple tracers (which will be discussed later), but the principles remain the same. The following two examples may help illustrate this point. Baffles, especially horizontal baffles, allow one to modify the complex flow in a fluid bed such that it approaches plug flow quite closely. Properly designed baffles achieve this by breaking up the bubbles and staging the bed, reducing solid mixing between top and bottom and thereby also reducing recirculation of gas either adsorbed on the catalyst or entrained in the dense phase inside particle clusters. An example of this effect and its measurement by a tracer experiment is given in Fig. 18 [for details, see Overcashier et al., 1959].

Figure 19 gives an example of the study of gas distributors in fluid beds. In Fig. 19 some of the f(t) plots are double peaked, which clearly indicates a maldistribution. Here both f(t) and the escape probability $\Lambda(t)$ clearly indicate that the design of the distributor is unsatisfactory and leads to a maldistribution in the overall reactor, which can be corrected by a proper design. This example illustrates the importance of a proper base case in comparing alternatives. Fluid beds, especially short fluid beds, have maldistributions of flow caused by the formation of larger bubbles. These are due to inherent hydrodynamic instabilities and cannot be prevented by the design of the distributor. In the cases shown in Fig. 19, some of the distributor designs give maldistributions which are much more pronounced compared to the best achievable case, which is a porous plate. Residence-time distributions cannot distinguish between these two phenomena and should always be considered only an additional, albeit important diagnostic tool.

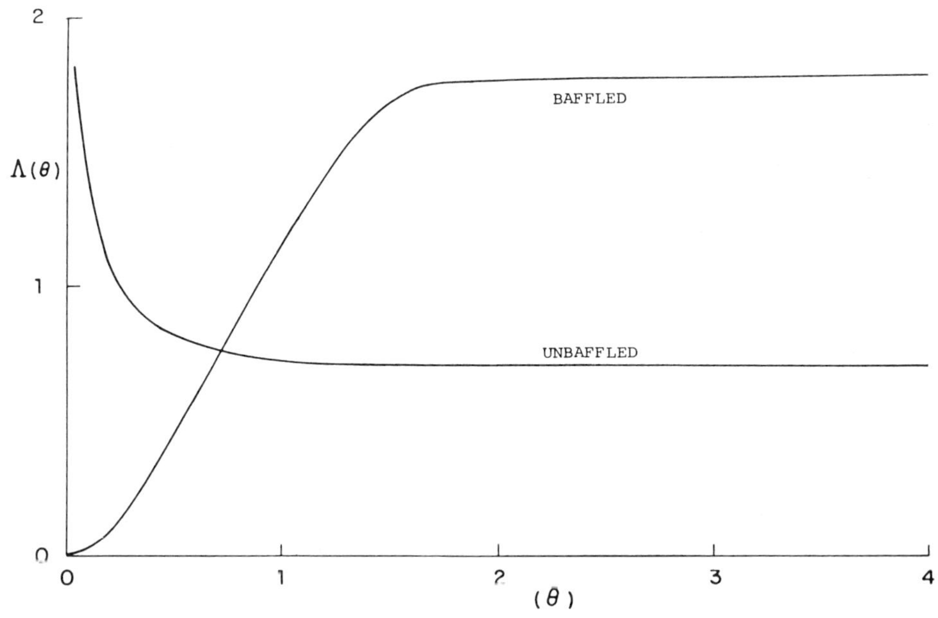

(a)

FIGURE 18 (a) Experimental residence-time distribution for gas flow in a fluidized-bed reactor (gas velocity = 1.6 ft/sec)—intensity function. (From Overcashier et al., 1959.) (b) Cumulative residence-time distribution for the flow in part (a).

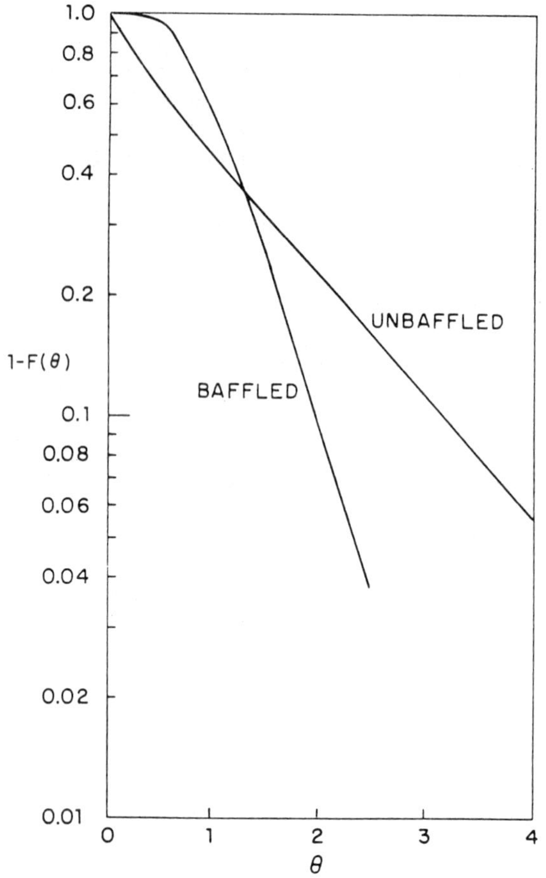

(b)

FIGURE 18 (Continued)

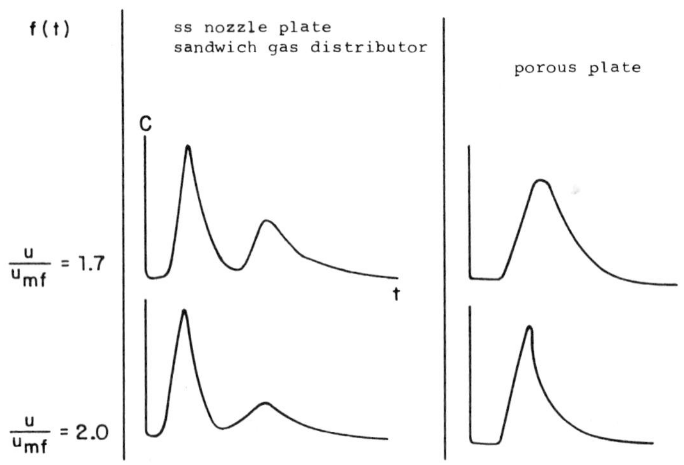

FIGURE 19 Residence-time distribution measured with different gas distributors in a fluid bed, as a function of fluidization indexes u/u_{mf}. (From Woodrow, 1978.)

Measurement of Flow Rates and Reactor Volume or Holdup

In hydrology and physiology, tracer experiments are widely used to measure flow rates and volumes [Weinstein and Dudukovic (1975)]. In reaction engineering it is possible in most cases to measure flow rates directly. Exceptions are reactors with internal circulation caused by density and pressure differences, where tracer experiments are sometimes useful to measure flow rates using Eq. (2). It is not a very accurate method and should be used only as a last resort.

In homogeneous reactors, the volume or holdup is almost always known. One can then compare the volume estimate from the residence-time distribution with the known volume. To do this, one has to estimate the contribution of the tail of either F(t) or f(t) to the first moment. If the volume estimated from the first moment is significantly smaller than the known volume, the vessel is not well perfused and has strongly stagnant regions. An example of this was shown in Fig. 13. In multiphase systems the relative volume and holdup of solid liquid and gas are sometimes hard to measure, and here tracer experiments can be valuable, despite the shortcomings mentioned above. An example of this use is presented later.

USE OF RESIDENCE-TIME DISTRIBUTIONS IN THE DESIGN OF HOMOGENEOUS REACTORS

One advantage of using the concept of a residence-time distribution in evaluating tracer experiments has already been discussed, namely, the physical insight into the properties of the system that is gained from the use of the escape probability $\Lambda(t)$. Just as important are the direct conceptual insights, which can help in understanding reactor design problems.

Tracer experiments are very useful in organizing and evaluating alternative reactor designs, as well as in designing and evaluating pilot plant experiments. This requires more than simple algorithmic uses of the distribution in which one tries to compute performance directly from a residence-time distribution. However, the conceptual framework which offers quidelines for scaleup and scale-down problems is generally worth the effort. The results in this chapter can be derived rigorously only for homogeneous reactors, which are a small fraction of the cases met by the design engineer. However, the guidelines for reactor design derived from them (in contrast to design algorithms) apply to heterogeneous reactors, as long as one understands the basic principles. The differences are discussed in a later section.

First-Order Isothermal Reactions

Consider the case of a first-order isothermal reaction system. There is a set of compounds A_j (j = 1, 2, . . ., n) which can undergo composition changes such that the rate of formation of a compound A_j is given by

$$r(A_j) = \sum_{i=1}^{n} (-k_{ij}A_j + k_{ji}A_i) \tag{23}$$

In other words, all reactions between two compounds A and B can be expressed by a set of reactions

$$r(A) = -k_1 A + \frac{k_1}{K_{AB}} B \tag{24}$$

where K_{AB} is the equilibrium constant of the reaction A $\Longleftrightarrow$ B.

If one starts with an inlet composite C_{jo}, the outlet composition of a plug-flow reactor is given (Wei and Prater, 1962) by the equation

$$C_j(t) = \alpha_{jo} + \sum_{r=1}^{n-1} \alpha_{jr} \exp(-\lambda_r t) \tag{25}$$

where C_j is the concentration of component j expressed as a mole fraction and λ_r are the (n − 1) nonzero eigenvalues of the system. α_{jo} and α_{jr} are constants that depend on the initial and equilibrium composition of the system, and t is the residence time in the homogeneous isothermal plug-flow reactor (or the reaction time in a homogeneous batch reactor).

Equation (23) and its solution [Eq. (25)] are completely equivalent to a problem in probability theory describing stochastic systems. Consider a large set of particles, which can change their state in a discrete way. Each particle is associated with a state A_j. The probability of a single particle being in a specific state A_j is p_j. If the number of particles in the set is very large, the probability of a particle to be in A_j is exactly equivalent to the mole fractions of particles A_j in the system, which in a first-order system is proportional to the concentration C_j. One can therefore look at C_j either as a concentration or as the probability of a single particle (or molecule) to be in state j. The mathematical proof is given in Shinnar et al. (1973). The reader unfamiliar with probability theory can interpret Eq. (25) as follows. The concentration C_{jo} of the particles entering the reactor expresses, for a single particle, the probability of being in state j. The reactor changes this probability into $C_j(t)$. Each particle behaves here independently from its neighbor and can undergo changes completely independent of any other molecule in the reactor. This is the physical meaning of a first-order reaction system.

Once this is understood, some important results in reaction engineering become clear without further mathematical derivation. Consider a homogeneous isothermal reactor which has a steady-state flow that can be characterized by a residence-time distribution f(t). The inlet concentration is C_{jo}. One can now perform the following thought experiment. An observer stands at the outlet of the reactor and collects all the particles and sorts them into bins according to their residence time. To do so, one has to divide the residence time into intervals Δt. If Δt is small, all particles in the bin, t to t + Δt, essentially have the same residence time, t, in the system. As each molecule in the bin behaves independently of its neighbors, the concentration in the bin would be the same as at the outlet of a homogeneous isothermal plug-flow reactor with residence time t.

$$p_j(t) = C_j(t) = \alpha_{jo} + \sum_{r=1}^{n-1} \alpha_{jr} \exp(-\lambda_r t) \tag{26}$$

The average concentration at the outlet of the reactor $\langle C_j \rangle$ can be obtained by averaging over all the bins:

$$\langle C_j \rangle = \int_0^{\infty} f(t) \left[\alpha_{jo} + \sum_{r=1}^{n-1} \alpha_{jr} \exp(-\lambda_r t) \right] dt \tag{27}$$

which is equal to

$$\langle C_j \rangle = C_{jo} + \sum_{r=1}^{n-1} \alpha_{jr} \hat{f}(\lambda_r) \tag{28}$$

where $\hat{f}(\lambda_r)$ is the Laplace transform of f(t) with λ_r substituted for the transform variable (s). Equation (28) is equivalent to the outlet of a hypothetical reactor described in Fig. 20a, which consists of a set of parallel tubular flow reactors with different residence times t. The fraction of fluid entering a tube with a residence time between t and t + Δt is f(t) Δt. In practice, a real reactor would hardly look like Fig. 20a, but there is one case that is equivalent. Consider the case where the reactor feed consists of separate droplets encapsulated by impermeable membranes which have a residence-time distribution F(t) in the reactor (it is equivalent to the reactor in Fig. 20a). The individual droplets in suspension polymerization behave this way. In a mixed reactor this is true only in the limiting case called segregated flow. In a first-order system, by definition, each molecule can be considered such a separate droplet.

The well-known case of an ideally stirred tank will illustrate Eq. (27). Consider the irreversible reaction A $\rightarrow$ B. In the case of a plug-flow reactor, $C_A(t) = C_{Ao}e^{-kt}$ and for a stirred-tank reactor, a mass balance will give

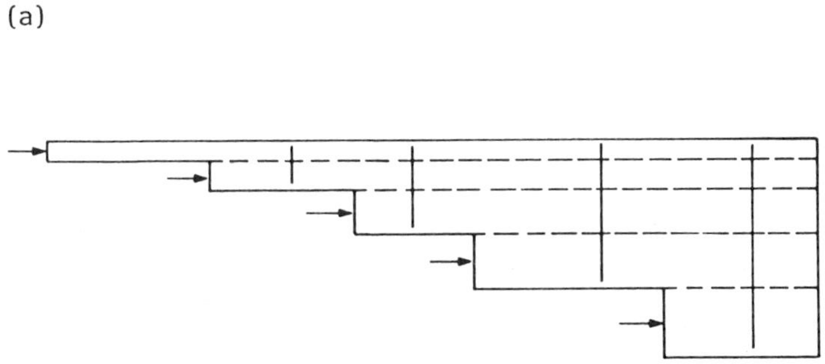

FIGURE 20 Alternative flow models for Zwietering method of bounding the conversion for a reaction of nth order: (a) segregated flow; (b) maximum mixedness.

$$QC_{A0} - QC_A - VkC_A = 0$$

or (29)

$$C_A = \frac{C_{A0}}{1 + Vk/Q} = \frac{C_{A0}}{1 + k\tau}$$

For a stirred tank, f(t) is here $(1/\tau)e^{-t/\tau}$ and therefore according to Eq. (27),

$$\langle C_A \rangle = \int_0^\infty C_{A0} e^{-kt} \frac{1}{\tau} e^{-t/\tau} \, dt = \frac{C_{A0}}{1 + k\tau} \tag{30}$$

In the mass balance [Eq. (26)], it was assumed that the concentration in the stirred tank is uniform at any point in the tank. For a linear system, a much weaker assumption is sufficient, namely, that for residence-time distribution is exponential or that the escape probability is constant. Residence-time distributions are linear property of the nonlinear model of the transport process in the reactor, and first-order reactions depend only on this linear property of the flow.

In reality very few reactions are truly linear, but it is a sufficiently good approximation for many systems. Many reactions are also pseudolinear in the same sense that while the reaction rates contain nonlinear terms such as a Langmuir expression, they can be written in a form

$$r_j = \phi(C_{jo}) K_{ij} C_i \tag{31}$$

where K_{ij} is a matrix of linear reaction rate coefficients and $\phi(C_{jo})$ is a nonlinear function of inlet concentration and pressure that can be taken out in front of the reaction rate matrix. Although this is seldom completely correct, it is often a good approximation. For all those systems residence-time distributions are an important modeling tool.

Deviation of the Residence-Time Distribution from Kinetic Experiments

One can also use Eq. (28) in the reverse way. If one has a complex reaction system in which the eigenvalues λ_r change over a significantly wide range, we can use it to reconstruct $\hat{f}(s)$. Assume that the values of λ_r have been measured from a batch experiment or a plug-flow reactor, and one now has experimental values for all the concentrations C_j at the outlet of an industrial reactor. One can then compute the values of λ_r and use them to estimate $\hat{f}(\lambda_r)$ for the industrial reactor. One then has several values of the Laplace transform $\hat{f}(s)$ from which a reconstruction of $\hat{f}(s)$ can be attempted. If the λ_r are widely spaced, the reconstruction is quite good. This approach will be discussed in more detail later when we deal with contact-time distribution in heterogeneous reactors.

Optimal Reactor Concepts for First-Order Reactions

Isothermal Reactors

The concepts derived in the preceding section lead to some important conclusions with regard to choosing a reactor configuration. Specifically, they provide guidelines for cases in which a plug-flow reactor is the preferred choice and for cases in which other reactor configurations will be preferable. The principle that a first-order reaction system is completely equivalent to the case where single molecules undergo changes from one state to other states, with defined probabilities, simplifies many design problems. Many optimal problems that are very difficult to deal with analytically become intuitively clear if one understands the foregoing principle and applies it appropriately.

In the preceding section it was shown that in the case of a first-order isothermal reaction system of arbitrary complexity, the hypothetical reactor in Fig. 20a will give identical results to any reactor having the same residence-time distribution. The outlet concentration is therefore the average obtained from several plug-flow reactors with different residence times. Once one recognizes that all one needs to do is average over a number of plug-flow reactors, optimization is simple. Consider, for example, the case where $\langle C_j(t)\rangle$ for a specific value of t has a desirable optimum minimum or maximum concentration of $\langle C_j\rangle$. Then assume that one has found a reactor with a residence-time distribution f(t) that gives an optimum yield of C_j. In this reactor the outlet concentration $\langle C_j\rangle$ is simply the average of all the different plug-flow reactors. There is a simple algebraic theorem that says that in any set of numbers with average X, there is at least one member of the set equal or larger than X and at least one member equal or smaller than X. If f(t) is a plug flow, one has only a single plug-flow reactor with $t = Ef(t) = \tau$ and which has a unique $C_j = \langle C_j\rangle$. But if f(t) is not a plug flow, one has in Fig. 20a different tubes with different outlet concentration. One of these reactors has an outlet concentration equal or larger than the average concentration $\langle C_j\rangle$ and in one of these reactors C_j is smaller (or larger) than $\langle C_j\rangle$. Therefore, depending on whether one wishes to maximize or minimize $\langle C_j\rangle$, one can always find a tubular reactor in Fig. 20 with a uniform residence time which is as good or better than the reactor with the arbitrary residence time f(t). This is a simple proof that for first-order systems a plug-flow reactor is optimal in the sense that one can design a plug-flow reactor with a yield or selectively as high or higher than any isothermal reactor with arbitrary residence-time distribution. It results in a very simple but powerful conceptual tool for reactor design. In all cases where the system is such that one can obtain the outlet concentration simply by averaging over the bins of different residence times, then, for any reaction system which has the property that there is an optimum yield of one compound, a plug-flow reactor is the best choice. If, for other reasons a simple plug-flow reactor is not feasible or desirable, one wants a design in which the residence time distribution approaches plug flow as closely as possible. For example, F(t) for a series of stirred tanks approaches plug flow if the number of stirred tanks is reasonably large. One can judge the required number of tanks simply by the allowable deviation from plug flow. Staged systems in general have a similar property and one can use the same approach to estimate the number of stages required.

For averaging to apply, the reaction does not have to be strictly first order. All one requires is that the reaction behavior of a molecule is not affected by the nature of its neighbors, as discussed in the preceding section. Otherwise, one needs to know not only how long a molecule has been in the system, but also who its neighbors were. If the outcome is strongly affected by the nature of its neighbors, a residence-time distribution cannot provide an answer for the outcome. Nor can one show that plug flow is optimum, although one can still get some guidelines which will be discussed in a later section.

In a similar way, if one looks at the conversion of a first-order simple reaction A<=> B with reaction rate

$$r_A = -k_1 a + k_2 b \tag{32}$$

it is given by

$$a - a_{eq} = (a_0 - a_{eq})e^{-(k_1+k_2)t} \tag{32a}$$

This is a convex function of t. Again, the outlet concentration of a reactor with residence-time distribution f(t) is obtained by averaging over the bins, which really means averaging $e^{-(k_1 + k_2)t}$ over f(t). It is easy to see that a plug-flow reactor with the same average residence time t will have a lower $a - a_{eq}$ and hence higher conversion than a reactor in which the bins have different t. (Had the function been concave, the opposite would have been true.) For high converison in a first-order isothermal system, a plug-flow reactor will have the lowest residence time. Again, one can evaluate the permissible deviation by some very simple models (such as a series of stirred tanks) and judge the design by measuring the deviation from plug flow via a tracer experiment. For small deviations the coefficient of variation γ^2 of f(t) is a good criterion. In the literature the Péclet number is often used instead and is approximately equal to $2/\gamma^2$. More accurately, it is given by the expression (see Shinnar and Naor, 1967).

$$\gamma^2 = \frac{2[e^{-Pe} - (1 - Pe)]}{Pe^2} \tag{33}$$

which for large Pe approaches

$$\gamma^2 = \frac{2}{Pe} \qquad (Pe \to \infty) \tag{33a}$$

But one should be careful not to use Péclet numbers smaller than 10, because in most cases this is meaningless, as the residence-time distribution will not fit a one-dimensional diffusion model.

In fact that the results are not sensitive to the form of the deviation from plug flow is illustrated in Fig. 21, where two cases are plotted. The

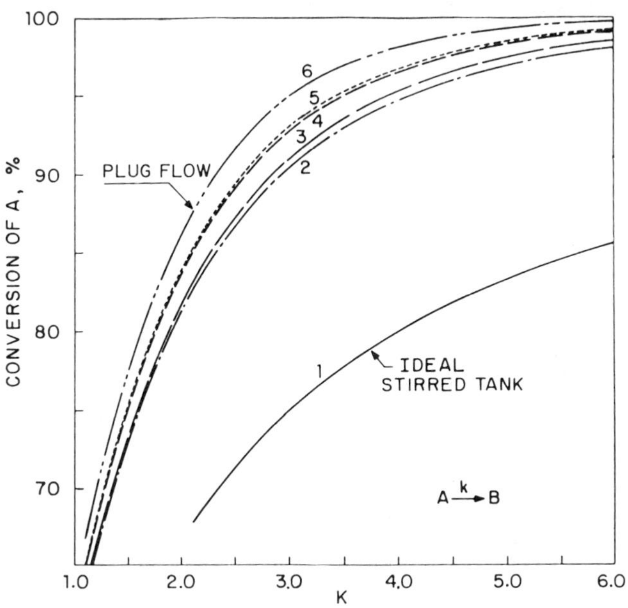

(a)

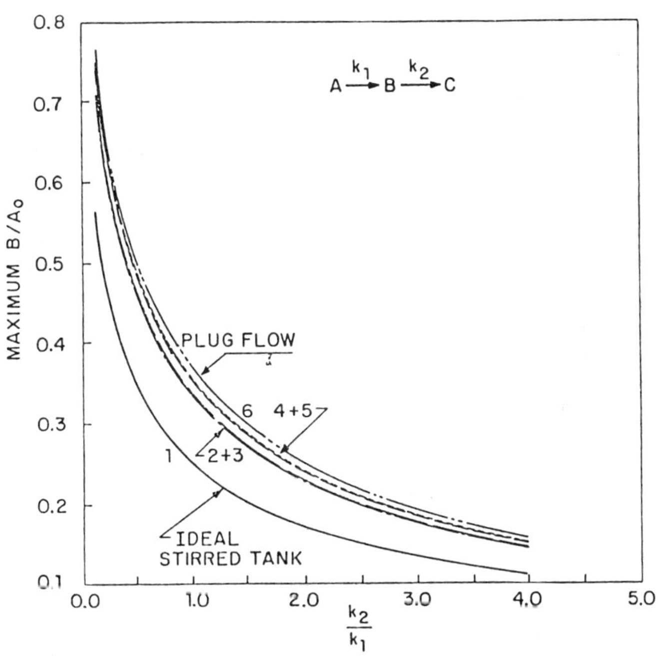

(b)

FIGURE 21 (a) Conversion of a first-order irreversible reaction for different residence-time distributions. (b) Selectivity of a consecutive first-order reaction for different residence-time distributuons. 1. Mixed flow; 2, five CSTRs in series; 3, one-dimensional flow, Pe = 8.87; 4, 10 CSTRs in series; 5, one-dimensional flow, Pe = 20; 6, plug flow.

first illustrates an irreversible first-order reaction at high conversion, and the second shows the selectivity of a consecutive first-order reaction. For both cases, the limiting cases of plug flow and a stirred tank are given and two models with equal Péclet number are given. One is a network of equal stirred tanks in series and the other is a one-dimensional diffusion system with a Péclet number chosen such that γ^2 is equal for both cases. For $\gamma^2 = 0.1$, the results are so close that it is impossible to plot the difference. For $\gamma^2 = 0.2$ (Péclet number 8.9 or 5 stirred tanks in series) the difference is already recognizable but still small. One should not misinterpret this statement. The results are always sensitive only to the variance of f(t). In this case only models for which f(t) has a reasonable similar form were compared. For such cases all models with equal γ^2 will give similar results, provided that γ^2 is sufficiently small. For computation it is therefore advantageous to use simpler models based on stirred tanks instead of diffusion, especially since none of the models represents the actual physics accurately.

Nonisothermal Reactors

The arguments in the preceding section can be easily extended to a nonisothermal reactor, provided that the temperature profile is imposed from the outside and is independent of the reactions. For example, consider an arbitrary network of stirred tanks, each of which is homogeneous and has a uniform temprature imposed from the outside, although each reactor may have a different temperature. The reactors are interconnected in an arbitrary manner. There could be several inlets and outlets. Again, each entering molecule is equipped with a clock, but this time a more complex clock is chosen that not only records the time spent in a specific tank, but also records each step in a specific tank in the order in which it occurs. If one looks at each clock, one finds a set of times $(t_1, t_2, t_5, t_1, t_2, \ldots)$ identifying each successive stay in tank i until it leaves the system. One again collects them into bins, but this time one insists that not only the total residence time t is constant in each bin, but that all molecules in a bin have an identical clock history. Thus in each bin not only are the separate stays in each tank of equal length, but they occur in the same sequence.

In each stirred tank the reactions are first order with fixed reaction rate, determined by the temperature of the tank. This means that if one knows the state of a molecule (or the concentration of each compound) entering the tank and the length of stay in the tank, the probability distribution at the time it leaves the tank is known or can be computed. This means that if one knows the whole history of a molecule in a bin, one knows the probability of it being in each one of the permissible states, which in reaction terms means that one knows the concentration of the different compounds in the bin. For those who are somewhat uncomfortable with probability theory, this concept can be translated as follows. Follow a blob of molecules that has an equal history. As they travel through each tank, the concentration of the blob changes in each tank in the same way it would in a plug-flow reactor with the same temperature and residence time. There is now only a cascade of plug-flow reactors with different residence times and temperatures (or reaction rates). Each bin represents a unique sectional plug-flow reactor of this type (see Fig. 22). It was shown previously that each stirred tank can be presented in such a way that molecules with residence time t can be lumped together and treated as if they would be in a plug-flow reactor of the same residence time t. The average outlet

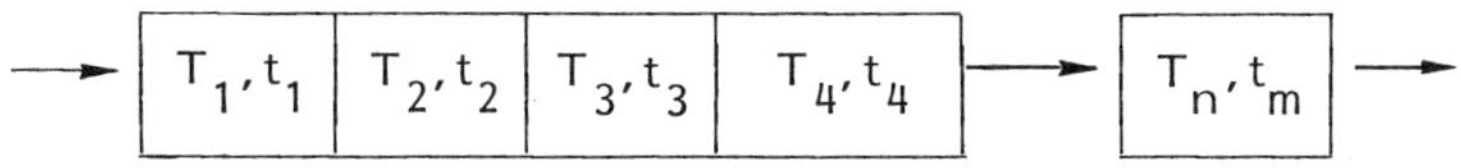

FIGURE 22 Equivalent representation of a particle history in a nonisothermal reactor. Sectionally uniform plug-flow reactor. T_i, temperature of each section; t_i, residence time of each section.

concentration is now the average of all those sectional plug-flow reactors and it is clear that if one is allowed to average them, one can always find a single bin (or a plug-flow reaction with a special temperature and time sequence), which yields a result as good or better than the network of stirred tanks.

This proof is not limited to networks of stirred tanks. One can take any arbitrary flow and represent it by such a network if one makes the stirred tanks small enough. This is therefore true for any arbitrary flow system provided that:

1. The reaction rates are locally first order in concentration.
2. The temperature profile is independent of the reactions and imposed from the outside.

If condition 2 is not fulfilled, as for example in an adiabatic reactor, one can no longer average. The temperature of a tank will depend on its concentration, and therefore the reaction rates affecting a single molecule will depend on the nature of the other molecules in the tank, which violate one condition for averaging. However, as many reactions are approximately first order, averaging provides a simple guideline for reactor design. Even more important, it lets one understand and recognize the exceptions to the rule, in the cases where plug-flow reactors are not optimal. These include, for example, all cases in which it is desirable to maximize the contact between products and reactants.

Bounding Methods for Nonlinear Reactions

For a first-order system, knowledge of the residence-time distribution completely determines the product composition in a premixed isothermal homogeneous reactor. For nonlinear reactions this is not enough, as one needs to know not only the history of a molecule, but also the property and state of the molecules in its close neighborhood. However, for a certain class of nonlinear reactions wehre the reaction rate is of the type

$$r_A = -ka^n \tag{34}$$

Zwietering (1959) has shown that one can get rigorous bounds on the possible conversions directly from f(t). Consider again a stirred tank with the simple irreversible reaction A → B, but this time let the reaction rate be second order. If the tank is ideally stirred, a mass balance will give

$$QC_{A_o} - QC_Z - VkC_A^2 = 0 \tag{35}$$

and

$$C_A = C_{A0}\left[\sqrt{\frac{\beta}{2}\left(\frac{\beta}{2} + 2\right)} - \frac{\beta}{2}\right] \qquad \beta = \frac{1}{kC_{A0}t} \tag{36}$$

One can also look at the other extreme idealized case mentioned before, namely, that the entering fluid enters as separate encapsulated drops that stay together (which Zwietering terms segregated flow). This case is equivalent to the reaction in Fig. 13. In each tubular reactor the outlet concentration will be

$$\frac{C_A}{C_{A0}} = \frac{1}{1 + C_{A0}kt} \tag{37}$$

and the average outlet concentration will be simply the average over f(t), or

$$\langle C_A\rangle = C_{A0}\beta \exp(\beta)\mathrm{Ei}(\beta) \qquad \beta = \frac{1}{kC_{A0}t} \qquad E_i = \int_0^\infty \frac{e^{-U}}{U}\, 0/u \tag{38}$$

where Ei is the exponential integral. Note that the two cases lead to different results. However, if $kC_{A0}\,t$ is not very large, the difference is small (see Fig. 23).

The reason for the difference is that for a second-order reaction the probability of a molecule of substance A reacting depends on its probability of being close to another molecule of A. In the segregated case, this probability is larger than in the mixed case, as in the latter the entering molecule is immediately exposed to products that reduce C_A in its surroundings. Knowledge of the residence-time distribution alone is here insufficient to compute C_A, as one needs to know what happened not only to the molecule but to its neighbors. However, for this simple case of second-order reaction, Zwietering was able to show that ideal mixing and segregated flow conditions are extremes and that all other possible mixing situations for which f(t) is the same will result in an outlet concentration such that

$$C_A(\text{ideal mixing}) > C_A(\text{segregated flow}) \tag{39}$$

For n = 1/2 or

$$r_A = -kC_A^{1/2} \tag{40}$$

the limits become

$$C_A(\text{ideal mixing}) = C_{A0}\left[1 + 2\left(\frac{kt}{2\sqrt{C_{A0}}}\right)^2 - 2\left(\frac{kt}{2\sqrt{C_{A0}}}\right)\sqrt{1 + \left(\frac{kt}{2\sqrt{C_{A0}}}\right)^2}\right] \tag{41}$$

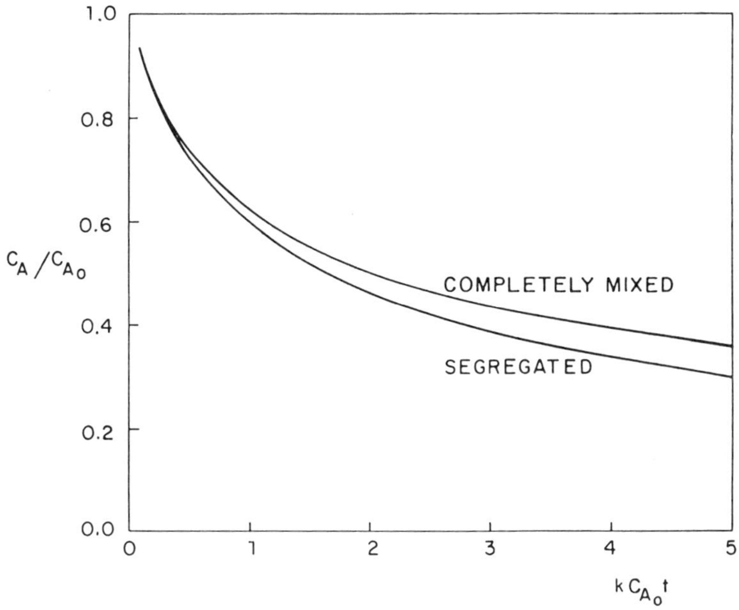

FIGURE 23 Conversion of a second-order irreversible reaction (A − B) in a stirred tank as a function of time for segregated and completely mixed flow.

and

$$C_A(\text{segregated flow}) = C_{A0}\left[1 + 2\left(\frac{kt}{2\sqrt{C_{A0}}}\right)^2 - 2\left(\frac{kt}{2\sqrt{C_{A0}}}\right)\right] \qquad (42)$$

Note that C_A for the ideal stirred tank is less than C_A for segregated flow. Here the presence of molecules of A reduces the reaction rate and mixing with products improves conversion. However, one has to be very careful with such arguments when applied to reactions of fractional order. They arise normally from Langmuir-type reaction rate expressions. Expressing them as a fractional order is correct only over a limited range of concentrations. There is a high probability that in integrating the results for Eq. (40), one may go outside the range where this is applicable, and therefore one must verify that the final result is not sensitive to this mistake. The results for a nonlinear reaction therefore depend not only on f(t) but on a more complete description of the mixing phenomena which Zwietering terms *micromixing*.

Swietering has also shown that these results can be extended to any arbitrary residence-time distribution as long as the reaction is a simple reaction of order n. In that case C_A/C_{A0} will always be bounded by two extreme cases: the case of segregated flow given in Fig. 20a and another case defined as maximum mixedness, which is illustrated in Fig. 20b. Whereas in Fig. 20a each tubular reactor is separate, in Fig. 20b, there is intermixing between them and the feed inlets are arranged such that the expected future life is constant for all molecules in a given cross section. Physically, one could realize the arrangement of Fig. 20b by designing a plug flow reactor in which additional feed is introduced at different points along the tube. For a reaction of order n the two limits or bounds on conversion are segregated flow and maximum mixedness. For the case of segregated flow,

$$\frac{\langle C_A \rangle}{C_{A0}} = \int_0^\infty [1 + (n - 1)R]^{1/1-n} f(t)\, dt \tag{43}$$

$$R = \frac{-kt}{C_{A0}^{1-n}}$$

For the case of maximum mixedness, $\langle C_A \rangle / C_{A0}$ has to be obtained by solving the differential equation

$$\frac{dC_A}{d\lambda} = -kC_A^n + \Lambda(\lambda)(C_A - C_{A0}) \tag{44}$$

where λ is an integration variable and $\Lambda(\lambda)$ is the escape probability as defined by Eq. (8). Equation (44) can be integrated numerically, although for certain values of n and certain values of $\Lambda(\lambda)$ it has analytical solutions. Here one sees another useful property of the escape probability and why one wants to measure the residence-time distribution in such a way that $\Lambda(t)$ can be computed reliably. For reactors close to plug-flow reactors, the bounds are very close and one does not have to worry about micromixing which is important mainly in stirred-tank reactors.

Strictly speaking, the results of Eq. (43) apply only to simple reaction of order n, in homogeneous ideally stirred reactors, which is a rather rare design problem. However, the concepts of bounding the outlet composition of a reactor is an essential tool of reaction engineering. It is really a form of performing a sensitivity analysis on a design. Consider the example of a stirred tank. A priori one knowns that this is an idealized model that cannot be fully realized. However, one can approach it in the sense that the time a droplet of feed spends in the reactor before it becomes totally mixed is very small compared to the total residence time (Evangelista et al., 1969b; Shinnar, 1961). One also knows that during scaleup the residence time stays constant, whereas the mixing time always increases (Evangelista et al., 1969a,b; Shinnar, 1961; Weinstein and Dudkovic, 1975). For a second-order single reaction, Zwietering's results show that one probably does not need to worry about this, as the completely mixed case is the worst (least desirable) case. There are many other cases for which this is not true. One wants to know how important such deviations from complete mixing are. To answer this, the two limiting cases for any arbitrary reaction system are determined. If the reactions are complex and nonlinear, these two limiting cases (segregated flow and complete mixing) will not give rigorous bounds but will indicate if the system is sensitive to mixing and what penalties are incurred for incomplete mixing. In scaleup, one cannot achieve segregated flow, so the case of complete mixing must be satisfactory or the design must be changed. If the system is sensitive to mixing, one can estimate the permissible mixing time by using models for incomplete mixing that simulate small deviations from incomplete mixing (Evangelista et al., 1969a,b; Belevi et al., 1981).

If, in a stirred-tank reactor, segregated flow leads to improved results, a plug-flow reactor will normally be even better. If one requires the stirring, one can approach plug flow by putting several such stirred tanks in series. Experiments that demonstrate the sensitivity to deviations in mixing can also be performed in a pilot plant. For example, one can design a pilot plant similar to the one shown in Fig. 24 despite the fact that the final reactor is completely mixed. There have been a number of models proposed that look at micromixing for more complex flow conditions over a wider range of micromixing [see, e.g., Weinstein and Adler (1967); Ng and Rippin (1965)]. These models, which are reviewed in detail in Nauman (1981), are more in the nature of learning models than design models, and are therefore outside the scope of this chapter.

Similar considerations apply to packed-bed and fluid-bed reactors. For nonlinear reactions the only safe way to scale them up is to approach plug flow as closely as possible. Again, one wants to know how closely plug flow must be approached and what the penalty is for small deviations. One can again perform experiments which experimentally check the sensitivity of the reaction system to deviations from plug flow and design the reactor accordingly. Here residence-time distributions are directly useful, since for small deviations from plug flow, one can estimate the deviation from plug flow from the tracer response by measuring γ^2. A simple model will give reasonably reliable bounds as to the possible impact of such small deviations, even if the reactions are highly nonlinear. If the sensitivity is large, one should take a safety coefficient with respect to the allowable coefficient of variation. This is illustrated in Fig. 25, where the effect of small deviations from plug flow on several reactions are compared. One is simply the irreversible reaction $A \rightarrow B$ of second order, and the other is a nonlinear case very sensitive to mixing. In both cases the results are indistinguishable, and for $\gamma^2 = 0.2$ (five stirred tanks in series, or a Péclet number of 10) the results are reasonably close. In both cases we also give the limiting cases of plug flow and a stirred tank to indicate the sensitivity to back mixing. The comments made with regard to Fig. 22 apply here

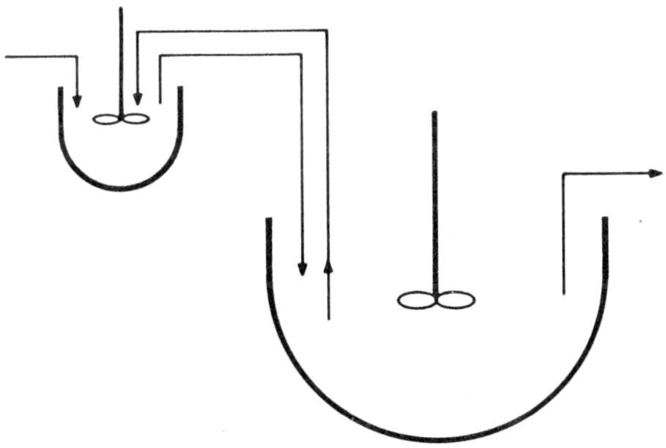

FIGURE 24 Simulation of finite mixing in a pilot plant.

equally well. The results are not unique functions of γ^2. However, for small deviations from plug flow, the results are not model sensitive as long as the residence-time distributions are reasonably similar. For simulation purposes, a model based on stirred tanks is therefore preferable, as it is much easier to compute. The fact that γ^2 alone is not sufficient to characterize the system unless γ^2 is very small is illustrated in Fig. 25c where two other models with $\gamma^2 = 0.1$ are compared to a series of stirred tanks. For a reaction sensitive to mixing, such as the one shown in Fig. 25b, the differences are very large and a much smaller γ^2 is necessary for them to become negligible. Such reactions therefore require extra care in scaleup and a close approach to plug flow (or a stirred tank).

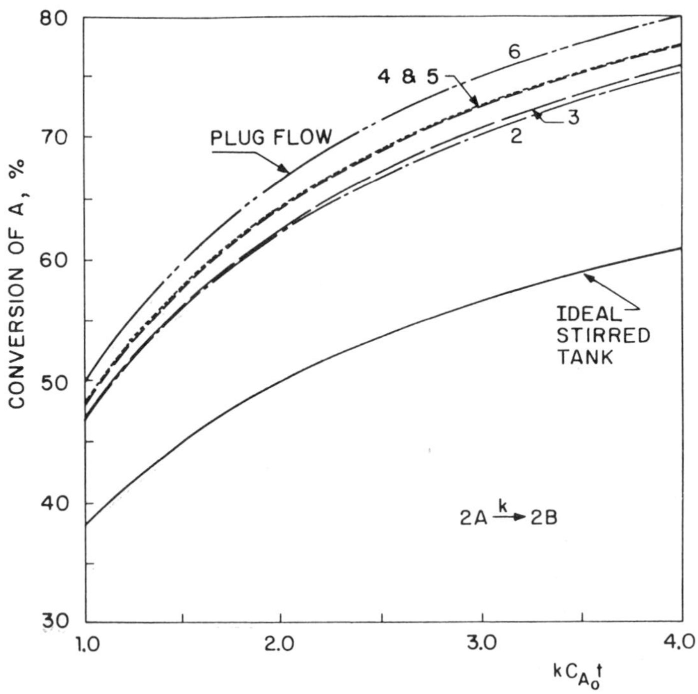

(a)

FIGURE 25 (a) Conversion of a second-order irreversible reaction for different flow models. (b) Comparison of product yield in a consecutive nonlinear reaction for different flow models. Reaction A→ B → C first order; reaction A + C → D second order. Inlet concentration of A, 1.0 (dimensionless units). 1, Mixed flow; 2, five CSTRs in series; 3, one-dimensional flow, Pe = 8.87; 4, 10 CSTRs in series; 5, one-dimensional flow, Pe = 20; 6, plug flow. (c) Same as in part (b) for different flow models:

(I) → [ ] → [∞] → (II) 10 stirred tanks in series (III) → [∞] → [ ] →

$V_{PF} = 0.69\ V, V_{ST} = 0.31.$

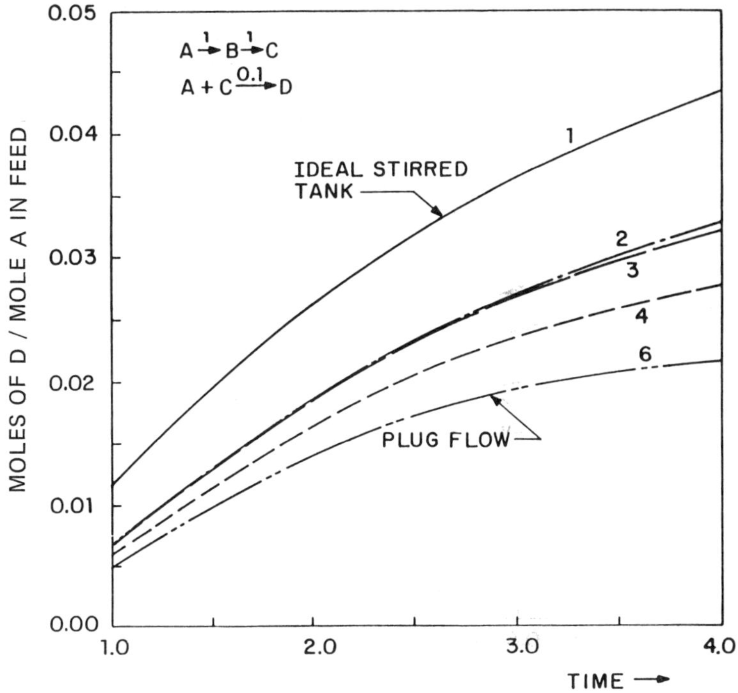

(b)

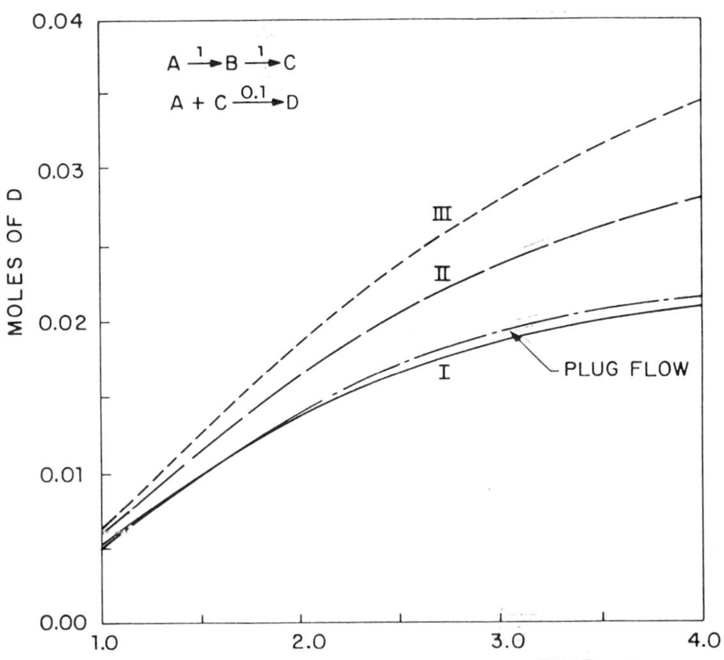

(c)

FIGURE 25 (Continued)

There is another area in which the concept of a residence-time distribution is useful in organizing ones thinking and experimentation. Thinking in terms of residence times and joint probabilities is helpful in deciding what reactor types to choose. For all linear and pseudolinear systems, isothermal as well as nonisothermal, plug-flow reactors are always preferable. The case where backmixing or a stirred tank has a positive influence is when there are complex interactions between products and reactants. Examples are autocatalytic reactions, autothermic reactions (heat can be considered a product), or reactions of the types

$$A \rightarrow B \rightarrow C$$

$$A + C \rightarrow D$$

If D is desirable, a mixed reactor is preferable, as can be seen from Fig. 25. If D is undesirable, a plug-flow reactor is preferable. An example of this reaction discussed in Renick et al. (1982):

Methanol → olefines → aromatics + paraffins

Aromatics + methanol → durene

If Durene formation is to be avoided, the reactor should be as close as possible to plug flow. This result can be derived without reference to residence-time distribution, but thinking in terms of residence time provides a useful framework.

There are a number of cases in which the desirable configuration is neither plug flow nor a stirred tank. Combustion is an example. But here one can often break down the optimal or desired flow system into configurations composed of well-defined flows. If, on the other hand, the flow is complex, residence-time distributions may provide diagnostic tests, but one can neither model nor scale such cases solely on the basis of residence-time distributions.

It should be emphasized that at the present state of knowledge, there is no way that one can reliably scale up a reactor if the reactor is sensitive to the form of f(t) and the reactor does not approach one of the well-defined ideal cases.

HETEROGENEOUS REACTORS AND MULTIPHASE SYSTEMS

Properties of Tracer Experiments in Multiphase Systems

Residence-time distributions provide a powerful method for the analysis of homogeneous isothermal reactors. Unfortunately, very few real reactors used are truly homogeneous. Those few that are seldom provide serious difficulties in design or scaleup. The majority of the reactors used are heterogeneous, either gas-solid, gas-liquid, or triple phase. Catalytic reactors are almost always heterogeneous, even in homogeneous catalysis, where most of industriral reactors deal with gas-liquid reactions.

If the use of residence-time distributions was restricted to homogeneous reactors, their usefulness would be rather limited. Luckily, most of the concepts derived in the previous sections can with some care, be applied to heterogeneous reactors. One must, however, be careful to take into account the heterogeneous nature of the system. This rules out many of the algorithmic uses of residence-time distributions. One cannot simply measure a residence-time distribution and predict the performance of a complex reactor. Even the concept of a residence-time distribution is not simple any more, as different reactants may have totally different residence-time distributions. In a heterogeneous reaction one cannot measure reaction rates and residence-time distributions independently and reliably predict reactor performance even for first-order systems. But the main uses focused on in the preceding section remain.

1. The intensity function is still a useful tool to diagnose maldistributions in the flow and provide clues to the nature of the flows in the system. Of special importance is that it allows one to check how closely a design approaches plug flow (or an ideally stirred tank). It is also useful in cold-flow models, where it can be used to test the efficiency of flow distributions or baffles. The examples given earlier were from heterogeneous reactors.
2. Residence-time distributions still provide a powerful conceptual tool for design and scaleup. Although often one cannot get accurate predictions, one can derive a better understanding of the scaleup problem.
3. Residence-time distributions in multiphase systems allow measurement of the holdup of each phase.

There are some important differences between tracer experiments in homogeneous and heterogeneous systems. It is important to understand these differences not only in evaluating tracer experiments, but also because they are related to design and scaleup. In a heterogeneous system, different compounds may have strongly different residence-time distributions. In a homogeneous reactor the difference in residence-time distributions between different compounds is, in most cases, negligible. This is especially so if the mixing process is turbulent. The only exceptions are slow laminar flows, where molecular diffusion will be important. Thus in most homogeneous reactors the nature of the tracer compounds is of small importance. On the other hand, in heterogeneous systems, there can be large differences between the response of different tracers, and therefore between the response of a tracer and that of a reactant. Even different compounds in the feed or different products may have different residence-time distributions. An example of this is shown in Fig. 26 (see also Orth and Schügerl, 1972). Figure 26 illustrates data taken from a packed-bed Fischer-Tropsch reactor in which CO and H_2 were reacted over a Ni catalyst. The reactor was operated at steady state and a step input of CO containing ^{13}C was introduced together with a step input of helium. The helium concentration was kept very small so as not to disturb the steady state. The response of the ^{13}C in the CO is plug flow with a longer residence time, which shows that the CO is adsorbed but that the interchange of the adsorbed CO with the CO in the gas phase is very fast. On the other hand, both thc CH_4 and C_3H_6 product behave like a stirred tank in series with a plug flow. This indicated not only strong adsorption but that, in this case, the interchange between the adsorbed phase and the gas phase for the product is slow and has a characteristic time similar to the residence time. The results could also have been the reverse (slow interchange of the reactant and fast interchange of the product). In a reacting system the concept of a residence-time distribution itself becomes rather complex to define.

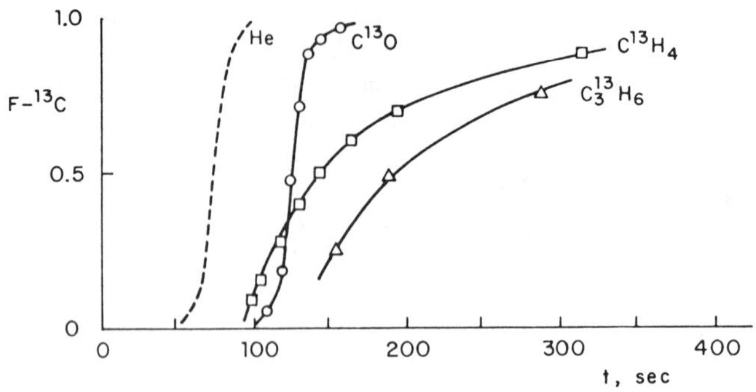

FIGURE 26 Response of a packed-bed Fischer-Tropsch reactor, fed with CO and H_2, to a step input of ^{13}CO and He. Reactor kept at steady state. (From Biloen et al., 1983.)

One can always define it with respect to specific elements such as C in Fig. 26 (or molecular combinations that do not change such as benzene rings), but the only way to measure this would be by means of proper isotope markers. In Fig. 26 the tracer response is given separately for ^{13}CO and $^{13}CH_4$. From the original data one could have measured a tracer response for ^{13}C, which would give a residence-time distribution for carbon atoms. It is a proper residence-time distribution, containing valuable information, but its interpretation is more difficult. It is not only a function of the flow pattern but also a function of conversion (or catalyst reactivity). To measure maldistribution in flow, a nonreactive tracer is often preferable.

The average residence time of a single tracer is not necessarily equal to total reactor volume divided by the flow rate. It is therefore hard to estimate independently the average residence time, which due to adsorption or similar phenomena, can vary by a factor of 5 or more between different tracers. Furthermore, not all tracers give linear responses. This makes the choice of a proper tracer much more critical than in a homogeneous system. Residence-time distributions are by definition linear functions. At steady state all compounds in a reactor have a measurable residence-time distribution regardless of the nonlinearity of the flow, nonlinear adsorption phenomena, and so on. If one can properly mark a compound (e.g., using an isotope), the tracer experiment will result in a residence-time distribution. But if one introduces a tracer that is not present in the system, it could behave in a nonlinear way, and the result could be very dependent on the amount of tracer introduced. Furthermore, one does not always want a tracer that behaves similar to the reactant. For example, in some cases where one is studying reactor internals in a packed bed, nonadsorbing tracers are preferable, as they are more sensitive to a maldistribution in the flow. This can be seen from Fig. 26. A helium tracer would have given a much better indication of a maldistribution in gas flow than using ^{13}CO and measuring the total amount of ^{13}C leaving. The proper choice of tracers is discussed in more detail in the next section.

Residence-time distributions cannot be used directly to predict outlet composition even for linear reaction systems. Reaction rates in catalytic reactors are measured and reported on the basis of either reactor volume, catalyst volume, or catalyst weight. In a homogeneous system this has the same time scale as the residence-time distribution. In a heterogeneous system this relationship is not valid(due to adsorption, absorption, different solubilities, etc.), and the difference in time scale can be a factor of 5 or more. The ratio is different for different reactants. Consider, for example, a packed-bed reactor. The estimates of kinetics are based in most cases on overall residence times of mole reactant/lb catalyst per hour. The real residence-time distribution of the gas is in most cases unknown because the holdup of the gas depends on adsorption. If one could really tag a reactant one would not necessarily get plug flow in a tubular laboratory reactor (see Fig. 26). However, the kinetic data underlying the scaleup assumed plug flow in the lab reactor. The rate expressions are really not molecular kinetic expressions describing real molecular events but represent overall relations allowing one to scale up a lab reactor to an industrial packed bed.

This approach works very well even if the reactant is strongly adsorbing and the reactions are highly nonlinear. The fact that for a highly adsorbing reactant the real residence time of a molecule is, in reality, far from plug flow, and different for all reactants, does not affect the usefulness of such overall kinetic relations for steady-state design. It may, however, strongly affect the dynamic behavior of the reactor and one has to be careful when using rate expressions obtained from steady-state experiments to describe the dynamic behavior of a reactor (Shinnar, 1978; Biloen et al., 1983). This also prevents one from using such methods as the boundary methods of Zwietering, unless once again one is dealing with small perturbations.

In catalytic reactors, reactivity can change with position even in isothermal reactors. In some catalytic reactions, especially at high temperatures, reactions occur in the gas phase as well as at the catalytic surface. There is therefore no unique correlation between residence time and the state of the molecule, which is a problem already encountered in the case of the nonisothermal reactor. While a residence-time distribution can give important information about the flow, one would need additional information to predict the outlet composition, even if the reactor is isothermal and first order.

The concept of a residence-time distribution is still helpful in the sense outlined in the beginning, as long as one is careful to recognize the differences. This will be discussed in detail at the end of the section. First, some techniques that are especially applicable to heterogeneous reactors will be discussed, namely, multiple tracer experiments and contact-time distributions.

Multiple Tracer Experiments

As in heterogeneous systems, different compounds may have different residence times. Independent knowledge of the residence-time distribution of several tracer compounds gives a better description of the system. If the feed contains different compounds, one may wish to use markers for each of them to study the system. One can also learn a considerable amount about the system by using a number of different tracers, even if they are completely different from the reactant fluid (Shinnar et al., 1972; Nauman, 1981; Orth and Schügerl, 1972). However, one is really measuring the residence-time distribution of the specific tracer, which is not necessarily representative of the system. If several tracers of strongly varying properties are utilized, the residence-time distributions of the compounds in the system can be bracketed. One should note, however, that even if one uses an isotope of a reactant, one still cannot call this a residence-time distribution of the system, as in a heterogeneous system different reactants may have totally different residence-time distributions.

One does not always want a tracer that behaves like the reactant fluid. Consider, for example, a packed-bed reactor. To measure maldistributions of the gas flow, one wants a tracer that does not diffuse into the catalyst and does not get adsorbed. In this case the transport processes inside the catalyst particles and adsorption processes are irrelevant for scaleup or reactor performance in the sense that one deals with scaleup of a pilot plant with identical catalyst particles. The main problem one needs to consider is that mass transport processes to and from the particle to not change as the linear velocity in the large reactor increases. Tracer experiments are not sensitive enough to check this property, but there are other methods to ensure this. The main use of a tracer experiment is to test that there are no maldistributions in the overall flow. An example of this was shown in Fig. 16. In a packed bed, if there are maldistributions, one can use the result of a tracer experiment with a nonadsorbing tracer to estimate their impact on conversion and product distribution. In trickle beds one must worry about both liquid and gas distribution, and would therefore want two tracers, one for each phase. In all these cases, $\Lambda(t)$ is the most sensitive way to check for a maldistribution, but in a trickle bed one must be careful. Strong deviations from plug flow can occur due to fluid dynamic effects and one therefore needs a good reference case (Mills and Dudukovic, 1981). In fluid-bed reactors the situation is more complex. If the solid phase is not stationary but is fed and removed from the bed (as in the treatment of solids in an FCC cracker), one may want to use a solid tracer to check if the solid has the desired residence-time distribution. In a single stage one probably wants complete mixing of the solids or a constant $\Lambda(t)$, whereas in a multistage bed and for treatment of solids, one wants the residence-time distribution for the solid to correspond to a series of stirred tanks, or to approach plug flow.

In some trickle beds, part of the feed evaporates. In this case, to study the flow properly, it is recommended that both an evaporating and nonevaporating fraction of the liquid be tagged. An example of such a multiple tracer experiment is shown in Fig. 27, which illustrates results obtained in a hydrotreater for fuel oil (Snow, 1983). Three tracers are used. The C_8 compound enters in the gas phase. The C_{16} compound represents a compound that under the conditions of this hydrotreater is present in both phases, whereas the C_{32} compound stays mainly in the liquid phase. Tracer results are given for two reactors operating under similar conditions. Reactor A contains a 1/8-in. extrudate hydrotreating catalyst, whereas reactor B operates with a 1/16-in. extrudate. Reactor B showed a better performance at high conversion, allowing a 50% higher throughput at 99% conversion. The results give a good illustration of the advantage of using proper multiple tracers. The most interesting information in Fig. 27 is not the form of the individual tracer response but the relative position of the individual tracer responses. In a trickle bed, one would expect that the residence time of the lqiuid is larger than that of the gas. The results bear this out, as the average residence time of the C_{32} response is larger in both reactors then in the C_8 tracer. In reactor B the C_{32} tracer appears 3 min after the C_8 tracer. However, looking at Fig. 27a one notices immediately

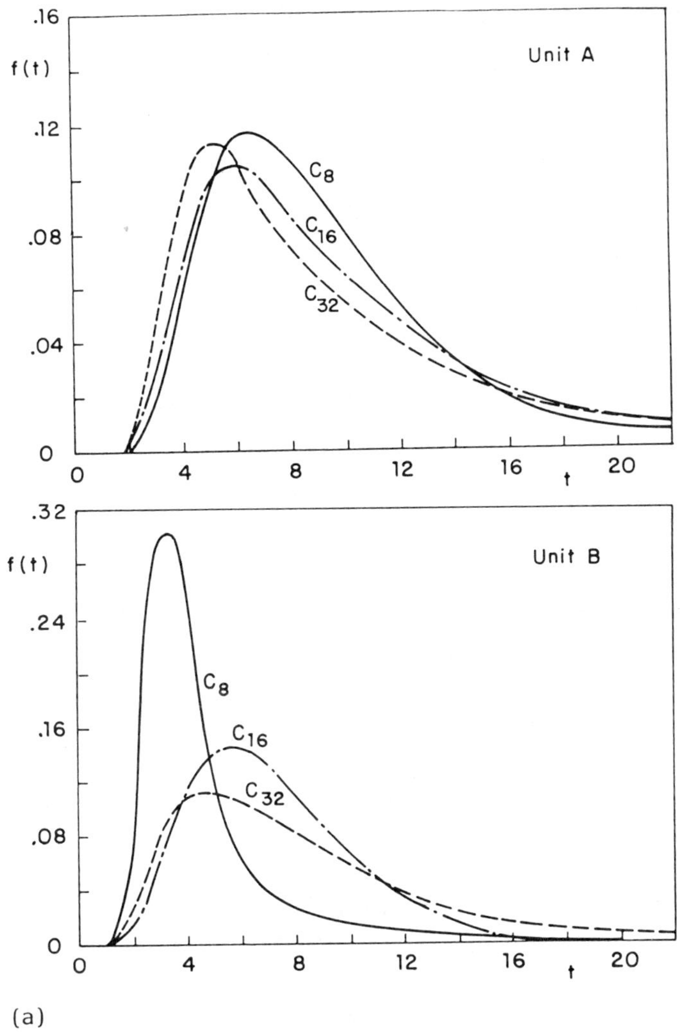

FIGURE 27 Multiple tracer experiments in two industrial fuel oil hydrotreaters. (Private communication with A. I. Snow, Arco Petroleum Co., 1983.) Three compounds tagged with ^{13}C introduced simultaneously in feed (pulse experiment) and measured at reactor outlet. (a) Residence-time density function; (b) intensity function (escape probability) of unit A; (c) intensity function (escape probability) of unit B; (d) cumulative residence-time distribution (unit A); (e) cumulative residence-time distribution (unit B).

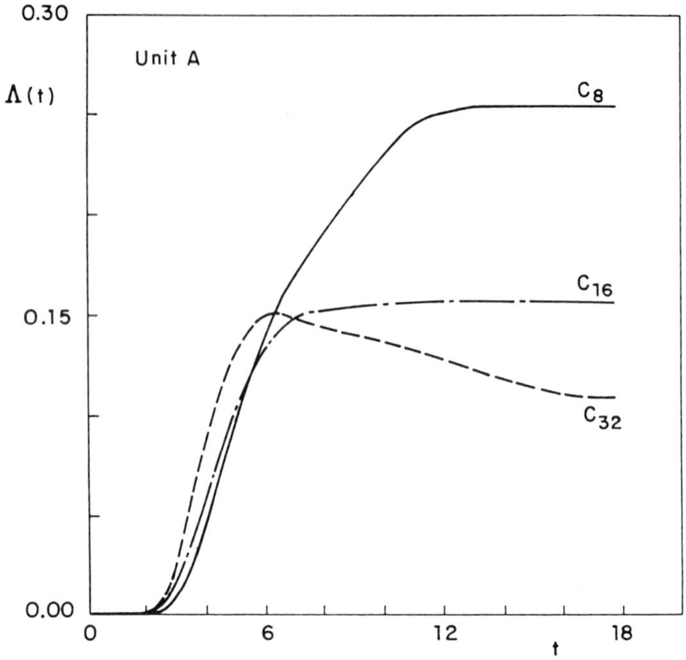

(b)

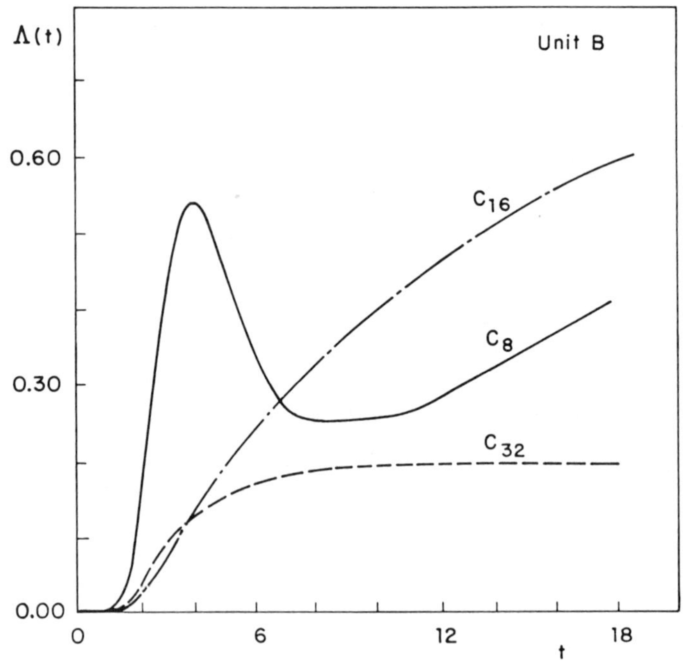

(c)

FIGURE 27 (Continued)

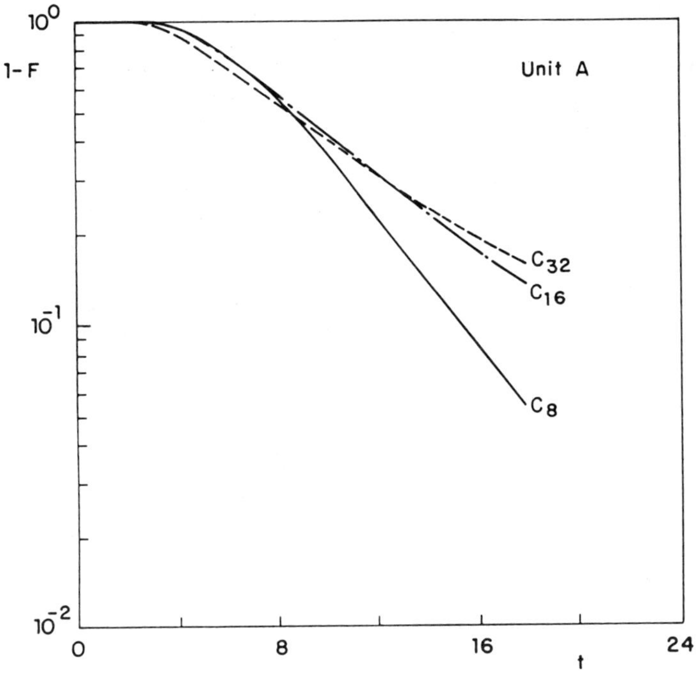

(d)

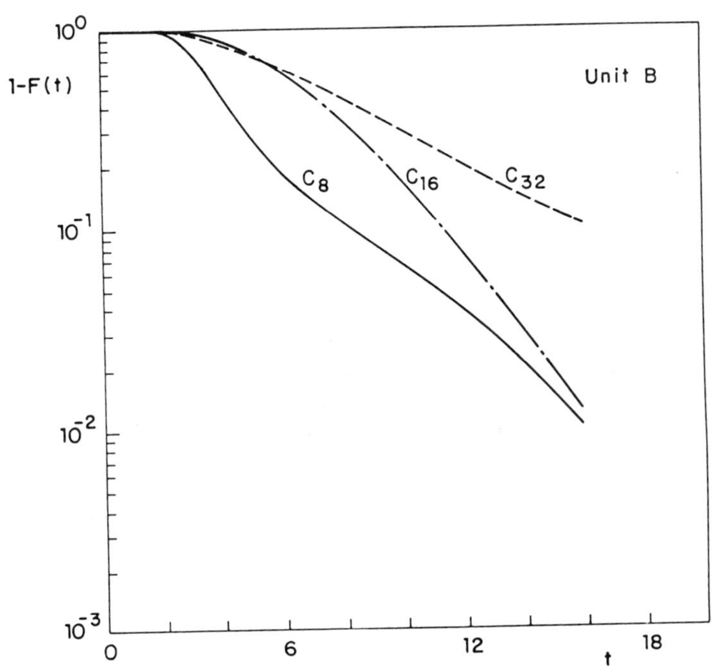

(e)

FIGURE 27 (Continued)

that the C_{32} tracer in reactor A appears about 1 min before the C_8 tracer. This shows a significant difference between the two reactors and shows that something is wrong with reactor A which explains its lower performance. If one examines in detail at the escape probability (Fig. 27c and d), one also notes that the C_{32} response shows a strong maldistribution in reactor A and good liquid distribution in reactor B. Interestingly, reactor B has a maldistribution in the gas flow (C_8) which is not evident in reactor A. But it is apparently not significant enough to cause problems.

These tracer results contain some very interesting information about what happens inside these reactors, and having simultaneous information from both a liquid and gas tracer greatly adds to their value. However, the interpretation of these results is in no way unique, as the maldistribution of the C_{32} response in reactor A could be due to several causes. It could be caused by a mechanical problem resulting in a nonuniform liquid distribution, or by inherent flow problems, or by the different catalyst size. Some reactor modeling could indicate which of these explanations is more likely.

It is important to note that in such a multiphase system, there is no way to deduce from a single tracer experiment if a maldistribution such as the C_{32} response in reactor A (Fig. 27) is caused by a bad distributor or by bypass phenomena inherent in the flow regime of this specific reactor or by phenomena inside the catalyst particle. Trickle beds, by their nature, do not scale equally well in different flow regimes, which are determined by various flow parameters, such as gas velocity and density, liquid properties, and liquid-to-gas ratio. This problem was already encountered in the discussion of the fluid bed, and it is important to realize this. However, often one has two reactors operating under similar conditions and, in such a case, a difference in tracer response clearly indicates a mechanical problem.

In studying the gas flow in a fluid bed, one faces a problem not encountered in packed beds. If the reactant is adsorbed, it will move with the solid. In an isothermal or adiabatic packed bed, if a pilot plant and a large plant have the same residence-time distribution for a nonadsorbing tracer, one would expect them to behave similarly. In a fluid bed one cannot make the same assumption, as the mixing processes of the solid and the adsorbed reactant with it are different for the two cases.

The only time that one can reliably scale a fluid bed and predict its performance from a pilot plant is where in both cases the gas flow approaches plug flow as closely as possible. One way to achieve this is by using baffles in the fluid bed. Baffles here have a double purpose. They prevent bypassing of gas due to large bubbles by breaking up the bubbles, and reduce solid circulation, thereby also reducing the backmixing of gas by moment of reactants adsorbed on the catalyst.

In cases where one wants to test whether the flow distributor operates well, a nonadsorbing tracer is a more sensitive tool (see Fig. 19). When there are baffles in a fluid bed, the problem becomes more complex. A good case can be made that an adsorbing tracer is more sensitive, as it shows if adsorbed tracer is recycled by the catalyst to the bottom of the reactor. Therefore, if a baffled fluid bed approaches plug flow with a strongly adsorbing rracer, it gives one better confidence in the design. However, one has to be careful to check that the tracer behaves linearly. To check this, one always needs comparative results from a packed-bed reactor. SF_6 has been reported to have desirable properties, and its adsorption can be changed in a cold-flow model over a wide range, as the adsorption equilibrium is a function of the water content of the air. If a properly baffled design gives a residence-time distribution with a high Péclet number (low coefficient of variation) for both a nonadsorbing and an adsorbing tracer, this improves the confidence in the scalability of the design.

Understanding the effect of adsorption on the residence-time distribution is helpful in another way. It explains why different reactions may have different sensitivity to scaleup in the same fluid bed. To understand this concept, one must first consider the evaluation of tracer experiments with adsorptive tracers in a more rigorous way. The approach used is described in detail in Shinnar et al. (1972). Consider a packed bed or a chromatographic column in which a tracer can be adsorbed inside a porous particle. One can now perform a tracer experiment with a nonadsorbing tracer which can also diffuse into the porous particles and has the same coefficient of diffusion inside the particles. For simplicity a pulse tracer experiment will be used. The density function of the nonadsorbing tracer is designated f_0 and that of the adsorbing tracer f_a. Also assume that the adsorption equilibrium is linear and the amount of tracer adsorbed at equilibrium per unit volume of C_0 gas phase is dC_0, where C_0 is the concentration in the gas phase. Consider the case where ideal plug-flow conditiond exist. Then $f_0(t)$ for the nonadsorbing tracer is a delta function at $t = \tau$. One can then measure $f_a(t)$ for different values of τ and define a function $\psi_a(t)$.

$$\psi_a(t) = f_a\,(t - \tau) \tag{45a}$$

which gives the sojourn-time distribution of the adsorbing tracer in the adsorbed state. The simplest case is if the tracer adsorbs in a way that the rate of adsorption and desorption is very fast. In this case

$$\psi_a(t) = \delta\,(\alpha\tau) \tag{45b}$$

and for the general case where α is a constant, that gives the ratio of the residence time in the adsorbed phase to that in the continuous phase.

$$f_a(t) = f_0[(\alpha + 1)t] \tag{45c}$$

If adsorption and desorption have finite rates, Eq (45c) will not hold. However, if the bed has uniform properties and the flow in the gas phase is plug flow, $\psi_a(t)$ must be a unique function of τ in the continuous phase. One can further show that for various values of τ

$$\psi_a(t/\tau_1 + \tau_2) = \int_0^t \psi_{ap}(\theta/\tau_1)\psi_{ap}([1 - \theta]/\tau_2)d\theta \tag{46a}$$

where $\psi_{ap}(t/\tau_1)$ is the function $\psi_{ap}(t)$ given that τ for the continuous phase is τ_1. The subscript p was added to indicate that the flow in the continuous phase is plug flow. If Eq. (46) holds, one can also show that the Laplace transfer of $\psi_a(t)$ must have the form [Shinnar et al., 1972)

$$\hat{\psi}_{ap}\left(\frac{s}{\tau}\right) = e^{-\tau\hat{\rho}(s)} \tag{46b}$$

and therefore

$$\hat{\psi}_{ap}\left(\frac{s}{\tau_1 + \tau_2}\right) = e^{-(\tau_1+\tau_2)\hat{\rho}(s)} \tag{46c}$$

$\hat{\rho}(s)$ therefore characterizes the adsorption process or in a more general way the transport processes in an outer phase not reachable for the tracer use for f_0.

In Eq. (45c) it was assumed that $\hat{\rho}(s) = \alpha s$. If one assumes a finite rate of adsorption and desorption, then in the example given $\hat{\rho}(s)$ becomes

$$\hat{\rho}(s) = \frac{\alpha s}{1 + (\alpha/\underline{\lambda})s} \tag{47a}$$

where $\underline{\lambda}$ is a normalized exchange coefficient. For understanding the general concepts, one does not need more complex forms of $\hat{\rho}(s)$. The reader should consult Shinnar et al. (1972) for more details. If $\hat{\rho}(s)$ is given by Eq. (47a), then

$$\hat{\psi}_{ap}(s) = e^{-\alpha\tau s/[1+(\alpha/\underline{\lambda})s]} \tag{47b}$$

which is a well-known function used in chromatography. To derive it, one must not only assume that $f_0(t)$ is a plug flow, but also assume that the catalyst bed has uniform properties and that the catalyst is stationary. This means that any molecule that got adsorbed returned to the same place it left. In the limiting case where $\underline{\lambda} \to \infty$ [Eq. (45c)], the last assumption is not needed. However, the assumption of uniformity is still necessary.

If the bed is uniform in its properties and each particle returns to the same place in the gas phase, Eq. (46) can be generalized to the case where $f_0(t)$ is not plug flow. For any such case with arbitrary $f_0(t)$,

$$\hat{\psi}_a(t) = \int_0^t f_0(\tau)\psi_{ap}\,\frac{t}{\tau}\,d\tau \tag{48a}$$

$$f_a(t) = \int_0^t f_0(\tau)\psi_{ap}\left(\frac{t-\tau}{\tau}\right)d\tau \tag{48b}$$

and

$$\hat{\psi}_a(s) = \hat{f}_0(\hat{\rho}(s)) \tag{49a}$$

where $\hat{\rho}(s)$ has been substituted for the variable s in $\hat{f}(s)$. Similarly,

$$\hat{f}_a(s) = \hat{f}_0(s + \hat{\rho}(s)) \tag{49b}$$

Note that in the previous simplified case,

*
$$\hat{f}_a(s) = \hat{f}_0 \left[s + \frac{\alpha s}{1 + (\alpha/\underline{\lambda})s} \right] \tag{49c}$$

The results of Eq. (49c) are experimentally testable, as one can measure $\alpha s/(1 + \alpha s/\underline{\lambda})$ in a packed-bed reactor under conditions approaching plug flow. However, there are strong restrictions on its applicability. Equation (49) also allows computation of $\hat{\psi}_a(s)$ and therefore of $\psi_a(t)$ from measurements of $f_a(t)$ and $f_0(t)$ in other cases where the adsorbed particle returns to the same place. The only other case where this direct computation is possible is when there is no correlation between the total residence time of a particle in the adsorbed phase and its residence time in the gas phase. In this case

$$f_a(t) = f_0(t) * \psi_a(t) \tag{50}$$

where * is again the convolution integral. Unfortunately, this is a rather unrealistic case for a catalytic reactor. In real cases when can Eq. (49) be used? One case is a packed bed with back diffusion. The other is a stirred tank reactor in which both the continuous and gas phase are mixed (such as in a Berty reactor). In that case all molecules in the continuous phase have equivalent location and therefore Eq. (49) holds. There are additional important implications that can be deduced from Eqs. (48) and (49). Consider the special case of a packed-bed reactor in which the flow in the internal phase is plug flow or $\hat{f}_0(s) = e^{-\tau_0 s}$. Further assume that all reactants and products have the same rate of adsorption and desorption and also the same adsorption equilibrium. Therefore, $\psi_a(t)$ is constant for all compounds. Finally, assume that a first-order complex reaction occurs in the adsorbed phase. In that case one can write an equivalent of Eq. (28) for the product distribution:

*
$$\langle C_j \rangle = \alpha_{j0} + \sum_{r=1}^{n-1} \alpha_{jr} \hat{\psi}_a(\lambda_r) \tag{51}$$

which for a packed-bed reactor can be written

*
$$\langle C_j \rangle = \alpha_{jo} + \sum_{r=1}^{n-1} \alpha_{jr} e^{-\tau(\hat{\rho}(\lambda_r))} \tag{51a}$$

In standard kinetic practice one neglects adsorption processes and simply assumes that the reactor is plug flow and one writes an equivalent of Eq. (26):

$$P_j(t) = C_j(t) = \alpha_{j0} + \sum_{r=1}^{n-1} \alpha_{jr} e^{-\tau_\rho \lambda_r^*} \tag{52}$$

where λ_r^* are eigenvalues based on phenomenological rate constants. τ_p is a time constant that has the dimensions of time but does not refer to any real measurable time. It is expressed in lb feed/lb catalyst per hour (or lb mol feed/lb catalyst per hour). Note that

$$\tau_0 \hat{\rho}(\lambda_r) = \tau_p \lambda_r^* \tag{53}$$

Equation (53) implies that it is always possible to describe a first-order reaction occurring in the adsorbed phase of a catalyst by assuming that the reactor is a plug-flow reactor. The reaction rates will not relate to any real reaction rates at the catalyst surface, but this is not important in phenomenological rate equations in reactor design. Normally, the kineticist is interested in eliminating from λ_r^* transport resistances that deal with transport to and from the catalyst particles and diffusion in the macropores, as well as back diffusion in the reactor. This can be achieved by assuming that the catalyst particles are small enough and that the microreactor used is long enough [Silverstein and Shinnar, 1975).

In this special case, one can then estimate the conversion of an isothermal packed-bed reactor in which $f_0(t)$ deviates from plug flow (either due to diffusional resistances or due to maldistributions), by measuring $f_0(t)$ with an inert tracer and then computing the effect of the deviation from plug flow just as one did in a homogeneous reactor. This is the result of deriving Eq. (53) from Eq. (51). One must, however, note that $f_0(t)$ has a different time scale than Eq. (53). One must therefore nondimensionalize $f_0(t)$ and scale it such that the new first moment is the same as that used in computing λ_r^*. This is particularly important, as kineticists often use different definitions for τ_p in computing λ_r^*.

Equation (49) to (53) also imply that for a first-order reaction, one can predict the behavior of a packed bed by getting kinetic measurements in a completely mixed reactor such as a Berty reactor, and vice versa. One cannot, however, predict the outcome of any reactor for which Eq. (49) does not hold, such as a fluid bed in which the catalyst moves. If the movement of the catalyst is slow compared to the adsorption-desorption processes of all reactants, one can get approximate predictions. For example, if the fluid bed is staged by vertical baffles, it approaches a series of stirred tanks or, for practical purposes, the reactants return to the same place. The outcome can then be predicted using a packed-bed reactor (or another fluid bed close to plug flow) as a model.

The relationship between Eqs. (51) and (52) was derived for the case where $\psi_a(t)$ is equal for all reactants. This is not a very likely case but was used for didactic reasons. However, the principle that one can predict the outcome of a first-order reaction in a reactor for which Eq. (49) holds by measuring $f_0(t)$ and the overall kinetics in a packed-bed reactor is valid for arbitrary sojourn-time distributions in the adsorbed phase, even when they are different for each compound. The concept therefore has much broader applicability. To prove the validity of the more general principle, one can use an argument similar to the one used earlier when we discussed optimal reactor concepts for nonisothermal reactors.

For any first-order reaction one can derive a phenomenological equation such as Eq. (52). Kineticists define a first-order reaction by this condition as they do not measure real reaction rates in the adsorbed phase. Consider such a case in which $\psi_a(t)$ is different for all compounds. The average sojourn-time distribution of the molecules will now depend on their state and therefore on the reactivity of the catalyst. This is included in λ^*. Once again, the hypothetical observer stands at the outlet of a plug-flow reactor and collects the molecules. The molecules are now sorted into bins. The molecules in each bin have a unique sojourn time θ in the continuous phase. In each bin, given θ one can compute $p_j(\theta)$, the probability of being in state j given θ. As the number of molecules is large, $p_j(\theta)$ is equal to $C_j(\theta)$. Averaging over all the histories yields $\langle p_j \rangle$ or $\langle C_j \rangle$ or the outlet concentration. Note that if the conditions leading to Eq. (49) hold [namely, the catalyst has uniform properties and a molecule always returns to the same place in the space of $f_0(t)$], then $\langle C_j \rangle$ is a unique function of the residence time t of the molecule in the nonadsorbed phase. If one looks at molecules with different sojourn times in the nonadsorbed phase, their probability of being in state j should be the same as that of a molecule passing a plug-flow reactor with residence time t in the nonadsorbed phase. The previous arguments should therefore hold regardless of the fact that all compounds have different adsorption desorption behavior as long as the system is linear and fulfills the basic assumptions.

The previous discussion leads to the disconcerting conclusion that in order to predict the behavior of a first-order reaction, one does not want a true residence-time distribution. A more accurate measure will be given by a nonadosrbing tracer that measures $f_0(t)$. Thus the choice of tracer can be crucial in predicting the outcome even of a first-order reaction. There are cases where the "true" residence-time distribution will not correctly predict the outcome of even a first-order reaction. This has important implications for the scaleup of first-order reactions. It was shown that there is a fundamental similarity between a mixed reactor and a plug-flow reactor and that one can always scale up from one to the other regardless of how the reactants behave in the adsorbed phase. However, one cannot apply the kinetic data obtained from either one to predict the outcome in a fluid-bed reactor. This has important implications for the scaleup of first-order reactions, as it indicates a fundamental similarity between a mixed reactor and a plug-flow reactor in terms of arbitrary behavior in the adsorbed phase.

An important exception is the limiting case of a fluid bed which is reasonably uniform and approximates plug flow. This will occur when either $\underline{\lambda}$ is very large or the mixing of the solid is slow compared to the time scale of $\underline{\lambda}$. ($\tau_{mixing} > 1/\underline{\lambda}$). Otherwise, $\psi_a(t)$ would depend on both $\underline{\lambda}$ and solid mixing. Unbaffled fluid beds in the bubbling range are not very uniform and are often modeled by a two-phase model. Here, Eq. (49) does not apply even if $\underline{\lambda}$ is large. One can, however, use an adsorbing tracer with a large value of $\underline{\lambda}$ to test how close the bed approaches the desired condition, namely, that the bed has no bypass of gas that does not contact the solid and approaches plug flow. Adsorbing tracers are an effective tool to test whether the system approaches plug flow or not.

In a packed-bed reactor one can measure $\underline{\lambda}$. In fact, if one uses tracers with lower values of $\underline{\lambda}$ and still gets a good approximation to plug flow, one can then get an estimate of the solid backmixing. To estimate its impact on the reaction, one would have to bound λ for all the reactants and products, which can in principle be done by using tracer methods similar to those used in Fig. 26.

There is one essential point to remember in testing for maldistribution of flow or for the effectiveness of internals in fluid beds and trickle beds. This is the need for a good comparison case of a well-performing reactor. If the residence-time distribution obtained by a tracer experiment deviates strongly from plug flow, it does not necessarily indicate a maldistribution of flow due to bad design. It could be a fluid dynamic instability phenomenon which is common in fluid beds and which can also occur in large trickle-bed reactors. In trickle-bed reactors it can be due to diffusion and adsorption phenomena in the liquid phase. Input output tracer experiments cannot distinguish between the two cases. What they can do is provide comparisons between two reactors and give clues to the effectiveness of corrective measures such as baffles, which is a very valuable tool properly used.

Measurement of Holdup in Multiphase Systems

In multiphase systems the separate holdup of each phase may be unknown. If the system is simply solid gas or liquid gas, the holdup of the solid or liquid might be obtained by a pressure difference. The holdup of the gas is more difficult to measure, as it may dissolve in the liquid or adsorb on a solid. Here tracer experiments can be very helpful. One has to be careful with the choice of the tracer. For a solid tracer it is important to find particles that do not separate and have the same size distribution and density as the solid studied. For best results the solid should be impregnated with the tracer when possible. If the liquid does not adsorb or evaporate almost any tracer will do, but if it does, it is best to tag the liquid or part of it. Choice of a proper tracer from the gas phase presents similar problems. If the gas adsorbs or is soluble in the liquid, one needs a tracer with similar properties or preferably an isotope of the real compound. This is sometimes the only direct way to measure holdup. When a suitable tracer is available, two further requirements must be met.

1. Accurate measurement of total feed rate for the flow of the phase measured
2. Accurate measurement of the amount of tracer introduced

In a pulse experiment one should always carefully check that the integral $\int_0^\infty QC(t)$ is equal to the amount of tracer introduced with an accuracy better than 95%. In this case, one should never use Eq. (13). The first moment of f(t) will then give an estimate of the holdup. In a step experiment one can use $\int_0^\infty (1 - F)\, dt$ to estimate the holdup. The estimate of the holdup obtained is one that includes adsorption phenomena. Therefore, if the tracer is not an isotope of the real flow compound, one has to be very careful in its interpretation.

The example given in Fig. 10 can serve to illustrate these problems. It is based on data supplied to the author by Exxon (Mills and Dudukovic, 1981) from a study of the Exxon Donor Solvent Reactor. The tracer experiments here were performed to measure holdup of the three phases. The system illustrated is a three-phase system. The reacting gas is a mixture of hydrogen and hydrocarbons, and the liquid is a mixture of dissolved coal liquids and recycled solvent. The solid phase is ash and coal particles. The tracer used to measure gas holdup was an isotope of argon which dissolves in the coal liquid, and it is this experiment that is shown in Fig. 10. The solubility of the tracer at the reactor temperature and pressure can be measured and, at equilibrium, the mole fraction of argon in the gas is five times the mole fraction of argon in the liquid.

In Fig. 10 the Laplace transform of the residence-time distribution computed from the tracer response was plotted. The first moment can be obtained from it by estimating the tangent of $\hat{f}(s)$ as $s \to 0$. This is relatively easy here, as the flow is close to plug flow and therefore $\ln[\hat{f}(s)]$ versus (s) is almost a straight line (see Fig. 10b), the slope of which is equal to the first moment. The variance has to be estimated from the second derivative near zero. This is not easy, but if the coefficient of variation is small, one can do it by model fitting using

$$\hat{f}(s) = \frac{1}{(1 + \tau s/n)^n} \tag{19c}$$

where n in here is a curve-fitting parameter. This approach was discussed earlier. There are several problems caused by the fact that the gas tracer is soluble in the liquid. What one really wants to know is not the holdup of the tracer, but an estimate of the gas volume in the reactor. If the coefficient of variation γ is large, there is no way that one can reliably estimate the gas holdup. Fortunately, in this particular case, γ is small and the estimate could be obtained. The small γ could be due to several reasons. Either the gas-liquid mass transfer coefficient is small or it is very large. If the mass transfer coefficient is small, adsorption in the liquid phase has only a very small impact on $C_2(t)$, in the range of accurate measurements because it mainly affects $C_2(t)$ for $t \to \infty$ (or $s \to 0$). The coefficient of variation is then in reality large, but this is not appreciated when computed from the measured form of f(s). One can then get a large error in the estimate of μ and γ^2. In the example given, the mass transfer coefficient could be measured independently and this alternative could be ruled out. It is therefore reasonable to assume that the mass transfer is fast, which allows one to assume that the gas adsorbed in the liquid phase is in equilibrium with the gas phase. If one knows the liquid holdup and the solubility of the gas, a condition for the effects of absorption can be made. Note that if γ^2 was not very small, there is no reliable way to make any such correction, and it would have to be difficult to estimate the volume of the gas phase in the reactor.

CONTACT-TIME DISTRIBUTIONS

The term *contact-time distribution* has been used in the literature in two ways. One is a sojourn-type distribution in the active phase which is obtained by using a slightly adsorbing tracer as well as a nonadsorbing inert tracer. If one designates f(t) of the adsorbing tracer as $f_a(t)$ and of the nonadsorbing tracer as $f_0(t)$, then, for pure plug flow, one can define the density function of a sojourn-time distribution in the adsorbed phase as

$$\psi_a(t) = f_a(t - \tau_0)$$

where τ_0 is the first moment of $f_0(t)$. This distribution was discussed in the preceding section, where it was shown that the function $\psi_a(t)$ can be measured not only for plug flow but also for a limited number of other cases. However, although this distribution is useful for reactor modeling and for diagnosing maldistributions, it is not directly useful for estimating the behavior of a first-order reaction.

The kinetic relations assume plug flow in a packed bed, whereas tracer experiments with a proper isotope tracer will not necessarily give plug flow (see Fig. 26). A close examination of Fig. 26 will demonstrate that one cannot use the real residence-time distribution of the reactants to reliably predict the behavior of a heterogeneous reactor. In Fig. 26 the overall flow is plug flow. However, the residence time of the ^{13}C tracer is not plug flow and has a large coefficient of variation. For other reactants the difference could even be larger. If one applies the kinetic relations which are based on the assumption of plug flow to a model based on the actual residence-time distribution, one would obtain very erroneous results. More meaningful results can be obtained by using the data from the helium tracer. The helium tracer has a residence-time distribution similar to $f_0(t)$ but dissimilar to the true overall residence-time distribution of any of the reactants. If the actual reactor is a fluid bed, one can no longer rely on the results from a helium tracer, nor can one rely on the true residence-time distribution. However, if the true residence-time distribution for both the fluid bed and the packed bed are available and they are very close to each other, one could deduce that the reaction should scale up safely. In all other cases, one cannot directly apply the information.

A more useful definition of contact-time distribution is based on the relation between the product distribution of first-order reactions and the residence-time distribution of the system. Earlier it was shown that knowledge of the product distribution defines the Laplace transform of the residence-time distribution density function [Eq. (28)]. For a first-order irreversible reaction $A \rightarrow B$ with a reaction rate $r(A) = -ka$, the outlet concentration in a homogeneous system

$$\frac{a}{a_f} = \hat{f}(k) \tag{54}$$

where $\hat{f}(k)$ is the Laplace transform of $f(t)$ with the reaction rate k substituted for s. For a first-order reaction occurring in a heterogeneous reactor, one can use Eq. (28) to define the distribution $\underline{\hat{\psi}}$ by

* $$\langle C_{jp} \rangle = \alpha_{j0} + \sum_{r=1}^{n-1} \alpha_{jr} \underline{\hat{\psi}}(\lambda_r) \tag{55}$$

which is an equivalent of Eq. (28). This distribution, originally proposed by Orcutt (Orcutt et al., 1962) and Glasser (Glasser et al., 1973), is not a real distribution of sojourn times but has the same dimensions and properties. Regardless of the real sojourn-time distribution, λ_r is defined and measured by assuming that $\underline{\hat{\psi}}(\lambda_r)$ is plug flow for a packed-bed reactor.

For a complex first-order reaction occurring in a packed-bed reactor, one can determine λ_r for each r. If the reaction then occurs in a different reactor such as a fluid bed, one can obtain measurements of C_j. The information from C_j and the λ_r from the packed-bed reactor gives estimates for several values of $\underline{\hat{\psi}}(\lambda_r)$. To obtain the complete function $\underline{\hat{\psi}}(\lambda_r)$, one would have to use curve-fitting functions. However, in practice this is often not needed, as sufficient information is available from the linear values of $\underline{\hat{\psi}}(\lambda_r)$. In this case, λ_r corresponds to the Laplace transform variable of $\underline{\psi}(t)$ even though $\underline{\psi}(t)$ is an artificial constraint defined by Eq. (55) with no physical meaning.

Another way of measuring $\underline{\hat{\psi}}(\lambda_r)$ is to vary k in Eq. (54). In a catalytic reactor, this can be done by either changing the temperature of the reactor or the reactivity of the catalyst. One can then use an equivalent of Eq. (54)

$$\frac{a}{a_t} = \hat{\underline{\psi}}(k) \tag{56}$$

to define $\hat{\underline{\psi}}$. It has the same meaning as the definition in Eq. (55) with k replacing λ_r. But it may not have the identical numerical form. Changing temperature or reactivity may change the complex transport process as well as the adsorption and desorption rates inside the catalyst particles. Therefore, the values of $\hat{\underline{\psi}}(k)$ obtained in this way will not be identical to those obtained using $\hat{\underline{\psi}}(\lambda_r)$ as defined in Eq. (55). However, the form will be similar.

Assume that one measures $\hat{\underline{\psi}}(k)$ or $\hat{\underline{\psi}}(\lambda_r)$ in a packed bed and then repeats these experiments in a fluidized bed keeping the space velocity constant, and measures the conversion for each (k). The highest conversion possible at each (k) is the same as in a plug-flow reactor. If, for any (k), the conversion is higher [or $\hat{f}(k)$ smaller], it follows that either the reaction is not first order, or the reactor is not isothermal, or the initial small packed-bed reactor was diffusion controlled. If the conversions for all values of (k) are equal to that of the packed bed, the fluid is essentially described by a plug flow. If not, the conversion at each point is a measure of the way the fluid bed deviates from the packed bed. If a/a_0 for the two cases is plotted vs. (k), one can interpret the curves as plots of $\hat{\underline{\psi}}(k)$ for the two reactors. $\hat{\underline{\psi}}(k)$ for the packed bed is by definition plug flow. Much can be learned from the nature of the plot. In Fig. 28 an experimental contact-time distribution for a relatively short bubbling fluid bed is given and compared with a plug-flow reactor and a stirred tank. Note that $\hat{\underline{\psi}}(k)$ of the fluid bed is larger than $\hat{\underline{\psi}}(k)$ of the packed-bed reactor (plug-flow case) for all values of (k). At large values of (k), $\hat{\underline{\psi}}(k)$, for the fluid bed, is very flat and, within experimental accuracy, almost parallel. This indicates that part of the material essentially bypasses the reactor. In fact, a simple bypass model

$$\hat{\underline{\psi}}(k) = a \exp(-k\tau) + b \tag{57}$$

fits the experimental results quite well. Figure 28 also gives $\hat{\underline{\psi}}(k)$ for a stirred tank. Note that it fits the experimental data less well. $\hat{\underline{\psi}}(k)$ is larger than that

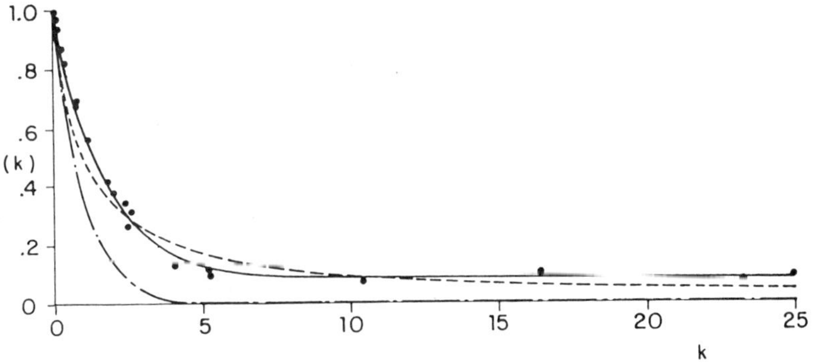

FIGURE 28 Experimental contact-time distribution for a fluid-bed (Orcutt et al., 1962). Comparison with various models: — — — — —, stirred-tank reactor; — - — - — plug-flow reactor; ———— plug-flow reactor allowing for bypass and stagnancy (experimental). (From Glasser et al., 1973.) $\hat{\underline{\psi}}(k)$ is measured here by conversion of ozone in fluid bed over an iron catalyst. k is varied by dilating the catalyst with inerts.

of a stirred tank for small and large values of (k). For intermediate range $\hat{\underline{\psi}}(k)$ is smaller than for a stirred tank, which is typical of some bypass phenomena.

The $\hat{\underline{\psi}}(k)$ for the fluid bed shown in Fig. 27 has another interesting property that is common to most fluid beds. The slope of $\hat{\underline{\psi}}(k)$ near $k \to 0$ is smaller than for a plug-flow reactor. (The only way to know this is to measure the reaction in a packed bed with the same catalyst under isothermal conditions.) This indicates that not only does the bed have a bypass in the bubble phase, but that part of the catalyst phase is not well contacted. In the author's experience properly baffled (or turbulent) fluid beds do not show any bypass in the gas phase but will share the problem that the dense catalyst phase is not perfused. Such baffled beds behave like near-plug-flow reactors with a smaller effective catalyst volume (about 50 to 70% of real catalyst holdup). [A possible explanation is given in Krambeck et al. (1967).]

If one could really measure the slope of $\hat{\underline{\psi}}(k)$ for very small (k), it would approach τ as $k \to 0$. This is the same problem that was illustrated in Fig. 13 when measuring the tail of F(t). If a region of the reactor is not well perfused, τ will be underestimated. If an independent estimate of τ is available, underestimating τ using this technique is evidence that the reactor is not well perfused. For the case shown in Fig. 13, one could estimate τ from the volume of the reactor and the feed rate. In the case of $\hat{\underline{\psi}}(k)$, the only way to estimate the first moment is by using a packed-bed reactor for comparison. Note that $\hat{\underline{\psi}}(k)$ here yields some valuable information directly applicable to reactor design. One can compare the information obtained from Fig. 27 with that obtained by tracer experiments in similar fluid beds (see Fig. 17). Figure 17 for a similar bed also clearly indicates a bypass phenomenon. The escape probability $\Lambda(t)$ measured from the tracer response of a nonadsorbing tracer shows a clear maximum, indicating a bypass or stagnancy.

The method shown in Fig. 17 has one significant advantage. It allows one to plot $\Lambda(t)$ which gives direct physical insight into the nature of the transport process. There is as yet no equivalent way to extract equivalent information directly from inspection of $\hat{\underline{\psi}}(k)$. If f(t) is measured, computation of $\hat{f}(s)$ or $\hat{f}(k)$ is straightforward and relatively easy. The opposite is not true. $\hat{\underline{\psi}}(k)$ is hard to invert accurately, and therefore it is hard to obtain an accurate estimate of the escape probability $\Lambda(t)$ from $\hat{\underline{\psi}}(k)$.

There are ways of inverting $\hat{\underline{\psi}}(k)$ either by numerical inversion or by fitting it with a theoretical model, but both methods are inaccurate. However, $\hat{\underline{\psi}}(k)$ is a valuable tool since it contains the information in a form that allows conclusions about the impact of the mixing processes on first-order chemical reactions. Note that the time scale of $\hat{\underline{\psi}}(k)$ does not refer to any real sojourn time. $\hat{\underline{\psi}}(k)$ is really a linear operator that changes the product distribution of a linear reaction system in a packed-bed reactor to that of fluid bed (or other) reactor. The properties of this linear operator are conceptually similar to a residence-time distribution and one can derive from it similar intuitive deductions about the nature of the transport processes in the reactor. In most heterogeneous catalytic reactions one desires plug flow in the reactor itself. $\hat{\underline{\psi}}(k)$ is therefore a useful tool for checking internals in a cold flow model or verifying how closely the reactor approaches plug flow, provided that one can find a suitable reaction. Both ozone and hydrogen peroxide decomposition have been proposed (Nauman and Buffham, 1983) and used.

The main advantage of using the $\hat{\underline{\psi}}(k)$ as defined here compared to $\psi_a(t)$ defined in the preceding section is that $\hat{\underline{\psi}}(k)$ is measurable in any heterogeneous catalytic reactor with a first-order reaction. On the other hand, $\psi_a(t)$ is uniquely measurable only in these cases where the absorbed tracer particles return to the same place, in other words, where adsorption does not change the residence-time distribution in the continuous phase. As this condition is often not met, $\hat{\psi}(k)$ provides a useful alternative.

Unfortunately, $\hat{\psi}(k)$ suffers from the same problem as $\psi_a(t)$. That is, although it is useful in modeling, it cannot be used for scaleup, as the $\hat{\psi}(k)$ depends on adsorption behavior and would be different for different first-order reactions. One must therefore be careful when applying $\hat{\psi}(k)$ to actual scaleup problems. It will, in a cold flow model, give good indications of bypass phenomena. There is, however, no guarantee that the real reactant will have the same behavior as reactions used in the cold flow model. $\hat{\psi}(k)$ is useful in studying maldistributions of flow in actual industrial reactors. If an industrial reactor does not perform up to expectations, one can often obtain several values of $\hat{\psi}(k)$ by changing the temperature and thereby (k). One can then compare the effect of changing (k) in the industrial reactor to pilot plant results or similar industrial reactors. From this one can then infer $\hat{\psi}(k)$ and possible reasons for the poor performance. This is a powerful diagnostic tool, complementing or substituting for tracer studies in industrial reactors. One can get similar results even if the reaction is nonlinear and complex involving several compounds. One starts with the assumption of a plug-flow reactor, estimates a set of rate constants, and then obtains expressions completely equivalent to Eqs. (16) and (20). In this case, unlike the homogeneous case, one cannot use f(t) or $\psi(t)$ to compute the outlet composition. The outlet composition can still be used to estimate $\hat{\psi}(k)$, as $\hat{\psi}(k)$ is an operator transforming the results of a plug-flow reactor into our reactor. Note that reliable design requires that $\hat{\psi}(k)$ be close to plug flow.

An especially interesting case is one of consecutive reactions with equal reaction rates. Consider, for example, the case

$$A \xrightarrow{k_1} B \xrightarrow{k_1} C \xrightarrow{k_1} D$$

The solution of this case is given in Table 2. If the values of the reaction rates are equal, one gets an estimate not only of $\hat{\psi}(k)$ but also of the higher derivatives of $\hat{\psi}(k)$ or $(d^n\hat{\psi}(k)/dk^n)_{k=k_i}$. Even if the values of (k) are different, the values of C_j give a good approximation of the derivatives as long as the differences are small. If estimates of the derivatives are available, $\hat{\psi}(k)$ can be reconstructed from a Taylor series expansion around (k) (see Glasset et al., 1973). It is hard to use such a reaction in a cold flow model. However, if an industrial reactor has a consecutive reaction, one can compare the product distribution to that of a packed-bed reactor. If significant deviations are found, this allows some conclusions as to the nature of the flow. A good base case, preferably an isothermal packed-bed reactor, is essential as well as a well-performing reactor of a similar type. One does not always want to reconstruct $\hat{\psi}(k)$ (although this is possible). The nature and magnitude of the deviations give some immediate qualitative information about the nature and magnitude of the deviation from plug flow, which can be interpreted by looking at some simple flow models and styding their effect on product distribution. An example is given in Table 3. In these examples, the overall residence time τ is kept constant for all cases. The reaction rate k is varied for each flow model so as to keep the outlet concentration of A (A/A_0) constant. This is done to make the example more consistent with actual practice. In many cases, catalyst activity changes and is compensated for

TABLE 2 Dependence of the Product Distribution of a First-Order Consecutive Reaction on the Residence-Time or Contact-Time Distribution[a]

	Reaction:
	$k_1 \neq k_2 \neq k_3$
Plug-flow reactor	Reactor with residence-time distribution f(t)
$\frac{A}{A_0} = e^{-k_1\tau}$	$\frac{A}{A_0} = \hat{f}(k_1)$
$\frac{B}{A_0} = k_1 \sum_{i=1}^{2} \frac{e^{-k_i\tau}}{\prod_{j=1,2;\, j\neq 1} (k_j - k_i)}$	$\frac{B}{A_0} = k_1 \sum_{i=1}^{2} \frac{\hat{f}(k_1)}{\prod_{j=1,2;\, j\neq i} (k_j - k_i)}$
$\frac{C}{A_0} = k_1 k_2 \sum_{i=1,3} \frac{e^{-k_i\tau}}{\prod_{j=1,3;\, j\neq i} (k_j - k_i)}$	$\frac{C}{A_0} = k_1 k_2 \sum_{i=1}^{3} \frac{\hat{f}(k_1)}{\prod_{j=1,3;\, j\neq i} (k_j - k_i)}$
$\frac{D}{A_0} = 1 - A - B - C$	$\frac{D}{A_0} = 1 - A - B - C$
$\frac{D}{A_0} = 1 - \sum_{i=1,3} e^{-k_i\tau} \prod_{j=1,3;\, j\neq i} \frac{k_j}{k_j - k_i}$	$\frac{D}{A_0} = 1 - \sum_{i=1}^{3} f(k_1) \prod_{j=1,3;\, j\neq i} \frac{k_j}{k_j - k_i}$

[a]For contact-time distribution, substitute (k) for $\hat{f}(k)$.

by changing the reactor temperature. In practice one would therefore compare the results of the industrial reactor to the pilot plant at constant A/A_0. One does the same if the conversion is reduced due to a maldistribution in the flow. Here the concentration of various products of a consecutive reaction, which are a function of the residence-time distribution, can clearly show if the reduced conversion is due to a change in catalyst activity or due to a maldistribution in flow (provided, of course, that one can confirm in the pilot plant that catalyst reactivity does not strongly affect selectivity). The results for the case of consecutive reactions $A \rightarrow B \rightarrow C \rightarrow D$ where all reaction rates are equal are given in Table 3. The results in this example are very sensitive to the residence-time and contact-time distributions.

$A \xrightarrow{k_1} B \xrightarrow{k_2} C \xrightarrow{k_3} D$	
$k_1 = k_2 = k_3 = k$	
Plug-flow reactor	Reactor with residence-time distribution f(t)
$\frac{A}{A_0} = e^{-k\tau}$	$\frac{A}{A_0} = \hat{f}(k_1)$
$\frac{B}{A_0} = k\tau e^{-k\tau}$	$\frac{B}{A_0} = -k_1 \left.\frac{d\hat{f}(k)}{dk}\right\vert_{k=k_1}$
$\frac{C}{A_0} = \frac{k^2\tau^2}{2!} e^{-k\tau}$	$\frac{C}{A_0} = \frac{k_1^2}{2!} \left.\frac{d^2\hat{f}(k)}{dk^2}\right\vert_{k=k_1}$
$\frac{D}{A_0} = 1 - A - B - C$	$\frac{D}{A_0} = 1 - A - B - C$
$\frac{D}{A_0} = 1 - \left[\sum_{i=0}^{2} \frac{(k\tau)^i}{i!}\right] e^{-k\tau}$	

APPLICATIONS OF CONTACT- AND RESIDENCE-TIME DISTRIBUTIONS TO THE DESIGN OF HETEROGENEOUS REACTORS

Earlier, rigorous concepts as to how residence-time distributions affect the conversion and contact-time distribution in isothermal homogeneous reactors were discussed. Unfortunately, very few reactions are truly homogeneous and fewer are isothermal. In the previous sections it was shown that these principles cannot be applied to heterogeneous reactors in straightforward way. However, the main applications are still valid if one is careful to recognize the special features of residence-time distributions in heterogeneous

TABLE 3 Dependence of the Product Distribution of a First-Order Consecutive Reaction on Residence-Time (Contact-Time) Distribution: Numerical Example

$$A \xrightarrow{k_1} B \xrightarrow{k_2} C \xrightarrow{k_3} D$$

$$k_1 = k_2 = k_3 = k$$

Flow model	Average residence time (h)	K (h^{-1})	A/A_0	B/B_0	C/C_0	D/D_0
1. Plug flow	1.0	1.0	0.368	0.368	0.184	0.08
2. Ten stirred tanks in series	1.0	1.05	0.368	0.35	0.183	0.099
3. 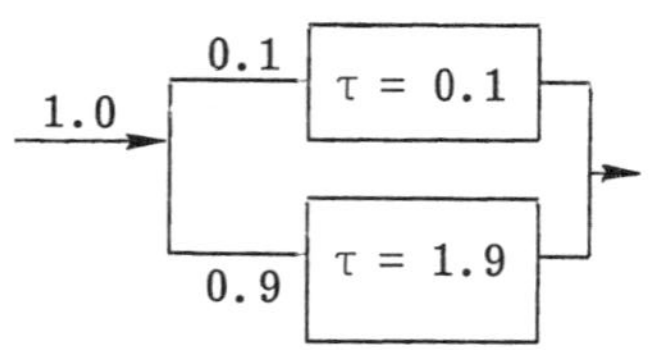	1.0	1.068	0.368	0.336	0.192	0.104
4. 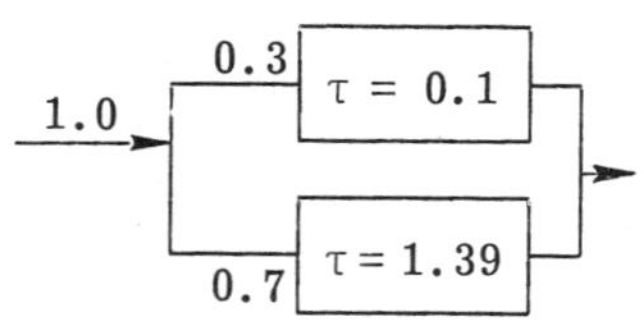	1.0	1.36	0.368	0.236	0.191	0.205
5. 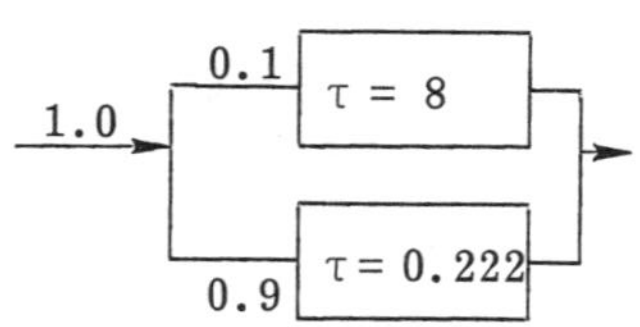	1.0	4.03	0.368	0.33	0.147	0.156

reactors. In this section the concepts presented earlier will be extended to the more general case of heterogeneous reactors.

In the present state of knowledge, rigorous algorithmic use of residence-time distributions to predict product distribution in heterogeneous reactors is very restricted, even for first-order reactions. There is no real good algorithmic equivalent of Zwietering's method.

Fortunately, the concept of bounding the outlet composition is still valid. Utilizing different models, one can still evaluate whether the reaction performance is sensitive to the actual flow model. The residence-time and contact-time distributions can then be used to place constraints on the range of flow models that one should consider. If the reactor is close to plug flow, one can evaluate bounds by measuring the deviation from plug flow by means of a tracer experiment. This works well only if the deviation from plug flow is small.

The conceptual uses of residence-time distributions given earlier are still valid. Plug-flow reactors or staged reactors approaching plug flow are optimal for all first-order reactions and for all reactions in which contact between products and reactants should be minimized. In fact, the example mentioned earlier was from catalytic heterogeneous reactors (Belevi et al., 1981). Also noteworthy is the conceptual use of the contact-time distribution $\hat{\underline{\psi}}(k)$ in reactor design and scaleup. Recall that $\hat{\underline{\psi}}(k)$ was really an operator that showed quantitatively the way the transport process in the reactor modify the product distribution as compared to a packed-bed reactor. Its time scale was the reaction rate used to predict the results of a packed-bed reactor, which is simply the overall reaction rate per unit volume or per unit weight of catalyst. It therefore has the same time scale as the inverse of the space velocity.

In a packed-bed reactor with a good flow distribution, it is possible to keep both $\hat{\underline{\psi}}(k)$ and $f_0(t)$, the residence-time distribution in the gas phase, constant. However, during scaleup, this is not automatically guaranteed. $f_0(t)$ will normally stay constant and close to the plug flow if the microreactor or pilot plant is sufficiently long [Silverstein and Shinnar, 1975). However, $\hat{\underline{\psi}}(k)$ will stay constant only if there is no mass transfer resistance between the gas phase and the catalyst particle. Otherwise, increasing reactor length will change the mass transfer coefficient, which will change $\hat{\underline{\psi}}(t)$ and therefore $\hat{\underline{\psi}}(k)$.

Another interesting conceptual conclusion can be derived from the concept of a contact-time distribution. It can be seen from Eqs. (48) to (50) that, for certain classes of systems, $\psi_a(t)$ and $\hat{\psi}_a(s)$ can be uniquely derived from two sets of information.

1. The residence-time distribution $f_0(t)$ in the continuous phase
2. The residence-time distribution $f_a(t)$ of an adsorbing tracer in a plug-flow reactor

Consider a first-order reaction that occurs in the adsorbed phase. As each of the different reactants has a different adsorption behavior (its adsorptive equilibrium may be different, as well as the rate at which it adsorbs or desorbs), each species has a different $\underline{\psi}(t)$ or $\hat{\underline{\psi}}(s)$. To properly model the system, one would need an exact tracer (isotope) for each reactant. However, for the special case described, one could predict $\underline{\psi}(t)$ or $\hat{\underline{\psi}}(s)$ for each of the species, given the tracer results from a plug-flow reactor and $f_0(t)$, which is common to all species. If one follows a molecule undergoing a first-order reaction, one can conceptually measure for each molecule a total sojourn time in the adsorbed phase. To compute its residence-time distribution, one must also know the reaction rates in order to determine how much time it spends in each of the various states it can assume. This is similar to the concept used earlier to deal with first-order reactions in nonisothermal homogeneous system.

The above is equivalent to changing the nature of $\hat{\rho}(s)$ in Eq. (49a). Although one cannot make this computation without detailed knowledge of adsorption kinetics, one can use the concept to derive an importan criterion in reaction engineering: namely, that for all cases for which Eqs. (48) to (50) apply, one can use the pseudokinetics obtained from a packed-bed reactor and apply it to all cases, using $f_0(t)$ to compute the results. This allows one to deal with dispersion and recycling in packed beds. It also shows that for a first-order reaction, the result of a stirred-tank catalytic reactor is directly scalable to a packed-bed reactor, and vice versa, as a stirred tank is one of the few cases with moving catalyst for which Eq.(48) applies. This scaling is possilbe despite the fact that one is really measuring a pseudokinetic first-order reaction rate based on space velocity and not the actual kinetics based on real residence times.

This concept can be extended qualitatively to second-order reactions in a catalytic reactor. In reactors of this type, micromixing is not important. However, it is often important to know whether the product distribution can be predicted reasonalby from $f_0(t)$. For cases where Eq. (48) applies, the Berty reactor provides a simple test. If a stirred-tank catalytic reactor yields kinetic relations that predict well the behavior of a packed-bed reactor, then nonlinear interactions between products and reactants are not important and $f_0(t)$ can be used to predict the effect of dispersion in scaleup.

The comparison above is often important in the design of complex reactors. It is therefore important to understand the principles by which Eqs. (48) and (49) were derived. The main assumptions were:

1. The system is spatially uniform.
2. An adsorbed particle (or a particle diffusing into a catalyst particle) always returns to the same place [or to a place equivalent to it in terms of $f_0(t)$].

These assumptions are not always valid. Nevertheless, there are many cases where one wants to predict the performance of a heterogeneous reactor which is not a packed bed from either a small packed-bed reactor or a pilot plant. This applies to fluid beds as well as transport reactors such as a riser cracker. If one compares such a reactor to a packed bed, it is impossible to keep both $f_0(t)$ and $\hat{\underline{\psi}}_0(k)$ constant. For example, the first moment of $f_0(t)$ will be much larger for the riser. If the reaction occurs only at the solid surface of this catalyst, this should not matter. But in many complex reaction systems catalytic surface reactions compete and interact with gas-phase reactions. The catalyst or solid may also act as an inhibitor for the gas-phase reactions. In some catalytic reactions, intermediates formed at the surface will either diffuse through the gas phase or react in the gas phase. One should therefore be careful in such cases, and evaluate the possible impact of changes in $f_0(t)$ relative to $\psi_a(t)$.

Rigorously, it is not enough that $f_0(t)$ and $\psi_a(t)$ stay constant. The distribution of single stays in either phase should also stay constant. If one cannot assume this to be true, one can experimentally test the sensitivity of the reactions to variations in $f_0(t)$, keeping $\psi_a(t)$ constant. This technique is a simple and powerful conceptual tool for organizing one's thinking about the pitfalls of a given scaleup problem. Consider, for example, a catalytic fluid-bed cracker in which the cracking really occurs in the dilute phase riser. At low temperatures, gas-phase reactions are slow and the results correlate very well with those obtained in a dense-phase bed at the same space velocity. At higher temperatures, the gas-phase reactions become significant and the correlation is poorer. As these gas-phase reactions are undesirable, it is important to reduce the average residence time of $f_0(t)$, keeping the space velocity constant.

Another important example is the case of a dense-phase catalytic fluid bed. Equations (48) and (49) do not apply rigorously in a fluid bed for two reasons:

1. The gas phase is nonuniform, as part of the gas is in bubbles and part in a dense phase around the catalyst.
2. The solid circulation and adsorbed reactants circulate with it. The sojourn-time distribution at each reactant depends on the rate at which it adsorbs and desorbs from the catalyst.

One can, however, design a fluid bed such that Eq. (48) approximately applies. For example, one can see to it that $f_0(t)$ approaches plug flow. One can also use an adsorbing tracer with fast exchange and design the reactor such that $f_a(t)$ approaches plug flow. That would occur only if all gas particles are adsorbed and desorbed many times. This then approaches the condition of spatial uniformity despite the pressure of bubbles. One can achieve this either by working in the turbulent fluidization regime or by using properly designed baffles. Designing the turbulent flow regime only helps condition 1. The second condition can be approached only by baffling or staging the bed. One can always approach plug flow by a sufficient number of stages. If the catalyst circulation between the stages is slow compared to the rate of adsorption and desorption of reactants, one approaches condition 2, namely, that each molecule always returns to a place such that its residence time in the continuous phase is not changed.

These arguments allow one to understand on theoretical grounds the importance of baffles in scaleup. Baffles give added confidence to the scaleup even if the fluidization is turbulent and a tracer experiment shows that one approaches plug flow without baffles. This is so because solid mixing in a fluid bed always increases during scaleup even when the pilot plant is reasonably large. It is important that the baffles really reduce solid mixing and divide the solid catalyst phase into stages. Vertical baffles, which are often used, do not achieve this goal and therefore do not provide the needed safety net for scaleup which can be achieved using horizontal baffles.

We will conclude with one final example. Recently, flash hydrogenation of coal, in which coal is devolatilized and partially gasified in the presence of hydrogen at high pressure, has received considerable attention. A simplified kinetic scheme of the reaction is

$$\text{Coal} \xrightarrow[(1)]{H_2} \text{heavy liquids} \xrightarrow[(2)]{H_2} \text{BTX} \xrightarrow[(3)]{H_2} \text{methane}$$

Steps 2 and 3 can occur in the gas phase and are also catalyzed by the coal ash as well as by the char. As this is a consecutive reaction, plug flow of the gas is required for good selectivity, and it is also important to control the residence time in a narrow optimal range. Kinetic and thermodynamic considerations require the reactor to be as isothermal as possible, and the easiest way to achieve this is by backmixing of solids. It has therefore been suggested that fluid-bed reactors or risers with high solid circulation would have significant advantages for such a design. However, all present experimental data were obtained in current dilute phase reactors in which the solid/gas ratio is an order of magnitude lower than would prevail in a riser with solid recirculation. There is no way to estimate reliably from data in a dilute-phase cocurrent reactor what would happen in riser with higher concentration of solids. One cannot keep residence time and contact time constant simultaneously. For any design estimates one would need data from a system in which both contact time and residence time are similar to the desired design. Or one would need evidence that one of these two has no impact on the reaction.

These simple examples should be sufficient to illustrate the varied applications of contact- and residence-time distributions to the design of heterogeneous reactor. They provide guidelines for dealing with the difficult problems of reactor scaleup in complex reactors.

NOTATION

a	concentration of compound A
A_j	name of compound j
A(t)	future life distribution of particle inside the reactor fraction of particles inside the reactor that will leave it before time t
C_A	concentration of compound A in product stream
C_{Af}	concentration of product f in feed stream
C(t)	concentration of tracer in exit stream
$C_f(t)$	concentration of tracer in feed stream
$E_f(t^n)$	expected value of t^n or $\int_0^\infty t^n f(t)\,dt$
$\hat{f}(s)$	Laplace transform of f(t)
F(t)	residence-time distribution: fraction of particle entering at time zero that leave the reactor before time t
f(t)	density function of residence time distribution: fraction of particles exiting between t and t + Δt = f(t) Δt
g(t)	density function of G(t)
G(t)	internal age distribution: fraction of particles which are in the reactor at a given time of observation that have been inside the reactor less than time t
k	kinetic rate constant
k_{ij}	rate constant of reaction $A_i \rightarrow A_j$

K_{ij}	equilibrium constant of reaction $A_i \rightarrow A_j$
m	amount of tracer introduced in pulse experiment
p_j	probability of molecule or particle to be in state j
Pe	Péclet number
Q	volumetric flow rate
t	time
V	volume of reactor
V_f	variance of density function $f(t) = \int_0^\infty (t - \tau)^2 f(t)\, dt$

Greek Letters

α_{jr}	constants in Eq. (25)
$\gamma = \frac{\sqrt{V_F}}{\tau}$	coefficient of variation of f(t)
λ_r	eigenvalues of first-order kinetic systems [see Eq. (25)]
$\Lambda(t)$	escape probability or intensity function of residence-time distribution
$\mu_r =$	$E_f(t^r)$ = rth moment of f(t)
τ	average residence time in homogeneous reactor = V/Q
$\psi(k)$	Laplace transform of contact-time distribution
$\psi_a(t)$	sojourn-time distribution density function of stay in the adsorbed phase [see Eq. (46)]

REFERENCES

Belevi, H., J. R. Bourne, and P. Rys, Chemical selectivities disguised by mass diffusion, Helv. Chim. Acta, *64*, 1618 (1981).

Biloen, P., J. N. Helle, F. G. A. van der Berg, and W. M. H. Sachfler, Isotopic transients in Fischer Tropsch reactions, J. Catal., (1983, in press).

Danckwerts, P. V., Continuous flow systems, Chem. Eng. Sci., *2*, 1 (1953).

Evangelista, J. J., S. Katz, and R. Shinnar, The effect of imperfect mixing on stirred combustion reactors, 12th Symp. (Int.) Combust., Combustion Institute, Pittsburg, Pa., (1969a), p. 901.

Evangelista, J. J., R. Shinnar, and S. Katz, Scale-up criteria for stirred tank reactors, AIChE J., *15*(6), 843 (1969b).

Froment, G. F. and K. B. Bischoff, *Chemical Reactor Analysis and Design*, Wiley, New York, (1979).

Gilliland, E. R. and E. A. Mason, Gas mixing in beds of fluidized solids, Ind. Eng. Chem., *44*, 218 (1952).

Glasser, D., S. Katz, and R. Shinnar, The measurement and interpretation of contact time distributions for catalytic reactor characterization, Ind. Eng. Chem. Fundam., *12*, 165 (1973).

Hiester, N. K and T. Vermeulen, Saturation performance of ion-exchange and adsorption columns, Chem. Eng. Prog., *48*, 505 (1952).

Himmelblau, B. M. and K. B. Bischoff, *Process Analysis and Simulation*, Wiley, New York (1968).

Krambeck, F. J., S. Katz, and R. Shinnar, Stochastic models for fluidized beds, Chem. Eng. Sci., *24*, 1497 (1967).

Krambeck, F. J., S. Katz, and R. Shinnar, Interpretation of tracer experiments in systems with fluctuating throughput, Ind. Eng. Chem. Fundam., *8*, 431 (1969).

Lehman, J. and K. Schügerl, Investigation of gas mixing and gas distributor performance in fluidized beds, Chem. Eng. J., *15*, 91 (1978).

Levenspiel, O., and K. B. Bischoff, Patterns of flow in chemical process vessels, Adv. Chem. Eng., *4*, 95 (1963).

Mills, P. L., and M. P. Dudukovic, Evaluation of Liquid-solid contacting in trickle-bed reactors by tracer methods, AIChE J., *27*, 893 (1981).

Murphree, E. V., J. Voorhies, Jr., and F. X. Mayer, Application of contacting studies to the analysis of reactor performance, Ind. Eng. Chem. Process Des. Dev., *3*, 381 (1964).

Naor, P. and R. Shinnar, Representation and evaluation of residence time distributions, Ind. Eng. Chem. Fundam., *2*, 278 (1963).

Naor, P., R. Shinnar, and S. Katz, Indeterminacy in the estimation of flowrate and transport functions from tracer experiments in closed circulation, Int. J. Eng. Sci., *10*, 1153 (1972).

Nauman, E. B., Residence time distributions and micromixing, Chem. Eng. Commun., *8*(53) (1981).

Nauman, E. B. and B. A. Buffham, *Mixing in Continuous Flow Systems*, Wiley, New York (1983).

Ng, D. Y. C., and D. W. T. Rippin, The effect of incomplete mixing on conversion in homogeneous reactions, Proc. 3rd Eur. Symp. Chem. React. Eng., Pergamon Press, Oxford (1965), pp. 161–165.

Orcutt, J. C., J. F. Davidson, and R. L. Pigford, Reaction time distributions in fluidized catalytic reactors, AIChE Chem. Eng. Prog. Symp. Ser., *38*, 1 (1962).

Orth, P. and K. Schügerl, Distribution of residence times and contact times in packed bed reactors: influence of the chemical reaction, Chem. Eng. Sci., *27*, 497 (1972).

Overcashier, R. H., D. B. Todd, and R. B. Olney, Some effects of baffles on a fluidized system, AIChE J., *5*, 54 (1959).

Østergaard, K. and M. L. Michelsen, On the use of the imperfect tracer pulse method for determination of hold-up and axial mixing, Can. J. Chem. Eng., *47*, 107 (1969).

Pennick, J. E., W. Lee, and J. Maziuk, Development of the methanol to gasoline (MTG) Process, 7th Int. Meet. React. Eng., Boston (1982).

Petho, A. and R. D. Noble, eds., *Residence Time Distribution Theory in Chemical Engineering*, Verlag Chemie, Weinheim, West Germany (1982).

Seinfeld, J. and L. Lapidus, *Process Modeling, Estimation, and Identification*, Prentice-Hall, Englewood Cliffs, N.J. (1974).

Shinnar, R., On the behavior of liquid dispersions in mixing vessels, J. Fluid Dyn., *10*, Pt. 2, 259 (1961).

Shinnar, R. and P. Naor, Residence time distributions in systems with internal reflux, Chem. Eng. Sci., *22*, 1369 (1967).

Shinnar, R., P. Naor, and S. Katz, Interpretation and evaluation of multiple tracer experiments, Chem. Eng. Sci., *27*, 1627 (1972).

Shinnar, R., D. Glasser, and S. Katz, First order kinetics in continuous reactors, Chem. Eng. Sci., *28*, 617 (1973).
Shinnar, R., Tracer experiments in reactor design, 2nd Conf. Physico-chem. Hydrodyn., Oxford (1977).
Shinnar, R. Process control research: an evaluation of present status and research needs, ACS Symp. Ser., *72*, 1 (1978).
Silverstein, J., and R. Shinnar, Design of fixed bed catalytic microreactors, Ind. Eng. Chem. Process Des. Dev., *14*, 127 (1975).
Snow, A. I., private communication, ARCO Petroleum Co. (May 1983).
Tarmy, B., private communication, Tracer experiments performed on the Exxon EDS Pilot Plant (March 1983).
Wei, J. and C. D. Prater, Structure and analysis of complex chemical reactions, Adv. Catal., *13*, (1962).
Weinstein, H. and R. J. Adler, Micromixing effects in continuous chemical reactors, Chem. Eng. Sci., *22*, 65 (1967).
Weinstein, H. and M. P. Dudukovic, Tracer methods in the circulation, in *Topics in Transport Phenomena*, R. Gutfinger, ed., Hemisphere, New York (1975).
Woodrow, P. T., Use of intensity function representation of residence time variability to understand and improve performance of industrial reactors, ACS Symp. Ser., *65*, 571 (1978).
Zvirin, Y. and R. Shinnar, Int. J. Multiphase Flow., *2*(5/6), 495 (1975).
Zwietering, T. N., The degree of mixing in continuous flow systems, Chem. Eng. Sci., *11*, 1 (1959).

3
Catalytic Surfaces and Catalyst Characterization Methods

W. NICHOLAS DELGASS *Purdue University, West Lafayette, Indiana*

EDUARDO E. WOLF *University of Notre Dame, Notre Dame, Indiana*

INTRODUCTION

Catalysis is a complex phenomenon in which a multitude of interesting variables intervene. A review of the principles and factors involved in determining catalytic activity is beyond the scope of this chapter. We have focused on the characterization of catalytic properties to enhance our understanding of the underlying processes. Emphasis is given to the modern surface analysis techniques, which can provide new insights for interpreting catalytic phenomena. Such methods, which were often first applied to study model surfaces, are now being used to analyze the more complex surfaces found on technical heterogeneous catalysts. Two types of characterization techniques are discussed in this chapter: (a) physical characterization, and (b) physicochemical characterization. The first type refers to total surface area and pore size distribution. Total area is relevant in determining the contact between the catalytic agent and reactants. Porosity is significant in controlling the transport of the catalytic agent into the support during catalyst preparation, as well as the transport of reactants and products between the bulk fluid phase and the active sites during reaction. Since surface area and pore volume are related, a balance between these two variables is a key factor in catalyst design. The BET adsorption method to measure total surface area and methods to measure pore size distribution are discussed in the following two sections.

Other important variables that characterize a supported catalyst are the area of the active catalytic component (i.e., metal area in the case of a supported metal catalyst) and the crystallite size distribution of the active phase. Many reactions are sensitive to the crystallite size, and they are termed *structure-sensitive* or *demanding reactions*. Various factors are involved in this structure sensitivity. The occurrence of a particular crystal phase, the ratio of edge to corner sites, the structure of the surface, and the stability of overlayers can all depend on crystallite size. A distinctive chemical feature of crystallites in the size region 0.5 to 5 nm is the low coordination number of the surface atoms. Later sections describe selective chemisorption of gases as a method to measure total metal area and x-ray diffraction and transmission electron microscopy as methods to determine average crystallite size and size distribution.

Although the methods mentioned previously have been known for many years, their development for chemical analysis of surfaces is more recent. As knowledge of the principles of operation, capabilities, and limitations of surface analysis methods expands, the interpretation of catalytic phenomena will become accordingly more sophisticated. Ultimately, this new knowledge of surface chemical behavior will support the search for and design of new catalysts. Knowledge of the composition of a surface is a prerequisite for understanding its chemistry. X-ray photoelectron spectroscopy (XPS), Auger electron spectroscopy (AES), and ion scattering spectroscopy (ISS) provide quantitative analysis of surfaces or thin surface layers. This information can reveal surface enrichment in alloys, the surface concentration of promoters in multicomponent supported catalysts, and the composition of the surface of mixed oxides. These methods can also detect impurities and poisons and, in general, follow changes in the catalyst surface after exposure to the reaction environment. Oxidation state and phase identification come most readily from XPS and, for selected elements, from the Mössbauer effect. The current frontier in surface analysis is the measurement of local surface composition and structure by application of (a) extended x-ray absorption of fine structure (EXAFS), and (b) secondary ion mass spectroscopy (SIMS). We discuss these analytical tools and magnetic resonance in relation to the determination of surface composition and structure.

The distribution of a catalytic material, or a poison within a support, can be a key variable in catalyst performance. Slight changes in procedure during catalyst preparation may introduce significant changes in the distribution of catalytic materials inside the pellets. We discuss the use of electron microprobes and electron microscopes equipped with analytical capabilities to assess the distribution of materials inside pellets.

To complete the chemical characterization of a surface, one must examine its interaction with adsorbing and reacting gases. We discuss methods by which the effects of gas-solid contact on the composition and chemistry of the solid surface can be examined. The influence of the surface on the adsorbed molecules is usually probed with vibrational spectroscopy (IR or Raman) and can be investigated with a magnetic resonance (NMR or ESR). Another measure of gas-surface interaction is the strength of the adsorbate bond. Sabatier suggested long ago that this bond must be strong enough to perturb the molecule but not so strong as to form an unreactive surface compound. Temperature-programmed desorption and reaction methods for studying surface bonding are discussed, along with vibrational spectroscopic techniques for examining the adsorbate molecules themselves. These adsorbate studies can serve either as a direct probe of the reaction itself or as an additional means of chemical characterization of the catalytic surface.

The techniques reviewed in this chapter are those best developed and most commonly applied or most promising at this time. The presentation emphasizes the essential concepts and features of each technique and major applications in catalysis. We have attempted to avoid excessive use of jargon to keep the discussions understandable to the nonspecialist. References to more detailed descriptions are given in each section. We note, finally, that thorough characterization of catalysts often requires use of combinations of techniques chosen to give complementary information, and that characterization of the catalyst surface encompasses only half of the information necessary to fully understand a catalytic system. The other half is the evaluation of the elementary kinetic steps of the reaction which are the manifestations of the influence of the catalyst surface on the reacting molecules.

TOTAL SURFACE AREA: BET METHOD

There are several techniques to estimate the total surface area and pore size distribution of porous materials. Various review articles concerning such measurements have been published (Emmett, 1954; Innes, 1968). The BET method is, however, the most common technique employed today for such measurements; consequently, it is the only technique discussed here. The reader is directed elsewhere in reference to other techniques (Innes, 1968).

BET Adsorption

Many physical adsorption isotherms exhibit S shapes which are inconsistent with the Langmuir isotherm based on monolayer adsorption-desorption equilibrium. Brunauer, Emmett, and Teller (1938) proposed an explanation based on the assumption that multilayer adsorption could occur. The BET model also postulates surface homogeneity and no intermolecular interaction between adsorbed species. The most common form of the BET adsortpion isotherm also assumes that an infinite number of multilayers can be adsorbed. The resulting linearized BET isotherm has the form

$$\frac{P}{V_{ads}(P_0 - P)} = \frac{1}{V_m C} + \frac{C - 1}{V_m C}\frac{P}{P_0} \tag{1}$$

where V_{ads} is the volume at STP occupied by molecules adsorbed at a pressure, P, V_m is the volume corresponding to monolayer coverage, C is a constant, and P_0 is the saturation vapor pressure of the adsorbate over a plane surface. A plot of $P/V_{ads}(P_0 - P)$ versus P/P_0 is a straight line with slope $S = (C - 1)/V_mC$ and intercept $I = 1/V_mC$. Knowing S and I permits calculation of V_m and therefore the number of gas molecules adsorbed in a monolayer, which, when multiplied by the cross-sectional area of the adsorbate, gives the total surface area of the solid. The following cross-sectional areas (in square angstroms per molecule) give reasonably self-consistent total areas: $N_2 = 16.2$, $O_2 = 14.1$, $Ar = 13.8$, $Kr = 19.4$ (Adamson, 1976). N_2 is the preferred adsorbate, but Ar and Kr are also commonly used.

The most common experimental configurations used to measure BET isotherms (Innes, 1968; Everett and Otterwill, 1970) are (a) the classical volumetric static apparatus used by Emmett, in which pressure associated with incremental volumes of gas is measured; (b) a recirculating batch flow system with helium dilutent; and (c) a pulse chromatographic dynamic flow system.

Several commercial systems based on static or dynamic techniques are available (Everett and Otterwill, 1970). The experimental objective in every case is to determine V_{ads} as a function of the adsorbate pressure. These values are then substituted into Eq. (1) so that the intercept and the slope of the linearized BET isotherm can be determined. The region of best fit for the isotherm is in the P/P_0 range 0.05 to 0.3.

Continuous and automatic systems have also been designed and differ from batch systems in that the gas is added continuously to the system at a very low rate instead of in large increments. Continuous addition permits automation, which, when coupled with microcomputers, provides direct calculation of surface area and pore size distribution. Major factors in all BET measurements are proper degassing of the sample and equilibration. Other techniques used for calculation of total surface area, such as gravimetric techniques, selective adsorption from liquids, and small-angle x-ray scattering, are briefly summarized by Innes (1968).

PORE SIZE AND DISTRIBUTION

The distribution of pore sizes is one of the important characteristics of supported catalysts since it is related to the value of the effective transport coefficients, and consequently it can affect activity, selectivity, and rates of deactivation. Catalyst pellets have a complex pore structure which presents a wide distribution of sizes. Pores between 10 and 100 Å are referred to as *micropores*, those between 1000 Å and 10 μm are called *macropores*, and those of intermediate size are called *mesopores*. The micropores are usually characteristic of the support porosity, whereas macropores can originate in the interparticle space created during formation of the pellets. A material exhibiting both micropores and macropores is described as having a *bimodal pore distribution*. Detailed discussions of pore structure and its determination have been presented in various textbooks and monographs (Everett and Otterwill, 1970; Gregg and Singh, 1967; Parfitt and Singh, 1976).

A first-order approximation or estimate of the average pore size can be performed by assuming that all pores are nonintersecting cylinders of uniform length L and radius r_p; then

$$\frac{\text{Pore surface}}{\text{Pore volume}} = \frac{S}{V_p} = \frac{2}{r_p} \qquad (2)$$

or

$$r_p = \frac{2V_p}{S}$$

where V_p is the total pore volume and S the total surface area.

For materials with a distribution of pore sizes, it is necessary to know the pore volume at various pore sizes. The derivative of the pore volume curve with respect to the radius gives the pore size distribution. The two most common techniques used to determine pore volumes are the BJH method (Barrett et al., 1951), based on the use of either adsorption isotherms or porosimetry.

Pore Size Distribution from Physical Adsorption Isotherms

If a physical adsorption isotherm is extended to the region in which P/P_0 is about 1, a rapid increase of V_{ads} occurs due to condensation of the adsorbate on the pore walls. Conversely, a decrease in pressure results in evaporation of the adsorbed or condensed liquid. Kelvin derived the following relation between the vapor pressure reduction over a liquid contained in a cylindrical capillary and the radius r:

$$\ln \frac{P}{P_0} = -\frac{2V\sigma}{rRT} \cos \phi \qquad (3)$$

where P_0 is the saturation pressure, σ the surface tension of the liquid, ϕ the contact angle (for simplicity assumed to be 0°), and V the molar volume of the liquid. The Kelvin equation can be used in principle to calculate r for any corresponding value of P/P_0. However, several complications arise: (a) a hysteresis loop is often observed in adsorption isotherms; (b) the pores are not cylindrical and have varying radii; and (c) an adsorbate film of varying thickness decreases the effective Kelvin radius. A discussion of the various theories that have been proposed to account for the factors cited above is given by Gregg and Singh (1967).

A detailed calculation procedure, known as the *BJH method*, has been reported by Barrett, Joyner, and Halenda (1951). The method consists of stepwise calculations based on the data obtained from the desorption branch of the isotherm about the region where $P/P_0 \simeq 1$. The calculations yield a cumulative pore volume versus pore radius which is then differentiated with respect to the radius to obtain the pore size distribution.

Mercury Porosimeter

In this case the pore volume is measured directly by forcing a nonwetting liquid, mercury, into the pores. A force balance between external pressure and surface tension in a cylindrical pore gives

$$r = \frac{2\sigma \cos \theta}{p} \qquad (4)$$

or for Hg,

$$r = \frac{7500}{p}$$

Equation (4) allows direct calculation of the pore radius at a given pressure. After plotting the cumulative pore volume, which equals the volume of Hg forced into the pores, versus pore radius, one can obtain the pore size distribution by differentiation with respect to r.

With proper experimentation, both N_2 desorption and Hg porosimetry generally agree well. The N_2 desorption method is adequate for small pores (2 to 30 nm), whereas large pores are preferentially measured by low pressure Hg porosimetry. With some materials high-pressure Hg penetration modifies the pore structure by breaking pores, which can lead to spurious results.

DETERMINATION OF METAL AREA BY SELECTIVE CHEMISORPTION OF GASES

The most common method employed to determine metal area is selective chemisorption of gases. The method has been discussed in review articles and textbooks (Whyte, 1973; Anderson, 1975). It consists of measuring the volume of gas adsorbed at monolayer coverage as in the case of BET adsorption; however, gases that chemisorb selectively on the surface under study are used instead of those that physisorb. Table 1 lists a number of metals and gases that form chemisorbed monolayers. Since for selected gases chemisorption is irreversible, fast, and usually single layer, monolayer coverage is determined from the plateau of the isotherm, extrapolated to zero pressure.

A simple stoichiometry is required to relate the number of gas molecules adsorbed with the number of surface atoms. For a given metal, an average site density can be calculated for the crystallographic planes exposed. From the site density, an area per surface atom (S_i) is calculated. For example, for Pt the densities of the (100) and (110) planes are 1.31×10^{15} sites/cm^2 and 0.93×10^{15} sites/cm^2 or 1.12×10^{15} sites/cm^2 on the average (50–50 distribution). This corresponds to an average area of 8.9 Å^2/atom. The total metal area per gram of catalyst (S_M) is then calculated as

$$S_M = \frac{S_i \nu V_{ads}(6.023 \times 10^{23})}{22,400 \text{ m}} \tag{5}$$

where ν is the adsorption stoichiometry (surface atoms/number of adsorbed molecules, V_{ads} the volume of gas adsorbed at monolayer coverage and standard temperature pressure, and m the mass of catalyst used.

TABLE 1 Selective Chemisorption of Gases on Metals[a]

Metals	Gases						
	O_2	C_2H_2	C_2H_4	CO	H_2	CO_2	N_2
Ti, Zr, Hf, V, Nb, Ta, Cr, Mo, W, Fe, Ru, Os	+	+	+	+	+	+	+
Ni, Co	+	+	+	+	+	+	–
Rh, Pd, Pt, Ir	+	+	+	+	+	–	–
Mn, Cu	+	+	+	+	±	–	–
Al, Au	+	+	+	+	–	–	–
Li, Na, K	+	+	–	–	–	–	–
Mg, Ag, Zn, Cd, In, Si, Ge, Sn, Pb, As, Sb, Bi	+	–	–	–	–	–	–

[a]+, Strong chemisorption occurs; ± chemisorption is weak; –, chemisorption is unobservable.

The metal fraction exposed, or dispersion, is calculated as

$$D = \frac{\text{metal surface atoms}}{\text{total metal atoms}} = \frac{\nu\mu_i M}{mw} \times 10^{-6} \qquad (6)$$

where M is the molecular weight of the metal, w the fractional metal loading, and μ_i the uptake in micromoles. The average crystallite size is calculated by assuming that crystallites are of the same regular uniform shape (i.e., crystallites are often assumed to be cubes with five faces exposed and the sixth being in contact with the support). For this geometry

$$d = \frac{5w}{S_M \rho_M} \qquad (7)$$

where ρ_M is the density and d the average crystallite size.

Spenadel and Boudart (1960) demonstrated that H_2 chemisorption on Pt black (10 torr, 25°C) agreed well with BET data. Agreement between H_2 chemisorption and x-ray diffraction results was also obtained in the case of a Pt supported on alumina catalyst. A variation of the direct chemisorption technique is titration of preadsorbed O_2 with H_2 (Gruber, 1962). Benson and Boudart (1965) proposed that the titration stoichiometry is

$$PtO + \frac{3}{2} H_2 \rightarrow Pt{-}H + H_2O$$

which allows for the uptake of three hydrogen atoms per Pt atom, thus increasing the sensitivity of the technique. Mears and Hansford (1967) found that a different $Pt{-}H_2$ stoichiometry appeared to be applicable to their results. The dilemma was resolved by Wilson and Hall (1970), who found that the stoichiometry of oxygen chemisorption depended on crystallite size. For this reason, titration techniques are not used with samples having very small sizes.

The use of CO chemisorption has also been proposed to determine metal surface areas (Hughes et al., 1962). However, the relative contribution of the linear and bridged forms of CO must first be assessed to establish the proper stoichiometry. Other gases, such as NO, have been used to evaluate the metal surface area of Ru (Ramamoorthy and Gonzalez, 1979).

The experimental techniques to measure V_{ads} have not been mentioned above, but in most references a static or volumetric technique similar to the one used in BET absorption has been used. A flow pulse techniuqe that permits rapid determination of V_{ads} has been described by several authors (Gruber, 1962; Benesi et al., 1971; Huasen and Gruber, 1971). The effect of sample volume, amount of catalyst, pretreatment, and so on, on the value obtained for V_{ads} using the pulse technique has been reported. Pulse results agree well with the static techniques if adsorption is very *fast* and *irreversible*. Reversible adsorption introduces tailing on the emerging peaks, which can lead to errors (Wanke et al., 1979). A stringent requirement of the pulse technique is the use of clean ultrahigh-purity gases and various traps to eliminate trace impurities. Sarkany and Gonzalez (1982b) have discussed conditions for proper use of the pulse technique, in particular, linearity of the TC cell response and the effect of reversible adsorption. Appropriately used, the flow technique gives good and expedient results and the apparatus is easy to build. For a new catalyst it is advisable first to establish the operating conditions in which reproducibility with the static or other techniques is attained.

DETERMINATION OF AVERAGE CRYSTALLITE SIZE AND CRYSTALLITE SIZE DISTRIBUTION

X-ray Diffraction

X rays have often been used in catalytic work in relation to the estimation of crystallite size and to obtain information on the bulk crystallographic structure of catalytic materials. A number of monographs exist on x-ray diffraction (Klug and Alexander, 1974; Cohen and Schwartz, 1977; Cullity, 1978; Schultz, 1982) and the subject is so extensive that it can be treated here only in basic terms.

Nature of X rays

X rays are electromagnetic radiation with wavelengths in the angstrom range. Consequently, they can penetrate matter and therefore are especially well suited to probe the structure of solids. X rays are commonly generated by bombarding a solid (target) with high-energy electrons to create inner-shell electron vacancies. Two types of x rays are thus generated: (a) a continuous spectrum, and (b) a characteristic line spectrum. The continuous spectrum arises from the deceleration of the incoming high-speed electrons by the target. Line spectra occur when an electron from a high-energy orbital fills a low-energy vacancy in the inner electron orbitals. The line spectrum is characteristic of the emitting material and use used for x-ray spectroscopy, which will be discussed in detail later. The peaks of the line spectra are designated by the orbital (shell) into which the electrons fall (K, L, M, N) and a Greek letter indicating the original orbital (α, β, etc). Due to the characteristic well-defined energy, the x-ray lines are used as a quasimonochromatic x-ray source in diffraction equipment.

X-ray Diffraction

Diffraction occurs when a wave interferes with an array of scattering centers, causing the outcoming waves to reinforce each other (constructive interference) or to be out of phase and cancel each other (destructive interference) (see Fig. 1a). X rays are scattered by the electrons of the irradiated matter. Elastically scattered x rays have the same frequency as the incoming x rays (coherent scattering); the opposite is true for inelastic scattering (incoherent scattering). Elastically scattered x rays, which are important in diffraction, can be depicted as reflected by the scattering atom.

Bragg used the reflection analogy to explain x-ray diffraction. Figure 1b shows a monochromatic x-ray beam of wavelength λ directed on two successive planes of a crystal. The reflected beam (i.e., scattered)

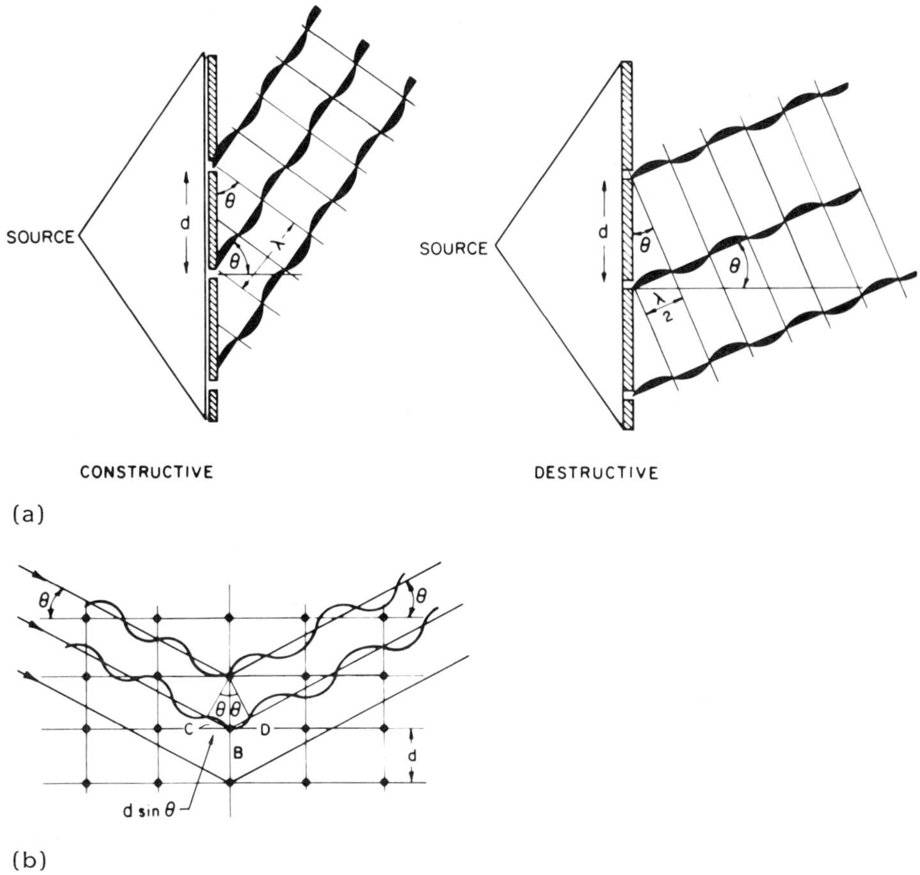

FIGURE 1 Schematic representation of (a) constructive (in phase) and destructive (out of phase) interference when a wave motion is scattered by slits, and (b) Bragg's reflection analogy for x-ray diffraction showing constructive interference.

and associated wavelengths are also shown. If the reflected rays are to be in phase, the path difference of the two successive incoming rays, CB + BD = 2d sin θ, must be equal to an integral number of wavelengths, that is,

$$2d \sin \theta = N\lambda \tag{8}$$

The equation above, known as *Bragg's law*, relates the d spacing of the crystal with the angle of incidence and wavelength of the incoming x rays. N is known as the order of the reflection. Equation (8) indicates that measuring the intensity of the diffracted beam as a function of the incident angle gives a diffraction pattern which is characteristic of the crystallographic structure of the irradiated sample. Such measurements, which constitute one of the many methods of XRD analysis, are most easily attained with an x-ray diffractometer.

An x-ray diffractometer consists of a circular table with a stationary x-ray source and a moving detector located in the circumference of the table with the sample set at the center. The moving detector, usually a proportional counter, records the intensity of the reflected beam as a function of the reflected angle 2θ. A straightforward application of XRD is the use of the XRD patterns to identify the various phases existing in a material. Diffraction patterns of powders are compiled in the *Powder Diffraction File Search Manual* (Berry, 1975).

Crystallite Size Determination from XRD Line Broadening

Diffraction lines should, in principle, be very narrow; however, when the crystallite size of a polycrystalline material falls below 1000 Å, broadening of the diffraction lines is observed. Instrumental limitations and lattice strain can also cause line broadening.

Line broadening due to particle size only arises because of incomplete destructive interference. When the incident beam is slightly off the Bragg angle, the reflected ray which would nullify it originates from N planes inside the crystal. If the crystal is smaller than about 1000 Å, most planes do not have their destructive counterparts N planes away, so that finite diffracted lines will be observed at the off-Bragg angle setting, causing line broadening. Various measures of the peak broadening are used, the most common being the width at half peak intensity $\beta_{\frac{1}{2}}$, and the integrated breadth, $\beta_i = \int_{\theta_1}^{\theta_2} I d\theta / I_{max}$ where I is the observed intensity of the diffracted beam.

The relation between line broadening and crystallite size for a stress free material, known as the *Scherrer formula*, is

$$\langle L \rangle_{hkl} = \frac{K\lambda}{\beta_{hkl} \cos\theta} \tag{9}$$

where K is a constant that depends on the definition of β used and the crystal geometry. The constant K varies from 0.98 to 1.39, but because of experimental uncertainties, the constant is often set equal to 1. When β is defined as the integral breadth, $\langle L \rangle$ is a volume average crystallite size.

In the absence of lattice stress, broadening also occurs due to experimental limitations such as nonparallel and nonmonochromatic irradiation. Other experimental factors also contribute to broadening, even if the specimen is of "infinite size" (i.e., larger than 1000 Å). Consequently, for the correct estimation of crystallite size using Scherrer's formula, the experimental contribution to line broadening must be separated to obtain the pure diffraction broadening. Although various procedures exist, two are outlined here: (a) a simplified procedure, and (b) an accurate Fourier transform method.

Simplified Procedure: In this case the sample is mixed with a standard that has particle sizes greater than 1000 Å and produces a diffraction line near the line of the sample under analysis. Alternatively, diffraction patterns of the sample and of another sample of the same material containing crystallites larger than 1000 Å can be used. In either case, if B is the width of the sample containing small crystallites and b the width of the standard sample with large crystallites (>1000 Å), the breadth or broadening of the pure diffraction pattern β is estimated by

$$\beta^2 = B^2 - b^2 \tag{10}$$

or

$$\beta = B - b \tag{11}$$

The first correction is derived by assuming that the shape of the peak is gaussian, whereas the second one corresponds to a Cauchy profile. Here β would be either $\beta_{\frac{1}{2}}$ or β_i. Over the years, the gaussian approximation has been widely used, although some evidence indicates that the pure diffraction profile is better approximated by the Cauchy form when a distribution of particle size exists (Klug and Alexander, 1974). However, either correction can be used. It must be emphasized that because the shape of the profiles must be assumed, the simplified procedures give speedily results of good relative accuracy but have restricted absolute significance.

Fourier Transform Method: The theory of this method is beyond the scope of this outline and the reader is referred to the specialized literature (Klug and Alexander, 1974; Cohen and Schwartz, 1977). The experimental determinations are similar to those outlined above, but the pure diffraction profile is calculated from the Fourier synthesis of the sample profile, h(ε), and a reference profile obtained with a reference sample of infinite size, g(ε), where ε is the angular change through the peak.

The Fourier transform coefficients of the true diffraction profiles are calculated from the real and imaginary Fourier series calculated from the intensities of sample and reference profiles measured at various angular intervals of the broadened peaks. Computation procedures are given Ganeson et al. (1978) and Sashital et al. (1977). The crystallite size can be calculated using the breadth of the reconstructed profile and Scherrer's formula or directly using the Fourier coefficients. Warren and Averbach (1950) related the crystallite size to the first derivative of the Fourier coefficients with respect to the Fourier harmonic. Furthermore, a plot of the second derivative of the Fourier coefficient is proportional to the distribution of crystallite sizes. Figure 2 shows results of Sashital et al. (1977) obtained using the Fourier transform method, including strain and imperfection corrections. Table 2 summarizes the average crystallite size for the various crystal planes along with dispersion results obtained from hydrogen chemisorption techniques. It can be seen that the Fourier transform calculations agree well with the chemisorption results.

Transmission Electron Microscopy

The most common use of an electron microscopy (EM) in catalysis is to measure the distribution of crystallite sizes as well as the morphology of supported or unsupported active materials. However, new analytical electron microscopes, which combine EM with spectroscopic analysis, also reveal information about the chemical nature of the crystallites. To understand the possibilities of EM beyond that of taking pictures, it is necessary to have a basic understanding of EM. A detailed description of the subject is presented in various textbooks (Heidenreich, 1964; Hawkes, 1972) and review articles (Beer, 1980; Schmidt et al., 1982).

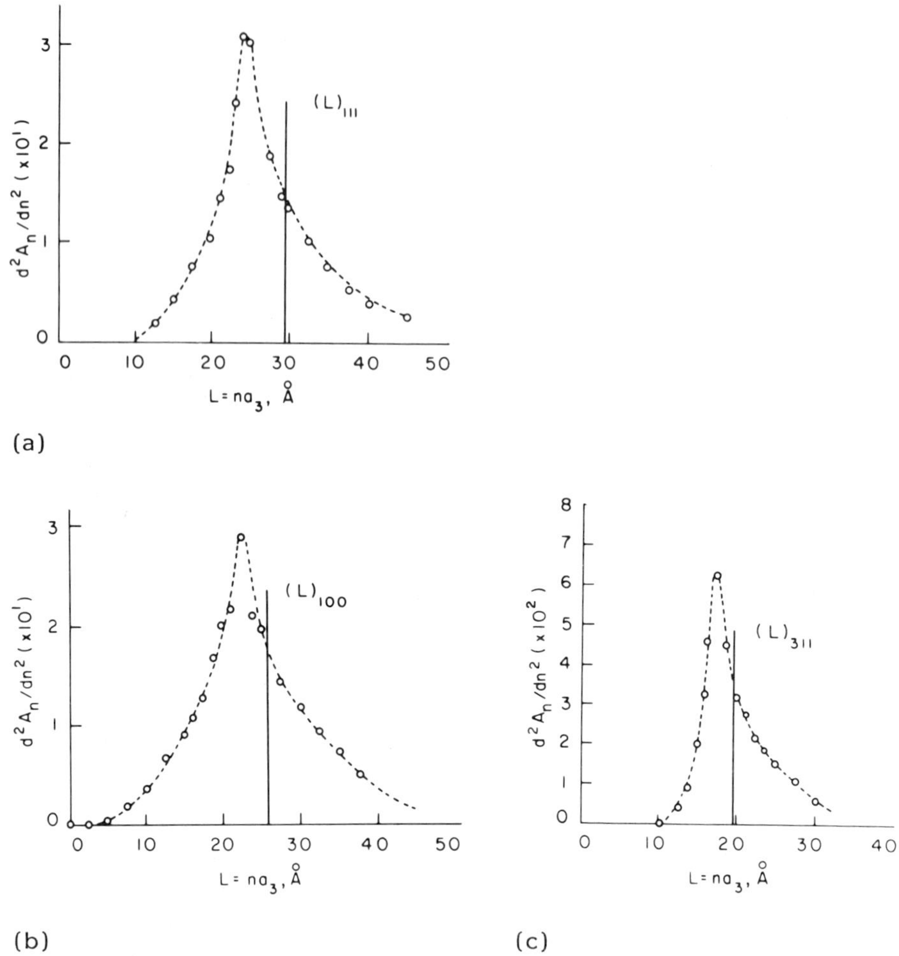

FIGURE 2 Particle size distribution for a catalyst containing 40% Pt on SiO_2 prepared by impregnation of chloroplatinic acid: (a) <111> direction; (b) <100> direction; (c) <311> direction.

An electron microscope differs from an optical microscope in that the electron beam replaces the light beam and electromagnetic lenses replace the optical lenses. The use of an electron beam improves the resolution of the microscope, which is proportional to the wavelength λ of the incident beam, in accord with (Heidenreich, 1964)

$$x = 0.6C_S^{1/4}\lambda^{3/4} \tag{12}$$

where x is the resolution (i.e., the minimum size that can be distinguished in the microscope) and C_S is the spherical aberration of the incident beam. Relation (12) illustrates the advantage of using an electron beam (λ = 0.5 to 10 Å) instead of light (λ = 4000 to 7000 Å).

TABLE 2 Particle Sizes $\langle L \rangle_{hkl}$ and Dispersion

Catalyst[a]	$\langle L \rangle_{111}$ (Å)	$\langle L \rangle_{100}$ (Å)	$\langle L \rangle_{110}$ (Å)	$\langle L \rangle_{311}$ (Å)	Dx (%)	Dh (%)
7.1-SiO_2-PtCl-S						
(1)	131	82	108	81	10	7.1
(2)	111	75	—	—	12	7.1
21.5-SiO_2-IonX-L						
(1)	49	43	45	47	25	21.5
(2)	48	43	45	47	25	—
27-SiO_2-IonX-S						
(1)	43	39	42	31	27	27.3
(2)	40	39	39	39	29	—
40-SiO_2-PtCl-S						
(1)	30	26	22	20	44	39.8
(2)	25	23	—	—	47	—

[a]7.1, 21.5, 27, and 40 indicate percent of Pt loading on SiO_2. PtCl and IonX indicate preparation via chloroplatinic acid impregnation and ion exchange, S and L refer to SiO_2 with small (120 to 140) and large (70 to 80) mesh size. Dx is the dispersion obtained from x-ray line broadening, and Dh is the value obtained from hydrogen chemisorption.

Two fundamental models of operation are used in EM: transmission electron microscopy (TEM) and scanning electron microscopy (SEM). A third, hybrid mode, scanning transmission electron microscopy (STEM), is also used. Figure 3 depicts the schematics of the various modes of operation together with the fundamental components involved in each case.

In transmission electron microscopy (TEM) the electron beam, generated by a heated filament (electron gun), passes through two electrodes and a condenser lens. Parallel rays thus created impinge on the specimen, where they are scattered as a result of the spatially variable refractive index. Scattered rays from the same point in the specimen are brought to the same point in the image formed by the objective lens. The overall effect is equivalent to transmisssion of primary electrons through the sample. Electrons are also diffracted (as in the case of x rays), thus producing rays which are slightly off the angle of rays formed by transmitted electrons. Thus, by selecting the proper aperture, an image of the transmitted (bright field) or diffracted electrons (dark field) is obtained after the electrons pass through intermediate and projector lenses and impinge on a fluorescent screen or a photographic plate. A typical conventional electron microscope operates at 100 kV, a vacuum of 10^{-5} to 10^{-6} torr, and is capable of 4.5-Å point-to-point resolution and mangification of 300,000. Dedicated high-resolution electron microscopes are capable of resolution of 2.5 Å with magnifications of 800,000 to 1,000,000.

An important aspect of TEM is that a three-dimensional sample yields a two-dimensional TEM image which might be difficult to interpret. Furthermore, the contrast between the support and the active component might

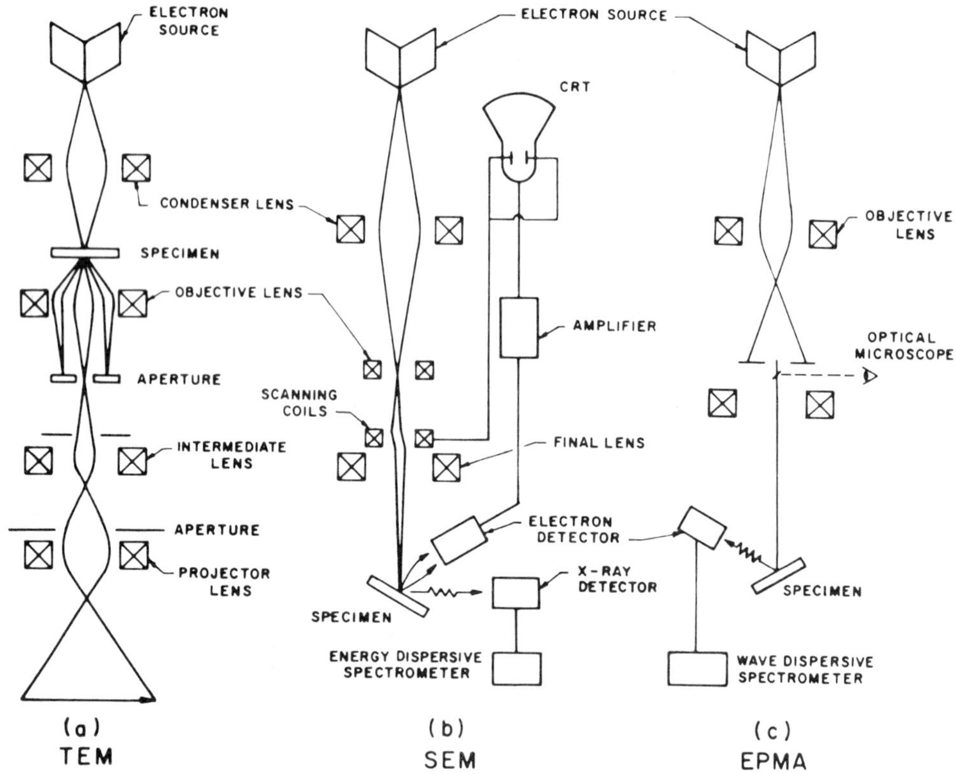

FIGURE 3 Schematics of (a) transmission electron microscope (TEM): (b) scanning electron microscope (SEM), and (c) electron probe microanalyzer (EPMA).

not be sufficient to distinguish the metal crystallites. In this case, dark-field images are useful since electrons diffract preferentially from the crystallites, thus producing brighter spots from them compared to the weak diffraction from the nearly amorphous support.

Supported Catalysts

Figure 4 shows bright- and dark-field images of a supported 5% Pt/SiO_2 catalyst. The sample is prepared by mortar grinding the catalyst particles (200 mesh) to ultrafine powder, which is then dispersed in a liquid (water, alcohol). A droplet of the solution is then deposited on an EM grid. The liquid evaporates, leaving on the grid microscopic particles which are thin enough for electron transmission. The micrographs shown in Fig. 4a were obtained with a 100 JEOL STEM operating at a magnification of 400,000. In this case, the contrast is relatively low because of the presence of the support. Counting and sizing the crystallites in both the bright- and dark-field images is often difficult, and many micrographs are required to obtain a statistically meaningful size distribution. A major limitation here can be the presence of small crystallites that cannot be seen under the conditions used. Furthermore, in some cases the diffraction pattern of the metal cannot be distinguished from that of the support (Schmidt et al.,

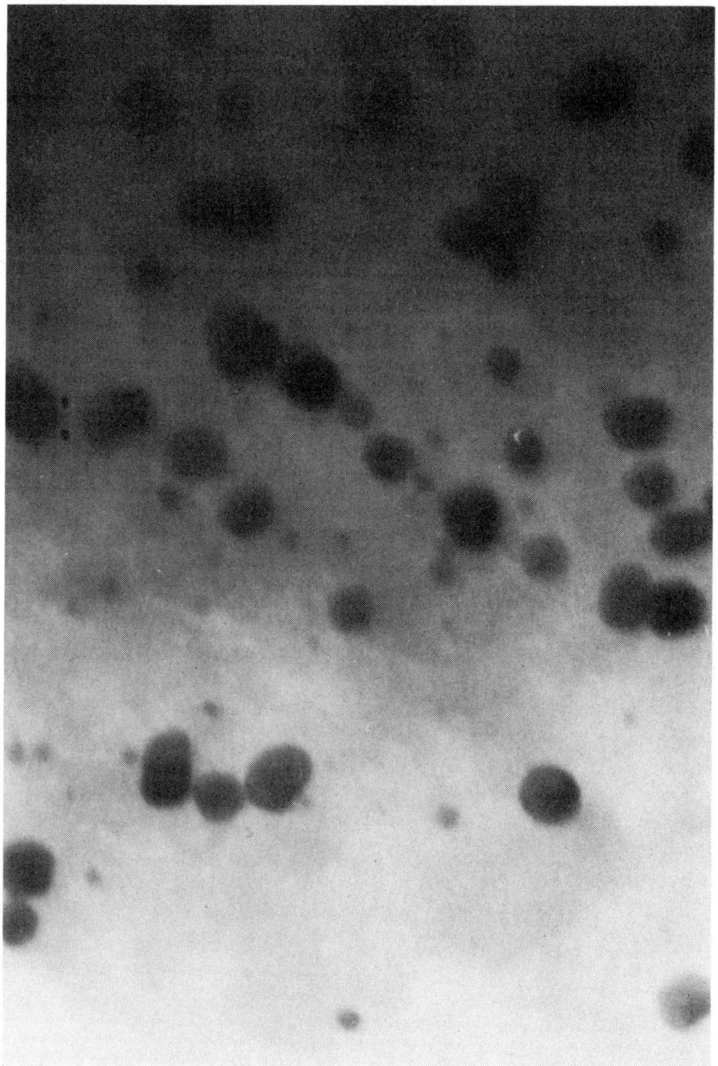

(a)

FIGURE 4 TEM photographs, all at magnification 400,000; 1 cm = 250 Å: (a) 5% Pt/SiO_2; (b) model sample, Pt sputtered on SiO_2 film; and (c) model sample, Pt impregnated on SiO_2 film with chloroplatinic acid. All catalysts were reduced in H_2 at 300°C.

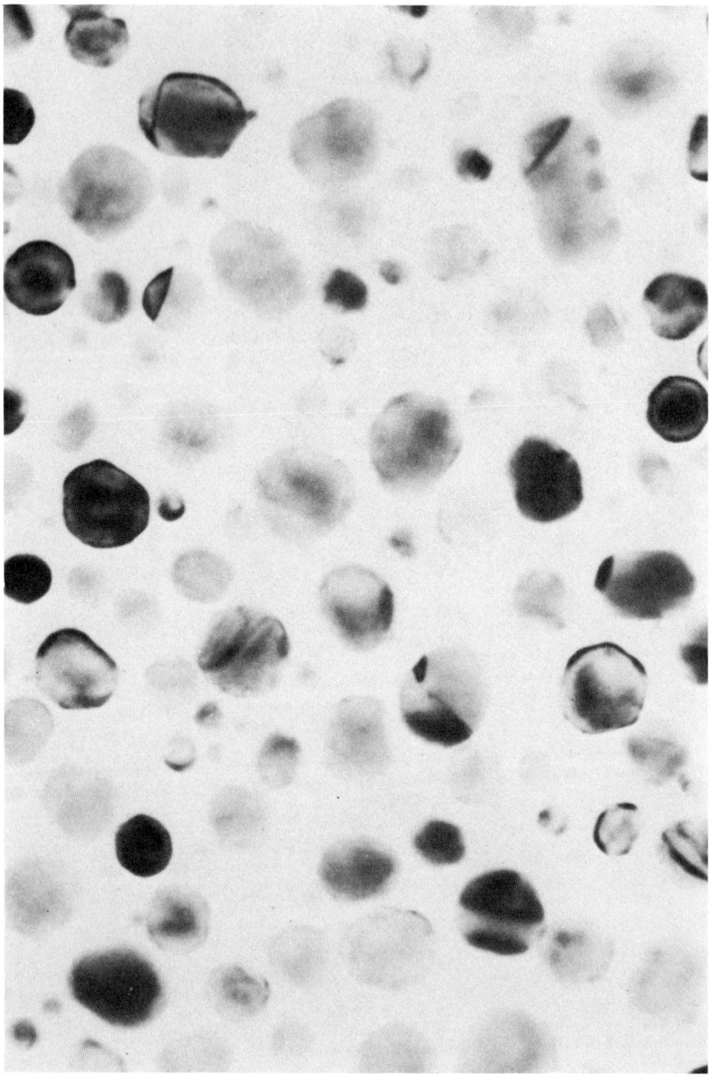

(b)

FIGURE 4 (Continued)

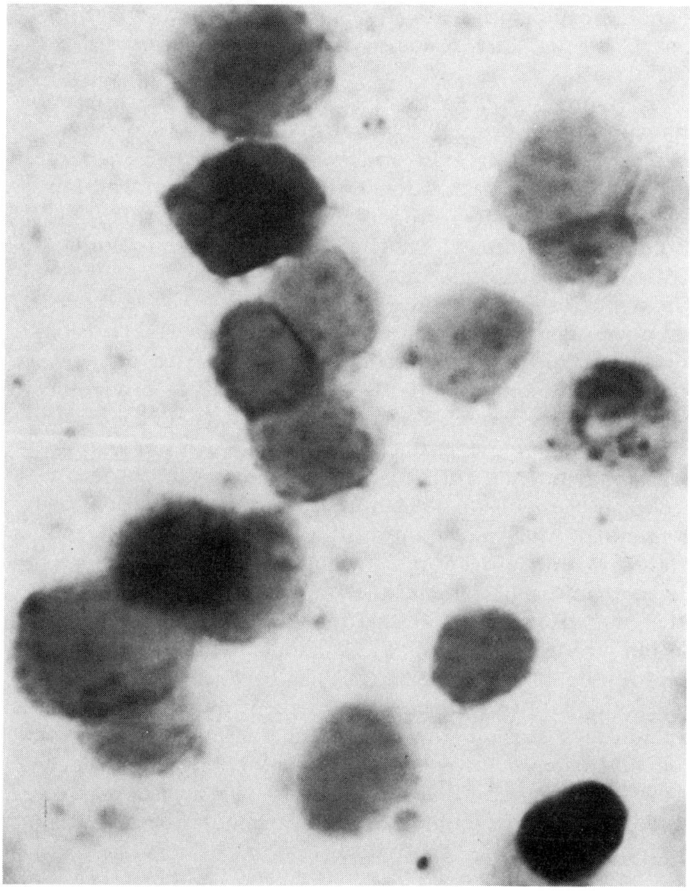

(c)

FIGURE 4 (Continued)

1982). It is always recommended that one check the electron diffraction pattern to determine the relative role of the support and active metal in dark-field studies. Charging of the sample by the electron beam is a common problem with catalysts in which the support is an insulator. Charging can be reduced by coating the sample with a carbon or gold film. Higher metal loadings (5 to 10%) and higher beam voltages result in better contrasts; however, sample damage can be caused by the electron beam. Sample preparation procedures are given by Baker et al. (1979) Glassl et al. (1980) report artifacts introduced due to sample preparation. The particle size distribution is obtained by the method described by Underwood (1970) or by means of an automatic sizing and counting device.

Model Samples

Schmidt and coworkers (Schmidt et al., 1982), among others, have discussed the advantages of using model samples in which the catalyst is deposited on a planar nonporous substrate of amorphous silica. A thin flake of silica is prepared by vacuum deposition of silica followed by vacuum deposition of the metals (e.g., Ni, Rh) onto the substrate. This technique avoids particle overlap and produces high contrast between the metal particles and the support. Figure 4b shows TEM results obtained by sputtering Pt onto a model silicon grid and Fig. 4c shows similar results obtained by impregnating the grid with H_2PtCL_6 and then reducing it in H_2 at 400°C. The disadvantage of the technique is that the total active area is low for reaction studies. Furthermore, the morphology of the particles deposited on the model, nonporous support, may be different than when the metal is deposited on porous silica.

Although TEM has been used extensively to study catalysts, in conventional operation it views samples in a vacuum environment, pre or post reaction. A technique that allows one to use TEM in a reactive environment is controlled atmosphere transmission electron microscopy (CATEM). Details of the CATEM technique have been documented in a review article by R. T. K. Baker (1979), who has introduced the technique to catalytic researchers. In CATEM, the specimen is mounted in a gas reaction cell which is fixed to the translational manipulator of the sample. With this arrangement it is possible to operate at gas pressures of the order of several torr within the cell and at temperatures up to 1300°C. This control of sample environment requires some sacrifice of resolution. Changes occurring during reaction are recorded by a TV video system. Baker's excellent paper (1979) illustrates the use of CATEM in studying gas-solid systems where the reaction produces carbon deposits or during gasification of the deposits, which induces motion of small metal particles and changes in the structure of the specimen. Extraordinary movies have been produced showing the growth of carbon filaments as well as the movements of metal particles during carbon gasification. Such information has revealed the complexity of the dynamics of gas-solid interactions.

COMPOSITION AND STRUCTURE OF SURFACES AND SOLIDS

Since the 1960s, several techniques have been developed to characterize surfaces. A great deal of effort has been spent on the measurement of basic properties of model surfaces (such as single crystals) to establish the theory and validity of each surface analysis method. These studies have led, in some cases, to interpretation of catalytic activity in terms of fundamental properties and have demonstrated the potential of these techniques in studies of technological interest. The discussion that follows is presented in the context of the application of these techniques in the interpretation of surface processes related to catalysis. Theoretical aspects are treated only on the most basic terms needed for interpretation of data. The reader is referred to the specialized literature for further detail concerning theory and for detailed discussion of equipment and operation.

Most of the techniques now commercially available are based on the excitation of the sample by an incoming beam of particles or electromagnetic energy. Following such excitation are de-excitation processes, which result in the emission of electrons or photons of discrete energy which can then be measured by a suitable energy analyzer. Figure 5 shows the schematics

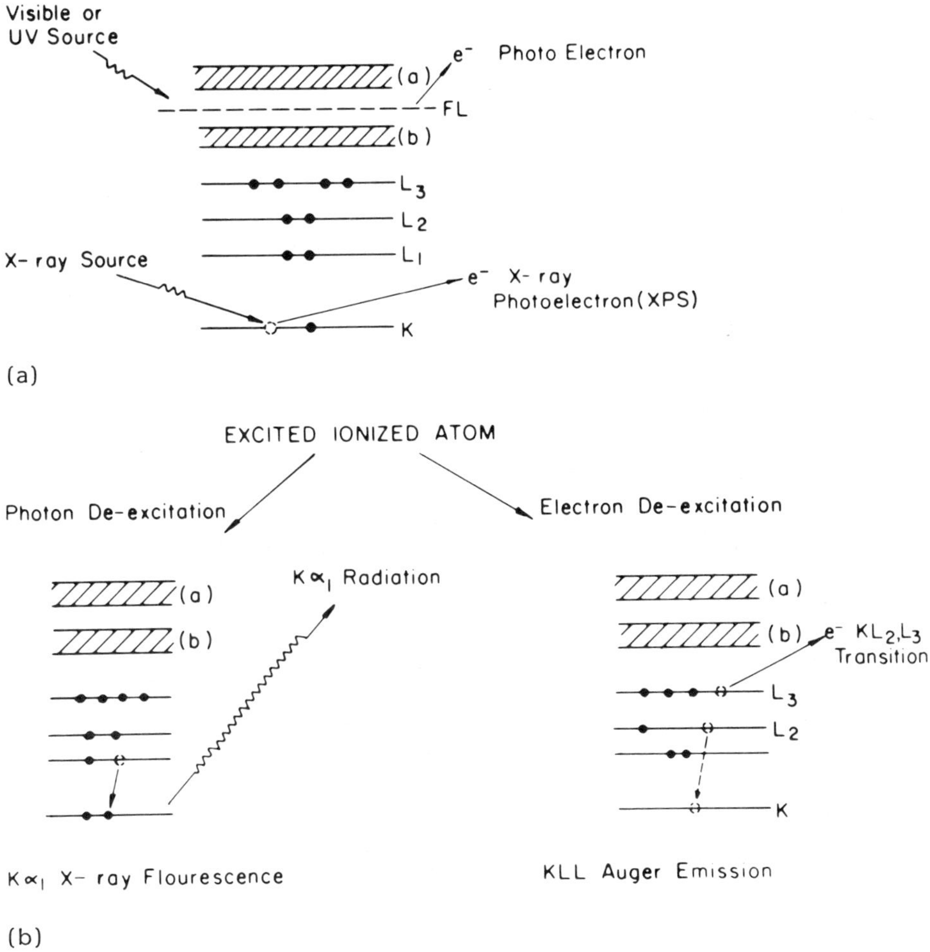

FIGURE 5 Schematics of (a) photon excitation and de-excitation by photoelectron emission, and (b) de-excitation processes leading to x-ray and Auger electron emission

of the two excitation and de-excitation processes most commonly used for surface analysis, together with the name and acronym of the analysis technique (XPS-AES). While several of the techniques discussed in this section yield similar information, the type of instrumentation and data interpretation varies. For the sake of clarity, each technique is discussed separately; however, there are many aspects that are common or complementary among them. One such common factor is the need for using ultrahigh vacuum (UHV). At 10^{-6} torr, about 10^{15} molecules of gas strike 1 cm^2 of the surface per second. At 10^{-9} torr, monolayer coverage occurs in about 15 min if every molecule that hits sticks; thus a clean surface can be attained only at UHV conditions.

The first technique described below is XPS (also known as ESCA, electron spectroscopy for chemical analysis) which corresponds to the case where the excitation source is x-ray photons, and the result is electron emission.

X-ray Photoelectron Spectroscopy

X-ray photoelectron spectroscopy (XPS) was developed in the early 1960s by Kai Siegbahn (Siegbahn et al., 1967) to characterize solids, liquids, and gases. Because the technique probes the top atomic layers of a solid, it has become one of the main tools for surface analysis. Several books, monographs (Siegbahn et al., 1967; Carlson, 1967; Ertl and Kuppers, 1979; Briggs, 1977; Delgass et al., 1979), and review articles (Palmberg, 1975; Hercules and Carver, 1974; Delgass et al., 1970; Hercules and Hercules, 1974) have been published describing theory and applications, with a few articles focusing on applied heterogeneous catalysts (Hercules and Carver, 1974; Delgass et al., 1970; Bhasin, 1981).

Photoelectron Emission (PE) or Photoionization

While the detailed physics of electron emission by photoexcitation, is quite complex, the basic principle is straightforward and has been known for decades. Emission of electrons occurs when matter is exposed to or excited by electromagnetic radiation or photons. If the photons have energies in the ultraviolet range, they cause ejection of the less tightly bound valence electrons that emerge at discrete energies [ultraviolet photoelectron spectroscopy (UPS)]. If the incoming photons are x rays of suitable energies (1 to 3 keV or soft x rays), ejection of inner or core electrons occurs. If the energy of the incoming photons is known, analysis of the energy of the emitted electrons permits calculation of core electron binding energies. Since the electron binding energies are quantized, the emitted electrons have discrete energies that reflect the electronic structure of the parent atom and thus identify the irradiated element.

Energy Relationships

The basic energy relationship occurring during photoionization was first proposed by Einstein. It states that the energy of the incoming photons, $h\nu$, equals the kinetic energy of the ejected electron (E_k) plus its binding energy, E_b (or photoionization energy):

$$h\nu = E_b + E_k \tag{13}$$

In practice, relation (13) must be modified to account for additional potentials existing in the spectrometer. Figure 6 shows the energies of electron binding measurements in solids. Once the electron is free from its core level (E_b) and reaches the Fermi level, it must spend some energy to free itself from the solid matrix (work function ϕ_s). The remainder energy appears as a kinetic energy E_k which is slightly changed to E_k' by the contact potential between the sample and the spectrometer. Since the Fermi level of the spectrometer (ϕ_{sp}) is equal to that of the sample because they are in electrical contact, one can write

$$h\nu = E_b + \phi_{sp} + E_k' \tag{14}$$

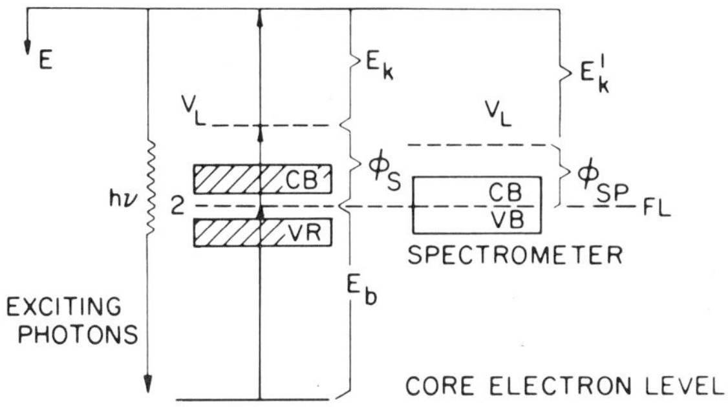

(a)

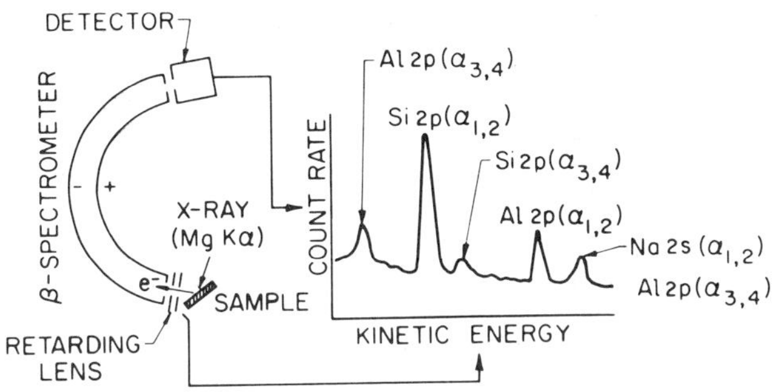

(b)

FIGURE 6 (a) Energetics of electron binding in solids. hν, excitation photon energy; E_b, electron binding energy; ϕ_s, sample work function; E_k, kinetic energy as ejected from the sample; E'_k, kinetic energy as measured by the spectrometer; FL, Fermi level; CB and VB, conduction and valence bands. (b) Schematic of basic components of an x-ray photoelectron spectrometer with data for a Y zeolite. (From Delgass et al., 1979.)

In most applications, relative binding energies or changes in their values are the relevant information. For each measurement, it is only necessary that ϕ_{sp} remain constant; conversely, the instrument can be calibrated using as a reference an element of known binding energy (Au or C, for example).

When the sample is irradiated, electrons emitted from various levels, and therefore possessing various kinetic energies, are emitted. A photoelectron spectrum consists of the number of photoelectrons emitted at a given energy level, N(E), versus their kinetic or binding energy (see Fig. 6b). From the spectral lines or peaks, which are designated in accord with the energy level from which electrons are ejected, E_b is calculated using Eq. (14) and the values compared with theoretical or experimental values (Wagner et al., 1978; Fadley et al., 1968). Thus XPS can be used for chemical identification of elements, but more significantly it can be used for chemical analysis of surfaces because of its surface sensitivity. Although, in principle, a single spectral line would be observed when a core electron is ejected, often one finds additional lines of lower intensity known as *satellites*. One type of satellite is the "shake-up" line. This occurs when the energy of the x-ray photon causes both photoemission and excitation of a valence electron to a low-lying bound state. A second common satellite line is a discrete energy loss line caused, for example, when the photoelectron excites a plasmon oscillation in the electrons in the sample.

Surface Sensitivity

Although x rays can penetrate solids far beyond the surface, the mean free path of electrons with energies between 10 and 1000 eV is of the order of 10 to 30 Å. Consequently, only electrons of atoms located near the surface obey the energy conversion equation [Eq. (14)] and are analyzed by the spectrometer. The relation between mean free path (or escape depth) of electrons as a function of energy is shown in Fig. 7. It can be seen that the mean free path is affected by the substrate, and for most elements the escape depth of photoelectrons comes from the top one to three atomic layers near the surface. The ability to select electrons ejected near the surface is what makes XPS a surface-sensitive technique.

Another important characteristic of XPS is its ability to produce an adequate spectrum even though the absolute amount of material present is of the order of 10^{-6} to 10^{-8} g. Fractions of monolayer as small as 0.01 can be detected in favorable cases.

Chemical Shifts

While the electron binding energies of particular elements are sufficiently invariant to provide chemical identification of the element, small but measurable shifts in binding energies occur when the XPS spectra of various compounds of the same element are measured (Siegbahn et al., 1967). The observed

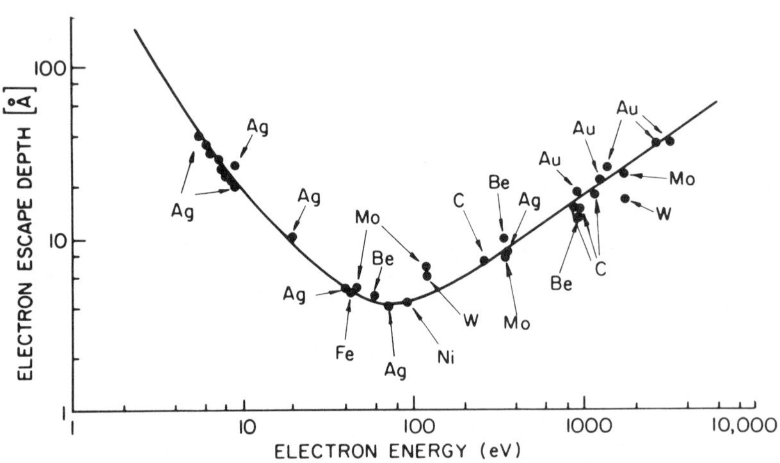

FIGURE 7 Mean free path or escape depth of electrons as obtained from Auger and photoelectron spectroscopy data. (From Delgass et al., 1979.)

shifts in E_b, or chemical shifts, can be explained in terms of changes in electrostatic forces which participate in electron binding (Siegbahn et al., 1967; Fadley et al., 1968). A core electron is subjected to attractive nuclear forces which are proportional to the atomic number and are also affected by the presence of the outer valence electrons. If an electron is removed from the outer shell, the core electrons feel an increased attractive force from the nucleus and therefore E_b increases. Conversely, a gain of an outer electron has the opposite effect.

Although various theories have been proposed to obtain a priori information on chemical shifts, not all the contributions to shifts on solid samples can be evaluated accurately. However, sufficient data have now been compiled to make use of chemical shifts as an indicator of the oxidation state of many elements (Wagner et al., 1978).

Instrumentation

The majority of the x-ray photoelectron spectrometers consist of the following five basic components: (a) source, (b) sample mount, (c) electron energy analyzer, (d) detector, and (e) data acquisition system. The source is an x-ray tube of conventional design consisting of a heated cathode or electron gun from which electrons are accelerated to a cooled anode or target. Al and Mg with K_α lines located at 1486.6 and 1253.6 eV are the most commonly used anodes. The sample region must be maintained at pressures lower than 10^{-9} torr to avoid contamination of the sample. This restriction is more severe than the 10^{-5} torr maximum pressure in the spectrometer required to minimize electron energy loss on collision with gaseous molecules.

Improvements in electron energy analyzers led to the development of XPS. Although various types of analyzers have been used, the most common types in use today are electrostatic analyzers (also known as β-spectrometers or monochromators). They consist of two concentric bodies (spheres, cylinders) onto which a voltage is superimposed. Electrons entering the space between the bodies follow trajectories that depend on electron energy. Thus for a given voltage only those electrons which have a specific kinetic energy will reach the detector. Figure 8 shows the schematics of an XPS spectrometer with a cylindrical mirror analyzer (CMA). The CMA consists of two concentric cylinders with a slit permitting electrons with energies proportional to the applied voltage to reach the detector. Generally, the analyzer is set at 100 eV and the photoelectrons are retarded by a negative potential applied to a lens system in front of the energy analyzer. The complete spectrum is produced by sweeping the retarding voltage. A data acquisition system is used to collect electron counts at the detector as a function of electron kinetic energy. Accumulation of counts during repetitive scanning improves the signal-to-noise ratio. The spectral lines are often superimposed on a raising background which occurs due to secondary and inelastically scattered electrons. Powdered samples can be spread uniformly into a self-adhesive material which is attached to the mount or pressed into self-supporting wafers. Charging of the sample is not as severe as in the case of excitation by electron impact and is usually compensated by flooding the sample region with low-energy electrons. For this reason, XPS is the preferred surface analysis technique for supported catalysts.

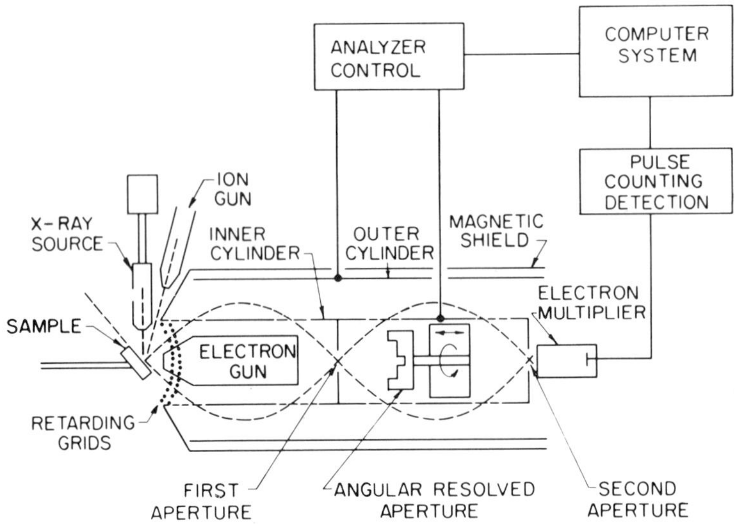

FIGURE 8 Schematic representation of a combined ESCA-Auger-SAM system using a double-pass cylindrical mirror analyzer (PHI model 550). (From Wagner et al., 1978.)

Applications

Applications of XPS are numerous, and many articles have been published in various areas of materials science, surface science, and catalysis. The XPS chapter by Delgass et al (1979) presents and discusses several examples of XPS applications. Rather than enumerating the many examples published in the literature, the major areas of application related to catalytic activity are discussed below in terms of qualitative and quantitative analysis

Qualitative Analysis

A straightforward application is a survey of elements existing on a catalyst. This is particularly important in connection with catalyst poisoning by impurities contained in the feed, for example, sulfur compounds or metals (Bhasin, 1981).

Chemical shifts provide an important clue about the oxidation state of the active component of a catalyst and its modification by pretreatment, reaction environment, and support. Several studies are reported by Delgass et al. (1979) in which chemical shifts have been used to determine the oxidation state of W supported on SiO_2 and Al_2O_3, Cr supported on SiO_2, and Co-Mo/Al_2O_3 hydrodesulfurization catalysts. Other studies on supported metals include Pt/Al_2O_3, Pt Au/C, NiO, FeV_2O_3, CuO/Al_2O_3, and Rh/C, among many other examples. The spectra of a 2% Pd supported on a ZSM-5 zeolite illustrating the use of chemical shifts arising from different oxidation states of Pd are shown in Fig. 9.

Quantitative Analysis

The areas under the XPS peaks carry quantitative information; however, accurate interpretation requires a structural model of the surface layer.

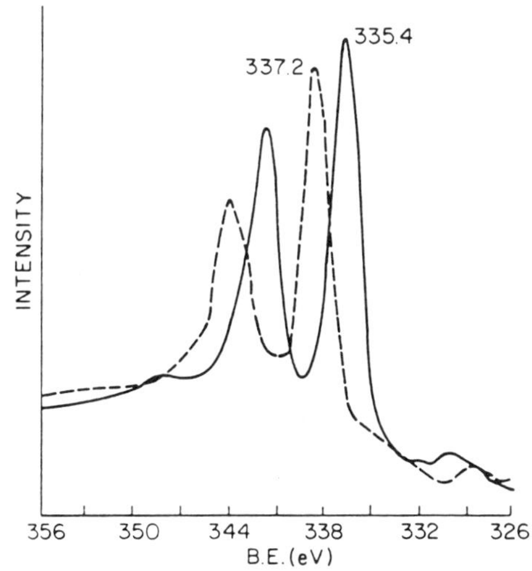

FIGURE 9 XPS spectra of Pd 3d transition of a 2% Pd/Na-ZSM-5 catalyst. Dashed line: oxidized; solid line: freshly reduced. (From Saha and Wolf, 1984.)

For a homogeneous solid with a flat surface, the following relation has been derived (Carlson, 1975):

$$I = K\sigma n\lambda F \tag{15}$$

where K is a spectrometer constant, σ the photoelectron cross section for a given shell, n the concentration of atoms undergoing photoelectron ejection, F the x-ray flux, and λ the electron mean free path for inelastic collision. Other models considering angular energy dependence have also been proposed (Fraser et al., 1973). In the case of porous materials such as supported metal catalysts, the model leading to the integrated intensity should consider metal crystallite size distribution, distribution of the metal within the pore structure, and the effect of surface roughness on an angle dependence. Some simplified models for quantitative analysis of supported catalyst have been proposed (Angevine et al., 1977). Many of the various factors involved in the expression of the integrated intensity cancel out when the ratios of intensities of two key components are compared. For example, for a completely uniformly and monatomic dispersed material, the ratio $I_{metal}/I_{support}$ derived from the integrated areas corrected for cross section should be proportional to the ratio of the number of metal atoms per unit area of support (Fraser et al., 1973). If the intensity ratio is higher than expected from the metal surface density, it indicates migration of the metal to the external surface, and if it is lower, it suggests particle growth or preferential deposition in small pores not easily seen from the exterior of the particle. Another case wherein relative comparison can give quantitative information is in analysis of metallic or polymetallic catalysts, where the integrated intensity ratios, corrected for photoelectron cross section, can be compared with the atomic ratio of the metals as loaded into the catalysts. Deviation from a 1:1 correspondence would suggest surface enrichment of one metal in preference to the other (Liao and Wolf, 1982).

In a few cases it has been possible to correlate the catalytic activity with the intensity of a particular XPS signal. Brinen and Melera (1972) correlated the activity of a Rh/C catalyst with the ratio of Rh in oxidized to metal form. Several studies of adsorbate metal interaction in clean metal surfaces have been published in connection with this application.

In closing this brief introduction of the utilization of XPS to study catalysis, the main applications are summarized. XPS is a useful tool in qualitative analysis of surfaces to identify contaminants and promoters, and to determine the oxidation state of metals and examine metal-support interactions in supported catalysts. Accurate absolute quantitative analysis requires a model of the surface layer, a difficult task with supported catalysts. However, relative corrected integrated intensity ratios are a useful clue to understanding surface enrichment and surface composition in polymetallic catalysts.

Nearly all catalyst applications require careful handling of the sample, including in situ pretreatment in a preparation chamber attached to the spectrometer. A comparison of XPS with other surface analysis techniques (Powell, 1982) concludes that XPS analysis causes the least damage to the sample during analysis and is the technique for which the most extensive procedures for qualitative and quantitative analysis have been reported.

Auger Electron Spectroscopy

Another tool now routinely used for surface analysis is Auger electron spectroscopy (AES). While the application of XPS to characterize catalysts has now become well established, AES has been a useful tool in applications involving flat surfaces such as single crystals. For this reason, less space will be devoted to AES than to XPS. However, since most of these techniques are in constant development, the potential applications of AES to catalysis are still quite open. As in the case of XPS, the principles of AES are briefly described, but emphasis is given to the possible applications of AES to heterogeneous catalysis. The reader is referred to the various review articles and books (Palmberg, 1975; Hercules, 1978; Chang, 1974; Davis et al., 1976; McGuire, 1979; Bhasin, 1976) on the subject for a more detailed theoretical discussion.

Auger Effect

Auger electron spectroscopy (AES) is named after its discoverer, Pierre Auger, who in 1925 discovered in a Wilson chamber that a two-electron emission occurred when argon atoms were excited by x rays. One of the electrons was a photoelectron ejected as described in Fig. 5a in the case of photoelectron emission. The second electron, known as the Auger electron, originates from another deexcitation process, which competes with photoelectron emission. When a primary excitation process, such as x ray or electron bombardment, causes the creation of a vacancy in an inner electron shell, for example, the K shell, deexcitation of the ionized atom can occur via the dropping of an electron from an outer level (i.e., L_1) to the vacated inner level. The excess energy is equilibrated by ejection of the second electron from an upper energy level (i.e., L_2); this Auger transition is then known as KL_1L_2. The kinetic energy of the ejected Auger electron is approximately equal to $E_K - E_{L_1} - E_{L_2}$ and is characteristic of the irradiated element. Moreover, the energy of the Auger transitions is independent of the energy of the incident beam. The number of Auger transitions possessed by atom increases with the atomic number of the element. An atom generally possesses one or two strong Auger transitions.

Whereas Auger electrons can be generated by using both x rays and electron beams, the latter mode of operation has become the most common type used in AES because the capability of electron beams to be focused in a very small surface area (10 to 100 μm in diameter) provides better spatial resolution than do x rays. X-ray excitation is only used in conjunction with XPS analysis. An Auger spectrum consists of a plot of the number of electrons emitted, N(E), as a function of energy. Auger emission occurs accompanied by secondary and scattered primary electrons; consequently, Auger transitions are not very prominent in such N(E) versus E plots. To enhance the Auger signal and remove the background caused by other electrons, the derivative dN(E)/dE is usually plotted as a function of E. Figure 10 shows both integral and differential spectra for Ag. The minimum of the differential signal is usually used to characterize the energy of the Auger transitions. The Auger spectra of all the elements (except H_2 and He) have been obtained and compiled (Davis et al., 1976).

Surface Sensitivity

Similar arguments given for XPS photoelectron escape depth are also valid for AES. The literature indicates that for Auger electrons with energies between 50 and 2000 eV, the escape depth is of the order of 4 to 20 Å (Palmberg, 1975). For most elements with Auger transitions below 1000 eV, the escape depth is only a few atomic layers. The greater surface sensitivity sometimes cited for AES versus XPS results because many elements have Auger lines at 100 eV kinetic energy, the minimum in Fig. 7.

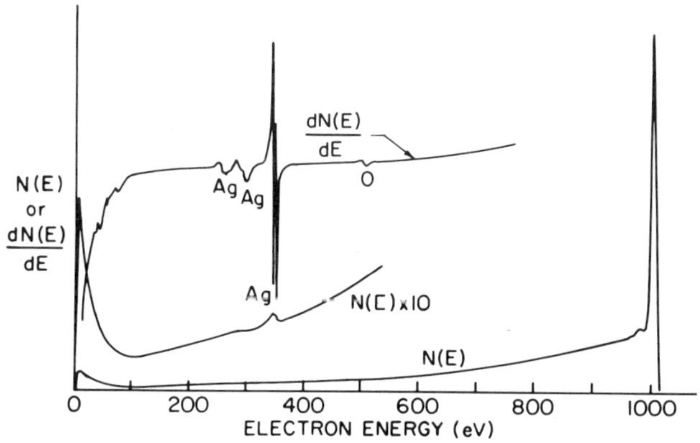

FIGURE 10 Secondary electron energy distribution curves from a silver sample: (A) N(E) versus E; (b) 10 × N(e) versus E; (c) dN(E)/dE versus E. (From Bhasin, 1976.)

Chemical Shifts

As in the case of XPS, chemical shifts have also been observed in AES. However, their interpretation is not well understood at present. Chemical shifts of many light elements ($Z < 40$) have been observed and have been related to oxidation state. In general, Auger lines involving only core electrons are easiest to interpret. In some cases the shape of the Auger peaks is also affected by structural factors; for example, carbide and graphitic carbon peaks deposited on stainless steel are strikingly different (Siegbahn et al., 1967). Such lines are coupled to the valence level and thus carry detailed bonding information. Complex Auger line shapes can also be used to fingerprint the identity of adsorbed molecules (Grant and Hooker, 1976).

Scanning Auger

The ability to focus an electron beam to a very fine diameter (ca. 5 μm) which can be scanned across a surface as in scanning electron microscopy is the basis for scanning Auger electron spectroscopy. Most AES instruments are equipped with low-resolution SEM (2000× magnification) capabilities to select an area for analysis. Furthermore, by locking the voltage of the energy analyzer (which is of the same type as in XPS), the detector only receives electrons coming from the selected element. This capability permits one to obtain a map of the element as the electron beam scans a selected area on the sample under analysis. This capability is referred to as a scanning auger microprobe (SAM) and is particularly useful in microelectronics.

Depth Profiling

SAM systems can be equipped with an Ar ion gun that can be used to impact (sputter) the surface with high-energy ions that remove atoms from the top surface layer. If the electron beam is also directed to the same area, it is possible to analyze the composition of the freshly exposed sublayer as a function of depth (depth profiling). This capability is important for detection of selective deposition of impurities, formation of surface layers, and, in general, to detect compositional changes with depth into the material (subsurface analysis).

Instrumentation

The basic components of an Auger electron spectrometer are those described for XPS except that an electron gun replaces the x-ray source with the optional addition of the ion gun for sputtering or etching, and an additional electron gun to obtain SEM images of the sample. Since the acquisition of a spectrum requires energy analysis of electrons, the same type of energy analyzer used in XPS (cylindrical mirror) can be used for AES. In fact, some manufacturers offer systems that can do both AES and XPS as well as other types of analysis (UPS, ISS, etc.).

As pointed out earlier, solid flat surfaces are ideal for AES analysis. In the case of powders, they must first be pressed into the shape of a wafer or a flat plate. Degassing and drying the sample is necessary to retain the UHV (10^{-8} to 10^{-9} torr) vacuum required in the analysis chamber. Unfortunately, most catalysts are supported on insulating oxides (Al_2O_3, SiO_2) which result in severe surface charging. Although some remedies for surface charging can be adopted, the problem is often so severe as to limit the applicability of this technique with technical catalysts.

Applications

In qualitative analysis, AES is used to survey possible contaminants and poisons that might affect the catalyst activity. The poisoning of a copper catalyst by lead, and of a $Pd/\gamma\text{-}Al_2O_3$ catalyst by iron, has been reported by Bhasin (1976). AES is used routinely in single-crystal studies to measure the cleanliness of such surfaces, in particular, to determine carbon and oxygen contamination, surface composition, and the presence of adsorbed layers (Chang, 1974).

As in XPS, quantitative analysis of powders is complicated by surface roughness and heterogeneity. Further complications arise from crystallite size distribution and nonuniform distribution of the metals in the support. Approximate analysis can be done in terms of relative intensities between the sample and a known standard. The atomic concentration of an element C_i in the sample is given approximately by

$$C_i = C_{i_{Std}} \frac{I_i}{I_{i_{Std}}} \qquad (16)$$

where I_i and $I_{i_{Std}}$ are the peak-to-peak Auger currents of the sample and standard, respectively, and $C_{i_{Std}}$ the surface concentration of the standard. Relation (16) assumes that escape depth from the sample and standard are similar and that the standard is homogeneously distributed in the sample. Auger sensitivity factors with respect to the sensitivity to oxygen have been given in the literature (Palmberg, 1975). These factors must be used when the concentrations of different elements are compared. Although many papers describing the use of AES in single-crystal studies have been published [about 300 are listed by Chang (1974)], papers in relation to applications of AES to heterogeneous catalysis are limited, presumably due to charging problems. A comparison of the various spectroscopic techniques for surface analysis and results compiled by the ASTM committee for surface analysis, have recently been published (Powell et al., 1979; 1982).

In summary, AES renders information similar to XPS, and similar problems can be studied with this technique. AES has a major advantage in lateral spatial resolution and sublayer analysis that is not yet available in XPS. As a disadvantage, AES presents serious charging problems with heterogeneous catalysts supported on insulating materials and chemical shifts are not as easily interpreted in AES as in XPS.

Ion Scattering Spectroscopy

In ion scattering spectroscopy (ISS), measurement of the energy of primary ions scattered by the solid sample provides the data from which sample composition is calculated. At medium and high energies (10^2 to 10^3 keV), ions penetrate deep into the solid. The elastic collision with an atomic core is governed by the mass of that atom, while interactions between the ion and electrons in the sample cause a loss of ion energy which is proportional to the distance traveled in the solid. At these energies, the scattering cross section is governed by simple nuclear potentials and the experiment is called *Rutherford scattering*. At energies from 0.5 to 10 keV, scattering cross sections and neutralization probabilities are so high that ions are scattered only by the first one or two atomic layers. Thus low-energy ion scattering is a true surface technique. With proper calibration, peaks in the scattered ion spectrum identify masses and amounts of elements present in the surface. This is a relatively new technique to catalysis, but in combination with ion etching offers both quantitative and structural analysis of complex surfaces.

Principles

The two important parameters in low-energy ISS (LEISS) are the energy and intensity of the scattered ions. Calculation of the mass from the energy of the peak is straightforward. Quantitative interpretation of intensity is more difficult.

The energy of the scattered primary ion is governed by momentum and kinetic energy conservation in the two-body collision between the primary ion and a single atom in the solid sample. The energy E_1 of the ion scattered at angle θ_L is given by

$$\frac{E_1}{E_0} = \left[\frac{m_1^2}{(m_1 + m_2)^2}\right]\left[\cos \theta_L + \left(\frac{m_2^2}{m_1^2} - \sin^2\theta_L\right)^{1/2}\right]^2 \tag{17}$$

where m_2 is the target atom mass, E_0 the initial primary ion energy, and m_1 its mass (Buck, 1975). At $\theta_L = 90°$, a common experimental value, Eq. (17) reduces to

$$\frac{E_1}{E_0} = \frac{m_2 - m_1}{m_2 + m_1} \tag{18}$$

The simplicity of this analysis is possible, in part, because of the relatively high energy of the collision. Multiple scattering effects do occur, but theory and experiment show that the binary collision mechanism dominates the scattering of the particles that remain ionized (Poate and Buck, 1976) and, except for possible resolution problems, Eq. (17) or (18) provides assignment of masses to the corresponding peaks. As is clear from these equations, m_1 can approach m_2 to improve resolution.

To be quantitative about surface composition, one needs either a theoretical description of the collision cross section and the ion neutralization probability along the ion path or a method to calibrate these effects empirically. The well-defined Rutherford scattering law, which describes the higher-energy cross sections, is complicated in the 1-keV region by the fact that the nuclear potentials are partially screened by the electrons. A Thomas-Fermi potential is often used to account for this screening effect (Poate and Buck, 1976), but others are also used (Taglauer and Heiland, 1976).

The ion neutralization process is equally important and difficult to model. The extra time spent by a low-energy ion in scattering from subsurface levels appears to be sufficient to allow substantial neutralization, preventing its detection. The charge transfer process depends on ion energy and identity as well as on substrate material and target atom. At present there is not a theoretical method that predicts these effects reliably. Thus one must turn to experimental calibration methods.

If the relation between surface composition and ISS signal is linear for a given primary ion and energy, calibration is straightforward. Linear relationships have been found for alloys (Smith, 1971) and for sulfur and oxygen on nickel (Taglauer and Heiland, 1974, 1975). Surface properties such as work function can be expected to affect ion neutralization, however, and nonlinearities in calibration functions have been observed (Niehus and Bauer, 1975). In practice, some of the most useful information is obtained by observing ion ratios as a function of the number of surface layers removed by ion etching. Changes in ion ratio are taken to indicate changes in relative concentration (Taglauer and Heiland, 1975). This approach assumes that any changes in cross section or neutralization probability accompanying chemical changes from layer to layer cancel out in the ion yield ratio.

Experiment

The essential components of the ISS instrument, shown in Fig. 11, are the ion beam, the sample, and the energy analyzer, all enclosed in a UHV scattering chamber. The 0.5- to 5-keV He, Ne, or Ar beams used for low-energy ISS are typically formed by electron impact on the gas at 5×10^{-6} to 10^{-3} torr (Buck, 1975). The ion gun is differentially pumped, delivers a beam with a stable ion current of 10 to 200 nA, a spot size on the order of 1 mm in diameter, and an energy spread of the order of 1 eV. The beam may also be mass filtered. The scattering chamber is a typical ultrahigh-vacuum bell jar with a base pressure of $\leqslant 10^{-10}$ torr and operating pressure $<10^{-8}$ torr. As with other particle spectroscopies, an attached catalyst pretreatment chamber is especially useful. The sample is mounted in the scattering chamber on a manipulator to facilitate other analytical experiments in the same instrument and possible angular-dependent studies. Insulator samples can accumulate a positive charge during

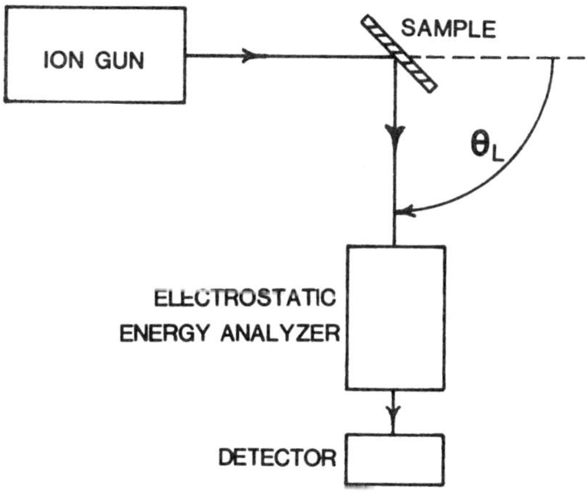

FIGURE 11 Schematic of an ion scattering spectrometer.

an experiment. This charge is removed by the electron flooding technique familiar in XPS and SIMS. Ion etching of the surface can be done with the primary beam itself or, preferably, with a second higher current beam. In any case it is important to avoid edge effects when substantial etching is done. Generally, one monitors only a small central portion of the etched region.

Electrostatic ion energy analyzers are used in low-energy scattering. Energy resolution $\Delta E/E$ on the order of 1% is typical. Thus Auger or XPS electron monochrometers are candidates for this task, provided that collimation can establish an unique scattering angle θ_L.

Applications

The power of this method is its ability to give quantitative or at least semiquantitative analysis of the uppermost layer of a surface. Its surface sensitivity surpasses both XPS and AES, and this makes the tool especially well suited to studies of phenomena such as surface enrichment in alloys and the structure of the upper layers of more complex surfaces. The work of Brongersma et al. (1978b) nicely illustrates the use of low-energy ISS for surface enrichment studies. They show strong surface enrichment in Cu for both CuNi and CuPt and illustrate effects of temperature in annealing and quenching surface compositions. Analysis of the data confirms the efficacy of calibration by the pure compounds and shows the necessity for truly clean surfaces since adsorbed layers can differentially shadow metal atoms. More direct applications in catalysis have emphasized mixed oxides and supported oxides. ISS analysis of the transition metal cation content of spinel surfaces has been found by Shelef et al. (1975) to be in good quantitative agreement with titration of the surface by NO adsorption. Studies of supported CoMo (Delannay et al., 1980a,b) and NiMo (Knozinger et al., 1981; Canosa Rodrigo et al., 1981; Abart et al., 1982; Jeziorowski et al., 1983) oxides on alumina offer excellent examples of the ability of ISS, coupled with ion etching, to examine surface structure in these hydrosulfurization catalyst precursors. For the NiMo system, Mo/Al and Ni/Mo intensity ratios as a function of etching time indicate essentially single-layer coverage of Al_2O_3 by Mo with Ni concentrated at the Mo-Al interface after calcination at 870 K (Jeziorowski et al., 1983). The ISS spectra shown in Fig. 12 show a low Al intensity at the outset, indicating that the Al_2O_3 surface is covered, and a maximum in the Ni/Mo intensity ratio with sputtering time, indicating the enhanced concentration of Ni below the surface. The order of addition of Ni and Mo was also found to affect the position of the Ni (Abart et al., 1982). If Ni was impregnated on the Al_2O_3 first, then the Ni/Mo intensity increased with etching time, showing Ni to be enhanced near the Mo/Al_2O_3 interface. If Ni was added after Mo, its concentration was maximum at the outermost surface of the catalyst. Applications to catalysis also include studies of Bi_2MoO_6 (Brongersma et al., 1978a) and individual metals of Al_2O_3 (Wu and Hercules, 1979; Chin and Hercules, 1982a,b; Zingg et al., 1980). The technique is relatively new, but progress is sufficient to mark the method as worthy of continued attention.

The features of this method that make it attractive for catalysis also apply in surface science. Here ordered or even just flat surfaces allow a more suitable use of the shadowing or blocking effects caused by the relative positions of atoms on the surface. The relative intensities of O and C, for example, show CO to bond carbon down on Ni (Smith, 1967).

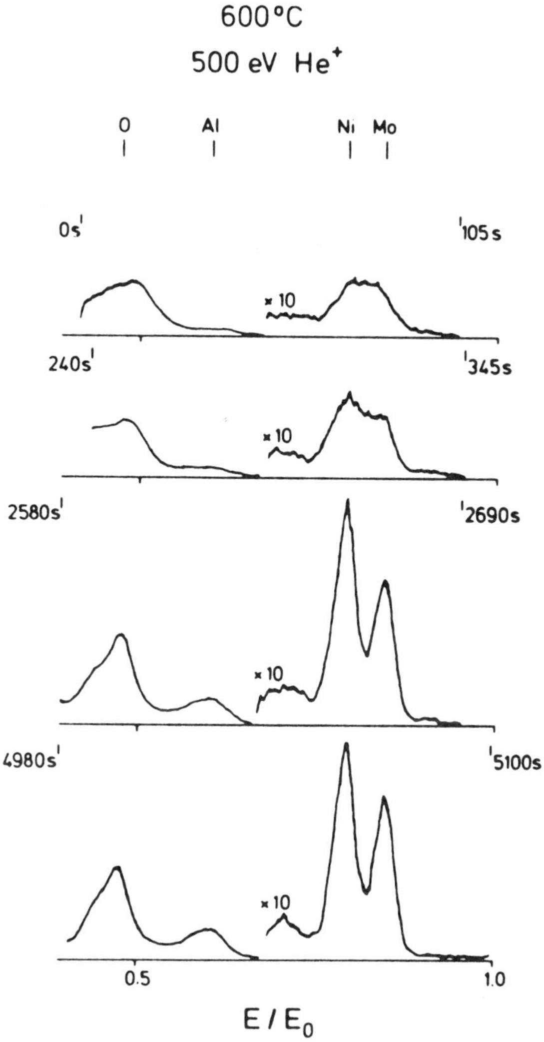

FIGURE 12 Ion scattering spectra of Ni3Mo12/Al_2O_3 as a function of increasing bombardment time. (From Jeziorowski et al., 1983.)

Quantitative analysis of shadowing effects on intensity as a function of azimuthal angle have also been used to determine the structure of ordered adlayers (Heiland and Taglauer, 1972).

Secondary Ion Mass Spectrometry

Secondary ion mass spectrometry (SIMS) is a particle spectroscopy using 200 to 5000-eV Ar^+ primary ions to bombard the sample surface and cause ejection of secondary ions which are detected by mass spectrometry. Figure 13 is a schematic representation of the experiment. The pressure in the spectrometer must be low enough to avoid significant ion/molecule collisions. Normal operating pressures vary from 10^{-6} to 10^{-11} torr.

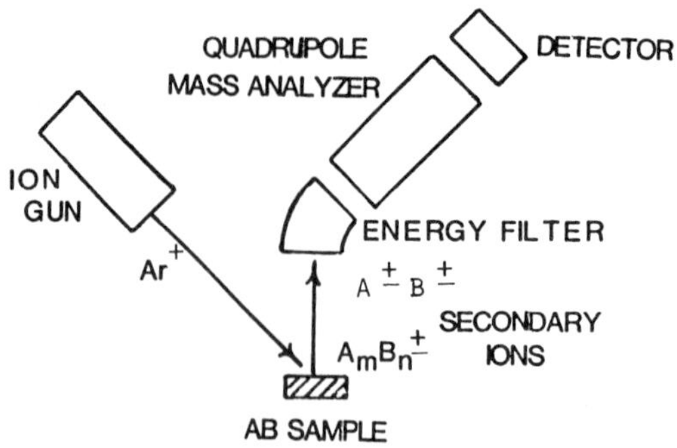

FIGURE 13 Schematic of a secondary ion mass spectrometer.

Materials scientists and engineers have for some time used dynamic (high primary ion flux) SIMS for depth profiling in metals and semiconductors (McHugh, 1975; Werner, 1975). Introduction of the static mode (Benninghoven, 1973), in which low primary ion fluxes cause removal of only a fraction of a monolayer during the experiment, has made SIMS a relatively recent entrant as a surface science and catalytic tool. The unique feature of this method is that the combinations of elements that appear in the cluster ions recorded in the spectra are representative of local environments on the surface. Thus SIMS stands with EXAFS as one of the few methods with the potential to provide proximity, local composition, and ultimately, local structure information.

Principles

The primary ion brings a large amount of energy to a small surface region and starts a collision cascade that affects hundreds of atoms and can cause ejection of as many as 10 or more atoms from the surface. The energy spectrum of the secondary ion peaks at about 2 eV, so only a small fraction of the primary energy finds its way to the secondary ions that comprise the SIMS spectrum. There are, of course, some high-energy secondary and primary ions, but these are filtered out and not detected. A thermodynamic model of the SIMS process appears to be successful in quantitative analysis of dynamic SIMS (Anderson and Hinthorne, 1973). A complete, quantitative foundation for static SIMS has not yet emerged, but many characteristics of the process are known. The secondary ion yield has both kinematic components, which account for energy transfer in the collision cascade, and electronic components, which govern the probability that an ejected particle will be ionized or remain ionized until detected.

The kinematic effects have been modeled by classical dynamic calculations which include all pairwise interactions in an integration of the equations of motion describing the collision of the primary ion with the solid (Harrison et al., 1978; Winograd and Garrison, 1980). These calculations show that for metals and metal/adsorbate systems, cluster ions form primarily over the surface by recombination of atoms that were close neighbors but not necessarily contiguous (Garrison et al., 1978). Strongly bonded species such as CO or C_6H_6 are shown to be ejected intact, however (Garrison et al., 1980; Garrison, 1980). The clearest demonstration of strong electronic influences on the SIMS spectrum is the observation that positive monomer ion yields from a metal can change by more than three orders of magnitude when oxygen is chemisorbed on the surface. The presence of oxygen raises the work function, favoring positive charge emission and retarding electrons. The importance of initial ion charge (Day et al., 1980) and electron exchange in the near-surface region (Lin and Garrison, 1982) has also been demonstrated.

Perhaps the most important issue concerning SIMS is the degree to which structural information is preserved in the spectrum. Extensive work on molecular solids shows that when intramolecular bonding is stronger than bonding between molecules, emission of intact molecules is the rule rather than the exception (Day et al., 1980). Spectra of materials such as quaternary amines show intact emission and some gas-phase ion fragmentation but little evidence for molecular damage during emission (Day et al., 1979). Highly ionic solids can give large ions (Barlak et al., 1982), while recombination over the surface is the dominant cluster ion mechanism for metals (Garrison et al., 1978). The utility of the SIMS method for identification of molecular species and element proximity on surface will continue to increase as the reliability of structural interpretation advances. Significant advances in understanding and control of the ionization process will be needed before the method can be truly quantitative.

Experiment

A SIMS spectrometer consists of an ion gun, sample, secondary ion energy filter, mass analyzer, and detector, all of which are enclosed in an ultra-high-vacuum chamber and controlled by a digital data acquisition system. The ion gun is differentially pumped and can deliver ions at energies from a few hundred eV to about 5000 eV at currents ranging from 10^{-10} to 10^{-5} A. The beam diameter at the sample is of the order of 1 mm. In some cases the beam is mass filtered and may be rastered over the surface. The sample is mounted on a manipulator. It is not uncommon to have other analytical devices, (e.g., AES, XPS) in the chamber and the sample is moved from one analysis position to another on the manipulator. Ability to heat and cool the sample in the analysis position and to dose adsorbing gases on it is highly desirable. For catalyst studies, a pretreatment chamber designed for clean transfer to and from the main chamber is a necessity (e.g., Ott et al., 1979). Catalysts can then be pretreated or reacted just prior to analysis. Insulating samples such as supported catalysts often charge up during an experiment. Such charging seriously degrades the spectrum and must be avoided. Electron-beam charge compensation is often an acceptable solution (Witmaack, 1979; Reuter et al., 1980). In some cases charging can also be avoided by proper sample preparation. Burnishing a very thin sample layer onto a metal foil, mixing the sample or with a conductor such as metal powder or graphite, scratching the surface of a thick coating to expose some of the metal backing plate will have all been successful (Day et al., 1980).

The prefilter conditions the ions for optimum mass analysis, usually by a quadrupole filter. In fast ion bombardment (FAB), the primary ion beam is neutralized so that the sample can ride at a high potential and a high-resolution magnetic mass analyzer can be used (Barber et al., 1981). Secondary ion mass spectra are often intense enough to be collected in an analog mode, but capability for pulse counting is necessary for the study of weak signals. The time needed to collect a spectrum depends on many factors but is generally of order 1 min to 1 hr.

Added features especially useful in surface science applications are energy selection of the secondary ions and capability of azimuthal and polar angle resolution of the secondary ions (Winograd and Garrison, 1980). In catalytic experiments, good charge compensation and convenience and effectiveness of sample pretreatment are prime considerations.

Applications

Since this technique is still developing, SIMS applications are seldom routine. The strengths of the method are (a) high sensitivity (ca. 10^{-6} monolayer in the highly favorable case of alkalai metals) and restriction to the uppermost surface layer, (b) ability to detect hydrogen, (c) sensitivity to isotopes, (d) detection of adsorbed molecular species, (e) indication of element proximity on a multicomponent surface, and (f) the possibility of structural analysis. The main weaknesses of the method are the difficulty of quantitization and the characteristic of most particle spectroscopies that they require low ambient pressure and see only the external surface of a porous catalyst.

Applications of SIMS to catalysis have been relatively few, but $FeRu^+$ ions have indicated the formation of FeRu alloys on SiO_2 (Delgass et al., 1982). For cobalt on alumina catalysts, $AlCo^+$ ions have shown the intimate mixing and low coverage of Co on the Al_2O_3 at low Co loadings (Chin and Hercules, 1982b). These ions were replaced by CoO_2^+, characteristic of Co_3O_4, when the Al_2O_3 surface was heavily loaded with Co. Molecular species on catalyst surfaces have shown up as a variety of C_2 to C_6 carboxylates in the negative-ion spectrum of a commercial silver catalyst used for ethylene oxidation (Benninghoven, 1973) and as the protonated molecular ion for isobutene and pyridine adsorbed on Si and Al oxides (Schumacher, 1983). Structural analysis of oxides, potentially a very interesting area, has begun with evaluation of MgCuAl oxides (Barber et al., 1976) and Fe aluminates (Buhl and Preisinger, 1975). The spectra have been interpreted as indicating oxide structure. The potential of the method for analysis of heterogenized homogeneous catalysts have been demonstrated by the observation of structurally significant secondary ions from transition metal complexes (Pierce et al., 1982).

Studies on well-defined metal surfaces show the utility of the method for surface reaction studies. Drechsler et al. (1979) have used SIMS to identify NH as the most abundant surface intermediate in NH_3 decomposition on Fe(110). Other studies identify surface species present during decomposition of formic acid on nickel (Mohri et al., 1978) and during flash desorption of C_2H_4 from Ru(001) (Lauderback and Delgass, 1984). SIMS has also confirmed the ethylidyne structure for ethylene adsorbed on Pt (111) (Creighton and White, 1983). Figure 14 shows direct SIMS observation of associative and dissociative adsorption of CO on Ru(001) (Lauderback and Delgass, 1983). The molecular state is indicated by $RuCO^+$ and Ru_2CO^+ ions. Finally, the angular-dependent SIMS measurements of Winograd and coworkers for adsorbate/metal crystal systems show strong angle-dependent variations in SIMS intensity characteristic of surface symmetry (Gibbs et al., 1982). These measurements, together with the classical dynamic calculations, suggest new SIMS methods for detailed surface structure determinations (Winograd, 1981).

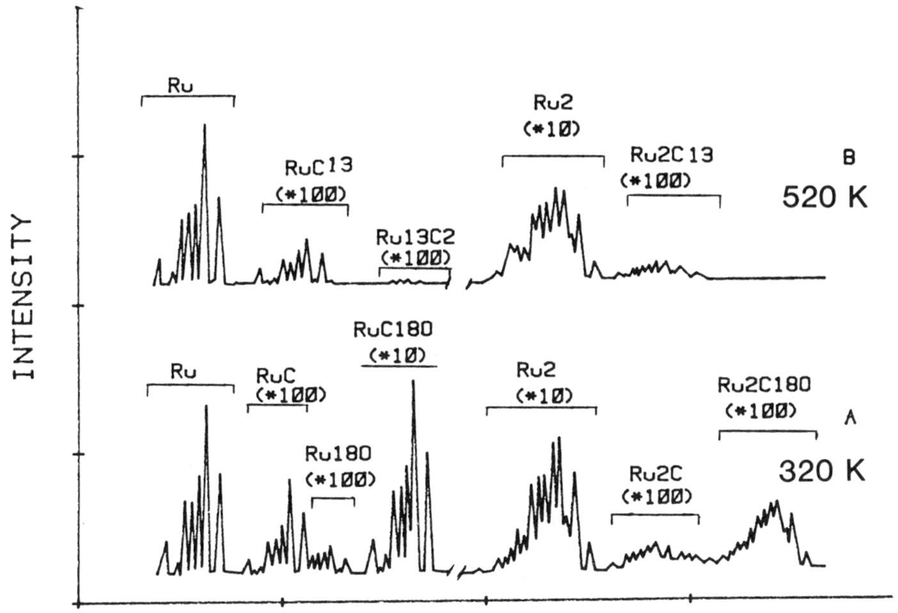

FIGURE 14 Positive-ion SIMS spectra of CO on Ru(001) after (A) 9.0L $C^{18}O$ at 320 K, and (B) 480L ^{13}CO at 520 K. (From Lauderback and Delgass, 1983.)

Extended X-ray Absorption Fine Structure

Active sites on a catalyst surface are defined by local structure. The average surface composition of a catalyst, as determined by XPS, Auger electron spectroscopy, or ISS, is information important in differentiation of catalysts and in following changes occurring during activation and deactivation. These tools can only imply local structure, however. A more direct measure is needed. Extended x-ray absorption fine structure (EXAFS), although not the complete answer to this local structure need, offers information on atomic arrangements in catalysts. The method has several important characteristics. It is element specific (i.e., one can individually examine x-ray absorption edges characteristic of the various elements in a multicomponent catalyst) and does not require crystallinity. Analysis of the extended fine structure of these edges gives the interatomic distances, to ca. 0.01 Å for subsequent shells; the occupation of the shells (coordination number), with a current accuracy not better than ±20%; and the mean-square displacement of the atom in question (Joyner, 1980). Although a relatively new and not yet fully developed technique, EXAFS has shown low coordination numbers but relatively unchanged bond distances for highly dispersed supported metals (Via et al., 1979), confirmed the Cu surface coating model for supported CuRu clusters (Sinfelt et al., 1980), and shown sulfided $Mo(Al_2O_3$ and $CoMoAl_2O_3)$ to contain small domain MoS_2 structures (Clausen et al., 1981a). In the United States, experiments are best done at the synchrotron radiation facilities at Stanford, Cornell, or Brookhaven (Rowe, 1981).

Principles

EXAFS begins with the measurement of transmitted x-ray intensity as a function of frequency. From these data one can calculate the x-ray absorption cross section as a function of energy. Each time the energy reaches the threshold for photoemission of a core electron in the sample material, the absorption cross section increases dramatically. Thus a broad scan of energy, as shown in Fig. 15, produces an absorption spectrum with steps or "edges" characteristic of the binding energies of core electrons in the sample. Except for accidental overlaps, each edge corresponds to a particular element. Close examination of a given edge shows that for several hundred eV to the high-energy side of the steps, the absorption initially oscillates with energy. The oscillations beginning 50 to 100 eV above the edge, the extended fine structure, result from the constructive and destructive interference between the outgoing photoelectron wave and its component backscattered from near neighbors. A pebble dropped among reeds in a still pond produces a classical analog of this effect.

Because of its origin, the detailed structure of the EXAFS spectrum contains information on the number and location of the neighbors surrounding a particular type of atom. Extraction of this information is complex but tractable and still developing (Teo and Joy, 1981; Hayes and Boyce, 1982; Einstein, 1982; Sandstrom and Lytle, 1979). First the nonoscillatory part of the absorption is removed by curve fitting and the EXAFS function x(k) is formed from the equation

$$x(k) = \frac{\mu - \mu_0}{\mu_0} \tag{19}$$

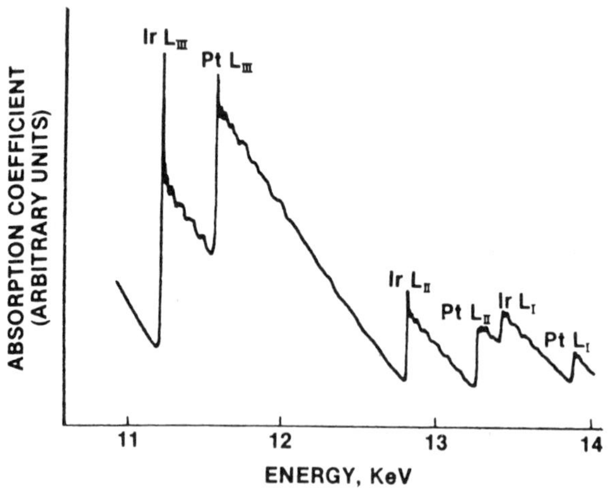

FIGURE 15 X-ray absorption spectrum at 100 K in the region of the L absorption edges of Ir and Pt for a PtIr-SiO_2 catalyst containing 10 wt % of each metal. (From Sinfelt et al., 1982.)

where $k = (2mE)^{\frac{1}{2}}/\hbar$, m is the mass of an electron h is Planck's constant by 2π, E is the kinetic energy of the photoelectron, μ the natural log of the ratio of the incident and transmitted x-ray photon flux, and μ_0 the absorption coefficient of the free atom. Theories based on single scattering of the photoelectron by atoms in the coordination shells surrounding the absorbing atom give

$$x(k) = \sum_j A_j(k) \sin[2kr_j + \delta_j(k)] \quad (20)$$

where the summation is over the coordination shells, r_j is the radial distance from the absorbing atom to atoms in the j shell and $\delta_j(k)$ the total phase shift for emission and scattering, $A_j(k)$ is an amplitude function that depends on the number of atoms in a given shell; r_j, a backscattering amplitude; and the root-mean-square deviation about r_j. With proper accounting for phase shifts and the k = 0 point, Fourier transformation of a k^n weighting of x(k) (Stern et al., 1975), produces a radial structure function that reflects the positions and number of atoms for each coordination shell as shown in Fig. 16. These are the data from which initial structural assignments are made. One method of refinement of structural parameters is to retransform a small region of radial space (say, the nearest-neighbor-shell) back into k space. The process acts as a Fourier filter which removes all but the nearest-neighbor contribution to the x(k) function. Nonlinear least-squares techniques can then be used to fit the structural parameters to this function (Via et al., 1981).

This discussion shows the origin of the EXAFS function and the radial structure function. Further details of the theory and data reduction procedure are given in the references cited above, but a few additional comments about the technique are included here. The EXAFs phenomenon results from backscattering of the excited photoelectron. X-ray absorption is one of several ways to detect the occurrence of this effect. Alternative ways of detecting x-ray absorption include x-ray fluorescence and Auger electron measurements. Stimulation of core electron emission may also be by electron bombardment (Hayes and Boyce, 1982; Einstein, 1982). The low mean free path of electrons in solids offers opportunity for EXAFS analysis accentuating surface structure (Landman and Adams, 1976; Einstein, 1982). Among the avenues opened by the use of electron stimulation is the observation of extended electron energy loss fine structure (EXELFS) in an electron microscope. This experiment gives both local structure and high spatial resolution over a sample (Leapman et al., 1981).

Experimental Methods

The alternative approaches mentioned above may eventually provide routine access to EXAFS-type information. X-ray absorption measurements can be made with conventional x-ray sources (Knapp et al., 1978; del Cueto and Shevchik, 1977) but counting times are generally long compared to those obtained with the intense beams available from synchroton radiation (Hayes and Boyce, 1982). Since absorption must be scanned as a function of energy, a broad continuous band of radiation is needed and the sharp emission lines of the conventional x-ray source must be avoided. A synchrotron or storage ring produces this radiation by the curved orbit of a beam of high-energy electrons. This, or the Bremstrahlung

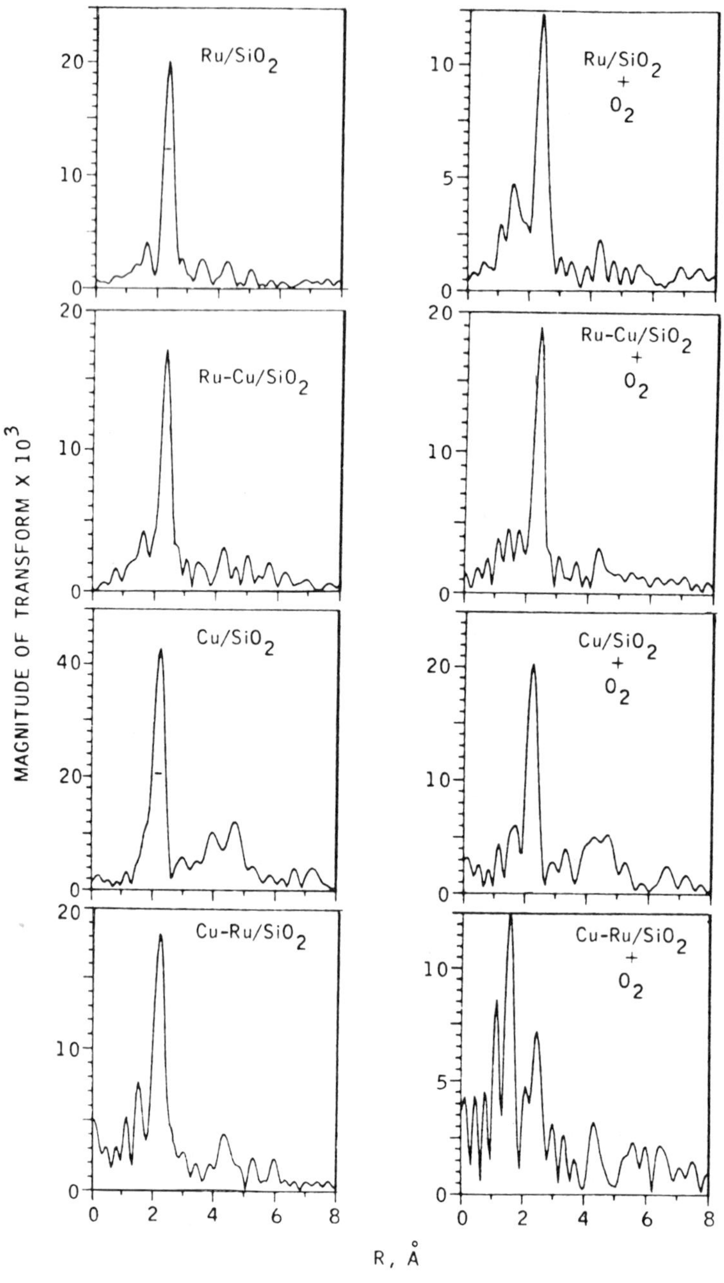

FIGURE 16 Effect of exposure of silica-supported ruthenium, ruthenium-copper, and copper catalysts to oxygen (1% oxygen in helium) at room temperature on the properties of the catalysts as shown by Fourier transforms of EXAFS data at 100 K for the catalysts in the absence and presence of oxygen. The transforms in the upper half of the figure are for the K absorption edge of ruthenium. Those in the lower half are for the Cu K edge. (From Sinfelt et al., 1980.)

background radiation of an x-ray tube, is passed through a monochromator and then through an incident flux detector, the sample, and a transmitted flux detector. X-ray energy is scanned with the monochromator. The raw data, transmitted intensity ratio as a function of energy, are then treated as described in the preceding section. As with any work on catalysts, proper cells for pretreating the sample and maintaining its environment during the measurement are essential (Via et al., 1979). Samples are often maintained at low temperatures (80 to 100 K) to decrease the mean-square displacement of the atoms and improve resolution.

Applications

EXAFS structural studies of interest in catalysis have ranged from examination of supported metal carbonyl clusters (Besson et al., 1980) to analysis of transition metal coordination in zeolites (Morrison et al., 1980). To illustrate the method, we focus on applications to supported metals and HDS catalysts.

In a series of papers, Sinfelt, Via, and Lytle report EXAFS of supported metals and bimetallic clusters. Studies of OS, Ir, and Pt on silica and alumina (Via et al., 1979) utilized samples with 1 wt % metal loadings and percent metal exposed (dispersion) of 70 to 100% and compared results to values for the pure metals. Nearest-neighbor distances for the supported metals were within 0.02 Å of the pure metal values. Coordination numbers for the nearest-neighbor shell were 7 to 10 for the catalysts, lower than the bulk value of 12 because of the high dispersion of the metals. The high surface-to-volume ratio of the metals also caused high root-mean-square displacement values for the catalysts compared to the bulk metals. In general, the close agreement between the interatomic distances for the catalysts and the bulk metal speaks against structural rearrangement of the small metal clusters (Burton, 1974). Distortions in structure have been observed, however. Moraweck et al. (1979) report Pt nearest-neighbor distances shortened by 0.12 Å for 12-Å Pt particles in Y zeolite. Interestingly, this value returned to the Pt-Pt bulk distance after adsorption of hydrogen. Joyner (1980), in discussing the data of Via et al. (1979), notes that the radial structure functions for Pt and Ir on silica are consistent with three-dimensional face-centered-cubic particles, but Ir and especially Pt on Al_2O_3 appear to have variations in peak intensities beyond the first neighbor distance that are best described by rafts with (100) orientation.

EXAFS of supported bimetallic clusters has included the RuCu (Sinfelt et al., 1980) OsCu (Sinfelt et al., 1981), PtIr (Sinfelt et al., 1982), and RhCu (Meitzner et al., 1983) systems. Here a new dimension is added to the power of the technique because EXAFS of edges corresponding to each element can be compared to analyze element proximity. For both RuCu and OsCu, where the metals are essentially immiscible, copper coats the group VIII metal. Ru(Os) in these catalysts has interactomic distances corresponding primarily to Ru(Os) neighbors. The Cu, on the other hand, is coordinated extensively to Ru(Os) as well as to Cu. Figure 16 shows the effects of O_2 on the transform of the weighted EXAFS function for both Cu and Ru. Note that the data are taken at low temperature to improve resolution. The intensities of lines below 2 Å are suppressed in this data because of the restricted region of k space used for the transform. Thus the interpretation of the results is based mainly on the major metal-metal nearest-neighbor peak. For Ru/SiO_2, oxygen exposure at room temperature significantly reduces the number of Ru-Ru neighbors. This peak is essentially unchanged after oxygen exposure of the $RuCu/SiO_2$ catalysts. Thus Cu shields Ru from oxidation as expected from the Cu coating model. The copper data in the bottom four spectra of Fig. 16 show that Cu in Cu/SiO_2 is only partially oxidized by O_2 (the core of the particle is protected by an oxide skin) but that Cu in the thin layer on Ru in $CuRu/SiO_2$ is completely altered by oxygen exposure. The RhCu and PtIr systems also show surface segregation phenomena but indicate more intermixing of the elements, in accord with the greater miscibility of these systems. For a 50-Å metal particles in $PtIr/SiO_2$, x-ray diffraction is consistent with homogeneous alloy formation, but EXAFS shows the particles to consist of Pt-rich and Ir-rich regions.

The structure of supported cobalt-molybdenum hydrodesulfurization catalysts has been a topic of interest approached by many analytical techniques. Recent EXAFS studies have contributed significant new information. Clausen et al. (1981b) have used the Mo K edge to examine the local structure around Mo in sulfided Mo/AlO_3 and $CoMo/Al_2O_3$ catalysts. Comparison of results for well-crystallized MoS_2 and the two catalysts shows that Co has little effect on the Mo environment, the first coordination shell of Mo is essentially the same in all three materials, and the second coordination shell occurs at the same distance for the three samples but its population is significantly lower for the catalysts. MoS_2 shows structure at high interatomic distances, but the catalysts do not. The authors interpret these data to indicate that Mo has the same sulfur coordination in the catalysts as in MoS_2 and that the catalytic species is not an oxysulfide. The low amplitude of the peaks corresponding to larger interatomic distances is interpreted as indicating very small MoS_2 crystallites (on the order of 10 Å). Studies of EXAFS following the Co K edge for the sulfided $CoMo/Al_2O_3$ catalysts show that Co is surrounded by sulfur, but the lack of any more distant peaks suggests a highly disordered environment and that Co is located at the surface of the small MoS_2 crystallites (Clausen et al., 1981a). Finally, we note that EXAFS of Mo has also been used to confirm an MoS_2 structure for a propene methathesis catalyst prepared by reaction of $Mo_2(\pi\text{-}C_3H_5)_4$ with the OH groups of $\gamma\text{-}Al_2O_3$ (Sato et al., 1982).

At this time, catalytic materials most amenable to EXAFS study are the ones in which the element of intest is in a single environment. The structure and composition of the first coordination sphere is the information most readily attained. Developing experimental methods will undoubtedly widen the scope of this technique, while improved analysis increases the sophistication of interpretation and allows solution of increasingly complex structural problems.

Mössbauer Spectroscopy

The Mössbauer effect is a nuclear gamma-ray resonance with such high precision that spectral features reflect the chemical state of the corresponding atom. The requirement of a relatively low-lying nuclear excited state restricts the number of isotopes convenient and possible for use in catalytic experiments to those shown in Table 3. Isotopic abundances indicate the fraction of the naturally occurring element which is the Mössbauer active isotope. When the technique applies, it is a powerful method for

TABLE 3

Mössbauer isotope	Parent[a]	Parent half-life[b]	Mössbauer λ-ray energy (keV)	Natural abundance (%)
^{57}Fe	^{57}CO	270d	14.4	2.2
^{119}Sn	^{119m}Sn	245d	23.9	8.6
^{151}Eu	^{151}Sm	93y	21.6	47.8
^{121}Sb	^{121m}Sn	75y	37.2	57.2
^{125}Te	^{125}I	57d	35.5	7.0
^{127}I	^{127m}Te	105d	57.6	100
^{197}Au	^{197}Pt	20h	77.3	100
^{99}Ru	^{99}Rh	16d	90	12.7
^{193}Ir	^{193}Os	32h	73	62.7
^{195}Pt	^{195}Au	192d	98.8	33.8
^{83}Kr	^{83}Rb	83d	9.3	11.6
^{181}Ta	^{181}W	140d	6.25	100

[a]Other parents are sometimes available.
[b]d, day; y, year; h, hour.

measuring oxidation state, local symmetry, magnetic ordering, and crystallite size effects for specific Mössbauer elements bound to or within solid matrices. For isotopes not requiring low temperature in order to observe a spectrum, the technique can be applied in situ to catalysts operating in the reaction environment. Applications have included use of ^{57}Co, ^{119}Sn, ^{151}Eu, ^{121}Sb, ^{197}Au, and ^{99}Ru, but by far the largest number of studies have been on ^{57}Fe. The reader is referred to Dumesic and Topsøe (1976) and Delgass et al. (1979) for detailed reviews of catalytic applications, and to Wertheim (1964) and Greenwood and Gibb (1971) for reference books on the Mössbauer effect itself.

Principles

The Mössbauer effect can be thought of as the nuclear equivalent of optical spectroscopy. The source of radiation is generally a metastable isotope which decays to the nuclear excited state of the Mössbauer isotope. Decay from this excited state to the ground state emits the Mössbauer γ ray. Resonant absorption can then occur if this γ ray impinges on a nucleus of the Mössbauer isotope in its ground state. Since the γ-ray energies are high (10^4 to 10^5 eV), the Mössbauer atoms must be bound to a solid to prevent the recoil "kick" from destroying the resonance. The area under a Mössbauer absorption line is a measure of the probability of "recoil-free" emission and absorption, which in turn is related to the coupling of the Mössbauer atom to the solid. This relation is complex (Debrunner and Frauenfelder, 1971), but spectral areas can be used as indicators of changes in bonding of a Mössbauer atom to a catalyst matrix (Garten et al., 1970). Since the width of the Mössbauer line is of order 10^{-9} eV, the precision of the measurement is so high that it reveals the very small perturbations made on the nuclear energy levels by the atomic electrons. These nuclear hyperfine interactions project chemical information to the nucleus and hence to the Mössbauer spectrum.

Three spectral parameters are of particular interest. The isomer shift (IS), measured in simple cases by the centroid of the spectrum (Kalvius, 1971), indicates relative electron density at the nucleus. Only s and relativistic p electrons contribute directly to this term; other electrons affect its value because of shielding. For example, isomer shift between high spin Fe^{3+} and Fe^{2+} is about 0.8 mm/s or about three line widths. Fe^{3+} has a higher electron density at the nucleus than does Fe^{2+} because it has only five 3d electrons while Fe^{2+} has six to shield the 1s, 2s, and 3s electrons. Covalent bonding can complicate interpretation of isomer shifts of ^{57}Fe because the 4s character of the bond increases the electron density at the nucleus while non-s character decreases it. Nevertheless, assignment of oxidation state for iron in oxides and other ionic compounds is usually straightforward.

The second spectral parameter is quadrupole splitting (QS). This effect removes the magnitude but not the sign degeneracy of the nuclear spin states. In ^{57}Fe, for example, the $I = 3/2$ excited state is split into $I_z = \pm 3/2$ and $I_z = \pm 1/2$ states. The ground state, with $I = 1/2$, is unsplit. A quadrupole split spectrum for ^{57}Fe in an isotropic powered sample has two lines of equal density (Greenwood and Gibb, 1971). The size of the splitting depends on the nuclear quadrupole moment (a constant) and the electric field gradient (EFG) at the nucleus. The EFG is a measure of the symmetry of the electronic environment of the atom. It is zero for cubic and tetrahedral symmetry, but increases in magnitude for more highly asymmetric arrangements. There are two important contributions to the EFG. The lattice component arises from the surroundings of the atom. The electronic component is a result of the valence election configuration of the atom itself. High-spin Fe^{3+} has an electron EFG of zero. The extra d electron on Fe^{2+} causes a strong, temperature-dependent quadrupole splitting when ^{57}Fe is in a low-symmetry environment (Wertheim, 1964). The combined effects of isomer shift and quadrupole splitting are shown in Fig. 17, which illustrates oxidation/reduction of Fe/Y-zeolite (Garten et al., 1970). Spectrum 17a shows two ferrous states, with spectra corresponding to a large outer and smaller inner doublet. Spectrum 17b corresponds to Fe^{3+} and spectrum 17d to ferrous ions whose bonding to the zeolite has been weakened by hydration.

When the net electron spin of an atom is greater than zero and is fixed in space longer than 10^{-7} s, the nucleus feels an effective magnetic field and all the (2I + 1) nuclear sublevels are split. For ^{57}Fe, the selection rules allow six transitions with relative intensity 3:2:1:1:2:3 for an isotopic powder sample (Greenwood and Gibb, 1971). The six-line, magnetic dipole splitting is characteristic of iron in any magnetically ordered material. The splitting of the outer lines is proportional to the effective magnetic field at the nucleus, a parameter often useful for phase identification (Topsøe et al., 1973). Because the effective magnetic field is measured at the atomic site, antiferro and ferri- as well as ferromagnetic states are detected. Different crystallographic sites in a given

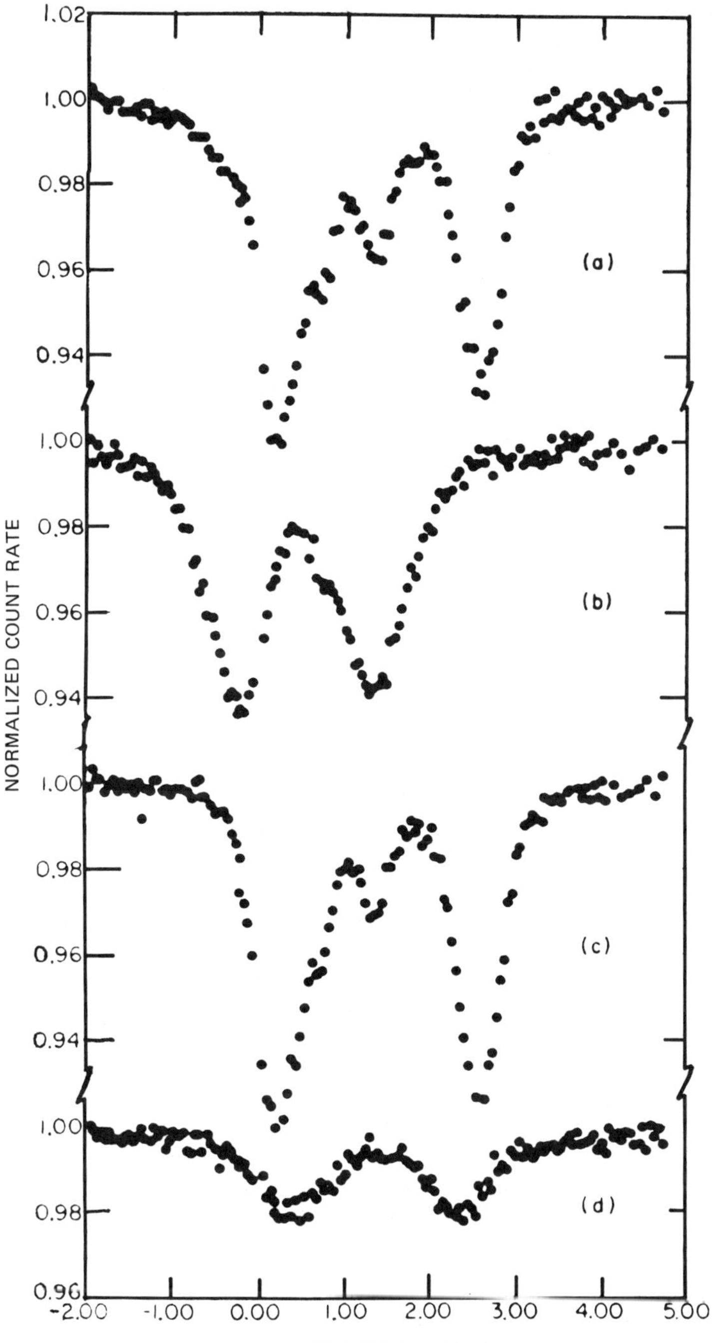

FIGURE 17 Oxidation/reduction of iron Y zeolite: (a) Fe^{2+}-Y dehydrated at 673 K; (b) oxidized at 673 K; (c) reduced in H_2 at 673 K; (d) exposed to 15 torr H_2O. All spectra at room temperature and on the same sample. (From Garten et al., 1970.)

material can also have different effective magnetic fields. This is the case for χ-carbide ($Fe_{2.5}C$) where three different sites produce 18 overlapping lines. When the volume of a magnetically ordered crystallite becomes small enough, thermal energy, kT, may be sufficient to flip the entire coupled spin system fast enough to cancel out the magnetic field at the nucleus. The Mössbauer effect shows this conversion from a magnetic to a superparamagnetic state when particle diameters are of the order 30 to 150 Å, depending on the material and the temperature of the measurement (Morup et al., 1980).

Experiment

In the typical Mössbauer experiment, γ rays pass from the source, through the catalyst absorber wafer, to the detector. The source is mounted on an electromechanical drive which imparts a back-and-forth motion with a carefully programmed velocity along the γ-ray path. The velocity causes a Doppler shift in the energy of the γ ray and is thus the means of scanning energy. A periodic triangular velocity versus time function is common. The Doppler shift corresponding to velocities ranging from −10 to +10 mm/s is sufficient to scan the full range of ^{57}Fe shifts and splittings. The detected count rate (number of counts in a fixed dwell time) is collected in a multichannel analyzer or minicomputer whose repetitive sweep is synchronized so that each channel corresponds to a given source velocity. By convention, spectra are plotted as percent transmission versus velocity, with positive velocity indicating motion of the source toward the absorber.

Sources must have a strong, narrow, single emission line. They consist of the parent of the Mössbauer isotope well bonded in a cubic matrix and are available commercially. ^{57}Co diffused into a Rh foil is typical for ^{57}Fe experiments. The detectors are usually proportional or scintillation counters, but solid-state devices are also used. Full systems, including the drive, alignment track, detector, data accumulation, and output device, and all associated electronics are commercially available. Absorber cells for control of the catalyst environment are also available, but are often custom made (Delgass et al., 1976; Shen et al., 1981; Clausen et al., 1979). The cell is a vacuum-tight enclosure with facility to heat and cool the catalyst wafer in a selected gas environment and Be windows for transmission of the γ rays. Special dewars are necessary if both source and absorber must be cooled to liquid-helium temperature, as for ^{99}Ru (Clausen and Good, 1977).

In the transmission mode discussed above, the optimum sample size is 10^{18} atoms of ^{57}Fe atoms per cm^2 for each line in the spectrum. A 5/8-in.-diameter wafer typically contains 0.1 to 0.8 g of catalyst. This value gives signals in the transmission range 90 to 99%, big enough to see, but not so big as to cause excessive broadening due to reabsorption (Margulies and Ehrman, 1961). Iron loading below 1 wt % usually require use of iron enriched in ^{57}Fe, available from Oak Ridge National Laboratories. The time required to obtain a spectrum depends on the strength of the source and varies from 10 min to 24 h, with 4 to 6 h being common. Transient effects are best followed by operating the spectrometer in the constant-velocity mode and following the intensity of a given peak as a function of time (Raupp and Delgass, 1979).

An alternative to the approach just described is the source mode. Here the catalyst must be made with the radioactive parent isotope. It is then placed in a reaction/pretreatment cell and used as the γ-ray source while the drive moves a standard single-line absorber to scan energy and generate the spectrum (Topsøe et al., 1981). Surface sensitivity of the Mössbauer effect can also be enhanced by using the backscattering geometry. Phillips et al. (1980) have used this method to follow the decomposition of $Fe(CO)_5$ on oriented graphite foils.

Applications

The Mössbauer effect is a seasoned tool in catalysis research and accounts for advances in the understanding of bulk solids (Topsøe et al., 1973), supported metals and alloys (Garten, 1976), oxides (Lund and Dumesic, 1982), sulfides (Topsøe et al., 1981), carbides (Amelse et al., 1978; Niemantsverdriet et al., 1980), catalyst genesis and activation (Boudart et al., 1977), oxidation-reduction behavior (Garten et al., 1970), support interactions (Tatarchuk and Dumesic, 1981), and the dynamics of some catalytic processes (Raupp and Delgass, 1979).

In deciding what information the Mössbauer effect can provide about a catalyst, one must bear in mind that the radiation penetrates the sample completely and, generally, all Mössbauer atoms in the sample contribute to the spectrum. As the fraction exposed (dispersion) of the phase containing the Mössbauer atom increases, the Mössbauer effect becomes more sensitive to the surface. Surface sensitivity is probably best checked experimentally by observing the spectral perturbations caused by adsorbed gases.

A few specific examples of Mössbauer studies are outlined here. Garten (1976), Garten and Sinfelt (1980), and Vannice et al. (1978), have shown that iron forms bimetallic clusters on SiO_2 and Al_2O_3 with a variety of metals, including Pd, Pt, Ir, and Ru. Lund and Dumesic (1982), have observed a support interaction between Fe and SiO_2 that involves SiO_2 migration over the iron phase, and Tatarchuk and Dumesic (1981), have reported one between Fe and TiO_2 involving dissolution of Fe in the TiO_2 phase. The role of Sn in a supported PtSn steam dehydrogenation catalyst has been carefully detailed by Gray and Farha (1976). The recent ^{57}Co source experiments of Topsøe et al. (1981) use the Mössbauer parameters to identify a mixed CoMo sulfide as the active site for hydrodesulfurization. It is of interest in this study that ^{57}Fe-doped into CoMo catalysts does not probe the correct cobalt chemistry (Clausen et al., 1978).

Figure 18 illustrates information obtained on supported iron-cobalt alloy catalysts by Stanfield and Delgass (1981). Spectra A to C show nearly complete reduction of iron since paramagnetic Fe^{2+} or Fe^{3+} states would contribute extra spectral intensity in the region 0 to 2 mm. The change in splitting of the outer lines shows that the effective magnetic field follows the expected composition variation and that iron-cobalt alloys with the nominal composition have been formed successfully. Other spectra show that improper pretreatment can completely prevent alloy formation. Spectra D to F show that increasing Co concentration supresses carburization of the catalyst during Fischer-Tropsch synthesis. The complex spectra of the carbides lie between -4 and $+4$ mm/s. Constant-velocity data taken in situ at the velocity of the rightmost Fe^0 line can give a dynamic trace of the amount of carburization as a function of time.

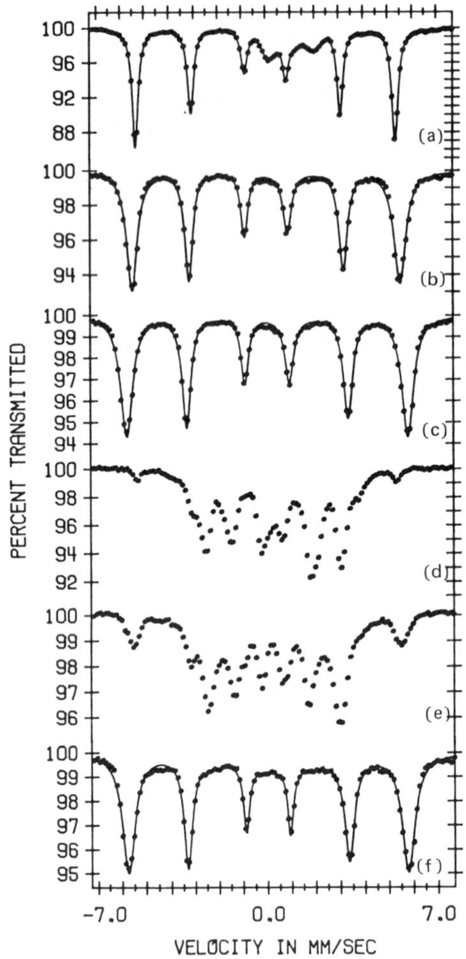

FIGURE 18 Room-temperature Mössbauer spectra of reduced and reacted vacuum-dried catalysts: (A) $10Fe/SiO_2$, reduced 8 h; (B) 8.95 Fe1.05Co/SiO_2, reduced 8 h; (C) 7.4Fe2.6Co/SiO_2, reduced 8 h; (D) $10Fe/SiO_2$, reacted 6 h; (E) 8.95Fe1.05Co/SiO_2, reacted 6 h; (F) 7.4Fe2.6Co/SiO_2, reacted 6 h. (From Stanfield and Delgass, 1981.)

Magnetic Resonance

Nuclear magnetic resonance (NMR) accurately measures the properties of a nuclear spin system. To an even greater extent than with the Mössbauer effect, these nuclear properties can reflect detailed chemical structure and bonding information. Electron spin resonance (ESR) measures the energies of the valence electron spin system directly and thus provides a sensitive measure of the chemical state of species having a net electron spin. In both measurements an applied magnetic field orients the spins, and transitions between spin states are observed by absorption of electromagnetic radiation. NMR absorptions are in the radio-frequency region, while ESR is in the microwave.

The high sensitivity of these two magnetic resonance techniques to subtle chemical changes makes the methods at once potentially powerful probes of chemical reactivity and techniques demanding thorough understanding of the underlying quantum mechanical effects in order to extract the full measure of information available. ESR has a long history of applications in catalysis. It has been successful in studies of the environment of paramagnetic transition metal cations in catalysts, in observing the chemical properties of electrons trapped in defects in oxide catalysts, and using the paramagnetic probe molecule NO, in examining the interactions of adsorbed molecules with surfaces (Lunsford, 1972, 1979a). Recently, ESR has been particularly important in identifying and characterizing O^- and other oxygen anions on catalytic surfaces (Che and Tench, 1982) and in examining mobility of adsorbed species (Howe, 1982).

NMR has made significant contributions to catalysis research both in the identification of chemical species and in measurements of species mobility (Lunsford, 1979b). Magic angle spinning (Andrew, 1981) and other recent advances in solid-state NMR make this technique one of the most promising for future catalysis research. High-resolution ^{29}Si NMR, for example, can quantitatively measure the numbers of Si ions with zero, one, two, three, or four Al next-nearest neighbors and thus has provided new understanding of Si-Al ordering in zeolites (Ramdas et al., 1981; Magi et al., 1981; Nagy et al., 1981). A complete review of magnetic resonance is well beyond the scope of this chapter. Our main purpose is to point out some of the most recent developments in NMR that suggest that an investment of time to understand magnetic resonance tools may be worthwhile.

Principles

The magnetic resonance effect is easiest to visualize for a spin-1/2 system. In the presence of a magnetic field, the magnetic moment associated with the spin will precess about the magnetic field direction. This motion is quantized and for a spin 1/2 system only two z components of the spin vector are allowed, the +1/2 and the −1/2 states. The energy of the spin system in the presence of the magnetic field is proportional to the product of the size of the field and the z component of the spin. Thus, as the magnetic field increases, the energy separation between the +1/2 and −1/2 states increases proportionately. At a given magnetic field strength, resonant absorption of energy will occur when the frequency ν of the applied electromagnetic radiation is such that $h\nu$ is equal to the energy separation between the +1/2 and −1/2 states, as shown in Fig. 19. In real systems spins are not isolated in free space, but reside in atomic or nuclear environments which are generally not isotropic and can provide a variety of perturbations that affect the energy of the system. In general, one must write down a spin Hamiltonian that describes the appropriate interactions and use it in fitting the data obtained.

In ESR the proportionality between the electron spin and the magnetic field is indicated by the g tensor. This tensor has three principal values, which deviate from the free electron value because of the interaction between the electron spin and orbital motion. Account must also be taken of the fact that the powder samples typical of catalysts have crystallites at all possible orientations with respect to the applied magnetic field (Lunsford, 1972). Hyperfine interactions, the coupling of electron and nuclear spins, cause additional splittings in ESR spectra and can be

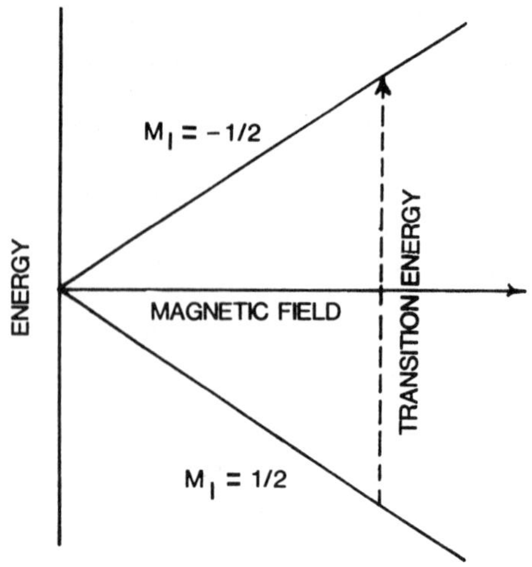

FIGURE 19 Splitting of a spin-1/2 nucleus in a magnetic field.

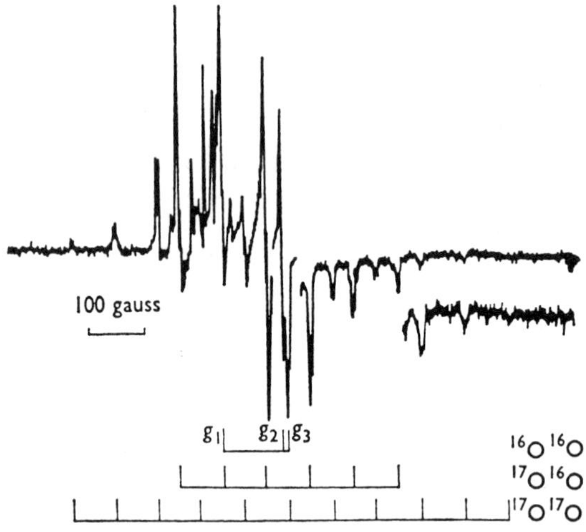

FIGURE 20 First derivative ESR spectrum of O_2 on the surface of high-surface-area MgO. The field increases from left to right and the gain has been reduced fivefold for the central line; a small portion of the high-field spectrum is overmodulated to show the outermost lines. For clarity no attempt has been made in this diagram to insert levels for the lines around g_1. (From Tench and Holroyd, 1968.)

helpful in defining the environment of an electron spin. As shown in Fig. 20, the isotope ^{17}O, with a nuclear spin of 5/2 shows 6 line splittings for $^{17}O^{16}O$ but 11 line splittings for $^{17}O_2^-$ (Tench and Holroyd, 1968). Note that the spectrum is displayed as the derivative of the energy absorption curve, as is standard in ESR spectroscopy.

NMR has sensitivity to chemical bonding through shielding effects. The more free electrons around the nucleus of interest are to circulate in response to the applied magnetic field, the more they decrease the strength of the field felt by the nuclear spin. The range of chemical shifts induced by this shielding is generally large with respect to NMR line widths in liquids, giving the method high chemical sensitivity. Spin-spin couplings between nonequivalent nuclei also occur, giving NMR further capability for defining the chemical environment of a given type of nucleus. In liquids, many interactions are averaged to zero because of the rapid motion of the molecules.

In solids, however, relaxation times are long and direct-dipole coupling of the spins and anisotropy of the chemical shift tensor both tend to broaden lines to such an extent that meaningful chemical shift information cannot be obtained from a standard spectrum. Analysis of these two interactions shows that if the sample is spinning on an axis 54°44' to the magnetic field direction, the static components of the dipole-dipole interaction and the chemical shift anisotropy both become zero (Andrew, 1981). Sidebands, whose positions depend on rotational frequency, still appear but do not obscure the spectrum. This technique, called *magic angle spinning*, dramatically improves the resolution of NMR of solids (Andrew, 1981). Further improvements can be made with spin decoupling, cross polarization (Yannoni, 1982), and multiple pulse techniques (Taylor et al., 1980; Ryan et al., 1980). In the past, NMR has been a powerful tool for chemical structure analysis. The potential of the new NMR methods is to bring such detailed structural information to the study of catalytic solids.

NMR Applications

Magic angle spinning NMR has provided a variety of new opportunities for studying the detailed structure of zeolites. As shown in Fig. 21, ^{29}Si NMR clearly resolves Si with zero to four Al neighbors bonded through oxygen bridges. In this work by Ramdas et al. (1981), the distribution of Al neighbors around Si was studied as a function of Si/Al ratio for synthetic faujasites. They found that Loewenstein's rule forbidding Al atoms from occupying neighboring tetrahedral sites was obeyed. As the Si/Al ratio approached 1, the NMR spectrum became a single line corresponding four Al neighbors around Si. By analyzing the relative intensities of lines as a function Si/Al ratio, the authors constructed an ordering scheme for Al placement in the faujasite lattice. The onset of para positioning for Al in the hexagonal rings facing the supercages was found to begin at a Si/Al ratio of about 1.4. The simulation in Figure 21 is based on this structural model. It is interesting to note that Loewenstien's rule appears to be broken for zeolite A (Bursill et al., 1980). Thomas and coworkers have also used solid-state NMR to show the similarity of ZSM-5 and Silicalite (Fyfe et al., 1982). Nagy et al. (1981) have observed that the chemical shifts for ZSM-5 and similar zeolites are more negative than those of the faujasites and have ascribed that difference to the presence of the five-membered rings. In another study,

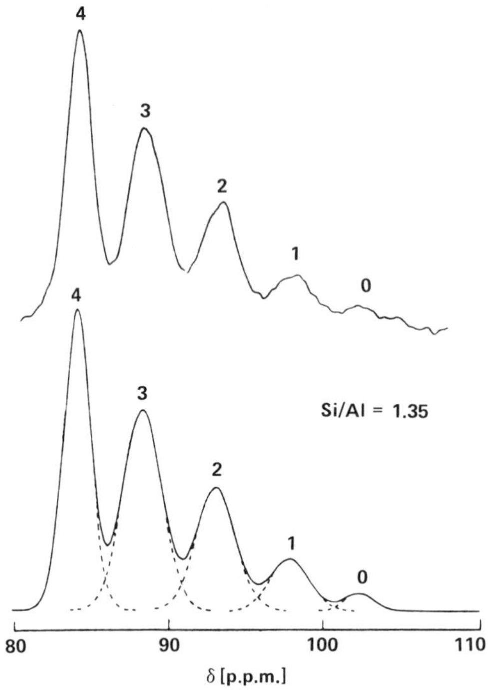

FIGURE 21 Observed (at 79.6 MHz) (upper) and simulated (lower) high-resolution ^{29}Si NMR spectra with magic angle spinning for a Si/Al ratio of 1.35 in synthetic faujasite. (From Ramdas et al., 1981.)

Nagy et al. (1982), used cross polarization (coupling of the proton and ^{29}Si resonances) to identify silanol groups in the spectrum of ZSM-5.

Although magic angle spinning is certainly important to applications of solid-state NMR to catalysts, it is not always necessary. Slichter (1981) has reported a very interesting set of ^{195}Pt NMR experiments on Pt/Al_2O_3 samples supplied by Sinfelt. The data were obtained by the spin echo technique. The lines observed were very broad, about 4.5 kG, the range of chemical shift between diamagnetic Pt compounds and Pt metal. The large shift of metallic Pt, called the *Knight shift*, arises from the polarization of the spins of the conduction electrons. Analysis of the peak shape as a function of Pt dispersion (fraction exposed) shows that the amount of diamagnetic Pt is directly related to dispersion. The spectra were interpreted to indicate that adsorption of gases on the surface ties up the otherwise free electrons of the metal and effectively demetallizes the surface atoms. Direct observation of such fundamentally important phenomena can have a great impact on our understanding of catalysis.

Fourier transform ^{13}C-NMR of species adsorbed on catalyst surfaces has also been effective. Nagy et al. (1982) have been able to follow the isomerization of 1-butene to 2-butenes on a tin- antimony oxide catalyst. Gay (1974) observed a variety of molecules adsorbed on silica and examined their mobility. Cirillo et al. (1980) have used proton NMR to identify three types of low-temperature H_2 adsorption on alumina and molybdenum/alumina catalysts.

DISTRIBUTION OF CATALYTIC MATERIALS WITHIN CATALYST PELLETS: ELECTRON MICROPROBE, SCANNING ELECTRON MICROSCOPY, AND ANALYTICAL MICROSCOPY

An important variable in the performance of a catalyst is the distribution of catalytic and noncatalytic materials inside the pellets. The latter is significant in the case of catalyst poisoning due to metal deposition. The main tools now available to provide such analysis are the electron probe microanalyzer (EPMA), scanning electron microscope (SEM), and scanning transmission electron microscope (STEM). The principle of operation in all three cases is the same, but the EPMA is designed to give best resolution for chemical analysis, whereas the electron microscopes are designed to give high magnification and good spatial and depth resolution, with chemical analysis being an added capability. The SEM produces two-dimensional images from electrons backscattered from the material instead of from electrons transmitted through the material, as is the case in TEM. A STEM is a TEM with scanning capabilities and can also be used for chemical analysis. The synopsis that follows will briefly describe the principle of x-ray emission used for chemical analysis in these instruments and then present a description of scanning electron microscopy. A brief comparison of the EPMA and SEM is also presented. Finally, applications to catalysis are described.

Principles of X-ray Emission or X-ray Fluorescence

Chemical microanalysis is carried out by analysis of the energy of x rays emitted when a sample is excited by electron impact or x rays. The phenomenon has already been referred to in previous sections in connection with sources for x-ray diffraction and photoelectron emission. When x rays are used as the excitation source, the technique is called *x-ray fluorescence analysis*. Such instruments are used primarily for chemical analysis of bulk samples because x rays are difficult to focus and therefore the spatial resolution is not as good as when electron impact is used as the excitation source. Electron beams can be focused to diameters ranging from 25 to 1000 Å. Furthermore, since electron beams are the primary excitation source in electron microscopy, x-ray analysis constitutes a natural additional capability for such instruments.

Figure 5b shows that when an electron is removed from an inner shell of an atom, creating a vacancy, it may be filled by an electron from an outer shell. The energy released by the transition may be emitted as an x-ray photon. Deexcitation by Auger electron emission can also occur. Since the energy levels are quantized, it follows that the energy of the emitted x-ray photon is also a discrete quantity and is characteristic of the element from which it was emitted. Consequently, knowledge of the emitted x-ray energy can be used for chemical characterization of the element. If the vacancy occurred in the K level and the transition occurs from the L level, the photon is designated as K_α (K_{α_1} and K_{α_2} exist due to differences in spin). Such emission gives rise to characteristic lines superimposed on a continuous spectrum. The characteristic lines produce quasi-monochromatic x rays which are used as the source in XRD. When x rays are used as an excitation source (x-ray fluorescence), no continuous spectrum is observed, and the background is very low, which is especially suited for energy analysis.

The additional processes resulting from electron impact are shown schematically in Figure 22. Secondary and transmitted electrons are used to produce images in SEM and TEM, respectively. Auger electrons are used to obtain surface chemical information. Energy analysis of the characteristic x rays also provides such information. Also shown in Fig. 22 is the depth of escape of the various deexcitation processes. It can be seen that the escape depth of x rays is on the order of micrometers; therefore, they do not contain surface information (as in the case of Auger electrons). X-ray emission is generated in a volume of about 1 μm^3, which makes x-ray analysis a microanalytical tool.

Two methods are used to analyze the energy of the emitted x rays: (a) wavelength dispersive spectrometry (WDS), and (b) energy dispersive spectrometry (EDS). In the former method, the energy of the x ray is obtained from knowing its wavelength ($E = h\nu = hc/\lambda$). λ is determined using a crystal of known crystal spacing which diffracts (disperses) the incoming x rays, allowing only one wavelength element to reach the detector. The principle is the same as described in XRD, and Bragg's law is used to determine λ. Energy dispersive analysis (EDS) is accomplished by using a semiconductor crystal that absorbs x rays, thereby exciting the crystal's electrons into the conduction band, producing an electric pulse proportional to the energy of the x rays. These pulses are then counted and sorted in a multichannel analyzer, and a display of intensity of the signal versus its energy gives the desired spectrum (see Fig. 23a). The advantages and disadvantages of these two types of x-ray energy analyzers are summarized in Table 4.

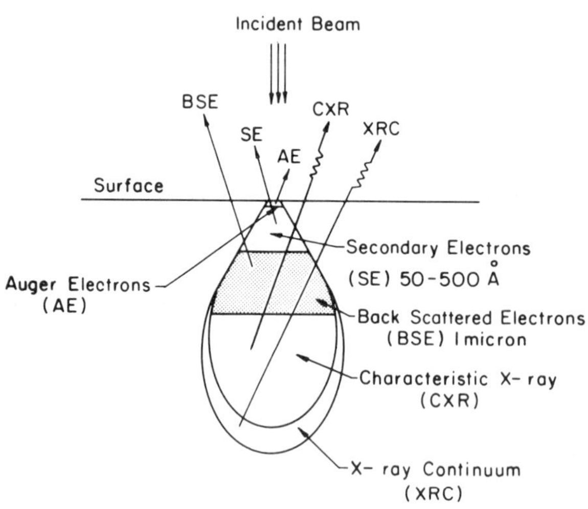

FIGURE 22 Schematic of various signals originated by electron impact, showing the corresponding volume of primary excitation.

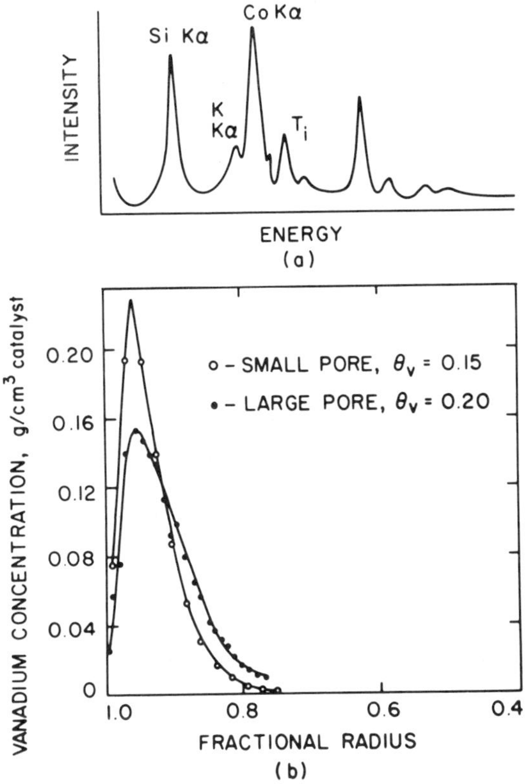

FIGURE 23 (a) X-ray spectrum of a sample containing Si, K, Co, and Ti obtained by EDS. (b) Vanadium concentration profiles on a 1/16" extrudate catalyst during hydrotreating of a heavy oil residuum. Effect of pore diameter. (From Tamm et al., 1979.)

TABLE 4 Comparison of EDS and WDS

Advantages of energy dispersive spectroscopy
Compact; low cost
Rapid (qualitative analysis)
Simultaneous multielement analysis of the full x-ray spectrum (Na upward)
Display of the entire spectrum in digital format
High collection efficiency
Low sensitivity to geometric effects
Lack of higher-order lines which are generated in crystal diffraction
Digitally produced outputs for element line scans and distribution maps
Advantages of wavelength-dispersive spectroscopy
Higher inherent element separation (resolution)
High count rate on individual elements
Analysis can be highly quantitative
Better peak-to-background ratios
Analysis of a wide range of elements (Be to U)
Higher sensitivity
Operates at room temperature

Instrumentation

A brief discussion of the major components of the various instruments used in electron probe microanalysis and SEM is given below; several books (Postek et al., 1980; Goldstein and Yakowitz, 1975; Thornton, 1968) and review articles (Sargent and Embury, 1976; Johari and Samudra, 1974) on SEM, and EPMA (Heinrich, 1980; Purdy and Anderson, 1976) are available for a more detailed description.

Figure 3 shows the schematics of a TEM, an electron probe microanalyzer (EPMA), and a SEM. These instruments are equipped with a system to produce and focus an electron beam and with a system for detection and analysis of x rays. The EPMA is designed primarily for x-ray analysis, with an electron-beam intensity which optimizes the x-ray yield and a low-magnification optical microscope for viewing the sample. The SEM, on the other hand, is designed to obtain images which, under ideal conditions, can be magnified in the 20× to 100,000× range with resolutions of 70 to 100 Å. Because of the greater versatility of the SEM, it is described in more detail in the following paragraphs.

The SEM differs from the TEM in that images are produced from secondary electrons generated by the electron beam. Thus the sample need not be transparent to electron transmission as in TEM. Electron impact creates a volume of primary excitation (see Fig. 22) from which elastically scattered (backscattered) and inelastically scattered electrons or secondary electrons are emitted. Secondary electrons have low energy (less than 50 eV), and therefore only those electrons near the surface can be emitted and detected. Secondary and backscattered electrons are emitted outward in all directions. An electron detector system is used to collect both types of emitted electrons and produce a signal proportional to their intensity. The detection of secondary electrons depends on the sample topography. Electrons emitted from the specimen surfaces facing the detector will show a bright face, whereas those emitted from surfaces facing away from the detector will not be detected, creating a shadowing effect. Furthermore, as the name indicates, an SEM is equipped with deflecting coils which permit scanning the electron beam over the sample in a raster mode. The raster pattern of the primary electron beam is synchronized with the scanning pattern of a cathode ray tube (CRT) used to display the detector signal so that a two-dimensional image of the scanned area is produced. Images are produced in the CRT or a TV, and permanent records can be obtained directly from the CRT using an instant camera or by videotape recording from a TV.

Of the two types of x-ray energy analyzers, WDS is the preferred mode of analysis from the standpoint of x-ray spectroscopy and is used exclusively in x-ray spectrometers as well as on EPMA instruments. WDS analyzes one wavelength element at a time and requires mechanical components to rotate the crystal as well as electrical components. Crystal alignment is quite important, and WDS is difficult to use with rough surfaces. EDS analyzers are used in most SEMs and in some electron probes. EDS provides a faster and easier survey analysis of the elements but at lower energy resolution and lower sensitivity, and it does not cover the entire spectrum, especially of light elements such as carbon. Clearly, for versatility one would like to have an SEM capable of using high beam intensities such as in EPMA and equipped with both WDS and EDS analyzers. For high-resolution microscopy work or x-ray spectroscopy work, separate instruments are preferred. Figure 23a shows the x-ray spectra of one spot on the sample. However, the beam can be scanned over the sample with the analyzer fixed on a certain energy so that only x rays corresponding to a specific element are displayed. This mode of operation produces an x-ray map of the element in question over a line (line scan) or area under analysis. Another way of obtaining elemental distribution is to move the sample under the fixed electron beam. This technique is particularly useful when analyzing catalyst pellets because it permits one to scan the entire width of the pellet.

For electron microanalysis, the sample must have a flat surface. For powders, the sample is first embedded onto a plastic medium, which, after drying, is ground to reveal the cross section of its interior. Conductive coatings (C, Au) are evaporated onto the surface of nonconductive materials (Purdy and Anderson, 1976). For SEM, the sample is mounted on a small stub coated with adhesive which holds the particles as received or after fracturing to reveal the interior.

Applications

Specific applications of SEM and EPMA to catalysis are discussed by Sargent and Embury (1976) and Purdy and Anderson (1976). SEM is used in general for physical characterization of catalyst particles, such as particle size, shape, surface structure and porosity, makeup of aggregates, and dispersion of one phase into another. The practical magnification and resolution of the SEM is relatively low compared with the TEM and STEM, which are preferentially used for examining crystallites in the size range 1 to 10 mm.

X-ray emission capabilities of the electron microprobe or SEM instruments are used to obtain bulk chemical information about the materials present in the sample. The most important application is to measure the distribution of catalytic materials or impurities over the cross section of catalyst pellets. A classical example is the effect of catalyst preparation in the distribution of the active catalyst inside pellets. The metal profiles inside the pellets can be followed as a function of preparation conditions. Depending on the method used, the catalyst can be deposited uniformly inside the pellet, or it can concentrate toward the external face of the pellet. Extension of this application will increase our understanding of catalyst preparation.

Another classical application is the study of metal deposition occurring during dehydrodesulfurization of heavy residuum (Tamm et al., 1979). Figure 23b shows the V content of a catalyst as a function of position inside the pellet, as obtained using an EPMA. The poisoning of catalysts due to impurities is an area of study in which EPMA and SEM are particularly useful tools.

ADSORBED SPECIES ON SURFACES

Infrared Spectroscopy

Unlike the surface analysis techniques based on electron spectroscopy, infrared (IR) spectroscopy is a well-seasoned tool for catalytic research, as witnessed by the various books (Little, 1976; Hair, 1967; Clark and Hester, 1974–1982) and review articles (Delgass et al., 1979; Eischens and Pliskin, 1958) published on the subject. IR spectroscopy was first developed to study the chemical structure of bulk compounds, but it was soon realized that it can also be used to gain structural information about species adsorbed on solid surfaces. The books by Hair (1967) and Little (1976) contain a good introductory description of IR together with many applications. The review article by Eichens and Pliskin (1958) has withstood the test of time and remains one of the most quoted references. The more recent review articles (Delgass et al., 1979; Basila, 1968; Blyholder, 1968; Pritchard and Catterick, 1976) emphasize newer developments. The description given here covers only the most fundamental aspects of IR spectroscopy and its major applications to catalytic research.

IR spectroscopy gives qualitative information about the way adsorbed molecules are bonded to surfaces, as well as structural information on solids. It can also be used to measure the amount of materials adsorbed, and in some cases it can be used to study the rate at which certain surface processes occur.

Principles of Vibrational Spectroscopy

When an atom or molecule is placed in an electromagnetic field, there is a transfer of energy between the field and the quantized energy levels of the molecules. When energy is adsorbed, measurement of the transmitted energy versus the frequency of the radiation produces a spectrum unique to the excited molecule. If the frequency of the incoming radiation is in the IR range, the spectrum is due to transitions in vibrational and rotational energy levels. In each electronic energy level there is a series of vibrational transitions, and for each vibrational level there is a series of rotational transitions. Transition in electronic energy levels occurs when the excitations have frequencies in the UV or visible range. Selection rules govern the permissible transitions among the various energy levels.

Diatomic Molecule: Vibrational spectra arise from the motion of one atom relative to another. While a detailed group theory has been developed to describe the phenomena, a simple explanation in which a diatomic molecule is modeled as an harmonic oscillator is often adequate to describe the first-order processes involved.

The energy and frequency of an electromagnetic wave are related by $E = h\nu$, where h is Planck's constant. Since $\lambda\nu = c$, where c is the speed of light, it has become a common practice to report spectra in terms of frequency instead of λ. The unit most commonly used is the wave number $= 10^4/\lambda$ (micrometers), which has units of cm^{-1}. The IR spectrum is divided into near IR, 10,000 to 5,000 cm^{-1} (λ: 1 to 2 μm); mid IR 5000 to 200 cm^{-1} (λ: 2 to 50 μm), and far IR, 200 to 10 cm^{-1} (λ: 50 to 1000 μm).

Solving the classical mechanical equations describing the oscillations of two masses, m_1, m_2, attached to a perfect spring which obeys Hooke's law, one can calculate the change in potential energy associated with the oscillations as

$$U = \frac{kr^2}{2} \tag{21}$$

where k is the force constant that measures the stiffness of the spring and r is the displacement from the center of mass. The motion of this system is sinusoidal and the frequency is given by $\nu = (2\pi)^{-1}\sqrt{k/\mu}$, where $\mu = m_1m_2/(m_1 + m_2)$ is the reduced mass. The simple result of the harmonic oscillator has immediate practical applications in the interpretation of spectra. Knowing ν, one can calculate the force constant k; conversely, changes in ν upon adsorption reflect weakening (lower ν) or strengthening (higher ν) of the bond.

Another application of the result obtained for the simple diatomic oscillator is in connection with changes in frequency due to isotopic substitution. If the force constant remains the same, then upon isotopic substitution of species with reduced mass μ_1 and a frequency ν_1 by an isotope of reduced mass μ_2, a frequency change predicted by

$$\frac{\nu_2}{\nu_1} = \left(\frac{\mu_1}{\mu_2}\right)^{1/2} \tag{22}$$

should be observed. Failure to observe a frequency change indicates that the assumed species or group is not involved in the vibration being observed, and another structure must be postulated [see Delgass et al. (1979) and Hair (1967) for illustrations of this application]. The potential energy diagram described by Eq. (21) is similar to the lower region of the energy diagram of a real diatomic molecule. For this reason, the quantized form of the harmonic oscillator analogy is adequate for a simple description of vibrational spectroscopy.

To account for quantum effects, the potential energy function of the harmonic oscillator is substituted into Schrödinger's wave equation. The solution of this equation gives the energy of the quantized vibration as

$$E = h\nu\left(v + \frac{1}{2}\right) \tag{23}$$

where v is the vibrational quantum number. Certain selection rules are used to ascertain the probability of transitions between energy levels. For harmonic oscillators, only transitions between $\Delta v = \pm 1$ are permissible. Since at room temperature most molecules are in the ground state, vibrational transitions between ground state and the first energy level are usually observed. Another important selection rule requires that a change in dipole moment with vibration must exist in order to couple the vibration to the electromagnetic radiation. In the absence of such a dipole moment change, as in the case of linear diatomic molecules containing the same atom (i.e., N_2), no vibrational energy transitions are observed. A corollary of the foregoing selection rule is that IR inactive molecules may give rise to dipoles when they are adsorbed on a surface and therefore exhibit an IR spectrum.

Rotational transitions often occur simultaneously with vibrational transitions. The combined vibrational-rotational energy relationship is of the form

$$E = h\nu_0 + \frac{h^2}{4\pi^2 I} m \tag{24}$$

where I is the moment of inertia of the molecule and m is a rotational quantum number that can have values ±1, ±2, and so on. The center of the vibrational-rotational band is at ν_0; thus the second term in Eq. (24)

defines the rotational fine structure. Depending on whether m is positive or negative, two branches, known as P and R branches, are produced in the high- and low-frequency sides of ν_0. Another branch, known as the Q branch, may appear in the vicinity of ν_0 due to changes in angular momentum about the axis joining the nuclei or due to unpaired electrons. Figure 24 shows, in addition to adsorbed CO, the high-resolution absorbance spectrum of gaseous CO, depicting the R and P branches resolved into specific lines due to rotational transitions. Lower resistance spectra will show only two broad bands about ν_0. For CO adsorbed on a metal surface, the rotational fine structure disappears and give rise to a single band.

Polyatomic Molecules: Diatomic molecules exhibit only one vibrational mode, but as more atoms are added to a molecule, additional vibrations are possible. For a nonlinear polyatomic molecule containing n atoms, 3n − 6 vibrations are possible (3n − 5 for linear molecules). For these, n − 1 can be defined as stretching (opposite movement of atoms) and 2n − 5 are bending or deformation vibrations (atoms move at an angle to each other). In some cases, the vibrations correspond to a similar type

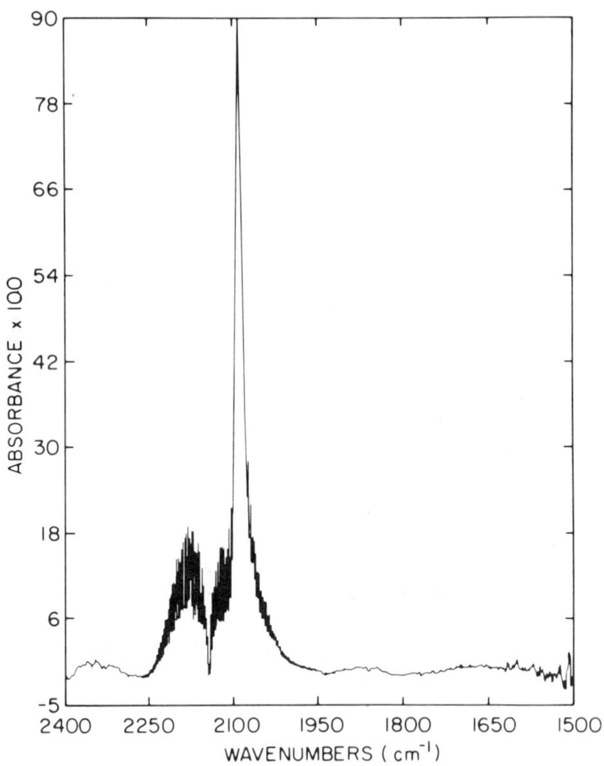

FIGURE 24 Infrared spectrum of CO in the gas plase and adsorbed on a 5% Pt/SiO_2 catalysts. The data were collected at a resolution of 2 cm^{-1} wave number in a single scan in a FTIR instrument. The fine structure superimposed into the broad peaks (P,R branches) represents rotational transitions. The sharp peak at 2070 cm^{-1} corresponds to adsorbed CO.

of motion and can be distinguished only because they are perpendicular to each other (degenerate vibrations). Hair (1967) discusses in detail the number of possible vibrations in polyatomic molecules.

Group Frequencies: While any vibrational mode involves all atoms in a molecule, much of the motion is dominated by a few atoms or group of atoms. Thus, whenever a characteristic combination of atoms is present on a compound, the same characteristic frequency appears more or less independent of the structure of the rest of the molecule. Group frequencies permit one to identify structural groups from the IR spectra. Group frequencies often involve terminal atoms of small mass such as C—H,O—H, N—H, or internal atoms connected by bonds which have a distinctive force constant. Extensive tabulations and discussions regarding the use of group frequencies for chemical identification are available (Alpert et al., 1970).

The identification of the spectra of adsorbed species often consists of matching the surface spectra with those of known compounds. Furthermore, infrared bands that are forbidden for molecules in the gas phase on grounds of symmetry may be possible when the molecule is adsorbed and perturbed by surface forces. The frequency of the adsorbed species shifts with surface coverage, coadsorption of other molecules, particle size, or, in general, with changes in the electronic environment of the metal-adsorbate interaction.

Band Intensities: If radiation of intensity I_0 impinges on a media of thickness d containing a concentration of molecules C, the intensity of the transmitted radiation is given by the *Beer-Lambert law*,

$$A = \ln \frac{I_0}{I} = \varepsilon C d \tag{25}$$

where ε is the extinction coefficient and A the absorbance. For adsorbed molecules, the extinction coefficient can be larger or smaller than in the gas phase (Little, 1976). If ε is constant with coverage, there is a linear relation between absorbance and coverage which can be used for quantitative analysis. The best approach to measuring extinction coefficients is to measure independently the amount of gas adsorbed along with the IR absorbance. Recent approaches have used a pulse technique (Sarkany and Gonzalez, 1982a) and mass spectroscopy (Savatsky and Bell, 1982) for evaluation of the amount adsorbed. If the band width of the adsorption peak changes, it is recommended to use an integrated absorbance over the full frequency range.

Band intensity increases with increasing metal loading and sample thickness. However, scattering and absorption by the support sets an upper limit to the amount of metal and material that can be used. Scattering is proportional to the cube of particle size and inversely proportional to the fourth power of the wavelength. It has been found that good transmission results are obtained if particle sizes are less than 1 μm. Thus, a choice of support with a particle size of <1 μm will facilitate IR studies.

Experimental

In the past, the majority of IR studies have been carried out using dispersive IR instruments operating in a transmission mode. However, Fourier transform infrared (FTIR) spectroscopy offers a number of advantages over dispersive instruments, in particular in catalytic applications. Reflection spectroscopy (specular, diffuse) is also available to study opaque samples and is especially suited to use with flat, opaque surfaces (foils, single crystals) in combination with other spectroscopic techniques. The description that follows compares the various modes of IR utilization and summarizes briefly advantages and disadvantages.

Dispersive Versus Fourier Transform IR: The classical IR instrument is a dispersive instrument consisting of a source producing a reference and a sample beam, a sample chamber, an optical system to combine the beams and separate the various frequencies, and a detector. Frequency discrimination depends on dispersion of the radiation by a grating system. A brief but clear description of dispersion instruments is presented by Bell (1980) and is available from the technical literature for commercial instruments. Fourier transform instruments operate on an entirely different principle to separate the frequencies. A single beam of IR radiation from the source is sent into a Michelson interferometer, which, by the displacement of a mirror, produces an interferogram containing all the frequencies within the desired IR spectrum. The details and principles of FTIR spectroscopy are presented in Griffith's (1975) book on the subject and are also summarized in Bell's (1980) review article. The information contained in the interferogram is converted into a frequency versus absorbance or transmittance spectrum by using a Fourier transform inversion. The transformation is carried out by a dedicated minicomputer which also operates the spectrometre and is used to display spectra. FTIR has two major advantages over grating spectrometers. First, the interferometer observes all spectral resolution elements simultaneously, whereas a monochrometer spends only a fraction of the total time observing each resolution element. Consequently, under identical conditions, the FT spectrometer can acquire spectra much faster than the dispersive spectrometer (this is known as *Felgett's advantage*). Conversely, for the same time of data collection, the FT spectrometer can produce spectra of much better signal-to-noise ratio. The second fundamental advantage of an FT spectrometer over a grating spectrometer is the increased optical thoughput of an interferometer compared to a grating monochrometer (*Jacquinot's advantage*). Griffith (1975) has presented a quantitative comparison between dispersive and FTS instruments which has been summarized by Bell (1980). Furthermore, FTS data acquisition and computer data storage capability provide an additional advantage over dispersive IR instruments, although it must be pointed out that similar computer capabilities are now being added to dispersive units. The main disadvantage of FTS instruments is their higher cost; however, new models with restricted capabilities are now being introduced in the market at prices that are comparable to those of dispersive units. In general, the higher IR throughput and smaller data collection time required for FTIR instruments make them especially suited for catalytic studies.

Transmission Versus Reflection and Emission: One of the most widely used methods of IR spectroscopy is the transmission mode, where in the IR beam is transmitted through the sample, which must be partially transparent to IR radiation. Many oxides, such as silica, alumina, zeolites, and stoichiometric zinc oxide, are IR transparent. The sample is pressed in the form of a thin disk (0.3 to 0.1 mm thick) with a density between 10 and 100 mg/ cm^2. The preparation of the wafer is often the key to successful experimentation. Although specific values depend on the material in question, there are certain general guidelines. The disk is pressed in a die which must have highly polished surfaces to prevent the disk from adhering to the die surfaces. In some cases it is convenient to condition the powders with a certain degree of moisture and temperature prior to pressing. Transparency increases when the crystallites of the powders are very small (silica and alumina prepared by hydrolysis of the chlorides in a H_2-O_2 flame produces small particles). An open pore structure (i.e., low die pressure) is desirable to facilitate transport of reactants inside the disk. In the case of supported catalysts, metal loading should be such that crystallite size should not exceed 100 Å. Pt can be loaded from 5 to 10% on silica (Cabosil) or alumina aerogel and still give a good absorbance spectra. The support should not absorb in the same frequency as the adsorbate. Other methods of sample preparation include evaporation of the metal onto an IR window, forming a thin film; however, the sensitivity is much less than when the metal is highly dispersed in a porous support (Pritchard and Catterick, 1968). The books of Little (1976) and Hair (1967) contain details of disk preparation, and the IR chapter in the Delgass et al. (1979) book also contains useful information in this regard.

In transmission experiments, the disk is placed in a cell containing special IR windows (CaF_2, sapphire, etc.) and equipped to meet the pressure and temperature requirements of the experiment. Several cells designs are reproduced by Little (1976) and Hair (1967). Newer designs for transient experiments consist of small-volume cells capable of rapid response to changes in feed concentration and temperature (Savatsky and Bell, 1982; Kaul and Wolf, 1984).

When the solid is completely opaque to IR, the pressed salt method can be used. The opaque solid is first ground and mixed with a halide transparent to IR (KBr, CsI) and can then be pressed into a self-supporting disk. Depending on the material, dilutions of 1/100 to 1/400 sample per halide are used. The disadvantage of this technique is that access to the surface by gases is lost and therefore adsorption or desorption studies are no longer possible. However, the technique is useful when analysis of the structure of the catalysts rather than the adsorbate-surface interaction is sought. Other methods that have been used are deposition on a liquid matrix and in a solid argon matrix (Pritchard and Catterick, 1976).

Reflectance techniques have been used with opaque samples that do not permit transmission. Specular reflection has been used in the case of metal mirrors such as single crystals or foils and, in principle, provides an attractive combination with other spectroscopic techniques involving flat surfaces. Diffuse reflectance spectroscopy (DRS) is used when the reflecting surface is rough and the incident radiation is reflected (scattered) diffusely. Such radiation must be collected by a hemispherical or elliptical reflector and focused into the detector. Klier (1980) summarizes the literature, theory, and application of DRS to the study of adsorbed species on catalysts.

New developments in vibrational spectroscopy (excluding IR excitation) are reviewed in the monograph by Bell and Hair (1980). They include electron energy loss spectroscopy (EELS), which is useful with single crystals or metallic foils; inelastic electron tunneling spectroscopy (IETS), used to study metal-insulator junctions; and neutron scattering spectroscopy (NSS).

Applications

Rather than an extensive listing of the many IR studies published in the literature, the major classes of applications in connection with catalysts are summarized below. The reader is referred to the several review papers for specific applications.

Bulk Structures: Studies of the structure of silica surfaces and acid oxide surfaces, including alumina, silica-alumina, and molecular sieves, are summarized by Hair (1967). A classical example of the use of IR in structural analysis is the work of Flanigan et al. (1971), who, using KBr-zeolite disks, compared the spectra of various A, X, and Y zeolites and hydroxysodalite and assigned IR bands to specific structural groups on each zeolite. In more recent studies (Pichat et al., 1974), the degree of crystallinity of the zeolite has been related to the IR bands occurring near 560 to 380 cm^{-1}. A relation between degree of crystallinity (as obtained from IR) and crystal size (obtained from XRD) has also been attained.

IR Studies of Adsorbed Molecules: A great deal of the IR work relevant to catalysis has been done with adsorbed molecules. The review of Eichens and Pliskin (1958) summarizes studies of hydrocarbons (ethylene, ethane, acetylene, olefins) on nickel and CO chemisorption on Pd, Pt, iron, and rhodium. Much of this work consisted of structural assigment of the observed IR bands to elucidate the type of bonding involved. In the case of CO chemisorption, IR can be used to detect the number of metal atoms bonded to CO by the observation of frequencies corresponding to linear CO (CO/M = 1), bridged CO (CO/M = 1/2, or two CO per metal (CO/Rh = 2). Observation of frequency shifts with surface coverage, crystallite size, and so on, are also important in the elucidation of the CO-metal interaction. IR studies of NO adsorbed on metals and oxides have also been used as a probe to elucidate the type of sites existing on supported catalysts (Peri, 1974).

Significant frequency shifts have been observed when hydrogen, from hydrogen-containing molecules, is bonded to the OH groups of silica. Such frequency shifts are interpreted in terms of charge transfer or changes in the OH dipole due to the presence of the adsorbed molecule (Basila, 1968).

One of the classical examples in the use of IR spectroscopy to examine the nature of active sites is the differentiation of Brønsted or proton donor acid sites and Lewis or electron acceptor acid sites. Ammonia (or nitrogen-containing molecules gives absorption bands near 3333 and 1639 cm^{-1} if in the NH_3 configuration (Lewis acid) and near 3125 and 1428 cm^{-1} if the NH_4^+ form (Brønsted acid). The frequencies of the coordinatively bonded molecules usually shift with coverage, which is interpreted in terms of the strength of the acid sites. Basila (1968) presents a review of IR studies of the acidity of silica-alumina and zeolites, together with frequency assignments for ammonia and pyridine.

It has already been mentioned that isotopic substitution can be used to elucidate the structure of surface complexes; H_2 on ZnO, ethylene-oxygen on Ag, and propylene on ZnO are a few examples discussed in some detail in the book by Delgass et al. (1979).

Infrared Kinetic Studies: If the rate of reaction can be correlated with the rate of formation or disappearance of an IR band, it can be concluded that such species is a reaction intermediate. If the reaction is carried out in a batch cell under transient conditions, and if the IR spectrometer is capable of measuring spectra in the time scale of the reaction experiments, transient IR experiments premit one to determine the kinetics of surface species and the reaction mechanism. The use of transient techniques in catalysis, although proposed some time ago by Tamaru (1964), has been the subject of various review articles (Tamaru, 1964, 1978; Bennett, 1976; Kobayashi and Kobayashi, 1974).

A combination of IR transient studies with isotopic labeling has been reported by Kokes et al. (1972) for 1-butene isomerization on ZnO. Band assignment was first resolved by isotopic substitution; then the IR bands were followed as a function of time and compared with gas-phase kinetic data. Simultaneous measurement of surface and gas phase composition, together with isotopic substitution, have also been carried out by Tamaru (1978) during the water-gas shift reaction catalyzed by ZnO.

The use of fast FTIR spectrometers has expanded the use of transient studies to catalytic systems more active than ZnO. Bennett (1982) describes examples of the use of pulses, steps, and cycled feed to study transient catalysis using IR spectroscopy. The reactions of carbon monoxide on Pt and iron have been studied using transient IR. The effect of transport processes and dynamics of chemisorption have been analyzed by Ho and Hegedus (1982). The dynamics of three-way catalysts have been studied by Herz (1982) using IR diode laser spectroscopy. Studies of self-sustained oscillations using FTIR have been carried out by Kaul and Wolf (1985). In most applications, the time constant associated with the transient surface processes must be of the order of that for IR data collection (about 0.1 to 1 s for FTIR spectrometers). Time-resolved FTIR spectroscopy has also been used to measure fast chemical reactions (Durana and Mantz, 1969); however, the technique has not yet been applied to surface studies.

In closing the synopsis on the fundamentals and applications of IR spectroscopy to catalysis, it is clear that this is a mature and necessary tool in catalytic research. New frontiers are now being opened with the use of transient techniques. The development of reflection and new vibrational spectroscopies will be useful in fundamental studies using flat or opaque surfaces.

Raman Spectroscopy

Raman spectroscopy has much in common with infrared spectroscopy, but since it records vibrational spectra from scattered rather than absorbed radiation, it gives slightly different and often complementary information. The availability of intense, monochromatic laser light sources compensates for the low scattering cross sections, and commercial spectrometers are capable of recording Raman spectra of catalytic materials and adsorbed species. The main advantage of this technique is that absorption by the solid matrix does not interfere with the low-frequency bands. Thus vibrational modes of the catalyst itself and of the bond between the adsorbate and the catalyst are accessible by Raman spectroscopy. Since the effect requires a change in polarizability with vibration rather that a change in dipole moment, spectra show strong peaks for nonionic bonds and symmetric vibrations. The method has been particularly successful for the study of hydrodesulfurization catalysts containing molybdenum. It has also been used to study adsorbed aromatic molecules and electrochemical surfaces. Theory, experimental considerations, and examples of applications are discussed below. Review articles by Hendra (1974), Cooney et al. (1975), Egerton and Hardin (1975), and Haller (1979) provide further information.

Principles

Electromagnetic radiation induces a dipole moment in a molecule. The size of the dipole moment is proportional to the magnitude of the electric field. The proportionality constant is the molecular polarizability. If the distribution of charge in the molecule changes during a particular vibration, the polarizability will also vary with the vibrational motion, and the vibration will be Raman active. Analysis of the interaction between the oscillating electric field of the exciting radiation and the induced dipole oscillating with the vibrational motion reveals three main scattering terms (Haller, 1979). The first is the elastic or Rayleigh scattering, which occurs at the frequency of the exciting radiation. The second and third terms, 10^{-9} times the intensity of the Rayleigh scattering, represent inelastic scattering in which energy is exchanged between the radiation and the molecule. These terms have frequencies $\nu_0 - \nu_m$ and $\nu_0 + \nu_m$, where ν_0 is the frequency of the exciting radiation and ν_m is the frequency of the molecular vibration. They are called the *Stokes* and *anti-Stokes Raman lines*, respectively (Chantry, 1971). The intensities of these lines depend on the change in polarizability with the vibration and, for symmetric vibrations, can be related to bond order (Hendra and Stratton, 1969). For molecules with a center of symmetry, vibrations that are symmetric about the center of symmetry change the polarizability and are Raman active, as shown for CO_2 in Fig. 25. Asymmetric vibrations of such molecules are Raman inactive. These sensitivities are exactly opposite to those governing IR activity for such molecules, but IR and Raman activity need not be mutually exclusive for more complex molecular symmetries. The vibrational selection rules are the same for Raman and IR (i.e., $\Delta U = \pm 1$ for harmonic oscillators).

A special case of Raman spectroscopy must be mentioned in a discussion of catalyst surfaces. Jeanmarie and Van Duyne (1977) and Albrecht and Creighton (1977) reported that pyridine adsorbed on silver in an electrochemical cell showed scattering cross sections 10^5 to 10^6 higher than for pyridine in solution. This phenomenon of surface-enhanced Raman spectroscopy (SERS) is still not completely understood (Furtak and Reyes, 1980; Van Duyne, 1979; Aravind and Metiu, 1983), but for Ag, Cu, and Au, the metals for which it has been observed, it offers a signal enhancement which makes study of molecules adsorbed on foils and other low-surface-area samples straightforward. Applications have been made in electrochemistry (Birke et al., 1983) and should also be possible in catalysis. The technique is particularly attractive because, with an optical multichannel analyzer, surface spectra can be recorded every 20 msec (Weaver et al., 1983).

Experimental Methods

A Raman spectrometer consists of a light source, the sample scattering chamber, a monochrometer, and a detector. A laser light source is essential for surface work. Argon ion (514.5 nm and 488.0 nm), helium-neon (632.8 nm), and krypton ion (647.1 nm) lasers are in current use. This light is highly collimated or passed through an interference filter to remove background radiation accompanying the laser line.

The light scattered from the sample can be viewed at any angle, but 0°, 90°, or 180° are commonly used. Cells are typically glass with an optically flat window. Because of the relative high power of the laser light, care must be taken to avoid sample decomposition or excessive heating by the exciting radiation. The light can be scanned across the sample (Zimmerer and Kiefer, 1974) or the sample can be rotated (Cheng et al., 1980) to avoid this problem.

To obtain the Raman spectrum, the scattered intensity must be measured as a function of frequency. A monochrometer accomplishes this

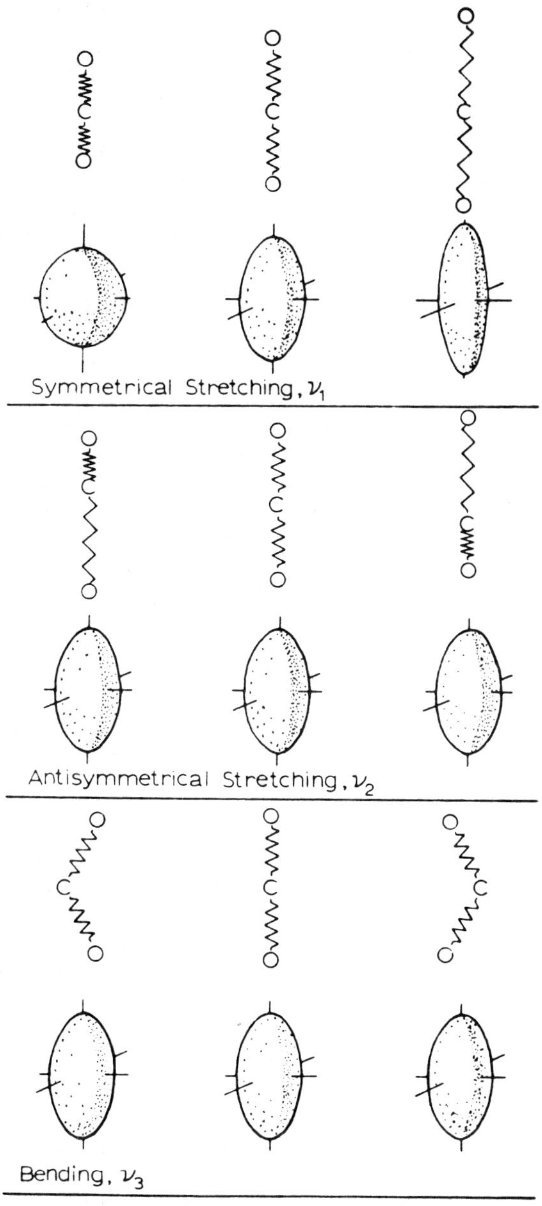

FIGURE 25 Polarizability changes during the vibration of carbon dioxide (not to scale). (From Tobias, 1967.)

task. The light passing through the monochrometer system at a given energy is recorded by counting photons with a photoelectric detector.

In some cases, species present in catalyst samples are excited by the laser light and emit a fluorescence spectrum that can obscure weak Raman lines. When organic impurities are the source of this problem, the fluorescence background can often be removed by heating the sample in O_2 (Egerton et al., 1974) or treating it with hydrogen peroxide or concentrated nitric acid (Adams et al., 1976). Other fluorescence rejection schemes use time resolution to eliminate the relatively long lifetime fluorescence signal (Van Duyne et al., 1974). Note that fluorescence radiation is easily identified by the lack of dependence of its frequency on the frequency of the exciting radiation.

Applications

Raman spectroscopy applies to the study of vibrations within the catalyst itself and to those of adsorbed molecules at the surface. The utility of the method for studying catalysts themselves is well illustrated by the growing literature on the genesis, precursors, and sulfided forms of molybdenum-containing hydrodesulfurization catalysts.

Molybdenum is usually added to the alumina support by impregnation with an aqueous solution of $(NH_4)_6Mo_7O_{24}$ to the point of incipient wetness. As summarized by Wang and Hall (1980), the metal deposition process is complex. At high pH, the molybdenum species in solution is monometric MoO_4^{2-}, while the polymolybdate species $Mo_7O_{24}^{6-}$ predominates at low pH. Furthermore, the local pH in the pores can be modified by the buffer capacity of the alumina and by the evolution of ammonia. Spectra of both wet (Fig. 26) and dried catalysts prepared at carefully controlled pH of 8.6, 3.9, and 1.0 show the ability of the Raman spectroscopy to follow the molybdenum chemistry (Wang and Hall, 1980). At pH 8.6, peaks at 326 cm^{-1} and 905 cm^{-1} identify the major species as MoO_4^{2-}. Bands at 215 cm^{-1} and 365 cm^{-1} show $Mo_7O_{24}^{6-}$ to dominate at pH 3.9 and 1.0.

Raman spectra also yield much information about calcined or oxidized catalysts. Zingg et al. (1980) have used the chracteristic spectra of MoO_3 and $Al_2(MoO_4)_3$ to show the appearance of MoO_3 when molybdenum supported on alumina exceeds monolayer coverage and that increased calcination temperature or duration converts MoO_3 to $Al_2(MoO_4)$ or another tetrahedral Mo species. A number of authors have reported on the interactions of Mo with alumina (Brown et al., 1977a,b; Jeziorowski and Knozinger, 1979; and Dufresne et al., 1981). In general increasing Mo coverage of the surface increases the two-dimensional aggregation of octahedral Mo species, as indicated by increasing frequency of the Raman bands. The aggregated Mo species interacting with alumina support has a characteristic Raman line in the 950-cm^{-1} region. Details of the effects of addition of Ni and Co to MoO/Al_2O_3 catalysts have also been elaborated (Cheng and Schrader, 1979a; Jeziorowoski and Knozinger, 1980; Knozinger et al., 1981).

Schrader and Cheng (1982) reported that sulfiding of Mo/Al_2O_3 in 10% H_2S/H_2 at 400°C produces a sample with a Raman spectrum characteristic of MoS_2, although slightly broadened and shifted. Studies by these authors of the effects of Mo loading, sulfiding temperature, and susceptibility of the sulfide to reoxidation show MoO_3 to be difficult to sulfide at low temperature, isolated Mo species to be the most difficult to sulfide, and polymolybdate species to sulfide most easily but to produce vacancies caused by reduced Mo. Brown et al. (1977b) and Payen et al. (1982) also

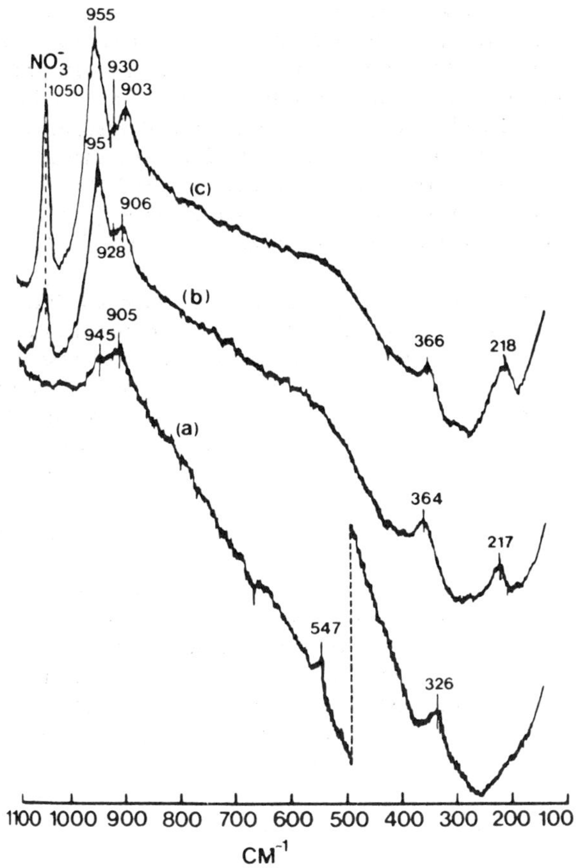

FIGURE 26 Raman spectra of wet catalysts prepared by impregnation of alumina with $(NH_4)_6Mo_7O_{24}$ at pH (a) 8.6, (b) 3.9, and (c) 1.0. (From Wang and Hall, 1980.)

report MoS_2 spectra from sulfided catalysts. The latter paper demonstrates the spatial resolution of the Raman microprobe, which takes advantage of the ca. 1 μm spot size of the laser radiation.

The Raman literature on adsorbed molecules is not large, but a number of interesting findings have been reported. Studies of molecules with aromatic rings offer good examples of the utility of the Raman measurement. Raman spectra of benzene adsorbed on silica show only slight peak shifts from the spectrum for liquid benzene, but the intensity of the 994-cm^{-1} ring breathing band is significantly reduced. Egerton et al. (1974) interpret this result as a decrease in polarization of the π electrons and an indication of π bonding of the silica surface. Surprisingly, thiophene shows similar π interaction with surface SiOH groups by loss of intensity in the ring breathing vibration at 834 cm^{-1} (Tam et al., 1975). Pyridine, on the other hand, indicates heteroatom interaction with a silica surface by a frequency shift in the ring breathing mode from 922 cm^{-1} in the liquid to 1007 cm^{-1} at low coverage, where hydrogen bonding dominates (Hendra et al., 1971; Egerton et al., 1974). In addition, these studies report coordinative bonding of pyridine to Lewis sites on alumina. Pyridine adsorption on hydrodesulfurization catalysts gives Raman spectra characteristic of physisorbed species and a specific chemisorbed species with a peak in the region 1008 to 1014 cm^{-1} (Cheng and Schrader, 1979b; Payen et al., 1982).

Raman spectroscopy has also elucidated surface reactions of adsorbed molecules. Kuiper et al. (1973) have combined Raman and IR results to show that benzaldehyde is converted to benzoate and benzyl alcoholate on an alumina surface. Turner et al. (1976) used Raman spectroscopy to observe the isomerization of 1-hexene to 2-hexene on alumina activated at 1200 K.

Most of the applications of surface-enhanced Raman spectroscopy have been in an electrochemical environment, but it is clear that the technique does apply to the interaction of gases with clean metals which show the effect. SERS spectra of carbon monoxide (Wood and Klein, 1980), ethylene (Moskovits and Dilella, 1980), and water (Pockrand, 1982) adsorbed on silver at low temperature have all been reported. For ethylene on silver, the Raman spectra contain bands for vibrational modes that are Raman inactive in the gas phase. Thus adsorption alters the symmetry of the molecule. Moscovits and Dilella interpret the band intensities and shifts as suggesting that both ethylene and propylene are π bonded to the silver surface.

Thus we see that, in general, Raman spectroscopy has been effective in identifying surface phases and coordination. It can reflect details of adsorbate bonding and gas-surface reaction, and in the surface-enhanced mode, it can give this information on submonolayers of adsorbates on flat surfaces. Applications to catalysis have been significant, but this is a technique in the process of being developed for catalysis research. Increasing use can be expected.

Temperature-Programmed Desorption and Temperature-Programmed Reaction

Although many experiments in catalysis demand isothermal conditions, the unsteady state produced by a rapid, programmed change in temperature can yield a wealth of information about a catalytic system. Perhaps the simplest version of this experiment is what is often called flash desorption spectroscopy (FDS) or thermal desorption spectroscopy in the surface science literature (King, 1975). In this experiment, gas is irreversibly adsorbed on a foil or single-crystal surface at low temperature. The temperature of the sample is then raised linearly at a typical rate of 10 K/sec. The desorption spectrum, usually recorded by a mass spectrometer in the vacuum system, shows peaks at temperatures related to the heats of adsorption of the various adsorbed states. Temperature-programmed desorption (TDS) (Cvetanovic and Amenomiya, 1972) is essentially the same experiment on a catalyst sample, but extra care must be taken to account for possible influence of heat transfer, diffusion, or readsorption of the gas in the porous sample. Temperature-programming rates are usually slower in TPD and most experiments are done in a carrier gas instead of a vacuum. The concept of using temperature to scan the energies of interaction of molecules with surfaces is a useful one and has appeared in many variations, including temperature-programmed reduction (Hurst et al., 1982), temperature-programmed oxidation (Miura et al., 1975), and temperature-programmed reaction (TPR) (Falconer and Schwarz, 1983).

In all of these techniques, the appearance of more than one peak in the spectrum indicates heterogeneity in the reaction energetics of the surface or the adlayer. Quantitative interpretation of these results can be difficult (Gorte, 1982), but detailed analysis of data is possible (Falconer and Schwarz, 1983) and even fingerprinting of such complex phenomena on catalysts can be very helpful in differentiating and developing new catalysts.

Principles

In flash desorption spectroscopy, molecules desorb from the surface directly and are pumped from the vacuum system. If the ratio of pumping speed to system volume is high and the heating rate is not excessive, the desorption rate is proportional to the pressure in the system or to the mass spectrometer signal above background. In the absence of readsorption

$$\frac{d\theta}{dT} = \frac{-k_{d,0}\theta^n}{\beta} \exp\left(\frac{-E_d}{RT}\right) \tag{26}$$

where θ is the fractional coverage of the surface, n the order of the desorption, β the heating rate dT/dt, $k_{d,0}$ the preexponential factor of the desorption rate constant, E_d the activation energy for desorption, T the absolute temperature, and R the gas constant. In general, $k_{d,0}$, E_d, and even n can be a function of coverage. If these factors are constant and the measured signal is proportional to $d\theta/dt$, then setting the derivative of Eq. (26) equal to zero produces the following relation describing the peak maximum:

$$E_d = \frac{k_{d,0}\theta_m^{n-1} nRT_m^2}{\beta} \exp\left(\frac{-E_d}{RT_m}\right) \tag{27}$$

Here θ_m and T_m refer to values at the peak maximum. Note that for first-order reactions T_m does not depend on θ, and E_d can be calculated if β and $k_{d,0}$ are known. A rough estimate for E_d (kcal/g mol) is given by $T_m/16$ (Madix, 1980).

At high carrier gas flows and in the absence of readsorption and diffusion limitations, TPD is equivalent to FDS. Increasing degrees of sophistication must be added to the analysis when coverage dependence of the parameters, readsorption, or diffusion limitations apply. As outlined succinctly by Falconer and Schwartz (1983), variation of heating rate and initial coverage; partial integrations of curves to produce desorption rate isotherms; detailed analyses of peak width, shape, and skewness; and direct curve fitting provide a variety of approaches for maximizing the return from TPD experiments. Intrusion of transport and readsorption problems cannot always be avoided, but recent analyses have provided quantitative guidelines for optimizing the design of the experiment (Herz et al., 1982; Lee and Schwartz, 1982; Gorte, 1982). Even under ideal experimental conditions, fast interconversion between various surface states can hide information from view by temperature-programmed methods. Nevertheless, this approach creates a useful family of tools for characterizing gas-surface energetics and surface reactivity.

Experimental Methods

TPD/TPR began as a simple and inexpensive experiment. Purified carrier gas or reactant passes through one side of a thermal conductivity cell, over the catalyst, and through the other side of the cell. In the absence of desorption or reaction, the thermal conductivity cell is balanced and gives a zero reading. When a linear temperature program is applied to the furnace around the sample, the temperature of the small catalyst charge is recorded by a delicate thermocouple placed in the bed and a desorption or reaction event is recorded by the imbalance in the thermal conductivity cell, which is mounted near the reactor exit to minimize the delay time. Higher sensitivity can be gained by using a flame ionization detector, but a more flexible apparatus is created when the detector can identify the composition of the reactor effluent. Since analysis must be done at the speed of the temperature program, a mass spectrometer is the most versatile detector and quadrupole mass spectrometers are often used. Accuracy and fast response require a well-designed inlet system connecting the atmospheric pressure reactor effluent to the low-pressure environment needed for ionization and mass detection. Further details of experimental design and operation are discussed by Falconer and Schwarz (1983).

Applications

Temperature-programmed techniques have been widely applied to catalysts. Falconer and Schwarz (1983) cite over 100 references. Our purpose here is to illustrate temperature-programmed desorption, reduction, reoxidation, and reaction with a few representative examples.

Temperature Programmed Desorption: Anderson et al. (1979) have examined H_2 desorption from Pt as a function of metal dispersion (fraction exposed) on different supports and of alloying with Au on silica. Broadening of the H_2 desorption profile with increasing dispersion was attributed to an increase in the number of high-energy sites at high dispersion. Alloying of Pt with Au decreased the amount of H_2 desorption but not the peak position, showing that Au acted only as an inert diluent in the Pt surface. For some metals, adsorption of hydrogen can be activated. H_2 TPD as a function of adsorption temperature has revealed strong activated adsorption effects for a variety of cobalt catalysts (Zowtiak et al., 1983).

CO shows complex, multipeaked desorption spectra for many metals. In studies of the effects of sulfur poisoning on CO desorption from Ni/SiO_2, Zagli and Falconer (1983) showed that sulfur preferentially removes the highest-energy CO adsorption sites. In an interesting contrast to the findings for H_2 desorption, Foger and Anderson (1979) showed that alloying Au with Pt changed the CO desorption spectrum significantly even though pure Au does not adsorb CO (see Fig. 27).

Temperature-Programmed Reduction: This experiment is based on the premise that the reducibility of metals on a catalyst surface can be a sensitive function of the surface chemical environment (Holm and Clark, 1968). Reducibility is typically measured by H_2 consumption during a linear temperature program. In the first application of this method to supported catalysts, Robertson et al. (1975) elucidated the conditions favoring alloying of Cu and Ni on SiO_2. Precalcination tended to maintain separate phases with identifiable reduction peaks for each metal. After direct reduction and intervening calcination, temperature-programmed reduction showed one

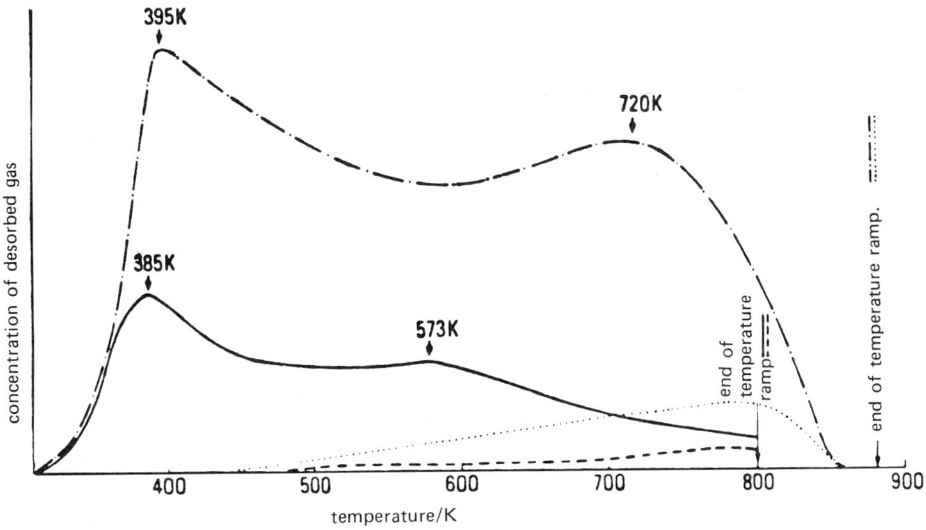

FIGURE 27 Temperature-programmed desorption profiles for carbon monoxide from platinum-gold/aerosil catalysts. Catalyst Pt15, Au85: _____, CO; ----, CO_2, both components by mass spectrometry. Catalyst Pt76 Au24 —·—·—, CO; ..., CO_2; both components by mass spectrometry. Samples: platinum-gold/aerosol (Pt15, Au85), 0.43 g. Platinum-gold/aerosol (Pt76, Au24), 0.43 g. (From Foger and Anderson, 1979.)

main reduction peak, corresponding to rereduction of both Cu and Ni. Interactions of Pt with Ge and Re on γ-Al_2O_3 have also been studied (McNicol, 1980). A calcined 0.375 wt % Pt 0.25 wt % Ge/γ-Al_2O_3 reforming catalyst showed Pt^{4+} to Pt^0 reduction at 553 K and Ge^{4+} to Ge^{2+} reduction at 873 K. Oxidation-state changes were calculated from the H_2 uptake per metal atom. Rereduction after reoxidation showed both Pt and Ge reduction occurring below 553 K and only a small residual peak at 873 K. H_2 chemisorption after interruption of the TPR cycle showed that the H_2 uptake on Pt was diminished in concert with the onset of Ge reduction. This result was interpreted as a coating of the Pt surface by Ge.

Temperature-Programmed Oxidation: Partial oxidation over mixed-oxide catalysts often involves an oxidation-reduction sequence in the surface layer. Temperature-programmed reaction is applicable to the study of the surface redox properties and has been used to advantage to study reoxidation of prereduced surfaces. Miura et al. (1975) and Uda et al. (1980) found two reoxidation peaks for bismuth molybdate and associated the higher temperature (600 K) peak with catalyst activity. In the latter study, Auger electron spectroscopy was used to assign the low-temperature peak to Mo^{4+} to Mo^{6+} conversion and a partial reoxidation of the bismuth. The high-temperature reoxidation, necessary for conversion of propylene to acrolein, completed the conversion of bismuth to Bi^{3+}.

The lability of sulfur in hydrodesulfurization catalysts is also amenable to temperature-programmed studies. Nag et al. (1980) have shown a correspondence between activity and decreasing temperature of H_2S appearance in an H_2-He stream passing over a sulfided catalyst.

Temperature-Programmed Reaction: These studies can include decomposition of adsorbed gases or surface complexes, reaction of adsorbates or reactive surface layers with gas-phase reactants, and reaction of coadsorbed species.

Falconer and Wise (1976) used temperature-programmed methods to study hydrazine and ammonia decomposition on Ir/Al_2O_3. The similarities in nitrogen evolution from the N_2H_2 and NH_3 dosed surfaces suggested that the N—N bonds are broken in N_2H_2 decomposition. Preadsorbed oxygen was found to shift the N_2 desorption peak dramatically.

CO/H_2 reactions have been a focal point for many TPR studies. Zagli et al. (1979) examined the temperature response of coadsorbed CO and H_2; and H_2 reaction with preadsorbed CO on supported nickel catalysts. Methane desorbed in a temperature range much narrower than that for CO desorption. Water desorbed in the same temperature range. CO_2 adsorption was found to be activated and dissociative on Ni/SiO_2. Temperature programming of adsorbed CO_2 in hydrogen gave methanation behavior identical to that for CO (Falconer and Zagli, 1980).

McCarty and Wise (1979) deposited carbon on Ni/Al_2O_3 by CO disproportionation at 550 K and studied its temperature-programmed reaction in hydrogen. This method is termed *temperature-programmed surface reaction* (TPSR). They observed methane production at room temperature as well as in the region expected from the CO/H_2 results. Even more carbon reactive at room temperature was found in TPSR on Ru/SiO_2 (Low and Bell, 1979). Both studies showed that thermal aging of the carbon reduces its reactivity.

Wolf and Kaul (1985) have developed a technique that combines temperature and concentration programming with transient FTIR spectroscopy. With its added control and observation of important variables, this technique has proven to be a powerful tool for studying catalytic reactions under transient conditions.

REFERENCES

Abart, J. E. Delgado, G. Ertl, H. Jeziorowski, H. Knozinger, N. Thiele, X. Zh. Wang, and E. Taglauer, Surface structure and reduction behavior of $NiO-MoO_3/Al_2O_3$ catalysts, Appl. Catal., *2*, 155 (1982).

Adams, D. M., I. R. Gardner, and N. W. Parkyns, Raman spectra of molybdenum absorbed on porous silica and of pyridine on supported nickel oxide, J. Catal., *45*, 145 (1976).

Adamson, A. W., *Physical Chemistry of Surfaces*, Wiley-Interscience, New York, (1976).

Albrecht, M. G. and J. A. Creighton, Anomalously intense Raman spectra of pyridine at a silver electrode, J. Am. Chem. Soc., *99*, 5215 (1977).

Alpert, N. L., W. E. Keiser, and H. A. Szymanski, *IR Theory and Practice of Infrared Spectroscopy*, Plenum Press, New York (1970).

Amelse, J. A., J. B. Butt, and L. H. Schwartz, Carburization of supported iron synthesis catalysts, J. Phys. Chem., *82*, 558 (1978).

Anderson, J. R., *Structure of Metallic Catalysts*, Acedemic Press, New York (1975).

Anderson, C. A. and J. R. Hinthorne, Thermodynamic approach to the quantitative interpretation of sputtered ion mass spectra, Anal. Chem., *45*, 1421 (1973).

Anderson, J. R., K. Foger, and R. J. Breakspere, Adsorption and temperature-programmed desorption of hydrogen with dispersed platinum and platinum-gold catalysts, J. Catal., *57*, 458 (1979).

Angevine, P. J., J. C. Vartuli, and W. N. Delgass, Dispersion and uniformity of supported catalysts by x-ray photoelectron spectroscopy, Proc. 6th Int Congr. Catal. Vol. 2 (1977), p. 611.

Aravind, P. K., and H. Meitiu, The effects of the interaction between resonances in the electromagnetic response of a sphere-plane structure: applications to surface enhanced spectroscopy, Surf. Sci., *124*, 506 (1983).

Baker, R. T. K., In-situ electron microscopy studies of catalyst particle behavior, Catal. Rev., *19*, 161 (1979).

Baker, R. T. K., E. B. Prestridge, and R. L. Garten, Electron microscopy of supported metal particles: I. Behavior of Pt titanium dioxide, aluminum oxide, silicon oxide and carbon, J. Catal., *56*, 390 (1979).

Barber, M. P. K. Sharpe, and J. C. Vickerman, Studies of the surface behavior of oxide catalysts by secondary ion mass spectrometry (SIMS): the surface composition of copper containing spinel catalysts and their precursors, J. Catal., *41*, 240 (1976).

Barber, M., R. S. Bordoli, R. D. Sedgewick, and A. N. Tyler, Fast atom bombardment of solids (F.A.B.): a new ion source for mass spectrometry, J. Chem. Soc. Chem. Commun., 7, 325 (1981).

Barlak, T. M., J. R. Wyatt, R. J. Colton, J. J. Decorpo, and J. E. Campana, Secondary ion mass spectrometry of metal halides: 2. Evidence for structure in alkali iodide clusters, J. Am. Chem. Soc., *104*, 1212 (1982).

Barrett, E. P., L. G. Joyner, and P. C. Halenda, The determination of pore volume and area distributions in porous substances: I. Computation from nitrogen isotherms, J. Am. Chem. Soc., *73*, 373 (1951).

Basila, M. R., Infrared spectra of adsorbed molecules, Appl. Spectrosc. Rev., *1*, 289 (1968).

Beer, M., Electron microscopy, Anal. Chem. Rev., *52*, 40R (1980).

Bell, A. T., Applications of Fourier transform infrared spectroscopy to studies of adsorbed species, in *Vibrational Spectroscopies of Adsorbed Molecules*, A. T. Bell and M. L. Hair, eds., ACS Symp. Ser., *137*, 13 (1980).

Benesi, H. A., L. T. Atkins, and R. R. Mosley, Rapid measurement of hydrogen chemisorption by supported catalytic metals, J. Catal., *23*, 211 (1971).

Bennett, C. O., The transient method and elementary steps in heterogeneous catalysis, Catal. Rev. Sci., Eng., *13*, 121 (1976).

Benninghoven, A., Surface investigations of solids by the statical method of secondary ion mass spectroscopy (SIMS), Surf. Sci., *35*, 427 (1973).

Benson, J. E. and M. Boudart, Hydrogen-oxygen titration method for the measurement of supported platinum surfaces areas, J. Catal., *4*, 704 (1965).

Berry, L. G., ed., *Powder Diffraction File, Search Manual, Inorganic Compounds*, Joint Committee on Powder Diffraction Standards, Swarthmore, PA. (1975).

Besson, B., B. Moraweck, A. K. Smith, J. M. Basset, R. Psaro, A. Fusi, and R. Ugo, IR and EXAFS characterization of a supported osmium cluster carbonyl, J. Chem. Soc. Chem. Commun., 569 (1980).

Bhasin, M. M., Auger electron spectroscopy a passing curiosity or a practical surface analysis tool for heterogeneous catalysis, in *Catalysis in Organic Synthesis*, Academic Press, New York (1976), p. 49.

Bhasin, M. M., Applications of Auger and ESCA spectroscopies, Chem. Eng. Prog., *77*, 60 (1981).

Birke, R. L., J. R. Lombardi, and L. A. Sanchez, Surface enhanced Raman spectroscopy: a new technique for studying interfacial phenomena, Adv. Chem. Ser., American Chemical Society, Washington, D.C. (1982), Vol. 201, p. 69.

Blyholder, G., in *Experimental Methods in Catalytic Research*, Vol. 1, R. B. Anderson, ed., Academic Press, New York (1968), p. 323.

Boudart, M., J. A. Dumesic, and H. Topsøe, Surface catalytic and magnetic properties of small iron particles. The effect of chemisorption of hydrogen on magnetic anisotropy, Proc. Natl. Acad. Sci., USA, *74*, 806 (1977).

Briggs, D., *Handbook of X-Ray and Ultraviolet Photoelectron Spectroscopy*, Heyden, London (1977).

Brinen, J. S. and A. J. Melera, Electron spectroscopy of chemical analyses (ESCA) starches on catalysts. Rhodium on charcoal, J. Phys. Chem., *76*, 2525 (1972).

Brongersma, H. H., L. C. M. Bierens, and G. C. J. van der Ligt, Applications of low-energy ion scattering, in *Material Characterization Using Ion Beams*, (J. P. Thomas and A. Cachard, eds.), Plenum Press, New York (1978a), p. 65.

Brongersma, H. H., M. J. Sparnaay, and T. M. Buck, Surface segregation in Cu-Ni and Cu-Pt alloys: a comparison of low energy ion scattering results with theory, Surf. Sci., *71*, 657 (1978b).

Brown, F. R., L. E. Makovsky, and K. H. Rhee, Raman spectra of supported molybdena catalysts: I. Oxide catalysts, J. Catal., *50*, 162 (1977a).

Brown, F. R., L. E. Makovski, and K. H. Rhee, Raman spectra of supported molybdena catalysts, J. Catal., *50*, 385 (1977b).

Brunauer, S., P. H. Emmett, and E. Teller, Adsorption of gases in multimolecular layers, J. Am. Chem. Soc., *60*, 309 (1938).

Buck, T. M., Low energy ion scattering spectrometry, in *Methods of Surface Analysis*, A. W. Czanderna, ed., Elsevier, New York (1975), p. 75.

Buhl, R. and A. Preisinger, Crystal structures and their secondary ion mass spectra, Surf. Sci., *47*, 344 (1975).

Bursill, L. A., E. A. Lodge, and J. M. Thomas, Zeolitic structures as revealed by high resolution electron microscopy, Nature, *286*, 111 (1980).

Burton, J. J., Structure and properties of microcrystalline catalysts, Catal. Rev., *9*, 209 (1974).

Canosa Rodrigo, B., H. Jeziorowski, H. Knozinger, X. Zh. Wang, and E. Taglauer, Elemental distribution in NiMo/γ—Al_2O_3 catalysts as studied by ion scattering spectrometry—influence of calcination time, Bull. Soc. Chem. Belg., *90*, 1339 (1981).

Carlson, T. A., *Photoelectron and Auger Spectroscopy*, Plenum Press, New York (1975).

Chang, C. C., Analytical Auger spectroscopy, in *Characterization of Solid Surfaces*, P. F. Kane and G. B. Larrabee, eds., Plenum Press, New York (1974), p. 509.

Chantry, G. W., Polarizability theory of the Raman effect, in *The Raman Effect*, Vol. 1, A. Anderson, ed., Marcel Dekker, New York (1971).

Che, W. and A. J. Tench, Characterization and reactivity of mononuclear oxygen species on oxide surfaces, Adv. Catal., *31*, 77 (1982).

Cheng, C. P. and G. L. Schrader, Characterization of supported molybdate catalysts during preparation using laser-Raman spectroscopy, J. Catal., *60*, 276 (1979a).

Cheng, C. P. and G. L. Schrader, Raman spectra of pyridine adsorbed on supported molybdate catalysts, Spectrosc. Lett., *12*, 857 (1979b).

Cheng, C. P., J. D. Ludowise, and G. L. Schrader, Controlled-atmosphere rotating cell for in situ studies of catalysts using laser Raman spectroscopy, Appl. Spectrosc., *34*, 146 (1980).

Chin, R. L. and D. M. Hercules, The influence of zinc on the surface properties of cobalt-alumina catalysts, J. Catal., *74*, 121 (1982a).

Chin, R. L and D. M. Hercules, Surface spectroscopic characterization of cobalt-alumina catalysts, J. Phys. Chem., *86*, 360 (1982b).

Cirillo, A. C., Jr., J. M. Dereppe, and W. K. Hall, Studies of the hydrogen held by solids: XXV. The proton resonance from H_2 adsorbed on alumina and molybdena-alumina catalysts, J. Catal., *61*, 170 (1980).

Clark, R. J. H, and R. E. Hester, eds., *Advances in Infrared and Raman Spectroscopy*, Vols. 1–8, Heyden, London (1974–1982).

Clausen, B. S., S. Morup, P. Nielsen, N. Thrane, and H. Topsøe, Mössbauer in situ cell with variable temperature and applied magnetic field capabilities, J. Phys. E. Sci. Instrum., *12*, 439 (1979).

Clausen, B. S., B. Lengler, R. Candia, J. Als-Nielsen, and H. Topxøe, EXAFS studies of calcined and sulfided Co-Mo Hyds catalysts, Bull Soc. Chem. Belg., *90*, 1249 (1981a).

Clausen, B. S., H. Topsøe, R. Candia,J. Villadsen, B. Lengler, J. Als-Nielsen, and F. Christensen, Extended x-ray absorption fine structure study of Co-Mo hydrodesulfurization catalysts, J. Phys. Chem., *85*, 3868 (1981b).

Clausen, C. A. and M. C. Good, A Mössbauer study of automotive emission control catalysts, J. Catal., *46*, 58 (1977).

Cohen, J. B. and L. H. Schwartz, *Diffraction from Materials*, Academic Press, New York (1977).

Cooney, R. P., G. Curthoys, and N. T. Tam, Laser Raman spectroscopy and its application to the study of adsorbed species, Adv. Catal., *24*, 293 (1975).

Creighton, J. R. and J. M. White, A SIMS study of the dehydrogenation of ethylene on Pt(111), Surf. Sci., *129*, 327 (1983).

Cullity, B. D., *Elements of X-Ray Diffraction*, Addison-Wesley, Reading, Mass. (1978).

Cvetanic, R. J. and Y. Amenomiya, A temperature programmed desorption technique for investigation of practical catalysts, Catal. Rev., *6*, 21 (1972).

Davis, L. E., N. C. MacDonald, P. W. Palmberg, G. E. Riach, and R. E. Weber, *Handbook of Auger Electron Spectroscopy*, 2nd ed., Physical Electronics Industries, Eden Prairie, Minn. (1976).

Day, R. J., S. E. Unger, and R. G. Cooks, Ionization of quaternary nitrogen compounds by secondar ion mass spectrometry, J. Am. Chem. Soc., *101*, 501 (1979).

Day, R. J., S. E. Unger, and R. G. Cooks, Molecular secondary ion mass spectrometry, Anal. Chem., *52*, 557A (1980).

Debrunner, P. G. and H. Frauenfelder, Introduction to the Mössbauer effect, in *Introduction to Mössbauer Spectroscopy*, L. May, ed., Plenum Press, New York (1971), p. 1.

Delannay, F., E. N. Haeussler, and B. Delmon, Etude par spectrométrie diffusion ionique de la structure de la bi-couche d'oxide de molybdène et de cobalt à la surface du support des catalyseurs d'hydrodésulfurization, Bull. Soc. Chim. Belg., *89*, 255 (1980a).

Delannay, F., E. N. Haeussler, and B. Delmon, The surface structure of the oxide precursor of CoMo/γ-Al_2O_3 hydrodesulfurization catalysts, J. Catal., *66*, 469 (1980b).

del Cueto, J. A. and N. J. Shevchik, EXAFS of Cu obtained with a conventional x-ray source, J. Phys. F, 7, L215 (1977).

Delgass, W. N., T. R. Hughes, and C. S. Fadley, X-ray photoelectron spectroscopy: a tool for research in catalysis, Catal. Rev., *4*, 179 (1970).

Delgass, W. N., L.-Y. Chen, and G. Vogel, Absorber cell for Mössbauer spectroscopy of heterogeneous catalysts at reaction conditions, Rev. Sci. Instrum., *47*, 968 (1976).

Delgass, W. N., G. L. Haller, R. Kellerman, and J. H. Lunsford, *Spectroscopy in Heterogeneous Catalysis*, Academic Press, New York (1979), p. 132.

Delgass, W. N., L. L. Lauderback, and D. G. Taylor, SIMS of reactive surfaces, in *Chemistry and Physics of Solid Surfaces*, Vol. 4, R. Vanselow and R. Howe, eds., Springer-Verlag, New York (1982), p. 51.

Dreschler, M., H. Hoinkes, H. Kaarman, H. Wilsch, G. Ertl, and M. Weiss, Interaction of NH_3 with Fe(110): identification of surface species by means of secondary ion mass spectroscopy (SIMS), Appl. Surf. Sci., *3*, 217 (1979).

Dufresne, P., E. Payen, J. Grimbolt, and J. P. Bonnelle, Study of Ni-Mo-γ-Al_2O_3 catalysts by x-ray photoelectron and Raman spectroscopy. Comparison with Co-Mo-γ-Al_2O_3 catalysts, J. Phys. Chem., *85*, 2344 (1981).

Dumesic, J. A. and H. Topsøe, Mössbauer spectroscopy applications to heterogeneous catalysis, Adv. Catal., *26*, 121 (1976).

Durana, J. F. and A. W. Mantz, Laboratory studies of reacting and transient systems, in *Fourier Transform Infrared Spectroscopy—Application to Chemical Systems*, Vol. 2, J. B. Ferraro and L. J. Basile, eds., Academic Press, New York (1969), p. 1.

Egerton, T. A. and A. H. Hardin, The application of Raman spectroscopy to surface chemical states, Catal. Rev. Sci. Eng., *11*, 71 (1975).

Egerton, A. A., A. H. Hardin, Y. Kozerouski, and N. Sheppard, Reduction of fluorescence from high area oxides of the silica, γ-alumina, silica-alumina and Y-zeolite types and Raman spectra for a series of molecules absorbed of these surfaces, J. Catal., *32*, 343 (1974).

Eichens, R. P. and W. A. Pliskin, The infrared spectra of adsorbed molecules, Adv. Catal., *10*, 1 (1958).

Einstein, T. L., Extended absorption fine structure analysis of surface structure, Appl. Surf. Sci., *11/12*, 42 (1982).

Emmett, P. H., *Catalysis*, Vol. 1, Reinhold, New York (1954).

Ertl, G. and J. Kuppers, *Low Energy Electrons and Surface Chemistry*, Verlag Chemie, Weinheim, West Germany (1979).

Everett, P. H. and R. H. Otterwill, eds., *Surface Area Determination*, Butterworth, London (1970).

Fadley, C. S., S. B. M. Hagstrom, M. P. Klein, and D. A. Shirley, Chemical effects on core-electron binding energies in iodine and europium, J. Chem. Phys., *48*, 3779 (1968).

Falconer, J. L. and J. A. Schwarz, Temperature programmed desorption and reaction: applications to supported catalysts, Catal. Rev. Sci. Eng., *25*, 141 (1983).

Falconer, J. L. and H. Wise, Temperature programmed desorption spectroscopy of N_2H_4 decompsition on Al_2O_3-supported Ir catalyst, J. Catal., *43*, 220 (1976).

Falconer, J. L. and A. E. Zagli, Absorption and methanation of carbon dioxide on a nickel/silica catalyst, J. Catal., *62*, 280 (1980).

Flanigan, E. D., H. Khatami, and H. A. Szymanski, in *Molecular Sieve Zeolites*, Vol. 1 American Chemical Society, Washington, D.C. (1971), p. 201.

Foger, K. and J. R. Anderson, Temperature programmed desorption of carbon monoxide adsorbed on supported platinum catalysts, Appl. Surf. Sci., *2*, 335 (1979).

Fraser, W. A., J. V. Florio, W. N. Delgass, and W. D. Roberson, Surface sensitivity and angular dependence of x-ray photoelectron spectra, Surf. Sci., *36*, 661 (1973).

Furtak, T. E. and J. Reyes, A critical analysis of theoretical models for the giant Raman effect from adsorbed molecules, Surf. Sci., *93*, 351 (1980).

Fyfe, C. A., G. C. Gobbi, J. Klinowski, J. M. Thomas, and S. Ramdas, Resolving crystallographically distinct tetrahedral sites in silicalite and ZSM-5 by solid state NMR, Nature, *296*, 530 (1982).

Ganeson, P., H. K. Kuo, A. Saavedra, and R. J. De Angelis, Particle size distribution function of supported metal catalysts by x-ray diffraction, J. Catal., *52*, 310 (1978).

Garrison, B. J., Mechanism of ejection of organic molecules from surfaces by keV ion bombardment, J. Am. Chem. Soc., *102*, 6553 (1980).

Garrison, B. J., N. Winograd, and D. E. Harrison, Jr., Atomic and molecular ejection from ion-bombarded reacted single-crystal surfaces. Oxygen on Cu(100), Phys. Rev. B, *18*, 600 (1978).

Garrison, B. J., N. Winograd, and D. E. Harrison, Jr., Mechanisms of CO ejection from ion bombarded single crystal surfaces, J. Chem. Phys., *73*, 3473 (1980).

Garten, R. L., Mössbauer spectroscopy of supported bimetallic catalysts, in *Mössbauer Effect Methodology*, Vol. 10, I. J. Gruversman and C. W. Seidel, eds., Plenum Press, New York (1976), p. 69.

Garten, R. L. and J. H. Sinfelt, Structure of Pt-Ir catalysts: Mössbauer spectroscopy studies employing ^{57}Fe as a probe, J. Catal., *62*, 127 (1980).

Garten, R. L., W. N. Delgass, and M. Boudart, A Mössbauer spectroscopic study of the reversible oxidation of ferrous ions in Y-zeolite, J. Catal., *18*, 90 (1970).

Gay, I. D., Carbon-13 nuclear magnetic resonance spectroscopy of molecules adsorbed on silica, J. Phys. Chem., *78*, 38 (1974).

Gibbs, R. A., S. P. Holland, K. E. Foley, B. J. Garrison, and N. Winograd, Energy and angle resolved SIMS studies of CO on Ni(001), J. Chem. Phys., *76*, 684 (1982).

Glassl, H., R. Kramer, and K. Hayek, Electron microscopy of Pt/γAl_2O_3 model catalysts: artifacts in sample preparation, J. Catal., *63*, 167 (1980).

Goldstein, J. I and H. Yakowitz, *Practical Scanning Electron Microscopy*, Plenum Press, New York (1975).

Gorte, R., Design parameters for temperature programmed desorption from porous catalysts, J. Catal., *75*, 164 (1982).

Grant, J. T. and M. P. Hooker, Oxygen KLL Auger spectra from O_2 and CO_2 adsorbed on Ni, Solid State Commun., *19*, 111 (1976).

Gray, P. R. and F. E. Farha, A Mössbauer investigation of a platinum-tin parafin dehydrogenation catalyst, in *Mössbauer Effect Methodology*, Vol. 10, I. J. Gruverman and C. W. Seidel, eds., Plenum Press, New York (1976), p. 47.

Greenwood, N. N. and T. C. Gibb, *Mössbauer Spectroscopy*, Chapman & Hall, London (1971).

Gregg, S. J. and K. S. L. Singh, *Adsorption Surface Area and Porosity*, Academic Press, New York (1967).

Griffith, P. R., *Fourier Transform Infrared Spectroscopy*, Wiley, New York (1975).

Gruber, H. L., Chemisorption studies on supported platinum, J. Phys. Chem., *66*, 48 (1962).

Hair, M. L., *Infrared Spectroscopy in Surface Chemistry*, Wiley-Interscience, New York (1967).

Haller, G. L., in *Heterogeneous Catalysis*, W. N. Delgass, G. L. Haller, R. Kellerman, and J. H. Lunsford, Academic Press, New York (1979), p. 58.

Harrison, D. E., Jr., P. W. Kelley, B. J. Garrison, and N. Winograd, Low energy ion impact phenomena on single crystal surfaces, Surf. Sci., *76*, 311 (1978).

Hausen A., and H. L. Gruber, Gas chromatographic study of hydrogen adsorption on supported metals, J. Catal., *20*, 97 (1971).

Hawkes, P. W., *Electron Optics and Electron Microscopy*, Taylor & Francis, London (1972).

Hayes, T. M. and J. B. Boyce, Extended x-ray absorption fine structure spectroscopy, Solid State Phys., *37*, 173 (1982).

Heidenreich, R. D., *Fundamentals of Transmission Electron Microscopy*, Wiley-Interscience, New York (1964).

Heiland, W. and E. Taglauer, Investigation of surface topography of oxygen on nickel single crystals by helium ion backscattering, J. Vac. Sci. Technol., *9*, 620 (1972).

Heinrich, K. F. J., *Electron Beam X-Ray Microanalysis*, Van Nostrand Reinhold, New York (1980).

Hendra, P. J., Raman spectroscopy at surfaces, Spex Speaker, *19*, 1 (1974).

Hendra, P. J and P. M. Stratton, Laser-Raman spectroscopy, Chem. Rev., *69*, 325 (1969).

Hendra P. J., J. A. Horder, and E. J. Loader, The laser-Raman spectrum of pyridine adsorbed on oxide surfaces, J. Chem. Soc., A, 1766 (1971).

Hercules, D. M., Challenges in surface science, Anal. Chem., *50*, 734A (1978).

Hercules, D. M. and J. C. Carver, Electron spectroscopy: x-ray and electron excitation, Anal. Chem., *46*, 133R (1974).

Hercules, S. H and D. M. Hercules, Surface characterization by electron spectroscopy for chemical analysis (ESCA), in *Characterization of Solid Surfaces,* P. F. Kane and G. B. Larrabee, eds., Plenum Press, New York (1974), p. 307.

Herz, R., in *Catalysis Under Transient Conditions,* A. T. Bell and L. Hegedus, eds., ACS Symp. Ser., *178,* 59 (1982).

Herz, R. K., J. P. Krela, and S. P. Marin, Adsorption effects during temperature programmed desorption of carbon monoxide from supported platinum, J. Catal., *73,* 66 (1982).

Ho, S. and L. Hegedus, Dynamics of high temperature carbon monoxide chemisorption on Pt alumina by fast-response IR spectroscopy, in *Catalysis under Transient Conditions,* A. T. Bell and L. Hegedus, eds., ACS Symp. Ser., *178,* 79 (1982).

Holm, V. C. F. and A. Clark, Reduction studies of supported metal oxide catalysts, J. Catal., *11,* 305 (1968).

Howe, R. F., EPR spectroscopy in surface chemistry: Recent developments, Adv. Colloid Interface Sci., *18,* 1 (1982).

Hughes, T. R., R. J. Houston, and R. P. Sieg, Flow adsorption method for catalyst metal surface measurements, Ind. Eng. Chem. Prod. Res. Dev., *1,* 95 (1962).

Hurst, N. W., S. J. Gentry, A. Jones, and B. D. McNicol, Temperature programmed reduction, Catal. Rev. Sci. Eng., *24,* 233 (1982).

Innes, W. B., in *Experimental Methods in Catalytic Research,* R. B. Anderson, ed., Academic Press, New York, (1968).

Jeanmarie, D. J. and R. P. Van Duyne, Surface Raman spectroelectrochemistry: Part I. Heterocydic aromatic and aliphatic amines adsorbed on the anodized silver electrode, J. Electroanal. Chem., *84,* 1 (1977).

Jeziorowski, H. and H. Knozinger, Raman and ultraviolet spectroscopic characterization of molybdena on alumina catalysts, J. Phys. Chem., *83,* 1166 (1979).

Jeziorowski, H. and H. Knozinger, Interaction between nickel and molybdena in alumina-supported catalysts as studied by Raman spectroscopy, Appl. Surf. Sci., *5,* 335 (1980).

Jeziorowski, H., H. Knozinger, E. Taglauer, and C. Vogdt, Low energy ion scattering study of Ni-Mo-Al_2O_3 catalysts, J. Catal., *80,* 286 (1983).

Johari, O. and A. V. Samudra, Scanning electron microscopy, in *Characterization of Solid Surface,* P. F. Kane and G. B. Larrabee, eds., Plenum Press, New York (1974), p. 107.

Joyner, R. W., Extended x-ray absorption fine structure studies on supported catalysts, in *Characterization of Catalysts,* J. M. Thomas and R. M. Lambert, eds., Wiley, New York (1980), p. 237.

Kalvius, G. M., Evaluation of Isomer Shifts, Proc. Conf. Hyperfine Interact. Exited Nuclei, Vol. 2, (1971), p. 523.

Kaul, D. J. and E. E. Wolf, Transient FTIR studies on surface dynamics: Part II. Self sustained oscillations during oxidation on Pd/SiO_2 catalyst, J. Catal., *91,* 216 (1985).

Kaul, D. J. and E. E. Wolf, Transient FTIR studies of surface reaction dynamics: Part I. CO oxidation on Pt/SiO_2, J. Catal., *89,* 348 (1984).

King, D. A., Thermal desorption from metal surfaces: A review, Surf. Sci., *47,* 384 (1975).

Klier, K., in *Vibrational Spectroscopies for Adsorbed Species,* A. T. Bell and M. L. Hair, eds., ACS Symp. Ser., *137,* 141 (1980).

Klug, H. P. and L. E. Alexander, *X-Ray Diffraction Procedures*, Wiley-Interscience, New York (1974).

Knapp, G. S., H. Chen, and T. E. Klippert, Development of a laboratory EXAFS facility, Rev. Sci. Instrum., *49*, 1658 (1978).

Knozinger, H., H. Jeziorowski, and E. Taglauer, Characterization of NiO-MoO_3-Al_2O_3 catalysts, *Proceedings, International Congress on Catalysis, 7th (Tokyo, 1980)*, Elsevier, Amsterdam (1981), p. 604.

Kobayashi, H. and M. Kobayashi, Transient response method in heterogeneous catalysis, Catal. Rev. Sci. Eng., *10*, 139 (1974).

Kokes, R. J., A. L. Dent, C. C. Chang, and L. T. Dixon, Infrared studies of isotopic effects for H_2 adsorption on zinc oxide, J. Am. Chem. Soc., *94*, 4429 (1972).

Kuiper, A. E. T., J. Medema, and J. J. Van Bokhoven, Infrared and raman spectra of benzaldehyde absorbed on alumina, J. Catal., *29*, 40 (1973).

Landman, U. and D. C. Adams, Extended x-ray absorption fine structure-auger process for surface structure analysis: theoretical consideration of a proposed experiment, Proc. Natl. Acad. Sci. USA, *73*, 2250 (1976).

Lauderback, L. L. and W. N. Delgass, SIMS studies of the structure and reactivity of carbon on Ru(001), ACS Petrol. Div. Prepr., *28*, 1273 (1983).

Lauderback, L. L. and W. N. Delgass, A SIMS study of the C_2H_4/Ru(001) interaction, ACS Symp. Ser., *248*, 21 (1984).

Leapman, R. D., L. A. Grunes, P. L. Fejes, and J. Silcox, Extended core edge fine structure in electron energy loss spectra, in *EXAFS Spectroscopy: Techniques and Applications*, B. K. Teo and D. C. Joy, eds., Plenum Press, New York (1981), p. 217.

Lee, P. I. and J. A. Schwartz, Adsorption desorption kinetics of H_2 from supported nickel catalysts, J. Catal., *73*, 272 (1982).

Liao, P. C. and E. E. Wolf, CO oxidation activity and XPS studies of Pt-Cu/γ-Al_2O_3 bimetallic catalysts, J. Catal., *74*, 307 (1982).

Little, L. H., *Infrared Spectra of Adsorbed Species*, Academic Press, New York (1976).

Low, G. G. and A. T. Bell, Studies of CO desorption and reaction with H_2 on alumina-supported Ru, J. Catal., *57*, 397 (1979).

Lund, C. R. F. and J. A. Dumesic, Strong oxide-oxide interations in silica-supported magnetite catalysts, J. Catal., *76*, 93 (1982).

Lunsford, J. H., Electron spin resonance in catalysis, Adv. Catal., *22*, 265 (1972).

Lunsford, J. H., Electron spin resonance spectroscopy, in *Spectroscopy in Heterogeneous Catalysis*, W. N. Delgass, G. L. Haller, R. Kellerman, and J. H. Lunsford, Academic Press, New York (1979a), p. 183.

Lunsford, J. H., Nuclear magnetic resonance spectroscopy, in *Spectroscopy in Heterogeneous Catalysis*, W. N. Delgass, G. L. Haller, R. Kellerman, and J. H. Lunsford, Academic Press, New York (1979), p. 236.

Madix, R. J., Reaction kinetics and mechanism on metal single crystal surfaces, Adv. Catal., *29*, 1 (1980).

Margulies, S. and J. R. Ehrman, Transmission and line broadening of resonance radiation incident on a resonance absorber, Nucl. Instrum. Methods, *12*, 131 (1961).

McCarty, J. G. and H. Wise, Hydrogenation of surface carbon on alumina-supported nickel, J. Catal., *57*, 406 (1979).

McGuire, G. E., *Auger Electron Spectroscopy Reference Manual*, Plenum Press, New York (1979).
McHugh, J. A., Secondary ion mass spectrometry, in *Methods of Surface Analysis*, A. W. Czanderna, ed., Elsevier, New York (1975), p. 223.
McNicol, B. D., Catalyst characterization by temperature and voltage programming, in *Characterization of Catalysts*, J. M. Thomas and R. M. Lambert, eds., Wiley, New York (1980), p. 141.
Mears, D. E. and R. C. Hansford, The stoichiometry for hydrogen titration of oxygen on supported platinum, J. Catal., *9*, 125 (1967).
Meitzner, G., G. H. Via, F. W. Lytle, and J. H. Sinfelt, Structure of Bimetallic clusters. Extended x-ray absorption fine structure (EXAFS) studies of Rh-Cu clusters, J. Chem. Phys., *78*, 882 (1983).
Miura, H., Y. Morikawa, and T. Shirasaki, Studies on the reduction-reoxidation of bismuth molybdate catalysts by temperature programmed reoxidation method, J. Catal., *39*, 22 (1975).
Mohri, M., H. Kakibayashi, K. Watanabe, and T. Yamashina, Application of SIMS-FDS-AES combined system to surface studies: adsorption and desorption of formic acid over clean surface of nickel, Appl. Surf. Sci., *1*, 170 (1978).
Morrison, T. I., A. H. Reis, Jr., E. Gebet, L. E. Iton, G. D. Stucky, and S. L. Suib, EXAFS as a probe of local coordination environments of transition metal ions in polycrystalline zeolite catalysts. Co(II) in hydrated zeolite-A and Y, J. Chem. Phys., *72*, 6276 (1980).
Morup, S., J. A. Dumesic, and H. Topsøe, Magnetic microcrystals, in *Applications of Mössbauer Spectroscopy*, Vol. 2 (R. L. Cohen, ed., Academic Press, New York (1980), p. 1.
Moskovits, M. and D. P. Dilella, Enhanced Raman spectra of ethylene and propylene adsorbed on silver, Chem. Phys. Lett., *73*, 500 (1980).
Nagy, J. B., J.-P. Gilson, and E. G. Derouane, Evidence for secondary building unit effects on the solid state ^{29}Si NMR. Resonance of silicon in zeolitic structures, J. Chem. Soc., Chem. Commun., 1129 (1981).
Nagy, J. B., A. Abou-Kais, M. Guelton, J. Harmel, and E. G. Derouane, Isomerization of 1-butene adsorbed on mixed tin-antimony oxide catalysts, A ^{13}C-NMR study, J. Catal., *73*, 1 (1982).
Niehus, H. and E. Bauer, Quantitative aspects of ion scattering spectroscopy (ISS), Surf. Sci., *47*, 222 (1975).
Niemantsverdriet, J. W., A. M. van der Kraan, W. L. van Dijk, and H. S. van der Baan, The behavior of metallic iron catalysts during Fischer-Tropsch synthesis studied with Mössbauer spectroscopy x-ray diffraction, carbon content determination and reaction kinetic measurements, J. Phys. Chem., *84*, 3863 (1980).
Ott, G. L., T. Fleisch, and W. N. Delgass, Fischer-Tropsch synthesis over freshly reduced iron ruthenium alloys, J. Catal., *60*, 394 (1979).
Palmberg, P. W., Auger electron spectroscopy, in *Electron Spectroscopy*, D. A. Shirley, ed., North Holland, Amsterdam (1975), p. 835.
Parfitt, G. D. and S. L. Singh, *Characterization of Powder Surfaces*, Academic Press, New York (1976).
Payen, E., M. C. Dhamelincourt, P. Dhamelincourt, J. Grimbolt, and J. P. Bonnelle, Study of Co (or Ni)-Mo oxide phase transformation and hydrodesulfurization catalysts by Raman microprobe equipped with new cells, Appl. Spectrosc., *36*, 30 (1982).
Peri, J. B., Infrared study of nitric oxide and carbon monoxide adsorbed on chromia/alumina, J. Phys. Chem., *78*, 588 (1974)

Phillips, J., B. Clausen, and J. A. Dumesic, Iron pentacarbonyl decomposition over grafoil. Production of small metallic iron particles, J. Phys. Chem., *84*, 1814 (1980).

Pichat, P., R. Beumont, and P. Barthomeuf, Infrared structural study of aluminum-deficient Y zeolite, J. Chem. Soc. Faraday Trans, I, *70*, 1402 (1974).

Pierce, J. L. and R. A. Walton, Characterization of nickel complexes on oxide supports by secondary ion mass spectrometry, J. Catal., *81*, 375 (1983).

Pierce, J. L., K. L. Busch, R. G. Cooks, and R. A. Walton, Desorption ionization mass spectrometry: secondary ion and laser desorption spectra of transition-metal complexes of B-diketones, Inorg. Chem., *21*, 2597 (1982).

Poate, J. M. and T. M. Buck, Ion scattering spectroscopy, in *Experimental Methods in Catalytic Research*, Vol. 3, R. B. Anderson and P. T. Dawson, eds., Academic Press, New York (1976), p. 175.

Pockrand, I., A Raman vibrational study of water adsorption on silver, Surf. Sci., *122*, L569 (1982).

Postek, M. T., K. S. Howard, A. H. Johnson, and K. L. McMichael, *Scanning Electron Microscopy, A Student's Handbook*, Ladd Research Industries, Inc., Boston, Mass. (1980).

Powell, C. J., Comparison of ESCA with other surface analysis techniques, in *Applied Electron Spectroscopy for Chemical Analysis*, H. Windlaw and F. Ho, eds., Wiley, New York (1982), p. 19.

Powell, C. T., N. E. Erickson, and T. E. Madey, Results of a joint Auger/ESCA round robin sponsored by ASTM Committee E-42 on surface analysis: Part I. ESCA results, J. Electron Spectrosc. Related Phenom., *17*, 361 (1979).

Powell, C. J., N. E. Erickson, and T. E. Madey, Results of a joint Auger/ESCA round robin sponsored by ASTM Committee E-42 on surface analysis: Part II. Auger results, J. Electron Spectrosc. Related Phenom., *25*, 87 (1982).

Pritchard, J. and T. Catterick, Infrared spectroscopy, in *Experimental Methods in Catalytic Research*, Vol. 3, R. B. Anderson and P. T. Dawson, eds., Academic Press, New York (1976), p. 281.

Purdy, G. R. and R. B. Anderson, The use of the electron probe microanalyzer in catalysis, in *Experimental Methods in Catalytic Research*, Vol. 2, R. B. Anderson and P. T. Dawson, eds., Academic Press, New York (1976), p. 95.

Ramamoorthy, P. and R. D. Gonzalez, Surface characterization of supported Pt-Ru bimetallic clusters using IR spectroscopy, J. Catal., *58*, 188 (1979).

Ramdas, S., J. M. Thomas, J. Klinoski, C. A. Fyfe, and J. S. Hartman, Ordering of aluminum and silicon in synthetic faujasites, Nature, *292*, 228 (1981).

Raupp, G. B. and W. N. Delgass, Mössbauer investigation of supported Fe catalysts: III. In situ kinetics and spectroscopy during Fischer-Tropsch synthesis, J. Catal., *58*, 361 (1979).

Reuter, W., M. L. Yu, M. A. Frisch, and M. B. Small, Charging effects in the secondary ion mass spectrometric analysis of targets containing low-conductivity regions, J. Appl. Phys., *51*, 850 (1980).

Robertson, S. D., B. D. McNicol, J. H. de Baas, S. C. Kloet, and J. W. Jenkins, Determination of reducibility and identification of alloying in copper-nickel-on-silica catalysts by temperature programmed reduction, J. Catal., *37*, 424 (1975).

Rowe, E. M., Synchrotron radiation facilities in the United States, Phys. Today, *34*, 28 (May 1981).

Ryan, L. M., R. E. Taylor, A. J. Paff, and B. C. Gerstein, An experimental study of resolution of proton chemical shifts in solids: Combined multiple pulse NMR and magic angle spinning, J. Chem. Phys., *72*, 508 (1980).

Saha, N. S. and E. E. Wolf, CO hydrogenation activity and XPS studies on Pd supported on ZSM-5 and Y-zeolites, J. Appl. Catal., *13*, 101 (1984).

Sandstrom, D. R. and F. W. Lytle, Developments in extended x-ray absorption. Fine structure applied to chemical systems, Annu. Rev. Phys. Chem., *30*, 215 (1979).

Sargent, C. M. and J. D. Embury, Transmission and scanning electron microscopy, in *Experimental Methods in Catalytic Research*, Vol. 2, R. B. Anderson and P. T. Dawson, eds., Academic Press, New York (1976), p. 140.

Sarkany, J. and R. Gonzalez, The measurement of extinction coefficients using a new infrared micropulse reactor technique, adsorption of CO on silica supported Pt, Appl. Spectrosc., *36*, 320 (1982a).

Sarkany, J. and R. Gonzalez, On the use of the dynamic pulse method to measure metal surface areas, J. Catal., *76*, 75 (1982b).

Sashital, S. R., J. B. Cohen, R. L. Burwell, and J. B. Butt, Characterization of the gel and platinum particles by x-ray diffraction, J. Catal., *50*, 479 (1977).

Sato, Y., Y. Iwasawa, and H. Kuroda, EXAFS study of highly active Mo_2 catalyst, Chem. Lett., 1101 (1982).

Savatsky, B. J. and A. T. Bell, Nitric oxide reduction by hydrogen over rhodium using transient response techniques, in *Catalysis Under Transient Conditions*, A. T. Bell and L. Hegedus, eds., ACS Symp., *178*, 105 (1982).

Schmidt, L., T. Wang, and A. Vazquez, Electron microscopy of catalyst particles, Ultramicroscopy, *8*, 175 (1982).

Schrader, G. L. and C. P. Cheng, In situ laser Raman spectroscopy of the sulfiding of Mo-γ-Al_2O_3 catalysts, J. Catal., *80*, 369 (1983).

Schultz, J. M., *Diffraction for material scientists*, Prentice-Hall, Englewood Cliffs, N.J. (1982).

Schumacher, P. J., Secondary ion mass spectrometry of clean and hydrocarbon treated cracking catalysts, M.S. Thesis, Purdue University (1983).

Shelef, M., M. A. Z. Wheeler, and H. C. Yao, Ion scattering spectra from spinel surfaces, Surf. Sci., *47*, 697 (1975).

Shen, W. N., J. A. Dumesic, and C. G. Hill, Jr., Mössbauer spectroscopy cell for in situ catalyst characterization and reaction kinetic studies at high pressures, Rev. Sci. Instrum., *52*, 858 (1981).

Siegbahn, K., et al., ESCA, *Atomic Molecular and Solid State Structure by Means of Electron Spectroscopy*, Almqvist and Wiksells, Uppsala (1967).

Sinfelt, J. H., G. H. Via, and F. W. Lytle, Structure of bimetallic clusters. Extended x-ray absorption fine structure (EXAFS) studies of Ru-Cu clusters, J. Chem. Phys., *72*, 4832 (1980).

Sinfelt, J. H., G. H. Via, F. W. Lytle, and R. B. Gregor, Structure of bimetallic clusters. Extended x-ray absorption fine structure (EXAFS) studies of Os-Cu clusters, J. Chem. Phys., *75*, 5527 (1981).

Sinflet, H., G. H. Via, and F. W. Lytle, Structure of bimetallic clusters, extended x-ray absorption fine structure (EXAFS) studies of Pt-Ir clusters, J. Chem. Phys., *76*, 2779 (1982).

Slichter, C. P., NMR study of platinum catalysts, Surf. Sci., *106*, 382 (1981).

Smith, D. P., Scattering of low-energy noble gas ions from metal surfaces, J. Appl. Phys., *38*, 340 (1967).

Smith, D. P., Analysis of surface composition with low-energy backscattered ions, Surf. Sci., *25*, 171 (1971).

Spenadel, L. and M. Boudart, Dispersion of platinum on supported catalysts, J. Am. Chem. Soc., *64*, 204 (1960).

Stanfield, R. M. and W. N. Delgass, Mössbauer spectroscopy of supported Fe-Co alloy catalysts for Fischer-Tropsch synthesis, J. Catal., *72*, 37 (1981).

Taglauer, E. and W. Heiland, Direct comparison of low-energy ion backscattering with Auger electron spectroscopy in the analysis of S adsorbed on Ni, Appl. Phys. Lett., *24*, 437 (1974).

Taglauer, E. and W. Heiland, Low energy ion scattering and Auger electron spectroscopy of clean nickel surfaces and adsorbed layers, Surf. Sci., *42*, 234 (1975).

Taglauer, E. and W. Heiland, Surface analysis with low energy ion scattering, Appl. Phys., *9*, 261 (1976).

Tam, N. T., R. P. Cooney, and G. Curthoys, Laser Raman spectra of thiophene absorbed on silica, J. Colloid Interface Sci., *51*, 340 (1975).

Tamaru, K., Adsorption measurements during surface catalysis, Adv. Catal., *15*, 65 (1964).

Tamaru, K., *Dynamic Heterogeneous Catalysis*, Academic Press, London (1978).

Tamm, P. W., H. F. Harnsberger, and A. G. Bridge, The effect of feed metals on catalyst aging in hydroprocessing residium, 72nd AIChE Meet., San Francisco (1979).

Tatarchuk, B. J. and J. A. Dumesic, Physical characterization of Fe/TiO_2 model supported catalysts: II. Electron spectroscopic studies of reduction behavior, J. Catal., *70*, 323 (1981).

Taylor, R. E., M. M. Silva Crawford, B. C. Gerstein, Self diffusion of hydrogen in $H_{1.65}MoO_3$, J. Catal., *62*, 401 (1980).

Tench, A. J. and P. Holroyd, The identification of O_2^- adsorbed on magnesium oxide, Chem. Commun., 471 (1981).

Teo, B. K. and D. C. Joy, eds., *EXAFS Spectroscopy: Techniques and Applications*, Plenum Press, New York (1981).

Thornton, P. R., *Scanning Electron Microscopy*, Chapman & Hall, London (1968).

Tobias, R. S., Raman spectroscopy in inorganic chemistry, J. Chem. Educ., *44*, 2 (1967).

Topsøe, H., J. A. Dumesic, and M. Boudart, Alumina as a textural promoter of iron synthetic ammonia catalysts, J. Catal., *28*, 477 (1973).

Topsøe, H., B. S. Clausen, R. Candia, C. Wivel, and S. Mørup, In situ Mössbauer emission spectroscopy studies of unsupported and supported sulfided Co-Mo hydrodesulfurization catalysts. Evidence for and nature of a Co-Mo-S-phase, J. Catal., *68*, 433 (1981).

Turner, I. D. M., S. O. Paul, E. Reid, and P. J. Hendra, Laser-Raman study of the isomerization of olefins over alumina, J. Chem. Soc., Faraday Trans. I, *72*, 2829 (1976).

Uda, T., T. T. Lin, and G. W. Keulks, The catalytic oxidation of propylene VII. The use of temperature programmed reoxidation to characterize γ-bismuth molybdate, J. Catal., *62*, 26 (1980).

Underwood, E. E., *Quantitative Stereology*, Addison-Wesley, Reading, Mass. (1970).

Van Duyne, R. P., Laser excitation of Raman scattering from adsorbed molecules on electrode surfaces, in *Chemical and Biochemical Applications of Lasers*, Academic Press, New York (1979), p. 10.

Van Duyne, R. P., D. L. Jeanmarie, and D. F. Shriver, Mode-locked laser Raman spectroscopy—a new technique for the rejection of interfering background luminescence signals, Anal. Chem., *46*, 213 (1974).

Vannice, M. A., Y.-L. Lam, and R. L. Garten, CO hydrogenation over well-characterized FeRu alloys, Am. Chem. Soc. Div. Petrol. Chem. Prepr., *23*, 495 (1978).

Venkataraman, B., Time resolved election spin resonance spectroscopy, Curr. Sci., *51*, 397 (1982).

Via, G. H., J. H. Sinfelt, and F. W. Lytle, Extended x-ray absorption fine structure (EXAFS) of dispersed metal catalysts, J. Chem. Phys., *71*, 690 (1979).

Via, G. H., J. H. Sinfelt, and F. W. Lytle, EXAFS studies of supported metal catalysts, in *EXAFS Spectroscopy: Techniques and Applications*, B. K. Teo and D. C. Joy, eds., Plenum Press, New York (1981), p. 159.

Wagner, C. D., W. M. Riggs, L. E. Davis, J. F. Moulder and G. E. Muilenberg, eds., *Handbook of X-Ray Photoelectron Spectroscopy*, Perkin Elmer Corporation, Eden Prairie, Minn. (1978).

Wang, L. and W. K. Hall, On the genesis of molybdena-alumina catalyst, J. Catal., *66*, 251 (1980).

Wanke, S., B. K. Lotochinski, and H. C. Sidwell, Hydrogen adsorption measurements by the dynamic pulse method, Prepr. Can. Symp. Catal., *6*, 78 (1979).

Warren, B. E. and B. L. Averbach, The effect of cold work distortion on x-ray patterns, J. Appl. Phys., *21*, 595 (1950).

Weaver, M. J., F. Barz, J. G. Gordon, and M. R. Philpott, Surface enhanced Raman spectroscopy of electrochemically characterized surfaces: potential dependence of Raman spectra for thiocyanate at silver electrodes, Surf. Sci., *125*, 409 (1983).

Werner, H. W., The use of secondary ion mass spectrometry in surface analysis, Surf. Sci., *47*, 301 (1975).

Wertheim, G. K., *Mössbauer Effect, Principles and Applications*, Academic Press, New York (1964).

Whyte, T. E., Metal particle size determination of supported metal catalysts, Catal. Rev., *8*, 117 (1973).

Wilson, G. R. and W. K. Hall, Study of hydrogen held by solids, J. Catal., *17*, 190 (1970).

Winograd, N., Characterization of solids and surfaces using ion beams and mass spectrometry, in *Progress in Solid-State Chemistry*, G. M. Rosenblatt and W. L. Worrell, eds., Pergamon Press, Elmsford, N.Y. (1981).

Winograd, N. and B. J. Garrison, Surface structure determinations with ion beams, Acc. Chem. Res., *13*, 406 (1980).

Wittmaack, K., Primary-ion charge compensation in SIMS analysis of insulators, J. Appl. Phys., *50*, 493 (1979).

Wood, T. H., and M. V. Klein, Studies of the mechanism of enhanced Raman scattering in ultrahigh vacuum, Solid State Commun., *35*, 263 (1980).

Wu, M. and D. M. Hercules, Studies of supported nickel catalysts by x-ray photoelectron and ion scattering spectroscopies, J. Phys. Chem., *83*, 2003 (1979).

Yannoni, C. S., High-resolution NMR in solids: the CPMAS experiments, Acc. Chem. Res., *15*, 201 (1982).

Zagli, A. E., J. L. Falconer, and C. A. Keenan, Methanation of supported nickel catalysts using temperature programmed heating, J. Catal., *56*, 453 (1979).

Zagli, A. E. and J. L. Falconer, unpublished results (1983).

Zimmerer, N. and W. Kiefer, Rotating surface scanning technique for Raman spectroscopy, Appl. Spectrosc., *28*, 279 (1974).

Zingg, D. S., L. E. Makovsky, R. E. Tischer, F. R. Brown, and D. M. Hercules, A surface spectroscopic study of molybdenum-alumina catalysts using x-ray photoelectron, ion-scattering and Raman spectroscopies, J. Phys. Chem., *84*, 2898 (1980).

Zowtiak, J. M., G. D. Weatherbee, and C. H. Bartholomew, Activated Adsorption of H_2 on cobalt and effects of support thereon, J. Catal., *82*, 230 (1983).

4 Diffusion-Reaction Interactions in Catalyst Pellets

DAN LUSS *University of Houston, Houston, Texas*

INTRODUCTION

The reactants within a porous catalytic pellet are consumed by chemical reactions and replenished by transport processes. Competitive kinetic and transport rate processes affect the observable rate and selectivity of a catalyst. The engineering implications of this interaction between the diffusional transport and chemical reactions were pointed out first by Damköhler (1937), Zeldovich (1939), Thiele (1939), Wheeler (1951), and Weisz and Prater (1954). The most comprehensive review of the subject is the monograph by Aris (1975).

A porous medium consists of a complex matrix of many interconnected channels of varying diameters and lengths, through which the reactant species have to diffuse in order to reach the active catalytic sites. Obviously, accurate geometric models of a porous structure cannot be developed. A proper analysis of the performance of a catalyst requires use of a reasonable, simplified model which incorporates the essential chemical and physical properties of the pellet. This is usually accomplished by use of a single-phase pseudohomogeneous model. This crude representation eliminates the distinction between the solid and fluid phases within the pellet and simplifies considerably the prediction of most properties which are of engineering importance.

We describe here methods of predicting the impact of the coupling between the chemical and transport rate processes on the activity and selectivity of porous catalytic pellets. This information is essential for the rational selection, testing, development and application of catalysts. We start with a review of the rate of transport within and to porous materials.

TRANSPORT WITHIN AND TO POROUS MEDIA

Transport of Gases Within Porous Media

There exists at present no flux relation with a completely sound theoretical basis capable of describing transport of gases in porous media over a wide range of pore sizes and pressures. A comprehensive review of the subject can be found in the monograph by Jackson (1977). Existing transport models, which relate the flux to local gradients in the compositions, pressure, and temperature, assume that a porous medium can be described either as a matrix of interconnected channels or as an assembly of stationary dust particles dispersed in the gas.

Flux models for capillary networks are naturally based on known transport laws for a single capillary. Molecular-wall collisions is the mechanism responsible for transport within an isothermal, isobasic, narrow capillary, whose diameter is small in comparison with the mean free path of the gaseous species. The flux relation for this Knudsen flow is

$$N_A = -\frac{D_{KA}}{RT} \text{ grad } p_A \tag{1}$$

$$D_{KA} = \frac{2r_p}{3}\left(\frac{8RT}{\pi M_A}\right)^{0.5} = \frac{2}{3} r_p u_A \tag{2}$$

where p_A is the partial pressure of species A and u_A is the corresponding average molecular velocity.

Gaseous diffusion is the dominant transport mechanism in a wide capillary with a diameter much larger than the mean-free-path length. The corresponding isobaric, isothermal fluxes satisfy the Stefan-Maxwell relations

$$-\frac{p}{RT} \text{ grad } x_i = \sum_{\substack{i=1 \\ i \neq j}}^{n} \frac{x_j N_i - x_i N_j}{D_{ij}} \tag{3}$$

where D_{ij} is the binary gaseous diffusion coefficient. According to the simple kinetic theory of gases, the self-diffusion coefficient is obtained as

$$D_{ii} = \frac{\lambda u_i}{3} \tag{4}$$

Very accurate predictions of binary diffusion coefficients can be obtained from the Chapman-Enskog formula, which can be found in standard text and reference books (e.g., Bird et al., 1960, pp. 504–513; Perry, 1973, Chap. 3, pp. 230–234; Reid et al., 1977, pp. 548–564). n Stefan-Maxwell relations may be written for an n-component mixture, but only n − 1 of these are independent.

The viscous flow of a pure component substance through a circular capillary satisfies the relation

$$N = -\frac{r_p^2 p}{8\mu RT} \frac{dp}{dz} \tag{5}$$

where μ is the viscosity of the gas.

A porous media may be considered to be a complicated matrix of interconnected channels. It is reasonable to speculate that if all the channels are small compared with the mean-free-path length, the local averaged flux vector, which describes the flow per unit total cross section, is related to the interstitial local average concentration gradient in the following way:

$$\underline{N}_A = -\frac{D^e_{KA}}{RT} \text{ grad } p_A \tag{6}$$

The effective Knudsen diffusivity is defined as

$$D^e_{KA} = \frac{\varepsilon}{\tau}\frac{2\langle r_p\rangle}{3}\left(\frac{8RT}{\pi M_A}\right)^{0.5} \triangleq B_2\left(\frac{8RT}{\pi M_A}\right)^{0.5} \tag{7}$$

ε_p is the pellet void fraction, and $\langle r_p\rangle$, the mean pore diameter, is computed as

$$\langle r_p\rangle = \int_0^\infty r_p f[\varepsilon(r_p)]\, dr_p \approx \frac{2\varepsilon}{\rho_b S_g} \tag{8}$$

where $f[\varepsilon(r_p)]\, dr_p$ is the fraction of the void volume with pores of radius r_p to $r_p + dr_p$. The tortuosity factor, τ, accounts for the tortous path of the molecules in the porous medium and the varying pore cross section. This empirical factor depends mainly on the porous material, its porosity, and the method of preparation. The value of τ is usually in the range 2 to 7. Wohlfahrt (1982) presented correlations of τ for several types of porous catalysts. The data for catalysts prepared from compacted powders were correlated as

$$\tau = (2 - \varepsilon)^{m+1}\varepsilon^{1-m} \tag{9}$$

where m depends on the porous material. Typical data and values of the parameter m for several catalysts are described in Fig. 1. Additional data were reported by Probst and Wholfahrt (1979). Feng and Stewart (1973) predicted that τ should be about 3 for isotropic systems with no dead-end branches.

It is reasonable to assume that the fluxes in a porous media, with pores that are wide compared to the mean free path length, still satisfy the Stefan-Maxwell relations if the binary diffusion coefficients are replaced by effective diffusion coefficients

$$D^e_{ij} = \frac{\varepsilon}{\tau}D_{ij} = B_1 D_{ij} \tag{10}$$

Omata and Brown (1972) pointed out that the tortuosity factor is not necessarily the same for both Knudsen and molecular diffusion. However, without detailed measurements, this is a necessary assumption.

Similarly, it is reasonable to expect that the forced flow through a porous media satisfies the relation

$$N = -\frac{B_0 p}{\mu RT}\frac{dp}{dz} \tag{11}$$

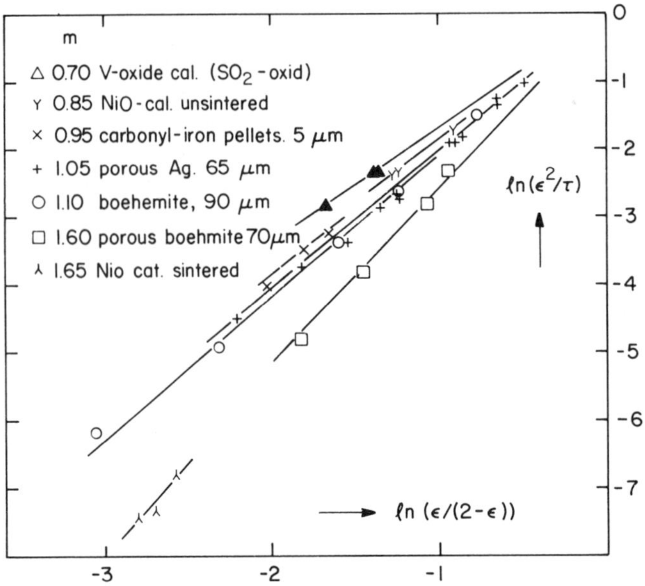

FIGURE 1 Dependence of the tortuosity factor on the porosity for several materials. (From Wohlfahrt, 1982.)

where the permeability of the medium, B_0, depends on the geometry and structure of the pores.

A method of finding the flux relations in the intermediate situations, where the three transport mechanisms may be of comparable importance, was proposed by Mason and Evans (1969). They suggested that the rate of momentum transfer by collision with the walls and among molecules should be combined additively. This leads to a set of n independent equations for the diffusive fluxes N_i^D:

$$-\frac{1}{RT}\,\text{grad}\;p_i = \frac{N_i^D}{D_{ki}^e} + \sum_{\substack{j=1\\ i=j}}^{n} \frac{x_j N_i^D - x_i N_j^D}{D_{ij}^e} \tag{12}$$

In the special case of a binary mixture, Eq. (12) reduces to

$$N_A^D = -\frac{D_A^e}{RT}\,\text{grad}\;p_A \tag{13}$$

where

$$\frac{1}{D_A^e} = \frac{1}{D_{KA}^e} + \frac{1-(1+\alpha)x_A}{D_{AB}^e} \tag{14}$$

and

$$\alpha = \frac{N_B}{N_A} \tag{15}$$

Solving the n equations Eq. (12) gives

$$N_i^D = -\frac{1}{RT}\sum_j^n G_{ij} \text{ grad } p_i \tag{16}$$

where the elements of G_{ij} are those of the matrix F^{-1}. The entries of F are

$$F_{ij} = -\frac{x_i}{D_{ij}^e} \qquad \text{for } i \neq j$$

$$F_{ii} = \frac{1}{D_{ki}^e} + \sum_{\substack{j=1 \\ i\neq j}}^{n} \frac{x_j}{D_{ij}^e} \qquad \text{for } i = j \tag{17}$$

Mason and Evans assumed that the total flux is the sum of the diffusive flux and viscous flow:

$$N_i = N_i^D + N_i^V \tag{18}$$

where

$$N_i^V = -\frac{x_i B_0 p}{\mu RT} \text{ grad } p \tag{19}$$

Substitution of Eqs. (18) and (19) in Eq. (12) gives a relation for the total fluxes:

$$-\frac{1}{RT} \text{ grad } p_i = \frac{N_i}{D_{ki}^e} + \sum_{\substack{j=1 \\ i\neq j}}^{n} \frac{x_j N_i - x_i N_j}{D_{ij}^e} + \frac{x_i B_0 p}{D_{ki}^e \mu RT} \text{ grad } p \tag{20}$$

An electrical analog of the additivity procedure proposed by Mason and Evans is shown in Fig. 2. It is remarkable that identical additivity rules are obtained by a rigorous analysis based on the dusty gas model (Mason et al., 1967).

Figure 3 enables a rapid rough estimate of the dependence of the diffusional regime on the average pore size and operating pressure. For example, when the pore size is about 100 Å, Knudsen flow is a dominant mechanism for pressures below 5×10^5 Pa, while gaseous diffusion is the main contributor for pressures above 5×10^6 Pa.

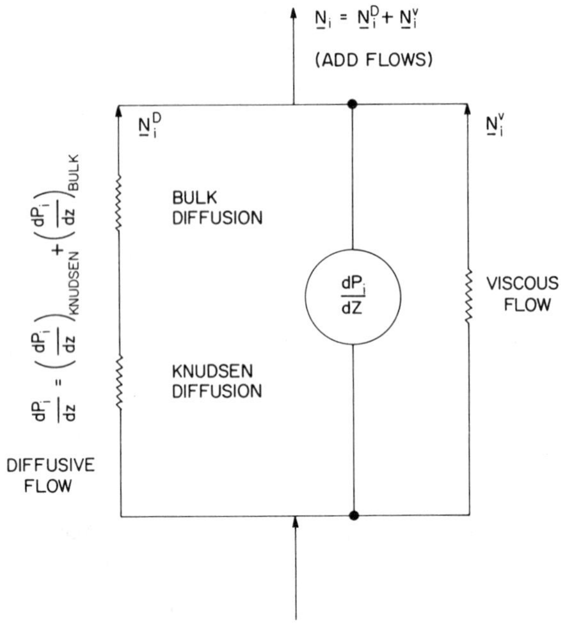

FIGURE 2 Electrical analog for the additivity rules for the gaseous fluxes according to the dusty-gas model. (From Mason et al., 1967.)

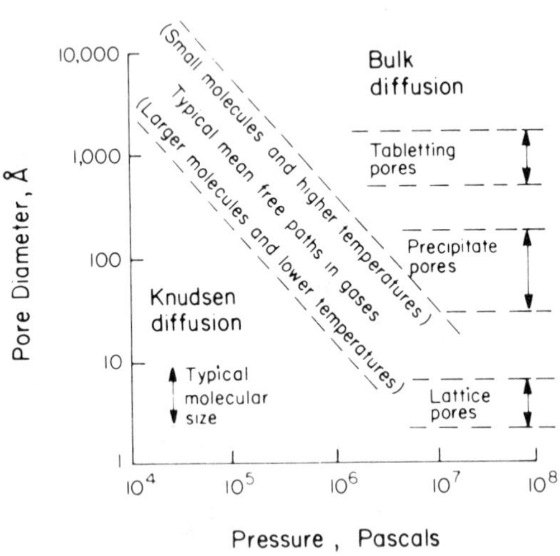

FIGURE 3 Transition between bulk and Knudsen diffusion for typical catalysts. (From Andrew, 1981.)

The isobaric flux relations are usually not applicable under reaction conditions. The reason is that the set of isobaric fluxes through a porous medium has to satisfy the Grahm relation,

$$\sum_{i=1}^{n} N_i \sqrt{M_i} = 0 \tag{21}$$

which is usually inconsistent with the stoichiometric requirement of

$$\sum_{i=1}^{n} N_i M_i = 0 \tag{22}$$

Consequently, a pressure gradient usually exists in a porous pellet in which a chemical reaction occurs. Although this pressure gradient may be small, it may have an important influence on the species fluxes.

A very general transport model was formulated by Feng and Stewart (1973), who assumed that a porous medium may be represented by a closely interconnected network of cylindrical pores and that the dusty-gas model describes the flux in each pore with D^e_{KA}, D^e_{AB}, and B_0 having values appropriate for a straight cylindrical pore of radius r_p. Their model predicts that if the pores are isotropically oriented, the tortuosity factor is independent of pore size and surface diffusion is negligible, then

$$N_i = -\frac{\varepsilon}{\tau RT}\sum_{j=1}^{n} g_{ij}(r_p)\,\text{grad}\ p_j - \frac{x_i\langle r_p^2\rangle p}{8\mu RT}\,\text{grad}\ p \tag{23}$$

where

$$g_{ij}(r_p) = \int_0^\infty G_{ij}(r_p) f[\varepsilon(r_p)]\, dr_p$$

$$\langle r_p^2\rangle = \int_0^\infty r_p^2 f[\varepsilon(r_p)]\, dr_p \tag{24}$$

and $G_{ij}(r_p)$ are the entries of the matrix G for a cylindrical capillary of radius r_p. Feng et al. (1974) have shown that the integrals in Eq. (24) may be described rather closely by a two-term approximation even for pellets with a wide pore size distribution. This avoids the need to compute G_{ij} for a wide range of r_p and simplifies considerably the application of this model. A major advantage of this model is that it incorporates information about the structure of the porous medium into the determination of the empirical transport coefficients. A comparison of the predictions of this model with experimental data was presented by Feng et al. (1974).

For isobaric binary systems Eq. (23) reduces to Eq. (13) with

$$D^e_A = \frac{\varepsilon}{\tau}\int_0^\infty \left[\frac{1}{D_{KA}} + \frac{1-(1+\alpha)x_A}{D_{AB}}\right]^{-1} f[\varepsilon(r_p)]\, dr_p \tag{25}$$

It is sometimes desired to compare the flux relations of the dusty-gas model with those obtained using a pseudo-Knudsen diffusion coefficient. Schneider (1976) suggested that when a single chemical reaction

$$\sum_{i=1}^{k} \nu_i A_i = 0 \tag{26}$$

occurs in a porous pellet, in which the total pressure gradient is negligible, the following approximation should be used:

$$\frac{1}{D_i^e} = \frac{\nu_i}{D_{ki}^e} + \sum_{j=1}^{k} \frac{\nu_i x_j - \nu_j x_i}{D_{ij}^e} + \nu_i \sum_{j=k+1}^{n} \frac{x_i}{D_{ij}^e} \qquad i = 1, 2, \ldots, k$$

$$\frac{1}{D_i^e} = -x_i \sum_{j=1}^{k} \frac{\nu_j}{D_{ij}^e} \qquad i = k+1, \ldots, n \tag{27}$$

Wakao and Smith (1962) developed a completely predictive model for the binary, isobaric flux in compressed powder catalysts which have a bimodal pore size distribution. They refer to the small pores within the powder particles as micropores and to the interstices between the particles as macropores. Wakao and Smith assume that the one-dimensional flux consists of three contributions: (a) flux through macropores having a radius of r_a, (b) flux through micropores which have a radius of r_i, and (c) flux through a series of micropores and macropores. They denoted the Knudsen coefficient and void fraction corresponding to the macro and micropores by D_{KA_a}, D_{KA_i}, ε_a, and ε_i, respectively. Their model predicts that

$$D_A^e = \varepsilon_a^2 D_a + (1 - \varepsilon_a)^2 D_i + 4\varepsilon_a(1 - \varepsilon_a)\left[\frac{1}{D_a} + \frac{1}{D_i}\right]^{-1} \tag{28}$$

where

$$D_a = \frac{D_{AB}}{1 - \alpha x_A + D_{AB}/D_{KA_a}} \qquad D_i = \frac{D_{AB}\varepsilon_i^2/(1 - \varepsilon_a)^2}{1 - \alpha x_A + D_{AB}/D_{KA_i}} \tag{29}$$

The main advantage of this model is that it contains no adjustable parameters. Its main deficiency is that it is not clear how to incorporate into the model pressure gradients (which are usually present in reactive systems) or how to generalize it for multicomponent mixtures.

Several steady-state and transient experimental techniques have been developed for determining the effective diffusivity of porous pellets and of the parameters appearing in the transport models. The most commonly used steady-state method is that in which either a cylindrical (Wicke and Kallenbach, 1941) or spherical (Weisz, 1957) pellet in a special cell is exposed to two gas streams having the same pressure and temperature but different compositions. The net diffusive flux is determined from the change in the composition of the gas streams.

For a binary mixture the one-dimensional flux satisfies the relation (Satterfield, 1970, p. 44).

$$N_A = \frac{pD^e_{AB}}{RTL(1 + \alpha)} \ln \frac{1 - (1 + \alpha)x_A(L) + D^e_{AB}/D^e_{KA}}{1 - (1 + \alpha)x_A(0) + D^e_{AB}/D^e_{KA}} \tag{30}$$

where

$$\alpha = \frac{N_B}{N_A} = -\left(\frac{M_A}{MB}\right)^{0.5} \tag{31}$$

Noting that

$$\frac{D^e_{AB}}{D^e_{KA}} = Hp \tag{32}$$

the values of D^e_{KA} and D^e_{AB} are determined by measuring the flux at several pressures. This technique is very sensitive to leaks next to the wall and this prevents its use at high temperatures.

The most widely used transient technique is the pulse response of either a chromatographic column (Cerro and Smith, 1970; Haynes and Sarma, 1973) or a single pellet (Suzuki and Smith, 1972; Dogu and Smith, 1976). The main advantages of this method are that it can be applied to particles of arbitrary shapes, it accounts for the contribution of dead-end pores, and it can be used to measure the diffusivity at high temperatures and pressures. Its main disadvantage is that the interpretation of the data is based on simplifying assumptions which are not always valid.

The chromatographic technique consists of measuring the first and second moments of a tracer that has been injected into a packed bed of porous pellets at several different flow rates of the inert carrier gas. The theory used to interpret the measurements is described in detail by Cerro and Smith (1970). The accuracy of this method is higher for adsorbable gases and improves with increasing particle size and bed length and with decreasing effective dispersion in the bed.

Experimental measurements and theoretical predictions of gaseous transport through porous media are usually dominated by the flux through large-diameter pores. However, most catalytic sites are present in micropores, and it is questionable whether the effective diffusivity is adequate for characterizing the transport under reaction conditions.

Surface Diffusion

Most gases are adsorbed to some extent on solid surfaces. The migration of the adsorbed species is referred to as *surface diffusion* and is particularly important for porous materials with large surface area and for strongly adsorbed species. The surface diffusion fluxes combine additively with the gaseous diffusive fluxes and therefore can be computed separately. The surface diffusive flux may be determined from a Wicke-Kallenbach diffusion cell operating at low pressures, so that the gaseous diffusion is of the

Knudsen type. A separate measurement with a nonadsorbing gas is used to determine the contribution of this Knudsen flow to the total flux.

Assuming that the surface and bulk concentration satisfy the equilibrium relation

$$f(C_A, C_B, \ldots, C_n, T^s) = C_A^s \tag{33}$$

the surface diffusion flux may be expressed as

$$-N_A^s = \frac{D_A^s \rho_b S_g}{\tau} \text{ grad } C_A^s = \frac{D_A^s \rho_b S_g}{\tau} \sum_{i=1}^{n} \frac{\partial f}{\partial C_i} \text{ grad } C_i \tag{34}$$

For sufficiently low gas concentration, the adsorption equilibrium approaches linearity and Eq. (34) simplies to

$$-N_A^s = \frac{D_A^s \rho_b S_g}{\tau} K_A(T) \text{ grad } C_A \tag{35}$$

where the adsorption equilibrium constant, $K_A(T)$, usually decreases with increasing temperatures according to an Arrhenius temperature dependence

$$K_A(T) = K_A(T_0) \exp\left[\frac{E_A^a}{R}\left(\frac{1}{T} - \frac{1}{T_0}\right)\right] \tag{36}$$

For low surface coverage the diffusivity may be computed from the site hopping mechanism,

$$D_A^s = \frac{\lambda^2}{2\tau^*} = \frac{\lambda^2 \nu}{2} \exp\left(-\frac{E_A^d}{RT}\right) \tag{37}$$

where λ is the average hopping distance, τ^* the lingering time, and ν the vibrational frequency of the adsorbed molecules. Combining Eqs. (35), (36), and (37) gives

$$-N_A^s = \frac{D_A^s(T_0) \rho_b S_g}{\tau} K_A(T_0) \exp\left[\frac{E_A^a - E_A^d}{R}\left(\frac{1}{T} - \frac{1}{T_0}\right)\right] \text{grad } C_A \tag{38}$$

Thus the surface diffusive flux increases with decreasing pore size (increasing S_g) and may either increase or decrease with temperature, depending on the sign of $E_A^a - E_A^d$.

The surface diffusivity usually depends on surface coverage and hence on the pressure and composition of the gas phase. Higashi et al. (1963) predicted that

$$D^s(\theta) = \frac{D^s(\theta = 0)}{1 - \theta} \tag{39}$$

where θ is the fractional surface coverage. This relation properly predicts an increase in the diffusivity with surface coverage, but is not expected to be valid for $\theta > 0.9$. Yang et al. (1973) derived a modified relation for high values of θ. Sladek et al. (1974) have shown that surface diffusivity depends on the strength of the adsorption and can be correlated with the differential heat of adsorption.

The surface diffusivity of physically adsorbed molecules at room temperature is in the range 10^{-3} to 10^{-6} cm^2/s. Measured values were reported by Barrer (1967, p. 557), Reed and Butt (1971), Schneider and Smith (1968), and Sladek et al. (1974). In most gas-solid reaction surface diffusion is not an important factor.

Diffusion in Zeolites

Many important industrial catalysts consist of small aluminosilicate (zeolite) crystals incorporated in a porous matrix. These crystals have narrow (5 to 10 Å diameter) structured channels through which the reactants and products have to migrate. The diameter of the pores is often close to the molecular dimension of the reacting species. Therefore, small differences in the shape of a molecule can have a large impact on the migration rate, and this can be utilized to create large differences in the catalytic selectivity of the catalyst. The diffusional processes in zeolites can affect significantly the selectivity of catalysts as only molecules which are smaller than a critical size can enter the pores and only products smaller than this size can leave the pores. Moreover, the restricted size of the cavities prevent formation of large intermediates. Consequently, by a proper design of the pore size it is possible to engineer a shape selective catalyst which will produce a desired species at a concentration which exceeds significantly the equilibrium concentration. This will occur if the desired product can leave the cavity, at which it is being formed, at a much faster rate than the other products. For example, when xylene is produced by the alkylation of toluene with methanol on a conventional catalyst, three different isomers of xylene are obtained. When the same reaction is carried out over certain zeolites, the product is essentially pure *para*-xylene, as the effective diffusivities of the ortho and meta isomers are smaller than those of the para isomer by three orders of magnitude.

It is customary to assume that the rate of migration in zeolites can be described by an activated diffusional process so that

$$N_A = -D_A^e \text{ grad } C_A \tag{40}$$

even though this is probably not the proper functional dependence. The reason is that the movement in the narrow pores of the zeolite is definitely not random, as it should be for a diffusional process, as the potential energy of a migrating molecule must be influenced by the attractive forces of the lattice at least when they pass through the openings between the cavities. Moreover, in the case of counterdiffusion the flux in one direction may be strongly hindered by that in the opposite direction, even though this effect is not predicted by Eq. (40).

The values of the effective diffusivity are usually determined by gravimetric sorption experiments [see, e.g., Gorring (1973)]. Riekart (1971) presented a theoretical analysis of this method and its pitfalls. Eberly

(1969) used the chromatographic technique to measure the diffusivities of various gases in zeolites. A review of the various methods used to determine the effective diffusivity in zeolites was presented by Eberley (1976).

The reported values of the effective diffusivities are in the wide range 10^{-6} to 10^{-14} cm^2/s. The temperature dependence of these diffusivities is stronger than that for bulk diffusion and is usually described by an Arrhenius temperature dependence. Goring (1973) found that the activation energy for diffusion of various hydrocarbon species varied in the range 10 to 60 kJ/mol.

There exists at present no reliable method for predicting the effective diffusivity in zeolites. The difficulties involved in the development of such a theory are brought out by Fig. 4, which describes the dependence of the diffusivity in a potassium T zeolite on the carbon chain length of normal paraffins. The experiments by Goring (1973) show that an *n*-dodecane molecule moves through the zeolite lattice at about 140 times the velocity of an *n*-octane molecule, despite the difference in size. Thus the interactions of the molecules with the zeolite lattice and cations must be the major factor affecting the diffusion, while molecule-molecule interactions are likely to be weak and determine mainly if a given path is free or blocked.

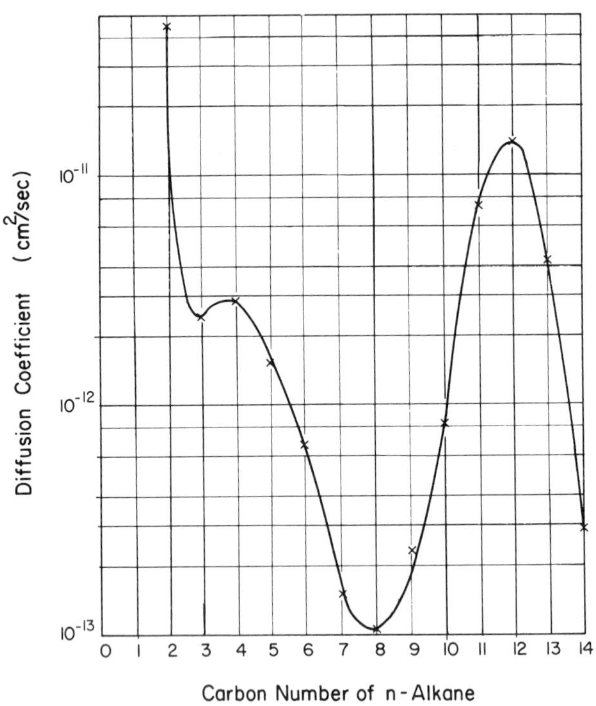

FIGURE 4 Diffusion coefficients of *n*-alkanes in potassium T zeolite at 300°C. (From Goring, 1973.)

Diffusion in Liquid-Filled Pores

The transport in liquid-filled pores is usually described by a Fickian diffusion relation with an effective diffusion coefficient

$$D_{AB}^{e} = \frac{\varepsilon D_{AB}}{\tau} \tag{41}$$

The bulk binary diffusion coefficient is computed by one of the standard available correlations (Reid et al., 1977, pp. 567–595). Surface diffusion may become important with strongly adsorbing molecules, and this contribution has then to be added to that in the pores. Komiyama and Smith (1974) described a case in which surface diffusion was the main contribution to the species flux.

In certain applications the sizes of the diffusing molecules and the pores are of the same order of magnitude. The interaction between the walls and the molecules tends to decrease the effective diffusion coefficient, which has to be computed by the equation

$$D_{AB}^{e} = \frac{\varepsilon D_{AB} K_p K_r}{\tau} \tag{42}$$

The equilibrium partition factor, K_p, is the ratio between the concentration of species A inside and outside the pore and depends on λ, the ratio of pore to molecule diameter. For small solvent molecules, K_p satisfies the relation (Ferry, 1936)

$$K_p = (1 - \lambda)^2 \tag{43}$$

When either the solute or the solvent adsorb on the surface K_p may be much smaller than the value predicted by Eq. (43). For example, Satterfield et al. (1973) measured a partition factor of 1/183 for λ of 0.38.

The resistance caused by drag exerted on the moving molecules by the walls is expressed through the drag coefficient,

$$K_r = (1 - \lambda)^{m+2} \tag{44}$$

where m is between 0 and 2. Combining Eqs. (42), (43), and (44) gives

$$D_{AB}^{e} = \frac{\varepsilon D_{AB}(1 - \lambda)^{4+m}}{\tau} \tag{45}$$

The values of K_p and K_r are strongly affected by the shape of the molecule and any interaction between the surface and either the solute or the solvent. Thus Eq. (45) is just a rough approximation for the influence of λ on the restricted diffusion coefficient. Recent experimental measurements and correlations of restricted diffusion coefficients were presented by Chantong and Massoth (1983) and Baltus and Anderson (1983).

Effective Thermal Conductivity

Intraparticle temperature gradients may have an important influence on the activity and selectivity of porous catalysts. A small number of experimental measurements of the effective thermal conductivity of pellets have been reported in the literature and a compilation of most of them can be found in Satterfield (1980, pp. 344–348). The data indicate that the effective conductivity is usually rather insensitive to changes in the porosity, pore size distribution, and the impregnated metal. A comparison of different methods for determining the effective thermal conductivity was presented by Hoffman et al. (1978).

At present there exists no general theory capable of predicting the effective conductivity of porous pellets as the mechanism of heat transport is rather complex. Butt (1965) generalized the random pore model of Wakao and Smith (1962) and developed a theory that is capable of correlating the effect of changes in the gas temperature and porosity on the thermal conductivity. Fortunately, the temperature profile within most catalytic pellets is rather uniform, so that the uncertainty concerning the value of the conductivity does not have an important impact on our ability to predict the performance of porous catalysts.

Mass and Heat Transfer Coefficient

A mass transfer coefficient is used to compute the flux of species between the ambient fluid and the surface of the catalyst. For a binary system

$$N_A = \frac{k_c}{\chi} C(x_{Ab} - x_{As}) \tag{46}$$

where χ is the bulk-flow correction factor, which accounts for the influence of the convective flux. This factor is equal to unity for equimolar counterdiffusion and is close to unity for dilute mixtures. In applications the factor is often assumed to be unity. Its exact value is given by the relation (Geankopolis, 1972, p. 254)

$$\chi = \frac{(R_N - x_A)_{LM}}{R_N} \tag{47}$$

where

$$R_N = \frac{N_A}{N_A + N_B} \tag{48}$$

and $(R_N - x_A)_{LM}$ is the ln mean of $R_N - x_{As}$ and $R_N - x_{Ab}$. The special case of diffusion through a stagnant film Eq. (47) reduces to

$$\chi = (1 - x_A)_{LM} \tag{49}$$

The prediction of mass transfer coefficients for multicomponent mixtures is quite complex, as the coefficients depend on the concentrations of the various species in the mixture. In general, approximate methods must be used in this case, and a survey of the subject is presented in the monograph by Cussler (1976, Chap. 9).

The mass and heat transfer coefficients between the pellet and the ambient fluid are determined by empirical correlations, as the processes are too complex to analyze in detail. It is customary to correlate the data in terms of j factors, which depend on the Reynolds number, the geometry of the pellet, and the boundary conditions. These factors are defined as

$$j_D = \frac{k_c}{u}\left(\frac{\mu}{\rho D_{AB}}\right)^{2/3} \tag{50}$$

$$j_H = \frac{h}{\mu \rho c_p}\left(\frac{c_p \mu}{\lambda_f}\right)^{2/3} \tag{51}$$

Dwivedi and Upadhay (1977) found that the following relations represented well most of the available data about mass transfer between a fluid and pellets in packed beds:

$$\varepsilon_b j_D = \begin{cases} 1.1068\text{Re}^{-0.72} & \text{for Re} < 10 \quad (52) \\ 0.4548\text{Re}^{-0.4069} & \text{for Re} > 10 \quad (53) \end{cases}$$

where

$$\text{Re} = \frac{u \rho d_p}{\mu} \tag{54}$$

The heat flux to a particle satisfies the equation

$$Q = h(T_b - T_s) \tag{55}$$

This relation is useful for predicting the temperature difference between the surface of a pellet and the ambient fluid. The value of the heat transfer coefficient is computed from the j_H factor for all cases in which the heat exchange by radiation can be neglected. Experimental data show that the values of j_H and j_D are essentially equal (Gupta and Thodos, 1962).

IMPACT OF MASS TRANSPORT LIMITATIONS

Determination of the Effectiveness Factor

The reactants within a porous catalytic pellet are consumed by the reaction and replenished by diffusion. When the time constant of the diffusion is small relative to that of the reaction, the concentration of the reactant is rather uniform throughout the pellet. However, when the time constant of the chemical reaction is small compared to that of diffusion, the concentration of the reactant within the pellet is expected to be lower than that close to the exterior surface, so that the average reaction rate (per unit volume of catalyst) depends on the diffusion-reaction interaction. Thus we expect the observable (or net) reaction rate to depend on the ratio between the time constants of the chemical reaction and the diffusion within the pellet.

The overall impact of the diffusional resistances is usually assessed by computing the effectiveness factor, η, which is defined as the ratio between the observed reaction rate to that obtained when neither concentration nor temperature gradients exist between the ambient phase and any point within the pellet. Clearly, the larger the deviation of the effectiveness factor from unity, the more important is the influence of the transport resistances. A comprehensive survey of the literature about the effectiveness factor can be found in the monograph by Aris (1975).

We illustrate the procedure of computing the effectiveness factor by considering the case of a first-order, isothermal, irreversible $A \rightarrow B$ reaction occurring within a spherical catalytic pellet. For the sake of compactness, we denote from now on by C and D^e the concentration and effective diffusivity of species A. The concentration of the reactant satisfies the equation

$$\frac{D^e}{r^2}\frac{d}{dr}\left(r^2\frac{dC}{dr}\right) - kC_1 = 0 \tag{56}$$

$$C = C_s \qquad \text{at } r = R \tag{57}$$

$$\frac{dC}{dr} = 0 \qquad \text{at } r = 0 \tag{58}$$

Solving Eq. (56) subject to Eqs. (57) and (58) we get

$$\frac{C}{C_s} = \frac{\sinh \phi\zeta}{\zeta \sinh \phi} \tag{59}$$

where

$$\zeta = \frac{r}{R} \qquad \phi = R\sqrt{\frac{k}{D^e}} \tag{60}$$

The parameter ϕ is defined as the Thiele modulus and is the square root of the ratio between the time canstant for diffusion ($\tau_D = R^2/D^e$) and that for the chemical reaction [$\tau_R = C_{As}/r(C_{As}) = 1/k$]. The Thiele modulus is independent of the surface concentration of the reactant only in the special case of a first-order reaction. The inset in Fig. 5 describes the concentration profiles for several values of the Thiele modulus. For ϕ smaller than about 0.3 ($\tau_D < 0.1\tau_R$) the concentration throughout the pellet is close to that at the surface. This region is referred to as the *kinetic-controlled regime*. For large ϕ ($\tau_D > 10\tau_R$) the concentration decreases exponentially with distance from the surface. We refer to these conditions as the *diffusion-controlled* or *diffusion-limited regime*. Here the reaction rate in the interior of the catalyst is rather low compared to that at the exterior shell. The region close to $\phi = 1$ or $\tau_D = \tau_R$ is referred to as the intermediate region.

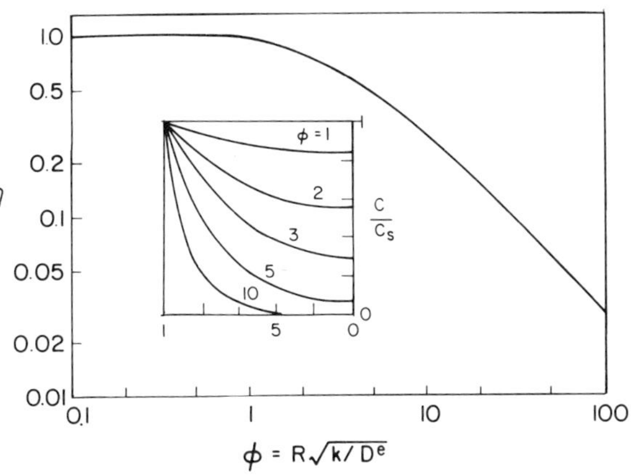

FIGURE 5 Dependence of the effectiveness factor on the Thiele modulus for a first-order reaction in a spherical pellet. Inset describes corresponding concentration profiles.

The effectiveness factor for this case is given by the expression

$$\eta = \frac{3}{\phi^2}(\phi \coth \phi - 1) \tag{61}$$

The logarithmic graph of η versus ϕ is shown in Fig. 5. As expected, for small ϕ the impact of diffusion is small and η is close to unity. For large ϕ diffusional limitations reduce significantly the observed reaction rate and η approaches asymptotically $3/\phi$.

The effectiveness factor has been computed for pellets of various geometrical shapes (Aris, 1975, Chap. 3). Fortunately, a rapid estimate of the value of η for particles of complex geometries can be obtained by capitalizing on the fact that if V_p/S_x is used as the characteristic length of a generalized Thiele modulus Λ, then the graph of η versus Λ is rather insensitive to changes in the geometry (Aris, 1957). Thus the graph for the spherical pellet can be used to predict the value of η for pellets of other geometries. The adequacy of this approximation is illustrated by Fig. 6, which compares the η versus Λ graph for a first-order reaction in a sphere with those for a semi-infinite slab and cylinder.

Similarly, by use of a properly normalized Thiele modulus,

$$\Lambda = \frac{V_p r(C_s)}{S_x}\left[2D^e \int_{C_{eq}}^{C_s} r(C)\, dC\right]^{-0.5} \tag{62}$$

one obtains a graph of η versus ϕ which is independent of the form of the kinetics rate expression for either large or small Λ (Aris, 1965; Bischoff, 1965; Petersen, 1965). For an nth-order reaction, the normalized Thiele modulus is

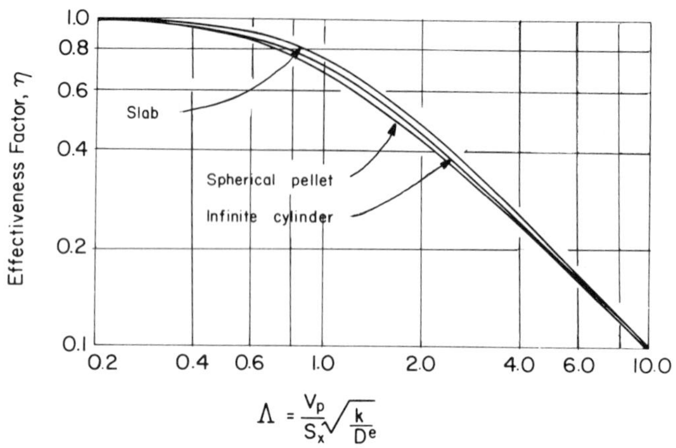

FIGURE 6 Dependence of the effectiveness factors for slab, cylindrical, and spherical pellets on the generalized Thiele modulus. (From Petersen, 1965.)

$$\Lambda = \frac{V_p}{S_x} \sqrt{\frac{(n + 1)kC_s^{n-1}}{2D^e}} \tag{63}$$

and Fig. 7 shows that the η versus ϕ graph for a first-order reaction approximates closely those of reactions with different orders.

When the reaction rate is a nonmonotonic function of the concentration of the limiting reactant, such as the Langmuir-Hinshelwood rate expression

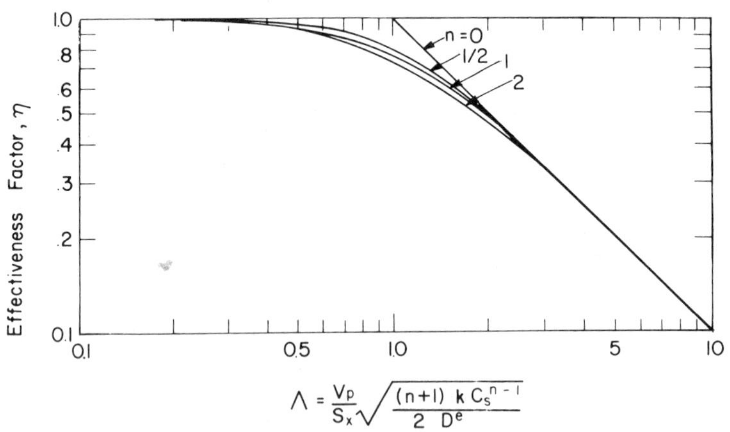

FIGURE 7 Dependence of the effectiveness factor on the generalized Thiele modulus for several nth-order reactions.

$$r(C) = \frac{kC}{(1 + KC)^2} \tag{64}$$

the effectiveness factor may exceed unity for some intermediate values of the Thiele modulus (Roberts and Satterfield, 1966). This occurs when the surface concentration exceeds the value corresponding to the maximum rate. In these cases a reduction in the concentration of the reactant by the diffusional limitations can increase the reaction rate over that corresponding to the surface concentration and the corresponding η may exceed unity. For certain rate expressions the interaction between transport and chemical rate processes can lead to steady-state multiplicity over a bounded range of the Thiele modulus, so that several values of the effectiveness factor may be obtained for certain kinetic parameters and pellet dimensions. This behavior is rather uncommon and it can be proven (Luss, 1971) that it cannot occur for any rate expression that satisfies the condition

$$(C_s - C)\left(-\frac{dr}{dC}\right) < r(C) \qquad \text{for all } 0 < C < C_s \tag{65}$$

where r(C) is the rate of consumption of the reactant A. For example, condition (65) is satisfied by the rate expression defined by Eq. (64) for all $C < C_s$ if and only if $KC_s < 8$. Clearly, condition (65) is satisfied for any rate expression that is monotonically increasing function of the concentration of the reactant.

Many repetitive calculations of the effectiveness factor need to be carried out in computer simultations of packed-bed reactors and parameter estimation of rival rate expressions. Several methods have been developed for getting a rapid approximation of η. These include the single-point collocation method of Stewart and Villadsen (1969) and several methods that capitalize on knowledge of the asymptotic value of η (Churchill, 1977; Wedel and Luss, 1980; Gottifredi et al., 1981a,b).

Several investigators computed the effectiveness factor using the dusty-gas model to describe the transport within the pellet (e.g., Abed and Rinker, 1973; Wong and Denny, 1975). It is of interest to know the difference between the results obtained using this model and that obtained using a pseudo-Knudsen diffusion model. Unfortunately, this difference depends on the value chosen for the effective diffusion coefficient and there exists no unambigious way of predicting its value. One possible scheme is the use of Eq. (27). The few available computations for gaseous mixtures seem to indicate that use of an effective Knudsen diffusion coefficient usually gives a reasonable approximation to the effectiveness factor computed by the dusty-gas model. This is probably due to the fact that diffusion coefficients of most gases are of similar magnitude. However, this point requires further analysis.

Many catalytic reactions involve a change in the number of moles and hence in volume. Rigorous calculations of the effectiveness factor for such cases are rather cumbersome, as they require use of the dusty-gas flux relations to account for the changes in both the composition and total pressure within the pellet. Calculations by Kehoe and Aris (1973) indicate that ignoring the momentum balance and total pressure gradient does not usually introduce a large error in the value of the effectiveness factor.

Weekman and Goring (1965) computed the effectiveness factor for symmetric pellets in which the reaction $A \rightarrow mB$ occurs, ignoring the influence of the pressure gradient on the transport within the pellet. This simplified model enables a rapid computation of the ratio between the effectiveness factor that accounts for the volume changes (η') to that which ignores it (η). Typical results for a second-order reaction are shown in Fig. 8. They indicate that when the reaction decreases the number of moles ($m < 1$), the bulk flow toward the center of the pellet increases the effectiveness factor above that obtained when the volume change is not taken into account (i.e., $\eta' > \eta$). On the other hand, when the reaction increases the number of moles ($m > 1$), the outward bulk flow causes η' to be smaller than η. The influence of the change in volume is decreased by the presence of inerts (i.e., by a decrease in x_{As}, the mole fraction of the reactant at the surface).

When the influence of external mass transfer is not negligible, the boundary condition on the surface of the catalyst becomes

$$k_c(C_b - C_s) = D^e\left(\frac{\partial C}{\partial \underline{n}}\right)_s \tag{66}$$

where $\underline{n}$ is a vector normal to the surface. For a symmetric pellet the surface concentration is uniform and it can be shown that for a first-order reaction (Aris, 1975)

$$\frac{1}{\eta} = \frac{1}{\eta_D(\Lambda)} + \frac{\Lambda^2}{Bi_m} \tag{67}$$

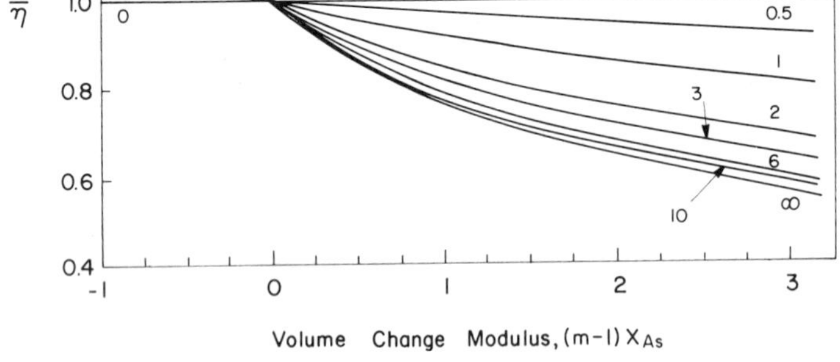

FIGURE 8 Dependence of the ratio η'/η on the volume change modulus for a second-order reaction. (From Weekman and Goring, 1965.)

where

$$Bi_m = \frac{k_c}{D^e} \frac{V_p}{S_x} \tag{68}$$

and $\eta_D(\Lambda)$ is the effectiveness factor for the case of negligible external mass transfer resistance (i.e., when Bi_m is infinite). It follows that the effectiveness factor is the sum of an internal resistance, which is the reciprocal of η in the absence of an external mass transfer resistance, and an external transport resistance of Λ^2/Bi_m. Note that in the presence of a mass transfer resistance, the asymptotic value of η for large Λ is Bi_m/Λ^2 rather than $1/\Lambda$.

Aris (1975, Sec. 3.9) has proven the remarkable result that Eq. (67) is valid for reactions with arbitrary kinetic expressions for large values of the Thiele modulus. Yortsos and Tsotsis (1982) have generalized this result for pellets with nonuniform catalytic activity.

In most practical problems Bi_m exceeds 10. Thus, according to Eq. (67), the value of η may be affected by the external mass transfer resistance only when η_D is considerably smaller than unity. This observation led to the development of a rule of thumb that it is possible to encounter cases in which the external mass transfer resistance is negligible while the internal diffusional resistance is appreciable, but the converse does not occur. Note that this conclusion would not be valid if Bi_m could attain small values.

In trickle-bed reactors one has often to compute the effectiveness factor of a liquid-filled catalytic pellet, of which only a fraction f of the exterior surface is wetted by the liquid. Tan and Smith (1980) and Mills and Dudukovic (1980) have shown that the corresponding effectiveness factor for a first-order reaction can be approximated by the relation

$$\eta = \frac{fC_L/C_L^*}{1/\eta_D(\Lambda) + \Lambda^2/Bi_w} + \frac{1 - f}{1/\eta_D(\Lambda) + \Lambda^2/Bi_d} \tag{69}$$

where C_L is the concentration of the reactant in the bulk fluid, C_L^* the liquid phase concentration of the reactant in equilibrium with the bulk gas concentration, and Bi_w and Bi_D the mass transfer Biot numbers for the wet and dry parts of the pellet, respectively. Herskowitz and Smith (1983) presented a comprehensive review of the influence of wetting on the effectiveness factor of liquid-filled catalysts.

Diffusion Disguise of Intrinsic Kinetics

Kinetic expressions are based on experiments in which the rate is measured as a function of the concentration of the reactants and the temperature. If the influence of the diffusional resistance on the observed rate is not taken into account, erroneous or disguised rate expressions are obtained.

The temperature dependence of the rate constant can usually be described by the Arrhenius relation

$$k(T) = k(T_0) \exp\left[\frac{E}{R}\left(\frac{1}{T_0} - \frac{1}{T}\right)\right] \tag{70}$$

When the rate expression is of a separable form, that is,

$$r(T,C) = k(T)f(C) \tag{71}$$

the apparent or observed activation energy is determined from the slope of the natural logarithm of the measured rate at a specific conversion against the reciprocal of the absolute temperature, that is,

$$E_a = -R \frac{d \ln r_m}{dT^{-1}} \tag{72}$$

The measured reaction rate satisfies the relation

$$r_m = k(T)f(C)\eta \tag{73}$$

Differentiation of Eq. (73) and substitution in Eq. (72) gives

$$E_a = E + 0.5 \frac{d \ln \eta}{d \ln \phi} \left(E + R \frac{d \ln D^e}{dT^{-1}} \right) \tag{74}$$

In the kinetic-controlled regime (small ϕ), the effectiveness factor is essentially independent of the Thiele modulus and according to Eq. (74) the apparent and intrinsic activation energies are equal. In the diffusion-controlled regime ($\phi > 3$), $d \ln \eta / d \ln \phi \to -1$, so that

$$E_a = 0.5E - 0.5R \frac{d \ln D_A^e}{dT^{-1}} \tag{75}$$

The effective diffusivity is proportional to T^m, so that the last term on the left-hand side of Eq. (75) is equal to 0.5mRT, which is usually very small in comparison to 0.5E. We conclude that the apparent activation energy in the diffusion-controlled regime is about half its intrinsic value. When the external mass transfer resistance is not negligible, $d \ln \eta / d \ln \phi \to -2$ for large ϕ, so that

$$E_a = -R \frac{d \ln D^e}{dT^{-1}} \tag{76}$$

In this case the apparent activation energy is very small and is independent of the intrinsic activation energy.

We conclude that transport intrusions mask and decrease the observed activation energy. The reason for this effect is that increasing the temperature decreases the time constant for the chemical reaction and the effectiveness factor. This, in turn, decreases the sensitivity of the reaction rate to changes in the temperature. This behavior is clearly illustrated in Fig. 9, which describes the logarithm of the rate of cumene cracking at various temperatures for catalysts of different sizes (Weisz and Prater, 1954). The

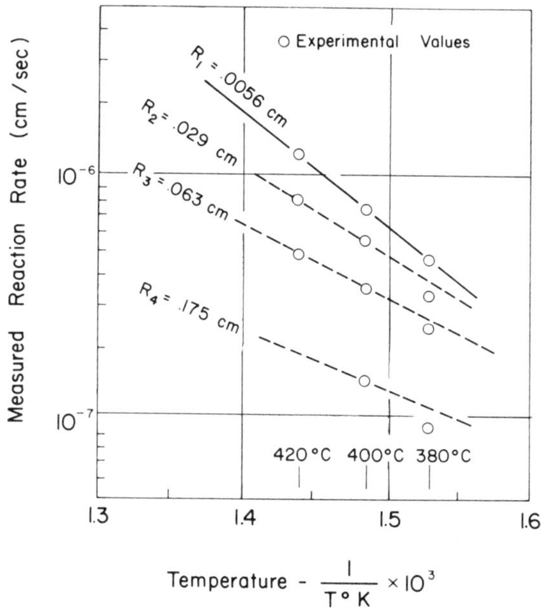

FIGURE 9 Influence of temperature and particle size on the cracking rate of cumene on SiO_2—Al_2O_3 catalyst. (From Weisz and Prater, 1954.)

data show that the slope of the graphs, or equivalently the apparent activation energy, decreases by a factor of about 2 as the particle's radius increases from 0.0056 cm to 0.175 cm.

Diffusional intrusions may also disguise the functional form of the rate expression. Consider as an illustration the case of an nth-order reaction for which the external resistance is negligible. Here

$$r_m = -\frac{1}{V_c}\frac{dn}{dt} = kC^n\eta(\Lambda) \tag{77}$$

The slope of the logarithmic graph of the observed rate against the concentration of the reactant is defined as the apparent reaction order n':

$$n' = \frac{d \log r_m}{d \log C} = n + \frac{1}{2}\frac{d \log \eta}{d \log \Lambda}\left(n - 1 - \frac{d \log D^e}{d \log C}\right) \tag{78}$$

It follows that the apparent and true orders are equal if the experiments are carried out in the kinetic-controlled regime in which η is essentially independent of Λ. In the diffusion-controlled regime the asymptotic slope of d log η/d log Λ is -1 and the apparent order depends on the relation between the concentration of the limiting reactant and the effective diffusivity. When the changes in the concentration of the reactant do not affect D^e, Eq. (78) predicts that

$$n' = \frac{n + 1}{2} \tag{79}$$

Thus the apparent and true orders are different for all but first-order reactions. The diffusional intrusions disguise the true order and shift it toward unity. For example, a second-order reaction behaves as if it is of order 1.5.

When the changes in the reactant's concentration are accomplished by varying the total pressure of a gaseous mixture and if gaseous bulk diffusion is the dominant mechanism, then

$$\frac{d \log D^e}{d \log C} = \frac{d \log D^e}{d \log p} = -1 \tag{80}$$

Substituting Eq. (80) into Eq. (78) we get that for large Λ,

$$n' = \frac{n}{2} \tag{81}$$

Thus in this case diffusional intrusions disguise the order of all but zero-order reactions.

We conclude that it is essential to account properly for the influence of diffusion to avoid pitfalls in the analysis and modeling of kinetic data.

Experimental Detection and Determination of Diffusional Limitations

Several experimental methods exist for detecting diffusional limitations and predicting the effectiveness factor. The most direct (but lengthy) method consists of determining the isothermal reaction rate for pellets of different sizes at a fixed mixture composition. The experiments are continued until pellets of two different sizes have the same activity, which is assumed to be the intrinsic one. The effectiveness factor for larger pellets is the ratio between their activity and the intrinsic one. The experiments can be carried out either in a differential reactor or a well-stirred heterogeneous reactor or a recycle reactor.

This test is meaningful only if all the pellets have the same intrinsic activity. This can be accomplished by crushing large particles to smaller sizes. This method may lead to pitfalls when the pellets have a nonuniform activity profile. For example, if a catalyst has a very narrow active exterior shell and an inert core, cutting the pellets does not affect the observed activity even if diffusion limits the rate in the exterior shell.

A more efficient method of determining η is based on the fact that if an isothermal reaction is carried out in spherical pellets of two different sizes, then

$$\frac{\Lambda_1}{\Lambda_2} = \frac{R_1}{R_2} \tag{82}$$

It follows from Eq. (61) that for a first-order reaction the ratio between the two observed rates satisfies the relation

$$\frac{(r_m)_1}{(r_m)_2} = \frac{\eta_1}{\eta_2} = \left(\frac{\Lambda_2}{\Lambda_1}\right)^2 \frac{3\Lambda_1 \coth 3\Lambda_1 - 1}{3\Lambda_2 \coth 3\Lambda_2 - 1} \tag{83}$$

The two equations (82) and (83) can be solved for Λ_1 and Λ_2, which in turn can be used to compute η_1 and η_2. The method is very sensitive to experimental errors if both experiments are conducted in the diffusion-controlled regime [i.e., when $(r_m)_1/(r_m)_2 \sim R_2/R_1$]. In this case additional experiments with smaller pellets need to be carried out. The advantage of this method is that, in principle, two experiments suffice to determine the effectiveness factor. However, to minimize the influence of inherent experimental errors, it is preferable to conduct experiments with three or four different particle sizes and determine the Thiele modulus and effectiveness factor that fit best all the experimental data.

Equation (83) is rigorously correct only for a first-order reaction. However, when the generalized definition of the Thiele modulus is used [Eq. (62)], the effectiveness factor is a rather insensitive function of the kinetic expression if the rate is a monotonic function of the reactant's concentration. Thus this method gives a reasonable estimate of η for many non-first-order reactions.

Weisz and Prater (1954, p. 167) have shown that the effectiveness factor for an nth-order reaction could always be related to the parameter

$$\Phi \triangleq \left(\frac{3V_p}{S_x}\right)^2 \frac{r_m}{D^e C_s} = \left(\frac{3V_p}{S_x}\right)^2 \frac{1}{D^e} \Bigg/ \left(\frac{C_s}{r_m}\right) \triangleq \frac{\tau_D}{\tau_R(\text{obs})} \tag{84}$$

which contains only measurable quantities. They noted that for reactions of order zero, one, and two, η exceeds 0.95 if Φ is smaller than 6, 0.6, and 0.3, respectively. Thus they suggested that $\Phi < 1$ be used as a rapid diagnostic for the absence of important diffusional limitations. The major advantage of this criterion is that it does not require any information about the rate expression or the kinetic parameters. Unfortunately, this useful criterion fails for rate expressions that are very sensitive to changes in C_s. Using singular perturbations (Wedel and Luss, 1980), it can be shown that η deviates from unity by less than 5% for a spherical pellet if

$$\Phi < 0.75/R'(1) \tag{85}$$

where $R(C_A/C_{A_s})$ is a dimensionless normalized rate [i.e., $R(1) = 1$] and R' is its derivative with respect to the dimensionless concentration C_A/C_{A_s}. Note that for a first-order reaction, $R'(1) = 1$ and Eq. (85) is essentially equal to the Weisz-Prater criterion.

Horak (1967) suggested that $\eta > 0.9$ if

$$\Phi < 6\xi \tag{86}$$

where ξ is the conversion for which the reaction rate is 0.9 of $r(C_s)$. For example, ξ is 0.1 for a first-order reaction, so that according to Eq. (86), $\eta > 0.9$ if $\Phi < 0.6$, which is in close agreement with the Weisz-Prater criterion (85). The agreement between the two criteria is not always close, as illustrated by the reaction

$$CO_2 + C \rightarrow 2CO$$

the rate of which satisfies the equation

$$r = \frac{kP_{CO_2}}{1 + 4P_{CO_2} + 50{,}000P_{CO}} \tag{87}$$

For a feed of pure carbon dioxide at 1 atm, ξ is equal to 0.55×10^{-5} and Eq. (86) predicts that $\eta > 0.9$ only if $\Phi < 3.3 \times 10^{-5}$, while Eq. (85) predicts that $\eta > 0.95$ only if $\Phi < 3.75 \times 10^{-5}$. Both predictions are much smaller than the value of unity predicted by the Weisz-Prater diagnostic.

We conclude that (a) if Φ is larger than unity, diffusional limitations definitely exist; (b) if Φ is much smaller than unity, diffusional intrusions are absent; and (c) if Φ is slightly smaller than unity, knowledge of the form of the rate expression is needed to decide if diffusional limitations affect the observed rate.

The presence of diffusional intrusions may sometimes be detected by changing the effective diffusion coefficient. This may be accomplished when gaseous diffusion dominates by using reaction mixtures containing a large excess of an inert diluent. If the same rate is obtained for mixtures with inerts of widely different molecular weights, and hence binary diffusion coefficients, it can be assumed that intraparticle diffusion is not limiting the rate. This test cannot be used for nonisothermal reactions since the difference in the inerts leads to differences in the heat transfer coefficients, and the difference in the pellet temperatures may mask the results.

Koros and Nowak (1967) proposed to use as a diagnostic test for the presence of diffusional limitations the fact that the reaction rate can be proportional to the number of catalytic sites only in the kinetic regime. They suggested to prepare catalysts with different number of active sites but similar diffusional properties either by pelletizing mixtures of catalyst particles with inert powder or by inpregnating the catalyst with different amount of the active component. Certain reactions are structure sensitive, so that their intrinsic activity depends on the crystallite sizes. Thus a failure to satisfy this test does not always imply the presence of diffusional limitations. A detailed discussion and illustration of the application of this test was presented by Madon and Boudart (1982).

The external mass transfer resistance is assumed to be negligible if the conversion does not change upon an increase in the linear velocity in an isothermal packed-bed reactor the length of which is adjusted to give a constant residence time. In nonisothermal cases thermal gradients may mask the results of this test.

In gas-liquid slurry reactors the test consists of checking the constancy of the rate as the agitation is increased. Chambers and Boudart (1966) pointed out that at the low Reynolds numbers encountered in many laboratory reactors, the transport coefficients may be rather insensitive to changes in the flow rate, and this diagnostic test may lead to erroneous conclusions. Figure 10 illustrates the application of this diagnostic test to the oxidation of SO_2 using catalyst particles of two different sizes and beds of two different lengths. The external mass transfer resistance is negligible in these cases, as changes in the velocity do not affect the conversion. Pore diffusion limitations are present at 470°C, as the conversion depends on the size of the pellets.

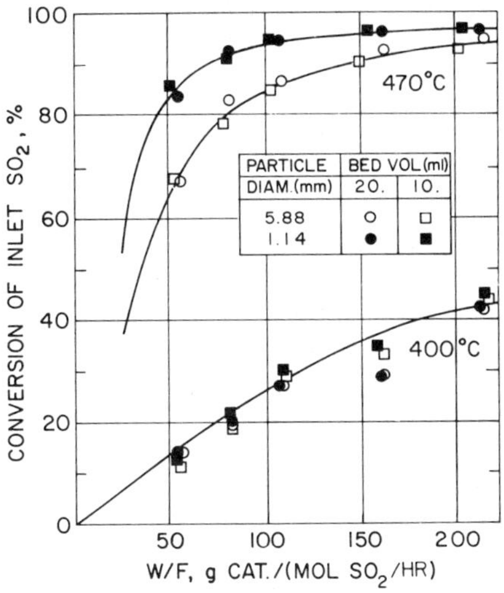

FIGURE 10 Influence of linear velocity and particle size on the conversion of SO_2. (From Dowden and Bridger, 1957.)

Mears (1971b) has shown that for an nth-order reaction the interphase concentration gradient can be ignored if

$$\Phi^* \triangleq \frac{r_m}{C_b k_c} \frac{V_p}{S_x} < 0.05 \tag{88}$$

It is sometimes difficult to predict the value of the mass transfer coefficient at the low velocities used in laboratory reactors. This value may be determined from a mass-transfer-limited reaction rate measured at elevated temperatures.

Influence of Diffusion on Multireaction Networks

A comprehensive analysis of the diffusion-reaction interaction for all possible reaction networks and kinetics is obviously an impossible task. Following Wheeler (1951), we examine the impact of diffusion limitations on the yield of a desired product for three typical isothermal two-reaction networks and illustrate how the knowledge of the single-reaction case may be exploited to predict the qualitative features of more complex reaction networks.

Consider first two parallel, independent, first-order reactions of which only the first is desirable:

$$A \xrightarrow{k_1} P_1$$

$$B \xrightarrow{k_2} P_2$$

An example of such a case is the hydrogenation of olefins in the presence of aromatic compounds. We define the selectivity as

$$S \triangleq \frac{r_A}{r_B} = \frac{k_1 A_b \eta_1}{k_2 B_b \eta_2} \tag{89}$$

In any useful catalyst $k_1 \gg k_2$, so that the time constant for the fast reaction is much smaller than that for the second reaction and the rate of the desired reaction is affected by the diffusional limitations much more than is the rate of the undesired reaction. In the absence of diffusional limitations $\eta_1 = \eta_2 = 1$, so that

$$S_0 = \frac{r_A}{r_B} = \frac{k_1 A_b}{k_2 B_b} \tag{90}$$

When diffusion limits both reactions,

$$\frac{\eta_1}{\eta_2} = \frac{\phi_2}{\phi_1} = \sqrt{\frac{k_2}{k_1}} \tag{91}$$

and the selectivity is given by the relation

$$S = \frac{A_b}{B_b} \sqrt{\frac{k_1}{k_2}} = S_0 \sqrt{\frac{k_2}{k_1}} \tag{92}$$

Thus the diffusional resistance taxes the desired reaction at a higher rate than it does the undesired reaction. We conclude that it is essential to minimize the diffusional resistance to get a high selectivity for this reaction network.

Next consider two consecutive first-order reactions

$$A_1 \xrightarrow{k_1} A_2 \xrightarrow{k_2} A_3$$

The species balances are described by the differential equations

$$\nabla^2 C_1 = \Lambda_1^2 C_1 \tag{93}$$

$$\nabla^2 C_2 = \Lambda_2^2 C_2 - \frac{\Lambda_1^2 C_1}{\delta} \tag{94}$$

where

$$\Lambda_i = \frac{V_p}{S_x} \sqrt{\frac{k_i}{D_i^e}} \qquad i = 1, 2 \tag{95}$$

$$\delta = \frac{D_2^e}{D_1^e}$$

The corresponding boundary conditions are

$$C_i = C_{is} \qquad i = 1, 2 \tag{96}$$

Solution of Eqs. (93) and (94) gives the point ratio of the two fluxes (i.e., the differential yield of component 2) as

$$-\delta \left.\frac{dC_2}{dC_1}\right|_s = -\frac{dC_{2s}}{dC_{1s}} = \frac{\delta s}{\delta s - 1} - \frac{\delta \eta_2 \Lambda_2^2}{\eta_1 \Lambda_1^2}\left(\frac{C_{2s}}{C_{1s}} + \frac{s}{\delta s - 1}\right) \tag{97}$$

where

$$s = \frac{k_1}{k_2} = \frac{\Lambda_1^2}{\delta \Lambda_2^2} \tag{98}$$

and η_i is the effectiveness factor for a single first-order reaction having a Thiele modulus of Λ_i.

When diffusional resistances are negligible, $\eta_i \to 1$ and Eq. (97) predicts that

$$-\frac{dC_{2s}}{dC_{1s}} = 1 - \frac{C_{2s}}{sC_{1s}} \tag{99}$$

and as expected the differential yield is independent of the diffusional processes. However, in the diffusion-controlled regime, $\eta_i \Lambda_i^2 \to \Lambda_i$ (for i = 1, 2) and the point yield approaches asymptotically the value

$$-\frac{dC_{2s}}{dC_{1s}} = \frac{\sqrt{s\delta}}{\sqrt{s\delta} + 1} - \frac{C_{2s}}{C_{1s}}\sqrt{\frac{\delta}{s}} \tag{100}$$

For all useful catalysts $s \gg 1$ and the asymptotic yield in the diffusion-controlled regime is much lower than that obtained in the kinetic regime.

Equation (97) indicates that the differential yield depends on the local C_{2s}/C_{1s} ratio. The concentration of the desired intermediate at the exit of an isothermal plug-flow reactor is found by integration of Eq. (97) to be

$$\frac{C_2(L)}{C_1(O)} = \frac{s(P - \delta)}{(\delta s - 1)(P - 1)}\left\{\left[\frac{C_1(L)}{C_1(O)}\right]^P - \frac{C_1(L)}{C_1(O)}\right\} + \frac{C_2(O)}{C_1(O)}\left[\frac{C_1(L)}{C_1(O)}\right]^P \tag{101}$$

where

$$P = \frac{\delta \eta_2 \Lambda_2^2}{\eta_1 \Lambda_1^2} \tag{102}$$

P is equal to 1/s when the diffusional resistances are negligible and to $\sqrt{\delta / s}$ in the diffusion-controlled regime. An experimental illustration of the shift with increasing diffusional resistance of the integral yield (at a fixed conversion level) from one asymptote to another is shown in Fig. 11 for the oxidehydrogenation reaction

Butene → butadiene → CO_2 + cracked products

By differentiation of Eq. (101) we find that the maximal yield of C_2 is obtained when

$$\left[\frac{C_1(L)}{C_1(0)}\right]^{P-1} = \frac{\dfrac{s}{P(\delta s - 1)} \dfrac{P - \delta}{P - 1}}{\dfrac{C_2(0)}{C_1(0)} + \dfrac{s}{\delta s - 1} \dfrac{P - \delta}{P - 1}} \tag{103}$$

Note that when $\delta = 1$, the expressions for $C_2(L)/C_1(0)$ in Eqs. (101) and (103) simplify considerably. An increase in the diffusional resistances (increasing Λ_i) decreases the conversion at which the maximal yield is obtained and reduces its value, as illustrated by the example shown in Fig. 12. Thus it is not possible to compensate for loss of yield due to diffusional limitations by increasing the conversion or equivalently the reactor

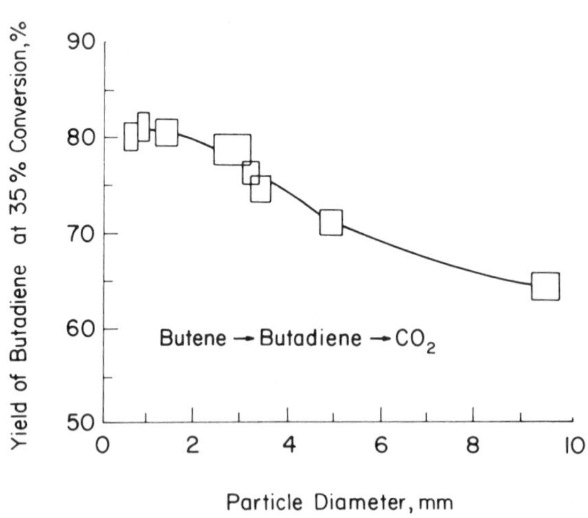

FIGURE 11 Dependence of the yield of butadiene on iron oxide catalyst particle size at 620°C. (From Voge and Morgan, 1972.)

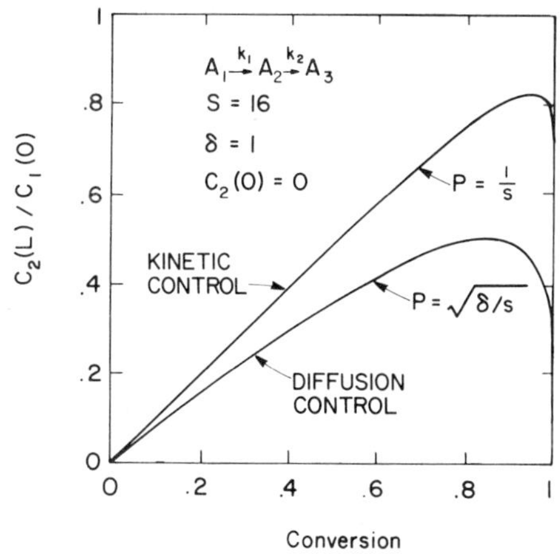

FIGURE 12 Dependence of the yield of the desired product on the conversion in the kinetic and diffusion-controlled regimes. (From Carberry, 1962.)

length. We conclude that the impact of mass transport resistance on the selectivity may be much more critical than its effect on the conversion.

When the interparticle mass transfer resistance is not negligible, the yield for symmetric pellets may be expressed in terms of the effectiveness factor for a single reaction. The yield is found by integration of Eq. (104), which is a simple modification of Eq. (97) (Carberry, 1962):

$$-\frac{dC_{2s}}{dC_{1s}} = \frac{\delta s}{\delta s - 1} - \frac{\delta \eta_2 \Lambda_2^2}{\eta_1 \Lambda_1^2} \left\{ \frac{C_{2s}}{C_{1s}} + \frac{s}{\delta s - 1} \left[1 + \frac{\eta_1 \Lambda_1^2}{Bi_{m1}} \left(\frac{Bi_{m1}}{Bi_{m2}} - 1 \right) \right] \right\} \tag{104}$$

Van de Vusse (1966) has presented a similar analysis for cases in which the first reaction is of order n while the second is of order m, where n, m = 0, 1, 2.

The analysis above indicates that the yield for two consecutive first-order reactions can be expressed as a function of the effectiveness factor for a single first-order reaction. Wei (1962) was the first to develop a scheme for predicting the yield for a general network of first-order reactions in terms of the effectiveness factor for a single reaction. A different approach to the solution of the same problem was presented by Aris (1975, p. 345).

Consider now the two simultaneous reactions

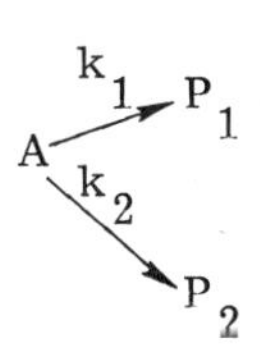

(desired, of order n)

(undesired, of order m)

occurring in a porous catalyst. As an analog to Eq. (90), we define the selectivity to be the ratio between the rate of formation of the desired to the undesired product:

$$S = \frac{\int_{V_p} k_1 A^n \, dv}{\int_{V_p} k_2 A^m \, dv} = S_0 \frac{\int_{V_p} (A/A_s)^n \, dv}{\int_{V_p} (A/A_s)^m \, dv} \tag{105}$$

where S_0 the intrinsic selectivity is defined as

$$S_0 = \frac{k_1}{k_2}(A_s)^{n-m} \tag{106}$$

When both reactions are of the same order, the ratio between the rate of formation of P_1 and P_2 is independent of the concentration of A, and the selectivity is equal to k_1/k_2. In this case the diffusional resistances affect only the rate of consumption of the reactant.

When the two reactions are of different orders, the selectivity is equal to the intrinsic one only when no diffusional limitations exist (i.e., when the concentration of the reactant is uniform throughout the pellet). However, when diffusional limitations exist, we expect them to tax at a higher rate the reaction with the higher order. Thus diffusional resistances can either increase or decrease the selectivity, depending on the sign of $m - n$. The intuitive guess is verified by Eq. (105), which shows that the selectivity is enhanced by diffusional limitations if the order of the desired reaction is less than that of the undesired one. The converse is true when n is larger than m.

It is not possible to obtain a general, explicit expression for the selectivity of this reaction network. Roberts (1972) derived some explicit asymptotic expressions for the selectivity and the interested reader is referred to that paper.

The analysis of the three two-reaction networks illustrates that it is often possible to predict the qualitative influence of diffusion on the yield of a desired product from knowledge of its impact in the single-reaction case. This ability to predict the qualitative influence of various variables on the yield is very useful. Obviously, a quantitative prediction of the yield requires a numerical solution of the governing species balances. Numerical procedures for solving these equations are described in Chapter 1.

In certain molecular sieve catalysts (zeolites) the effective diffusion coefficients of some species may differ by several orders of magnitude from those of the other. In these cases diffusional limitations may have a profound influence on the product distribution and on the yield of the desired product. For example, the xylene mixture produced by alkylation of toluene on a zeolite ZCM-5 catalyst can have *para*-xylene far in excess of the bulk gas-phase equilibrium concentration. Wei (1982) has analyzed this selectivity enhancement and showed that it was caused by the large difference in the diffusivity between the *para*-xylene and the *ortho*- and *meta*-xylenes.

INFLUENCE OF TEMPERATURE EFFECTS

Heat release or consumption by chemical reactions may lead to intraparticle and interphase temperature gradients, which can have an important impact on the activity and selectivity of catalytic pellets. The impact of these temperature gradients may be more pronounced than that of the concentration gradients.

When a single chemical reaction occurs in a symmetric catalytic pellet, the maximal intraparticle temperature Max(T) and the surface temperature T_s must satisfy the relations (Lee and Luss, 1969; Carberry, 1976, p. 231)

$$\frac{\text{Max}(T) - T_s}{T_b} < \beta_b(1 - \Phi^*) \tag{107}$$

$$\frac{T_s - T_b}{T_b} = \beta^*\Phi^* \tag{108}$$

where

$$\beta_b = \frac{(-\Delta H)D^e C_b}{\lambda^e T_b} \qquad \beta^* = \frac{(-\Delta H)k_c C_b}{hT_b}$$
$$\Gamma = \frac{\beta^*}{\beta_b} = \frac{k_c}{h}\frac{\lambda^e}{D^e} = \frac{Bi_m}{Bi_h} \qquad Bi_h = \frac{V_p}{S_x}\frac{h}{\lambda^e} \tag{109}$$

Note that all the terms on the right-hand side of Eqs. (107) and (108) are observable quantities. When $\Phi^* = 1$, Max(T) $\sim T_s$ and $T_s \to T_b(1 + \beta^*)$ (i.e., the pellet has a uniform temperature which exceeds by $\beta^* T_b$ that of the bulk phase). When $\Phi^* \to 0$, Max(T) $< T_b(1 + \beta_b)$ and $T_s \sim T_b$ and most of the temperature rise is within the pellet. Dividing Eq. (108) by Eq. (107) gives a bound on the ratio between the inter- and intraphase temperature gradients,

$$\frac{T_s - T_b}{\text{Max}(T) - T_s} \geqslant \frac{\Gamma\Phi^*}{1 - \Phi^*} \tag{110}$$

In practice the anticipated values of β_b are in the range 10^{-3} to 0.3, while those of Γ are in the range 10 to 10^4 for gas-solid systems and 10^{-4} to 10^{-1} for liquid-solid systems (Carberry, 1976, p. 230). It follows from Eqs (110) and (109) that in gas-solid systems a large temperature difference may exist between the surface of the pellet and the bulk gas. For example, Maymo and Smith (1966) measured a 300°C interphase temperature gradient during the oxidation of hydrogen over a platinum catalyst. In liquid-solid systems the interphase temperature gradient is usually small because of the low value of Γ, and most of the thermal resistance is within the catalytic pellet. This intrapellet temperature gradient is usually small, as the value of β_b is rather small for liquid-solid systems.

The foregoing criteria are most useful for estimating whether any of the temperature gradients are negligible. This information is very useful, as the analysis of a limiting case in which either the intrapellet or the interphase temperature gradients are negligible is much simpler than that of the general case.

Single-Reaction Case

Interphase temperature gradients are usually the dominant ones for gas-solid reactions. We examine here the impact of the interphase gradients on an nth-order catalytic reaction A → B occurring in a porous pellet in which the intraparticle gradients are negligible. The corresponding species and energy balances are

$$S_x k_c(C_b - C_s) = V_p r(C_s, T_s) \tag{111}$$

$$S_x h(T_s - T_b) = (-\Delta H) V_p r(C_s, T_s) \tag{112}$$

Combining the two equations gives

$$\frac{C_s}{C_b} = \frac{1 + \beta^* - y_s}{\beta^*} \tag{113}$$

where

$$y_s = \frac{T_s}{T_b} \tag{114}$$

Substitution of Eq. (113) into Eq. (112) gives a single equation describing the steady-state temperature. For example, for a first-order reaction with an Arrhenius temperature dependence the dimensionless steady-state temperature satisfies the equation

$$y_s - 1 = \mathrm{Da}(1 + \beta^* - y_s)X(y_s) \stackrel{\Delta}{=} f(y_s) \tag{115}$$

where

$$\mathrm{Da} = \frac{k(T_b)V_p}{k_c S_x} \qquad X(y_s) = \exp\left[\gamma\left(1 - \frac{1}{y_s}\right)\right] \qquad \gamma = \frac{E}{RT_b} \tag{116}$$

We define the effectiveness factor to be the ratio between the observed rate and that obtained at bulk fluid conditions. Thus if intraparticle gradients are absent,

$$\eta = \frac{r(C_s, T_s)}{r(C_b, T_b)} = \frac{r_m}{r(C_b, T_b)} \tag{117}$$

To illustrate the calculation of η, we consider the case of an irreversible nth-order reaction for which Eq. (117) becomes

$$\eta = \left(\frac{C_s}{C_b}\right)^n X(y_s) = \left(\frac{1 + \beta^* - y_s}{\beta^*}\right)^n X(y_s) \tag{118}$$

Equation (111) may be rewritten as

$$\frac{C_s}{C_b} = 1 - \Phi^* \tag{119}$$

Substitution of this result and Eq. (108) into Eq. (118) gives

$$\eta = (1 - \Phi^*)^n \exp\left[\gamma\left(1 - \frac{1}{1 + \beta^*\Phi^*}\right)\right] \tag{120}$$

Figure 13 describes the dependence of the effectiveness factor for a first-order reaction on the observable parameter Φ^*. It is apparent that in certain cases the effectiveness factor exceeds unity, as the increase in the reaction rate constant due to the interphase temperature gradient overcompensates for the decrease in the rate due to the concentration gradient. This surprising effect is encountered mainly in highly exothermic reactions.

The external transport resistances in laboratory and pilot plant packed-bed reactors are larger than those in industrial reactors, as the former operate usually at a much lower linear velocity. Thus it is essential to account properly for the impact of temperature gradients in the analysis of kinetic data and reactor design.

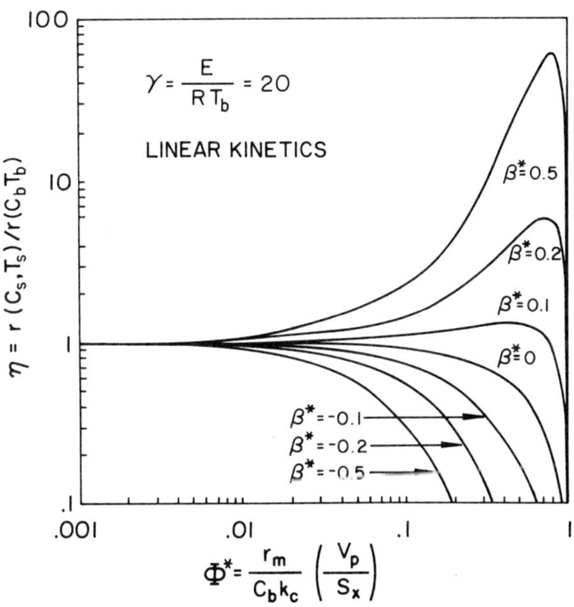

FIGURE 13 Dependence of the external effectiveness factor on Φ^* and β^*. (From Carberry and Kulkarni, 1973.)

The interaction between the external gradients and the intrinsic kinetics may lead to the existence of multiple steady-state solutions for the same set of parameters. For example, if for a first-order reaction

$$\gamma > 4\left(1 + \frac{1}{\beta^*}\right) \tag{121}$$

three steady-state solutions exist for all Damköhler numbers in the region

$$\frac{y_{min-1}}{f(y_{min})} > Da > \frac{y_{max-1}}{f(y_{max})} \tag{122}$$

where

$$y_{max,min} = \frac{\gamma(2 + \beta^*) \pm \sqrt{\gamma\beta^*[\gamma\beta^* - 4(1 + \beta^*)]}}{2(1 + \beta^*)} \tag{123}$$

Similar criteria may be derived for reactions with different rate expressions (Chang and Calo, 1979; Tsotsis and Schmitz, 1979; Leib and Luss, 1981; Luss, 1980).

The existence of steady-state multiplicity may lead to a hysteresis in the reaction rate or surface temperature when one of the operating conditions, such as the ambient concentration or temperature, is slowly changed. Figure 14 illustrates this behavior for the oxidation of butane on a platinum wire (Cardoso and Luss, 1969). The gas temperatures at which a sudden increase or decrease in the wire's temperature occurs upon a slight change

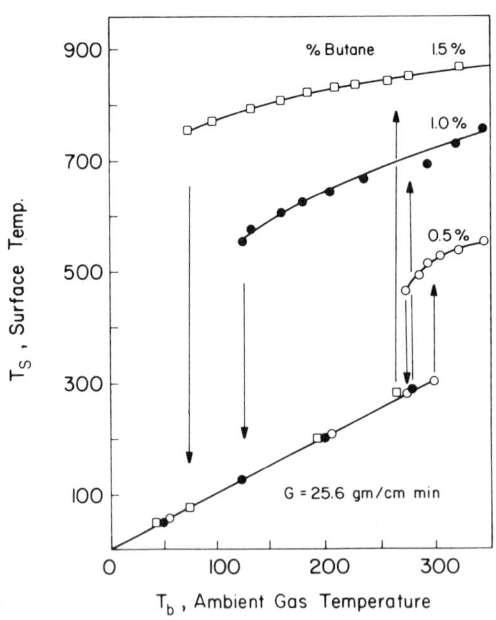

FIGURE 14 Dependence of catalyst wire surface temperatures on gas temperatures and butane concentrations. (From Cardoso and Luss, 1969.)

in the ambient conditions are referred to as the *ignition* and *extinction temperatures*, respectively.

We examine next the more common case in which the pellet temperature is uniform but different from the ambient one, while the intraparticle concentration gradients are not negligible. The species and energy balances for an nth-order reaction are

$$S_x k_c (C_b - C_s) = \int_{V_p} k(T_s) C^n \, dv \tag{124}$$

$$S_x h(T_s - T_b) = (-\Delta H) \int_{V_p} k(T_s) C^n \, dv \tag{125}$$

Combining Eqs. (124) and (125) gives

$$\frac{C_s}{C_b} = \frac{1 + \beta^* - y_s}{\beta^*} \tag{126}$$

Substitution of Eq. (126) into Eq. (125) and rearrangement gives the following equation for the dimensionless surface temperature:

$$y_s - 1 = \beta^* \mathrm{Da} X(y_s) \left(\frac{1 + \beta^* - y_s}{\beta^*}\right)^n \eta_i[\Lambda(y_s)] \tag{127}$$

where

$$\mathrm{Da} = \frac{V_p}{S_x} \frac{k(T_b) C_b^{n-1}}{k_c} \qquad \Lambda^2(1) = \left(\frac{V_p}{S_x}\right)^2 \frac{k(T_b) C_b^{n-1}}{D_e}$$
$$\Lambda^2(y_s) = \Lambda^2(1) X(y_s) \left(\frac{1 + \beta^* - y_s}{\beta^*}\right)^{n-1} \tag{128}$$

and $\eta_i(\Lambda)$ is the effectiveness factor for an nth-order isothermal reaction. Equation (127) can be solved for y_s and the corresponding Λ and η_i.

An analysis of this model indicates that most of its qualitative features are similar to those of the lumped-parameter model Eq. (115). The overall effectiveness factor is the ratio between the observed to ambient reaction rates,

$$\eta = \frac{k(T_s) C_s^n \eta_i}{k(T_b) C_b^n} = X(y_s) \left(\frac{1 + \beta^* - y_s}{\beta^*}\right)^n \eta_i \tag{129}$$

and its value may exceed unity. The main influence of the pore diffusion limitations is to reduce η_i from the value predicted in their absence.

The model predicts that steady-state multiplicity may exist in certain cases. It is difficult to derive simple, explicit criteria predicting the parameters for which the uniqueness-multiplicity transition occurs, but it is relatively easy to obtain bounds on the parameters for which either uniqueness or multiplicity exists. For example, for a first-order reaction a unique solution exists for all Da and $\Lambda(1)$ if (Van den Bosch and Luss, 1977)

$$\gamma < 4\left(1 + \frac{1}{\beta^*}\right) \tag{130}$$

while multiplicity exists for some Da and $\Lambda(1)$ if

$$\gamma > 8\left(1 + \frac{1}{\beta^*}\right) \tag{131}$$

Sharper bounds may be obtained for specific $\Lambda(1)$ and similar criteria may be derived for nth-order reactions. Studies by Nielsen and Villadsen (1984) and Hu et al. (1985a,b) have revealed that this model predicts that up to five steady-state solutions exist for some nth-order reactions and set of parameters. In contrast, the lumped model predicts that at most three solutions can exist for any nth-order reaction.

When the interphase concentration and temperature gradients are negligible, the steady-state species and energy balances are

$$D^e \nabla^2 C - r(C,T) = 0 \tag{132}$$

$$\lambda^e \nabla^2 T + (-\Delta H) r(C,T) = 0 \tag{133}$$

subject to the boundary conditions

$$C = C_s \quad \text{and} \quad T = T_s \qquad \text{on the surface} \tag{134}$$

Combining the two equations gives

$$\frac{C}{C_s} = \frac{1 + \beta - y}{\beta} \tag{135}$$

where

$$\beta = \frac{(-\Delta H) C_s D^e}{\lambda^e T_s} \qquad y = \frac{T}{T_s} \tag{136}$$

Thus the steady state is the solution of a single equation of the form

$$\nabla^2 y + \beta\phi^2 f(y) = 0 \tag{137}$$

where

$$\phi^2 = R^2 \frac{r(C_s, T_s)}{C_s D^e} \tag{138}$$

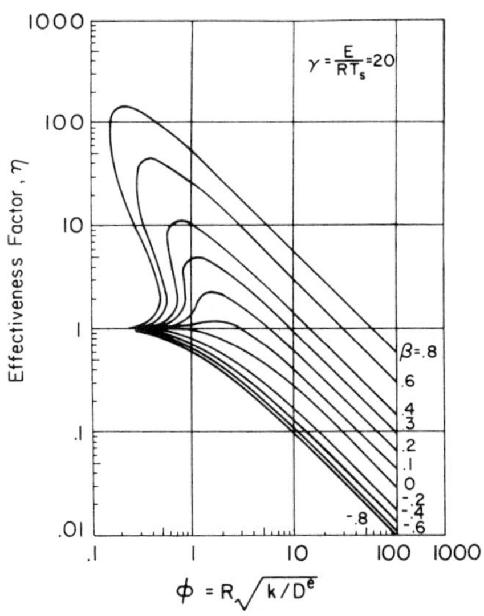

FIGURE 15 Nonisothermal effectiveness factors for a first-order reaction in a spherical pellet. (From Weisz and Hicks, 1962.)

$$f(y) = \frac{r(C,T)}{r(C_s,T_s)} = \frac{r(y,T_s)}{r(1,T_s)} \tag{139}$$

The corresponding effectiveness factor is

$$\eta = \frac{1}{V_p} \int_{V_p} f(y)\, dv \tag{140}$$

Numerical calculations of the effectiveness factor for a first-order reaction in a sphere by Weisz and Hicks (1962) (Fig. 15) indicate that the effectiveness factor may exceed unity, due to the competing influences of the intraparticle temperature and the concentration gradients. Moreover, steady-state multiplicity may be caused by the interaction of the diffusion-reaction rate processes.

Luss (1968) has shown that a sufficient condition for the existence of a unique solution for all ϕ^2 is that

$$(y - 1)\frac{d \ln f(y)}{dy} < 1 \qquad \text{for all } 1 < y < 1 + \beta \tag{141}$$

For the case of a first-order reaction, criterion (141) guarantees uniqueness for all ϕ if

$$\gamma < 4\left(1 + \frac{1}{\beta}\right) \tag{142}$$

This criterion is usually satisfied, as β is most cases rather small. Methods of bounding the values of ϕ for which multiplicity exists were presented by Luss (1971).

When the temperature of the pellet is uniform, no more than five steady-state solutions are obtained for a first-order reaction. However, an arbitrary large number of solutions may be found for a first-order reaction occurring in a spherical pellet with intraparticle temperature and concentration gradients (Villadsen and Ivanov, 1978). This behavior is found for extremely high values of γ and β which are not encountered in practice (Copelowitz and Aris, 1970). When both the interphase and intraparticle gradients have to be accounted for, one needs to solve the two equations (132) and (133) subject to the boundary conditions

$$k_c(C_b - C_s) = D^e \underline{n} \nabla C \tag{143}$$

$$h(T_b - T_s) = \lambda^e \underline{n} \nabla T \tag{144}$$

The behavioral features of this case are similar to those found for the limiting models discussed before (i.e., the effectiveness factor may exceed unity and multiplicity occurs for certain values of the parameters). The interaction between the interphase and intraparticle gradients may cause the appearance of two distinct regions of multiplicity, as shown in Fig. 16. The two multiplicity regions may overlap, giving five steady-state solutions.

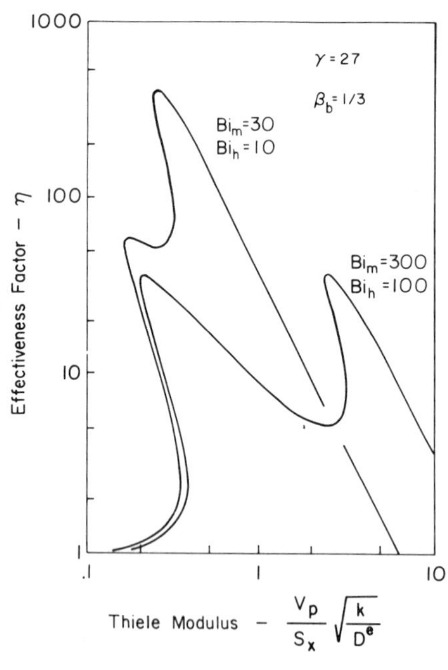

FIGURE 16 Effectiveness factor as a function of the Thiele modulus for a first-order reaction in a slab catalyst. (From Aris and Hatfield, 1968.)

It is rather difficult to derive exact criteria predicting explicitly the set of parameters for which multiplicity exists in this case. Sufficient conditions for uniqueness were presented by Jackson (1972).

The analysis above indicates that either intra- or interphase temperature gradients may cause the effectiveness factor to exceed unity and lead to steady-state multiplicity. Estimation of the magnitude of these gradients is useful for deciding which of the limiting models should be used in a specific application.

Multireaction Networks

Temperature gradients may have a significant influence on the yield of a desired product in a multireaction network. Whenever the activation energies of the various reactions are not equal, thermal gradients affect the ratio among the rates of the different reactions and hence the intrinsic yield. The temperature gradients also affect the ratio among the time constants for diffusion and reaction and hence the influence of diffusion on the yield.

It is impossible to analyze the behavior of all reaction networks and rate expressions. Thus we shall discuss just two typical two-reaction networks and comment on the behavior of more intricate networks.

We start by examining a case of two consecutive reactions, $A_1 \to A_2 \to A_3$. To simplify the analysis, we ignore intraparticle concentration and temperature gradients and assume that only interphase gradients exist. The corresponding steady-state species and energy balances are

$$S_x k_{c1}(C_b - C_{1s}) = V_p k_1 C_{1s} \tag{145}$$

$$S_x k_{c2}(C_{2s} - C_{2b}) = V_p(k_1 C_{1s} - k_2 C_{2s}) \tag{146}$$

$$S_x h(T_s - T_b) = V_p[(-\Delta H_1)k_1 C_{1s} + (-\Delta H_2)k_2 C_{2s}] \tag{147}$$

The three equations may be combined to give the following equation for the dimensionless temperature:

$$y_s - 1 = \frac{\beta_1^* Da_1}{1 + Da_1}\left(1 + \frac{\sigma Da_2}{1 + Da_2}\right) + \frac{\beta_2^* Da_2}{1 + Da_2} \tag{148}$$

where

$$Da_i = \frac{k_i V_p}{k_{ci} S_x} \qquad \beta_i^* = \frac{(-\Delta H_i) k_{ci} C_{ib}}{h T_b}$$

$$\sigma = \frac{\Delta H_2}{\Delta H_1} \tag{149}$$

Equation (148) is a highly nonlinear function of y_s, as each of the Damköhler numbers includes a temperature-dependent rate constant.

The corresponding yield is given by the equation (Cassiere and Carberry 1973)

$$-\frac{dB}{dA} = \frac{k_1C_{1s} - k_2C_{2s}}{k_1C_{1s}} = \eta_2^* - \frac{\eta_2^*k_2}{\eta_1^*k_1}\frac{C_{2b}}{C_{1b}} \tag{150}$$

where

$$\eta_i^* = \frac{1}{1 + Da_i} \qquad i = 1,\ 2$$

When both Damköhler numbers are very small ($Da_i \ll 1$), the external effectiveness factors η_i^* are close to unity and Eq. (150) becomes

$$-\frac{dB}{dA} = 1 - \frac{k_2C_{2b}}{k_1C_{1b}} \tag{151}$$

In this case the temperature gradients affect the yield only by changing the ratio k_1/k_2.. When $E_1 > E_2$ the temperature gradients enhance the yield, while the converse occurs if $E_2 > E_1$. When the Damköhler numbers are not small, temperature gradients affect the rate constants as well as the diffusional limitations.

An analysis of Eq. (150) indicates that it may have up to five different solutions, even though at most three exist when only one reaction occurs. Balakotaiah and Luss (1982a) have proven that multiplicity can exist for some Damköhler numbers if and only if at least one of the two reactions is exothermic and the set γ_i, β_i, and σ satisfies at least one of five conditions given in that work. Moreover, they showed how to determine the Damköhler number for which multiplicity exists. The interested reader is referred to that work.

When multiple steady states exist in this system, the bifurcation diagrams, which describe the yield or surface temperature as a function of a slowly changing state variable, may be much more intricate than those found for the single-reaction case. Moreover, the number of different types of bifurcation diagrams tends to increase rapidly with the number of the reactions (Balakotaiah and Luss, 1982b, 1983b; Harold and Luss, 1985). Figure 17 describes a case in which two hysteresis loops are completely nested within a large one. Obviously, it would be easy to miss the internal nested loop in an experimental study, unless there exists some theoretical guidance about its existence. The existence of these "exotic" bifurcation diagrams points out the need for a thorough analysis of the impact of thermal gradients, to avoid pitfalls in the design and operation of the reactor.

When the influence of intraparticle concentration gradients is taken into account, the equations that describe the dimensionless steady-state temperature becomes more intricate (McGreavy and Thornton, 1970) and their analysis is more tedious. We shall not discuss this case here.

Intraparticle temperature and concentration gradients may have an important influence on the yield of the desired intermediate A_2 for this reaction network. Butt (1966) has shown that even when the activation energies of both reactions are equal, the intraparticle gradients may affect the yield when both reactions are exothermic. In some extreme cases, the temperature gradients may even change the sign of the yield, so that instead of producing the desired species the reaction consumes it. When $E_1 = E_2$ the temperature gradients do not affect the ratio between the two

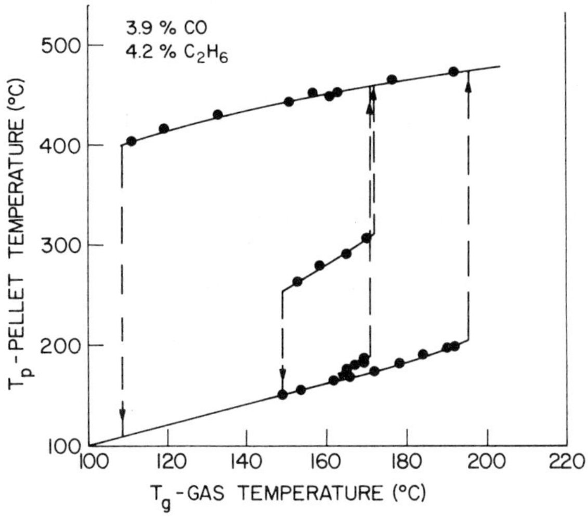

FIGURE 17 Dependence of the temperature of a catalytic pellet on the gas temperature during the simultaneous oxidation of CO and C_2H_6. (From Harold and Luss, 1982.)

rate constants k_1 and k_2. However, the increase in the values of these rate constants with rising temperatures increases the deleterious impact of the concentration gradients, which tax the desired fast reaction ($A_1 \rightarrow A_2$) more than they do the slower undesired reaction ($A_2 \rightarrow A_3$). On the other hand, the yield may be improved by thermal gradients when $E_1 = E_2$ for endothermic reactions. This is, however, accompanied by a large decrease in the reaction rate.

When the activation energy of the desired reaction is larger than that of the undesired one ($E_1 > E_2$), thermal gradients tend to improve the yield for exothermic reactions, as the increased ratio of k_2/k_1 overcompensates for the larger diffusional resistances. The inverse occurs when $E_2 > E_1$.

Consider now a single reactant that is consumed by two simultaneous first-order reactions. When intraparticle gradients can be ignored, the local selectivity is defined as

$$S = \frac{r_1}{r_2} = \frac{k_1(T_s)}{k_2(T_s)} \tag{152}$$

so that the impact of the temperature gradients depends on its influence on the ratio $k_1(T_s)/k_2(T_s)$. Thus the selectivity is improved by thermal gradients if the reaction is exothermic and $E_1 > E_2$ or if the reaction is endothermic and $E_2 > E_1$.

The interaction between the chemical reaction and the interpellet temperature gradients leads to steady-state multiplicity for some Damköhler numbers in this case if at least one of the following three conditions is satisfied (Michelsen, 1977):

$$\gamma_1 > 4\left(1 + \frac{1}{\beta^*}\right) \tag{153}$$

$$\gamma_1 > 4\left(1 + \frac{1}{\beta_2^*}\right) \tag{154}$$

$$(\gamma_2 - \gamma_1)(\beta_2^* - \beta_1^*) > 4(1 + \beta_1^*)(1 + \beta_2^*) \tag{155}$$

The last condition [Eq. (155)] may be satisfied even if both reactions are endothermic—a rather surprising result in view of the fact that multiplicity cannot occur for a single, first-order endothermic reaction.

When intraparticle concentration and temperature gradients exist, the species and energy balances can be solved numerically and the selectivity is computed from

$$S = \frac{\int_{V_p} k_1(T)C\, dv}{\int_{V_p} k_2(T)C\, dv} \tag{156}$$

When the intraparticle resistances are small ($\phi \ll 1$), the selectivity approaches asymptotically the value of $k_1(T_s)/k_2(T_s)$. For large diffusional resistances the selectivity approaches a different asymptotic value which depends on the reaction heats and activation energies of both reactions. Figure 18 illustrates some typical results obtained by Østergaard (1965) for two exothermic reactions. As expected, the asymptotic selectivity in this case is enhanced by temperature gradients when the activation energy of the desired reaction exceeds that of the undesired one (i.e., $E_1/E_2 > 1$).

The impact of the thermal gradients on the intrinsic selectivity of any multireaction network can be determined rapidly by use of the external resistances model. When all the reactions are either exothermic or endothermic,

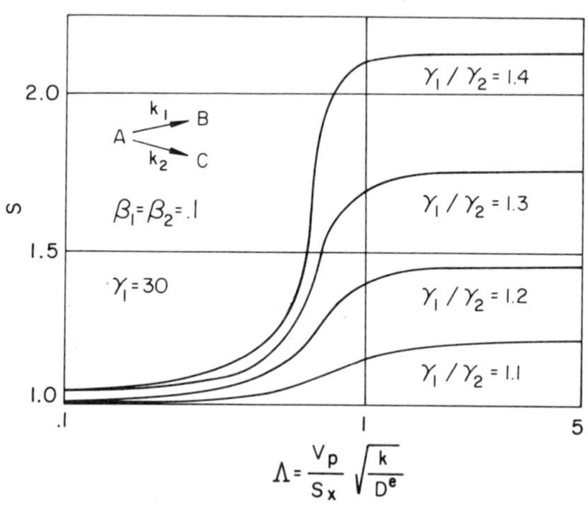

FIGURE 18 Dependence of the yield for two simultaneous first-order reactions in a slab on the Thiele modulus. (From Østergaard, 1965.)

the qualitative effect of the intraparticle temperature and concentration gradients can usually be predicted from knowledge of the influence of the diffusional resistance in the single-reaction case. The information about the impact of the interphase thermal gradients on the intrinsic selectivity and of the intraparticle diffusional limitations can be used to estimate the qualitative effect of temperature gradients in multireaction systems. Obviously, the quantitative prediction of the influence of thermal gradients requires a numerical solution of a set of nonlinear differential equations that describe the system. Balakotaiah and Luss (1983a) have devised a procedure for getting sufficient multiplicity criteria for networks of many irreversible first-order reactions.

Diagnostic Tests for Temperature Gradients

Criteria (107) and (108) enable a rapid prediction of the magnitude of the temperature gradients based on observable quantities. Several criteria have been proposed for predicting whether thermal resistances have an important influence on the intrinsic reaction rate. Obviously, the impact of temperature gradients depends on the exothermicity of the reaction, the sensitivity of the reaction rate to changes in the temperature, the rate and the transport coefficients.

Anderson (1963) predicted that intraparticle temperature gradients cause the observed rate to deviate by less than 5% from the value obtained under isothermal conditions if

$$\Phi \beta \gamma < 1 \tag{157}$$

Kubota and Yamanaka (1969) predicted that for positive-order reactions the combined effect of concentration and intraparticle temperature gradients is less than 5% if

$$\Phi < \frac{1}{n - \beta\gamma} \tag{158}$$

When $n - \beta\gamma$ is close to zero, the perturbation approach used to derive Eq. (158) is no longer applicable. In this case the asymptotic expansion of the effectiveness factor for large values of ϕ (Petersen, 1965) can be used to derive a different criterion. Mears (1971b) suggested using a value of 13 for the right-hand side of Eq. (158) when both n and $\gamma\beta$ are close to unity. Criterion (85) can also be used in this case to predict when the intraparticle gradients do not affect the rate by more than 5%.

The major resistance to heat transfer is in the boundary layer for gas-solid reactions. Mears (1971a) has shown that interphase temperature gradients cause less than a 5% deviation in the rate if

$$\Phi^* \gamma_b |\beta^*| < 0.05 \tag{159}$$

A comparison of criteria (157) and (159) indicates that interphase gradients limit the rate before the intraparticle gradients do if

$$Bi_a = \frac{h}{\lambda^e} \frac{V_p}{S_x} < 2.2 \tag{160}$$

This condition is usually satisfied in laboratory reactors due to the low flow rates and heat transfer coefficients.

A comparison of criteria (88) and (159) indicates that the external temperature gradients affect the rate before the external mass transfer does if

$$|\beta^*|\gamma_b > 1 \tag{161}$$

This condition is satisfied by many highly exothermic or endothermic reactions.

Mears (1971b) has shown that for an nth-order reaction, the combined intraparticle-interphase gradients affect the rate by less than 5% if

$$\Phi < \frac{1 + \Phi^*\beta^*\gamma_b}{|n - \gamma_b\beta_b|(1 + n\Phi^*)} \tag{162}$$

This criterion reduces to Eq. (158) when the external resistances are negligible ($\Phi \ll 1$).

Although the diagnostic tests are very useful, the estimation of some of the relevant parameters may be difficult at times. In such cases direct experimental tests may have to be carried out to check for the impact of the thermal gradients.

A common test for external gradient is checking for the influence of the flow rate on the conversion at a fixed space velocity. If no effect is found, it is concluded that the external resistances are not important. Unfortunately radial temperature gradients usually affect the conversion in an integral packed-bed reactor more than do the interphase gradients. Thus this test does not provide a conclusive answer as to whether the interphase resistance is important. A direct examination of the influence of the interphase gradients can be carried out either in a well-stirred heterogeneous catalytic reactor, such as the Berty reactor, or in a recycle reactor in which radial temperature gradients can be eliminated.

ACKNOWLEDGMENTS

I wish to thank John Villadsen, Arvind Varma, and Stig Wedel for helpful comments.

NOTATION

Bi_a	Biot number for heat transfer $kV_p/(\lambda^e S_x)$
Bi_m	Biot number for mass transfer $k_c V_p/(D^e S_x)$
B_0, B_1, B_2	structural parameters
c_p	specific heat
C_{eq}	concentration corresponding to chemical equilibrium
C_j	molar concentration of species j
C_L	concentration of reactant in bulk fluid

C_L^*	liquid-phase concentration in equilibrium with bulk gas concentration
d_p	diameter of particle
D_{ij}	binary molecular diffusivity
D_K	Knudsen diffusion coefficient
D^e	effective diffusivity
D^s	surface diffusivity
Da	Damköhler number, defined by Eq. (116)
E	activation energy
E_a	apparent activation energy
E_A^a	activation energy for adsorption of species A
E_A^d	activation energy for desportion of species A
f	fraction of wetted exterior surface of pellet
$f(C_A)$	surface adsorption isotherms
f(r)	pore size density function
F_{ij}	element of a matrix defined by Eq. (17)
g_{ij}	element of matrix defined by Eq. (24)
G_{ij}	element of the matrix F^{-1}
h	heat transfer coefficient
H	proportionality coefficient, defined by Eq. (32)
j_D	j factor for mass transfer, defined by Eq. (50)
j_H	j factor for heat transfer, defined by Eq. (51)
k	reaction rate constant
k_c	mass transfer coefficient
K_A	adsorption equilibrium constant
K_p	equilibrium partition factor
K_r	increased drag factor
L	half thickness of slab
m	exponent in correlation
M	molecular weight
n	reaction order; number of moles
n'	apparent reaction order
$\underline{n}$	normal outward unit vector
N_i	total molar flux of species i
N_i^D	diffusive flux of species i
N_i^V	viscous molar flux of species i

p	pressure
P	quantity defined by Eq. (102)
r	radial position
r_m	measured reaction rate
r_p	pore diameter
r(C)	reaction rate
R	particle radius; universal gas constant
R_N	ratio of molar fluxes, defined by Eq. (48)
Re	Reynolds number defined by Eq. (54)
s	ratio of rate constants k_1/k_2
S	selectivity
S_g	surface area per unit weight of catalyst
S_x	external surface area of catalyst
t	time
T	temperature
T_s	surface temperature
u	bulk fluid velocity
u_i	average molecular velocity of species i
V_c	volume of catalyst
V_p	volume of pellet
x_A	mole fraction of A
y	dimensionless temperature, defined by Eq. (136)
y_{max}	quantity defined by Eq. (123)
y_{min}	quantity defined by Eq. (123)
y_s	dimensionless surface temperature, defined by Eq. (114)
z	length coordinate

Greek Letters

α	ratio of molar fluxes, defined by Eq. (15)
β	Prater number, defined by Eq. (136)
β_b	Prater number, defined by Eq. (109)
β^*	Prater number, defined by Eq. (109)
γ	Arrhenius number, defined by Eq. (116)
Γ	ratio of Biot numbers, defined by Eq. (109)
δ	ratio of effective diffusivities, D_2^e/D_1^e
$-\Delta H$	heat of reaction

ε	porosity of pellet
ε_b	bed voidage fraction
ζ	dimensionless length, r/R
η	effectiveness factor
η^*	external effectiveness factor = 1/(1 + Da)
θ	fraction of surface covered
λ	ratio of molecule to pore diameter; average hopping distance
λ_f	thermal conductivity of fluid
λ^e	effective thermal conductivity
Λ	normalized Thiele modulus, defined by Eq. (62)
μ	viscosity
ν	vibrational frequency
ν_i	stoichiometric coefficient
ξ	conversion at which rate is 0.9 of initial rate
ρ	density
ρ_b	bulk catalyst density
σ	ratio of heats of reaction = $\Delta H_2/\Delta H_1$
τ	tortuosity factor for
τ^*	lingering time
ϕ	Thiele modulus with R as characteristic length
Φ	observable modulus defined by Eq. (84)
Φ^*	observable modulus defined by Eq. (88)
χ	bulk-flow correction factor

REFERENCES

Abed, R., and R. G. Rinker, Reaction with mole changes in porous catalysts in the molecular, transition, and Knudsen regimes, AIChE, J., *19*, 618 (1973).

Anderson, J. B., A criterion for isothermal behavior of a catalyst pellet, Chem. Eng. Sci., *18*, 147 (1963).

Andrew, S. P. S., Theory and practice of the formulations of heterogeneous catalysts, Chem. Eng. Sci., *36*, 1431 (1981).

Aris, R., On shape factors for irregular particles, Chem. Eng. Sci., *6*, 262 (1957).

Aris, R., A normalization for the Thiele modulus, Ind. Eng. Chem. Fundam., *4*, 227 (1965).

Aris, R., *The Mathematical Theory of Diffusion and Reaction in Permeable Catalysts*, Vol. 1: *The Theory of the Steady State*, Clarendon Press, Oxford (1975).

Balakotaiah, V. and D. Luss, Exact steady state multiplicity criteria for two consecutive or parallel reactions in lumped-parameter systems, Chem. Eng. Sci., *37*, 433 (1982a)

Balakotaiah, V. and D. Luss, Structure of the steady-state solutions of lumped parameter chemically reacting systems, Chem. Eng. Sci., *37*, 1161 (1982b)

Balakotaiah, V. and D. Luss, Multiplicity criteria for multi-reaction networks, AIChE J., *29*, 552 (1983a)

Balkotaiah, V. and D. Luss, Multiplicity features of reacting systems; dependence of the steady states of a CSTR on the residence time, Chem. Eng. Sci., *38*, 1709 (1983b).

Baltus, R. E. and J. L. Anderson, Hindered diffusion of asphaltines through microporous membranes, Chem. Eng. Sci., *12*, 1959 (1983).

Barrer, R. M., Surface and volume flow in porous media, in *The Solid Gas Interface*, Vol. 2, E. A. Flood, ed., Marcel Dekker, New York (1967).

Bird, R. B., W. E. Stewart, and E. N. Lightfoot, *Transport Phenomena*, Wiley, New York (1960).

Bischoff, K. R., Effectiveness factors for general reaction rate forms, AIChE, J., *11*, 351 (1965).

Butt, J. B., Thermal conductivity of porous catalysts, AIChE J., *11*, 106 (1965).

Butt, J. B., Internal mass and heat transport effects on catalytic behavior-activity, selectivity and yield in complex reaction systems, Chem. Eng. Sci., *21*, 275 (1966).

Carberry, J. J., Mass diffusion and isothermal catalytic selectivity, Chem. Eng. Sci., *17*, 675 (1962).

Carberry, J. J., *Chemical and Catalytic Reaction Engineering*, McGraw-Hill, New York (1976).

Carberry, J. J and A. A. Kulkarni, The non-isothermal catalytic effectiveness factor for monolith supported catalysts, J. Catal., *31*, 41 (1973).

Cardoso, M. A. and D. Luss, Stability of catalytic wires, Chem. Eng. Sci., *24*, 1699 (1969),

Cassiere, G. and J. J. Carberry, The interphase catalytic effectiveness factor-activity, yield and non-isothermality, Chem. Eng. Educ., *7*(1), 22 (1973).

Cerro, R. P. and J. M. Smith, Chromatography of nonadsorbable gases, AIChE J., *16*, 1034 (1970).

Chambers, R. and M. Boudart, Lack of dependence of conversion on flow rate in catalytic studies, J. Catal., *6*, 141 (1966).

Chang, H. C. and J. M. Calo, Exact criteria for uniqueness and multiplicity of an nth order chemical reaction via a catastrophe theory approach. Chem. Eng. Sci., *34*, 285 (1979).

Chantong, A. and F. E. Massoth, Restrictive diffusion in aluminas, AIChE J., *29*, 725 (1983).

Churchill, S. W., A generalized expression for the effectiveness factor of porous catalyst pellets, AIChE J., *23*, 208 (1977).

Copelowitz, I. and R. Aris, Finding and conjectures concerning the multiplicity of solutions, Chem. Eng. Sci., *25*, 906 (1970).

Cussler, E. L., *Multicomponent Diffusion*, Elsevier, Amsterdam (1976).

Damköhler, G., Einfluss von Diffusion, Stromung und Wärmetransport auf die Ausbeute bei chemisch-technischen Reaktionen, Chemieingenieur, *3*, 359 (1937).

Dogu, G and J. M. Smith, Rate parameters from dynamic experiments with single catalyst pellets, Chem. Eng. Sci., *31*, 123 (1976).
Dowden, D. A. and G. W. Bridger, The testing of heterogeneous catalysts, Adv. Catal., *9*, 669 (1957).
Dwivedi, P. N. and S. N. Upadhyay, Particle-fluid mass transfer in fixed and fluidized beds, Ind. Eng. Chem. Process Des. Dev., *16*, 157 (1977).
Eberley, P. E., Jr., Diffusion studies in zeolites and selected solids by gas chromatographic techniques, Ind. Eng. Chem. Fundam., *8*, 25 (1969).
Eberley, P. E., Jr., Diffusion in zeolites, in *Zeolite Chemistry and Catalysis*, J. A. Rabo, ed., ACS Monogr., *171*, 392 (1976).
Feng, C. E. and W. E. Stewart, Practical models for isothermal diffusion and flow of gases in porous solids, Ind. Eng. Chem. Fundam., *12*, 143 (1973).
Feng, C. V., V. V. Kostov, and W. E. Stewart, Multicomponent diffusion of gases in porous solids. Model and experiments, Ind. Eng. Chem. Fundam., *13*, 5 (1974).
Ferry, J. D., Statistical evaluation of sieve constant in ultra-filtration, J. Gen. Physiol., *20*, 95 (1936).
Geankopolis, C. J., *Mass Transport Phenomena*, Holt, Rinehart, and Winston, New York (1972).
Goring, R. L., Diffusion of Normal paraffins in zeolite T. Occurrence of window effect, J. Catal., *31*, 13 (1973).
Gottifredi, J. C., E. E. Gonzo, and O. D. Quiroga, Analytical expression for single reaction with arbitrary kinetics, geometry and activity distribution, Chem. Eng. Sci., *36*, 713 (1981a).
Gottifredi, J. C., E. E. Gonzo, and O. D. Quiroga, Analytical expressions for single reaction with arbitrary kinetics slab geometry, Chem. Eng. Sci., *36*, 705 (1981b).
Gupta, S. and G. Thodos, Mass and heat transfer in the flow of fluids through fixed and fluidized beds of spherical particles, AIChE J., *8*, 608 (1962).
Harold, M. P. and D. Luss, An experimental study of steady-state multiplicity of two parallel reactions, AIChE Los Angeles Meet. (1982).
Harold, M. P. and D. Luss, An experimental study of steady state multiplicity features of two parallel catalytic reactions, Chem. Eng. Sci., *40*:39 (1985).
Hatfield, B. and R. Aris, Combined effects of internal and external diffusion in the non-isotermal case, Chem. Eng. Sci., *24*, 1213 (1969).
Haynes, W. H., Jr., and P. N. Sarma, A model for the application of gas chromatography to measurements of diffusion in bidisperse structured catalysts, AIChE J., *19*, 1043 (1973).
Herskowitz, M. and J. Smith, Trickle-bed reactors: a review, AIChE J., *29*, 1 (1983).
Higashi, K., H. Ito, and J. Oishi, Surface diffusion phenomena in gaseous diffusion: 1. Surface diffusion of pure gas, J. Atom. Energy Soc. Jpn., *5*, 846 (1963).
Hofmann, U., G. Emig, and H. Hofmann, Comparison of different determination metods for effective thermal conductivity of porous catalysts, ACS Symp. Ser., *65*, 189 (1978).
Horak, J., Criteria for evaluation of the effect of the diffusion of substances on heat transfer inside a catalyst particle on the reaction rate, Int. Chem. Eng., *7*, 475 (1967).

Hu, R., V. Balakotaiah, and D. Luss, Multiplicity features of porous catalytic pellet: 1. Single zeroth order reaction case, Chem. Eng. Sci., *40*:589 (1985a).

Hu, R., V. Balakotaiah, and D. Luss, Multiplicity features of porous catalytic pellets: 1. Influence of reaction order and pellet geometry, Chem. Eng. Sci., *40*:599 (1985b).

Jackson, R., Some uniqueness conditions for the symmetric steady state of a catalyst with surface resistances, Chem. Eng. Sci., *27*, 2205 (1972).

Jackson, R., *Transport in Porous Catalysts*, Elsevier, Amsterdam, 1977.

Kehoe, J. P. G. and R. Aris, Internal pressure and forced flow for reactions with volume change, Chem. Eng. Sci., *28*, 2094 (1973).

Komiyama, H. and J. M. Smith, Surface diffusion in liquid filled pores, AIChE J., *20*, 1110 (1974).

Koros, R. M. and E. J. Nowak, A diagnostic test of the kinetic regime in packed bed reactor, Chem. Eng. Sci., *22*, 470 (1967).

Kubota, H. and Y. Yamanaka, Remarks on approximate estimation of catalyst effectiveness factor, J. Chem. Eng. Jpn., *2*, 238 (1969).

Lee, J. C. M. and D. Luss, Maximum temperature rise inside catalytic pellets, Ind. Eng. Chem. Fundam., *8*, 596 (1969).

Leib, T. M. and D. Luss, Exact uniqueness and multiplicity criteria for an n-th order reaction in a CSTR, Chem. Eng. Sci., *36*, 210 (1981).

Luss, D., Sufficient conditions for uniqueness of the steady-state solutions in distributed parameter systems, Chem. Eng. Sci., *23*, 1249 (1968).

Luss, D. Uniqueness criteria for lumped and distributed parameter chemically reacting systems, Chem. Eng. Sci., *26*, 1713 (1971).

Luss, D., Steady-state multiplicity and uniqueness criteria for chemically reacting systems, in *Dynamics and Modeling of Reactive Systems*, W. E. Stewart, W. H. Ray, and C. C. Conley, eds., Academic Press, New York (1980).

Madon, R. J. and M. Boudart, Experimental criterion for the absence of artifact in the measurement of rates of heterogeneous catalytic reactions, Ind. Eng. Chem. Fundam., *21*, 438 (1982).

Mason, E. A., and R. B. Evans III, Graham's law: simple demonstrations of gases in motion, J. Chem. Educ., *46*, 358 (1969).

Mason, E. A., A. P. Malinauskas, and R. B. Evans, III, Flow and diffusion of gases in porous media, J. Chem. Phys., *46*, 3199 (1967).

Maymo, J. A. and J. M. Smith, Catalytic oxidation of hydrogen-intrapellet heat and mass transfer, AIChE, J., *12*, 845 (1966).

McGreavy, C. and J. M. Thornton, Prediction of the effectiveness factor and selectivity for highly exothermic reactions, Chem. Eng. J., *25*, 303 (1970).

Mears, D. E., Diagnostic criteria for heat transport limitations in fixed bed reactors, J. Catal., *20*, 127 (1971a).

Mears, D. E., Tests for transport limitations in experimenal catalytic reactors, Ind. Eng. Chem. Process Des. Dev., *10*, 541 (1971b).

Michelsen, M. L., Steady-state multiplicity of lumped parameter systems with two parallel reactions, Chem. Eng. Sci., *32*, 454 (1977).

Mills, P. L. and M. P. Dudukovic, Analysis of catalyst effectiveness in trickle bed reactors processing volatile or nonvolatile reactants, Chem. Eng. Sci., *35*, 2267 (1980).

Nielsen, P. H. and J. Villadsen, An analysis of the multiplicity pattern of models for simultaneous diffusion, chemical reaction and adsorption, Chem. Eng. Sci., *40*, 571 (1985).

Omata, H. and L. F. Brown, Using the dusty gas diffusion equation in catalyst pores smaller than 50 A radius, AIChE J., *18*, 967 (1972).

Østergaard, K., The influence of intraparticle heat and mass diffusion on the selectivity of parallel heterogeneous catalytic reactions, Proc. 3rd Int. Congr. Catal., North-Holland, Amsterdam, (1965, p. 1348).

Perry, R. H., and C. H. Chilton, *Chemical Engineers Handbook*, 5th ed., McGraw-Hill, New York (1973).

Petersen, E. E., *Chemical Reaction Analysis*, Prentice-Hall, Englewood Cliffs, N. J. (1965).

Probst, K. and K. Wohlfart, Empirische Abschutzung effektiver diffusions koeffizienten in porösen Systemen, Chem. Ing. Tech., *51*, 737 (1979).

Reed, E. M., and J. B. Butt, Surface diffusion of single sorbates at low and intermediate surface coverage, J. Phys. Chem., *75*, 133 (1971).

Reid, R. C., J. M. Prausnitz, and T. K. Sherwood, *The Properties of Gases and Liquids*, 3rd ed., McGraw-Hill, New York (1977).

Reikert, L., Rates of sorption and diffusion of hydrocarbons in zeolites, AIChE J., *17*, 446 (1971).

Roberts, G. W., The selectivity of porous catalysts: parallel reactions, Chem. Eng. Sci., *27*, 1409 (1972).

Roberts, G. W. and C. N. Satterfield, Effectiveness factor for porous catalysts. Langmuir-Hinshelwood kinetic expressions for bimolecular surface reactions, Ind. Eng. Chem. Fundam., *5*, 317 (1966).

Satterfield, C. N., *Mass Transfer in Heterogeneous Catalysis*, MIT Press, Cambridge, Mass. (1970).

Satterfield, C. N., *Heterogeneous Catalysis in Practice*, McGraw-Hill, New York (1980).

Satterfield, C. N., C. K. Colton, and W. H. Pitcher, Jr., Restricted diffusion in liquid within fine pores, AIChE J., *19*, 628 (1973).

Schneider, P., Catalyst effectiveness for multicomponent reactions with concentration dependent diffusivities, Chem. Eng. Commun., *2*, 155 (1976).

Schneider, P. and J. M. Smith, Chromatographic study of surface diffusion, AIChE J., *14*, 886 (1968).

Sladek, K. J., E. R. Gilliland, and R. F. Baddour, Diffusion on surfaces. Correlation of diffusivities of physically and chemically adsorbed species, Ind. Eng. Chem. Fund., *13*, 100 (1974).

Stewart, W. E. and J. Villadsen, General calculation of multiple steady states and the effectiveness factors for porous catalysts, AIChE J., *15*, 28 (1969).

Suzuki, M., and J. M. Smith, Dynamics of diffusion and adsorption in a single catalyst pellet, AIChE J., *18*, 326 (1972).

Tan, C. S. and J. M. Smith, Catalyst particle effectiveness with unsymmetrical boundary conditions, Chem. Eng. Sci., *35*, 1601 (1980).

Thiele, W. E., Relation between catalytic activity and size of particle, Ind. Eng. Chem., *31*, 916 (1939).

Tsotsis, T. T., and R. A. Schmitz, Exact uniqueness and multiplicity criteria for a positive order Arrhenius reaction in a lumped system, Chem. Eng. Sci., *34*, 135 (1979).

Van den Bosch, B., and D. Luss, Uniqueness and multiplicity criteria for an n-th order chemical reaction, Chem. Eng. Sci., *32*, 203 (1977).

Van de Vusse, J. G., Consecutive reactions in heterogeneous systems—influence of order of reaction on selectivity, Chem. Eng. Sci., *21*, 645 (1966).

Villadsen, J. and E. Ivanov, A note on the multiplicity of the Weisz-Hicks problem for an arbitrary geometry-factor, Chem. Eng. Sci., *33*, 41 (1978).

Voge, H. H. and C. Z. Morgan, Effect of catalyst particle size on selectivity in butene dehydrogenation, Ind. Eng. Chem. Process. Des. Dev., *11*, 454 (1972).

Wakao, N. and J. M. Smith, Diffusion in catalysts pellets, Chem. Eng. Sci., *17*, 825 (1962).

Wedel, S. and D. Luss, A rational approximation of the effectiveness factor, Chem. Eng. Commun., *7*, 245 (1980).

Weekman, V. W., Jr. and R. L. Goring, Influence of volume changes on gas phase reactions in porous catalysts, J. Catal., *4*, 260 (1965).

Wei, J., Intraparticle diffusion effects in complex systems of first-order reactions, J. Catal., *1*, 526 (1962).

Wei, J., A mathematical theory of enhanced para xylene selectivity in molecular sieve catalysts, J. Catal., *76*, 433 (1982).

Weisz, P. B., Diffusivity of porous particles. Measurements and significance for internal reaction velocities, Z. Phys. Chem., *11*, 1 (1957).

Weisz, P. B. and J. S. Hicks, The behavior of porous catalyst particles in view of internal mass and heat diffusion effects, Chem. Eng. Sci., *17*, 265 (1962).

Weisz, P. B. and C. D. Prater, Interpretation of measurements in experimental catalysis, Adv. Catal., *6*, 144 (1954).

Wheeler, A., Reaction rates and selectivity in catalyst pores, Adv. Catal., *3*, 249 (1951).

Wicke, E. and R. Kallenbach, Die Oberflächendiffusion von Kohlendioxyde in aktiven Kohlen, Kolloid-Z., *97*, 135 (1941).

Wohlfahrt, K., The design of catalyst pellets, Chem. Eng. Sci., *37*, 283 (1982).

Wong, R. L. and V. E. Denny, Diffusion, flow and heterogeneous reactions of ternary mixtures in porous catalytic media, Chem. Eng. Sci., *30*, 709 (1975).

Yang, R. T., J. F. Fenn, and G. L. Haller, Modification of the Higashi model for surface diffusion, AIChE, J., *19*, 1052 (1973).

Yortsos, Y. C. and T. T. Tsotsis, On the relationship between the effectiveness factor for the Robin and Dirichlet problems for a catalyst with uniform catalytic activity. The case of generalized isothermal and nonisothermal kinetics, Chem. Eng. Sci., *37*, 1436 (1982).

Zeldovich, Y. B., On the theory of reactions on powders and porous substances, Acta Phys. Chim. USSR, *10*, 583 (1939).

5
Gas-Solid Noncatalytic Reactions

L. K. DORAISWAMY and B. D. KULKARNI *National Chemical Laboratory, Pune, India*

INTRODUCTION

Reactions between gases and solids play a major role in the materials processing industries, encompassing a broad range of operations, such as extractive metallurgy, combustion of solid fuels, environmental control, energy generation, and catalyst manufacture and regeneration. Several examples of gas-solid reactions of industrial importance are cited in Table 1. The breadth and diversity of systems encountered in gas-solid reactions, as evident from the table, suggests that any attempt at formulating a unifying theme for the design of gas-solid reactors would not meet with much success. However, one can recognize a number of common steps in the overall design of gas-solid reactions, a combination of which may be used in the development of models and formulation of design strategies for these systems. It is in this spirit that the present chapter attempts to thread together the otherwise diverse systems.

The foremost consideration in the analysis and design of these systems is the mode of contact. The number of reactor types (or modes of contact) is very large in the chemical industry. Even for the same operation, for example, calcination of lime, different types of reactors (e.g., rotary kiln, flat hearth, fluidized bed, moving bed) are used. A similar situation prevails in many other gas-solid reactions involving calcination, decomposition, roasting, and so on. The fixed-, fluidized-, and moving-bed techniques appear to be most commonly employed and have received greater attention from both the theoretical and practical points of view. The other modes of contact, less commonly employed, include horizontal moving bed, pneumatic conveyor, rotating cylinder, and flat hearth furnace. Although a priori selection of the mode of contact is difficult, a choice will have to be made at as early a stage as possible—certainly before any significant effort is put into the detailed design.

This makes it necessary to undertake an experimental program with a view to obtaining some qualitative and semiquantitative information regarding the system. The tools of mathematical modeling should be invoked at this

TABLE 1 Examples of Gas-Solid Reactions

System	Type of reactor	Remarks	References
1. Reduction of hematite	Fixed bed	Consecutive reaction of solids	Spitzer et al. (1968)
	Fixed bed		Ohmi et al. (1978)
	Moving bed		Tsay et al. (1976)
	Multistage fluid bed		Doheim (1973)
	Fluid bed		Ahner and Feinman (1964)
2. Coal gasification	Moving bed	Complex reaction	Yoon et al. (1979)
	Fluidized bed		Ubhayakar et al. (1977); Yoshida and Kunii (1974)
3. Selective chlorination of ilmenite	Fluidized bed	Reaction of two solids with the same gas	Doraiswamy et al. (1959); Fuwa et al. (1978)
Regeneration of coked catalyst			Ramachandran et al. (1975)

4. Reduction of iron ore with CO and/or H_2	Moving bed	Reaction of two gases with same solid	Tsay et al. (1976)
	Fixed bed		Osman et al. (1966); Barner et al. (1963)
5. Oxidation of ZnS	Fluid bed	Simple reaction	Natesan and Philbrook (1969); Yoshida and Wen (1972)
	Multistage fluid bed		Fukunaka et al. (1976)
	Fixed bed		Beveridge and Kawamura (1974)
6. Reduction of metal chlorides with hydrogen		Homogeneous-heterogeneous reaction	Rao (1981)
7. Reduction of cobalt sulfide in presence of CaO		Gaseous product of first reaction reacting with a second solid	Fahim et al. (1978)
Decomposition of potassium benzoate			Gokhale et al. (1975); Kulkarni and Doraiswamy (1980a)
8. Simultaneous oxidation and sulfation of cuprous oxide		Solid product of first reaction reacting with a second gas	Bourgeois et al. (1974)

stage to further substantiate the final choice. Focusing on the phenomena taking place in the reactor (i.e., the reaction, and transfer of heat, mass, and momentum), one can reduce the apparent diversity to a small number of models or basic reactor types. The modeling and design of reactors is then based on the equations describing these phenomena.

After a tentative choice of the mode of contact is made, the next immediate problem concerns the particles—their physical and chemical interactions with the surroundings. The variability involved here could be from one extreme set of conditions to another. The particles may be packed, as in a fixed-bed reactor, or completely mixed or moving in plug flow, as in a fluidized- or moving-bed reactor. The gas phase could be in plug flow or in a state of complete mixing. Intermediate stages of mixing are also possible. Additionally, the chemical interactions could involve simple single-step reactions or multistep reactions with or without changes in the volume of the phases and heat effects. The particles themselves could be porous or nonporous or may become porous during the reaction; they may also change their size, and significant alterations in the structure might accompany the chemical reaction. No single model can account for all the possible events listed. In fact, different classes of models, with one more suitable than the others for a given situation, can be developed. Clearly, the first step in the rational synthesis of a design strategy is a full understanding of the particle behavior.

SINGLE-PARTICLE STUDIES

Basic Steps in Gas-Solid Reactions

Despite the bewildering complexities that can occur in the study of a single particle, the formulation and accounting of these complexities can be simplified if one adheres to and understands the basic steps involved. Figure 1 sketches a solid particle reacting with a gas species to give products of reaction. The typical elemental steps involved include gas-phase mass transfer, diffusion inside the pores, adsorption of reactants, and reaction and desorption of products. Although these steps occur in succession, any one or more of these might be rate limiting.

The rate of reaction for the particle as a whole for a slowly reacting system will be influenced by the surface kinetics (the intrinsic rate). For somewhat faster reactions, pore diffusion intrudes to limit the rate. For still faster reactions with exothermic effects, the temperature gradient across the particle dimension or the gas film may become the limiting factor. In the extreme case of very rapid reactions, mass transfer across the fluid interface may become the controlling factor. It is necessary to demarcate these controlling regimes—for in a given situation when, say, gas film mass transfer is controlling, it would be wasteful to show undue concern for the detailed reaction mechanism. These considerations are common to catalytic and noncatalytic gas-solid reactions.

An important point that differentiates gas-solid reactions from their catalytic counterparts is the fact that here the solid is also involved as a reactant. The continuous consumption of the solid reactant leads to inevitable structural changes during the reaction, and the system as a whole is always in a transient state. An important implication of this is that even the controlling regime might continually change with time for the same particle.

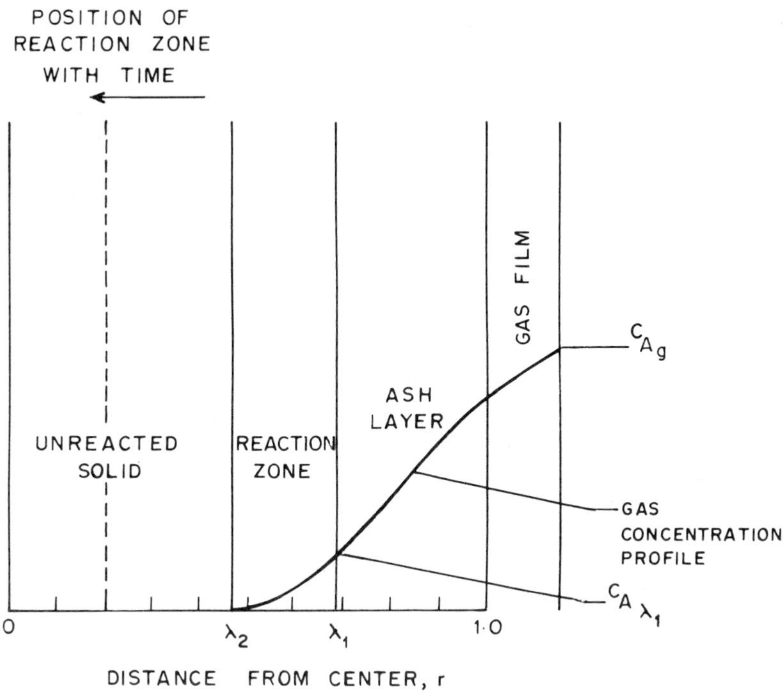

FIGURE 1 Basic steps when a solid particle reacts with a gas species.

The problem, complex as it is in the case of a single particle, assumes further dimensions when a reactor is involved. The particle, influenced by its own microenvironment, determines the controlling regime (and therefore the net rate of production per pellet), which at a given instant of time is different for different particles. This problem, although considered in greater detail later, here follows logically from a consideration of the basic steps in gas-solid reactions.

The rates of reaction when gas film mass transfer is controlling can easily be obtained from a knowledge of the transport coefficient across the fluid-solid interface. Several correlations for predicting the transport coefficients for a fixed or moving solids and a moving gas stream have been reported in the literature. These correlations should be used with some caution, for many of them are valid for restricted operating conditions. The recommended equations for estimating the mass transfer coefficient are given in a later section.

In the presence of pore diffusion effects, it is customary to invoke the laws of molecular diffusion and assume that they are applicable to a porous medium with an effective diffusion parameter. This equivalence is at best an approximation, and the total flux balance should include, besides the diffusive flux, the contributions from the pressure gradient and bulk flow due to diffusion. Conventionally, these contributions are neglected and perhaps justifiably so. However, in certain instances, such as when multicomponent diffusion is involved, they should be taken into account. The concept of effective diffusivity as applied here (as distinct from gas-solid catalytic reactions) also needs reconsideration. In most gas-solid systems, the morphology of the solid changes due to the presence of reaction, temperature gradient, and so on. The typical changes that may accompany a reaction include pore shrinkage or closure, swelling, softening, or cracking of the particle. The temperature may affect the sintering of particles, whereby the density of contact is increased. In view of these influences, the effective diffusivity would be modified continuously with the progress of reaction.

Basic Models

The relevance of the basic steps in gas-solid reactions has brought out a number of features that are more or less unique to these types of reactions. It is necessary now to model these features such that a further insight into the behavior of these systems can be gained using numerical experiments. As already mentioned, it is obviously not possible for a single model to incorporate all the features of gas-solid reactions. In fact, different classes of models have been postulated and used to describe these systems. In the present section we briefly summarize these basic models, and in the next present a state-of-the-art discussion of the extension of these basic models to different situations of greater practical relevance. A comprehensive review has recently been published on the modeling of gas-solid reactions by Ramachandran and Doraiswamy (1982b) (see also Doraiswamy and Sharma, 1983).

Sharp Interface Model

The sharp interface model (SIM) is one of the earliest models used and is well described in the standard textbooks on chemical reaction engineering. The model is mainly applicable to nonporous or relatively less porous solids and assumes that the reaction occurs at a sharp interface that separates the reacted outer shell (or ash layer) and the unreacted inner core of solid. The steps involved are diffusion through the gas film, diffusion through the ash layer, and reaction at the interface. The respective rates of these steps in mol/s (for a single pellet) are:

Diffusion through gas film:

$$R_{Ag} = 4\pi R^2 k_g (C_{Ab} - C_{As}) \tag{1}$$

Diffusion through ash layer:

$$R_{Ad} = \frac{4\pi R R_i D_e}{R - R_i} (C_{As} - C_{Ai}) \tag{2}$$

Chemical reaction at the interface:

$$R_{Ac} = 4\pi R_i^2 k_s C_{Ai} \tag{3}$$

The rate of reaction per pellet at any time t accounting for all the three resistances is obtained simply by summing Eqs. (1) to (3) as

$$\frac{1}{R_A} = \frac{1}{R_{Ag}} + \frac{1}{R_{Ad}} + \frac{1}{R_{Ac}} \qquad (4a)$$

or

$$R_A = \left(\frac{1}{4\pi R^2 k_g} + \frac{R - R_i}{4\pi R R_i D_e} + \frac{1}{4\pi R_i^2 k_s} \right)^{-1} C_{Ab} \qquad (4b)$$

This rate of consumption of A can be related to the conversion of B through a stoichiometric balance on solid B as

$$-\frac{d}{dt}\left(\frac{4}{3} \pi R_i^3 \frac{\rho_B}{M_B} \right) = \frac{\nu_B}{\nu_A} R_A \qquad (5)$$

Integration of this equation subject to the initial radius (=R) at t = 0 leads to the conversion-time relationship,

$$x + \frac{Rk_g}{2D_e} [1 - 3(1 - x)^{2/3} + 2(1 - x)]$$

$$+ \frac{3k_g}{k_s} [1 - (1 - x)^{1/3}] = \frac{3k_g C_{Ab} M_B \nu_B}{\nu_A \rho_B R} t \qquad (6)$$

The time required for complete conversion can be obtained by setting x = 1. The equation can also be written in terms of the time required for complete conversion if any one of the three basic steps were to control the rate as

$$t = t_{film} + t_{ash} + t_{reaction}$$

$$= f_0(x)\tau_f + f_1(x)\tau_a + f_2(x)\tau_r \qquad (7)$$

where τ_f, τ_a, and τ_r, represent, respectively, the time required for complete conversion if film transfer, ash diffusion, or reaction alone were to control the rate. The functions f_0, f_1, and f_2 and the various τ's assume different forms for different geometries of the particle and are defined in Table 2 together with other relevant parameters.

It is evident from the relationships present in the table that the time required for a given conversion x depends on τ, which in turn depends on (a) the radius of the particle R, (b) the concentration in the gas phase C_{Ag}, (c) the molal density of solid ρ_B, and (d) the stoichiometric coefficient ν_B, in addition to k_g, D_e, or k_s, depending on whether film, ash diffusion, or reaction controls.

It may be noted that the dependence of τ on pellet size is different for controlling regimes. Thus it is first order in R for reaction control, second order for ash diffusion control, and 1.5 to 2 order for film diffusion control. By varying the size of the particle it is therefore easy to delineate the controlling regime.

TABLE 2 Time-Conversion Relationships for SIM for Different Particle Geometries[a]

$$A(g) + \nu_B B(s) \rightarrow R(g) + S(s)$$

Controlling regime	Flat plate	Cylinder	Sphere	τ
Film diffusion, $f_0(x) =$	x	x	x	$\frac{\rho_B R}{\nu_B(\beta + 1)k_g C_{Ag}}$
Ash diffusion, $f_1(x) =$	x^2	$x + (1 - x)\ln(1 - x)$	$1 - (1 - x)^{2/3} + 2(1 - x)$	$\frac{\rho_B R^2}{2\nu_B(\beta + 1)D_e C_{Ag}}$
Reaction, $f_2(x) =$	x	$1 - (1 - x)^{1/2}$	$1 - (1 - x)^{1/3}$	$\frac{\rho_B R}{\nu_B k_S C_{Ag}^n}$

[a]The conversion $x = 1 - (r/R)^{\beta+1}$, where $\beta = 0$, 1, and 2 for flat plate, cylinder, and sphere, respectively.

Further, for pellets of different sizes, different controlling mechanisms can prevail. At the start of the reaction there is no ash layer present. Also, unless the reaction is extremely rapid, film diffusion is normally not limiting. The system therefore begins with the rate controlled by the surface reaction. With progress in time, sufficient ash layer builds up and the system passes to ash diffusion control.

The dependence of τ on the gas-phase concentration, while not very important for a single particle, assumes great significance for particles in a reactor. As already mentioned, in a reactor the gas-phase concentration is usually not constant but varies with position. Thus, even for particles of the same dimensions, we can have different controlling regimes when they are placed in a reactor. This point is discussed further later.

An alternative approach to the analysis of gas-solid reactions is to express the results in analogy with catalytic systems, in terms of an effectiveness factor. Such an approach has been used by Ishida and Wen (1968), and for a general nth-order reaction, the effectiveness factor can be simply expressed as

$$\eta = \frac{R_A}{4\pi R^2 k_s C_{Ab}^n} \tag{8}$$

In view of the changing rate R_A, the effectiveness factor is also changing with time and the following general implicit equation can be derived:

$$\eta = 1 - R_i^2 Da \frac{1}{Sh} + \frac{1 - R_i}{R_i} \tag{9}$$

where $Sh = k_g R/D_e$ and $Da = k_s R C_{Ab}^{n-1} / D_e$. The movement of the interface R_i with time appearing in Eq. (9) can be obtained by integrating Eq. (5), where the term R_A on the right-hand side can be replaced by R_A given by Eq. (8).

The case considered thus far assumes linear kinetics. It is possible for a reaction to follow fractional-order kinetics with respect to the gaseous reactant. Typical examples are the reduction of Cr_2O_3 by hydrogen (Chu and Rahmel, 1979) and oxidation of ZnS (Cannon and Denbigh, 1957). In many instances the reaction rate may also be represented by Hougen-Watson (H-W) kinetics. In general, in the presence of such nonlinear kinetics, no analytical expression for the conversion-time relations can be obtained. Numerical integration procedures for power law kinetics using SIM have been proposed by Sohn and Szekely (1972) and for H-W kinetics by Ramachandran (1982). The case of zero-order reaction has been reported by Simonsson (1979). A zero-order reaction typically exhibits sharp transition between the kinetic and diffusion control regimes. The combination of first- and zero-order kinetics has been analyzed by Kam et al. (1976) for the case for oxidation of carbon.

SIM is a phenomenological model, and in the extreme case of reaction control or ash diffusion control requires only one parameter—the relevant time factor τ—to describe the system. The model predicts total conversion of solid in a finite time and is well suited for many practical systems. But it cannot account for such features as the leveling off of conversion of the S-shaped nature of the conversion-time curve.

Volume Reaction Model

When the solid is porous, the gas can penetrate the solid and the reaction may now be assumed to take place all over the volume of the pellet rather than at a sharp interface. In general, the rate of reaction at the interior points would be lower than at the surface due to the concentration gradient. In a special situation when no diffusion resistance exists, the reaction occurs uniformly all through the particle, leading to the so-called *homogeneous model*. The rate of reaction per unit volume of the pellet is represented as

$$R_A = kC_A^m C_B^n$$

where m and n refer to the orders with respect to the gas and solid, respectively. A distinctive feature of the volume reaction model is that the rate of chemical reaction depends on the activity and concentration of the solid reactant. This means that only for certain values of n ($n < 1$) can the solid reactant be completely reacted at any point in a particle and its concentration reduced to zero.

The general conservation equations for the gaseous and solid reactant species for the volume reaction model illustrated in Fig. 2 can be written as

$$\nabla^2 \hat{C}_A = \phi^2 \hat{C}_A^m \hat{C}_B^n \tag{10}$$

$$-\frac{d\hat{C}_B}{d\hat{t}} = \hat{C}_A^m \hat{C}_B^n \tag{11}$$

where

$$\phi = \left[\frac{kC_{Ab}^{m-1} C_{B0}^n R^2}{D_e}\right]^{1/2}$$

and

$$\hat{t} = \frac{\nu_B k}{\nu_A} C_{Ab}^m C_{B0}^{n-1} t$$

The boundary conditions required for the solution of the model are

$$t = 0,\ \hat{R} = 1:\quad \hat{C}_B = 1 \tag{12}$$

$$t > 0,\ \hat{R} = 1:\quad \frac{d\hat{C}_A}{d\hat{R}} = Sh(1 - \hat{C}_A) \quad \text{or} \quad \hat{C}_A = 1 \tag{13}$$

$$t > 0,\ \hat{R} = 0:\quad \frac{d\hat{C}_A}{d\hat{R}} = 0 \tag{14}$$

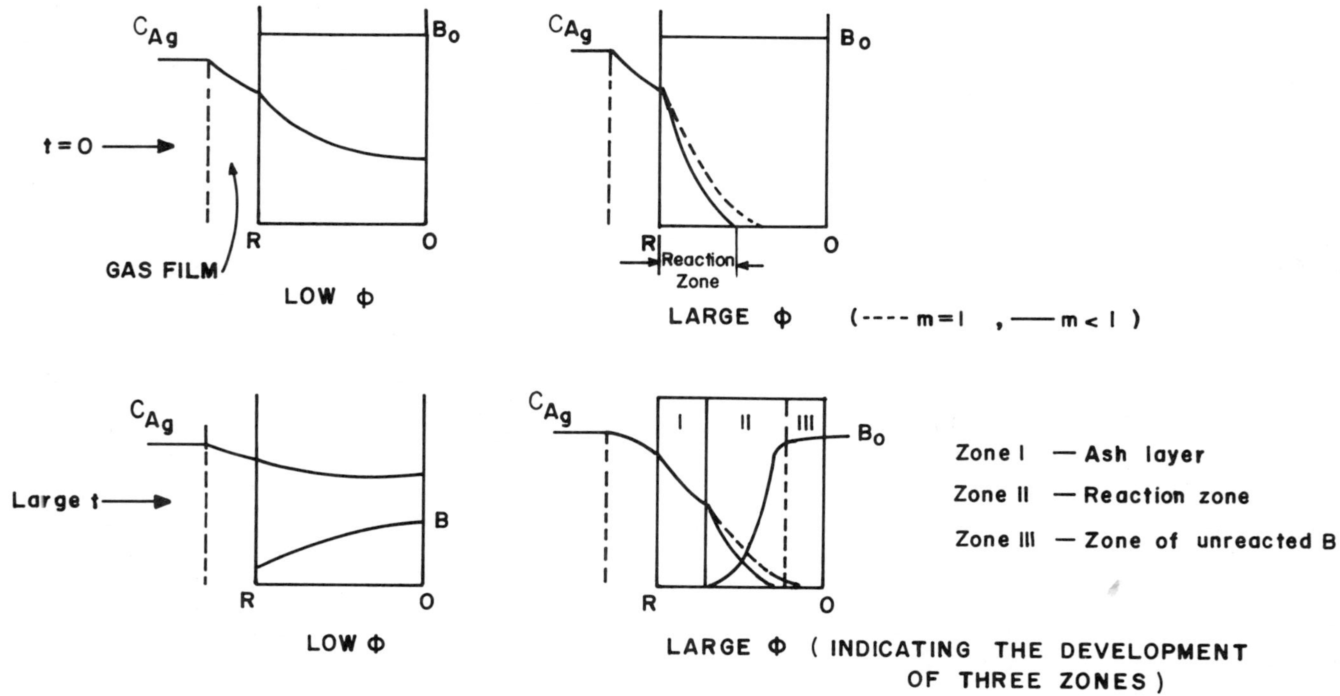

FIGURE 2 Concentration profiles for the homogeneous model for various Thiele moduli.

No analytical solution to this set of equations for arbitrary values of m and n seems possible, and recourse to numerical methods is necessary. However, for certain simplified cases analytical solutions can be obtained. Thus for the case of low values of ϕ the concentration profile within the pellet will be nearly uniform and integration of Eq. (11) for n = 0 and 1 would lead to the following simple conversion-time relations:

$$x = \begin{cases} 1 - \exp(-\hat{t}) & \text{for } n = 1 \quad (15) \\ \hat{t} & \text{for } n = 0 \quad (16) \end{cases}$$

At higher values of ϕ ($\phi > 0.2$) the concentration profile within the pellet cannot any more be neglected and both Eqs. (10) and (11) need to be solved simultaneously. An especially important case from the practical standpoint is when the order with respect to the gaseous component is unity (n = 1). In this situation it is possible to combine Eqs. (10) and (11) into a single equation by defining a cumulative gas concentration as

$$\psi = \int_0^{\hat{t}} \hat{C}_A \, d\hat{t} \tag{17}$$

The transformed equation for the special case of m = n = 1 can be written as

$$\nabla^2\psi = \phi^2[1 - \exp(-\psi)] \tag{18}$$

with boundary conditions $\psi(1) = \hat{t}$ and $(d\psi/d\hat{R})_{\hat{R}=0} = 0$. This transformation, suggested by Del Borghi et al. (1976) and Dudukovic and Lamba (1976), is extremely useful for systems with no structural variations and for power-law-type kinetics. This case (m = 1, n = 1) has also been analyzed by Dudukovic and Lamba (1978a); approximate analytical solutions to the problem have been proposed by Ramachandran and Kulkarni (1980) and by Wen and Wu (1976).

Some of the useful time-conversion relations are presented in Table 3. The case of m = 0, n = 1 has been analyzed by Dudukovic and Lamba (1978b); m being equal to zero in this case, the gas concentration can drop to zero within the pellet, depending on the value of ϕ. In fact, a critical value of $\phi_{cr} = 6\sqrt{Sh}/(2 + Sh)^{1/2}$ can be deduced for this case. For $\phi < \phi_{cr}$, the concentration would be finite all over the pellet and the simple relation given by Eq. (15) describes the conversion-time behavior. For $\phi > \phi_{cr}$, the governing equations are applicable only in the region $\hat{R}_i < \hat{R} < 1$, and solutions to this case are provided by Dudukovic and Lamba (1978a). A comparative study of the effects of reation order on the conversion-time relationship has been presented by these authors and their results are reproduced in Fig. 3. It is seen from the figure that the differences are not very significant.

Other analyses employing different values for n and m have also been presented in the literature. The case of m = 1, n = 0 has been analyzed by Ausman and Watson (1962) and by Ishida and Wen (1968). In view of the zero-order dependency on the solid concentration the local conversion of the solid in the pellet reaches completion. This leads to distinct zones,

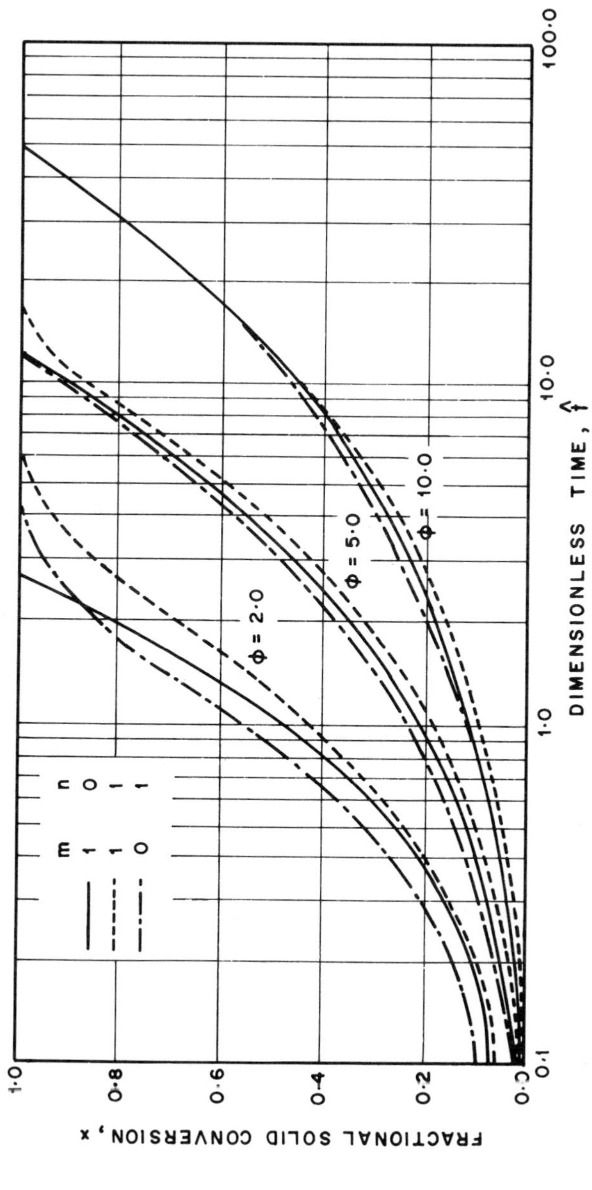

FIGURE 3 Effect of reaction orders on x versus $\hat{t}$ plots based on volume reaction model for slab geometry; $Bi_m = \infty$. (From Dudukovic and Lamba, 1978b.)

TABLE 3 Conservation Equations for Heat and Mass for Volume Reaction and Particle-Pellet Models

Model	Governing equations
Volume reaction model	$N_1 \dfrac{\partial \hat{C}_A}{\partial t} = \dfrac{1}{\hat{R}^2} \dfrac{\partial}{\partial \hat{R}} \left(\hat{R}^2 \dfrac{\partial \hat{C}_A}{\partial \hat{R}} \right) - \phi^2 \hat{C}_A^m \hat{C}_B^n \exp \left[\gamma \left(1 - \dfrac{1}{\hat{T}} \right) \right]$
	$N_2 \dfrac{\partial \hat{T}}{\partial \hat{t}} = \dfrac{\partial \hat{T}}{\partial \hat{R}^2} + \dfrac{2}{\hat{R}} \dfrac{\partial \hat{T}}{\partial \hat{R}} + \beta_m \phi^2 \hat{C}_A^m \hat{C}_B^n \exp \left[\gamma \left(1 - \dfrac{1}{\hat{T}} \right) \right]$
	$\dfrac{\partial \hat{C}_B}{\partial \hat{t}} = - \hat{C}_A^m \hat{C}_B^n \exp \left[\gamma \left(1 - \dfrac{1}{\hat{T}} \right) \right]$
	$\hat{R} = 1: \quad \dfrac{\partial \hat{C}_A}{\partial \hat{R}} = Sh(1 - \hat{C}_A); \quad \dfrac{\partial \hat{T}}{\partial \hat{R}} = Nu(1 - \hat{T})$
	$\hat{R} = 0: \quad \dfrac{\partial \hat{C}_a}{\partial \hat{R}} = \dfrac{\partial \hat{T}}{\partial \hat{R}} = 0$
	$\hat{t} = 0: \quad \hat{C}_A = 0, \ \hat{T} = 1, \ \hat{C}_B = 1$
	where
	$N_1 = \dfrac{k_v C_{Ab}^m C_{Bs}^{n-1} f_c R^2 t}{D_{eB}}$

	$N_2 = \dfrac{\rho_s C_{ps} k_v C_{Ab}^m C_{Bs}^{n-1} f_0 R^2 t}{k'_e}$
	$\beta_m = \dfrac{(-\Delta H) D_{eB} C_{Ab}}{k'_e T_b}$
Particle-pellet model	$N_1^1 \dfrac{\partial \hat{C}_A}{\partial \hat{t}} = \dfrac{\partial^2 \hat{C}_A}{\partial \hat{R}^2} + \dfrac{2}{\hat{R}} \dfrac{\partial \hat{C}_A}{\partial \hat{R}} - \phi^2 \hat{r}_{Gi}^2 \hat{C}_A a'$
	$N_2^1 \dfrac{\partial \hat{T}}{\partial \hat{t}} = \dfrac{\partial^2 \hat{T}}{\partial \hat{R}^2} + \dfrac{2}{\hat{R}} \dfrac{\partial \hat{T}}{\partial \hat{R}} + \beta_m \phi^2 \hat{r}_{Gi}^2 \hat{C}_A a'$
	$\dfrac{\partial \hat{r}_{Gi}}{\partial \hat{t}} = -\hat{C}_A a'$
	where
	$a' = \dfrac{\exp[\gamma(1 - 1/\hat{T})]}{1 + Da_G \hat{r}_{Gi}(1 - \hat{r}_{Gi}) \exp[\gamma(1 - 1/\hat{T})]}$
	$N_1^1 = \dfrac{\nu f_c R^2 M_B C_{Ab} k_s}{D_e \rho_\beta r_{G0}}$
	$N_2^1 = \dfrac{\nu R^2 \rho_s C_{ps} M_B C_{Ab} k_s}{k'_e r_{G0}}$

[a] The conversion $x = 1 - (r/R)^{\beta+1}$, where $\beta = 0$, 1, and 2 for flat plate, cylinder, and sphere, respectively.

comprising a product zone nearer the surface and a reaction zone within the pellet.

Ishida and Wen (1968) divided the total reaction time into two periods (i.e., the constant-rate period and falling-rate period), in analogy with the drying process. The time-conversion relations for the two periods are

$$x = \frac{3(\phi \coth \phi - 1)\hat{t}}{\phi^2[1 + (1/\mathrm{Sh})(\phi \coth \phi - 1)]} \quad \text{(constant-rate period)} \tag{19}$$

and

$$x = 1 - \hat{R}_i^3 + \frac{3\hat{R}_i}{\phi^2}[\phi\hat{R}_i \coth(\phi\hat{R}_i)] \quad \text{(falling-rate period)} \tag{20}$$

The conversion at the end of constant-rate period is obtained by substituting $\hat{t} = \hat{t}_{cr}$ in Eq. (19), where $\hat{t}_{cr}$ is the time required for completion of the constant-rate period is given by

$$\hat{t}_{cr} = 1 + \frac{\phi^2}{6}\left(1 + \frac{2}{\mathrm{Sh}}\right) - \left(1 - \frac{D_{eB}}{D_{eS}}\right) \ln \frac{\sinh \phi}{\phi} \tag{21}$$

Equation (20) relates the conversion in the falling-rate period to the dimensionless interface position in the particle. It is possible to obtain a relationship between the interface position and time, which is given implicitly as

$$\hat{t} = 1 - \left(1 - \frac{D_{eB}}{D_{eS}} \ln \frac{\hat{R}_i \sinh \phi}{\sinh \phi\hat{R}_i}\right) + \frac{\phi^2}{6}(1 - \hat{R}_i)^2(1 + 2\hat{R}_i)$$

$$+ \frac{\phi^2}{3\mathrm{Sh}}(1 - \hat{R}_i^3) + \left[\frac{D_{eB}}{D_{eS}}(1 - \hat{R}_i) + \frac{\hat{R}_i}{\mathrm{Sh}}\right]\left[\phi \coth(\phi\hat{R}_i) - 1\right] \tag{22}$$

In the two-zone models described above, when the rate of reaction is very rapid, the concentration of the gas species drops very sharply in the reaction zone. The reaction zone can then be further subdivided into two zones, comprising the actual reaction zone where the bulk of the reaction occurs and a core of completely unreacted solid. This has led to the development of so-called *zone models* (Bowen and Cheng, 1969; Mantri et al., 1976). A common feature of these models is the existence of a reaction zone of thickness ΔY which is a function of the Thiele modulus. Mantri et al. (1976) have prepared a plot of ΔY versus ϕ, which is reproduced in Fig. 4. Such a plot is useful in parameter estimation. The width of the reaction zone (ΔY) can be independently determined from measurements using electron probe microanalysis (see Prasannan and Doraiswamy, 1982). For extremely large values of ϕ, the reaction zone width becomes very narrow, leading to the sharp interface model.

Ramachandran and Doraiswamy (1982a) considered the case where the reaction is zero order with respect to both gaseous and solid reactant

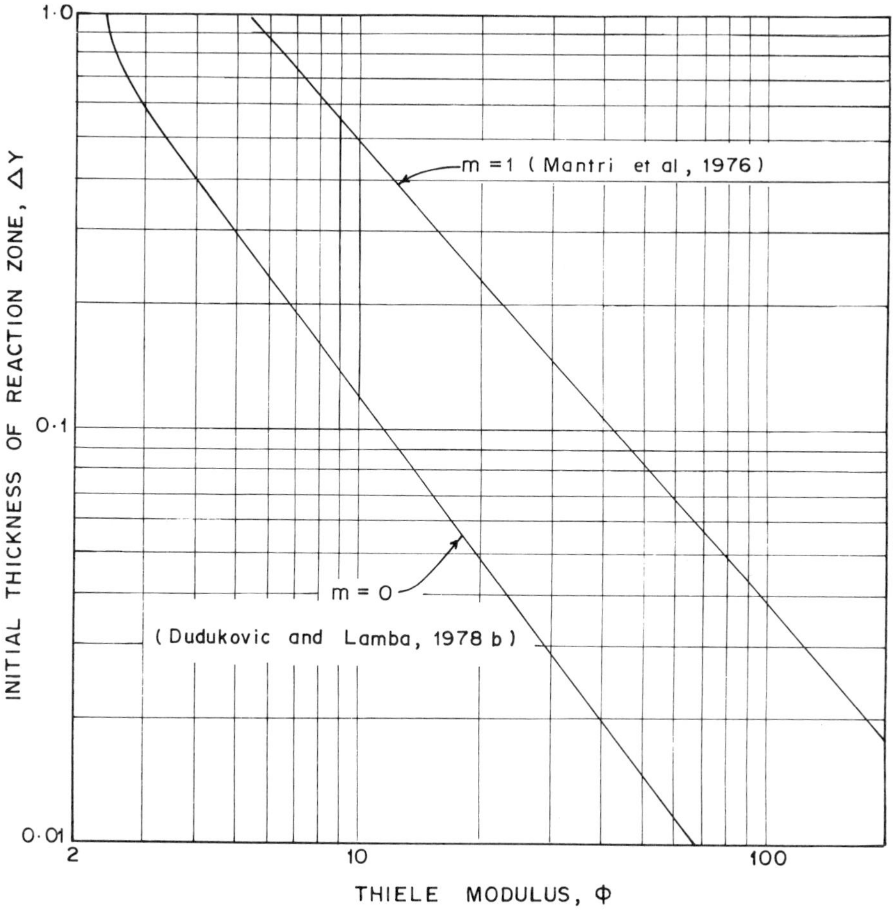

FIGURE 4 Reaction zone thickness as a function of the Thiele modulus for the volume reaction model. (From Mantri et al., 1976.)

species. For finite concentrations of gas within the pellet (no severe diffusional limitations), this case leads to the simple solution

$$x = \hat{t} \tag{23}$$

When significant resistance to the diffusion of A exists, a case (depending on the value of ϕ) may arise where the concentration of A actually drops to zero within the pellet. A simple criterion for this situation to occur (as already mentioned) is given by

$$\phi_{cr} \leqslant \frac{6}{(1 + 2/Sh)^{1/2}} \tag{24}$$

For values of $\phi > \phi_{cr}$, the governing diffusion equation has to be solved for the zone where the actual reaction occurs and the position of this zone shifts with time. Once the zone is formed, the reaction process occurs within this zone until the solid there gets completely exhausted. The reaction then jumps to another zone, where the phenomenon continues. In this case, therefore, we have a physical picture of where the reaction zone jumps from one position to another, and Ramachandran and Doraiswamy (1982a,b) have termed this the *jumping zone model*. The jumping zone concept is illustrated in Fig. 5.

Grain (Particle-Pellet) Model

In the basic grain model, also referred to as the *particle-pellet model*, it is assumed that each pellet consists of several grains of solid. The grains are spherical and of the same size. It is further assumed that each grain reacts according to SIM and that the size of the grain does not change with reaction (thereby implying no voidage change in the pellet). A diffusion resistance for the gaseous species exists in the interparticle spaces within the pellet. The basic features of the model are sketched in Fig. 6.

The mathematical formulation of the model requires consideration of the rate processes within an individual grain, and the overall mass balance for the gaseous reactant in the pellet and its stoichiometric relationship with the extent of solid consumed. As far as the individual grain is concerned, the rates of diffusion through the reacted portion of the grain and of reaction at the interface can be obtained in analogy with Eqs. (2) and (3) as

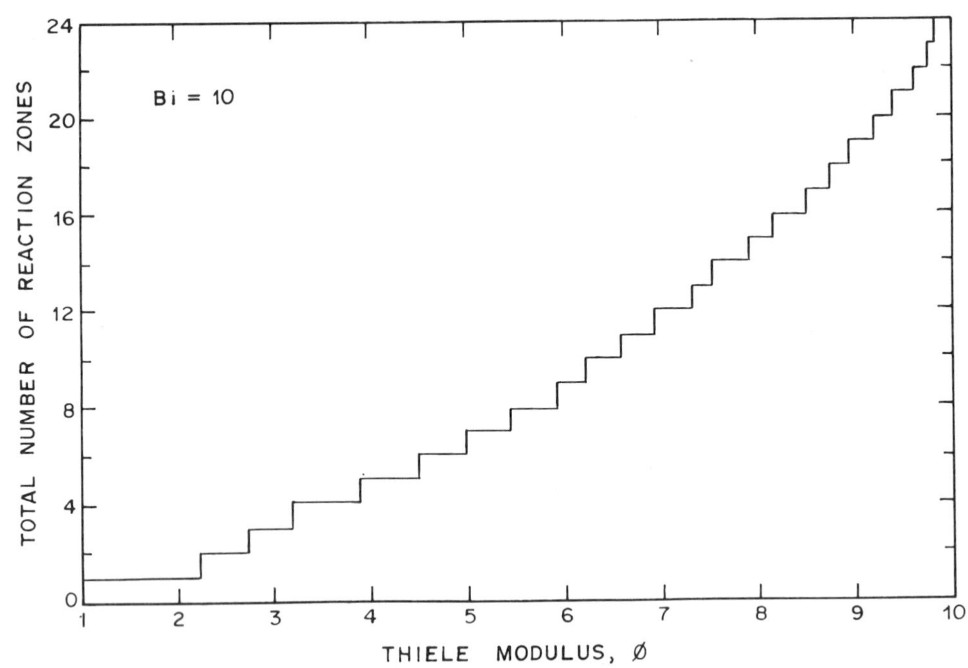

FIGURE 5 Number of reaction zone that develop in the pellet for various values of ϕ.

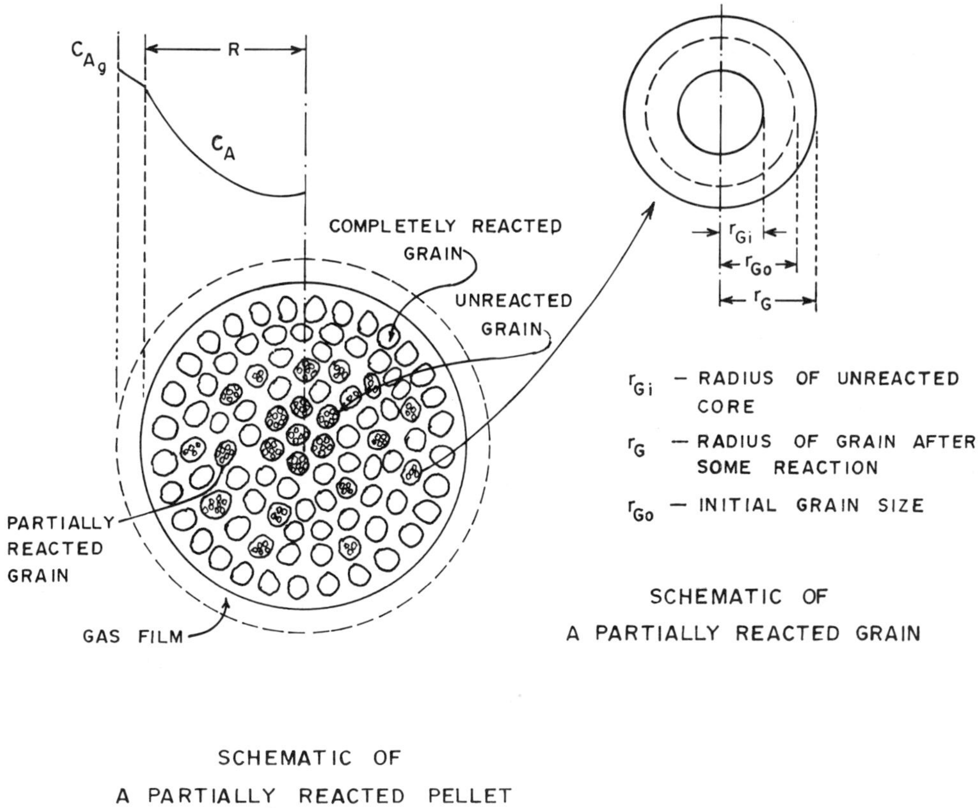

FIGURE 6 Schematic representation of the grain size model.

$$R_{GA} = \frac{4\pi D_{eG} r_{Gi} r_{G0}}{r_{Gi} - r_{G0}} (C_{As} - C_{Ai}) \tag{25}$$

$$R_{GA} = 4\pi r_{Gi}^2 k_s C_{Ai} \tag{26}$$

Eliminating the unknown concentration C_{Ai} at the interface, we obtain the overall rate per unit grain as

$$R_{GA} = \frac{4\pi r_{Gi}^2 k_s C_A}{(1 + r_{Gi} k_s / D_{eG})(1 - r_{Gi}/r_{G0})} \tag{27}$$

Once the rate of reaction for the individual grain is known, we can proceed to write the overall pellet equation as

$$\nabla^2 C_A - R_{GA} \frac{3(1 - f)}{4\pi r_{G0}^3} \tag{28}$$

where the term in parentheses refers to the total number of grains in the pellet volume. The term R_{GA} involves a knowledge of the interfacial position r_{Gi} within each grain, which is a function of both the time and position within the pellet. To evaluate this, a stoichiometric balance on the solid reactant B can be written in analogy with Eq. (5) for an individual grain as

$$-\frac{d}{dt}\left[\frac{4}{3}\pi r_{Gi}^3\frac{\rho_B}{M_B}\right] = R_{GA} \tag{29}$$

The term C_A appearing in R_{GA} in this equation fixes the position of the individual grain in the pellet.

The appropriate boundary conditions to the problem are

$$\begin{aligned} r = R: &\quad D_e\frac{dC_A}{dr} = k_g(C_{Ab} - C_{As}) \\ r = 0: &\quad \frac{dC_A}{dr} = 0 \\ t = 0: &\quad r_{Gi} = r_{G0} \end{aligned} \tag{30}$$

In general, this set of equations requires numerical solution.

A mathematical analysis of the particle-pellet model has been presented by Calvelo and Smith (1970) and Szekely and Evans (1971a,b) for a simple isothermal first-order reaction. An approximate analytical solution has also been proposed by Evans and Ranade (1980). The constant-size grain model has been generalized by Szekely and Propster (1975) to include grain size distribution. The case of nonlinear kinetics has been proposed by Sohn and Szekely (1973a,b) for reactions following the H-W rate law.

The microscopic visualization of the solid as consisting of grains requires two parameters to be known for describing the model. Considering the physical features of the model, the two characteristic times involved are τ_g, the time required for complete conversion of the grain in the C_A environment, and τ_a, the time for complete conversion of the particle by diffusion if the grain conversion process is extremely fast. In the limiting case of τ_g controlling, the diffusional resistance for the gas within the particle is negligible and the simple homogeneous model results. The individual grains could follow the shrinking core model with ash diffusion or reaction controlling. In view of the processes within the grain determining the system behavior, the conversion-time relationship will be independent of the pellet dimensions. On the other extreme, when diffusion within the pellet controls, one would observe shrinking core behavior with ash diffusion control, and typically (see Table 2) the system behavior($\tau \alpha R^2$) will be dependent on the pellet dimensions.

In the intermediate region where both τ_g and τ_d are of the same order of magnitude (note that τ_g is not a specific number but has a distribution that is commensurate with the C_A profile in the pellet, which in turn is dictated by τ_d), one could expect the pellet behavior to lie within the confines of the limiting cases of shrinking core with reaction and ash diffusion controlling processes.

The model is useful in cases where pellets are formed by compaction of particles of very fine sizes. This is not so in naturally occurring minerals. in which case fictitious grains will have to be invoked to apply the model. Also the model, in its simple form, does not explain the earlier quoted S-shaped behavior and leveling off of conversion.

Crackling Core Model

Park and Levenspiel (1975) proposed the so-called *crackling core model* (Fig. 7) to account for the S-shaped behavior of x versus t plots observed in some systems. In this model the reaction is assumed to occur in two steps

A(nonporous) → A (porous)

A(porous) → Product

The first stage could represent a simple physical process or may involve a porous intermediate, formed out of reaction, that further undergoes reaction according to the second stage. The reduction of hematite and manganese oxide (De Bruijin et al., 1980) and uranium oxide (Le Page and Fane, 1974) with hydrogen follows such a sequence:

$$Fe_3O_4 \xrightarrow{+H_2} Fe_2O_3 \xrightarrow{+H_2} FeO \xrightarrow{+H_2} Fe$$

$$MnO_2 \rightarrow Mn_2O_3 \rightarrow Mn_3O_4 \rightarrow MnO$$

$$UO_3 \rightarrow U_3O_8 \rightarrow UO_2$$

The crackling core model has been used successfully in the interpretation of experimental data for these systems.

The model envisages two stages, each of which is characterized by a certain characteristic time, and the model therefore involves at least two parameters. As before, τ_g represents the time required for complete conversion of the grain, and τ_c represents the time required for the original nonporous particle to become completely porous. In the limiting case of fast crackling ($\tau_g \ll \tau_c$), the controlling processes shift over to individual grains and one observes a simple homogeneous model. The individual grains

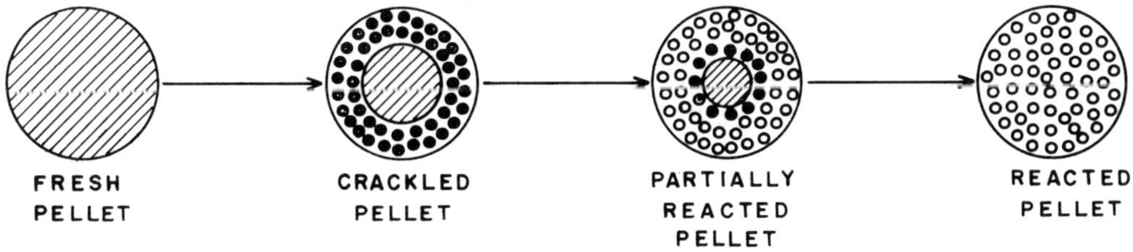

FIGURE 7 Basic stages in a crackling core model.

follow the shrinking core model with either reaction or ash diffusion controlling. In view of the processes within the grain controlling, the conversion-time relationship becomes independent of the pellet dimensions. In the other extreme of slow crackling, the processes within the grain (in the porous part of the pellet) get over much faster, and for the pellet as a whole one observes shrinking core behavior with reaction control. Typical of such situations (see Table 2), the system time constant shows linear dependence on the size of the pellet.

The study of the limiting cases as observed here makes possible some comparisons with the earlier particle-pellet model. The case of $\tau_g \gg \tau_c$ is identical to the case of $\tau_g \gg \tau_d$ in the particle-pellet model. For the reverse case $\tau_g \ll \tau_c$, however, SIM with reaction control prevails as against SIM with ash diffusion for the particle-pellet model. In view of the different size dependencies in these regimes, studies with different sizes of particles can afford a distinction between the two types of models.

In the intermediate regime where both τ_g and τ_c are of comparable magnitude, the model yields an algebraic equation for the conversion-time relationship which can readily be used. The basic feature of this model is that it predicts the S-shaped curve observed experimentally in some systems. Also, in the event that $\tau_g \to \infty$, the crackling proceeds faster, rendering the pellet porous and taking it to some intermediate level of conversion (x_I). Subsequent conversion to 100% cannot, however, be observed in view of large τ_g. The model also explains the leveling off of conversion below 100%, as observed in some systems.

The simple crackling core model could be extended, for example, by removing the assumption of no diffusion resistance to gas inside the porous part of the pellet. Also, one might consider a volume reaction model in the porous part of the pellet rather than a shrinking core model for the individual grain.

Nucleation Model

An alternative approach to provide an explanation for the observed sigmoid behavior in some gas-solid systems has been attempted in a class of models which are generally referred to as *nucleation models*. Nucleation effects are often significant in systems such as reduction of metallic oxides. In these systems the process proceeds with the generation of metallic nuclei, which subsequently grow and finally overlap. The rates of these individual processes strongly affect the conversion-time behavior of the system. In the event that the nuclei generation rate is faster, the whole surface gets covered with the metallic phase and the reaction proceeds in a topochemical manner. On the other hand, for a slow generation rate, the metal-oxide interface is irregular and different considerations prevail in estimating the conversion-time relations.

An empirical model that relates the conversion to the time has been proposed by Avrami (1940):

$$x = 1 - \exp(-Ct^N) \tag{31}$$

C and N in this equation are constants which can be obtained by fitting experimental data. The following three-parameter equation in differential form has been proposed by Erofeev (1961):

$$\frac{dx}{dt} = kx^a(1 - x)^b \tag{32}$$

and seems to provide adequate fit to experimental data (Neuburg, 1970).

In a more rational approach to nucleation, Ruckenstein and Vavanellos (1975) considered the existence of germ nuclei in the form of impurities present and the embryos of the new solid phase. In the process the germ nucleus is transformed into a growth nucleus, leading finally to product formation. The growth process is treated as an activated process, and employing the modified Avrami model, the following conversion-time relation may be obtained:

$$x = \beta k^3_{growth} N_{n,o} k_{nf} \int_0^1 \exp(-k_{nf} t_{nf})(t - t_{nf})^3 [1 - x(t_{nf})] dt_{nf} \tag{33}$$

where $N_{n,o}$ is the number of germ nuclei per unit solid volume at the beginning, and k_{nf} and t_{nf} are the rate constant and time required for nucleus formation, respectively.

Bhatia and Perlmutter (1980) have also analyzed this problem using the population balance approach. The possibility of both homogeneous and heterogeneous nucleation around a germ nucleus is considered and the basic Avarami model has been modified to include a term for the initial volume of a growth nucleus.

Some experimental data on the formation and growth of nuclei for the oxidation of copper iodate and reduction of wustite have been provided by Neuburg (1970) and El-Rahaiby and Rao (1979). Some aspects of the theory of nucleation have also been discussed (Rao, 1979).

Extensions of the Basic Model

The preceding section dealt with simple basic models. In the present section we extend these models by considering several complexities that are likely to be present in practical situations. The complexities that can be considered are those due to the effect of bulk diffusion, existence of pressure gradients in the pellets, nonisothermal effects, and effects due to structural changes. The effect of structural changes, being of much greater practical relevance, will be considered at some length at the end of the section.

Bulk-Flow Effects

Detailed formulations accounting for the effect of bulk flow have been presented by a number of authors (e.g., Beveridge and Goldie, 1968; Gower, 1971; Sohn and Sohn, 1980) for the sharp interface model. The bulk flow within the solid assumes greater importance for reactions with volume change, and typical examples of industrial gas-solid reactions with volume change are

$$C(s) + CO_2(g) = 2CO(g)$$

$$Ni(s) + 4CO(g) = Ni(CO)_4(g)$$

$$FeCl_2(s) + H_2(g) = Fe(s) + 2HCl$$

The conservation equation for the mass of reactant in situations of this type should include a term for bulk flow and can be represented in analogy with the gas-solid catalytic systems (Weekman and Gorring, 1965) in dimensionless form as

$$\frac{d}{d\hat{R}}\left(\hat{R}^{\beta}\frac{d\hat{C}_A}{d\hat{R}}\frac{1}{1+\theta\hat{C}_A}\right)=0 \tag{34}$$

with the boundary conditions

$$\hat{R}=1:\quad \hat{C}_A=1 \tag{35}$$

$$\hat{R}=\hat{R}_1:\quad \frac{d\hat{C}_A}{d\hat{R}}=2(\beta+1)\phi^2C_A(1+\theta\hat{C}_A) \tag{36}$$

β is a constant, and ϕ and θ represent, respectively, the Thiele modulus and volume change modulus in this equation, and are defined as

$$\phi=\frac{k}{2(\beta+1)D_e}\,\frac{(\beta+1)V_p}{A_p} \tag{37}$$

$$\theta=\left(\frac{\nu_C}{\nu_A}-1\right)C_{Ab} \tag{38}$$

The concentration profile for the gaseous reactant as obtained from the solution of Eq. (34) can be related to the consumption of solid as follows:

$$\frac{d\hat{R}_i}{d\hat{t}}=-\hat{C}_{Ai} \tag{39}$$

where

$$\hat{t}=\frac{kC_tC_{Ab}}{\rho_B}\,\frac{A_p}{(\beta+1)V_p}t \tag{40}$$

Bearing in mind the relationship between conversion and interfacial position,

$$x=1-\left(\frac{R_i}{R}\right)^{\beta+1} \tag{41}$$

we can calculate the conversion-time relationship. Sohn and Sohn (1980) have obtained the following asymptotic solution for the conversion-time behavior:

$$\frac{\ln(1+\theta)}{\theta}\frac{\hat{t}}{\phi^2} = 1 - \frac{(\beta+1)(1-x)^{2/(\beta+1)} - 2(1-x)}{\beta - 1} = p(x) \tag{42}$$

In general, the net effect of volume change is to change the time required to attain a given extent of conversion by a factor $\theta/\ln(1+\theta)$. It is also possible to incorporate the effect of external mass transfer in the presence of volume change effects. The conversion-time relationship can now be obtained as

$$\hat{t} = [1 - (1-x)^{1/(\beta+1)}] + \phi^2\left[\frac{\theta}{\ln(1+\theta)}p(x) + \frac{2(1+\theta)}{Sh}x\right] \tag{43}$$

In the event that nonisothermal conditions prevail, significant pressure gradients can exist within the pellet and the simple analysis above is no longer applicable. In fact, there is a need now to modify the conventional Fickian law of diffusion, and more sophisticated models for diffusion in porous media (such as the dusty-gas model of Mason et al. (1969) may have to be invoked. (This has not been attempted so far.)

Even in instances when the bulk-flow effects are not present (no mole change), pressure gradients may arise due to temperature gradients. In the presence of pressure gradients it has been shown (Wong et al., 1976; Hite and Jackson, 1977) that the conventional Weisz-Prater temperature holds true only in the case of Knudsen diffusion. Significant alterations occur when bulk diffusion prevails. In fact, an additional term $(\beta_m^2/4)$ is included in the equation for maximum temperature rise where β_m refers to the thermicity factor. Pressure gradients can thus have a significant effect on the conversion-time relations, depending on the regime of diffusion. In the case of gas-solid reactions, the presence of structural effects continually alters the diffusional regimes. The effects of pressure gradients cannot therefore be ignored all through the course of conversion. The influence of pressure gradients in gas-solid reactions following SIM has been examined by Deb Roy and Abraham (1974) and Turkdogan et al. (1973). Detailed numerical solutions have been obtained to the pressure gradients that arise because of Knudsen flow in the ash layer with small pores and due to the reactant gas having a diffusivity different from that of the product.

Nonisothermal Effects

When the chemical reaction is accompanied by evolution of heat, temperature gradients may exist within the pellet. The local reaction rate in the interior of the particle can thus be higher, and the higher inside temperature can cause some difficulties, such as sintering. Additionally, depending on the heat generation/removal rate, the system may exhibit multiplicity behavior.

In the analysis of nonisothermal SIM, equations describing heat transfer through the gas film and ash layer and heat generation due to reaction at the interface can be written in a manner analogous to the mass transfer process. Such a formulation affords an easy analysis of the nonisothermal effects. The approach, however, presumes the validity of the pseudo-steady-state assumption, which while justifiable for mass transfer processes, is not a good approximation for heat transfer processes.

Luss and Amundson (1969) have provided an analysis that incorporates the transient heat accumulation term and their solution can be represented as

$$\hat{T}_i - 1 = \frac{2}{3} P_1 \sum_{n=1}^{\infty} \frac{\sin \mu \hat{R}_i}{\hat{R}_i} f(\mu) \int_{\hat{R}_i}^{1} \sin(\mu\xi)\hat{R}_i \exp[-\mu^2 P_2 f(\xi)]d\xi \quad (44)$$

where μ's are the positive roots of the transcendental equation

$$\mu \cot \mu + Nu - 1 = 0 \quad (45)$$

and the various other parameters are defined as follows:

$$f(\xi) = \frac{\xi - \hat{R}_i}{Da \exp[\gamma(1 - 1/\hat{T}_i)]} + \frac{1}{2}(\xi^2 - \hat{R}_i^2) + \frac{1}{3}\left(\frac{1}{Sh} - 1\right)(\xi^3 - \hat{R}_i^3) \quad (46)$$

$$P_1 = \frac{3(-\Delta H)\nu\rho_B}{M_B \rho_S C_{ps} T_b} \qquad P_2 = \frac{\rho_B k'_e}{M_B \rho_s C_{ps} D_e C_{Ab}} \quad (47)$$

$$f(\mu) = \frac{(Nu - 1)^2 + \mu^2}{Nu(Nu - 1) + \mu^2} \quad (48)$$

Equation (44) gives the interface temperature in terms of the interfacial position $\hat{R}_i$, which can be obtained from

$$\frac{d\hat{R}_i}{d\hat{t}} = \eta \quad (49)$$

where η refers to the effectiveness factor under nonisothermal conditions.

For the specific cases of coke regeneration and reaction of carbon with oxygen and steam, model equations based on SIM have also been developed by Shettigar and Hughes (1972) and Rehmat and Saxena (1980). The problem of thermal instability in gas-solid systems has been discussed by Cannon and Denbigh (1957). Shen and Smith (1963), Aris (1967), Ishida and Wen (1968), and Wen and Wang (1970) for SIM. The phenomenon of thermal instability leads to sudden transitions in the rate-controlling steps and has an important bearing on reactor operation, as runaway conditions may occur.

The nonisothermal SIM is well suited for describing decomposition reactions (Narsimhan, 1961; Campbell et al., 1970; Hills, 1968). For a decomposition reaction

$$A(s) \rightarrow B(s) + P(g)$$

the application of the phase rule suggests one degree of freedom. Each temperature T_d therefore has a fixed value of partial pressure p_d of the product gas P, and once this value is reached, the reaction starts and the front moves inside. The process is controlled either by heat or gas diffusion through the product layer and typically yields SIM behavior.

In the event that heat transfer through the ash layer controls, the reaction interface stays isothermal and the equation for SIM with ash diffusion control and D_e replaced by k_e' (in the definition of τ in Table 2) represents the conversion-time behavior. Where gas diffusion through the ash layer is controlling, the variation of D_e with temperature ($D_e \propto T^{1/2}$ in the knudsen regime: $D_e \propto T^{1.5-2}$ in the bulk diffusion regime) should be appropriately accounted for (Mu and Perlmutter, 1980).

Nonisothermal effects have also been analyzed for the volume reaction model by Shettigar and Hughes (1972), and for the particle-pellet model by Calvelo and Smith (1970). The latter analysis is based on the pseudo-steady-state hypothesis, which is not valid for heat transfer. The governing transient equations for heat and mass for the volume and grain models are presented in Table 4 together with the relevant boundary conditions. An analysis of the complete transients for the particle-pellet model indicates a shift in the temperature maximum from the surface to the center with progress of time (Sampath et al., 1975). Nonisothermal effects along with structural variations have not yet been analyzed.

Effects of Structural Variations

The earlier parts of this section were concerned with the extension of the basic models to incorporate certain modifications resulting from the nonlinear nature of chemical reaction, pressure gradients and bulk flow within the pellet. Here we shall reexamine these models with a view to incorporating the structural changes that occur within the solid during reaction. The main structural changes are due to chemical reaction and sintering. The differences in the molal volumes of the reactant and product lead to variations in the particle voidage and effective diffusivity during the course of reaction. To account for these effects it is necessary to formulate a relationship between the voidage, diffusivity, and extent of conversion. A large number of such relationships to account for these effects have been proposed using different models, and the more important of these are summarized in Table 5. The incorporation of such effects in the basic models give rise to conversion-time relationships which show a sigmoid behavior observed experimentally. Also, where the porosity at the surface of the pellet becomes zero (pore closure), the governing equations predict incomplete conversion. Such behavior has, in fact, been observed in some experimental systems. Hartman and Coughlin (1974, 1976) and Georgakis et al. (1979) have used a modified grain model to describe pore closure due to grain swelling in the sulfation of lime.

In a departure from models based on pore behavior, Garza and Dudukovic (1982a,b) have proposed a variable-size grain model for systems with structural changes. Incorporating the diffusivity variations with porosity for systems with no pore closure, the model first calculates the time required for complete conversion of the surface layer of the solid. This product layer then moves inside, toward the center. In systems where pore closure occurs, the model evaluates the time required for pore closure and obtains the concentration profile of the unreacted solid at this instant. Further processes inside the pellet are described by SIM with the incorporation of a solid concentration profile. Approximate analytical solutions have

TABLE 4 Typical Diffusivity and Porosity Variations due to Structural Changes in Pellet

Number	Relationship	Model	References
1	$f_c = f_{co} + \alpha_1 \left(1 - \frac{B}{B_0}\right)$	Volume reaction model	Fan et al. (1977)
2	$\frac{D_e}{D_{e_0}} = \alpha_2 \left\{ \exp\left[1 - \frac{1}{(B/B_0)^{\alpha_3}}\right] + \alpha_2 - 1 \right\}^{-1}$ where α_1, α_2, and α_3 are constants and B refers to solid concentration	Volume reaction model	Fan et al. (1977)
3	$f_c = f_{co} + (1 - f_{co}) \frac{V_B + V_I}{V_B - \nu_s V_s}$ $D_e = D_{eo} \frac{f_c}{f_{co}}$ where V_B and V_s refer to molar volumes of species and B and s per mole of solid. V_I refers to molar volume of inert per mole of solid B.		Levenspiel (1979)
4	$\frac{1 - f_c}{1 - f_{co}} = \left(\frac{r_G}{r_{Go}}\right)^3$ where r_G refers to the radius of the grain $\frac{D_e}{D_{eo}} = \frac{f_c}{f_{co}}$	Particle-pellet model	Ramachandran and Smith (1977a)

<table>
<tr>
<td></td>
<td>Eventually, with progress of time, the pore will be completely blocked. The time required for pore closure is obtained as
$$= \frac{1}{Bi_p}\left[\frac{Z_\nu}{Z_\nu - 1}\,\frac{1}{(1 - f_{co})^{2/3}}\left(1 + \frac{Z_\nu}{1 - Z_\nu}\,f_{co}\right) - \frac{1}{Z_\nu - 1}\,\frac{1}{(1 - f_{co})^{2/3}}\right] + 1 - \frac{1}{(1 - f_{co})^{1/3}}\left(1 + \frac{Z_\nu}{1 - Z_\nu}\,f_{co}\right)$$
where $Z_\nu = \dfrac{\nu_C \rho_B M_C}{\nu_B \rho_C (1 - f_c) M_B}$</td>
<td>Particle-pellet model</td>
<td>Georgakis et al. (1979)</td>
</tr>
<tr>
<td>5</td>
<td>$$\frac{D_e}{D_{eo}} = 1 - \frac{1 - f_{co}}{f_{co}}(\gamma - 1)(1 - \hat{r}_{Gi}^{\,3})$$
where γ refers to the ratio of the volume of solid product formed per unit volume of reactant consumed and $\hat{r}_{Gi}$ refers to the dimensionless position of the shrinking core in the grain. The time required for pore closure is obtained as
$$\tau = 1 - \tilde{\eta} + \frac{Da_g}{2}\left\{1 - \tilde{\eta}^2 + \frac{[\gamma + (1 - \gamma)\tilde{\eta}^3]^{2/3}}{1 - \gamma}\right\}$$
where
$$\tilde{\eta} = \left[1 + \frac{f_{co}}{(1 - f_{co})(1 - \gamma)}\right]^{1/3}$$</td>
<td>Variable grain model</td>
<td>Garza and Dudukovic (1982)</td>
</tr>
</table>

TABLE 5 Some Useful Time-Conversion Relationships

Model	System description	Time conversion	References
Volume reaction model	Reaction first order in gas and solid; spherical pellet	$\ln\left[\frac{1-x}{0.699} - \frac{\exp(-\hat{t})}{2.33}\right] + \hat{t} = \frac{\phi^2}{10.5}\left[1 - \frac{1-x}{0.699} + \frac{\exp(-\hat{t})}{2.33}\right]$	Kulkarni and Ramachandran (1980)
Particle-pellet model	First-order reaction	$\hat{t} = 1 - (1-x)^{1/3} - \frac{\phi_1^2}{18}[1 - 3(1-x)^{2/3} + 2(1-x)] + (0.21x - 0.31x^2)\left(1 + \frac{\phi_1^2}{18}\right)\exp\left[-0.9\left(\ln\frac{\phi_1^2}{19.44}\right)^2\right]$ where $\hat{t} = \frac{\nu_B t M_B A_g k_s(T_0)}{\rho_B r_{G0}}$ and $\phi_1 = \left[\frac{3(1-f_{c0})k_s(T_0)R^2}{r_{G0}D_e}\right]^{1/2}$	Evans and Ranade (1980)
	Uniform grains of unchanging size; both pellets and grains can be of arbitrary geometry	$\hat{t} = g_{Fg}(x) + \frac{Da_g}{2F_g} f_{Fg}(x) + \frac{\phi_p^2}{2F_pF_g} f_{F_p}(x)$ where subscripts g and p refer to grain and pellet respectively and F to the shape factor and takes values of 1, 2, or 3 for infinite slab, cylinder, and sphere respectively.	Szekely et al. (1976)

		$g_{Fg(X)} = 1 - (1 - x)^{1/F_g}$ $f_{F_i}(x) = \frac{F_i}{f_i - 2}\left[1 - (1 - x)^{2/F_i} - \frac{2}{F_i}x\right]$ where $i = g$ or p	
	Uniform grains of varying sizes; both pellet and grain can have arbitrary geometry	$\hat{t} = gF_g(x) + \frac{Da_g}{2F_g} P_{F_g}(x) + \frac{\phi_p^2}{2F_pF_g} f_{f_p}(x)$ where gFg(x) as above and $P_{F_g}(x) = F_g\left\{\frac{1}{F_g - 2}[1 - (1 - x)^{2/F_g}] + \frac{1 - [\gamma + (1 - \gamma)(1 - x)]^{2/F_g}}{(F_g - 2)(\gamma - 1)}\right\}$	Garza and Dudukovic (1982)
Random pore model	Takes account of structural variations through population balance approach; kinetic regime	$x = 1 - \left(1 - \frac{k_sA_g^mt}{R}\right)^3 \exp\left[-\frac{k_sA_g^mS_0t}{1 - f_{c0}}\left(1 + \frac{\psi}{4}\frac{k_sA_g^mS_0t}{1 - f_{c0}}\right)\right]$ where $\psi = \frac{4L_0(1 - f_{c0})}{S_0^2} \qquad \hat{t} = \frac{k_sA_g^mS_0t}{1 - f_{c0}}$ This expression can also be expressed as $\frac{dx}{d\hat{t}} = (1 - x)[1 - \psi \ln(1 - x)]^{1/2}$	Bhatia and Perlmutter (1980)

TABLE 5 (Continued)

Model	System description	Time-conversion	References
	Diffusion control regime	$\frac{dx}{d\hat{t}} = \frac{a(1-x)[1-\psi \ln(1-x)]^{1/2}}{1 + \frac{2k_s \rho_b (1-f_{c0})}{\nu_B M_B D_{eG} S_0} \frac{Z_\nu}{\psi}[\sqrt{1-\psi \ln(1-\psi)} - 1]}$	Bhatia and Perlmutter (1981a,b)
SCM	Gasification reaction	$x = \frac{f_{c0}}{1-f_{c0}}\left[\left(1 + \frac{k_s A_g^m t}{r_{p0}}\right) \frac{G' - 1 - (k_s A_g^m t)/r_{po}}{G' - 1} - 1\right]$ where G' is a solution of $\frac{4}{27} f_{c0} G'^3 - G' + 1 = 0$	Szekely et al. (1976)
Random capillary model	Gasification reaction	$x = 1 - \exp[-2\pi(N_0 \nu^2 t^2 + 2N_1 \nu t)]$ where ν is the reaction velocity defined as the rate of change of pore radius with time, and N_0 and N_1 are structural parameters $N_0 = \int_{S_{min}}^{S_{max}} p(S_0) dS_0$ $N_1 = \int_{S_{min}}^{S_{max}} S_0 p(S_0) dS_0$ where $p(S_0)$ is the density distribution based on the initial pore size distribution, and S is the pore size	Gavalas (1980)

been developed by Garza and Dudukovic (1982a,b) and are included in Table 5 together with the solutions for other models.

A new class of models incorporating structural features such as pore size distribution have come up in recent years. The simplest of such models is the single-pore model of Ramachandran and Smith (1977a) and Chrostowski and Georgakis (1978). The model of Ramachandran and Smith, sketched in Fig. 8, analyzes the changes taking place in a single pore which is supposed to be representative of the overall pellet. The single pore is assumed to be cylindrical with the concentric ring of solid B associated with it. Depending on the molal volumes of the reactant and product, the pore size increases, decreases, or stays constant with progress of reaction. The model yields a simple conversion-time relationship and requires a knowledge of the average structural properties of the solid (pore radius, length, radius of associated solid). The simple model has been extended to incorporate the effect of bulk flow and reversibility of reaction (Ulrichson and Mahoney, 1980).

The use of average structural properties leads to mathematically tractable models that are normally adequate for most gas-solid systems. More rigorous models, however, should include the distribution and variations in the structural properties with reaction. Thus most solids have an initial pore size distribution that evolves during the reaction due to pore intersection and coalescence. The net reaction surface is therefore a variable, and any rigorous structural model should account for this variation. Some of the recent models that account for this will now be discussed.

The model of Christman and Edgar (1980) assumes the applicability of a single-pore model at a local point r in the pellet. The rate of reaction thus obtained is averaged over the existing pore size distribution, which is then used in the mass balance equation together with the average values of other parameters (effective diffusivity, rate constant, etc.). The mass balance equation is supplemented by an equation describing the evolution of pore size distribution with the progress of reaction, and Christman and Edgar (1980) modeled this process using a population balance approach. The key feature of this model is that it gives the variation of pore size distribution with time and position. This additional information can be used for model verification and selection. A similar approach based on population balance has been introduced by Gavalas (1980).

The model of Simons and Rawlins (1980) assumes that each pore within the solid ends up at the surface of the pellet, where a distribution of pore radii can exist. The model calculates the flux for a pore of radius r_p and averages it over the surface pore size distribution to obtain an average flux of the gaseous reactant at the surface. This is then related to the conversion of solid B. The model, though simple, does not account for the variation of reaction surface with time.

Perhaps the most realistic model is the random pore model of Bhatia and Perlmutter (1980, 1981b), which assumes that the actual reaction surface of solid B is the result of the random overlapping of a set of cylindrical surfaces. The development of the reaction surface as envisaged in this model is sketched in Fig. 9. The model proceeds with calculation of the actual reaction surface and then the conversion-time relationship in terms of the intrinsic structural properties of the solid.

The following equations for the conversion-time behavior for the cases of kinetics and ash diffusion control have been obtained:

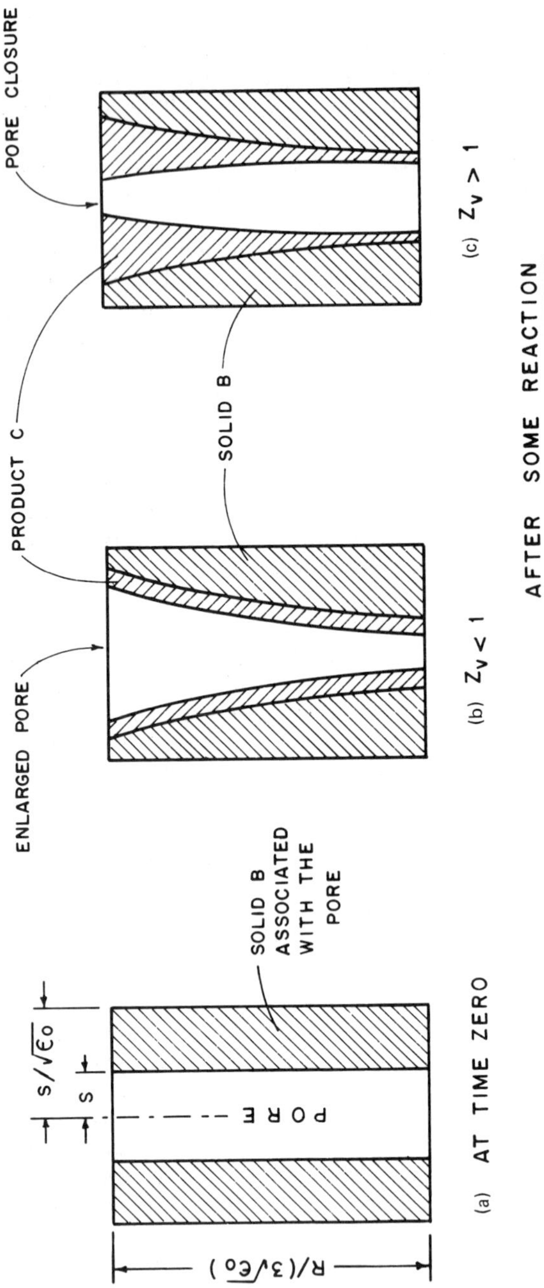

FIGURE 8 Schematic representation of the single-pore model.

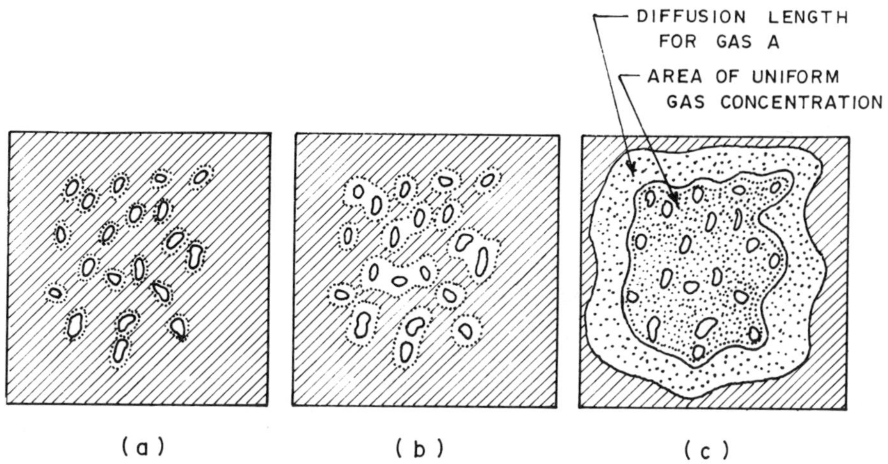

FIGURE 9 Development of reaction surface according to the random pore mode. The shaded area represents unreacted solid B, the dashed area the product layer. (a) Early stage showing product layer around each pore; (b) intermediate stage, showing some overlapping reaction surfaces; (c) later stage, showing full development of product layer and reaction surface for the particular view chosen. (From Bhatia and Perlmutter, 1981b.)

$$\frac{dx}{d\hat{t}} = (1 - x)[1 - \psi \ln(1 - x)]^{1/2} \tag{50}$$

and

$$\frac{dx}{d\hat{t}} = \frac{\hat{C}_A(1 - x)[1 - \psi \ln(1 - x)]^{1/2}}{1 + \beta' Z/\psi\{[1 - \psi \ln(1 - x)]^{1/2} - 1\}} \tag{51}$$

where

$$\psi = \frac{4\pi L_0(1 - f_0)}{S_0^2} \quad \text{and} \quad \hat{t} = \frac{k_s C_{Ag}^m S_0 t}{1 - f_0} \tag{52}$$

The parameter β' characterizes the diffusional resistance to the flow of the gaseous species in the product layer. For $\psi = 0$, Eq. (50) reduces to the volume reaction model with $n = 1$, and for $\psi \to 1$ the behavior closely approaches the grain model. It is apparent, therefore, that the concept of reaction order with respect to solids is closely related to the structure of the solid, and the random pore model for the first time provides a rational meaning to the order of reaction with respect to solids. The relation between ψ and n is graphically illustrated in Fig. 10.

Several distributions for the parameter ψ can be assumed, and Bhatia and Perlmutter (1981a) have shown that a uniform pore size leads to the lowest reactivity of solid. Also, for a given pore size distribution, an optimal structure exists for highest activity of solid.

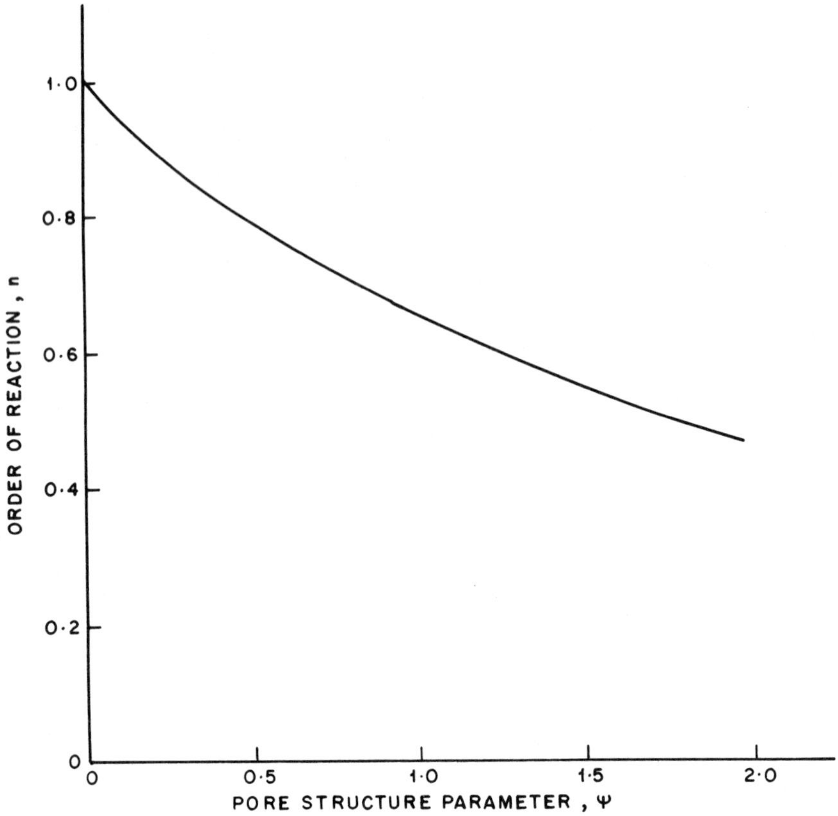

FIGURE 10 Relationship between the structural parameter of the random pore model and the order of reaction with respect to solid in the volume reaction model. (From Bhatia and Perlmutter, 1980.)

The model of Bhatia and Perlmutter (1980, 1981a,b), discussed above, considers the pores and reaction surface areas to be equal. Although this seems to be a good approximation at low levels of conversion, it is less accurate at high conversion levels. In a subsequent paper, Bhatia and Perlmutter (1983) provided generalization in terms of moving pore and reaction surface, and for the case of no intraparticle and boundary layer diffusional resistances, obtained the following relation:

$$\hat{t} = \frac{dx_1}{S^*(x_1)} + \frac{\beta'}{2}\int_0^x \int_0^{x_1} \left[\frac{1}{S^{*2}(x')} + \frac{Z_v - 1}{S_p^{*2}(x')} \right] dx'\, dx_1 \qquad (53)$$

For $\beta' \to 0$ and ∞, this equation reduces to the cases of kinetic control and product layer diffusion control. In this equation S^* and S_p^* refer to the dimensionless reaction surface area and pore surface area, respectively, and these quantities can be appropriately evaluated for different models of the internal structure of the solid. Thus for the grain model,

$$S^*(x) = (1 - x)^m \tag{54}$$

$$S_p^*(x) = [1 + (Z_v - 1)x]^m \tag{55}$$

while for the random pore model,

$$S^*(x) = (1 - x)\sqrt{1 - \psi \ln(1 - x)} \tag{56}$$

$$S_p^*(x) = [1 + (Z_v - 1)x]\sqrt{1 - \psi \ln[1 + (Z_v - 1)x]} \tag{57}$$

Substituting these expressions for S^* and S_p^* in Eq. (53) and integrating leads to the desired conversion-time relationship. A comparison of predictions obtained using the grain model and the random pore model for a set of parameter values is illustrated in Fig. 11.

Sintering

The temperature variations within the pellet in the presence of an exothermic reaction result in sintering, which causes a decrease in the effective diffusivity of the pellet. Alternatively, the grain size increases and the specific surface area of the grain decreases. Simple empirical laws such as exponential decay of diffusivity with reaction (Evans et al., 1973) or first-order decay in surface area (Ranade and Harrison, 1979, 1981) can be used to account for sintering. The primary influence of sintering is to decrease pore intersection. This causes a decrease in porosity and an increase in the tortuosity factor of the pellet. These effects have been taken into account by Kim and Smith (1974). Chan and Smith (1976), and Ramachandran and Smith (1977b). The combined effect of chemical reaction and sintering on the diffusivity in the pellet has been expressed by the relation

$$D_e = \left(\frac{1}{D_e} + \frac{1}{D_k}\right)^{-1} \frac{1}{F(f_p)}\left[1 - (1 - f_0)\left(\frac{r_{Gi}}{r_{G0}}\right)^3\right]^2 (1 - f_p) \tag{58}$$

where f_p represents the fraction of pores removed. Assuming the rate of removal of pores to be porportional to their availability, the following simple equation for the variation of f_p with time results:

$$\frac{df_p}{dt} = k_p(1 - f_p) \tag{59}$$

where k_p, analogous to the conventional rate constant, represents the rate constant for pore removal.

General Formulation of Model Equations

In the previous sections we described the basic models and considered their extensions to include more complex cases, such as bulk flow, nonisothermicity, and so on. In the present section we develop a general mathematical formulation of these models, including a few case studies.

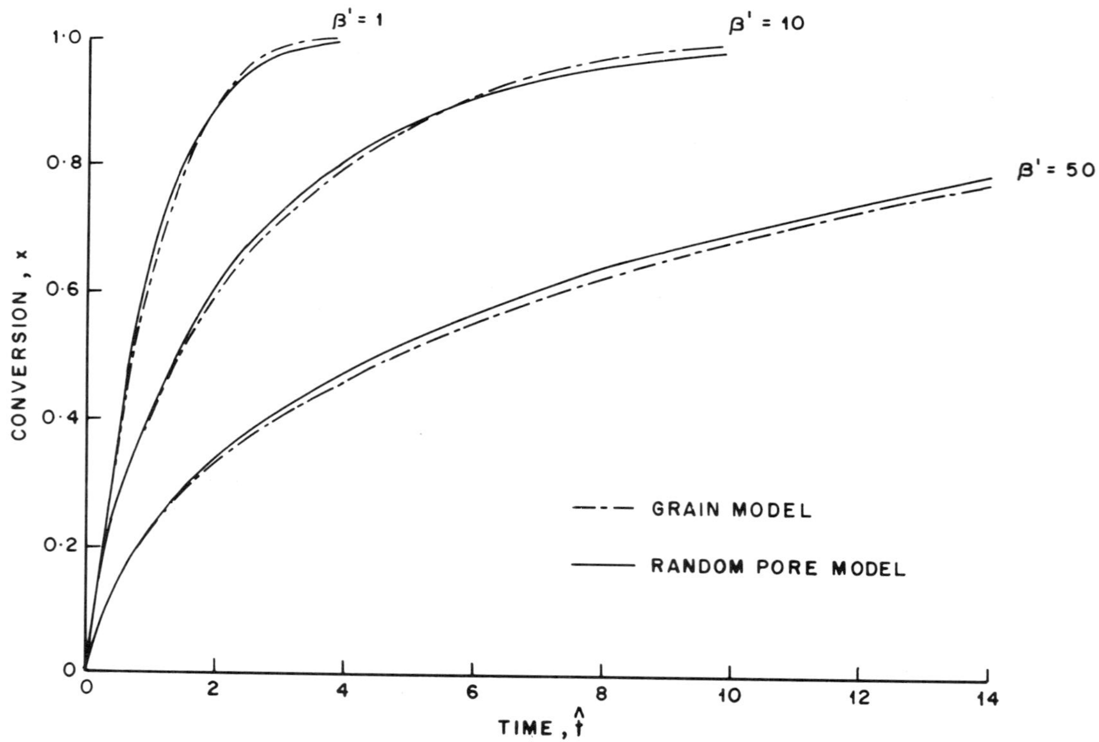

FIGURE 11 Comparison of conversion predictions for several models for various values of β'. Parameters are $\psi = Z_V = 1$, $\phi = 0$.

It is preferable to begin with the volume reaction model, which is described by Eqs. (10) to (14). For a reaction first order in the gaseous component, these equations can be rewritten as

$$\frac{1}{\hat{R}^{\beta}} \frac{\partial}{\partial \hat{R}} \left(\alpha \hat{R}^{\beta} \frac{\partial \hat{C}_A}{\partial \hat{R}} \right) = \phi^2 \frac{\partial x}{\partial \hat{t}} \tag{60}$$

$$\frac{dx}{d\hat{t}} = \frac{\hat{C}_A}{F(x)} \tag{61}$$

where α refers to the diffusivity ratio D_e/D_{e0}, x the conversion, ϕ the Thiele modulus based on the initial diffusivity, and β the geometric factor. Following the general technique developed by Del Borghi et al. (1976), the cumulative gas concentration can be written as

$$\psi = \int_0^{\hat{t}} \hat{C}_A \, d\hat{t} = \int_0^x f(x) \, dx \tag{62}$$

and Eqs. (60) and (61) then transformed into

$$\nabla^2 \psi = \int_0^x \frac{\phi^2}{\alpha} \, dx = k(x) \tag{63}$$

Substituting for ψ from Eq. (62) we obtain

$$f(x) \frac{d^2}{d\hat{R}^2} + \beta \frac{f(x)}{\hat{R}} \frac{dx}{\hat{R}} + f'(x)\left(\frac{dx}{d\hat{R}}\right)^2 - k(x) = 0 \tag{64}$$

Equation (64) is an ordinary second-order differential equation of second degree and its solution directly gives the conversion profiles. The equation is sufficiently general, since several functional forms [f(x)] suitable to different models can be chosen. Typical forms of f(x) for some of the more frequently used models are presented in Table 6. Prasannan et al. (1982) has employed a collocation procedure to solve this equation, and the results obtained by them using the functional forms of f(x) corresponding to the particle-pellet model, the particle-pellet model with structural variations, and the nucleation model are reproduced in Fig. 12.

ANALYSIS OF REACTORS

We have so far been concerned with the description of the single-particle system, and several models attempting to describe qualitatively and semiquantitatively the complexities in real systems have been discussed. Newer models—many of them extensions of the basic model—are constantly appearing in the literature and efforts to quantitatively and realistically describe the systems continue. Despite these efforts, the single-particle behavior may continue to involve uncertainties, and the situation becomes even more

TABLE 6 Functional Forms for F(x) for Different Gas-Solid Reaction Models

Number	Functional form f(x)	Reaction model
1	$(1 - x)^{-n}$	Volume reaction model
2	$-1 - \frac{(1 - x)^{1/3} + (1 - x)^{2/3}}{Sh}$	Grain model
3	$-1 - \frac{(1 - x)^{1/3}}{Sh} + \frac{(1 - x)^{2/3}}{Sh[Z_\nu + (1 - Z_\nu)(1 - x)]^{1/3}}$	Grain model with structural variations
4	$\frac{1}{n} \frac{1}{1 - x} \ln\left(\frac{1}{1 - x}\right)^{(1-n)/n}$	Nucleation model

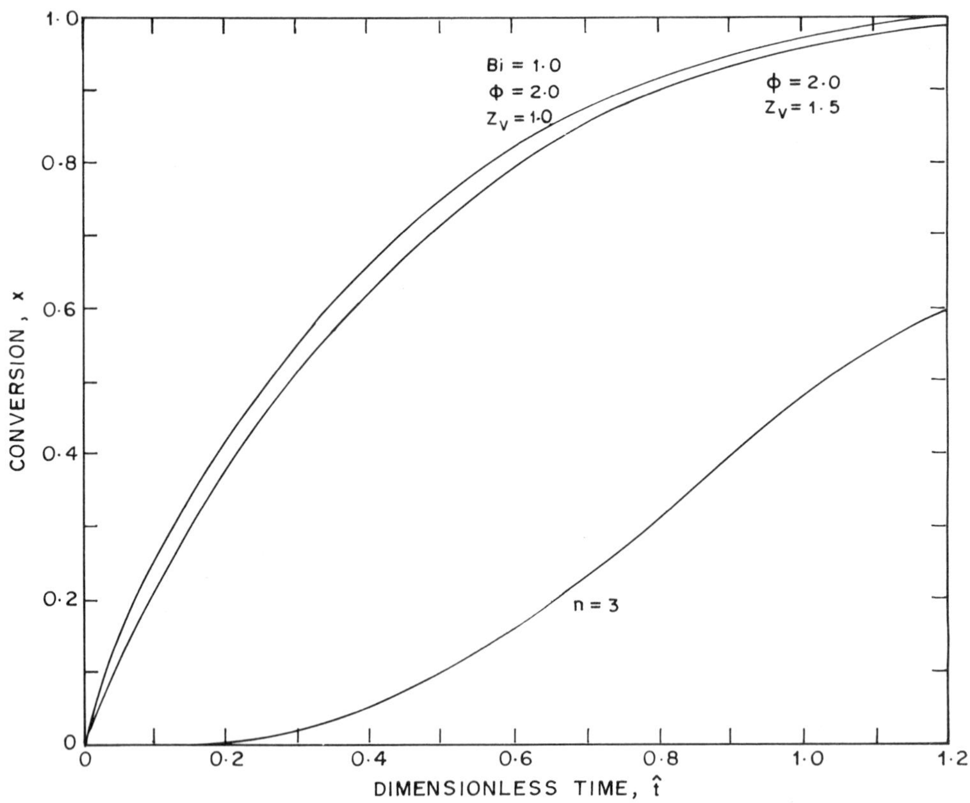

FIGURE 12 Solid conversion versus dimensionless reaction time for particle pellet model with and without structural changes and for nucleation model.

difficult when particle assemblies, as in the case of reactors, are concerned. The common types of reactors employed in gas-solid noncatalytic systems are packed-bed, fluidized-bed, and transport-line reactors.

The additional complexities that need to be accounted for in the analysis of reactors arise out of the fluid flow aspect of the problem. In addition to the heat and mass transport processes within the pellet, one needs now to consider the transport processes between the pellets. Also, specific to each mode of operation, the design criterion may depend quite critically on the pressure drop-gas flow relationship as in the packed-bed, or on the minimum fluidization velocity as in the fluidized-bed, or on the terminal velocities of particles as in the transport-line reactors. The solids feed to these reactors may be of constant size and shape or may involve a size distribution. None of these effects manifest themselves in single-particle studies but become important considerations in the design of reactors. We present below a discussion of each of these reactor types.

Packed-Bed Reactors

Many gas-solid noncatalytic reactions are carried out in packed-bed reactors, some important examples being the roasting and sintering of ores, reduction of metal oxides, production of light weight aggregates, and incineration of solid wastes. A typical sketch of a packed-bed reactor is shown in Fig. 13. The conventional packed bed contains a charge of solid reactant and the gas passes through the voids in the bed. The reaction between the gas and the solid leads to the formation of a reaction front that gradually moves along the packed height. This mode of contact is batch with respect to the solid and is unsuitable for large-scale particle applications. In many instances it is preferable to have solids on a moving grate supplied with gas blown through the bed on the grate (Fig. 13b). The principal correspondence between these two modes of contact is that the time for the reaction front in Fig. 13a can be identified with L/u in Fig. 13b, providing justification for the work done on conventional transient fixed-bed systems (Szekely, et al., 1976).

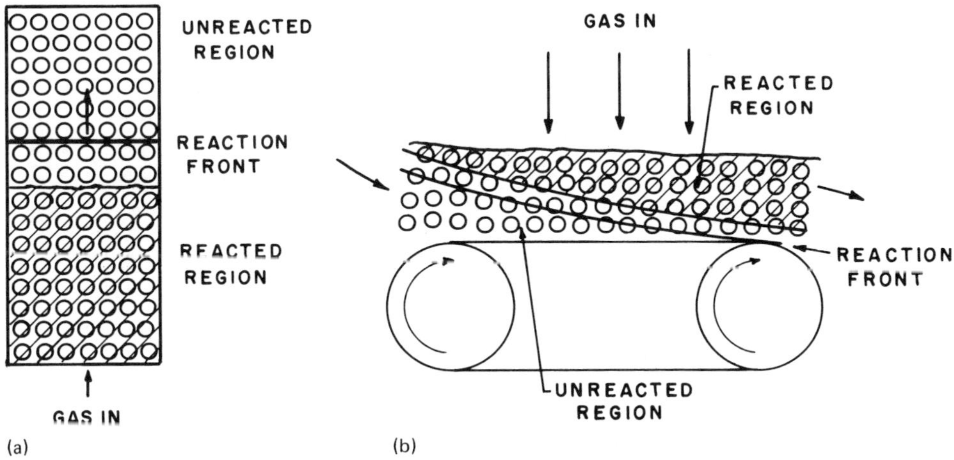

FIGURE 13 Gas-solid contacting in a fixed-bed arrangement: (a) packed bed; (b) packed bed with cross-flow contacting.

In the packed bed as sketched in Fig. 13a, several nonidealities can exist. The flow can deviate from the ideal because of radial variation in velocity and mixing effects brought about by the presence of pellets. Also, the temperature across the cross section of the packed bed in practical systems may not be uniform. These problems lead to a consideration of the variation of system properties in the axial and radial directions. One way of accounting for these situations is by superimposing an effective transport mechanism on the overall transport by plug flow. The fluxes due to these additional transport mechanisms are described by the Fick's law of diffusion for mass and the Fourier law of conduction for heat transfer with effective system properties. Although this way of representing dispersion in packed reactors appears quite sound, the resulting equations are invariably rather cumbersome in that they involve solution of two-point boundary value problems. Also, this class of models predicts that the injected pulse of tracer spreads at an infinite velocity in all directions, which has never been observed experimentally. Chiefly due to these disadvantages, two other classes of models have been proposed for describing the packed bed.

In so-called *finite-stage models* the bed is considered as being composed of a system of interconnected mixing cells of roughly pellet diemnsions. The first quantitative study using such a model appeared in a paper by Kramers and Alberda (1953). Deans and Lapidus (1960a,b) developed a mathematical model describing mixing characteristics in fixed beds for both reactive and nonreactive systems. McGuire and Lapidus (1965) analyzed the transient behavior of packed-bed reactors. The review by Hlaváček and Votruba (1977) covers most of the applications of finite-stage models, and it appears that practically no application to gas-solid noncatalytic systems exists.

The finite-stage models are attractive if only a few stages are involved; however, for a rigorous description, it is often necessary to consider a large number of stages—when significant computational advantage is lost. To overcome this difficulty Hinduja et al. (1980) have developed a cross-flow model, which divides the void region in the bed into stagnant and flowing parts. The stagnant region corresponds to the wakes of packing elements, and dispersion is introduced by assuming that there is exchange of fluid between the stagnant and flowing regions throughout the bed. The physical situation gives rise to first-order initial value differential equations, as against recurrence relations in the mixing cell models and second-order differential equations in the dispersion models. None of these alternative models have yet been applied to gas-solid noncatalytic systems. In the present section, therefore, we shall be mainly concerned with dispersion models.

Before presenting details of the dispersion models, some comments on their suitability for noncatalytic reactions are in order. The section on single-pellet studies has amply brought out the fact that the behavior of a single pellet is characterized by changes in the controlling regime during the course of its conversion. In the simplest case of SIM, even if one starts with a reaction-controlled system, the behavior eventually becomes diffusion controlled, suggesting that unlike in the case of nondecaying catalytic systems, the effectiveness factor for the pellet is a function of time. This makes it necessary to consider individual pellets in the reactor explicitly. In other words, the use of pseudohomogeneous models becomes inappropriate. Of necessity, therefore, we have to use the one- or two-dimensional heterogeneous models. These models require a knowledge of intraparticle and interphase transport properties. In view of the nonisotropic nature of the packed bed, the interphase transport properties are generally different in the axial and radial directions. The models, in general, are therefore more complex and do not yield analytical solutions even in simple cases.

In what follows we present a general framework for the development of these models by considering the one-dimensional heterogeneous model. As far as the single pellet is considered, it will be assumed that the process can be described by the grain model.

Formulation of the Model

The first step in the formulation of the model is writing the mass and energy balance equations for the external field surrounding the pellets. These equations are coupled to the mass and energy balance equations for the individual pellets, where the general chemical reaction

$$A(g) + B(s) \rightleftharpoons C(g) + D(s) \qquad (\Delta H < 0)$$

occurs. It is seen from the reaction stoichiometry that there is no change in moles during the reaction, and therefore bulk-flow effects within the pellet can be ignored. To simplify the analysis, it is further assumed that the physical properties of the solid, such as thermal conductivity, heat capacity, molar density, and so on, are constant. The chemical reaction is a simple first-order reversible reaction occurring at a sharp interface within each grain in the pellet and transport resistance through the ash layer in the individual grains is negligible. The nonisothermal nature of the reaction does not bring about sintering or agglomeration of solid particles and the initial physical structure remains unaltered. Also, the solid temperature is assumed to be uniform throughout the pellet. A rigorous proof for the isothermality of pellets has been given by Carberry (1976).

In the external field it is assumed that axial diffusion terms are negligible compared to convective flow terms, the concentration and temperature are constant in the radial direction, and the pressure drop across the bed is negligible. These assumptions allow the use of the simple one-dimensional heterogeneous model. The nonisothermicity of reaction affects the linear velocity and density of gas along the bed length, and these variations cannot be ignored. Also, the heat generated in the reactor is lost through the sensible heat carried by the gas leaving the system and by convection at the outside surface of the reactor wall.

These assumptions simplify the formulation of the mathematical model although the model becomes somewhat more restrictive. Omitting the detailed material and energy balances, one could write the dimensionless equations as follows:

Fluid Phase: Reactant:

$$\frac{f_B A_4 W_2}{A_2} \frac{\partial}{\partial \tau}\left(\frac{y'_A}{\theta}\right) + \frac{\partial}{\partial z}\left(\frac{\alpha y'_A}{\theta}\right) + 3(1 - f_B)A_4 W_2\left(\frac{y'_A}{\theta} - \frac{y_A}{\theta^*}\bigg|^{\hat{R}=1}\right) = 0 \tag{65}$$

Product:

$$\frac{f_B A_4 W_2}{A_2}\frac{\partial}{\partial \tau}\left(\frac{y'_c}{\theta}\right) + \frac{\partial}{\partial z}\left(\frac{\alpha y'_C}{\theta}\right) - 3(1 - f_B)A_4W_2\left(\frac{y'_A}{\theta} - \frac{y_A}{\theta^*}\bigg|^{\hat{R}=1}\right) = 0 \tag{66}$$

The initial and boundary conditions associated with these equations expressed in dimensionless form are:

Initial conditions:

$$y'_A(z, \tau \leqslant 0) = 0 \tag{67}$$

$$y'_C(z, \tau \leqslant 0) = 1.0 \tag{68}$$

$$\alpha(z, \tau \leqslant 0) = 1.0 \tag{69}$$

$$\theta(z, \tau \leqslant 0) = 1.0 \tag{70}$$

Boundary conditions:

$$y'_A(z = 0, \tau > 0) = \left[y'_{Ao} + \frac{3A_4(1 - f_B)y_A}{\theta^*}\bigg|^{\hat{R}=1}\right] \bigg/ \left[1 + \frac{3_A 4(1 - f_B)}{\theta}\right] \tag{71}$$

$$y'_c(0,\tau) = 1 - y'_A(0,\tau) \tag{72}$$

$$\alpha(0,\tau) = \theta(0,\tau) \tag{73}$$

$$= \frac{\theta^* A_3(1 - f_B) + 1}{1 + A_3(1 - f_B)} \tag{74}$$

Equations (71) to (74) are obtained by carrying out a molar balance of gaseous reactant A, an overall molar balance, and an energy balance of the gaseous mixture at the entrance to the packed-bed reactor.

Pellet Equations:

$$\frac{\nu}{\hat{R}^2}\frac{\partial}{\partial \hat{R}}\left(\hat{R}^2 \frac{\partial y_A}{\partial \hat{R}}\right) - 3(1 - f_p)\hat{r}_{Gi}^2 A_1 W_1 k^* \left(y_A - \frac{y_C}{K}\right) = 0 \tag{75}$$

Product:

$$\frac{\nu'}{\hat{R}^2}\frac{\partial}{\partial \hat{R}}\hat{R}^2\left(\frac{\partial y_c}{\partial \hat{R}}\right) + 3(1 - f_p)\hat{r}_{Gi}^2 A_1 W_1 k^* \left(y_A - \frac{y_c}{K}\right) = 0 \tag{76}$$

where the dimensionless radial position of the reaction front inside each grain is given by

$$\frac{\partial \hat{r}_{Gi}}{\partial \tau} = \frac{A_1 W_1 \gamma_1 k^*}{\theta^*}\left(y_A - \frac{y_C}{K}\right) \tag{77}$$

and the initial and boundary conditions associated with the internal field problem are given in dimensionless form as follows:

Initial conditions:

$$y_A(z, \hat{R}, \tau \leqslant 0) = 0 \tag{78}$$

$$y_C(z, \hat{R}, \tau \leqslant 0) = 0 \tag{79}$$

$$\hat{r}_{Gi}(z, \hat{R}, \tau \leqslant 0) = 1 \tag{80}$$

Boundary conditions:

$$\frac{\partial y_A}{\partial \hat{R}}(z, 0, \tau) = 0 \tag{81}$$

$$\frac{\partial y_C}{\partial \hat{R}}(z, 0, \tau) = 0 \tag{82}$$

$$\frac{\partial y_A}{\partial \hat{R}}(z, 1, \tau) = \frac{A_2}{\nu}\left[\frac{\theta^*}{\theta} y'_A - y_A(z, 1, \tau)\right] \tag{83}$$

$$\frac{\partial y_C}{\partial \hat{R}}(z, 1, \tau) = \frac{\nu}{\nu'} \frac{\partial y_A}{\partial \hat{R}}(z, 1, \tau) \tag{84}$$

Energy Balance Equations: Fluid phase:

$$f_B \frac{\partial \theta}{\partial \tau} = \theta \frac{\partial \alpha}{\partial z} - \alpha \frac{\partial \theta}{\partial z} \tag{85}$$

The change of superficial gas velocity with respect to the axial position is obtained as

$$\frac{\partial \alpha}{\partial z} = 3(1 - f_B) A_3 W_2 (\theta^* - \theta) + 2A_5 W_3 (\theta - \theta_c) \tag{86}$$

where

$$\alpha(0, \tau) = \theta(0, \tau) \tag{87}$$

The change of temperature of the reacting pellets at any particular axial position with respect to time in dimensionless form can be obtained as

$$\frac{\partial \theta^*}{\partial \tau} = A_6 R - \frac{3}{1 - f_p} A_7 (\theta^* - \theta) \tag{88}$$

with the initial condition

$$\theta^*(z, \tau \leqslant 0) = 1 \tag{89}$$

The parameter *R* defines the extent of reaction and is given by

$$R = -9\int_0^1 [\hat{R}\hat{r}_{Gi}^2] \frac{\partial \hat{r}_{Gi}}{\partial \tau} d\hat{R} \tag{90}$$

Equations (65) to (90) completely define the system subject to the assumptions stated earlier. No analytical solution to this set of equations exists and numerical methods have to be employed. The model involves formulation of several dimensionless groups which express the rate of one process in relation to another. For large and small values of these parameters, the system approaches one or the other asymptotic limit. In what follows we shall qualitatively analyze the influence of these dimensionless groups.

The dimensionless group A_6 $(= -\Delta H/C'_{ps}T_0)$ involves the heat of reaction. In the event that A_6 approaches zero, the system behavior tends to be isothermal. Even for finite values of A_6, isothermal behavior can be approached if A_5 defines the convective heat transfer through the wall of the reactor in relation to the axial heat transfer through the gas phase in the bed. The parameter A_4 is a measure of the relative rate of mass transfer across the gas-solid interface to the rate of mass transfer through the gas phase in the bed. For large values of A_4, the flow through the bed becomes controlling, and in the limiting case, the gaseous reactant gets completely consumed at some axial distance in the bed. The sharp interface thus formed moves along the axial coordinate with time. In the other extreme of very low values of A_4, the transfer across the fluid-solid interface takes control. More strictly, at low values of A_4, the controlling regime is shifted from the external field to the internal field.

The internal field is specified in terms of two other dimensionless groups. The group A_1 measures the relative rate of the chemical reaction in relation to the internal gas diffusion, while the group A_2 measures the relative importance of external gas film mass transfer to internal diffusion. For a specified value of A_4 depending on the value of A_2, the bed behavior approaches either gas-film control ($A_2 \to 0$) or a stage where the entire bed behaves as a single pellet with a uniform environment of gas around ($A_2 \to$ large value). The parameter A_1, for a specified value of A_4, determines whether reaction control or diffusion control would prevail in the system.

The advantages of a skeleton analysis of this type is that it provides asymptotic solutions to the general model as well as guidelines for the behavior of the model. Thus, for isothermal systems ($A_6 = 0$) with a very small value of the parameter A_4, it is seen that any of the three mechanisms (gas film, internal diffusion, or reaction) in the internal field will govern the reactor behavior. These equations can be solved analytically to obtain the following expressions for the extent of reaction (ϕ^*) in the bed.

Diffusion Control

In view of constant concentration of gas everywhere in the bed, the pellet and the bed becomes indistinguishable from each other and the extent of reaction in the bed is obtained as

$$\phi^* = 1 - Z_1^3 \tag{91}$$

where Z_1 refers to the dimensionless position of the reaction front within the pellet and is obtained as

$$2Z_1^3 - 3Z_1^2 + 1 = \tau_D = \frac{6\gamma_1 \tau}{1 - f_p} \tag{92}$$

The mole fraction of gaseous reactant within the pellet is given by

$$1 - y_A = \frac{1 - \hat{R}}{1 - Z_1} \frac{Z_1}{\hat{R}} \tag{93}$$

Reaction control: The extent of reaction in the bed is obtained as

$$\phi^* = 1 - \hat{r}_{Gi}^3 \tag{94}$$

where $\hat{r}_{Gi}$ refers to the dimensionless position of the reaction front within each grain and is obtained as

$$\hat{r}_{Gi} = (1 - A_1 W_1 \gamma_1 \tau)^3 \tag{95}$$

Gas film control: The extent of reaction in the bed is given by

$$\phi^* = \frac{3\gamma_1 A_2}{1 - f_p} \tau \tag{96}$$

In the other extreme case, where the gas flow through the bed becomes controlling in relation to its transfer across the gas-solid interface, the controlling regime lies in the external field. This was the case analyzed by Evans and Song (1974), who considered isothermal and nondispersive conditions in the reactor and spherical pellets with no external mass transfer resistance. For this case the gas concentration varies with axial position in the bed and each pellet is subjected to a different environment. A reaction modulus σ analogous to the reaction modulus A_1 was defined, and their numerical results indicate that after an initial period the plot of the extent of reaction against position in the bed shows a sigmoid shape which subsequently travels through the bed at constant velocity. The sigmoid shape corresponds to the reaction zone width, and the material lying outside this zone is either completely reacted or unreacted. For higher values of σ the reaction width increases and for sufficiently large values can cover the entire bed length. The simple model of Evans and Song (1974) has been extended further by Ranade and Evans (1980), who additionally considered the nonisothermal nature of the bed. The variation of velocity and gas properties with the temperature profile in the bed, as formulated in the general equations here, has, however, not be accounted for. The authors used the model to describe the reduction of commercial iron ore (taconite) with hydrogen, and generally satisfactory simulation was reported. A nonisothermal packed bed has also been analyzed by Beveridge and Kawamura (1974). These authors also neglected the axial and radial mixing in the bed; the behavior of the pellet was accounted for by defining an effectiveness factor. The model has been applied to the roasting of zinc sulfide, and the experimental and calculated results show reasonable success of the model.

The incorporation of axial and radial dispersion of heat and mass in the system leads to considerable complications. While the case of axial dispersion of heat and mass has been treated by Sampath et al. (1975), the case of radial variations does not seem to have been attempted in the literature. In the model of Sampath et al. (1975) the governing equations, which now include axial Peclet numbers for heat and mass, were simplified using orthogonal collocation. The authors applied this model to the regeneration of coked catalyst in a fixed bed. The results of this analysis predicted the maximum temperature that could be attained in the reactor, and a number of operating strategies to keep this temperature within permissible limits were suggested as a consequence of model simulations.

The preceding section discussed some of the basic models for the packed-bed noncatalytic reactor. Despite the simplifying assumptions made, the models are unable to present a general qualitative picture of the bed; indeed, most of these models were developed with a view to obtaining additional information for specific systems. In real packed beds there exist additional complexities, such as porosity and velocity distributions. In a randomly packed bed it has been experimentally observed that the porosity near the wall differs from the average value (Pillai, 1977; Korolev et al., 1971; Benenati and Brosilow, 1962). All these authors report an oscillatory behavior of porosity near the wall. The significant data on porosity variations have been correlated by Chandrasekhar and Vortmeyer (1979) as

$$f_B(r) = \bar{f}_B \left\{ 1 + \exp\left[3 \frac{d_t}{2d_p} \left(\frac{2r}{d_t} - 1 \right) \right] \right\} \tag{97}$$

This equation is adequate for most practical purposes.

The variation of void fraction near the walls leads to a lower conversion in the wall region than within the bed. In the presence of an exothermic reaction, where the local heat liberated depends on the extent of conversion, the use of average void fraction in the bed can therefore lead to results significantly different from those actually observed. It is desirable in such a situation to employ the initial void distribution as given by Eq. (97). In many gas-solid noncatalytic reactions where changes in particle size occur as a result of reaction, it is evident that this initial distribution will vary and more sophisticated models employing the evolution of porosity distribution in packed beds with progress of reaction would become necessary. Although no attempts to incorporate any of these features in models for gas-solid noncatalytic reactors are evident as of now, they should form part of any further development programs in this area.

Another area of research, where practically no information exists, relates to the dynamic response of packed-bed reactors to minor disturbances in the feed composition, temperatures, and so on. It is known that in a reactor the heat flow, fluid flow, and mass flow transients are coupled and are interacting with each other. A small disturbance sometimes can cause significant variations in the speeds of propagation of concentration and temperature fronts in the bed. As is known for catalyst reactors (Mehta et al., 1981), this can lead to transient high-temperature peaks which can sinter or damage the bed. It is necessary to know a priori when such a phenomenon can occur, and predictive criteria for the purpose are essential. Admittedly, the heterogeneous nature of the packed-bed noncatalytic reactor does not make it possible to obtain such a criterion for quick use in practice. However, attempts to obtain simple criteria using alternative descriptions (such as cell models or cross-flow models) for packed bed would be noteworthy.

Conventional packed-bed reactors, although suitable for most isothermal systems, pose problems for highly exothermic or endothermic reactions. Additionally, these reactors do not offer any advantage from the viewpoint of solids handling. It is preferable in such instances to employ fluidized-bed reactors, where a greater case in the handling of solids and uniform bed temperature can be realized. The fluid-bed reactors, however, are unsuitable for systems where the possibility of agglomeration of bed material exists. Also, in view of the nearly complete mixing of solids in these reactors, high conversion of solids would require longer residence times. The moving-bed technique has an advantage here, since solids flow more or less in a plug-flow manner. However, for highly exothermic reactions the occurrence of hot spots, and so on, could be a drawback to moving-bed operations. In a subsequent two sections we briefly describe the modeling aspects of these reactors.

Moving-Bed Reactors

A number of contacting arrangements can be envisaged for moving bed systems:

1. Cocurrent flow of gas and solid
2. Countercurrent flow of gas and solid, as in a shaft furnace
3. Cross flow of gas and solid, as in sinter bed operations

In most of these modes of contact the gas and the solids flow pattern is closer to plug flow; however, deviations from ideality are expected in certain situations and can easily be accommodated. In many situations the uptake rate of gas by the solids is marginal in comparison with the net flow, resulting in an uniform environment of gas all around the bed. In such situations the conversion of solids leaving the reactor is the same as that of a single particle exposed to a gas for a time equal to the residence time of the particle in the reactor.

For the case of varying gas compositions, the model equations under isothermal conditions can be formulated as follows:

$$\frac{D_s A_c \rho_s (1 - f_B)(d^2x/dl^2)}{M_s} - \frac{F_s \rho_s (dx/dl)}{M_s} - R_A A_c = 0 \tag{98}$$

$$\frac{D_g A_c \rho_g f_B (d^2C_A/dl^2)}{M} - \frac{F_g \rho_g (dC_A/dl)}{M} = R_A A_c \tag{99}$$

where D_s and D_g refers to the effective axial dispersion coefficients of the solid gas phases, F_s and F_g the volumetric flow rates, A_c the cross-sectional area, x the conversion of solid, and C_A the gas-phase concentration. The equations can be put in the dimensionless form as

$$\frac{1}{Pe_s}\frac{d^2x}{dz^2} - \frac{dx}{dz} + \frac{f_B}{1 - f_B}\frac{\rho_g}{\rho_s}\frac{F_g}{F_s}\left(\frac{1}{Pe_g}\frac{d^2C}{dz^2} - \frac{dC}{dz}\right) = 0 \tag{100}$$

$$\frac{1}{Pe_g}\frac{d^2C}{dz^2} - \frac{dC}{dz} - \frac{A_c Z(1 - f_B)}{F_g}\frac{\rho_g}{\rho_s}\frac{dx}{d\tau} = 0 \tag{101}$$

These equations are accompanied by inlet and outlet boundary conditions given by

$$x = x_0 + \frac{1}{Pe_s}\frac{dx}{dz}\bigg|_{z=0} \tag{102}$$

$$C = C_0 + \frac{1}{Pe_g}\frac{dC}{dz}\bigg|_{z=0} \tag{103}$$

$$\frac{dx}{dz} = 0 = \frac{dC}{dz}\bigg|_{z=1} \tag{104}$$

The model as formulated assumes isothermal steady-state conditions, uniform conditions in the radial direction, negligible mass transfer across gas-solid interface, and cocurrent movement of the gas and solid phases. With appropriate changes in the boundary conditions and sign of the flow term according to the direction of flow, the system can be generalized to include countercurrent operation.

The set of equations (100) to (104) has been solved by Hartman et al. (1979) to obtain the concentration profiles for the gas and solid reactants. The specific system considered by the authors was removal of SO_2 from flue gases using calcium sulfate and sodium sulfide particles.

The case of countercurrent movement of gas and solid phases with no dispersion in the phases has been analyzed by Evans and Song (1974), and illustrative results for conversion as a function of various operating parameters were presented. The consideration of nonisothermality of operation requires additional equations describing heat balances in the gas and solid phases. Assuming plug-flow conditions, these equations can be easily formulated as

$$\frac{F_g \rho_g C_{pg}}{MA_c}\frac{dT_g}{dl} = ha(T_s - T_g) \tag{105}$$

The equation suggests that the temperature change in the axial direction is due only to the effect of heat transfer from the pellets. The heat balance on the solid phase yields

$$(1 - f_B)\overline{\rho_s C_{ps}} \frac{dT_s}{dl} = ha(T_g - T_s) + (-\Delta H)R_A \tag{106}$$

where $\overline{\rho_s C_{ps}}$ refers to the combined properties of solid reactants and products and R_A refers to the reaction rate in appropriate units. The equation assumes that the temperature change in the axial direction occurs as a result of heat generated in the pellet due to chemical reaction. In view of the large thermal conductivity of solids, isothermal conditions within the pellet can be assumed. The set of equations needs additional conditions, which can be defined as

$$T_g = T_{g0} \qquad l = 0 \tag{107}$$

$$T_s = T_{s0} \qquad l = 0 \text{ (cocurrent)} \tag{108}$$

$$T_s = T_{s0} \qquad l = L \text{ (countercurrent)} \tag{109}$$

The case of nonisothermal countercurrent moving-bed reactions has been analyzed by Tsay et al. (1976), who applied this model to the reduction of hematite with hydrogen and carbon monoxide. The simulation results indicate that an optimum gaseous feed composition exists for certain range of operating variables. It is possible to incorporate additional features in the model, such as a side stream injection (to adjust temperature and composition in the bed), corrections for maldistribution of gas and solid flow (Yagi and Szekely, 1979), and so on.

It is necessary for larger particles to incorporate the diffusive processes within the particle. Conventional analyses ignore the mass and heat transfer processes within the particle. Also, for exothermic reactions the axial and more particularly radial dispersion effects should be included. A complete analysis incorporating these effects is lacking. The dispersion phenomenon present in the bed is especially important for exothermic or more specifically nonlinear processes, as it establishes the feedback necessary for the occurrence of nonunique behavior of the bed. The existence of multiplicity and instability has been established for catalytic reactors. The literature on noncatalytic systems is relatively scarce and more efforts in this direction would be fruitful.

Fluidized-Bed Reactors

Fluidized-bed reactors represent a multiphase system consisting of two, or more correctly, three phases. As a prerequisite to the design of these reactors, it is necessary to know the gas flow through the constituent phases as well as their volume fractions, extent of mixing, and rates of interphase mass transfer. Several models depicting the behavior of the fluid bed are available in the literature, and the present treatment presupposes a knowledge of these essentials. Relationships exist between the properties of the bed and the diameter of bubbles (which grow along the bed due to coalescence), expansion of the bed, and gas transfer from the emulsion phase. The predictive equations for obtaining these properties of the bed are listed in the section on estimation of parameter values.

The chief features of the fluid bed as used for the noncatalytic reactor differ from those for the conventional catalytic reactor. The main departures arise due to different considerations associated with the solid phase. In noncatalytic systems the solid acts as a reactant and therefore its consumption during the reaction should be accounted for in contrast to catalytic systems. The reaction brings about changes in the size and shape of these particles which affect the fluidization behavior. Nondecaying catalytic systems can be operated in the batch mode with respect to the solid, but the consequences of reaction and solids consumption often require that the noncatalytic systems be operated on a continuous basis. The solids feed to these reactors would generally contain a size distribution. Besides its influence on the fluidization behavior, the size distribution has a direct effect on the residence-time distribution and therefore on the conversion of the solid reactant. In catalytic systems, the gas mixing and interchange coefficients play an important role in deciding the conversion in the reactor. In noncatalytic reactions, the degree of backmixing and residence time of particles are additional important factors. Conventional catalytic reactors operate with sufficiently fine particles so that processes within the particles can often be ignored. In noncatalytic reactions with size distribution of solids, pseudohomogeneous conditions cannot always be assumed and explicit recognition of the processes within the catalyst particle becomes necessary. In the discussion that follows we attempt to progressively incorporate some of these complexities.

Constant Gas-Phase Environment

In this section we analyze the case where the gas-phase concentration is constant throughout the reactor. Such a case arises for situations where a large excess of reactant gas is available in the system. The analysis for conversion now depends mainly on the effect of solids mixing. The mixing pattern of the solids in most reactors can be assumed to be close to backmixing, with the residence-time distribution of the solids given by the expression

$$E(t) = \frac{1}{\bar{t}(R_j)} \exp\left[-\frac{t}{\bar{t}(R_j)}\right] \tag{110}$$

where $\bar{t}(R_j)$ is the average residence time for particles of size R_j and $E(t)dt$ represents the fraction of the particles that have a residence time between t and t + dt. This information on residence-time distribution can be combined with a suitable model for computing the conversion in a single particle, to yield the average conversion of solids in the reactor:

$$\bar{x} = \sum_{0}^{R_M} \frac{F_s(R)_j}{F_s} \int_0^{\infty} x(R)_j \frac{1}{\bar{t}(R)_j} \exp\left[-\frac{t}{\bar{t}(R)_j}\right] dt \tag{111}$$

where R_M is the maximum size in the feed and $F_s(R)_j$ is the quantity of particles of size R_j entering with the feed.

However, the analysis ignores any loss of particles due to elutriation. We shall now consider this effect.

Effect of Elutriation

Elutriation rates from fluidized beds depend on many factors: size distribution of particles; particle characteristics, such as shape and attrition behavior; mode of operation of the reactor; location and type of internals; feedboard height; and so on. Often, many of these factors are interrelated, and it is customary to express the rate of elutriation on the basis of an elutriation constant E* as follows:

$$\text{Rate of removal of solids of size } R_j = E^* \begin{pmatrix}\text{Weight of that size}\\ \text{of solid in the bed}\end{pmatrix}$$

or

$$\text{Rate of removal of solids of size } R_j \text{ per unit bed area} = E_s^* \begin{pmatrix}\text{Fraction of bed weight}\\ \text{consisting of size } R_j\end{pmatrix}$$

or

$$-\frac{1}{A_c}\frac{d(R)_i}{dt} = E_s^* \frac{W(R)_j}{W} \tag{112}$$

where E_s^* may now be viewed as a specific elutriation constant. The two quantities E* and E_s^* are related simply as follows:

$$E^* = E_s^* \frac{A_c}{W} \tag{113}$$

The following general equation, based on the experimental data for batch operations with binary mixtures, has been suggested by Wen and Hashinger (1960):

$$\frac{E_s^*}{g(u - u_t)} = (1.7 \times 10^{-5}) \left(\frac{u - u_t}{gd_p}\right)^{0.5} \left(\frac{d_p u_t}{\mu}\right)^{0.725} \left(\frac{\rho_s - \rho_g}{\rho_g}\right)^{1.15} \left(\frac{u - u_t}{u_t}\right)^{0.1} \tag{114}$$

The average residence time of solids of size R_j in a fluidized bed can be predicted in terms of the elutriation constant (Levenspiel, 1979):

$$\bar{t}(R_j) = \left[\frac{F_1}{W} + E^*(R_i)\right]^{-1} \tag{115}$$

Knowing $\bar{t}(R_j)$, we can use Eq. (115) to predict the solids conversion in the presence of elutriation.

Other studies on elutriation include those of Sycheva and Donat (1974), Tanaka and Shinohara (1972), Gontanev and Paulin (1975), Lange et al. (1977), and Horio et al. (1980).

Models Based on Particle Growth or Shrinkage

As has been pointed out earlier, change in particle size is a basic feature of certain gas-solid reactions. In the previous sections, the effect of particle size distribution was included by defining an average residence time for each particle size, leading finally to Eq. (111). An alternative approach is to account for change in particle size by assuming a simple topochemical model for particle growth or shrinkage as a first approximation. Equations can then be developed for the quantity of material leaving the reactor as well as for particle size distribution in the existing stream.

The rate of change of particle radius can be represented for most practical cases by a simple linear model or an inverse shrinkage model as

$$\frac{dR}{dt} = -k = \text{constant} \tag{116}$$

$$\frac{dR}{dt} = -\frac{k}{R} \tag{117}$$

The parameter k in the first case is related to the rate constant k_s for the SIM by

$$k = \frac{\nu k_s C_{Ai} M_B}{\rho_B} \tag{118}$$

The latter case arises for a gas-film-controlled reaction (e.g., oxidation of carbon).

The equations to predict changing particle size can be utilized to calculate the quantity of material of a given size ($F_{s,1}$) leaving the reactor. For a system with feed $F_{s,0}$ having a size distribution $\breve{\phi}(R)$, an exit stream $F_{s,1}$ with a size distribution $\breve{\phi}_1(R)$, an elutriation stream $F_{s,2}$, and an arbitrary rate law for the changing particle size, a material balance on solids of size between R and (R + dR) yields

$$\begin{pmatrix}\text{Solids}\\\text{in}\\\text{feed}\\\text{(kg/s)}\end{pmatrix} - \begin{pmatrix}\text{solids}\\\text{in}\\\text{outflow}\\\text{(kg/s)}\end{pmatrix} - \begin{pmatrix}\text{solids leaving}\\\text{in elutriation}\\\text{stream}\\\text{(kg/s)}\end{pmatrix} - \begin{pmatrix}\text{growth of}\\\text{solids in}\\\text{and out}\\\text{of the}\\\text{interval}\\\text{(kg/s)}\end{pmatrix} + \begin{pmatrix}\text{mass}\\\text{increase}\\\text{of solid}\\\text{within the}\\\text{interval}\\\text{(kg/s)}\end{pmatrix} = 0 \tag{119}$$

On the assumption of complete backmixing of the solid, this equation can be written as

$$\begin{aligned} &F_{s,0}\breve{\phi}_0(R)\,dR - F_{s,1}\breve{\phi}_1(R)\,dR - WE^*(R)\breve{\phi}_1(R)\,dR \\ &- W\frac{d}{dR}[r(R)\breve{\phi}_1(R)\,dR] + \frac{(SF)W}{R}\,r(R)\breve{\phi}_1(R)\,dR = 0 \end{aligned} \tag{120}$$

The terms E*(R), (SF), and r(R) represent, respectively, the elutriation constant defined by Eq. (113), the shape factor of the particles, and the rate of change of particle size. Equation (120), valid for a particular size range, can be supplemented by a balance over all sizes to give

$$F_{s,2} + F_{s,1} - F_{s,0} = (SF)W \int_{RM}^{R} \frac{\breve{\phi}(R)r(R)}{R}\, dR \tag{121}$$

Equation (121) assumes positive or negative values depending on whether the particle grows or shrinks. Equations (120) and (121) can be rearranged to obtain the outflow stream in terms of the input stream as

$$\frac{W}{F_{s,0}} = \int_{R_M}^{R} \frac{R^3}{r(R)}\, I \left[\int_{R_M}^{R} \frac{\breve{\phi}_0(R)\, dR}{R^3 I} \right] dR \tag{122}$$

The corresponding outflow size distribution is given by

$$\breve{\phi}_1(R) = \frac{F_{s,0} R^3}{Wr(R)}\, I \left[\int_{R_M}^{R} \frac{\breve{\phi}_0(R)\, dR}{R^3 I} \right] \tag{123}$$

In these equations R_M represents the smallest feed size for a growing particle or largest size for a shrinking particle, and I is an integral defined as

$$I = \exp \left[\int_{R_M}^{R} \frac{F_{s,1}/W + E^*(R)}{r(R)}\, dR \right] \tag{124}$$

The set of Eqs. (122) to (124) cannot in general be solved analytically. However, they represent a total generalization with respect to feed size distribution, reaction kinetics, and presence of an elutriation stream. For specific situations, such as constant size feed, linear kinetics, or no elutriation, these equations can be simplified to obtain analytical solutions. Levenspiel (1979) has summarized some of these analytical solutions.

Analysis for Systems with Varying Gas-Phase Composition

The assumption of known and constant gas environment in the reactor simplified the treatment considerably, since an analysis based on the solid phase alone could be used to describe the bed behavior. Many practical systems using large particle beds fluidized with a large excess of gas conform to this situation. In many other systems (such as vigorously fluidized beds of fine particles), however, the composition of gas varies with position in the reactor. A simple analysis as above would then be inapplicable. In these instances, it is necessary to include separate equations for the bubble and cloud-wake phases.

A number of models to depict the behavior of the bed are available, and some of those have been extended to gas-solid noncatalytic reactors. Thus the two-phase model of Davidson and Harrison (1963) has been used by Campbell and Davidson (1975) to analyze the data on the combustion of carbon particles for short periods of combustion in a batch reactor. The model has also been used and considerably extended by Amundson (see Bukur et al., 1977). Tigrel and Pyle (1971) have used this model for the not-too-different problem of catalyst deactivation. Kunii and Levenspiel (1969) and Kato and Wen (1969) have extended their models to gas-solid noncatalytic systems. A general model for noncatalytic beds should take account of variation in both size and density of particles. No model that takes account of all the realities in the bed is however available. A particularly useful model that takes account of some of the complexities in practical systems has been suggested by Chen and Saxena (1978).

Based on a more rigorous description of the bed, Chen and Saxena have shown that the solids in the emulsion phase may have a net downward, stationary, or net upward movement, depending on the rate of solids feed. As a result, other parameters, such as bubble volume fraction and rise velocity, become dependent on the solids feed rate. A detailed solids population balance has been formulated and a computer simulation program has been presented by these authors. The results show that the extent of gas bypassing through bubbles is reduced when the solids feed rate is increased.

These authors have also proposed a criterion to predict the direction of solids movement in the emulsion phase. An illustrative plot of this criterion is reporduced in Fig. 14. The quantities F_{c1} and F_{c2} in this figure are defined as follows:

$$F_{c1} = \frac{\rho_s A_c (1 - f_{mf}) s_{wb}}{1 + s_{wb} f_{mf}} u - u_{mf} \frac{\rho_s A_c s_{wb} (1 - f_{mf})(1 - \delta - s_{wb}\delta)}{1 + s_{wb} f_{mf}} - \frac{s_{wb}\delta}{1 - \delta} \frac{dW_r}{dt} \tag{125}$$

and

$$F_{c1} = F_{c1} + \frac{\delta s_{wb} - 1 + \delta}{1 - \delta} \frac{dW_r}{dt} \tag{126}$$

where W_r is the weight of the solids in the reactor and ρ the average particle density over the entire bed.

If $F_0 < F_{c1}$, the solids move down in the emulsion, whereas for $F_0 > F_{c2}$, them move upward.

Modeling of Shallow-Bed Reactors

The modeling of gas-solid noncatalytic fluid-bed reactors followed the developments in catalytic systems and in most cases these models were derived as their extensions. Adequate for qualitative purposes, these models are really applicable only for tall beds ($L/d_t > 1$). Most industrial fluid beds, however, operate with small L/d_t (<1), and the use of these models becomes restrictive. The essential differences between the two types of beds arise due to the changed flow profiles. Thus in tall beds it is justifiable to assume no radial variations of concentration and temperature. The measurements on lateral mixing of solids in shallow beds, however, indicate a much lower value than the axial mixing coefficients. This implies that the lateral variations cannot generally be ignored. Thus it is likely that strong temperature and concentration gradients in the radial direction

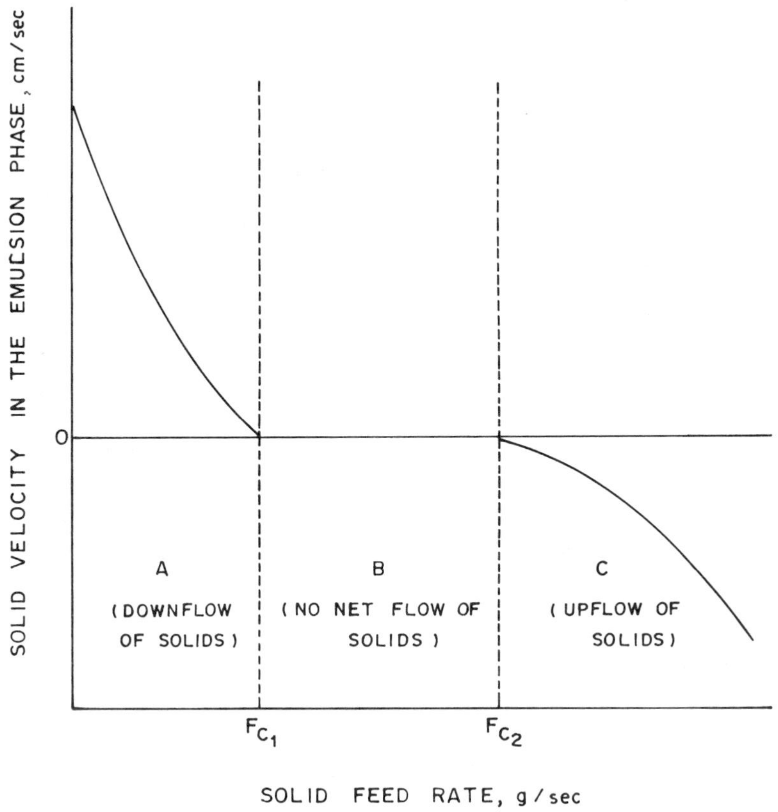

FIGURE 14 Classification of fluidizations according to the solids feed rate.

exist in shallow fluidized beds (Highley and Merrick, 1971; Chavarie, 1973; Fan et al., 1979; Fan and Chang, 1980; Grace, 1980). The remarks, generally applicable to gas-solid catalytic and noncatalytic systems, are much more pertinent to the latter case in view of the solids being involved as a reactant. A rigorous model for the gas-solid fluid-bed reactor therefore should account for the radial dispersion of gas and solid reactant and for temperature distribution. These considerations would lead to a far more complex formulation, which may be difficult to use practically. However, simplifying assumptions, suitable to each case, can be made to get the qualitative behavior of the bed, which will be far more realistic than that obtained using the models for tall beds.

Thus one might consider the applicability of the simple two-phase theory with the emulsion phase at minimum fluidization conditions and completely mixed—in this case, only in the axial direction. The bubble phase would be considered to move in plug flow. The solids could be introduced at the center or at the bottom and overflow occurs at the farther end. The solid and gas feed temperatures can be identical or might differ, depending on the specific case. The formulation of the model requires accounting of the gas in the bubble phase. In the interest of generality, an overall heat balance to take care of nonisothermal situations can also be readily written.

Emulsion Phase

Solid reactant: The material balance on solid S in the emulsion phase over the control volume can be written as

$$\begin{pmatrix}\text{Accumulation}\\ \text{of solid B}\end{pmatrix} = \begin{pmatrix}\text{net rate of}\\ \text{B in by}\\ \text{convection}\end{pmatrix} - \begin{pmatrix}\text{net rate of}\\ \text{B in by}\\ \text{dispersion}\end{pmatrix} - \begin{pmatrix}\text{rate of}\\ \text{disappearance}\\ \text{of solid B}\\ \text{due to}\\ \text{reaction}\end{pmatrix} + \begin{pmatrix}\text{solid B}\\ \text{in feed}\\ \text{stream}\end{pmatrix}$$

or in equivalent mathematical form as

$$\frac{\partial C_B}{\partial t} = \frac{1}{x^\beta}\frac{\partial}{\partial x}\left(x^\beta D_{eB}\frac{\partial C_B}{\partial x}\right) - r_B + \frac{F}{H(1-\delta)}$$

where C_B is the concentration of solid B, D_{eB} the effective diffusivity of solid species B, r_B the rate of consumption of solid B, F the molal feed rate of solid per unit area of feeder, H the height of bed, and δ the bubble fraction in the bed.

Gaseous reactant: The material balance on the gaseous component A in the emulsion phase can be written as

$$\begin{pmatrix}\text{Accumulation}\\ \text{of gaseous}\\ \text{species A}\end{pmatrix} = \begin{pmatrix}\text{net rate}\\ \text{of A}\\ \text{in by}\\ \text{diffusion}\end{pmatrix} + \begin{pmatrix}\text{net rate of}\\ \text{A in by}\\ \text{convection}\end{pmatrix} + \begin{pmatrix}\text{exchange}\\ \text{of gas}\\ \text{with the}\\ \text{bubble}\\ \text{phase}\end{pmatrix} - \begin{pmatrix}\text{consumption}\\ \text{due to}\\ \text{reaction}\end{pmatrix}$$

or in equivalent mathematical form as

$$f_{mf}\frac{\partial C_{Ae}}{\partial t} = \frac{1}{x^\beta}\frac{\partial}{\partial x}\left(x^\beta D_{eA}\frac{\partial C_{Ae}}{\partial x}\right) + \frac{U_{mf}}{H(1-\delta)}(C_{A0} - C_{Ae}) + \frac{\delta}{1-\delta}\frac{1}{H}\int_0^H K_{be}(C_{Ab} - C_{Ae})\,dl - r_A$$

Bubble Phase: A material balance on the gaseous component A in the bubble phase can be written in accordance with Davidson's model as

$$\begin{pmatrix}\text{Accumulation}\\ \text{of A}\end{pmatrix} = \begin{pmatrix}\text{net rate of A}\\ \text{by convection}\end{pmatrix} - \begin{pmatrix}\text{exchange to emulsion}\\ \text{phase}\end{pmatrix}$$

or in equivalent mathematical form as

$$\frac{\partial C_{Ab}}{\partial t} = -u_b\frac{\partial C_{Ab}}{\partial l} - K_{be}(C_{Ab} - C_{Ae}) \tag{127}$$

The overall energy balance for the controlled volume can be written as

Accumulation energy	=	net rate of heat energy in by conduction	−	net rate of heat energy from solids and gas	+	heat generation due to reaction

or in equivalent mathematical notation as

$$\overline{\rho C_p} \frac{\partial T}{\partial t} = \frac{1}{x^{\beta}} \frac{\partial}{\partial x} \left(k'_e x^{\beta} \frac{\partial T}{\partial t} \right) + \frac{u \rho C_p}{H} (T_{g0} - T)$$

$$+ \frac{F M_B C_{ps}}{H} (T_{B0} - T) + (1 - \delta)(-\Delta H) r_A \qquad (128)$$

The set of Eqs. (127) and (128) describe the fluid-bed behavior for a shallow bed. In the form given, these equations are quite complex and no analytical solution is possible. Chang et al. (1982) have solved these equations under steady-state conditions on the assumption of isothermality of operation and no radial variation of concentration of gaseous component. Also, the reaction rates r_A and r_B are related through stoichiometry, and zero-order and first-order dependence on solid concentration was considered. The analytical results obtained by the authors are summarized in Table 7 and should be useful in a quick estimation of reactor performance.

The general model developed above has been applied to the case of roasting of zinc sulfide, with the following general conclusions: The lateral

TABLE 7 Exit Concentration of Solid from the Bed

Reaction	Solid concentration
Zero order	$(\hat{C}_B)_{exit} = \frac{\phi_2}{2\phi_3} \hat{x}_f^2 - \frac{\phi_1 \hat{C}_{Ae}^n}{2\phi_3}$
First order	$(\hat{C}_B)_{exit} = \frac{\phi_2 I_1(p\hat{x}_f)[I_0(p)K_1(p) + I_1(p)K_0(p)]}{p^2[I_0(p\hat{x}_f) + I_1(p\hat{x}_f)k_0(p\hat{x}_f)][pI_1(p) + \phi_3 I_0(p)]}$
	where n is order with respect to gaseous reactant, $\hat{x}_f$ the dimensionless position of the feeder point, and p, ϕ_2, and ϕ_3 are constants defined as
	$p = (\phi_1 \hat{C}_{Ae})^{1/2} \qquad \phi_1 = \frac{\nu_B k L^2 C_{A0}^n C_{B0}^{n-1}}{D_{eB}}$
	$\phi_2 = \frac{F L^2}{(1 - \delta) H D_{eB} C_{B0}} \qquad \phi_3 = \frac{F L}{D_{eD} C_{D0}(1 - \delta)}$

mixing of solids has a profound influence on the conversion of solids for beds with smaller bubble dimensions. As such a situation is common in shallow fluidized beds, the influence of lateral mixing should not be overlooked. In beds with large-diameter bubbles—where the controlling mechanism lies across the bubble/emulsion interface—lateral mixing of solids has relatively little influence on bed performance.

Staged Fluidized-Bed Reactors

Frequently, from the standpoint of better contact between the gas and the solid, it becomes necessary to stage the fluid-bed reactor. A three-stage fluidized bed together with relevant streams is shown in Fig. 15. The solids in each bed can be considered to be completely mixed. The bed weight, average residence time of solids, and residence-time distribution (RTD) of solids may differ from bed to bed.

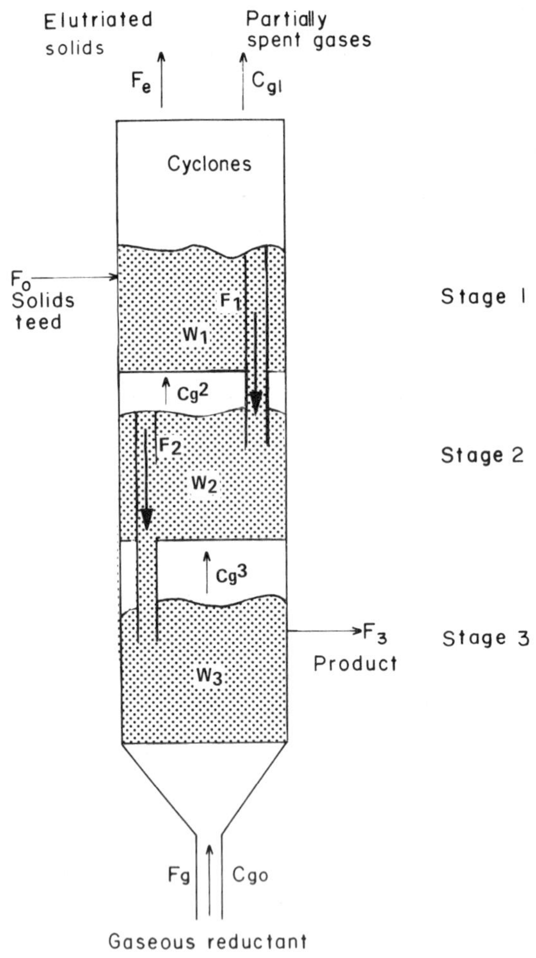

FIGURE 15 Schematic representation of a three-stage fluidized reduction unit.

The RTD function for any particular size of feed solids d_p is given by Eq. (110), where the average residence time of solids of size d_p is given by

$$\bar{t}(d_p) = \frac{1}{F_1/W_1 + E(d_p)[1 - \eta(D_p)]} \tag{129}$$

$E(d_p)$ in this equation is the elutriation constant appropriately defined taking account of the parameters in Figure 15 as

$$E(d_p) = \frac{F_1 S_e(d_p)}{W_1 S_1(d_p)} \tag{130}$$

and S_1 and S_e are defined by the equations

$$F_1 S_1(d_p) = \frac{\alpha_1 F_0 S_0(d_p)}{1 + (W_1/F_1)E(d_p)} \tag{131}$$

and

$$F_1 = \alpha_1 F_0 \int_{(d_p)}^{(d_p)_{max}} \frac{S_0(D_p)\, d(d_p)}{1 + (W_1/F_1)E(d_p)}$$

since

$$\int_{(d_p)_{min}}^{(d_p)_{max}} S(d_p)\, d(d_p) = 1 \tag{132}$$

$\eta(d_p)$ in Eq. (129) represents the cyclone efficiency.

For an assumed conversion-time relationship, the unconverted fraction of solids of size d_p is given by Eq. (111). Since the bed is composed of many size fractions, this equation has to be integrated over the size distribution to obtain mass conversion of solid in the stage. The mass conversion in the stage can be obtained as

$$1 - \bar{x} = 1 - 3\gamma_i + 6\gamma_i^2 - 6\gamma_i^3(1 - e^{-1/\gamma_i}) \tag{133}$$

where

$$\gamma_i = \frac{\bar{t}_i(d_p)}{\tau_i(d_p)} \tag{134}$$

and $\bar{t}_i$ and τ_i represent, respectively, the average residence time of solid of size d_p and the time required for its complete conversion in the ith stage. It is important to bear in mind that the value of τ_i will depend on the local concentration of gas in the stage. The gas phase could either have a constant composition through the stage or may be varying. Suitable balance equations in the gas phase should be written for this purpose.

TABLE 8 Recommended Equations for Effective Diffusivity and Flow Permeability

Number	Parameter	Recommended equation	References
1	Effective diffusivity		
	Molecular diffusivity	$D_{Ae} = D_{AB} \frac{f_c}{\tau_f}$	
		where	
		$D_{ij} = \frac{0.0018583T^{3/2}}{P\sigma_{ij}^2\Omega_{ij}} \left(\frac{1}{M_i} + \frac{1}{M_j}\right)^{1/2}$	
		M = molecular weight of species i, j	
		σ, Ω = constants in Lennard-Jones potential energy function	
		τ_f = tortuosity factor	
		S_g = total surface area at porous solid	

	Knudsen diffusivity	$D_{Ak} = \frac{19{,}400 f_c^2}{\tau_f S_g \rho_s} \left(- \frac{T}{M} \right)^{1/2}$	
	Effective diffusivity in transition region	$\frac{1}{D_e} = \frac{1}{D_{Ae}} + \frac{1}{D_{Ake}}$	
2	Flow permeability	$\frac{k_m}{k_{m0}} = \left[1 + \left(\frac{0.14}{KN} \right)^{6/5} \right]^{5/6}$ where KN, the modified Knudsen number, is defined as $KN = \left(\frac{\pi R_g T}{8M_A} \right)^{1/2} \left(\frac{1 - f_c}{f_c} \right) \frac{\mu S}{P}$ where μ refers to the viscosity of the gas and S is the specific surface	Asaeda and Toei (1979)

TABLE 9 Recommended Equations for Estimating Transport Coefficients

Coefficient	Recommended equation	Reference
Fluid-particle transport coefficient		
Mass transfer coefficient	$Sh = 2 + 1.1Sc^{1/3}Re^{0.6}$ $3 < Re < 10{,}000$	Wakao and Funazkri (1978)
Heat transfer coefficient	$Nu = 2 + 1.8Re^{0.5}Pr^{1/3}$ $Re > 100$ $Nu = 0.36Re^{0.94}$ $Re < 100$	Glidden and Crafield (1970)
General correlation for fluidized beds	$f_f J_h = \frac{2.876}{Re} + \frac{0.3023}{Re^{0.35}}$	Gupta et al. (1974)
Bed-to-wall heat transfer coefficient	$h_{bw} = 35.8(k'_g)^{0.6} d_p^{-0.36} \rho_s^{0.2}$ (MKS units)	Botterill (1975)
Particle-to-horizontal stages and tubes	$h = \frac{7.2k'_g(1 - f_f)^{2/3}}{d_p} + 26.6U^{0.2}C_{pg}\rho_g d_p$	Zabrodsky et al. (1981)

Interchange coefficients in fluid beds		
Bubble-emulsion mass and heat transfer coefficient	$k_{be} = 4.5 \dfrac{U_{mf}}{d_b} + 5.85 \left(\dfrac{D_b g^{1/2}}{d_b^{5/2}}\right)^{1/2}$	
	$h_{be} = 4.5 \dfrac{U_{mf} \rho_g C_{pg}}{d_b} + 5.85 \left(\dfrac{k'_g C_{pg} \rho_g g^{1/2}}{d_b^{5/2}}\right)^{1/2}$	Kunii and Levenspiel (1969)
Cloud-emulsion mass and heat transfer coefficient	$k_{ce} = 6.75 \left(\dfrac{f_{mf} D_e U_b}{d_b^3}\right)^{1/2}$	Mori and Wen (1975)
	$h_{ce} = 6.75 \left(\dfrac{f_{mf} k'_{eg} C_{pg} \rho_g U_b}{d_b^3}\right)^{1/2}$	

TABLE 10 Recommended Equations for Computing Basic Parameters of the Fluidized Bed

Parameter	Equation	Reference
1. Minimum fluidization velocity	$\frac{g(\rho_s - \rho_g)d_p}{\rho_g U_{mf}^2} = 2.4 \times 10^5 \left[\frac{\mu^2}{P_g g(\rho_s - \rho_g)d_p^3}\right]^{0.85} \left(\frac{\rho_s}{\rho_g}\right)^{0.13} + 33.7$	Broadhurst and Becker (1975)
	$0.01 < Re_{mf} < 1000, \quad 500 < \rho_s/\rho_g < 50{,}000$	
	$1 < \frac{P_s g(\rho_s - \rho_g)\, d_p^3}{\mu^2} < 10$	
2. Bubble diameter	$d_b = d_{bm} - (d_{bm} - d_{b0}) \exp\left(-\frac{0.3\ Z}{d_t}\right)$	
	where	
	$d_{bm} = 0.652[A_c(U_0 - U_{mf})]^{0.4}$	

	$d_{b0} = 0.347 \left[\dfrac{A_c(U_0 - U_{mf})}{n_0}\right]^{0.4}$		
3. Bubble velocity	$U_b = U_0 - U_{mf} + 0.711\sqrt{gd_b}$		
4. Bubble fraction in bed	$\delta = \dfrac{U_0 - U_{mf}}{U_b}$		
5. Cloud fraction in bed	$\delta_c = \dfrac{\delta \alpha_b}{\alpha_b - 1} \qquad \alpha_b = \dfrac{f_{mf}U_b}{U_{mf}}$		
6. Solid mixing rate	$= (U_0 - U_{mf})A_c f_m (1 - f_{mf})_s$	$0.067 < f_m < 0.3$	Rajan and Wen (1980)
7. Attrition rate	$= \dfrac{K(U_0 - U_{mf})}{M}$	$2.73 \times 10^{-8} < K < 9.11 \times 10^{-8}$	Merrick and Highley (1974)

PARAMETER ESTIMATION OF THE MODELS

The preceding section was concerned with models for gas-solid noncatalytic reactors. These models were formulated in terms of dimensionless groups which consist of the physical quantities and structural parameters of the system. The present section is devoted to the estimation of these properties. Some of the physical quantities can be measured directly or are readily available in the literature. On the other hand, certain other properties, such as effective diffusivity and heat and mass transfer coefficients, have to be obtained using proven relationships. Several of these parameters are not specific to gas-solid noncatalytic reactors, but occur very often in operations of gaseous mass transfer in porous materials such as catalytic reaction, adsorption-desorption, drying, and so on. Consequently, a large volume of literature exists on the estimation of these properties under a variety of conditions. We present in Tables 8 to 10 the recommended equations for these properties for use in design.

NOTATION

a	interfacial area for transfer, cm^2/cm^3
A	general notation for gaseous reactant species
A_1	dimensionless group defined as $k_s R/D_e$
A_2	dimensionless group defined as $k_g R/D_e$
A_3	dimensionless group defined as $hR_g T_0/PC_p U_0$
A_4	dimensionless group defined as k_g/U_0
A_5	dimensionless group defined as $h_w R_g T_0/U_0 PC_p$
A_6	dimensionless group defined as $-\Delta H/C'_{ps} T_0$
A_7	dimensionless group defined as $hR/D_e C'_{ps} T_0$
A_c	cross sectional area of the bed, cm^2
A_p	area of pellet, cm^2
B	general notation for the solid reactant species
C	dimensionless concentration of gas in the bed
C_A, C_B	concentration of species A, B, mol/cm^3
C_{Ab}	concentration of gaseous species A in the bulk or in the bubble phase, mol/cm^3
C_{Ae}	concentration of gaseous species A in emulsion phase, mol/cm^3
C_{Ai}	concentration of gaseous species A at the interface, mol/cm^3
C_{As}	concentration of gaseous species A at the surface, mol/cm^3
C_{B0}	initial concentration of species b, mol/cm^3

C_p	specific heat of gaseous species, cal/g•s
C_{ps}	specific heat of solid, cal/g•s
C'_{ps}	molar specific heat of solid, cal/mol•s
$\overline{C}_p$	average specific heat of gaseous stream, cal/g•s
$\hat{C}_A$, $\hat{C}_B$	dimensionless concentration of species A, B
D_e	effective diffusivity of gaseous species A in the solid, cm^2/s
D_{eB}	effective diffusivity of gaseous species A in the solid B, cm^2/s
D_{eG}	effective diffusivity of gaseous species A in the grain, cm^2/s
D_{eS}	effective diffusivity of gaseous species A in the solid S, cm^2/s
D_g	effective axial dispersion coefficient of gas in the bed, cm^2/s
D_k	Knudsen diffusivity, cm^2/s
D_s	effective axial dispersion coefficient of solid in bed, cm^2/s
Da	Damköhler number
E	activation energy, cal/mol
E^*, E^*_s	elutriation constant
E(t)	residence-time distribution function defined by Eq. (110)
f, f_c	pellet voidage
f_0	initial porosity
$f_0(x)$, $f_1(x)$, $f_2(x)$	functions of conversion defined in Table 2
f_B	void fraction in bed
f_{mf}	voidage at minimum fluidization
f_p	fraction of pores removed
F_{c1}, F_{c2}	parameters defined by Eq. (125)
F_g	volumetric flow rate of gas phase, cm^3/s
F_s	volumetric flow rate of solid, cm^3/s
$F_{s,0}$, $F_{s,1}$, $F_{s,2}$	volumetric feed rate of the entering, leaving, and elutriating streams in fluid-bed reactor, cm3/s
$F(F_p)$	function of fraction of pores removed
h	heat transfer coefficient, $cal/s{\cdot}cm^2{\cdot}°C$
h_w	heat transfer coefficient at the wall, $cal/s{\cdot}cm^2{\cdot}°C$
H	total height of the bed, cm

k	general notation for rate constant, $(cm^3/mol)^{n-1}$ s^{-1}
k'_e	effective thermal conductivity of solid, cal/s·cm·°C
k_g	phenomenological mass transfer coefficient, cm/s
k_p	rate constant for pore removal
k_s	surface reaction rate constant, cm/s
k^*	dimensionless reaction rate constant, $k/k(T_0)$
K	equilibrium constant
K_{be}	interchange coefficient across bubble-emulsion phase, cm/s
m	general notation for order of reaction with respect to solid species
M_B	molecular weight of species B
n	general notation for order of reaction with respect to gaseous species
p	partial pressure of species
P	pressure: also used to denote product species
P_d	decomposition pressure
r_G	radial position in the grain, cm
r_{Gi}	radial position of interface in the grain, cm
r_{G0}	initial radius of the grain, cm
R	radius of pellet, cm
R_A	general notation for rate of reaction, mol/s
R_{Ac}	rate of chemical reaction at the interface, mol/s
R_{Ad}	rate of diffusion through ash layer, mol/s
R_{Ag}	rate of diffusion through gas film, mol/s
R_g	gas constant
R_{GA}	rate of reaction per grain, mol/cm^3
R_i	radius of the interface, cm
R_j	radius of the ith particle
$\hat{R}$	dimensionless radial position
$\hat{R}_i$	dimensionless radial position of the interface
$\mathcal{R}$	extent of reaction parameter defined by Eq. (90)
s_{wb}	volume fraction of solids in wake per unit volume of bubble
S^*_p	dimensionless pore surface area
S^*	dimensionless reaction surface area

SF	shape factor of solid
t	time
$t(R_j)$	average residence time for particle of size R_j
$\hat{t}$	dimensionless time as variedly defined in text
$\hat{t}_{cr}$	dimensionless time defined by Eq. (21)
T	temperature variable
T_b, T_0	temperature in the bulk
T_{B0}, T_{g0}	inlet temperature of solid and gaseous streams
T_g	temperature in the gas phase
T_i	temperature at the interface
T_s	temperature at the solid surface
$\hat{T}_i$	dimensionless temperature at the interface
U	superficial velocity in the bed, cm/s
U_0	initial velocity in the bed, cm/s
U_b	bubble rise velocity, cm/s
U_{mf}	minimum fluidization velocity, cm/s
V_p	volume of the pellet, cm^3
W	weight of feed
W_1	dimensionless ratio of the pellet size to grain size, R/r_{G0}
W_2	dimensionless ratio of reactor length to pellet size, L/R
W_3	dimensionless ratio of reactor length to reactor radius, $2L/d_b$
y_A	mole fraction of gaseous reactant within the pellet
y'_A	mole fraction of gaseous reactant in the gas mixture entering the bed
y_c	mole fraction of gaseous product within the pellet
y'_c	mole fraction of gaseous product in the gas mixture entering the bed
z	dimensionless axial distance in the reactor
Z_1	defined by Eq. (92)

Greek Letters

α	diffusivity ratio, D_e/D_{e0}: also refers to dimensionless superficial gas velocity in Eq. (65)
β	shape factor equal to 0, 1, and 2 for slab, cylinder, and sphere, respectively

β_m	exothermicity factor
γ	Arrhenius parameter, defined as E/R_gT_b
γ_1	molar density of gas to that of the solid
δ	volume fraction of bubbles in the bed
ΔH	heat of reaction, cal/g mol
η	effectiveness factor as defined by Eq. (8)
θ	dimensionless gas temperature in the bed, T_g/T_0
θ_c	dimensionless coolent temperature, T_c/T_0
θ^*	dimensionless solid temperature in the bed, T_s/T_0
μ	roots of transcendental equation Eq. (45)
ν	dimensionless diffusivity of gaseous $D_{eA}/D_{eA,0}$
ν_A, ν_B	stoichiometric coefficients
ν'	dimensionless diffusivity of gaseous product D_e/D_{e0}
ρ_B	density of solid species B, g/cm^3
σ	reaction modulus
τ	dimensionless time relative to the process in the pellet, defined as D_et/R^2
τ_a	time required for complete conversion if ash diffusion alone controls
τ_c	the original nonporous material to become completely porous
τ_d	time required for complete conversion for diffusion-controlled processes
τ_D	dimensionless time relative to the internal diffusion controlled process
τ_g	time required for complete conversion of grain
τ_f	time required for complete conversion if film diffusion alone controls
τ_r	time required for complete conversion if reaction alone controls
ϕ^*	Thiele modulus defined in Eq. (11)
ϕ	extent of reaction in the bed
ϕ_1, ϕ_2	parameters defined in Table 7
ϕ_{cr}	critical values of Thiele modulus
$\check{\phi}_0(R)$, $\check{\phi}_1(R)$	size distribution of particles in the entering and leaving streams, respectively

REFERENCES

Ahner, W. D. and J. Feinman, Kinetics of hydrogen reduction of ferrous oxide in a batch fluidized bed: effects of mass velocity and pressure, AIChE J., *10*, 652 (1964).

Aris, R., Transition between regimes in gas-solid reactions, Ind. Eng. Chem. Fundam., *6*, 315 (1967).

Asaeda, M. and R. Toei, Correlation of gaseous flow and diffusional permeabilities in porous bodies, J. Chem. Eng. Jpn., *12*, 214 (1979).

Ausman, J. M. and C. C. Watson, Mass transfer in a catalyst pellet during regeneration, Chem. Eng. Sci., *17*, 323 (1962).

Avrami, M., Kinetics of phase change: II. Transformation-time relations for random distribution of nuclei, J. Chem. Phys., *8*, 212 (1940)

Barner, H. E., F. S. Manning and W. O. Philbrook, Theoretical analysis of hydrogen reduction of hematite in a fixed bed, Trans. Met. Soc. AIME, *227*, 897 (1963).

Benenati, R. E. and C. B. Brosilow, Void fraction distribution in beds of spheres, AIChE J., *8*, 359 (1962).

Beveridge, G. S. G. and P. J. Goldie, Effectiveness factors and instability in noncatalytic gas-solid reactions: the effect of solid heat capacity, Chem. Eng. Sci., *23*, 912 (1968).

Beveridge, G. S. G. and M. Kawamura, in *Process in Pyrometallurgy*, M. J. Jones, Ed., Institute of Mineral Meterology, London (1974).

Bhatia, S. K. and D. D. Perlmutter, A random pore model for fluid-solid reactions: I. Isothermal kinetic control, AIChE J., *26*, 379 (1980).

Bhatia, S. K. and D. D. Perlmutter, The effect of pore structure on fluid-solid reactions: application to the SO_2-lime reaction, AIChE J., *27*, 226 (1981a).

Bhatia, S. K. and D. D. Perlmutter, A random pore model for fluid-solid reactions: II. Diffusion and transport effects, AIChE J., *27*, 247 (1981b).

Bhatia, S. K. and D. D. Permutter, Effect of the product layer on the kinetics of the CO_2-lime reaction, AIChE J., *29*, 79 (1983).

Botterill, J. S. M., *Fluid Bed Heat Transfer,* Academic Press, London (1975).

Bourgeois, S. V., Jr., F. R. Groves, Jr., and A. H. Weke, Analysis of fixed bed sorption: flue gas desulfurization, AIChE J., *20*, 93 (1974).

Bowen, J. H. and C. K. Chen, A diffuse interface model for fluid-solid reaction, Chem. Eng. Sci., *24*, 1829 (1969).

Broadhurst, T. E. and H. A. Becker, Onset of fluidization and slugging in beds of uniform particles, AIChE J., *21*, 238 (1975).

Bukur, D., H. S. Caram, and N. R. Amundson, Some model studies of fluidized bed reactors, in *Chemical Reactor Theory,* L. Lapidus and N. R. Amundson, eds., Prentice-Hall, Englewood Cliffs, N.J. (1977).

Calvelo, A. and J. M. Smith, Intrapellet transport in gas-solid non-catalytic reactions, Chem. Proc., No. 3, 1 (1970).

Campbell, E. K. and J. F. Davidson, in *Fluidization Technology,* D. F. Keairns, ed., Vol. 2, Hemisphere, Washington, 285 (1975).

Campbell, R. R., A. W. D. Hills, and A. Paulin, Transport properties of porous lime and their influence on the decomposition of porous compacts of calcium carbonate, Chem. Eng. Sci., *25*, 929 (1970).

Cannon, K. J. and K. J. Denbigh, Studies on gas-solid reactions: I, The oxidation of zinc sulfide, Chem. Eng. Sci., *6*, 145 (1957).

Carberry, J. J., *Chemical and Catalytic Reaction Engineering*, McGraw-Hill, New York (1976).

Chan, S. F. and J. M. Smith, Effect of sintering on diffusion in porous solids, Indian Chem. Eng., *18*, 42 (1976).

Chandrashekar, B. C. and D. Vortmeyer, Warme-Stoff., *12*, 105 (1979).

Chang, C. C., L. T. Fan, and S. X. Rong, Modeling of shallow fluidized bed reactors: 1. Analytical studies of a simplified model, Can. J. Chem. Eng., *60*, 272 (1982).

Chavarie, C., Chemical reaction and interphase mass transfer in gas fluidized bed, Ph.D. thesis, McGill University (1973).

Chen, T. P. and S. C. Saxena, A mechanistic model applicable to coal combustion in fluidized beds, AIChE Symp. Ser., *74*(176) 149 (1978).

Christman, P. G. and T. F. Edgar, Distributed pore size model for the sulfation of lime stone, AIChE Meet., Chicago (1980).

Chrostowski, J. W. and C. Georgakis, Pore plugging model for gas-solid reactors, ACS Symp. Ser., *65* (Chem. React. Eng. Houston) 225 (1978).

Chu, W. F. and A. Rahmel, The kinetics of reduction of chromium oxide by hydrogen, Met. Trans. B, *10*, 401 (1979).

Davidson, J. F. and D. Harrison, *Fluidized Particles*, Cambridge University Press, Cambridge (1963).

Deans, H. A. and L. Lapidus, Computational model for predicting and correlating the behavior of fixed bed reactors: derivation of model for nonreactive system, AIChE J., *6*, 656 (1960a).

Deans, H. A. and L. Lapidus, A computational model for predicting and correlating the behavior of fixed bed reactors: extension to chemically reacting systems, AIChE J., *6*, 663 (1960b).

De Bruijn, T. J. W., T. H. Soeravidjaya, W. A. Jong, and P. J. Van Den Berg, Modeling of the reduction of maganese oxides with hydrogen, Chem. Eng. Sci., *35*, 1591 (1980).

Deb Roy, T. and K. P. Abraham, An analysis of the pressure build-up inside a reacting pellet during gas-solid reactions, Met. Trans., *5*, 349 (1974).

Del Borghi, M., J. C. Dunn. and K. B. Bishoff, A technique for solution of the equations for fluid-solid reactions with diffusion, Chem. Eng. Sci., *31*, 1065 (1976).

Doheim, M. A., A mathematical model for iron ore reduction: continuous multistage fluidized systems with solid if size distribution and elutriation occurs, J. Appl. Chem. Biotechnol., *23*, 375 (1975).

Doraiswamy, L. K. and M. M. Sharma, *Heterogeneous Reactions—Analysis Examples and Reactor Design*, Vol. 1, Wiley, New York (1983).

Doraiswamy, L. K., H. C. Bijawat, and M. V. Kunte, Chlorination of ilmenite in a fluidized bed, Chem. Eng. Prog., *55*, 80 (1959).

Dudukovic, M. P. and H. S. Lamba, Solution of moving boundary problems for gas-solid noncatalytic reactions by orthogonal collocation, Chem. Eng. Sci., *33*, 303 (1978a).

Dudkovic, M. P. and H. S. Lamba, A zone model for reactions of solid particles with strongly adsorbing species, Chem. Eng. Sci., *33*, 471 (1978b).

El-Rahaiby, S. K. and Y. K. Rao, The kinetics of reduction of iron oxides at moderate temperatures, Met. Trans. B, *10*, 257 (1979).

Eroffev, B. V., Proc. 7th Int. Symp. Reactivity of Solids (1961), p. 553.
Evans, J. W. and M. G. Ranade, The grain model for reaction between a gas and a porous solid—a refined approximate solution to the equations Chem. Eng. Sci., *35*, 1261 (1980).
Evans, J. W. and S. Song, Application of porous pellet model to fixed, moving and fluidized bed gas-solid reactors, Ind. Eng. Chem. Process Des. Dev., *13*, 146 (1974).
Evans, J. W., J. Szekely, W. H. Ray, and Y. K. Chaung, On the optimum temperature progression for irreversible non-catalytic gas-solid reactions, Chem. Eng. Sci., *28*, 683 (1973).
Fahim, M. A., N. Wakao, and J. D. Ford, Gas-solid reactions: I. A grain-cell model for complex reactions, Can. J. Chem. Eng., *56*, 725 (1978).
Fan, L. T. and C. C. Chang, Modeling and simulation of dynamic and steady state characteristics of shallow fluidized bed combustions, 2nd Congr. North American Continent, Las Vegas (Aug. 24–29, 1980).
Fan, L. S., K. Miyanami, and L. T. Fan, Transients in isothermal fluid-solid reaction systems—modeling of the sigmoidal-conversion-time behaviour of a gas-solid reaction, Chem. Eng. J., *13*, 13 (1977).
Fan, L. T., K. Tojo, and C. C. Chang, Modeling of shallow fluidized bed combustion of coal particles, Ind. Eng. Chem. Process Des. Dev., *18*, 333 (1979).
Fukunaka, Y., T. Monta, Z. Asaki, and Y. Kondo, Oxidation of zinc sulfide in a fluidized bed, Met. Trans. B, *7*, 707 (1976).
Fuwa, A., E. Kimura, and S. Fukushima, Kinetics of iron chlorination of roasted illmenite ore, Fe_2TiO_5 in a fluidized bed reactor, Met. Trans. B, *1*, 643 (1978).
Garza-Garza, O. and M. P. Dudukovic, Solution of models for gas-solid noncatalytic reactions by orthogonal collocation on finite elements with moving boundary, Chem. Eng. Sci., *6*, 131 (1982a).
Garza-Garza, O. and M. P. Dudukovic, A variable rise model for gas-solid reactions with structural changes, Chem. Eng., J., *24*, 35 (1982b).
Gavalas, G. R., A random capillary model with application to char gasification at chemically-controlled rates, AIChE J., *26*, 577 (1980).
Georgakis, C., C. W. Chang, and J. Szekely, A changing grain size model for gas-solid reactions, Chem. Eng. Sci., *34*, 1072 (1979).
Glidden, B. J. and R. R. Crafield, Gas particle heat transfer coefficients in packed beds at Reynolds numbers between 2 and 100, Br. Chem. Eng., *15*, 481 (1970).
Gokhale, M. V., A. T. Naik, L. K. Doraiswamy, Gas-solid reactions: an unusual observation in the disproportionation of potassium benzoate to terephthalate, Chem. Eng. Sci., *30*, 1409 (1975).
Gontaney, V., and A. Paulin, Rud.-Metal. Zb., *4*, 379 (1975).
Gower, R. C., A determination of the transport-limited reaction rate for a gas-solid reaction forming a porous reaction product, Ph.D. thesis, Lehigh University (1971).
Grace, J. R., Fluidized bed reactor modeling—an overview, 2nd Chem. Congr. North American Continent, Las Vegas (Aug. 24–29, 1980).
Gupta, S. N., R. B. Chaube, and S. N. Upadhyay, Fluid-particle heat transfer in fixed and fluidized beds, Chem. Eng. Sci., *29*, 839 (1974).
Hartman, M. and R. W. Coughlin, Reaction of sulfur dioxide with lime-stone and the influence of pore structure, Ind. Eng. Chem. Process Des. Dev., *13*, 248 (1974).
Hartman, M. and R. W. Coughlin, Reaction of sulphur dioxide with limestone and the grain model, AIChE J., *22*, 490 (1976).

Hartman, M., J. Hejna, and Z. Beran, Application of the reaction kinetics and dispersion model to gas-solid reactors for removal of SO_2 from flue gas, Chem. Eng. Sci., *34*, 475 (1979).

Highley, J. and D. Merrick, The effect of the spacing between solid feed points on the performance of a large scale fluidized bed reactor, AIChE Symp. Ser. *67*, 219 (1971).

Hills, A. W. D., The mechanism of the thermal decomposition of calcium carbonate, Chem. Eng. Sci., *23*, 297 (1968).

Hinduja, M. J., S. Sundaresan, and R. Jackson, Crossflow model of dispersion in packed bed reactors, AIChE J., *26*, 274 (1980).

Hite, R. H. and R. Jackson, Pressure gradients in porous catalyst pellets in the intermediate diffusion regime, Chem. Eng. Sci., *32*, 703 (1977).

Hlaváček, V., and J. Votruba, Steady-state operation of fixed bed reactors and monolithic structures, in *Chemical Reactor Theory*, L. Lapidus and N. R. Amundson, eds., Prentice-Hall, Englewood Cliffs, N.J. (1977).

Horio, M., A. Taki, Y. S. Hsei, and I. Muchi, Elutriation and particle transport through the freeboard of a gas-solid fluidized bed, in *Fluidization*, T. R. Grace, and D. L. Keairns, eds. (1980), p. 509.

Ishida, M. and C. Y. Wen, Comparison of kinetic and diffusional models for solid-gas reactions, AIChE J., *14*, 311 (1968).

Kato, K. and C. Y. Wen, Bubble assemblage model for fluidized bed catalytic reactors, Chem. Eng. Sci., *24*, 1351 (1969).

Kim, K. K. and J. M. Smith, Diffusion in nickel-oxide pellets—effects of sintering and reduction, AIChE J., *20*, 670 (1974).

Korolev, V. N., N. I. Syromyatnikov, and E. M. Tolmachev, Inzh.-Fiz. Zh., *6*, 973 (1971).

Kramers, H. and G. Alberda, Frequency response analysis of continuous flow systems, Chem. Eng. Sci., *2*, 173 (1953).

Kulkarni, B. D. and L. K. Doraiswamy, A delayed diffusion model for unusual reaction (disproportionation of potassium benzoate to terephthalate) involving an expanding reaction zone, Chem. Eng. Sci., *33*, 817 (1980a).

Kulkarni, B. D. and L. K. Doraiswamy, Estimation of effective transport properties in packed bed reactors, Catal. Rev. Sci. Eng., *22*, 431 (1980b).

Kulkarni, B. D. and P. A. Ramachandran, An approximate analytical solution to some problems in gas-solid noncatalytic reactions, Ind. Eng. Chem. Process. Des. Dev., *19*, 467 (1980).

Kunii, D. and O. Levenspiel, *Fluidization Engineering*, Wiley, New York (1969).

Lange, J. F., Y. Martinie, and M. A. Bergougnon, J. Powder Bulk Solids Technol., *1*, 15 (1977).

Le Page, A. H. and A. G. Fane, The kinetics of hydrogen reduction of UO_3 and U_3O_8 derived from ammonium diuranate, J. Inorg. Nucl. Chem., *36*, 87 (1974).

Levenspiel, O., *Chemical Reaction Omnibook*, Oregon State University Press, Corvallis, Oreg. (1979).

Luss, D. and N. R. Amundson, Maximum temperature rise in gas-solid reactions, AIChE J., *15*, 194 (1969).

Mantri, V. B., A. N. Gokarn, and L. K. Doraiswamy, Analysis of gas-solid reactions: formulation of a general model, Chem. Eng. Sci., *31*, 779 (1976).

Mason, E. A., A. P. Malinauskas, and R. B. Evans, III, Flow and diffusion in gases in porous media, J. Chem. Phys., *46*, 3199 (1967).

McGuire, H. L. and L. Lapidus, On the stability of a detailed packed bed reactor, AIChE J., *11*, 85 (1965).

Mehta, P. S., W. N. Sams, and D. Luss, Wrong way behavior of packed bed reactors—the pseudo homogeneous model, AIChE J., *27*, 234 (1981).

Merrick, D. and J. Highley, Particle size reduction and elutriation in a fluidized bed process, AIChE Symp. Ser., *70* (137), 366 (1974).

Mori, S. and C. Y. Wen, Estimation of bubble diameters in gaseous fluidized beds, AIChE J., *21*, 109 (1975).

Mu, J. and D. D. Perlmutter, An equilibrium interface model for solid decomposition, Chem. Eng. Sci., *35*, 1645 (1980).

Narsimhan, G., Thermal decomposition of calcium carbonate, Chem. Eng. Sci., *16*, 7 (1961).

Natesan, K. and W. O. Philbrook, Mathematical model for temperature variation within a particle undergoing reaction with application to roasting of zinc-sulphide, Trans. Met. Soc. AIME, *245*, 1417 (1969).

Neuburg, H. J., Kinetics of cuprous iodide oxidation with air, Ind. Eng. Chem. Process. Des. Dev., *9*, 285 (1970).

Ohmi, M., T. Usui, Y. Minamide, and M. Natio, Proc. 3rd Int. Congr. Iron Steel Conf. (1978), p. 472.

Osman, M. A., F. S. Manning, and W. D. Philbrook, Reduction of single particles and packed beds of hematite with carbon monoxide, AIChE J., *12*, 685 (1966).

Park, J. and O. Levenspiel, The crackling core model for the multistep reaction of solid particles, Chem. Eng. Sci., *32*, 233 (1975).

Pillai, K. K., Voidage variation at the wall of a apcked bed of spheres, Chem. Eng. Sci., *32*, 59 (1977).

Prasannan, P. C. and L. K. Doraiswamy, Gas-solid reactions: experimental evaluation of the zone model, Chem. Eng. Sci. *37*, 925 (1982).

Prasannan, P. C., P. A. Ramachandran, and L. K. Doraiswamy, Deocmposition reactions: effect of bulk flow and heat capacity of the solid, Chem. Eng. J., *40*, 1251 (1985).

Rajan, R. R. and C. Y. Wen, A comprehensive model for fluidized bed coal combustors, AIChE J., *26*, 642 (1980).

Ramachandran, P. A., Analysis of non-catalytic reactions following langmuir-Hinshelwood kinetics, Chem. Eng. J., *23*, 223 (1982).

Ramachandran, P. A. and L. K. Doraiswamy, Analysis of gas-solid reactions with zero order dependency on both gas and solid: concept of jumping reaction zones, AIChE J., *28*, 881 (1982).

Ramachandran, P. A. and L. K. Doraiswamy, Gas-Solid reactions— a review, AIChE J., *28*, 898 (1982b).

Ramachandran, P. A. and B. D. Kulkarni, Approximate analytical solutions to gas-solid problems, Ind. Eng. Chem. Process Des. Dev., *19*, 717 (1980).

Ramachandran, P. A. and J. M. Smith, Effect of sintering and porosity changes on the rate of gas-solid reactions, Chem. Eng. J., *4*, 137 (1977a).

Ramachandran, P. A. and J. M. Smith, A single-pore model for gas-solid noncatalytic reactions, AIChE J., *23*, 353 (1977b).

Ramachandran, P. A., M. H. Rashid, and R. Hughes, A model for coke oxidation from catalyst pellets in the initial burning period, Chem. Eng. Sci., *30*, 1391 (1975).

Ranade, M. G. and J. M. Evans, The reaction between a gas and a solid in a nonisothermal packed: simulation and experiment, Ind. Eng. Chem. Process Des. Dev., *19*, 118 (1980).
Ranade, P. V. and D. P. Harrison, The grain model applied to solids with varying structural properties, Chem. Eng. Sci., *34*, 427 (1979).
Ranade, P. V. and D. P. Harrison, The variable property grain model applied to the zinc oxide-hydrogen sulfide reaction, Chem. Eng. Sci., *36*, 1079 (1981).
Rao, Y. K., Mechanism and the intrinsic rates of reduction of metallic oxides, Met. Trans. B, *10*, 243 (1979).
Rao, Y. K., Dual-reaction model for hydrogen reduction of metallic chlorides, Trans. Inst. Min. Metall., *90*, C47 (1981).
Rehmat, A. and S. C. Saxena, Agglomeration of ash in fluidized bed gasification of coal by steam-oxygen (air) mixture, Ind. Eng. Chem. Process Des. Dev., *19*, 223 (1980).
Ruckenstein, E. and T. Vavanellos, Kinetics of solid phase reactions, AIChE J., *21*, 756 (1975).
Sampath, B. S., P. A. Ramachandran, and R. Hughes, Modeling of non-catalytic gas-solid reactions: I. Transient analysis of the particle-pellet model, Chem. Eng. Sci., *30*, 125 (1975).
Shen, J. and J. M. Smith, Diffusional effects in gas-solid reactions, Ind. Eng. Chem. Fundam., *4*, 9293 (1963).
Shettigar, U. R. and R. Hughes, Prediction of transient temperature distribution in gas-solid reactions, Chem. Eng. J., *3*, 93 (1972).
Simons, G. A. and W. T. Rawlins, Reaction of sulfur dioxide and hydrogen sulfide with calcined limestone, Ind. Eng. Chem. Process Des. Dev., *19*, 565 (1980).
Simonsson, D., Reduction of fluouride by reaction with limestone and particles in a fixed bed, Ind. Eng. Chem. Process. Des. Dev., *18*, 288 (1979).
Sohn, H. Y. and H. J. Sohn, The effect of bulk flow due to volume change in the gas phase on gas-solid reactions: initially nonporous solids, Ind. Eng. Chem. Process Des. Dev., *19*, 237 (1980)
Sohn, H. Y. and J. Szekely, The effect of reaction order in non-catalytic gas-solid reactions, Can. J. Chem. Eng., *50*, 674 (1972).
Sohn, H. Y. and J. Szekely, Reactions between solids through gaseous intermediates: I. Reactions controlled by chemical kinetics, Chem. Eng. Sci., *28*, 1789 (1973a).
Sohn, H. Y. and J. Szekely, A structural model for gas-solid reactions with a moving boundary: IV. Langmuir-Hinshelwood kinetics, Chem. Eng. Sci., *28*, 1169 (1973b).
Spitzer, R. H., F. S. Manning, and W. O. Philbrook, Simulation of topochemical reduction of hematite via intermediate oxides in an isothermal countercurrent reactor, Trans. Met. Soc. AIME, *242*, 618 (1968).
Sycheva, T. N. and E. V. Donat, Khim. Prom., *6*, 459 (1974).
Szekely, J. and J. W. Evans, Studies in gas-solid reactions: Part I. A structural model for the reaction of porous oxides with a reducing gas. Met. Trans., *2*, 1691 (1971a).
Szekely, J. and J. W. Evans, Studies in gas-solid reactions: Part II. An experimental study of nickel oxide reduction with hydrogen, Met. Trans., *2*, 1699 (1971b).

Szekely, J. and M. Propster, A structural model for gas-solid reactions with a moving boundary: VI. The effect of grain size distribution on the conversion of porous solids, Chem. Eng. Sci., *30*, 1049 (1975).

Szekely, J., J. W. Evans, and H. Y. Sohn, *Gas-Solid Reactions*, Academic Press, New York (1976).

Tanaka, I. and H. Shinohara, Mem. Fac. Eng. Kyushu Univ., *32*, 117 (1972).

Tigrel, A. Z. and D. L. Pyle, A model for a fluidized bed catalytic cracker, Chem. Eng. Sci., *26*, 133 (1971).

Tsay, Q. T., W. H. Ray, and J. Szekely, The modeling of hematite reactions with hydrogen plus carbon monoxide mixture, AIChE J., *22*, 1064 (1976).

Turkdogan, E. T., R. G. Olsson, H. A. Wriedt, and L. S. Darken, Calcination of lime stone, SME Trans. AIME, *254*, 9 (1973).

Ubhayakar, S. K., D. B. Stickler, and B. E. Gannon, Fuel, *56*, 281 (1977).

Ulrichson, D. L. and D. J. Mahoney, Pore closure model for gas-solid reactions: the effect of bulk flow and reversibility, Chem. Eng. Sci., *35*, 567 (1980).

Wakao, N. and T. Funazkri, Effect of fluid dispersion coefficients on particle to fluid mass transfer coefficients in packed beds, Chem. Eng. Sci., *33*, 1375 (1978).

Weekman, V. W., Jr. and R. L. Gorring, Influence of volume change on gas phase reactions in porous catalysts, J. Catal., *4*, 260 (1965).

Wen, C. Y. and S. C. Wang, Thermal and diffusional effects in noncatalytic solid-gas reactions, Ind. Eng. Chem., *62*(8), 30 (1970).

Wen, C. Y. and N. T. Wu, An analysis of slow reaction in a porous particle, AIChE J., *22*, 1012 (1976).

Wong, R. L., G. L. Hubbard, and V. E. Denny, Effects of temperature and pressure gradients on catalyst pellet effectiveness factors: I. Chem. Eng. Sci., *31*, 541 (1976).

Yagi, J. and J. Szekely, Effects of gas and solids maldistribution on the performance of moving bed reactors, reduction of iron oxide pellets with hydrogen, AIChE J., *25*, 800 (1979).

Yoon, H., J. Wei, and M. M. Denn, Feasible operating regions for moving bed coal gasification, Ind. Eng. Chem. Process Des. Dev., *18*, 306 (1979).

Yoshida, K. and D. Kunii, Complex reaction in fluidized beds-simulation of gasification, J. Chem. Eng. Jpn., *7*, 34 (1974).

Yoshida, K. and C. Y. Wen, Noncatalytic solid-gas reactions in fluidized-bed reactors, Adv. Chem. Ser., *109*, 138 (1972).

Zabrodsky, S. S., Y. G. Epanov, D. M. Galershtein, S. C. Saxena, and A. K. Kolar, Heat transfer in a large particle fluidized bed with immersed in-line and staggered bundles of horizontal smooth tables, Int. J. Heat Mass. Transfer, *24*, 571 (1981).

6
Design of Fixed-Bed Gas-Solid Catalytic Reactors

GILBERT F. FROMENT *Rijksuniversiteit, Gent, Belgium*

HANNS P. K. HOFMANN *Universität Erlangen—Nürnberg, Erlangen, Federal Republic of Germany*

INTRODUCTION

At present, the majority of commercial gas-phase catalytic processes are carried out in fixed-bed reactors. Important exceptions, for which fluidized-bed operation is preferred—either to continuously regenerate the catalyst or to improve heat removal—are the catalytic cracking of gas-oil, the syntheses of acrylonitrile and ethylene dichloride, and naphthalene oxidation into phthalic anhydride.

Fixed-bed reactors are preferred because of simpler technology and ease of operation. This is particularly true for adiabatic reactors. When the process has a pronounced heat effect, simple adiabatic operation is not always applicable, however. With very exothermic reactions the temperature rise may become excessive and harm the selectivity or the catalyst, shift the equilibrium to the wrong side, or cause safety problems. Very endothermic reactions, on the other hand, may be extinguished when carried out in a single adiabatic reactor. This is why multibed adiabatic reactors with interstage heat exchange, direct or indirect, internal or external, are frequently encountered. Examples are ammonia, methanol, and sulfuric acid synthesis; carbon monoxide conversion; and catalytic reforming of naphtha. The flow in such reactors is generally axial, but when a fraction of the effluent is recycled, it may be advantageous to use radial flow and to spread the catalyst out over a larger surface to reduce the pressure drop.

With multibed adiabatic reactors, interstage injection with the reacting gases is possible only between the beds. The temperature profile is of the sawtooth type and usually suboptimal. Theoretically, only continuous heat exchange permits a close approximation of the optimum temperature profile, but the heat exchange coefficients required to obtain this often take values beyond those that can be reached in practice.

Continuous heat exchange is achieved in two types of fixed-bed reactor. In the first, the catalyst is packed inside a number of parallel tubes surrounded by a heat-exchanging fluid. In the second, the tubes through which the heat-exchanging fluid flows are inserted into a massive bed of catalyst. The first type is referred to in this chapter as a *multitubular fixed-bed reactor* and is encountered, for example, in *o*-xylene oxidation to phthalic anhydride, steam reforming, and Fischer-Tropsch synthesis. The second type is referred to as a *fixed-bed reactor with internal heat exchange* and is encountered in ammonia and methanol synthesis.

In the strongly exothermic *o*-xylene oxidation, the tubes have an internal diameter of only 2.5 cm. They are filled with catalyst and cooled by means of circulating molten salt. One vessel may contain 20,000 tubes. For endothermic steam reforming the tubes are inserted into a gas-fired furnace. In ammonia synthesis a bed of 1.8 m internal diameter for a daily production of 200 metric tons, which is rather low by today's standards, may contain 750 cooling tubes. The feed flows through these tubes and is preheated by the reacting gases before entering the catalyst bed. The process can be operated in an autothermal way.

Exothermic equilibrium reactions can reach higher conversions at lower temperature levels, because of more favorable equilibrium conditions. At low temperatures the reaction rate is too low, however, so that excessive amounts of catalyst would be required. Nonisothermal operation is preferred in this case and the optimum temperature profile decreases toward the exit. Such an optimum profile is better approximated in a multitubular reactor than in a multibed reactor, at the expense of a considerably more complicated internal design and flow patterns, however. The effluent temperature may even be sufficient for partially preheating the feed and for producing high-pressure steam, as is done in ammonia synthesis. To avoid heat losses, but also for safety reasons, the boiler is preferably inserted into the high-pressure vessel. High-pressure reactors may reach a considerable level of sophistication.

Another reason for imposing a certain temperature profile may be the reaction sequence. Take, for example, a consecutive reaction scheme consisting of two steps. When the second step has an activation energy that is higher than that of the first step and the intermediate product is the desired one, the selectivity for this intermediate is favored by a decreasing temperature profile.

Fixed-bed reactors can be scaled up in several ways. Similarity theory is of little use in this matter. It is general practice to proceed from the laboratory to the commercial scale gradually, involving at least two intermediate scales: the pilot and demonstration units. Even large pilots cannot be expected to behave exactly like the ultimate commercial reactor, so that further insight is sometimes required. Mockups are useful for providing information on hydrodynamics, for example, on the redistribution over a downstream stage of a multibed reactor of gases coming from an external heat exchanger, a problem encountered in sulfuric acid synthesis. To date it has been inconceivable to proceed with reactor scaleup without at least a certain degree of modeling. The modeling can be fundamental and rely on separate observation of various aspects of the process, tied together and cast into a set of equations simulating the reactor, or it can be semi-empirical, thereby relying on pilot experimentation for tuning so-called "effective" parameters of the model.

The approach followed in this chapter is fundamental. Obstacles encountered in such an approach are complex hydrodynamics and the lack of sufficiently detailed and accurate chemical kinetics (a chronic problem with complex reactions and complex feedstocks). Modern aspects of kinetic analysis of complex porcesses have been discussed in recent years by

Froment (1975, 1976b), Froment and Hosten (1981), Hosten and Froment (1984), and Hoffmann and Hofmann (1976a,b). Additional difficulties arise when internal diffusion is limiting. It is generally too complicated—if not impossible—to derive both the kinetic equation and the diffusional parameters simultaneously from the usual kinetic experimentation. The intrinsic kinetics should be obtained first and in the absence of any diffusional limitations. The diffusional parameters are determined in a second stage by experimentation with particles of larger size (Dumez and Froment, 1976) or by purely physical measurements using, for example, the Wicke-Kallenbach cell or the gas chromatographic technique (De Deken et al., 1982). Other complications arise from catalyst deactivation by coking, poisoning, and sintering. Several reviews deal with this phenomenon (Froment, 1976a, 1980a, 1982; Butt, 1978, 1980, 1982; Delmon and Grange, 1980; Pommersheim and Dixit, 1981).

It should be emphasized here that the main deficiencies in the modeling of fixed-bed catalytic processes originate from incomplete and inaccurate chemical kinetics. Unfortunately, this is insufficiently recognized by those who have the background to contribute to this area: the chemical engineers. After this brief discussion of qualitative design aspects and of problem areas encountered in the quantitative design based on the fundamental approach, the latter will now be dealt with in detail.

CLASSIFICATION OF MODELS

Table 1 shows the classification of models for fixed-bed tubular reactor models proposed by Froment (1972a, 1972b, 1979) and widely accepted since. The classification contains two categories: the pseudohomogeneous and heterogeneous models. The latter distinguish between temperatures and concentrations in the bulk gas phase and those inside, or at the surface of the solid. The former do not. Each category contains two- and one-dimensional models, to account, in more or less detail, for gradients on the reactor scale. The basic model, AI, postualtes plug flow through the bed. It is one-dimensional because it assumes uniformity in a section perpendicular to flow, except in a very thin layer near the wall, when heat is exchanged with the surroundings. A first complication, leading to model AII , is to account for deviations from plug flow by superimposing mixing or "dispersion" in the axial direction (i.e., in the flow direction). When dispersion is also considered in the radial direction (i.e., in a cross section), because of the occurrence of temperature gradients, the model becomes two-dimensional.

TABLE 1 Classification of Models for Fixed-Bed Tubular Reactors

		Pseudohomogeneous $T = T_S;\ C = C_S$		Heterogeneous $T \neq T_S;\ C \neq C_S$
One-dimensional	AI	Basic, ideal	BI	+ interfacial gradients
	AII	+ axial dispersion	BII	+ intraparticle gradients
Two-dimensional	AIII	+ radial dispersion	BIII	+ radial dispersion

The first model of the heterogeneous category is derived from the basic pseudohomogeneous model AI by considering temperature and concentration gradients over the film surrounding the catalyst particles. The second model, BII, still one-dimensional, adds gradients inside the catalyst particle to the preceding model. Finally model BIII considers axial and radial gradients in the reactor as well as inter- and intraparticle local gradients.

The problem is now that of quantitatively expressing these phenomena, particularly the dispersion. This is generally done in terms of a continuum-type model, but so-called *cell models*, accounting for the two-phase nature of the reactor, have also been used. In continuum models the overall flow is considered to follow parallel streamlines, although not necessarily at a uniform velocity. Any mixing caused by the presence of the packing is modeled in terms of a Fickian, diffusion-type mechanism. This requires the flux of a component between two points to result from a large number of steps, each independent of the preceding one, as in a random walk. The flux resulting from such a mechanism is proportional to the gradient, just as in molecular or turbulent diffusion. The proportionality factor between the flux and the gradient is lumped with the molecular and turbulent diffusivities into an "effective" diffusivity or dispersion coefficient. Whereas the basic model AI leads to first-order differential equations (ordinary for steady state, partial for transient operation), the models including dispersion lead to second-order differential equations.

For nonisothermal situations, heat transfer has to be included in the model. In the basic pseudohomogeneous model, AI, heat is considered to be transferred by the overall convection only. When axial and/or radial dispersion is added, yielding models AII or AIII, heat is also transferred by effective conduction. This is a mode of heat transfer occurring in a pseudosolid that lumps the fluid and solid and therefore the mechanisms associated with each phase.

In heterogeneous models, heat and mass transfer from the bulk to the catalyst surface is described in terms of convective mechanisms. Transport mechanisms inside the particle are expressed in terms of effective diffusion and conduction, with the appropriate effective diffusivities and conductivities. In model BIII the dispersion in the bed is accounted for in a way analogous to that used in the pseudohomogeneous equivalent, AIII, but the mechanisms are separated into those occurring in the fluid and solid phases, respectively. Because of their nature, continuum models, based on effective transport mechanisms and leading to differential equations, are adequate only from a certain bed depth and a certain ratio of bed to particle diameter onward.

The first cell models were one-dimensional and consisted simply of a series of completely mixed tanks, each associated with a single particle. They lead to algebraic equations for the steady state and to ordinary differential equations for transient situations. Unlike continuum models, these cell models cannot generate backmixing. This feature was only introduced later by the backflow cell models. Although the properties of this type of cell model were thoroughly analyzed (Roemer and Durbin, 1967), their application seems to be lagging behind. However, Deckwer (1974) has applied them to nonisothermal reactors.

Cell models contain the property of infinite speed of propagation of a signal. To avoid this, a time delay would have to be introduced between the cells, but this has not been done so far (Sundaresan et al., 1980). Any sophistication of the model, such as allowing for a coupling between the particles by radiation or by considering a cell to contain the back half of a particle and the front half of a particle in the next row, destroys the simplicity of the computation.

Cell models can also be used for heterogeneous situations, when, for example, heat transfer resistance in the external film and mass transfer resistance inside the particle have to be accounted for. Two-dimensional cell models were introduced by Deans and Lapidus (1960). In these models the cells of a given row are connected to the next, downstream row, which is offset one-half cell width, thus causing radial mixing. Generally, heat is considered to be transferred through the fluid phase only. This has been shown to lead to serious errors, however (De Wasch and Froment, 1971). Kunii and Furusawa (1972) have accounted for all possible heat transfer mechanisms in a fixed bed, according to the lines set by Yagi and Kunii (1957) and Kunii and Smith (1960), to be discussed further in this chapter, but the computational advantage of the original cell models with respect to the continuum models is then completely lost.

The equivalence between the predictions of continuum and cell models has been studied extensively for both one- and two-dimensional cases (Aris and Amundson, 1957; Olbrich et al., 1966; Sundaresan et al., 1980; Varma, 1980). It should not be forgotten that the cell models were developed for mathematical convenience. In spite of their seemingly closer relation to reality, they are still far from true hydrodynamic models. The cell size, for example, is not the void fraction between particles: it is calculated to match the results of residence-time-distribution experiments. To achieve this, the cell size in the radial direction has to be 0.9 times the particle diameter.

The models used in this chapter are exclusively of the continuum type. With present-day numerical techniques and computers, integration of differential equations no longer presents great challenges. The methods available for this will be mentioned briefly in the appropriate sections.

MODEL EQUATIONS AND THEIR APPLICATIONS

Pseudohomogeneous Models

Basic Model with Plug Flow

General Model Equations: In this and the following sections the model equations will be written for constant density of the fluid phase, constant number of moles, and for one reaction only, to avoid obscuring the essential features. More complicated features will be dealt with, when necessary, in specific examples.

The model equations for a nonisothermal, nonisobaric reactor with axial flow, operating in the steady state, may be written

$$u_s \frac{dC}{dz} + \rho_B r_A = 0 \tag{1}$$

$$u_s \rho_g c_p \frac{dT}{dz} + 4\frac{\alpha_i}{d_t}(T - T_w) - (-\Delta H)\rho_B r_A = 0 \tag{2}$$

$$-\frac{dp_t}{dz} = \frac{f\rho_g u_s^2}{g d_p} \tag{3}$$

with initial conditions: at $z = 0$, $C = C_0$, $T = T_0$, and $p_t = p_{t0}$. The coupling of Eqs. (1) and (2) is evident. The coupling of these with Eq. (3) results from the total pressure dependence of the rate and the variation of ρ_g with temperature and the number of moles. The set of equations (1) to (3) is easily integrated using Runge-Kutta techniques.

In Eq. (2) T_w is the internal wall temperature and α_i is a heat transfer coefficient between the bed and the internal wall for which Leva (1948) proposed the following correlations. When the reaction mixture is heated from the wall,

$$\frac{\alpha_i d_t}{\lambda_g} = 0.813 \left(\frac{d_p G}{\mu}\right)^{0.9} e^{-6(d_p/d_t)} \tag{4}$$

when it is cooled from the wall

$$\frac{\alpha_i d_t}{\lambda_g} = 3.50 \left(\frac{d_p G}{\mu}\right)^{0.7} e^{-4.6(d_p/d_t)} \tag{5}$$

The ratio d_p/d_t determines the void fraction, particularly in the vicinity of the wall and therefore the flow conditions in that region and is introduced in an empirical way. The range of application of these equations is limited, however, since they predict a zero value of α_i at zero Reynolds number $d_p G/\mu$, thereby ignoring the contribution of the solid phase and of natural convection.

De Wasch and Froment (1972), who measured heat transfer over a range of Reynolds numbers extending from 30 to 1000, correlated their results as follows:

$$\frac{\alpha_i d_t}{\lambda_g} = \frac{\alpha_i^\circ d_p}{\lambda_g} + 0.033 \frac{c_p \mu}{\lambda_g} \frac{d_p G}{\mu} \tag{6}$$

where $\alpha_i^\circ = 2.44\lambda_{er}^\circ/d_t^{4/3}$ and λ_{er}° is calculated from a correlation to be presented in the section on the two-dimensional pseudohomogeneous model.

Equation (3) is the pressure-drop equation for turbulent flow, to be used in conjunction with the friction factor correlations discussed below. The friction factor f is generally correlated in the following form versus the void fraction and the Reynolds number:

$$f = \frac{1-\varepsilon}{\varepsilon^3}\left(a + b\,\frac{1-\varepsilon}{Re}\right) \tag{7}$$

Hicks (1970) proposed a different form of correlation for spheres:

$$f = 6.8\,\frac{(1-\varepsilon)^{1.2}}{\varepsilon^3}\,Re^{-0.2} \qquad \text{for } 300 < \frac{Re}{1-\varepsilon} < 60{,}000 \tag{8}$$

which implies that a and b are not true constants, in agreement with Tallmadge (1970).

TABLE 2 Values of Parameters in Ergun-Type Friction Factor Correlations

Reference	a	b	Flow range
Ergun (1952)	1.75	150	$\frac{Re}{1-\varepsilon} < 500$
Handley and Heggs (1968)	1.24	368	$1000 < \frac{Re}{1-\varepsilon} < 5000$
McDonald et al. (1979)	1.8 (smooth particles) 4.0 (rough)	180	$\frac{Re}{1-\varepsilon} < 10{,}000$
Mehta and Hawley (1969)	$1.75\left[1 + \frac{4}{6}\frac{d_p}{(1-\varepsilon)d_t}\right]$	$150\left(1 + \frac{4}{6}\frac{d_p}{(1-\varepsilon)d_t}\right)^2$	
Tallmadge (1970)	1.75	$4.2\ Re^{5/6}$	$0.1 < \frac{Re}{1-\varepsilon} < 10{,}000$

For nonspherical geometries the particle diameter is replaced by an equivalent diameter,

$$d'_p = 6 \frac{V_p}{S_p}$$

where V_p and S_p are the volume and surface area of the particle. With hollow rings, however, this correction factor is not accurate enough. Brauer (1957) introduced the following groups to relate the results for hollow rings to cylinders:

$$d_p = 6 \frac{V_{cyl}}{S_{cyl}} E^n \tag{9}$$

with

$$E = \frac{V_p S_{cyl}}{S_p V_{cyl}} \quad \text{and} \quad n = \frac{d_t/d_i}{(\varepsilon d_t^2/d_i^2)^{0.4} + 0.010(\varepsilon d_t^2/d_i^2)^{0.75}}$$

according to Reichelt and Blasz (1971). The void fraction ε in packed beds of spheres has been correlated by Haughey and Beveridge (1969) in the following way:

$$\varepsilon = 0.38 + 0.073 \left[1 + \frac{(d_t/d_p - 2)^2}{(d_t/d_p)^2}\right] \tag{10}$$

The basic pseudohomogeneous model is applied in the following sections to a number of typical fixed-bed reactor problems encountered in industrial practice. These applications are intended as illustrations of the fundamental approach advocated in this chapter.

The various aspects of reactor design and operation illustrated here may require, in specific cases, more appropriate models. Even then the simulation by means of this relatively simple model will be a valuable guide for the selection of the ultimate higher-level model, because it will reveal essential features about the process and the reactor behavior. The criteria for the escalation to the higher-level models listed in Table 1 are given in each corresponding section.

Application to Adiabatic and Multibed Adiabatic Reactors: Equations (1) to (3) are also applicable to adiabatic reactors by putting α_i equal to zero in Eq. (2). A useful relation, valid in isobaric adiabatic reactors, is obtained by combining Eqs. (1) and (2):

$$\Delta C = \frac{\rho_g c_p}{-\Delta H} \Delta T \tag{11}$$

Douglas and Eagleton (1962) have integrated Eqs. (1) and (2) (with $\alpha_i = 0$) for zero-, first-, and second-order kinetics and expressed the results in terms of exponential integrals, which are tabulated functions.

As mentioned earlier, it is of importance to reduce the pressure drop through the catalyst bed, in particular when a fraction of the effluent has to be recycled, as in catalytic reforming or ammonia synthesis. This requires rather shallow bed arrangements, but care has to be taken to achieve a good distribution of the feed. When this condition is satisfied, it is evident from the continuity equation that for isothermal conditions at least and for equal total pressure, the same conversion will be obtained when W/F_{A_0} in the shallow bed equals that in the deep bed. To avoid impractical reactor configurations and dimensions, radial flow is sometimes substituted for axial flow. The question then rises if centrifugal or centripetal flow is preferable. It follows from the detailed theoretical study of Balakotaiah and Luss (1981) that, for plug-flow, isothermal, and isobaric operation, the conversion is independent of the flow direction.

Marin and Froment (1982) recently compared axial with both centrifugal and centripetal radial flow in the catalytic reforming of a C_6 stream, under conditions and in configurations suggested by industrial catalytic reforming. For radial flow the continuity equations for the components are written as follows. The continuity equations for a feed component i, when the axis is oriented in the flow direction and the origin of r is at the inlet of the catalyst bed, are written

$$\frac{d(u_s C_i r)}{dr} = \sum_j \alpha_{ij} \rho_B r_j \tag{12}$$

where j is one of the reactions in which component i is involved. Introducing the conversion

$$x_i = \frac{F'_{i_0} R_0 - F'_i r}{F'_{i_0} R_0}$$

with $F'_i = u_s C_i$, leads to

$$\frac{dx_i}{dr} = r \frac{\sum_j \alpha_{ij} \rho_B r_j}{u_{s0} \rho_g R_0}$$

for both centrifugal and centripetal flow. The radial axis is directed in the flow direction and R_0 is the radial distance, measured from the reactor vertical axis onward, where the feed enters the bed.

The energy equation can be written

$$\frac{dT}{dr} = \frac{r \sum_j \alpha_{ij} (-\Delta H) \rho_B r_j}{c_p u_{s_0} \rho_g R_0} \tag{13}$$

The pressure drop is still calculated from Eq. (3) with f as given in Eq. (7), but with u_s in the Reynolds number varying with position. The boundary conditions are obvious.

TABLE 3 Simulation of Reactors with Axial and with Radial Flow for the Reforming of a C_6 Stream

	Makeup feed rates (kg/h)	Feed rate recycle
C_1-C_4		67,808
Dimethylbutanes	10,164	
Methylpentanes	17,279	
n-Hexane	40,656	
Methylcyclopentane	20,845	
Benzene	11,056	
Hydrogen		9,986
	100,000	77,794

Total hydrocarbon feed rate: 167,808 kg/h
Molar ratio recycle/HC feed: 6.00
Molar ratio H_2/HC feed: 4.20
Weight of catalyst: 63,000 kg

	Axial flow	Centripetal
Bed diameter internal (m)	8	2.00
Bed diameter external (m)	8	5.04
Bed height (m)	2.02	6.03
Inlet temperature (°C)	495	495
Exit temperature (°C)	460.5	459.01
Inlet pressure (bar)	15	15
Exit pressure (bar)	14.96	14.98
Effluent flow rates, C_1-C_4 (kg/h)	70,343	70,192
Dimethylbutanes	7,599	7,659
Methylpentanes	26,763	26,454
n-Hexane	34,275	34,611
Methylcyclopentane	6,300	6,672
Benzene	21,765	21,478
Hydrogen	10,684	10,667

The simulation results are given in Table 3. The selected radial configuration leads to a pressure drop which is slightly lower than the axial, yields the same product distribution, and looks acceptable from a practical point of view.

For several reasons already referred to, multibed operation is required when the temperature effects of the process are too pronounced. The distribution of the catalyst over the various beds and the inlet temperatures to each of these are design variables. Well-established rigorous techniques such as dynamic programming (Bellman, 1957; Aris, 1960) can be used to derive optimal values for these variables. Examples have been published for ammonia and sulfuric acid synthesis and CO conversion by the water-gas shift reaction. Figure 1 shows the result for SO_3 synthesis using Capelli's rate equation (Froment and Bischoff, 1979). Analogous results and a comparison of several optimization techniques were presented by Chartrand and Crowe (1969).

A conversion versus temperature diagram is a convenient way to represent the operation of a reactor for the case of a single reaction. It does not, of course, contain any information as to where in the reactor a given set (x,T) is reached. For an adiabatic bed the relation (x,T) is obtained from the equivalent of Eq. (11). In Fig. 1, the curve Γ_e represents the equilibrium line, the curve $\Gamma_{\lambda m}$ the locus of optimum reaction rates in an adiabatic bed, and the curves Γ_1, Γ_2, Γ_3 and Γ'_1. Γ'_2, Γ'_3 are loci of

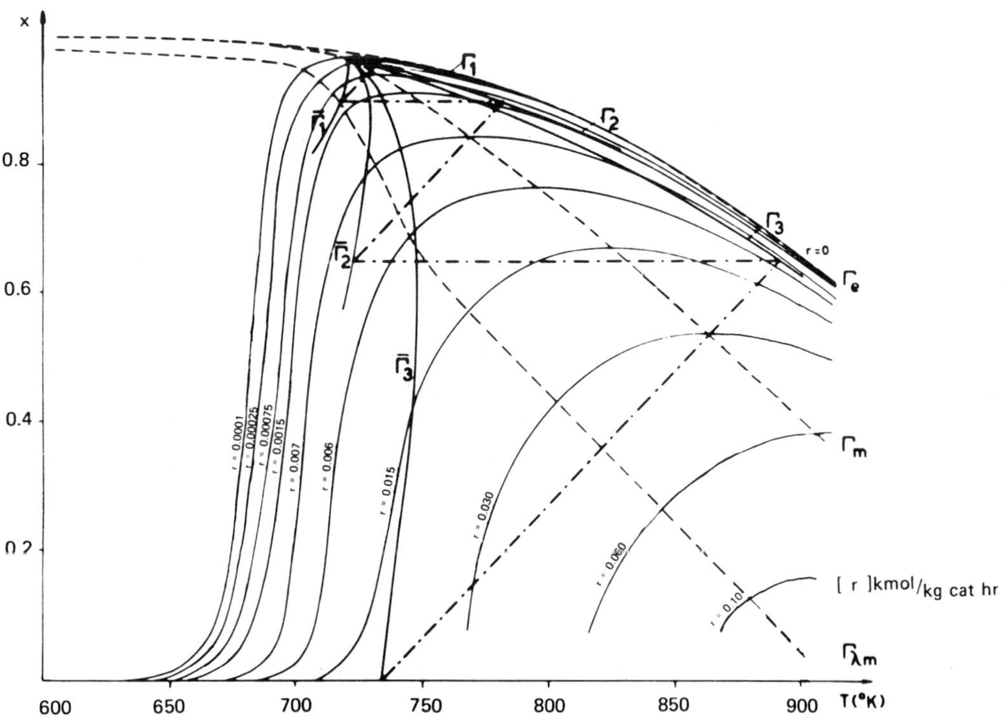

FIGURE 1 Optimal conversion versus temperature trajectories in SO_2 oxidation. (After Froment and Bischoff, 1979.)

optimum inlet and outlet conditions for each of the three beds. The profit function includes the benefit resulting from the conversion and the penalty associated with the cost of catalyst and operation. To achieve optimal operation with an exothermic equilibrium reaction, the temperature has to be lowered as the conversion increases. The multibed adiabatic reactor permits only a sawtooth temperature profile. Optimal reaction rate in each point can only be obtained by following the Γ_m curve, and this can only be attempted in a reactor with continuous heat exchange, either of the multitubular type or of the internal heat exchanger type, as mentioned earlier.

Tubular Reactors with Continuous Heat Exchange: Multitubular reactors with catalyst inside the tubes are generally modeled by considering only one tube, supposed to be representative for all of them. This is generally justified when the heat-exchanging medium is either well stirred or flows along or across the tubes in a single pass. When, however, the shell side of the bundle is baffled and the heat transfer medium flows through it in several passes, the complete bundle has to be considered, to account for the different conditions in each tube and for the interaction between tubes. The latter case will not be considered here; let it suffice to refer to work by McGreavy and Dunbobbin (1976, 1978).

One problem that is of particular importance when exothermic reactions are carried out in reactors of the type considered here is the occurrence of temperature peaks, called *hot spots*. The operating conditions, such as the feed concentration and temperature, or the coolant temperature and the tube diameter, have to be chosen judiciously to keep the hot spot within limits. What this limit is can be decided on by external constraints such as safety or catalyst stability, or by what Van Welsenaere and Froment (1970) called *intrinsic* constraints or criteria, related to the nature and the parameters of the reaction and to the operating conditions.

Bilous and Amundson (1956) were among the first to draw attention to the sensitivity of temperature profiles to the properties of the reaction or to the operating conditions. Barkelew (1959) derived, from an extensive set of simulations, simple intrinsic rules that are useful in early stages of design to select ranges of safe operation and to set limits on the geometry. Unfortunately, the results were obtained for a distorted temperature dependence of the rate coefficient. The problem was taken up later, with the Arrhenius dependence of the rate coefficient, by Dente and Collina (1964), Marek et al. (1969) (also Hlaváček, 1970), and Van Welsenaere and Froment (1970). The latter compared the various results in a diagram represented in Fig. 2.

The approach followed by Van Welsenaere and Froment has been extended to first-order reactions carried out in the presence of interfacial temperature gradients and intraparticle concentration gradients (model BII) by Rayadhyaksha et al. (1975). Recently, Morbidelli and Varma (1982) derived simple criteria for all positive-order reactions. The procedure for obtaining the critical values is explicit but requires a (single) numerical integration of a differential equation. All other parameters being equal, runaway is more likely the lower the reaction order.

The detailed design or simulation of a reactor requires numerical integration to generate the various profiles inside the tubes. An example of one aspect of the design is given in Fig. 3, which shows reactor temperature profiles in axial direction and their sensitivity with respect to the salt bath temperature in *o*-xylene oxidation over a V_2O_5 catalyst (Froment, 1967).

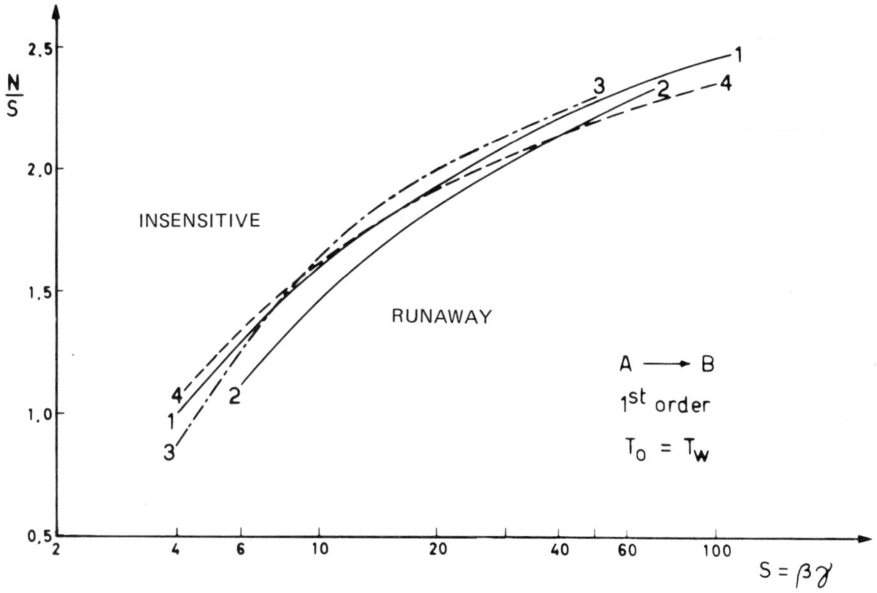

FIGURE 2 Comparison of criteria for parametric sensitivity and runaway based on one-dimensional pseudohomogeneous model with plug flow. (After Froment, 1974.)

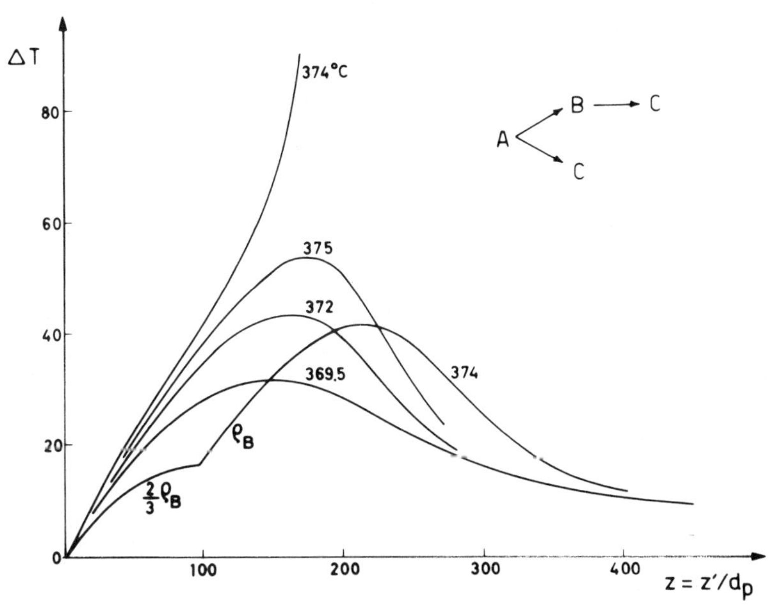

FIGURE 3 Parametric sensitivity in *o*-xylene air oxidation. Effect of salt bath temperature. Curve at 374°C with diluted bed. (After Froment, 1967.)

The figure shows how, beyond a certain salt bath temperature, runaway leading to an almost adiabatic temperature rise occurs. The phenomenon illustrated here is parametric sensitivity, not instability *sensu strictu*. The latter cannot be generated by this type of reactor model when the wall temperature is constant.

The parametric sensitivity of a multitubular reactor with coolant in cocurrent flow and varying in temperature was recently studied by Soria Lopez et al. (1981). In certain cases it is important to optimize the temperature profile when a sequence of reactions with different temperature dependencies is taking place or when the process is reversible and exothermic, as in ammonia, methanol, and SO_3 synthesis. Again this is best illustrated in a conversion versus temperature diagram such as the one presented in Fig. 1 for SO_3 synthesis. In this diagram the Γ_m curve represents the sets (x,T) associated with maximum rates, so that following this set (x,T) in the reactor ensures maximum conversion with a minimum amount of catalyst. The temperature profile that achieves this can be obtained through the use of Pontryagin's (1969) maximum principle. Murase et al. (1970) performed such calculations for an ammonia-synthesis reactor, but the practical realization is not so simple, because unrealistic heat transfer rates are required. Optimal temperature profiles may also be obtained by appropriate dilution of the catalyst bed with inert particles (Butt and Weng, 1976; Froment, 1971; Narsimhan, 1976).

Autothermal operation becomes possible when the heat of reaction is sufficient to heat the feed to the reaction temperature. In modern multibed adiabatic ammonia synthesis reactors the effluent and the feed exchange heat in an external heat exchanger. Internal heat exchangers whereby heat is exchanged between feed and reacting gases are also encountered, again in ammonia synthesis. The equations for the simulation and design of such reactors are given below.

For autothermal operation in a multibed adiabatic reactor for a single reaction with key component A (Fig. 4):

Catalyst bed:

$$\frac{dx_A}{dz} = \frac{\pi d_t^2}{4F_{A_0}} \rho_B r_A$$

$$\frac{dT}{dz} = \frac{(-\Delta H)\rho_B r_A}{u_s \rho_g c_p} \qquad (14)$$

External heat exchanger:

$$\frac{dT_1}{dz'} = -4 \frac{U}{d_t} \frac{T - T_1}{(u_s \rho_g c_p)_1}$$

$$(u_s \rho_g c_p)_1 \, dT_1 = -(u_s \rho_g c_p)_2 \, dT_2$$

Boundary conditions:

Catalyst bed: $x_A(0) = x_i$

$T(0) = T_1(L)$

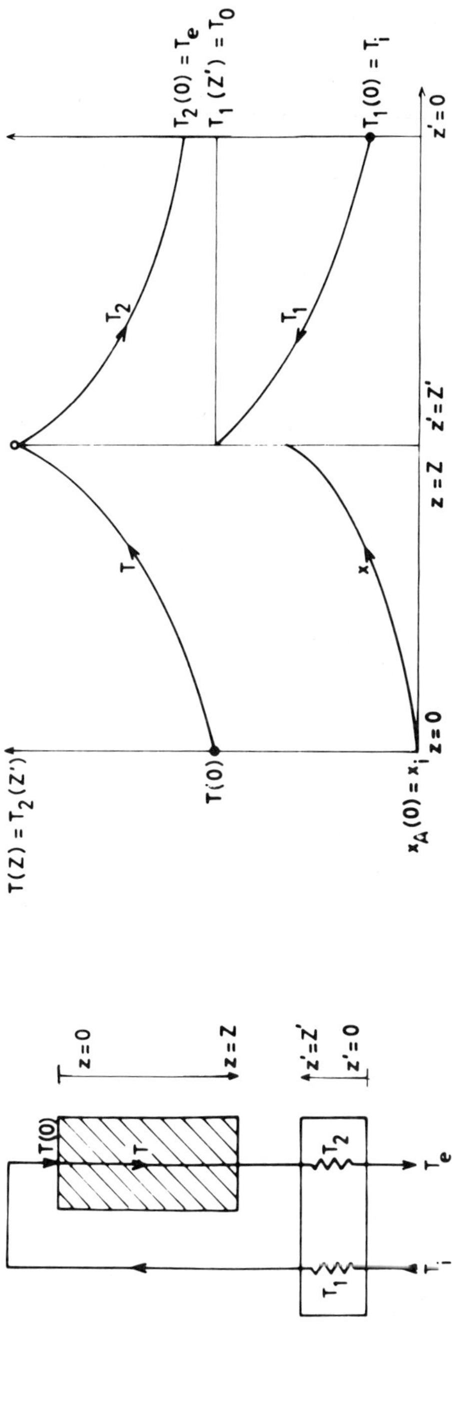

FIGURE 4 Autothermal operation of an adiabatic reactor with external heat exchanger.

Heat exchanger: $T_1(0) = T_i$

$T_2(0) = T_e$

$T_2(Z') = T(Z)$

For autothermal operation in a multitubular reactor (Fig. 5):

$$\frac{dx_A}{dz} = \frac{\pi d_t^2}{4F_{A0}} \rho_B r_A$$

$$\frac{dT}{dz} = \frac{(-\Delta H)\rho_B r_A}{u_s \rho_g c_p} - 4 \frac{U}{d_t} \frac{T - T_1}{u_s \rho_g c_p} \tag{15}$$

$$\frac{dT_1}{dz} = -4 \frac{U}{d_t} \frac{T - T_1}{u_s \rho_g c_p}$$

Boundary conditions:

$x_A = x_i$ at $z = 0$

$T = T_1$ $z = 0$

$T_1 = T_i$ $z = Z$

Notice that U is an overall heat transfer coefficient, not to be confused with α_i in Eq. (2).

The feedback of heat introduces a feature hitherto not encountered in this chapter, namely the possibility of multiple steady states. This

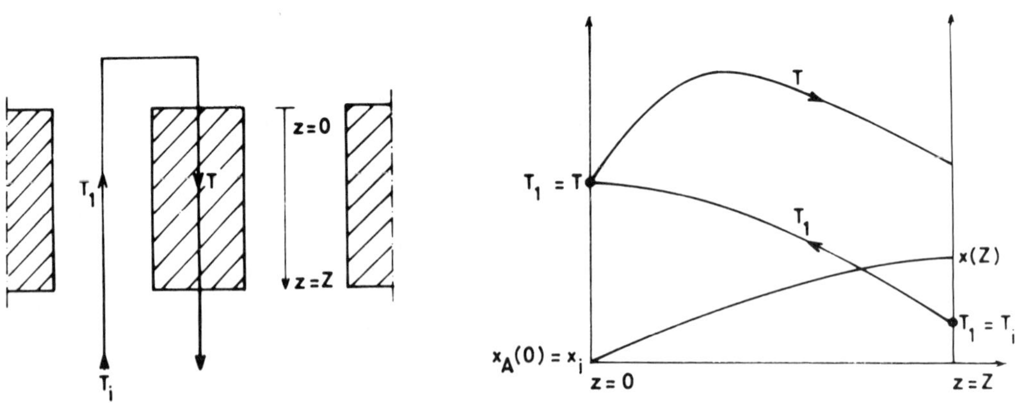

FIGURE 5 Autothermal operation of a multitubular reactor.

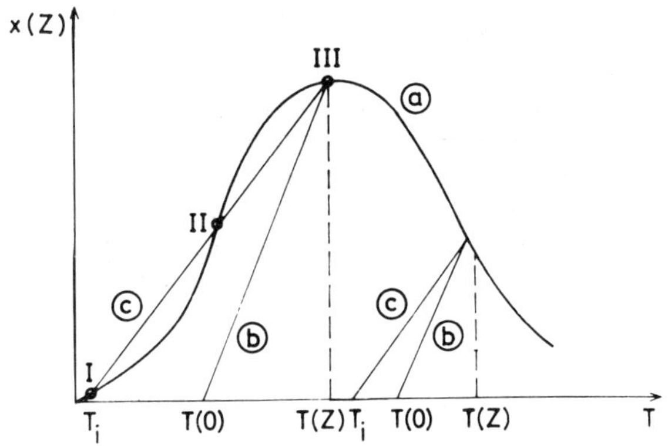

FIGURE 6 Steady-state operating points in an autothermal multibed adiabatic reactor.

phenomenon is readily understood on the basis of a now classical diagram (Liljenroth, 1918; Van Heerden, 1953) shown in Fig. 6.

In an x-T plane the sigmoidal curve is a measure of the heat produced by an exothermic equilibrium reaction, while the straight line is a measure of the amount of heat exchanged between, for example, the feed to an adiabatic bed and the effluent. The slope of this straight line is given by the ratio $\lambda/[1 + (U\pi d_t/mc_p)L]$. For a certain range of inlet temperatures and for certain values of this slope, therefore of operating conditions and configuration, three intersections (i.e., three steady states are possible) instead of one. The intersection leading to an intermediate temperature and conversion corresponds to an unstable steady state, the upper and the lower to stable steady states. The usual operation of an ammonia-synthesis reactor is represented by a straight line nearly tangent to the sigmoidal curve. Slight variations in the feed conditions may therefore cause extinction of the reactor, with the associated drastic cooling and possible mechanical damage to the catalyst and reactor. Consequently, autothermal reactors can run away either because of parametric sensitivity, already encountered in the simple tubular reactor without feedback, or because of instabilities *sensu strictu*, associated with the existence of multiple steady states.

Evidently, computer simulations permit locating ranges of operation in which such undesirable situations could develop. Yet it is always convenient to have an analytical criterion for this purpose. Such a criterion was developed by Inoue (1978) for nth-order reactions:

$N > \beta_r \gamma - n$ for irreversible reactions

$N > \beta_r \gamma - \kappa n$ with $\kappa = \dfrac{1 + K}{K}$ for reversible reactions

where

$$N = \frac{4U}{d_t} \frac{C_{A_0}}{c_p \rho_g r_A(T_0) \rho_B} \tag{16}$$

and β_r is the dimensionless adiabatic temperature rise in the reactor. There is no risk of instability with nth-order irreversible reactions when $\beta_r\gamma < 8$ and for first-order reversible reactions when $\beta_r\gamma < 8\kappa$.

The optimal operating conditions for autothermal reactions are close to the blow-off conditions, beyond which an external heat supply is required. For first-order reactions, irreversible or reversible, Ampaya and Rinker (1977a,b) related the critical feed temperature, T_F, leading to blow-off and the critical bed inlet temperature, T_0, to the operating and design parameters. At blow-off the following equation is valid for a first-order irreversible reaction:

$$(\gamma - 1)\tau \exp(\gamma) = 1 \tag{17}$$

where

$$\tau = \frac{2U \, \Delta T_{ad} R'}{d_t c_p M_m k \rho_B p_t}$$

The required degree of preheating then follows from

$$\frac{T_0 - T_f}{\Delta T_{ad}} = \frac{4UL}{u_s c_p \rho_g d_t} - \frac{1}{\gamma - 1} \frac{T_0}{\Delta T_{ad}} \tag{18}$$

When transients have to be accounted for, the time derivative also enters into the models. Most dynamic studies have been carried out by means of the one-dimensional model with plug flow [Eqs. (1) to (3)]. Even then the computational requirements become important. The dynamics of fixed-bed reactors were discussed by Ray (1972). An interesting case of transients was studied analytically by Mehta et al. (1981)—so-called wrong-way behavior—a reduction of the feed temperature may lead to a transient temperature rise because of the different speeds of propagation of concentration and temperature disturbances in the bed. The cold feed cools the upstream section of the reactor and decreases the reaction rates and therefore the conversion in that region. The less converted cold fluid then contacts hot catalyst, thus leading to a substantial temperature rise. Again, the key parameters determining the magnitude of the response are the adiabatic temperature rise, the activation energy, the heat transfer capacity, the coolant temperatures, the magnitude of the disturbance, and the reactor length. For a first-order reaction carried out in an adiabatic reactor and for a zero-order reaction in a cooled reactor, simple expressions can be derived for the maximum temperature rise. For the first case the expression is as follows:

$$\begin{aligned}(\beta_r + \omega) \exp\left(\frac{\gamma}{\beta_r + \omega}\right) - \omega \exp\left(\frac{\gamma}{\omega}\right) + \gamma\left[E_i\left(\frac{\gamma}{\omega}\right) - E_i\left(\frac{\gamma}{\beta_r + \omega}\right)\right] \\ = \frac{T}{T_0} \exp\left(\frac{\gamma T_0}{T}\right) - \exp \gamma + \gamma\left[E_i(\gamma) - E_i\left(\frac{\gamma T_0}{T}\right)\right]\end{aligned} \tag{19}$$

where ω represents the ratio of the temperature of the feed after and before the disturbance.

For dynamic situations, however, it may be necessary to distinguish between gas and solid, because of the difference in heat capacity of the two phases. This will be illustrated later when we deal with the one-dimensional heterogeneous model with interfacial gradients. The cyclic variation of the feed concentration or temperature has been shown, in certain cases, to be beneficial for the conversion and selectivity but has not been adopted yet in industrial practice. Reviews on this subject have been presented by Schädlich et al. (1983) and Renken (1982).

Axial Dispersion Model

It is clear that the flow conditions in fixed-bed reactors deviate to a certain extent from plug flow. Since it is impossible—or too complicated—to rigorously describe the true flow pattern, the deviation in mass and heat transport is described in terms of diffusion—and conduction—like mechanisms that operate in superposition on transport by plug flow. The fluxes of these mechanisms are expressed by Fick- and Fourier-type equations. Consequently, a temperature peak will generate a heat flux in the upstream direction in addition to that caused by radiation. The magnitude of this flux depends on the gradient and on the effective conductivity appearing in the Fourier-type law. Hiby (1963) has shown that this effect would extend to only about one particle diameter and according to Wicke (1960, 1975) and Sundaresan et al. (1980), it would be overestimated by the correlations that are presently available for the model parameters.

The reactor model equations are as follows, with the assumptions and restrictions already made:

$$(D_{ea})_s \frac{d^2C}{dz^2} - u_s \frac{dC}{dz} - \rho_B r_A = 0 \tag{20}$$

$$\lambda_{ea} \frac{d^2T}{dz^2} - u_s \rho_g c_p \frac{dT}{dz} - 4 \frac{\alpha_i}{d_t}(T - T_W) + (-\Delta H)\rho_B r_A = 0 \tag{21}$$

and the pressure-drop equation (3). The boundary conditions are

$$u_s(C_0 - C) = (D_{ea})_s \frac{dC}{dz} \qquad \text{at } z = 0$$

$$u_s \rho_g c_p (T_0 - T) = -\lambda_{ea} \frac{dT}{dz}$$

$$\frac{dC}{dz} = \frac{dT}{dz} = 0 \qquad \text{at } z = L$$

These boundary conditions have been extensively discussed (Wicke, 1975). Mathematically, they lead to a two-point boundary value problem, for which an appropriate integration algorithm is recommended.

Each term in Eq. (20) has dimensions $kmol/m_r^3 \cdot h$ and $(D_{ea})_s$ is based on the superficial flow velocity. Notice, however, that the effective axial dispersion coefficient is obtained from correlations Pe_{ma} versus Re in which the Péclet group usually contains u_i. These correlations therefore yield

$(D_{ea})_i = (D_{ea})_s/\varepsilon$. For gases, Pe_{ma} approaches a value of 2 at the Re numbers usually encountered in commercial fixed-bed reactors (Froment and Bischoff, 1979, p. 527). The effective axial thermal conductivity, λ_{ea}, is obtained from correlations Pe_{ha} versus Re in which the former contains u_s. These correlations show considerable spread, however (Dixon and Cresswell, 1979). Yagi et al. (1960) derived the following correlation from their experimental data at Re < 20:

$$\frac{\lambda_{ea}}{\lambda_g} = \frac{\lambda^o_{ea}}{\lambda_g} + \psi_a \mathrm{RePr} \qquad (22)$$

with $\psi_a = 1/Pe_{ma}$ of the order of 0.75 and λ^o_{ea}/λ_g of the order of 8 to 10, which is close to values measured for λ^o_{er}/λ_g, to be discussed later. This would lead to a value for

$$Pe_{ha} = \frac{u_s \rho_g c_p d_p}{\lambda_{ea}}$$

of the order of 2. A theoretical derivation by Bischoff (1962), based on axial mass dispersion and radial velocity profiles in packed beds, leads to a dependence on Reynolds, which is more complicated than that of Eq. (22). Extensive experimental data of Votruba et al. (1972), extending to Re = 150 and higher, lead to considerably lower Pe_{ha}, of the order of 0.5 (i.e., λ_{ea} roughly four times higher than those of Yagi et al.).

Axial dispersion effects may be neglected when the following criteria, derived by Young and Finlayson (1973), are satisfied. For situations in which the rate is monotonically decreasing with bed length (isothermal operation, adiabatic operation of an endothermic reaction, exothermic reaction with excessive cooling), the criteria are set up for inlet conditions.

Axial dispersion effects are negligible at the inlet when

$$\frac{r_{A_0} \rho_B d_p}{u_s C_0} \ll Pe_{ma} \qquad (23)$$

and

$$\frac{(-\Delta H) r_{A_0} \rho_B d_p}{(T_0 - T_w) u_s \rho_g c_p} \ll Pe_{ha}$$

Even when effects are noticeable at the inlet they are not necessarily felt at the exit when total or equilibrium conversion is achieved inside the reactor, of course, but intermediate states may be affected. When the rate of reaction is maximum at some intermediate position (e.g., because of a hot spot) the criteria above are inadequate. Evidently, the steepness of the temperature peak or of the concentration gradient determines whether or not axial dispersion has to be accounted for, but generally, a priori criteria are not available. Again according to Young and Finlayson (1973), axial dispersion is negligible when

$$\max\left|\frac{dx}{d(z/d_p)}\right| \ll Pe_{ma}$$
$$\max\left|\frac{dT}{d(z/d_p)}\right| \ll Pe_{ha} \qquad (24)$$

The gradients can be approximated from a simulation based on the one-dimensional pseudohomogeneous model with plug flow, AI. In industrial reactors with moderate temperature peaks, the flow velocities and reactor length are such that axial dispersion effects may be neglected. In their simulation of the TVA ammonium synthesis reactor, Baddour et al. (1965) observed that axial dispersion of heat altered the steady-state temperature profile by less than 0.6°C from that computed by means of the basic model, AI. Van Welsenaere and Froment (1970) simulated hydrocarbon oxidation close to runaway conditions, with a hot spot of 50°C and a conversion of some 70% in the first meter. Even for such a severe situation the criteria (23) and (24) were satisfied. More drastic cases are generally not realistic and have to be avoided for reasons of selectivity.

The structure of the model equations (20) and (21) introduces the possibility of multiple steady states. Consider, for illustrative purposes, the adiabatic version of these equations and let $\lambda_{ea} = (D_{ea})_s$. Combining the equations leads to

$$\frac{d^2T}{dz'^2} - Pe'_{ma}\frac{dT}{dz'} + f(T) = 0 \qquad (25)$$

where

$$z' = \frac{z}{d_p} \qquad Pe'_{ma} = \frac{u_i L}{(D_{ea})_i}$$

$$f(T) = \frac{k(T_0)L^2}{(D_{ea})_s}\rho_B(T_{ad} - T)\exp\left(\frac{E}{R'T_0}\right)\left(1 - \frac{T_0}{T}\right)$$

This equation may lead to a monotonic relation between T(Z) and T_0, yielding unique steady-state profiles or, for a certain range of parameter values, to a relation as shown in Fig. 7.

With such a relation three steady-state profiles are possible for inlet temperatures $(T_0)_3 < T_0 < (T_0)_1$. Of the three values of exit temperatures, T(Z), corresponding to a T_0 inside that range, only the outer two are stable and can be observed without external action. The runaway associated with multiplicity of steady states has different properties from that generated by parametric sensitivity, discussed under the basic plug-flow model, AI.

Indeed, as T_0 is increased along branch A, the exit temperature, T(Z), increases progressively up to $T(Z)_1$. For a value slightly exceeding $(T_0)_1$, the exit temperature T(Z) jumps to $T(Z)_2$. Notice the discontinuity in T(Z): whereas in the parametric sensitivity case any exit temperature can be achieved, in the present case the range $T(Z)_1 < T(Z) < T(Z)_2$ is

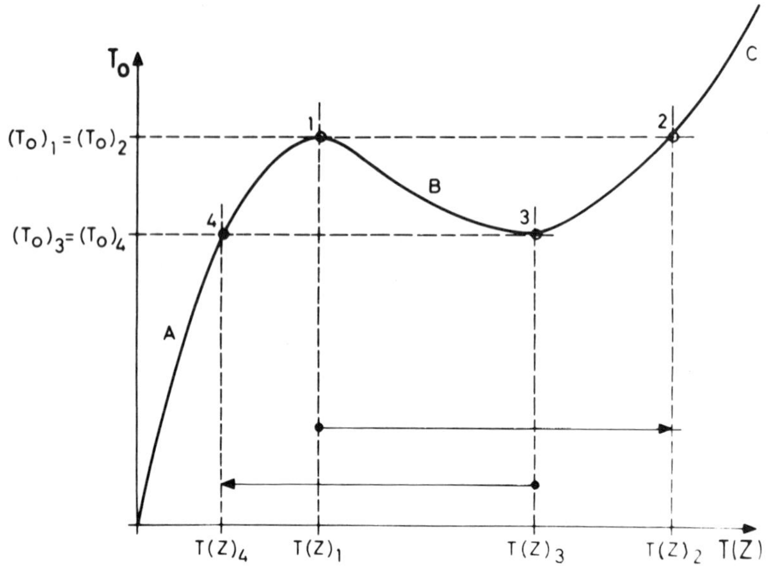

FIGURE 7 Relation of feed and exit temperatures in an adiabatic fixed-bed reactor with axial mixing leading to the possibility of three steady-state profiles. (From Froment, 1980b.)

excluded for the C_0 and flow rate that were selected. This is clearly an instability in the restricted sense. Indeed, if the inlet temperature is lowered from $(T_0)_2$ onward, T(Z) will progressively decrease along branch C until a value $T(Z)_3$ is reached. If T(Z) is further reduced, T(Z) will jump downward to $T(Z)_4$, a value that differs from $T(Z)_3$. The temperature history exhibits a hysteresis, which is typical for an unstable system. Several criteria for detecting multiplicity have been derived, either through an analytical approach or through numerical computations. It is clear from Fig. 7 that a necessary and sufficient condition for uniqueness is the absence of a hump, or in mathematical terms that no bifurcation occurs. The hump is not a sufficient condition for multiplicity, however, since this is possible for a certain range of T_0 values only. In short, multiplicity would be possible only for strongly exothermic processes with a very high activation energy, carried out in reactors with considerable influence of axial mixing.

According to Hlaváček and Hofmann (1970), for first-order reactions multiplicity would be possible as soon as

$$\beta_r \gamma > \frac{4\gamma}{\gamma - 4}$$

where β_r is the dimensionless adiabatic temperature rise in the reactor, $(-\Delta H)C_0/\rho_g c_p T_0$, and γ the dimensionless activation energy, E/RT_0. Figure 8 shows that even when this condition is satisfied, multiplicity can occur only when Pe_{ma} is smaller than a certain value, depending on the ratio

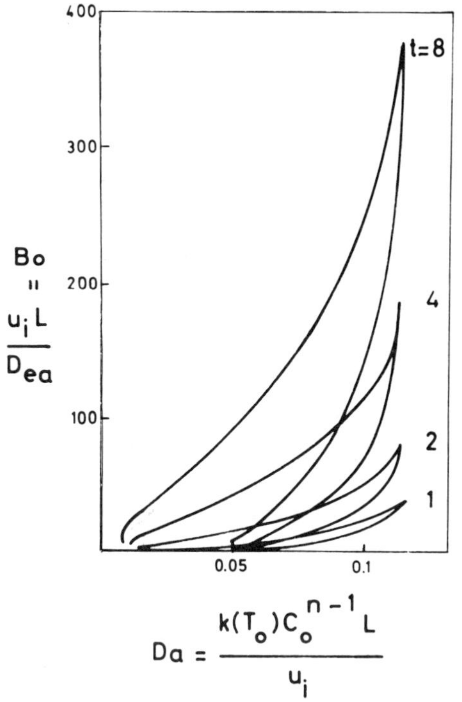

FIGURE 8 Areas of multiplicity of steady-state profiles in an adiabatic tubular reactor with axial mixing. $\beta_r\gamma = 10$; $\gamma = 20$; $t = \lambda_{ea}/D_{ea}\rho_g c_p$. (From Hlaváček and Votruba, 1977.)

$$\frac{Pe_{ma}}{Pe_{ha}} = \frac{\lambda_{ea}}{\rho_g c_p (D_{ea})_s}$$

and between two values of the first Damköhler number, $k(T_0)L/u_i$.

The value of β_r of 0.5 is exceeded in ammonia-methanol and oxy synthesis. Values of $\lambda_{ea}/(D_{ea})_s\rho_g c_p$ range from 1 to 4. Even for the maximum value, a value of Pe'_{ma} of 200, corresponding to a bed depth of roughly 100 particles, would exclude multiplicity. Industrial reactors typically lead to Pe'_{ma} of 600 to 2000. These values were confirmed in a very extensive analysis of this problem by Jensen and Ray (1982), who concluded further that oscillations caused by axial dispersion are totally unlikely in fixed-bed reactors.

It should be added that multiplicity of steady states has been observed in laboratory tubular reactors, for some extreme conditions. Even if axial dispersion is relatively more important in laboratory reactors, it now looks as if other phenomena are responsible for this behavior: intraparticle gradients, a particular form of rate equation, or surface phenomena on the catalyst. These causes will be dealt with in later sections.

Two-Dimensional Pseudohomogeneous Model

When reactions with a pronounced heat effect, such as ethylene, benzene, xylene, and naphthalene oxidation, are carried out in multitubular reactors, radial temperature and consequent concentration gradients develop. It then becomes necessary to extend the one-dimensional model AI to account for these gradients, caused by finite rates of heat and mass transfer in radial direction. In fixed beds the resistance to heat transfer is not limited to a very thin layer in the immediate vicinity of the wall, but extends throughout the cross section. The flux in radial direction of heat and mass is superimposed on the flux in axial direction by plug flow and is again expressed in terms of effective diffusion and conduction, obeying Fick's and Fourier's law. The effective dispersion coefficient contains contributions from molecular and turbulent diffusion and from the dispersion caused by the catalyst particles. The effective thermal conductivity contains various contributions, such as conduction in the solid and in the fluid, radiation between gas-filled voids and between solids, convection from the fluid to the solid, and conduction through stagnant fluid in the vicinity of contact surfaces between solid particles. It has been derived from experiments that the effective thermal conductivity λ_{er} is relatively constant in the core of the bed, but rapidly decreases toward the wall—as if supplementary resistance operates close to this boundary. This is not surprising: it is known that the packing—and therefore the flow characteristics—in that region are quite different from those in the core. Rather than using point values, λ_{er} has generally been considered to be constant up to the wall and a second parameter, accounting for the heat transfer at the wall, α_w, has been introduced. It is defined by

$$-\lambda_{er}\frac{\partial T}{\partial r} = \alpha_w(T - T_w) \qquad \text{at } r = R \tag{26}$$

This wall heat transfer coefficient should not be confused with the heat transfer coefficient, α_i, of the one-dimensional model, which lumps the resistance to heat transfer completely in the film near the wall.

Since in industrial reactors the flow rate is generally high, the contribution of axial effective diffusion can be neglected. The model equations can then be written, for steady state, in an isobaric reactor and no change in number of moles:

$$(D_{er})_s\left(\frac{\partial^2 C}{\partial r^2} + \frac{1}{r}\frac{\partial C}{\partial r}\right) - u_s\frac{\partial C}{\partial z} - r_A\rho_B = 0 \tag{27}$$

$$\lambda_{er}\left(\frac{\partial^2 T}{\partial r^2} + \frac{1}{r}\frac{\partial T}{\partial r}\right) - u_s\rho_g c_p\frac{\partial T}{\partial z} + (-\Delta H)r_A\rho_B = 0 \tag{28}$$

with boundary conditions

$$\begin{aligned} &C = C_0,\ T = T_0 && \text{at } z = 0,\ 0 < r < R \\ &\frac{\partial C}{\partial r} = 0 && \text{at } r = 0 \text{ and } r = R \end{aligned} \tag{29}$$

$$\frac{\partial T}{\partial r} = 0 \qquad \text{at } r = 0$$

$$\frac{\partial T}{\partial r} = -\frac{\alpha_w}{\lambda_{er}}(T - T_w) \qquad \text{at } r = R \tag{29}$$

Notice that axial effective diffusion has been neglected in this model for the reasons explained in the preceding section.

Like D_{ea}, the effective radial diffusivity is obtained from correlations Pe_{mr} versus Re. In Pe_{mr} the velocity is the interstitial, so that $(D_{er})_i$ is derived from this. Again, $(D_{er})_s = \varepsilon(D_{er})_i$. Correlations for D_{er} have been derived from experiments. According to Fahien and Smith (1955), for Re > 10,

$$\frac{Pe_{mr}}{1 + 19.4(d_p/d_t)^2} = 10$$

The denominator contains the ratio d_p/d_t, to account for the radial nonhomogeneity of the packing.

For heat transfer De Wasch and Froment (1972) obtained the following correlation for air and a V_2O_5-TiO_2 catalyst:

$$\lambda_{er} = \lambda^\circ_{er} + \frac{0.0025}{1 + 46(d_p/d_t)^2}\,Re \tag{30}$$

$$\alpha_w = \alpha^\circ_w + \frac{0.0115 d_t}{d_p}\,Re \tag{31}$$

Yagi and Kunii (1957) and Kunii and Smith (1960) developed models for calculating λ_{er} and α_w from fundamentals. In these models the flux by effective conduction is considered to consist of two contributions: the first static, independent of flow conditions, the second dynamic,

$$\lambda_{er} = \lambda^\circ_{er} + \lambda^t_{er} \tag{32}$$

The model for the calculation of λ°_{er} contains the mechanisms mentioned above. Zehner and Schlünder (1970) developed a slightly different concept yielding the following equation:

$$\frac{\lambda^\circ_{er}}{\lambda_g} = (1 - \sqrt{1-\varepsilon})\left(1 + \varepsilon\,\frac{\alpha_{rs} d_p}{\lambda_g}\right) + \frac{2\sqrt{1-\varepsilon}}{1 + [(\alpha_{rs} d_p/\lambda_g) - B](\lambda_g/\lambda_s)}\,\theta \tag{33}$$

with

$$\theta = \left\{ \frac{\{1 + [(\alpha_{rs} d_p/\lambda_g) - 1](\lambda_g/\lambda_s)\}B}{\{1 + [(\alpha_{rs} d_p/\lambda_g) - B](\lambda_g/\lambda_s)\}^2} \ln \frac{1 + \alpha_{rs} d_p/\lambda_s}{B(\lambda_g/\lambda_s)} - \frac{B - 1}{1 + [(\alpha_{rs} d_p/\lambda_g) - B](\lambda_g/\lambda_s)} + \frac{B + 1}{2B}\left(\frac{\alpha_{rs} d_p}{\lambda_g} - B\right)\right\}$$

where $B = b[(1 - \varepsilon)/\varepsilon]^{10/9}$ with b = 1.25 for spheres and 2.5 for cylinders and Raschig rings.

The dynamic contribution λ_{er}^t is obtained from

$$\frac{\lambda_{er}^t}{\lambda_g} = \psi_r \mathrm{PrRe} \tag{34}$$

where $\psi_r = 1/\mathrm{Pe}_{mr}$. From the heat transfer experiments of De Wasch and Froment (1972), the following correlation for ψ_r can be derived:

$$\psi_r = \frac{0.14}{1 + 46(d_p/d_t)^2} \tag{35}$$

whereas Yagi and Kunii (1957) obtained values between 0.10 and 0.14 Yagi and Wakao (1959) and Yagi and Kunii (1960) also derived the following correlation for α_w and developed a model along the lines adopted for the modeling of λ_{er} to predict this parameter:

$$\frac{\alpha_w d_p}{\lambda_g} = \frac{\alpha_w^\circ d_p}{\lambda_g} + \psi_w \mathrm{PrRe} \tag{36}$$

where $\psi_w = 0.041$. Values of $\alpha_w^\circ d_p/\lambda_g$ in cylindrical beds are of the order 0f 3.0, depending on the solid, its size, and the ratio d_t/d_p. Hennecke and Schlünder (1973) also presented a correlation for α_w. based on an extensive set of experimental data. Spechia et al. (1980) have presented new data and correlations, both for λ_{er}^t and α_w. The numerical problems associated with the solution of sets of second-order nonlinear partial differential equations have been thoroughly investigated and various techniques, such as the implicit Crank-Nicolson scheme (Grosjean and Froment, 1962) and orthogonal collocation (Finlayson, 1971), have been used and compared with one another (Mihail and Jordache, 1976). Computer programs for their application are available (Hoffmann et al. (1977); Lerou and Froment, 1982; Dente et al. (1972). Yet for situations that are very demanding in computational efforts, such as those involving transients, it may still be of interest to use a model with a structure requiring less computation, but containing, nevertheless, certain features of the two-dimensional model. An "equivalent" one-dimensional model has been developed that aims at reproducing the radially averaged concentrations and temperatures generated by the two-dimensional model. By matching temperature profiles in the absence of reaction, Froment (1961) derived the following relation from which the parameters α_i of the equivalent one-dimensional model can be calculated:

$$\frac{1}{\alpha_i} = \frac{1}{\alpha_w} + \frac{d_t}{8\lambda_{er}} \qquad (37)$$

Slightly different but analogous relations were derived by Crider and Foss (1965), Hlaváček (1970), and Finlayson (1971). A comparison of both model predictions was made by Froment (1967) for *o*-xylene oxidation. For severe but still realistic operating conditions, the deviations between both models become important. And, of course, the values in the axis may still significantly differ from the radially averaged value, necessitating a more detailed model.

Examples of simulations based on the two-dimensional pseudohomogeneous model have been presented by Froment (1967), Hawthorn et al. (1968), Carberry and White (1969) and Garcia-Ochoa et al. (1981), among others. These simulations reveal that the radial profiles are not very sensitive to the Péclet number for radial mass transfer, but that, on the other hand, for severe conditions, λ_{er} and α_w have to be known with great accuracy. Figure 9 shows measured and computed temperatures in a cross section halfway between two tubular reactors for the endothermic dehydrogenation of methylcyclohexane on a Pt/Al_2O_3 catalyst (Hawthorn et al., 1968). The parameters in the two-dimensional model were independently determined. The reactor with internal tube diameter 1.65 cm leads to a temperature difference between the wall and the axis of 34°C. This ΔT is made up of two

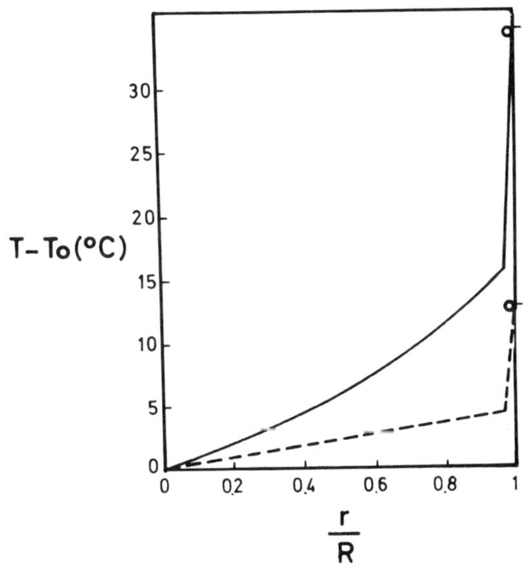

FIGURE 9 Measured and computed radial temperature profiles in a methylcyclohexane dehydrogenation reactor. Solid line: I.D. tube, 1.65 cm; $\alpha_w d_t/2\lambda_{er}$ = 2.9; dashed line: I.D. tube, 0.71 cm; $\alpha_w d_t/2\lambda_{er}$ = 1.1. (From Hawthorn et al., 1968.)

nearly equal parts and the reaction rate is significantly varying with radial position, while the temperature profile is deviating from parabolic. In the reactor with internal tube diameter 0.71 cm, the ΔT is 13°C, with nearly two-third accounted for by the wall film resistance; the reaction rate is nearly uniform in the cross section and the temperature profile is very nearly parabolic. In this case a one-dimensional model would be sufficient to successfully simulate the reactor, provided that the heat transfer coefficient α_i would be adapted to account for the influence of the radial temperature profile, as is done in Eq. (36).

Hofmann and coworkers (Hofmann, 1979) studied reactions with widely different heat effects in a sophisticated bench-scale tubular reactor allowing axial and radial profiles of conversion and temperature to be measured. This work again emphasizes the high degree of accuracy required in the modeling of reaction and heat transfer under severe conditions. The expressions for the static contributions to λ_{er} and α_w contain a number of parameters which are not always available with a sufficient accuracy: even the apparently trivial solid conductivity, λ_s, is a problem, in particular with complex catalysts (Hoffmann et al., 1978).

Discrepancies between observed and predicted parameter values are more likely also at low d_t/d_p, for which the reported data show a wide scatter and for which deviations from plug flow should no longer be described by a Fickian mechanism. Recently, a non-Fickian model was developed by Chang (1982), but much remains to be done in this area.

Little has been done also to account for the radial variation in the velocity caused by the nonuniform void fraction. The latter has a value of 1 at the wall and decreases in an oscillating way to a rather constant value around the central axis. The following empirical correlation was derived for the void fraction in a bed of spheres by Martin (1978) from the experimental data of Benenati and Brosilow (1962):

$$\varepsilon(r') = \begin{cases} \varepsilon_{min} + (1 - \varepsilon_{min})r'^2 & -1 \leqslant r' < 0 \\ \varepsilon_\infty + (\varepsilon_{min} - \varepsilon_\infty)e^{-0.25r'} \cos \dfrac{\pi}{a} r' & r' \geqslant 0 \end{cases}$$

where $r' = 2(r/d_p) - 1$ and $a = 0.816$ for $d_t/d_p = \infty$ and 0.876 for $d_t/d_p = 20.3$. ε_{min} occurs at a distance of $(1/2)d_p$ from the wall, while ε_∞ refers to the value in the axis. For beds of spheres ε_{min} equals 0.2 and ε_∞ roughly 0.4, depending on d_t/d_p. For a d_t/d_p of 8, Schwartz and Smith (1953) measured velocities near the wall twice as high as those in the axis. Accounting for the point values of the flow velocity into the treatment of their radial heat transfer data, Lerou and Froment (1977) came to values for λ_{er} very close to those predicted by the correlation (30), in which the dependence on the Reynolds number was arrived at through the variation of the global flow rate.

Figure 10 shows a comparison of radial temperature profiles in an *o*-xylene oxidation reactor, computed on the basis of uniform flow velocity and a flow velocity distribution induced by the local void fraction (Lerou and Froment, 1977).

Again it is clear that considerable sophistication is required in the modeling of demanding situations. It is unrealistic to believe that gross

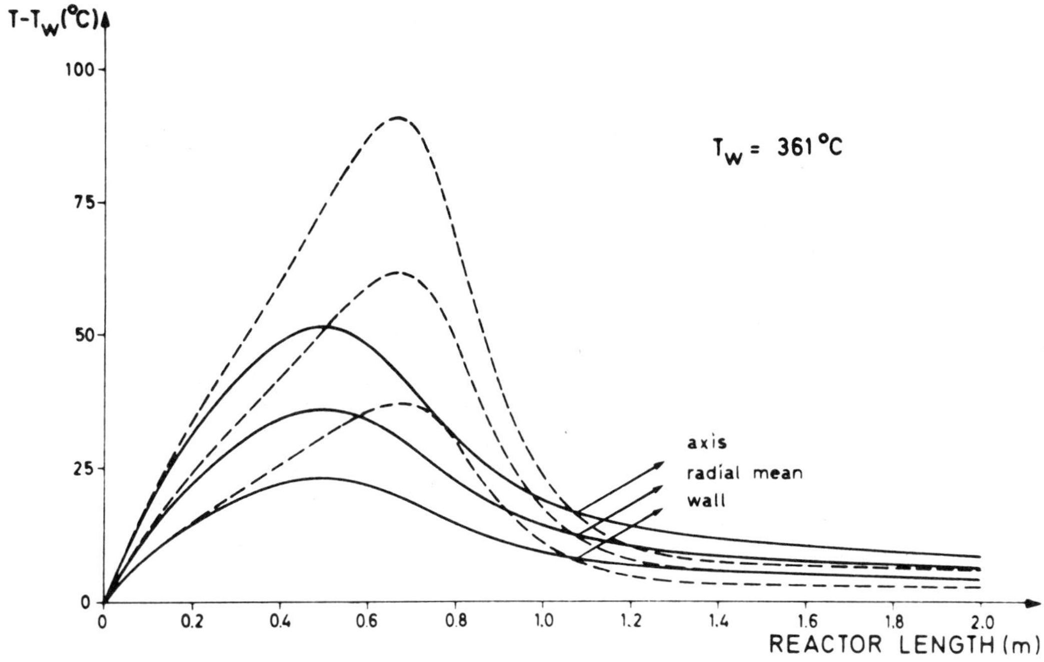

FIGURE 10 Effect of flow velocity profile on radially averaged temperature, and conversion profiles in a tubular reactor for *o*-xylene oxidation. d_p = 0.0045 m; d_t = 0.025 m; T = 361°C. Dotted line, model with radial velocity profile; solid line, model with uniform velocity. (After Lerou and Froment, 1977.)

models would be sufficient for the modeling of a reactor, just because it is industrial and geometrically relatively simple. To decide whether or not a two-dimensional model is required, Mears (1971a) proposed to evaluate the group

$$\frac{|\Delta H|\, r_A(T)\rho_B d_t^2 E}{4\lambda_{er} R' T_w^2}\left(1 + \frac{8}{\alpha_w d_p/\lambda_{er}}\, \frac{d_p}{d_t}\right) \tag{37}$$

at the hot spot predicted by the one-dimensional simulation. If the value of the group exceeds 0.4, the simulation should be extended to the pseudo-homogeneous two-dimensional model. The idea behind the criterion is that if the cross section is to be uniform in temperature, the radially averaged rate in the hot-spot location should not deviate from the rate taken at the wall temperature by more than 5%. For exothermic reactions the criterion is somewhat optimistic, since the temperature predicted by the one-dimensional model at a given bed depth is lower than the radial average of the two-dimensional temperature profile.

Clearly, there are cases where there is no justification for not using a two-dimensional model in designing or simulating a reactor. And from the work of Hofmann and coworkers (Hofmann, 1979), this would seem to be true for kinetic analysis also, when pronounced hot spots are inevitable. Finally, it should be added that the mathematical structure of the reactor model as presented here cannot generate multiple steady states (Young and Finlayson, 1974). The parametric sensitivity of the model has been investigated numerically along the Barkelew approach by Agnew and Potter (1966).

Heterogeneous Models

Plug-Flow Model with Interfacial Gradients

Model Equations and Correlations for the Parameters: When the transfer of heat and mass from the bulk gas phase to the catalyst involves significant gradients in the vicinity of the surface of the particle, the model commonly used to describe such a situation in a fixed-bed reactor is of the heterogeneous type. It is written, in the steady state, for a single reaction, constant density, and isobaric operation:

Fluid phase:

$$-u_s \frac{dC}{dz} = k_g a_v (C - C_s^s) \tag{38}$$

$$u_s \rho_g c_p \frac{dT}{dz} + \frac{4\alpha_i}{d_t}(T - T_w) = h_f a_v (T_s^s - T) \tag{39}$$

Solid phase:

$$k_g a_v (C - C_s^s) = \rho_B r_A \tag{40}$$

$$h_f a_v (T_s^s - T) = (-\Delta H)\rho_B r_A \tag{41}$$

with initial conditions $C = C_0$, $T = T_0$ at $z = 0$. The algebraic equations (40) and (41) are first iteratively solved, yielding the values of C_s^s and T_s^s which enter into the right-hand sides of the differential equations for the fluid. These are then solved by a Runge-Kutta routine. Notice that this model does not explicity contain any axial coupling between the solid phase; heat and mass are transferred in the axial direction through the gas phase only.

The correlations for the mass and heat transfer coefficients have been reviewed by Schlünder (1978). A frequently used correlation for k_g is that of Petrovic and Thodos (1968):

$$\varepsilon\, Sh = 0.357 Re^{0.641} Sc^{0.33} \qquad 3 \leqslant Re \leqslant 2000 \tag{42}$$

whereas for heat transfer Whitaker (1972) proposed

$$\varepsilon \text{ Nu} = (1 - \varepsilon) \left[0.5\left(\frac{\text{Re}}{1 - \varepsilon}\right)^{0.5} + 0.2\left(\frac{\text{Re}}{1 - \varepsilon}\right)^{0.66} \right] \text{Pr}^{0.33} \tag{43}$$

and Gelbin et al. (1976):

$$\varepsilon \text{ Nu} = 1.00 \left(\frac{\text{Re}}{\varepsilon}\right)^{0.563} \text{Pr}^{0.33} \tag{44}$$

Evidently, these transfer coefficients between bulk gas and a bed of particles differ from those between a gas and a single particle, but Martin (1978) and Gnielinski (1982) expressed Nu in terms of that for a single particle, Nu_{sp}, which is in turn first related to laminar and turbulent contributions:

$$\text{Nu}_{sp} = 2 + \sqrt{\text{Nu}_l^2 + \text{Nu}_t^2} \tag{45}$$

with

$$\text{Nu}_l = 0.664 \sqrt[3]{\text{Pr}} \sqrt{\text{Re}'}$$

and

$$\text{Nu}_t = \frac{0.037\text{Re}'^{0.8}\text{Pr}}{1 + 2.443\text{Re}'^{-0.1}(\text{Pr}^{0.66} - 1)} \qquad \text{and} \qquad \text{Re}' \frac{u_s \rho_g \sqrt{S_p/\pi}}{\mu \varepsilon}$$

The Nusselt number for the packed bed is then calculated from

$$\text{Nu} = f_a \text{Nu}_{sp} \tag{46}$$

with the shape factor, f_a = 1.0 for spheres, 1.6 for cylinders, 2.3 for Berl-Saddles, and 2.1 for Raschig rings. Relation (46) and the shape factors also hold for the Sherwood number. Gnielinski's correlation fits a large number of experimental data of various authors, except at low values of Pe_h = PrRe. Martin (1978) and Schlünder (1978) explained this by the error introduced by assuming uniform radial flow velocity in the deviation of h_f from the experimental data and derived an asymptotic relation valid when $\text{Pe}_h \to 0$.

The fluid-solid heat transfer coefficient derived from experiments in packed beds heated from the wall is really a conglomerate resulting from several mechanisms. Three contributions are considered by Balakrishnan and Pei (1979): the true fluid to solid heat transfer coefficient, the particle-to-particle heat transfer coefficient, and the coefficient for heat transfer through the solid. Using microwaves to heat the solid enabled Bhattacharyya and Pei (1975) to measure the true fluid to solid heat transfer coefficient. As mentioned already, the model equations (38) to (41) do not contain any coupling between particles. Therefore, for fixed beds exchanging heat through the wall, the use of the global h_f, obtained from the correlations (43), (44) or (46) is consistent with the model, although it is not necessarily the most accurate way of dealing with the phenomenon. Interfacial gradients develop when the rate of reaction is high, the heat effect pronounced, and the flow velocity relatively low The most likely gradient is a temperature gradient.

Mears (1971b) derived the following criterion for neglecting interfacial concentration gradients in the presence of an irreversible reaction with order n:

$$\frac{r_A \rho_S d_p}{2Ck_g} < \frac{0.15}{n} \tag{47}$$

For a less than 5% deviation in the rate equation, due to interfacial temperature gradients alone, he derived (1971a)

$$\frac{(-\Delta H) r_A \rho_S d_p}{2h_f T} < 0.15 \frac{R'T}{E} \tag{48}$$

Interfacial gradients are generally not of concern in industrial reactors, since the flow velocity is high. In ammonia synthesis Baddour et al. (1965) simulated a ΔT of 2.3°C between gas and solid at the top of the reactor, where the rate is maximum and 0.4°C at the outlet. In methanol synthesis, Cappelli et al. (1972) calculated a ΔT of maximum 1.5°C. In phthalic anhydride synthesis. Froment and Bischoff (1979) calculated a ΔT of maximum 3°C for normal operating conditions and in steam reforming of natural gas, De Deken et al. (1982) obtained less than 4°C. There are exceptions, however, for example, when a component of the catalyst itself is involved in the reaction, as in catalyst reoxidation or catalyst regeneration by burning off the coke formed by side reactions.

Example of Application: An illustration of the need for heterogeneous models with interfacial gradients is provided by Hatcher et al. (1978). The case dealt with is the reoxidation of a nickel catalyst in the secondary reformer of an ammonia plant, prior to opening the reactor for maintenance. During normal operation the gases coming from the primary reformer are fed at 1100°C and exit at 900°C. The reactor is adiabatic. When it has to be opened, the first step in stabilizing the catalyst is to circulate steam through the bed until the exit has cooled down to 250°C. During the cooling period the steam strips hydrogen from the catalyst support, but it does not remove hydrogen adsorbed on the nickel surface. At this point, nitrogen, containing small amounts of oxygen only, is fed to the reactor. The oxygen reacts with the adsorbed hydrogen and also partially oxidizes the nickel catalyst. Both reactions are very exothermic and a thermal wave travels through the bed. The reactor simulation equations have to account for the non-steady-state nature of the operation. The continuity equation for the oxygen in the gas phase may be written

$$\frac{GR'T}{M_m p_t a_v k_g}\left(\frac{\partial p_A}{\partial z} + \frac{\varepsilon \rho_g}{G}\frac{\partial p_A}{\partial t}\right) = p_{A_s} - p_A \tag{49}$$

The energy equation for the gas phase is

$$\frac{Gc_p}{a_v h_f}\left(\frac{\partial T}{\partial z} + \frac{\varepsilon \rho_g}{G}\frac{\partial T}{\partial t}\right) = T_s - T \tag{50}$$

The continuity equation for oxygen in the solid phase is

$$\frac{k_g}{R'T}(p_A - p_{A_s}) = \frac{d_p}{6}\rho_s\left[\frac{1}{2} r_{H_2} + \frac{1}{2} r_{Ni}\right] \tag{51}$$

where the term in brackets is simply r_{O_2}.

Since the rate of hydrogen oxidation is instantaneous and the phenomenon is therefore mass-transfer limited, the oxygen present on the surface is essentially zero as long as there is hydrogen on the surface, so that

$$\frac{k_g}{R'T} p_A = \frac{d_p}{12}\rho_s \frac{\partial C_{H_2}}{\partial t} \tag{52}$$

with C_{H_2} is the concentration of adsorbed hydrogen, in kmol/kg cat. The energy equation for the solid phase is written

$$h_f(T_s - T) + \frac{d_p}{6} c_s \rho_s \frac{\partial T_s}{\partial t} = \frac{d_p}{6}\rho_s[(-\Delta H_{Ni})r_{Ni} + (-\Delta H_{H_2})r_{H_2}] \tag{53}$$

The initial and boundary conditions are

$$t = 0,\ z \geqslant 0:\quad T_s = T = T_0,\ p_{A_s} = p_A = 0,\ C_{H_2} = C^\circ_{H_2},\ C_{Ni} = C^\circ_{Ni}$$

$$t \neq 0,\ z = 0:\quad p_A = p_{A_0},\ T = T_0$$

The rate equation for nickel reoxidation, derived from bench-scale experimentation, was written as

$$r_{Ni} = \frac{k'K_A p_A C_{Ni}^2}{1 + K_A p_A} \tag{54}$$

with r_{Ni} in kmol/kg cat·h, C_{Ni} in kmol/kg catalyst, and k' in kmol/kg catalyst·h. The simulation was carried out for a reactor with a diameter of 1.8 m and a length of 2.5 m, operated adiabatically. The reoxidation gas was fed at 250°C and at an absolute pressure of 1.5 bar. The oxygen partial pressure in the feed was 0.075 atm. Its mass flow velocity amounted to 1342.8 kg/m^2·h. The equivalent catalyst diameter was 0.014 m and the catalyst bed density 950 kg/m^3 reactor. The initial Ni concentration was 0.00624 kmol/kg cat, and the hydrogen concentration, 0.00093 kmol/kg cat. The set of equations was integrated along the characteristics by means of a Runge-Kutta-Gill routine. The results are represented in the Figs. 11 and 12.

The simulation illustrates the combined effect of too high an oxygen content and too low a flow velocity, which led to the fusion of the catalyst from a bed depth of 1.46 m onward. This requires a temperature exceeding 1400°C. Figure 11 shows how the simulation predicts a temperature of 1400°C, reached between 1.2 and 1.3 m, in excellent agreement with the observation. Figure 12 illustrates that the difference between gas and solid temperatures can reach 100°C. Clearly, a reactor model accounting for interfacial gradients is required here. Notice from Fig. 12 how the solid temperature exceeds the gas temperature at positions already passed by the thermal wave and how the difference rapidly increases on the back slope of the wave. Near the peak of the wave the difference becomes smaller. In front of the wave the gas temperature exceeds the catalyst temperature. It is also observed that after a certain time the wave travels at a nearly constant rate. Analogous equations are used to simulate the regeneration of coked catalysts and similar results are obtained (Olson et al., 1968).

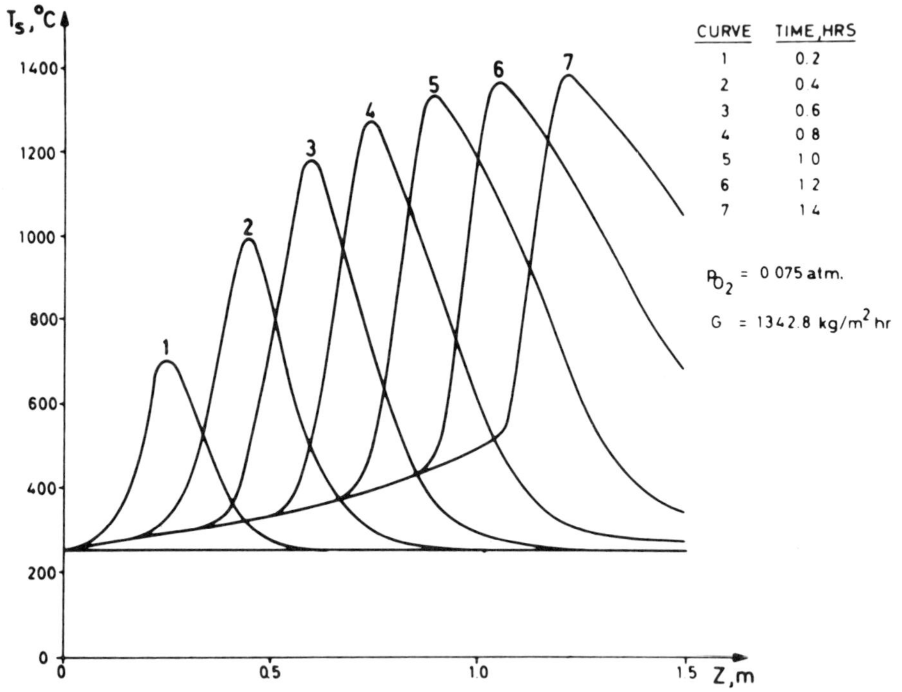

(a)

FIGURE 11 Reoxidation of a steam reforming catalyst in an adiabatic secondary reformer, simulated by means of the heterogeneous one-dimensional model with interfacial gradients. (a) Solid temperature profiles; (b) Nickel and hydrogen concentration profiles. (After Hatcher et al., 1978.)

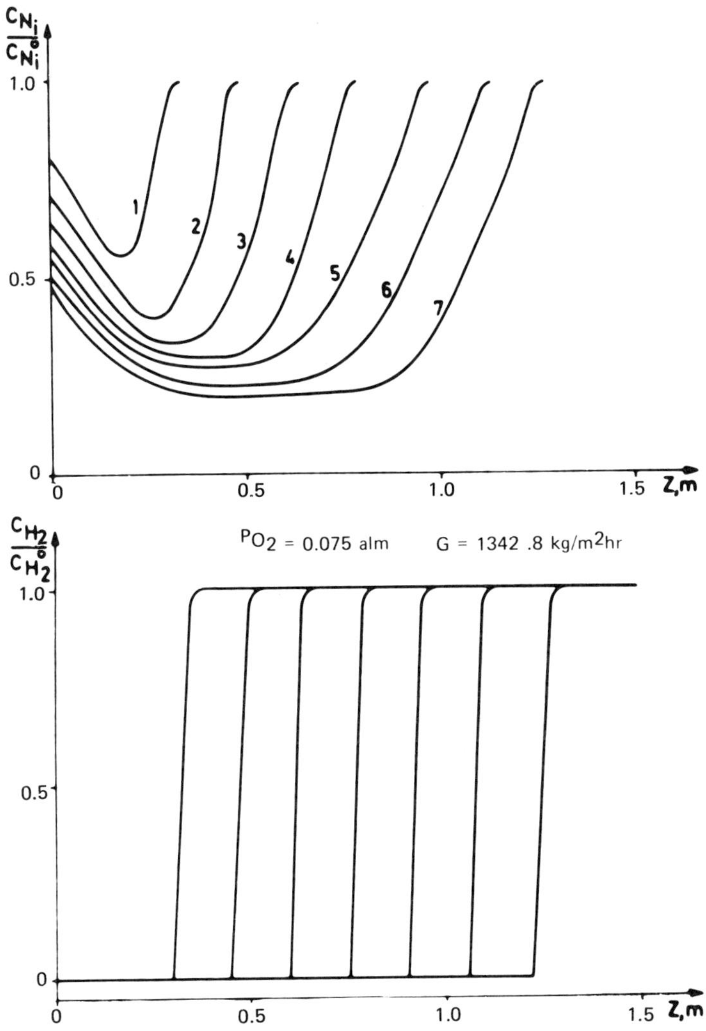

(b)

FIGURE 11 (Continued)

Multiplicity of Steady States: Ever since the work of Wicke (1960), Liu and Amundson (1962) and Liu et al. (1963), there has been intensive academic interest in multiplicity of steady states generated by the existence of gradients at the fluid solid interface in exothermic processes. The purpose of the present section is to describe this phenomenon in simple terms and to evaluate whether it is likely to occur or not in industrial tubular reactions.

Consider a single reaction and a catalyst particle with uniform temperature T_s and concentration C_s of a key component. The bulk fluid surrounding it is at a temperature T and concentration C. At steady state the heat

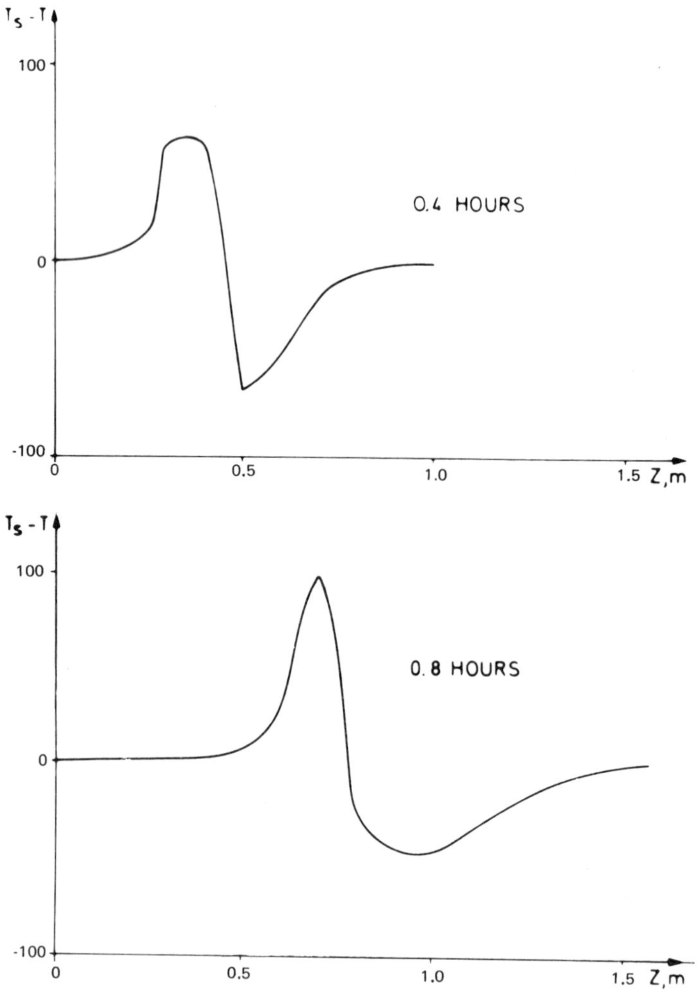

FIGURE 12 Simulated temperature difference between solid and gas in the reoxidation of a steam reforming catalyst. (After Hatcher et al., 1978.)

generated by reaction $Q_R = (-\Delta H) r_A \rho_B$ has to equal the heat transferred from the particle to the bulk fluid, $Q_T = h_f a_v (T_s - T)$. It can be seen from Fig. 13 that Q_R has a sigmoidal shape when plotted versus T_s, whereas Q_T leads to a straight line. When the intersection of Q_r and Q_T occurs in A, the corresponding values of T_s are low, whereas intersection in C requires a higher fluid temperature and leads to a high temperature for the solid. For a certain range of T, three intersections are possible, the outer two corresponding to stable, the intermediate to unstable operation. Which one of the two stable steady states is selected by the system depends on the fluid temperature and on the previous history.

Suppose that the system operates in a point on branch A. As soon as the gas temperature is increased above a value leading to the solid temperature $(T_s)_1$, the latter will jump to a value $(T_s)_2$. For the history just

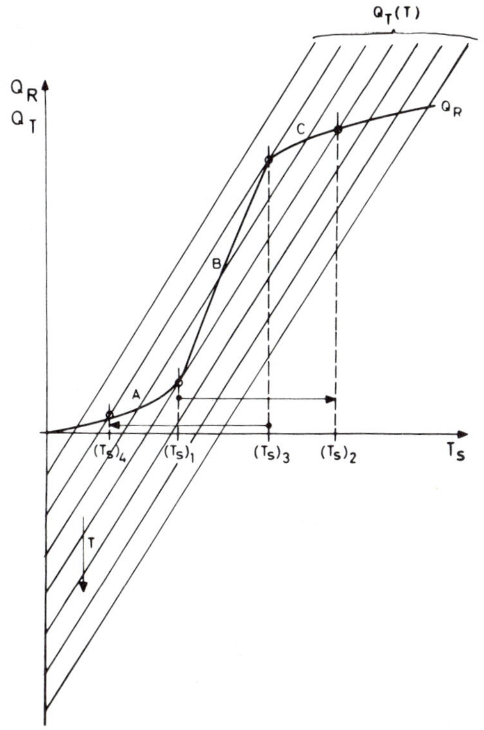

FIGURE 13 Heat production and transfer rates for various gas and catalyst temperatures.

described, the points in the range $(T_S)_1$, $(T_S)_2$ do not correspond to a steady state for T_S. If the particle is initially at $(T_S)_2$ and its temperature is slightly lowered below $(T_S)_3$, the temperature will drop to $(T_S)_4$ and extinction will occur. With this history the range $(T_S)_3$ to $(T_S)_4$ does not correspond to a steady state for the solid temperature. The hysteresis involved in the description above is shown in Fig. 14.

Extending this reasoning from a single particle to a fixed-bed reactor leads to the conclusion that the concentration and temperature profile in a reactor—even without feed-effluent heat exchange—would not only depend on the feed conditions and the wall temperature, but also on the temperature profile at the time the feed is initially admitted to the reactor.

Figure 15 shows simulated temperature profiles in an adiabatic reactor with interfacial gradients (Liu and Amundson, 1962a,b). Figure 15a relates to a situation in which multiplicity is not possible and in which runaway would result solely from parametric sensitivity. Upon removal of the perturbation causing the runaway, the original profile would be recovered. Any intermediate temperature profile would be possible with an appropriate selection of operating conditions. Figure 15b, on the other hand, relates to multiplicity of steady states and true instability. Notice the much higher feed concentration required to achieve this. In case A the initial bed temperature is lower than 393°C. Up to 0.27 m the particles are in the lower steady state. The temperature then rises very steeply over a few

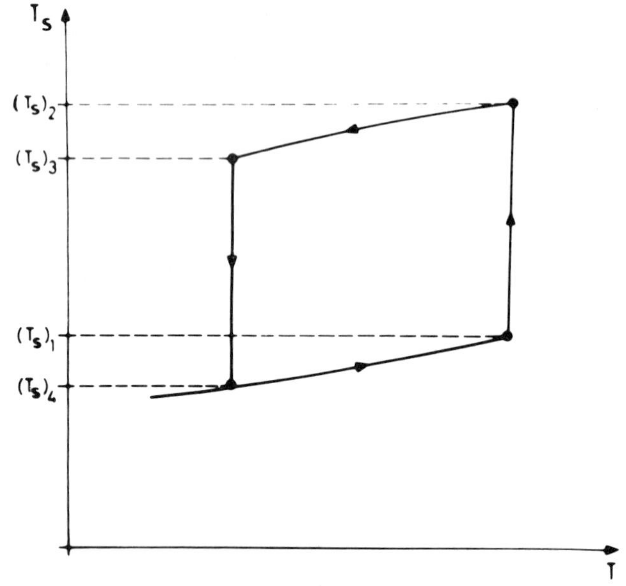

FIGURE 14 Temperature hysteresis in catalyst temperature upon variations in bulk gas temperature for the case of multiple steady state.

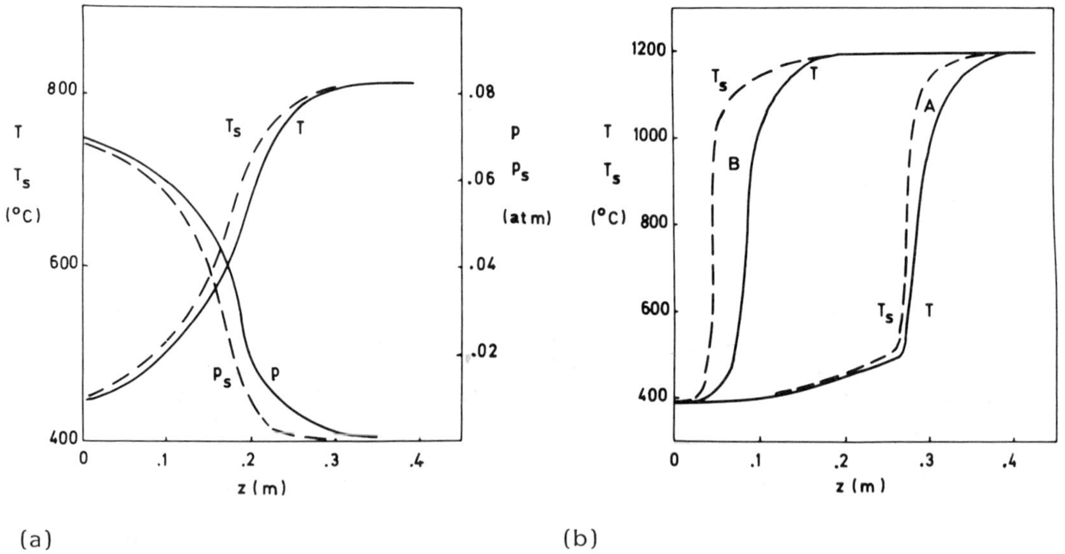

FIGURE 15 Simulated temperature profiles in an adiabatic tubular reactor with interfacial gradients for a first-order reaction. (a) Nonunique steady-state case: $p_0 = 0.15$ atm, $T_0 = 393°C$; initial T_s for A: 393°C; for B: 560°C. (b) Unique steady state; $p_0 = 0.007$ atm, $T_0 = 449°C$. (After Lee and Amundson, 1962.)

layers of catalyst to reach the upper steady state. In case B the initial bed temperature is 560°C. The feed is rapidly heated from 393°C onward and it takes only a few centimeters to ignite the reactor to the upper steady state.

Experimentally, the characteristics associated with Fig. 15 would have to be checked to distinguish between the causes of runaway: parametric sensitivity or multiplicity of steady states. Similar, although less drastic behavior would be calculated for orders higher than one. For the more general Hougen-Watson-type rate equations, which account for adsorption of the reacting components, the results may differ in some aspects (Cardoso and Luss, 1969).

The analysis of the multiplicity of steady states caused by interfacial gradients proceeds along the lines described earlier. A priori criteria for testing multiplicity associated with interfacial gradients were reviewed by Luss (1976). Briefly, for exothermic reactions uniqueness is guaranteed in a cross section in the reactor with bulk conditions (C,T) when

$$\frac{\beta_f \gamma}{1 + \beta_f} \leqslant f(\beta_f, n)$$

For orders $0 \leqslant n \leqslant 2$, the function $f(\beta_f, n)$ is bounded between 1 (for $n = 0$ and $\beta_f = 0$) and 5.82 (for $n = 2$ and $\beta_f = 0$). For large β_f the function value asymptotically approaches 4, for all n. For n = 1 the functions $f(\beta_f, n)$ equals 4, for all β_f. When $\beta_f \gamma / 1 + \beta_f$ exceeds the value of $f(\beta_f, n)$, multiplicity is possible for some values of the dimensionless parameter

$$\alpha = \frac{V_p k(T) \rho_s C^{n-1}}{S_p k_g \beta_f^{n-1}}$$

(Van den Bosch and Luss, 1977).

Recently, Tsotsis et al. (1982) presented criteria for Hougen-Watson-type rate equations. In this case multiplicity may also occur with endothermic reactions and for isothermal situations ($\beta_f = 0$). In ammonia, methanol, and oxo synthesis, values of $\beta_f \gamma / 1 + \beta_f$ exceeding 10 may be encountered. Yet for realistic operating conditions, ΔT of 2.5°C and less were simulated by Baddour et al. (1965) and Capelli et al. (1972) in ammonia and methanol synthesis. Such low ΔT values cannot generate multiplicity.

It has already been mentioned that the model [Eqs. (38) to (41)] does not contain any axial coupling between the particles, a serious simplification when steep gradients occur over a few layers of catalyst particles only. Eigenberger (1972a,b) extended the model to account for this and found a significantly different behavior in the range of multiplicity. The backward flow of heat prevents some parts of the reactor to be in the lower steady state and others in the high steady state, which would be possible if there were no coupling. If the reactor is in the high steady state at the exit, it will be so at the inlet, too, so that the temperature peak is right at the entrance. Vanderveen et al. (1968) analyzed such a situation by means of a cell model.

Plug-Flow Model with Interfacial and Intraparticle Gradients

Model Equations and Parameters: Many industrial processes make use of catalysts and operating conditions leading to intraparticle gradients. In the present section a rather general model will be given first, before entering into the possibility of simplifying it. The equations can be written accounting for the already mentioned assumptions:

Fluid phase:

$$-u_s \frac{dC}{dz} = k_g a_v (C - C_s^s) \quad (55)$$

$$u_s \rho_g c_p \frac{dT}{dz} + 4 \frac{\alpha_i}{d_t}(T - T_w) = h_f a_v (T_s^s - T) \quad (56)$$

Solid phase, with spherical particles:

$$\frac{D_e}{\xi^2} \frac{d}{d\xi}\left(\xi^2 \frac{dC_s}{d\xi}\right) - \rho_s r_A(C_s, T_s) = 0 \quad (57)$$

$$\frac{\lambda_e}{\xi^2} \frac{d}{d\xi}\left(\xi^2 \frac{dT_s}{d\xi}\right) + \rho_s(-\Delta H) r_A(C_s, T_s) = 0 \quad (58)$$

with boundary conditions

$$z = 0: \quad C = C_0, \; T = T_0$$

$$\xi = 0: \quad \frac{dC_s}{d\xi} = \frac{dT_s}{d\xi} = 0$$

$$\xi = \frac{d_p}{2}: \quad k_g(C_s^s - C) = -D_e \frac{dC_s}{d\xi} \quad (59)$$

$$h_f(T_s^s - T) = -\lambda_e \frac{dT_s}{d\xi}$$

The effective diffusivity D_e is usually related to the molecular and Knudsen diffusion inside the pores of the catalyst by means of the Bosanquet relation:

$$D_e = \frac{\varepsilon_s}{\tau(1/D_m + 1/D_K)} \quad (60)$$

which is strictly speaking valid for equimolar countercurrent diffusion in a binary mixture only.

Bulk flow due to a pressure gradient or surface diffusion can generally be neglected. The internal void fraction of the catalyst, ε_s, enters into this formula because the diffusion evidently occurs in the pores only and the tortuosity, τ, because these are randomly oriented (i.e., not along the normal from the surface to the center). D_e can be measured using the steady-state Wicke-Kallenbach method (1941) or the transient gas chromatographic technique, which also accounts for the contribution of dead and pores (Van Deemter et al., 1956).

Probst and Wohlfahrt (1979) presented empirical methods for estimating D_e. Several pore structure models have been proposed for calculating D_e (or τ), since reliable correlations are available for D_m and D_K and since ε_s is easily measured. No such models are available for λ_e, except that of Butt (1965), based on the simple pore model of Wakao and Smith (1962), but temperature effects inside the particle are generally negligible when the reaction does not involve any component of the solid. The numerical effort introduced by the integration of the system of second-order (non-linear) differential equations for the solid phase in each of the nodes used in the integration of the fluid-phase equations is substantial, in particular for complex reactions, even with modern collocation methods.

It has been shown, both by calculations and experimentally, that intraparticle temperature gradients are generally negligible, so that Eq. (58) can be dropped. Also, when there are external gradients, these are generally only temperature gradients. The system of equations then reduces to:

Fluid phase:

$$-u_s \frac{dC}{dz} = -D_e a_v \left.\frac{dC}{d\xi}\right|_{\xi=d_p/2} \tag{61}$$

$$u_s \rho_g c_p \frac{dT}{dz} + 4\frac{\alpha_i}{d_t}(T - T_w) = h_f a_v (T_s^s - T) \tag{62}$$

Solid phase:

$$\frac{D_e}{\xi^2}\frac{d}{d\xi}\left(\xi^2 \frac{dC_s}{d\xi}\right) - \rho_s r_A(C_s, T_s) = 0 \tag{63}$$

with boundary conditions for the solid:

$$\text{At } \xi = \frac{d_p}{2}: \quad -D_e \frac{dC}{d\xi} = \rho_s r_A(C_s^s, T_s^s)$$

$$h_f(T_s^s - T) = -(-\Delta H) D_e \frac{dC}{d\xi} \tag{64}$$

$$\text{At } \xi = 0: \quad \frac{dC_s}{d\xi} = 0$$

and the usual boundary conditions for the fluid phase. Clearly, the problem is considerably further simplified for first-order reactions, for which an analytical solution for Eq. (63) is possible. The result is generally cast into the following convenient form:

$$\eta = \frac{1}{\phi}\,\frac{3\phi \coth 3\phi - 1}{3\phi} \tag{65}$$

where η, the effectiveness factor, is the ratio of the actual rate to that in the absence of diffusional limitations, that is, when the surface conditions prevail throughout the particle and

$$\phi = \frac{V_p}{S_p}\sqrt{\frac{k\rho_s}{D_e}}$$

is a modified Thiele modulus, which, like the original Thiele modulus, contains the rate coefficient and the effective diffusivity and therefore varies with position in the reactor whenever there is a temperature profile. The use of ϕ leads to a unique relation between η and ϕ, whatever the shape of the particles, for both low and high values of the modulus and to differences between cylinders and spheres smaller than 15% for intermediate values, as shown in Fig. 16.

The η concept allows replacing Eq. (57) [or Eq. (63)], and when it is included, Eq. (58), by the algebraic equations

$$k_g a_v (C - C_s^S) = \eta \rho_B r_A (C_s^S, T_s^S) \tag{66}$$

$$h_f a_v (T_s^S - T) = \eta(-\Delta H)\rho_B r_A (C_s^S, T_s^S) \tag{67}$$

When the particle is nonisothermal, η may exceed 1. It cannot be obtained analytically, but Eq. (71) can still be used for the sake of uniformity and abbreviation of the equations.

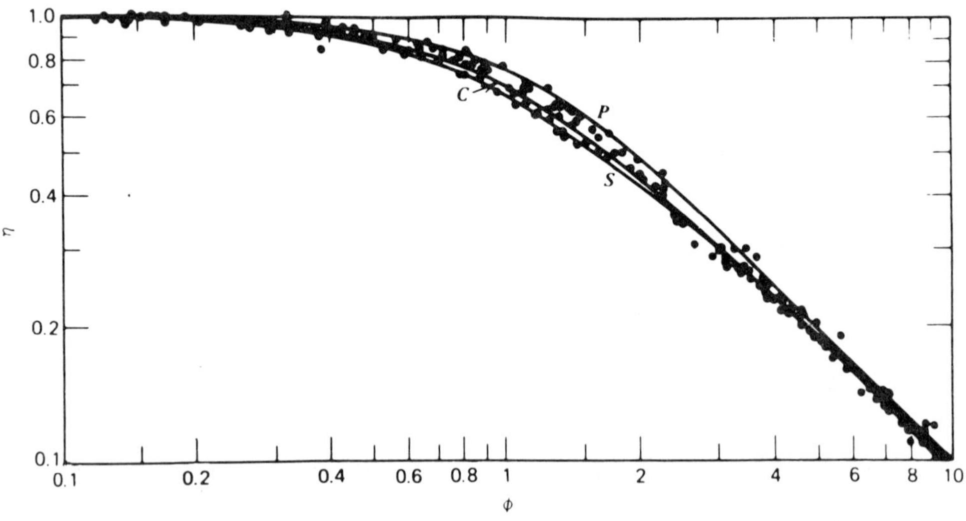

FIGURE 16 Effectiveness factor versus Thiele modulus $\phi = (V_p/S_p)\sqrt{k\rho_s/D_e}$. S, slab; P, sphere; C, cylinder.

In the treatment above, the rate r_A is taken at surface condition (C_s^s, T_s^s). Sometimes it is referred to the bulk-phase conditions, $r_A(C,T)$. In other words, the external concentration gradient is included into the η value. The analytical expression for the isothermal effectiveness factor for first-order reaction then becomes

$$\eta_G = \frac{3\ \mathrm{Sh}}{\phi^2} \frac{\phi \cosh \phi - \sinh \phi}{\phi \cosh \phi + (\mathrm{Sh} - 1)\sinh \phi} \tag{68}$$

where $\mathrm{Sh} = k_g \delta / D_e$* and δ is the film thickness.

The use of η_G further reduces the system Eqs. (55), (56), (66), and (67) into

$$-u_s \frac{dC}{dz} = \eta_G \rho_B r_A(C,T) \tag{69}$$

$$u_s \rho_g c_p \frac{dT}{dz} + 4 \frac{\alpha_i}{d_t}(T - T_w) = \eta_G \rho_B r_A(C,T) \tag{70}$$

The structure of these model equations is identical to that of the basic pseudohomogeneous model, AI .

For orders different from one, η and η_G cannot be obtained analytically. In those cases Eqs. (71) and (74) can still be used, but the η concept does not introduce any gain any more, since obtaining η numerically in each integration node is not any different from solving Eqs. (57) and (58). This is why approximate solutions for the relation η versus ϕ have been worked out by Aris (1965), Bischoff (1965), and Petersen (1965b), based on the introduction of

$$\phi = \frac{V_p r_A(C_s^s, T_s^s)\rho_s}{\sqrt{2}\ S_p} \left[\int_{C_{s_0}}^{C_s^s} D_e r_A(C_s^s, T_s^s)\rho_s \, dC \right]^{-1/2} \tag{71}$$

The use of this "generalized modulus" extends the application of Eq. (65) to any shape and to any type of rate equation, even the Hougen-Watson form, but still only for a single reaction. Equation (71) also shows that for orders different from one, the surface concentration enters into the modulus, so that η is a point value, varying in the reactor, even when the latter is isothermal. Evidently, the use of η does not yield the concentration profile inside the particles, but this is generally not of concern in the reactor design stage. When it is of importance to have information on intraparticle gradients—for example, when coke or/and poison are

*The ratio of a convective coefficient to an internal diffusivity or conductivity is more properly termed a Biot number for mass and heat.

deposited inside the catalyst and an insight is required into the relation between the catalyst structure and penetration depth of coke or poison, as exemplified by work of Summers and Hegedus and co-workers in automobile exhaust purification (Summers and Hegedus, 1979)—the required calculations can be disconnected from the reactor design and can be carried out for surface values generated in the reactor simulation. Various criteria have been derived for evaluating, on the basis of observables, whether or not it is necessary to account for intraparticle gradients. A rather extensive set of proposed criteria is given by Butt (1980). A few only are given here. For the absence of concentration gradients in an isothermal particle—in other words, to have a value of $\eta > 0.95$—Weisz and Prater (1959) proposed

$$\frac{r_A \rho_s d_p^2}{4D_e C_s^s} < \begin{cases} 6.0 & \text{for zero-order reactions} \\ 0.6 & \text{for first-order reactions} \\ 0.3 & \text{for second-order reactions} \end{cases} \tag{72}$$

in which r_A is the observed rate.

For first-order reactions, this criterion is slightly more conservative than that previously proposed by Weisz and Prater (1953) and requiring the left-hand side to be smaller than 1. For the absence of temperature gradients (i.e., the observed rate differs by no more than 5% from the rate prevailing in an isothermal particle), Anderson (1963) derived

$$\frac{|\Delta H| r_A \rho_s d_p^2}{4\lambda_e T_s^s} < \frac{T_s^s R'}{E} \tag{73}$$

or

$$\frac{r_A \rho_s d_p^2}{D_e C_s^s} < \frac{1}{\beta_p \gamma}$$

with β_p and γ taken at surface conditions C_s^s and T_s^s. For the absence of combined intraparticle and interfacial gradients ($\eta = 1 \pm 0.05$), Mears (1971a,b) proposed

$$\frac{r_A \rho_s d_p^2}{4D_e C} < \frac{1 + 0.33\gamma[(-\Delta H) r_A \rho_s d_p / 2h_f T]}{n - \beta_p \gamma[1 + 0.33n(r_A \rho_s d_p / 2k_g C)]} \tag{74}$$

with β_p and γ evaluated at bulk gas conditions, C and T.

Multiplicity of Steady States: In recent years great attention has been paid to the mathematical features of the solid-phase equations (57) and (58). For certain ranges of the parameter values, more than one steady-state profile would be possible for given geometry and operating conditions. Again the approach used to derive criteria for uniqueness or multiplicity in a cross section of the bed is the one mentioned earlier. The condition for uniqueness arrived at by Luss (1977) and by Van der Bosch and Luss (1977) for a first-order irreversible reaction as regards a particle with concentration and temperature gradients and a particle having internal concentration gradients and interfacial concentration and temperature gradients can be written in the same form, $\beta\gamma/(1 + \beta) < 4$, with $\beta = \beta_p$ in the first and $\beta = \varepsilon_s \beta_f$ in the second case. If this condition is violated, multiplicity may exist for a certain range of values of the Thiele modulus in the first case and of the Sherwood number, $\varepsilon_s(k_g/D_e)(V_p/S_p)$, in the second case. In the second case no simple analytical expressions can be given for these ranges. Internal concentration gradients reduce the region in parameter space for which multiplicity of steady states is possible, as compared to the situation in which only interfacial gradients occur. In isothermal particles, multiplicity would still be possible with rate equations that are not monotonic with respect to the conversion (e.g, Hougen-Watson-type rate equations with very high values for the adsorption coefficients of one of the reactants).

The rate equations for air oxidation of CO or H_2 on platinum/alumina are examples of this. Multiplicity has indeed been observed in laboratory studies of these reactions (Smith, 1975; Hegedus et al., 1977). Whether this is really caused by the interaction between reaction and interfacial and intraparticle gradients or by some different phenomenon has not been settled unequivocally, however. According to Keil and Wicke (1980), the multiplicity observed in their work on CO oxidation is kinetically induced (i.e., would result from an interaction between reaction and adsorption). Eigenberger (1976, 1978) has modeled this interpretation and its consequences. A realistic review of the area of multiplicity of steady states has been presented by Schmitz (1975). It should be stressed that multiplicity has not been reported yet for industrial operation of tubular reactors without feedback.

Examples of Application: There are very few examples of design studies which explicitly include internal diffusion limitations into the model. The first reason for this is that very few kinetic studies of catalytic processes went into sufficient depth to permit a quantitative expression for the intraparticle diffusional limitations to be derived. The second reason is the computational effort involved in the integration of the full set of equations (55) to (58). What has generally been done is to use an "effective" rate equation, based on bulk gas phase conditions only and assumed to account implicitly for these effects, so that the reactor model is reduced to the pseudohomogeneous base case AI. The accuracy of such an "effective"—in reality distorted—rate equation may be seriously questioned, however.

In an optimization study of 1-butene dehydrogenation on a chromia-alumina catalyst, requiring a large number of reactor simulations, Dumez and Froment (1976, 1977) circumvented the necessity of integrating the solid-phase equations (57) and (58) in a far more rigorous way. Having determined actual reaction rates, r_E, for various particle sizes, they plotted these versus the true, intrinsic rates, r_H, obtained with finely crushed catalyst under corresponding bulk gas phase compositions. A unique curve was derived which could be expressed as

$$r_E = 0.0034 + \sqrt{0.00685(r_H + 0.0017)} \qquad (75)$$

When both sides of this equation are divided by r_H, the resulting equation asymptotically behaves like the relation between η and ϕ [Eq. (65)], since

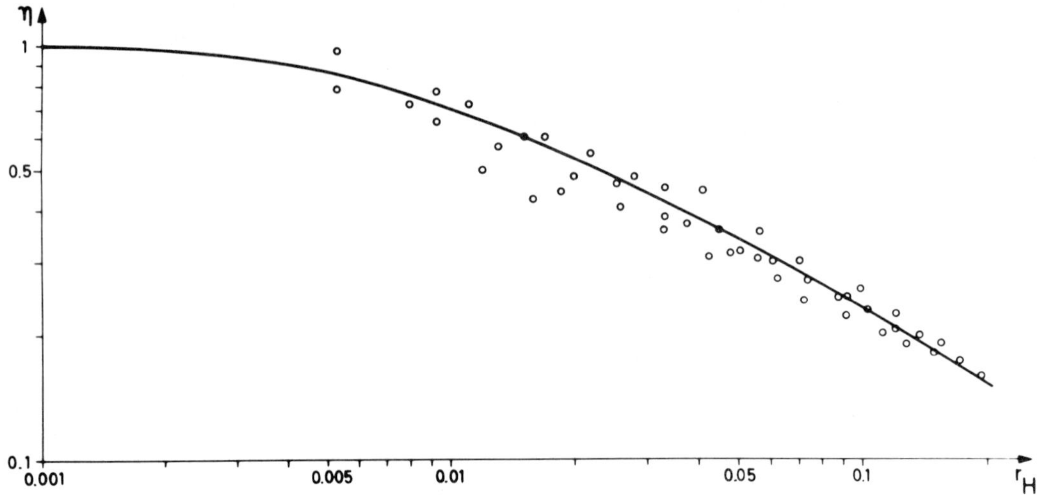

FIGURE 17 Effectiveness factor versus intrinsic rate of 1-butene dehydrogenation. (After Dumez and Froment, 1977.)

r_E tends toward r_H as the latter goes to zero and becomes proportional to $\sqrt{r_H}$ as r_H tends to infinity. The relation $\eta = r_E/r_H$ versus r_H is shown in Fig. 17.

Dumez and Froment (1976) also considered the full set of equations (55) to (58). They derived a value of 5 for τ from rate measurements at various sizes and the knowledge of the intrinsic rate. With D_m and D_K calculated from published correlations and ε_s determined by physical measurements, D_e was calculated from Eq. (60). The model equations (57) and (58) were integrated in each node used in the integration of the fluid field equation by collocation. In their work on steam reforming of natural gas on a nickel-alumina catalyst, De Deken et al. (1982) obtained a τ of 4.5 from the gas chromatographic technique, using finely crushed catalyst as a column material. The effectiveness factor was very low with this very active catalyst: on the order of 0.02 to 0.04. The reaction was confined to a very thin layer near the surface, but nevertheless still representing an internal area of about 1000 times the external surface area.

The model used in the reactor simulation considered interfacial temperature and intraparticle concentration gradients, leading to equations (61) to (63). Two continuity equations were required to define the composition of the reaction mixture. The rate equations were of the Hougen-Watson type and the generalized modulus concept [Eq. (71)] was used. Strictly speaking, this concept is limited to a single reaction, but an approximate relation could be established between the two independent reactions determining the process. Figure 18 shows the variation of η_{CH_4} and η_{CO} along the reactor for typical operating condition. The discontinuity at a bed depth of 6 m results from the use of a different catalyst size farther downstream. Notice that both η are essentially zero at the inlet, where the reaction rates are extremely high. The ΔT over the external film surrounding the catalyst is very small: 4°C up to 6 m, 3°C from there onward, and 2°C

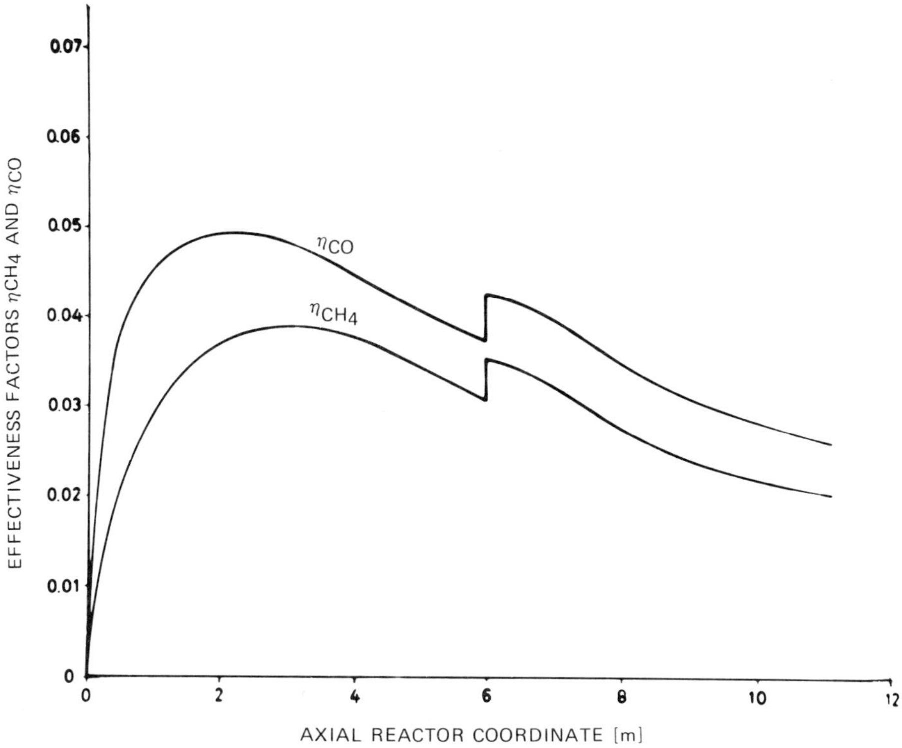

FIGURE 18 Steam reforming of natural gas. Variation of the effectiveness factor for methane conversion and for CO production with bed depth. (After De Deken et al., 1982.)

and less from 10 m onward. Neglecting this ΔT would further simplify the set of model equations without significantly modifying the results.

Two-Dimensional Heterogeneous Model

The model equations, considering axial and radial gradients in the reactor, interfacial and intraparticle concentration, and temperature gradients, can be written, when use is made of the effectiveness factor η, either in its analytical expression or just formally, to abbreviate the system of equations (De Wasch and Froment, 1971):

Fluid phase:

$$(D_{er})_s \left(\frac{\partial^2 C}{\partial r^2} + \frac{1}{r} \frac{\partial C}{\partial r} \right) - u_s \frac{\partial C}{\partial z} = k_g a_v (C - C_s^s) \tag{76}$$

$$\lambda_{er}^f \left(\frac{\partial^2 T}{\partial r^2} + \frac{1}{r} \frac{\partial T}{\partial r} \right) - u_s \rho_g c_p \frac{\partial T}{\partial z} = h_f a_v (T - T_s^s) \tag{77}$$

Solid phase:

$$k_g a_v (C - C_s^s) = \eta \rho_B r_A \tag{78}$$

$$h_f a_v (T_s^s - T) = \eta(-\Delta H)\rho_B r_A + \lambda_{er}^s \left(\frac{\partial^2 T}{\partial r^2} + \frac{1}{r}\frac{\partial T}{\partial r} \right) \tag{79}$$

with boundary conditions

$$C = C_0 \qquad \text{at } z = 0, \text{ all } r$$

$$T = T_0$$

$$\frac{\partial C}{\partial r} = 0 \qquad \text{at } r = 0, \text{ all } z$$

$$\frac{\partial T}{\partial r} = \frac{\partial T_s}{\partial r} = 0$$

$$\frac{\partial C}{\partial r} = 0$$

$$\alpha_w^f (T_w - T) = \lambda_{er}^f \frac{\partial T}{\partial r} \qquad \text{at } r = R, \text{ all } z$$

$$\alpha_w^s (T_w - T_s) = \lambda_{er}^s \frac{\partial T_s}{\partial r}$$

The second term in the right-hand size of Eq. (79) expresses heat transfer in radial direction involving both the solid and fluid phases. The two-phase character of the model is also reflected in the boundary condition for heat transfer at the wall. The model parameters are now D_{er}, λ_{er}^f, λ_{er}^s, α_w^f, and α_w^s.

The effective conductivity and wall heat transfer coefficient concept, used in the two-dimensional pseudohomogeneous model (Yagi and Kunii, 1957; Kunii and Smith, 1960; Zehner and Schlünder, 1970), is not directly applicable here. In this concept the heat transfer mechanisms are lumped according to static and dynamic contributions (Re = 0 and ≠ 0, respectively) and not according to phase, as was done by Singer and Wilhelm (1950). However, Yagi and Kunii (1957) considered the following heat transfer mechanisms, grouped here according to phase:

Fluid phase:

1. Conduction
2. Convection
3. Transfer from fluid to solid
4. Radiation from void to void

Solid phase:

1. Conduction
2. Conduction from particle to particle through contact surfaces
3. Conduction through the stagnant film surrounding the contact surfaces
4. Radiation from particle to particle

Combining these mechanisms in the appropriate way and expressing them in terms of the heat transfer formulas lead to (De Wasch and Froment, 1971)

$$\lambda^f_{er} = \varepsilon(\lambda_g + \beta d_p \alpha_{rv} + \rho_g c_p D_{er}) \tag{80}$$

$$\lambda^s_{er} = \frac{\beta(1 - \varepsilon)}{(\gamma/\lambda_s) + \{1/[(\lambda_g/\phi) + h_f d_p + \alpha_{rs} d_p]\}} \tag{81}$$

where β is a coefficient comprised between 0.9 and 1.0, α_{rv} and α_{rs} are radiation coefficients between voids and solids, respectively, and ϕ can be calculated from a formula presented by Kunii and Smith (1960). Based on the model of Yagi and Kunii (1960) for wall heat transfer, the following relations can be written for α^f_w and α^s_w, provided that T and T_s are not too different:

$$\alpha^f_w = \frac{\lambda^f_{er}}{\lambda_{er}} \alpha_w \tag{82}$$

$$\alpha^s_w = \frac{\lambda^s_{er}}{\lambda_{er}} \alpha_w \tag{83}$$

Dixon and Cresswell (1979) discussed experimental heat transfer correlations and derived relations between the pseudohomogeneous parameters λ_{er} and α_w and the parameters of the heterogeneous model, λ^f_{er}, λ^s_{er}, a^f_w, and a^s_w. De Wasch and Froment (1971) simulated a hydrocarbon oxidation reactor and obtained the radial temperature profiles shown in Fig. 19, for conditions whereby $\eta = 1$. The ΔT between solid and fluid does not exceed 2°C, although the reaction is very exothermic. Figure 20 compares radial mean temperatures as a function of bed depth, calculated by various models discussed in this chapter and for typical industrial operating conditions.

Curves 1 and 2 were obtained by means of the one-dimensional models, pseudohomogeneous (AI) and heterogeneous, but with $\eta = 1$ (BI), curves 3 and 4, by the two-dimensional models AIII and BIII, respectively. Curve 5, which predicts an excessive hot spot, was obtained by means of a model neglecting heat transfer through the solid. For the conditions of the example, some 25% of the radial heat flux occurs through the solid phase.

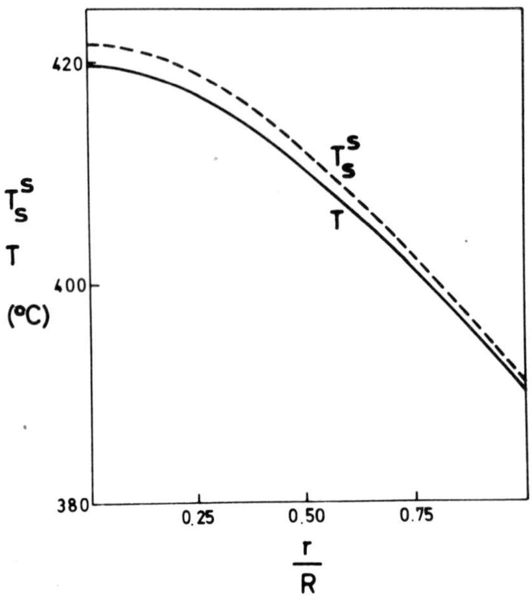

FIGURE 19 Simulation of hydrocarbon oxidation by means of two-dimensional heterogeneous model. Gas and solid radial temperature profiles. (After De Wasch and Froment, 1971.)

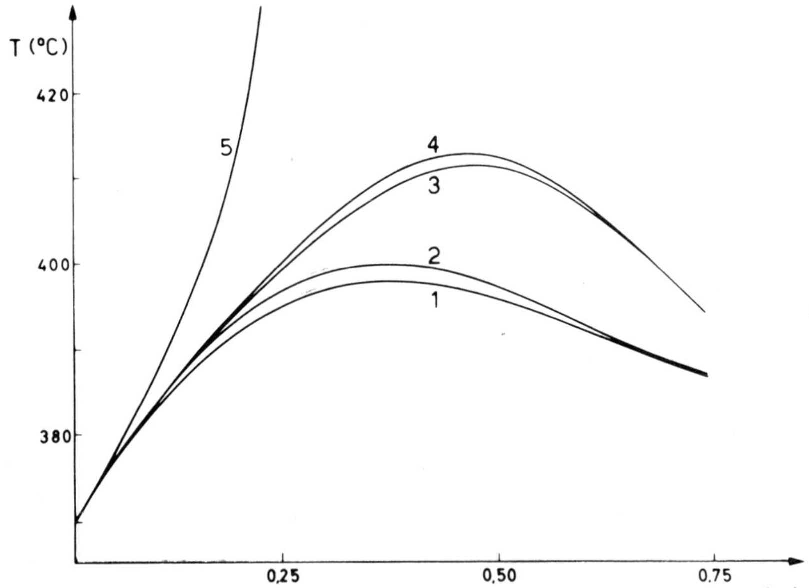

FIGURE 20 Hydrocarbon oxidation. Simulation by means of various models: 1, basic pseudohomogeneous one-dimensional model; 2, one-dimensional heterogeneous model with interfacial gradients; 3, pseudohomogeneous two-dimensional model; 4, two-dimensional heterogeneous model; 5, two-dimensional heterogeneous model with heat transfer in radial direction through gas phase only. (After De Wasch and Froment, 1971.)

The heterogeneous nature of the case dealt with here results from interfacial gradients, but again these were shown to be negligible. De Deken et al. (1982) used a two-dimensional model to simulate a process in which the interfacial gradients could be neglected, but not the intraparticle concentration gradient—the steam reforming dealt with earlier.

Figure 21 shows radial temperature profiles at various bed depths. The maximum difference in process gas temperature between the centerline and the inner tube wall amounts to 33°C, which is not excessive and permits the radially averaged temperature to be simulated accurately by means of the corresponding one-dimensional model with a heat transfer coefficient calculated from Eq. (37).

Note that the model is really a heterogeneous model, but that since there are no interfacial gradients and since the intraparticle concentration gradient is accounted for by means of the effectiveness factor concept, its structure is that of the pseudohomogeneous two-dimensional model AIII,

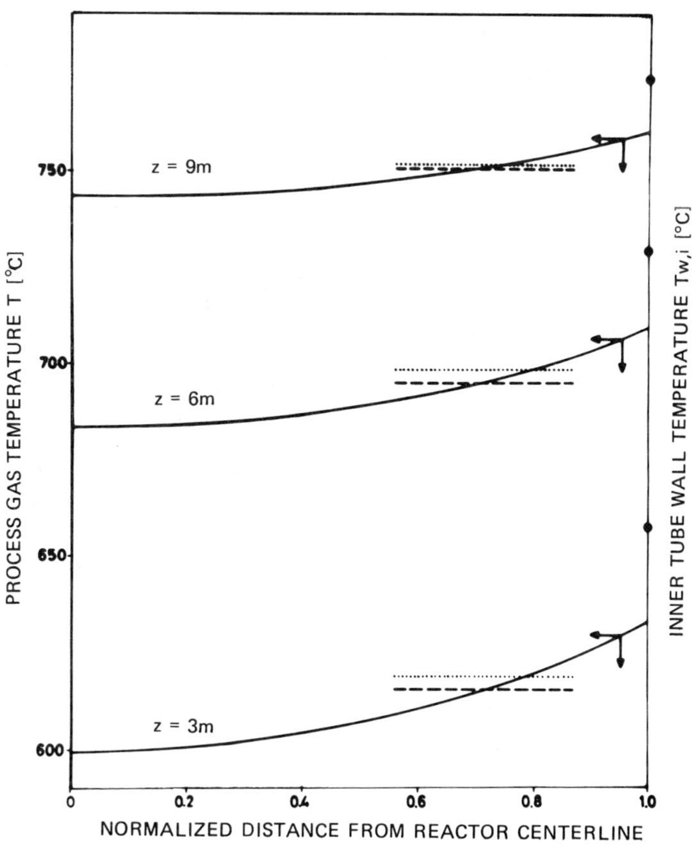

FIGURE 21 Steam reforming of natural gas. Simulation of radial temperature profiles at various bed depths by means of two-dimensional heterogeneous model with intraparticle concentration gradient only. Comparison of radially averaged values with prediction of temperature by means of one-dimensional model with intraparticle concentration gradient. (After De Deken et al., 1982.)

with η multiplying the rate term $r_A \rho_B$. Again because there are no interfacial gradients and because the particles are isothermal, the radial heat transfer parameters are those of the pseudohomogeneous model, λ_{er} and α_w.

SPECIAL CASE: THE MONOLITH CONVERTER

The monolith converter consists of an array of ducts with catalyst coating on the walls. It is used for control of automotive emission and is being considered for waste abatement. The discussion of reactor design and operation in the previous sections was oriented exclusively toward reactors for production processes. The catalytic muffler does not fall in this category, but since there may be a potential use for monolith converters in other areas and since its simulation makes use of reactor models described in this chapter, it is discussed briefly below, mainly from the simulation point of view.

More detailed discussions were published by Wei (1975), who reviewed some of the reaction engineering aspects and early models for CO oxidation. Lee and Aris (1976) also reviewed the features of possible models, whereas Hegedus and coworkers (Hegedus et al., 1980) significantly contributed to the mathematical modeling of the phenomena at the level of the catalyst particle, in particular of instabilities and oscillations. Below, models are given for the comparison of the performance of a monolith converter, along the lines set by Young and Finlayson (1974). They are written, for the sake of generality, for transient operation, which is also the most typical type of operation for the catalytic muffler. Since the transient response is determined mainly by the thermal response of the solid, all other time derivatives are set equal to zero. The reaction considered is CO oxidation. Diffusional limitations in the thin catalytic layer is neglected, although it could—formally, at least—easily be incorporated into the model equations by means of the effectiveness factor η. A cell comprising a cylindrical duct and surrounding solid is chosen. The operation of the cell is adiabatic; that is, the flow rate is the same over all ducts and the solid temperature at any cross section of the monolith is uniform. Axial conduction in the solid is neglected, as is axial diffusion in the fluid since the Péclet number, based on the length of the duct, exceeds 500. The first model dealt with by Young and Finlayson is a one-dimensional heterogeneous model with plug flow and interfacial gradients (BI in the terminology used here). The continuity equation for CO is written, in terms of mol fractions,

$$u \frac{\partial y}{\partial z} = 4 \frac{k_g}{d_t}(y_s - y) \tag{84}$$

and the energy equation for the fluid,

$$u_s \rho_g c_{p_g} \frac{\partial T}{\partial z} = 4 \frac{h_f}{d_t}(T_s - T) \tag{85}$$

At the wall surface, where the reaction occurs (and with the rate of disappearance of CO taken to be positive),

$$k_g \frac{\rho_g}{M_m}(y - y_s) = r_{CO} \tag{86}$$

$$(-\Delta H) r_{CO} = \frac{\rho_s c_{p_s}}{S_s} \frac{\partial T_s}{\partial t} + h_f (T_s - T) \tag{87}$$

where S_s represents the surface area per volume of solid in the monolith cell. The boundary and initial conditions are

$$\begin{aligned} T &= T_0(t) \quad \text{at } z = 0 \\ y &= y_0(t) \\ T_s &= T_{s_i}(z) \quad \text{at } t = 0 \end{aligned} \tag{88}$$

The second model is of the two-dimensional type, but with laminar flow, since the Reynolds number in the duct ranges from 20 to 400. Continuity equation for CO:

$$D_m \frac{1}{r}\frac{\partial}{\partial r}\left(r \frac{\partial y}{\partial r}\right) = u_{ax}\left[1 - \left(\frac{r}{R}\right)^2\right]\frac{\partial y}{\partial z} \tag{89}$$

Energy equation for the fluid:

$$\lambda_g \frac{1}{r}\frac{\partial}{\partial r}\left(r \frac{\partial T}{\partial r}\right) = u_{ax}\left[1 - \left(\frac{r}{R}\right)^2\right]\rho_g c_{p_g} \frac{\partial T}{\partial z} \tag{90}$$

and at the wall surface:

$$- D_m \frac{\rho_g}{M_m} \left.\frac{\partial y}{\partial r}\right|_{r=R} = r_{CO}$$

$$\lambda_g \left.\frac{\partial T}{\partial r}\right|_{r=R} = (-\Delta H) r_{CO} + \frac{\rho_s c_{p_s}}{S_s}\frac{\partial T_s}{\partial t} \tag{91}$$

where r_{CO} is expressed in kmol/h and unit wall surface area. The boundary and initial conditions are

$$\begin{aligned} T_s &= T \quad \text{at } r = R, \text{ all } z \\ T_s &= T_{s_i} \quad \text{at } t = 0 \\ T &= T_0(t) \quad \text{at } z = 0 \\ y &= y_0(t) \end{aligned}$$

The one-dimensional model may have multiple steady states, but the two-dimensional may not. Figure 22 shows that the lower steady-state profiles of T_s and T of the one-dimensional model are obtained when the monolith is heated from an initial low temperature, while the higher steady state profiles are reached when the converter is cooled from some initial high temperature. For intermediate initial temperatures the final steady state may lie between those shown. The unique T_s and T profiles of the two-dimensional model are closer to those corresponding to the lower steady state of the one-dimensional model.

The two-dimensional model is definitely a more realistic model of the monolith, but it is computationally much more demanding. This is why it is of interest to investigate whether or not it is possible to approximate it by means of the one-dimensional model—overlooking for a while the multiplicity of steady states associated with the latter. The curves related to the one-dimensional model in Fig. 22 were calculated for constant values of the Sherwood and Nusselt numbers, $k_g d_t / D_m$ and $h_f d_t / \lambda_g$, respectively.

To come to an exact fit, however, these numbers have to continuously vary along the reactor. The Nu number, for example, would have to be calculated, for each axial position, from the temperature gradient at the wall, generated by the two-dimensional model:

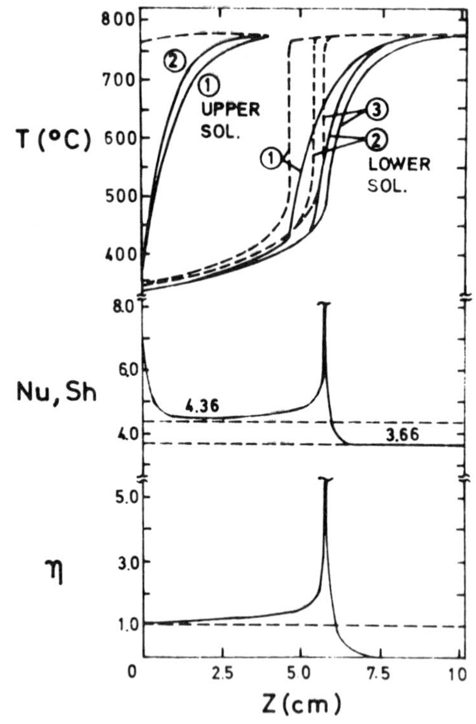

FIGURE 22 One- and two-dimensional simulation of monolith converter. Solid line, one-dimensional heterogeneous model; dashed line, two-dimensional model. (After Young and Finlayson, 1974.)

$$\mathrm{Nu} = \frac{2(\partial T/\partial r)\big|_{r=R}}{T_s - \overline{T}} \tag{92}$$

where $\overline{T}$ is the cup mixing temperature of the fluid. The Sherwood number would be obtained in an analogous way. Figure 22 also shows how Nu and Sh, taken to be identical in this case, vary with axial position. Also shown is the ratio of the reaction rate at wall conditions to that at mixing cup conditions. This ratio varies strongly with position and becomes infinite when the gas reaches the wall temperature, in the reaction zone, after light off.

To avoid the two-dimensional simulation as a prerequisite for one-dimensional calculations, Heck et al. (1976) approximated the Nu profile before the discontinuity by the value for pure heat transfer with constant heat flux and that beyond the discontinuity by the value corresponding to constant wall temperature. Sundaram and Froment (1979, 1980) derived the local Nu for laminar flow and in the absence of reaction from a large number of two-dimensional simulations. They came to the following correlations:

$$\mathrm{Nu} = 3.655 + 0.1450Z^{-0.778}\exp(-7.158Z) \tag{93}$$

The asymptotic value of Nu of 3.655 completely agrees with the experimental value.

For turbulent flow, Whitaker (1972) obtained the following correlation:

$$\mathrm{Nu} = \mathrm{Nu}_\infty + a_1\exp(a_2Z)Z^{0.743} \tag{94}$$

with

$$a_1 = 1.045 \times 10^{-2}\mathrm{Re}^{0.296}$$

$$a_2 = -1.37 \times 10^{2}\mathrm{Re}^{0.992}$$

where Nu_∞ is the asymptotic Nu number (e.g., $2.28 \times 10^{-2}\mathrm{Re}^{0.8}$, in agreement with the Dittus-Boelter correlation). Sundaram and Fromont (1979, 1980) also showed that Nu depends strongly on the properties of the reaction. For endothermic reactions and both laminar and turbulent flow they correlated the Nusselt profile versus the dimensionless adiabatic temperature rise, the Lewis number, and the dimensionless activation energy. The discontinuity in the ratio (92) for exothermic reactions at $T = T_s$ precludes a correlation that would be as accurate as that given for endothermic reactions.

The preceding discussion once again illustrates the importance of a representative model, not only to come to the highest possible accuracy, but also to avoid the prediction of a behavior that may not be real at all. The question of whether or not multiplicity is possible in a monolith converter remains largely unanswered, even if Hlaváček and Votruba (1974) presented experimental evidence for it. This has been explained in terms of axial heat transfer in the wall. It is more likely, however, to result from phenomena at the catalyst surface instead of transport phenomena. Surface phenomena causing multiplicity would be reflected by more than one rate-determining step and rate equations more complicated than those used in the work of Young and Finlayson. Multiplicity would then, of course, also become possible with the two-dimensional model.

CONCLUSION

In this chapter the design or simulation of fixed-bed reactors is based on fundamental models. These combine the hydrodynamic and heat transfer phenomena with the chemical reaction aspects by means of a set of mathematical equations of varying complexity. With the present computational facilities and numerical techniques, the solution of these equations is no longer an obstacle to the application of these models to the real problems and configurations encountered in industrial operation.

Evidently, no model will ever represent the reactor behavior in an absolutely rigorous way. The equations contain a number of parameters, some of which are "effective" in the sense that they lump the effects of phenomena that would otherwise have to be expressed by means of additional equations, to be solved simultaneously in the reactor simulation. Admittedly, considerable improvement and extension of existing correlations is required for severe reaction conditions or unusual reactor configurations. Indeed, for convenience, correlations for these parameters were generally obtained from purely physical experiments, possibly without sufficient testing under conditions similar to those encountered in the presence of reaction. Under reaction conditions the parameter values may then turn out to be inaccurate, even when the underlying description of the phenomenon is correct. But the parameters should not be tuned arbitrarily on the basis of experimental testing of the reactor model, to avoid inconsistency and violation of their physicochemical background.

The models developed here are all of the continuum type. This dispute between the advocates of continuum and cell models is mainly of academic interest, except perhaps for very severe reaction conditions. An escalation of model sophistication is presented and criteria are given for the selection of the appropriate model. Which model is "appropriate" depends in the first place on the reactor, the operating conditions, and the reactions, but also on the required accuracy and on the availability and precision of the required fundamental information. Some of the models presented here are a bare minimum for a realistic description of the reactor operation, others may look too refined and of academic interest only. Nevertheless, it is safe to predict that what may seem to be too refined today may well be the minimum level of modeling tomorrow, particularly for demanding conditions. Models will certainly continue to be refined, but realism should be a guide in this development.

It should never be forgotten, however, that advanced reactor modeling requires first, accurate kinetics, not only on fresh, but also on gradually deactivating catalysts. In the past this has been the main obstacle to the fundamental approach in reactor design. It will continue to be a serious challenge in the future, because it requires a considerable experimental effort of a team with both broad scientific and technical background which is seldom put together at the right time (i.e., at an early stage of development).

NOTATION

a_v	external particle surface area per unit reactor volume, m_p^2/m_r^3
C	gas phase concentration, $kmol/m_g^3$
C_0	concentration of A in feed, $kmol/m^3$
c_p	specific heat of gas, kJ/kg·K
c_{p_s}	specific heat of gas inside the solid, kJ/kg·K
c_s	specific heat of solid, kJ/kg·K
C_s	concentration inside catalyst, $kmol/m_g^3$
C_s^s	surface concentration, $kmol/m_g^3$
d_i	internal diameter of catalyst ring, m_p
d_p	particle diameter, m_p
d_t	internal tube diameter, m_r
D_e	effective diffusivity inside catalyst, $m_g^3/m_p \cdot s$
D_K	Knudsen diffusivity, m^2/s
D_m	molecular diffusivity, m^2/s
$(D_{ea})s$	effective diffusivity in axial bed direction based on superfacial flow velocity, $m_g^3/m_r \cdot s$
$(D_{er})_s$	effective diffusivity in radial bed direction based on superfacial flow velocity, $m_g^3/m_r \cdot s$
E	activation energy, kJ/kmol or $m^3 \cdot atm/kmol \cdot K$
E_i	exponential integral
f	friction factor
F_{A_0}	molar feed rate of component A, kmol·A/s
g	acceleration of gravity, m/s^2
G	superficial mass flow velocity, $kg/m_r^2 s$
G_i	superficial mass flow velocity of component i, kg $i/m_r^2 \cdot s$
h_f	gas-to-solid heat transfer coefficient in a fixed bed, $kJ/m_p^2 \cdot s \cdot K$
k_0	reaction rate coefficient; for first order, e.g., $m^3 g/kg$ cat·s
k_g	gas-to-solid mass transfer coefficient in a fixed bed, $m_g^3/m_p^2 \cdot s$
K	equilibrium constant,
K_A	adsorption constant of component A, 1/atm
L	total length of reactor, m_r
m	mass flow rate, kg/s
M_m	molecular weight of mixture, kg/kmol
n	order of reaction
Nu	Nusselt number $(h_f d_t/\lambda_g)$

p_A	partial pressure of component A, atm
p_t	total pressure, atm
Pe_{hr}	Péclet number for radial effective diffusion, $[d_p u_i / D_{er}]$
Pe_{ma}	Péclet number for axial effective diffusion in fixed bed, $[d_p u_i / (D_{ea})_i]$
Pe'_{ma}	Péclet number for axial effective diffusion in fixed bed, $[d_p / L / (D_{ea})_i]$
Pr	Prandtl number $(c_p \mu / \lambda_g)$
r	radial position, m
r_A	rate of reaction of A per unit catalyst weight, kmol/kg cat·s
r_E	effective rate of dehydrogenation of 1-butene, kmol/kg cat·s
r_H	intrinsic rate of dehydrogenation of 1-butene, kmol/kg cat·s
R	tube radius, m
R'	gas constant, kJ/kmol·K or m^3·atm/kmol·K
Re	Reynolds number $(d_p G / \mu)$
S_{cyl}	external surface area of cylinder with same external diameter and height as the ring-shaped catalyst, m_p^2
S_p	surface area of a particle, m^2
S_s	surface area per volume of solid in monolith cell, m^2/m^3
Sh	Sherwood number $(k_g d_t / D_m)$
t	time, s
T, T_1	process gas temperature, K
T_2	coolant temperature, K
T_0	feed temperature of gas, K
T_s	solid temperature, K
T_s^s	solid surface temperature, K
u_{ax}	velocity in the axis of monolith duct, m/s
u_i	interstitial flow velocity, m_r/s
u_s	superficial flow velocity, m_g^3/m_r^2·s
U	overall heat transfer coefficient, kJ/m^2·s·K
V_{cyl}	volume of cylinder with same external diameter and height as ring-shaped catalyst, m_p^3
V_p	particle volume, m_p^3
x	conversion
y	mole fraction
z, z'	axial distance in reactor (respectively, heat exchanger), m_r
Z, Z'	length of reactor (respectively, heat exchanger), m_r

Greek Letters

α_i	bed-to-wall heat transfer coefficient in one-dimensional model, $kJ/m^2 \cdot s \cdot K$
α_{ij}	stoichiometric coefficient of component i in reaction j
α_{rs}	radiation coefficient between particles, $kJ/m^2 \cdot s \cdot K$
α_{rv}	radiation coefficient between voids, $kJ/m^2 \cdot s \cdot K$
α_w	wall heat transfer coefficient in two-dimensional model, $kJ/m^2 \cdot s \cdot K$
α_w^o	static contribution to α_w, $kJ/m^2 \cdot s \cdot K$
β_f	dimensionless T_{ad} in a film surrounding a particle $[k_g(-\Delta H)C/h_f T]$
β_p	dimensionless T_{ad} in a particle $[D_e C_s^s(-\Delta H)/\lambda_e T_s^s]$
β_r	dimensionless adiabatic temperature rise in a reactor $[(-\Delta H)C_0/\rho_g c_p T_0]$
γ	dimensionless activation energy $(E/RT_0, E/RT, E/RT_s^s)$
$(-\Delta H)$	heat of reaction, kJ/kmol
ΔT_{ad}	adiabatic temperature rise, K
ε	void fraction of bed, m_g^3/m_r^3
ε_s	void fraction of particle, m_g^3/m_p^3
ζ	a group $(M_m L/u_s \rho_g)$, $m_r^3 \cdot s/kmol$
η	catalyst effectiveness factor based on surface conditions (C_s^s, T_s^s)
η_G	catalyst effectiveness factor based on bulk gas conditions (C,T)
λ_e	effective conductivity inside a particle, $kJ/m \cdot s \cdot K$
λ_{ea}	effective axial conductivity of bed + gas, $kJ/m \cdot s \cdot K$
λ_{ea}^o	static contribution to the effective axial conductivity $kJ/m \cdot s \cdot K$
λ_{er}	effective radial conductivity of bed + gas, $kJ/m \cdot s \cdot K$
λ_{er}^o	static contribution to the effective radial conductivity, $kJ/m \cdot s \cdot K$
λ_{er}^f	effective radial conductivity for gas phase, $kJ/m \cdot s \cdot K$
λ_{er}^s	effective radial conductivity for solid phase, $kJ/m \cdot s \cdot K$
λ_g	gas conductivity, $kJ/m \cdot s \cdot K$
μ	dynamic viscosity, $kg/m \cdot s$
ξ	inside particle, oriented toward surface coordinate, m_p
ρ_B	bulk density of bed, kg cat/m_r^3
ρ_g	gas phase density, kg/m_g^3
ρ_s	catalyst density, kg cat/m_p^3
τ	tortuosity
ϕ	Thiele modulus $[= (V_p/S_p)(k/D_e)]$

ψ_a $1/Pe_{ma}$

ψ_r $1/Pe_{mr}$

REFERENCES

Agnew, J. B. and O. E. Potter, The design of stable packed tubular reactors for highly exothermic reactions, Trans. Inst. Chem. Eng., *44*, T216 (1966).

Ampaya, J. P. and R. G. Rinker, Autothermal reactors with internal heat exchange: I. Numerical and experimental studies in the multiple steady-state region, Chem. Eng. Sci., *32*, 1327 (1977a).

Ampaya, J. P. and R. G. Rinker, Design correlations for autothermal reactors with internal countercurrent heat exchange, Ind. Eng. Chem. Process Des. Dev., *16*, 63 (1977b).

Amundson, N. R., Mathematical models of fixed bed reactors, Diskussionstagung Bunsen-Gesellschaft und Dechema, Köningstein, FGR, Ber. Bunsenges., *74* (1970).

Anderson, J. B., A criterion for isothermal behaviour of a catalyst pellet, Chem. Eng. Sci., *18*, 147 (1963).

Aris, R., On shape factors for irregular particles, Chem. Eng. Sci., *6*, 262 (1957).

Aris, R., *The Optimal Design of Chemical Reactors*, Academic Press, New York (1960).

Aris, R., A normalization for the Thiele modulus, Ind. Eng. Chem. Fundam., *4*, 227 (1965).

Aris, R. and N. R. Amundson, Some remarks on longitudinal mixing or diffusion in fixed beds, AIChE J., *3*, 280 (1957).

Baddour, R. F., P. L. T. Brian, B. A. Logeais, and J. P. Eymery, Steady state simulation of ammonia synthesis convertor, Chem. Eng. Sci., *20*, 281 (1965).

Balakotaiah, V. and D. Luss, Effect of flow direction on conversion in isothermal radial flow fixed-bed reactors, AIChE J., *27*, 442 (1981).

Balakrishnan, A. R. and P. C. T. Pei, Heat transfer in gas-solid packed bed systems: 1. A critical review; 2. The conduction mode; 3. Overall heat transfer rates in adiabatic beds, Ind. Eng. Chem. Process Des. Dev., *18*, 31, 40, 47 (1979).

Barkelew, C. H., Stability of chemical reactors, Chem. Eng. Prog. Symp. Ser., *55*(25), 38 (1959).

Bellman, R., *Dynamic Programming*, Princeton University Press, Princeton, N.J. (1957).

Benenati, R. F. and C. B. Brosilow, Void fraction distribution in beds of spheres, AIChE J., *3*, 359 (1962).

Bhattacharyya, D., and D. C. T. Pei, Heat transfer in fixed-bed gas-solid systems, Chem. Eng. Sci., *30*, 293 (1975).

Bilous, O. and N. R. Amundson, Chemical reactor stability and sensitivity, AIChE J., *2*, 117 (1956).

Bischoff, K. B., Axial thermal conductivity in packed beds, Can. J. Chem. Eng., *40*, 161 (1962).

Bischoff, K. B., Effectiveness factors for general reaction rate forms, AIChE J., *11*, 351 (1965).

Brauer, M., Druckverlust in Füllkörpersäulen bei Einphasen-strömung, Chem. Ing. Tech., *29*, 785 (1957).

Butt, J. B., Thermal conductivity of porous catalysts, AIChE J., *11*, 106 (1965).

Butt, J. B., *Reaction Kinetics and Reactor Design*, Prentice-Hall, Englewood Cliffs, N.J. (1980).

Butt, J. B., in *Catalyst Deactivation*, B. Delmon and G. F. Froment, eds., Elsevier, Amsterdam (1982).

Butt, J. B. and R. M. Billimoria, Catalyst deactivation, Chem. React. Eng. Rev.—Houston, ACS Symp. Ser., *72*, D. Luss and V. W. Weekman, eds., ACS, Washington, D.C. (1978), p. 288.

Butt, J. B. and M. S. Weng, Some problems in the operation and deactivation of stability-limited fixed bed catalytic reactors, Chem. React. Eng., Proc. 4th Int. Symp., Heidelberg, 1976, Dechema (1976), p. VII–310.

Calderbank, P. H., A. Caldwell, and G. Ross, The diluted catalyst fixed bed reactor for exothermic catalytic reactions, Proc. 4th Eur. Symp. Chem. React. Eng., Pergamon Press, Elmsford, N.Y. (1968).

Capelli, A., A. Collina, and M. Dente, Mathematical model for simulating behavior of Fauser-Montecatini industrial reactors for methanol synthesis, Ind. Eng. Chem. Process Des. Dev., *11*, 184 (1972).

Carberry, J. J., and D. White, On the role of transport phenomena in catalytic reactor behavior. Digital simulation of naphthalene oxidation over V_2O_5, Ind. Eng. Chem., *61*, 27 (1969).

Cardoso, M. A. A. and D. Luss, Stability of catalytic wires, Chem. Eng. Sci., *24*, 1699 (1969).

Chang, H. C., A non-Fickian model of packed bed reactors, AIChE J., *28*, 208 (1982).

Chartrand, G. and C. M. Crowe, The optimization of the catalytic oxidation of sulphur dioxide in a multi-bed adiabatic reactor, Can. J. Chem. Eng. *47*, 296 (1969).

Crider, J. E. and A. S. Foss, Effective wall heat transfer coefficients and thermal resistances in mathematical models of packed beds, AIChE J., *11*, 1012 (1965).

Deans, H. A. and L. Lapidus, A computation model for predicting and correlating the behavior of fixed-bed reactors: I. Derivation of model for nonreactive systems; II. Extension to chemically reactive systems, AIChE J., *6*, 656, 663 (1960).

Deckwer, W. D., The backflow cell model applied to nonisothermal reactors, Chem. Eng. J., *8*, 135 (1974).

De Deken, J. C., E. F. Devos, and G. F. Froment, Steam reforming of natural gas. Intrinsic kinetics, diffusional influences and reactor design, Proc. ISCRE7, ACS, Washington, D.C. (1982).

Delmon, P. and P. Grange, in *Catalyst Deactivation. Solid State Chemical Phenomena in Aging and Deactivation of Catalysts*, B. Delmon and G. F. Froment, eds., Elsevier, Amsterdam (1980).

Dente, M. and M. Collina, Behavior of tubular reactors with regard to sensitivity, Chim. Int. (Italy), *46*, 752 (1964).

Dente, M., G. Biardi, and E. Ranzi, Mathematical model for tubular gas-solid catalytic reactors, Chem. React. Eng., Proc. 5th. Eur./2nd. Int. Symp. Chem. React. Eng., Amsterdam, Elsevier, Amsterdam, (1972), p. B.5–13.

De Wasch, A. P. and G. F. Froment, A two-dimensional heterogeneous model for fixed bed catalytic reactors, Chem. Eng. Sci., *26*, 629 (1971).

De Wasch, A. P. and G. F. Froment, Heat transfer in packed beds, Chem. Eng. Sci., *27*, 567 (1972).

Dixon, A. G. and D. L. Cresswell, Theoretical prediction of effective heat transfer parameters in packed beds. Reply to comments by Vortmeyer and Berninger on Theoretical Predictions. . ., AIChE J., *25*, 663 (1979); *28*, 511 (1982).

Douglas, J. M. and L. C. Eagleton, Analytical solutions for some adiabatic reactor problems, Ind. Eng. Chem. Fundam., *1*, 116 (1962).

Dumez, F. J. and G. F. Froment, Dehydrogenation of 1-butene into butadiene. Kinetic catalyst coking and reactor design, Ind. Eng. Chem. Process Des. Dev., *15*, 291 (1976).

Dumez, F. J. and G. F. Froment, The problem of intraparticle mass transfer limitations in reactor design. A simplified approach, Chem. Eng. Sci., *32*, 974 (1977).

Eigenberger, G., On the dynamic behavior of the catalytic fixed-bed reactor in the region of multiple steady states: I. The influence of heat conduction in two-phase models; II. The influence of the boundary conditions in the catalyst phase, Chem. Eng. Sci., *27*, 1909, 1917 (1972).

Eigenberger, G., Kinetic instabilities in catalytic reactions. A modeling approach, Chem. React. Eng., Proc. 4th Int. Symp., Heidelberg, 1976, Dechema (1976), p. VII–290.

Eigenberger, G., Kinetic instabilities in heterogeneously catalyzed reactions: I. Rate multiplicity with Langmuir-type kinetics; II. Oscillatory instabilities with Langmuir-type kinetics, Chem. Eng. Sci., *33*, 1255, 1269 (1978).

Ergun, S., Fluid flow through packed columns, Chem. Eng. Progr., *48*(2), 89 (1952).

Fahien, R. W. and J. M. Smith, Mass transfer in packed beds, AIChE J., *1*, 25 (1955).

Finlayson, B. A., Packed bed reactor analysis by orthogonal collocation, Chem. Eng. Sci., *26*, 1081 (1971).

Froment, G. F., Design of fixed bed catalytic reactors based on effective transport models, Chem. Eng. Sci., *7*, 29 (1961).

Froment, G. F., Fixed bed catalytic reactors. Current design status, Ind. Eng. Chem., *59*(2), 18 (1967).

Froment, G. F., Some aspects of the design of fixed bed reactors for hydrocarbon oxidation, Period. Polytech. (Budapest), *15*, 219 (1971).

Froment, G. F., Analysis and design of fixed bed catalytic reactors. A state of the art review, Proc. 1st Int. Symp. Chem. React. Eng., Adv. Chem. Ser., ACS, Washington, D.C. (1972a).

Froment, G. F., Fixed bed reactors—steady state conditions, Proc. 5th Eur. Symp. Chem. React. Eng., Elsevier, Amsterdam (1972b).

Froment, G. F., Fixed bed catalytic reactors. Technological and fundamental aspects, Chem Ing. Tech., *26*, 374 (1974).

Froment, G. F., Model discrimination and parameter estimation in heterogeneous catalysis, AIChE J., *21*, 1041 (1975).

Froment, G. F., Catalyst deactivation by coking, Proc. 6th Int. Congr. Catal., Vol. 1, Chemical Society, London (1976a), p. 10.

Froment, G. F., Kinetic analysis of reactions catalyzed by solids, Chem. React. Eng., Proc. 4th Int. Symp., Heidelberg, 1976, Dechema (1976b).

Froment, G. F., in *Catalyst Deactivation*, B. Delmon and G. F. Froment, eds., Elsevier, Amsterdam (1980a).

Froment, G. F., Hot spots and runaway in fixed bed tubular reactors, in *Chemistry and Chemical Engineering of Catalystic Processes*, NATO Adv. Study Inst. Ser., R. Prins and G. C. A. Schuit, eds., Sijthoff en Noordhoff, Alphen aan den Rijn, The Netherlands (1980b), p. 535.

Froment, G. F., A rigorous formulation of the effect of coke formation on catalyst activity, in *Progress in Catalyst Deactivation*, J. L. Figueiredo, ed., Nato Adv. Study Inst. Ser. E54, Nÿhoff, Dordrecht, The Netherlands (1982).

Froment, G. F. and K. B. Bischoff, *Chemical Reactor Analysis and Design*, Wiley, New York (1979).

Froment, G. F. and L. H. Hosten, *Catalysis, Science and Technology*, Vol. 2, J. R. Anderson and M. Boudart, eds., Springer-Verlag, Berlin (1981).

Garcia-Ochoa, F., M. Leuz, G. Emig, and H. Hofmann,, Reaktions technische Untersüchüng der katalytischen Dehydrierüng von Cyclohexanol zü Cyclohexanon, Chem. Ztg., *105*, 349 (1981).

Gelbin, D., K. H. Radeke, B. Rosahl, and W. Stein, Stoff- und Wärmetransport in Schüttschichte: Teil II. Wärmetransport und axiale Wärmeleitung, Int. J. Heat Transfer, *19*, 987 (1976).

Gnielinski, V., Berechnung des Wärme- und Stoffaustauchs in durchströmte ruhenden Schüttungen, Verfahrenstechnik, *16*, 36 (1982).

Grosjean, C. C. and G. F. Froment, Computation of axial and radial profiles of temperature and conversion in chemical reactors, Meded. Vlaam. Akad., *24*(1) (1962).

Handley, D. and P. J. Heggs, Momentum and heat transfer mechanisms in regular shaped packings, Trans. Inst. Chem. Eng., *46*, T251 (1968).

Hatcher, W., L. Viville, and G. F. Froment, The reoxidation of a nickel reforming catalyst. Kinetic analysis and fixed bed simulation, Ind. Eng. Chem. Process Des. Dev., *17*, 491 (1978).

Haughey, D. P. and S. G. Beveridge, Structural properties of packed beds. A Review, Can. J. Chem. Eng., *47*, 130 (1969).

Hawthorn, R. D., G. H. Ackerman, and A. C. Nixon, A mathematical model of a packed-bed heat-exchanger reactor for dehydrogenation of methylcyclohexane. Comparison of predictions with experimental results, AIChE J., *14*, 69 (1968).

Heck, R. H., J. Wei, and J. R. Katzer, Mathematical modeling of monolith catalysts, AIChE J., *22*, 477 (1976).

Hegedus, L. L., Temperature excursions in monoliths, AIChE J., *21*, 849 (1975).

Hegedus, L. L., S. H. Oh, and K. Baron, Multiple steady states in an isothermal integral reactor: The catalytic oxidation of carbon monoxide over Pt-alumina, AIChE J., *23*, 632 (1977).

Hegedus, L. L., C. C. Chang, D. J. McEwan, and E. M. Sloan, Response of catalyst surface concentrations to forced concentration oscillations in the gas phase. The NO, CO, O_2 system over Pt-alumina, Ind. Eng. Chem. Fundam., *19*, 367 (1980).

Hennecke, F. W. and E. U. Schlünder, Wärmeübergang in beheizten oder gekuhlten Rohren mit Schuttungen aus Kugeln, Zylinder und Raschig-Ringen, Chem. Ing. Tech., *45*, 277 (1973).

Hiby, J. W., Longitudinal and transverse mixing during single-phase flow through granular beds, in *Interaction Between Fluids and Particles*, Inst. Chem. Eng., London (1963).

Hicks, R. E., Pressure drop in packed beds of spheres, Ind. Eng. Chem. Fundam., *9*, 500 (1970).

Hlaváček, V., Aspects in Design of packed catalytic reactors, Ind. Eng. Chem., *62*, 8 (1970).

Hlaváček, V. and H. Hofmann, Steady state axial heat and mass transfer in tubular reactors. An analysis of the uniqueness of solution, Chem. Eng. Sci., *25*, 173, 187 (1970).

Hlaváček, V. and J. Votruba, Experimental study of multiple steady state in adiabatic catalytic systems, Chem. React. Eng. II, Adv. Chem. Ser., *133*, H. Hulburt, Ed., ACS, Washington, D.C. (1974).

Hlaváček, V. and J. Votruba, Steady-state operation of fixed-bed reactors and monolithic structures, in *Chemical Reactor Theory. A Review*, L. Lapidus and N. R. Amundson, eds., Prentice-Hall, Englewood Cliffs, N.J. (1977), p. 314.

Hlaváček, V., M. Marek, and T. M. John, Modelling of chemical reactors: XII, Collect. Czech. Chem. Commun., *34*, 3868 (1969).

Hofmann, H., Standardprogramme für die Optimierung chemischetechnologischer Probleme, Dechema-Monogr., *53*, 143 (1965).

Hofmann, H., Probleme bei der mathematischen Modellierung von Schüttschict-Reaktoren, Chem. Ing. Tech., *46*, 236 (1974).

Hofmann, H., Forschritte bei der Modellerierung von Festbettreaktoren, Chem. Ing. Tech., *51*, 257 (1979).

Hoffmann, U. and H. Hofmann, Kinetik von Reaktionssystemen mit vielen Komponenten und vielen Reaktionen sowie deren verinfachende Behandlung: I, Chem. Ing. Tech., *48*, 465 (1976a).

Hoffmann, U. and H. Hofmann, Kinetik von Reaktionsystemen mit vielen Komponenten und vielen Reaktionen sowie deren vereinfachende Behandlung: II, Chem. Ing. Tech., *48*, 630 (1976b).

Hoffmann, U., G. Emig, and H. Hofmann, Comparison of different methods for effective thermal conductivity of porous catalysts, ACS Symp. Ser., *65*, 189 (1978).

Hosten, L. H. and G. F. Froment, Kinetic modelling of complex reactions, in *Recent Advances in the Engineering Analysis of Chemically Reacting Systems*, L. K. Doraiswamy, ed., Wiley Eastern Limited, India (1984).

Inoue, H., Static thermal stability of an autothermal reactor, Chem. Eng. J. Jpn., *11*, 40 (1978).

Jensen, K. F. and H. W. Ray, The bifurcation behavior of tubular reactors, Chem. Eng. Sci., *37*, 199 (1982).

Keil, W. and E. Wicke, Über die kinetischen Instabilitäten bie der CO-oxidation tion an Platin-Katalysatoren, Ber. Bunsenges, Phys. Chem., *84*, 377 (1980).

Kunii, D. and J. M. Smith, Heat transfer characteristics of porous rocks, AIChE J., *6*, 71 (1960).

Kunii, D. and T. Furusawa, Prediction of hot-spot location in catalytic fixed bed reactors, Chem. Eng. J., *4*, 268 (1972).

Lee, S. T. and R. Aris, Modeling the monolith: some methodological considerations, Chem. React. Eng., Proc. 4th Int. Symp., Heidelberg, Dechema (1976), p. VI–232.

Lee, H. H. and J. Butt, Heterogeneous catalytic reactors undergoing chemical deactivation, AIChE J., *28*, 405, 411 (1982).

Lerou, J. J. and G. F. Froment, Velocity, temperature and conversion profiles in fixed bed catalytic reactors, Chem. Eng. Sci., *32*, 853 (1977).

Lerou, J. J. and G. F. Froment, Fixed bed reactor design and simulation, in *Computer-Aided Process Plant Design*, M. E. Leesley, ed., Gulf Publishing Company, Houston, Tex. (1982).

Leva, M., Cooling of gases through packed tubes, Ind. Eng. Chem., *40*, 747 (1948).

Liljenroth, F. G., Starting and stability phenomena of ammonia-oxidation and similar reactions, Chem. Met. Eng., *19*, 287 (1918).

Liu, S. L. and N. R. Amundson, Stability of adiabatic packed bed reactors. An elementary treatment, Ind. Eng. Chem. Fundam., *1*, 200 (1962a).

Liu, S. L., R. Aris, and N. R. Amundson, Stability of non adiabatic packed bed reactors, Ind. Eng. Chem. Fundam., *2*, 12 (1963).

Luss, D., Interactions between transport phenomena and chemical rate processes, Chem. React. Eng., Proc. 4th Int./6th Eur. Symp., Heidelberg, Dechema (1976), p. 487.

Luss, D., Steady state and dynamic behavior of a single catalytic pellet, in *Chemical Reactor Theory. A Review*, L. Lapidus and N. R. Amundson, eds., Prentice-Hall, Englewood Cliffs, N.J., (1977), p. 191.

Marek, M., V. Hlaváček, and T. M. John, Modelling of chemical reactors: XI. Nonadiabatic non-isothermal catalytic packed bed reactor. An analysis of a two-dimensional model, Coll. Czech. Chem. Commun., *34*, 3664 (1969).

Marin, G. B. and G. F. Froment, Reforming of C_6-hydrocarbons on a Pt-Al_2O_3 catalyst, Chem. Eng. Sci., *37*, 759 (1982).

Martin, H., Low Peclet number particle-to-fluid heat and mass transfer in packed beds, Chem. Eng. Sci., *33*, 913 (1978).

McDonald, I. F., M. S. El Sayed, K. Mow, and F. A. Dullien, Flow through porous media, Ind. Eng. Chem. Fundam., *18*, 199 (1979).

McGreavy, C. and B. R. Dunbobbin, Some aspects of the design of multitubular fixed bed reactors for highly exothermic reactions, Chem. React. Eng., Proc. 4th Int. Symp., Heidelberg, 1976, Dechema (1976), p. III–96.

McGreavy, C., and B. R. Dunbobbin, Operational flexibility consideration in the design of multitubular reactors, Chem. React. Eng.—Houston, ACS Symp. Ser., *65*, V. W. Weekman and D. Luss, eds., ACS, Washington, D.C. (1978).

Mears, D. E., Diagnostic criteria for heat transport limitations in fixed bed reactors, J. Catal., *20*, 127 (1971a).

Mears, D. E., Tests for transport limitations in experimental catalytic reactors, Ind. Eng. Chem. Process Des. Dev., *10*, 541 (1971b).

Mehta, D. and M. C. Hawley, Wall effect in packed columns, Ind. Eng. Chem. Process Des. Dev., *8*, 280 (1969).

Mehta, P. S., W. N. Sams, and D. Luss, Wrong-way behavior of packed bed reactors: I. The pseudo-homogeneous model, AIChE J., *27*, 234 (1981).

Mihail, R. and C. Jordache, Performances of some numerical techniques used for simulation of fixed bed catalytic reactors, Chem. Eng. Sci., *31*, 83 (1976).

Morbidelli, M. and A. Varma, Parametric sensitivity and runaway in tubular reactors, AIChE J., *28*, 705 (1982).

Murase, A., H. L. Roberts, and A. O. Converse, Optimal thermal design of an autothermal ammonia synthesis reactor, Ind. Eng. Chem. Process Des. Dev., *9*, 503 (1970).

Narsimhan, G., Catalyst dilution as a means to establish an optimum temperature profile, Ind. Eng. Chem. Process Des. Dev., *15*, 302 (1976).

Oh, S. H., L. L. Hegedus, and J. C. Cavendish, Mathematical modeling of catalytic converter lightoff, AIChE J., *26*, 935 (1980).

Olbrich, W. E., J. B. Agnew, and O. E. Potter, Dispersion in packed beds and the cell model, Trans. Inst. Chem. Eng., *44*, T207 (1966).

Olson, K. E., D. Luss, and N. R. Amundson, Regeneration of adiabatic fixed beds, Ind. Eng. Chem. Process Des. Dev., *7*, 96 (1968).

Petersen, E. E., *Chemical Reaction Analysis*, Prentice-Hall, Englewood Cliffs, N.J. (1965a).

Petersen, E. E., A general criterion for diffusion influenced chemical reactions in porous solids, Chem. Eng. Sci., *20*, 587 (1965b).

Petrovic, L. F. and G. Thodos, Mass transfer in the flow of gases through packed beds, Ind. Eng. Chem. Fundam., *7*, 274 (1968).

Pommersheim, J. M. and R. S. Dixit, Fixed bed reactors with deactivating catalysts, ACS Symp. Ser., *168*, 367 (1981).

Pontryagin, L. S., V. G. Boltryanski, V. Gamkrelidze, and E. I. Mrschenko, *Mathematische Theorie Optimaler Prozesse*, R. Oldenbourg, Munich, West Germany (1969).

Probst, K. and K. Wohlfahrt, Empirische Abschätzung effektiver Diffusionskoeffizienten in porösen Systemen, Chem. Ing. Tech., *51*, 737 (1979).

Ray, H. W., Catalyst particle and fixed bed reactor dynamics. A review, Proc. 5th Eur. Symp. Chem. React. Eng., Elsevier, Amsterdam (1972).

Rayadhyaksha, R. A., R. Vasudeva, and L. K. Doraiswamy, Parametric sensitivity in fixed bed reactors, Chem. Eng. Sci., *30*, 1399 (1975).

Reichelt, W. and E. Blasz, Strömungstechnische Untersuchungen an mit Rashig-ringen gefüllten Füllkörperrohren und -Säulen, Chem. Ing. Tech., *43*, 949 (1971).

Renken, A., Instationäre Prozesführung kontinuierlicher Reaktoren, Chem. Ing. Tech., *54*, 571 (1982).

Roemer, H. H. and L. D. Durbin, Transient response and moment analysis for backflow cell models for systems with longitudinal mixing, Ind. Eng. Chem. Fundam., *6*, 120 (1967).

Schädlich, K., U. Hoffmann, and H. Hofmann, Periodical operation of processes and evaluation of conversion improvements, Chem. Eng. Sci., *38*, 1375 (1983).

Schlünder, E. U., Transport Phenomena in Packed Bed Reactors, Chem. React. Eng. Rev.—Houston, ACS Symp. Ser., *72* (1978).

Schmitz, R., Multiplicity, Stability and Sensitivity, Chem. React. Eng. Rev., Adv. Chem. Ser., *148*, H. Hulburt, ed., ACS, Washington, D.C. (1975), p. 156.

Schwartz, C. E. and J. M. Smith, Flow distribution in packed beds, Ind. Eng. Chem., *45*, 1209 (1953).

Singer, E., and E. H. Wilhelm, Heat transfer in packed beds. Analytical solutions and design methods, Chem. Eng. Prog., *46*, 343 (1950).

Smith, T. G., J. Zahradnik, and J. J. Carberry, Non-isothermal inter-intraphase effectiveness factors for negative order kinetics—CO-oxidation over Pt, Chem. Eng. Sci., *30*, 763 (1975).

Soria Lopez, A., H. I. de Lasa, and J. A. Porras, Parametric sensitivity of a fixed bed catalytic reactor. Cooling fluid influence, Chem. Eng. Sci., *36*, 285 (1981).

Spechia, V., G. Baldi, and S. Sicardi, Heat transfer in packed bed reactors with one-phase flow, Chem. Eng. Commun., *4*, 361 (1980).

Summers, J. C. and L. L. Hegedus, Modes of catalyst deactivation in stoichiometric automobile exhaust, Ind. Eng. Chem. Prod. Res. Dev., *18*, 318 (1979).

Sundaram, K. M. and G. F. Froment, A comparison of simulation models for empty tubular reactors, Chem. Eng. Sci., *34*, 117 (1979).
Sundaram, K. M. and G. F. Froment, Two dimensional model for the simulation of tubular reactors for thermal cracking, Chem. Eng. Sci., *35*, 364 (1980).
Sundaresan, S., N. R. Amundson, and R. Aris, Observations on fixed-bed dispersion models. The role of the interstitial fluid, AIChE J., *26*, 529 (1980).
Tallmadge, J. A., Packed bed pressure drop—An extension to higher Reynolds numbers, AIChE J., *16*, 1092 (1970).
Tsotsis, T. T., A. E. Haderi, and R. A. Schmitz, Exact uniqueness and multiplicity criteria for a class of lumped reaction systems, Chem. Eng. Sci., *37*, 1235 (1982).
Van Deemter, J. J., F. J. Zuiderweg, and A. Klinkenberg, Longitudinal diffusion and resistance to mass transfer as causes of nonideality in chromatography, Chem. Eng. Sci., *5*, 271 (1956).
Van den Bosch, B., and D. Luss, Uniqueness and multiplicity criteria for an nth order chemical reaction, Chem. Eng. Sci., *32*, 203 (1977).
Vanderveen, J. W., D. Luss, and N. R. Amundson, Stability of packed-bed reactors: effect of flow variations and coupling between the particles, AIChE J., *14*, 636 (1968).
Van Heerden, C., Autothermic processes. Properties and reactor design, Ind. Eng. Chem., *45*, 1242 (1953).
Van Welsenaere, R. J. and G. F. Froment, Parametric sensitivity and runaway in fixed bed catalytic reactors, Chem. Eng. Sci., *25*, 1503 (1970).
Varma, A., On the number and stability of steady states of a sequence of continuous-flow stirred tank reactors, Ind. Eng. Chem. Fundam., *19*, 316 (1980).
Vortmeyer, D. and R. Berninger, Comments on the paper, a Theoretical prediction of effective heat transfer parameters in packed beds by A. Dixon and D. L. Cresswell, AIChE J., *25*, 663 (1979); AIChE J., *28*, 508 (1982).
Votruba, J., V. Hlaváček, and M. Marek, Packed bed axial thermal conductivity, Chem. Eng. Sci., *27*, 1845 (1972).
Wakao, N. and J. M. Smith, Diffusion in catalyst pellets, Chem. Eng. Sci., *17*, 825 (1962).
Wei, J., The Catalytic Muffler, Chem. React. Eng. Rev., Adv. Chem. Ser. *148*, H. Hulburt, ed., ACS, Washington, D.C. (1975).
Weisz, P. B., and C. D. Prater, Interpretation of measurements in experimental catalysis, Adv. Catal., *6*, 143 (1959).
Whitaker, S., Forced connective heat transfer correlations for flow in pipes past flat plates, single cylinders, single spheres and for flow in packed beds and tube bundles, AIChE J., *18*, 361 (1972).
Wicke, E., Transportprozesse und chemische Reaktion bei Umsetzungen Gasförmig-Fest, Acta Tech. Chim. Acad. Naz. Lincei-Varese (1960).
Wicke, E., Physical Processes in Chemical Reactor Engineering, Chem. React. Eng. Rev., p. 75, Adv. Chem. Ser., *148*, H. Hulburt, ed., ACS, Washington, D.C. (1975).
Wicke, E. and R. Kallenbach, Die Oberflächendiffusion von Kohlendioxyd in aktive Kohle, Kolloid-Z., *97*, 135 (1941).
Yagi, S. and D. Kunii, Studies on effective thermal conductivities in packed beds, AIChE J., *3*, 373 (1957).

Yagi, S. and D. Kunii, Studies on heat transfer near wall surface in packed beds, AIChE J., *6*, 97 (1960).
Yagi, S. and N. Wakao, Heat and mass transfer from wall to fluid in packed beds, AIChE J., *5*, 79 (1959).
Yagi, S., D. Kunii, and N. Wakao, Studies on axial effective thermal conductivity in packed beds, AIChE J., *6*, 543 (1960).
Young, L. C. and B. A. Finlayson, Axial dispersion in non-isothermal packed bed chemical reactors, Ind. Eng. Chem. Fundam., *12*, 412 (1973).
Young, L. C. and B. A. Finlayson, Mathematical modeling of the monolith convertor, Chem. React. Eng. II, Adv. Chem. Ser., *133*, ACS, Washington, D.C. (1974).
Zehner, P. and E. U. Schlünder, Die effektive Wärmeleitfähigkeit durchströmter Kugelschüttungen bei mäszigen und hohen Temperaturen, Chem. Ing. Tech., *42*, 933 (1970); *45*, 272 (1973).

7
Fluidized-Bed Reactors

PETER N. ROWE and JOHN G. YATES *University College London, London, England*

INTRODUCTION

Industrial reactors in which solid particles are maintained in a fluidized state by an upward-flowing gas stream were first applied on a wide scale in the 1940s in the fluidized catalytic cracking (FCC) process for the production of gasoline, although a more localized use of the technique had been practiced in Germany 10 years earlier in the Winkler coal gasification process. The main aim of those who developed the FCC process was to devise a method of engineering the catalytic cracking reaction of gas oil, discovered by Houdry, without using fixed-bed reactors that not only suffered from considerable operational problems but also were strongly protected by patents (Jahnig et al., 1980). The success of the development work on the FCC process may be judged from the short time span of four years between its initial conception and the startup of full-scale plant and it is apparent with hindsight that a major reason for this success was the eminent suitability of the cracking process for operation in fluidized-bed reactors. For example, the relative ease of transfer of the finely powdered catalyst between reactor and regenerator overcame the difficulty of cyclic operation of fixed beds and at the same time solved the problem of transferring heat from the exothermic regeneration to the endothermic cracking reaction. Furthermore, the high throughputs that it was possible to maintain gave the FCC process an enormous advantage over its competitors and it is not surprising that some 40 years later it remains one of the most important and elegant catalytic processes in all industry.

Subsequent developments in the field of industrial fluidized-bed reactors, however, did not always result in the same degree of success for the processes to which they were applied and it rapidly became clear that more needed to be understood about the basic physics of gas-solid fluidization before large-scale reactors could with confidence be designed or scaled up from smaller units, and during the last three decades with this aim in view a great deal of research has been carried out on the fundamentals of the subject.

The type of information required for the design of a fluidized-bed reactor is not markedly different from that needed for any other gas-solid reactor and relates to chemical kinetics and the mode and extent of contact between the gas and solid phases. The way in which this information is assembled constitutes the reactor model. Consider the general scheme shown in Fig. 1, where gas enters at the base and rises through the bed at a superficial velocity U, in excess of the minimum, U_{mf}, required just to fluidize the particles. Under these conditions it is generally observed that the bed divides into two phases and each accommodates a proportion of the total flow (Fig. 1b). The interstitial emulsion-phase flow Q_i passes at a velocity close to U_{mf} through the space between the fluidized particles, while the excess, Q_b, flows in the form of rapidly moving voids or "bubbles" which are essentially free of particles and grow by coalescence as they rise through the bed. Gas flowing interstitially is in intimate contact with particles and, if the chemical kinetics are favorable, the efficiency of the chemical conversion in this phase will be high. Bubble-phase gas, however, does not make good contact with solids, so reaction here will be inefficient. There is some exchange between the two phases, but the extent of this, Q_E, is determined mainly by the size of the bubbles and decreases as they become larger.

The extent to which a reactant gas is converted to products as it passes through the bed is therefore determined by the following factors:

1. The division of flow between the two phases
2. The contacting pattern of gas and solids within each phase

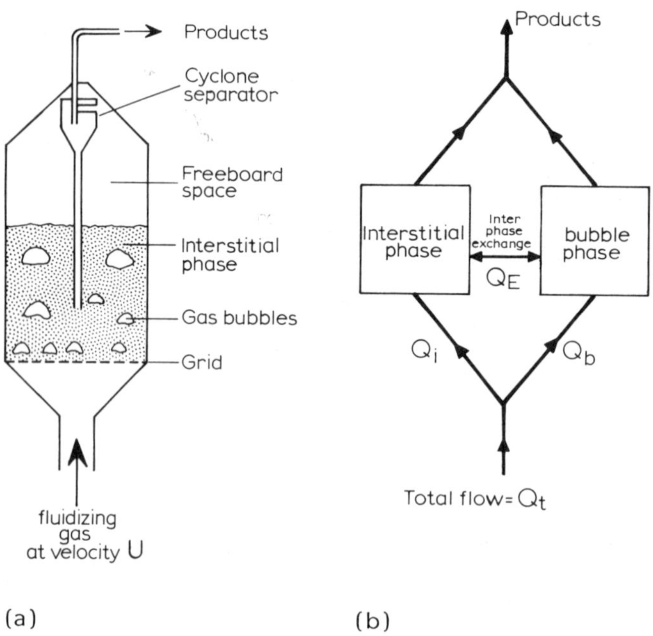

FIGURE 1 Flow distribution in a fluidized bed. (a) Cross section through a fluidized bed. (b) Schematic of gas flow division.

3. Interphase exchange
4. Chemical kinetics

The three hydrodynamic factors are discussed in detail below, but first it will be useful to consider some of the more basic features of gas-fluidized beds that have important implications for their use as chemical reactors.

BASIC FEATURES OF GAS-SOLID FLUIDIZATION

Minimum Fluidization Velocity

A bed of particles is normally load bearing because any impressed force is ultimately resisted by static friction between grains in contact and is spread over a volume of the whole. Thus one can walk over dry sand, although sinking a little at each step as the grains rearrange to provide frictional resistance. If a fluid, liquid, or gas, is passed upward through a particle bed, relative motion causes an upward drag force on each particle, effectively reducing its apparent weight. At a critical velocity drag force equals gravitational force, each particle is individually supported and they no longer rest upon one other. This condition can be written

$$C_D \frac{\pi}{4} d_p^2 \frac{1}{2} \rho_f U^2 = \frac{\pi}{6} d_p^3 (\rho_s - \rho_f) g \tag{1}$$

for uniformly sized spherical particles. In this condition there is no static friction between particles and the whole ceases to be load bearing. It behaves like a fluid with shear forces resisted essentially by the viscosity of the interstitial fluid—hence the origin of the descriptive name *fluidized bed*. The fluid velocity at which this condition is first reached is called the minimum fluidization velocity U_{mf}. It is conventional and convenient to express this as a superficial velocity,

$$U_{mf} = \frac{Q_{mf}}{A} \tag{2}$$

where A is the cross-sectional area of the bed.

Light bodies will float on fluidized particles and heavy ones sink. The bed surface remains level although subject to wave motion and splashing, just as a liquid. The whole can be readily stirred and particles will flow through orifices in the side or bottom of the containing vessel. The above applies whether the fluid is a liquid or a gas, but the effects are more dramatic and initially surprising in the latter case.

The essential condition of Eq. (1) can be expressed differently by considering the bed as a whole rather than individual particles. The bed becomes fluidized when the upward flow of fluid produces a pressure drop equivalent to the bed weight. In a parallel-sided vessel,

$$H(\rho_s - \rho_f)(1 - \varepsilon)g = \Delta p \tag{3}$$

is the necessary condition for fluidization. Generally, in gas-fluidized systems $\rho_s \gg \rho_f$, and the latter can ususally be neglected.

The bulk density of dry powder in a vessel of constant cross-sectional area is given by

$$\rho_{bulk} = \frac{M}{HA} \tag{4}$$

and bulk density is related to the individual particle density by

$$\rho_{bulk} = (1 - \varepsilon)\rho_s \tag{5}$$

so that

$$H(1 - \varepsilon) = \frac{M}{A\rho_s} \tag{6}$$

from which it is seen that the product $H(1 - \varepsilon)$ in Eq. (3) is constant. Equation (6) also shows how bed voidage and its variation is calculated from the measurement of bed height. A difficulty arises in measuring particle density ρ_s when the particles are porous, as is commonly the case with materials such as catalysts (Knight et al., 1980; Knight and Rowe, 1980). The density required for this purpose is the ratio of particle mass to the volume enclosing all solid material together with gas contained in all pores and cavities (i.e., gas that is not part of the flowing stream).

Up to the point of fluidization the bed behaves as a packed bed and pressure drop can be calculated from well-known semiempirical equations. For fine particles where the local Reynolds number is small, the Carman-Kozeny equation is suitable and can be written for minimum fluidization conditions (Richardson, 1971)

$$U_{mf} = \frac{\varepsilon_{mf}^3}{5(1 - \varepsilon_{mf})^2} \frac{\Delta p}{S^2 \mu H_{mf}} \tag{7}$$

where S is the particle specific surface. Combining this with Eq. (3) yields

$$U_{mf} = \frac{\varepsilon_{mf}^3}{5(1 - \varepsilon_{mf})^2} \frac{(\rho_s - \rho_f)g}{S^2 \mu} \tag{8}$$

For uniformly sized spherical particles $S = 6/d_p$ and $\varepsilon_{mf} = 0.4$, so

$$U_{mf} = \frac{0.0006 d_p^2 (\rho_s - \rho_f) g}{\mu} \tag{9}$$

It is useful from this to remember the way in which U_{mf} varies with size, density, and viscosity, but it is not an accurate predictor for most powders of practical interest. It should be noted that when considering a fluidized-bed chemical reactor, fluid density and viscosity and the particle density will generally be fixed by chemical and thermodynamic considerations. U_{mf} will then be determined solely by particle size to which it is quite sensitive, at least with finely divided particles.

A more general pressure-drop equation applicable to larger particles and Reynolds numbers is that due to Ergun (1952), which can be written for the present purpose

$$\frac{\Delta p}{H_{mf}} = 150\left[\frac{(1-\varepsilon_{mf})^2}{\varepsilon_{mf}^3}\right]\left(\frac{\mu U_{mf}}{d_p^2}\right) + 1.75\left[\frac{(1-\varepsilon_{mf})}{\varepsilon_{mf}}\right]\left(\frac{\rho_f U_{mf}^2}{d_p}\right) \qquad (10)$$

The two terms on the right-hand side correspond, respectively, to the viscous or skin friction component and the turbulent or form drag contribution and thus span a wide Reynolds number range. The first term is the same as Eq. (7), although with a slightly different constant. Substituting from Eq. (3) and rearranging yields

$$\frac{\rho_f(\rho_s - \rho_f)gd_p^3}{\mu^2} = 150\left[\frac{(1-\varepsilon_{mf})}{\varepsilon_{mf}^3}\right]\left(\frac{U_{mf}d_p\rho_f}{\mu}\right) + \frac{1.75}{\varepsilon_{mf}^3}\left(\frac{U_{mf}d_p\rho_f}{\mu}\right)^2 \qquad (11)$$

Defining

$$Re_{mf} = \frac{U_{mf}d_p\rho_f}{\mu} \qquad (12)$$

and Galileo number,

$$Ga = \frac{\rho_f(\rho_s - \rho_f)gd_p^3}{\mu^2} \qquad (13)$$

we have

$$Ga = 150\left(\frac{1-\varepsilon_{mf}}{\varepsilon_{mf}^3}\right) Re_{mf} + 1.75\left(\frac{1}{\varepsilon_{mf}^3}\right) Re_{mf}^2 \qquad (14)$$

Thus Re_{mf} (containing the required U_{mf}) is quadratic in Ga, which is a combination of the physical properties of particles and fluid. For the particular case of uniform spheres, $\varepsilon_{mf} = 0.4$:

$$Re_{mf}^2 + 51.4\,Re_{mf} - 0.0366\,Ga = 0 \qquad (15)$$

or

$$Re_{mf} = 25.7[(1 + 5.53 \times 10^{-5}\,Ga)^{1/2} - 1] \qquad (16)$$

Equation (14) is quite a good and reliable predictor, although a little clumsy to use. Most powders of practical interest include a range of particle sizes in which case the surface mean diameter should be used. Nonspherical particles can be characterized by the diameter of a sphere with the same specific surface, but it must be kept in mind that particles differing greatly from spheres in overall shape (e.g, plates and needles) may not be readily fluidized.

A purely empirical correlation due to Leva (1959) becomes, in SI units,

$$U_{mf} = \frac{7.9 \times 10^{-3} d_p^{1.82} (\rho_s - \rho_f)^{0.94}}{\mu^{0.88}} \tag{17}$$

and is valid for $Re_{mf} < 10$. At higher values the result of Eq. (17) is multiplied by the correction factor shown in Fig. 2. Leva's equation is simple to use and reasonably reliable but not as good as Eq. (14).

A little insight comes from considering the experiments of Rowe (1961). The drag force was measured on an isolated spherical particle held in a fluid stream and the result expressed as drag coefficient varying with Reynolds number. Similar measurements were made under identical conditions except that the observed spherical particle was closely surrounded by similar particles also held stationary. The results are summarized in Fig. 3, from which it is seen that at a given Reynolds number drag coefficient is increased by a factor of about 70 when an isolated particle is surrounded by close neighbors.

If a particle is to be held stationary in an upward-flowing stream, the conditions of Eq. (1) must apply whether the particle is isolated or one within a fluidized bed. The product $C_D U^2$ must be the same for these two cases:

$$C_{D_I} U_T^2 = C_{D_{mf}} U_{mf}^2 \tag{18}$$

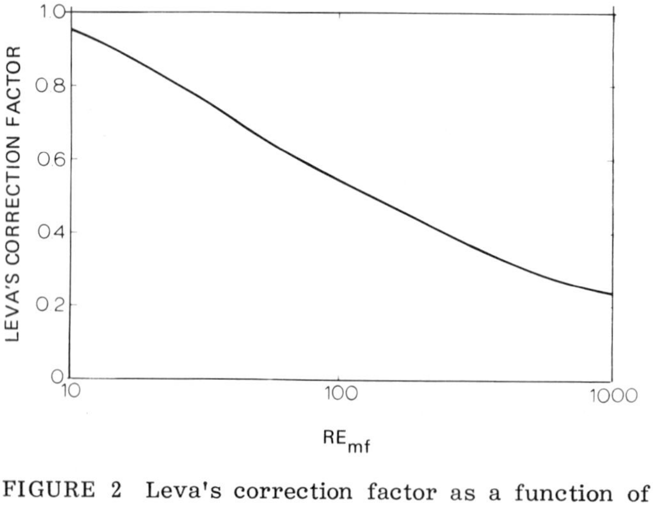

FIGURE 2 Leva's correction factor as a function of Reynolds number at minimum fluidization.

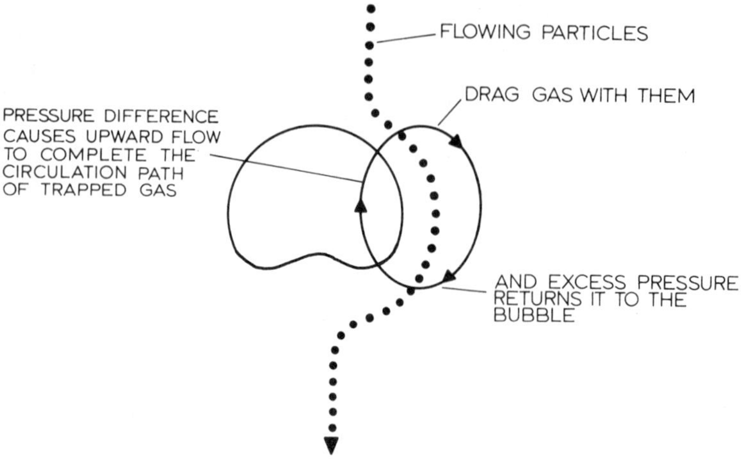

FIGURE 8 Formation of gas cloud around fast-moving bubble.

is actually in contact with particles. Only this fraction of gas can react at any instant, although since the whole is circulating, each element will have a limited period of reaction time. Furthermore, since a necessary condition for this hydrodynamic situation is that $U_B > U_i$, gas forming the cloud has a shorter residence time than interstitial gas. Plainly these limitations have important consequences in reaction engineering and the mechanism described above forms a basis for reactor modeling.

It is evident that the two velocities, U_B and U_i, are important parameters in determining the nature of gas flow through the bed. They are independent of each other. U_B depends only on bubble size, as shown by Eq. (38). U_i is proportional to U_{mf}, which in turn depends on the gas and particle properties as given by Eq. (9), (14), or (17). For small values of U_{mf}, when clouds are likely to form, interstitial gas velocity depends primarily on particle size, roughly as d_p^2. The important ratio U_B/U_i is then proportional to $d_B^{1/2}/d_p^2$. It should be noted that conventionally, U_{mf} is expressed as a superficial velocity

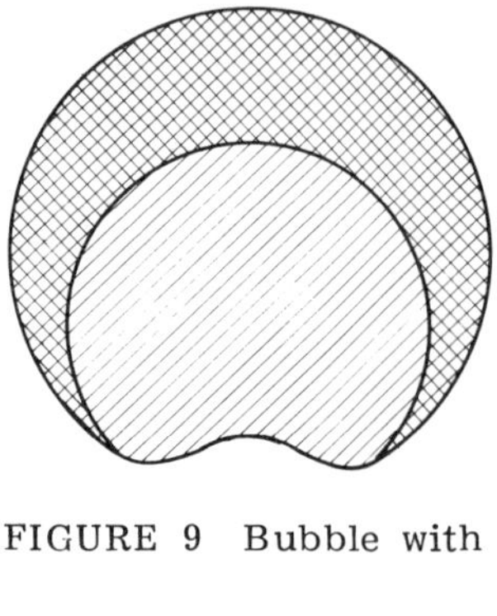

FIGURE 9 Bubble with captive cloud.

$$U_{mf} = \frac{Q_{mf}}{A} \tag{41}$$

and therefore when the simple two-phase theory applies,

$$U_i = \frac{U_{mf}}{\varepsilon_{mf}} \tag{42}$$

Clouds form around the bubbles whenever $U_B > U_i$, that is, when the ratio

$$\alpha = \frac{U_B}{U_i} > 1 \tag{43}$$

A simple but elegant theory (Davidson and Harrison, 1963) treats the bubbles as perfect circles around which concentric clouds form the diameter, being given by

$$\frac{d_c}{d_B} = \left(\frac{\alpha + 1}{\alpha - 1}\right)^{1/2} \tag{44}$$

for a two-dimensional or disklike bubble (the only case observable experimentally) or

$$\frac{d_c}{d_B} = \left(\frac{\alpha + 2}{\alpha - 1}\right)^{1/3} \tag{45}$$

for three-dimensional or spherical bubbles. These equations are shown graphically in Fig. 10. From this it is seen that when $\alpha < 1$, the cloud is of infinite extent, or in other words, is open to through-flow.

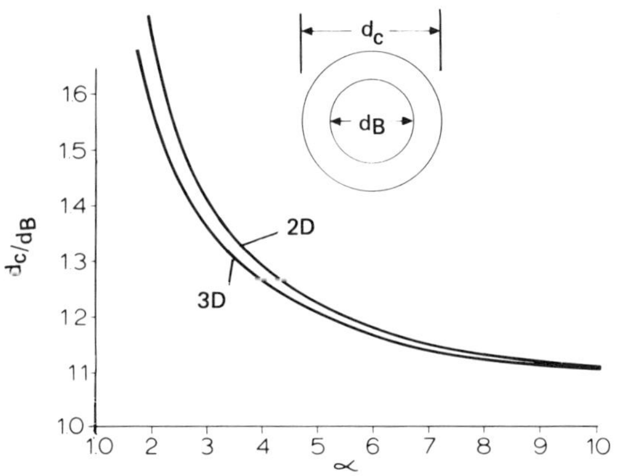

FIGURE 10 Ratio of cloud to bubble diameter as a function of the dimensionless parameter α.

Equation (45) gives an estimate that is useful in calculating contacting efficiency in reactor modeling, but a better estimate is given by the semi-empirical equation (Partridge and Rowe, 1966)

$$\frac{V_c}{V_B} - 1 = \frac{1.17}{\alpha - 1} \tag{46}$$

In gas fluidized beds of particles where $\bar{d}_p \lessdot 100\ \mu m$, values of the velocity ratio are commonly in the rage $10 < \alpha < 100$. From Eq. (46) it is then seen that the cloud volume is between 13 and 1% greater than bubble volume. That is, only between 13 and 1% of gas in the bubble or cloud phase is in contact with solid particles at any instant. This is very poor contacting and explains the performance of reactors that operate under such ill-chosen conditions.

Bubbles usually form initially at the distributor plane and unless there is change in the volumetric gas flow with height, the bubble population changes thereafter only by coalescence and splitting. There is, however, an initial or formation zone in which this is not true.

Distributor and Bubble Formation

In an industrial reactor the distributor plate usually contains a number of regularly spaced holes which may or may not be capped in some way to prevent solids falling through on shut down and to help distribute the flow uniformly. Gas velocity through individual holes can be considerable (typically tens of m/sec), yet under all realistic conditions gas enters the bed from them in the form of bubbles (Rowe et al., 1979). The gas momentum is effectively zero and even with horizontal entry, bubbles rise vertically from the source. The mode of entry is very similar to the way bubbles form when gas is blown into a liquid.

In one important respect gas entering a fluidized bed differs from that entering a liquid. With increasing flow rate the frequency remains approximately constant at eight per second and the size of bubble at detachment increases. Into a liquid the reverse is true at least at modest flow rates.

This is an important and useful fact, as it enables initial bubble size to be calculated. If it is assumed that the total gas flow is equally divided between the holes, the bubble volume at detachment follows from the observed frequency. However, this is subject to the correction mentioned below.

From x-ray observation it is seen that the size of bubble at detachment multiplied by observed frequency accounts for only about one-half of the known flow through the distributor orifice. That is, the bubbles are only about half the expected volume. These newly formed bubbles continue to grow after detachment and by the time their centers are about two diameters above the hole, they are of a size and frequency to account for all the flow as depicted in Fig. 11. It is now seen that the initial bubble size is given by

$$d_{B(0)} = \left(\frac{3Q_0}{8\pi}\right)^{1/3} \tag{47}$$

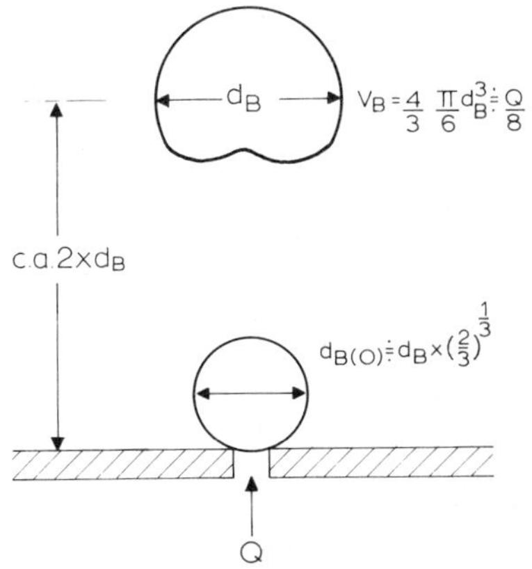

FIGURE 11 Bubble formation at an orifice.

where Q_0 is the flow through a single orifice in m^3/s. At the moment of detachment the bubble is spherical but quickly develops a wake, as shown in Fig. 11.

If for a proposed design of distributor the calculation of Eq. (47) gives an impossibly small diameter (e.g., less than the hole size or markedly less than 100 particle diameters), it is evident that the design is unsatisfactory and not all holes will operate. Similarly, if the bubble size and hole spacing is such that gross overlap occurs, the holes will not operate independently. These considerations, together with the well-known empirical rule that pressure drop across the plate should be about one-third of that across the bed to ensure stability, are essential guides to successful distributor design.

The facts described above can be explained in a semiqualitative way that is important in modeling chemical reactor behavior in the shallow distributor zone. Unlike a bubble forming in a liquid, one in a fluidized bed can grow only by gas flowing perpendicularly across the bubble boundary. This is the only way in which a force can be applied to individual particles to cause them to move and so extend the boundary. After time t the still attached bubble may have grown to diameter d_B, as shown in Figure 12,

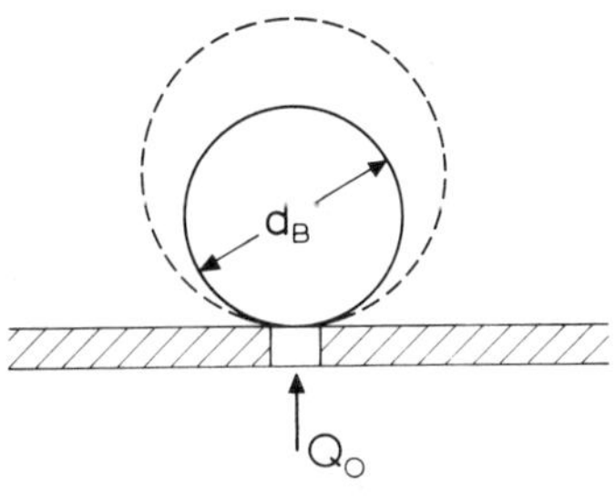

FIGURE 12 Region of gas penetration during bubble formation.

but during this time the gas has advanced as far as the dashed line indicated. The volume within the outer boundary is Qt plus the volume occupied by particles in the eccentric annular region.

This annular region now contains excess gas, and therefore the local powder voidage, ε, must increase. Such increase is unstable and evidently when the bubble detaches and begins to rise through the unstable expanded region, it relaxes and the excess gas pours back into the bubble, almost certainly through its base. The experimental evidence shows that this is completed by the time the bubble has traveled little more than one diameter. Since eight bubbles form each second, it is a very rapid sequence of events and hydrodynamic equilibrium is achieved in a time of order 10^{-1} s.

Referring again to Fig. 11, the upper bubble is now stable and will have formed a cloud around itself if the velocity ratio is appropriate. Half the gas in this cloud and bubble first passed into the dense phase and then returned to form a stable bubble. This brief but efficient contact between gas and particles can lead to good reactor performance in this shallow layer, and a model for this is developed later in this chapter.

Bubble Coalescence and Average Size

Bubbles coalesce readily for although the fluidized powder has a property equivalent to surface tension, it is small. The rate of coalescence depends essentially on the closeness or concentration of bubbles. From this it is possible to postulate the way in which bubble concentration will change with height, and combining this with the assumption that the bubble flow remains constant, bubble size can be estimated. The empirical constant is fitted from the data, leading to (Rowe, 1976)

$$\bar{d}_B = \frac{(U - U_{mf})^{1/2}(h + h_0)^{3/4}}{g^{1/4}} \tag{48}$$

h_0 in Eq. (48) arises as a constant of integration and its physical meaning is clear by reference to Fig. 13. It is that imaginary distance beneath the distributor where the bubbles would be of zero size. It is related to initial bubble diameter by

$$h_0 = \frac{d_{B(0)}^{4/3} g^{1/3}}{(U - U_{mf})^{2/3}} \tag{49}$$

and so can be estimated from Eq. (47). In practice, h_0 is usually only a few centimeters and can often be neglected when estimating bubble size in beds of reasonable depth.

For bubbles to achieve an average size at a given height and flow rate, it is obvious that the containing vessel must be big enough. If not, the bed will slug and Eq. (48) is useful in indicating when this condition will begin, an important point in reactor modeling (Rowe, 1978a).

Another expression for bubble size that is based on slightly different reasoning is (Darton et al., 1977)

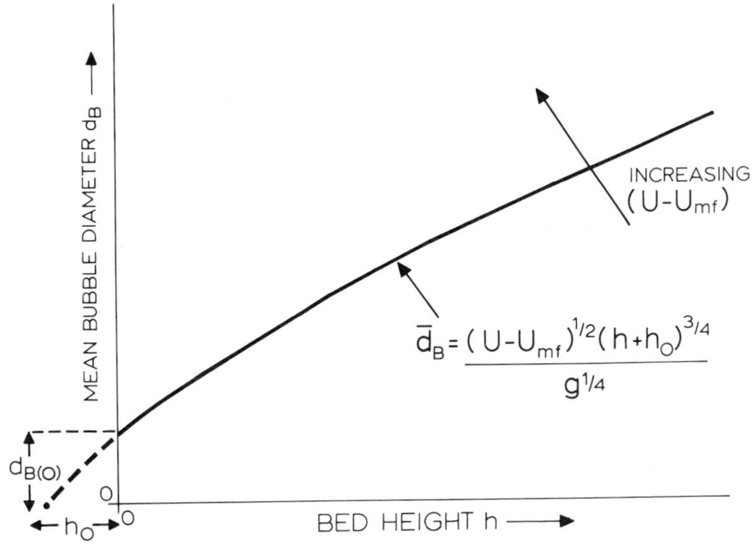

FIGURE 13 Variation of bubble diameter with bed height.

$$d_e = \frac{0.54(h + 4\sqrt{A_0})^{0.8}(U - U_{mf})^{0.4}}{g^{0.2}} \tag{50}$$

where d_e is the equivalent sphere diameter $[d_e = (3/4)^{1/3} d_B]$ and A_0 is the catchment area for a bubble stream at the distributor plate. The two equations (48) and (50) are very similar numerically and, within the limits of observational error, describe the data equally well.

Bubbles also split by instabilities that develop as "fingers" from the roof which may grow rapidly enough to divide the bubble usually into two unequal parts. It is quite common for the two parts subsequently to coalesce. On average, small daughter bubbles persist as individuals for some time, so that the size distribution becomes positive skewed with an excess of small sizes. A well-designed distributor with uniform holes produces initially a more or less mono-sized distribution, but by coalescence and splitting this develops toward an exponential distribution with height which is equivalent to time.

Within a region where there are many bubbles and frequent coalescence, average bubble velocity is increased by the accelerations accompanying coalescence. It has already been pointed out that even with isolated and undisturbed bubbles, there is a high error variance in estimating velocity from bubble size. This variance is increased when many bubbles are present, so that from the data it is not clear that there is any correlation at all. It is, however, apparent that velocity increases with gas flow rate, and a commonly used expression that takes this into account is

$$U_B = \sqrt{\frac{gd_B}{2}} + (U - U_{mf}) \tag{51}$$

which is probably satisfactory for most reaction engineering purposes.

Equation (48) can be expressed as average bubble volume using the bubble shape and proportions given by Eq. (39) and (40). If the two-phase theory applies as given by Eq. (36), the volumetric flow is constant and this can be written

$$Q_B = n\overline{V}_B \tag{52}$$

where n is the frequency with which bubbles cross a horizontal plane through the bed. Combining these equations gives an expression for this frequency,

$$\frac{n}{A} = \frac{8g^{3/4}}{\pi(U - U_{mf})^{1/2}(h + h_0)^{9/4}} \tag{53}$$

Bubble concentration is related to frequency by

$$N = \frac{n}{A\overline{U}_B} \tag{54}$$

$$= \frac{8g^{3/4}}{\pi\overline{U}_B(U - U_{mf})^{1/2}(h + h_0)^{9/4}} \tag{55}$$

from which it is seen that bubble concentration falls rapidly with increasing height. A large industrial-scale reactor operated at a fairly high gas velocity may bubble furiously near the bottom but appear almost quiescent at the top. Under these conditions bubbles can be treated as isolated in much of the upper part of the reactor.

The question is commonly asked: Is there a maximum bubble size? But it is an academic question without much practical importance. Equation (48) implies growth without limit and, while it is based on coalescence frequency observed on a fairly small scale, on an industrial scale it does fit the few data there are. Certainly, bubbles of diameter in the region of 1 m have been reported from more than one source. Rate of coalescence falls off rapidly with concentration so that rate of growth declines with height, and it requires an impossibly large and deep bed to observe bubbles much more than 1 m in diameter. It is inconceivable that close bubbles should fail to coalesce just because the distributor plane is far below. Certainly, big bubbles tend to break up, but then the fragments all too commonly recoalesce. Evidence and logic suggest there is no upper limit, and the possibility of experimental proof seems remote.

On the contrary, there is surely a lower size limit. A bubble of particle size is meaningless. The bubble has reality only when particles can flow around is as a pseudofluid. This suggests a lower limit of order $10^2 \times d_p$. There is no experimental evidence for bubbles less than about 1 cm in diameter, but observation would be difficult.

Particle Mixing and Segregation

The transport of a wake and streamline motion of particles causes a largely predictable displacement of material. This can be demonstrated by preparing

a bed initially with, say, black particles forming the lower half and white ones above with a horizontal interface between (Rowe et al., 1965b).

Figure 14 indicates the kind of displacement that occurs as first one and then successive bubbles disturb the initial arrangement. Each bubble carries wake material to the surface, spilling some enroute while streamline motion draws up a spout or drift profile. Repetition complicates the pattern and the whole quickly becomes well mixed. If bubbles tend to rise in the center, the overall motion is a convection pattern, up the center and down the sides. This reverses if bubbles rise near the walls and more complicated patterns occur in large beds.

It is important to realize from these demonstrations that particle mixing is caused solely by displacement by the bubbles. Without bubbles there can be no mixing. There is no mechanism equivalent to molecular diffusion whereby neighboring particles may jostle and exchange positions—except in the wake. In spite of appearances to the contrary, particle movement is relatively orderly and predicable to a limited degree.

It follows from this that the rate of particle mixing depends on the bubbling rate. It is directly related to the excess gas flow rate $(U - U_{mf})$, as seen from Eq. (35). In most realistic circumstances, the bubbling rate is such that mixing occurs rapidly, and in the majority of cases it is appropriate to treat the particles as perfectly mixed. The circulation rate can be estimated roughly (Rowe, 1973) from

$$t_c = \frac{Ah_{wf}}{0.6\,Q_B[1 - (Q_B/AU_B)} \qquad (56)$$

where t_c is a measure of the time taken for a particle to make a circuit of the bed. If in a chemical reactor where the particles are changing composition the rate of chemical change is small compared with circulation rate, the solids system can justifiably be treated as perfectly mixed, and the above is a simple test to check the correctness of the assumption. The oversimplified nature of Eq. (56) still leaves it quite suitable for this purpose.

Uneven gas distribution quickly leads to convection patterns of solids motion, and the type of pattern is not difficult to imagine with the foregoing mechanism in mind. Sometimes this is done deliberately to induce a

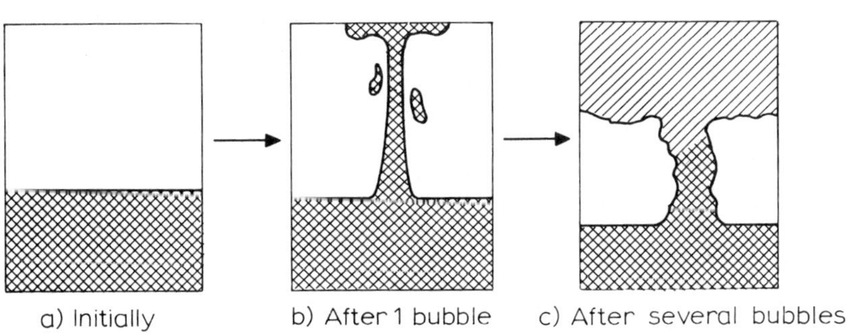

FIGURE 14 Particle mixing caused by rising bubbles.

desired circulation. As it is not easy to ensure uniform distribution over a large area, it may occur involuntarily and unwanted in industrial reactors.

Bulk movement of particles in some kind of circulation pattern will affect bubble velocity and processes such as coalescence. Equations such as (48) and (51) must be used with caution in reactors where appreciable convection is known or suspected to occur.

If the powder to be fluidized contains grains of different types, segregation may occur. Shape differences seem to be unimportant and the only relevant differences between particles are size and density. Of these, density is much the more important. In contrast with mechanical powder mixers, fluidized beds will tolerate a very wide range in particle size without appreciable segregation.

Minimum fluidization velocity varies with mean particle size, so that if the size composition of a mixture changes, so does U_{mf}. If fines are elutriated or if agglomeration occurs, U_{mf} increases and at a constant gas flow rate the vigor of fluidization decreases. This may eventually cause segregation because, as explained above, the rate of mixing depends on the bubbling rate, which declines with increasing U_{mf} according to Eq. (36).

Again in contrast with mechanical mixers, particles of different density can readily segregate in a fluidized bed. While there is usually a spectrum of particle sizes in powders of industrial interest, there are rarely more than two or three different densities and most commonly only two. The majority of research studies are confined to binary systems.

If two powders made up of particles differing only in density are well mixed and then gently fluidized, the dense component settles to the bottom as jetsam. It forms a pure layer, but a small portion remains uniformly mixed in the less dense flotsam above it. With increasing gas velocity (bubbling rate), the proportion of jetsam dispersed in the flotsam increases, and eventually it may be possible to achieve an overall uniform mix. This is illustrated graphically in Fig. 15.

The ratio $M = x/\bar{x}$ as defined by Fig. 15 is a simple mixing index, zero for complete segregation and unity for perfect mixing. It is a function of velocity, of density ratio, and to a small extent, of size ratio. A simple empirical relationship is (Rowe and Nienow, 1976)

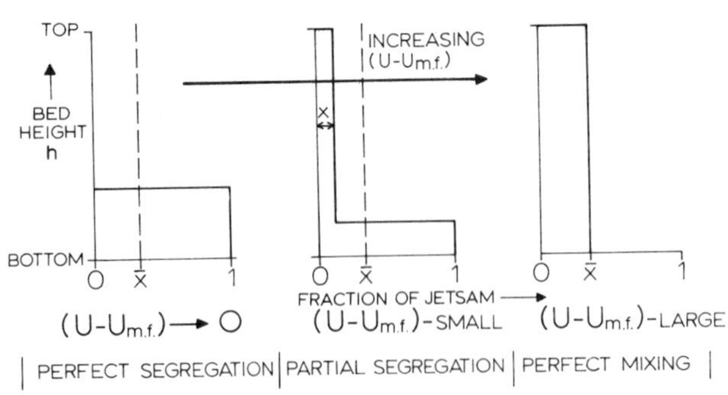

FIGURE 15 Segregation patterns for binary mixtures of solids.

$$M = f(U) \left(\frac{\rho_J}{\rho_F}\right)^{5/2} \left(\frac{d_1}{d_2}\right)^{1/2} \tag{57}$$

where $(d_1/d_2) > 1$. This shows quantitatively the relative importance of density and size difference. The variation with velocity, $f(U)$, is complicated by the above-mentioned fact that U_{mf} varies with composition. It is of the double exponential form shown in Fig. 16. For each binary system there is a critical velocity U_{T0} below which the system rapidly segregates and above which it rapidly mixes. An empirical equation from which this can be estimated is

$$\frac{U_{T0}}{(U_{mf})_F} = \left[\frac{(U_{mf})_J}{(U_{mf})_F}\right]^{1.2} + 0.9\left(\frac{\rho_J}{\rho_F} - 1\right)^{1.1} \left(\frac{d_1}{d_2}\right)^{0.7} - 2.2\bar{x}^{1/2}(H^*) \tag{58}$$

where H^* is a reduced bed height defined by

$$H^* = 1 - \exp\left(-\frac{H}{D}\right) \tag{59}$$

Like mixing segregation is brought about by the bubbles. A bubble passing through a region of jetsam enables it to descend through neighboring flotsam. Similarly, the bubbles cause mixing by lifting jetsam in their wakes, carrying it to the upper part of the bed. The two processes are in competition, but mixing increases more rapidly than does segregation.

From the point of view of chemical reaction engineering, interest lies in knowing the distribution of material in cases where segregation may occur. Complete mixing is often desired and the equations above give a guide as to whether or not this will occur.

Heat and Mass Transfer

Two quite different and independent heat transfer coefficients apply to a fluidized bed. $h_{G/S}$ refers to heat transfer between gas and solid particles

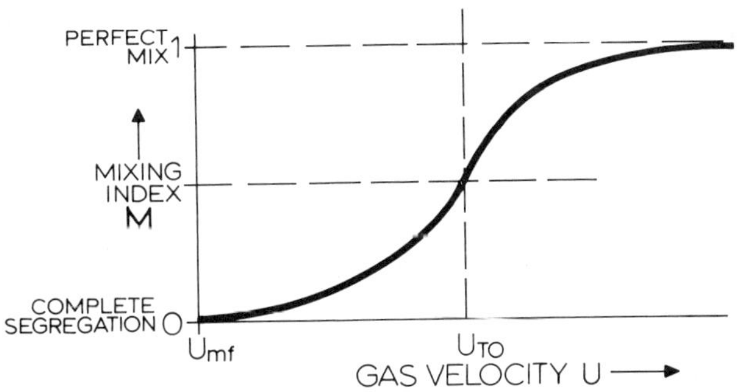

FIGURE 16 Variation of mixing index with gas velocity.

and $h_{B/W}$ refers to heat transfer between the bed and its containing walls or immersed surfaces.

Heat transfer between gas and particles is usually very rapid but only because of the enormous surface area presented by finely divided solids and because of their large heat capacity on a volume basis compared with that of the gas. The transfer coefficient $h_{G/S}$ is small, near its lower limiting value, because the local Reynolds number is small, as explained earlier. The local number can be estimated from (Rowe and Claxton, 1965)

$$Nu = h_{G/S} \frac{d_p}{k} \tag{60}$$

$$= A + BPr^{1/3} Re^n \tag{61}$$

where

$$A = \frac{2}{1 - (1 - \varepsilon)^{1/3}} \tag{62}$$

$$B = \frac{2}{3\varepsilon} \tag{63}$$

and

$$(2 - 3n)/(3n - 1) = 4.65Re^{-0.28} \tag{64}$$

Consider the case where fluidizing gas enters the bed at a different temperature from the particles. Assume that the particles are uniform spheres and are perfectly mixed by bubble disturbance as described above. Further assuming no heat loss and that the solids have a much greater heat capacity per unit volume than gas, it is readily shown that the depth of bed over which 90% of the gas change in temperature occurs is given by

$$(\Delta h)_{90\%} = \frac{0.38UC_G\rho_G d_p}{h_{G/S}} \tag{65}$$

If this is evaluated for a typical gas-solid fluidized system with $h_{G/S}$ calculated from Eq. (61), it is seen that this layer is only a few particle diameters thick. That is, virtually all heat transfer occurs in a very shallow bottom layer. Experimental attempts to detect a temperature gradient at the bottom of the bed invariably fail, which is in accordance with Eq. (65). This follows from three simple facts: the particles are well mixed, they have much larger volumetric heat capacity than gas, and they offer a lot of surface area for heat transfer.

It is well known that a vigorously bubbling gas fluidized bed is remarkably uniform in temperature even when highly energetic reactions are occurring. This all follows from what is written above and is not a consquence of a high transfer coefficient when this is considered on an individual particle basis.

Heat transfer from bed to walls is also generally very high, and this is expressed by a high value of the coefficient $h_{B/W}$ when, as is logical, it is defined in terms of wall area. The magnitude of the coefficient is usually comparable with that for heat transfer to a boiling liquid, and often the

overall coefficient is limited by conditions on the other side of the wall. Numerical values depend on particle and gas properties in a complicated way that has been well documented (Botterill, 1975).

The reason for high heat transfer rates is easily explained. It follows from the fact that particles are well mixed and of uniform temperature. The temperature profile across a section of the bed is flat, and therefore there is a very steep gradient or driving force against the containing walls. Another way of saying this is that hot particles from the center are continually brought to the cold walls and then moved on before they have time to cool appreciably. The requirement for good heat transfer is simply that the particles keep moving.

The foregoing also applies to surfaces immersed in the bed, but difficulties can arise when these interfere with fluidization, the rise of bubbles, and the consequential movement of particles. It is easy in these general terms to understand why a submerged heat exchanger may not perform well, but improvement depends on engineering empiricism with this model kept in mind.

There is no realistic equivalent of mass transfer between surfaces and the bed but an obvious one between the particles and the gas. On an individual particle basis this is closely analogous to heat transfer and Eqs. (60) and (61) simply become

$$\mathrm{Sh} = k_{G/S} \frac{d_p}{2} \tag{66}$$

$$= A + B\mathrm{Sc}^{1/3}\mathrm{Re}^n \tag{67}$$

with A, B, and n defined as before by Eqs. (62) to (64).

There is very little more to say about mass transfer, although the simpleminded often seek an overall coefficient that will yield conversion in a reactor from facts such as the inlet conditions and the bed temperature. There is no simple answer of this kind. Overall conversion follows from an integration of all the histories experienced by elements of gas in their passage through the bed, in bubbles, in interstitial flow, or as part of a cloud. The local reaction rate will depend on local concentration and on the transfer coefficient given by Eq. (67). But this is the basis of the reactor models dealt with later in this chapter.

Fine Powders

Fine powders behave differently from coarser ones in ways that are important for reaction engineering, and some of the previous description requires modification. Fine powder in this context means Geldart's group A powders (Geldart, 1973), which are roughly defined as lying in the range

$$\begin{aligned} (\rho_s - \rho_f)d_p^2 &> 10^6 \\ (\rho_s - \rho_f)d_p &< 2 \times 10^5 \end{aligned} \tag{68}$$

when density is expressed in $\mathrm{kg/m^3}$ and particle diameter in μm. Thus solids of density 10^3 $\mathrm{kg/m^3}$ show group A behavior when in the range

$30 < d_p < 200$ μm. Below this size they are very difficult to fluidize uniformly and tend to channel (group C), and above it they behave as hitherto described (group B). The classification is approximate at best, depends also on other poorly understood particle properties, and there is a gradation of behavior and not abrupt change. Most powders clearly show group A behavior when $\bar{d}_p < 100$ μm.

The distinguishing feature of group A powders is that they expand uniformly when fluidized at low gas velocities and do not begin to bubble until a higher velocity, U_{mb}, is reached. For group A powders $U_{mb}/U_{mf} > 1$. This is illustrated by Fig. 17, which shows how bed height changes with gas velocity.

The uniformly expanded state is quite stable. There is no hysteresis and the same height is observed whether gas velocity has increased or decreased. The first appearance of bubbles is a breakdown of this stability, and observations around U_{mb} are not very reproducible, making it difficult to determine with precision.

Behavior in the region $U_{mf} < U < U_{mb}$ is closely akin to that of liquid fluidized beds as described briefly earlier. The height or bed voidage varies with velocity as Eq. (31) except that the constant of proportionality is less than U_T. As a corollary, the exponent m is greater than observed with liquids at the same Reynolds number.

With decreasing average particle size this group A behavior becomes more marked. Bed height can double before bubbling begins when $U_{mb}/U_{mf} \doteqdot 3$. Expansion is increased by the addition of fines to the powder, which has the effect of reducing mean particle size $\bar{d}_p$.

Many catalyst powders prepared for use in fluidized-bed reactors behave as group A and fluid-bed cracking catalyst (FCC) is a good example. Expansion of the dense phase and increased interstitial flow above U_{mf} has important consequences in reactor behavior and modeling. Since this occurs only with fine powders where interstitial flow is necessarily small, increase in this very effective phase can increase reactor performance markedly.

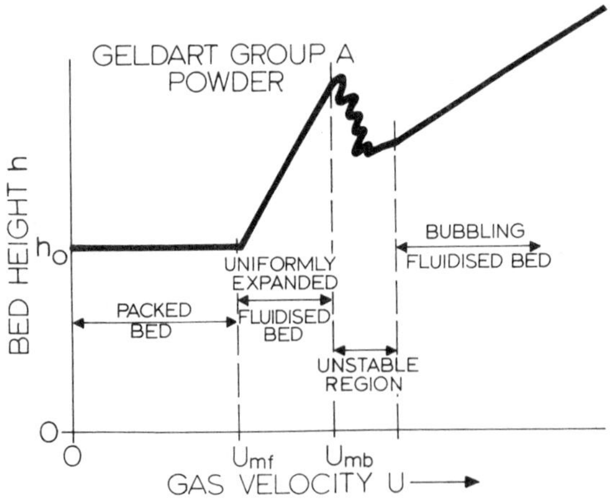

FIGURE 17 Variation of bed height with gas velocity for group A powders.

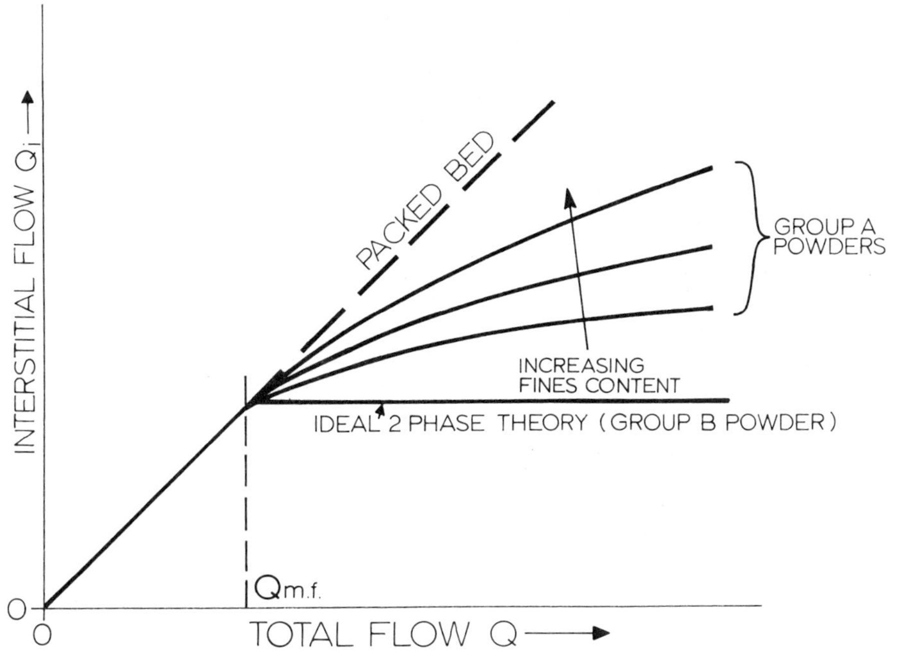

FIGURE 18 Variation of interstitial flow with total flow as a function of fines content for group A powders.

Although observation is difficult, there is evidence that the dense phase remains expanded after bubbling begins and more gas than Q_{mf} flows interstitially in a vigorously fluidized bed of group A powder (Rowe et al., 1978). Figure 18 shows how interstitial flow increases for group A powders as the fines content is increased ($\bar{d}_p$ decreased). Only limited data are so far available, and it is not yet possible to quantify the information of Fig. 18 for a previously untested powder. It should be noted that interstitial flow will decrease and reactor performance will consequently fall as fines are elutriated from the bed. Such is commonly observed by plant operators.

Effect of Increasing Pressure

Most fluidized-bed reactors operate at or a little above atmospheric pressure, but there are cases when it is required to operate at higher pressures. Fluidized coal combustion under pressure, for example, gives improved efficiency because of the reduced volume of the plant for a given duty and because of the opportunity to recover further energy by expanding hot gases through a turbine (Roberts et al., 1980). For this application, pressures up to 20 bar are feasible. There are, additionally, chemical processes where it is thermodynamically desirable to operate at high pressure.

At the same mass flow rate it is obvious that velocity will decrease with increasing pressure, so there will be fewer and smaller bubbles with an increasing proportion of interstitial flow. It is of more interest to compare

behavior at the same volumetric flow rate as pressure increases and this is the condition in all that follows.

It is commonly reported by plant operators that fluidization is "smoother" as pressure increases, and this is said to be because the bubbles are smaller. The evidence for this is very slender because of the difficulty of observation, but the conclusion is widely accepted. Data from research observations under pressure are scarce, again because of the difficulty of observation, but what little evidence there is points to the conclusion that proportionately more gas flows interstitially as pressure increases (Rowe and MacGillivray, 1980).

The ratio U_{mb}/U_{mf} for class A powders does increase with increasing pressure but not dramatically, about 25% for a fivefold increase in pressure (Rowe et al., 1982). Certainly, some group B powders show group A behavior as pressure increases, and this is not predicted by the criteria of Eq. (68). An alumina powder of $\bar{d}_p$ = 450 μm that is typically group B at NTP began to show group A behavior at pressures as low as 2 bar and showed it markedly at higher pressures (Rowe et al., 1983). X-ray ciné photographs of this powder fluidized with nitrogen show that as expected, bubble size decreases with pressure and the average volume is halved at around 60 bar. Again, this is not a dramatic change. Bubble frequency increases and the visible bubble flow varies approximately in proportion with $A(U - U_{mb})$. The bubble velocity coefficient $\sqrt{U_B/(1/2)gd_B}$ increases by about 50% at 30 bar and thereafter changes little.

It should be noted that U_{mf} falls with increasing pressure as predicted by Eq. (14). Terminal falling velocity, U_T, also falls with pressure. The change can be considerable, and for the 450 μm alumina referred to above, U_{mf} falls to one third of its atmospheric value at about 60 bar. U_T falls to about one-fifth over the same range. The change increases with particle size. Equation (14) fits the presently available data very well up to pressures of 100 bar.

Change of U_{mf} with pressure means that at constant volumetric flow rate the bed will be fluidized at a greater multiple of U_{mf} as pressure increases. The development of group A behavior means that a larger proportion of gas will flow interstitially, and the net result may be a reduction in bubble flow with consequently smaller bubbles. These changes would improve reactor performance.

With the 450-μm alumina (the only material so far to be studied systematically in this way), dramatic changes do occur at pressures greater than about 80 bar. As this pressure is approached, the wake obtrudes more deeply into the bubble and they begin to break up by wake reaching to the roof, a mode of breakup never observed at lower pressures. At 80 bar the bubbles are no longer clearly observable, and it may be that the system has changed toward particulate fluidization.

Entrainment and Elutriation

When fluidized beds are operated at gas velocities in excess of the minimum fluidization value, the eruption of gas bubbles at the bed surface causes particles to be carried upward into the freeboard region. It is generally accepted that these particles originate from the bubble wake region (George and Grace, 1978) and that the height to which they are carried above the surface before falling back is a function of their size and density as well as of the velocity of the fluidizing gas.

Under conditions of operation in which the superficial gas velocity exceeds the terminal fall velocity of the bed particles, entrainment leads to elutriation and particles are carried out of the system completely. Thus there is usually a concentration gradient in the freeboard, with the larger denser particles being carried some way above the surface before returning to it, and the smaller, lighter particles being removed from the system. The height above the surface at which the particle concentration becomes constant is known as the *transport disengaging height* (TDH).

It is clearly important from a design point of view to be able to predict how much bed material will be elutriated for a given set of operating conditions, since this has a strong bearing on cyclone and filter specification. A basic assumption generally made is that entrainment is a first-order process and the rate of elutriation of particles in a particular size range is directly proportional to the mass fraction x_i of those particles in the bed:

$$-\frac{1}{A}\frac{d}{dt}(x_i M) = K^* x_i \tag{69}$$

where A is the bed area, M the bed mass, and K* ($kg/m^2 \cdot s$) a first-order rate constant. Then for an approximately constant bed mass,

$$\frac{x_i}{x_{i0}} = \exp\left(-\frac{K^* A t}{M}\right) \tag{70}$$

where x_{i0} is the initial mass fraction of elutriable particles. A great deal of work has been carried out to establish relationships between K* and the properties of fluidizing gas and bed particles, and as a result a number of correlations have been devised and some are listed in Table 4. It must be

TABLE 4 Some Published Correlations for Elutriation from Fluidized Beds

Yagi and Aochi (1955):

$$\frac{K^* g d_p^2}{\mu(U_0 - U_{ts})^2} = 0.0015 Re_t^{0.6} + 0.01 Re_t^{1.2}$$

Zenz and Weil (1958):

$$\frac{K^*}{\rho_g U_0} = \begin{cases} 39.1 \times 10^2 \left(\dfrac{U_0^2}{g d_p \rho_s^2}\right)^{1.87} & \dfrac{U_0^2}{g d_p \rho_s^2} \leqslant 581.8 \times 10^{-3} \\[2ex] 7.02 \times 10^3 \left(\dfrac{U_0^2}{g d_p \rho_s^2}\right)^{1.15} & \dfrac{U_0^2}{g d_p \rho_s^2} \geqslant 581.8 \times 10^{-3} \end{cases}$$

TABLE 4 (Continued)

Wen and Hasinger (1960):

$$\frac{K^*}{\rho_g(U_0 - U_{ts})} = 1.52 \times 10^{-5} \left[\frac{(U_0 - U_{ts})^2}{gd_p}\right]^{0.5} Re_t^{0.725} \left(\frac{\rho_s - \rho_g}{\rho_g}\right)^{1.15}$$

Tanaka et al. (1972):

$$\frac{K^*}{\rho_g(U_0 - U_{ts})} = 4.6 \times 10^{-2} \left[\frac{(U_0 - U_{ts})^2}{gd_p}\right]^{0.5} Re_t^{0.3} \left(\frac{\rho_s - \rho_g}{\rho_g}\right)^{0.15}$$

Merrick and Highley (1974):

$$\frac{K^*}{\rho_g U_0} = A + 130 \exp\left[-10.4\left(\frac{U_{ts}}{U_0}\right)^{0.5} \left(\frac{U_{mf}}{U_0 - U_{mf}}\right)^{0.25}\right]$$

Geldart et al. (1979):

$$\frac{K^*}{\rho_e U_0} = 23.7 \exp\left[-5.4\left(\frac{U_{ts}}{U_0}\right)\right]$$

$\rho_e = \rho_g + \Sigma\rho_i$ (ρ_i = solid loading of ith size fraction in exit gas)

Colakyan et al. (1979):

$$K^* = 33\left(1 - \frac{U_{ts}}{U_0}\right)^2$$

Bachovchin et al. (1979):

$$K^* = 3.35 \times 10^{-5}\left(\frac{U_0}{\sqrt{\bar{d}_p g}}\right)^{4.67} \left(\frac{\rho_g}{\rho_s}\right)^{1.62} \left(\frac{\mu}{\bar{d}_p}\right) \left(\frac{D_c\sqrt{X_s}}{\bar{d}_p}\right)^{1.15}$$

where

$\bar{d}_p$ = average size of particle elutriated

X_s = fraction of fines at the bed surface

Lin et al. (1980):

$$\frac{K^*}{\rho_g U_0} = 9.43 \times 10^{-4}\left(\frac{U_0^2}{gd_p}\right)^{1.65}$$

TABLE 5 Experimental Conditions for Entrainment or Elutriation of Fines from Fluidized Beds (from Chen and Wen, 1982)

Investigator	Experiments	Column diameter, D_c bed height, L; freeboard, H (m)	Gas	Distributor
Yagi and Aochi (1955)	Elutriation two and multicomponents, batch operations	$D_c = 0.052-0.071$ $L = 0.031-0.12$	Air	Fixed bed of steel balls
Zenz and Weil (1960)	Entrainment of FCC catalyst, steady-state operation	$D_c = 0.051 \times 0.61$ $H = 0.254-2.80$	Air, mean grid pressure 8 atm	
Wen and Hashinger (1960)	Elutriation, two and multi-components, batch operation	$D_c = 0.067$ $L + H = 2.08, 1.83$	Air, He	Filter cloth
Tanaka et al. (1972)	Elutriation, two components, batch and continuous	$D_c = 0.067$ $L + H = 1.80$	Air	Perforated plate
Merrick and Highley (1974)	Elutriation, steady-state operation	$D_c = 0.90 \times 0.45$ $L = 0.61-1.22$ $L + H = 3.96$	Air	—
Geldart et al. (1979)	Elutriation, multicomponents, batch operation	$D_c = 0.076$ $L = 0.35-0.45$ $L + H = 3.80$	Air	Filter paper covered by wire mesh
Colakyan et al. (1979)	Elutriation, multicomponents, batch operation	$D_c = 0.90 \times 0.90$ $L + H = 6.30$	Air	Perforated plate
Bachovchin et al. (1979)	Entrainment, multicomponents, steady-state operation	$D_c - 0.1524$ $L = 0.23-0.25$ $H = 0.75-4.00$	Air	Perforated plate
Lin et al. (1980)	Entrainment and elutriation, multicomponents, steady-state operation	$D_c = 0.60 \times 0.60$ $L = 0.25$ $H = 0.63-3.27$	Air	Grid

Particles	Size of fines to be entrained (μm)	Size of coarse solids in parent bed (μm)	Initial weight fraction of fines (%)	Superficial gas velocity (m/sec)	Internals
Sand, glass seed, refractory	85–500	310–1640	5–20	0.92–1.62	None
FCC catalyst with size distribution	20–150	—	—	0.30–0.72	None
Glass spheres, coal powder	40–140	100–280	6–10	0.22–1.32	None
Glass beads, sand, stainless balls, lead balls, Neobeads	60–800	141–2300	—	1.28–2.70	None
Coal ash with size distribution	0–1400	1400–3170	—	0.61–2.44	None
Sand, shot, alumina	38–327	150–355	5–75	0.60–3.00	None
Sand with size distribution	37–356	356–2360	10	0.90–3.60	With or without immersed heat transfer tubes
Sand	22–180	Coarse: 443–1095; Medium: 22–1095	10–20	0.61–1.25	None
Sand/char with size distribution	0–125 (char)	125–419	0.01–1	0.10–0.30	None

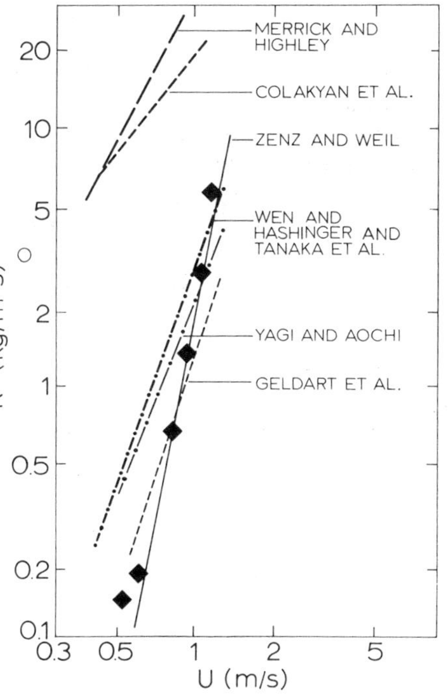

FIGURE 19 Specific elutriation rate constant versus gas velocity for 64-μm size fraction. (From George and Grace, 1981.)

emphasized, however, that owing to the physical complexity of the elutriation process, these correlations are frequently inapplicable outside the range of conditions for which they were determined. Furthermore, there is evidence that wall effects can have an influence on particle carryover, so that bed diameter should be included in a correlation as an important variable; this is seldom done, however. Table 5 indicates the range of conditions, particle size, and so on, for which the previously listed correlations were obtained.

A number of investigators have compared the elutriation correlations with their own experimental results and in many cases have found poor agreement (see, e.g., Gugnoni and Zenz, 1980). George and Grace (1981), however, found the Zenz-Weil correlation to fit their data quite well over a range of conditions, whereas the relationships of Merrick and Highley (1972) and Colakyan et al. (1979) were generally poor (Fig. 19). It would thus seem that a mechanistic model independent of empirical measurements would be preferable for predicting elutriation rates under any foreseeable circumstances, but despite some attempts to establish such a model (see Chen and Wen, 1982; George and Grace, 1981), progress so far has been limited.

INTERPHASE GAS EXCHANGE

It is generally accepted that gas bubbles in fluidized beds are devoid of particles and that except in the cloud and wake regions, the opportunity

for gas-solid interaction is limited. As a result of the "porous" nature of the bubble-cloud boundary, however, gas exchange can occur between the unreactive bubble phase and the reactive dense phase, and this can lead to a significant increase in the extent of conversion in a fluidized-bed reactor. The precise mechanism of the gas-exchange process has been the object of much speculation, and a number of methods have been proposed for calculating its magnitude under given conditions. There is, however, a second mechanism by means of which bubble gas can contact bed particles, and this is during the coalescence process when gas "leaks" from a trailing to a leading bubble through the intervening dense phase. It will be convenient first to examine the proposed transfer mechanisms for single bubbles and then to consider what modifications need to be made to take account of bubble coalescence.

Gas Exchange from Single Bubbles

Chavarie and Grace (1976) identified three "classical" models of mass transfer from single bubbles that have been discussed in the literature. The first is based on the assumption that transfer is controlled by diffusion across the boundary between the cloud and interstitial phase, and was developed principally by Rowe et al. (1966, 1971), Toei et al. (1969), and Chiba and Kobayashi (1970). The Rowe approach, for example, assumes gas in the bubble and cloud to be completely mixed and the transfer rate to be a function of a single coefficient k_{gc} (m/s):

$$\frac{-dV_b \cdot C_{bc}}{dt} = k_{gc} S_{ci} (C_{bc} - C_i) \quad \text{mol/s} \tag{71}$$

where V_{bc} is the volume of the bubble plus cloud, S_{ci} the surface area of the cloud-dense phase interface, and C_{bc} and C_i the concentrations of gas in the bubble-cloud and interstitial phases, respectively. The gas-solid system is then modeled on the basis of transfer from a drop of one immiscible liquid rising through another, the drop diameter being equal to the cloud diameter d_c. The Sherwood number, Sh_c, for mass transfer from a sphere of uniform composition in the appropriate range of Reynolds number ($30 < Re_c < 2000$) is given, according to Rowe et al. (1965a), by the expression

$$Sh_c = \frac{k_{gc} d_c}{D} = 2 + 0.69\, Sc^{1/3} Re_c^{1/2} \tag{72}$$

where the Schmidt number,

$$Sc = \frac{\mu}{\rho D} \tag{73}$$

and the Reynolds number,

$$Re_c = \frac{U_R d_c \rho}{\mu} \tag{74}$$

with D being the molecular diffusivity of the gas and U_R the relative velocity between bubble and interstitial gas: $U_R = U_b - U_i$. The bubble

cloud-to-interstitial phase exchange rate (volume of gas per unit volume of bubble per second) is then

$$Q_{bci} = \frac{k_{gc}\pi d_c^2 \varepsilon}{V_{bc}} = \frac{3.9\varepsilon D\, Sh_c}{V_{bc}^{2/3}} \tag{75}$$

A second approach to interphase gas exchange was explored by Davidson and Harrison (1963), who disregarded the bubble cloud and considered transfer to occur directly between bubble and interstitial phases by a combination of convective and diffusive flow. The molar gas exchange rate is

$$-\frac{dV_b C_b}{dt} = (q + k_{bi}S_{bi})(C_b - C_i) \tag{76}$$

where the convective term, q, is given for a spherical bubble by

$$q = \frac{3\pi}{4} U_{mf} d_b^2 \tag{77}$$

d_b being the bubble diameter, V_b the bubble volume, and S_{bi} the interphase surface area.

The mass transfer coefficient k_{bi} is evaluated on the basis of the analogy with diffusion from a gas bubble rising in a liquid:

$$k_{bi} = 0.975 D^{1/2} d_b^{-1/4} g^{1/4} \tag{78}$$

The overall rate of exchange is then given by

$$Q_{bi} = \frac{q + k_{bi}S_{bi}}{V_b} \tag{79}$$

$$= 4.5\left(\frac{U_{mf}}{d_b}\right) + 5.85\left(\frac{D^{1/2}g^{1/4}}{d_b^{5/4}}\right) \tag{80}$$

The third "classical" model is that of Kunii and Levenspiel (1969), who combined elements of the foregoing two methods and considered the exchange process to take place in two stages. The first stage involves transfer from the bubble to the cloud region and is described in terms of a coefficient Q_{bc} identical with the bubble-to-interstitial phase coefficient Q_{bi} of Davidson and Harrison. In the second stage gas is considered to be transferred from cloud to interstitial phase by a diffusive process, described by the coefficient Q_{ci}, which may be evaluated on the basis of the penetration theory of Higbie (1935):

$$Q_{ci} = 6.78\left(\frac{\varepsilon_{mf} D U_b}{d_b^3}\right)^{1/2} \tag{81}$$

The two transfer processes are assumed to occur in series, so that the overall flow rate of gas from bubble to interstitial phase, Q_{bi}, is given by

$$\frac{1}{Q_{bi}} = \frac{1}{Q_{bc}} + \frac{1}{Q_{ci}} \tag{82}$$

Although the Kunii-Levenspiel model assumes separate resistances to flow at the bubble-cloud interface and at the cloud-dense phase boundary, in practice the latter is normally the dominant factor, so the model predictions are not very different from those derived from the purely diffusive models. This is shown by the values given in Table 6, where exchange coefficients from the three models are calculated for a system in which a single spherical bubble of ozone ($D_G = 2 \times 10^{-5}$ m^2/s) with a diameter of 0.10 m is assumed to be injected into a bed of particles of $U_{mf} = 0.053$ m/s incipiently fluidized with air.

It is apparent that the model of Davidson and Harrison in which the convective through-flow term is dominant over the diffusion term gives a value an order of magnitude greater than those derived on the basis of the mainly diffusive models.

Charvarie and Grace (1976) and Sit and Grace (1978) carried out gas exchange experiments in a two-dimensional bed into which bubbles of air containing low concentrations of ozone were injected. They found values of overall transfer coefficients, corrected for unsteady bubble growth and entrance effects, that were somewhere between those predicted by the through-flow- and diffusion-dominated models. They concluded that for

TABLE 6 Predicted Interphase Gas Exchange Coefficients for Ozone Bubbles in an Air-Fluidized Bed

Model	Exchange coefficient	Calculated value (s^{-1})
Partridge and Rowe (1966)	$Q_{bci} = \frac{3.9\varepsilon D\ Sh_c}{V_{bc}^{2/3}}$	0.245
Davidson and Harrison (1963)	$Q_{bi} = 4.5\left(\frac{U_{mf}}{d_b}\right) + \left(\frac{D^{1/2}g^{1/4}}{d_b^{5/4}}\right)$	3.21
Kunii and Levenspiel (1969)	$Q_{bi} = \frac{Q_{bc}Q_{ci}}{Q_{bc} + Q_{ci}}$	0.466

Source: After Charvarie and Grace (1976).

single bubbles the best agreement with experiment was shown by the model of Murray (1965, 1966) (which ascribes a modest value to the through-flow), modified on the basis of penetration theory to give an overall transfer coefficient for spherical bubbles of

$$Q_{bi} = \frac{1.5U_{mf}}{d_b} + \frac{12}{d_b^{3/2}} \frac{(D\varepsilon_{mf}U_b)^{1/2}}{\pi} \tag{83}$$

The value of Q_{bi} from Eq. (83) calculated for the ozone-air system considered in Table 6 is 1.34 s^{-1}.

Gas Exchange in Systems of Multiple Bubbles

As already mentioned, interphase gas exchange can increase considerably during the process of bubble coalescence, so that in vigorously bubbling beds the extent of gas-solid contact leading to chemical reaction can be expected to be greater than would be predicted on the basis of the single-bubble models discussed in the preceding section. Single-bubble models are nevertheless useful in suggesting a basic mechanism of transfer into which the effects of coalescence may subsequently be incorporated.

There is as yet no convincing theory to account for all the complexities of bubble coalescence. It is known, however, that the process normally proceeds by a trailing bubble accelerating into the rear of a leading one and producing a single bubble whose volume is greater than the sum of the volumes of the original pair. In a continuation of the study mentioned earlier, Sit and Grace (1981) investigated the interphase transfer of ozone during coalescence in a two-dimensional bed. From ciné film of the process, they divided coalescence into four periods:

1. An approach period between the time of bubble formation and the point of contact of the nose of the trailer with the wake region of the leader
2. An encroachment period when the nose of the trailer passed through the wake region and touched the lower edge of the leader
3. A period during which coalescence proceeded
4. A postcoalescence period

They found the greatest change in ozone concentration to occur during the approach and encroachment periods and that 77% of the gas transfer occurred within the first 50% of the total time span. Furthermore, they estimated the mass transfer coefficient of the leading bubble to be two to three times greater during the approach period than when the bubble was in isolation.

On the basis of these observations, Sit and Grace proposed that during coalescence the convective through-flow process is increased while the diffusive processes are little affected. They thus proposed applying an empirical enhancement factor of 1.8 to the through-flow term in the expression for Q_{bi} based on the Murray theory [Eq. (83)]. However, since from visual observation 40% of the bubbles are at any instant in the early stages of coalescence, a more realistic estimate of the enhancement factor was

suggested to be $(1.8 \times 0.4 + 0.6) = 1.32$, giving the modified expression for the exchange coefficient:

$$\overline{Q}_{bi} = \frac{2U_{mf}}{\overline{d}_b} + \frac{12}{\overline{d}_b^{3/2}} \frac{(D\varepsilon_{mf}\overline{U}_b)^{1/2}}{\pi} \qquad (84)$$

where $\overline{U}_b$ is the mean velocity of bubbles in the bed and $\overline{d}_b$ their mean diameter. Equation (84) appears to give the best estimate of the interphase transfer coefficient in freely bubbling, three-dimensional beds of nonabsorbing particles.

REACTOR MODELS

A large number of fluidized-bed reactor models have been described in the literature during the last three decades, and although most of them are designed for application to gas-solid catalytic reactions, suitable modifications for noncatalytic processes are relatively easily incorporated (Kunii and Levenspiel, 1969). The essential feature of all the models is the description given of the bed hydrodynamics, and since this involves the interplay of a number of physical variables, the number of possible permutations and combinations and hence the number of models, is quite large. Several reviews have been published and these should be referred to for detailed comparisons (Pyle, 1970; Grace, 1971; Rowe, 1972; Yates, 1975; Horio and Wen, 1977; van Swaaij, 1978). In general, the starting point of any model is a consideration of the division of flow between the dense phase and the lean or bubble phase, and in the majority of cases the two-phase theory is assumed to apply (i.e., it is assumed that all gas flowing into the bed in excess of that required for minimum fluidization passes through in the form of bubbles). It is then necessary to consider the extent of conversion in each phase, the amount of transfer between phases as gas flows through, and finally how the two streams are combined at the bed surface to give the outlet reactant concentration. In addition to the events taking place in the bed itself under conditions of steady two-phase flow, it is necessary for completeness to examine the possiblity of chemical reaction occurring in both the grid zone, where fresh gas enters the bed and bubbles are in embryonic form, and in the freeboard region above the bed surface, where high concentrations of active entrained particles can be found. We will, however, examine first some of the purely two-phase models and defer a consideration of the effect of these end zones.

The models to be considered have been chosen to represent three approaches to the problem of reactor design. The first (Davidson and Harrison, 1963) models bed performance from first principles and uses the minimum amount of empirical information. The second (Kunii and Levenspiel, 1969), while based on two-phase theory, imposes an operating condition on the bed (an artifically controlled bubble size) that enables a relatively simple solution to be obtained. The third model (due to the Shell company) incorporates theory with a large amount of empirical information obtained from pilot plant experiments and is a fairly typical example of industrial methodology.

Model of Davidson and Harrison (1963)

This was one of the earliest models to be based on the two-phase theory, and although its basic assumptions represent an oversimplification of the physics of fluidization, it has been successfully applied in a number of cases. The assumptions on which the model is based are as follows:

1. Gas bubbles are evenly distributed throughout the bed and are of equal size.
2. Reaction occurs only in the dense phase, where gas flowing at U_{mf} is either completely mixed or in plug flow.
3. Gas is exchanged between the bubble and dense phases by the convection-diffusion mechanism discussed earlier.

A material balance on reactant A, with initial concentration C_{A_0}, leaving the interstitial phase in the case where this is assumed to be completely mixed gives the concentration, C_{A_i}, in terms of three dimensionless parameters, X the interphase exchange term, β the bubble fraction, and k' the reaction kinetic term:

$$C_{A_i} = \frac{C_{A_0}(1 - e^{-X})}{1 - \beta e^{-X} + k'} \tag{85}$$

where

$$X = \frac{Q_{bi}H}{U_b V_b} \tag{86}$$

$$\beta = 1 - \frac{U_{mf}}{U} \tag{87}$$

and

$$k' = \frac{kH_{mf}}{U} \tag{88}$$

Here H is the expanded bed height, H_{mf} the height at minimum fluidization, U_b the bubble velocity, V_b the bubble volume, and k a first-order reaction rate constant. The interphase transfer coefficient Q_{bi} is given by Eq. (80) and this leads to an expression for X in terms of the bubble diameter d_b:

$$X = \frac{H}{U_b}\left[4.50\left(\frac{U_{mf}}{d_p}\right) + 5.85\left(\frac{D^{1/2}g^{1/4}}{d_b^{5/4}}\right)\right] \tag{89}$$

For reactions involving mixtures of gases, Davidson and Harrison recommended the use of the methods of Arnold or Hirschfelder (Reid and Sherwood, 1958) for the estimation of the gas-phase diffusion coefficients, *D*.

A separate mass balance on the bubble phase leads to

$$C_{A_b} = C_{A_i} + (C_{A_0} - C_{A_i})e^{-X} \tag{90}$$

and the overall outlet concentration of A, C_A, is then found by substituting for C_{A_i} and C_{A_b} from Eqs. (85) and (90) in

$$C_A = \beta C_{A_b} + (1 - \beta)C_{A_i} \tag{91}$$

giving the overall fraction of reactant unconverted:

$$\frac{C_A}{C_{A_0}} = \beta e^{-X} + \frac{(1 - \beta e^{-X})^2}{1 - \beta e^{-X} + k'} \tag{92}$$

Inspection of Eq. (92) shows that for large values of the interphase exchange term X, the model, as would be expected, reduces to that of a CSTR. Furthermore, Eq. (92) predicts that conversion can never be complete even for very fast values of k':

$$\frac{C_A}{C_{A_0}} \rightarrow \beta e^{-X}$$

as a result of the bypassing action of bubble gas.

If the dense-phase gas is assumed to be in plug flow, mass balance equations on both phases lead to the following expression for the fraction of reactant leaving the bed unconverted:

$$\frac{C_A}{C_{A_0}} = \frac{1}{m_1 - m_2}\left[m_1 e^{-m_2 H}\left(\frac{1 - m_2 H U_{mf}}{XU}\right) - m_2 e^{-m_1 H}\left(\frac{1 - m_1 H U_{mf}}{XU}\right)\right] \tag{93}$$

m_1 and m_2 being found from

$$2H(1 - \beta)m = (X + k') \pm [(X + k')^2 - 4k'X(1 - \beta)]^{1/2} \tag{94}$$

where $m = m_1$ with the positive sign and $m = m_2$ with the negative sign. In both forms of the model the variable playing the most important role is the bubble diameter, d_b, since it determines, among other things, the extent of interphase gas transfer. In the earlier publications an average value of d_b for the whole bed was estimated from the reaction data of other workers, but more recently, Darton (1979) has proposed a method whereby the variation in bubble diameter with bed height may be allowed for. The method is based on a deterministic model of bubble growth by stagewise coalescence and leads to the following expression for the interphase transfer term:

$$X = 32.0 \left\{ \frac{\varepsilon_{mf}^{1/2} D^{1/2} g^{1/10}}{(U - U_{mf})^{7/10} A_0^{1/5}} \left[1 - \left(\frac{4A_0^{1/2}}{H + 4A_0^{1/2}} \right)^{2/5} \right] \right\} \tag{95}$$

In this equation A_0 is the area of distributor plate per orifice and H is the expanded bed height estimated by iteration from

$$\frac{H}{H_{mf}} = 1 + 2.0 \left\{ \frac{(U - U_{mf})^{4/5}}{(gH_{mf})^{2/5}} \left[\left(\frac{4A_0^{1/2}}{H_{mf}} + \frac{H}{H_{mf}} \right)^{3/5} - \left(\frac{4A_0}{H_{mf}} \right)^{3/5} \right] \right\} \tag{96}$$

The use of Eq. (95) in conjunction with the "dense phase completely mixed" equation (92) was shown by Darton to give reasonably good predictions of the conversions measured by Fryer and Potter (1976) using a 0.229-m-diameter reactor fitted with a bubble cap distributor with 61 caps, giving $A_0 = 675\ \text{mm}^2$.

Model of Kunii and Levenspiel (1969)

This model is designed specifically for beds that are vigorously bubbling but not slugging and in which the bubble size is controlled at a uniform value throughout by the use of internals. As with the foregoing model of Davidson and Harrison, the average bubble diameter is the single most important parameter since it determines the bubble velocity by Eq. (51) and the extent of interphase gas exchange by Eq. (80). Conversion in the interstitial phase at gas velocities in excess of about $3U_{mf}$ is considered to be negligible, the great majority of gas flowing as bubbles. The bubble gas is assumed to be in plug flow and the fraction of reactant emerging unconverted at the bed surface is

$$\frac{C_A}{C_{A_0}} = \exp(-k_f) \tag{97}$$

where k_f is an effective overall rate constant given by

$$k_f = \frac{Hk_r}{U_b} \left\{ \gamma_b + \frac{1}{\dfrac{k_r}{Q_{bc}} + \dfrac{1}{\gamma_c + [1/(k_r/Q_{ci}) + 1/\gamma_i]}} \right\} \tag{98}$$

Here k_r is the first-order reaction rate constant, Q_{bc} and Q_{ci} are the interphase exchange rates defined in Eqs. (8) ($Q_{bi} \equiv Q_{bc}$) and (81), and the γ terms represent the ratios of solids dispersed in bubbles, γ_b, clouds, γ_c, and interstitial phase, γ_i, to the total volume of bubbles in the bed:

$$f_b(\gamma_b + \gamma_c + \gamma_i) = (1 - \varepsilon_{mf})(1 - f_b) \tag{99}$$

f_b is the fraction of bed occupied by bubbles, a term also dependent on bubble diameter:

$$f_b = \frac{U - U_{mf}}{U_b} \tag{100}$$

The degree of gas-solid contact, and hence the extent of reaction, depends strongly on the values of the γ terms, and the model in its simplest form takes $\gamma_b = 0$ while γ_c is calculated from

$$\gamma_c = (1 - \varepsilon_{mf}) \left[\frac{3U_{mf}/\varepsilon_{mf}}{0.711(gd_b)^{1/2} - U_{mf}/\varepsilon_{mf}} + \frac{V_w}{V_b} \right] \tag{101}$$

where the ratio of bubble volume V_b to wake volume V_w is found from the empirical correlation of Rowe and Partridge (1965). The value of γ_i is found by substitution of γ_b and γ_c in Eq. (99). The Kunii-Levenspiel model was found by Chavarie and Grace (1975) to show the best agreement of all the several models they tested against experimental results of the ozone decomposition reaction in a large two-dimensional fluidized bed, although they found some of its simplifying assumptions to be self-compensating.

A much more sophisticated treatment of the same basic model has been developed by Potter and his co-workers who included a consideration of reaction in the dense phase of the bed and incorporated equations to allow for the change in bubble size with bed height. Their so-called *counter-current back mixing model* has been well reviewed by Potter (1978).

Good for summary

Shell Model

The two models discussed earlier although firmly based on the two-phase theory of fluidization, do incorporate empirical relationships derived from experiments with laboratory-scale equipment, and this has caused the validity of their application to large industrial units to be questioned. A number of different approaches to the modeling of large beds have therefore been explored, and although most remain hidden behind the cloak of industrial secrecy, one such approach has been discussed fairly fully in the open literature (van Swaaij and Zuiderweg, 1972, 1973; de Vries et al., 1972). The Shell company, faced with designing of fluidized bed reactor for their solid-catalysed hydrogen chloride oxidation process, adopted a combined theoretical and empirical attack on the problem in which data from a number of beds of increasing diameter were used to obtain scaleup correlations. The theoretical basis was the two-phase theory model of van Deemter (1967), which assumes that:

1. Bubble gas is in plug flow.
2. Gas through-flow in the dense phase is negligible.
3. Gas mixes in the dense phase by a process of eddy diffusion.

As with the Davidson-Harrison model, mass balances on reactant gas flowing into the two phases leads to an equation for the bubble phase:

$$\frac{\partial C_{Ab}}{\partial \xi} + N_\alpha(C_{A_b} - C_{A_i}) = 0 \qquad (102)$$

$$N_\alpha(C_{A_i} - C_{A_b}) - \frac{1}{N_E}\frac{\partial^2 C_{A_b}}{\partial \xi^2} + N_r C_{Ai} = 0 \qquad (103)$$

Here ξ is the dimensionless bed height (h/H), and N_α, N_E, and N_r are dimensionless groups defined as follows:

1. The number of interphase transfer units, $N_\alpha = k'_g H/U$, where k'_g is a mass transfer coefficient
2. The number of interstitial phase mixing units, $N_E = UH/f_i E$ where f_i is the volume fraction of bed occupied by interstitial gas and E is a mixing coefficient
3. The number of reaction units, $N_r = k'H/U$, where k' is an effective rate constant for dense phase

The two hydrodynamic groups N_α and N_E were found from the results of tracer experiments in beds 0.3, 0.6, and 1.5 m in diameter. Thus from the residence-time distribution of tracer gas, the following expression for the height of a transfer unit, H_α, was obtained:

$$H_\alpha = \left(1.8 - \frac{1.06}{D^{1/4}}\right)\left(3.5 - \frac{2.5}{H^{1/4}}\right) \qquad (104)$$

Then $N_\alpha = H/H_\alpha$. Equation (104) was predicted to be valid for the microspheroidal catalyst in beds with heights and diameters of up to 10 m. The term N_E was found from values of E measured in experiments with labeled solid particles in pilot-scale beds up to 3 m in diameter; E is plotted as a function of bed diameter in Fig. 20, and values of N_r and N_α are given as functions of solids fines content and gas velocity in Table 7. The dimensions of the full-scale reactor were calculated in the following way. Bed diameter was found in a straightforward way from known stoichiometry of the process and a realistic value of the fluidizing gas velocity. An estimated figure for bed height was put into Eq. (104) and a value of H_α, and hence N_α, calculated. The latter, along with the empirical N_E and N_r terms, were then substituted into the solutions of the two simultaneous equations (102) and (103) and an overall reactant conversion calculated. The difference between this calculated conversion and the required value was then adjusted by choosing a second value of the bed height and repeating the calculation. The method, although costly in terms of pilot plant work was claimed to result in a satisfactory final design.

Model for Reaction in the Distributor Region

As mentioned earlier, there is experimental evidence to show that up to the point of detachment of a bubble from an orifice in the distributor of a fluidized bed, the volumetric flow of gas through the orifice is greater than that leading to the formation of a visible bubble, and furthermore

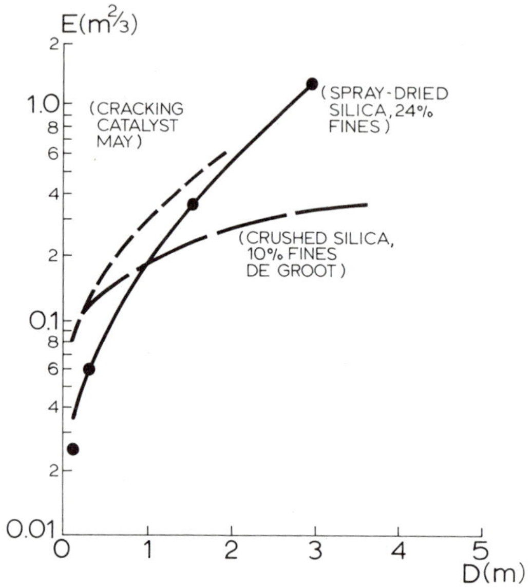

FIGURE 20 Mixing coefficient as a function of bed diameter. (From de Vries et al., 1972.)

that equalization of flow is observed only after the bubble has risen some distance into the bed. The sequence of events is illustrated in Figs. 11 and 12, and the conclusion that is drawn from the visual evidence is that during the period up to the time of detachment of the bubble, t_d, the unaccounted or invisible gas has passed into the dense phase, causing the particles to expand to a void fraction ε in excess of ε_{mf}. The time t_d is clearly related to the frequency of bubble formation n (Hz) by

$$t_d = \frac{1}{n} \tag{105}$$

TABLE 7 Empirical Parameters Used in Shell Chlorine Process Model

Percent fines	u (m/s)	Conversion relative to equilibrium	H (m)	N_r	N_α	H_α (m)
20	0.2	95.7	10.0	15	4.0	2.50
17	0.22	95.0	10.4	20	3.5	2.97
12	0.2	93.5	9.7	33	3.0	3.26
7	0.2	91.0	10.0	34	2.56	3.86

Source: de Vries et al. (1972).

and during this period the volume of gas entering the dense phase and reacting is $Q_0t_d(1 - x_0)$, where x_0 is the fraction of flow that forms the visible bubble. Further, if we assume that the gas entering the dense phase undergoes a pseudohomogeneous first-order reaction with the particles at a rate given by

$$-r_a = (1 - \varepsilon)k_s C_A \tag{106}$$

where k_s is the first-order rate constant, the fraction of reactant A remaining unconverted after time t is

$$\frac{C_A}{C_{A_0}} = \exp[-(1 - \varepsilon)k_s t] \tag{107}$$

Now if the time taken for the bubble gas to reach equilibrium (i.e., the time for the bubble to form completely) is t_e and if the mean residence time of the gas in the expanded dense phase is $0.5t_e$, then Eq. (107) becomes

$$\frac{C_A}{C_{A0}} = \exp[-0.5(1 - \varepsilon)k_s t_e] \tag{108}$$

When the bubble has formed completely, the gas it contains will be made up partly of that which has been in contact with the dense phase and partly of that which has not, so that at equilibrium the fraction of A remaining unreacted inside the bubble will be

$$\frac{C_{Ab}}{C_{A0}} = \frac{Q_0t_dx_0 + Q_0t_d(1 - x_0)(C_A/C_{A0})}{Q_0t_d} \tag{109}$$
$$= x_0 + (1 - x_0)\exp[-0.5(1 - \varepsilon)k_s t_e]$$

Assuming that a value of the rate constant k_s is available from auxiliary experiments in, for example, a fixed-bed reactor, the bubble-phase concentration of reactant gas may be calculated once the three remaining parameters x_0, ε, and t_e are known.

In the present work the proportion of incoming gas appearing as a visible bubble at the orifice ($x_0 = 0.36$) and at a position 25 cm above the orifice ($x_{25} = 0.79$) have been measured, so by a linear extrapolation from these two values the height at which $x = 1.0$, and hence at which the bubble gas has reached equilibrium, may be found. If this height is H_e, the time for the bubble to reach equilibrium is clearly

$$t_e = \frac{H_e}{U_b} \tag{110}$$

so that if the bubble velocity is known, the residence time of gas in the expanded phase may be calculated. Now the velocity of a bubble in a

fluidized bed is normally expressed in terms of its diameter, d_b, by the Davies-Taylor relationship

$$U_b = \left(\frac{gd_b}{2}\right)^{1/2} \tag{111}$$

the observed coefficient of proportionality being unity. As approximately one-fourth of the volume of the sphere centered on the bubble is occupied by bubble wake, the volume of gas forming the bubble is

$$V_b = \frac{3}{4}\frac{\pi}{6} d_b^3 \tag{112}$$

$$= x_0 \frac{Q_{0r}}{n_b} \tag{113}$$

Hence

$$d_b = \left(\frac{8}{\pi}\right)^{1/3} V_b^{1/3} \tag{114}$$

$$U_b = \left(\frac{g}{2}\right)^{1/2}\left(\frac{8}{\pi}\right)^{1/6} V_b^{1/6} \tag{115}$$

$$= 0.259V_b^{1/6} \tag{116}$$

and combining Eqs. (110), (113), and (116), we have

$$t_e = 3.86H_e \left(\frac{n_b}{Q_0}\right)^{1/6} \tag{117}$$

The value of the third parameter in Eq. (109), the voidage of the expanded dense phase, ε, is at present unknown, but reasonable values in the range 0.4 to 0.7 may be assumed for the purpose of an illustrative calculation. Combining Eqs. (109) and (117), the fraction of reactant A remaining unreacted in the bubble phase at a height H_e above the distributor is

$$\frac{C_{Ab}}{C_{A0}} = x_0 + (1 - x_0) \exp\left[-1.93(1 - \varepsilon)k_s H_e \left(\frac{n_b}{Q_0}\right)^{1/6}\right] \tag{118}$$

Example The value of H_e calculated from the foregoing measurements of visible bubble volume at two bed heights (0 and 0.25 m above the orifice) is 0.375 m. Other values used in the calculation are as follows:

$x_0 = 0.36$

$n_b = 7$ Hz

$$Q_0 = 5 \times 10^{-4}\ m^3/s$$

$$k_s = 1,\ 5,\ 10\ s^{-1}$$

$$\varepsilon = 0.4,\ 0.5,\ 0.6,\ 0.7$$

Table 8 gives the calculated values of C_{Ab}/C_{A0} at height H_e above the distributor and it is clear that, particularly in the case of fast reactions, the model predicts considerable conversion in the distributor zone.

Reactions in the Freeboard

When bubbles burst at the surface of a fluidized bed, the solid particles released into the freeboard can cause further reaction to take place with any gas that has passed through the bed unreacted. Reaction in the freeboard region is not subject to the same constraints as that which occurs in the bed itself, and it can give rise to severe temperature excursions in the gas space above the surface, leading to damage of both bed structures and catalyst particles.

It is well known, for example, that uncontrolled reactions in the freeboard of fluidized-bed phthalic anhydride plants were the cause of considerable problems, as also were afterburning reactions in catalytic cracker regenerators.

Yates and Rowe (1977) developed a simple model of a catalytic reaction based on the assumption that the freeboard contained perfectly mixed, equally dispersed particles derived from bubble wakes. The physical form of the model is illustrated in Fig. 21. Assuming the two-phase theory to apply to the fluidized bed and the fraction of wake particles ejected above the surface to be f, the rate of particle ejection is

$$\frac{(1 - \varepsilon_{mf})fQ_b}{3} = \frac{f(U - U_{mf})A(1 - \varepsilon_{mf})}{3} \tag{119}$$

TABLE 8 Calculated Fraction Unconverted Leaving Distributor Zone

	$\frac{C_{Ab}}{C_{A0}}$		
ε	$k_s = 1\ s^{-1}$	$k_s = 5\ s^{-1}$	$k_s = 10\ s^{-1}$
0.40	0.876	0.577	0.434
0.50	0.895	0.620	0.466
0.60	0.914	0.672	0.512
0.70	0.934	0.733	0.577

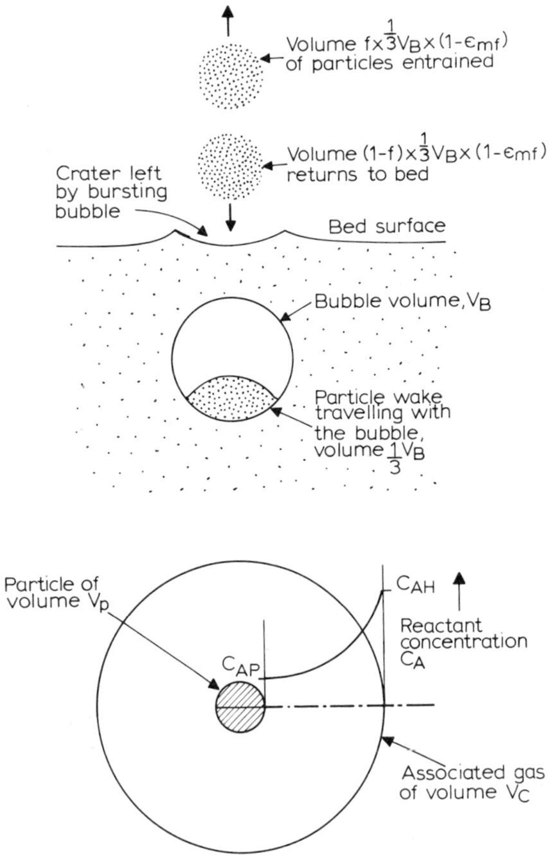

FIGURE 21 Basis of freeboard reaction model. (From Yates and Rowe, 1977.)

where Q_B is the volumetric flow rate of bubble gas and A is the bed cross-sectional area. The total holdup of particles in the freeboard, V, is

$$V = \left(\frac{H_F}{U - U_T} \right) f \frac{(U - U_{mf})A(1 - \varepsilon_{mf})}{3} \tag{120}$$

where H_F is the freeboard height and U_T is the particle terminal fall velocity. The freeboard voidage is then

$$\varepsilon_f = 1 - \left[\frac{f(1 - \varepsilon_{mf})}{3} \left(\frac{U - U_{mf}}{U - U_T} \right) \right] \tag{121}$$

Now if the particles are equally spaced, the volume V_c of the gas cell surrounding each particle is given by

$$V_c = \frac{3V_p}{f(1 - \varepsilon_{mf})}\left(\frac{U - U_{mf}}{U - U_T}\right) \tag{122}$$

where V_p is the particle volume. The rate of change of concentration of reactant A within a cell is a function of the particle surface area A_p, the concentration driving force ($C_{Ah} - C_{Ap}$), and a mass transfer coefficient h_M:

$$-\frac{dC_A}{dt} = \frac{h_M A_p}{V_c}(C_{Ah} - C_{Ap}) \tag{123}$$

while the rate of change of C_A at the particle surface is determined by the rate of chemical reaction. If the freeboard gas is assumed to be in plug flow, the equations lead to the following expression for the fraction of reactant unconverted at any height h above the bed surface:

$$\frac{C_{Ah}}{C_{As}} = \exp\left\{-\frac{fh}{3(U - U_T)\left[\dfrac{d_p}{6h_M(1 - \varepsilon_{mf})} + \dfrac{1}{k}\right]}\right\}$$

where C_{As} is the reactant concentration at the bed surface and k is an assumed first-order reaction rate constant. The corresponding equation for mixed flow of gas in the freeboard is

$$\frac{C_A}{C_{As}} = \frac{1}{\left\{1 + \dfrac{fH_F}{3(U - U_T)\left[\dfrac{d_p}{6h_M(1 - \varepsilon_{mf})} + \dfrac{1}{k}\right]}\right\}} \tag{125}$$

The mass transfer coefficient h_M may be found from the expression of Rowe et al. (1965a):

$$Sh = 2 + 0.69Sc^{1/3}Re_T^{1/2} \tag{126}$$

where the Sherwood number,

$$Sh = \frac{h_M d_p}{D} \tag{127}$$

the Schmidt number,

$$Sc = \frac{\mu}{\rho D} \tag{128}$$

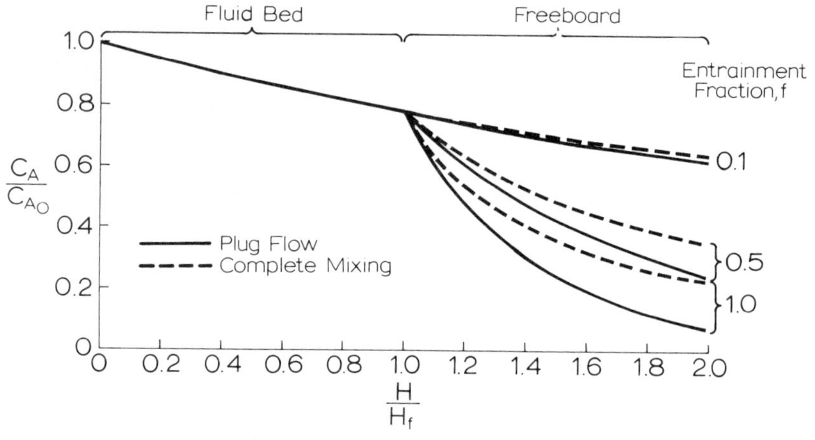

FIGURE 22 Conversion versus bed height for $k = 5\ s^{-1}$.

and the Reynolds number,

$$Re_T = \frac{U_T d_p \rho}{\mu} \tag{129}$$

The terminal fall velocity may be obtained from graphical correlations between Re_T and the velocity independent group $2d_p\rho g(\rho_s - \rho_g)/3\mu^2$ (Kunii and Levenspiel, 1969), and hence the Sherwood number and h_M may be calculated from Eqs. (128) and (129).

Solutions to the equations were obtained for various combinations of the parameters, and it was shown that under some circumstances a greater level of conversion can be obtained in the freeboard than in the fluidized bed itself (Fig. 22). Experimental support for these conclusions has been provided by Miyauchi and Furusaki (1974) and Furusaki et al. (1976), who have also developed theoretical models of the effect.

Conclusions

It would seem from the discussion presented in this section that three zones need to be considered in the design of a fluidized-bed reactor: First, the entry zone near the distributor where fresh reactant gas first comes into contact with bed particles and where bubbles are in the early stages of development; second, the fluidized bed itself, where in general reaction is controlled by mass transfer between the bubble and interstitial phases; and third, the freeboard zone above the bed surface. The necessary design equations for each zone are quite different and must be examined thoroughly for the full range of potential operating variables before the overall performance of a reactor can be predicted with confidence.

NOTATION

A c/s area of bed, m^2

C_A concentration of reactant A, mol/m^3

C_D drag coefficient (single particle) $= F/\frac{\pi}{4}d_p^2\ \frac{1}{2}\rho_f u^2$

C_G specific heat of gas, J/mol·K

$\mathcal{D}$ molecular diffusivity, m^2/s

D bed diameter, m

d_p particle diameter, μm

f_B volume fraction bubble holdup

F drag force on a single particle, N

g acceleration of gravity, m/s^2

h height above distributor, m

$h_{B/w}$ heat transfer coefficient: bed to wall, $W/m^2 \cdot K$

$h_{G/s}$ heat transfer coefficient: gas to solids, $W/m^2 \cdot K$

H total bed height above distributor, m

K thermal conductivity, W/m·K

K a constant

$K_{G/s}$ mass transfer coefficient, m/s

m exponent in Eq. (31)

M total mass of bed, kg

M simple mixing index $= X/\overline{x}$ (Fig. 13)

n bubble frequency crossing horizontal plane, s^{-1}

n exponent in Eq. (45)

N bubble number concentration, m^{-3}

p absolute pressure, N/m^2

Q volumetric flow rate, m^2/s

S particle external surface per unit volume, m^{-1}

U superficial fluid velocity, m/s

V_B bubble volume, m^3

Greek Letters

α velocity ratio (U_B/U_i)

Δp pressure drop across the bed, N/m^2

ε bed voidage

μ fluid viscosity, $N \cdot s/m^2$

ρ_{bulk} powder bulk density, kg/m^3

ρ_f fluid density, kg/m^3

ρ_s particle density, kg/m^3

Special Numbers

Nu	Nusselt number ($h_{G/s}d_p/K$)
Pr	Prandtl number ($C_G\mu/K$)
Re	Reynolds number ($Ud_p\rho/\mu$)
Sc	Schmidt number ($\mu/\rho D$)
Sh	Sherwood number ($K_{G/s}d_p/D$)

Subscripts

B	bubble
C	cloud
D	distributor plane
f	fluid
F	flotsam
G	gas
H	plane at height h
i	interstitial
I	isolated
J	jetsam
mf	minimum fluidization
mb	minimum bubbling
s	solid (particle)
T	terminal falling
W	wake

REFERENCES

Bachovchin, D. M., J. M. Beer, and A. F. Sarofim, Elutriation from fluidized beds, AIChE 72nd Meet., San Francisco (1979).

Baeyens, J. and D. Geldart, Predictive calculations of flow parameters in gas fluidized beds and fluidization behaviour of various powders, Int. Symp. Fluidization Appl., Toulouse (1973), p. 263.

Botterill, J. S. M., *Fluid-Bed Heat Transfer*, Academic Press, London (1975).

Chavarie, C. and J. R. Grace, Performance analysis of a fluidized-bed reactor, Ind. Eng. Chem. Fundam., *14*, 75, 79, 86 (1975).

Chavarie, C. and J. R. Grace, Interphase mass transfer in a gas-fluidized bed, Chem. Eng. Sci., *31*, 741 (1976).

Chiba, T. and H. Kobayashi, Gas exchange between the bubble and emulsion phases in gas-solid fluidized beds, Chem. Eng. Sci., *25*, 1375 (1970).

Colakyan, M., N. Catipovic, G. Jovanovic, and T. Fitzgerald, Elutriation from fluidized beds with immersed heat transfer tubes, AIChE 72nd Meet., San Francisco (1979).

Darton, R. C., A bubble growth theory of fluidized bed reactors, Trans. Inst. Chem. Eng., *57*, 134 (1979).

Darton, R. C., R. D. La Nauze, J. F. Davidson, and D. Harrison, Bubble growth due to coalescence in fluidized beds, Trans. Inst. Chem. Eng., *55*, 274 (1977).

Davidson, J. F. and D. Harrison, *Fluidized Particles*, Cambridge University Press, Cambridge (1963).

Davidson, J. F., D. Harrison, R. C. Darton, and R. D. La Nauze, *Chemical Reactor Theory*, Prentice-Hall, Englewood Cliffs, N.J. (1977).

Davies, R. M. and G. I. Taylor, The mechanics of large bubbles rising through extended liquids and through liquids in tubes, Proc. R. Soc., *A200*, 375 (1950).

de Vries, R. J., W. P. M. van Swaaij, G. Mantovani, and A. Heijkoop, Design criteria and performance of the commercial reactor for the Shell chlorine process, Proc. 2nd Int. Symp. Chem. React. Eng., Amsterdam (1972), p. B9–59.

Ergun, S., Fluid flow through packed columns, Chem. Eng. Prog., *48*, 89 (1952).

Fryer, C. and O. E. Potter, Experimental investigation of models for fluidized bed catalytic reactors, AIChE J., *22*, 38 (1976).

Furusaki, S., T. Kikuchi, and T. Miyauchi, Axial distribution of reactivity inside a fluid-bed contactor, AIChE J., *22*, 354 (1976).

Geldart, D., Types of gas fluidization, Power Technol., 7, 285 (1973).

Geldart, D., J. Cullinan, S. Georghiades, D. Gilvray, and D. J. Pope, The effect of fines on entrainment from gas fluidized beds, Trans. Inst. Chem. Eng., *57*, 269 (1979).

George, S. E. and J. R. Grace, Entrainment of particles from aggregative fluidized beds, AIChE Symp. Ser., *74*(176), 67 (1978).

George, S. E. and J. R. Grace, Entrainment of particles from a pilot scale fluidized bed, Can. J. Chem. Eng., *59*, 279 (1981).

Grace, J. R., An evaluation of models for fluidized-bed reactors, AIChE Symp. Ser., *67*(116), 159 (1971).

Gugnoni, R. J. and F. A. Zenz, Particle entrainment from bubbling fluidized beds, in *Fluidization*, Plenum Press, New York (1980), p. 501.

Higbie, R., The rate of absorption of a pure gas into a still liquid during short periods of exposure, Trans. AIChE, *31*, 365 (1935).

Horio, M. and C. Y. Wen, An assessment of fluidized-bed modeling, AIChE Symp. Ser., *73*(161), 9 (1977).

Jahnig, C. E., D. L. Campbell, and H. Z. Martin, History of fluidised solids development at Exxon, in *Fluidization*, Plenum Press, New York (1980), p. 3.

Knight, M. J. and P. N. Rowe, On measuring the density of fine particles by pycnometry, Chem. Eng. Sci., *35*, 997 (1980).

Knight, M. J., P. N. Rowe, H. J. MacGillivray, and D. J. Cheesman, On measuring the density of finely divided but porous particles such as fluid bed cracking catalyst, Trans. Inst. Chem. Eng., *58*, 203 (1980).

Kunii, D. and O. Levenspiel, *Fluidization Engineering*, Wiley, New York (1969).

Leva, M., *Fluidization*, McGraw-Hill, New York, (1959).

Lin, L., J. T. Sears, and C. Y. Wen, Elutriation and attrition of char from a large fluidized bed, Powder Technol., *27*, 105 (1980).
Merrick, D., and J. Highley, Particle size reduction and elutriation in a fluidized-bed process, AIChE, Symp. Ser., *70*(137), 366 (1974).
Miyauchi, T. and S. Furusaki, Relative contribution of variables affecting the reaction in fluid bed contactors, AIChE J., *20*, 1087 (1974).
Murray, J. D., On the mathematics of fluidization, J. Fluid Mech., *22*, 57 (1965).
Murray, J. D., Mathematical aspects of bubble motion in fluidized beds, Chem. Eng. Prog. Symp. Ser., *62*(62), 57 (1966).
Partridge, B. A. and P. N. Rowe, Chemical reaction in a bubbling gas-fluidized bed, Trans. Inst. Chem. Eng., *44*, 335 (1966).
Potter, O. E., Modeling fluidized-bed reactors, Catal. Rev.-Sci. Eng., *17*, 155 (1978).
Pyle, D. L., Fluidized-bed reactors: a review, Proc. 1st Int. Symp. Chem. React. Eng., Washington, D.C. (1970), p. 106.
Reid, R. C. and T. K. Sherwood, *The Properties of Liquids and Gases*, McGraw-Hill, New York, (1958).
Richardson, J. F., Incipient fluidization and particulate systems, in *Fluidization*, J. F. Davidson and D. Harrison, eds., Academic Press, London (1971).
Richardson, J. F. and W. N. Zaki, Sedimentation and fluidization, Trans. Inst. Chem. Eng., *32*, 35 (1954).
Rietema, K., Application of mechanical stress theory to fluidization, Int. Symp. Fluidization, Eindhoven (1967), p. 154.
Roberts, A. G., P. Raven, R. N. Phillips, S. N. Barker, K. K. Pillai, and P. Wood, PFBC—a 1000 hour programme, in *Fluidized Combustion*, Inst. Energy Symp. Ser., No. 4, 3.1.1 (1980).
Rowe, P. N., Drag forces in a hydraulic model of a fluidized bed: Part II, Trans, Inst. Chem. Eng., *39*, 178 (1961).
Rowe, P. N. Experimental properties of bubbles, in *Fluidization*, J. F. Davidson and D. Harrison, eds., Academic Press, London, (1971), Chap. 4.
Rowe, P. N., Fluidised bed reactors, Proc. 2nd Int. Symp. Chem. React. Eng., Amsterdam (1972), p. A9.
Rowe, P. N. Estimation of solids circulation rate in a bubbling fluidised bed, Chem. Eng. Sci., *28*, 979 (1973).
Rowe, P. N., Prediction of bubble size in a gas fluidised bed, Chem. Eng. Sci., *31*, 285 (1976).
Rowe, P. N., Some simple considerations in the design of a fluidised bed chemical reactor, Chem. Ind., 424 (1978a).
Rowe, P. N., The two phase theory and fluidised bed reactor models, Proc. 5th Int. Symp. Chem. React. Eng., Houston, ACS Symp. Ser., *65*, 436 (1978b).
Rowe, P. N. and K. T. Claxton, Heat and mass transfer from a single sphere to fluid flowing through an array, Trans. Inst. Chem. Eng., *43*, 321 (1965).
Rowe, P. N. and H. J. MacGillivray, A preliminary study of the effect of pressure on a bubbling gas fluidised bed, in *Fluidized Combustion*, Inst. Energy Symp. Ser., No. 4, 3.1.1 (1980).
Rowe, P. N. and A. W. Nienow, Particle mixing and segregation in gas fluidised beds, Powder Technol., *15*, 141 (1976).

Rowe, P. N. and B. A. Partridge, An x-ray study of bubbles in fluidised beds, Trans. Inst. Chem. Eng., *43*, 157 (1965).

Rowe, P. N., K. T. Claxton, and J. B. Lewis, Heat and mass transfer from a single sphere in an extensive flowing fluid, Trans, Inst. Chem. Eng., *43*, 14 (1965a).

Rowe, P. N., B. A. Partridge, A. G. Cheyney, G. A. Henwood, and E. Lyall, The mechanism of solids mixing in fluidised beds, Trans. Inst. Chem. Eng., *43*, 271 (1965b).

Rowe, P. N., T. J. Evans, and J. C. Middleton, Transfer of gas between bubbles and dense phase in a two-dimensional fluidised bed, Chem. Eng. Sci., *26*, 1943 (1971).

Rowe, P. N., L. Santoro, and J. G. Yates, Division of gas between bubble and interstitial phases in fluidised beds of fine powders, Chem. Eng. Sci., *33*, 133 (1978).

Rowe, P. N., H. J. MacGillvray, and D. J. Cheesman, Gas discharge from an orifice into a gas fluidised bed, Trans. Inst. Chem. Eng., *57*, 194 (1979).

Rowe, P. N., P. U. Foscolo, A. C. Hoffman, and J. G. Yates, Fine powders fluidised at low velocity at pressures up to 20 bar with gases of different viscosity, Chem. Eng. Sci., *37*, 1115 (1982).

Rowe, P. N., P. U. Foscolo, A. C. Hoffman, and J. G. Yates, X-ray observations of gas fluidised beds under pressure, in *Fluidization IV*, D. Kunii and R. Toei, eds., Engineering Foundation, New York (1984), p. 53.

Sit, S. P. and J. R. Grace, Interphase mass transfer in an aggregative fluidised bed, Chem. Eng. Sci., *33*, 1115 (1978).

Sit, S. P. and J. R. Grace, Effect of bubble interaction on interphase mass transfer in gas-fluidized beds, Chem. Eng. Sci., *36*, 327 (1981).

Tanaka, I., H. Shinohara, H. Hirousue, and Y. Tanaka, Elutriation of fines from fluidized bed, J. Chem. Eng. Jpn., *5*, 51 (1972).

Toei, R., R. Matsuno, H. Miyigawa, K. Nishitani, and Y. Komagawa, Gas interchange between a bubble and the continuous phase in gas-solid fluidized beds, Int. Chem. Eng., *9*, 358 (1969).

van Swaaij, W. P. M., The design of gas-solids and related reactors, Proc. 5th Int. Symp. Chem. React. Eng., Houston (1978).

van Swaaij, W. P. M. and F. J. Zuiderweg, Investigation of ozone decomposition in fluidized beds on the basis of a two-phase model, Proc. 2nd Int. Symp. Chem. React. Eng., Amsterdam (1972), p. B9–25.

van Swaaij, W. P. M. and F. J. Zuiderweg, The design of gas-solids fluidized beds: prediction of chemical conversion, Proc. Int. Symp. Fluidization Appl., Toulouse (1973), p. 454.

van Deemter, J. J., The countercurrent flow model of a gas-solids fluidized bed, Int. Symp. Fluidization, Eindhoven (1967), p. 334.

Wen, C. Y. and L. H. Chen, Fluidized bed phenomena—entrainment and elutriation, AIChE J., *28*, 117 (1982).

Wen, C. Y. and R. F. Hashinger, Elutriation of solid particles from a dense phase fluidized bed, AIChE J., *6*, 220 (1960).

Yagi, S. and T. Aochi, Elutriation of particles from a batch fluidized bed, Soc. Chem. Eng. Japan, Spring Meet. (1955).

Yates, J. G., Fluidized bed reactors, Chem. Engr., 671 (1975).

Yates, J. G. and P. N. Rowe, A model for chemical reaction in the freeboard region above a fluidized bed, Trans. Inst. Chem. Eng., *55*, 137 (1977).

Zenz, F. A. and N. A. Weil, A theoretical-empirical approach to the mechanism of particle entrainment of fluidized beds, AIChE J., *4*, 472 (1958).

8
Coal Gasification Reactors

MORTON M. DENN *University of California, Berkeley, California*

REUEL SHINNAR *The City College of the City University of New York, New York, New York*

INTRODUCTION

Coal gasification reactors are devices that contact coal with reactive gases, usually oxygen and steam, to produce gaseous products, generally CO and hydrogen or methane. Proposed or existing configurations include fluidized and countercurrent moving (nonfluidized) beds and entrained flow reactors; product compositions depend on the coal type and the reactor configuration.

The *dry ash free* composition of a typical eastern United States bituminous coal is roughly $CH_{0.8}O_{0.1}$. In excess of 40 wt% of this material will volatilize with slow heating, yielding a product containing tars, oils, water, methane, hydrogen, CO, CO_2, and other gaseous products. More of the coal may volatilize with very rapid heating. The porous residue, or *char*, consists mostly of carbon and inorganic ash.

Typical proximate and ultimate analyses of Illinois No. 6 bituminous coal, an eastern U.S. coal which is used subsequently for illustrative purposes, are shown in Table 1. The following primary reactions can be expected to take place between the carbon in the char and the gaseous reactants:

Oxidation:

$$C + \frac{1}{2}O_2 \rightleftarrows CO \qquad \Delta H = -26.4 \tag{1}$$

$$C + O_2 \rightleftarrows CO_2 \qquad \Delta H = -94.2 \tag{2}$$

Gasification:

$$C + H_2O \rightleftarrows CO + H_2 \qquad \Delta H = +32.2 \tag{3}$$

$$C + CO_2 \rightleftarrows 2CO \qquad \Delta H = +41.4 \tag{4}$$

$$C + 2H_2 \rightleftarrows CH_4 \qquad \Delta H = -20.2 \tag{5}$$

$$C + H_2O \rightleftarrows \frac{1}{2}CH_4 + \frac{1}{2}CO_2 \qquad \Delta H = +1.4 \tag{6}$$

TABLE 1 Typical Analysis of Illinois No. 6 Coal

	Wt %
Proximate	
Ash	9.6
Moisture	4.2
Volatile matter	34.2
Fixed carbon	52.0
	Wt % dry ash free basis
Ultimate	
C	77.3
H	5.9
N	1.4
S	4.3
O	11.1

In addition, three important gas-phase reactions may take place:

Water-gas shift: $$H_2O + CO \rightleftarrows H_2 + CO_2 \qquad \Delta H = -9.2 \tag{7}$$

Methanation $$3H_2 + CO \rightleftarrows H_2O + CH_4 \qquad \Delta H = -52.4 \tag{8}$$

Oxidation: $$H_2 + \frac{1}{2} O_2 \rightleftarrows H_2O \qquad \Delta H = -58.6 \tag{9}$$

Reactions (7) and (8) can also take place between the same species in the devolatilization products; in addition, the volatiles are combustible, and the higher-molecular-weight constituents might crack. The catalytic activity of the ash, which differs from coal to coal, is likely to affect the rates of both the heterogeneous and gas-phase reactions.

Many other reactions between the major species might also occur, but other reactions of possible interest are derivable from this basic set. The methanation reaction

$$C + \frac{2}{3} H_2O \rightleftarrows \frac{1}{3} CH_4 + \frac{2}{3} CO$$

is simply the sum of one-third reaction (5) and two-thirds reaction (3), for example. Only four reactions are, in fact, independent for the seven chemical species made up of three elements, but the remaining five are included here because of their possible importance. Reactions (1) to (3) and (5) and (2) to (5) and (7) form useful independent sets.

USES OF MEDIUM-BTU GAS

This chapter is concerned mainly with medium-btu gasifiers, a generic name for gasifiers in which the gasification medium is a mixture of steam

and oxygen. Air-blown gasifiers, which are normally an integrated part of power plants, will not be treated here. The gas obtained from medium-btu gasifiers has several uses, each of which requires a different gas composition and hence affects design considerations.

Fuel Gas Production

Medium-btu gas is a clean fuel that can directly substitute for natural gas in existing boilers and many other industrial applications and is also suitable for combined-cycle power plants. The disadvantage of medium-btu gas for fuel gas use is that it is expensive to pump over long distances.

For fuel gas uses the major criterion is the lower heating value, which is approximately proportional to the number of moles of $CO + 0.85H_2 + 2.85CH_4$ produced per mole of carbon or coal. A high CO content is therefore preferable, and the shift reaction (7) is undesirable. (It is still common practice to use the higher heating value, which is proportional to $CO + H_2 + 3CH_4$, and we shall sometimes follow this practice. Efficiency based on lower heating value is close in value for coal conversion processes to the efficiency based on available free energy.)

SNG

Although interest in pure SNG production has presently subsided, it was the main driving force for many recent developments in gasification technology. There are two criteria that distinguish SNG production from fuel gas and syngas gasifiers, based on the fact that the total yield of methane per mole of coal is proportional to $CO + H_2 + 4CH_4$:

1. Maximizing CH_4 in the outlet of the gasifier reduces gas cleanup and methanation cost.
2. The ratio of H_2/CO has to be larger than unity, and higher ratios are preferable. Otherwise, the gas has to be shifted.

Syngas

CO and H_2 are used in the manufacture of many chemicals. They can also be converted to transportation fuels, either via a Fischer-Tropsch process or via methanol and further conversion of methanol by zeolites. For each use a special H_2/CO ratio is required. In all such cases any methane formed in the gasifier is a by-product that has to be separated from the product. This by-product methane can either be sold or reformed to syngas.

The need to separate methane does not necessarily imply that for syngas it is always desirable to minimize methane production. As we will show later, gasifiers with high thermal efficiency always produce significant amounts of methane, and eliminating methane production imposes severe penalties on the design. These penalties therefore have to be weighed against the cost of separating the methane from the final product. The penalty is very low in the production of methanol and in some Fischer-Tropsch processes, but is significant for many chemicals. The separation penalty also depends on the scale of the plant and the value of the methane by-product.

In a synfuel plant that is coproducing SNG and liquid fuels, the total clean fuel has a higher heating value that is proportional to $CO + H_2 + 4CH_4$,

and it is interesting that this is almost independent of the fuel consumption as long as we deal with pure hydrocarbons.

GASIFICATION AND PARTIAL COMBUSTION

It is useful to draw a distinction between *gasification* and *partial combustion*, since they correspond to different modes of reactor operation. By partial combustion we mean a process in which the oxygen supplied to the gasifier is sufficient to convert all carbon to CO. In that case, the role of steam in the feed is simply to shift a portion of the product to hydrogen. In contrast, steam is required for conversion of at least a portion of the carbon in gasification.

Consider a coal with a composition CH_aO_b; a is typically 0.8; b is typically 0.1 for an eastern U.S. bituminous coal and 0.2 for a western U.S. subbituminous coal. Linear combination of the basis reactions (1) to (3) and (5) leads to the overall equation

$$CH_aO_b + RO_2 + R(Su + Sp)H_2O \rightarrow (2 - 2R - SuR - b - 2m)CO + (2R + SuR + b + m - 1)CO_2 + \left(SuR + \frac{a}{2} - 2m\right)H_2 + mCH_4 + SpRH_2O \qquad (10)$$

R is the molar ratio of oxygen in the feed to carbon in the coal. Su + Sp is the feed steam-to-oxygen ratio; Su denotes steam utilized for gasification or the shift reaction, while Sp denotes unreacted steam that leaves with the product gas. m is the molar production rate of methane, either through reaction or through volatile evolution.

This overall reaction system has an important invariant: the total molar production of syngas (CO plus H_2) plus four times the methane production per mole of carbon feed. We denote this invariant by I:

$$I = \frac{CO + H_2 + 4CH_4}{C} = 2 + \frac{a}{2} - b - 2R \qquad (11)$$

For a given coal (a and b fixed) and essentially complete carbon conversion, this combination of products is uniquely fixed by the oxygen-to-carbon feed ratio and is independent of gasifier type. An alternative form when there is no CO_2 in the feed, based only on product gas mole fractions, is

$$I = \frac{CO + H_2 + 4CH_4}{CO + CO_2 + CH_4} \qquad (12)$$

This relation must be modified slightly if carbon conversion is not complete or if the feed contains CO_2.

The reactor operates as a partial combuster when Eq. (10) can be balanced without the need to utilize steam or feed CO_2 to effect complete carbon conversion. For purposes of our discussion here, it is most

convenient to neglect methane formation. In that case the criterion for partial combustion is

$$R \geqslant \frac{1 - b}{2}$$

The minimum oxygen rate for partial combustion, which we denote R_c, is then

$$R_c = \frac{1 - b}{2} \tag{13}$$

[We could attempt to be more precise in the definition of R_c by accounting for methane produced by devolatilization and considering the maximum amount of carbon that could be gasified by devolatilization in the presence of H_2. To do so would make the discussion dependent on reactor configuration. Equation (13) is conservative in that a reactor that operates as a partial combuster under this definition is a partial combuster under any more stringent definition, and it is thus useful for reactor classification.]

We wish to emphasize that the definition of R_c is strictly stoichiometric. If $R > R_c$, we could in principle reach the final composition by first totally gasifying the coal with oxygen and then reacting the gaseous products with steam. The fact that in reality the oxygen-coal reaction may lead to CO_2 and the coal will later react with CO_2 and H_2O is not relevant here. The value of R_c will typically be between 0.40 and 0.45. The kinetic processes occurring in the reactor are fundamentally different depending on whether R is greater or less than R_c; in the former case, carbon conversion will be largely by oxidation, and the steam is simply used to shift the product, while direct steam (and perhaps CO_2) gasification of carbon must occur in the latter case. Combination of Eqs. (11) and (13) leads to the following equality, based solely on coal and product gas compositions:

$$R_c - R = \frac{1}{2}\left[I - \left(1 - \frac{a}{2}\right)\right] \tag{14}$$

It is particularly important that pilot studies be carried out in the same $R_c - R$ regime as the planned commercial process, to ensure that scaleup occurs under comparable kinetic conditions. We shall return to this point in the discussion of specific gasifers.

The gross molar heating values of hydrogen and CO are nearly the same, and approximately one-third that of methane. Thus the gross product heating value per mole of carbon fed is approximately proportional to 3m plus the sum of the coefficients of CO and H_2 in Eq. (10); we donate this sum as h:

$$h = 1 + \frac{a}{2} - 2(R_c - R) = I - m \tag{15}$$

The significance of methane formation in a reactor for the production of syngas is illustrated in Eq. (15). I is fixed entirely by the oxygen-to-carbon ratio. For a given amount of oxygen, production of methane is detrimental to the heating value of fuel or syngas, and the methane formation is of value only if it reduces the amount of oxygen required. As we

shall see when we discuss stoichiometry in more detail, formation of methane significantly reduces oxygen requirements because reaction (6) has a much lower heat of reaction than reaction (3).

Finally, as we shall discuss further below, overall efficiency is greatly affected by the cost of feed steam and oxygen, with 1 mol of oxygen costing about four times as much as 1 mol of steam. This cost is proportional to R(4 + Su + Sp), and economic optimization of a syngas process includes making this quantity small while making $R_c - R$ large. Minimizing R might be of less importance in the overall optimization than minimizing Sp, the unused steam appearing in the product. It readily follows from the requirement that CO and CO_2 have positive stoichiometric coefficients in Eq. (10) that the steam requirement for gasification and shift (without methane formation) is bounded as follows:

$$\frac{2}{R}(R_c - R) \leqslant \text{Su} \leqslant \frac{2}{R}\left(R_c - R + \frac{1}{2}\right) \tag{16}$$

Comparison of the actual feed steam-to-oxygen ratio to the upper bound in Eq. (16) provides a minimum estimate of the excess steam. Excess steam is required either to moderate temperature or due to inherent thermodynamic limitations.

GASIFIER TYPES

There are four broad classes of reactor configurations: moving (or "fixed") bed, entrained flow, fluidized bed, and molten bath. The last of these is fundamentally different from the first three and is not discussed here. Numerous designs have been proposed within each class, and we will focus here on generally applicable principles and a few representative designs that are in or near commercial use. There is a complete compilation of reactor descriptions, operating ranges, and operating histories in Nowacki (1981), and very useful descriptions in Bissett (1978) and the proceedings of an EPRI conference on synfuels (EPRI, 1981). We focus here on reactor design principles as applicable to coal gasifiers.

Countercurrent Moving Bed

The prototypical moving-bed gasifier is the Lurgi dry ash pressurized gasifier, as shown schematically in Fig. 1, which was first put into commercial operation in 1936. The antecedents of this reactor can be traced to nonpressurized town gas producers that have long been in use. Coal with a particle size range of about 6 to 30 mm (the reactor cannot accept coal fines that are formed during grinding) is fed through a lock hopper and moves under gravity countercurrent to the rising gas stream. Steam and oxygen, air, or enriched air are fed through the bottom and distributed through the rotating grate, which is used to remove powdered ash. A similar reactor configuration developed at the U.S. Department of Energy Morgantown Energy Technology Center and implemented as well in the General Electric GEGAS reactor does not have a water-cooled wall but does have a stirrer that can be moved through the coal bed. A typical effluent composition from gasification of Illinois No. 6 bituminous coal with steam and oxygen in a Lurgi reactor is shown in Table 2. Illinois No. 6 is a low reactivity,

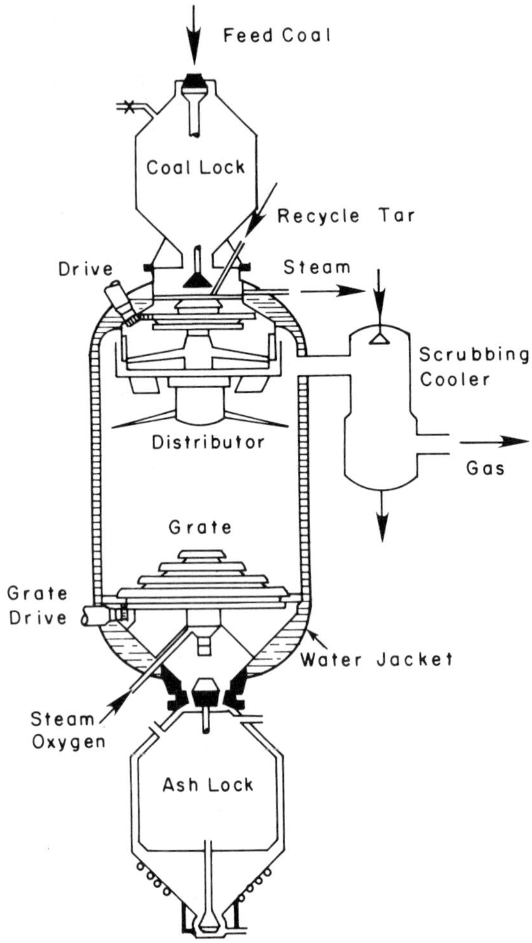

FIGURE 1 Schematic of dry ash Lurgi pressurized gasifier.

TABLE 2 Effluent Composition for Oxygen-Blown Pressurized Gasification of Illinois No. 6 Coal

	Dry ash Lurgi (Woodall-Duckham Ltd., 1974)	Texaco (McDaniel, 1981)
H_2	39.1	37.5
CO	17.3	39.5
CO_2	31.2	21.5
CH_4	9.4	—
C_2H_6	0.7	—
H_sS + COS	1.1	1.1
Inert	1.2	0.4

caking coal, and the choice of this particular example is for comparative purposes only. The current applications of this type of gasifier, such as at SASOL, are limited to reactive, noncaking coals, although successful gasification of caking coals has been reported in stirrer-equipped reactors.

The moving-bed gasifier can be conceived as operating as shown in Fig. 2. The coal first contacts hot product gases and is dried. The volatiles are then driven off by relatively slow heating. Next, the char contacts a hot, oxygen-free gas, enabling the endothermic gasification reactions (3) and (4) to occur. Finally, the remaining char enters a region of oxygen-rich gas, allowing combustion to take place.

There are two important consequences of this reactor configuration. First, the volatiles are released in such a way that they cannot react with oxygen. Thus the methane, CO, and hydrogen from the volatiles appears in the product gas, as does the tar. The condensible liquids are generally

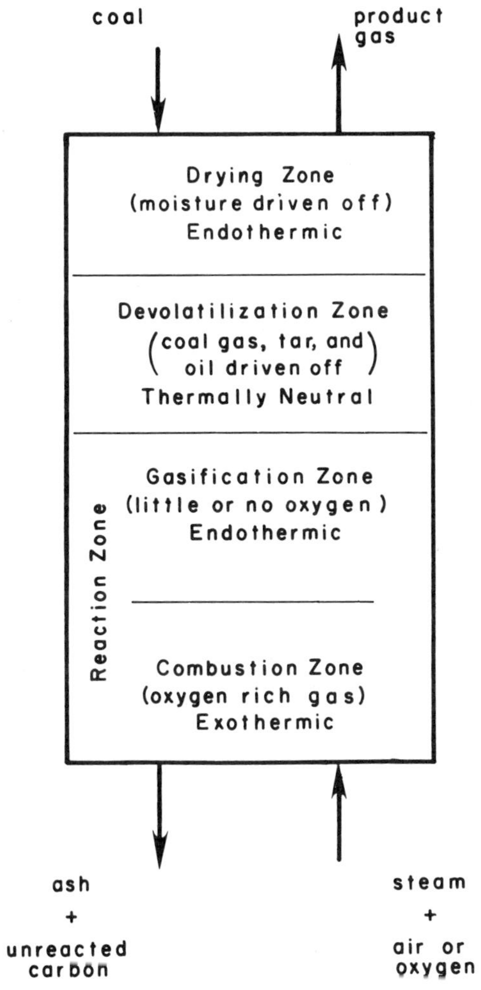

FIGURE 2 Schematic of moving-bed gasifier.

undesirable in the product gas and must be separated by quenching. (The wastewater treatment can be quite expensive, particularly in a small installation. In larger installations the tars could in principle be recovered as a valuable byproduct. One of the most important challenges in gasification research is to overcome this problem and devolatilize the coal without tar formation. This might be done by decomposing the tar in a catalytic reactor before condensing the steam.) Devolatilization is the primary source of methane in this reactor, since direct methanation of char is relatively slow, although some product methane undoubtedly comes from one or more of the methane-forming reactions. Second, good control of the maximum temperature in the bed is essential to ensure that the inorganic ash remains in a dry state for removal by the grate and does not become sticky and agglomerate. The ash softening temperature is approximately 1200°C (but highly dependent on coal type), and large amounts of excess steam (typically 7 mol of steam per mole of oxygen) are required to keep the temperature in the combustion region below this value. The excess steam is the primary reason for the relative thermal inefficiency of this type of gasifier. The excess steam requirement for high-reactivity coals is less than for low-reactivity coals, which partly explains why the gasifier is used mainly for the former. The data in Tables 1 and 2 lead to a value of $R_c - R$ of 0.13 for the Lurgi with eastern coal (and a higher value if methane formation is included in the definition of R_c); thus this reactor clearly operates in the gasification regime.

The excess steam can be reduced in principle if the grate is replaced with a hearth and the steam-to-oxygen ratio is reduced to a level that allows the maximum temperature in the combustion region to rise above the point at which ash is a freely running liquid; this temperature is typically about 1450°C but is highly dependent on coal type and fluxing agent. A schematic of the British Gas Corporation/Lurgi slagging gasifier is shown in Fig. 3. This system has not yet been commercialized. Reported data (Scott, 1981) for the slagger for gasification of a Pittsburg No. 8 bituminous coal with a molar steam-to-oxygen ratio of between 1.1 and 1.2 show considerably more CO and less hydrogen than in the dry ash Lurgi at nearly the same pressure (Woodall-Duckham Ltd., 1974), and little CO_2. This is to be expected from the equilibrium of the water-gas shift reaction (7), in view of the substantially reduced steam content. Less methane was produced in the slagger. The slagger has a value of $R_c - R$ that is close to zero, but slightly in the gasification regime. The moving beds are the only reactors in commercial or near-commercial operation that are true gasifiers in the sense defined above ($R_c - R > 0$).

Entrained Flow Reactor

Entrained flow reactors are devices in which coal particles that are typically about 100 μm in diameter are fed cocurrently with the steam and the oxidant. The coal may be transported either in a gas stream or in a water slurry; in the latter case the suspending water provides the steam for gasification. Bissett (1978) has provided a comprehensive review of the characteristics of entrained flow reactors.

Two entrained flow gasifiers have reached major commercial status. The Koppers-Totzek atmospheric reactor, shown schematically in Fig. 4, was first put into operation in 1952. The coal is fed through opposing jets. The steam-to-oxygen ratio is less than unity, and the combustion zone is

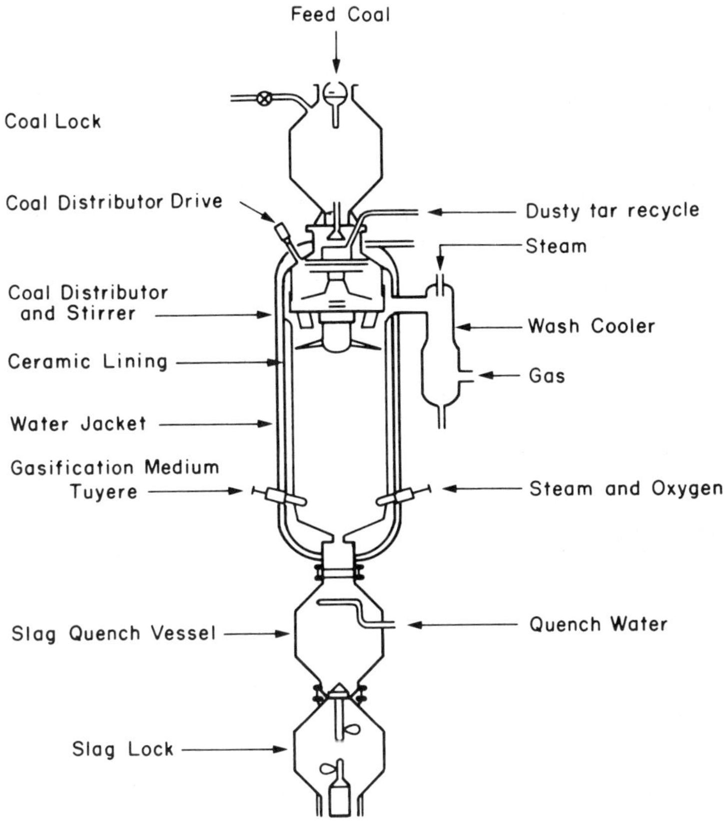

FIGURE 3 Schematic of slagging pressurized gasifier.

above the slagging temperature. A pressurized gasifier of this type is under development by Shell and Koppers.

The Texaco pressurized gasification reactor, shown schematically in Fig. 5, is a modification of an established reactor for the partial combustion of crude oil residual. The first major commercial installations were under way at the end of 1982, although Bissett (1978) gives a detailed description of a semicommercial-scale Texaco gasifier that operated from 1956 to 1958. The coal is fed in a water slurry, which has been reported to contain up to 70% solids by weight. The result is a steam-to-oxygen ratio of close to unity, and the ash slags. Pilot plant data for gasification of Illinois No. 6 coal are shown in Table 2. These data illustrate the fundamental difference between the Texaco and Lurgi gasifiers. The value of $R_c - R$ for the Texaco gasifier is -0.09, and the reactor operates as a partial combustor. [The same result is obtained for the data of Dillingham et al. (1982). In both cases the value of R computed from Eqs. (11) and (12) is 0.52, indicating excellent carbon material balance closure.] The Koppers-Totzek and Shell-Koppers gasifiers both also operate with slightly negative values of $R_c - R$, based on data compiled in Shinnar and Kuo (1978).

Devolatilization takes place in these entrained flow reactors in a region that is oxygen-rich. Thus combustion of the volatiles occurs, and there is

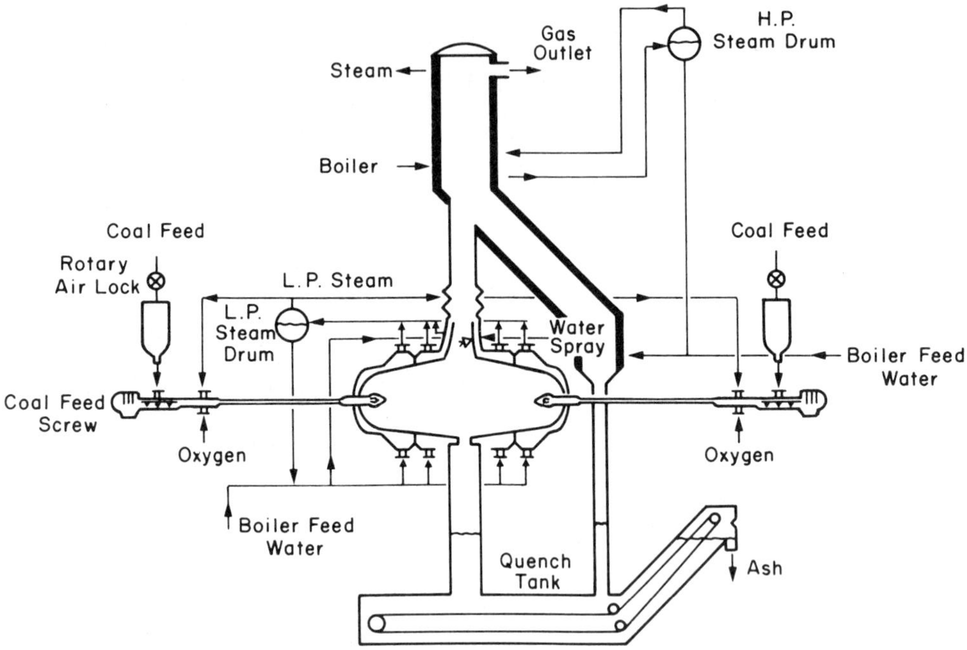

FIGURE 4 Schematic of Koppers-Totzek reactor.

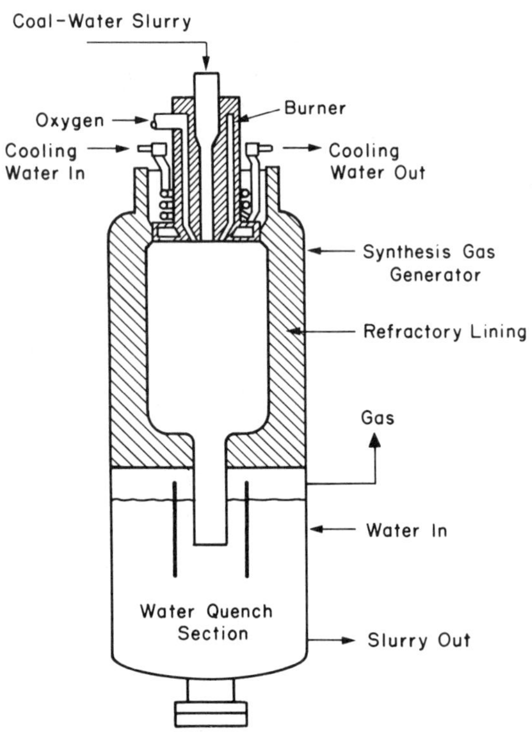

FIGURE 5 Schematic of Texaco pressurized gasifier.

little or no methane or condensible liquid in the product gas. The gas need not be quenched prior to heat exchange. An entrained flow reactor configuration that causes devolatilization to occur after combustion, and hence maximizes methane production, is embodied in the design of the high-pressure Bi-Gas reactor, but it has never been successfully carried beyond the pilot stage.

Fluidized Bed

No pressurized fluidized-bed gasifier is as well developed as the examples cited above of other gasifier configurations, and it is therfore appropriate to discuss here some of the problems faced in the design and scaleup of a fluid-bed gasifier.

The only economically available fluid-bed gasifier is the atmospheric pressure Winkler, which was developed in the 1920s for German Braunkohle. The reactor operates only with very reactive coals, and carbon conversion is limited to 90%. Typical effluent data are shown in Table 3. Ruhr Chemie has developed a pressurized version that will operate at 30 atm, and a demonstration plant is planned. The published estimates claim that overall performance characteristics will be identical to the old Winkler, but with 95% carbon conversion.

A fluidized bed can in principle be staged, and it therefore has a potential advantage over a true countercurrent reactor. Conditions in the devolatilization zone can be adjusted independently of those in the reaction zone in order to crack tars and maximize methane production; the temperature profile in a countercurrent reactor cannot be controlled, and the residence time in the devolatilization zone is quite short. There are also several inherent disadvantages relative to a reactor like the slagger. One is that the gasification zone temperature is limited by the need to prevent agglomeration. No good data exist on the maximum permissible temperature. The second is that a significant amount of carbon must be maintained in the gasification zone in order to achieve reasonable reaction rates. If this

TABLE 3 Effluent Data for Oxygen-Blown Gasification of German Braukohle in a Winkler Gasifier at 2 atm and 1700°C

H_2	38.44
CO	33.36
CO_2	21.4
CH_4	1.8
$R = O_2/C$	0.96
Steam/O_2	1.52
Cold gas efficiency	74.7
Net efficiency	56.8

zone is mixed, the ash leaving the reactor will contain excess carbon; this is the reason for the limited conversion in the Winkler.

There are two further problems that affect the Winkler and all other fluid-bed gasifiers under development:

1. Some of the partially converted coal disintegrates into fines. Fines are difficult to convert, as shown in Chapter 5, and they are swept from the gasifier and tend to adhere to the wall of the cyclone. If fed back to the bed, buildup of fines could reduce the bed density. Reducing the load on the cyclone requires a low bed velocity that is high enough to prevent ash agglomeration. This problem is solved in the Winkler by adding oxygen near the top of the bed, raising the temperature of the dilute phase and preferentially combusting the fines. (There are no detailed data to indicate if the fines really combust, or if the combustion is in the gas phase and the fines convert because the increased temperature.) The net result is that the Winkler operates as a partial combuster, with $R < R_c$.

2. The combustion reaction near the oxygen inlet is much faster than can be dissipated either by the gasification reaction or by the mixing of the fluid bed. The method used to date to overcome this fast reaction and prevent agglomeration and high temperatures is to dilute the oxygen with either steam or cold recycle gas to act as a heat sink. The Higas and Synthane reactors developed in the United States required as much steam as a diluent as the dry ash Lurgi, eliminating most potential economic advantage. An alternative way to overcome the problem is to increase the mixing intensity near the nozzle.

Recent American development work has concentrated on two pilot plants, the UGas gasifier of IGT and the Westinghouse fluid-bed reactor. Both are currently conceived as single-stage gasifiers (e.g., Schwartz et al., 1982) in that the coal is fed directly to the fluid bed and there is no separate zone for devolatilization. They therefore produce less methane than a gasifier designed to have a separate devolatilization zone with independent temperature control, although not all of the methane formed by devolatilization is reformed. The major achievement of these programs has been the development of an ash agglomerating zone at the bottom of the gasifier. The ash agglomerates grow much larger than coal particles and hence separate out at the bottom of the bed and can be removed selectively. A high fluidization velocity is required to prevent agglomeration of the coal particles, especially in caking coals; this permits a reasonably high coal concentration to be maintained in the bed, while removing ash at the bottom having a much lower carbon content.

Both the UGas and Westinghouse reactors have demonstrated the feasibility of ash agglomeration and segregation. The Westinghouse has operated thus far with coal conversions below 90%, and usually below 85%, mainly because of carbon loss due to imperfect fines recovery. The Westinghouse has also operated with large heat losses, both because of nonadiabatic operation and heat losses in the fines recycle. Pilot data for $R - R_c$ are therefore quite negative (−0.2), although Westinghouse claims that the commercial gasifier is projected to operate with a positive value of $R - R_c$ comparable to that of the dry ash Lurgi. The UGas has operated until now only at low pressure (ca. 3 atm), but claims to have achieved 95% coal conversion. Results of both gasifiers were significantly better for reactive coals than for bituminous coals, as would be expected from simple kinetic considerations.

STOICHIOMETRIC ANALYSIS

Some useful principles regarding gasifier operation can be obtained by examining overall stoichiometric constraints and energy balances under idealized conditions. We will consider only the gasification of char, which is taken as consisting entirely of carbon. Any effect of the presence of volatiles can be added afterward in this analysis.

Reactions of char to form methane in the absence of a suitable catalyst are generally much slower kinetically than combustion or steam and CO_2 gasification. It therefore suffices in the first approximation to neglect any methanation and to remove reaction (5) from the set of reactions under consideration. Reactions (1) to (3) then form a basis from which to construct all other reactions to form CO, CO_2, and hydrogen from carbon, water, and oxygen.

The three species fed to the reactor are conveniently represented in Fig. 6 as the vertices of a triangular diagram on which each point represents the mole fraction of reactants. The feed compositions for reactions (1) to (3) are shown on the axes. It is readily established that incomplete carbon conversion will occur for starting compositions that lie above the line connecting C + (1/2)O_2 and C + H_2O, while incomplete oxygen conversion will occur for compositions below the line connecting C + O_2 to the H_2O vertex. A gasifier would operate ideally with no excess steam. It can be shown by linear combination of the basis reactions [twice (3) plus (2) minus twice (1)] that steam must appear in the effluent for starting compositions that lie below the line connecting C + O_2 with the reaction

$$C + 2H_2O \rightleftarrows CO_2 + 2H_2 \qquad \Delta H = +23.0 \tag{17}$$

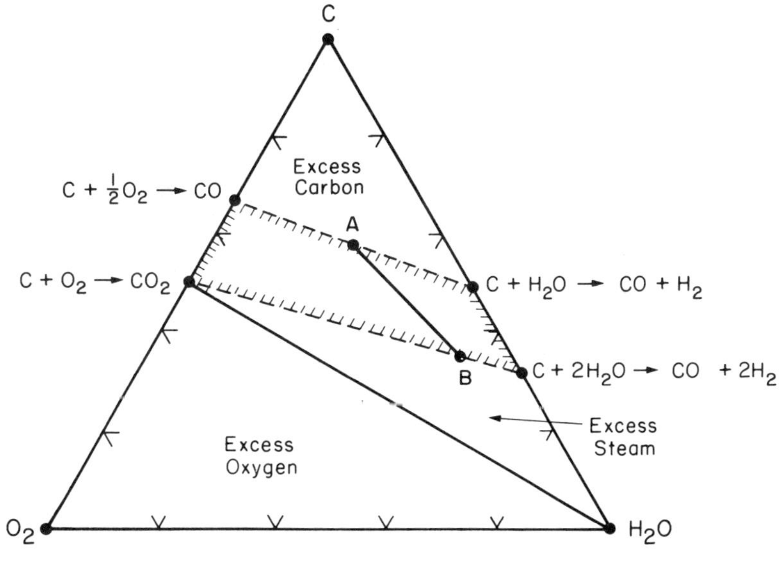

FIGURE 6 Triangular diagram for reactor feed conditions.

[Some steam will, of course, appear in the effluent for starting formulations above this line, since the analysis assumes complete conversion and neglects any possible chemical equilibria, including that of the water-gas shift reaction (7).] The shaded trapezium area thus represents the most desirable area for gasifier operation.

The operating range can be further restricted by the condition of autothermal operation (overall thermal neutrality): the exothermic reactions must produce just sufficient reaction enthalpy to drive the endothermic reactions (assuming that an external heating source, such as a nuclear reactor, is not to be used). We suppose that the inlet and outlet streams for char gasification are at the same temperature, which we will take to be 700°C for this discussion. Let X_1, X_2, and $1 - X_1 - X_2$ represent the fractional conversion of carbon through each of the basis reactions (1), (2), and (3), respectively. Thermal neutrality is then represented by

$$(-26.4)X_1 + (-94.2)X_2 + (+32.2)(1 - X_1 - X_2) = 0$$

and the thermal efficiency based on the lower heating value (LHV) of the feed char at 700°F will be 100%. The overall reaction for thermal neutrality is the one-parameter family

$$C + (0.255 + 0.036X_1)O_2 + (0.745 - 0.536X_1)H_2O \rightarrow$$

$$(0.745 + 0.464X_1)CO + (0.255 - 0.464X_1)CO_2 + (0.795 - 0.536X_1)H_2$$

or, in terms of the basis reactions,

$$X_1\left(C + \frac{1}{2}O_2 \rightarrow CO\right)$$

$$(0.255 - 0.464X_1)(C + O_2 \rightarrow CO_2)$$

$$(0.745 - 0.536X_1)(C + H_2O \rightarrow CO + H_2)$$

This reaction family lies along line AB in Fig. 6. Point A corresponds to $X_1 = 0.55$, with reactions (1) and (3) in the following proportions:

$$C + \frac{1}{2}O_2 \rightarrow CO$$

$$0.82C + 0.82H_2O \rightarrow 0.82CO + 0.82H_2$$

X_1 cannot exceed 0.55, or else CO_2 will be required as a feed. Point B corresponds to $X_1 = -1.61$ and represents the minimum value for which it is possible in principle to produce a product gas without steam. It is most convenient to express point B in terms of reactions (2) and (17) as

$$C + O_2 \rightarrow CO_2$$

$$4.09C + 8.17H_2O \rightarrow 0.82CO + 0.82H_2$$

Values of X_1 below −1.61 would require CO as a feed in this simplified analysis. The H_2/CO ratio in the product gas varies from 0.45 at point A to infinity (no CO) at point B. Conversely, point A has no CO_2 in the product, while the H_2/CO_2 ratio at point B is 2.

This simple stoichiometric analysis, which neglects all kinetic and equilibrium limitations, is generally instructive but is misleading in one important regard: the thermal efficiency of all points along line AB based on the lower heating value at 700°F is the same. A more realistic picture is obtained by accounting for the energy required to product the feed steam and oxygen at the assumed pressure of 400 psi. It can be shown (Shinnar and Kuo, 1978) that production of 1 mol of oxygen at 700°F and 400 psi requires the same energy as 4.1 mol of steam at the same conditions. The overall reactions at points A and B are, respectively,

$$A: \quad C + 0.275O_2 + 0.45H_2O \rightarrow CO + 0.45H_2$$

$$B: \quad C + 0.195O_2 + 1.61H_2O \rightarrow CO_2 + 1.16H_2$$

Point A requires the energy equivalent of 1.58 mol of steam to convert one mol of carbon, while point B requires the equivalent of 2.41 mol of steam. There is thus a net energy loss equal to the production of 0.83 mol of steam in passing from point A to point B, which is equivalent to 9% based on the lower heating value of reactants. When the energy required to prepare steam and oxygen is taken into account, the efficiencies for the idealized processes at points A and B drop to 81% and 72%, respectively. Line AB can be parameterized by the feed steam-to oxygen ratio; the computed efficiency as a function of steam-to-oxygen ratio for autothermal operation is shown in Fig. 7.

An equivalent analysis can be carried out by assuming that methanation will take place and that the reaction products are only CO, CO_2, and CH_4. This limit is unrealistic in the absence of the appropriate catalysts, but helps in defining the performance that might be obtained. Combinations of the

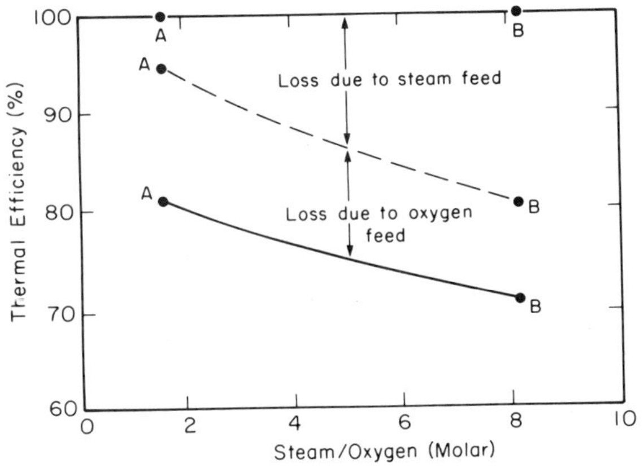

FIGURE 7 Efficiency as a function of molar steam to-oxygen ratio. (From Shinnar and Kuo, 1978.)

TABLE 4 Thermal Efficiency of Limiting Cases in Char Gasifiers with Stoichiometric and Energy Constraints[a]

	Point A	Point B	Point E	Point F
Overall reaction	$C + 0.275O_2 + 0.45H_2O \rightarrow CO + 0.45H_2$	$C + 0.195O_2 + 1.61H_2O \rightarrow CO_2 + 1.61H_2$	$C + 0.015O_2 + 0.985H_2O \rightarrow 0.4925CH_4 + 0.5075CO_2$	$C + 0.18O_2 + 0.428H_2O \rightarrow 0.213CH_4 + 0.788CO$
Steam-to-oxygen ratio	1.6	8.2	64.5	2.4
Steam requirements [lb/MMBtu (LHV) gas]	47.8	170.9	104.7	45.4
Steam (MBtu/lb mol carbon)	9.2	32.7	20.0	8.7
Oxygen requirements [scf/MMBtu gas (scf/mscf)]	611 (190)	411 (124)	35 (10)	405 (126)
Oxygen (MBtu/lb mol carbon)	23.0	14.9	1.3	15.0
Thermal efficiency	81	72	87	86

[a]Conditions: feed and outlet temperature, 700°F; no heat recovery from products below 700°F; Btu requirements of feeds, 220 Btu/scf oxygen, 1130 Btu/lb steam, both at 700°F and 400 psia.
Source: Shinnar and Kuo (1978).

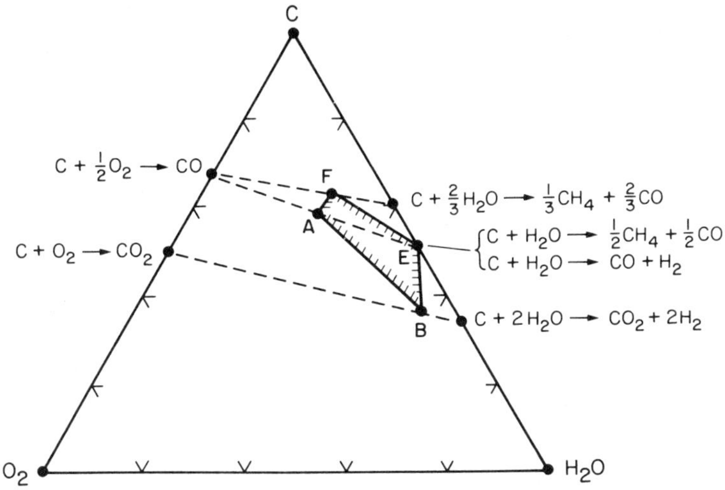

FIGURE 8 Region of autothermal operation with methane formation.

two limiting analysis leads to the operating diagram shown in Fig. 8. The quadrilateral ABEF bounds the autothermal region and lies within the region of possible complete utilization of carbon and steam. The characteristics of the corner points are summarized in Table 4. If methanation could be carried out within the gasifier, it is clear that operation should be along line EF. Kinetic constraints, to be discussed subsequently, currently preclude this option, however. Thus it is clear that the desirable region for operation of a gasification reactor is close to point A.

The assumption in this analysis that all reactions go to completion is not as restrictive as it might appear. The oxygen and steam utilized to convert carbon must satisfy the constraints outlined here, which thus define upper bounds on efficiency. Equilibrium, kinetic, and transport limitations provide information regarding excess steam. Oxygen utilization will always be essentially complete.

THERMODYNAMIC AND PROCESS CONSTRAINTS

The net efficiency of a gasification process depends strongly on the energy required to prepare the feed, which is approximately proportional to the moles of steam plus 4.1 times the moles of oxygen (Shinnar and Kuo, 1978). The ratio of irrecoverable energy used in preparing the feed to the lower heating value of the product gas is therefore a useful measure of gasifier efficiency; we denote this ratio as E_L:

$$E_L = \frac{H_2O + 4.1O_2}{CO + 0.85H_2 + 2.85CH_4} \tag{18}$$

The lower E_L, the higher the net efficiency. (For an SNG plant we would replace the denominator with $CO + H_2 + 4CH_4$.) Points E and F in Table 4

are clearly superior to points A and B in having lower values of E_L, but there are several reasons why this stoichiometric limit cannot be approached.

The first is a process constraint. The calculations in Table 4 assume that product gas can be heat exchanged with the feeds and only consider the heat requirements of chemical reactions. The actual heat requirements are greater. Gasification requires high temperatures (>1500°F) without a catalyst. It is impractical to heat exchange a high-temperature product stream, for two reasons:

1. Heat transfer at high pressure from streams containing H_2S is limited by material constraints to temperatures below 1100°F.
2. Gasifier product streams often contain tars and solids which will coke and cover heat exchanger surfaces.

In present commercial gasifiers the only way in which heat is fed back from the product gas to the gasifier feed is by countercurrent gasifiers. This exchange recovers a large fraction of the sensible heat of the product gases and hence gives an inherent advantage to countercurrent gasifiers, since they require less heat to be supplied to the gasification zone and therefore less oxygen. In a cocurrent or well-mixed single-stage gasifier the only way in which the sensible heat of the product gases can be recovered is a process steam. Since the steam can also be produced directly from a combustion of coal, we suffer the penalty incurred in separating the incremental oxygen that is required to supply the heat to raise the product to gasifier temperature.

The second reason that points E and F are inaccessible with current gasifiers is the thermodynamic consequence of the kinetic rates of the different gasification reactions. These points require methane formation by reaction (6), where no H_2 is produced. Reactions (3), (4), and (8) are all faster than reaction (6) at normal gasification conditions, and reaction (7) is generally much faster than all the others. Thus, while the equilibrium conversion of reaction (6) is high at high temperatures, the other product gases will also be present in the reactor and the global equilibrium must be taken into account. Methane yield at global equilibrium is low at temperatures above 1100°F, except at very high pressures (Shinnar et al., 1982). At low temperatures, where methane is the dominant product, steam conversion is approximately 50%, and the increased steam requirement removes any advantage over point A (at least for fuel gas).

It is important to emphasize that the inability to obtain a high methane yield at high temperatures is not a rigorous thermodynamic constraint but is a consequence of the fact that the primary gasification reaction is $C + H_2O \rightarrow CO + H_2$. Methane is formed mainly by consecutive processes involving reactions (5) and (8); in neither case could the product exceed the equilibrium composition. If, on the other hand, one could find a catalyst that promoted reaction (6) or some other reaction (e.g., $1.5C + H_2O \rightarrow 0.5CH_4 + CO$) via a surface mechanism that does not first require formation of molecular hydrogen, there is no thermodynamic reason why methane yield consistent with the equilibrium yields of these latter reactions could not be approached. Present catalysts, such as potassium carbonate, do not allow this, at least not as presently applied, and very high pressure operation would only present means of obtaining a high methane yield with high steam conversion.

E_L is shown as a function of temperature in Figs. 9 and 10 under equilibrium conditions for an eastern and western U.S. coal, respectively,

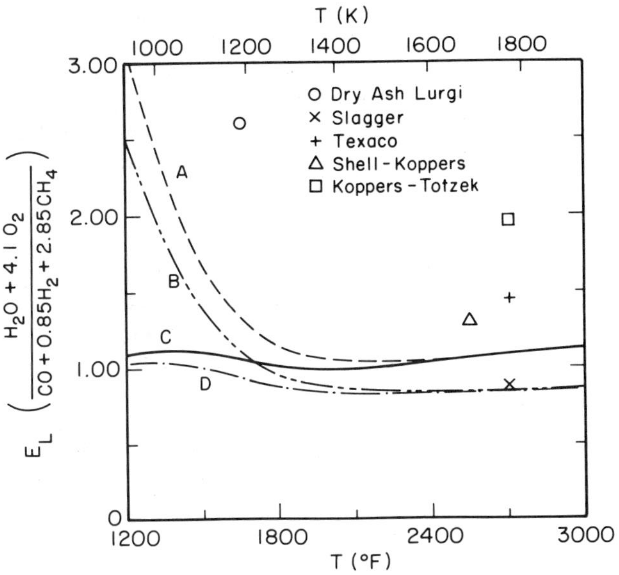

FIGURE 9 E_L as a function of gasifier temperature for an eastern U.S. coal. A, well-mixed reactor, no methane formation; B, countercurrent reactor, no methane formation; C, well-mixed reactor, methane at global equilibrium; D, countercurrent reactor, methane at global equilibrium.

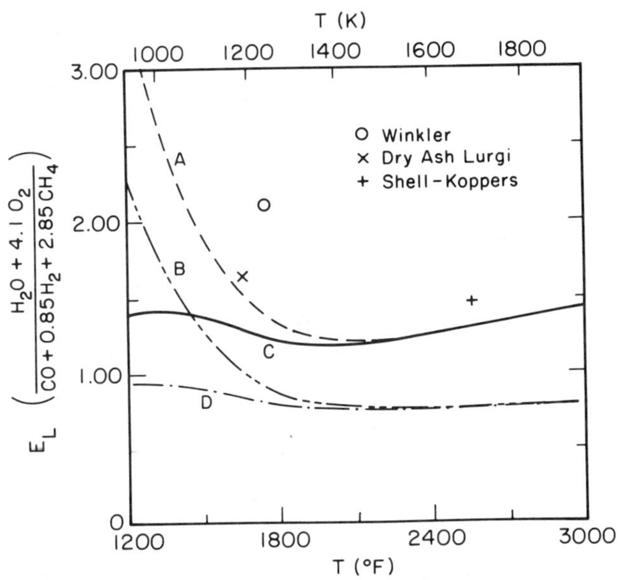

FIGURE 10 E_L as a function of gasifier temperature for a Western U.S. coal. Symbols as in Figure 9.

at 400 psi with steam and oxygen fed at 1100°F and coal at ambient temperature. There are four calculations: the reactor is assumed to be either well mixed or countercurrent, and methane is either formed and is at the global equilibrium composition or is not formed at all. Gases formed during devolatilization in the countercurrent gasifier do not affect the equilibrium, while in the well-mixed or cocurrent gasifier they reduce the maximum steam conversion. There appears to be a broad minimum in all cases around 200°F, corresponding to optimal thermal efficiency for fuel gas. The presence or absence of methane has little effect on the location of the minimum, nor does the gasifier configuration. E_L increases at lower temperatures, but the increase is much less steep for the cases with methane production. If one had to operate at lower temperatures for process reasons, methane formation would be essential for high efficiency.

The actual values of E_L for several gasifiers operating on eastern coal are also plotted in Fig. 9. We note that the slagger approaches to theoretical minimum quite closely. (The devolatilization compositions used for the countercurrent calculations were derived from data on the slagger, so the countercurrent reactor curve must pass near the slagger point. This has no bearing on the discussion of relative reactor efficiencies.) The Texaco gasifier, by contrast, lies well above the thermodynamic equilibrium line, in part because the reactor uses a water slurry as a feed and requires excess oxygen to supply heat for vaporization. The Shell-Koppers and Koppers-Totzek cocurrent gasifiers, which use steam instead of water, also lie above the optimum; the short residence time apparently precludes attainment of equilibrium. E_L for the dry ash Lurgi is also very large, despite the fact that it is countercurrent. This reactor is not suitable for eastern U.S. coal because of the high steam-to-oxygen ratio required to keep the ash below the fusion temperature.

As shown in Fig. 10, the dry ash Lurgi performs much better on western coal, where the more rapid endothermic gasification reactions aid in moderating the temperature and reduce the excess steam requirement. There are no data available for the slagging moving bed, and none of the commercial or semicommercial reactors have achieved results as good as those of the slagger for eastern coal. The Texaco gasifier is not suitable for these coals at present because the heating value is too low to permit efficient slurrying with water.

Calculations of R under the same four assumptions are shown in Figs. 11 and 12 for eastern and western coals, respectively, together with R_c and data for the same gasifiers. R never reaches R_c for any of the calculations, regardless of CH_4 formation, and there is no optimum near the minimum of E_L. The minimum of E_L is a consequence of the fact that R increases with temperature, whereas steam requirements decrease. R is approximately equal to R_c for the Texaco, Shell-Koppers, Koppers-Totzek, and Winkler gasifiers. The reason for this high value of R is kinetic and is not a thermodynamic constraint; these reactors use oxygen for partial combustion because steam gasification is too slow under their process conditions. This is easiest to see for the Texaco and rather puzzling for the fluid bed, but we lack any reliable data to explain the latter. R is close to the equilibrium limit for the slagger and is still low for the dry ash Lurgi; the moving beds are therefore the only gasifiers that operate presently at true gasification conditions, and the slagger is the only gasifier that operates close to the thermodynamic optimum.

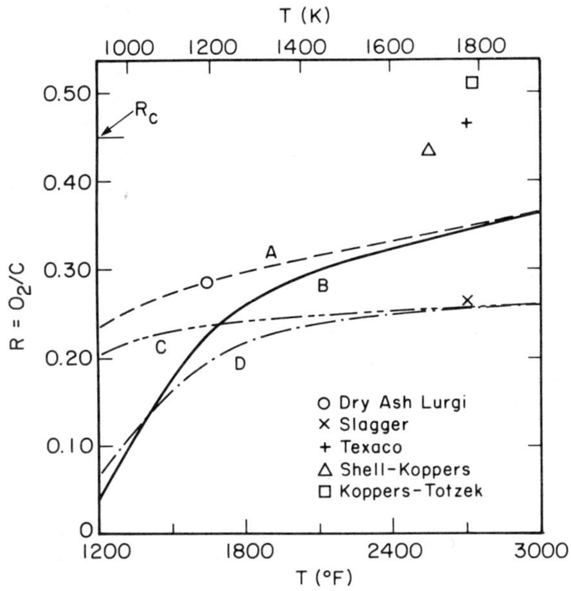

FIGURE 11 R as a function of gasifier temperature for an eastern U.S. coal. Symbols as in Fig. 9.

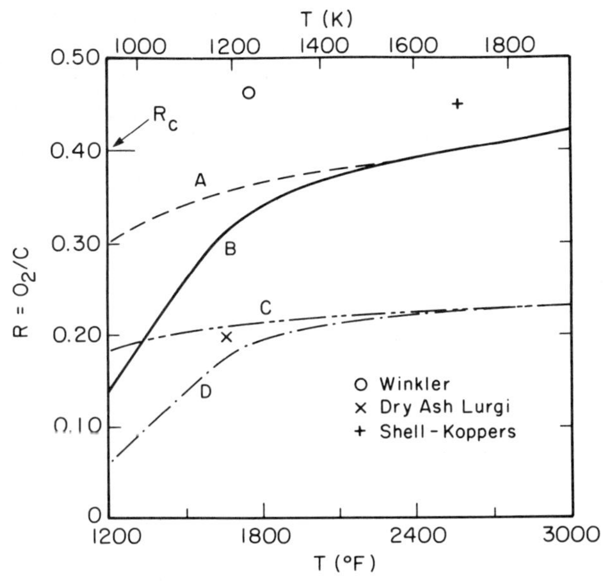

FIGURE 12 R as a function of gasifier temperature for a western U.S. coal. Symbols as in Fig. 9.

It is important to note that methane formation by itself is not always beneficial, but depends on where it occurs. If methane is formed in the gasification zone it reduces the heat requirement, and therefore R, and also improves steam conversion. If it is formed in the low-temperature zone of a countercurrent gasifier, the value depends on the mechanism. Methane formation by devolatilization or by reaction of H_2 and steam with tar and coal are beneficial, since they reduce the char feed to the gasifier. The heating value of methane formed from these reactions is larger than the heating value of the hydrogen formed in the gasification zone (Shinnar et al., 1982). On the other hand, the impact is negative if methane is formed by reaction of CO or CO_2 with H_2, since the reaction is highly exothermic and the heat of reaction is not fed back to the gasification zone. The heating value of the methane formed in this way is lower than that of the syngas reacted. Such methane formation can occur in the lower temperature zone since fresh iron oxide and other ash constituents are methanation catalysts which are later poisoned by the H_2S. This problem also causes difficulties in the evaluation of pilot plant results. Steel and stainless steel catalyze methanation as well as shift, and can therefore lead to outlet compositions which contain more CH_4 and H_2 than would be obtained in a commercial gasifier where surface effects are much smaller.

KINETICS-FREE CALCULATIONS

It is often possible to estimate reactor performance by the use of *kinetics-free* calculations; such calculations assume that key reactions occur infinitely fast, and hence go to completion or to thermodynamic equilibrium, or that they do not occur at all. (Assumptions are thus made both about relative kinetic rates and about residence times, so such calculations are only apparently free of kinetics.) The procedure is the same for all reactor types, but the detailed calculations differ slightly depending on whether or not the volatiles are released in an oxygen-rich environment. Thus it is convenient to consider co- and countercurrent reactors separately for illustrative purposes.

Countercurrent Moving Bed

Consider the moving-bed reactor, shown schematically in Figs. 1 to 3. We will focus on the reaction zone, below the regions where devolatilization and drying take place. Direct hydrogenation to form methane is usually slow and will initially be neglected in this elementary analysis. Oxygen reactivity is assumed to be infinitely fast, and oxygen attack on the fixed carbon is irreversible and is completed before any other reaction begins. Steam gasification, reaction (3), is taken to be sufficiently fast to reach equilibrium or completion at the top of the reaction zone. The water-gas shift, reaction (7), is taken to be instantaneous and always at equilibrium. Reactions (1), (3), and (7) form a fundamental set for overall balances without methanation, and are therefore the only reactions that need be considered.

Two general types of behavior can exist, depending on relative feed rates. If the coal feed rate is too large to utilize all the steam, reactions (3) and (7) [and hence also reaction (4)] will be in equilibrium at the top of the reaction zone and there will be unreacted carbon in the ash at the bottom of the reactor. Reduction in the carbon feed rate relative to steam

will lead to reduced amounts of carbon in the ash, while maintaining the equilibria at the top. The carbon in the ash will reach zero at a critical carbon rate. At carbon feed rates below this critical value there will be an excess of steam and the carbon-steam reaction can no longer be at equilibrium at the top of the gasification zone, although the equilibrium of reaction (7) will be maintained. (This behavior is contrary to the intuition of many people and is a consequence of the countercurrent operation.)

The overall mass and energy balances and the equilibrium equations are sufficient to define the compositions and temperatures of the effluents (with the additional assumption that the solid and gas temperatures are equal). The calculation therefore requires only the solution of a set of algebraic equations. The fixed carbon-to-ash ratio must be given, since the ash is included in the energy balance, and we use a ratio of 5:1 as typical of Illinois No. 6 coal. The calculations shown in Fig. 13 from Yoon et al. (1979a) are for a gas feed temperature of 700°F and an operating pressure of 25 atm. The solid lines are lines of constant fraction of unreacted carbon in the ash. The line of zero unreacted carbon represents the case in which feed rates of carbon, steam, and oxygen are such that the carbon-steam reaction reaches equilibrium at the top of the reaction zone without unreacted carbon at the bottom; this line, and all points below, are characterized by complete utilization of carbon, first by oxygen and then by steam.

There is a noticeable change in slope of the lines of constant unreacted carbon at a steam-to-oxygen molar ratio of about 1.2. To the left of the change in slope the equilibrium of the carbon-steam reaction is far to the

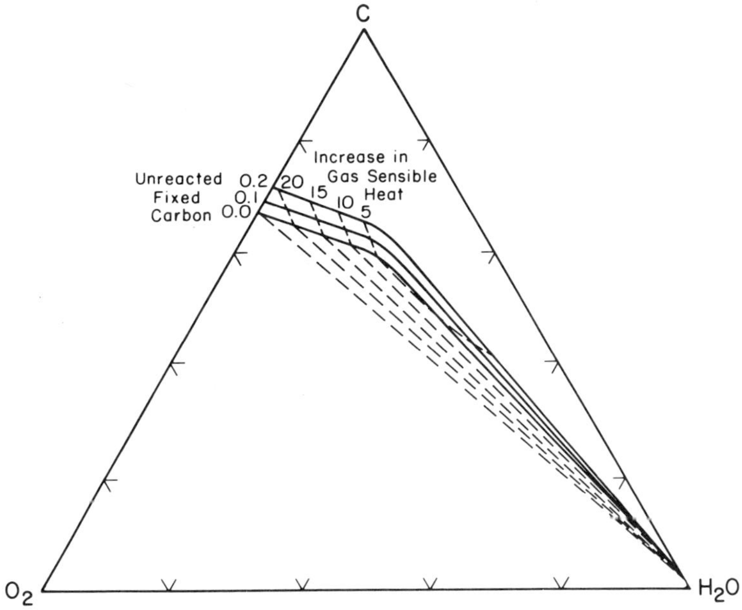

FIGURE 13 Lines of constant fraction of unreacted fixed carbon (solid), and increase in gas sensible heat (dashed, kcal/mol fixed carbon) for a countercurrent moving-bed reactor. (After Yoon et al., 1979a.)

right, and this reaction as well as oxidation can be taken to be essentially irreversible. The line of complete carbon conversion in this region is thus very close to the line connecting reactions (1) and (3) in Figs. 6 and 8. The region to the right of the break represents an excess of steam. The excess steam lowers the temperature, and sufficient excess steams brings the peak temperature below the ash melting point and allows operation of a dry ash reactor. The lines of constant carbon in the ash in this region approximate lines of constant oxygen-to-carbon molar feed ratios.

The dashed lines in Fig. 13 are lines of constant increase in gas sensible heat in kcal/g mol of fixed carbon. Autothermal operation for char gasification lies to the right of the line of complete carbon conversion. A coal containing 10% moisture will require about 5 kcal/g mol of carbon for heating and drying; if we presume that the gas exit temperature must be at least 700°F to prevent condensation of tars, operation in the region to the right of the 5-kcal/mol line is excluded because of the need to add heat to the gasifier. Operating points to the left of the 5-kcal line will represent an appreciable loss of energy unless the sensible heat can be conveniently recovered downstream, which will be difficult in a gas stream containing condensibles that will foul a heat exchanger. The intersection of the 5-kcal/mol sensible heat line with the complete carbon conversion line thus represents the most efficient operating point. (This is slightly to the left of point A in Fig. 6, which was determined on the basis of a dry, hot char feed.) This intersection occurs near a steam-to-oxygen ratio of 1.2, which, as noted above, is typical of the feed to the BCG/Lurgi slagger. The complete carbon conversion line is close to the 10-kcal/mol sensible heat line at a steam-to-oxygen ratio of about 7, which is typical of a dry ash Lurgi reactor; this represents an additional loss of efficiency of about 5%. The net thermal efficiency (Shinnar and Kuo, 1978) is shown as a function of steam-to-oxygen ratio at complete carbon conversion in Fig. 14. (This calculation

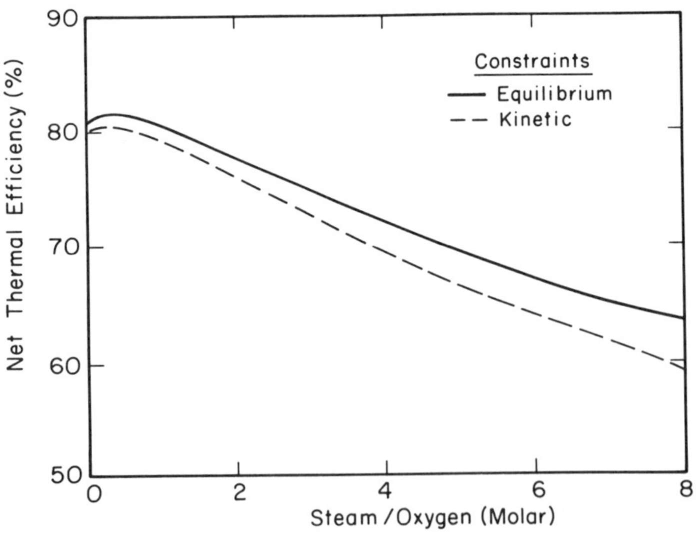

FIGURE 14 Thermal efficiency for a countercurrent moving-bed reactor as a function of molar steam-to-oxygen ratio. (From Shinnar and Kuo, 1978.)

is at 400 psi.) The dashed line is for an equilibrium constant that is only 10% of the true value as a means of attempting to account for kinetic effects. This approach to kinetic and transport constraints in a moving-bed gasifier has been used by Gumz (1950) and Woodmansee (1976).

One further useful conclusion can be drawn from this analysis. The steam gasification reaction (3) is not equimolar in gaseous species, so the equilibrium will be pressure dependent. The effect of decreasing the pressure is to make the reverse reaction more unfavorable, and hence to move the break point on the line of complete carbon conversion to the right along the line connecting the reactions (1) and (3) in Fig. 6. Since all points to the left of the break correspond to essentially complete utilization of steam, it can be concluded that slagging moving-bed gasification will be insensitive to decreasing pressure; this phenomenon has been observed by Ellman et al. (1977) in pilot plant studies. Conversely, at pressures substantially higher than 370 psi, the breakpoint would move sufficiently far to the left to cause excessive unreacted carbon.

There is no conceptual or computational difficulty in repeating these calculations with reaction (5) included in the basis in order to account for methanation. The line of complete carbon conversion with methanation equilibrium is shown in Fig. 15, together with the corresponding line without methanation. The two lines converge at steam-to-oxygen ratios of order unity, which is the region of slagger operation. The high temperatures here drive the equilibrium far to the left and suppress methane formation. Thus, for slagging operation the methanation reactions need not be considered. At the higher steam-to-oxygen ratios and lower temperatures characteristic of dry ash moving-bed reactors, however, methanation is thermodynamically favorable. As noted in Table 4 and in the preceding section, formation of methane in the reaction zone increases the efficiency. This is because methanation is exothermic and enables gasification to be carried out with less utilization of oxygen.

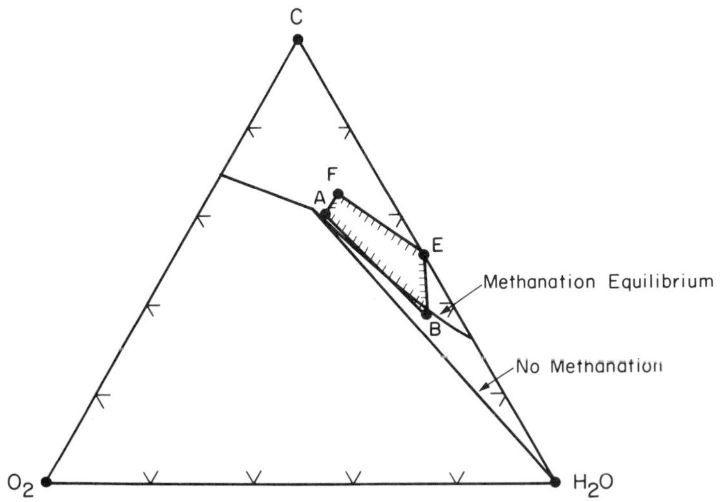

FIGURE 15 Lines of complete carbon conversion with and without methanation for a countercurrent moving-bed reactor. (After Yoon et al., 1979a.)

There are competing thermodynamic effects with pressure at the high steam-to-oxygen rations in dry ash operation. Methanation is favored by increased pressure, but we have already noted that increased pressure will lower the efficiency of steam gasification. This observation is consistent with the preliminary report from the Ruhr 100 dry ash Lurgi gasifier (Lohman and Langhoff, 1981; Lohmann, 1982), which is intended for operation at up to 100 atm. A pressure increase from 25 to 50 atm (with a change in ash content of the coal) showed an increase in raw gas methane from 9.1 to 13.5% (but little change in other hydrocarbons), with decreases in hydrogen and CO from 39.1 to 37.9% and 16.1 to 15.0%, respectively, and an increase in raw gas heating value of 13%. There was also a 5% increase in carbon conversion and a 5% decrease in oxygen consumption. Thus the optimal pressure will depend on the reactivity of the coal. Reactivity of methanation reactions seems generally to be low at intermediate pressure; the methane yields shown in Table 5 for moving-bed gasifiers operating at 30 atm and less are not significantly greater than would be expected solely from the volatiles. The methane yield from the gasification of coke probably represents the maximum methane obtainable in the reaction zone (adjusted for pressure if compared to the other data). The slightly higher yields in the dry ash reactor relative to the slagger may reflect the lower temperatures in the gasification zone of the former, which are thus more favorable to methane formation. The increased methane production at 50 atm is accompanied by a decrease in oxygen utilization corresponding roughly to a trade-off between the exothermic reaction (5) and reactions (1) and (2) in a 3:1 ratio, and methane formation in the reaction zone (rather than a change in the devolatilization process) is a plausible explanation for the increased methane production.

The trapezemium of autothermal operation in the reaction zone (ABEF) is also included in Fig. 15. The neighborhood of line EF, which is carbon conversion to methane, CO, and CO_2, and contains the highest efficiencies, is quite far removed from the line of complete carbon conversion with methane equilibrium. This region around line EF therefore appears not to be obtainable. Selective catalysts that provide a kinetic pathway approaching global equilibrium by a trajectory passing through the neighborhood of line EF would be required, as discussed by Shinnar et al. (1982).

The analysis above also pinpoints the deficiencies of a kinetics-free analysis. It fits the slagger quite well and explains certain trends for the dry ash Lurgi, but it misses the main feature that controls the design of the dry ash reactor. The steam-to-oxygen ratio in a dry ash Lurgi is controlled by the relative selection rates of combustion (which is diffusion controlled) and reaction (3). Nonreactive coals therefore require a much higher steam-to-oxygen ratio, as discussed below in the section on modeling. In the slagger this is of no consequence, and therefore the assumption that combustion is infinitely fast has no effect on the steam requirements. It is always important in any design or development effort to understand clearly what is the critical constraint, and the case above is a good example of how constraints can shift by changes in design.

Cocurrent Entrained Flow

An analogous "kinetics-free" analysis can be carried out for any reactor configuration. We do so here, with less detail, for the cocurrent entrained flow

TABLE 5 Methane Production of Various Moving-Bed Gasifiers

Gasifier	Coal	Steam/oxygen	Pressure (atm)	scf Methane per lb dry ash free coal[a]
Dry ash Lurgi	Wyoming (western U.S.)	7.5	27	4.2
Dry ash Lurgi	Illinois No. 6 (eastern U.S.)	8.5	30	3.85
Dry ash Ruhr-100 Lurgi	German Long Flame	7.7	25	4.0
Dry ash Ruhr-100 Lurgi	German Long Flame	6.5	50	5.4
Grand Forks Slagger	Lignite	1.0	27	2.1
BCG/Lurgi Slagger	Scottish Frances	1.3	30	2.65
Solihull Slagger	Avenue No. 2 coke	1.17	20	0.3

[a] 1 scf C_nH_m converted to equivalent scf CH_4 by ratio of heats of combustion.

Source: Compiled by Shinnar and Kuo (1978), except data on the Solihull slagger, which are from Lacey (1969), and the Ruhr-100 gasifier, which are from Lohmann and Langhoff (1981).

reactor without char recycle, such as the Texaco gasifier (Fig. 5). The volatiles are driven off in an oxygen-rich environment in this type of reactor, and combustion of the volatiles is expected to occur even before oxygen attack on the remaining fixed carbon. The volatiles must therefore be included in the material and energy balances, and the graphical constructions will be specific to each coal.

It is easiest for comparison to use the *fixed carbon* in the coal as the basis for calculations on the triangular diagram. The lines defining the broad bounds of the operating region are thus shifted somewhat from those in Fig. 6. The corresponding lines connecting reactions (1) and (3) and reaction (2) and the H_2O vertex are drawn in Fig. 16 for a coal with the analysis in Table 1.

The pulverized coal is fed to the Texaco gasifier in a water slurry. The maximum slurry concentration is limited by current technology to about 70% solids by weight. This limit is shown as a straight line emanating from the O_2 vertex. The gasifier must operate to the right of this line. A gasifier in which the coal feed is transported solely by gas could operate anywhere to the right of the C-O_2 axis. The 50% solids slurry line, which is probably the practical lower limit for operation, is also shown in Fig. 16.

The effluent temperature and composition can be estimated by making assumptions about relative kinetic rates and the available residence time. Reaction (1) is assumed to occur first, and to go to completion. Since such reactors always operate with slightly more oxygen than required to utilize all of the carbon through reaction (1), the excess oxygen is assumed to react to form CO_2. The residence time is then assumed to be sufficiently long to achieve equilibrium in the gas phase. The reactor operates in the slagging regime, and methanation is not expected at the high temperatures, so

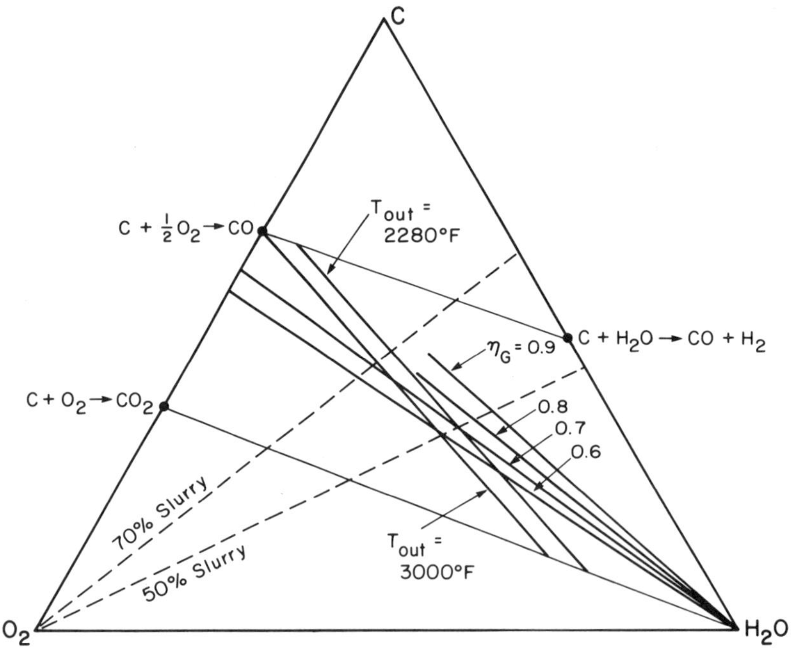

FIGURE 16 Operating diagram for a Texaco entrained flow gasifier, Illinois No. 6 coal. (After Denn and Wei, 1982.)

reaction (5) need not be included in the basis. The calculations shown here, from Denn and Wei (1982), were carried out for an operating pressure of 620 psi and an oxygen feed temperature of 300°F.

Lines of constant effluent temperatures of 2280 and 3000°F are shown on Fig. 16. The former is the minimum temperature that will ensure slagging, while the latter probably represents an absolute upper bound on the system temperature because of materials limitations. The effluent composition is summarized in lines of constant efficiency, from 60 to 90%. Efficiency in this figure is defined as the ratio of the *higher*, or gross heating value of the product gas to the higher heating value of the coal. As shown in Eq. (15) (with $m = 0$), lines of constant higher heating value are lines of constant carbon-to-oxygen ratio, increasing in efficiency with decreasing oxygen (and hence with decreasing effluent CO_2). It is clear that for a given slurry concentration, the maximum efficiency will be obtained by operating at the lowest permissible effluent temperature. The maximum obtainable gross efficiency, without taking any kinetic constraints into account, is slightly over 0.8 with the limitation of a 70% slurry. This must be reduced when the maximum temperature in the reaction zone is incorporated in the analysis.

It is important to note that all present cocurrent gasifiers operate in the partial combustion regime ($R - R_c > 0$). This can be predicted from thermodynamic considerations for a Texaco gasifier with a coal-water slurry containing 30% water if we assume a minimum operating temperature of 2700°F. It follows from Figs 11 and 12, on the other hand, that $R - R_c$ need not be positive for a cocurrent gasifier with dry feed, and yet all published data for such gasifiers are also in the partial combustion regime. This is a consequence of the use of sufficient steam with the coal feed to require a high oxygen-to-coal ratio in order to reach the operating temperature. The steam is used here only as a coolant and not as a reactant for the gasification reactions, and it serves only to shift the product gas. This seems to indicate that the steam-carbon reaction is too slow even at these high temperatures for significant reaction to occur in the short residence time in such reactors. The assumption of equilibrium in a cocurrent reactor seems therefore to be justified only if $R - R_c > 0$, and even then it is questionable since it gives possible operating regimes but does not predict the operating conditions. It is very important to understand this limitation when we deal with kinetic models.

DETAILED REACTOR MODELING

The stoichiometric and kinetics-free calculations define useful bounds on reactor operating regions, but reactor operability is ultimately determined by the detailed temperature, composition, and ash-state profiles. We briefly describe in this section the approaches to such detailed modeling and the problems that must be dealt with. Detailed models do not seem to have been utilized in the initial design of any of the existing operating systems, and their major use will undoubtedly lie in the development of process understanding for control, feedstock change, and perhaps design modifications for increased efficiency (e.g., Lu and Denn, 1984).

Before undertaking a discussion of detailed reactor modeling, it is well to say a few words about model validation. Most data have been obtained on reactor effluents, and effluent conditions are extremely insensitive to the details of the model. Data are generally obtained under successful operating

conditions, which means nearly complete carbon conversion. Given the facts that combustion and water-gas shift are fast and that complete carbon conversion implies a sufficient residence time for any necessary steam or CO_2 gasification, the effluent is essentially fixed by feed conditions. (It is only a slight overstatement to insist that it is difficult to construct a model of a successful gasifier that will *not* predict the effluent correctly. Many of the models in the literature, especially for high-temperature gasifiers, are simply complex ways to do thermodynamic calculations.) Temperature and composition profiles would be sensitive tests, but only limited and incomplete data are available from pilot- and full-scale units, and these do not seem to have been used in any modeling study.

Reaction Rates

All detailed reactor models require a description of the solid-gas reactions, using models of varying degrees of detail to describe the physicochemical phenomena in the neighborhood of a coal or char particle. The kinetic data in the literature indicate considerable variation with coal rank in reactivity to steam and CO_2, but less variability in the rate of oxygen attack. It is generally agreed that reactions (3) and (4), steam and CO_2 gasification, proceed at roughly the same rate, and considerably slower than oxidation.

Some published data on steam gasification rates are shown in Fig. 17. The data represent a range of coals, pressures, and steam and hydrogen mole fractions, so absolute rate magnitudes are not directly comparable, but each line is at constant gas composition and pressure. The variability in reported activation energies is a cause for concern and must affect any modeling study. Lupa and Kliesch (1979) have compiled and evaluated the available rate data. Typical data showing relative magnitudes of the several gas-solid reaction rates in a constant environment are plotted in Fig. 18; the combustion data are from Sergent and Smith (1973) and the gasification data from Gibson and Euker (1975). The diffusion limits are based on a shell-progressive model for a 10-mm-diameter particle at 2000°F and 50% carbon conversion, which is typical of conditions in the combustion zone of a moving bed gasifier. Gasification may be comparable in rate to combustion under these conditions.

One of the important, still unresolved differences in the kinetic expressions employed in moving-bed models is whether to include reaction (9), the formation of water from hydrogen and oxygen. This free-radical reaction would be expected to occur very rapidly in the combustion zone, but there may be sufficient char and ash surface present to quench the reaction. The presence or absence of gas-phase oxidation has little effect on reactor effluent properties, but it will greatly affect the computed maximum temperature and hence the range of operability.

Moving-Bed Models

Moving-bed reactor models are of two basic types: homogeneous and heterogeneous. The equations for the former take the gas and solid temperatures to be equal, and average over the two phases, while the latter treat gas and solid phases separately. Steady-state multiplicity and limit cycles have been observed only with heterogeneous models. The major differences between models are in the choice of reactions to be considered, sources of kinetic data, details of single-particle models, treatment of devolatilization,

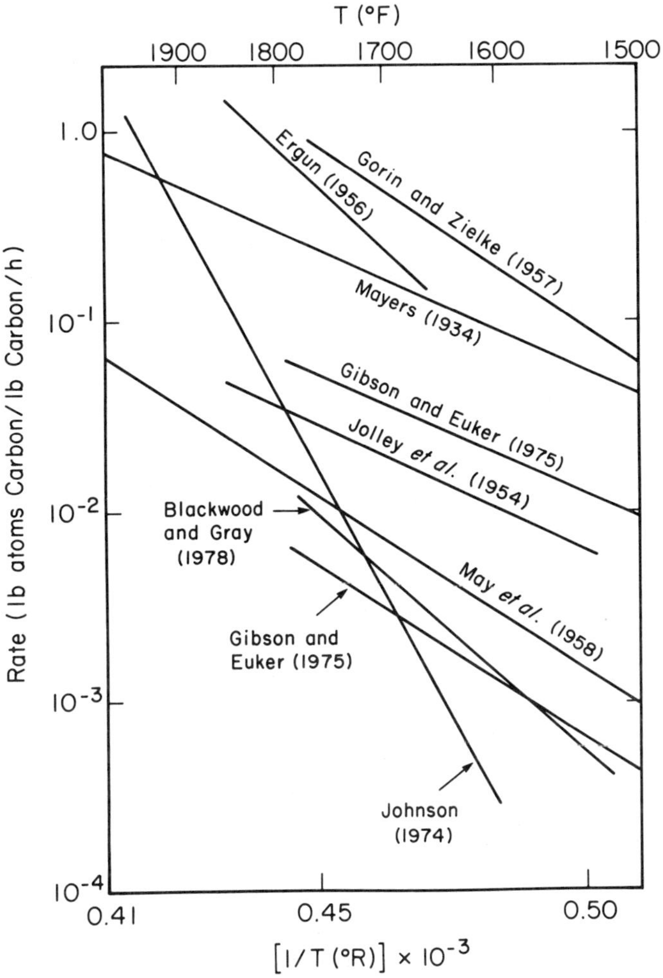

FIGURE 17 Published data on steam gasification rates.

and inclusion of radial variations and heat transfer. All models assume a single particle size and plug flow of the gas. The published moving-bed models are summarized in Table 6.* A one-dimensional, adiabatic analysis is adequate to establish the important operating features of a full-scale gasifier with wall cooling at normal throughput, since the penetration distance of the wall thermal boundary layer can be estimated to be only a small

*Stillman's (1979, 1981) model contains an error in the energy balance which results in computations equivalent to replacing the heat capacity c_p by $c_p(1 + d \ln c_p/d \ln T)$. This introduces a position-dependent error in the heat capacity that can exceed 50% in the gas phase in the region of steepest temperature profiles. The error is unlikely to change the qualitative results published using the model, but detailed steady-state and transient profiles are unreliable.

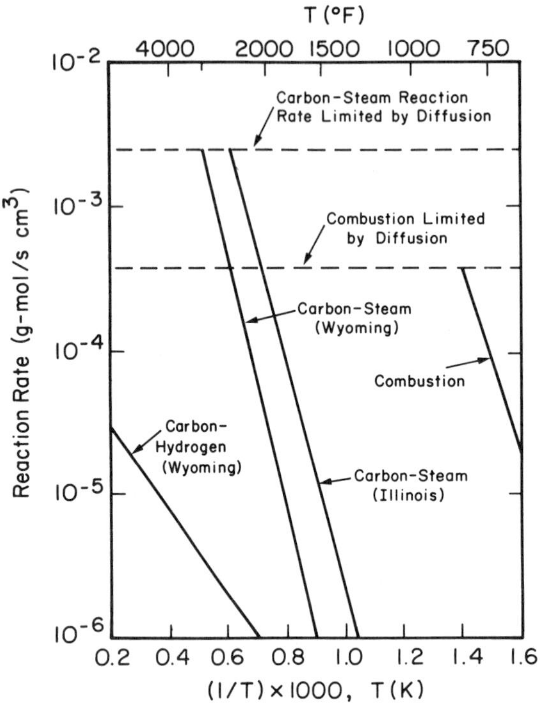

FIGURE 18 Rates of combustion and gasification reactions.

fraction of the diameter (Yoon et al., 1978). Wall cooling and radial variation will be important in a small-diameter pilot reactor or in a full-scale reactor at low throughput or in a banked (hot standby) state.

All moving-bed models that have been compared with plant effluent data have shown generally good agreement. Table 7 show results of two simulations of the Westfield test on Illinois No. 6 coal in an oxygen-blown Lurgi gasifier. The difference between the two simulations seems to be the result of different means of treating volatile evolution and reaction, which is the source of nearly all the methane in both cases, and is not a consequence of the fact that one model is homogeneous and one is heterogeneous. The effluents all satisfy the invariant relation defined by Eq. (12) (adjusted to account for the H_2S and C_2H_6). Yoon's calculations agree with the plant data to within the uncertainty of the oxygen plant purity, but we have already noted that effluent properties are not a reliable means of model validation. The computed maximum temperatures are at the upper end of the dry ash range, which is where the plant would be expected to operate, but the maximum temperature is very sensitive to the assumptions made regarding the CO/CO_2 selectivity of oxidation (Denn et al., 1979), while the effluent is insensitive to this parameter. Reliable model validation would require detailed axial profiles, which are extremely difficult to obtain in a pressurized moving bed. Profile data of Hebden et al. (1954) are notable, and have apparently not been used. Woodmansee and Floess (1975) have reported data on the inside wall temperature profile in the GEGAS reactor, but these data may not be representative of the interior.

TABLE 6 Published Models of Moving-Bed Gasifiers

References	Steady-state or transient	Homogeneous or heterogeneous	One-dimensional or radial dependence	Detailed single-particle model	Computer code available
Amundson and Arri (1978)	S-S	Hetero	1-D	Yes	—
Biba et al. (1976, 1978)	S-S	Hetero	1-D	No	—
Caram and Fuentes (1982)	S-S	Hetero	1-D	No	—
Cho and Joseph (1981)	S-S	Hetero	1-D	No	—
Cwiklinski et al. (1981)	S-S	Hetero	1-D	Yes	—
Daniel (1982)	T	Homo	1-D	No	—
Desai and Wen (1978)	S-S	Homo	1-D	No	DOE
Kim and Joseph (1982)	T	Hetero	1-D	No	—
Stillman (1979, 1981)	S-S and T	Hetero	1-D	No	IBM
Yoon et al. (1978, 1979b)	S-S and pseudo-S-S	Homo	1-D + core/boundary layer	No	EPRI
Yu et al. (1982, 1983)	S-S and T	Homo	2-D	No	EPRI

TABLE 7 Simulation of Pressurized Lurgi Gasification of Illinois No. 6 Coal at Westfield Using Models of Yoon et al. (1978) and Cho and Joseph (1981)[a]

	Yoon et al.	Cho and Joseph	Westfield plant data[b]
CO_2	32.9	27.5	31.2
CO	13.8	19.5	17.3
H_2	42.4	44.4	39.1
CH_4	8.6	6.5	9.4
C_2H_6	0.7	—	0.7
H_2S	1.0	2.1	1.1
Inert	0.5	—	1.2

[a]Steam/oxygen molar ratio = 9.6, fixed carbon/oxygen molar ratio = 2.73.

[b]Plant data are from Woodall-Duckham Ltd. (1974).

Typical temperature profiles computed with the model of Yu et al. (1983), which includes Yoon's model as a special case, are shown in Fig. 19 for gasification of Illinois No. 6 coal in an air-blown 12-ft-diameter dry ash Lurgi gasifier. The cold zone near the wall is the source of nearly all the unreacted carbon. The effect of blast temperature is shown in Fig. 20. There is a critical blast temperature below which carbon conversion effectively ceases. The model is probably inaccurate in predicting the precise

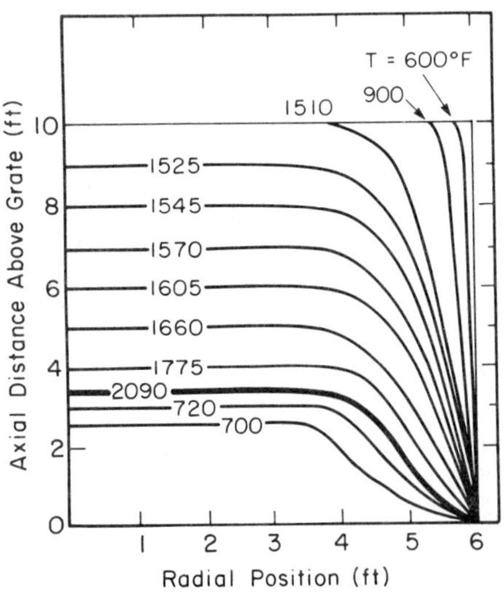

FIGURE 19 Computed temperature contours for a dry ash Lurgi gasifier. (From Yu et al., 1983.)

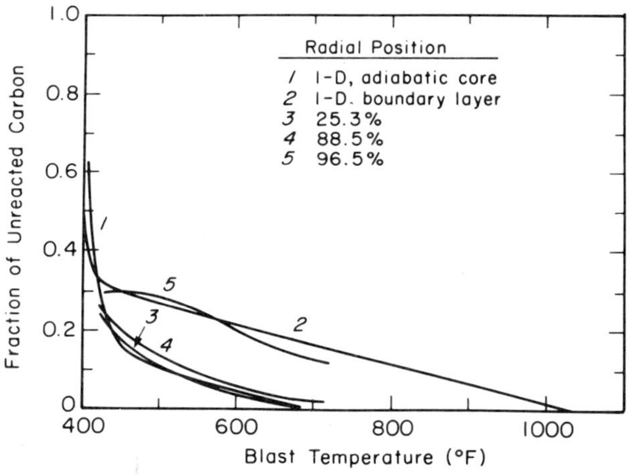

FIGURE 20 Computed radial distribution of unreacted carbon at the bottom of a dry ash Lurgi gasifier as a function of blast temperature. (From Yu et al., 1983.)

value of this critical temperature, since the finite heat transfer rate between the blast gas and the ash is not taken into account in the homogeneous model, but the effect is certainly real and important. There is a detailed discussion of the interactions between wall cooling and blast temperature in Yu et al. (1982, 1983).

The dashed line in Fig. 21 is the locus of optimal feed conditions computed by Yoon et al. for Illinois No. 6 coal; it lies very close to the line in Fig. 13 computed with the kinetics-free approach for complete carbon conversion and no methanation. Computed lines of maximum temperature for oxygen-blown gasification of Illinois No. 6 coal are also plotted in Fig. 21. This completes the definition of operability regions. The dry ash reactor must operate to the right of the 1175°C contour, but as close to that line as possible. The point denoted "L" on Fig. 21 represents the operating point of the Westfield test of Illinois No. 6 coal in a dry ash Lurgi reactor (Woodall-Duckham Ltd., 1974), and it is consistent with the reactor model calculations. A slagging reactor must operate to the left of the ash melting temperature and, as noted previously, ideally close to point A. Point S is the reported operating point in a BCG/Lurgi slagger at Westfield (Scott, 1981) for a British Rossington coal that is believed to be similar to Illinois No. 6. The steam-to-oxygen ratio is essentially the same as at point A, and corresponds to complete steam utilization, but the carbon-to-blast gas ratio is lower than would be expected from the model. The model predicts more unreacted carbon than is observed in the slagger, probably because of an inadequate description of the region near the hearth, and this is one possible explanation of the difference in feed ratios.

As a final note on this aspect of modeling, it is useful to return briefly to the kinetics-free method of analysis. Table 8 shows a comparison of the gas properties above the reaction zone, but prior to devolatilization, as computed with the detailed model and by the kinetics-free method, assuming complete carbon conversion and no methanation with the latter. The

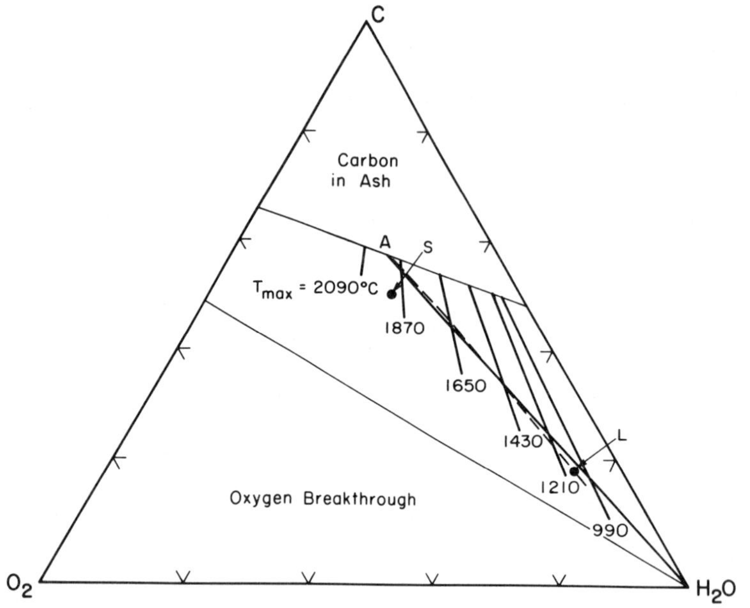

FIGURE 21 Operating region and locus of optimal feed conditions computed for dry ash Lurgi reactor. (After Denn and Wei, 1982.)

TABLE 8 Comparison of Detailed and Kinetics-Free Model Calculations for Oxygen-Blown Gasification of Illinois and Wyoming Coals in a Dry Ash Lurgi Gasifier

	Illinois coal		Wyoming coal	
	Detailed model	Kinetics-free	Detailed model	Kinetics-free
H_2O	0.68	0.68	0.49	0.45
H_2	0.32	0.32	0.45	0.55
CO	0.09	0.09	0.22	0.25
CO_2	0.22	0.22	0.30	0.30
CH_4	0.002	0	0.03	0
Exit temperature (°F)	1441	1437	1383	1320
H_2/CO	3.43	3.52	2.07	2.21
Maximum temperature (°F)	2072	2004	2361	2362

Source: Denn et al. (1979).

maximum temperature is estimated from the adiabatic temperature rise of combustion followed by water-gas shift equilibrium, assuming complete utilization of oxygen and equal amounts of CO and CO_2 as combustion products. The detailed model calculations assume the same selectivity and an initial particle size of 20 mm. All of the results, including the estimate of maximum temperature, are close. Agreement is better with the low-activity Illinois coal then with the high-activity Wyoming coal, since the latter allows for more methanation. The kinetics-free approach is clearly an effective means of estimating process performance *if* the feed conditions necessary for complete carbon conversion are known. (The maximum temperature estimated in this way will differ significantly from the kinetics-free calculation only if the particle size is sufficiently large that the oxidation rate is limited by mass transfer and becomes comparable to the steam gasification rate, or if large amounts of methane are formed.)

There are two time scales that are relevant in the transient response of a moving-bed gasifier. The first is of the order of the gas residence time, which is approximately 1 min in the dry ash reactor and 10 s in the slagger. The response to gas load changes over this time scale is only a consequence of changing amounts of the hot gas effecting devolatilization and drying. The temperature profile in the reaction zone is unchanged, so the kinetic processes remain the same and the steady-state model of the reaction zone suffices. This short time scale is the important one for changes in gas demand in a power cycle, and the response can be computed from an overall energy and mass balance on the drying and devolatilization zones.

There is a second time scale, of order hours, which corresponds to the movement of a thermal wave through the bed. A pseudo-steady-state analysis can be used to follow the wave propagation through the bed following a load change, and to compute the effluent properties from the steady-state model for the given profile, as done by Yoon et al. (1979b). This approach is not adequate for the large changes associated with startup from, or turndown to, a significantly reduced throughput or a hot banked state.

The most important processing result from the full transient model of Yu et al. (1982) is that the long-term transient response of a dry ash gasifier is approximately first order, with a time constant that depends on the operating level but (based on limited calculations) seems to be independent of the magnitude of the change. Typical computed transients for an air-blown dry ash Lurgi with Illinois No. 6 coal are shown in Fig. 22. The time constant for turndown is about 6 h, while the time constant for a load increase is about 3 h. The transient time for startup is shorter than for turndown because an increase in the gas flux increases the flame velocity. Published experimental data on transients in dry ash gasifiers are not available, but the long time constants are consistent with the known operating characteristics of these reactors. Yu et al. used the dynamic model to study feed policies during turndown to maintain the combustion zone in a fixed position, and concluded that a step change to the new steady-state ratio is effective.

The nature of the transient in a slagger is fundamentally different, since the low steam-to-oxygen ratio and hot slag tap burner gas always keep the combustion zone at the hearth. Thus a throughput change does not cause a thermal wave, and the major dynamic response is in the region close to the wall, where the greater relative importance of heat loss decreases the conversion. Response times for load increases and decreases are about six times as fast as in the dry ash reactor, and effluent compositions are essentially unchanged. The major transient is the heat transfer between

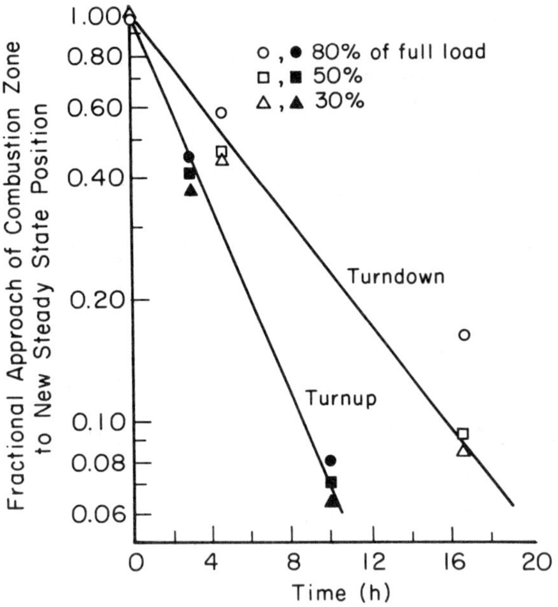

FIGURE 22 Computed transient response of a dry ash Lurgi gasifier to load changes. (After Yu et al., 1982.)

the bed and the water jacket. Slagger tests at Westfield (Scott, 1981) show stable operation with large load changes effected over minutes, but the insensitivity of the effluent, even under transient operation, makes such experiments inadequate for testing a dynamic model.

Entrained Flow Models

Entrained flow reactor models fall into three general categories: two-compartment, multicompartment, and two-dimensional turbulence. The published descriptions are summarized in Table 9. Most models represent commercial proprietary codes and are noted by the corporate or institutional name as well as the authors of descriptive papers.

The two compartment models simply assume that the gas hydrodynamics can be approximated by a well-stirred region immediately downstream of the nozzle, followed by a plug-flow section, with plug flow throughout of the solid particles. Volatiles are assumed to evolve and burn in the well-stirred region. These models show good agreement with published conversion data for the Texaco gasifier, but such agreement is to be expected in light of the comments made previously about the factors that determine reactor effluents. There is little methane produced, so total syngas ($CO + H_2$) is fixed entirely by the feed oxygen-to-carbon ratio. The steam serves only to shift the product and determine the CO/H_2 ratio; prediction of the effluent temperature in order to compute the equilibrium of reaction (7) is thus the main result of the model, and the temperature prediction can be "tuned" by small changes in the assumptions about the size of the well-stirred region and the heat loss in the plug-flow region.

TABLE 9 Entrained Flow Gasifier Models

Model	Reference	Type
Avco-Everett	Ubhayakar et al. (1977)	Multicompartment
Bringham Young 1-DICOG	Smoot et al. (1979), Hebden et al. (1981)	Multicompartment
Brigham Young PCGC-2	Hebden et al. (1981), Martin (1982)	Two-dimensional
Delaware	Denn and Wei (1982), Olsen (1981)	Two-compartment
Jaycor	Ghate and Martin (1982)	Two-dimensional
Physical Sciences	Ghate and Martin (1982)	Multicompartment
S-Cubed	Schneyer et al. (1982)	Two-dimensional
Texaco	Lupa and Kliesch (1979), Lupa (1982)	Two-compartment
West Virginia	Wen and Chuang (1979)	Two-compartment

The prediction of the maximum temperature is an important model output, since this value must lie within the bounds established by the ash slagging temperature and the materials of construction. The region comprising maximum temperatures in the range 2800° to 3000°F computed from the Delaware model (Denn and Wei, 1982) is added to the triangular operating diagram in Fig. 23. The shaded region is the feasible operating range for a slurry-fed, oxygen-blown gasifier with Illinois No. 6 coal, according to the model calculations, bounded by a maximum temperature of 3000°F, an effluent gas temperature of 2280°F, and a slurry in the range 50 to 70% solids by weight. The maximum temperature prediction is very sensitive to the assumed size of the recirculating zone, however, which depends on the unknown jet configuration and velocity. In addition, the assumption of a single well-stirred region must be a poor approximation to the temperature profile from the jet core to the wall and should be expected to predict too low a maximum. Lupa and Kliesch (1979) note that reactor temperatures (at an unstated location) predicted by their model differ from plant measurements by several hundred degrees and are systematically low. (The raw data are proprietary and have not been published.) The transient model predictions (Lupa, 1982) are generally good with regard to total syngas production, but they deviate from the component data and fail to predict the "wrong-way" dynamics observed in some of the reported runs.

The multicompartment models attempt to describe the physical processes on a finer spatial scale, including kinetic models of devolatilization, but still use macroscopic approximations to the gas and solid dynamics. The two-dimensional models solve full continuum descriptions of the momentum, mass, and energy transport, including detailed kinetic and turbulence (mixing-length or k-ε) models. A considerable data base, including detailed profiles of the type computed in the two-dimensional models, is not being developed for model testing on small entrained reactors (Hebden et al., 1981;

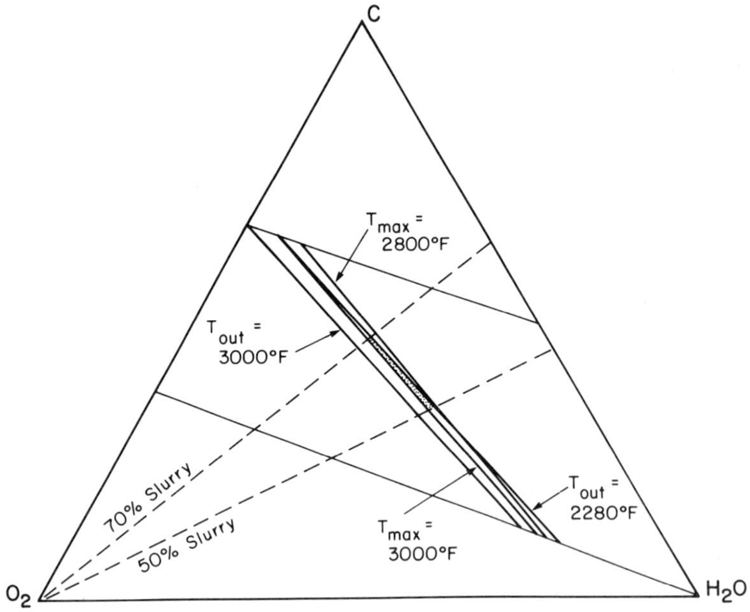

FIGURE 23 Operability region computed for a Texaco entrained flow gasifier. (After Denn and Wei, 1982.)

Smoot, 1982; Solomon and Hamblen, 1982), but applicability of these complex models to the analysis of full-scale reactors has not been demonstrated in the open literature.

Fluidized-Bed Models

Several steady-state models of fluidized char gasification have been published using conventional fluidization models: perfectly mixed, Davidson-Harrison, and Kunii-Levenspiel. The most complete such model is that of Caram and Amundson (1979); other, less general treatments include those of Yoshida and Kunii (1974) and Weimer and Clough (1980). A model briefly described by Finson (1982) seems similar in concept to that of Caram and Amundson. Caram and Amundson show good agreement with effluent data from a small char gasifier, but the model is not applicable to an industrial or demonstration scale configuration. Finson shows poor agreement with UGas and Westinghouse reactor data.

A number of "first principle" models to follow detailed two-phase fluid mechanics, as well as the chemical reactions, are under development and are described in Ghate and Martin (1982). Although these detailed hydrodynamic models seem to show the flow characteristics associated with fluidized beds, the published descriptions indicate that they are still far from providing an engineering tool for dealing with the problems described earlier regarding fluidized gasifier design. The most difficult problem is near the oxygen nozzle. This is a highly nonisothermal zone in which performance depends strongly on the relative rates of combustion and solid mixing around the nozzle. Both rates depend on the properties of the coal and the

state of the char, and an adequate description of this region is beyond current modeling capabilities. The only reliable way to develop a gasifier is therefore to have a pilot plant in which the mixing zone is large enough to be duplicated in the commercial plant by multiplication. Hence both the nozzle and the region around each oxygen nozzle must be kept constant to allow reliable scaleup.

A second problem is that the kinetic properties of the char depend on particle history, and the kinetics measured in a fixed or small fluid bed might not be applicable to a large bed; the reactivity of a particle alternately exposed to oxidizing and reducing atmospheres is not necessarily the same as one kept in a constant environment. Finally, the ash agglomeration process is not sufficiently understood to be modeled reliably or scaled to different operating conditions.

REFERENCES

Amundson, N. R. and L. E. Arri, Char gasification in a countercurrent reactor, AIChE J., *24*, 87 (1978).

Biba, V., E. Klose, J. Malecha, and J. Macak, Mathematisches Modell zur Kohlevergasung unter Druck, Energietechnik, *26*(1), 28; *26*(2), 26 (1976).

Biba, V., J. Macak, E. Klose, and E. Malecha, Mathematical model for the gasification of coal under pressure, Ind. Eng. Chem. Process Des. Dev., *17*, 92 (1978).

Bissett, L. A., *An Engineering Assessment of Entrainment Gasification*, MERC/RI-78/2, U.S. Department of Energy, Morgantown, W.Va. (1978).

Caram, H. and N. R. Amundson, Fluidized bed gasification reactor modeling, Ind. Eng. Chem. Process Des. Dev., *18*, 80, 96 (1979).

Caram, H. and H. S. Fuentes, Simplified model for a countercurrent char gasifier, Ind. Eng. Chem. Fundam., *21*, 464 (1982).

Cho. Y. S. and B. Joseph, Heterogeneous model for moving-bed coal gasification reactors, Ind. Eng. Chem. Process Des. Dev., *20*, 314 (1981).

Cwiklinski, R. R., J. Wei, and M. M. Denn, A heterogeneous moving bed gasifier model, AIChE Detroit Natl. Meet. (Aug. 1981). See also Denn (1982).

Daniel, K. J., Fixed bed gasifier modeling, appendix to R. R. Priestley, *Moving-Bed Gasification-Combined-Cycle Control Study*, Vol. 2, EPRI AP-1840, Vol. 2, Electric Power Research Institute, Palo Alto, Calif. (1982).

Denn, M. M., Steady-state and dynamic simulation of moving-bed gasifiers, in *Coal Gasification Modeling Workshop Proceedings*, M. Ghate and J. W. Martin, eds., DOE/METC/82-24/UC-90c, U.S. Department of Energy, Morgantown, W.Va. (1982), p. 391.

Denn, M. M. and J. Wei, Operability constraints for coal gasification reactors, Proc. Joint Meet. Chem. Eng., CIESC/AIChE, Beijing, China (1982), p. 631.

Denn, M. M., W.-C. Yu, and J. Wei, Parameter sensitivity and kinetics-free modeling of moving bed coal gasifiers, Ind. Eng. Chem. Fundam., *18*, 286 (1979).

Desai, P. R. and C. Y. Wen, *Computer Modeling of Morgantown Energy Research Center's Fixed Bed Gasifier*, MERC/CR-79/3, U.S. Dept. of Energy, Morgantown, W.Va. (1978).

Dillingham, E. W., G. N. Richter, and W. G. Schlinger, Enriched air and oxygen gasification of Illinois No. 6 coal in a Texaco coal gasification unit, in *Proceedings of the First Annual EPRI Contractors' Conference on Coal Gasification*, EPRI AP-2394, Electric Power Research Institute Palo Alto, Calif. (1982), p. 1-1.

Ellman, R. C., B. C. Johnson, H. H. Schobert, L. E. Paulson, and M. M. Fegley, Current status of studies in slagging fixed-bed gasification at the Grand Forks Energy Research Center, 1977 Lignite Symp. Grand Forks, N.Dak. (1977).

EPRI, *Conference Proceedings: Synthetic Fuels—Status and Directions*, 2 vol., EPRI WS-79-238, Electric Power Research Institute, Palo Alto, Calif. (1981).

EPRI, *Proceedings of the First Annual EPRI Contractors' Conference on Coal Gasification*, EPRI AP-2394, Electric Power Research Institute, Palo Alto, Calif. (1982).

Finson, M. L. PSI model for fluidized-bed coal gasification, in *Coal Gasification Modeling Workshop Proceedings*, M. Ghate and J. W. Martin, eds., DOE/METC/82-24/UC-90c, U.S. Department of Energy, Morgantown, W. Va. (1982), p. 58.

Ghate, M. and J. W. Martin, *Coal Gasification Modeling Workshop Proceedings*, DOE/METC/82-24/UC-90c, U.S. Dept. of Energy, Morgantown, W.Va. (1982).

Gibson, M. A. and C. A. Euker, Jr., Mathematical modeling of fludized bed coal gasification, AIChE Annu. Meet., Los Angeles (Nov. 1975).

Gumz, W., *Gas Producers and Blast Furnaces*, Wiley, New York, (1950).

Hebden, D., R. F. Edge, and K. W. Foley, *Investigations with a Small Pressure Gasifier: Part I. The Study of the Reactions Occurring in the Fuel Bed*, Gas Council Research Communication GC14, The Gas Council, Birmingham, United Kingdom (1954).

Hebden, P. O., L. D. Smoot, and P. J. Smith, *Prediction and Measurement of Optimum Operating Conditions for Entrained Flow Coal Gasification Processes*, 3 vols., DOE/MC/14380-1210, U.S. Dept. of Energy, Morgantown, W.Va. (1981).

Kim, M. and B. Joseph, The dynamic behavior of moving-bed coal gasifiers, in *Coal Gasification Modeling Workshop Proceedings*, M. Ghate and J. W. Martin, eds., DOE/METC/82-24/UC-90c, U.S. Dept. of Energy, Morgantown, W.Va. (1982), p. 352.

Lacey, J. A., Gasification of coal in a slagging pressure gasifier, Adv. Chem., 69, 31 (1969).

Lohmann, C., Kohlevergasung-Erfahrungen mit der versuchsanlage Dorsten (LURGI-RUHR 100), Gas- Wasserfach, Gas-Erdgas, 123, 287 (1982).

Lohmann, C. and J. Langhoff, "Ruhr 100"—development work on the pressurized gasifier in Dorsten, in *Conference Proceedings: Synthetic Fuels—Status and Directions*, 2 vol., EPRI WS-79-238, Electric Power Research Institute, Palo Alto, Calif. (1981), p. 23-1.

Lu, C. H. and M. M. Denn, Enhanced efficiency in dry ash moving bed gasifiers through a distributed feed, in *Frontiers in Chemical Reaction Enigineering*, Wiley Eastern, New Delhi (1983), p. 299.

Lupa, A. J., *Simulation of a Texaco Gasifier*, vol. 2: *An Unsteady State Model*, EPRI AF-1179, Vol. 2, Electric Power Research Institute, Palo Alto, Calif. (1982).

Lupa, A. J. and H. C. Kliesch, *Simulation of a Texaco Gasifier*, Vol. 1: *A Steady-State Model*, EPRI AF-1179, Vol. 1, Electric Power Research Institute, Palo Alto, Calif. (1979).

McDaniel, J., EPRI tests on Rukrkohle/Ruhrchemie's 165 ton per day Texaco coal gasification plant, in *Proceedings of the First Annual EPRI Contractors' Conference on Coal Gasification*, EPRI AP-2394, Electric Power Research Institute, Palo Alto, Calif. (1982), p. 2-1.

Nowacki, P., *Coal Gasification Processes*, Noyes Data Corp., Park Ridge, N.J. (1981).

Olsen, K. *Computer Modeling of the Texaco Entrained Flow Coal Gasifier*, B.Ch.E. (with Distinction) thesis, University of Delaware (1981).

Schneyer, G. P., J. L. Cook, D. H. Brownell, Jr., and T. R. Blake, *Dynamic Simulation of a Single-Stage Entrained Flow Coal Gasifier*, EPRI AP-2740, Electric Power Research Institute, Palo Alto, Calif. (1982).

Schwartz, C. W., L. K. Rath, and M. D. Frier, The Westinghouse Gasification process, Chem. Eng. Prog., 78 (5), 55 (1982).

Scott, J. E., *U.S. Coal Test Program on BCG/Lurgi Slagging Gasifier*, EPRI AP-1922, Electric Power Research Institute, Palo Alto, Calif. (1981).

Sergent, G. D. and I. W. Smith, Combustion rate of bituminous coal char in the temperature range 800 to 1700°K, Fuel, 52, 52 (1973).

Shinnar, R. and J. C. Kuo, *Gasifier Study for Mobil Coal to Gasoline Processes*, Final Report, FE-2766-13/UC-90D, U.S. Dept. of Energy, Germantown, Md. (1978).

Shinnar, R., G. Fortuna, and D. Shapira, Thermodynamic and kinetic constraints of catalytic synthetic natural gas processes, Ind. Eng. Chem. Process Des. Dev., 21, 728 (1982).

Smoot, L. D., Experimental data for code evaluation, in *Coal Gasification Modeling Workshop Proceedings*, M. Ghate and J. W. Martin, eds., DOE/METC/82-24/UC-90c, U.S. Dept. of Energy, Morgantown, W.Va. (1982), p. 292.

Smoot, L. D., P. O. Hedman, and P. J. Smith, *Mixing and Kenetic Processes in Pulverized Coal Combustors*, 2 vol., EPRI FP-1199, Electric Power Research Institute, Palo Alto, Calif. (1979).

Solomon, P. R. and D. G. Hamblen, Measurement of coal gasification reactions with on-line in-situ FTIR analysis, in *Coal Gasification Modeling Workshop Proceedings*, M. Ghate and J. W. Martin, eds., DOE/METC/82-24/UC-90c, U.S. Dept. of Energy, Morgantown, W.Va. (1982), p. 275.

Stillman, R., Simulation of a moving bed gasifier for a western coal, IBM J. Res. Dev., 23, 249 (1979).

Stillman, R., Moving bed coal gasifier dynamics using MOC and MOL techniques, ACS Symp. Ser., 186, 331 (1981).

Ubhayakar, S. K., D. B. Stickler, and R. E. Gannon, Modeling of entrained-bed pulverized coal gasifiers, Fuel, 56, 281 (1977).

Weimer, A. W. and D. E. Clough, Modeling of char particle/size conversion distributions in a fluidized bed gasifier: non-isothermal effects, Powder Technol., 27, 85 (1980).

Wen, C. Y. and T. Z. Chaung, Entrainment coal gasification modeling, Ind. Eng. Chem. Process Des. Dev., 18, 684 (1979).

Woodall-Duckhan, Ltd., *Trials of American Coals in a Lurgi Gasifier at Westfield, Scotland*, Res. and Dev. Rep. FE-105, ERDA, Washington, D.C. (1974).

Woodmansee, D. E., Modeling of fixed bed gas producer performance, Energy Commun., 2, 13 (1976).

Woodmansee, D. E. and J. K. Floess, Coal gasifiability evaluations in a one-foot diameter, fixed-bed gas producer, AIChE Natl. Meet., Houston, Tex. (1975).

Yoon, H., J. Wei and M. M. Denn, A Model for moving bed coal gasification reactors, AIChE J., 24, 885 (1978).

Yoon, H., J. Wei, and M. M. Denn, Feasible operating regions for moving bed coal gasification reactors, Ind. Eng. Chem. Process Des. Dev., 18, 306 (1979a).

Yoon, H., J. Wei, and M. M. Denn, Transient behavior of moving bed coal gasifiers, AIChE J., 25, 429 (1979b).

Yoshida, K. and D. Kunii, Complex reactions in fluidized beds—simulation of gasification, J. Chem. Eng. Jpn., 7, 34 (1974).

Yu, W. C., M. M. Denn, and J. Wei, Transient simulation of moving-bed coal gasifiers, ACS Symp. Series, 196, 359 (1982).

Yu, W. C., M. M. Denn, and J. Wei, Radial effects in moving bed coal gasifiers, Chem. Eng. Sci., 38, 1467 (1983).

9
Gas-Liquid Reactors

SERGIO CARRÀ and MASSIMO MORBIDELLI *Politecnico di Milano, Milan, Italy*

INDUSTRIAL REACTIONS AND TYPES OF REACTORS

Chemical reactions between a gas and a solute dissolved in a liquid are very common in industry. In Table 1 examples of some important processes performed in gas-liquid reactors are given. In such reactors the gaseous components are dissolved in the liquid, where they react with the other components. The fundamental analysis of these units is very complex as a consequence of the simultaneous occurrence of diffusion and chemical reaction. The hydrodynamic conditions of the system are also difficult to define.

The overall rate of the process is affected by the rates of mass transfer in the gas and liquid phases and by the rate of chemical reaction. Actually, the diffusivity in the gas phase is some orders of magnitude higher than the diffusivity in liquids. It follows that gas-side mass transfer resistance becomes significant only in the case of very fast chemical reactions.

According to the relative magnitude of mass transfer rate with respect to reaction rate, two extreme situations can be evidenced:

1. Instantaneous irreversible reaction in which the rate of the overall process is controlled by the diffusion rates of the reactants.
2. Slow reactions with uniform concentration of reactants in the whole volume of liquid, while the concentration of the dissolved gas is determined by phase equilibrium. The rate of the overall process is controlled by the rate of the chemical reaction.

Some cases of practical importance may be approximated by the mentioned asymptotic situations. However, for many important industrial processes the rate of mass transfer and chemical reactions are comparable. Therefore, both effects must be accounted for in the design of chemical reactors.

In a gas-liquid reaction, yield and selectivity are also affected by mass transfer, the nature of gas-liquid contacting, and residence-time distribution

TABLE 1 Examples of Processes Performed in Gas-Liquid Reactors

Absorption of acid gases	Absorption of SO_3 in dilute sulfuric acid
	Absorption of NO_2 in dilute nitric acid
	Removal of CO_2 and H_2S by absorption in alkaline liquids
Oxidation of organic compounds by oxygen or air	Oxidation of paraffins to acids
	Oxidation of *p*-xylene to terephthalic acid
	Oxidation of cyclohexane to cycloheanone
	Oxidation of acetaldehyde to acetic acid
	Oxidation of ethylene to acetaldehyde
Chlorination	Chlorination of benzene to chlorobenzene
	Chlorination of dodecane
	Chlorination of toluene to chlorotoluene
	Addition of chlorine to ethylene
Hydrogenation of organic compounds	Hydrogenation of olefins
	Hydrogenation of esters of fatty acids
Other reactions	Absorption of isobutylene by sulfuric acid
	Sulfation of alcohols by sulfur trioxide
	Polymerization of olefins in organic solvents

in the two phases. For instance, in a consecutive reaction network, the insufficient rate of mass transfer in the liquid may give rise to a decrease in selectivity with respect to an intermediate desired product. This situation may occur for example, in the chlorination of organic compounds (van de Vusse, 1966a,b). The oxidation of hydrocarbons is another gas-liquid reaction in which mass transport may plan an important role in defining the selectivity toward the desired product (Carrá and Santacesaria, 1980).

The design of a gas-liquid reactor proceeds in two steps:

1. The choice of a suitable type of reactor
2. The definition of reaction conditions and the design of the geometry parameters

To begin, we consider the first point. Some typical examples of gas-liquid reactors are illustrated in Fig. 1, and each will be considered in turn.

Stirred-Tank Reactors

These reactors, which are agitated mechanically, represent versatile equipment for dispersing gases. These units are the best for systems with

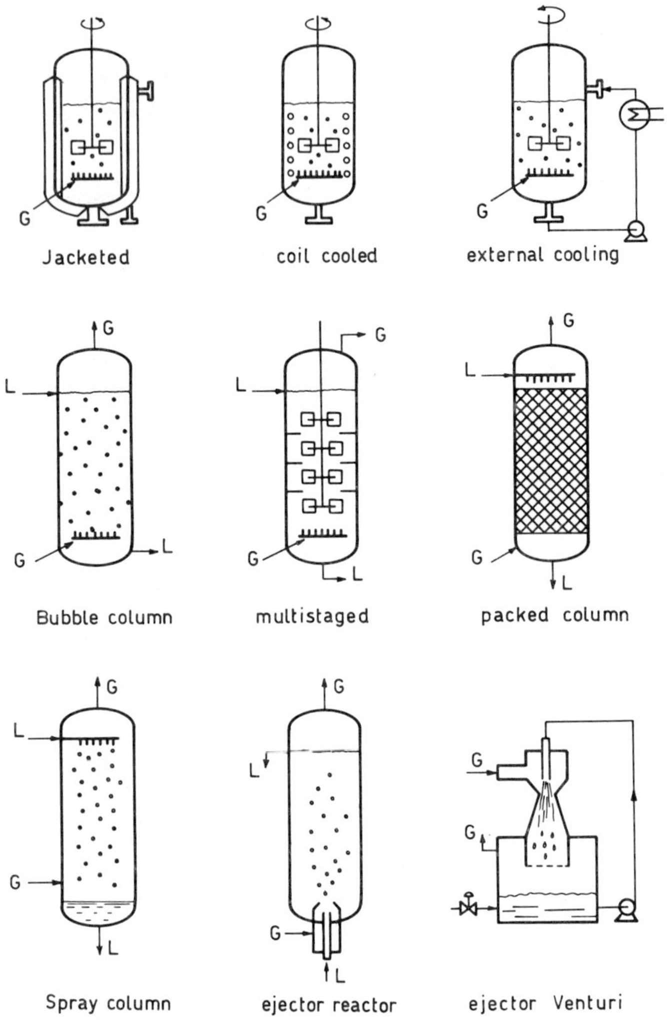

FIGURE 1 Schematic representation of some industrial gas-liquid reactors.

large heat effects and are particularly useful for performing slow reactions requiring high liquid holdup.

Bubble Columns

In these columns the gas is supplied at the bottom of a tower whose height is at least three times its diameter. Gas bubbles rise through the liquid; the bubbling reaction creates mixing. These units are employed for relatively slow reactions in which the key component is usually in the liquid phase. They are more economical than stirred tanks.

Packed Columns

In these columns liquid and vapor pass co- or countercurrent to each other through the passages created by the packing. They are very good for

service that requires low pressure drop and for corrosive service because of the wide selection of material of construction. These units are designed as absorbers and the reaction is viewed as an accelerated absorption.

Plate Columns (Plates: Perforated, with Bubble Caps or Valves)

Liquid and vapor follow separate paths between stages, and they contact each other on plates. In these units liquid holdup on plates is advantageous for slow reactions requiring long contact time.

Spray Columns

Liquid is sprayed with nozzles from the top of a column and the gas flows upward. The unit is empty, and it is required in the case of gas streams containing solid particles. A high liquid area is created near the nozzles, but drops coalesce and the area is rapidly reduced. For this reason these units are generally used for relatively easy absorption duty.

PHASE EQUILIBRIA AND DIFFUSIVITY OF GASES IN LIQUIDS

The design of gas-liquid reactors requires a knowledge of the distribution of involved components between phases in a state of thermodynamic equilibrium and the rate at which mass transfer occurs. Although complete theoretical predictions of the mentioned effects are not possible at present, some correlations have been proposed to evaluate their role in gas-liquid contactors. They are used only in the absence of experimental data and after careful analysis of the conditions of validity.

Phase Equilibria

Thermodynamic equilibrium conditions between the gas and liquid phase are conveniently expressed through a vaporization factor as

$$\frac{y_i}{x_i} = K_i(x,y,P,T) \tag{1}$$

Standard thermodynamics (Prausnitz, 1969) yields suitable expressions for the vaporization factor K_i. Two situations must be considered. The first refers to the case in which some components are present in the vapor phase at a temperature higher than their critical temperature (supercritical components). It occurs, for instance, when hydrogen or oxygen feed the gas-liquid reactor. In this case

$$K_i = \frac{H_i \gamma_i^*}{\phi_i P} \tag{2}$$

where ϕ_i is the fugacity coefficient of component i in the vapor phase. If the nonideal behavior of the gas phase is described through the virial equation of state, then

$$\ln \phi_i = 2\tilde{\rho}_g \sum_j y_i B_{ij} - \ln\left(\frac{P}{RT\tilde{\rho}_g}\right) \tag{3}$$

where B_{ij} is the second virial coefficient of the mixture components i and j, H_i the Henry constant, and γ_i the activity coefficient of superficial component i in the liquid phase. It must satisfy the unsummetrical normalization condition

$$\lim_{x_i \to 0} \gamma_i^* = 1 \tag{4}$$

The limiting Henry's constant can be predicted from nonideal solution behavior, in agreement with the theory of regular solutions (Hildebrand et al., 1970), through the relation

$$H_i = p_i^\circ \left[\exp \frac{\tilde{V}_i}{RT} (\delta_i - \bar{\delta})^2\right] \tag{5}$$

where p_i° is the extrapolated vapor pressure of component i, and $\tilde{V}_i$ is the molar volume of the solute gas, often assumed equal to the value of the liquid volume at 25°C; δ_i and $\bar{\delta}$ are the solubility parameters of the solute i and of the bulk liquid, respectively. The solubility parameter reflects the influence of forces between the molecules, through the cohesive energy density expressed by

$$\delta_i = \sqrt{\frac{\Delta\tilde{H}_{ev,i} - RT}{\tilde{V}_i}} \tag{6}$$

where $\Delta\tilde{H}_{ev,i}$ is the molar enthalpy of vaporization and $\tilde{V}_i$ is the molar volume of component i.

If the size of the molecules of the solute i are greatly different from those of the solvent B, Eq. (5) must be modified as follows:

$$H_i = p_i^\circ \exp\left[\frac{\tilde{V}_i}{RT} (\delta_i - \bar{\delta})^2 + \ln \frac{\tilde{V}_i}{\tilde{V}_B} + \left(1 - \frac{\tilde{V}_i}{\tilde{V}_B}\right)\right] \tag{7}$$

The preceding equations are satisfactory for solutions where both solvent and solute are nonpolar or for slightly polar solvents. The values of δ_i and $\bar{\delta}$ can be obtained either from Reid et al. (1977) or from Wilhem and Battino (1973). The solubility of some gases in water as a function of temperature is summarized in Fig. 2. An empirical correlation for nonpolar gas solubilities in alcohol-water solutions has been proposed by Tokunaga (1975) in terms of the alcohol volume fraction ϕ in liquid phase:

$$\ln H = \phi(1 - \phi)(A + B\phi + C\phi^2 + D\phi^3)$$

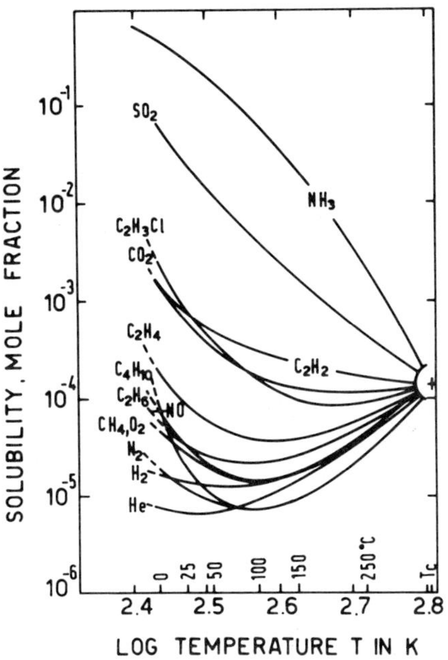

FIGURE 2 Solubility of gases in water. (From Hayduk and Laudie, 1974.)

where the empirical coefficients depend only on the nature of the alcohol and temperature, not on the nature of the solube gas (O_2, CO_2, N_2 or even CH_4).

At moderate pressure, $\phi_i \cong 1$ and the concentration of the gas in the liquid phase is small, so that K_i is simply given by the ratio between the Henry constant and the pressure.

The second situation mentioned refers to a mixture of subcritical components. This occurs in units in which a reaction is performed through distillation of volatile products. In this case,

$$K_i = \frac{f_i^\circ \gamma_i}{\phi_i P} \tag{9}$$

where f_i° is the fugacity of pure component i and γ_i is the activity coefficient in the liquid phase. It satisfies the symmetric normalization condition:

$$\lim_{x_i \to 1} \gamma_i = 1 \tag{10}$$

The influence of electrolytes on the solubility of gases can be estimated through the following equation proposed by van Krevelen and Hoftijzer (1948b):

$$\log_{10} \frac{H}{H_B} = hI \tag{11}$$

where H_B is the value of Henry's constant in pure solvent and I the ionic strength of the solution:

$$I = \frac{1}{2}\sum_i C_i Z_i^2 \tag{12}$$

C_i being the concentration of ions of valency Z_i; h is a salting coefficient that can be expressed through the sum of different contributions as

$$h = h_g + h_+ + h_- \tag{13}$$

h_g refers to the species of gas, h_+ and h_- to the species of positive and negative ions, respectively. The numerical values of these coefficients for some gases and ions are given in Tables 2 and 3.

In mixed electrolytes the following equation, suggested by Onda et al. (1970), can be applied:

$$\log_{10} \frac{H}{H_B} = \sum_j h_j I_j \tag{14}$$

TABLE 2 Salting Coefficients for Inorganic Ions (L/g mol)

Ion	h_+	Ion	h_-
H^+	−0.1110	Cl^-	0.3416
Na^+	−0.0183	Br^-	0.3310
K^+	−0.0362	NO_3^-	0.3230
NH_4^+	−0.0737	OH^-	0.3875
Li^+	−0.0416	HSO_3^-	0.3869
Mg^{2+}	−0.0568	HS^-	0.3718
Zn^{2+}	−0.0590	HCO_3^-	0.4286
Ca^{2+}	−0.0547	CO_3^{2-}	0.3754
Ba^{2+}	−0.0473	SO_4^{2-}	0.3446
Mn^{2+}	−0.0624	SO_3^{2-}	0.3275
Fe^{2+}	−0.0602	PO_4^{3-}	0.3265
Co^{2+}	−0.0532	$C_6H_5O^-$	0.4084
Ni^{2+}	−0.0520	MnO_4^-	0.2600
Cr^{3+}	−0.0986		

Source: Charpentier (1981).

TABLE 3 Salting Coefficient h_g for Gases (L/g mol)

T (°C)	H_2	O_2	CO_2	H_2S	NH_3	C_2H_2	C_2H_4	SO_2	N_2
0	—	−0.1653	−0.2110	—	—	—	—	—	—
10	−0.2170	—	—	—	—	—	—	—	—
15	−0.2197	−0.1786	−0.2222	—	—	−0.2124	−0.2003	—	—
20	−0.2132	−0.1771	—	—	—	—	—	—	—
25	−0.2115	−0.1892	−0.2277	−0.2551	−0.2394	−0.2240	−0.1951	−0.3154	−0.1904
35	—	—	—	—	—	—	—	−0.3122	—
40	—	—	−0.2327	—	—	—	—	—	—

Source: Charpentier (1981).

where I_j is the ionic strength attributable to the jth electrolytic species and h_j is a characteristic coefficient of that electrolyte.

Diffusivity in Liquids

Despite different theoretical treatments available for the description of diffusion in liquids, there are no satisfactory methods to predict the diffusivity coefficients. Therefore, prediction procedures must be applied only when experimental data are not available. Such prediction procedures can be divided in those suitable for nonelectrolytes and those suitable for electrolytes. Most studies have focused on the estimation of diffusivities in dilute solutions (Skelland, 1974). There are two theoretical approaches to the diffusional theory of nonelectrolytes. The first is known as *hydrodynamical theory*, and in it the drag of a large spherical molecule i, moving through a continuum of small solvent molecules j, is evaluated. The following equation is derived:

$$\frac{D_{ij}\mu_i}{T} = \frac{k_B}{6\pi r_i} \tag{15}$$

where r_i is the radius of the molecule i.

In the latter approach, the liquid is treated as a quasi-crystalline lattice in which holes are scattered. The following relation was proposed by Jost (1952):

$$\frac{D_{ij}\mu_j}{T} = f(\text{molecular volumes of mixture}) \tag{16}$$

The right-hand side of Eq. (16) has been semiempirically evaluated and a set of proposed relationships are summarized in Table 4.

The molecular volumes can be estimated by adding up the contributions of the atoms in the molecule as reported by Liley and Gambill (1973), pp. 232–234). The parameter ξ which appears in the Wilke-Chang equation (Eq. 1 in Table 4), is an association factor for the solvent. It is equal to 2.6 for water, 1.9 for methanol, 1.5 for ethanol, and 1 for unassociated solvents such as hydrocarbons, ethers, and so on. The uncertainity involved in assigning values to ξ for the new solvents has resulted in efforts to eliminate this factor, as in the other equations reported in the table.

The effect of concentration on diffusion coefficients can be accounted by means of the following equation, proposed by Vignes (1966):

$$(D_i\mu_{ij})_{\text{conc}} = (D^\circ_{ij}\mu_j)^{x_j}(D^\circ_{ji}\mu_i)^{x_i}\left(1 + \frac{d \ln \gamma_i}{d \ln x_i}\right) \tag{17}$$

D°_{ij} and D°_{ji} being the diffusion coefficients in dilute solution; μ_i, μ_j, and μ_{ij} the viscosities of pure components and of the solution, respectively; and γ_i the activity coefficient of the solute. Equation (17) can be applied to nonassociated solutions or to an associated solution with a constant degree of association.

In an electrolytic solution both cations and anions diffuse at the same rate, so that electrical neutrality of the solution is preserved. If complete

TABLE 4 Semiempirical Relationships for Diffusivity in Dilute Binary Solution of Nonelectrolytes

Equation	Restriction	Equation[a]
1	Exclude water as solute	$\frac{D_i \mu_{ij}}{T} = 7.4 \times 10^{-8} \frac{(\xi M_j)^{1/2}}{\tilde{V}_i^{0.6}}$
2	Exclude water as solute	$\frac{D_i \mu_{ij}}{T} = 8.2 \times 10^{-8} \frac{1 + (3\,\tilde{V}_j/\tilde{V}_i)^{2/3}}{\tilde{V}_i^{1/3}}$
3	General	$\frac{D_i \mu_{ij}}{T} = a10^{-8} M_j^{1/2}/(\tilde{V}_i \tilde{V}_j)^{1/3}$ being $a = 10$ for $(\tilde{V}_j/\tilde{V}_i) < 1.5$ $a = 8.5$ for $(\tilde{V}_i/\tilde{V}_j) > 1.5$
4	Organic solvents	$\frac{D_i \mu_j}{T} = 8.52 \times 10^{-8} \frac{1.40(\tilde{V}_j/\tilde{V}_i)^{1/3} + (\tilde{V}_j/\tilde{V}_i)}{\tilde{V}_j^{1/3}}$
5	Aqueous solutions only	$D_i = 14 \times 10^{-5} \mu_W^{-1.1} \tilde{V}_i^{-0.6}$

[a] μ = cP; D = cm^2/s; $\tilde{V}$ = cm^3/mol; T = K.

dissociation takes place, the diffusion coefficient can be predicted very accurately at infinite dilution by using the following equation proposed by Nerst (1888):

$$D_i^\circ = 8.931 \times 10^{-14} T \frac{\ell_+^\circ \ell_-^\circ}{\ell_+^\circ + \ell_-^\circ} \frac{Z_+ + Z_-}{Z_+ Z_-} \tag{18}$$

where D_i° is the diffusivity of electrolyte; ℓ_+° and ℓ_-° the cation and anion conductances at infinite dilution, respectively (mhos/equivalent); Z_+ and Z_- the absolute values of cation and anion valences. Some values of ℓ_+° and ℓ_-°, are givon by Liley and Gambill (1973, p. 235). For evaluating the diffusivities at other than infinite dilution, the following equation has been recommended by Gordon (1937):

Average error (%)				
Organic solvents	Water as solvent	Water as solute	Overall	Reference
27	11	up 250	10	Wilke and Chang (1955)
25	11	up 250	—	Scheibel (1954)
15	9	—	13.5	Reddy and Doraiswamy (1967)
18	—	26	18	
16	—	—	—	Othmer and Thakar (1953)
—	—	—	5	Lusis and Ratcliff (1968)

$$(D_i)_{conc} = D_i^\circ \frac{1}{C_j \bar{V}_j} \frac{\mu_j}{\mu_{ij}} \left(1 + m \frac{\partial \ln \gamma_+}{\partial m}\right) \tag{19}$$

m being the molality of the electrolyte, γ_+ the mean activity coefficient of ions based on molality, and $\bar{V}_j$ the partial molar volume of solvent.

MASS TRANSFER IN GAS-LIQUID CONTACTORS

Balance of a Reacting System

The mass balance equations of a gas-liquid reacting system are written in terms of phase equilibria, interphase mass transfer, and reaction rate data.

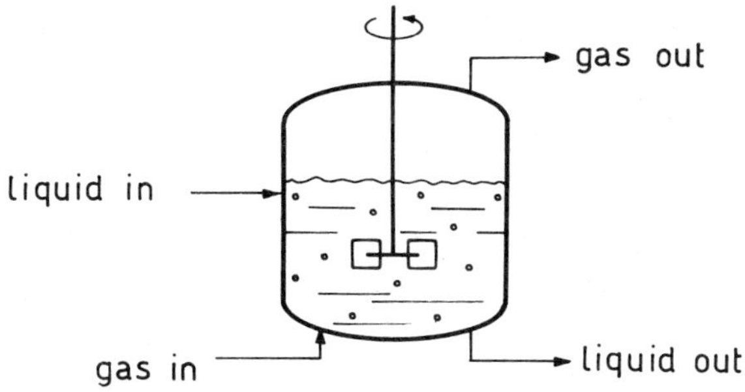

FIGURE 3 Schematic representation of a well-mixed gas-liquid unit.

The role of these factors may be evidenced by considering an idealized perfectly mixed, two-phase system, as illustrated in Fig. 3, in which a reaction occurs in the liquid phase. The mass transfer of the volatile components involves two steps. The first step is the transfer from the bulk gas phase to the gas-liquid interface; the second involves the transfer from the gas-liquid interface to the bulk liquid phase. Equilibrium conditions are assumed to exist at the interface. The mass transfer rate per unit interfacial area, N_i, of each component i is expressed as a function of the bulk-phase concentrations:

$$N_i = k_{\ell i}E_i(C^*_{\ell i} - C_{\ell i}) = k_{gi}(C_{gi} - C^*_{gi}) = K_{gi}(C_{gi} - H_i C_{\ell i}) \tag{20}$$

where

$$H_i = \frac{\tilde{\rho}_g K_i}{\tilde{\rho}_\ell} \tag{21a}$$

$$C^*_{gi} = H_i C^*_{\ell i} \tag{21b}$$

The asterisk indicates the gas-liquid interface; E_i is an *enhancement factor* which accounts for the influence of chemical reaction on mass transfer, as will be explained later.

Although more rigorous treatments are available (King, 1964), the usual combined resistance approach is employed to define the overall mass transfer coefficient K_{gi} in terms of gas-phase and liquid-phase resistances as follows:

$$\frac{1}{K_{gi}} = \frac{1}{k_{gi}} + \frac{H_i}{k_{\ell i}E_i} \tag{22}$$

In "bubble-type" gas-liquid reaction system, $1/k_g$ is generally small enough compared with $H_i/k_{\ell i}E_i$. The possible exception is the case in which $k_{\ell i}E_i$

is very large because of a fast reaction in the liquid phase. The mass balance of component i in the unit under consideration may be stated as

$$Q_g^f C_{gi}^f - Q_g C_{gi} = K_{gi} a(C_{gi} - H_i C_{\ell i}) V_R \tag{23}$$

In addition, the mass balance in the liquid phase is given by

$$Q_\ell^f C_{\ell i}^f - Q_\ell C_{\ell i} + K_{gi} a(C_{gi} - H_i C_{\ell i}) V_R = \varepsilon_\ell R_i V_R \tag{24}$$

where R_i is the overall rate of consumption of the ith component.

Effects of Chemical Reactions on Mass Transfer

The differential equation of change in the liquid phase arises from a mass balance on each component (Danckwerts, 1970; Astarita, 1967). It expresses the phenomenon of simultaneous diffusion and chemical reaction and may be written as

$$\frac{\partial C_i}{\partial t} = -\underset{\sim}{u} \cdot \nabla C_i + D_i \nabla^2 C_i - R_i \tag{25}$$

accumulation convection molecular transport reaction rate

If the bulk of the liquid is well mixed, the preceding equation needs to be integrated over the region near the gas-liquid interface. This implies an understanding of the fluid dynamics in such a region. The simplest model relies on the assumption that the velocity vector $\underset{\sim}{u}$ is parallel to the interface. In addition, the concentration gradient vector ∇C_i is assumed to be perpendicular to the interface. As a consequence of these assumptions, the convective term is dropped. This procedure is accomplished in classical film and penetration models. The models contain one parameter each, which can not be specified theoretically and which reflect the influence of the prevailing flow pattern (Sherwood et al., 1975). In the former model the parameter is the film depth δ, in which the resistance to mass transfer is confined. In the penetration model, the parameter is the mean lifetime $\bar{s}$ of the elements of fluid which are randomly replaced at the surface.

Both parameters are semiempirically evaluated from experimental values of the mass transfer coefficient k_ℓ without chemical reaction, as

$$\delta = \frac{D}{k_\ell} \qquad \bar{s} = \frac{k_\ell^2}{D} \tag{26}$$

Attempts have been made to identify the more reliable of the mentioned models. The general conclusion is that for describing the rate of gas absorption in the presence of chemical reaction, only small differences exist between the two models once the same value for the purely physical mass transfer coefficient k_ℓ is assumed. In fact, the accuracy of available

experimental data is frequently insufficient to discriminate between the two models, which do not usually differ by more than a few percent. Significant differences appear only when the gaseous and liquid reactants have greatly different diffusivities (Brian et al., 1961) or when a complex reaction scheme is involved (Tavares da Silva, 1974; Huang et al., 1980). It follows that the choice between the models is usually made on the basis of simplicity in describing the reacting system. For these reasons the film model has been more widely used and will be applied in the following. The dependence of the mass transfer coefficient on the diffusion coefficient predicted by the two models is different; the power 0.5, given by the penetration theory, is recommended.

The occurrence of a chemical reaction in the liquid phase affects the value of the mass transfer coefficient. This effect can be accounted for through the introduction of an *enhancement factor* E or a *reaction factor* E*. The former is given by the ratio between the mass transfer fluxes with and without reaction, in the same hydrodynamic conditions;

$$E_i = \frac{-D_i(dC_{fi}/ds)_{s=0}}{k_{\ell i}(C^*_{\ell i} - C_{\ell i})} \tag{27}$$

Instead, the reaction factor is defined as the ratio between the total mass flux of the component i under examination in the presence of a reaction and the hypothetical mass flux in the absence of reaction and with zero concentration of i in the liquid bulk:

$$E^*_i = \frac{-D_i(dC_{fi}/ds)_{s=0}}{k_{\ell i}C^*_{\ell i}} \tag{28}$$

The mathematical analysis of gas-liquid reactions is complex as a consequence of the interference of hydrodynamic and chemical factors. It follows that the possibility of obtaining analytical expressions of the mentioned factors is limited to cases of simplified reaction kinetics only.

The type of chemical system which has received the most attention is that in which the dissolved gas A undergoes an irreversible reaction with a reactant B dissolved in the liquid. The reaction is described by the following stoichiometric equation:

$$A + \nu_B B \rightarrow \text{products} \tag{29}$$

The film model assumes that even on a microscopic scale the process occurs in steady state, so that the term $(\partial C_i/\partial t)$ in Eq. (25) is dropped. It follows that the material balances for the two components are

$$\left.\begin{aligned} D_A \frac{d^2C_{fA}}{ds^2} - r &= 0 \\ D_B \frac{d^2C_{fB}}{ds^2} - \nu_B r &= 0 \end{aligned}\right\} \quad 0 \leqslant s \leqslant \delta \tag{30}$$

r being the reaction rate and s the distance in the liquid from the interface. D_A and D_B are the diffusivities of A and B in the liquid.

The boundary conditions are

$$\left.\begin{aligned} C_{fA} &= C^*_{\ell A} \\ \frac{dC_{fB}}{ds} &= 0 \end{aligned}\right\} \text{ at } s = 0 \tag{31}$$

$$\left.\begin{aligned} C_{fB} &= C_{\ell B} \\ C_{fA} &= C_{\ell A} \end{aligned}\right\} \text{ at } s = \delta \tag{32}$$

To obtain the bulk concentration of component A (i.e., $C_{\ell A}$), it must be taken into account that some of it reacts within the film and the rest is transferred across the film. For instance, in a well-mixed reactor, the following mass balance equation is obtained:

$$-aD_A \frac{dC_{fA}}{ds} = r(\varepsilon_\ell - a\delta) + \varepsilon_\ell C_{\ell A} \frac{Q_\ell}{V_R} \quad \text{at } s = \delta \tag{33}$$

This equation states that the transfer rate of dissolved gas to the bulk liquid must be equal to the amount reacted in the bulk plus that which leaves the reactor in the effluent stream. By integrating the preceding differential equations, the concentration profiles in the liquid film are obtained, and from there the values of the enhancement and reaction factors can be evaluated through Eqs. (27) and (28), respectively.

Enhancement Factor for Fast Reaction

If the reaction rate is high, it is justifiable to consider the reaction to be completed in the film. In this case the concentration of the absorbed gas in the bulk of the liquid is zero. Therefore, the boundary condition (32) can be simplified as follows:

$$C_{\ell A} = 0 \qquad \text{at } s = \delta \tag{34}$$

and the enhancement and reaction factors are then identical.

Although this situation is not fulfilled for many important gas-liquid reactions, it is worthwhile to consider it in some detail. Cases in which the reaction also occurs in the liquid bulk are discussed extensively in the next section.

The typical form of the concentration profiles of the reagents A and B of reaction (29) is illustrated in Fig. 4a. The reaction is fast and it occurs completely in the liquid film during the diffusion of component A. For reactors in which this situation occurs, high interfacial area is required, while the liquid holdup is not important. The use of packed columns is thus recommended. Let us consider, first, the case in which reaction (29) takes palce through a bimolecular reaction of first order with respect to both

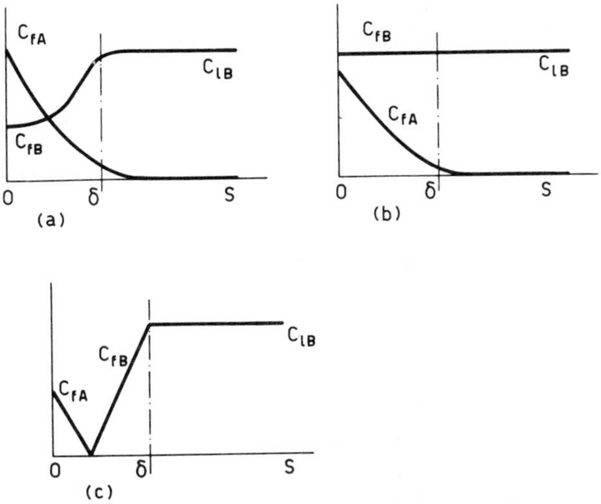

FIGURE 4 Concentration profile of gaseous and liquid reactants in the liquid phase: (a) fast reaction; (b) fast reaction with $C_{\ell B} \gg C^*_{\ell A}$; (c) instantaneous reaction.

reagents. In this case the enhancement factor can be expressed as a function of a dimensionless parameter known as the *Hatta number*:

$$M = \frac{\sqrt{D_A k C_{\ell B}}}{k_{\ell A}} \tag{35}$$

and of the concentration diffusion parameter:

$$Z_D = \frac{D_B}{\nu_B D_A} \frac{C_{\ell B}}{C_{\ell A}} \tag{36}$$

An approximate expression of the enhancement factor E has been proposed by van Krevelen and Hoftijzer (1948a). It was obtained through a linearization technique based on the assumption that the concentration value of B throughout the film is constant and equal to its value at the interface. With this assumption the film model equations can be integrated and the following implicit expression for the enhancement factor is obtained:

$$E_A = \frac{M[(E_{A,in} - E_A)/(E_{A,in} - 1)]^{1/2}}{\tanh \{M[(E_{a,in} - E_A)/(E_{A,in} - 1)]^{1/2}\}} \tag{37}$$

$E_{A,in}$ is the enhancement factor corresponding to an *instantaneous* reaction and is given by

$$E_{A,in} = Z_D + 1 = 1 + \frac{D_B}{\nu_B D_A} \frac{C_{\ell B}}{C_{\ell A}} \quad (38)$$

The behavior of the enhancement factor as a function of the parameters M and Z_D is given in Fig. 5. Two particular cases merit attention. The first occurs if the concentration of component B in the bulk liquid is much greater than $C^*_{\ell A}$, as shown in Fig. 4b. In this case the kinetics of the reaction becomes pseudo-first order with an apparent reaction rate constant given by $kC_{\ell B}$. An exact solution of Eq. (30) can thus be obtained, which gives the following expression of the enhancement factor:

$$E_A = \frac{M}{\tanh M} \quad (39)$$

If $4 < M < E_{A,in}/2$, the points representing the enhancement factor are very close to the diagonal in Fig. 5. The second case is the one in which the reaction rate is so high that the concentration of the reactant B is less than the solubility of the gas A, because it reacts instantaneously with component B. This situation occurs when $M > 10E_{A,in}$. In fact, it can be seen from Fig. 5 that when the preceding condition occurs, an increase of the Hatta number leads to a limiting value of $E_A = E_{A,in}$. Since the reaction is instantaneous, the liquid phase may be considered to be made up of two subsequent layers in which the resistance to the transformation is given by the diffusion of the reagents (Fig. 4c). In this case expression (38) for the enhancement factor is obtained.

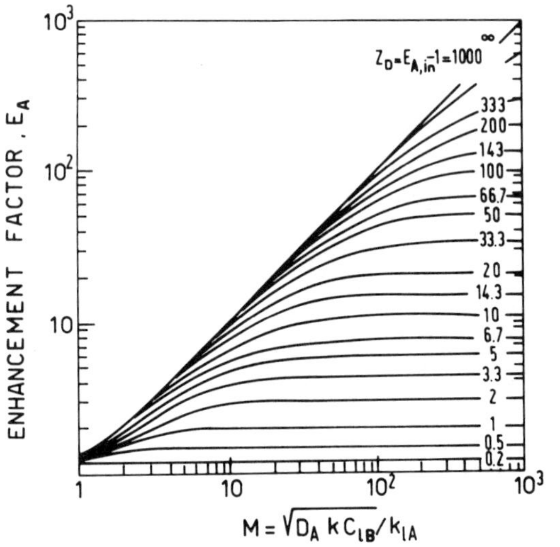

FIGURE 5 Enhancement factor for a fast bimolecular second-order reaction.

A more general case is that in which reaction (29) takes place through a bimolecular reaction of order m with respect to A and n with respect to B. This problem was studied by Hikita and Asai (1963), whose basic idea was to linearize the reaction rate with respect to the gaseous reactant concentration through the approximation

$$r = kC_A^m C_B^n \simeq kC_A^{*\,m-1} \frac{2}{m+1} C_B^n C_A \tag{40}$$

which is obviously exact in the case of m = 1. Introducing this approximation, the previously mentioned linearization technique can readily be applied (i.e., assuming C_B as constant within the film). This approach leads to the following analytical implicit expression for the enhancement factor:

$$E_A = \frac{M[(E_{A,in} - E_A)/(E_{A,in} - 1)]^{n/2}}{\tanh\{[(E_{A,in} - E_A)/(E_{A,in} - 1)]^{n/2}\}} \tag{41}$$

where the Hatta number in this case is defined as

$$M = \frac{1}{k_{\ell A}} \left[\left(\frac{2}{m+1}\right) kD_A C_{\ell A}^{*\,m-1} C_{\ell B}^n\right]^{1/2} \tag{42}$$

$E_{A,in}$ is still given by Eq. (38). The behavior of the enhancement factor as a function of the parameters is given in Fig. 6 for various values of n, with fixed m = 1. The accuracy of expression (40) is sufficient for engineering calculations. In Fig. 7, for a pseudo-mth-order reaction (i.e.,

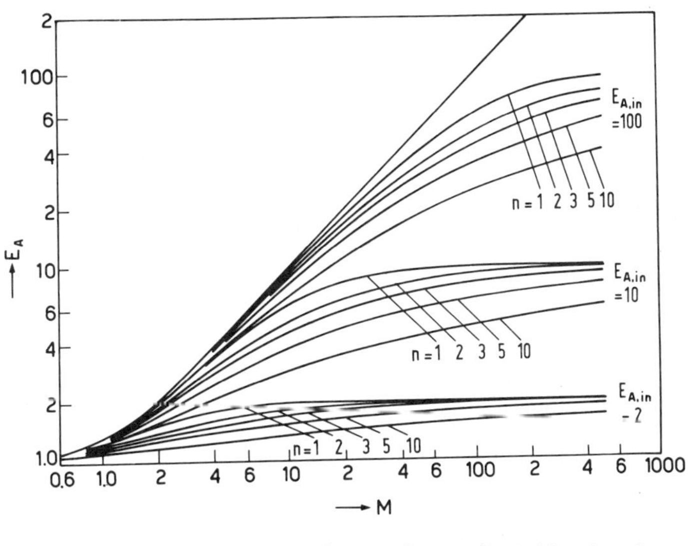

FIGURE 6 Enhancement factor for a fast bimolecular reaction of first order with respect to the gaseous reactant and nth order with respect to the liquid reactant.

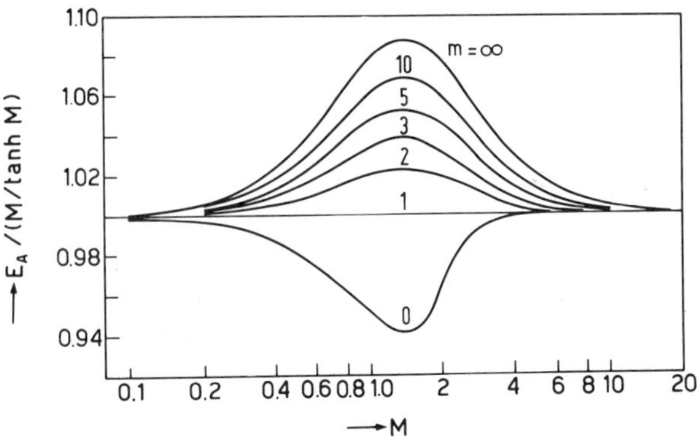

FIGURE 7 Ratio of the true enhancement factor numerically calculated, E_A to the approximate factor calculated by Eqs. (39) and (42), in the case of a pseudo mth-order reaction.

where C_B is constant) the ratio between the approximate and the numerically calculated enhancement factors is reported for various values of m as a function of the Hatta number. The same kind of accuracy is also exhibited by Eq. (40) in the case of reaction (29) of mth order with respect to A and nth order with respect to B.

It is worth mentioning here the large amount of literature devoted to the definition of *explicit* expressions of the enhancement factor under the assumption of complete gaseous reactant depletion in the liquid bulk, $C_{\ell A} = 0$. For reaction (29), first order with respect to both reactants, it has been reviewed by Wellek et al. (1978), who recommended the use of the expression proposed by Yeramian et al. (1970):

$$E_A = \frac{M^2}{2(E_{A,in} - 1)\tanh^2 M}\left\{\left[1 + \frac{4(E_{A,in} - 1)E_{A,in}\tanh^2 M}{M^2}\right]^2 - 1\right\} \qquad (43)$$

where M and $E_{A,in}$ are given by Eqs. (35) and (38), respectively. In the present section we have considered a particular situation for the gas-liquid reactor which allows the neglecting of the reaction in the liquid bulk, since this is not reached by the gaseous reactant. Deeper analysis of the problem requires the coupling of the film equations with those describing the reactor behavior. This is developed later and more general approximate expressions of the reaction factor are also given.

MEASUREMENTS OF INTERFACIAL AREAS AND MASS TRANSFER COEFFICIENTS

Gas holdup ε_g, interfacial area a, and mass transfer coefficients are important parameters determining the mass transfer rates in gas-liquid reactors. Different procedures have been proposed for their measurement, which can be classified in two categories: local measurements with physical techniques

and global measurements with chemical procedures. A detailed analysis of the procedures has been given by Charpentier (1981).

Physical Methods

The gas holdup is directly determined by measuring the height of the aerated liquid and that of the clear liquid without reaction (Fair et al., 1962; Marrucci and Nicodemo, 1967; Hughmark, 1967; Yoshida et al., 1970; Eissa and Schügerl, 1975; Yoshida and Akita, 1965; Miller, 1974). This procedure is rapid but not very accurate. A more precise technique, in which the gas holdup is obtained from measurements of the clear liquid height in the dispersion at successive manometric taps on the side of the froth container, has been proposed and employed by Burgess and Calderbank (1975a,b).

An electrical technique, proposed by Linek and Mayrhoferova (1969), can also be used. It involves the measurements of the surface elevation at certain selected points by means of an electrically conducting probe. The height is determined by the vertical position of the tip at which the sum of the contact times equals one-half of the measurement period. The γ-ray transmission technique has also been employed to determine the local gas holdup in gas-liquid contactors (Calderbank, 1958; Calderbank and Rennie, 1962; Bernard and Sargent, 1966; Hwa and Beckmann, 1960).

From the gas holdup, the specific interfacial area can be calculated as

$$a = \frac{6\varepsilon_g}{d_{sv}} \tag{44}$$

where d_{sv} is the volume surface mean diameter or Sauter mean diameter, defined as

$$d_{sv} = \frac{\sum_i n_i d_b^3}{\sum_i n_i d_b^2} \tag{45}$$

d_b is the diameter of the single bubble (or drop) and n_i the number of bubbles (or drops) of diameter d_b. The value of d_{sv} can be directly evaluated through a statistical analysis of high-speed photomicrographs performed in the dispersion (Calderbank and Rennie, 1962; Vermeulen et al., 1955; Porter et al., 1966; Akita and Yoshida, 1973; Ashley and Haselden, 1972).

The local interfacial contact area is directly evaluated by light transmission and reflection techniques. A parallel beam of light is passed through the dispersion and a photocell is placed at some distance from it (Vermeulen, 1955). Only the unscattered part of the incident parallel beam is recorded by the photocell. It has been shown by Calderbank (1958) that for bubbles the scattering cross section is equal to its projected area. The total interfacial area per unit volume of the dispersion is related to the intensity ratio of the radiation through the following equation:

$$\ln \frac{I_0}{I} = a \frac{\ell}{4} \tag{46}$$

where ℓ is the optical path length and I_0/I the intensity ratio between the incident beam of radiation and the transmitted beam.

Chemical Methods

The chemical methods are derived from the theory of mass transfer with chemical reaction, previously illustrated. A gas is absorbed into a liquid, where it reacts with a dissolved species. An excellent review of such methods has been published by Sharma and Danchwerts (1970). By choosing a reactant with adequate solubility and a suitable reaction rate, either the products $k_\ell a$, $k_g a$, or the interfacial area can be deduced from the overall rate of gas absorption.

The determination of the interfacial area relies on the fact that when the reaction between the two components in the liquid phase is mth order in A and nth order in B if

$$4 < M < \frac{E_{A,in}}{2} \tag{47}$$

where M is defined by Eq. (42), the rate of absorption is given by

$$r = a\left(\frac{2}{m+1} kD_A C^{*\,m+1}_{\ell A} C^n_{\ell B}\right)^{1/2} \tag{48}$$

This equation shows that the specific absorption rate is independent of hydrodynamic conditions. Therefore, if the solubility, the diffusivity, and the reaction kinetics are known, the specific interfacial area can be determined from the experimentally determined total absorption rate.

Interfacial areas determined from the physical and chemical methods may differ by more than 100%. A careful analysis reveals that the results of the photographic method and the sulfite oxidation method (a specific chemical method discussed later) show large variations at high gas velocities. For homogeneous bubbly flow, Schumpe and Deckwer (1980) have found that

$$a_{photo} = 1.35a_{sulfite} \tag{49}$$

For churn-turbulent flow, the difference is even larger. These results reveal that the photographic method is not very reliable.

The analytic determination of the absorption rates can be performed in two limiting situations, the transport or reaction-controlled cases. In the former case, which corresponds to instantaneous reaction regime, the chemical systems employed are indicated in Table 5. In the latter case, which corresponds to slow reaction regime, the employed chemical systems are shown in Table 6.

Of course, the interpretation of experimental data relies on the model employed to describe the gas-liquid dispersion, and customarily it is assumed that the liquid phase is perfectly mixed, while the gas phase is either in plug flow or perfectly mixed. This subject has been critically examined by Midoux et al. (1980), who proposed a flow model for gas dispersion into liquids inhibiting coalescence that seems more realistic than the plug-flow model.

TABLE 5 Chemical Systems Employed to Determine $k_\ell a$ in the Instantaneous Reaction Regime

Solute gas	Reactant	Reference
NH_3	H_2SO_4	Sharma and Danckwerts (1970)
SO_2, Cl_2, HCl	NaOH	Sharma and Danckwerts (1970)
H_2S, HCl, CO_2	Amines	Sharma and Danckwerts (1970)
O_2	$N_2S_2O_4$	Jhaveri and Sharma (1968)

Oxygen absorption into sodium sulfite in the presence of copper ions as catalyst has been widely used for the determination of absorption rates in packed columns, bubble columns, and mechanically agitated vessels.

The absorption occurs through the reaction

$$O_2 + 2Na_2SO_3 \rightarrow 2Na_2SO_4 \tag{50}$$

An extended review on its use as a model reaction of known kinetics in determining accurately mass transfer characteristics of gas-liquid contactors has been published by Linek and Vacek (1981).

TABLE 6 Chemical Systems Employed to Determine $k_\ell a$ in the Slow Reaction Regime

Solute gas	Reactant	Catalyst	References
CO_2	K_2CO_3	NaClO	Danckwerts and Sharma (1966) Danckwerts and Gillham (1966)
O_2 diluted with air	Cu Cl	—	Jhaveri and Sharma (1967)
O_2 diluted with air	Na_2SO_3	$CuSO_4$	De Waal and Okeson (1966) Wesselingh and Van't Hoog (1970) Astarita et al. (1964)
O_2 diluted with air	Na_2SO_3	$CoSO_4$	Linek (1966) Linek and Mayrhoferova (1970) Onda et al. (1972)
Butadiene	Maleic-anhydride	—	Neelakantan and Gehlawat (1982)

MODELING OF ISOTHERMAL GAS-LIQUID REACTORS

Similarly to most chemical reactors, reliable sizing and selection of optimal operating conditions for gas-liquid reactors require the use of suitable mathematical models. The hydrodynamics of industrial gas-liquid reactors is quite complex, due to the presence of two phases, one dispersed and the other continuous, in relative motion, in some cases through a stationary solid packing. In particular, mixing in both phases can significantly affect the reactor performance in terms of reactant conversion and product selectivity. Wen and Fan (1975) and Shah et al. (1978) have reviewed backmixing in empty and packed gas-liquid reactors.

The most common technique used to define the extent of backmixing in a given unit is first to measure the residence-time distribution (RTD) of a suitable tracer in each phase and then to fit the obtained data with an appropriate model. A number of such models have been proposed in the literature, and some of these are summarized in Table 7. They can be divided into two main groups: pseudocontinuous models and cell models, leading to differential and algebraic equations, respectively. Each model is characterized by one or more parameters, which are usually tuned in order to reproduce the experimental RTD data.

The simplest pseudocontinuous model is the axial dispersion model, where backmixing is characterized by a one-dimensional diffusional term, which also includes the effects of both radial dispersion and nonuniform

TABLE 7 Backmixing Models

	Adjustable parameters	Reference
Pseudocontinuous models		
Axial dispersion model	1	Levenspiel (1972)
Multistaged dispersion model	2	Nishiwaki and Kato (1974)
Modified mixing cell model	2	Deans (1963)
Cross-flow model	2	Hochman and Effron (1969)
Piston diffusion exchange model	3	van Swaaij et al. (1969)
Cell models		
Series of stirred-tank model	1	Dean and Lapidus (1960a,b)
Modified series of stirred-tank model	3	Bang and Cholette (1973)
Backflow model	Variable	Miyauchi and Vermeulen (1963)
Dispersion backflow model	Variable	Nishiwaki et al. (1973)

velocity profile. It is equivalent to the series of CSTRs models, where backmixing is again characterized by only one parameter: the number of tanks in the series. In the other models listed in Table 7, some more parameters are introduced in order into account the possible simultaneous presence of flow with axial dispersion, bypass, and a fraction of stagnant fluid. Although they give a better representation of experimental RTD data, when used to describe the kinetic performance of a gas-liquid reactor, they lead to complex mathematical equations, which can be of limited use for reactor simulation or design. Therefore, in the following we consider only the axial dispersion model for both the gas and liquid phases. Obviously, a model of this type can describe only macromixing situations intermediate between plug flow and perfect mixing. More complex situations can be examined by modeling the unit under examination as a combination of suitable smaller units, each described by the axial dispersion model, with different values of the dispersion coefficients. This model can then be regarded as the fundamental unit for building more complex models, such as those presented by Wolf and Resnick (1963) or those mentioned in Table 7.

We will first outline the most general form of the axial dispersion model, which can be used to simulate most gas-liquid reactors, provided that suitable values are given for the involved parameters. The CSTR model, which is also the limiting situation for infinite axial dispersion, is examined in detail due to its very frequent application. Two modes of operation will be considered: both phases continuous and batch for the liquid phase. As in most industrial applications, the chemical reactions are assumed to occur only in the liquid phase.

The mathematical model of a gas-liquid reactor is constituted of two interacting portions. The first describes the diffusion and reaction process within the liquid film, depending on the concentration values in the bulk of the two phases in relative motion. The second describes the time or space evolution of the bulk concentrations, taking into account the operating mode, hydrodynamics, and macromixing in each phase.

Liquid Film Model

Considering a generic ith component, its dimensionless steady-state mass balance in the liquid film, where NR reactions occur, is written as follows:

$$\xi_i \frac{d^2\phi_{fi}}{dx^2} = -Ha^2 \sum_{j=1}^{NR} \nu_{ij} p_j f_j \qquad x \in (0,1) \tag{51}$$

with boundary conditions (BCs)

$$x = 0: \quad \phi_{fi} = \phi^*_{\ell i} \tag{51a}$$

$$x = 1: \quad \phi_{fi} = \phi_{\ell i} \tag{51b}$$

where ϕ_{fi}, $\phi_{\ell i}$, and $\phi^*_{\ell i}$ indicate the dimensionless concentration of the ith component in the liquid film, bulk, and gas-liquid interface, respectively; the other quantities are defined as follows:

$$\xi_i = \frac{D_i}{D_A} \qquad \phi_{fi} = \frac{C_{fi}}{C^\circ_\ell} \qquad \phi_{\ell i} = \frac{C_i}{C^\circ_\ell} \qquad \phi^*_{\ell i} = \frac{C^*_{\ell i}}{C^\circ_\ell}$$

$$x = \frac{s}{\delta} \qquad Ha^2 = \frac{r^\circ_1 D_A}{k^2_{\ell A} C^\circ_\ell} \qquad p_j = \frac{r^\circ_j}{r^\circ_1} \tag{52}$$

$$f_j = \frac{r_j}{r^\circ_j} \qquad k_{\ell i} = \frac{D_i}{\delta}$$

where the subscript A indicates the gaseous reactant and the superscript ° indicates a reference condition, usually corresponding to the feed stream or the initial condition for continuous or batch operations, respectively.

It is worthwhile to recall that the last of equations (52) is a result of the film theory, but it is not verified in practice since from experimental measurements the purely physical mass transfer coefficient $k_{\ell i}$ appears to be proportional to the square root of the diffusion coefficient D_i.

In the solution of the liquid film model, it is only necessary to calculate the mass flux of the ith component across the gas-liquid film (x = 0) and the liquid film-liquid bulk (x = 1) interfaces. The first is given by

$$(N_i)_{s=0} = -D_i \left(\frac{dC_{fi}}{ds}\right)_{s=0} = k_{\ell i} E^*_i C^*_{\ell i} = k_{gi} (C_{gi} - C^*_{gi}) \tag{53}$$

where the reaction factor E^*_i, defined by Eq. (28), using Eqs. (52) reduces to

$$E^*_i = -\frac{1}{\phi_{\ell i}} \left(\frac{d\phi_{fi}}{dx}\right)_{x=0} \tag{54}$$

The last equality in Eq. (53) is based on also using the film theory for the gas-side mass transfer process, and in dimensionless form leads to

$$\phi^*_{gi} = \frac{\phi_{gi}}{1 + \gamma_i E^*_i / H^*_i} \tag{55}$$

where

$$\gamma_i = \frac{k_{\ell i} C^\circ_\ell}{k_{gi} C^\circ_g} \qquad H^*_i = \frac{C^\circ_\ell}{C^\circ_g} H_i \tag{56}$$

and the gas-liquid equilibrium at the interface [(Eq. (21b)] reduces to

$$\phi^*_{gi} = H^*_i \phi^*_{\ell i} \tag{57}$$

The material fluxes at the liquid boundaries can be calculated, using Eqs. (51), (53), and (55), through the relationships

$$(N_i)_{s=0} = \frac{k_{\ell i} C^\circ_\ell \phi_{gi} (E^*_i/H^*_i)}{1 + \gamma_i E^*_i/H^*_i} \tag{58}$$

$$(N_i)_{s=\delta} = k_{\ell i} C^\circ_\ell \left[\frac{\phi_{gi}(E^*_i/H^*_i)}{1 + \gamma_i E^*_i/H^*_i} + \frac{Ha^2}{\xi_i} \sum_{j=1}^{NR} \nu_{ij} p_j \int_0^1 f_j \, dx \right] \tag{59}$$

as a function of the bulk concentrations ϕ_{gi} and $\phi_{\ell i}$. These can be evaluated through the macroscopic balances examined later, depending on the reactor hydrodynamics and operation mode.

Axial Dispersion Model

Using the axial dispersion model for both phases, the isothermal, steady-state, dimensionless material balances for the ith component are, in the liquid phase,

$$\frac{1}{Pe_\ell} \frac{d^2 \phi_{\ell i}}{dz^2} + n^* \frac{d(v_\ell \phi_{\ell i})}{dz} + (J_i)_{x=1} + \alpha \frac{St_{\ell i}}{\xi_i} Ha^2 \sum_{j=1}^{NR} \nu_{ij} p_j f_j = 0$$

$$z \in (0,1) \tag{60}$$

and in the gas phase,

$$\frac{1}{Pe_g} \frac{d^2 \phi_{gi}}{dz^2} - \frac{d(v_g \phi_{gi})}{dz} - (J_i)_{x=0} = 0 \qquad z \in (0,1)$$

where besides those defined in Eqs. (52), the following dimensionless quantities have been introduced:

$$\phi_{gi} = \frac{C_{gi}}{C^\circ_g} \qquad z = \frac{y}{L} \qquad v_\ell = \frac{u_\ell}{u^\circ_\ell} \qquad v_g = \frac{u_g}{u^\circ_g}$$

$$St_{\ell i} = \frac{Lak_{\ell i}}{u^\circ_\ell} \qquad St_{gi} = \frac{Lak_{gi}}{u^\circ_g} \qquad Pe_\ell = \frac{u^\circ_\ell L}{E_\ell(1 - \varepsilon_g - a\delta)} \tag{61}$$

$$Pe_g = \frac{u^\circ_g L}{E_g \varepsilon_g} \qquad \alpha = \frac{1 - \varepsilon_g - a\delta}{a\delta}$$

$n^* = 1, -1$ indicates countercurrent and cocurrent flow, respectively. The dimensionless interfacial material fluxes are given, using Eqs. (55), (58), and (59), by

$$(J_i)_{x=0} = \frac{St_{gi} \phi_{gi} \gamma_i (E^*_i/H^*_i)}{1 + \gamma_i E^*_i/H^*_i} = St_{gi} \phi^*_{\ell 1} \gamma_i E^*_i \tag{63}$$

$$(J_i)_{x=1} = St_{\ell i}\left[\frac{\phi_{gi}(E_i^*/H_i^*)}{1+\gamma_i E_i^*/H_i^*} + \frac{Ha^2}{\xi_i}\sum_{j=1}^{NR}\nu_{ij}\, p_j \int_0^1 f_j\, dx\right] \tag{64}$$

When, as in Eqs. (60), the variation of the superficial velocities v_ℓ and v_g along the reactor axis must be taken into account, the equation of state of each phase should be introduced. However, for most usual applications the liquid superficial velocity can be assumed constant along the reactor axis, while for the gas phase the ideal gas law can be used:

$$C_{gi} = \frac{Py_i}{RT} \tag{65}$$

which introduced in Eqs. (61) and (63) leads to

$$-\frac{1}{Pe_g}\frac{d^2(P/P^\circ)}{dz^2} + \frac{d(v_g P/P^\circ)}{dz} + \sum_{i=1}^{NC} St_{gi}\,\phi_{gi}\,\frac{\gamma_i E_i^*/H_i^*}{1+\gamma_i E_i^*/H_i^*} = 0 \tag{66}$$

where P° is the pressure value at the reference conditions, which in this case are those relative to the feed stream (i.e., $P^\circ/RT = C_g^\circ = \Sigma_{i=1}^{NC} C_{gi}^f$). According to Deckwer (1976), the pressure gradient can significantly affect reactor behavior and should then be taken into account through the relationship

$$\frac{P}{P_T} = 1 + \bar{\alpha}(1 - z) \tag{67}$$

where $\bar{\alpha} = \rho_\ell g L(1 - \varepsilon_g)/P_T$ and P_T is the pressure at the reactor top. Note that, using Eq. (67), the first term in Eq. (66) vanishes.

The boundary conditions for Eqs. (60) and (61) depend on the reactor operation mode: cocurrent ($n^* = -1$) or countercurrent ($n^* = 1$):

Liquid phase:

$$z = \frac{1+n^*}{2}:\quad v_\ell^f \phi_{\ell i}^f = v_\ell \phi_{\ell i} - \frac{1}{Pe_\ell}\frac{d\phi_{\ell i}}{dz} \qquad v_\ell = v_\ell^f \tag{68a}$$

$$z = 1 - n^*:\quad \frac{d\phi_{\ell i}}{dz} = 0 \tag{68b}$$

Gas phase:

$$z = 0:\quad v_g^f \phi_{gi}^f = v_g \phi_{gi} - \frac{1}{Pe_g}\frac{d\phi_{gi}}{dz} \qquad v_g = v_g^f \tag{69a}$$

$$z = 1:\quad \frac{d\phi_{gi}}{dz} = 0 \tag{69b}$$

TABLE 8 Mass Balances for the CSTR Model

Liquid phase:

$$v_\ell \phi^f_{\ell i} = v_\ell \phi_{\ell i} - \alpha \frac{St_\ell}{\xi_i} Ha^2 \sum_{j=1}^{NR} \nu_{ij} p_j f_j - (J_i)_{x=1}$$

Gas phase:

$$v^f_g \phi^f_{gi} = v_g \phi_{gi} + (J_i)_{x=0}$$

$$v^f_g = v_g + \sum_{i=1}^{NC} (J_i)_{x=0}$$

Special cases of this model, particularly used in practice, are the CSTR and the PFR models, which are obtained assuming infinite ($Pe \to 0$) and zero ($Pe \to \infty$) axial dispersion coefficients, respectively. The equations for the CSTR model are summarized in Table 8, where constant pressure is assumed in the reactor. For the PFR model, the equations readily derive from those of the axial dispersion model, by simply deleting all the terms containing the Péclet numbers and the boundary conditions at the reactor outlet.

Semiflow Batch Model

A common operation mode for gas-liquid reactors is to continuously flow the gas phase through the liquid phase held inside the reactor. The liquid concentrations are then time dependent, but not space dependent, since the liquid can usually be considered well mixed. For the liquid phase, the following material balances can be written

$$\frac{d\phi_{\ell i}}{d\theta} = \alpha\, Ha^2 \sum_{j=1}^{NR} \nu_{ij} p_j f_j + \frac{\xi_i}{St_{\ell i}} (\bar{J}_i)_{x=1} = 0 \qquad \theta > 0 \tag{70}$$

with the initial conditions, $\phi_{\ell i} = \phi^{in}_{\ell i}$ at $\theta = 0$, and where

$$\theta = \frac{t k_{\ell A} a}{1 - \varepsilon_g - a\delta} \tag{71}$$

The total mass flux of the ith component entering the liquid bulk, $(\bar{J}_i)_{x=1}$, is given by

$$(\bar{J}_i)_{x=1} = \int_0^1 (J_i)_{x=1}\, dz \tag{72}$$

where $(J_i)_{x=1}$ is given by Eq. (64) as a function of the gas-phase composition, which for $Pe_g \neq 0$ depends on the reactor axis z according to Eq. (61).

Usually, pseudo-steady-state conditions can be assumed for the gaseous phase (Schaftlein and Russel, 1968), so that the mass balances in this phase are identical to those previously described for the axial dispersion model.

Variation of Interfacial Area and Gas Holdup

In all previous models it is assumed that the gas-liquid interfacial area and the gas holdup are constant along the reactor axis. Although experimental results do not fully confirm this assumption (Deckwer, 1977; Deckwer et al., 1978; Kölbel et al., 1972), sufficient data have not yet been collected to derive reliable correlations. However, in some applications (i.e., dead-end reactors) the variation of the gas flow rate due to absorption is such that changes in interfacial area and gas holdup cannot be neglected. To this end, Schaftlein and Russel (1968) and Shaikh and Varma (1983b) have suggested the following relationships:

$$\frac{a}{a^f} = \left(\frac{v_g}{v_g^f}\right)^{2/3} \tag{73a}$$

$$\frac{\varepsilon_g}{\varepsilon_g^f} = \frac{v_g}{v_g^f} \tag{73b}$$

where the ratio between the value of a (or ε_g) with and without (i.e., at feed conditions, $v_g = v_g^f$) absorption is related to the superficial gas velocity. These equations are based on the assumption that the dispersed gas phase is constituted by a swarm of spherical bubbles flowing with no breakage or coalescence and no shape variation during shrinkage.

ESTIMATION OF CHARACTERISTIC PARAMETERS

The use of mathematical models for the simulation of various types of industrial gas-liquid reactors requires a suitable evaluation procedure of all the parameters appearing in the model equations. In this section recommended semiempirical relationships for the three most widely used reactor types—sparged, packed, and mechanically agitated reactors—are reported.

Sparged Reactors

These are empty reactors where the gas-liquid mixing energy is predominantly introduced with the gas phase, which can be dispersed into the liquid phase through various sparger types: single orifices, pipes with holes, sieve and porous plates, and two-phase nozzle spargers of injector and ejector type (Mersmann, 1978; Fair, 1967a). Main advantages of these units are the absence of moving parts, eliminating the need for seals (less maintenance and cost); solids handling ability; large heat transfer rate per

unit reactor volume; and large values of the gas-liquid mass transfer coefficients. The disadvantages are backmixing, particularly in the liquid phase; high pressure drop (significant particularly at low pressure); and the reduction of interfacial area due to bubble coalescence, significant for $L/d_c > 12$.

Vertical bubble columns are the sparged gas-liquid reactors most widely used in industry. Some modifications have been introduced to improve their performance: sectionalized bubble columns (low backmixing and bubble coalescence), horizontal bubble columns (low pressure drop), downflow bubble columns (large residence time), and plate columns (large number of transfer units). General review articles on hydrodynamics and mass and heat transport have been reported by Fair et al. (1973), Pohorecki (1976), Stichlmair and Mersmann (1977), Charpentier (1981), and Zahradník et al. (1982) for plate columns, and Joshi and Sharma (1976, 1978) for horizontal bubble columns.

Various studies on the design and simulation of bubble columns have been reported in the literature (Shulman and Molstad, 1950; Sideman et al., 1966; Fair, 1967a,b; Sharma and Mashelkar, 1968; Cichy et al., 1969; Cichy and Russel, 1969; Mashelkar, 1970; Deckwer, 1977; Mersmann, 1978). Shah et al. (1982) have published an extensive survey on the characteristic parameters estimation for bubble columns.

These parameters are affected by the column regime, which changes by increasing the upflow gas velocity as follows:

1. Bubble flow, characterized by uniform flow of equally sized bubbles; it occurs approximately at $u_g \lesssim 0.05$ m/s (Fair, 1967a).
2. Churn turbulent flow, where large and small bubbles are present with different rise velocities leading to an unsteady flow; it is the most common in industrial units.
3. Slug flow, where the bubble occupies the entire column cross section; it occurs at large upflow gas velocity, with columns diameter, $d_c \lesssim 0.15$ m.

The a priori definition of the flow regime is difficult, since it is affected by many parameters, such as sparger type, physical properties, and velocity of the liquid phase. An approximate map of flow regimes is shown in Fig. 8.

The recommended empirical equations for the evaluation of the characteristic parameters of a bubble column are reported in Table 9. The volume-surface mean diameter of the swarm of bubbles, defined by Eq. (45) can be evaluated through Calderbank's (1967) correlation when gas is sparged by two-phase nozzles and the energy dissipation rate is given by $P_t/V_d = u_g \rho_\ell (1 - \varepsilon_g) g$. For less effective gas spargers, such as perforated plates and single orifices, the correlation proposed by Akita and Yoshida (1974) can be used. In sparged reactors the bubble size is controlled by the equilibrium between breakup and coalescence rates. For low-viscous aqueous solutions, in the bubbly flow regime, the bubble size stability diagram shown in Fig. 9 has been proposed by Berghams (1973). In it the critical Weber number, which corresponds to the mentioned equilibrium, separating the stable and unstable regions, is shown. Pressure values up to 1.6 MPa do not affect the bubble diameter if the gas velocity is corrected to take into account the pressure in the column (Kölbel et al., 1961).

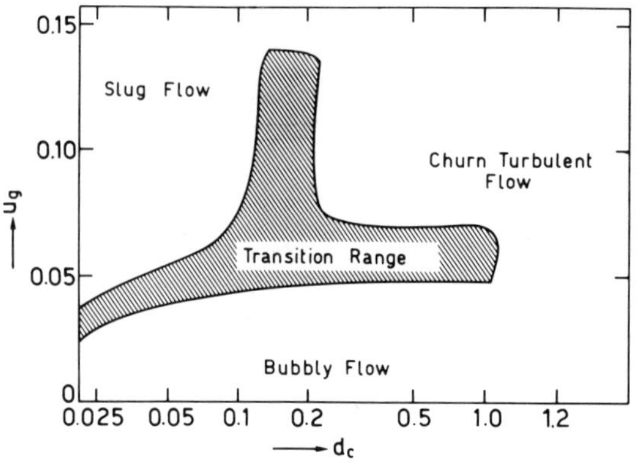

(a)

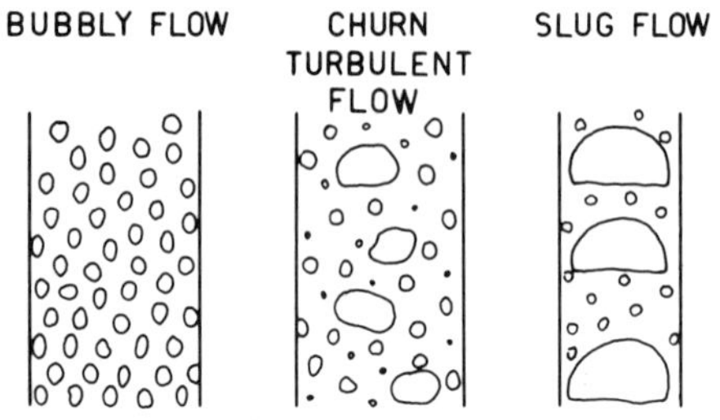

(b)

FIGURE 8 (a) Approximate dependency of flow regime on gas velocity u_g and column diameter d_c. (b) Schematic representation of flow regimes in bubble columns.

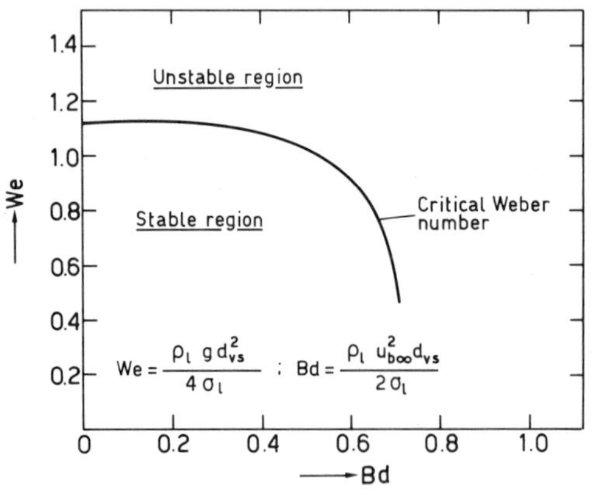

FIGURE 9 Stability diagram for low-viscous aqueous solutions in the bubbly flow regime.

TABLE 9 Semiempirical Expressions for Bubble Columns

Bubble diameter	Calderbank (1967)
$d_{vs} = C_1 \dfrac{\sigma_\ell^{0.6}}{(P_T/V_d)^{0.4}\rho_\ell^{0.2}} \varepsilon_g^n \left(\dfrac{\mu_g}{\mu_\ell}\right)^{0.25} + C_2$	
$C_1 = 2.25$; $n = 0.40$; $C_2 = 0$ for aqueous solutions of electrolytes	
$C_1 = 1.90$; $n = 0.65$; $C_2 = 0$ for aqueous solutions of alcohols	
$C_1 = 4.15$; $n = 0.50$; $C_2 = 0.0009$ m for pure liquids	
$d_{vs} = 26d_c \left(\dfrac{gd_c^2\rho_\ell}{\sigma_\ell}\right)^{-0.5} \left(\dfrac{gd_c^3\rho_\ell^2}{\mu_\ell^2}\right)^{-0.12} \left(\dfrac{u_g}{\sqrt{gd_c}}\right)^{-0.12}$	Akita and Yoshida (1974)
$0.07 < d_c < 0.6$ m; if $d_c > 0.6$ m, use $d_c = 0.6$ m; $u_g < 0.42$ m/s; $\varepsilon_g < 0.3$	
Terminal bubble rise velocity	Clift et al. (1978)
$Re_{b\infty} = M^{-0.149}(J - 0.857)$	
$J = 0.94H^{0.747}$ $(2 < H < 59.3)$	
$J = 3.42H^{0.441}$ $(H > 59.3)$	
$H = \frac{4}{3} E_0 M^{-0.149}(\mu_\ell/\mu_w)^{-0.14}$	

$$M = \frac{g\mu_\ell^4(\rho_\ell - \rho_g)}{\rho_\ell^2\sigma_\ell^3} < 10^{-3} \qquad Re_{b\infty} = \frac{d_{vs}u_{b\infty}\rho_\ell}{\mu_\ell} > 0.1$$

$$E_0 = \frac{g(\rho_\ell - \rho_g)d_{vs}^2}{\sigma_\ell} < 40$$

Gas holdup — Akita and Yoshida (1973)

$$\frac{\varepsilon_g}{(1 - \varepsilon_g)^4} = C\left(\frac{gd_c^2\rho_\ell}{\sigma_\ell}\right)^{1/8}\left(\frac{gd_c^3}{\nu_\ell^2}\right)^{1/12}\left(\frac{u_g}{\sqrt{gd_c}}\right)$$

C = 0.20 for nonelectrolyte solutions

C = 0.25 for electrolyte solutions

$0.15 < d_c < 0.6$ m; $u_g < 0.42$ m/s $\quad \varepsilon_g < 0.3$

Hikita et al. (1980)

$$\varepsilon_g = 0.672f\left(\frac{u_g\mu_\ell}{\sigma_\ell}\right)^{0.578}\left(\frac{\mu_\ell^4 g}{\rho_\ell\sigma_\ell^3}\right)^{-0.131}\left(\frac{\rho_g}{\rho_\ell}\right)^{0.062}\left(\frac{\mu_g}{\mu_\ell}\right)^{0.107}$$

f = 1 for pure liquid or nonelectrolyte solutions

$f = 10^{0.0414}$ for $I < 1$ kg ion/m^3

$f = 1.1$ for $I > 1$ kg ion/m^3

I = ionic strength of the solution

$d_c = 0.1$ m; $0.042 < u_g < 0.38$ m/s

TABLE 9 (Continued)

$$\varepsilon_g = \left[1 + 0.06855\,\frac{(1-\varepsilon^*)^{3.112}}{\varepsilon^{*0.395}}\left(\frac{\rho_\ell}{\rho_g}\right)^{0.0346}\left(\frac{\mu_\ell}{\mu_g}\right)^{0.254}\mathrm{Fr}^{0.36}\mathrm{We}^{0.543}\right]^{-1}$$	Friedel et al. (1980)
$$\varepsilon^* = \frac{u_g}{u_g + u_\ell}$$	
$$\mathrm{Fr} = \frac{F^2}{gd_c\rho_\ell^2}$$	
$$\mathrm{We} = \frac{F^2 d_c}{\sigma_\ell \rho_\ell}$$	
$$F = \rho_\ell u_\ell + \rho_g u_g$$	
$148 < F < 336$ kg/m^2·s; $0.003 < \varepsilon^* < 0.24$; $0.004 < \varepsilon_g < 0.3$	
$184 < \rho_\ell/\rho_g < 5340$; $37 < \mu_\ell/\mu_g < 2220$; $0.055 < \sigma_\ell < 0.074$ N/m	
Interfacial area per reactor volume	
$$a = \frac{1}{3d_c}\left(\frac{gd_c^2\rho_\ell}{\sigma_\ell}\right)^{0.5}\left(\frac{gd_c^3}{\nu_\ell^2}\right)^{0.1}\varepsilon_g^{1.13}$$	Akita and Yoshida (1974)
$0.07 < d_c < 0.6$ m; if $d_c > 0.6$ m, use $d_c = 0.6$ m; $u_g < 0.4$ m/s; $\varepsilon_g < 0.14$	

$$a = 48.7(u_g/\mu_{eff})^{0.51}$$ Schumpe and Deckwer (1981)

$d_c = 0.14$ m; $u_g > 0.02$ m/s

Volumetric liquid mass transfer coefficient

$$k_\ell a = 0.6 \frac{D_\ell}{d_c^2} \left(\frac{\mu_\ell}{\rho_\ell D_\ell}\right)^{0.5} \left(\frac{g d_c^2 \rho_\ell}{\sigma_\ell}\right)^{0.62} \left(\frac{g d_c^3 \rho_\ell^2}{\mu_\ell^2}\right)^{0.31} \varepsilon_g^{1.1}$$ Akita and Yoshida (1973)

$0.15 < d_c < 0.6$ m; if $d_c > 0.6$ m; use $d_c = 0.6$ m; $u_g < 0.33$ m/s; $\varepsilon_g < 0.3$

$$k_\ell a = b u_g^{0.8}$$ Deckwer et al. (1974)

Sparger	Liquid phase	b
Cross of nozzles (10^{-3}m), $d_c = 0.2$ m; $0.003 < u_g < 0.08$ m/s	Tap water	0.467
	Salt solutions	0.460
Sintered plates $0.1 < d_c < 0.15$ m; $0.003 < u_g < 0.08$ m/s	Tap water	1.174
	Salt solutions	1.445

$$k_\ell a = 0.00315 u_g^{0.59} \mu_{eff}^{-0.84}$$ Dekwer et al. (1982)

$u_g = 0.08$ m/s; $d_c = 0.14$ m

Liquid-side mass transfer coefficient — Calderbank and Moo-Young (1961)

$$\frac{k_\ell d_{vs}}{D_\ell} = 2 + 0.31 \left[\frac{d_{vs}^3(\rho_\ell - \rho_g)g}{\mu_\ell D_\ell}\right]^{1/3} ; \quad d_{vs} < 2.5 \times 10^{-3} \text{ m}$$

TABLE 9 (Continued)

$$\frac{k_\ell d_{vs}}{D_\ell} = 0.42\left(\frac{\mu_\ell}{\rho_\ell D_\ell}\right)^{1/2}\left[\frac{d_{vs}^3 \rho_\ell(\rho_\ell - \rho_g)g}{\mu_\ell^2}\right]^{1/3} \; ; \; d_{vs} > 2.5 \times 10^{-3} \text{ m}$$	
$$\frac{k_\ell d_{vs}}{D_\ell} = 2 + a\left(\frac{\mu_\ell}{\rho_\ell D_\ell}\right)^{0.546}\left(\frac{d_{vs} u_s \rho_\ell}{\mu_\ell}\right)^{0.779}\left(\frac{d_{vs} g^{1/3}}{D_\ell^{2/3}}\right)^{0.116}$$	Hughmark (1967)
a = 0.061 for single bubbles a = 0.0187 for swarms of bubbles $0.025 < d_c < 1.1$ m; $u_\ell < 0.12$ m/s; $u_s = u_g/\varepsilon_g - u_\ell/(1 - \varepsilon_g)$	
Gas-side mass transfer coefficient	Geddes (1946)
$$k_g = \frac{d_{vs}}{\tau_g(1 - y)}\left(0.083 + 6.58\,\frac{D_g \tau_g}{d_{vs}^2}\right)$$ $\tau_g = L/u_b$; u_b = bubble velocity y = gas mole fraction of diffusing component	
$$\frac{k_g d_{vs}}{D_g} = 3.29 \times 10^{-4}\left(\frac{d_{vs} u_g \rho_g}{\mu_g}\right)^{0.756}$$ $0.01 < d_c < 0.02$ m; $1 < u_g < 5$ m/s	Shilimkan and Stepanek (1978)

Liquid-phase axial dispersion coefficient	Deckwer et al. (1974)
$E_\ell = 0.678 d_c^{1.4} u_g^{0.3}$	
$d_c^{1.4} u_g^{0.3} < 400$	
$E_\ell = 0.33 d_c (u_c + u_\ell)$	Joshi (1980)
$u_c = 1.31 \left[g d_c \left(u_g - \frac{\varepsilon_g}{1 - \varepsilon_g} u_\ell - \varepsilon_g u_{b\infty} \right) \right]^{1/3}$	
$0.1 < d_c < 1.1$ m; $0.003 < u_g < 0.45$ m/s; $0.12 < u_\ell < 0.002$ m/s	
Gas-phase axial dispersion coefficient	Diboun and Schügerl (1967)
$E_g = 5 u_r d_c$	
u_r = relative velocity between the gas and liquid phase	
$E_g = 50 d_c^{1.5} (u_g/\varepsilon_g)^3$	Mangartz and Pilhofer (1980)
$E_g = 56.4 d_c^{1.33} (u_g/\varepsilon_g)^{3.56}$	Field and Davidson (1980)
$0.08 < d_c < 3$ m; $0.008 < u_g < 0.1$ m/s	
Heat transfer coefficient	Deckwer (1980)
$\frac{h_w}{\rho_\ell C_{p\ell} u_g} = 0.1 (\mathrm{ReFrPr}^2)^{-0.25}$	
$\mathrm{Re} = u_g d_{vs} \rho_\ell / \mu_\ell$; $\mathrm{Fr} = u_g^2 / g d_{vs}$; $\mathrm{Pr} = C_{p\ell} \mu_\ell / k_{T\ell}$	

TABLE 9 (Continued)

$0.03 < (\mathrm{ReFrPr}^2) < 7;\ 6 < \mathrm{Pr} < 985$	
$\dfrac{h_w}{\rho_\ell C_{p\ell} u_g} = 0.411 \left(\dfrac{u_g \mu_\ell}{\sigma_\ell}\right)^{-0.851} \left(\dfrac{\mu_\ell^4 g}{\rho_\ell \sigma_\ell^3}\right)^{0.308} \left(\dfrac{k_{T\ell}}{C_{p\ell}\mu_\ell}\right)^{2/3}$	Hikita et al. (1981a)
$5.4 \times 10^{-4} < u_g u_\ell/\sigma_\ell < 7.6 \times 10^{-2}$	
$4.9 < C_{p\ell}\mu_\ell/k_{T\ell} < 93$	
$7.7 \times 10^{-12} < \mu_\ell^4 g/\rho_\ell \sigma_\ell^3 < 1.6 \times 10^{-6}$	

The choice of one expression for the evaluation of the gas holdup is quite difficult, due to its sensitivity to the material system, including trace impurities. The gas holdup depends on the superficial gas velocity to the power 0.7 to 1.2 in the bubble flow regime, and 0.4 to 0.7 in the churn turbulent one. On the other hand, it is almost independent of pressure values up to 1.6 MPa (Kölbel et al., 1961), of the presence of internals (mainly for heat transfer), and of the column diameter for $d_c > 0.15$ m. It is then convenient, particularly for noncoalescing or non-Newtonian liquids, to evaluate ε_g through a simple experimental measurement in a laboratory-scale column ($d_c > 0.15$ m), with the desired material system and gas velocity. The most accurate correlations seem those proposed by Akita and Yoshida (1973) and Hikita et al. (1980), valid for single- or multinozzle spargers with diameter of about 1 mm or larger and for low-viscous ($\mu_\ell < 0.02$ Pa·s) and coalescing liquids. Such systems are characterized by the following condition (Mersmann, 1978):

$$-\frac{1}{\Delta\sigma}\frac{d\sigma_\ell}{dx} \leqslant 1 \qquad (74)$$

where x indicates the mole fraction of the component of lower surface tension, and $\Delta\sigma$ is the difference in the surface tension of the two components of the binary mixture. If condition (74) is violated, foaming occurs in the column and the gas holdup increases, as in the case of aqueous solutions of electrolytes (Deckwer et al., 1974). For downflow bubble columns the relationships proposed by Friedel et al. (1980) are recommended.

The gas-liquid interfacial area per unit reactor volume, a, can be calculated from Eq. (44) using the correlations mentioned for gas holdup and bubble size. In particular, the correlation proposed by Akita and Yoshida (1974) is recommended, which has been derived for less effective spargers (with one or more distributors) and can then be used in general at least for a conservative estimation. In the case of $0.14 < \varepsilon_g < 0.3$, the value of a should be calculated from Eq. (44) using the expressions for ε_g and d_{vs} proposed by the same authors and reported in Table 9. A more general correlation specifically tested for two-phase nozzle spargers is shown in Fig. 10 (Nagel et al., 1978). The direction of flow, up or down does not

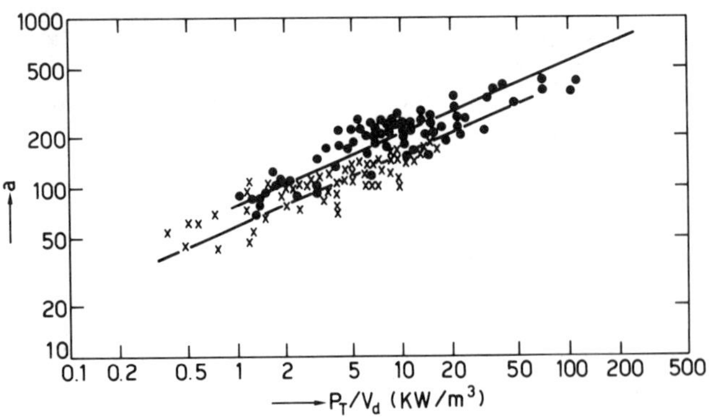

FIGURE 10 Specific interfacial areas in bubble columns; •, vertical pipe; x, horizontal pipe.

substantially affect the interfacial area in cocurrent flow (Shilimkan and Stepanek, 1978). Finally, for non-Newtonian highly viscous media and columns operating in the slug regime, as it is usually the case for such media at $u_g > 0.02$ m/s, Schumpe and Deckwer (1981) developed a useful correlation. The effective liquid viscosity μ_{eff} is calculated according to Nishikawa et al. (1977).

The volumetric liquid-side mass transfer coefficient depends on gas velocity, sparger type, and liquid physicochemical properties, while they are unaffected by the liquid velocity and the column diameter, above d_c = 0.15 m. Also, in this case the correlation proposed by Akita and Yoshida (1973) applies to less effective spargers, and the one proposed by Deckwer et al. (1982) to non-Newtonian highly viscous media. The $k_{\ell a}$ values obtained from Akita and Yoshida's equation are conservative values; much higher values can be obtained using sintered plates or two-phase nozzles for gas sparging [Deckwer (1977) suggests, as a rule of thumb, a correction factor of about 4 to 5]. This same equation can also be used for multistage columns. A more general expression is given by Deckwer et al. (1974), where the constant b is quite sensitive to sparger type and material system. This same expression can be used for downflow bubble columns (Herbrechtsmeier and Steiner, 1978).

Gas-side mass transfer resistance is usually negligible, and it becomes important only in the case of fast or instantaneous chemical reactions. Accurate correlations are not available; those reported in Table 9 are the most widely used. It is worthwhile reiterating that for all the parameters described up to this point, the effect of the column diameter d_c becomes insignificant for d_c values larger than 0.15 m for less viscous liquids, and larger than 0.30 m for highly viscous liquids ($\mu_\ell > 0.02$ Pa·s).

The liquid-phase axial dispersion coefficient E_ℓ depends on the gas velocity and column diameter and geometry, whereas it is independent of the liquid velocity, at least up to u_ℓ = 0.03 m/s. Rectangular columns exhibit larger backmixing than cylindrical ones (Striegel and Shah, 1977a). In a coiled column, E_ℓ depends on liquid velocity and approaches the same degree of axial mixing as in a single liquid-phase column for Reynolds number of the order of 400 to 2000. Information about backmixing in horizontal columns, in multistage columns partitioned with orifice and perforated plates, and in spray columns can be found in the review of Shah et al. (1978).

In nonisothermal bubble columns it is necessary to take into account the heat exchange with the surroundings. It is remarkable that for Newtonian liquids the heat transfer coefficients from the reactor wall and from immersed coils are almost identical. For non-Newtonian liquids, the latter is larger, and then it is advantageous to use coils. Also, the location of the heat transfer section along the column axis does not appear to affect the heat transfer coefficient (Hikita et al., 1981a).

Packed Columns

The presence of packings inside the column limits bubbles coalescence, leading to smaller bubbles, which rise with lower velocity. As a consequence, backmixing decreases in both phases, and interfacial area and gas holdup increase with respect to empty bubble columns. The pressure drop increases, but usually not significantly (Carleton et al., 1967). High-porosity packings, such as screen packings ($\varepsilon_p \simeq 0.90$), are particularly efficient

(Chen and Vallabh, 1970). Packed columns are specifically indicated when handling corrosive materials and when no significant heat exchange is required.

Two main flow patterns can be operated in a packed column: trickle flow, where the gas forms a continuous phase while the liquid phase flows downward as a film over the solid packing and its level is below the packing; and bubble flow, where the gas flows as bubbles through the liquid, whose level is now above the packing. A detailed description of the column hydrodynamics and the packing characteristics is reported by Fair et al. (1971) and Treybal (1968). In Table 10 approximate values of the void fraction ε_p and the total dry surface area per unit packed volume a_t are reported for some packings. The characteristic packing dimension is defined as follows:

$$d_p = \frac{6(1 - \varepsilon_p)}{a_t} \tag{75}$$

which is not normally the same as the nominal diameter d_n.

The recommended relationships for the evaluation of the characteristic parameter will be examined below in terms of the operation mode and the specific flow patterns. Useful surveys have recently been reported by Charpentier (1976, 1978) on mass transfer, by Shah et al. (1978) on backmixing in packed columns, and by Satterfield (1975) on trickle-bed reactors.

Countercurrent Packed Columns

These units are usually operated in the trickle-flow regime, that is, below the loading zone. For such situations the various involved parameters can be evaluated with the expressions reported in Table 11. Little information is available on columns operated countercurrently in the bubble-flow regime (Carleton et al., 1967; Sahay and Sharma, 1973).

The liquid holdup is given by the sum of two contributions: $\varepsilon_\ell = \varepsilon_\ell^s + \varepsilon_\ell^o$; the static holdup, ε_ℓ^s, given by the liquid retained by the drained packings, and the operational holdup, ε_ℓ^o. The values of the adjustable parameters appearing in the correlation developed by Shulman et al. (1955) are reported in Table 12 for some packings. For other types of packing the procedure proposed by Dombrowski and Brownell (1954) can be applied

The evaluation of the interfacial area per unit of packed volume, a_p, and the liquid-side mass transfer coefficient k_ℓ can be obtained, for the packings and operating conditions summarized in Table 10, from Figs. 11 and 12 as a function of the liquid superficial velocity. The gas velocity does not affect the reported curves. For situations not included in these figures, the relationships reported in Table 11 can be used. The values of the critical surface tension, σ_c for various materials, including polypropylene and polyethylene hydrophilized by treatment with chromosulfuric acid (Linek et al., 1977), are reported in Table 13. Note that the expression proposed by Onda et al. (1968) assumes that the gas-liquid interfacial area a_p identifies the packing wetted surface area a_w, which is not true when a portion of the liquid is dispersed as small droplets or rivulets, as for Pall rings (in this case the interfacial area may be underestimated by 50%).

TABLE 10 Packing Characteristics and Operating Conditions for the Data Shown in Figs. 11 and 12

	Packing						
	Type	Nominal diameter, d_n (in.)	Temperature (°C)	Column diameter d_c	Number of particles per unit of packed column (m^{-3})	Dry packing area, a_t (cm^{-1})	Void fraction, ε_p
A	Ceramic Intalox saddles	1/2	25	4 in.	630,000	4.7	0.78
B	Ceramic Pall rings	1/2	25	4 in.	360,000	4.2	—
C	Steel Pall rings	5/8	25	6 in.	220,000	3.5	0.93
D	Ceramic Raschig rings	3/8	30	10 and 20 cm	1,070,000	5.1	0.68
E	Ceramic Raschig rings	3/8	30	4.37 cm	980,000	—	—
F	Ceramic Intalox rings	1/2	30	10 cm	385,000	—	—
G	Ceramic Raschig rings	1/2	25	4 in.	370,000	3.8	0.64
H_1	Ceramic Pall rings	1	25	9 in.	49,000	2.2	—
H_2	Inox steel Pall rings	1	25	20 cm	49,000	2.0	0.94

H_3	Polypropylene Pall rings	1	25	20 cm	51,000	2.0	0.90
I_1	Ceramic Intalox saddles	1	25	9 in.	84,000	2.5	0.77
I_2	Ceramic Intalox saddles	1	25	20 cm	75,300	2.5	0.77
I_3	Polypropylene Intalox	1	25	20 cm	53,500	2.0	—
J	Ceramic Raschig rings	1	25	12 in.	—	2.03	—
K	Ceramic Raschig rings	1	25	9 in.	48,000	1.8	—
	Ceramic Raschig rings	1	25	20 cm	50,600	1.9	0.74
	PVC Raschig rings	1	25	20 cm	51,400	1.9	—
L	Ceramic Raschig rings	1 1/2	20	18 in.	14,000	1.3	0.71
M	Ceramic Intalox saddles	1 1/2	11–34	0.5 m	21,000	1.6	0.80
N	Ceramic Raschig rings	1 1/2	11–34	0.5 m	13,000	1.3	0.71
O	Polypropylene Pall rings	1 1/2	11–34	0.5 m	15,500	1.3	0.91
P	Polypropylene Intalox saddles	2	11–34	0.5 m	6,200	1.1	—

TABLE 11 Semiempirical Expressions for Countercurrent Packed Columns

Expression	Reference
Liquid holdup	
$\varepsilon_\ell = 6.28\left(\frac{\rho_\ell u_\ell d_p}{\mu_\ell}\right)^{0.676}\left(\frac{d_p^3 g\rho_\ell^2}{\mu_\ell^2}\right)^{-0.44} + \frac{3.8\times 10^{-4}}{d_p}$	Otake and Kunigita (1958)
$\varepsilon_\ell^s = \frac{C_1 \mu_\ell^{C_2}\sigma_\ell^{C_3}}{\rho_\ell^{0.37}}$	Shulman et al. (1955)
$\varepsilon_\ell^o = 2.2\left(\frac{\mu_\ell u_\ell}{g\rho_\ell d_n^2}\right)^{1/3} + 1.8\left(\frac{u_\ell^2}{g d_n}\right)^{1/2}$	Buchanan (1967)
d_n = nominal packing size (Table 10)	
Interfacial area per unit of packed volume	Onda et al. (1968)
$\frac{a_p}{a_t} = 1 - \exp\left[-1.45\left(\frac{\sigma_c}{\sigma_\ell}\right)^{0.75}\left(\frac{u_\ell\rho_\ell}{a_t\mu_\ell}\right)^{0.1}\left(\frac{u_\ell^2 a_t}{g}\right)^{-0.05}\left(\frac{u_\ell^2\rho_\ell}{\sigma_\ell a_t}\right)^{0.2}\right]$	
a_t = total dry area of the packing per unit of packed volume	
$\frac{a_p}{a_t} = 1.045\left(\frac{u_\ell\rho_\ell}{a_t\mu_\ell}\right)^{0.041}\left(\frac{u_\ell^2\rho_\ell}{\sigma_\ell a_t}\right)^{0.133}\left(\frac{\sigma_c}{\sigma_\ell}\right)^{0.182}$	Puranik and Vogelpohl (1974)

$0.25 < \rho_\ell u_\ell < 12\ \mathrm{kg/m^2 \cdot s}$; $10^{-2} < d_p < 0.0375\ \mathrm{m}$ $0.3 < \sigma_\ell/\sigma_c < 1.3$; $0.08 < a_p/a_t < 0.8$;	
Volumetric liquid mass transfer coefficient $k_\ell a_p = 0.0025\left(\frac{g\rho_\ell}{a_t\mu_\ell}\right)^{2/3}\left(\frac{g^2\rho_\ell}{\mu_\ell}\right)^{1/9}\left(\frac{\mu_\ell u_\ell^3 a_t^3}{g^2\rho_\ell}\right)^{1/4}\left(\frac{\mu_\ell}{\rho_\ell D_\ell}\right)^{-1/2}$ $0.1 < u_\ell\rho_\ell < 42\ \mathrm{kg/m^2 \cdot s}$; $0.015 < u_g\rho_g < 1.22\ \mathrm{kg/m^2 \cdot s}$; $0.006 < d_p < 0.05\ \mathrm{m}$; $0.06 < d_c < 0.5\ \mathrm{m}$	Mohunta et al. (1969)
Liquid-side mass transfer coefficient $k_\ell = 0.0051\left(\frac{\mu_\ell g}{\rho_\ell}\right)^{1/3}\left(\frac{u_\ell\rho_\ell}{a_w\mu_\ell}\right)^{2/3}\left(\frac{\mu_\ell}{\rho_\ell D_\ell}\right)^{-1/2}(a_t d_p)^{0.4}$ $0.0063 < d_p < 0.051\ \mathrm{m}$; $1 < (u_\ell\rho_\ell/a_w\mu_\ell) < 500$ $a_w = a_p$ (as given by the same authors in this table)	Onda et al. (1968)
Gas-side mass transfer coefficient $\frac{k_g}{a_t D_g} = f\left(\frac{\rho_g u_g}{a_t\mu_g}\right)^{0.7}\left(\frac{\mu_g}{\rho_g D_g}\right)^{1/3}(a_t d_p)^{-2}$ $f = 2.0$ for $d_p < 0.013\ \mathrm{m}$	Onda et al. (1968)

TABLE 11 (Continued)

$f = 5.2$ for $d_p > 0.013$ m	
$0.0063 < d_p < 0.051$ m; $5 < u_g \rho_g / a_t \mu_g < 500$	
Liquid-phase axial dispersion coefficient	Sater and Levenspiel (1966)
$(Pe'_\ell / \varepsilon_\ell) = 0.00758\ Re_\ell^{0.703}$	
$0.007 < d_p < 0.016$ m	
$Pe'_\ell = 836 Re_g^{-0.317} Re_\ell^{-2.01}$	Linek et al. (1978)
$2 \times 10^{-5} < Re_g^{-0.317} Re_\ell^{-2.01} < 2 \times 10^{-3}$	
$Pe'_\ell = u_\ell d_p / E_\ell$; $Re_\ell = \rho_\ell u_\ell d_p / \mu_\ell$; $Re_g = \rho_g u_g d_p / \mu_g$	
Gas-phase axial dispersion coefficient	De Maria and White (1960)
$Pe'_g / (\varepsilon_p - \varepsilon_\ell) = 2.4 Re_g^{-0.2} \times 10^{-(0.013 - 0.088\ d_p/d_c) Re_\ell}$	
$Pe'_g = 1.15 Re_g^{0.139 Re_\ell^{0.16}} \times 10^{-0.131\ Re_\ell^{0.385}}$	Linek et al. (1978)
$Pe'_g = u_g d_p / E_g$; $5 < Re_g < 50$; $Re_\ell < 100$	

TABLE 12 Values of the Adjustable Parameters Appearing in Shulman's Equation, Reported in Table 11, for Various Packings

Packing				
Type	Nominal diameter, d_n (in.)	C_1	C_2	C_3
Carbon Rashig rings	1	1.35	0.02	0.23
Ceramic Rashig rings	1	2.75	0.02	0.99
Ceramic Berl saddles	1	0.904	0.04	0.55

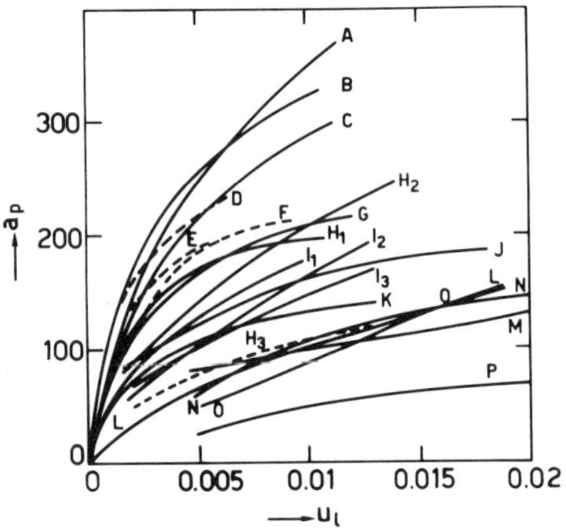

FIGURE 11 Interfacial area per unit packed volume, a_p in countercurrent packed columns for the packing illustrated in Table 10.

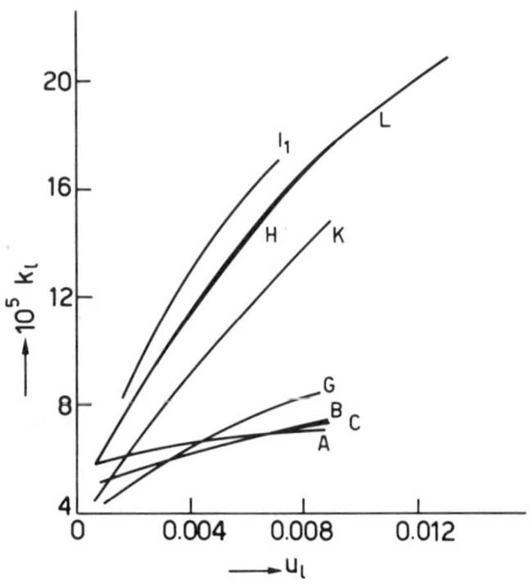

FIGURE 12 Liquid-side mass transfer coefficient k_ℓ in countercurrent packed columns for the packing illustrated in Table 10.

TABLE 13 Critical Surface Tension for Some Packing Materials

Material	$\sigma_C \times 10^3$ (N/m)
Carbon	56
Ceramic	61
Glass	73
Paraffin	20
Polyethylene	33
Polyvinyl chloride	40
Steel	75
Polypropylene and polyethylene hydrophylyzed	54

Axial dispersion is usually present in both phases, although it is larger in the liquid phase [roughly one order of magnitude in the Péclet number; Dunn et al. (1977)]. Reliable semiempirical expressions are not available for this parameter, due to discrepancies obtained using different experimental approaches (steady-state or transient) or different evaluation methods of the experimental response curves. In Table 11 the most widely used expressions are reported; the correlations proposed by Linek et al. (1978) appear particularly accurate, although they are limited to a 0.139-m-diameter column packed with 15-mm ceramic Raschig rings.

Cocurrent Packed Columns

In these units, since flooding cannot occur, unlimited flow rates can be used. So, unlike countercurrently operated columns, they do not exhibit capacity limitations and at a given gas and liquid flow rate, the pressure drop is lower. Obviously, due to reduced average interphase mass transfer driving force, their application is limited primarily to columns with fast chemical reactions. For small gas flow rates in a downward cocurrent column, the trickle or bubbly flow can be present, depending on the magnitude of the liquid flow rate. For increasing gas velocity, in both cases a transition to pulse flow, and subsequently to spray flow, takes place. Flow patterns and transitions from one form to another, as a function of the gas and liquid flow rates, are reported graphically by various authors, with reference to specific systems (Charpentier et al., 1969; Beimesch and Kessler, 1971; Charpentier and Favier, 1975). Trickle flow cannot be realized in upflow columns.

The recommended expressions for the evaluation of characteristic parameters are summarized in Table 14. In the expression developed by Larkins et al. (1961), which is not valid in the trickle flow regime, the pressure drop for each phase is calculated through the following Ergun-type equation:

TABLE 14 Semiempirical Expressions for Cocurrent Packed Columns

---

Liquid holdup and pressure drop — Larkins et al. (1961)

$$\log(\varepsilon_\ell/\varepsilon_p) = -0.744 + 0.525 \log x - 0.109(\log x)^2$$

$$\log \frac{\Delta P_{LG}}{\Delta P_L + \Delta P_G} = \frac{0.416}{(\log x)^2 + 0.666}$$

$$0.05 < x = (\Delta P_L/\Delta P_G)^{1/2} < 30$$

Liquid holdup — Bakos and Charpentier (1970)

$$\log(\varepsilon_\ell/\varepsilon_p) = a_1 + a_2 \log \bar{x} + a_3(\log \bar{x})^2$$

$$0.05 < \bar{x} = \left(\frac{u_\ell \rho_\ell}{u_g \rho_g}\right)^{1/2} \left(\frac{\Delta P_{LG}}{L \rho_g g} + 1\right)^{-1/2} < 100$$

	Rashig rings	Spheres	Pellets
a_1	−0.570	−0.280	−0.363
a_2	0.165	0.175	0.168
a_3	−0.095	−0.047	−0.043

$$d_p > 0.002 \text{ m}$$

Interfacial area per unit of packed volume

$$\frac{a_p}{a_t} = a_1 \bar{y}^{-a_2}$$

TABLE 14 (Continued)

$\bar{y} = \dfrac{\Delta P_{LG}}{L} \dfrac{\varepsilon_p}{a_t}$ Pa	
$\bar{y} > 12$ Pa, $a_1 = 0.25$, $a_2 = 0.5$	Gianetto et al. (1970)
$\bar{y} < 12$ Pa, $a_1 = 0.05$, $a_2 = 1.2$	Charpentier (1976)
$\varepsilon_p < 0.5$	
Volumetric liquid mass transfer coefficient	
$k_\ell a_p = a_1(\bar{E}_\ell D_\ell / D_w)^{a_2}$	
$\bar{E}_\ell = \dfrac{u_\ell \Delta P_{LG}}{L}$ W/m^3; $D_w = 2.4 \times 10^{-9}$ m^2/s	
$\bar{E}_\ell > 80$ W/m^3, $a_1 = 0.0173$, $a_2 = 0.5$	Reiss (1967)
$5 < \bar{E}_\ell < 100$ W/m^3, $a_1 = 0.0011$, $a_2 = 1.0$	Charpentier (1976)
$\bar{E}_\ell < 5$ W/m^3, $a_1 = 0.008$, $a_2 = 0$	

Volumetric gas mass transfer coefficient	Reiss (1967)
$k_g a_p = 2 + 0.07\,\overline{E}_g^{\,2/3}$	
$\overline{E}_g = \frac{\mathbf{u}_g\,\Delta P_{LG}}{L}\quad W/m^3$	
Liquid-phase axial dispersion coefficient	Co and Bibaud (1971)
$\frac{u_\ell d_p}{\varepsilon_\ell E_\ell} = 0.0165\left(\frac{u_\ell \rho_\ell d_p}{\mu_\ell}\right)^{0.4313}$	
$\frac{u_\ell d_p}{\varepsilon_\ell E_\ell} = 0.128\left(\frac{\rho_\ell u_\ell d_p}{\mu_\ell}\right)^{0.245}\left(\frac{\rho_g u_g d_p}{\mu_g}\right)^{-0.16}(a_t d_p)^{0.53}$	Stiegel and Shah (1977b)
Gas-phase axial dispersion coefficient	Hochman and Effron (1969)
$\frac{u_g d_p}{(\varepsilon_p - \varepsilon_g)E_g} = 1.8\left[\frac{u_g d_p \rho_g}{\mu_g(1 - \varepsilon_p)}\right]^{-0.7} \times 10^{-0.005\rho_\ell u_\ell d_p/\mu_\ell(1-\varepsilon_p)}$	

$$\frac{\Delta P}{L} \frac{\rho \varepsilon_p^3}{a_t^3 \mu^2} = \frac{\rho u}{a_t \mu} \left(h_1 + h_2 \frac{\rho u}{a_t \mu} \right) \tag{76}$$

where h_1 and h_2 are given in Table 15 for various packings. In the case of trickling liquid, ε_ℓ is given by the equation developed by Bakos and Charpentier (1970), and the pressure drop can be estimated from Eq. (76) by replacing the packing porosity ε_p with $(\varepsilon_p - \varepsilon_\ell)$ to account for the liquid film. Note that the relationships reported in Table 15 for the evaluation of pressure drop and liquid holdup are not valid for saddles and Pall rings; for these packings useful data are reported by Dodds et al. (1960), Wen et al. (1963), and Reiss (1967). Extensive data on downflow packed beds with foaming and nonfoaming materials are reported by Midoux et al. (1976). The same relationships proposed for downflow columns also give reasonably accurate results for upflow columns (Carleton et al., 1967).

The evaluation of mass transfer coefficients and effective interfacial areas is highly affected by the flow regime, which in turn depends on the energy dissipation rate. In the relationship proposed by Reiss (1967) for the evaluation of $k_\ell a_p$ in liquids more viscous than water, the complementary correction $D_\ell \mu_\ell^{0.8}$ = constant for constant temperature, should be applied. In the case of $k_g a_p$, the expression developed by Reiss (1967) is not valid for the trickle flow regime, where the same equation reported in Table 11 for the countercurrent case gives conservative results. Mitchell and Perona (1979) found, for high-porosity packings ($\varepsilon_p > 0.7$), interfacial area values significantly larger than those given by the relationship proposed by Gianetto et al. (1970) and Charpentier (1976). For upflow

TABLE 15 Values of Parameters h_1 and h_2 in Eq. (76) for Various Packings

Packing			
Type	Nominal diameter, d_n (in.)	h_1	h_2
Dumped	1/2	3.06	0.340
Stacked	1/2	2.33	0.039
Dumped	1	2.56	0.918
Stacked	1	3.58	0.0195
Stacked	3	3.39	0.018
Intalox	1	3.75	0.333
Cylinder	1/8	4.17	0.292
Sphere	3/8	3.28	0.167
Raschig ring	3/8	7.39	0.388

Source: Reiss (1967).

columns the $k_\ell a_p$ values are roughly 100% larger than those calculated using the relationships reported in Table 14 (Specchia et al., 1974); in the bubbly regime a conservative value, $k_\ell a_p = 0.15\ s^{-1}$ is recommended (Charpentier, 1976).

Axial dispersion in the liquid phase can be described using the relationships developed by Co and Bibaud (1971) and Stiegel and Shah (1977b) for trickle beds and bubble-packed columns, respectively. For the gas phase, the relationship developed by Hochman and Effron (1969) applies only to trickle flow [for $Re_\ell > 650$ the dispersion coefficient is independent of the liquid flow rate; Woodburn (1974)]. For bubble-packed columns no data are available on gas-phase axial dispersion, which can be expected to be smaller than in upacked bubble columns under equivalent flow conditions.

Mechanically Agitated Reactors

Stirred tanks are particularly suitable for viscous liquids, large liquid volumes, and low gas flow rates. Noticeable is the flexibility of these units with respect to the heat transfer rate and the liquid residence time. Principal drawbacks are the large backmixing in both phases and the high cost of the mechanical agitator, particularly in the case of highly corrosive liquids. As a rule of thumb with noncorrosive liquids, the mechanically agitated reactor is most economical when the overall reaction rate is five times larger than the mass transfer rate in a bubble column. If a large fraction of gas needs to be absorbed, stirred-tank reactors are not recommended. Recent reviews on these units have been reported by Calderbank (1967), Charpentier (1981), and Joshi et al. (1982).

In Table 16 a list of the most important agitator types is given, together with their main characteristics [u_{AM} and $\mu_{\ell M}$ indicate the maximum allowable values of the agitator tip velocity and of the liquid viscosity, respectively (Steiff et al., 1981)]. The most common agitator for dispersing gas into liquids is the six-flat-blade turbine, while a pitched-blade turbine is to be preferred when suspension of solid particles is required. Baffles in the vessel change the liquid flow patterns, improving the gas-liquid mass transfer. The baffling recommended is four 90° baffles, $d_R/12$ in width, offset from the vessel wall one-sixth of the baffle width. In a single impeller unit, values such as $H/d_R = 1$, $d_A/d_R = 0.4$ to 0.5, and $h_A/d_R = 0.33$ to 0.5 are recommended, where d_A and h_A indicate the agitator diameter and height off the tank bottom, respectively. Gas sparging can be realized with an inexpensive perforated plate (at least in the turbulent regime, where the bubble size is regulated by the coalescence and breakup phenomena). The sparger is located just below the agitator, with a diameter 0.8 to 1 of the impeller diameter. An even gas distribution, in the case of a sparge ring, can be obtained by arranging the size and number of holes so that the gas velocity through the holes is at least three times that through the pipe used as sparging ring. The use of a disk-mounted turbine is suitable, particularly at low agitation speed in order to prevent bubbles crossing a hub-mounted impeller instead of being dispersed. Further information on reactor design is given by Bates et al. (1966) and Nagata (1975).

In these units the gas-liquid dispersion is realized using not only the energy associated with the gas inlet flow rate, as in sparged reactors, but also a significant contribution is by mechanical agitation. The latter is largely predominant when the agitation speed exceeds a critical value N_0, which corresponds roughly to a tip velocity of about 2.25 m/s, and is given by

TABLE 16 Qualitative Information on the Application of the Most Common Agitators in Multiphase Systems[a]

Stirrer design	d_A/d_R	u_{AM} (m/s)	μ_{LM} (Pa·s)	Baffled	Heat transfer		Gas-liquid dispersion	Liquid-solid dispersion
					Wall	Coil		
Propeller	0.15–0.4	15	5	Yes	gg	g	gg	gg
Disk flat-blade turbine	0.2–0.45	15	10	Yes	gg	g	gg	g
Pitched-blade turbine	0.2–0.45	12	20	Yes	gg	gg	gg	gg
Impeller	0.5–0.7	12	20	Yes	gg	—	g	g
Paddle mixer	0.4–0.5	5	50	Yes	gg	gg	g	bg
Anchor impeller	0.9–0.98	5	50	No	gg	—	g	bg
Helical mixer	0.9–0.98	1	1000	No	gg	—	—	—

[a]gg, Highly suitable; g, suitable; bg, conditionally suitable.

$$N_0 d_A = \left(\frac{\sigma_\ell g}{\rho_\ell}\right)^{1/4} \left(a + b\,\frac{d_R}{d_A}\right) \tag{77}$$

where a and b depend on the agitator type: a = 1.22 and b = 1.25 for turbine impellers; a = 2.25 and b = 0.68 for two- and four-blade agitators (Westerterp et al., 1963). Further details on the hydrodynamics and flow patterns of agitated gas-liquid reactors, including the estimation of the agitator flooding conditions, are reported by Rushton and Bimbinet (1968), John (1971), Nagata (1975), Hicks and Gates (1976), Joshi et al. (1982), and Rietema (1982).

Another critical agitation speed, N_c, can be defined beyond which gas above the free liquid surface is entrapped by the liquid. It can be evaluated from the following relationship (Joshi et al., 1982):

$$N_c = 1.65\,\frac{d_R^{1.1}}{d_A^{1.98}}\,N_P^{-0.125}\left(\frac{\sigma_\ell g}{\rho_\ell}\right)^{0.19}\left(\frac{\mu_\ell}{\mu_g}\right)^{0.031}\left(\frac{w_A}{d_A}\right)^{0.625} \tag{78}$$

where N_p is the power number and w_A the impeller blade width.

Due to the strong dependence of mass and heat transfer parameters on liquid viscosity, surface tension, coalescence characteristics (such as solution ionic strength), and foam ability, it is very difficult to predict a priori these parameters (Elstner and Onken, 1981). Therefore, the semiempirical expressions reported in Table 17, developed primarily for the air-water system, should be used with caution.

The power absorbed by a mechanical stirrer in an unaerated liquid, P_0, can easily be calculated (Bates et al., 1966; Nagata, 1975). The reduced power consumption in gas-liquid systems can be evaluated according to the relationships reported in Table 17. The correlation developed by Hughmark (1980) is based on experimental data of various authors; it refers to flat-blade turbines, Newtonian fluids, and does not account for the power contribution of gas sparging. Further experimental results are reported by Hassan and Robinson (1977).

The correlations developed by Sridhar and Potter (1980a,b) for interfacial area and gas holdup constitute extensions of the corresponding correlations by Calderbank (1958), to account for larger gas flows and operating pressures beyond atmospheric. In such cases the contributions of the kinetic power P_k and the gas expansion power P_q are comparable to the mechanical agitation power P_g. At atmospheric pressure, $P_t \simeq P_g$ and $\rho_g \simeq \rho_a$, and these equations reduce to those developed by Calderbank (1958). Mean bubble diameters can be estimated using Eq. (44) and suitable correlations are reported in Table 17.

Specific interfacial area values are highly sensitive to the solution ionic strength. In fact, the inhibition of bubble coalescence by ionic solutes, even if present in very small amounts (Elstner and Onken, 1981), leads to larger values of a. The expressions proposed by Van't Riet (1979), valid for aqueous solution and for agitation speed beyond the critical value N_0, account for this effect.

The volumetric liquid mass transfer coefficient is not affected by the agitator type, location, and number and by the sparger type and position,

TABLE 17 Semiempirical Expressions for Mechanically Agitated Reactors

Impeller power consumption	Hughmark (1980)
$\frac{P_g}{P_0} = 0.10\left(\frac{Q_g}{NV_\ell}\right)^{-1/4}\left(\frac{N^2 d_A^4}{g w_A V_\ell^{2/3}}\right)^{-1/5}$	
$0.87 < \rho_\ell < 1.6\ kg/m^3;\ 0.0008 < \mu_\ell < 0.028\ Pa \cdot s;\ 0.025 < \sigma_\ell < 0.072\ N/m;$	
$0.1 < d_R < 1\ m;\ 0.25 < d_A/d_R < 0.46;\ 0.31 < P_g/P_0 < 0.8$	
Total power consumption	Gray et al. (1982)
$\frac{P_T}{N^3 d_A^5 \rho_\ell} = 0.75\left[\frac{g^2 h_A d_R}{(u_g/\varepsilon_g)^2 N^2 d_A^2}\right]^{1/4}\left(\frac{\sigma_\ell}{\sigma_w}\right)^{1/4}$	
h_A = agitator height off the tank bottom;	
$0.15 < d_R < 0.45\ m;\ 0.25 < d_A/d_R < 0.45;\ 0.75 < h_A/d_A < 1.5$	
Gas holdup	Hughmark (1980)[a]
$\varepsilon_g = 0.74\left(\frac{Q_g}{NV_\ell}\right)^{1/2}\left(\frac{N^2 d_A^4}{g w_A V_\ell^{2/3}}\right)^{1/2}\left(\frac{d_b N^2 d_A^4 \rho_\ell}{\sigma_\ell V_\ell^{2/3}}\right)^{1/4}$	
$d_b = 2.5 \times 10^{-3}\ m;\ 10^{-2} < V_\ell < 51\ m^3;\ u_g < 0.053\ m/s$	

$\varepsilon_g = \left(\frac{\varepsilon_g u_g}{u_b}\right)^{1/2} + 0.000216\left[\frac{(P_g/V_\ell)^{0.4}\rho_\ell^{0.2}}{\sigma_\ell^{0.6}}\right]\left(\frac{u_g}{u_b}\right)^{0.5}\left(\frac{P_T}{P_g}\right)\left(\frac{\rho_g}{\rho_a}\right)^{0.16}$	Sridhar and Potter (1980a)
$u_b = 0.265$ m/s; ρ_a = density of air at operating conditions	
$P_T = P_g + P_k + P_q$; $P_k = 0.5Q_g\rho_g u_s^2$;	
$P_q = \rho_\ell g h_L Q_g$; u_s = gas velocity at sparger exit	
Interfacial area per reactor volume	Hughmark (1980)[a]
$a = 1.38\left(\frac{g\rho_\ell}{\sigma_\ell}\right)^{1/2}\left(\frac{Q_g}{NV_\ell}\right)^{1/3}\left(\frac{N^2 d_A^4}{g w_A V_\ell^{2/3}}\right)^{0.592}\left(\frac{d_b N^2 d_A^4 \rho_\ell}{\sigma_\ell V_\ell^{2/3}}\right)^{0.187}$	
$d_b = 2.5 \times 10^{-3}$ m; $10^{-2} < V_\ell < 51$ m^3; $u_g < 0.053$ m/s	
$a = 1.44\left[\frac{(P_g/V_\ell)^{0.4}\rho_\ell^{0.2}}{\sigma_\ell^{0.6}}\right]\left(\frac{u_g}{u_b}\right)^{0.5}\left(\frac{P_T}{P_g}\right)\left(\frac{\rho_g}{\rho_a}\right)^{0.16}$	Sridhar and Potter (1980b)
symbols as for gas holdup correlation	
Volumetric liquid mass transfer coefficient	Chandrasekharan and Calderbank (1981)
$k_\ell a = \frac{0.0248}{d_R^4}\left(\frac{P_g}{V_\ell}\right)^{0.551} Q_g^{0.551/\sqrt{d_R}}$	

TABLE 17 (Continued)

$90 < P_g/V_\ell < 1700\ W/m^3$; $2.6 \times 10^{-3} < u_g < 1.8 \times 10^{-2}$ m/s; $0.3 < d_R < 1.2$ m	
$k_\ell a = b_0 \left(\frac{P_g}{V_\ell}\right)^m u_g^n$	Van't Riet (1979)
For pure water: $b_0 = 0.026$; $m = 0.4$; $n = 0.5$	
For ionic solutions: $b_0 = 0.002$; $m = 0.7$; $n = 0.2$	
$2 \times 10^{-3} < V_\ell < 4.4\ m^3$; $500 < P_g/V_\ell < 10{,}000\ W/m^3$	
Liquid-side mass transfer coefficient	Prasher and Willis (1973)
$k_\ell = 0.592 D_\ell^{1/2} (\bar{\varepsilon} \rho_\ell / \mu_\ell)^{1/4}$	
$\bar{\varepsilon} = \frac{8N^3 d_A^5 \phi}{d_R^2 h_L}$; $q = \frac{Q_g}{N d_A^3}$	
$\phi = 1 - 1.26q$ for $q < 0.035$	
$\phi = 0.62 - 1.85q$ for $q > 0.035$	
$d_A/d_R = 1/3$; $0.006 < \bar{\varepsilon} < 0.4\ m^2/s^3$	
Gas-side mass transfer coefficient	Newman (1931)
$k_g = 6.58 \left[\frac{D_g}{d_{vs}(1 - y)}\right]$	

y = gas mole fraction of diffusing component	
Liquid-phase axial dispersion coefficient $E_\ell = 0.116\left[\frac{(d_R + H)^2 N}{(d_R/d_A)^2}\right]$ H = clear liquid height; valid for disk turbine impeller	Joshi (1980)
Heat transfer coefficient $\frac{h_w d_R}{k_{T\ell}} = b\ \overline{Re}^m Pr^{0.33} Fr^{-0.1}\left(\frac{\mu_w}{\mu_\ell}\right)^{-0.14}$ $b = 1.35$; $m = 0.59$ for jacket $b = 0.87$; $m = 0.64$ for coil $\overline{Re} = \frac{d_A \rho_\ell}{\mu_\ell}(d_A N + 4u_g)$; $Pr = \frac{C_{p\ell}\mu_\ell}{k_{T\ell}}$ $Fr = \frac{N^2 d_A}{g}$; $\mu_w(\mu_c)$ = liquid viscosity at inside vessel (coil) wall temperature $10^3 < \overline{Re} < 5 \times 10^5$; $2 < Pr < 266$; $0.008 < Fr < 0.53$ $1.02 < \mu_w/\mu_\ell < 1.50$ for coil; $0.18 < \mu_w/\mu_\ell < 0.98$ for jacket	Rao and Murti (1973)

[a]These relationships have been corrected with respect to the original version according to Hughmark (1982).

as long as the indicative conditions on sparger diameter and agitator location previously mentioned are satisfied (Van't Riet, 1979). When operating at large flow rates, the gas sparging power contribution should also be considered together with mechanical power P_g in the evaluation of $k_\ell a$ (Botton et al., 1980). Extensions to systems other than O_2-water can be made as suggested by Fair (1967a):

$$(k_\ell a) = (k_\ell a)_{O_2\text{-water}} \left(\frac{D}{D_{O_2\text{-water}}}\right)^{1/2} \tag{79}$$

where $D_{O_2\text{-water}} = 2.4 \times 10^{-9}$ m^2/s. No general correlation is available for non-Newtonian liquids; experimental data reported by Höcker et al. (1981) can be useful. The correlations for gas- and liquid-side mass transfer coefficients are not very accurate; for the latter the relationship proposed by Calderbank and Moo-Young (1961) reported in Table 9, can be used.

The axial dispersion coefficients in both phases, E_g and E_ℓ, are greatly enhanced by mechanical agitation, so that the expressions reported for sparged reactors give in this case conservative estimates. Perfect mixing can reasonably be assumed at stirring speeds above N_0. Further information about liquid-phase backmixing in mechanically agitated contactors is reported by Sideman et al. (1966) for single-stage systems and by Sullivan and Treybal (1970) and Joshi (1980) for multistage systems.

In Table 17 semiempirical correlations for the evaluation of the heat transfer coefficients, for both the coil-aerated liquid and the wall-aerated liquid heat transfer, are reported. These relationships, obtained using a flat-blade turbine as agitator, are of general validity, since the type of

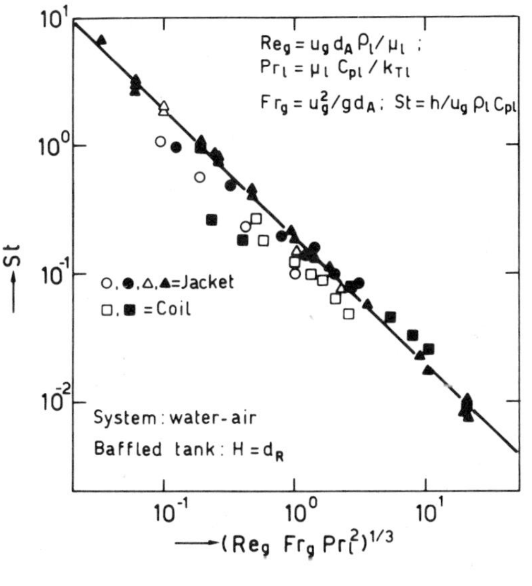

FIGURE 13 Correlation of heat transfer data in aerated stirred vessels.

impeller has little effect on heat transfer (Joshi et al., 1982). As in the case of mass transfer, the effect of gas sparging on the heat transfer coefficient decreases at higher agitation speed; in any case, it never causes variations with respect to the ungassed case (Uhl, 1966; Nagata, 1975) larger than about 25 to 30%. In Fig. 13 the data obtained by various authors for the air-water system at high stirring intensity ($Nd_A^2 \rho_\ell / \mu_\ell > 10^5$) are shown (Steiff et al., 1981).

APPROXIMATE SOLUTIONS FOR ISOTHERMAL MODELS

The complete model of a gas-liquid reactor is obtained by coupling the suitable macroscopic material balances with the liquid film model. The latter must then be solved for different values of the bulk concentrations ϕ_{gi} and $\phi_{\ell i}$, which vary along the reactor axis (steady-state axial dispersion model) or during time (semiflow batch model). The solution in general requires numerical techniques. Since boundary value differential equations are involved, the weighted residual methods (Finlayson, 1972; Villadsen and Michelsen, 1978) and the implicit difference methods, such as the one proposed by Lee (1968), are particularly convenient. Comments on their application to problems of this type are reported by Huang et al. (1980), Shah and Paraskos (1975), and Hc and Lee (1980) for the orthogonal collocation method; by Szeri et al. (1976) for the Galerkin method; and by Deckwer (1976) for the Lee method.

It is apparent that numerical solution of the complete models, although always feasible, is quite complex and time consuming. It is therefore useful to develop approximate solutions, valid only in limited situations, which can be applied for first-trial reactor design and optimization. Such solutions are developed below for the film and the reactor models separately.

Approximate Solutions of the Liquid Film Model

As already discussed, the only information relative to the film model that is required for the solution of the reactor model are the material fluxes at the interfaces $(J_i)_{x=0}$ (or E_i^*) and $(J_i)_{x=1}$ as a function of the bulk concentrations ϕ_{gi} and $\phi_{\ell i}$. Approximate relationships for $(J_i)_{x=0}$ and $(J_i)_{x=1}$ as functions of ϕ_{gi} and $\phi_{\ell i}$ for various kinetic schemes are reported in the following.

Some relationships of this type have already been reported, but they were all obtained assuming zero concentration in the liquid phase for the gaseous reactant (i.e., $\phi_{\ell A} = 0$), and therefore their use is confined to cases where the reactor is operating in the fast reaction regime, as discussed later. It is evident that this assumption depends on the reactor performance, which is obviously not known a priori, and therefore, if not used with caution it can lead to serious pitfalls in the simulation of gas-liquid reactors as shown by Shaikh and Varma (1983c).

Teramoto et al. (1969, 1973) modified the original expression of van Krevelen and Hoftijzer (1948a) for consecutive reactions to account for $\phi_{\ell A} \neq 0$ and the general reaction order, using the approximation (40) developed by Hikita and Asai (1963). This has been extended by Morbidelli et al. (1984b) to a complex scheme of consecutive reactions and compared with the exact numerical solution. The approximate solution is reported in Table 18 and applies to the following scheme of multiple irreversible reactions:

$$A + B_j \rightarrow B_{j+1} + P_j \qquad j = 1, 2, \ldots, NR \tag{80}$$

where the rate of the generic jth reaction is given by

$$r_j = k_j C_{fA}^{m_j} C_{fj}^{n_j} \tag{81}$$

The isothermal liquid film model equations are used, the only exception being the boundary condition (51a) for the liquid reactants (B_j and P_j), which, since only the gaseous reactant A is considered to be volatile, is replaced by $(d\phi_{fj}/dx)_{x=0} = 0$. In Fig. 14 the characteristic multistep behavior of the reaction factor E_A^* as a function of the Hatta number in the case of three reactions of first order with respect to both reactants (NR = 3, $m_j = n_j = 1$) is shown. Extensive comparison with the exact numerical solution for the case of a CSTR have been performed by Morbidelli et al. (1984b). The percentage error of the approximate solution never exceeds a few percent, except for multiple reactions characterized by small values of the selectivity, where it can be of the order of 20%. The approximate solution is exact for reactions of first order with respect to the gaseous reactant (m = 1) and zeroth order with respect to the liquid reactant (n = 0). For other reaction orders the accuracy of the approximate solution is more sensitive to deviations of m from unity than of n from zero. In conclusion, the approximate expressions reported in Table 18 are sufficiently accurate for reactor modeling. The only exception is for multiple reactions, where the gaseous reactant exhibits different reaction orders (i.e., $m_i \neq m_j$); in this case percentage errors of the order of 100% for the selectivity, also in the range where its value is not very small, can be found.

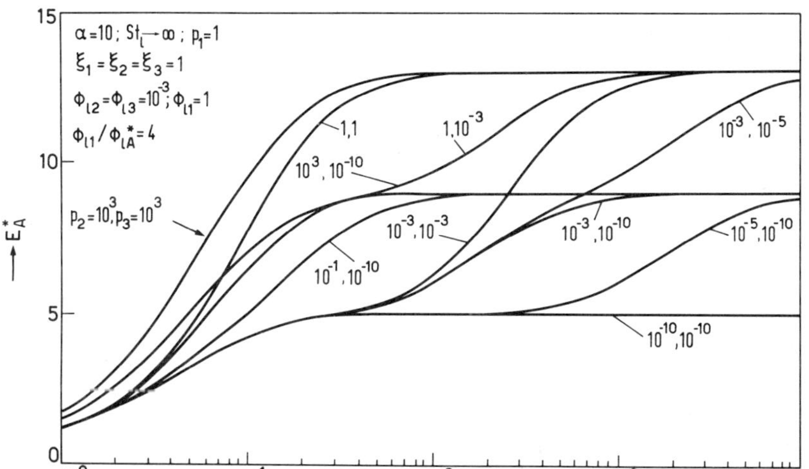

FIGURE 14 Behavior of the reaction factor, E_A^* as a function of the Hatta number Ha in the case of three consecutive bimolecular reactions of first order with respect to both reactants.

TABLE 18 Approximate Liquid Film Model Solution for the Kinetic Scheme (80)

Volatile reactant, A:

$$E_A = \frac{\gamma'}{\tanh \gamma'}\left(1 - \frac{\phi_{\ell A}/\phi^*_{\ell A}}{\cosh \gamma'}\right)$$

$$(J_A)_{x=0} = St_{gA}\frac{\gamma_A E^*_A/H^*_A}{1 + \gamma_A E^*_A/H^*_A}\phi_{gA}$$

$$(J_A)_{x=1} = St_{\ell A}\phi^*_{\ell A}\frac{\gamma'}{\tanh \gamma'}\left(\frac{1}{\cosh \gamma'} - \frac{\phi_{\ell A}}{\phi^*_{\ell A}}\right)$$

$$\phi^*_{\ell A} = \frac{\phi_{gA}/H^*_A}{1 + \gamma_A E^*_A/H^*_A}$$

Nonvolatile reactant, B_j:

$$E^*_j = 0;\ (J_j)_{x=0} = 0$$

$$(J_j)_{x=1} = -St_{\ell j}\frac{\Gamma_j}{\gamma' \tanh \gamma'}(\phi_{\ell A} + \phi^*_{\ell A})\left(1 - \frac{1}{\cosh \gamma'}\right)$$

$$\phi^*_{\ell j} = \phi_{\ell j} + \Gamma_j\left[\frac{\phi^*_{\ell A} - \phi_{\ell A}}{\gamma'^2} - \frac{\phi^*_{\ell A}}{\gamma'} + \frac{\phi^*_{\ell A} - \phi^*_{\ell A}\exp(-\gamma')}{\gamma' \operatorname{sh} \gamma'}\right]$$

where

$$\gamma'^2 = Ha^2\sum_{j=1}^{NR} p_j \frac{2}{m_j + 1}\phi^{*\,m_j - 1}_{\ell A}\phi^{*\,n_j}_{\ell j};$$

$$\Gamma_j = \frac{Ha^2}{\xi_j}\left(p_j\frac{2}{m_j + 1}\phi^{*\,m_j-1}_{\ell A}\phi^{*\,n_j}_{\ell j} - p_{j-1}\frac{2}{m_{j-1} + 1}\phi^{*\,m_{j-1}-1}_{\ell A}\phi^{*\,n_{j-1}}_{\ell j}\right)$$

Using a similar technique, Onda et al. (1972), Hikita et al. (1977), and Zarzycki et al. (1981) have developed approximate solutions for the kinetic schemes reported in Table 19. All the reacting components are considered nonvolatile, the only exception being the gaseous reactants A and A'; the boundary condition (51b) has also been used at the liquid film-liquid bulk interface (i.e., $\phi_{\ell A} \neq 0$). The approximate solutions are reported in Table 20 in the form of a system of nonlinear algebraic equations. The unknowns are, in addition to the reaction factors E^*_A and $E^*_{A'}$ (if A' is present), the dimensionless concentrations of all the nonvolatile reactants at the gas-liquid interface, $\phi^*_{\ell i}$. Comparison of the approximate values of the reaction factor with those obtained by numerical integration of the relevant equations, for various situations, have shown that the errors never exceed a few percent. It is remarkable that very similar results can be obtained for all the kinetic schemes in Table 19 using the penetration and the surface renewal

TABLE 19 Kinetic Schemes Considered in Table 20

Reaction	Reaction rate
(a) $A + \nu_B B \xrightarrow{k_1}$ product	$r_1 = k_1 C_A^m C_B^n$
(b) $A + \nu_B B \underset{k_2}{\overset{k_1}{\rightleftharpoons}} \nu_E E + \nu_F F$	$r_1 = k_1 C_A^m C_B^n$
	$r_2 = k_2 C_E^p C_F^q$
(c) $A + \nu_B B \xrightarrow{k_1} \nu_C C$ + product	$r_1 = k_1 C_A^m C_B^n$
$\nu_{A'} A' + \nu'_C C \xrightarrow{k_2} \nu_B B$	$r_2 = k_2 C_{A'}^p C_C^q$
(d) $A + \nu_{A'} A' \longrightarrow$ product	$r_1 = k_1 C_A C_{A'}$
(e) $A + \nu_B B \xrightarrow{k_1}$ product	$r_1 = k_1 C_A^m C_B^n$
$A + \nu'_C C \xrightarrow{k_2}$ product	$r_2 = k_2 C_A^p C_C^q$
(f) $A + \nu_B B \xrightarrow{k_1}$ product	$r_1 = k_1 C_A^m C_B^n$
$\nu_{A'} A' + \nu_B B \xrightarrow{k_2}$ product	$r_2 = k_2 C_{A'}^p C_B^q$

models (Onda et al., 1972). This confirms the substantial equivalence of the three models, already discussed.

Finally, it is worthwhile mentioning the study of Pangarkar (1974) and Shaikh and Varma (1983c) on the effect of liquid reactant volatility, of Karlsson and Bjerle (1980) on the effect of the gas-side mass transfer resistance, and various analyses of mass transfer enhancement in the case of complex kinetic schemes (Chandhari et al., 1975; Carmichael and Chang, 1980; Zarzycki et al., 1981; Astarita and Savage, 1982).

Another way to obtain approximate solutions for the liquid film model is to assume a specific operating regime for the reactor. According to Astarita (1967) and Danckwerts (1970), and considering, for example, the (m,n)th-order reaction (i.e., mth order with respect to A and nth order with respect to B):

$$A + \nu B \rightarrow \text{liquid products} \tag{82}$$

characterized by the following kinetic expression:

$$r = kC_A^m C_B^n \tag{83}$$

TABLE 20 Approximate Solution of the Film Model for the Kinetic Schemes Reported in Table 19

Reaction	Approximate solution	M_1 and M'_1
(a)	$E_A^* = \Phi$	$M\phi_{\ell B}^{*\,n/2}$
	$E_A^* = 1 - q_A + \dfrac{\xi_B}{\nu_B}\, q_B(1 - \phi_{\ell B}^*)$	
(b)	$E_A^* = \dfrac{(1 - q_A)(1 + \xi_E/\nu_E T) + (\xi_E/\nu_E)[(q_A/T) - q_B\phi_{\ell E}](1 - \operatorname{sech} M_1)}{1 + (\tanh M_1/M_1)\xi_E/\nu_E T}$	$M\phi_{\ell B}^{*\,n/2}\left(1 + \dfrac{\nu_E T}{\xi_E}\right)^{1/2}$
	$E_A^* - (1 - q_A) = \dfrac{\xi_B}{\nu_B}\, q_B(1 - \phi_{\ell B}^*) = \dfrac{\xi_E}{\nu_E}\, q_B(\phi_{\ell E}^* - \phi_{\ell E})$	
	$= \dfrac{\xi_F}{\nu_F}\, q_B(\phi_{\ell F}^* - \phi_{\ell F})$	
(c)	$E_A^* = \Phi;\ E_{A'}^* = \Phi'$	$M\phi_{\ell B}^{*\,n/2}$ and $M'\phi_{\ell C}^{*\,q/2}$
	$\Phi - (1 - q_A) = \dfrac{\xi_{A'}}{\nu_{A'}}\,\dfrac{\phi_{\ell A'}}{\phi_{\ell A}^*}[\Phi' - (1 - q_{A'})] + \dfrac{\xi_B}{\nu_B}\, q(1 - \phi_{\ell B}^*)$	
	$= \dfrac{\nu_{C'}\xi_{A'}\phi_{\ell A'}^*}{\nu_C\nu_{A'}\phi_{\ell A}^*}[\Phi' - (1 - q_{A'})] - \dfrac{\xi_C}{\nu_C}\, q_B(\phi_{\ell C} - \phi_{\ell C}^*)$	
(d)	$E_A^* = \dfrac{\cosh M_1 - q_A}{\sinh M_1}\, Ha_1$	$M'''\ S\ \dfrac{\nu_{A'}C_{\ell A}^*}{\xi_{A'}C_{\ell A'}^*}$

TABLE 20 (Continued)

Reaction	Approximate solution	M_1 and M'_1
	$E^*_{A'} = 1 + (E^*_A + q_A - 1)\dfrac{\nu_{A'}C^*_{\ell A}}{\xi_{A'}C^*_{\ell A'}} - q_{A'}$	
(e)	$E^*_A = \Phi$	
	$E^*_A = 1 - q_A + \dfrac{\xi_B}{\nu_B} q_B(1 - \phi^*_{\ell B}) + \dfrac{\xi_C}{\nu_C} q_B(\phi_{\ell C} - \phi^*_{\ell C})$	$M(\phi^{*n}_{\ell B} + V\phi^{*q}_{\ell C})$
	$\dfrac{\phi^*_{\ell C}}{\phi_{\ell C}} = \left(1 + \dfrac{\nu_C \xi_B}{\nu_B \xi_C} U \dfrac{1 - \phi^*_{\ell B}}{\phi^*_{\ell B}}\right)^{-1}$	
(f)	$E^*_A = \Phi;\ E^*_{A'} = \Phi'$	$M\phi^{*\,n/2}_{\ell B}$ and $M''\phi^{*\,q/2}_{\ell B}$
	$E^*_A = 1 - q_A + \dfrac{\xi_B}{\nu_B} q_B(1 - \phi^*_{\ell B}) - \dfrac{\xi_{A'}}{\nu_{A'}} \dfrac{\phi^*_{\ell A'}}{\phi^*_{\ell A}}[\Phi' - (1 - q_{A'})]$	

where

A, A' = gaseous reactant

B, C, E, F = liquid reactant, nonvolatile

$\phi^*_{\ell i} = C^*_{\ell i}/C_{\ell B}$; $\phi_{\ell i} = C_{\ell i}/C_{\ell B}$; $\xi_i = D_i/D_A$; $q_A = C_{\ell A}/C^*_{\ell A}$; $q_{A'} = C_{\ell A'}/C^*_{\ell A'}$;

$q_B = C_{\ell B}/C^*_{\ell A}$

$$\Phi = \frac{M_1}{\tanh M_1} (1 - q_A \text{ sech } M_1); \; \Phi' = \frac{M'_1}{\tanh M'_1}(1 - q_{A'} \text{ sech } M'_1)$$

$$T = \frac{k_2}{k_1} \frac{m+1}{p+1} C^{*\,(p+q-m-n)}_{\ell A} q_B^{\,(p+q-n-1)} \phi^{*\,(p-1)}_{\ell E} \phi^{*\,q}_{\ell F} / \phi^{*n}_{\ell B}$$

$$V = \frac{k_2}{k_1} C^{*\,(p+q-m-n)}_{\ell A} q_B^{\,(q-n)} \frac{m+1}{p+1}$$

$$M^2 = \frac{2}{m+1} k_1 D_A C^{*\,(m-1)}_{\ell A} C^n_{\ell B} / k^2_{\ell A}$$

$$M'^2 = \frac{2}{p+1} \nu_{A'} k_2 D_{A'} C^{*\,(p-1)}_{\ell A} C^q_{\ell C} / k^2_{\ell A'}$$

$$M''^2 = \frac{2}{p+1} \nu_{A'} k_2 D_{A'} C^{*\,(p-1)}_{\ell A'} C^q_{\ell B} / k^2_{\ell A'}$$

$$M'''^2 = k_1 D_A C^*_{\ell A'} / k^2_{\ell A}$$

$$U = V \phi^{*\,(q-1)}_{\ell C} / \phi^{*\,(n-1)}_{\ell B}$$

$$S = \frac{\xi_{A'} C^*_{\ell A'}}{C^*_{\ell A} \nu_{A'}} \left(1 - \frac{1 - q_{A'}}{3E^*_A}\right) + \frac{1 - q_A - E^*_A}{3E^*_A}$$

the following parameter can be introduced:

$$M^2 = \frac{2}{m+1} \frac{kC^{*m-1}_{\ell A} C^n_{\ell B} D_A}{k^2_{\ell A}} = \frac{2}{m+1} Ha^2 \phi^{*m-1}_{\ell A} \phi^n_{\ell B} \qquad (84)$$

which is the Hatta number defined at the liquid bulk conditions and represents the ratio of the characteristic times of the diffusion and reaction processes. Its value can then be used to define the reactor regime.

Slow Reaction Regime

In this case the reaction in the liquid film is negligible, and the reaction factor identifies with the purely physical diffusion value:

$$E^*_i = \frac{\phi^*_{\ell i} - \phi_{\ell i}}{\phi^*_{\ell i}} \qquad (85)$$

The material fluxes at the interfaces can then be calculated as follows:

$$(J_i)_{x=0} = St_{gi} \frac{\gamma_i/H^*_i}{1 + \gamma_i/H^*_i}(\phi_{gi} - \phi_{\ell i}H^*_i) = St_{gi}\gamma_i(\phi^*_{\ell i} - \phi_{\ell i}) \qquad (86)$$

$$(J_i)_{x=1} = St_{\ell i} \frac{1/H^*_i}{1 + \gamma_i/H^*_i}(\phi_{gi} - \phi_{\ell i}H^*_i) = St_{\ell i}(\phi^*_{\ell i} - \phi_{\ell i}) \qquad (87)$$

The gas-liquid interface concentration values are given by Eq. (57) and

$$\phi^*_{gi} = \frac{\phi_{gi} + \gamma_i \phi_{\ell i}}{1 + \gamma_i/H^*_i} \qquad (88)$$

Equations (86) to (88) represent the liquid film model solution under the condition

$$M^2 \ll 1 \qquad (89)$$

Such conditions has been further specified by Barona (1979) for a CSTR model with a (1,1)-th order reaction of type (82) as $M < 0.1$.

It is worth mentioning that two subregimes can be identified with the slow reaction regime depending, in addition to Eq. (89), on the following conditions:

Kinetic Subregime

$$\alpha M^2 \ll 1 \qquad (90)$$

In this case the chemical reaction is the rate-determining step, so that the gaseous reactant concentration is $\phi_{\ell A} = \phi^*_{\ell A}$, independent of the reaction

rate. The gas-liquid reactor then behaves as a homogeneous reactor, where only the liquid reactant concentrations change with time or position.

Diffusional Subregime

$$\alpha M^2 \gg 1 \tag{91}$$

In this case the gas-liquid mass transfer is the rate-determining step, so that the liquid concentration of the gaseous reactant vanishes. The total amount absorbed from the gas stream can then be calculated by solving the only gaseous phase mass balance. The interfacial material fluxes are easily evaluated from Eqs. (86) and (87) with $\phi_{\ell A} = 0$, leading to a linear expression for the sink term in the gas-phase mass balance of the gaseous reactant.

Fast Reaction Regime

In this case it is assumed that the reaction is completed within the liquid film, so that the liquid bulk concentration of the gaseous reactant vanishes, $\phi_{\ell A} = 0$. This simplification is of great utility in practice, since the liquid-phase mass balance of the gaseous reactant is not needed, and the approximate expressions for the reaction factor E_A^* reported earlier can now be used. This regime is established for

$$M^2 \gg 1 \tag{92}$$

In Figs. 15 and 16, a comparison between the two approximate solutions developed above and the complete liquid film model solution for two reactions of type (80) of first order with respect to both reactants (i.e., NR = 2, $m_j = n_j = 1$) in a CSTR is shown. It is apparent that the two approximations

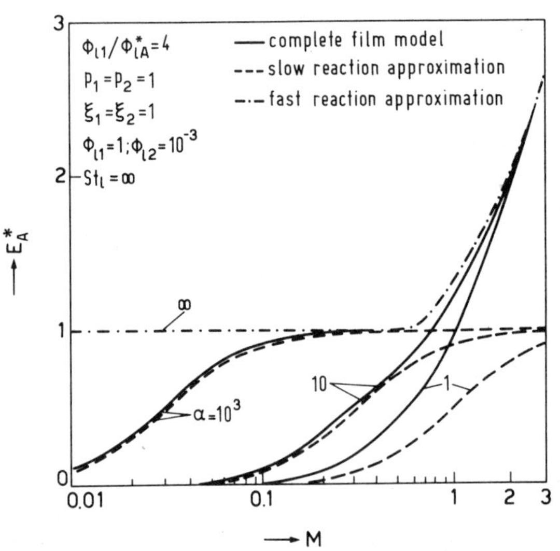

FIGURE 15 Comparison of the reaction factor, E_A^* calculated with the complete model and the slow and fast reaction approximations.

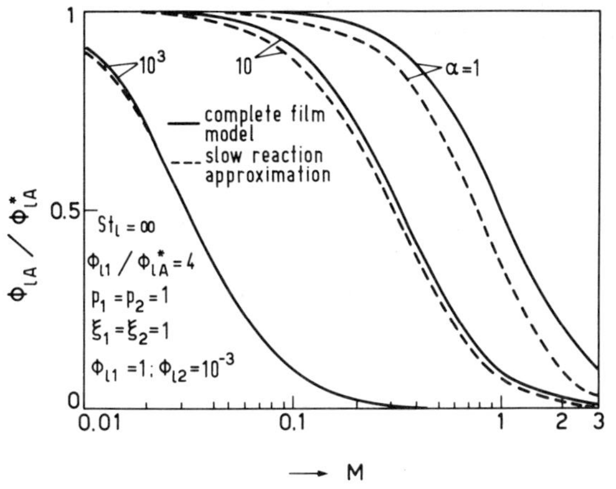

FIGURE 16 Comparison of the concentration ratio $\phi_{\ell A}/\phi_{\ell A}^*$ calculated with the complete model and the slow reaction approximation.

are quite accurate in the appropriate intervals of M values, while a lack of accuracy appears for both of them for intermediate M values, depending on the liquid holdup value α, defined by Eq. (62). It should be remarked that the value of M is not known a priori, since in general it depends on the reactor operating conditions through the bulk reactant concentrations (and then it is a function of the reactor axis in the steady-state axial dispersion model, and of time in the semiflow batch model). Therefore, care must be exercised to ensure that all possible values of M along the reactor axis or during time are such as to satisfy conditions (89) or (92). Obviously, this is not true for the approximate solutions reported in Tables 18 and 20, which can be applied for all reactor regimes.

The available solutions, valid for a specific reactor regime, and the possible model simplifications arising from limiting values of the dimensionless characteristic parameters are reviewed below for each reactor model reported earlier.

Approximate Solutions of the Axial Dispersion Model

The macromixing effects on reactor performance are taken into account in each phase separately through the introduction of a Péclet number. In Fig. 17 the effect of Pe_g on the reactor outlet conversion, in the case of a pseudo-first-order reaction in the slow regime, is shown. Extensive calculations for this system, performed by Pavlica and Olson (1970), have shown that the Péclet number affects the reactor conversion by no more than 10%, except in the interval $0.1 < Pe < 20$. Such a conclusion, valid for each phase independently, has been also verified in the fast reaction regime by Deckwer (1976). For $Pe < 0.1$ and $Pe > 20$, the CSTR and PFR models can be applied, respectively. Also, a PFR model can generally be assumed for the gaseous phase for small bubbles columns ($d_c < 0.5$ m)

Frequently, gas-liquid reactors can be described using a CSTR model for the liquid phase and a PFR or an axial dispersion model for the gaseous phase. In this case the reactor model can be obtained by suitably combining the appropriate equations of the models presented earlier. The only difference is that in the CSTR model the local material flux entering the

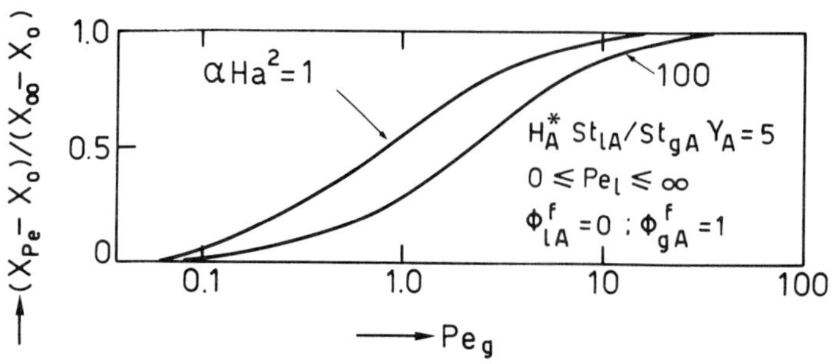

FIGURE 17 Effect of the gaseous phase Péclet number Pe_g on the outlet reactor conversion, in the case of a pseudo-first-order reaction in the slow reaction regime.

liquid bulk $(J_i)_{x=1}$ must be replaced by the average one, $(\bar{J}_i)_{x=1}$, defined by Eq. (72), as done in the semiflow batch model.

A useful simplification of the axial dispersion term has been proposed by Hlaváček and Hofmann (1970).

$$\frac{1}{Pe}\,\frac{d^2\phi}{dz^2} - \frac{d\phi}{dz} \cong -\left(\frac{\lambda^2}{Pe} + \frac{Pe}{4}\right)\phi \qquad (93)$$

where

$$tg\lambda = \frac{\lambda\ Pe}{\lambda^2 - Pe^2/4} \qquad (94)$$

Since such an approximation correctly approaches the CSTR model as $Pe \to 0$, while obviously it does not approach the PFR model as $Pe \to \infty$, its application should be limited to situations with relatively small Pe values.

Another important aspect of the axial dispersion model is the variation of the superficial gas velocity along the reactor axis. Gas expansion due to reduced pressure and gas shrinkage due to gas absorption cause opposite variations on the gas velocity and thus on the reactor performance. This is quantified in Fig. 18, where the effect of $\bar{\alpha}$ and the inlet gaseous reactant mole fraction y_A^f on the reactor conversion for a first-order reaction in the fast reaction regime is shown. From the results reported by Deckwer (1976), it can be concluded that the hydrostatic head variation can be neglected for $\bar{\alpha} < 0.1$ (in particular, for large pressure values, $P > 20$ atm), while the gas absorption effect can be neglected when $y_A^f < 0.2$.

Slow Reaction Regime

The model is given by Eqs. (60), (61), (66) to (69), (86), and (87), where the only nonlinearity is due to the gas velocity variation along the reactor axis and to the chemical reaction in the liquid bulk.

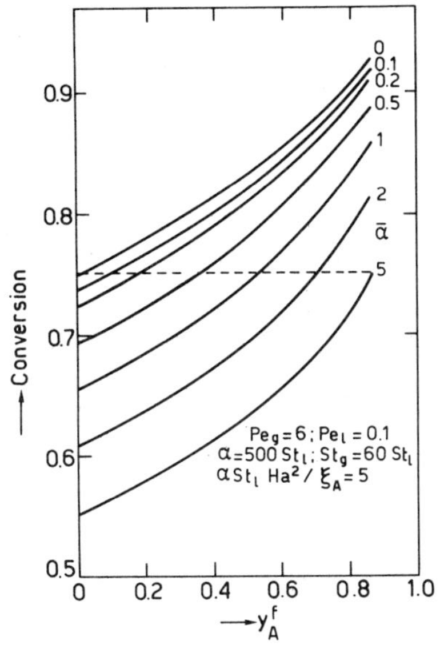

FIGURE 18 Influence of the hydrostatic head variation $\bar{\alpha}$ and the reactant mole fraction in the feed y_A^f on its conversion. The dashed line indicates the solution where the gas velocity is assumed constant.

For a reaction type (82) with the reaction rate given by Eq. (83), if the gas velocity is assumed constant ($v_g = v_g^f = 1$) and the kinetics of pseudo first order, the model becomes linear. The analytical solution can then be obtained by standard techniques (Pavlica and Olson, 1970). In Table 21 the solution obtained assuming a PFR model for both phases (Pe_g, $Pe_\ell \to \infty$) is reported (Cichy et al., 1969). In the same table the solution in the case of a gaseous phase constituted of pure reactant A is also reported.

Of particular practical interest is the case where the liquid phase is described by the CSTR model and the gaseous phase by the axial dispersion model. Then $\phi_{\ell i}$ is independent of the axial position, so the mass balances in the gaseous phase can be solved analytically. Those in the liquid phase require the introduction of the average fluxes $(\bar{J}_i)_{x=1}$ defined by Eq. (72). Since the reaction in the liquid film is negligible, $(N_i)_{s=\delta} = (N_i)_{s=0}$, which in dimensionless form can be rewritten as

$$(\bar{J}_i)_{x=1} = (\bar{J}_i)_{x=0} \frac{St_{\ell i}}{\gamma_i St_{gi}} \tag{95}$$

where $(\bar{J}_i)_{x=0}$ can be evaluated from Eq. (72), using Eqs. (61) and (69) as follows:

$$(\bar{J}_i)_{x=0} = v_g^f \phi_{gi}^f - v_g(1)\phi_{gi}(1) \tag{96}$$

TABLE 21 PFR Model with the Bimolecular Reaction (82) of Pseudo First Order ($v_g = 1$, $v_\ell = 1$)

$$\phi_{\ell A} = C_1 \frac{1 + \lambda_+}{H_A^*} e^{d\lambda_+ z} + C_2 \frac{1 + \lambda_-}{H_A^*} e^{d\lambda_- z}$$

$$\phi_{gA} = C_1 e^{d\lambda_+ z} + C_2 e^{d\lambda_- z}$$

$$C_2 = \phi_{gA}^f - C_1$$

$$C_1 = \begin{cases} \dfrac{\phi_{gA}^f (1 + \lambda_-) - H_A^* \phi_{\ell A}^f}{\lambda_- - \lambda_+} & \text{for } n^* = -1 \\ \dfrac{\phi_{GA}^f (1 + \lambda_+) e^{d\lambda_+} - H_A^* \phi_{\ell A}^f}{(1 + \lambda_+) e^{d\lambda_+} - (1 + \lambda_-) e^{d\lambda_-}} & \text{for } n^* = 1 \end{cases}$$

$$\lambda_\pm = \frac{-[q - n^*(p + 1) \pm (q^2 + p^2 + 2p + 1 + 2n^*pq - 2n^*q)^{1/2}}{2q}$$

$$b = u_g^o C_g^o/(u_\ell^o C_\ell^o); \quad q = 1/bH_A^*$$

$$p = q\alpha St_{\ell A} Ha^2/d; \quad d = St_{gA} \frac{\gamma_A/H_A^*}{1 + \gamma_A/H_A^*}$$

Case of gaseous phase of pure A:

$$\phi_{\ell A} = \frac{\phi_{\ell A}^f e^{\bar{\lambda} z} + bq\phi_{gA}^f (1 - e^{\bar{\lambda} z})}{p + 1}$$

$$\bar{\lambda} = \frac{d(p + 1)}{q}$$

TABLE 22 Liquid-Phase CSTR, Gaseous-Phase Axial Dispersion Model (v_g = 1)

$$\phi_{gi}(1) = \frac{4q(\phi_{gi}^f - b/d)e^{Pe_g/2}}{(1+q)^2 e^{Pe_g q/2} - (1-q)^2 e^{-Pe_g q/2}} + \frac{b}{d}$$

$$v_\ell \phi_{\ell i}^f = v_\ell \phi_{\ell i} - \alpha \frac{St_{\ell i}}{\xi_i} Ha^2 \sum_{j=1}^{NR} \nu_{ij} p_j f_j - \frac{St_{\ell i}}{\gamma_i St_{gi}} [\phi_{gi}^f - \phi_{gi}(1)]$$

For $Pe_g \to \infty$:

$$\phi_{gi}(1) = \left(\phi_{gi}^f - \frac{b}{d}\right) e^{-d} + \frac{b}{d}$$

$$d = St_{gi} \frac{\gamma_i/H_i^*}{1 + \gamma_i/H_i^*}$$

$$b = dH_i^* \phi_{\ell i}$$

$$q = \left(\frac{1 + 4d}{Pe_g}\right)^{1/2}$$

Assuming constant gas velocity (i.e., v_g = 1), the model reduces to the system of nonlinear algebraic equations reported in Table 22, which can be solved by standard techniques. Note that this procedure applies to all kinetic schemes and rate expressions, including reversible reactions. In the case of a nonvolatile component, the equations in Table 22 still hold with $H_i^* = 0$ and $\phi_{gi}^f = 0$, which lead to $\phi_{gi}(1) = 0$.

Fast Reaction Regime

In this case the mass balance of the gaseous reactant in the liquid phase is not necessary, since it is assumed that $\phi_{\ell A} = 0$. Under the conditions of constant velocity, $v_g = 1$, and reaction factor independent of the axial position, the mass balance of the gaseous reactant in the gas phase can be solved analytically as follows:

$$\frac{\phi_{gA}}{\phi_{gA}^f} = \frac{2(1+q)\exp[Pe_g(1-q)z/2] - 2(1-q)\exp[-Pe_g q + Pe_g(1+q)z/2]}{(1+q)^2 - (1-q)^2 \exp(-Pe_g q)} \tag{97}$$

where

$$q = \left(1 + 4\,\frac{St_{gA}}{Pe_g}\,\frac{\gamma_A E_A^*/H_A^*}{1 + \gamma_A E_A^*/H_A^*}\right)^{0.5} \tag{98}$$

The assumption E_A^* = const. is quite restrictive in practice, However, it is valid in the case of a reaction of type (82), first order with respect to A and nth order with respect to B, with component B nonvolatile and the liquid phase well mixed. Thus, according to Eq. (39), if $4 < M < E_{A,in}/2$, in the reactor factor is given by

$$E_A^* = M = \frac{(kD_A C_{\ell B}^n)^{0.5}}{k_{\ell A}} \tag{99}$$

which is constant along the reactor axis, because the liquid phase is well mixed, (i.e., $C_{\ell B}$ = const.). In the fast reaction regime, the gaseous reactant is depleted within the liquid film and therefore at steady state it follows that $\nu_B(N_A)_{s=0} = -(N_B)_{s=\delta}$, which in dimensionless form leads to

$$(\bar{J}_B)_{x=1} = -(\bar{J}_A)_{x=0}\,\frac{\nu_B\, St_{\ell A}}{\gamma_A\, St_{gA}} \tag{100}$$

where $(\bar{J}_A)_{x=0}$ can readily be evaluated, using Eqs. (61) and (69), as follows:

$$(\bar{J}_A)_{x=0} = v_g^f \phi_{gA}^f - v_g(1)\phi_{gA}(1) \tag{101}$$

The final system of nonlinear algebraic equations which constitutes the model is reported in Table 23.

With reference to the same reacting system, in the case of PFR model for the gaseous phase, the reactor model can be greatly simplified, although taking into account the variation of the superficial gas velocity (Deckwer, 1976). The final version of the model is reported in Table 24. Moreover, extensive calculations performed by Deckwer (1977) enable us to account for the effect of the gas-phase axial dispersion on the outlet reactor conversion. In Fig. 19 the ratio of the reactor lengths calculated using the axial dispersion model, L_{Pe} (Pe_g = finite), and the PFR model, L_∞ ($Pe_g \to \infty$), to obtain a given value of the outlet conversion is shown. The results are shown for various values of the Péclet number, Pe_g. It is remarkable that the hydrostatic head parameter, $\bar{\alpha}$, and the gaseous reactant mole fraction in the feed stream, y_A^f, although affecting the gas velocity appreciably, do not significantly alter the results of Fig. 19. That is, the effect of axial dispersion seems almost independent of the variation of the gas velocity along the reactor axis.

TABLE 23 Liquid-Phase CSTR, Gaseous-Phase Axial Dispersion Model, with the Bimolecular Reaction (82) of (1,n)-th Order ($v_g = 1$; B Nonvolatile)

$$\phi_{gA}(1) = \frac{4\phi^f_{gA} q e^{Pe_g/2}}{(1+q)^2 e^{Pe_g q/2} - (1-q)^2 e^{-Pe_g q/2}}$$

$$\phi_{\ell A} = 0$$

$$v_\ell \phi^f_{\ell B} = v_\ell \phi_{\ell B} + \frac{\nu_B}{\gamma_A} \frac{St_{\ell A}}{St_{gA}} [\phi^f_{gA} - \phi_{gA}(1)]$$

For $Pe_g \to \infty$:

$$\phi_{gA}(1) = \phi^f_{gA} e^{-d}$$

$$d = St_{gA} \frac{\gamma_A E^*_A/H^*_A}{1 + \gamma_A E^*_A/H^*_A} \;;\; q = \left(\frac{1+4d}{Pe_g}\right)^{1/2}$$

$$E^*_A = Ha\phi^{n/2}_{\ell B}$$

TABLE 24 Liquid-Phase CSTR, Gaseous-Phase PFR Model with the Bimolecular Reaction (82) of (1,n)th-Order (B Nonvolatile)

$$\phi^f_{gA}\xi - (1 - \phi^f_{gA}) \ln(1-\xi) = St_{gA} \frac{\gamma_A E^*_A/H^*_A}{1 + \gamma_A E^*_A/H^*_A} \frac{1 + 0.5\bar{\alpha}}{1 + \bar{\alpha}}$$

$$\xi = \frac{\phi^f_{gA} - (v_g \phi_{gA})_{z=1}}{\phi^f_{gA}} \;;\; v^f_g = 1$$

$$v_g(1) = \frac{(\bar{\alpha} + 1)(1 - \phi^f_{gA})}{1 - (\bar{\alpha} + 1)\phi_{gA}(1)}$$

$$\phi_{\ell A} = 0$$

$$v_\ell \phi^f_{\ell B} = v_\ell \phi_{\ell B} + \frac{\nu_B St_{\ell A}}{\gamma_A St_{gA}} \xi \phi^f_{gA}$$

$$E^*_A = Ha\phi^{n/2}_{\ell B}$$

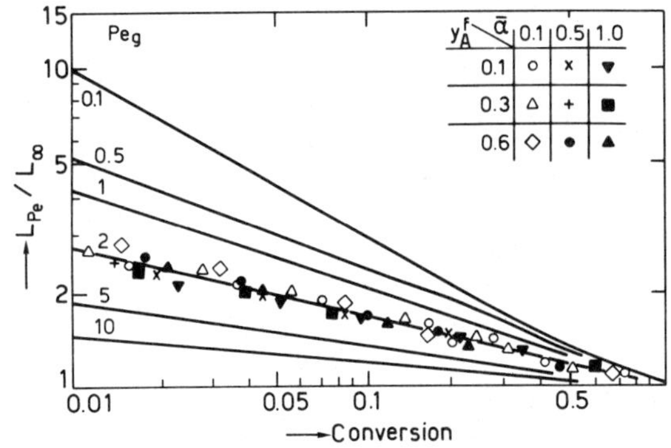

FIGURE 19 Influence of the gaseous-phase Péclet number Pe_g on the reactor length necessary to obtain a given conversion. The effect of $\bar{\alpha}$ and y_A^f is examined at $Pe_g = 2$.

The effect of the gaseous reactant order on reactor performance can be investigated through the relationships developed by Mhaskar (1974). They concern an (m,n)th-order reaction of type (82), with component B non-volatile and plug flow for both gaseous and liquid phases. Also in this case, assuming that $4 < M < E_{A,in}/2$, the reaction is given by

$$E_A^* = M = \left(\frac{2}{m+1} kD_A C_{\ell A}^{*m+1} C_{\ell B}^n\right)^{0.5} / k_{\ell A} \tag{102}$$

In Table 25 the analytical solutions obtained by Mhaskar (1974) for various values of the reaction orders m and n are reported, assuming plug flow in both phases (i.e., Pe_g, $Pe_\ell \to \infty$), negligible gas-side mass transfer resistance (i.e., $\gamma_A E_A^*/H_A^* \ll 1$), and constant superficial velocities (i.e., $v_\ell = v_g = 1$). Four situations have been considered: cocurrent flow ($n^* = -1$), countercurrent flow ($n^* = 1$), the case where the gaseous reactant concentration ϕ_{gA} does not vary appreciably along the reactor ($\phi_{gA} = 1$), and finally the case where the liquid reactant concentration $\phi_{\ell B}$ does not vary appreciably along the reactor ($\phi_{\ell B} = 1$). It is worth mentioning also the approximate solutions obtained by Szeri et al. (1976) for a (1,1)th-order reaction in the fast and instantaneous reaction regimes. Such solutions are valid in the range of large and small Péclet number and have been obtained using the perturbation and Galerkin methods, respectively.

Approximate Solutions of the CSTR Model

The equations that constitute the model are summarized in Table 8. Since these are algebraic equations, the problem here is only to simplify the liquid film model. In the case of consecutive reactions (80) of general order, the approximation reported in Table 18 can be used. The resulting nonlinear algebraic system can be solved through various methods, as reported by Hashimoto et al. (1968), Teramoto et al. (1973), Morbidelli et al. (1984), and Shaikh and Varma (1983b). In the last reference, a (1,1)th-order reaction is considered, taking into account the gas holdup and interfacial area variation

TABLE 25 PFR Model with the Bimolecular Reaction (82) of (m,n)-th Order (B Nonvolatile, $v_g = v_\ell = 1$, $\gamma_A E_A^*/H_A^* \ll 1$)

m	n	Cocurrent ($n^* = -1$)
1	2	$\phi_{gA} = d\Big/\left[1 - (1-d)\exp\left(\frac{dbz}{d-1}\right)\right]$ $(q \neq 1)$
1	2	$\phi_{gA} = 1/(bz + 1)$ $(q = 1)$
1	1	$\phi_{gA} = d + d\,\frac{[1 - \sqrt{q-1}\tan(bz\sqrt{q-1}/2)]^2}{[\sqrt{q-1} + \tan(bz\sqrt{q-1}/2)]^2}$ $(q >)$
1	1	$\phi_{gA} = 1/(1 + bz/2)^2$ $(q = 1)$
1	1	$\phi_{gA} = d - d\,\frac{[1 + \sqrt{1-q} + (1 - \sqrt{1-q})\exp(-bz\sqrt{1-q})]^2}{[1 + \sqrt{1-q} - (1 - \sqrt{1-q})\exp(-bz\sqrt{1-q})]^2}$ $(q < 1)$
1	1/2	$\ln\frac{\sqrt{q(\phi_{gA} - 1) + 1} - [4(q-1)(q\phi_{gA} - q + 1)]^{1/4} + \sqrt{q-1}}{\sqrt{q(\phi_{gA} - 1) + 1} + [4(q-1)(q\phi_{gA} - q + 1)]^{1/4} + \sqrt{q-1}}$ $+ 2\arctan\frac{[4(q-1)(q\phi_{gA} - q + 1)]^{1/4}}{\sqrt{q-1} - \sqrt{q(\phi_{gA} - 1) + 1}} = -bz\sqrt{2}\,(q-1)^{1/4}$ $+ \arctan\frac{[4(q-1)]^{1/4}}{\sqrt{q-1} - 1} + \ln\frac{1 - [4(q-1)]^{1/4} + \sqrt{q-1}}{1 + [4(q-1)]^{1/4} + \sqrt{q-1}}$ $(q > 1)$
1	1/2	$\phi_{gA} = \frac{1}{[(bz + 4)/4]^4}$ $(q = 1)$
1	1/2	$\ln\frac{[q(\phi_{gA} - 1) + 1]^{1/4} - (1-q)^{1/4}}{[q(\phi_{gA} - 1) + 1]^{1/4} + (1-q)^{1/4}} + 2\arctan\frac{[q(\phi_{gA} - 1) + 1]^{1/4}}{(1-q)^{1/4}}$ $= -bz(1-q)^{1/4} + \ln\frac{1 - (1-q)^{1/4}}{1 + (1-q)^{1/4}} + 2\arctan\frac{1}{(1-q)^{1/4}}$ $(q < 1)$
≠1	0	$\phi_{gA} = [1 + bz(m-1)/2]^{2/(1-m)}$
1	0	$\phi_{gA} = \exp(-bz)$

TABLE 25 (Continued)

m	n	Cocurrent ($n^* = -1$)
0	2	$\phi_{gA} = d\,\dfrac{\{1 + \sqrt{d} + (1 - \sqrt{d})\exp[-bz\sqrt{d}/(1-d)]\}^2}{\{1 + \sqrt{d} - (1 - \sqrt{d})\exp[-bz\sqrt{d}/(1-d)]\}^2}$ $(q > 1)$
0	2	$\phi_{gA} = \dfrac{1}{(1 + bz/2)^2}$ $(q = 1)$
0	2	$\phi_{gA} = -d\left[\dfrac{1 - \sqrt{-d}\tan(bqz\sqrt{-d}/2)}{\sqrt{-d} + \tan(bqz\sqrt{-d}/2)}\right]^2$ $(q < 1)$
0	1	$\sqrt{\phi_{gA}} = (1 + \sqrt{1-d})\exp(-bz(\sqrt{q}/2) - \sqrt{\phi_{gA} - d}$ $(q \neq 1)$
0	1	$\phi_{gA} = \exp(-bz)$ $(q = 1)$

The liquid concentration of B is given by

$$\phi_{\ell B} = 1 - q(1 - \phi_{gA})$$

m	n	Countercurrent ($n^* = 1$)
1	2	$\phi_{gA} = \dfrac{h}{q}\left\{1 - 1/\left[1 + \dfrac{q}{\phi_B^o}\exp(-bhz)\right]\right\}$
1	1	$\phi_{gA} = \dfrac{h}{q}\left\{1 - \dfrac{[\sqrt{h} - \sqrt{\phi_B^o} - (\sqrt{h} + \sqrt{\phi_B^o})\exp(b\sqrt{h}\,z)]^2}{[\sqrt{h} - \sqrt{\phi_B^o} + (\sqrt{h} + \sqrt{\phi_B^o})\exp(b\sqrt{h}\,z)]^2}\right\}$
1	1/2	$\ln\dfrac{h^{1/4} + (h - q\phi_{gA})^{1/4}}{h^{1/4} - (h - q\phi_{gA})^{1/4}} - 2\arctan\left(\dfrac{h - q\phi_{gA}}{h}\right)^{1/4} = bzh^{1/4} - 2\arctan\left(\dfrac{\phi^o_B}{h}\right)^{1/4} + \ln\dfrac{h^{1/4} + \phi_B^{o\,1/4}}{h^{1/4} - \phi_B^{o\,1/4}}$
≠1	0	$\phi_{gA} = [1 + bz(m-1)/2]^{2/(1-m)}$
1	0	$\phi_{gA} = \exp(-bz)$
0	2	$\phi_{gA} = \dfrac{h}{q}\,\dfrac{[\sqrt{h/q} - 1 - (\sqrt{h/q} + 1)\exp(-bz\sqrt{hq})]^2}{[\sqrt{h/q} - 1 + (\sqrt{h/q} + 1)\exp(-bz\sqrt{hq})]^2}$

TABLE 25 (Continued)

m	n	Cocurrent ($n^* = -1$)
0	1	$\phi_{gA} = \frac{h}{q}[\sqrt{q/h}\cos(bz\sqrt{q}/2) - \sqrt{1 - q/h}\sin(bz\sqrt{q}/2)]^2$

The liquid concentration of B is given by

$$\phi_{\ell B} = \phi_{\ell B}(0) + q(1 - \phi_{gA})$$

$$\phi_{\ell B}(0) = 1 - q[1 - \phi_{gA}(1)]$$

m	n	Cocurrent and pure A ($n^* = -1$; $\phi_{gA} = 1$)
Any	$\neq 2$	$\phi_{\ell B} = [1 + bqz(n - 2)/2]^{2/(2-n)}$
Any	2	$\phi_{\ell B} = \exp(-bqz)$

m	n	Negligible consumption ($\phi_{\ell B} = 1$)
$\neq 1$	Any	$\phi_{gA} = [1 + bz(m - 1)/2]^{2/(1-m)}$
1	Any	$\phi_{gA} = \exp(-bz)$

$$b = \frac{St_{gA}\gamma_A Ha}{H_A^{*(m+1)/2}}\sqrt{\frac{2}{m + 1}}; \quad q = \frac{\nu_B St_{\ell A}}{St_{gA}\gamma_A}$$

$$d = (1 - 1)/q; \; h = q + \phi_B^o; \; \phi_B^o = \phi_{\ell B}(0)$$

$$\phi_{gA}^f = 1; \; \phi_{\ell B}^f = 1$$

due to gas absorption according to Eqs. (73). In Figs. 20 and 21 the conversions of the gaseous and the liquid reactants, X_A and X_B, are shown as a function of the residence time, operating with fixed gas and liquid flow rate, respectively. The behavior represented can be explained considering the actual reactor regime, which is defined by the value of the parameter M defined by Eq. (84) (not by M_f, which is the M value calculated using the feed concentrations).

Similarly, for the kinetic schemes reported in Table 19, the approximate solutions of the film model listed in Table 20 can be used, thus reducing the reactor model to a system of algebraic equations. Some useful approximations for the model solution, based on the specific reactor regime, are given next.

Slow Reaction Regime

Assuming constant gas velocity ($v_g = 1$), the model equations reduce to those reported in Table 26. In general, this is a system of nonlinear equations with the liquid bulk concentrations $\phi_{\ell i}$ as unknowns.

Fast Reaction Regime

In the case of a (m,n)th-order irreversible reaction of type (82), assuming component B nonvolatile and $4 < M < E_{A,in}/2$, the reaction factor is given by Eq. (102). In this case the model equations can be rewritten as shown in Table 27.

Approximate Solutions of the Semiflow Batch Model

The mass balance in the liquid phase for these models is given by Eqs. (70) to (72). From comparison with the mass balance equations in the liquid phase of the CSTR model reported in Table 8, it appears that the two set of equations are identical, the only exception being that the accumulation term $d\phi_{\ell i}/d\theta$ in the batch model is replaced by the convective term $(v_\ell \phi_{\ell i} - v_\ell \phi_{\ell i}^f)\, \xi_i / St_{\ell i}$ in the CSTR model.

In semiflow batch models the gaseous phase is usually described by the steady-state axial dispersion model, with various levels of backmixing. Therefore, all the solutions reported in Tables 22 to 24, 26, and 27, where the CSTR model is assumed for the liquid phase and the axial dispersion or the CSTR model for the gaseous phase, can be used in this case as well. In particular, the solutions for the semiflow batch models are valid under the same conditions and are obtained from those reported in the above-mentioned tables, by simply replacing the term $(v_\ell \phi_{\ell i} - v_\ell \phi_{\ell i}^f)(\xi_i/St_{\ell i})$ by the derivative $d\phi_{\ell i}/d\theta$. Systems of ordinary differential equations, instead of algebraic equations, are then obtained. For illustrative purposes, the solution reported in Table 22 for the case of $M \ll 1$ and $v_g = 1$ is examined. In the case of an (m,n)th-order reaction of type (82) with A the only volatile component, the equations reported in Table 22 for the liquid mass balance reduce to

$$\frac{d\phi_{\ell A}}{d\theta} = -\alpha Ha^2\, \phi_{\ell A}^m \phi_{\ell B}^n + \frac{1}{\gamma_A\, St_{gA}} [\phi_{gA}^f - \phi_{gA}(1)] \tag{103}$$

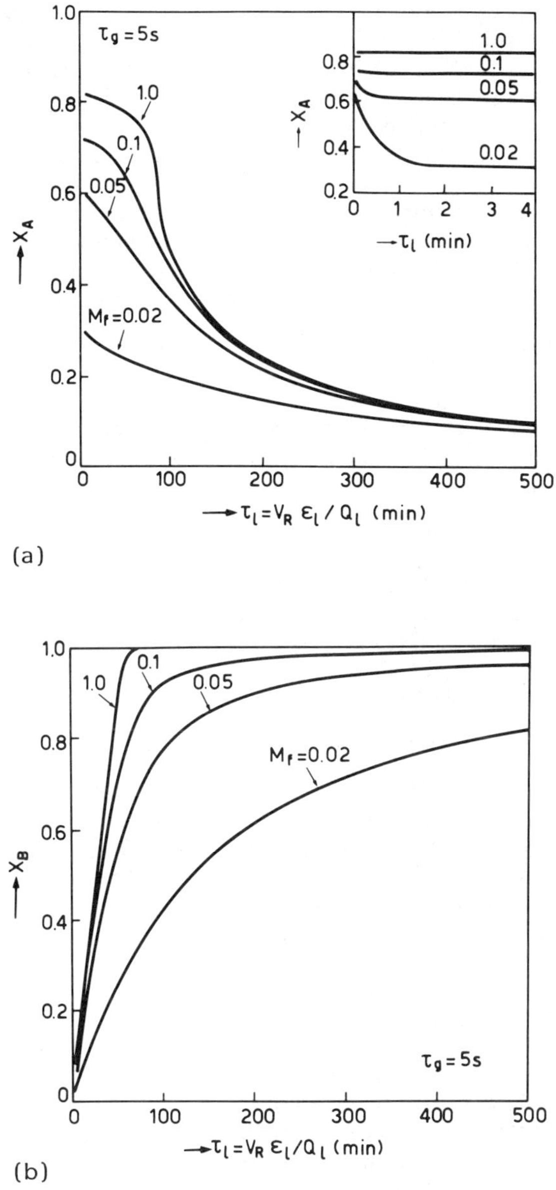

FIGURE 20 Conversion of the gaseous reactant X_A as a function of the liquid residence time τ_ℓ. (b) Conversion of the liquid reactant X_B as a function of the liquid residence time τ_ℓ. $M_f = (kD_B C^f_{\ell B})^{1/2} / k_{\ell A}$.

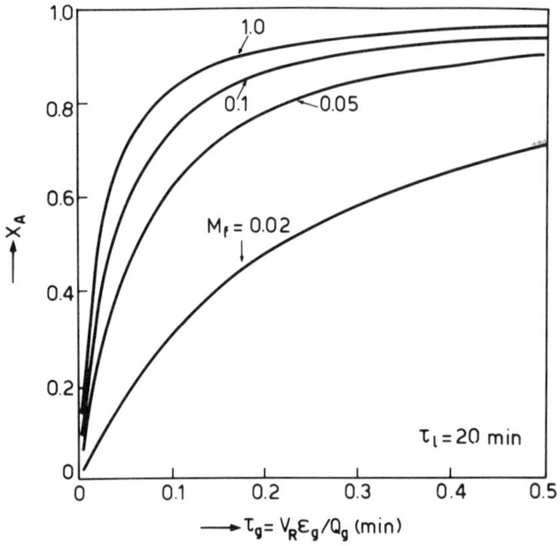

(a)

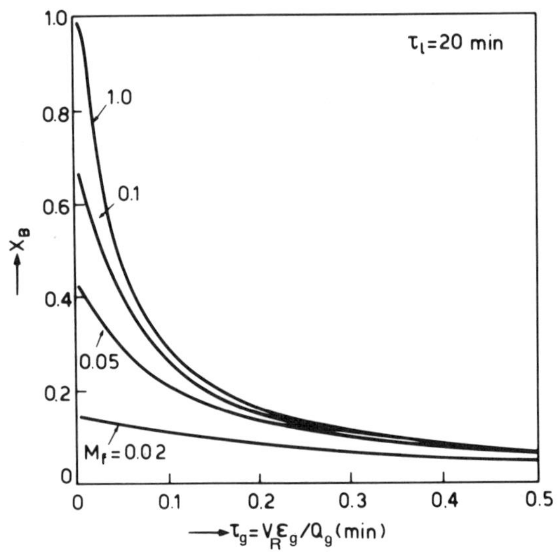

(b)

FIGURE 21 (a) Conversion of the gaseous reactant X_A as a function of the gas residence time, τ_g. (b) Conversion of the liquid reactant X_B as a function of the gase residence time τ_g. $M_f = (kD_A C^f_{\ell B})^{1/2}/k_{\ell A}$.

TABLE 26 CSTR Model in the Slow Reaction Regime with Constant Velocities ($v_g = v_\ell = 1$)

$$\phi_{gi} = \frac{\phi^f_{gi} + q_i H^*_i \phi_{\ell i}}{1 + q_i}$$

$$\phi^f_{\ell i} = \phi_{\ell i} - \alpha \frac{St_{\ell i}}{\xi_i} Ha^2 \sum_{j=1}^{NR} \nu_{ij} p_j f_j - \frac{St_{\ell i}}{\gamma_i\, St_{gi}} \left(\phi^f_{gi} - \phi_{gi}\right)$$

$$q_i = St_{gi} \frac{\gamma_i / H^*_i}{1 + \gamma_i / H^*_i}$$

$$v^f_g = 1; \; v^f_\ell = 1$$

TABLE 27 CSTR Model in the Fast Reaction Regime with the Bimolecular Reaction (82) of (m,n)th Order

$$(\phi^f_{gA} - \phi_{gA}) - St_{gA} \frac{\gamma_A E^*_A / H^*_A}{1 + \gamma_A\, E^*_A / H^*_A} \phi_{gA}(1 - \phi_{gA}) = 0$$

$$\phi^f_{\ell B} = \phi_{\ell B} + St_{\ell A} \nu_B \frac{E^*_A / H^*_A}{1 + \gamma_A E^*_A / H^*_A} \phi_{gA}$$

$$\phi^*_{\ell A} = \frac{\phi_{gA} / H^*_A}{1 + \gamma_A E^*_A / H^*_A}$$

$$E^*_A = Ha\left[\frac{2}{m + 1} \phi^{*\,m-1}_{\ell A} \phi^{*\,n}_{\ell B}\right]^{1/2}$$

$$v_g = 1 - St_{gA} \frac{\gamma_A E^*_A / H^*_A}{1 + \gamma_A E^*_A / H^*_A} \phi_{gA}$$

$$v_\ell = v^f_\ell = 1; \; v^f_g = 1$$

TABLE 28 Semiflow Batch Model in the Slow Reaction Regime, with the Bimolecular Reaction (82) of Pseudo First Order ($v_g = 1$)

$$\phi_{gA}(1) = h\phi_{gA}^{f} + \phi_{\ell A}H_A^*(1 - h)$$

$$\phi_{gB}(1) = 0$$

$$\phi_{\ell A} = \frac{a_0}{a_1} + \left(\phi_{\ell A}^{in} - \frac{a_0}{a_1}\right)\exp(-a_1\theta)$$

$$\phi_{\ell B} = \phi_{\ell B}^{in} - a_2\left\{\frac{a_0}{a_1}\theta + \left(\phi_{\ell A}^{in} - \frac{a_0}{a_1}\right)\frac{1}{a_1}[1 - \exp(-a_1\theta)]\right\}$$

where

$$h = 4qe^{Pe_g/2}/[(1 + q)^2e^{Pe_gq/2} - (1 - q)^2e^{-Pe_gq/2}]$$

[for $Pe_g \to \infty$, $h = \exp(-d)$]

$$a_0 = \frac{1}{\gamma_A St_{gA}}\phi_{gA}^{f}(1 - h)$$

$$a_1 = \alpha Ha^2 + a_0\frac{H_A^*}{\phi_{gA}^{f}}$$

$$a_2 = \alpha\nu_B Ha^2$$

$$d = St_{gA}\frac{\gamma_A/H_A^*}{1 + \gamma_A/H_A^*}$$

$$q = (1 + 4d/Pe_g)^{1/2}$$

$$\frac{d\phi_{\ell B}}{d\theta} = -\alpha\nu_B Ha^2\phi_{\ell A}^{m}\phi_{\ell B}^{n} \qquad (104)$$

while those for the gas mass balance are unchanged. In the case of pseudo-first-order kinetics (i.e., $m = 1$, $n = 0$), the analytical solution is possible and is reported in Table 28.

A useful approximation has been proposed by Barona and Prengle (1973), valid for slow reactions, such that $\alpha M^2 < 0.25$. It consists of evaluating $\phi_{\ell B}(\theta)$ from Eq. (104), assuming that $\phi_{\ell A} = \phi_{\ell A}^*$, and then calculating $\phi_{\ell A}$ from Eq. (103) where the time derivative is neglected (i.e., $d\phi_{\ell A}/d\theta = 0$).

TABLE 29 Semiflow Batch Model in the Slow Reaction Regime, with the Bimolecular Reaction (82) of (m,n)th Order ($v_g = 1$; $\alpha M^2 < 0.25$). Approximation of Barona and Prengle (1973)

$$\phi_{gA}(1) = h\phi^f_{gA} + \phi_{\ell A} H^*_A (1 - h)$$

$$\phi_{gB}(1) = 0$$

$$\phi_{\ell B} = [(\phi^{in}_{\ell B})^{1-n} + (n - 1)a_2 \phi^{*m}_{\ell A} \theta]^{1/(1-n)} \qquad (n \neq 1)$$

$$\phi_{\ell B} = \phi^{in}_{\ell B} \exp(-a_2 \phi^{*m}_{\ell A} \theta) \qquad (n = 1)$$

$$\alpha Ha^2 \phi^m_{\ell A} \phi^n_{\ell B} - \frac{1}{\gamma_A St_{gA}} [\phi^f_{gA} - \phi_{gA}(1)] = 0$$

$$\frac{1}{\gamma_A St_{gA}} [\phi^f_{gA} - \phi_{gA}(1)] = \phi^*_{\ell A} - \phi_{\ell A}$$

where the parameters h and a_2 are defined as in Table 28.

The gas-liquid interface mass flux can be obtained by noting from Eqs. (87), (95), and (96) that

$$(\bar{J}_A)_{x=1} = \frac{St_{\ell A}[\phi^f_{gA} - \phi_{gA}(1)]}{\gamma_A St_{gA}} = St_{\ell A}(\phi^*_{\ell A} - \phi_{\ell A}) \qquad (105)$$

Using these approximations, the model reduces to the system of nonlinear algebraic equations reported in Table 29. This solution is adequate for cases such as the chlorination of benzene to monochlorobenzene at 80°C and the auto-oxidation of *p*-toluic acid in acetic acid solution, catalyzed by cobalt and manganese bromide, where the reactions are (1,1)th order (Agnello and Williams, 1960; Groggins, 1958) and (0.5,1)th order (Ravens, 1959), respectively. In these cases the equations reported in Table 29 can easily be solved analytically.

NONISOTHERMAL GAS-LIQUID REACTORS

Modeling Nonisothermal Gas-Liquid Reactors

Gas-Liquid reactors are widely used for temperature-sensitive processes, such as oxidation, hydrogenation, alkylation, chlorination, sulfonation, and so on, which are all accompanied by large heat effects. In fact, owing to the high heat capacity of the liquid phase and the large dispersion prevailing

in gas-liquid reactors, the temperature is almost constant in the entire reactor volume and can easily be controlled. To this end, heat is usually removed through the vessel wall or through a coil immersed in the liquid volume. In both cases, the process-side heat transfer coefficient is much larger than in the absence of bubbles, due to the increased turbulence provoked by bubbles in the liquid film at the heating surface.

The joint effect of the heat of absorption and the heat of reaction can lead to significant temperature gradients within the liquid film, which can largely affect the total mass absorption rate and, in the case of a complex kinetic scheme, also the process selectivity.

Thermal effects in gas absorption without reaction have been studied by Chiang and Toor (1964) and Verma and Delancey (1975) by coupling the energy and mass balances in the surface renewal model. The latter considered the influence of heat of absorption, volume changes in the liquid phase, variable liquid density, and the pseudo-Dofour effect; they concluded that while for the accurate prediction of the surface temperature rise the first two mentioned phenomena must be taken into account, if only the interphase mass flow is to be evaluated, excellent estimates are obtained using the isothermal model. This is largely due to the equivalence between the enhancing effect of the liquid volume change and the depressing effect of the heat of absorption, which leads to larger superficial temperature and then to lower solubility of the gaseous reactant. This study, where gas-side resistances are always assumed negligible, has been verified experimentally for the systems ammonia-water and propane-decane.

If a chemical reaction is present, its thermal contribution must also be taken into account. While the heat of absorption is released only at the gas-liquid interface, the heat of reaction is released in the entire liquid film, and in the case of slow reaction, also in the liquid bulk. Thus the estimation of the temperature profile requires the solution of both energy and mass balances in the liquid film. An upper limit of the superficial temperature values has been reported by Danckwerts (1970), assuming that all the heat of reaction, as well as the heat of absorption, is liberated at the interface:

$$T^* - T_\ell = \frac{D_A(-\Delta\tilde{H}_a - \Delta\tilde{H}_R)}{k_{T\ell}k_{\ell A}} (N_A)_{s=0} \left(\frac{k_{T\ell}}{\rho_\ell C_{p\ell} D_A}\right)^{0.5} \tag{106}$$

Accurate analyses have been made in the case of a first-order reaction using both the penetration theory model [Danckwerts (1952, 1967) and Shah (1972) examined the effect of the temperature rise on reactant solubility, diffusion coefficient, and reaction rate constant] and the film model (Mann and Moyes, 1977). In the latter it was concluded that for pseudo-first-order kinetics in the fast reaction regime (i.e., $C_{\ell A} = 0$), the nonisothermal reaction factor, defined as the ratio between the absorption rate with reaction heat effects and the purely physical absorption rate with zero heat of solution, is given by

$$E_A^* = \frac{C_{\ell A}^*(T^*)\ \mathrm{Ha}^*}{C_{\ell A}^*(T_\ell)\tanh \mathrm{Ha}^*} \qquad \mathrm{Ha}^{*2} = \frac{D_A k(T^*)}{k_{\ell A}^2} \tag{107}$$

where T_ℓ and T^* indicate the liquid bulk and the gas-liquid interface temperature, respectively. The latter can be evaluated through the equation

$$h_\ell(T^* - T_\ell) = -(\Delta\tilde{H}_a + \Delta\tilde{H}_R)k_{\ell A}C^*(T_\ell)E_A^* \tag{108}$$

which must be coupled with a suitable expression of the gaseous reactant solubility as a function of temperature $C^*(T) = f(T)$.

As pointed out by Clegg and Mann (1969), from Eq. (107) it appears that the solubility decrease due to the interfacial temperature rise can reduce the interface absorption flows even below those obtained under conditions of purely physical isothermal absorption (i.e., $E_A^* < 1$). In these approaches the only heat effect on the surface temperature, and consequently on the gaseous reactant solubility, is taken into account. Therefore, such results can be applied with confidence to cases where heat effects are small or absorption rates as well as solubilities are low. If the converse is true, a deeper analysis of the problem is necessary.

The most complete treatment of nonisothermal gas-liquid mass transfer has been given by Tamir et al. (1975) and Tamir and Taitel (1975) using a penetration-type model for gas absorption into a laminar liquid stream. In it the solvent volatility is taken into account, together with the gas-side mass transfer resistance. A boundary layer model is assumed for the fluid mechanical behavior of the gas phase, and in the liquid phase the transverse bulk-flow contribution, ignored in previous works, is taken into account. This is found to affect the interface absorption flow only for low reaction rates or contact times (i.e., when the reaction contribution is small); under these conditions the decrease with respect to the isothermal case is limited, and the same situation depicted previously for purely physical absorption is obtained. At large reaction rates or contact times, the solvent evaporation and the gaseous environment effects are of the same order of magnitude as the heat effect, leading to serious decreases of the mass flux with respect to the isothermal case. The effect of heat conduction from the liquid film to the gaseous phase is negligible compared to other heat contributions. It is worth pointing out that for increasing reaction rates the interfacial absorption flow decreases, which is contrary to the trend obtained under isothermal conditions. In the case of absorption with chemical reaction of chlorine in toluene, experimentally investigated by Mann and Clegg (1975), this analysis shows that the effect of the gaseous environment is to decrease the total absorption flux of chlorine by 25% more than what is predicted using the simplified models mentioned previously, which take into account only the heat effect on the interfacial temperature.

The complete model of a nonisothermal gas-liquid reactor, including the detailed description of the nonisothermal liquid film, would be quite complicated, so, usually, several simplifications are introduced in order to model such units. In the first place, the gas- and liquid-phase temperatures are generally assumed to be identical at any given axial position. This assumption can be verified using the relationship for the heat transfer coefficient developed by Calderbank and Moo-Young (1961) for small bubbles ($d_{VS} <$ 0.0025 m, as is true in almost all industrial gas-liquid contactors):

$$\frac{h_\ell d_{vs}}{k_{T\ell}} = 2 + 0.31\left[\frac{d_{vs}^3(\rho_\ell - \rho_g)g}{\mu_\ell D_\ell}\right]^{1/3} \tag{109}$$

Only one energy balance is then required for these reactors, which using the axial dispersion model (assuming the gas and liquid velocities constant along the reactor axis, and the physical properties of the gas and liquid and the heats of reaction, absorption, and evaporation independent of temperature) is given by

$$E_T \frac{d^2T}{dy^2} - (u_g \rho_g C_{pg} + u_\ell \rho_\ell C_{p\ell}) \frac{dT}{dy} = Ua_c(T - T_c) + a \sum_{i=1}^{NC} (N_i)_{s=0}(-\Delta \bar{H}_i) - q_R \tag{110}$$

where on the right-hand side three contributions are present: heat exchange with heating or cooling devices (a_c is the heat exchange surface area of coils or jacket, per unit reactor volume), heat liberated or absorbed by evaporation or solution of the liquid [$(N_i)_{s=0} < 0$ and $\Delta\bar{H} = \Delta\tilde{H}_{ev}$] or the gaseous [$(N_i)_{s=0} > 0$ and $\Delta\bar{H} = \Delta\tilde{H}_a$] reactants, respectively, and finally, the heat of reaction per unit reactor volume. This last term, in the case of only one reaction with one gaseous reactant [see Eq. (82)], can be evaluated as follows:

$$q_R = \left[(N_A)_{s=0}a - u_\ell \frac{dC_{\ell A}}{dy}\right](-\Delta\tilde{H}_R) \tag{111}$$

The overall heat transfer coefficient can be calculated as usual through the following relationship:

$$\frac{1}{U} = \frac{1}{h_w} + \frac{s_c}{k_{TS}} + \frac{1}{h_c} \tag{112}$$

where the process-side heat transfer coefficient h_w can be evaluated as shown earlier, and the cooling (or heating) side heat transfer coefficient h_c using standard techniques (Kern, 1950).

Little information is available on the evaluation of effective axial thermal conductivity E_T. However, since the mechanisms for axial heat transfer are widely accepted to be responsible also for axial mass transfer (i.e., turbulent mixing due to liquid circulation, natural convection, and conduction), it is reasonable to assume equal effective axial diffusivity for mass and heat:

$$\frac{E_T}{\rho_d C_{pd}} = E_\ell \tag{113}$$

where the index d refers to the gas-liquid dispersion. Chen and McMillian (1982) have recently shown that Eq. (113) is reasonably accurate for batch bubble columns, while for packed columns it gives slightly lower values for the effective axial thermal conductivity, due to the additional heat conduction through the solid packings.

In many cases of practical interest, as shown by Fair et al. (1962) using a 1-m-diameter column with superficial gas velocities over 0.006 m/s, liquid

backmixing is quite extensive, leading to temperature uniformity. In this case ($E_T \to \infty$) the CSTR model can be applied for the energy balance:

$$(Q_g \rho_g C_{pg} + Q_\ell \rho_\ell C_{p\ell})(T^f - T) = Ua_c V_R (T - T_c) + aV_R \sum_{i=1}^{NC} (N_i)_{s=0}(-\Delta \bar{H}_i) - q_R V_R \tag{114}$$

where U is given by Eq. (112), and q_R, in the case of one reaction involving the only gaseous reactant A, can be evaluated as

$$q_R = \left[(N_A)_{s=0} a - \frac{Q_{\ell A} C_{\ell A}}{V_R} \right] (-\Delta \tilde{H}_R) \tag{115}$$

Models of this type have been used in several theoretical studies of nonisothermal gas-liquid reactors (Hoffman et al., 1975; Sharma et al., 1976; Raghuram et al., 1979; Huang and Varma, 1981c,d).

Finally, it is worth mentioning the shortcut and rigorous design calculation procedures for packed gas absorbers involving large heat effects proposed by Stockar and Wilke (1977a,b) and Feintuch and Treybal (1978) and applied to cases such as the absorption in aqueous solutions of air gaseous streams containing ammonia or acetone or hydrogen chloride.

Steady-State Multiplicity

The steady state obtained in isothermal backmixed gas-liquid reactors is usually unique and stable. This is not true for nonisothermal reactors, where the interaction between the chemical reaction, the heat and mass transport resistances, and the gaseous reactant solubility causes the occurrence of multiple steady states.

Experimentally, up to three steady states have been found by Ding et al. (1974) in a CSTR during chlorination of *n*-decane, while Hancock and Kenney (1972, 1977) showed unusual dynamic phenomena, such as sustained periodic oscillation, in a semiflow stirred-tank column during methanol chlorination, with hydrochloric acid in the presence of $ZnCl_2$ catalyst, to methyl chloride.

Earlier theoretical analyses of steady-state multiplicity and stability were presented by Schmitz and Amundson (1963a–d) and Luss and Amundson (1967) using reactor models which assume that the chemical reaction and the interphase mass transfer are two independent, noninteracting processes. Thus these analyses are restricted to the case of slow reaction. Hoffman et al. (1975) and Huang and Varma (1981d) have examined the steady-state multiplicity for the bimolecular reaction (82) of (1,1)th order in an adiabatic and a nonadiabatic reactor, respectively. In both cases the CSTR model for both phases and the energy balance (114) has been used, and the agreement with the experimental results reported by Ding et al. (1974) is satisfactory. It is concluded that an isola with up to five steady states can occur in an adiabatic CSTR, while in the corresponding nonadiabatic case (even with a small heat loss) only three steady states can exist. In the case of two consecutive reactions of the same kind, Sharma et al. (1976) have shown

the possible existence of up to seven steady states. Note that these are unique features of gas-liquid reactors, as in a single-phase adiabatic CSTR a maximum of three and five steady states can occur for one or two consecutive reactions, respectively. Also the occurrence of an isola is not possible in adiabatic reactors.

Analytically necessary and sufficient criteria for steady-state multiplicity, uniqueness, and local stability have been developed for CSTR reactors with pseudo-first-order reactions in the fast reaction regime (Raghuram and Shah, 1977; Raghuram et al., 1979; Huang and Varma, 1981b; Singh et al., 1982). Under these conditions, the occurrence of limit cycles is usually not possible. It is worthwhile mentioning the significant usefulness of steady-state multiplicity analysis as a tool for discriminating among rival models and for the accurate evaluation of some parameters of these models (Hoffman et al., 1975; Huang and Varma, 1981e). In the latter work surprising agreement between the multiplicity regions predicted by the pseudo-first-order reaction model in the fast reaction regime and the complete (1,1)th-order model was found.

Finally, Huang and Varma (1981c) and Aluko and Chang (1982) have examined the steady-state multiplicity of nonadiabatic bubble columns with fast pseudo-nth-order reactions, assuming the liquid phase well mixed, the gas phase in plug flow, and the energy balance in the form of Eq. (114). Every multiplicity pattern observed in a gas-liquid CSTR is also possible in a bubble column, and the possibility of multiplicity in the bubble column is always higher than that in the CSTR for equivalent system parameters. Analytic criteria are developed for the prediction of uniqueness and multiplicity of the steady states as a function of the system physicochemical parameters.

Stagewise Nonisothermal Models

Stagewise models are widely used in the simulation of nonisothermal gas-liquid contactors. Resorting to a stagewise, tray-tower analogy, they can describe not only plate columns but also packed columns. Following Carrà et al. (1979b), the model equations are summarized in Table 30. In each stage the gas-liquid system is assumed to be more or less close to the equilibrium conditions, depending on the value of the Murphree efficiency E_M. This parameter takes into account the possible influence of mass transfer resistances for the various components. In some cases, a similar approach can be used to describe the effect of the heat transfer resistance leading to different temperature values for the gas and liquid streams (Bourne et al., 1974a) on the jth stage:

$$T_{g,j} = T_{g,j-1} + E_{MT}(T_{\ell,j} - T_{g,j+1}) \tag{116}$$

Murphree efficiencies can be esitmated following several correlations, as reported by Stichlmair and Mersmann (1977) and Pohorecki (1976). However, since the model results can be quite sensitive to the value of such quantities, Bourne et al. (1974b) suggested determining the Murphree efficiency values as a function of the operating conditions through suitable measurements conducted on a single plate of the same design as those of the unit under consideration.

As it appears from Table 30, stagewise models lead to a system of nonlinear algebraic equations. Several numerical techniques are available for

TABLE 30 Stagewise Nonisothermal Model Equations

Material balances of the ith component on the kth stage:

$$F_k z_{i,k} + G_{k+1} y_{i,k+1} - G_k y_{i,k} + L_{k-1} x_{i,k-1} - L_k x_{i,k} + V_{\ell,k} \sum_{j=1}^{NR} \nu_{i,j} r_{j,k}$$

Equilibrium conditions for the ith component on the kth stage:

$$y_{i,k} = E_{M,k} K_{i,k} x_{i,k} + (1 - E_{M,k}) y_{i,k+1}$$

Enthalpy balance on the kth stage:

$$F_k \tilde{H}_k^f + G_{k+1} \tilde{H}_{g,k+1} - G_k \tilde{H}_{g,k} - L_k \tilde{H}_{\ell,k} + L_{k-1} \tilde{H}_{\ell,k-1} + V_{\ell,k} \sum_{j=1}^{NR} (-\Delta \tilde{H}_{R,j})\, r_{j,k} - q_k = 0$$

Overall material balance on the kth stage:

$$F_k + G_{k+1} - G_k + L_{k-1} - L_k + V_{\ell,k} \sum_{i=1}^{NC} \sum_{j=1}^{NR} \nu_{i,j} t_{j,k} = 0$$

Stoichiometric equation on the kth stage:

$$\sum_{j=1}^{NC} (x_{i,k} - y_{i,k}) = 0$$

the solution of these systems, and most of them have been reviewed by Hunek et al. (1976) and Holland (1981). The global Newton-Raphson method has been successfully used for the case of distillation columns with chemical reactions (Carrà et al., 1979b,d). On the other hand, a dynamic simulation method was found by Bourne et al. (1974a) to be particularly efficient in the case of gas absorbers.

The use of distillation columns with chemical reactions is quite common in chemical industry. Mayer and Wörz (1980) give useful suggestions on the optimal selection, from the standpoint of energy savings, of either a reaction column or a batch reactor mounted by a column, depending on the relative volatilities and the reaction rates of the compounds involved. Coupling distillation with chemical reaction is particularly useful when it is necessary to displace the chemical equilibrium in order to increase the desired product yield, as in esterification and transesterification reactions (Baratella et al., 1974). In the case of two consecutive reactions in the

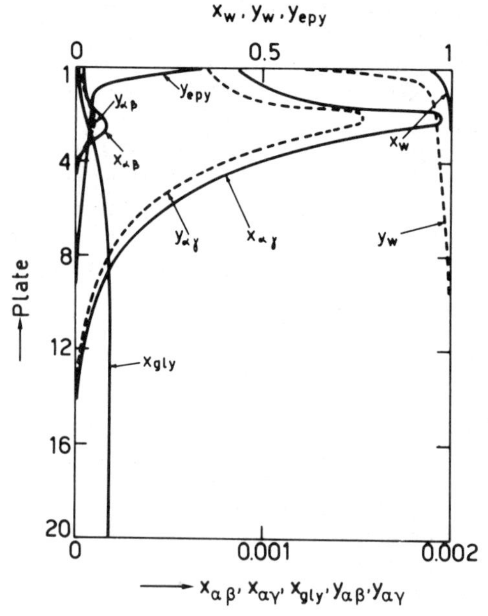

FIGURE 22 Profiles of some liquid-phase mole fractions x_i (solid lines) and some vapor-phase mole fractions y_i (dashed lines) along a column for the synthesis of epichlorohydrin.

liquid phase, where the intermediate desired product is highly volatile, the use of a distillation column reactor can significantly improve the process selectivity. This is the case in epichlorohydrin (Carrà et al., 1979a) and propylene oxide (Carrà et al., 1979c) synthesis by dehydrochlorination, which both follow the following simplified kinetic scheme: chlorohydrins → epoxide → hydrolysis products. In the case of epichlorohydrin synthesis

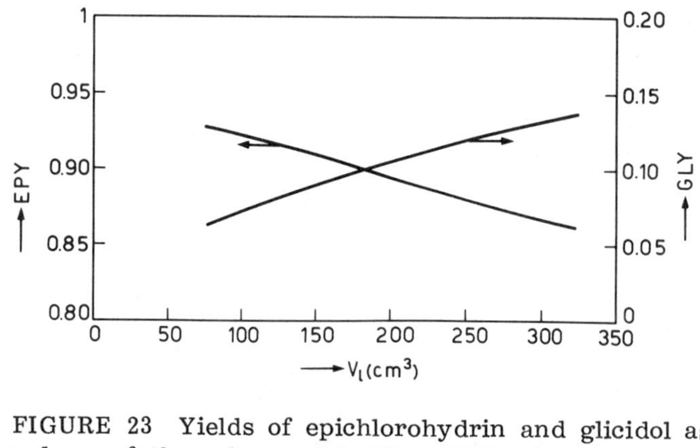

FIGURE 23 Yields of epichlorohydrin and glicidol as a function of the liquid volume of the column plate V_ℓ. EPY, (distilled epychlorohydrin)/(reacted dichlorohydrins); GLY, (glycidol produced)/(fed dichlorohydrins).

from a mixture of $\alpha\beta$- and $\alpha\gamma$-dichlorohydrins, the calculated concentration profiles along the column and yields to the main product and the by-product (glycidol) as a function of the liquid volume on the plate of a pilot column are shown in Figs. 22 and 23, respectively (Carrà et al., 1979b).

Finally, the application of stagewise models to the simulation of gas absorption in nonisothermal plate columns is worth mentioning. The model results have been verified accurately through comparison with experimental concentration and temperature profiles (including the column startup and shutdown transients) for the case of the air-ammonia-water systems in two different columns: one with five bubble cap plates and the other one with 11 sieve plates (Bourne et al., 1974a,b).

REACTOR CHOICE AND SCALEUP

Reactor selection for a given process is usually performed on the basis of previous experience or analogies with similar existing processes. In fact, especially in the first stages of the research, limited information is available; in gas-liquid reactors, several basic operations which could require different reactor types must often be performed simultaneously. Thus a rigorous selection procedure is not available, although some useful guidelines can be given. In Table 31 the characteristic values of the hydrodynamic and mass transfer parameters of various types of industrial reactors are reported. Such values, coupled with physical and kinetic information about the system under consideration, using the reactor models reported earlier, can lead to reactor selection. A procedure of this kind has been proposed by Nagel et al. (1978).

The main scope of a gas-liquid contactor is to produce suitably large values of the mass (and/or heat) transfer parameters k_g, k_ℓ, and a, using some kind of energy (gas sparging or mechanical agitation). Among these, only the mass transfer area can vary through several orders of magnitude depending on the choice of the engineering parameters, thus allowing the achievement of the desired reactor performance. Thus once the process requirements, in terms of mass transfer surface area, are known, the reactor choice can be made with the use of Fig. 24, where the surface area values attainable with various reactors, and the relative energy cost, are shown. The values in the figure are referred to the gas flow rate Q_g, since for a given reacting system, the depletion of the gaseous reactant concentration within the reactor is basically defined by the only parameter (a/Q_g) (Nagel et al., 1973, 1978).

In the case of mechanically agitated reactors, a useful procedure for the selection of commercially available turbine agitators has been presented by Hicks and Gates (1976). In Tables 32 and 33, the turbine prime-mover power and shaft speed, which uniquely identify a commercial turbine, are reported as a function of three parameters characterizing the gas-liquid dispersion process. These are the equivalent liquid volume $V_{eq} = (\rho_\ell/\rho_W)V_\ell$, the superficial gas velocity u_g°, and the scale of agitation index. The latter is evaluated using the gas-dispersion agitation scale reported in Table 34; where the index ranges from 1 to 10, and 0 indicates any condition producing a flooded impeller, never used in chemical processes. This procedure is valid for baffled reactors, designed according to the suggestions given earlier and equipped with a sparge ring and a six-flat-blade turbine whose prime-mover power is fully invested in the gassed conditions.

TABLE 31 Hydrodynamic and Mass Transfer Parameters Values in Industrial Reactors

Type of reactor	ε_ℓ	$10^4 k_g$ (mol/cm^2·s·atm)	$10^2 k_\ell$ (cm/s)	a (cm^{-1})	$10^2 k_\ell a$ (s^{-1})
Packed columns					
Countercurrent	2–25	0.03–2	0.4–2	0.1–3.5	0.04–7
Cocurrent	2–95	0.1–3	0.4–6	0.1–17	0.04–102
Plate columns					
Bubble cap	10–95	0.5–2	1–5	1–4	1–20
Sieve plates	10–95	0.5–6	1–20	1–2	1–40
Bubble columns	60–98	0.5–2	1–4	0.5–6	0.5–24
Packed bubble columns	60–98	0.5–2	1–4	0.5–3	0.5–12
Tube reactors					
Horizontal and coiled	5–95	0.5–4	1–10	0.5–7	0.5–70
Vertical	5–95	0.5–8	2–5	1–10	2–100
Spray columns	2–20	0.5–2	0.7–1.5	0.1–1	0.07–1.5
Mechanically agitated bubble reactors	20–95	—	0.3–4	1–20	0.3–80
Submerged and plunging jet	94–99	—	0.15–0.5	0.2–1.2	0.03–0.6
Hydrocyclone	70–93	—	10–30	0.2–0.5	2–15
Ejector reactor	—	—	—	1–20	—
Venturi	5–30	2–10	5–10	1.6–25	8–25

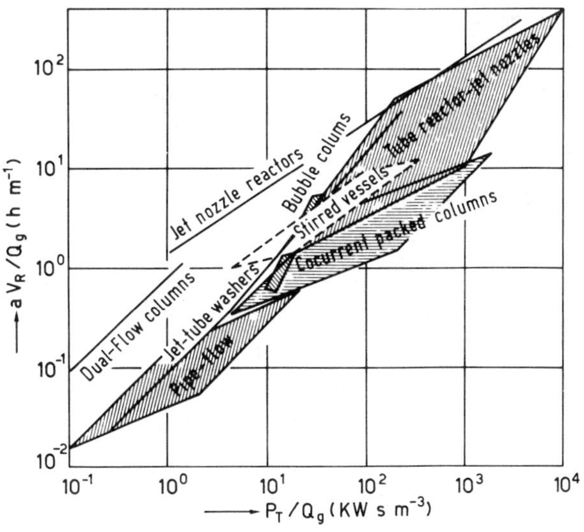

FIGURE 24 Approximate dependency of interfacial area on the power dissipation density for various types of reactors.

Therefore, for systems that are gassed at some times and ungassed at others, the agitator must be equipped with a two-speed motor to prevent dangerous overload when the gas flow ceases. Finally, the agitator diameter, expressed in meters, can be evaluated as

$$d_A = 8.128\left[\frac{Hp}{(P/P_0)N^3S_g}\right]^{1/5} \tag{117}$$

where Hp and N are the prime-mover power and the agitation speed reported in Tables 32 and 33 and expressed in hp and rpm, respectively; P/P_0 is evaluated as follows:

$$\ln\frac{P}{P_0} = \ln\left[1 + 235.86\left(\frac{Q_g}{Nd_A^3}\right)^{2.667}\right] - 9.155\left(\frac{Q_g}{Nd_A^3}\right)^{0.833} \tag{118}$$

In the case of multiple impeller systems the same procedure can be applied, as reported by Hicks and Gates (1976). It is worthwhile mentioning the alternative approach based on the Pfaudler agitation scale described by Barona (1979).

The reactor dimensioning can be performed using either a modeling or an experimental approach. In the first case, a first trial value of the geometric dimensions and the operating parameters (such as pressure, temperature, and feed stream characteristics) must be selected to initiate the procedure. The process specific parameters (i.e., physical and kinetic data of the system under examination) are then calculated or measured independently, and the self-adjusting parameters (relative to the hydrodynamics, mass,

TABLE 32 Prime-Mover Power (hp) and Agitator Speed (rpm) for Superficial Gas Velocity, $u_g^o = 0.021$ m/s

Scale of agitation	Equivalent volume, V_{eq} (gal)							
	750	1,500	3,000	5,000	10,000	20,000	50,000	75,000
1	1.5/125	2/56	5/84	7.5/68	15/56	30/45	100/56	150/45
				15/155	25/100	50/84	125/68	200/68
2	1.5/84	2/45	7.5/125	10/84	20/68	40/56	154/84	250/84
				7.5/45	14/45		200/125	
					30/125			
3	2/45	3/84	7.5/68	10/45	25/84	50/68	250/155	300/100
	2/84	3/68	5/45	10/56	25/56	75/125	125/45	350/125
	2/125	3/56	7.5/84		20/45	50/45		200/45
			5/56			50/56		
4	3/84	5/125	10/84	15/68	30/100	60/84	150/45	250/56
	3/68	5/84	10/100	20/100	40/155	75/100	200/68	300/68
		5/100	10/45	15/84	30/68	60/56	150/56	300/45
		5/45	10/56	20/68	40/84	75/68	200/45	
5	3/56	7.5/125	15/155	25/125	50/100	100/155	250/84	350/84
		7.5/155	15/68	25/84	50/68	100/100	250/56	400/100
		7.5/68	15/84	25/100	50/84	100/56		350/45
		7.5/84	15/45	25/56	50/45	100/68		400/56
6	5/125	10/84	20/100	30/155	60/125	125/125	300/100	500/68
	5/84	10/100	20/68	30/100	60/155	125/68	350/125	
	5/100		20/45	30/125	60/84	125/45	300/68	
				30/68	60/56		350/84	
7	7.5/155	10/56	25/125	40/155	75/190	150/155	400/100	600/84
			25/84	40/84	75/100	150/84	400/56	

TABLE 32 (Continued)

Scale of agitation	Equivalent volume, V_{eq} (gal)							
	750	1,500	3,000	5,000	10,000	20,000	50,000	75,000
7 (cont.)			25/100	40/100	75/125	150/100		
			25/56	40/56	75/68	150/45		
8	7.5/125	15/155	30/155	50/100	100/155	200/125	500/68	
	7.5/68	15/84	30/100	50/68	100/100	200/68		
	7.5/84		30/125	50/84	100/56	200/45		
				50/45	100/68			
9	10/100	15/68	30/68	60/125	125/125	250/155	600/84	
				60/155	125/68	250/84		
				60/84	125/45	250/56		
				60/56				
10	10/84	20/100	40/156	75/190	150/155	300/100		
	15/155	20/68	40/84	75/100	150/84	350/125		
	20/100			75/125	150/100	300/68		

TABLE 33 Prime-Mover Power (hp) and Agitator Speed (rpm) for Superficial Gas Velocity, $u_g^\circ = 0.061$ m/s

Scale of agitation	Equivalent volume, V_{eq} (gal)							
	750	1,500	3,000	5,000	10,000	20,000	50,000	75,000
1	1.5/84	3/56	7.5/68 10/100	10/45	30/68 60/155	50/45 60/56	150/45 250/84	
2	1.5/56	3/45	15/155	15/68 20/100	40/84	75/68 100/100	300/100 350/125	
3	2/84	5/100	10/84 7.5/45	25/125	75/190	125/125	200/56	
4	3/84 3/68	5/84 7.5/155 5/56	10/45 10/56	30/155 20/68 15/45 15/56	50/100 40/56 30/45	150/155 75/45	200/45	300/45
5	3/56	7.5/125 7.5/68 7.5/84	15/68 15/84 15/45 15/56	25/84 25/100 25/56	60/125 50/68 50/84 50/45	100/56 100/68	300/68	350/45 400/56
6	5/125 5/84 5/100	10/84 10/100	20/100 20/68	30/100 30/125 30/68 30/45	60/84 60/56 60/68	125/68 125/45	350/84 350/45 300/45	500/68
7	7.5/155	10/56	25/125 25/84 25/100 25/56	40/155 25/84 40/100 40/56	75/100 75/125 75/68 75/84	150/84 150/100 150/45 150/56	400/100 400/56	600/84

TABLE 33 (Continued)

Scale of agitation	Equivalent volume, V_{eq} (gal)							
	750	1,500	3,000	5,000	10,000	20,000	50,000	75,000
8	7.5/125 7.5/84	15/155 15/84	30/155 30/100 30/125	50/100 50/68 50/84 50/56	100/155 100/100 100/56 100/68	200/125 200/68 200/45	500/68	
9	10/100	15/68	30/68	60/125 60/155 60/84 60/56	125/125 125/68 125/45	250/155 250/84 250/56	600/84	
10	15/155	25/125 25/84	40/155 40/84	75/190 75/100 75/125	150/155 150/84 150/100	300/100 300/68		

TABLE 34 Scale of Agitation for Dispersing Gas into Liquids

Scale of agitation	Description
0	Agitation level of zero indicates a flooded impeller. Not recommended in CPI service.
1–2	Agitation levels 1–2 characterize applications in which the degree of gas dispersion is not critical. Agitators capable of scale level 2 will: Provide nonflooded impeller conditions for coarse dispersion of gas in the liquid system. Be typical of applications that are not mass transfer limited.
3–5	Agitation levels 3–5 characterize applications in which the degree of gas dispersion is moderate. Agitators capable of scale level 5 will: Drive fine bubbles completely to vessel wall. Provide recirculation of dispersed bubbles back into impeller.
6–10	Agitation levels 6–10 characterize critical gas-liquid reactors where rapid mass transfer is required. Agitators capable of scale level 10 will: Maximize interfacial area and recirculation of dispersed bubbles back into impeller.

and heat transfer) are calculated as shown earlier. All the values so obtained are fed to the suitable mathematical model, whose solution leads to the estimation of the reactor performance and then to its dimensioning and optimization (Deckwer, 1977). The most uncertain steps of this procedure concern the correct evaluation of the self-adjusting parameters, particularly for bubble columns and mechanically stirred contactors, and the related problem of the correct reactor model selection. Also, the complete physical (solubility and diffusivity of the solute in the reacting mixture) and kinetic (rate constant of all the involved reactions) characterization of the system can be difficult and uncertain.

An experimental procedure can overcome these difficulties, leading to more accurate results. In this case the equations reported earlier can be very useful to define the kind of dependence among the various parameters. Small-scale experiments are performed on the same gas-liquid system using a laboratory apparatus similar in shape, agitation, and contact time to the chosen industrial reactor. The scaling rule, which ensures the same specific interfacial area, is to maintain constant the total power input per unit liquid volume; that is, P_t/V_ℓ and u_g^f are constant in a mechanically agitated contactor, and $\rho_\ell g u_g$ is constant in a tubular sparged reactor (Reith, 1970). Note that using this scaleup criterion, the heat transfer coefficient h_w has roughly the same values in the model and in the prototype, since it decreases for increasing reactor diameter as $d_R^{-1/9}$. Therefore, on scaleup of agitated reactors with significant heat transfer, installation of

additional heating or cooling devices is usually necessary, since the contents increase with the cube and the heat exchange with the square of the reactor diameter (Steiff et al., 1981). Since the interfacial area (and then the reactor performance) increases for increasing values of the power input per unit liquid volume, and the operating costs are directly proportional to the power input, the reactor volume selection should be based on the minimum total cost.

A useful scaleup procedure, based on constant depletion of the reactant gaseous concentration, has been proposed by Nagel et al. (1973, 1978). It allows, through the use of suitable working diagrams, defining the optimal reactor volume using as a scaleup criterion a constant value of the quantity a/Q_g. An alternative scaleup procedure for bubble columns has been proposed by Kastanek et al. (1980).

An alternative experimental approach is the simulation of industrial units using laboratory-scale apparatuses, with the same gas-liquid system (so avoiding determination of kinetic, solubility, and diffusivity parameters) and a different mean of liquid agitation. It is based on the observation that if the purely physical mass transfer coefficients in the two units are equal, the absorption rates per unit interfacial area, also in the presence of the reaction, are identical. This fact appears quite reasonable noticing that the predicted effect of chemical reaction on the rate of absorption differs little whether the film or surface renewal-type models are used as a basis for calculations, provided that the same value of the purely physical mass transfer coefficient k_ℓ is assumed in all models (Danckwerts, 1970). When the absorption process involves a dilute solute in the gas phase (i.e., significant gas-phase transfer resistance is present), the gaseous mass transfer coefficient k_g must be the same in both units. Finally, for slow reactions that do not completely occur in the liquid film, the ratio a/ε_ℓ must be the same in both units, to account for the liquid bulk reaction contribution. So, in general, the three criteria for simulation (i.e., to reproduce the same absorption rate per unit interfacial area) are identical values of k_ℓ, k_g, and a/ε_ℓ in the industrial and laboratory absorber.

The procedure can then be outlined as follows:

1. Measurement or calculation of k_ℓ, k_g, and a/ε_ℓ in the industrial unit
2. Reproduction of the same values in the laboratory unit
3. Measurement of the interfacial mass flux in the laboratory unit, which is the same in the industrial unit

To accomplish the last two steps, it is necessary to use laboratory-scale absorbers where the interfacial area is known with good accuracy and the values of the three parameters k_ℓ, k_g, and a/ε_ℓ can be varied over a reasonably wide range of values. The first one is usually varied by mechanical liquid agitation, the second one by independent mechanical gas agitation, or simply by changing the relative gas-liquid velocity. The control of the third parameter requires more complex apparatus, such as a string of spheres with pools at the top (Alper and Danckwerts, 1976) or the modified stirred cell proposed by Levenspiel and Godfrey (1974). However, few applications of the third criterion are known in the literature, since usually only fast reactions have been examplined. In Table 35 and Fig. 25, the most important types of laboratory equipment have been summarized together with the relative values of the

TABLE 35 Values of the Characteristic Parameters of the Laborabory Models Sketched in Fig. 25

	Model							
		Wetted wall column						
Parameter	Laminar jet	Cylinder	Sphere	Cone	String of disks	Stirred cell	Rotating drum	Modified stirred cell
Contact time (s)	10^{-3}	10^{-1}	10^{-1}	2×10^{-1}	10^{-1}	6×10^{-2}	2×10^{-4}	2×10^{-2}
	10^{-1}	2	1	1	2	10	10^{-1}	10
Interfacial area, A (cm^2)	0.3	10	10		30	4	2	2
	1.0	100	40	80	360	80	100	80
k_ℓ ($cm \cdot s^{-1}$)	1.6×10^{-2}	3.6×10^{-3}	5×10^{-3}	5×10^{-3}	3.6×10^{-3}	1.6×10^{-3}	1.6×10^{-2}	2×10^{-3}
	1.6×10^{-1}	1.6×10^{-2}	1.6×10^{-2}	1.1×10^{-2}	1.6×10^{-2}	2.1×10^{-2}	3.56×10^{-1}	2×10^{-2}
$10^5 k_g$ ($mol \cdot cm^{-2}\ s^{-1}\ atm^{-1}$)	10	1	1	1	1	1	—	1
	40	9	9	9	25	15	—	15
A/V_ℓ (cm^{-1})	20	25	20	40	20	2×10^{-3}	100	2×10^{-3}
	80	60	60	70	60	5.4×10^{-1}	1250	10^{-1}

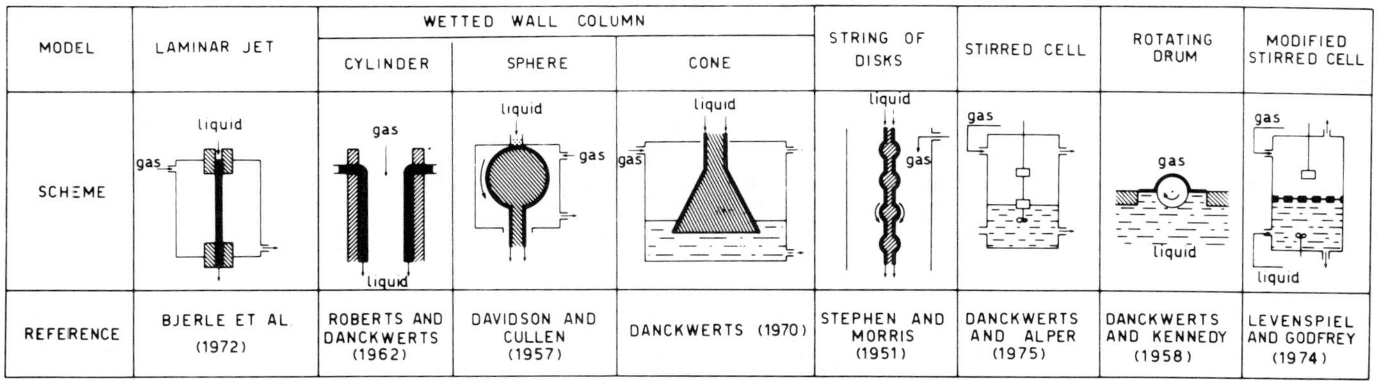

FIGURE 25 Schematic representation of some laboratory equipments.

relevant parameters. These, coupled with Table 31, can lead to the choice of the most suitable laboratory unit for the simulation of a given industrial unit (Charpentier and Laurent, 1974). Note that this equipment is usually employed for the determination of kinetic rate constants, diffusivities, and solubilities (Lauret et al., 1975).

A classic example of the procedure described above is the simulation of a packed column with a stirred cell for reactions occurring in liquid film (Danckwerts and Gillham, 1966; Charpentier, 1978; and references therein). In this case a mutual correspondence can be determined between the agitator speeds in the gas and liquid phases in the model unit (which define k_g and k_ℓ, respectively) and the gas and liquid superficial velocities in the packed column, respectively.

The interfacial mass flow values measured in the laboratory unit can be used to perform a differential or an integral simulation of an industrial tubular reactor. In the first case, the material flux is evaluated at the various concentration values present along the column, and reactor behavior is then predicted through a numerical integration (Danckwerts and Alper, 1975; Laurent and Charpentier, 1977; and references therein). In the case of multiple liquid or gaseous reactants or when the reaction contribution in the liquid bulk is significant, the differential simulation is not feasible, and only the integral one can be pursued. In this case the column is simulated as a whole, using the laboratory model unit described above. The global absorption rate in the model is measured at the same operating conditions, including the outlet and inlet concentration values, at which the industrial unit is operated. The global absorption rate in the industrial unit is then given by the product of the global absorption rate in the laboratory unit and the ratio between the liquid flow rates in the industrial and model units, respectively (Alper and Danckwerts, 1976). It is noticeable that both these types of simulation procedures exhibit, in all literature applications (Laurent and Charpentier, 1977), errors that never exceed 20%.

NOTATION

a	interfacial area per unit reactor volume
a_c	heat exchange surface per unit reactor volume
a_p	interfacial area per unit packed volume
a_t	total dry surface area of packing per unit packed volume
A, A'	gaseous reactant
B	liquid reactant
C	liquid reactant
C_i	concentration of the ith component
C_p	specific heat
d_A, d_c, d_R	diameter of agitator, columns, reactor
d_p, d_n	characteristic and nominal packing diameter
d_{vs}	Sauter mean bubble diameter, defined by Eq. (45)
D	diffusion coefficient

E, E*	enhancement and reaction factor, defined by Eqs. (27) and (28), respectively
E_g, E_ℓ	axial effective mass dispersion in the gas and liquid phases
E_{in}	reaction factor for an instantaneous reaction
E_M, E_{MT}	Murphree efficiency for mass and heat transfer
E_T	axial effective heat dispersion
f	fugacity
f	r/r°
F	feed molar flow rate
g	gravitational acceleration
G	gas molar flow rate
h_A	agitator height off the reactor bottom
h_c, h_w	coil and wall bed heat transfer coefficients
h_ℓ	gas-liquid heat transfer coefficient
H	interfacial equilibrium ratio, defined by Eq. (21a)
H	liquid height
H*	dimensionless Henry's constant, defined by Eq. (56)
$\tilde{H}$	molar enthalpy
$\mathcal{H}$	Henry's constant
Ha	Hatta number, defined by Eq. (52)
I	solution ionic strength, defined by Eq. (12)
J	dimensionless mass flux
k	reaction rate constant
k_B	Boltzmann's constant
k_g, k_ℓ	gas-side and liquid-side mass transfer coefficients
k_T	thermal conductivity
K	equilibrium ratio, defined by Eq. (1)
K_g	overall mass transfer coefficient, defined by Eq. (22)
L	reactor length
L	liquid molar flow rate
m	reaction order
m_i	molality of the ith component
M	Hatta number, defined by Eq. (84)
n	reaction order
n*	flow index; 1 = countercurrent, −1 = cocurrent
N	agitation speed

N_i	mass flux of the ith component (positive if leaving the gas phase)
N_p	power number
NC	component number
NR	reaction number
p_j	r_j°/r_1
p°	vapor pressure
P	pressure
P_0	power input in unaerated reactor
P_t, P_g	total and mechanical power input
Pe	Péclet number, defined by Eq. (62)
Q	volumetric flow rate
r	reaction rate
R	universal gas constant
Re	Reynolds number
$\mathcal{R}_i$	overall rate of consumption of the ith component
s	distance into the liquid phase
s_c	wall or coil width
S_g	specific gravity
St	Stanton number, defined by Eq. (62)
t	time
T	temperature
u	superficial velocity
U	overall heat transfer coefficient
v	dimensionless superficial velocity
V	volume
V_R	reactor volume
w_A	agitator blade width
x	s/δ
x_i	mole fraction of the ith component in the liquid phase
y	axial coordinate
y_1	mole fraction of the ith component in the gas phase
z	y/L
z_i	mole fraction of the ith component in the feed stream
Z_D	concentration diffusion parameter, defined by Eq. (36)

Greek Letters

α	ratio of the bulk liquid volume to film volume, defined by Eq. (62)
$\bar{\alpha}$	dimensionless parameter defined by Eq. (67)
γ	activity coefficient
γ	dimensionless parameter defined by Eq. (56)
δ	film thickness
$\Delta\tilde{H}_a$, $\Delta\tilde{H}_{ev}$, $\Delta\tilde{H}_R$	molar heat of absorption, evaporation and reaction (negative if exothermic)
ε_g, ε_ℓ	fractional gas and liquid holdup
ε_p	packing void fraction
θ	dimensionless time, defined by Eq. (71)
μ	viscosity
ν_B	stoichiometric coefficient of reactant B, positive
ν_{ij}	stoichiometric coefficient of the ith component in the jth reaction
ξ	D/D_A
ρ	density
σ	surface tension
ϕ	dimensionless concentration
ϕ	fugacity coefficient

Special Operators

∇	gradient operator
∇^2	Laplacian operator

Subscripts

d	dispersion
f	film
g	gas
i	component index
j	reaction index
k	stage index
ℓ	liquid
s	solid
w	water
~	vector

Superscripts

f feed

in initial

° reference condition (feed for continuous and initial for batch reactors)

~ molar

* surface

– partial molar

REFERENCES

Agnello, L. A. and W. H. Williams, Synthetic phenol, Ind. Eng. Chem., *52* (11), 894 (1960).

Akita, K. and F. Yoshida, Gas holdup and volumetric mass transfer coefficient in bubble columns, Ind. Eng. Chem. Process Des. Dev., *12*, 76 (1973).

Akita, K. and F. Yoshida, Bubble size, interfacial area, and liquid phase mass transfer coefficient in bubble columns, Ind. Eng. Chem. Process Des. Dev., *13*, 84 (1974).

Alper, E. and P. V. Danckwerts, Laboratory scale-model of a complete packed column absorber, Chem. Eng. Sci., *31*, 599 (1976).

Aluko, M. and H. C. Chang, Multiplicity, uniqueness and stability for an exothermic reaction in a non-adiabatic bubble column reactor, Chem. Eng. J., *24*, 151 (1982).

Ashley, M. J. and G. G. Haselden, Effectiveness of vapor-liquid contacting on a sieve plate, Trans. Inst. Chem. Eng., *50*, T119 (1972).

Astarita, G., *Mass Transfer with Chemical Reaction*, Elsevier, New York (1967).

Astarita, G. and D. W. Savage, Simultaneous absorption with reversible instantaneous chemical reaction, Chem. Eng. Sci., *37*, 677 (1982).

Astarita, G., G. Marrucci, and L. Coleti, Ossidazione catalitica del solfito di sodio: un metodo per la misura di aree interfaciali, Chim. Ind., *46*, 1021 (1964).

Bakos, M. and J. C. Charpentier, Taux de rétention pour des écoulements gaz-liquide à co-courrant vers le bas dans les colonnes à garnissage arrosé et noyé, Chem. Eng. Sci., *25*, 1822 (1970).

Bang, Y. V. and A. Cholette, Comparison du Modèle des réservoirs imparfaitement agités en série avec le modèle de dispersion axiale, Can. J. Chem. Eng., *51*, 149 (1973).

Baratella, P., S. Carrà, G. Giardini, and R. S. Santi, Multistage distillation with chemical reaction, Quad. Ing. Chim. Ital., *10*, 1 (1974).

Barona, N., Design CSTR reactors this way, Hydrocarbon Process., *58*(7), 179 (1979).

Barona, N. and H. W. Prengle, Jr., Design reactors this way for liquid-phase processes: Part 1. Types, design factors, selection and scale-up, Hydrocarbon Process., *52*(3), 63 (1973).

Bates, R. L., P. L. Fondy, and J. G. Fenic, Impeller characteristics and power, in *Mixing*, Vol. 1, V. W. Uhl and J. B Gray, eds., Academic Press, New York (1966), p. 112.

Beimesch, W. E. and D. P. Kessler, Liquid-gas distribution measurements in the pulsing regime of two-phase concurrent flow in packed beds, AIChE J., *17*, 1160 (1971).

Berghmans, J., Stability of gas bubbles rising in inviscid fluids, Chem. Eng. Sci., *28*, 2005 (1973).

Bernard, J. D. T. and R. W. H. Sargent, Hydrodinamic performance of a sieve-plate distillation column, Trans. Inst. Chem. Eng., *44*, T314 (1966).

Bjerle, I., S. Bengtsson, and K. Färnkvist, Absorption of SO_2 in $CaCO_3$-slurry in a laminar jet absorber, Chem. Eng. Sci., *27*, 1853 (1972).

Botton, R., D. Cosserat, and J. C. Charpentier, Operating zone and scale-up of mechanically stirred gas-liquid reactors, Chem. Eng. Sci., *35*, 82 (1980).

Bourne, J. R., U. V. Stockar, and G. C. Coggan, Gas absorption with heat effects: 1. A new computational method, Ind. Eng. Chem. Process Des. Dev., *13*, 115 (1974a).

Bourne, J. R., U. V. Stockar and G. C. Coggan, Gas absorption with heat effects: 2. Experimental results, Ind. Eng. Chem. Process Des. Dev., *13*, 124 (1974b).

Brian, P. L. T. and M. C. Beaverstock, Gas absorption accompanied by a two-step chemical reaction, Chem. Eng. Sci., *20*, 47 (1965).

Buchanan, J. E., Holdup in irrigated ring-packed towers below the loading point, Ind. Eng. Chem. Fundam., *6*, 400 (1967).

Burgess, J. M. and P. H. Calderbank, The measurement of bubble parameters in two-phase dispersions: I. The development of an improved probe technique, Chem. Eng. Sci., *30*, 743 (1975a).

Burgess, J. M. and P. H. Calderbank, The measurement of bubble parameters in two-phase dispersions: II. The structure of sieve tray froths, Chem. Eng. Sci., *30*, 1107 (1975b).

Calderbank, P. H., Physical rate process in industrial fermentation: Part I. The interfacial area in gas-liquid contacting with mechanical agitation, Trans. Inst. Chem. Eng., *36*, 443 (1958).

Calderbank, P. H., Mass transfer, in *Mixing*, Vol. 2, V. W. Uhl and J. B. Gray, eds., Academic Press, New York (1967), p. 2.

Calderbank, P. H. and M. B. Moo-Young, The continuous phase heat and mass transfer properties of dispersions, Chem. Eng. Sci., *16*, 39 (1961).

Calderbank, P. H. and J. Rennie, The physical properties of foams and froths formed on sieve-plates, Trans. Inst. Chem. Eng., *40*, 3 (1962).

Carleton, A. J., R. J. Flain, J. Rennie, and F. H. H. Valentin, Some properties of a packed bubble column, Chem. Eng. Sci., *22*, 1839 (1967).

Carmichael, G. R. and S. C. Chang, Mass Transfer accompanied by equilibrium and second-order irreversible reactions, Chem. Eng. Sci., *35*, 2459 (1980).

Carrà, S. and E. Santacesaria, Engineering aspects of gas-liquid catalytic reactions, Catal. Rev. Sci. Eng., *22*, 75 (1980).

Carrà, S., E. Santacesaria, M. Morbidelli, P. Schwarz, and C. Divo, Synthesis of epichlorohydrin by elimination of hydrogen chloride from chlorhydrins: 1. Kinetic aspects of the process, Ind. Eng. Chem. Process Des. Dev., *18*, 424 (1979a).

Carrà, S., E. Santacesaria, M. Morbidelli, P. Schwarz, and C. Divo, Synthesis of epichlorohydrin by elimination of hydrogen chloride from chlorohydrins: 2. Simulation of the reaction unit, Ind. Eng. Chem. Process. Des. Dev., *18*, 428 (1979b).

Carrà, S. E., Santacesaria, M. Morbidelli, and L. Cavalli, Synthesis of propylene oxide from propylene chlorhydrins: 1. Kientic aspects of the process, Chem. Eng. Sci., *34*, 1123 (1979c).

Carrà, S., M. Morbidelli, E. Santacesaria, and G. Buzzi, Synthesis of propylene oxide from propylene chloroydrins; 2. Modelling of the distillation with chemical reaction unit, Chem. Eng. Sci., *34*, 1133 (1979d).

Chandhari, R. V., B. D. Kulkarni, and L. K. Doraiswamy, Simultaneous absorption of two gases accompanied by a complex chemical reaction: approximate solutions, Chem. Eng. Sci., *30*, 945 (1975).

Chandrasekharan, K., and P. H. Calderbank, Further observations on the scale-up of aerated mixing vessels, Chem. Eng. Sci., *36*, 819 (1981).

Charpentier, J. C., Recent progress in two phase gas-liquid mass transfer in packed beds, Chem. Eng. J., *11*, 161 (1976).

Charpentier, J. C., Gas-liquid reactors, *Chemical Reaction Engineering Reviews*, D. Luss and V. W. Weekman, eds., ACS Symp. Ser., *72*, 223 (1978).

Charpentier, J. C., Mass transfer rates in gas-liquid absorbers and reactors, Adv. Chem. Eng., *11*, 1 (1981).

Charpentier, J. C. and M. Favier, Some liquid holdup experimental data in trickle-bed reactors for foaming and nonfoaming hydrocarbons, AIChE J., *21*, 1213 (1975).

Charpentier, J. C. and A. Laurent, Some considerations and recalls on the design of gas absorbers: laboratory apparatus to simulate absorber and to predict the effect of chemical reaction, AIChE J., *20*, 1029 (1974).

Charpentier, J. C., C. Prost, and P. Le Goff, Chute de pression pour des écoulements a co-courant dans les colonnes à garnissage arrosé: Comparison avec le garnissage Noyé, Chem. Eng. Sci., *24*, 1777 (1969).

Chen, B. H. and A. F. McMillian, Heat transfer and axial dispersion in batch bubble columns, Can. J. Chem. Eng., *60*, 436 (1982).

Chen, B. H. and R. Vallabh, Holdup and mass transfer in bubble columns containing screen cylinders, Ind. Eng. Chem. Process Des. Dev., *9*, 121 (1970).

Chiang, S. H. and H. L. Toor, Gas absorption accompanied by a large heat effect and volume change of the liquid phase, AIChE J., *10*, 398 (1964).

Cichy, P. T. and T. W. F. Russell, Two-phase reactor design: tubular reactors. Reactor model parameters, Ind. Eng. Chem., *61*(8), 15 (1969).

Cichy, P. T., J. S. Ultman, and T. W. F. Russell, Two-phase reactor design: tubular reactors. Reactor model development, Ind. Eng. Chem., *61*(8), 6 (1969).

Clegg, G. T. and R. Mann, A penetration model for gas absorbtion with first order chemical reaction accompanied by large heat effects, Chem. Eng. Sci., *24*, 321 (1969).

Clift, R., R. J. Grace, and M. E. Weber, *Bubbles, Drops, and Particles*, Academic Press, New York (1978), Chap. 7.

Co, P. and R. Bibaud, Longitudinal mixing of the liquid phase in packed columns with countercurrent two-phase flow, Can. J. Chem. Eng., *49*, 727 (1971).

Danckwerts, P. V., Temperature effects accompanying the absorption of gases in liquids, Appl. Sci. Res., *A3*, 385 (1952).

Danckwerts, P. V., Gas-absorption accompanied by first-order reaction: concentration of product, temperature-rise and depletion of reactant, Chem. Eng. Sci., *22*, 472 (1967).

Danckwerts, P. V., *Gas Liquid Reactions*, Mc-Graw Hill, New York (1970).

Danckwerts, P. V. and E. Alper, Design of gas absorbers: Part III. Laboratory "point" model of a packed column absorber, Trans. Inst. Chem. Eng., *53*, 34 (1975).

Danckwerts, P. V. and A. J. Gillham, The design of gas absorbers: 1. Method for predicting rates of absorption with chemical reaction in packed columns and tests with 1.5 in Rashig rings, Trans. Inst. Chem. Eng., *44*, T42 (1966).

Danckwerts, P. V. and A. M. Kennedy, The kinetics of absorption of carbon dioxide into neutral and alkaline solutions, Chem. Eng. Sci., *8*, 201 (1958).

Danckwerts, P. V. and M. M. Sharma, Absorption of carbon dioxide into solutions of alkalies and amines. Hydrogen sulfide and carbonyl sulfide, Chem. Eng. (London), CE244 (1966).

Davidson, J. F. and E. J. Cullen, The determination of diffusion coefficients for sparingly soluble gas in liquids, Trans. Inst. Chem. Eng., *35*, 51 (1957).

Deans, H. A., A mathematical model for disperison in the direction of flow in porous media, Soc. Petrol. Eng. J., *3*, 49 (1963).

Deans, H. A. and L. Lapidus, A computational model for predicting and correlating the behavior of fixed-bed reactors: I. Derivation of model for nonreactive systems, AIChE J., *6*, 656 (1960a).

Deans, H. A. and L. Lapidus, A computational model for predicting and correlating the behavior of fixed-bed reactors: II. Extension to chemically reactive systems, AIChE J., *6*, 663 (1960b).

Deckwer, W. D., Nonisobaric bubble columns with variable gas velocity, Chem. Eng. Sci., *31*, 309 (1976).

Deckwer, W. D., Absorption and reaction of isobutene in sulfuric acid: III. Considerations on the scale-up of bubble columns, Chem. Eng. Sci., *32*, 51 (1977).

Deckwer, W. D., Bubble column reactors: their modeling and dimensioning, Chem.-Ing.-Tech., *49*, 212 (1977), or Int. Chem. Eng., *19*, 21 (1979).

Deckwer, W. D., On the mechanism of heat transfer in bubble column reactors, Chem. Eng. Sci., *35*, 1341 (1980).

Deckwer, W. D., R. Burckhart, and G. Zoll, Mixing and mass transfer in tall bubble columns, Chem. Eng. Sci., *29*, 2177 (1974).

Deckwer, W. D., I. Adler, and A. Zaidi, A comprehensive study of CO_2 interphase mass transfer in vertical cocurrent and countercurrent gas-liquid flow, Can. J. Chem. Eng., *56*, 43 (1978).

Deckwer, W. D., K. Nguyen-Tien, A. Schumpe, and Y. Serpemen, Oxygen mass tranfer into aerated CMC solutions in a bubble column, Biotechnol. Bioeng., *24*, 461 (1982).

De Maria, K. and R. R. White, Transient response study of gas flowing through irrigated packing, AIChE J., *6*, 473 (1960).

De Waal, K. J. A., and J. C. Okeson, The oxidation of aqueous sodium sulphite solutions, Chem. Eng. Sci., *21*, 559 (1966).

Diboun, M. and K. Schügerl, Eine Blasensaule mit Gleichstrom von Wasser und Leift: I. Mischungsvorgänge in der Gasphase, Chem. Eng. Sci., *22*, 147 (1967)

Ding, S. Y., S. Sharma, and D. Luss, Steady-state multiplicity and control of the chlorination of liquid n-dcane in an adiabatic continuously stirred tank reactor, Ind. Eng. Chem. Fundam., *13*, 76 (1974).

Dodds, W. S., L. F. Stutzman, B. J. Sollami, and R. J. McCarter, Concurrent gas absorption mass transfer, AIChE J., *6*, 197 (1960).

Dombrowski, H. S. and L. E. Brownell, Residual equilibrium saturation of porous media, Ind. Eng. Chem., *46*, 1207 (1954).

Dunn, W. E., T. Vermeulen, C. R. Wilke, and T. T. Word, Longitudinal dispersion in packed gas-absorption columns, Ind. Eng. Chem. Fundam., *16*, 116 (1977).

Eissa, S. H. and K. Schügerl, Holdup and backmixing investigations in cocurrent and countercurrent bubble columns, Chem. Eng. Sci., *30*, 1251 (1975).

Elstner, F. and U. Onken, Effect of liquid phase properties on mass transfer in gas-liquid dispersions, Ger. Chem. Eng., *4*, 84 (1981).

Fair, J. R., Designing gas-sparged reactors, I, Chem. Eng., 67 (July 1967a).

Fair, J. R., Designing gas-sparged reactors, II, Chem. Eng., 207 (July 1967b).

Fair, J. R., A. J. Lambright, and J. W. Andersen, Heat transfer and gas holdup in a sparged contactor, Ind. Eng. Chem. Process. Des. Dev., *1*, 33 (1962).

Fair, J. R., D. E. Steinmeyer, W. R. Penney, and J. A. Brink, Liquid-gas systems, in *Chemical Engineers' Handbook*, 5th ed., R. H. Perry and C. H. Chilton, eds., McGraw-Hill, New York (1973), Chap. 18.

Feintuch, H. M. and R. E. Treybal, The design of adiabatic packed towers for gas absorption and stripping, Ind. Eng. Chem. Process. Des. Dev., *17*, 505 (1978).

Field, R. W. and J. F. Davidson, Axial dispersion in bubble columns, Trans. Inst. Chem. Eng., *58*, 228 (1980).

Finlayson, B. A., *The Method of Weighted Residuals and Variational Principles*, Academic Press, New York (1972).

Friedel, L., P. Herbrechtsmeier, and R. Steiner, Mean gas holdup in downflow bubble columns, Ger. Chem. Eng., *3*, 342 (1980).

Geddes, R. L., Local efficiencies of bubble plate fractionators, AIChE Trans. *42*, 79 (1946).

Gianetto, A., G. Baldi, and V. Specchia, Absorption in packed towers with cocurrent downward high-velocity flows: I. Interfacial areas, Quad. Ing. Chim., *6*, 125 (1970).

Gordon, A. R., The diffusion constant of an electrolyte, and its relation to concentration, J. Chem. Phys., *5*, 522 (1937).

Gray, D. J., R. E. Treybal, and S. M. Barnett, Mixing of single and two phase systems: power consumption of impellers, AIChE J., *28*, 195 (1982).

Gregory, G. A. and D. S. Scott, in *Cocurrent Gas-Liquid Flow*, E. Rhodes and D. S. Scott, eds., Plenum Press, New York (1969).

Groggins, P. H., *Unit Processes in Organic Synthesis*, McGraw-Hill, New York (1958), Chap. 6.

Hancock, M. D. and C. N. Kenney, Instabilities in two-phase reactors, Proc. 5th Eur./2nd Int. Symp. Chem. React. Eng., Elsevier, Amsterdam (1972), p. B.3–47.

Hancock, M. D. and C. N. Kenney, The stability and dynamics of a gas-liquid reactor, Chem. Eng. Sci., *32*, 629 (1977).

Hashimoto, K., M. Teramoto, T. Nagayasu, and S. Nagata, The effects of mass transfer on the selectivity of gas-liquid reactions, J. Chem. Eng. Jpn., *1*, 132 (1968).

Hassan, I. T. M., and C. W. Robinson, Stirred-tank mechanical power requirement and gas holdup in aerated aqueous phases, AIChE J., *23*, 48 (1977).

Hayduk, W. and H. Laudie, Prediction of diffusion coefficients for nonelectrolytes in dilute aqueous solutions, AIChE J., *20*, 611 (1974).

Herbrechtsmeier, P. and R. Steiner, Untersuchungen an einem Blasensäulen-Abstromreaktor, Chem. -Ing.-Tech., *50*, 944 (1978).

Hicks, R. W. and L. E. Gates, How to select turbine agitators for dispersing gas into liquids, Chem. Eng., p. 141 (July, 19, 1976).

Hikita, H. and S. Asai, Gas absorption with (m,n)-th order irreversible chemical reaction, Kagaku Kogaku, *11*, 823 (1963), or Int. Chem. Eng., *4*, 332 (1964).

Hikita, H., S. Asai, and H. Ishikawa, Simultaneous absorption of two gases which react between themselves in a liquid, Ind. Eng. Chem. Fundam., *16*, 215 (1977).

Hikita, H., S. Asai, K. Tanigawa, K. Segawa, and M. Kitao, Gas holdup in bubble column, Chem. Eng. J., *20*, 59 (1980).

Hikita, H., S. Asai, H. Kikukawa, T. Zalke, and M. Ohue, Heat transfer coefficient in bubble columns, Ind. Eng. Chem. Process Des. Dev., *20*, 540 (1981a).

Hikita, H., S. Asai, K. Tanigawa, K. Segawa, and M. Kitao, The volumetric liquid phase mass transfer coefficient in bubble columns, Chem. Eng. J., *22*, 61 (1981b).

Hildebrand, J. H., J. M. Prausnitz, and R. L. Scott, *Regular and Related Solutions*, D. Van Nostrand, Princeton, N.J. (1970).

Hlaváček, V., and H. Hofmann, Modeling of chemical reactors: XVI. Steady state axial heat and mass transfer in tubular reactors. An analysis of the uniqueness of solutions, Chem. Eng. Sci., *25*, 173 (1970).

Ho, S. P. and S. T. Lee, An analysis of complex gas-liquid reactions in an isothermal CSTR, Chem. Eng. Sci., *35*, 1139 (1980).

Hochman, J. M., and E. Effron, Two-phase concurrent downflow in packed beds, Ind. Eng. Chem. Fundam., *8*, 63 (1969).

Höcker, H., G. Langer, and U. Werner, Mass transfer in aerated Newtonian and non-Newtonian liquids in stirred reactors, Ger. Chem. Eng., *4*, 51 (1981).

Hoffman, L. A., S. Sharma, and D. Luss, Steady state multiplicity of adiabatic gas-liquid reactors: I. The single reaction case, AIChE J., *21*, 318 (1975).

Holland, C. D., *Fundamentals of Multicomponent Distillation*, McGraw-Hill, New York (1981), Chap. 8.

Huang, D. T. J. and A. Varma, Steady state and dynamic behavior of fast gas-liquid reactions in non-adiabatic CSTRs, Chem. Eng. J., *21*, 47 (1981b).

Huang, D. T. J. and A. Varma, Steady state multiplicity of a non-adiabatic bubble column with fast reactions, AIChE J., *27*, 111 (1981c).

Huang, D. T. J. and A. Varma, Steady state uniqueness and multiplicity of non-adiabatic gas-liquid CSTRs: Part I. Second-order reaction model, AIChE J., *27*, 481 (1981d).

Huang, D. T. J. and A. Varma, Steady state uniqueness and multiplicity of non-adiabatic gas-liquid CSTRs: Part II. Discrimination among rival reaction models, AIChE J., *27*, 489 (1981e).

Huang, D. T. J., J. J. Carberry, and A. Varma, Gas absorption with consecutive second-order reactions, AIChE J., *26*, 832 (1980).

Hughmark, G. A., Holdup and mass transfer in bubble columns, Ind. Eng. Chem. Process Des. Dev., *6*, 218 (1967).

Hughmark, G. A., Power requirements and interfacial area in gas-liquid turbine agitated systems, Ind. Eng. Chem. Process. Des. Dev., *19*, 638 (1980).

Hughmark, G. A., personal communication (1982).

Hunek, J., P. Földes, and J. Sawinsky, Methods for calculating reactive distillation, Mag. Kem. Lapja, *31*, 21 (1976), or Int. Chem. Eng., *19*, 248 (1979).

Hwa, C. S. and R. B. Beckmann, Radiological study of liquid holdup and flow distributions in packed gas-absorption columns, AIChE J., *6*, 359 (1960).

Jhaveri, A. S. and M. M. Sharma, Kinetics of absorption on oxygen in aqueous solutions of cuprous chloride, Chem. Eng. Sci., *22*, 1 (1967).

Jhaveri, A. S. and M. M. Sharma, Absorption of oxygen in aqueous alkaline solutions of sodium dithionite, Chem. Eng. Sci., *23*, 1 (1968).

John, O. G., Der Einfluss der Energiedichte auf den Stoffumsatz in Kontinuierlich Betriebenen, Begasten Rührbehältern, Chem.-Ing.-Tech., *43*, 342 (1971).

Joshi, J. B., Axial mixing in multiphase contactors. A unified correlation, Trans. Inst. Chem. Eng., *58*, 155 (1980).

Joshi, J. B. and M. M. Sharma, Mass transfer characteristics of horizontal sparged contactors, Trans. Inst. Chem. Eng., *54*, 42 (1976).

Joshi, J. B. and M. M. Sharma, Liquid phase backmixing in sparged contactors, Can. J. Chem. Eng., *56*, 116 (1978).

Joshi, J. B., A. B. Pandit, and M. M. Sharma, Mechanically agitated gas-liquid reactors, Chem. Eng. Sci., *37*, 813 (1982).

Jost, W., *Diffusion in Solids, Liquids and Gases*, Academic Press, New York (1952).

Karlsson, H. T. and I. Bjerle, A general explicit enhancement factor for irreversible gas-liquid reactions with gas-side resistance, Chem. Eng. Sci., *35*, 1005 (1980).

Kastanek, F., J. Zahradnik, M. Rylek, and J. Kratochvil, Scaling up of bubble reactors on basis of laboratory data, Chem. Eng. Sci., *35*, 456 (1980).

Kern, D. Q., *Process Heat Transfer*, McGraw-Hill Kogakusha, Tokyo (1950).

King, C. J., The additivity of individual phase resistances in mass transfer operations, AIChE J., *10*, 671 (1964).

Kölbel, H., E. Borchers, and H. Langemann, Grössenverteilung der Gasblasen in Blasensäulen, Chem.-Ing.-Tech., *33*, 668 (1961).

Kölbel, H., R. Beinhauer, and H. Langemann, Dynamische Messung des relativen Gasgehaltes in Blasensäulten mittels absorption von Röntgenstrahlen, Chem.-Ing.-Tech., *44*, 697 (1972).

Larkins, R. P., R. R. White, and D. W. Jeffrey, Two-phase concurrent flow in packed beds, AIChE J., *7*, 231 (1961).

Laurent, A. and J. C. Charpentier, Le Rôle et l'utilité des modèles expérimentaux de laboratoire dans la prévision des performances d'un réacteur gaz-liquide industriel. Exemples d'utilisation de modèles expérimentaux simulant le fonctionnement d'une colonne a garnissage, J. Chim. Phys., *74*, 1001 (1977).

Laurent, A., C. Prost, and J. C. Charpentier, Détermination par méthode chimque des aires interfaciales et des coefficients de transfert de matière dans les divers types d'absorbeurs et de reacteurs gaz-liquide, J. Chim. Phys., *72*, 236 (1975).

Lee, E. S., *Quasilinearization and Invariant Imbedding*, Academic Press, New York (1968).

Levenspiel, O., *Chemical Reaction Engineering*, Wiley, New York (1972).

Levenspiel, O. and J. H. Godfrey, A gradientless contactor for experimental study of interphase mass transfer with/without reaction, Chem. Eng. Sci., *29*, 1723 (1974).

Liley, P. E. and W. R. Gambill, Physical and chemical data, in *Chemical Engineer's Handbook*, 5th ed., R. H. Perry and C. H. Chilton, eds., McGraw-Hill, New York (1973), Chap. 3.

Linek, V., Bestimmung der Phasen-Grenzfläche in einem mit mechanischem Rührwerk versehenen Reaktor bei Gasdurchgang, Chem. Eng. Sci., *21*, 777 (1966).

Linek, V. and J. Mayrhoferova, The chemical method for the determination of the interfacial area. The influence of absorption rate on the holdup and on the interfacial area in a heterogeneous gas-liquid system, Chem. Eng. Sci., *24*, 481 (1969).

Linek, V. and J. Mayrhoferova, The kinetics of oxidation of aqueous sodium sulphite solution, Chem. Eng. Sci., *25*, 787 (1970).

Linek, V. and V. Vacek, Chemical engineering use of catalyzed sulfite oxidation kinetics for the determination of mass transfer characteristic of gas-liquid contactors, Chem. Eng. Sci., *36*, 1747 (1981).

Linek, V., Z. Krivsky, and P. Hudec, Effective interfacial area in plastic-packed absorption columns, Chem. Eng. Sci., *32*, 323 (1977).

Linek, V., P. Beneš, J. Sinkule, and Z. Křivsky, Simultaneous determination of mass transfer coefficient and of gas and liquid axial dispersions and holdups in packed absorption column by dynamic response method, Ind. Eng. Chem. Fundam., *17*, 298 (1978).

Lusis, M. A. and G. A. Ratcliff, Diffusion in binary liquid mixtures at infinite dilution, Can. J. Chem. Eng., *46*, 385 (1968).

Luss, D. and N. R. Amundson, An analysis of chemical reactor stability and control. Segregated two phase systems, Chem. Eng. Sci., *22*, 267 (1967).

Mangartz, K. H. and T. H. Pilhofer, Untersuchungen zur Gasphasendispersion in Blasensäulenreacktoren, Verfahrenstechnik (Mainz), *14*, 40 (1980).

Mann, R. and G. T. Clegg, Gas absorption with an unusual chemical reaction: The chlorination of toluene, Chem. Eng. Sci., *30*, 97 (1975).

Mann, R. and H. Moyes, Exothermic gas absorption with chemical reaction, AIChE J., *23*, 17 (1977).

Marrucci, G. and L. Nicodemo, Coalescence of gas bubbles in aqueous solutions of inorganic electrolytes, Chem. Eng. Sci., *32*, 1257 (1967).

Mashelkar, R. A., Bubble columns, Br. Chem. Eng., *15*, 1297 (1970).

Mayer, H. H. and O. Wörz, Selection of reactors for reactions with superposed distillation, Ger. Chem. Eng., *3*, 252 (1980).

Mersmann, A., Design and scaleup of bubble and spray columns, Ger. Chem. Eng., *1*, 1 (1978).

Mhaskar, R. D., Effect of backmixing on the performance of bubble column reactors, Chem. Eng. Sci., *29*, 897 (1974).

Midoux, N., M. Favier, and J. C. Charpentier, Flow pattern, pressure loss and liquid holdup data in gas-liquid downflow packed beds with foaming and nonfoaming hydrocarbons, J. Chem. Eng. Jpn., *9*, 350 (1976).

Midoux, N., A. Laurent, and J. C. Charpentier, Limits of the chemical methods for the determination of physical mass transfer parameters in mechanically agitated gas-liquid reactors, AIChE J., *26*, 157 (1980).

Miller, D. N., Scale-up of agitated vessels gas-liquid mass transfer, AIChE J., *20*, 445 (1974).

Mitchell, M. G. and J. J. Perona, Gas-liquid interfacial areas for high porosity tower packings in concurrent downward flow, Ind. Eng. Chem. Process Des. Dev., *18*, 316 (1979).

Miyauchi, T. and T. Vermeulen, Diffusion and back-flow models for two-phase axial dispersion, Ind. Eng. Chem. Fundam., *2*, 304 (1963).

Mohunta, D. M., A. S. Vaidyanathan, and G. S. Laddha, Effective interfacial areas in packed columns, Indian Eng. Chem., *11*, 39 (1969).

Morbidelli, M., A. Servida, and S. Carrà, Bimolecular reactions of general order in gas-liquid reactors with volatile reactants. Application to the case of autoxidation units, in *Frontiers in Chemical Reaction Engineering*, L. K. Doraiswamy and R. A. Mashelkar, eds., Wiley Eastern, New Deli (1984a).

Morbidelli, M., A. Servida, S. Carrà, and A. Varma, Consecutive bimolecular reactions of general order in gas-liquid reactors, in *Recent Advances in the Engineering Aspects of Chemically Reacting Systems*, L. K. Doraiswamy ed., Wiley Eastern, New Delhi (1984b).

Nagata S., *Mixing*, Halsted Press, Tokyo (1975).

Nagel, O., H. Kürten, and B. Hegner, Die Stoffaustauschfläche in Gas-Flüssügkeits Kontaktapparaten Auswahlkriterien und Uterlagen zur Vergrösserung, Chem.-Ing.-Tech., *45*, 913 (1973).

Nagel, O. H., B. Hegner, and H. Kürten, Criteria for the selection and design of gas-liquid reactors, Chem.-Ing.-Tech., *50*, 934(1978) and Int. Chem. Eng., *21*, 161 (1981).

Neelakantan, K. and J. K. Gehlawat, New chemical systems for the determination of liquid-side mass transfer coefficient and effective interfacial area in gas-liquid contractors, Chem. Eng. J., *24*, 1 (1982).

Nerst, W., Zur Kinetik der in Lösung bifindlichen Körper, Z. Phys. Chem., *2*, 613 (1888).

Newman, A. B., The drying of porous solids: diffusion calculations, AIChE Trans., *27*, 310 (1931).

Nishikawa, M., H. Kato, and K. Hashimoto, Heat transfer in aerated tower filled with non-Newtonian liquid, Ind. Eng. Chem. Process Des. Dev., *16*, 133 (1977).

Nishiwaki, A. and Y. Kato, Relation between the dispersion model and other models for first-order reactions, Can. J. Chem. Eng., *52*, 276 (1974).

Nishiwaki, A., T. Kago, and Y. Kato, Fraction of reactant unreacted of the first-order reaction in multistage reactors, Kagaku Kogaku, *37*, 960 (1973).

Onda, K., H. Takeuchi, and Y. Okumoto, Mass transfer coefficients between gas and liquid phases in packed columns, J. Chem. Eng. Jpn., *1*, 56 (1968).

Onda, K., E. Sada, T. Kobayashi, S. Kito, and K. Ito, Solubility of gases in aqueous solutions of mixed salts, J. Chem. Eng. Jpn., *3*, 137 (1970).

Onda, K., H. Takeuchi, and Y. Maeda, The absorption of oxygen into sodium sulphite solutions in a packed column, Chem. Eng. Sci., *27*, 449 (1972).

Otake, T. and E. Kunigita, Mixing characteristics of irrigated packed tower, Kagaku Kogaku, *22*, 144 (1958).

Othmer, D. F. and M. S. T akar, Correlating diffusion coefficients in liquids, Ind. Eng. Chem., *45*, 589 (1953).

Pangarkar, V. G., Gas absorption with reaction in a solution containing a volatile reactant, Chem. Eng. Sci., *29*, 877 (1974).

Pavlica, R. T. and J. H. Olson, Unified design method for continuous-contact mass transfer operations, Ind. Eng. Chem., *62*(12), 45 (1970).

Pohorecki, R., Mass transfer with chemical reaction during gas absorption on a sieve plate, Chem. Eng. Sci., *31*, 637 (1976).

Porter, K. E., M. B. King, and K. C. Varshney, Interfacial areas and liquid-film mass-transfer coefficients, Trans. Inst. Chem. Eng., *44*, T274 (1966).

Prasher, B. D. and G. B. Willis, Mass transfer in an agitated vessel, Ind. Eng. Chem. Process Des. Dev., *12*, 351 (1973).

Prausnitz, J. M., *Molecular Thermodynamics of Fluid-Phase Equilibria*, Prentice-Hall, Englewood Cliffs, N.J. (1969).

Puranik, S. S. and A. Vogelpohl, Effective interfacial area in irrigated packed columns, Chem. Eng. Sci., *29*, 501 (1974).

Raghuram, S. and Y. T. Shah, Criteria for unique and multiple steady states for a gas-liquid reaction in an adiabatic CSTR, Chem. Eng. J., *13*, 81 (1977).

Raghuram, S., Y. T. Shah, and J. W. Tierney, Multiple steady states in a gas-liquid reactor, Chem. Eng. J., *17*, 63 (1979).

Rao, K. B. and P. S. Murti, Heat transfer in mechanically agitated gas-liquid systems, Ind. Eng. Chem. Process Des. Dev., *12*, 190 (1973).

Ravens, D. A. S., The kinetics and mechanism of the auto-oxidation of p-toluic acid in acetic acid solution catalyzed by cobalt and manganese bromides, Trans. Faraday Soc., *55*, 1768 (1959).

Reddy, K. A. and L. K. Doraiswamy, Estimating liquid diffusivity, Ind. Eng. Chem. Fundam., *6*, 77 (1967).

Reid, R. C., J. M. Prausnitz, and T. K. Sherwood, *The Properties of Gases and Liquids*, Third ed., McGraw-Hill, New York (1977).

Reiss, L. P., Cocurrent gas-liquid contacting in packed columns, Ind. Eng. Chem. Process Des. Dev., *6*, 486 (1967).

Reith, T., Interfacial area and scaling up of gas-liquid contactors, Br. Chem. Eng., *15*, 1559 (1970).

Rietema, K., Science and technology of dispersed two-phase systems: I and II, Chem. Eng. Sci., *37*, 1125 (1982).

Roberts, D. and P. V. Danckwerts, Kinetics of CO_2 absorption in alkaline solutions: I. Transient absorption rates and catalysis by arsenite, Chem. Eng. Sci., *17*, 961 (1962).

Rushton, J. H. and J. J. Bimbinet, Holdup and flooding in air liquid mixing, Can. J. Chem. Eng., *46*, 16 (1968).

Sahay, B. N. and M. M. Sharma, Effective interfacial area and liquid and gas side mass transfer coefficients in a packed column, Chem. Eng. Sci., *28*, 41 (1973).

Sater, V. E. and P. Levenspiel, Two-phase flow in packed beds, Ind. Eng. Chem. Fundam., *5*, 86 (1966).

Satterfield, C. N., Trickle-bed reactors, AIChE J., *21*, 209 (1975).

Schaftlein, R. W. and T. W. F. Russel, Two-phase reactor design: tank-type reactors, Ind. Eng. Chem., *60*(5), 12 (1968).

Scheibel, E. G., Liquid diffusivities, Ind. Eng. Chem., *46*, 2007 (1954).

Schmitz, R. A. and N. R. Amundson, An analysis of chemical reactor stability and control: Va. Two-phase systems in physical equilibrium-1, Chem. Eng. Sci., *18*, 265 (1963a).

Schmitz, R. A. and N. R. Amundson, An analysis of chemical reactor stability and control: Vb. Two-phase gas-liquid and concentrated liquid-liquid reacting systems in physical equilibrium-2, Chem. Eng. Sci., *18*, 391 (1963b).

Schmitz, R. A. and N. R. Amundson, An analysis of chemical reactor stability and control: VI. Two-phase chemical reacting systems with heat and mass transfer resistances, Chem. Eng. Sci., *18*, 415 (1963c).

Schmitz, R. A. and N. R. Amundson, An analysis of chemical reactor stability and control: VII. Two-phase chemical reacting systems with fast reactions, Chem. Eng. Sci., *18*, 447 (1963d).

Schumpe, A., and W. D. Deckwer, Vergleich der Fotografischen und der Sulfit-Oxidation-Methode zur Phasengrenzflächen-Bestimmung in Blasensäulen, Chem.-Ing.-Tech., *52*, 468 (1980).

Schumpe, A. and W. D. Deckwer, Determination of interfacial areas in non-Newtonian gas-liquid flow, ACS Meet., Atlanta, Ca. March 30, (1981).

Shah, Y. T., Gas-liquid interface temperature rise in the case of temperature dependent physical, transport and reaction properties, Chem. Eng. Sci., *27*, 1469 (1972).

Shah, Y. T. and J. A. Paraskas, Approximate solutions to single and two-phases axial dispersion problems in isothermal tubular flow reactors, Chem. Eng. Sci., *30*, 465 (1975).

Shah, Y. T., G. J. Stiegel, and M. M. Sharma, Backmixing in gas-liquid reactors, AIChE J., *24*, 369 (1978).

Shah, Y. T., B. G. Kelkar, S. P. Godbole, and W. D. Deckwer, Design parameters estimation for bubble column reactors, AIChE J., *28*, 353 (1982).

Shaikh, A. A. and A. Varma, Gas absorption with chemical reaction: the case involving a volatile liquid reactant, Chem. Eng. Sci. (1983a, in press.)

Shaikh, A. A. and A. Varma, Studies in gas-liquid reactors: I. Steady-state behavior of isothermal CSTRs (1983b, submitted for publication.)

Shaikh, A. A. and A. Varma, Studies in gas-liquid reactors: II. On the use of explicit versus implicit enhancement factors. (1983c, submitted for publication.)

Sharma, M. M. and P. V. Danckwerts, Chemical methods of measuring interfacial area and mass transfer coefficients in two-fluid systems, Br. Chem. Eng., *15*, 522 (1970).

Sharma, M. M. and R. A. Mashelkar, Absorption with reaction in bubble columns, Inst. Chem. Eng. Symp. Ser., *28*, 10 (1968).

Sharma, S., L. A. Hoffman, and D. Luss, Steady state multiplicity of adiabatic gas-liquid reactors: II. The two consecutive reactions case, AIChE J., *22*, 324 (1976).

Sherwood, T. K., R. L. Pigford, and C. R. Wilke, *Mass Transfer*, McGraw-Hill, New York (1975).

Shilimkan, R. V. and J. B. Stepanek, Interfacial area in cocurrent gas-liquid upward flow in tubes of various size, Chem. Eng. Sci., *32*, 149 (1977).

Shilimkan, R. V. and J. B. Stepanek, Mass transfer in cocurrent gas-liquid flow: gas side mass transfer coefficients in upflow, interfacial areas and mass transfer coefficients in gas and liquid in downflow, Chem. Eng. Sci., *33*, 1675 (1978).

Shulman, H. L. and M. C. Molstad, Gas-bubble columns for gas-liquid contacting, Ind. Eng. Chem., *46*(6), 1058 (1950).
Shulman, H. L., C. F. Ullrich, N. Wells, and A. Z. Proulx, Holdup for aqueous and nonaqueous systems, AIChE J., *1*, 259 (1955).
Sideman, S., O. Hortacsu, and J. W. Fulton, Mass transfer in gas-liquid contacting systems, Ind. Eng. Chem., *58*, 32 (1966).
Singh, C. P. P., Y. T. Shah, and N. L. Carr, The effect of gas feed temperature on the steady state multiplicity of an adiabatic CSTR with a fast pseudo-first order reaction, Chem. Eng. J., *23*, 101 (1982).
Skelland, A. H. P., *Diffusional Mass Transfer*, Wiley, New York (1974).
Specchia, V., S. Sicardi, and A. Gianetto, Absorption in packed towers with cocurrent upward flow, AIChE J., *20*, 646 (1974).
Sridhar, T., and O. E. Potter, Gas holdup and bubble diameters in pressurized gas-liquid stirred vessels, Ind. Eng. Chem. Fundam., *19*, 21 (1980a).
Sridhar, T. and O. E. Potter, Interfacial areas in gas-liquid stirred vessels, Chem. Eng. Sci., *35*, 683 (1980b).
Steiff, A., R. Poggemann, and P. M. Weinspach, Heat transfer in agitated vessels with multiphase systems, Ger. Chem. Eng., *4*, 30 (1981).
Stephens, E. J. and G. A. Morris, Determination of liquid-film absorption coefficients, Chem. Eng. Prog., *47*, 232 (1951).
Stichlmair, J. and A. Mersmann, Dimensioning plate columns for absorption and rectifications, Chem.-Ing.-Tech., *49*, 106 (1977), or Int. Chem. Eng., *18*, 223 (1978).
Stiegel, G. J. and Y. T. Shah, Axial dispersion in a rectangular bubble column, Can. J. Chem. Eng., *55*, 3 (1977a).
Stiegel, G. J. and Y. T. Shah, Backmixing and liquid holdup in a gas-liquid cocurrent upflow packed column, Ind. Eng. Chem. Process Des. Dev., *16*, 37 (1977b).
Stockar, U. V. and C. R. Wilke, Rigorous and short-cut design calculations for gas absorption involving large heat effects: 1. A new computational method for packed gas absorbers, Ind. Eng. Chem. Process Des. Dev., *16*, 88 (1977a).
Stockar, U. V. and C. R. Wilke, Rigorous and short-cut design calculations for gas absorption involving large heat effects: 2. Rapid short cut design procedure for packed gas absorbers, Ind. Eng. Chem. Process Des. Dev., *16*, 94 (1977b).
Sullivan, G. A. and R. E. Treybal, Axial mixing and gas absorption in a mechanically agitated absorption tower, Chem. Eng. J., *1*, 302 (1970).
Szeri, A., Y. T. Shah, and A. Madgavkar, Axial dispersion in two phase co-current flow with fast and instantaneous reactions, Chem. Eng. Sci., *31*, 225 (1976).
Tamir, A., and Y. Taitel, The effect of the gas side resistance on absorption with chemical reaction from binary mixtures, Chem. Eng. Sci., *30*, 1477 (1975).
Tamir, A., P. V. Danckwerts, and P. D. Virkar, Penetration model for absorption with chemical reaction in the presence of heat generation, bulk flow and effects of the gaseous environment, Chem. Eng. Sci., *30*, 1243 (1975).
Tavares Da Silva, The penetration theory for gas absorption accompanied by a first order chemical reaction with a diffusing catalyst, Chem. Eng. Sci., *29*, 275 (1974).

Teramoto, M., T. Nagayasu, T. Matsui, K. Hashimoto, and S. Nagata, Selectivity of consecutive gas-liquid reactions, J. Chem. Eng. Jpn., *2*, 186 (1969).

Teramoto, M., K. Hashimoto, and S. Nagata, Effect of mass transfer on the selectivity of (m,n)-(p,q) order consecutive gas-liquid reactions, J. Chem. Eng. Jpn., *6*, 522 (1973).

Tokunaga, J., General correlation for nonpolar gas solubilities in aqueous alcohol solutions, J. Chem. Eng. Jpn., *8*, 7 (1975).

Treybal, R. E., *Mass-Transfer Operations*, 2nd ed., McGraw-Hill, New York (1968).

Uhl, V. W., Mechanically aided heat transfer, in *Mixing*, Vol. 1, V. W. Uhl and J. B. Gray, eds., Academic Press, New York (1966), p. 279.

Van de Vusse, J. G., Consecutive reactions in heterogeneous systems: I. The effect of mass transfer on selectivity, Chem. Eng. Sci., *21*, 631 (1966a).

Van de Vusse, J. G., Consecutive reactions in heterogeneous systems: I. Influence of order of reaction rates on selectivity, Chem. Eng. Sci., *21*, 645 (1966b).

van Krevelen, D. W., and P. J. Hoftijzer, Kinetics of gas-liquid reactions. General theory, Recl. Trav. Chim., *67*, 563 (1948a).

van Krevelen, D. W. and P. J. Hoftijzer, Sur la solubilité des gaz dans les solutions aqueouses, Int. Congr. Chim. Ind., 21st, Brussels, Special No., p. 168 (1948b).

van Swaaij, W. P. M., J. C. Charpentier, and J. Villermaux, Residence time distribution in the liquid phase of trickle flow in packed columns, Chem. Eng. Sci., *24*, 1083 (1969).

Van't Riet, K., Review of measuring methods and results in nonviscous gas-liquid mass transfer in stirred vessels, Ind. Eng. Chem. Process Des. Dev., *18*, 357 (1979).

Verma, S. L. and G. B. Delancey, Thermal effects in gas absorption, AIChE J., *21*, 96 (1975).

Vermeulen, T., G. M. Williams, and G. E. Langlois, Interfacial area in liquid-liquid and gas-liquid agitation, Chem. Eng. Prog., *51*, 85F (1955).

Vignes, A., Diffusion in binary solutions, Ind. Eng. Chem. Fundam., *5*, 189 (1966).

Villadsen, J., and M. L. Michelsen, *Solution of differential equation models by Polynomial Approximations*, Prentice-Hall, Englewood Cliffs, N.J. (1978).

Wellek, R. M., R. J. Brunson, and F. H. Law, Enhancement factors for gas-absorption with second-order irreversible chemical reaction, Can. J. Chem. Eng., *56*, 181 (1978).

Wen, C. Y. and L. T. Fan, *Models for flow systems and chemical reactors*, Chemical Processing and Engineering Monograph Series, Marcel Dekker, New York (1975).

Wen, C. Y., W. S. O'Brien, and L. T. Fan, Pressure drop through packed beds operated cocurrently, J. Chem. Eng. Data, *8*, 47 (1963).

Wesselingh, J. A. and A. C. Van't Hoog, Oxidation of aqueous sulfite solutions: model reaction for measurements in gas-liquid dispersions, Trans. Inst. Chem. Eng., *T69*, 48 (1970).

Westerterp, K. R., L. L. van Dierendonck, and J. A. de Kraa, Interfacial areas in agitated gas-liquid contactors, Chem. Eng. Sci., *18*, 157 (1963).

Wilhem, E. and R. Battino, Thermodynamic functions of solubilities of gases in liquids at 25°C, Chem. Rev., *73*, 1 (1973).

Wilke, C. R. and P. Chang, Correlation of diffusion coefficients in dilute solutions, AIChE J., *1*, 264 (1955).

Wolf, D. and W. Resnick, Residence time distribution in real systems, Ind. Eng. Chem. Fundam., *2*, 287 (1963).

Woodburn, E. T., Gas phase axial mixing at extremely high irrigation rates in a large packed absorption tower, AIChE J., *20*, 1003 (1974).

Yeramian, A. A., J. C. Gottifredi, and J. J. Ronco, Mass transfer with homogeneous second order irreversible reaction. A note for an explicit expression for the reaction factor, Chem. Eng. Sci., *25*, 1622 (1970).

Yoshida, F. and K. Akita, Performance of gas bubble columns: volumetric liquid-phase mass transfer coefficient and gas holdup, AIChE J., *11*, 9 (1965).

Yoshida, F., T. Yamane, and Y. Miyamoto, Oxygen absorption into oil-in-water emulsions, Ind. Eng. Chem. Process Des. Dev., *9*, 570 (1970).

Zahradnik, J., F. Kaštánek, and J. Kratochvíl, Hydrodynamics and mass transfer in uniformly aerated bubble columns reactors, Collect Czech. Chem. Commun., *47*, 262 (1982).

Zarzycki, R., S. Ledakowicz, and M. Starzak, Simultaneous absorption of two gases reacting between themselves in a liquid: II. Approximate solutions, Chem. Eng. Sci., *36*, 113 (1981).

10
Gas-Liquid-Solid Reactors

YATISH T. SHAH *University of Pittsburgh, Pittsburgh, Pennsylvania*

MAN MOHAN SHARMA *University of Bombay, Bombay, India*

INTRODUCTION: CLASSIFICATIONS: EXAMPLES OF INDUSTRIAL IMPORTANCE

Three-phase reactors, involving gas, liquid, and solid phases, are widely encountered in industrial practice. Several examples are illustrated in Table 1. Other examples are given by Østergaard (1968), Shah (1979), Shah et al. (1982), and Doraiswamy and Sharma (1983). As can be seen from this table, solid phase can be either reactant or product or it could be catalyst. In some cases we may have four phases, either through two immiscible liquid phases (either as reactants and/or products) or through two solid phases, one reactant or product and other catalyst. Although, three-phase systems are found in chemical, polymer, biochemical, and petroleum industries, they are best known in hydroprocessing operations.

Two main categories of three-phase reactors are fixed-bed reactors, which are normally used when catalyst particles are larger than 10^{-3} m, and slurry reactors, wherein the solid phase is in suspension. The most common type of fixed-bed reactor used in industry is the trickle-bed reactor, wherein liquid trickles over a fixed bed of catalyst while gas flows cocurrently downward as a continuous phase. Fixed-bed reactors are often operated in the bubble flow regime, wherein both gas and liquid flow cocurrently upward and gas is the dispersed phase. In some processes, such as for hydrodemetallization of high-metal-content heavy crude, the catalyst needs to be replaced frequently, and a moving fixed-bed reactor is used. In such reactors, a pulsating flow regime may prevail in the bottom part of the reactor. As pointed out by Østergaard (1968), in true slurry reactors, solid particles are very fine (generally less than 10^{-3} m) and gas flows mainly in and out of the reactor. In many cases, slurry reactors can be mechanically agitated. If solid particles are large such that they form a discrete phase within the reactor and if the liquid-solid mixture flows in and out of the reactor, the reactor is often called a *three-phase fluidized-bed reactor*. In the present chapter we concentrate on slurry reactors (with fine catalyst particles) with some allowance for slurry flow in and out of the reactor. Several types of fixed-bed and slurry reactors used in industrial practice are illustrated in Fig. 1.

TABLE 1 (a) Examples of Fluid-Fluid-Solid Systems: Solids Sparingly Soluble

Absorption of Co_2 in aqueous suspension of lime and $Ba(OH)_2$
Absorption of CO_2 in a suspension of BaS
Absorption of lean SO_2 in a slurry of $CaCo_3$
Absorption of lean H_2S, $COCl_2$, Cl_2, etc., in an aqueous suspension of lime
Absorption of CO in a suspension of lime in the manufacture of $(HCOO)_2Ca$

(b) Examples of Fluid-Fluid-Solid Systems: Solid Particles Insoluble in The Medium

Absorption of CO_2 in a suspension of MgO
Absorption of CO_2 in a suspension of CaS
Absorption of SO_2 in a suspension of $CaCO_3$
Chlorination of wood pulp
Hydrogenation of styrene-butadiene rubber latex and other polymeric substances suspended in solvents
Hydrogenation of sodium particles suspended in mineral oil to give NaH
Chlorination of suspended polyethylene (and PVC) particles in water
Manufacture of zinc dithionite by reaction between SO_2 and zinc particles suspended in water

(c) Examples of Fluid-Fluid-Solid (Catalytic) Reactions

System	Catalyst	Type of reactor
Hydrogenation of unsaturated fats	Raney nickel	SR
Hydrogenation of nitrobenzene sulfonic acid (potassium salt), nitrophenols, etc., in aqueous solutions	Pd, Pt, Ni, etc., based catalyst	SR
Hydrogenation and hydrocracking of α-cellulose (also powdered coal) to liquid and gaseous fuels	Ni catalyst	SR
Hydrogenation of nitriles	Ni-based catalyst	SR
Hydroisomerization of butene-1 to butene-2	Undisclosed	TBR: packed bubble column (up-flow)
Hydroformylation and hydrogenation of monoolefins, methyl sorbate, etc.	Nobel metal (Rh)-based catalyst	SR

TABLE 1 (Continued)

System	Catalyst	Type of reactor
Hydrogenation of xylose; hydrogenolysis of saccharides	Raney Nickel; Ru	SR
Fischer-Tropsh synthesis: conversion of $CO+H_2$ to CH_4, paraffins, etc.	Ni-MgO	SR
Conversion of $CO+H_2$ to methanol	Cu-Zn oxide	(Three-phase) fluidized-bed reactor
Reaction between nitrite and H_2 to give hydroxylamine salts; reaction between NH_4NO_3, H_3PO_4, and H2 to give hydroxylamine phosphate	Noble metal (Pd) on carbon	SR
Reduction of sodium chlorate in (diaphragm cell) aqueous caustic liquor with H_2	Not disclosed	TBR
Hydrogenation of iodine dissolved in aqueous II	Ru or Rh	SR
Reduction of uranyl(VI) to uranyl(IV) in aqueous solutions with H_2	PD based	TBR: packed bubble column reactor
Reduction of UF_5 dissolved in molten LiF-BeF_2-ThF_2 melt with H_2	Pt-based catalyst	SR
Deuterium exchange between H_2 and water	Pt on carbon	TBR
Air oxidation of ethanol to acetic acid; conversion of primary alcohols to the corresponding sodium salt of the acid	Pd-based catalyst, CdO, ZnO, etc.	TBR; SR
Oxidation of isobutylene glycol to α-hydroxyisobutyric acid	Pt on carbon	SR
Amination of monoethanolamine to ethylene diamine	Undisclosed	TBR
Isomerization of propylene oxide to allyl alcohol; rearrangement of oxirane compounds	Li_3PO_4 suspended in diphenyl; SiO_2-Al_2O_3	SR
Oxidation of glucose to K-gluconate	Pt on carbon	SR
Hydrogenation of 1-bromo, 2,4-dichlorobenzene to 1,3-dichlorobenzene	Pd on carbon	SR

TABLE 1 (Continued)

System	Catalyst	Type of reactor
Alkylation of benzene with propylene	Phosphoric acid	TBR
Reaction between isobutylene and aqueous formaldehyde (a step in the manufacture of synthetic isoprene)		TBR
Reaction between isobutylene and methanol to give methyl-*tert*-butyl ether	Ion-exchange resins	Multitubular packed-bed reactor
Epoxidation of ethylene to ethylene oxide	Silver oxide on silica gel suspended in dibutyl phthalate	SR
Epoxidation of propylene with *tert*-butyl hydroperoxide, ethyl benzene hydroperoxide, etc.; Epoxidation of styrene and α-olefins with cumene hydroperoxide	Mo sulfide/oxides	SR
Slurry polymerization	Organometallic compounds	SR
Oxidation of SO_2 to H_2SO_4 in water; absorption of H_2S from stack gases in alkaline solutions; osication of aqueous sodium sulfide (black liquor)	Active carbon	TBR, SR
Air oxidation of dilute formic acid and acetic acid solutions (treatment of waste streams)	Copper oxide-zinc oxide; ferric oxide	TBR
Hydrodesulphurization and hydrocracking of petroleum fractions	Molybdenum- and tungsten-based catalyst	TBR
Conversion of propylene to isopropanol	Tungstic oxide-based catalyst	TBR
Acrylamide from acrylonitrile	Copper chromite	TBR
Dehydrogenation of secondary alcohols to ketones	Ni or copper chrome	SR
Dehydrogenation of isoborneol to camphor	Cu-Ni-Mn	SR
Hydrogenation of polymeric substances	Raney Ni; Pd/C	SR

TABLE 1 (Continued)

System	Catalyst	Type of reactor
Hydrogenation of epoxides of styrene, α-olefins, etc., in the presence of water as a second liquid phase	Raney Ni	SR
Hydrogenation of 2-methyl nitrobenzene in conc. H_2SO_4 medium to 4-amino-3-methyl phenol	Pt/C	SR
Absorption of NO_x (at low ppm) in water	Patented alloy gauge packing	TBR
Coal liquefaction	Ni-Co-Mo	SR
Organic substitution reactions via phase transfer catalysis	Quarternary ammonium or phosphenium salts bound to cross-linked polystyrene resins, resin-bound linear polyethers, resins-supported oligooxyethlenes, poly(vinyl pyridine)-bound polymers	SR

The nature of the gas-liquid solid reaction may often be complex. The product of the reaction is usually soluble in the liquid phase but in several instances the product may be insoluble in the liquid phase. In slurry and fixed-bed reactors the liquid phase may be a reactant or may simply be a solvent. In a special case of the slurry reactor, the particles may simply be adsorbents. In some uncommon cases, such as demetallization of heavier petroleum fractions, the catalyst gets laden with the product (metals in the case of heavier petroleum fractions).

Solid-bound catalysts and reactants have gained importance in recent years and here both slurry and fixed-bed reactors are used. There is an incentive to use heterogenized-homogeneous catalysts, as the problem of the recovery of the catalyst can easily be handled using such catalysts. There are special reasons to immobilize a soluble reactant; the objective may be to achieve high selectivity or to avoid product contamination. In fact, ionic reactants exchanted on ion-exchange resins have also been considered in the literature. The solid particles may be reactant, which may be soluble/sparingly soluble/insoluble in the liquid phase, and if insoluble, may exist on the solid reactant as a product layer or as a discrete phase.

A variety of two-phase reactions involving a sparingly soluble reactant can be carried out successfully using the technique of phase transfer catalysis. Here the catalyst (PTC) forms a complex with the reactive

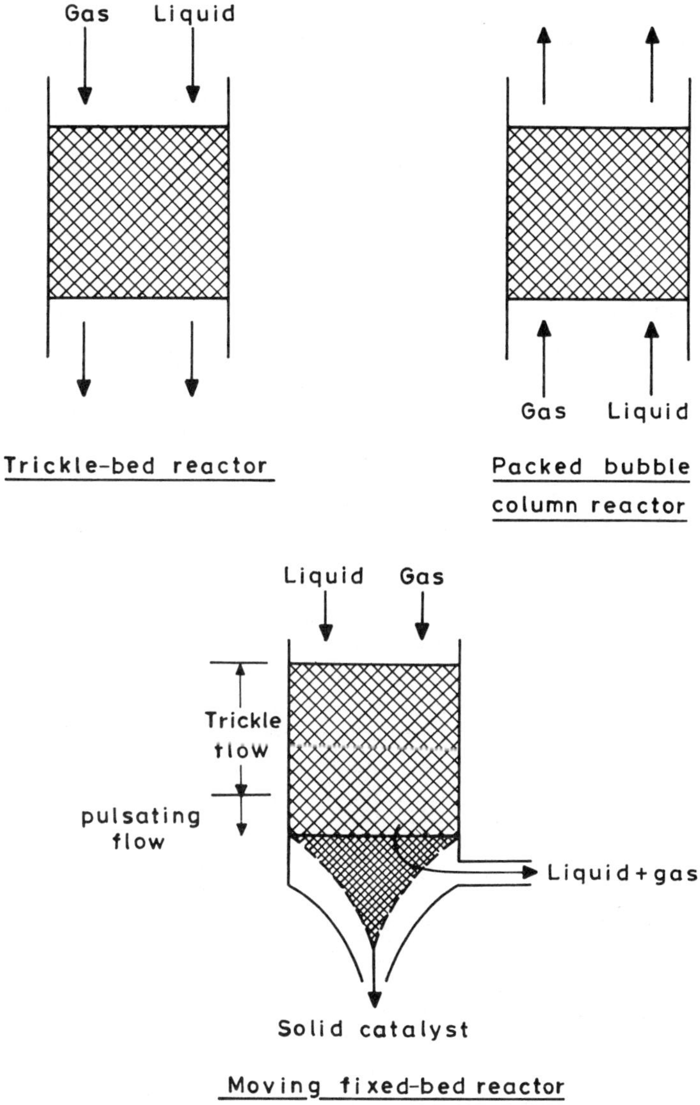

(a)

FIGURE 1 Various types of (a) fixed-bed reactors, and (b) slurry reactors.

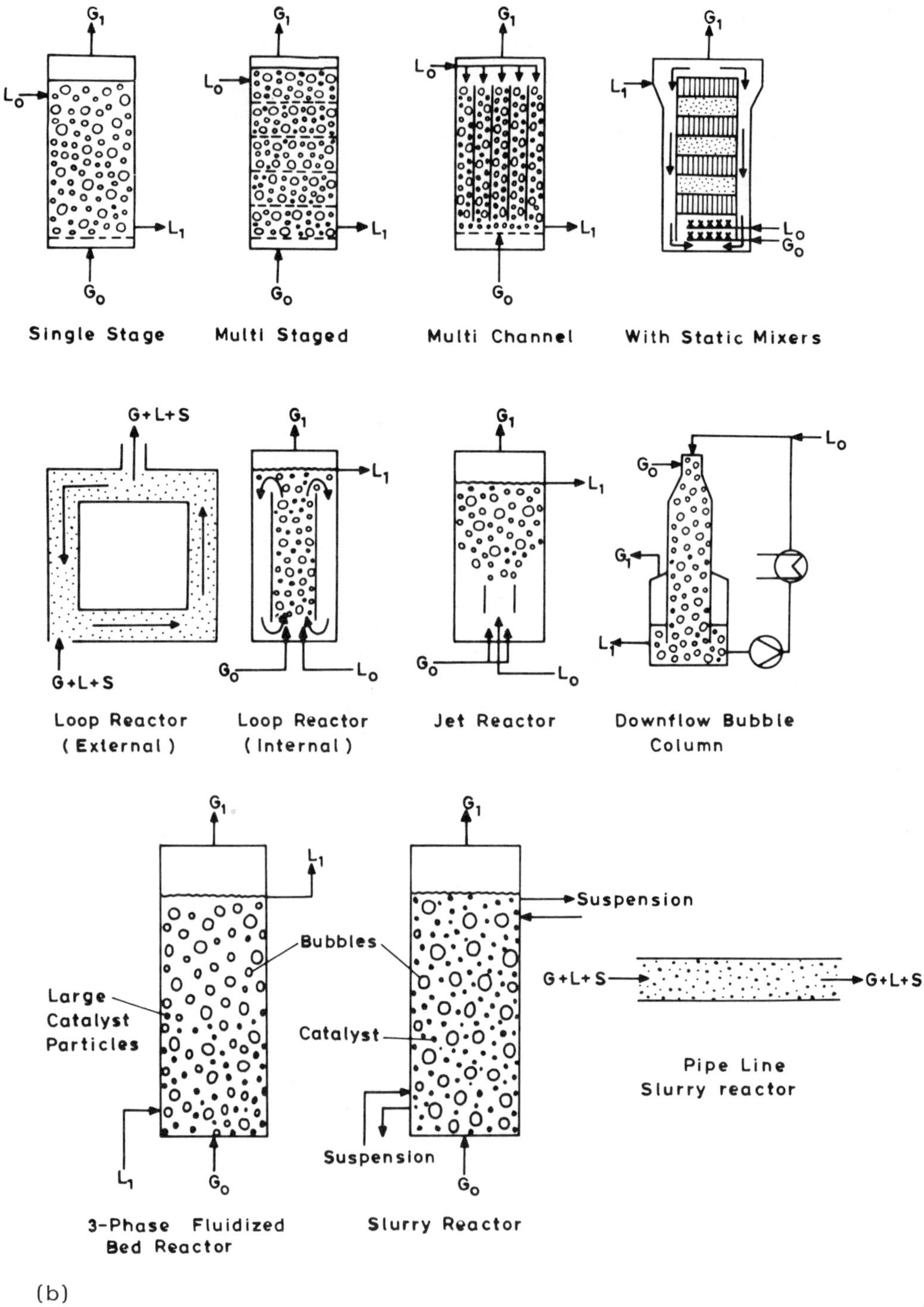

FIGURE 1 (Continued)

species from the aqueous or solid phase and transports it across the interface into the organic phase, where the reaction takes place. The main drawbacks of the technique involve the separation and recovery of the costly PTC, but these can probably be overcome by employing a solid-supported PTC. This type of catalysis was termed *triphase catalysis* by Regen (1975). The special feature of triphase catalysis is that the two reacting species are located in two different phases and the solid catalyst forms the third phase.

Finally, in recent years, three-phase electrochemical reactions have also gained some importance. Both slurry and fixed-bed reactors have been considered where gas-liquid-solid and liquid-liquid-solid systems are encountered, and some new designs have been suggested. In this chapter we attempt to cover briefly the main features of gas-liquid-solid reactors, with the principal emphasis on conventional fixed and slurry-bed reactors.

HYDRODYNAMICS, MIXING, AND TRANSPORT CHARACTERISTICS OF FIXED-BED REACTORS

Three-phase catalytic fixed-bed reactors are generally operated with downward or upward cocurrent flow of liquid and gas. The most predominant flow regime encountered in commercial operations is the trickle flow regime, in which gas and liquid flow cocurrently downward over a fixed bed of catalyst. As shown by Satterfield (1975), in many hydroprocessing operations, flow conditions are such that the reactor is operated at the boundary of trickle flow and pulsating flow regimes. The pulsating flow regime can also be achieved when gas and liquid flow cocurrently upward at high velocities. Some commercial reactors are operated under a bubble flow regime wherein gas and liquid flow cocurrently upward through a fixed bed of catalyst. Finally, hydrodemetallization of heavy crude is carried out in moving fixed-bed reactors with cocurrent downward gas-liquid flow due to rapid aging of the catalyst. The flow regime in such reactor can change from trickle flow at the top to pulsating flow at the bottom. The discussion presented below is therefore restricted to (a) trickle flow, (b) pulsating flow, (c) packed bubble flow, and (d) moving fixed-bed reactors.

Flow Regime Boundaries

In all work on two-phase downflow over packed beds before 1975, flow regimes were considered to be functions only of gas and liquid flow rates. The flow regime boundaries have been obtained in terms of superficial mass flow rates of air and water. The coordinates for the flow maps, showing flow regime boundaries, have been either G_G versus G_L (Weekman and Myers, 1964, 1965; Beimesch and Kessler 1971; Sato et al., 1973b; Satterfield, 1975; Chou et al., 1977; Charpentier et al. 1969; Hoffmann, 1978; Sicardi et al., 1979) or G_L versus G_G (Turpin and Huntington, 1967). Sato et al. (1973b) noted that the boundary between continuous and pulse gas flow moves to low liquid flow rates and returns to high liquid flow rates as the particle size is increased.

Charpentier and Favier (1975) introduced the effects of density, viscosity, and surface tension and the foaming nature of liquid of the flow regime boundaries and presented a flow map using Baker's coordinates for gas-liquid flow in an empty tube. Their results are shown in Fig. 2.

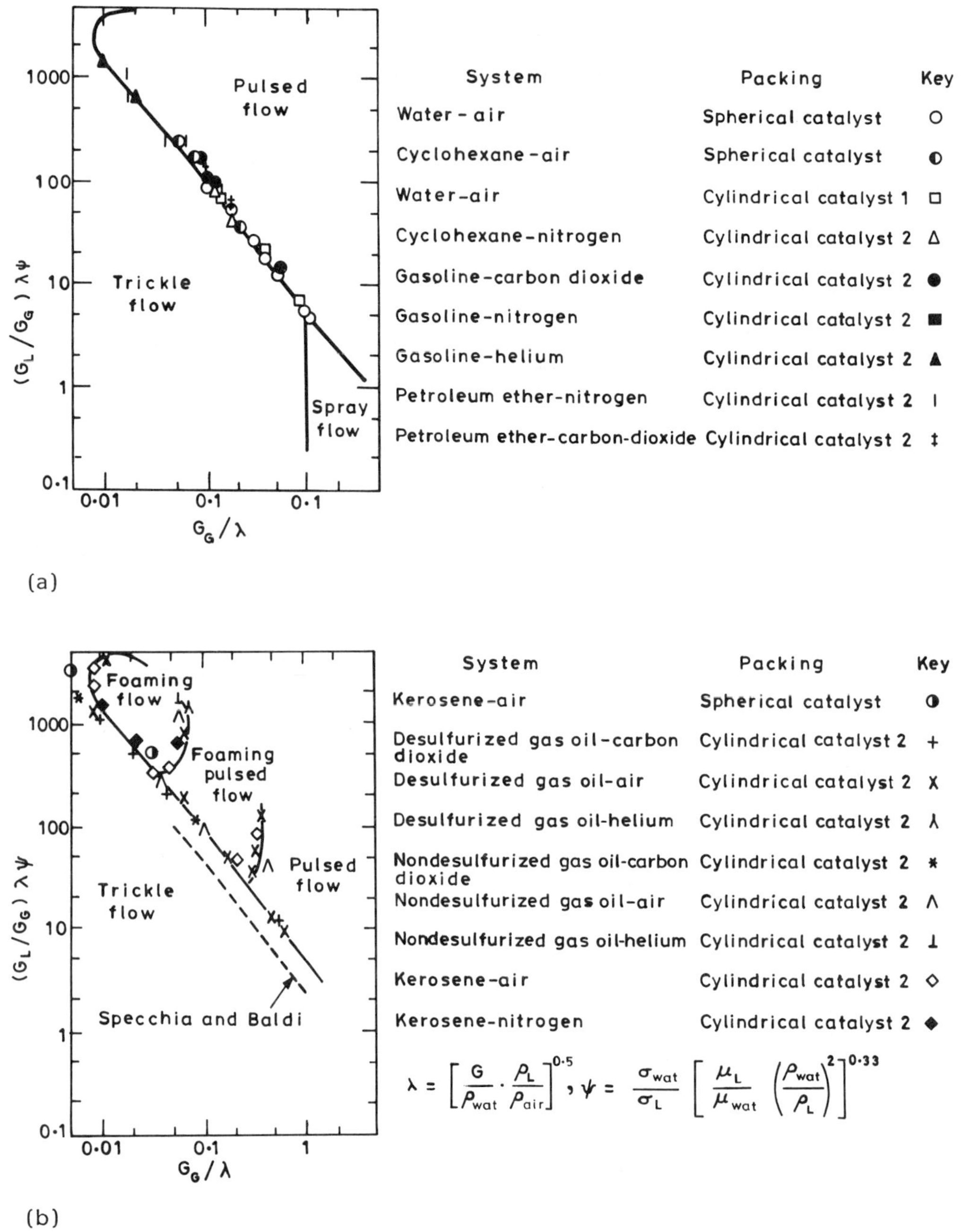

FIGURE 2 Flow-regime boundaries for (a) foaming and (b) nonfoaming liquids. (After Charpentier and Favier, 1975.)

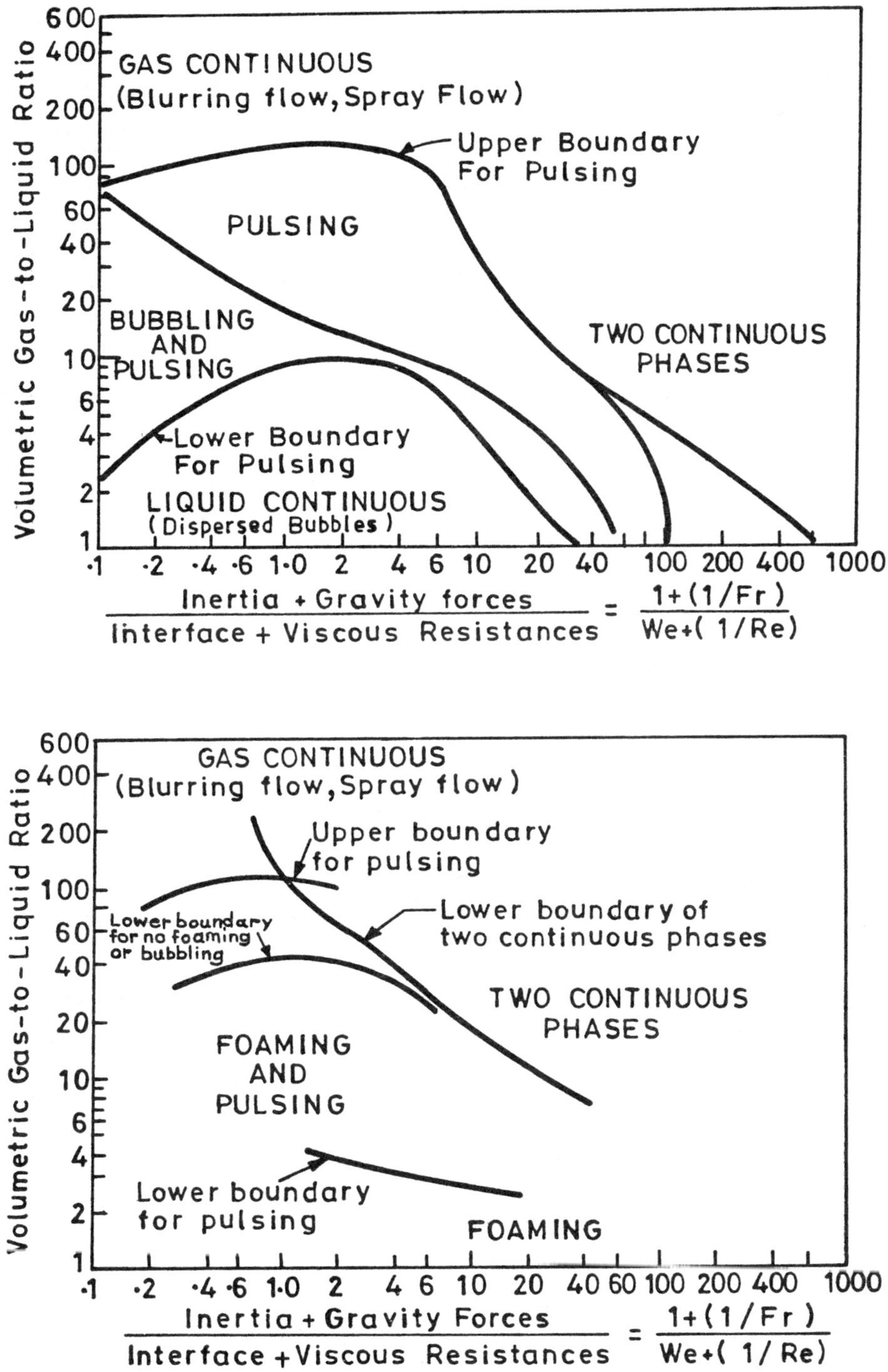

FIGURE 3 Flow maps: two-phase downflow through packed beds: (top) nonfoaming liquids; (bottom) foaming liquids. (From Shah et al., 1981.)

Talmor (1977) presented flow maps (Fig. 3) for foaming and nonfoaming liquids using a superficial volumetric gas-to-liquid ratio versus a ratio of driving force to resistance force. Clements and Halfacre (1979) showed that the data of Chou et al. (1977) for aqueous solution of alcohol, and their own data for air-isopropanol aqueous solution (0 to 48 wt %), were correlated by the flow maps of Talmor (1977) but not by the flow maps of Charpentier and Favier (1975). These workers also concluded that surface tension exerts the dominant influence over flow pattern transitions and foaming in the packed bed. A Talmor flow map has been used by Patil and Sharma (1981) for a tube packed with Pall rings and multifilament wire gauge packings.

The flow regime boundaries for cocurrent upflow over a fixed bed in terms of gas and liquid mass flow rates have been reported by several workers (Ford, 1960; Turpin and Huntington, 1967; Specchia et al., 1974; Sato et al., 1974; Pittsburgh Energy Research Center, 1976). The results of the latter study are shown in Fig. 4. Saada (1975) obtained data characterizing the flow regime for small particles (d_p < 0.2 cm). He also modified the coordinates of flow maps to include the effects of densities and viscosities of gas and liquid and bed porosity. In Saada's work, flow is divided into two regimes. At relatively high gas flow rates, the pores of the packing are considered to be traversed by both phases together, whereas at low

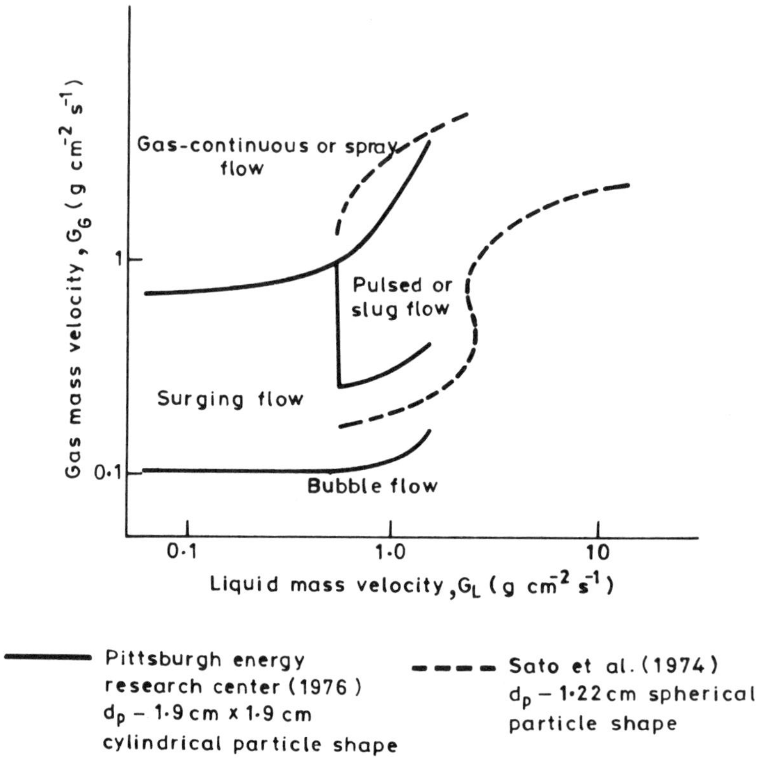

FIGURE 4 Flow regime boundaries of water-air upflow through a packed bed. (From Pittsburgh Energy Research Center, 1976.)

gas flow rates each phase is considered to flow seperately through the pores. The former is referred to as a *two-phase pore-flow regime* and the latter as a *single-phase pore-flow regime*. This flow map is shown in Fig. 5. Generalized flow maps such as those of Talmor (1977), for downflow over packed beds, are not available for the cocurrent upflow case.

Pressure Drop

Reported data on pressure drop in two-phase flow over a fixed bed cover various flow regimes and a large range of design and operating variables. However, most of these studies are for air-water systems. Other gas-liquid systems have also been investigated (Clements and Schmidt, 1976, 1980b; Larkins et al., 1961; Specchia and Baldi, 1977), but more data on hydrocarbon and foaming systems are needed. The data base of Larkin

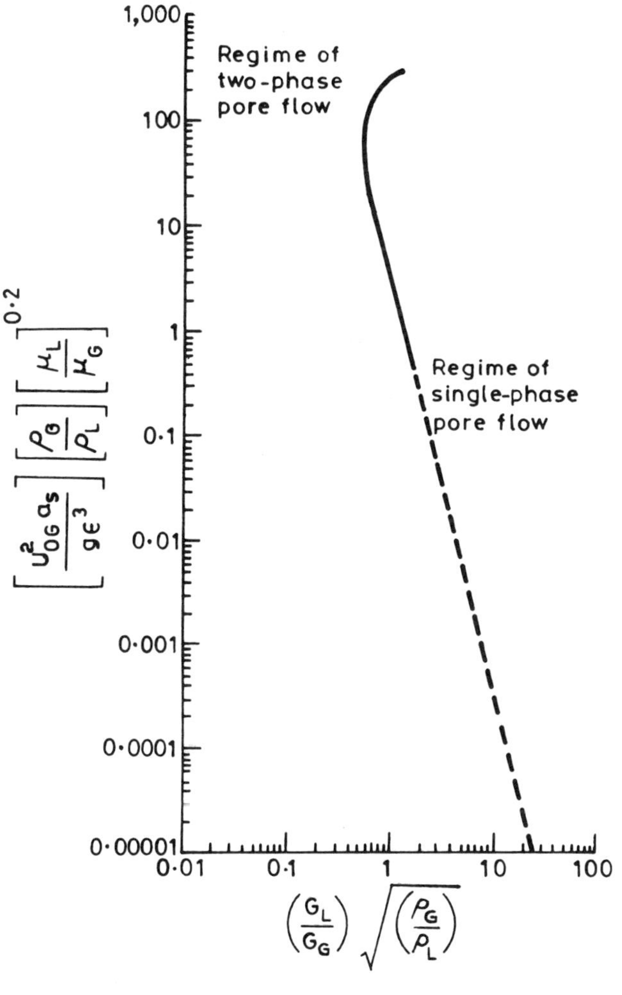

FIGURE 5 Correlation of flow transition by Saada (1975).

et al.'s (1961) correlation is relatively large and close to the conditions in a commercial reactor and therefore is more commonly used.

In the correlation of pressure-drop data for two-phase flow in packed beds, basically two approaches have been used. In one approach, the variables are lumped into various dimensionless groups, such as Reynolds numbers for gas and liquid, Weber numbers, and so on, and the pressure drop or friction factor is expressed in terms of these dimensionless groups. The use of Tallmadge or Lockhart-Martinelli parameters is included in this approach (Turpin and Huntington, 1967; Specchia and Baldi, 1977). The other approach is to correlate the two-phase pressure drop with the pressure drops of energy losses for individual phases in single-phase flows under identical conditions.

Ford (1960) measured pressure drop in packed beds for cocurrent upflow and identified the flow regimes based on the flow that takes place inside the pores: namely, *single-phase pore flow* and *two-phase pore flow*. He presented correlations having the form

$$\frac{1}{g\rho_L}\left(\frac{\Delta P}{\Delta z}\right)_{LG} = \alpha(Re_L)^x(Re_G)^y\left(\frac{\mu_L}{\mu_G}\right)^z \tag{1}$$

The values of α, x, y, and z depend on the type of flow regime. Different values of α, x, y, and z have accordingly been suggested for the two regimes. Saada (1975) modified the Ford correlation and replaced the μ_L/μ_G term by d_p/d_c (i.e., the ratio of the particle and column diameters).

A large number of data points obtained by the Pittsburgh Energy Research Center (PERC) (1976) in highly pulsed and spray flow regimes have been correlated by Tallmadge's correlation (1970). Sato et al. (1974), on the other hand, correlated their data with the Lockhard-Martinelli type of relation in all three flow regimes (i.e., bubble, pulsed, and spray).

The more widely used correlation of Larkins et al. (1961), for cocurrent downflow is based on the second approach. It correlates the overall two-phase energy loss for the gas and liquid passing through the reactor to the two individual single-phase energy losses as follows:

$$\log_{10}\frac{\delta_{LG}}{\delta_L + \delta_G} = \frac{K_1}{(\log_{10}X)^2 + K_2} \tag{2}$$

where δ_g is the gas-phase energy loss, δ_L the liquid-phase energy loss, δ_{LG} the two-phase energy loss, and $X = (\delta_L/\delta_G)^{1/2}$. The constants K_1 and K_2 have the values 0.416 and 0.66, respectively.

Larkins et al. (1961) defined the two-phase pressure gradient $(\Delta P/\Delta Z)_{LG}$ as follows:

$$\left(\frac{\Delta P}{\Delta Z}\right)_{LG} = \delta_{LG} - \rho_M \tag{3}$$

where

$$\rho_M = \varepsilon_L\rho_L + (1 - \varepsilon_L)\rho_G \tag{4}$$

Several other workers have used Eq. (3) directly (Abbot et al., 1967) or in modified form (Reiss, 1967; Sato et al., 1973a) for correlation of their data. Charpentier et al. (1969) derived pressure-drop correlations by comparing two-phase pressure loss ΔP_{LG} with that for single-phase flow, in terms of the energy-based dimensionless groups. Clements and Schmidt (1976, 1980a,b) combined the foregoing two approaches and gave the correlation for pressure drop in two-phase cocurrent downflow in terms of Reynolds numbers of gas and liquid phases, gas-phase Weber number, and pressure drop for the gas phase alone. Clements (1978) also gave a theoretical justification for this approach. Theoretical correlations for pressure drop under trickle flow conditions have been proposed by Hutton and Leung (1974) and Specchia and Baldi (1977).

It is important to note that in most of the studies reported, Reynolds numbers for liquid and gas phases are considered to be the controlling dimensionless groups. However, those studies that use systems other than air-water recognized the role of surface tension on the pressure drop and used either an empirical correction factor (Specchia and Baldi, 1977) or the dimensionless Weber number (Clements and Schmidt, 1976, 1980a,b). More data over a larger range of Weber numbers are needed.

Phase Holdups

Liquid Holdup

Total liquid holdup in a fixed-bed column is the sum of the operating and static holdups. *Static holdup*, the amount of liquid in the bed after the liquid inlet is shutt off and the column is allowed to drain, represents the liquid retained in the pore volume of the catalyst and its packing. *Operating holdup*, the liquid external to the catalyst particles, depends on the liquid and gas flow rates, their properties, and the nature of catalyst packing.

Static holdup is often correlated by Eotvos number, Eo ($=\rho_L g d_p^2/\sigma_L$, where d_p is the nominal particle diameter and g the gravitational acceleration). The correlation of Van Swaaij et al. (1969) indicates that smaller particle diameter and fluid density and larger surface tension give larger static liquid holdup. The correlation also indicates that a porous material gives a larger static liquid holdup than does a nonporous material.

Basically, two types of correlations for the dynamic or total liquid holdup are reported in the literature. Some investigators have correlated the liquid holdup directly with the liquid velocity and fluid properties by either dimensional or dimensionless relations. In more recent investigations, liquid holdup is correlated to the Lockhart-Martinelli parameter $\Delta P_L/\Delta P_G$ or its equivalent.

For cocurrent downflow, the dimensional correlations (Satterfield and Way, 1972; Jesser and Elgin, 1943) are very few. Most of the correlations are in the following dimensionless form (Clements and Schmidt, 1980a; Otake and Okada, 1963; Michell and Furzer, 1972; Buchanan, 1967; Gelbe, 1968; Mohunta and Laddha, 1965; Van Swaaij et al., 1969; Specchia and Baldi, 1977; Davidson et al., 1959):

$$\varepsilon_{dL} = \alpha(Rc_L)^{\beta}(Ga_L)^{\gamma}(a_s d_p)^{\eta} \tag{5}$$

The values of α, β, γ, and η obtained by different workers differ.

Many correlations using the Lockhart-Martinelli parameter or its equivalent are also reported in the literature (Abbott et al., 1967; Larkins et al., 1961; Sato et al., 1973a; Bakos and Charpentier, 1970; Midoux et al., 1976). The correlations of Midoux et al. (1976), in terms of energy based dimensionless groups, for nonfoaming and foaming liquids are recommended for use in the case of hydrocarbon liquids.

For cocurrent upflow a large amount of experimental data were obtained by the Pittsburgh Energy Research Center (1976). The average liquid holdup was found to be a very strong function of actual gas velocity and a relatively mild function of liquid velocity. Various dimensionless correlations are available for cocurrent upflow reactors. The modified correlation of Ford (1960) has been given by Saada (1975) in the form

$$\varepsilon_L = \alpha \left(\frac{Re_L}{Re_G} \right)^{\beta} \tag{6}$$

where α and β depend on the flow regime. Stiegel and Shah (1977b) proposed a correlation which is in a more general form:

$$\varepsilon_L = \alpha \, Re_L^{\beta} Re_G^{\gamma} (a_s d_s)^{\eta} \tag{7}$$

Sato et al. (1974) correlated their data in terms of the Lockhart-Martinelli parameter. A theoretical model of the liquid holdup for cocurrent upflow through a packed column was proposed by Hutton and Leung (1974). With the help of their model, these workers showed that cocurrent upflow operation gives higher liquid holdup than cocurrent downflow operation under the same gas and liquid flow conditions, a fact demonstrated experimentally by Turpin and Huntington (1967), and shown in Fig. 6.

The range of variables and physical properties examined in the literature can be roughly stated as $0.3 < Re_L < 3000$, $0.4 \text{ mm} < d_p < 50 \text{ mm}$,

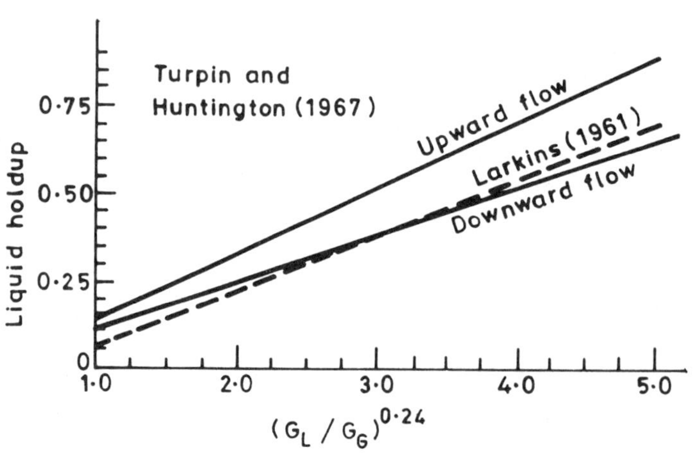

FIGURE 6 Liquid holdup for cocurrent upflow versus cocurrent downflow through a packed bed.

$2.6 < a_s d_p < 6.02$, and $14 < Ga_L < 320$ for particle shapes consisting of Raschig rings. Berl saddles, spheres, and irregular granules. Both foaming and nonfoaming liquids have been examined. Although there are significant discrepancies in the predictions of various correlations, qualitatively they all indicate that the liquid holdup under trickle flow conditions increases with liquid velocity and is essentially independent of gas flow rate. Although not completely clear, an increase in particle size appears to decrease the liquid holdup. An increase in Galileo number for the liquid also decreases the liquid holdup. These statements are generally verified by three theoretical models: those of Hutton and Lueng (1974), Reynier and Charpentier (1971), and Clements (1978). It is important to note that none of the controlling dimensionless groups (Re_L, Re_G, Ga_L, and $a_s d_p$) in liquid holdup correlations include a surface tension term (i.e., the liquid holdup is considered to be independent of surface tension). This conclusion needs to be verified in view of the observed dependence of flow regimes and pressure drop on surface tension.

Gas Holdup

In a packed-bed column, the gas holdup can be evaluated if the total liquid holdup and void volume of the column are known. However, in cases where the gas holdup is small, it is more desirable to evaluate it directly rather than through the difference. For cocurrent upflow in a bubble flow regime, Ford (1960) suggested the relation

$$\varepsilon_G = 21.2\left(\frac{Re_L}{Re_G}\right)^{0.2}\left(\frac{\mu_L}{\mu_G}\right)^{0.24} \tag{8}$$

whereas Achwal and Stepanek (1976) proposed two correlations for the bubble flow regime: one based on the homogeneous flow model and the other based on force balance on the column.

Axial Dispersion

Liquid Phase

While in most commercial trickle-bed operations, the liquid phase is believed to move in plug flow, in pilot-scale operations, the liquid phase can be backmixed. As shown in Fig. 7, the axial dispersion in the liquid phase under trickle flow conditions can be an order of magnitude higher than that in the single phase. For trickle flow, the correlation of Hochman and Effron (1969),

$$Pe_L = 0.042Re_L^{0.5} \tag{9}$$

where $Re_L = U_{oL}\rho_L d_p/\mu_L(1 - \varepsilon)$, is recommended. For the cocurrent upflow packed bubble column, the correlation shown in Fig. 8 is recommended. No similar correlation for the pulsating flow regime is presently available.

Gas Phase

Just as for the liquid phase, the axial dispersion in the gas phase under trickle flow conditions is higher than that under single-phase flow conditions. The correlation of Hochman and Effron (1969),

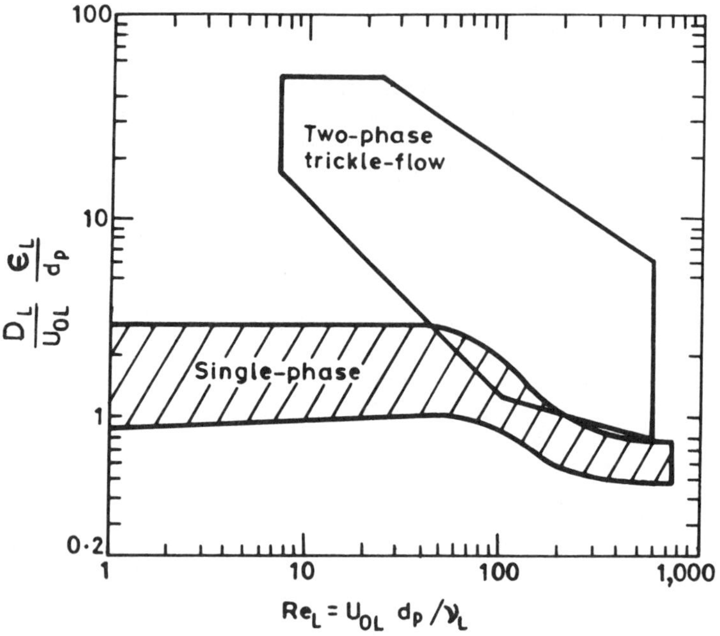

FIGURE 7 Pe and Re_L relations for single-phase and trickle flow. (After Hofmann, 1977.)

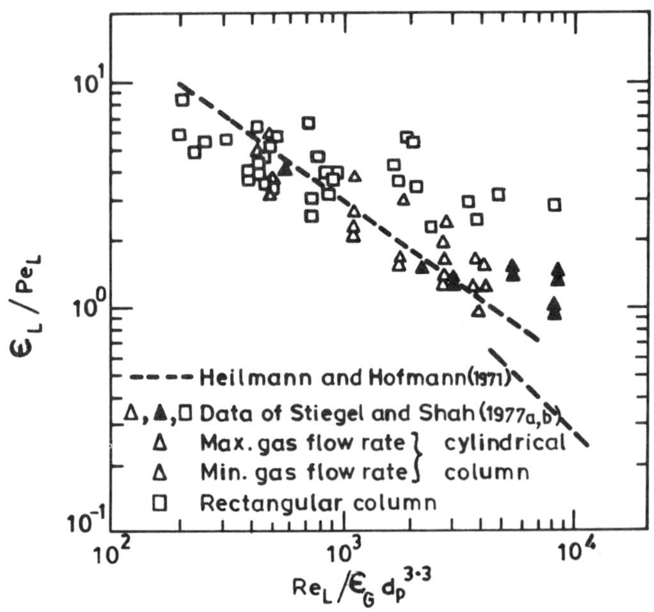

FIGURE 8 Heilmann-Hofmann correlation for backmixing in a cocurrent upflow packed bubble column. (After Stiegel and Shah, 1977a.)

$$Pe_G = 1.8Re_G^{-0.7}10^{-0.005Re_L} \tag{10}$$

where $Pe_G = U_G d_p/D_G$, $Re_G = G_G d_p/\mu_g(1 - \varepsilon)$, and $Re_L = G_L d_p/\mu_L(1 - \varepsilon)$ for trickle flow conditions, is recommended. No similar correlations for the bubble flow and pulsating flow regimes are presently available.

Mass Transfer

Gas-Liquid

Just as for liquid holdup, the correlations for the mass transfer coefficient ($K_L a_L$) are reported in two ways. Some investigators correlate $K_L a_L$ with liquid and gas velocities in dimensional (Gianetto et al., 1970; Goto and Smith, 1975; Goto et al., 1976) or dimensionless (Goto and Smith, 1975) forms, whereas others (Charpentier, 1976; Reiss, 1967; Satterfield, 1975) have presented energy correlations.

The dimensional correlations assume that $K_L a_L \propto U_L^{\bar{r}} U_G^s$, where the value of $\bar{r}$ and s depend on the types of packings (Shah, 1979). For very small gas and liquid velocities Goto and Smith (1975) and Goto et al. (1976) gave more accurate predictions. Charpentier (1976), Reiss (1967), Giannetto et al. (1970), and Specchia et al. (1974) gave relations between $K_L a_L$, a_L, and K_L and the energy parameter. Reiss (1967) gave a correlation $K_L a_L = 0.173\ [(\Delta P/\Delta Z)_{LG} U_{OL}]^{0.5}$, which was subsequently modified by Satterfield (1975) and later by Charpentier (1976) to account for the liquid properties. Specchia et al. (1974) gave energy correlations for both upflow and downflow, as shown in Figs. 9 to 11, and indicated that for the same value of energy parameters, upflow gives a better transfer coefficient and interfacial area than does downflow.

Better values of K_L are obtained for slower liquid velocities in upflow compared to downflow, presumably due to an increase in circulation inside

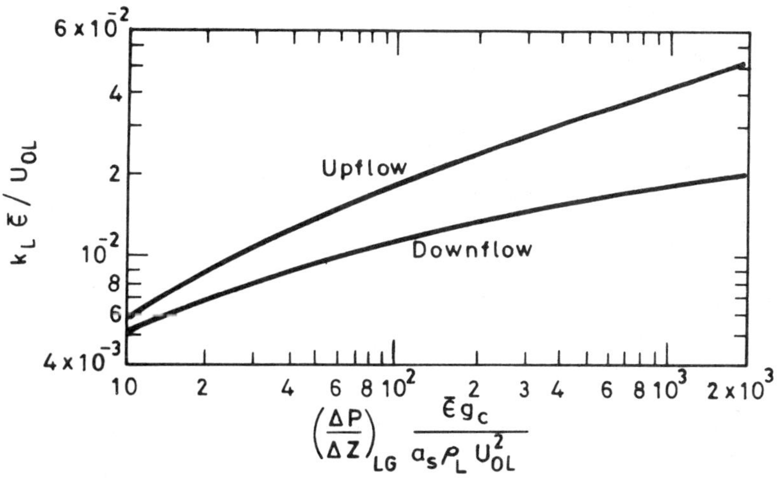

FIGURE 9 Energy correlations for k_L. (After Specchia et al., 1974.)

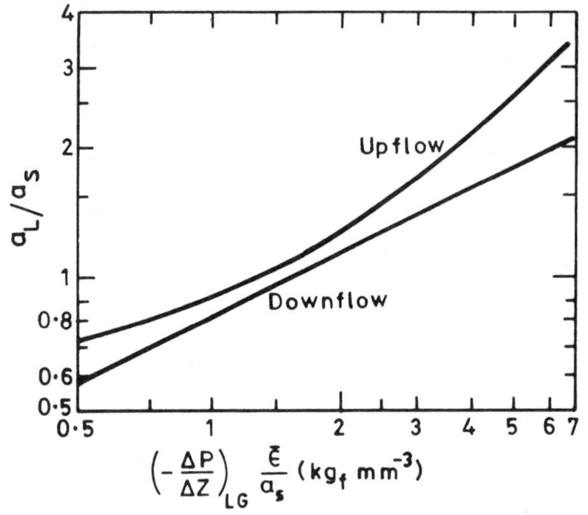

FIGURE 10 Energy correlations for a_L/a_s. (After Specchia et al., 1974.)

the liquid drops, caused among other things, by the greater slip velocity between the liquid and gas phases. Specchia et al. (1974) also showed that the upflow values of K_La_L are, on the average, two times that of downflow values in pulsed and spray flow regimes because the gravitational force leads to higher liquid holdup and pressure drop.

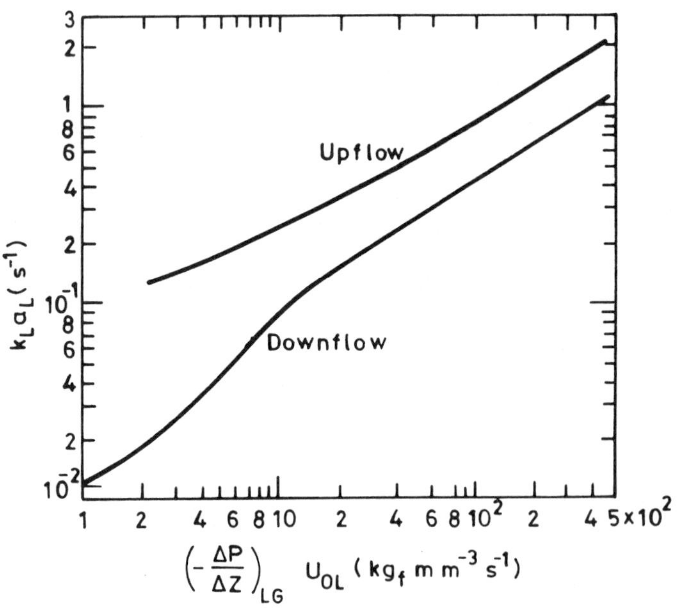

FIGURE 11 Energy correlations for k_La_L. (After Specchia et al., 1974.)

Charpentier (1976) suggested that if no reliable data or correlations for a given packing are available, as a conservative approximation, K_La_L can be taken to be 0.15 s^{-1} in the bubble flow regime. He also suggested that the gas-liquid interfacial area varies with about 0.5 power of the superficial gas velocity, regardless of the packing size and type, column diameter, and superficial liquid velocity. Morsi and Charpentier (1981) indicated that the mass transfer properties of hydrocarbon systems are substantially different from those of air-water systems. Also, the interfacial area (a_L) for hydrocarbon systems is much less than that for air-water systems. The effective interfacial area and the liquid-side mass transfer coefficient for trickle-bed reactors packed with a variety of packings and for organic liquids have also been measured by Mahajani and Sharma (1979, 1980). Future work must be directed toward systems other than air-water and under high-temperature and high-pressure conditions.

Liquid-Solid

Under trickle flow conditions, the liquid-solid mass transfer coefficient can be estimated using the correlation of Goto et al. (1975) (see Fig. 12) or that of Dharwadkar and Sylvester (1977). In the pulsating flow regime the correlation of Lemay et al. (1975) is recommended. For the packed bubble flow column, the correlation of Kirillov and Nasamanyan (1976) is recommended for large particles ($d_p > 3$ mm) and the correlation presented in Fig. 12 is recommended for small particles ($d_p < 3$ mm).

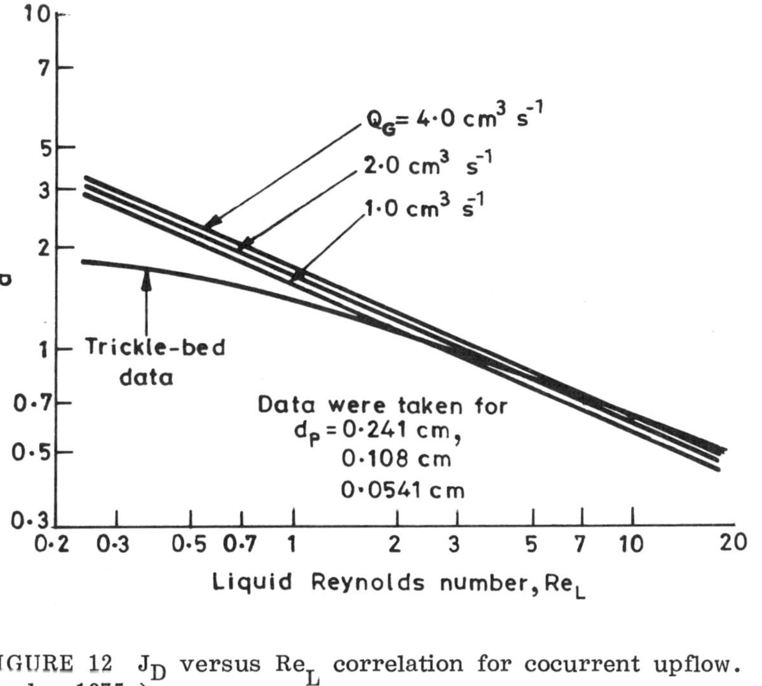

FIGURE 12 J_D versus Re_L correlation for cocurrent upflow. (After Goto et al., 1975.)

Heat Transfer

The data reported on heat transfer in trickle-bed reactors are sparse. Weekman and Myers (1965) measured wall-to-bed heat transfer coefficients in a cocurrent air-water downward flow through packed column. Their results are discussed by Shah (1979). A more recent work on heat transfer in trickle-bed reactors is that by Specchia and Baldi (1979). In the low-interaction regime, they proposed a correlation between Nusselt number and Prandtl number and a modified liquid Reynolds number. In the high-interaction regime, the heat transfer coefficient was found to be independent of gas flow rate and packing size and shape. Specchai and Baldi (1979) also obtained some results for the effective thermal conductivity of the bed in both low- and high-interaction regimes.

Scaleup Considerations

Experiments on the laboratory or pilot scale are invariably conducted with 25-mm-I.D. and 0.5- to 2-m-long reactors at about the same liquid hourly space velocity as in the industrial reactor. However, the latter reactor may have a height of 20 to 25 m. Thus the superficial liquid velocity in the pilot reactor may be merely about 10% of the industrial unit and one may have different flow regimes in the two cases; if this is the case, the reactor scaleup may be seriously jeopardized. We have already emphasized the importance of different flow regimes, in particular the different functional dependence of various design variables, on operating parameters. In fact, if the proper precautions are not taken, one may get a situation where external mass transfer resistance may become important or dominant in the pilot reactor but may be unimportant in the industrial reactor. There may also be different residence-time distributions for gas and liquid phases in the two cases. One may often benefit by replacing expensive catalyst with inexpensive support material of equivalent dimension when solid-liquid or gas-liquid film resistance is important.

General Remarks

Very little is known on the hydrodynamic, mixing, and transport characteristics of moving fixed-bed reactors. It is suggested that such reactors should be divided on the basis of the prevailing flow regime and approximate correlations for each flow regime should then be applied.

HYDRODYNAMIC, MIXING, AND TRANSPORT CHARACTERISTICS OF SLURRY REACTORS

As shown in Fig. 1, a variety of slurry reactors are used in industrial practice. Some of these are:

1. Gas sparged (including vibrating) reactor
2. Mechanically agitated (including gas inducing) reactor
3. Three-phase fluidized reactor
4. Transport reactor
5. Multitubular reactor
6. Loop reactor

7. Plate column reactor
8. Pipe reactor
9. Spouted bed reactor

Some novel distillation column reactors for MTBE and for some hydrogenation reactions, such as the conversion of isophorone to trimethylcyclohexanone and trimethylcyclohexanol, have also been reported (Smith and Huddleston, 1982; Schmitt, 1960).

In the following we discuss primarily reactor types 1 and 2. Some details are, however, applicable to other types of reactors, and these are noted.

Flow Regime Boundaries

In gas sparged, vertical bubble columns, flow regime maps have been described by a number of researchers and are discussed by Shah et al. (1982). As an approximate evaluation of the flow regime in a vertical slurry reactor, Fig. 13 can be used. It should be, noted, however, that the type of sparger used, liquid velocity, physicochemical properties of the liquid, and the presence of solid may also affect the flow regime boundaries. Wallis (1962, 1969) has suggested the use of flow regime charts based on the drift flux approach. Kara (1981) and Kelkar (1982) have recently shown that this approach can be applied to three-phase slurry reactor systems. The flow regime for gas-liquid flow in horizontal pipe has been discussed at great length by Govier and Aziz (1972). No similar flow maps for three-phase flow are presently available.

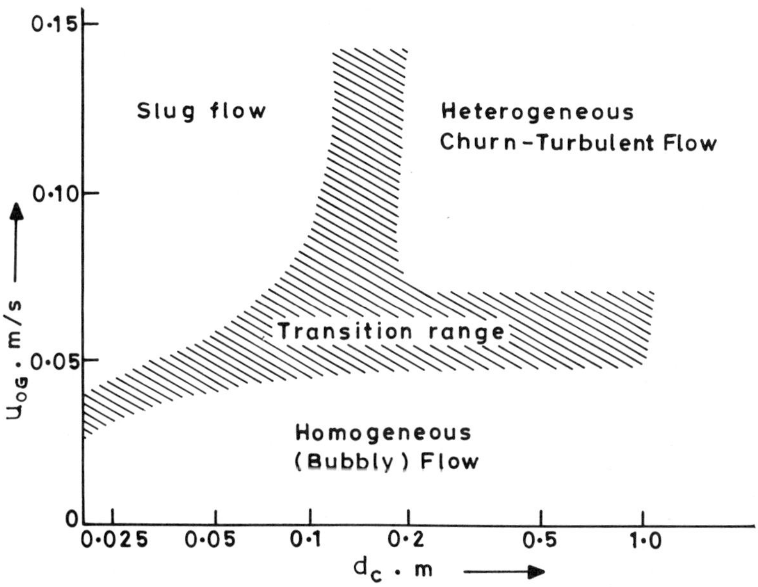

FIGURE 13 Approximate dependency of flow regime on gas velocity and column diameter (water and dilute aqueous solutions). (From Shah et al., 1982.)

Bubble Dynamics

Bubble size and its distribution and bubble rise velocity have a direct bearing on the performance of the bubble columns. A brief review of this subject has been given by Shah et al. (1981, 1982). When gas is sparged by single orifices or perforated or sintered plates, the following correlation of Akita and Yoshida (1974) can be used as first approximation for the estimation of average bubble size:

$$\frac{d_{bs}}{d_c} = 26\left(\frac{d_c^2 g \rho_L}{\sigma}\right)^{-0.5}\left(\frac{g d_c^3}{\nu_L^2}\right)^{-0.12}\left(\frac{u_{oG}}{g_c d}\right)^{-0.12} \tag{11}$$

Relation (11) has not been tested for three-phase systems. For a stirred vessel the correlation of Calderbank (1967),

$$d_{bs} = C\left[\frac{\sigma^{0.6}}{(P/V_D)^{0.4}\rho_L^{0.2}}\right]\varepsilon_G^n\left(\frac{\mu_G}{\mu_L}\right) \tag{12}$$

may be used as a first approximation. Here constants C and n depend on the stirrer type and the type of liquid phase. P/V_D represents the energy dissipation rate per unit volume of dispersion.

The bubble size distribution is often given in terms of large bubbles and small bubbles. The definition of large and small bubbles has been based on the dynamic bubble disengagement technique described by Sriram and Mann (1977), Schumpe (1981), and Vermeer and Krishna (1981), among others. Based on this technique, the fraction of gas throughput carried by large bubbles as a function of gas velocity obtained by Beinhauer (1971) is shown in Fig. 14. For slurry reactors, such a plot should depend on the

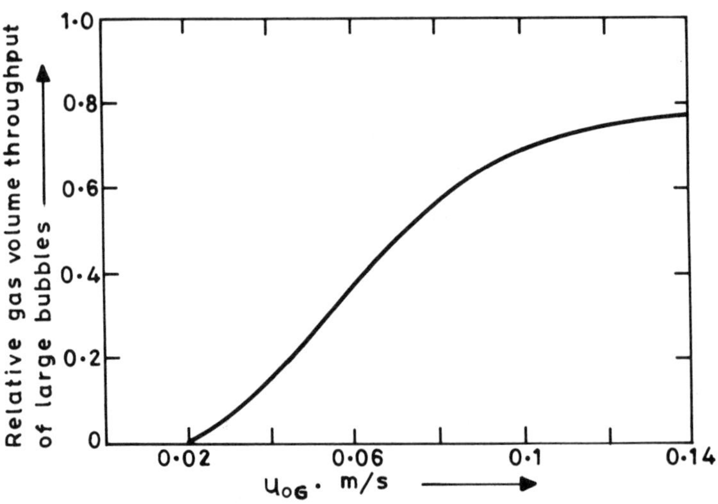

FIGURE 14 Fraction of volume throughput of large bubbles. (Calculated from Beinhauer, 1971.)

physical properties of liquid, solids particle size, nature, and concentration, a relationship not yet been developed in the literature.

Kara (1981) and Kelkar (1982) found that Zuber and Findlay's modification of drift flux theory predicted meaningful bubble size, bubble rise velocity, and radial distribution of gas holdup in three-phase slurry reactors. The bubble size evaluation for the gas-liquid flow in horizontal pipes is given by Govier and Aziz (1972). A similar evaluation of the mechanically agitated contactor is given by Joshi et al. (1982).

Minimum Gas and Liquid Velocities Required for Fluidization

In a three-phase system the particles form the fluidized phase, with the liquid as the fluidizing medium. The gas passing through the system imparts the requisite energy to the liquid to keep the particles suspended. The bed is fluidized when the superficial velocity past the particles is greater than their settling velocity. Kato (1963), Roy et al. (1964), Imafuku et al. (1968), Narayanan et al. (1969), and Begovich and Watson (1978) studied the minimum gas velocity required to suspend particles in a stagnant liquid medium. These works are reviewed by Shah (1979) and Shah et al. (1982).

Begovich and Watson (1978) measured the minimum gas and liquid velocities required to fluidize various types of solids as a function of particle size, particle density, and liquid viscosity. No effect of the initial bed height or column diameter was found. In mechanically agitated contactors, some minimum impeller speed is required to keep the particles in suspension. For liquid-solid suspensions (in the absence of gas), Zwietering (1958) proposed such a correlation. Arbiter et al. (1976) and Subbarao and Taneja (1979) have shown that the value of minimum speed is higher in the presence of gas then in the absence of gas. Smith and coworkers (Smith, 1980) have studied the effect of cavity formation, behind the impeller blades, on the impeller speeds required for suspension of solid particles in gas-liquid-solid systems. More discussion of this subject is given by Joshi et al. (1982).

Phase Holdup

The available correlations for the gas holdup in two- and three-phase bubble columns are reviewed by Shah et al. (1982). For slurry reactors, the gas holdup depends mainly on the superficial gas velocity and is often very sensitive to the physical properties of the liquid. The dependence of the gas holdup on gas velocity is generally of the form

$$\varepsilon_G \alpha U_{OG}^n \tag{13}$$

where $0.7 < n < 1.2$ for bubbly flow and $0.4 < n < 0.7$ for churn turbulent flow.

A large number of correlations for gas holdup have been reported in the literature, however, the large scatter in the reported data does not allow a single correlation. Some of the important correlations are reported by Shah et al. (1982). The large scatter (Schügerl et al., 1977; Bach and Pilhofer, 1978) is due mainly to the extreme sensitivity of the holdup to

the material system and to the trace impurities, which is not well understood. The easily available physical properties, such as density, viscosity, and surface tension, are not necessarily sufficient to expalin the scatter observed. The effect of gas properties on the gas holdup is shown to be small (Hikita et al., 1980). Kölbel et al. (1961) and Deckwer et al. (1980a) have also shown a very small effect of pressure on the gas holdup.

Jeffrey and Acrivos (1976) found that for particles as small as a few (<10) micrometers, suspensions behave macroscopically as non-Newtonian fluids. Therefore-gas holdup values involving non-Newtonian liquids might be useful in analyzing gas-slurry systems.

Gas holdup data in non-Newtonian media have been reported by Nishikawa et al. (1977), Buchholz et al. (1978), Nakanoh and Yoshida (1980), and Schumpe et al. (1981). Gas holdup values higher than predicted by the correlation of Akita and Yoshida (1973) were obtained. The effect of spargers was found to be insignificant in the slug flow regime.

Hikita et al. (1980) compared their data with those predicted from different correlations and attributed the disagreement between theories partly to the differences in the design of gas sparges. They concluded that at low gas velocities, the single-nozzle gas sparger gives lower values of gas holdup than does a multinozzle or a porous plate sparger. Zahradnik and Kastanek (1979) carried out experiments with sieve plates of different hole diameters and observed that the critical gas velocity corresponding to the onset of stable performance of a plate depends on the plate hole diameter. Numerous data for different kinds of spargers and various liquid media (synthetic fermentation media) have been reported by Schügerl et al. (1977) Oels et al. (1978). Hills (1974) and Ueyama et al. (1980) reported pronounced radial holdup profiles, and Kobayashi et al. (1970) proposed an empirical correlation to evaluate such a radial distribution.

Shah (1979) has critically reviewed the published information on individual phase holdups and has recommended the correlations of Kim et al. (1972, 1975) and Hovmand and Davidson (1968). More recently, Dhanuka and Stepanek (1978) measured and theoretically predicted the individual phase holdups in the 0.05-m-I.D. column. Begovich and Watson (1978) determined the individual phase holdups from 0.076- and 0.152-m-I.D. columns and developed correlations which cover their own data and those reported by Østergaard (1965), Østergaard and Thiesen (1966), Østergaard and Michelsen (1968), Michelsen and Østergaard (1970), Rigby and Capes (1970), Efremov and Vakhrusev (1970), Dakshinamurthy et al. (1971), Bhatia and Epstein (1974), and Kim et al. (1975).

Kato et al. (1972) and Kato and Nishiwaki (1972), Ying et al. (1980a), and Vasalos et al. (1980) investigated the effect of solids concentration and concluded that an increase in solids concentration generally decreases the gas holdup but that the effect becomes insignificant at high gas velocities (>0.1 m/s). Ying et al. (1980a) also applied the correlation of Akita and Yoshida (1973) to their data and concluded that the correlation of Akita and Yoshida is equally adequate for the three-phase system. Kojima and Asano (1980) measured the fractional gas holdup and average bubble diameter from 0.055- and 0.095-m-I.D. contactors. The particle size was varied in the range of 105 to 2000 μm. The liquid viscosity was varied from 1×10^{-3} to 18.8×10^{-3} Pa·s. The results were empirically correlated.

Liquid fluidized beds of small particles have been found to contract upon the introduction of gas (Stewart and Davidson, 1964; Østergaard and Thiesan, 1966; Kim et al., 1972), while the reverse trend occurs with large particles. The contraction can be explained by considering liquid wakes

behind the gas bubbles. The liquid in the wake moved at a much faster rate than the continuous liquid phase. As a result, the average velocity of the bulk liquid phase decreases and the bed contracts, causing an increase in the solid holdup. The large particles can cause a bubble breakup by virtue of their inertia and hence result in an expansion in fluidized beds. Bhatia et al. (1972) and Armstrong et al. (1976a) studied the effect of solids wettability on the behavior of three-phase fluidized beds of small and large particles, respectively. They have also qualitatively described the role of wakes and bubbles in such bed. Kim et al. (1975) have reported the existence of two distinct types of three-phase fluidization, which may be termed as *bubble coalescing* and *bubble disintegrating*. The former occurs when the particles are smaller than a critical size and the latter occurs when they are larger.

Axial solid concentration has been studied by Cova (1966), Imafuku et al. (1968), Farkas and Leblond (1969), Narayanan et al. (1969), Yamanaka et al. (1970), and Govindrao (1975). Cova (1966) reported that high gas and liquid velocities and high viscosities tend to give more uniform solid distribution. Imafuku et al. (1968) observed that the critical gas velocity for complete suspension depends mostly on the liquid flow near the gas distributor, while Govindrao (1975) pointed out that particle diameter and bed volume have a strong influence on the axial distribution of solids. As a rule of thumb, in a cocurrent column, particles less than 100 μm in diameter can form a pseudohomogeneous slurry, while particles greater than that will result in some axial solids distribution.

In multistage columns Schügerl et al. (1977) observed higher values of gas holdup than in a single-stage column. They also noted a strong effect of the distributor design in multistage columns for a noncoalescing medium. However, in an air-water medium, Zahradnik et al. (1974) and Zahradnik and Kastanek (1974) observed very little effect of the distributor design and the liquid velocity in multistage columns. Freedman and Davidson (1969) and Botton et al. (1978) carried out experiments with draught tubes and observed insignificant effects of the draught tube on the holdup values. Wieland (1978) reported that in a bubble column with an external loop, the holdup values are comparable to the values in a single-stage bubble column. As a rule of thumb, it can be concluded that the effect of internals on the gas holdup is negligible, and any correlation that fairly represents the values of holdup in a single-stage bubble column can be used to calculate the values in the presence of internals. As a first approximation the same conclusion should apply to three-phase systems. The phase holdups in mechanically agitated contactor is discussed by Joshi et al. (1982).

Axial Dispersion

Liquid Phase

The effects of suspended solid particles on liquid-phase backmixing have been studied by Schügerl (1967), Imafuku et al. (1968), Vail et al. (1968), Michelsen and Østergaard (1970), Kato et al. (1972), Østergaard (1978), and El-Temtamy et al. (1979). Shah et al. (1978) have critically reviewed the literature and outlined the present state of the art. The correlation reported by Kato et al. (1972) is suitable for cases where the particle size is relatively small. The experimental observations may be summarized as follows:

1. When the particles are very small and/or the difference between the densities of liquid and solid is small and/or the solid loading is relatively small, the extent of liquid-phase axial mixing in three-phase systems is practically the same as that for gas-liquid systems. The liquid-phase axial dispersion coefficient varies with column diameter and gas velocity, and under turbulent conditions it is practically independent of liquid velocity and particle diameter.
2. When the particle size is large and the density difference is high, the dispersion coefficient depends on the particle size and the superficial gas and liquid velocities (Michelsen and Østergaard, (1970).

Joshi (1980) has shown that the observation of Michelsen and Østergaard (1970) can be explained if the average liquid circulation velocity (V_c) is selected as the correlating parameter. When the rate of energy input is higher than that dissipated at the gas-liquid and liquid-solid interfaces, recirculation of liquid occurs. The data reported by Kato et al. (1972), Michelsen and Østergaard (1970), Østergaard (1978) and Vail et al. (1968) were analyzed by Joshi (1980). From an accuracy point of view, only those data were analyzed for which at least 20% of the input energy is dissipated in the liquid motion. Based on the analysis, the following equation with a standard deviation of 16% was obtained:

$$D_L = 0.29(V_c + V_{OL})d_c \tag{14}$$

Riquarts (1981) made use of another theoretical approach. In analogy to the well-known Pe = 2 relation for fixed beds, Riquarts assumed that this relation is also valid for bubble columns. The liquid-phase dispersion coefficient is given by

$$D_L = \frac{1}{2} u_{b\infty} d_b \tag{15}$$

and introducing appropriate expressions for the bubble rise velocity and bubble diameter, he arrived at the equation

$$Pe_L = 14.7\left(\frac{Fr_L^3}{Re_L}\right)^{1/8} \tag{16}$$

where

$$Fr_L = \frac{U_{OL}^2}{gd_c} \qquad Pe_L = \frac{U_{OL}d_c}{D_L} \qquad Re_L = \frac{U_{OL}d_c}{\nu_L} \tag{17}$$

Equation (16) describes measured data with accuracy similar to that of other correlations reported in the literature.

The effect of solid particles on the liquid dispersion coefficient is not clearly understood due to the lack of enough experimental evidence. At high solid concentrations, Ying et al. (1980b) observed large discrepencies between actual and theoretical values of dispersion coefficients predicted

with the help of different existing correlations. They also observed the effect of liquid velocity on dispersion coefficients. The observations are supported by Kara et al. (1982) and Kelkar (1982), who noted that the dispersion coefficients are lower at high liquid velocities than those observed in the absence of solids.

Solid Phase

Kato et al. (1972) studied solid-phase backmixing under a wide range of conditions. They found that for small particle sizes in small-diameter columns, the axial dispersion coefficient for solids is the same as that for the liquid. For relatively large particle sizes (up to 0.177 mm) the dispersion coefficient depends on the particle Reynolds number (Shah, 1979). More experimental investigations pertaining to solid-phase backmixing in the presence of large particles (>0.2 mm) are needed.

Gas Phase

The gas-phase backmixing in gas-liquid-solid systems have been measured by Schügerl (1967), Østergaard and Michelsen (1968), and Michelsen and Østergaard (1970). Schügerl (1967) reported that at low liquid velocities, the gas-phase Péclet number increases with the gas rate, but at high liquid velocities, the Péclet number shows a maximum with respect to gas rate. From Schügerl's data, the effect of liquid velocity on the gas-phase Péclet number is unclear, although at low velocities, the gas-phase Péclet number appears to decrease with increase in liquid velocity. Based on studies of many nonaqueous and aqueous systems, in 0.1- and 0.15-m-I.D. bubble columns, Mangartz and Pilhofer (1980) reported the following equation for the gas-phase dispersion coefficient:

$$D_G = 50d_c^{1.5}\left(\frac{U_{OG}}{\varepsilon_G}\right)^3 \tag{18}$$

where D_G is in m^2/s, d_c is in meters, U_{OG} in m/s, and ε_G is fractional gas holdup in the presence of solids. Field and Davidson (1980), who conducted measurements on gas-phase dispersion in a large-diameter (32 times that used by Pilhofer et al. (1978)] column, slightly modified Eq. (18) and proposed the empirical correlation

$$D_G = 56.4d_c^{1.33}\left(\frac{U_{OG}}{\varepsilon_G}\right)^{3.56} \tag{19}$$

Although the predictions of Eqs. (18) and (19) may differ up to 50%, both correlations are recommended due to lack of sufficient literature data. More data on three-phase systems are needed.

Mass Transfer

Gas-Liquid

Gas-liquid mass transfer in a bubble column has been reviewed extensively by Shah et al. (1982). In three-phase bubble column reactors, K_La_L can

be affected by the presence of solids. Investigations of various authors (Voyer and Miller, 1968; Slesser et al., 1968; Sharma and Mashelkar, 1968; Kato et al., 1973; Juvekar and Sharma, 1973; Sittig, 1977; Zlokarnik, 1979) indicate that the degree of influence of suspended particles on k_La_L depends on the particle concentration, the particle size, the liquid solid density difference, the geometrical sizes, and the operating conditions of the reactor (i.e., gas and liquid velocities). Nguyen-Tien and Deckwer (1981) showed that at high liquid velocities (U_{OL} = 0.093 m/s) and low gas velocities, the k_La_L values are slightly higher in the presence of solids. A small increase in k_La_L was also reported by Sittig (1977) for the solids (various plastic particles) concentrations below 15 wt % and particle sizes between 50 and 200 μm. With rising gas velocities and decreasing liquid velocities, the particle distribution becomes increasingly nonuniform and the k_La_L values are lower than the ones without the solids. The influence of the solid concentration on k_La_L depends largely on the liquid and gas velocities. Kato et al. (1973) have also shown that for higher solid concentration a steep decrease in k_La_L is found, which is caused by a decrease in a_L. It has been shown by Dhanuka and Stepanek (1980) that with an increase in particle size, k_La_L decreases because of a decrease in a. For the typical operating conditions prevailing in Fischer-Tropsch synthesis in the slurry phase, Zaidi et al. (1979) and Deckwer et al. (1980) have shown that the presence of solid particles (diameter less than 50 μm, concentrations of solids <16 wt %) has a negligible effect on k_La_L.

Weiland (1978) investigated mass transfer in a bubble column with an external loop. He reported an agreement of k_La_L data for water-O_2 with the data reported by Yoshida and Akita (1965) and Lin et al. (1976). In the absence of other data, these data should be used as a first approximation for three-phase systems.

Voigt et al. (1979, 1980) and Hecht et al. (1980) have studied mass transfer in a multistage bubble column for electrolyte solutions and highly viscous Newtonian and non-Newtonian (carboxymethyl cellulose and polyacrylate) solutions. They have recommended use of multistage columns for a coalescence promoting medium, although for non-Newtonian solutions better mass transfer can be achieved in single-stage bubble columns (Voigt et al., 1980). Hydrogenation of dissolved elastomeric materials may show pronounced non-Newtonian behavior (Falk, 1976). Mass transfer in such systems has been reviewed by Kumar et al. (1978) and Mashelkar and Chavan (1973). Fan et al. (1976) and Wang and Fan (1978) studies oxygen transfer in a bubble column packed with Koch motionless mixers. They reported a significant increase (as high as twofold) in k_La_L and dependence of k_La_L values with Koch mixers than the ones with sieve trays in bubble columns. Orazen and Erickson (1979) reported significantly larger k_La_L in the two-stage airlift tower for a gas velocity between 0.20 and 0.45 m/s. No difference in k_La_L was apparent at superficial gas velocities from 0.12 to 0.20 m/s. Lin et al. (1976) have shown that in tower cycling fermenters for volatile substrates, higher mass transfer rates can be achieved. Fujie et al. (1980) observed higher volumetric oxygen transfer coefficients in a downflow bubble column compared to those in conventional upflow systems.

The average liquid-side volumetric mass transfer coefficient depends on the average bubble diameter and their rise velocity. The bubble dynamics in a cocurrent-upflow, three-phase fluidized-bed system have been studied by Massimilla et al. (1959, 1961), Østergaard (1966), and Lee (1965).

Massimilla et al. (1961) found that the bubble size and the rate of bubble coalescence decreased with increasing fluidization intensity. These results were in agreement with a subsequent study of Østergaard (1966). Adlington and Thompson (1965) found that the gas-liquid interfacial area decreased with decreasing bed porosity and was less sensitive to changes in particle size. However, for large particles and large bubbles, Lee (1965) showed that the gas-liquid interfacial area increased with increasing distance away from the gas distributor.

The adosrption rate depends on the gas-liquid interfacial area, the gas residence time, and the gas-liquid mass transfer coefficient. It is the first factor that has an important and complex effect on the mass transfer rate in three-phase fluidized-bed systems. The study of Østergaard and Suchozebrski (1968) showed that the volumetric mass transfer coefficient (i.e., $K_L a_L$) increased with increasing gas flow rate but was uninfluenced by variations in the liquid velocity. Particle size was observed to have a pronounced effect on the adsorption rate. Østergaard and coworkers also showed that an increase in the superficial liquid velocity had no effect on the absorption rate in beds of 6-mm particles and in solid-free bubble columns. In beds of 1-mm particles, however, an increase in liquid velocity caused a marked increase in the volumetric mass transfer coefficient. These results agreed well with the fact that both the gas holdup and bubble size, in beds of 6-mm particles and in the solid-free bubble column, were independent of liquid velocity, whereas in beds of 1-mm particles (østergaard, 1966, 1971b) an increase in the liquid velocity caused a marked reduction in bubble size and an increase in gas holdup. The rate of bubble coalescence was higher at low liquid velocities than at high liquid velocities in beds of 1-mm particles (Østergaard, 1966). An increase in the gas velocity increases the volumetric mass transfer coefficient.

Shah (1979) reviewed the studies on mass transfer across the gas-liquid interface. The best up-to-date work is reported by Joosten et al. (1977). They showed that in the absence of solids, the volumetric mass transfer coefficient can be well correlated to the total power, dissipated by stirrer and gas, per unit volume. However, the correlation with power dissipated by the stirrer alone was poor. Joosten et al. (1977) also found that for a given stirrer power input and superficial gas velocity, the volumetric mass transfer coefficient first increased with the volume of solids added. Although this is in qualitative agreement with the observations of Slesser et al. (1968) and Chandrasekaran and Sharma (1977), unlike these authors, Joosten et al. (1977), found the increase to be rather small (between 10 and 20%). Furthermore, as more solids were added, the volumetric mass transfer coefficient remained constant at first and then started to decline at a specific concentration that depended on solid type and particle size. Joosten et al. (1977) explained the data based on the increase in the apparent viscosity of the slurry by the addition of solids. Mass transfer in mechanically agitated contactor is discussed further in Joshi et al. (1982).

Liquid-Solid

In three-phase bubble column slurry reactors, the mass transfer from bulk liquid to the solid surface can play an important role in the overall apparent reaction rate (Choudhari and Ramchandran, 1980). Liquid-solid mass transfer is based on the slip velocity (boundary layer) between two phases, and in general, leads to such expressions as

$$Sh = 2.0 + \alpha Sc^m Re^n \tag{20}$$

Frossling (1938) and Ranz and Marshall (1952) showed that for large values of Re, $\alpha = 0.6$, $m = 1/3$, and $n = 1/2$. Shah (1979) has discussed in detail the recommendations of various investigators (Friedlander, 1957, 1961; Harriot, 1962; Levich, 1962; Brian and Hales, 1969) for evaluating the particle Reynolds number

$$Re = \frac{u_r d_p}{\nu} \tag{21}$$

where u_r is the relative velocity between two phases. In the case of bubble columns where the solid particles are suspended, this velocity is difficult to determine. Kolmogoroff's theory of isotropic turbulence avoids this problem, as discussed by Shah et al. (1982).

Only a few experimental studies (Kamawura and Sasano, 1965; Sano et al., 1974; Sanger and Deckwer, 1981) have been carried out in three-phase vertical bubble columns. These investigations and the proposed correlations using Kolmogoroff's characteristic parameters are outlined by Shah et al. (1982). The data of Sano et al. (1974) covers a range of Schmidt numbers from 200 to 1400. Sänger and Deckwer (1981) have provided the data for Schmidt numbers ranging from 137 to 50,000. The measured data deviated at low Reynolds numbers because the assumption of isotropic turbulence is not justified, as discussed by Levins and Glastonbury (1972). Figure 15 illustrates that the data of these three investigators are consistent. The difference in three curves is due partly to

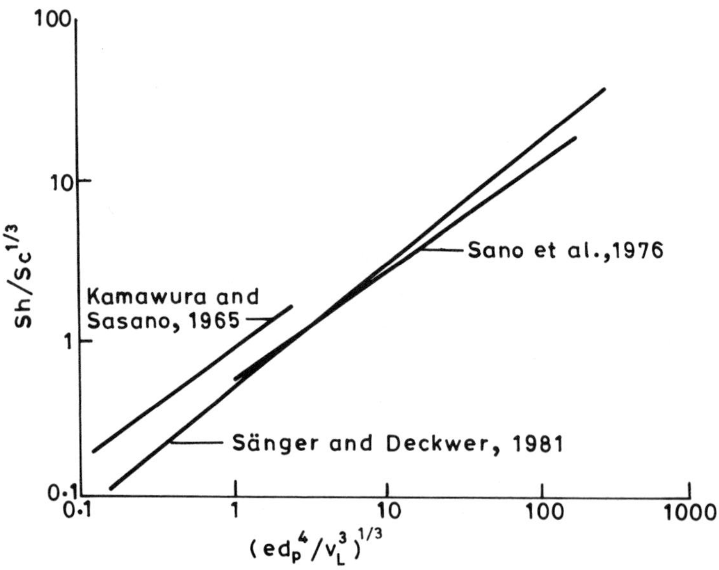

FIGURE 15 Correlations of liquid-solid mass transfer coefficients in bubble columns. (From Shah et al., 1982.)

uncertainties in the estimation of diffusivities. The figure also demonstrates the usefulness and applicability of the theory of isotropic turbulence. This is advantageous, as the energy dissipation can be easily measured in bubble columns. The liquid-solid mass transfer in bubble columns with and without internals have also been measured by Sharma and coworkers (Sharma, 1982) using Cu-dichromate system.

The equation by Sanger and Deckwer (1981) is recommended for the calculation of liquid-solid mass transfer since it covers a wide range of experimental parameters. More recommendations for mechanically agitated contactors are given by Joshi et al. (1982).

Heat Transfer

Heat transfer from the reactor wall and inserted rods and coils to the vertical flowing gas in liquid dispersion has been the subject of many investigations, and suitable correlations have been reported (Kölbel et al., 1958, 1960; Muller, 1958; Martins, 1959; Kast, 1963; Shaykhutdinov et al., 1971; Burkel, 1972; Mersmann, 1975; Hart, 1976; Steiff and Weinspach, 1978; Deckwer, 1980; Deckwer et al., 1980; Joshi et al., 1980; Kato et al., 1981). A summary of the published data collected on bubble columns and stirred vessels is presented by Steiff and Weinspach (1978).

Muller (1958), Kölbel et al. (1958, 1960), Viswanathan et al. (1965), Armstrong et al. (1976b), and Zaidi et al. (1979) measured heat transfer coefficients from the inserted heat source to the bed (gas-liquid-solid dispersion) or between the column wall and the bed. Analysis of the reported data indicate the following behavior of gas-liquid-solid systems:

1. The values of the heat transfer coefficient h_W are at least 10 to 100 times higher then that for single-phase (gas or liquid) flow, 0.9 to 1.3 times higher than for bubble columns, and 1.2 to 25 times higher than for gas-solid fluidized beds.
2. The values of h_W are practically independent of the column diameter and dimensions of heat transfer element.
3. In general, h_W increases with an increase in particle diameter up to 3 mm. However, for particle sizes larger than 3 mm, h_W becomes essentially independent of the particle diameter (Armstrong et al., 1976b).
4. When the solid concentration ε_S is increased from zero, at the beginning h_W increases with an increase in ε_S, reaches a maximum value at a certain value of ε_S, and then falls.
5. h_W varies as $U_{OG}^{0.25}$. Zaidi et al. (1979) have reported that above a U_{OG} of 0.1 m/s, h_W is independent of U_{OG} in a 0.1-m-I.D. column. This behavior was not observed by Armstrong et al. (1976b) up to a U_{OG} of 0.23 m/s in a 0.24-m-I.D. column.
6. The design of gas sprager has no effect on h_W.

For suspensions of Kieselguhr in water and oil, Kölbel et al. (1958, 1960) reported separate correlations for laminar and turbulent regions. Their results are discussed by Shah (1979). Vishwanathan et al. (1965) and Arstrong et al. (1976b) reported their experimental data in the form of graphs and correlations are not proposed.

Deckwer (1980) has analyzed the problem of heat transfer on the basis of Higbie's penetration theory together with Kolmogoroff's theory of isotropic turbulence and proposed the correlation

$$St = \alpha(ReFrPr^2)^{-1/4} \tag{22}$$

where

$$St = \frac{h_w}{\rho C_p U_{OG}} \qquad Re = \frac{U_{OG} d_{bs}}{\mu} \qquad Fr = \frac{U_{OG}^2}{g d_{bs}} \qquad Pr = \frac{C_p \mu}{K}$$

where k, ρ, C_p, and μ are the thermal conductivity, density, specific heat, and viscosity of the gas-liquid-solid system. These are calculated by the following equations:

$$\bar{k} = \bar{k}_L \frac{2\bar{k}_L + \bar{k}_s - 2\varepsilon_S(\bar{k}_L - \bar{k}_s)}{2\bar{k}_L + \bar{k}_s + \varepsilon_s(\bar{k}_L - \bar{k}_S)} \qquad \rho = \varepsilon_S \rho_S + \varepsilon_L \rho_L \tag{23}$$

$$\mu = \mu_L(1 + 4.5\varepsilon_S)$$

Zaidi et al. (1979) have shown that the data reported by Muller (1958), Kölbel et al. (1958, 1960), and Zaidi et al. (1979) can be correlated with Eq. (22) within 20%, and the value of α is reported to be 0.1.

For design purposes, Eq. (22) is recommended. However, a precaution is desirable for cases where no more than 80% of the input energy is dissipated at the gas-liquid and liquid-solid interfaces and at the column wall. Further Eq. (22) was proposed for slurry reactors, and therefore for the case of three-phase fluidized beds, U_{OG} may be replaced to get the form

$$h_w = 0.1\bar{k}^{-0.5} \rho^{0.75} C_p^{0.5} \mu^{-0.25} g^{0.25} \left[U_{OG} + U_{OL}\left(1 - \frac{\rho_L}{\varepsilon_S \rho_S + \varepsilon_L \rho_L}\right)\right]^{0.25} \tag{24}$$

Joshi et al. (1980) have proposed a new correlation based on the circulation cell model (Joshi, 1980). They reported that in some gas-liquid-solid systems, all the input energy due to the gas flow is dissipated at the gas-liquid and solid-liquid interface rather than in the liquid motion. By using the method of average recirculation velocity (Joshi, 1980), they proposed the following equation to calculate the wall bed heat transfer coefficient:

$$\frac{h_w d_c}{\bar{k}} = (0.48) \frac{d_C^{1.33} g^{0.33} (U_G - \varepsilon_G u_{b\infty}) \rho^{1/3}}{\mu_L} \left(\frac{C_\rho \mu_L}{\bar{k}}\right) \times \left(\frac{\mu_L}{\mu_w}\right)^{0.14} \tag{25}$$

This equation is based on the analogy between mechanically agitated contactors and bubble columns. Joshi et al. (1980) showed that the apparent

discrepancy between the deterministic (Joshi, 1980) and the stochastic (Deckwer, 1980) approach is immaterial and that both physical concepts are equally capable of representing transfer phenomena in the liquid phase of bubble columns and an be regarded as limiting cases of a more general model. This subject is discussed in more detail by Joshi et al. (1982).

Scaleup Considerations

Laboratory experiments are invariably made in 70- to 150-mm reactors, and in the case of gas sparged reactors, the height-to-diameter ratio may be 3:10. In large industrial reactors we often have a low height-to-diameter ratio of 1:2, and in the case of gas sparged reactors we also normally have lower values of superficial gas velocity. Even though the height-to-diameter ratio is low in industrial reactors, one may encounter problems associated with the hydrostatic head of the liquid and low values of superficial gas velocity. In particular, we have to ensure that the solid particles concentration is uniform throughout the reactor. In some cases we may encounter a substantial change in the superficial gas velocity from bottom to top, but such cases have received very little attention. When backmixing in the gas phase is relevant, one needs to ascertain this more closely. This problem may be particularly important in mechanically agitated reactors.

One must bear in mind that just as for the fixed-bed reactor, if expensive catalysts are used (say, those based on noble metals) and the resistance is essentially located in the solid-liquid film or gas-liquid film, one may benefit substantially by replacing a part of the catalyst with an equivalent support material (say, alumina or zeolite or activated carbon) but without the noble metal.

CALCULATIONS OF ABSORPTION/DESORPTION RATES

The calculation of absorption/desorption rates for a three-phase system when the solid phase is a catalyst is discussed at great length by Shah (1979), Shah and Sharma (1976), and Choudhary and Ramchandran (1980) and will not be repeated here. The film theory analysis illustrated in these references can be applied to more complex reaction systems. Here we restrict our discussion to novel reaction systems such as micellar catalysis used in liquid-liquid reactions, catalytic reactions in which catalyst particles are smaller than diffusion film thickness, and slurry reactors involving reactive solids. Some detailed mathematics of the last two cases is given by Choudhary and Ramchandran (1980).

Micellar Catalysis

Micellar catalysis has been shown to have a remarkable effect on the rate of a variety of liquid-liquid reactions. Micelles are aggregates of surfactants formed in aqueous solutions, where the apolar groups extend into the interior of the aggregate and the head groups are located at the interface with water. Typically, micelles have an average radii of 12–30 Å and contain 20 to 100 monomers. These micelles can solubilize large quantities of materials which otherwise have a low solubility in the aqueous phase.

Rate acceleration or inhibition of organic reactions in micellar solutions arises from the different rates of reaction of the substrate in the micellar medium and in the bulk phase and the distribution of substrates between these two phases. Typical reactions that have been studied in presence of micelles include hydrolysis reactions and nucleophilic substitutions. Fendler and Fendler (1975) give a good summary of the work done on micellar systems up to 1975. Most of these systems involved slow reactions where the mass transfer of the reactants is fast compared to the reaction rate and hence need not be considered. Menger and Portnoy (1967) developed a pseudophase model for unimolecular reaction systems. By assuming the total rate to be the sum of the rates of reactions occurring in the micellar and the aqueous pseudophases, they obtained an expression for the overall observed first-order rate constant:

$$k_{obs} = \frac{k_w + k_m K_s (D_n)}{1 + k_s (D_n)} \tag{26}$$

where k_w and k_m are the true first-order rate constants in the aqueous and micellar pseudophases, K_s the binding constant of the reactant to the micelle, and (D_n) the concentration of micellized surfactant.

The pseudophase model has also been applied to bimolecular reactions by considering the distribution of both reactants between the micellar and water phases (Bunton, 1979). For these systems the micellar effects may be due either to the increased concentration of reactants in the micellar medium (proximity effect) or to increased reactivity in the micellar medium over that in the aqueous phase (medium effect). Dilution effects may come into play when micellar concentrations are made very high because of the decreased probability of finding both the reactants in the same micelle, and this leads to reduced reaction rates.

Recently, Janakiram and Sharma (1982) have considered the case of slow and fast liquid-liquid reactions when mass transfer of reactant is important. The film theory applied to mass transfer with chemical reaction was used, wherein it was assumed that all the resistance to mass transfer existed in a thin film near the interface (called *diffusion films*). The reaction was assumed to occur in the aqueous phase only, which contained the micelles and the reactant B. The hydrophobic reactant A was assumed to have a low solubility in the aqueous phase. Because of their small size, the micelles are present in the film and as A diffuses through the film, it is taken up simultaneously by the micelles, where it reacts with the micellar bound B. The micelles thus steepen the concentration profile of A in the film and lead to higher mass transfer rates. By writing down transport equations for A and B in the film and considering micellar incorporation to be a parallel step, equations have been derived to predict mass transfer rates of A in micellar systems for different regimes (Janakiraman and Sharma, 1982).

Catalyst Particles Smaller than Film Thickness: Reaction Occurs in the Diffusion Film and Is Catalyzed by Solid Particles

We can conceive if situations where catalyst particles are sufficiently small in size (say, below 20 μm) to be smaller than the gas-liquid diffusion film

thickness. In such cases a parallel model of transport has to be considered. In the established practice of slurry reactors, the steps associated with gas-liquid and liquid-solid mass transfer are taken in series. Further, the indigenous reaction may be sufficiently fast to occur partially or wholly in the diffusion film, yet it is further amenable to catalysis by fine particles. Pal et al. (1982) have considered this situation. Earlier work in this area has been carried out by Deckwer, Alper, and coworkers. There is considerable practical significance for such cases, as not only may one get marked increase in the specific rate of absorption but one may also realize the desired selectivity. In the case of magnetic catalyst particles such as Fe, Ni, and CO, or those that can be magnetized, we have a powerful method of recovering fine catalyst particles efficiently via high-gradient magnetic separation. This subject merits further attention. A brief mathematical analysis is presented below. Since the reaction is fast enough to occur in the diffusion film wihtout a catalyst, it is logical to assume that in the presence of fine catalyst particles, the transport of dissolved gas to the catalyst particles will control the rate due to catalyst. For a reaction that is first order in the dissolved gas and dissolved reactive, species, the governing equations are

$$D_{MA} \frac{d^2C_A}{dz^2} = k_1C_A + k_sa_pC_A \tag{27}$$

where

$$k_1 = k_2C_B^\circ$$

The second term on the right-hand side of Eq. (27) accounts for the reaction in the film due to catalyst particles. A solution for the absorption rate of A with the boundary conditions

$$\begin{aligned} z = 0: &\quad C_A = C_A^*, \quad \frac{dC_B}{dz} = 0 \\ z = \delta_L: &\quad C_A = 0 \end{aligned} \tag{28}$$

can be given as

$$N_A = -D_{MA} \left.\frac{dC_A}{dz}\right|_{z=0} = \frac{C_A^*\sqrt{D_{MA}k_1 + D_{MA}k_sa_p}}{\tanh\left(\dfrac{\sqrt{D_{MA}k_1 + D_{MA}k_sa_p}}{k_1}\right)} \tag{29}$$

Here the contribution due to the parallel reaction in the film is unimportant as the particle is relatively very coarse, leading to low values of a_p and k_s (in the particle size range relevant here, k_s will be inversely proportional to the particle size).

When $D_{MA}k_1 + D_{MA}k_sa_p/k_1 > 3$, Eq. (29) can be expressed as

$$\frac{1}{D_{MA}}\left(\frac{\gamma_A^{1}}{C_A^*}\right)^2 - k_1 = k_sa_p = k_s \frac{6w}{\rho_pd_p} \tag{30}$$

For small d_p, where $k_s d_p / D_{MA} = 2$, a plot of $[(1/D_{MA})(N_A/C_A^*)^2 - k_1]$ against $D_{MA}w$ for a specified particle size should be a straight line through the origin having a slope of $12/\rho_p d_p^2$. Thus data pertaining to different temperatures can be accommodated in a single plot. Pal et al. (1982) have verified this theory for absorption of O_2 in aqueous solutions of sodium sulfide at elevated temperature and pressure with fine activated carbon particles as catalyst.

Slurry Reactors Involving Reactive Solids

In this category we can visualize two distinct situations: (a) particles are sparingly soluble in the medium and the reaction occurs between the dissolved gas and the dissolved reactive species obtained from the dissolution of solid particles (the product of the reaction may be soluble or insoluble in the medium), or (b) particles are insoluble in the medium (the product of the reaction may be soluble or insoluble in the medium).

Solid Particles Sparingly in the Medium

This situation is commonly encountered in practice and we can think of two distinctly different situations: (a) solid particles are larger than the thickness of the liquid film (this thickness is that of diffusion film within which the resistance to mass transfer is confined in the absence of chemical reactions, or (b) solid particles are smaller than the liquid film thickness.

Particles Larger Than the Liquid Film Thickness. It is quite easy to deal with this problem as the constituent steps; namely, diffusion of dissolved gas from the interface into the liquid phase and dissolution of solid particles in series (Fig. 16). We can predict the specific rate of mass transfer for any of the controlling regimes. It may well be that in view of the particle size and loading, the bulk liquid phase may be saturated with respect to the dissolved solid species. However, even if we have a "finite" slurry that causes bulk concentration to be lower than the saturation concentration, we can solve the problem analytically.

Particles Smaller Than the Liquid Film Thickness. Depending on the type of gas-liquid contactor, we may have a liquid film thickness (diffusion film thickness) in the range 2 to 40 μm. Thus if we have particles smaller than the film thickness, we can conceivably have the step of dissolution of solid particles in parallel to that of diffusion of dissolved gas from the gas-liquid interface to the bulk liquid phase. Depending on the relative rates of diffusion and chemical reaction, the entire amount of dissolved gas may be consumed in the film. Further, depending on the controlling regime, the concentration of dissolved reactive species in the liquid film may be uniform or may be zero close to the gas-liquid interface (see Figs. 17 and 18).

When the concentration of dissolved solid species is uniform in the film, the simultaneous dissolution of solid particles in the film will not affect the specific rate of mass transfer. However, if we have depletion of the reactive species, the simultaneous dissolution of solid particles can lead to a situation where the depletion is entirely overcome and concentration becomes uniform in the film.

The most exciting situation arises when the reaction is instantaneous and the concentration of both species, namely the dissolved gas and

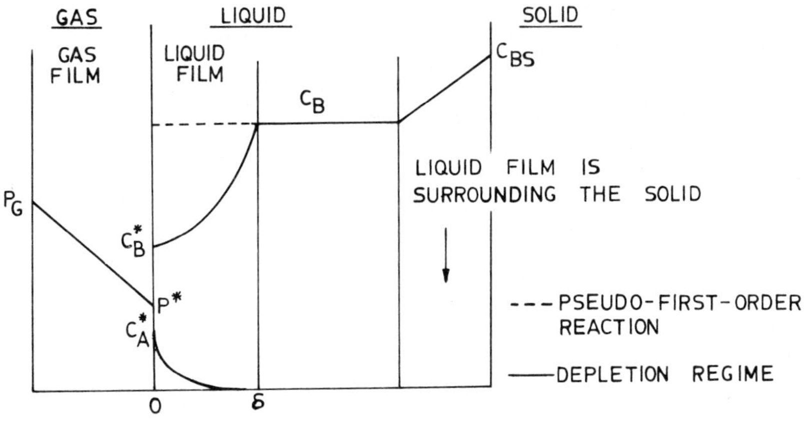

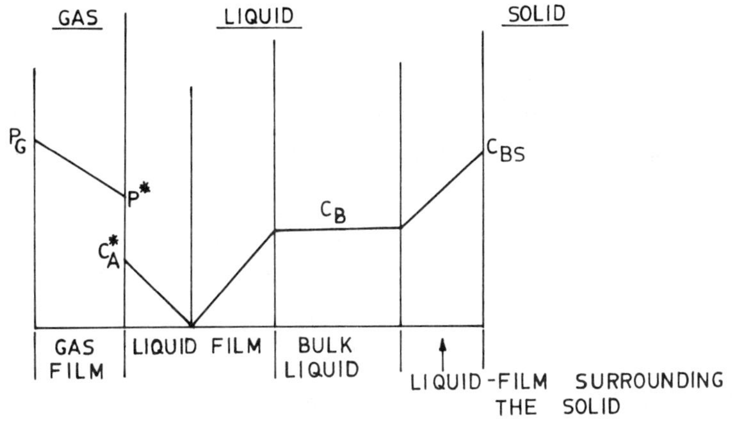

FIGURE 16 Concentration profiles based on the film theory for two gas-liquid-solid systems when the solid dissolution in the film is unimportant.

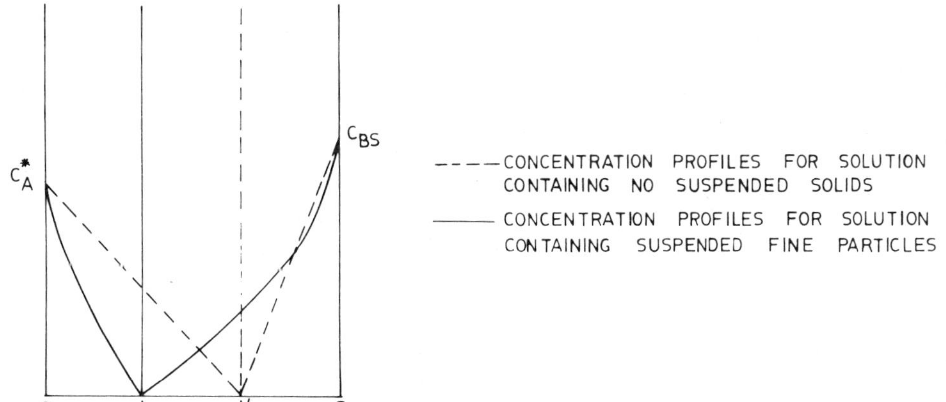

FIGURE 17 Concentration profiles for an instantaneous reaction when the solid dissolution in the film is important.

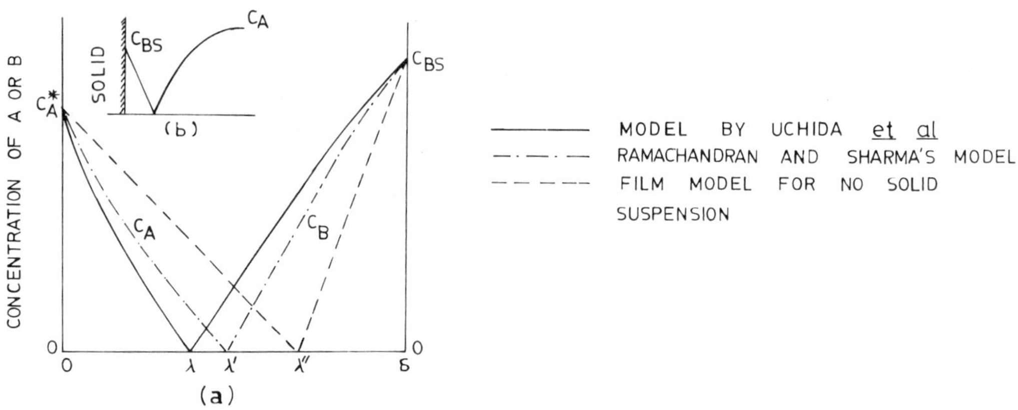

FIGURE 18 Concentration profiles of A and B (λ, λ', λ" are the main reaction planes for these models): (a) in liquid film near gas-liquid interface; (b) in the neighborhood of solid particles present in the region $z = 0$ to $z = \lambda$.

dissolved solid, is zero at a reaction plane that is very close to the gas-liquid interface. Here the concentration of dissolved species drops to zero from the bulk concentration, which may well be the same as the saturation concentration. The simultaneous dissolution of the solid particles in the film will obviously augment the flux of the dissolved solid species and hence the specific rate of absorption will increase. Ultimately, the reaction plane may coincide with the gas-liquid interface and we will get a very interesting situation where the specific rate of absorption will be proportional to the square root of the particle area. Since for very small particles the solid-liquid mass transfer coefficient is inversely proportional to the particle size (Sherwood number = 2), we can conceivably get a situation where the specific rate of mass transfer will be inversely proportional to the particle size.

The simultaneous dissolution of very fine solid particles in the film may lead to a situation where the "instantaneous" reaction-controlled mechanism (but having known value of rate constant) changes over to the case of depletion of reactants and eventually to fast pseudo-mth-order reaction-controlled mechanism.

We also think of situations where the interfacial concentration of the dissolved gas is comparable or even higher than the saturation concentration of the solid particles, and in such cases we will have significant enhancement in the rate of dissolution of solid particles in the film due to the instantaneous reaction (Fig. 18).

The mathematics associated with those cases is evaluated by Ramachandran and Sharma (1969). Sada and coworkers (1977, 1980) and Uchida and coworkers (1975) have also applied the theory to the absorption of SO_2 in slurries of $CaCO_3$ and $Ca(OH)_2$.

Solid Particles Insoluble in the Medium

Particles of Unchanging Size. Here the product is insoluble in the medium and stays with the particle, and the reactant has to diffuse through the product layer. Figure 19 shows typical concentration profiles.

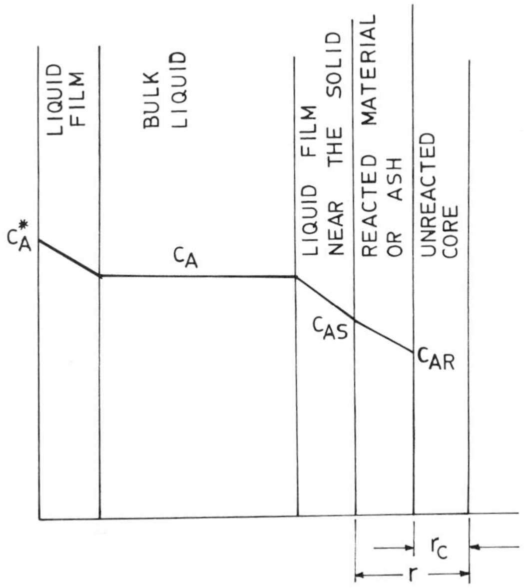

FIGURE 19 Concentration profiles for gas-liquid-solid systems where solids are reacting but insoluble in liquid; solids of constant size.

TABLE 2 Controlling Mechanisms for Constant Size Particles

Controlling mechanism	Controlling rate term		Remarks
Mass transfer from phase A to phase B	$k_L a_L$	$x_B \alpha t$	Hydrodynamic factors are very important
Solid-liquid mass transfer	k_s	$x_B \alpha t$	Hydrodynamic factors have some effect in the turbulent regime
Diffusion through solid	D_{ms}	$t \alpha [(1 - x_B)^{2/3} + (1 - x_B)]$	Hydrodynamic factors are likely to be unimportant, effect of temperature should be insignificant
Chemical reaction	k_R	$x_B \alpha t^3$	Effect of temperature very significant; no effect of hydrodynamic factors

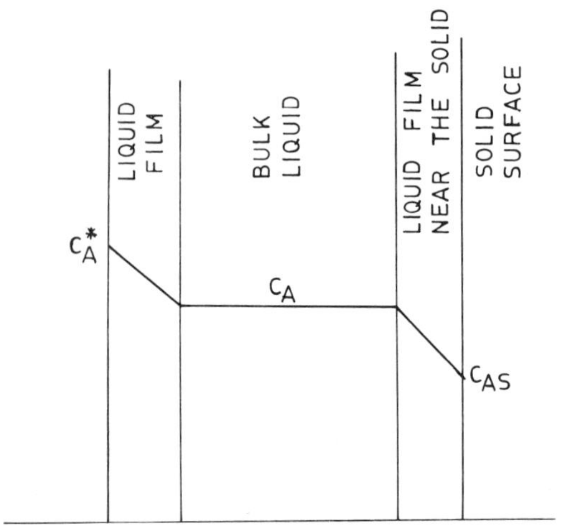

FIGURE 20 Concentration profiles for gas-liquid-solid systems where solids are reacting but insoluble in the liquid; solids of changing size.

It is possible to follow systematic procedures which will allow the controlling mechanism to be clearly discerned. Table 2 brings out the salient features of different controlling regimes.

Particles of Changing Size. Here the product of the reaction is soluble in the medium and hence the reactant shrinks. Figure 20 shows the concentration profiles and Table 3 shows the pertinent features of different controlling regimes.

Hydrometallurgy provides a variety of examples which fall in this category of reactions. Oxidative leaching of copper, cobalt, and silver are some typical examples. A novel case is that of the manufacture of alkali hydrides by hydrogenation of alkali metals suspended in a hydrocarbon solvent. Chlorination of wood pulp provides another example.

TABLE 3 Controlling Mechanisms for Particles of Changing Size

Controlling mechanism	Controlling rate term	Relation between time and fractional conversion	Remarks
Mass transfer from phase A to phase B	$k_L a_L$	$x_B \alpha t$	Hydrodynamic factors are very important
Mass transfer to solid particles	k_s	$x_B \alpha t$	Hydrodynamic factors have some effect in the turbulent regime
Chemical reaction at the surface	k_R	$x_B \alpha t^3$	Effect of temperature significant; no effect of hydrodynamic factors

MODELS FOR FIXED-BED CATALYTIC REACTORS

Isothermal Models

A generalized isothermal steady-state model for the fixed-bed catalytic reactor can be written as

$$D_{ij}\frac{d^2C_{ij}}{dz^2} - \frac{d(U_iC_{ij})}{dz} \pm T_{ij} \pm R_{ij} = 0 \qquad (31)$$

where the subscript i refers to the phase and j refers to the species. Equation (31) is a one-dimensional model which is generally applicable to the fixed-bed systems. If the reactor is to be operated under transient conditions, the right-hand side of Eq. (31) will have a time-dependent concentration term. In Eq. (31), D_{ij} is the axial dispersion coefficient for species j in phase i, U_i the superficial velocity of phase i, and T_{ij} the transfer in and/or out of species j from phase i to other phases. For example, for a gaseous reactant j in the liquid phase, T_{ij} can be expressed as

$$T_{ij} = K_La_L(C_{Gj} - C_{Lj}) - K_{sj}a_s(C_{Lj} - C_{sj}) \qquad (32)$$

Here i = L and the species is transferred in from the gas phase and transferred out to the solid catalyst phase. R_{ij} is the reaction rate of species in the phase i. The + sign indicates generation and the − sign indicates depletion. For the fixed-bed reactor, the first two terms on the left-hand side of Eq. (31) drop out. The first term on the left-hand side also drops out for the gas phase. If the gas velocities of each phase vary along the length of the reactor, this can be evaluated by taking a total material balance on each phase. In general, the total number of equations that need to be solved simultaneously equal the sum of component balances for each phase plus the total number of overall material balances. It should be noted that Eq. (31) assumes that the dispersion model applies to the reactor, an assumption generally found to be reasonable. Equation (31) is subjected to the boundary conditions

$$D_{ij}\frac{dC_{ij}}{dz} = U_i(C_{ij} - C_{ij}^{\circ}) \qquad \text{at } z = 0 \qquad (33)$$

and

$$\frac{dC_{ij}}{dz} = 0 \qquad \text{at } z = L \qquad (34)$$

where L is the length of the reactor. Various cases of this generalized model have been examined in the literature, and these are reviewed by Shah (1979).

In an evaluation of pilot-scale trickle-bed reactors the model described above is often not applicable due to flow anamolies such as incomplete catalyst wetting, insufficient liquid holdup, backmixing, and so on. Shah (1979) has reviewed at great length the reported modeling of pilot-scale hydroprocessing trickle-bed reactors.

Nonisothermal Models

Many industrial hydroprocessing reactors such as those for hydrodesulfurization and hydrocracking operate under adiabatic or close to adiabatic conditions. Since hydrodesulfurization, hydrocracking, and hydrogenation reactions are highly exothermic, the temperature rise in such reactors is frequently controlled by the injection of quench streams at one or more positions along the length of the reactor. These fixed-bed (under trickle flow conditions) reactors are usually operated under plug-flow conditions and the reactor temperature profile exhibits discontinuities at the position of the quenches. The maximum temperature rise in commercial trickle-bed reactors is controlled by the amount of nature of the quench and its position along the length of the reactor.

In commercial hydrodesulfurization reactor the catalyst bed deactivates nonuniformly due to the deposition of metals. The catalyst deactivation is usually counterbalanced by the increase in feed temperature in order to maintain the constant quality of the product. This process is continued until the temperature in any part of the reactor reaches the maximum allowable temperature based on the metallurgical limit. At this point, the reactor is shut down and the catalyst is regenerated. The time required to achieve this maximum allowable temperature is called the *reactor cycle life*. Shah et al. (1976) and Mhaskar et al. (1978) have pointed out that the reactor cycle life depends on the locations of quench positions (for one and two quenches) and that by properly choosing the quench positions, one can significantly improve the reactor cycle life. Abichandani et al. (1980) evaluated the problem for a more generalized nonadiabatic reactor. These studies illustrate that in industrial fixed-bed hydroprocessing reactors, catalyst activity can be maintained at a higher level over a longer period of time by properly controlling the temperature distribution in the reactor. The thermal behavior of a fixed-bed hydrocracking reactor with one or more quenches has been evaluated by Yan (1980).

Generally, commercial fixed-bed reactors have a large catalyst bed length-to-particle diameter ratio and they are operated under plug-flow conditions. The criteria for axial dispersion effects in adiabatic trickle-bed hydroprocessing reactors for (a) residual hydrodesulfurization, (b) hydrocracking of gas oils, and (c) denitrogenation of shale oils were derived by Shah and Paraskos (1975). These criteria indicate that at high conversions, an adiabatic reactor produces a larger axial dispersion effect than the isothermal operation. At low conversions, the opposite results are obtained. Some experimental verifications of these analytical criteria are needed.

MODELS FOR SLURRY REACTORS

Isothermal Models

A generalized model for the fixed-bed catalytic reactor described in the preceding section is applicable for both the gas and liquid phases in slurry reactors. If the solid particles are very fine (less than 100 μm), the liquid-solid slurry is often assumed to form pseudohomogeneous slurry and the separate material balances for the solid phase may not be required. A model of the slurry reactor wherein solid concentration gradient prevails within the reactor has been evaluated by Govindarao (1975) and discussed by Shah (1979). Some aspects of steady-state behavior of three-phase

fluidized-bed reactor as applied to coal liquefaction are discussed by Parulekar and Shah (1980) and Shah (1981). This model should be, in general, applicable to other three-phase systems.

The modeling of a three-phase slurry reactor with solid as a reactant is a more difficult task. Joshi et al. (1981) modeled the performance of an oxydesulfurization reactor wherein the sulfur in coal particles is oxidized to form sulfuric acid. When the solid phase dissolves or reacts substantially, its volume fraction along the reactor changes. This problem has not been evaluated in the literature.

The selectivity problem in the operation of a slurry reactor has also not been yet completely evaluated. For example, the liquid-phase half-hydrogenation of 2,4-dinitrotoluene (DNT) can lead to different isomer distribution depending on the presence or absence of mass transfer resistances (Acres and Cooper, 1972). Even more interesting is the case of obtaining stereospecific product depending on the presence/absence of pore diffusional resistance (Woerde et al., 1982). Selectivity is also very important in coal liquefaction and FT synthesis. These types of problems need to be evaluated in the future.

Nonisothermal Models

The steady-state thermal behavior of an adiabatic three-phase coal liquefaction reactor operated under slow and fast hydrogen consumption reaction regimes have been analyzed by Parulekar and Shah (1980, 1982). The theoretical calculations were based on the axial dispersion model, and the predictions of the model for steady-state temperature distribution in the presence as well as the absence of quench were found to agree well with those measured experimentally (Shah, 1981). It should be noted that unlike in plug-flow reactor, in a backmix reactor the temperature continuity at the quench location is maintained; and as shown by Krishnamurthy and Shah (1979) and Singh et al. (1981), the quench gas flattens the temperature profile in the entire backmixed reactor.

The multiple steady states in adiabatic reactors have been investigated by Singh et al. (1980). Nunez et al. (1982) studied the effects of preheater variables on the multiple steady state in an adiabatic coal liquefaction reactor. Shah and Singh (1981) presented some experimental data in a laboratory adiabatic reactor and showed that significant temperature excursions could occur under practical ranges of operations.

Parulekar et al. (1980) examined a first-order reaction in a three-phase bubble column reactor with the help of an axial dispersion model. For the extreme case of a continuous-stirred-tank reactor, these authors identified a fairly large region of five steady states for a nonzero concentration of the gaseous reactant in the liquid phase, a fact previously observed for an adiabatic gas-liquid CSTR (Raghuram and Shah, 1977). For the case of partially backmixed bubble column reactors, the authors derived criteria for the uniqueness of steady states.

SELECTION OF REACTORS

A variety of reactors have been used for gas-liquid-solid systems, and Satterfield (1970), Shah (1979), and Doraiswamy and Sharma (1983) have considered several aspects of these reactors. As indicated elsewhere, the main classification of reactors may be based on solid particles being very

fine when we have slurry reactors and solid particles being coarse when we usually adopt fixed-bed reactors. The selection of a reactor for a specified duty will depend on a number of factors, and Table 4 lists the important points.

Slurry Versus Fixed-Bed Reactors

The relative merits of slurry and fixed-bed reactors have been reported by several of the authors listed above. Slurry reactors operate as thermal flywheels and maintenance of isothermal conditions is relatively easy. The slurry can be pumped and regeneration can be conveniently accomplished. Further powdered catalysts are generally cheaper and fine catalyst particles whould show a negligible to small resistance to isothermal diffusion. Slurry reactors measure up very well with respect to the important points listed in Table 4. The main disadvantage of slurry reactors is associated with removal of the catalyst particles from the product stream. In case the products are volatile and are stripped under operating conditions (e.g., in Fischer-Tropsch operations for CH_4 and lower hydrocarbons), the question of catalyst removal does not arise except for purposes of catalyst regeneration/replacement. This disadvantage is liely to be relatively unimportant for catalyst particles, such as those based on Fe, Ni, Cr, and Co, which

TABLE 4 Gas-Liquid Solid Reactors: Checklist for Selection of Reactor

Residence time of liquid phase
Allowable pressure drop
Relative flow rates of gas and liquid
Scale of operation
Recycle of gas
Changes in type of feedstock
Countercurrent or cocurrent mode of operation
Corrosion
Foaming behavior
Rheological behavior
Micromixing
Selectivity
Heat removal/supply
Solid particles as reactant or catalyst
Particles fine or coarse
Product of reaction soluble in insoluble
Specification of solid product
Fouling of catalyst

are magnetic, or particles that can be magnetized, as the catalyst separation (HGMS). Recent reports indicate that many noble metal catalysts can be prepared on mangetized supports and that HGMS can even deal with submicron particles.

For very large scale operation for relatively cheap materials such as in hydroprocessing of petroleum fractions, slurry reactors may prove to be unwiedly and expensive. However, for vacuum residua or bottom fractions of heavy crude oils, which contain relatively very high amounts of organometallic compounds, fouling may be severe and slurry reactors have a clear edge over fixed-bed reactors.

In slurry reactors the liquid phase is essentially backmixed; the gas phase may be plug flow or partially backmixed. In fixed-bed reactors, depending on the regime of operation, we may approach plug-flow behavior for both phases.

In fixed-bed reactors, the trickle flow gives a relatively very low residence time for liquid, and slow to very slow reactions, which require a long residence time, cannot be conviently conducted. It is possible to overcome this problem to some extent by using packed bubble column reactors.

Fixed-bed reactors are not generally suited for application, where the feed consists of a slurry, or one of the products of reaction is solid, as this may be lead to excessive pressure drop and maloperation of the column.

Gas Sparged Versus Mechanically Agitated Slurry Reactors

Having decided to use slurry reactors, we can adopt different versions depending on circumstances. The process may call for relatively small amounts of gas, and in case gas sparged reactors are considered, the problem of catalyst suspension may become severe and we may be forced to opt for a mechanically agitated version. It is possible to use much higher flow rates of gas and to recycle the unreacted gas in case we are dealing with pure gas, which is the case in a variety of hydrogenations. In large reactors there is definite merit in opting for gas sparged reactors, as the problems of shaft seal, shaft stability, and so on, can be avoided. Values of the heat transfer coefficient are comparable in the two types of contactors. However, when the heat load is very high we can have a pump-around system as in a loop reactor. We also have the option of a heat exchanger of the multitubular reactor type, where the slurry passes through tubes together with the gas; heat transfer fluid is circulated through the shell. In gas sparged reactors, depending on the solids loading, particle size, and density, we may approach three-phase fluidized-bed behavior.

Mechanically agitated contactors can be operated using the "dead-end" system. Here, pure gas is introduced in the head space and the pressure is maintained at a specified level; no reaction product is volatile. Such a system is used extensively in hydrogenation of a variety of edible and nonedible unsaturated oils. This operation is very large worldwide; a few million tons per year are processed in this way. A variety of feedstocks are used and frequent changes in feedstock become necessary. Further, in the case of edible fats, the product has to meet fairly stringent requirements. For the foregoing reasons, the mechanically agitated contactor has remained a "workhorse" in this industry despite large-scale operation and other benefits of continuous operation in fixed-bed reactors with respect to heat recovery. Some improvement may be possible in mechanically agitated contactors by using a smaller impeller (size about 40% of that of the main impeller) at a distance from the interface roughly equal to its own diameter.

This arrangement should improve gas dispersion and hence gas-liquid interfacial area.

Recent work indicates that with better control strategies and faster analytical techniques it may become possible to use fixed-bed reactors for the hydrogenation of edible and nonedible oils.

WORKED EXAMPLE: MANUFACTURE OF HYDROXYLAMINE PHOSPHATE IN A SLURRY REACTOR

Design a continuous bubble column (slurry) reactor for the production of 100 metric ton per day of hydroxylamine phosphate (HYAM) based on the reduction of aqueous ammonium nitrate-phosphate buffer. A palladium-based catalyst will be employed.

Data

Temperature = 70°C

Pressure = 4 atm at the top
(solutions for pressures of 8, 10, 15, and 20 atm at the top will also be considered)

Superficial gas velocity at the top = 1 cm/s
(solutions for superficial gas velocities at the top of 0.5 and 2 cm/s will also be considered)

Concentration of NH_4NO_3 in feed = 20 wt %

Concentration of H_3PO_4 in feed = 23 wt %

Desired conversion of NH_4NO_3 to HYAM = 80%

Hydrogen rate at the bottom = 10% excess over the theoretical requirements

Henry's coefficient for hydrogen, $H = 4.63 \times 10^{-7}$ mol/cm^3·atm

Catalyst = 2% palladium on carbon

Concentration of active palladium, $C_{pd} = 1 \times 10^{-3}$ g/cm^3 of solution

Particle size of the catalyst = 8×10^{-3} cm

Density of catalyst = 0.6 g/cm^3

Diffusivity of dissolved hydrogen = 5.24×10^{-5} cm^2/s

Reaction

$$3H_2 + NO_3^- + 2\,H^+ \xrightarrow{\text{catalyst}} NH_3OH + 2H_2O$$

$$(H_3PO_4 + NH_2OH \longrightarrow NH_3OH^+ + H_2PO_4^-)$$

Solution

Rate of production of hydroxylamine phosphate

$$= \frac{100 \times 10^6}{24 \times 3600 \times 131}$$

$= 8.84$ mol/s

NH_4NO_3 required in the feed $= \frac{8.84}{0.8} = 11.05$ mol/s or 884 g/s

Concentration of NH_4NO_3 in the feed = 20 wt %

Feed rate = 884/0.2 = 4420 g/s

Hydrogen requirements = $8.84 \times 3 = 26.52$ mol/s

Hydrogen supplied is 10% over the theoretical requirements: $= 26.52 \times 1.1 = 29.17$ mol/s

Volumetric flow rate of hydrogen at the top

$$= 2.652 \times 0.1 \times 22.4 \times 10^3 \times \frac{273 + 70}{273} \times \frac{1}{4}$$

$$= 1.86 \times 10^4 \text{ cm}^3/\text{s}$$

Superficial gas velocity at the top = 1 cm/s

Cross sectional area of the column = 1.86×10^4 cm^2

Diameter of the column = 1.54 m

$$\text{Superficial liquid velocity} = \frac{4420}{1.22 \times 1.86 \times 10^4} = 0.194 \text{ cm/s}$$

Determination of Rate-Controlling Step in the HYAM Formation

The following steps are involved in the reaction between hydrogen and ammonium nitrate:

1. Transfer of hydrogen to the bulk liquid
2. Transfer of hydrogen and ammonium nitrate from bulk liquid to the catalyst surface
3. Reaction of hydrogen and ammonium nitrate on the catalyst surface

The relative importance of different resistances will be considered at the top of the column. The following assumptions have been made to simplify the problem:

1. The gas holdup and the (volumetric) liquid-side mass transfer coefficient are uniform throughout the reactor. (the value of ε is calculated by using an average value of V_G at the top and the bottom of the column.)
2. The vapor pressure of water over the solution has been ignored, as this is insignificant in practically all cases.
3. The catalyst particles are homogeneously dispersed throughout the reactor.
4. The liquid phase is completely backmixed.

Transfer of H_2 to the Bulk Liquid

If this is the rate-controlling step, the rate of absorption of hydrogen will be given by the equation

$$R'_A = k_L\underline{a}[A^*] \qquad (\text{units of } \underline{a} = \text{cm}^2/\text{cm}^3 \text{ of solution})$$

From the data of de Rooij et al. (1977), we have

$$k_L \ (\text{at } V_G = 1.0 \text{ cm/s}) = 2 \times 10^{-2} \text{ cm/s}$$

$$\frac{k_L}{d_B} = 0.1 \quad \text{or} \quad d_B = \frac{k_L}{0.9} = 2.22 \times 10^{-2} \text{ cm}$$

de Rooij et al. (1977) have reported some data on fractional gas holdup for the same system under consideration and the data can be correlated by the following equation in the range $1.0 < V_G < 10$, cm/s:

$$\varepsilon_G = \frac{1.5V_G}{30 + 2V_G}$$
$$= 0.0485 \ (\text{at } V_G = 1.0 \text{ cm/s}) \qquad (35)$$

Effective interfacial area:

$$\underline{a} = \frac{6}{d_B(1 - \varepsilon_G)}$$
$$= \frac{6 \times 0.0485}{2.22 \times 10^{-2} \times (1 - 0.0485)} \qquad (36)$$
$$= 13.77 \text{ cm}^2/\text{cm}^3 \text{ of solution}$$

(This value of $\underline{a}$ is high, but for the specific system under consideration it appears to be reasonable.)

$$R'_A = k_L\underline{a}[A^*]$$
$$= 2 \times 10^{-2} \times 13.77 \times 4 \times 4.63 \times 10^{-7} \qquad (37)$$
$$= 5.1 \times 10^{-7} \text{ mol/cm}^3 \text{ of solution}$$

Transfer of Hydrogen from Bulk Liquid to the Catalyst Surface

Sano et al. (1974) have given the following equation for the calculation of a true solid-liquid mass transfer coefficient.

$$\text{Sh} = \frac{k_{SL}d_p}{D_A}$$
$$= \left[2 + 0.4\left(\frac{Ed_p^4}{\upsilon^3}\right)^{1/4} \text{Sc}^{1/3}\right]\phi_c \qquad (38)$$

where

$$E = \text{rate of energy dissipation per unit mass of liquid}$$
$$= V_G \times g$$
$$= 981\ \text{cm}^2/\text{s}^3,\ \text{at the top}$$
$$d_p = \text{diameter of catalyst particle} = 8 \times 10^{-3}\ \text{cm}$$
$$\phi_c = \text{shape factor; here its value is assumed to be equal to 1}$$
$$\upsilon = \text{kinematic viscosity} = 4.2 \times 10^{-2}\ \text{cm}^2/\text{s}$$

Substitution in Eq. (38) gives

$$\text{Sh} = 2.02$$

and

$$k_{SL} = 1.32 \times 10^{-2}\ \text{cm/s}$$

Thus the Sherwood number is very close to 2 and this situation will be valid for other operating conditions considered in this example.

Concentration of catalyst:

$$w = \frac{1 \times 10^{-3}\ \text{g active Pd/cm}^3\ \text{of solution}}{2\ \text{g active Pd/100 g catalyst}}$$
$$= 0.05\ \text{g/cm}^3\ \text{solution}$$

(*Note*: wt% of Pd on C is 2 and concentration of active Pd is 1×10^{-3} g/cm^3 of solution.)

Solid-liquid effective interfacial area:

$$\underline{a}_p = \frac{6w}{\rho d_p} = \frac{6 \times 0.05}{0.6 \times 8 \times 10^{-3}} = 62.5\ \text{cm}^2/\text{cm}^3\ \text{solution} \tag{39}$$

$$k_{SL}\underline{a}_p = 0.825\ \text{s}^{-1}$$

For this step to be rate controlling, R'_A will be given by

$$R'_A = k_{SL}\underline{a}_p[A^*] = 0.825 \times 4 \times 4.63 \times 10^{-7} = 15.28 \times 10^{-7}\ \text{mol/(cm}^3\ \text{of solution)}\cdot\text{s} \tag{40}$$

Reaction of Hydrogen and Ammonium Nitrate on the Catalyst Surface

de Rooij et al. (1977) have given the following empirical equation for the rate of reaction of hydrogen in aqueous $NH_4NO_3-H_3PO_4$ buffer:

$$R'_A = ka_c C_{pd}[A^*]^{0.34} \tag{41}$$

where k is the reaction rate constant and is given by the following equation:

$$\begin{aligned} k &= 1.696 \times 10^4 \exp\left(\frac{-7674}{RT}\right) \\ &= 1.696 \times 10^4 \exp\left[\frac{-7674}{1.987} \times (273 + 70)\right] \\ &= 0.218 \frac{(\text{mol})^{0.66}(\text{cm})^{1.02}}{(\text{g active Pd})\cdot\text{s}} \end{aligned}$$

where

a_c = relative activity of the catalyst and is assumed to be equal to 0.8

C_{pd} = concentration of active palladium, 1×10^{-3} g active Pd/cm^3 of solution

$$\begin{aligned} R'_A &= 0.218 \times 0.8 \times 1 \times 10^{-3}(4 \times 4.63 \times 10^{-7})^{0.34} \\ &= 19.61 \times 10^{-7} \text{ mol/(cm}^3\text{ solution)}\cdot\text{s} \end{aligned}$$

From the individual values of R'_A above it can be seen that at the top of the column, all three steps contribute to the overall resistance. As we go toward the bottom of the column, the value of [A*] increases because of the increase in the static head of the liquid. Further, the absorption of gas is accompanied by considerable bubble shrinkage and the value of $\underline{a}$ is likely to decrease along the height of the column (as we go up the column). For the calculation of $\underline{a}$ it is assumed that Eqs. (35) and (36) are valid. Due to the two factors above, the rate of gas-liquid mass transfer increases as we go toward the bottom because of the increase in both [A*] and $\underline{a}$. For the case under consideration we have assumed, as stated earlier, that we have uniform values of ε and $k_L\underline{a}$ throughout the reactor.

Since, in bubble column slurry reactors the liquid phase is expected to be completely backmixed, the values of dissolved hydrogen concentration and the hydrogen concentration at the catalyst surface will be uniform throughout the contactor. As a result, the rate of solid-liquid mass transfer will be constant (the variations in the value of k_{SL} in the range of V_G encountered in the reactor were found to be small and hence the value of $k_{SL}\underline{a}_p$ has been assumed to be constant).

When all the three steps contribute to the overall resistance, the rate of dissolution of hydrogen in the liquid bulk is given by the equation

$$R'_A = k_L\underline{a}([A^*] - [A_0]) \tag{42}$$

The rate of transfer of hydrogen from bulk liquid to the catalyst surface is given by the equation

$$R'_A = k_{SL}\underline{a}_p([A_o] - [A_s]) \tag{43}$$

The rate of reaction of ammonium nitrate and hydrogen on the catalyst surface is given by the equation

$$R'_A = ka_c C_{pd}[A_s]^{0.34} \tag{44}$$

(The relative importance of surface reaction in relation to gas-liquid and liquid-solid mass transfer resistance will depend on the pressure, as the intrinsic kinetics shows 0.34 order in dissolved hydrogen.)

Since the liquid phase has been assumed to be completely backmixed, the values of $[A_0]$ and $[A_s]$ remain uniform throughout the contactor. Elimination of $[A_s]$ from Eqs. (43) and (44) gives

$$[A_0] = \frac{R'_A}{k_{SL}\underline{a}_p} + \left(\frac{R'_A}{ka_c C_{pd}}\right)^{1/0.34} \tag{45}$$

The overall material balance across the contactor gives

$$R'_A = \frac{G_i - G_o}{Sh_t(1 - \varepsilon)} \tag{46}$$

where h_t is the height of dispersion. Substitution of Eq. (46) in Eq. (45) gives $[A_0]$

$$[A_0] = \frac{G_i - G_o}{(1 - \varepsilon_G)Sh_t k_{SL} a_p} + \left[\frac{G_i - G_o}{Sh_y ka_c(1 - \varepsilon_G)C_{pd}}\right]^{1/0.34} \tag{47}$$

The material balance with respect to hydrogen over a differential height dh_d gives the following ordinary differential equation:

$$\frac{dG}{dh_d} = R'_A \times S(1 - \varepsilon) \tag{48}$$

$$= k_L\underline{a}([A^*] - [A_o]S(1 - \varepsilon_G) \tag{49}$$

where

$$[A^*] = H(P_T + \psi h_d)$$
$$\psi = \frac{dP}{dh_d} = \frac{(1 - \varepsilon_G)\rho}{1033} \tag{50}$$

Substitution of Eq. (50) in Eq. (49) takes the form

$$\frac{dG}{dh_d} = \alpha + \beta h \tag{51}$$

where

$$\alpha = k_{L}\underline{a}S(1 - \varepsilon_G)\{HP_T - [A_o]\} \tag{52}$$

$$\beta = k_{L}\underline{a}SH\psi(1 - \varepsilon_G) \tag{53}$$

and $[A_o]$ is given by Eq. (47). The boundary conditions are

At the top: $h = 0, \quad G = G_o = 2.652$ (54)

At the bottom: $h = h_t, \quad G = G_i = 29.17$ (55)

Integration of Eq. (51) gives

$$G_i - G_o = \alpha h_t + \frac{\beta h_t^2}{2} \tag{56}$$

or

$$h_t = \frac{-\alpha + \sqrt{\alpha^2 + 2\beta(G_i - G_o)}}{\beta} \tag{57}$$

Procedure for the Calculation of Height

The procedure is essentially trial and error:

1. Assume the height of the column and calculate $[A_o]$ using Eq. (47).
2. Calculate the value of $K_{L}\underline{a}$ from Eqs. (35) and (36) based on the average value of V_G.
3. Knowing $[A_o]$ and $k_{L}\underline{a}$, calculate α and β using Eqs. (52) and (53).
4. Calculate h_t from Eq. (57)
5. If the calculated value of h_t is greater/smaller than the initial guess, the next guess should be in between these two values. The numerical method of partial substitution may be used for quick convergence.

In case the reaction follows first-order kinetics, the height of dispersion can be given by the following analytical expression:

$$h_t = \frac{-\alpha \pm \sqrt{\alpha^2 + 2\gamma(G_i - G_o)(1 + \beta)}}{\gamma}$$

where

$$\alpha = k_{L}\underline{a}HP_TS(1 - \varepsilon_G)$$

$$\beta = k_{L}\underline{a}\left(\frac{1}{k_{SL}\underline{a}_p} + \frac{1}{ka_cC_{pd}}\right)$$

$$\gamma = k_{L}\underline{a}S(1 - \varepsilon_G)H\psi$$

$$k = \text{cm}^3/(\text{g active Pd})\cdot\text{s}$$

TABLE 5 Hydroxylamine Phosphate: Slurry Reactor Details for Different Operating Conditions

No.	Total pressure at the top (atm)	Superficial gas velocity at the top (cm/s)	Column diameter (cm)	Height of dispersion (cm)	Volume of dispersion (m^3)
1	4.0	0.5	218	1106	41.27
2	4.0	1.0	154	1878	33.35
3	4.0	2.0	109	3200	29.85
4	8.0	0.5	154	1241	23.15
5	8.0	1.0	109	2202	20.54
6	8.0	2.0	77.1	4303	20.07
7	10.0	0.5	138	1312	19.58
8	10.0	1.0	97.5	2397	17.90
9	10.0	2.0	68.9	4793	17.89
10	15.0	0.5	112.6	1493	14.86
11	15.0	1.0	79.6	2856	14.21
12	15.0	2.0	56.3	5913	14.71
13	20.0	0.5	97.5	1669	12.46
14	20.0	2.0	68.9	3280	12.24
15	20.0	2.0	48.7	6923	12.92

Height of dispersion = 1787 cm

Volume of dispersion = 33.35 m^3

The problem was reworked for several values of superficial gas velocity and total pressure at the top. The results are given in Table 5.

Due to complete backmixing in the liquid phase, the concentration of dissolved hydrogen remains uniform throughout the contactor. Therefore, for cases where there is substantial difference in the top and bottom pressures, desorption is likely to occur near the top of the column. It was found that desorption occurs in all cases for values of superficial gas velocity at the top of the column greater than 2 cm/s. We should avoid, as far as possible, designing columns for conditions that are conducive for desorption.

There is a remarkable effect of pressure on the volume of the contactor. Since hydrogen is usually available from steam reformers at higher pressures in the range 15 to 25 atm, it is useful to employ a pressure of 20 atm at the top of the contactor. The case of a superficial gas velocity of 0.5 cm/s at the op appears to be most reasonable. We may therefore opt for a 97.5-cm-diameter (say, 1.0 m) column with an effective dispersion height of 16.69 m (dispersion volume = 12.46 m^3). The case of a pressure

10 atm at the top of the contactor with a superficial gas velocity of 0.5 cm/s at the top also appears to be attractive with a column diameter of 1.38 m and an effective dispersion height of 13.12 m (dispersion volume = 19.58 m^3).

NOTATION

$\underline{a}$	gas-liquid interfacial area
a_p	particle surface area
a_s	particle surface area per unit volume in Eq. (5) and Figs. 9 to 11.
A^*, A_o, A_s	gas-liquid interface, bulk liquid and particle surface concentrations in worked example
C	concentration
C_p	specific heat
C^*	solubility
d_b	bubble diameter
d_c	column diameter
d_p	particle diameter
d_{vs}	average bubble size
D	axial dispersion coefficient
D_m	molecular diffusivity
D_{ms}	diffusion through solid product layer
D_n	concentration of micellized surfactant [Eq. (26)]
e	energy consumption in Fig. 15
F_r	Froude number, dimensionless [Eq. (17)]
g	gravitational constant
G	mass velocity
h_w	slurry wall heat transfer coefficient
j_D	j factor in Fig. 12
k_s	liquid-solid mass transfer coefficient
K_l, K_r	reactor rate constants
K_L	liquid-side gas-liquid mass transfer coefficient
$K_L$overall	gas-liquid mass transfer coefficient
K_s	binding constant in Eq. (26)
K_w, K_m, K_{obs}	various kinetic constants in Eq. (26)
$\overline{K}$	thermal conductivity
L	length of reactor
N_a	absorption rate of species A

p	Pressure
Pe	Péclet number, dimensionless
Pr	Prandtl number [Eq. (22)]
P/V_D	power per unit volume of dispersion
Q_v	volumetric gas flow rate
r	radius of particle
r_c	radius of unreacted core
Re	Reynolds number, dimensionless
Sc	Schmidt number ($\mu\rho/Dm$)
Sh	Sherwood number ($k_s d_p/D_m$)
St	Stanton number [Eq. (22)]
t	time
U	velocity
$U_{b\infty}$	bubble rise velocity
U_t	relative velocity between two phases
V_c	circulation velocity
w	particle loading
We	Weber number in Fig. 3
x_B	conversion in Tables 2 and 3
z	axial distance in the reactor or within liquid film

Greek Letters

δ	energy loss
δ_L	thickness of liquid film
μ	viscosity
ΔP	pressure drop
$\frac{\Delta P}{\Delta Z}$	pressure gradient
ε	volume traction
$\bar{\varepsilon}$	void fraction in the fixed bed
ε_{dL}	dynamic liquid holdup
λ, ψ	parameters in Fig. 2
$\bar{\lambda}$, $\bar{\lambda}'$, $\bar{\lambda}''$	reaction planes in Figs. 17 and 18
ν	kinematic viscosity
ρ	density
σ	surface tension

Subscripts

A	refers to species A
air	refers to air
B	refers to species B
G	gas phase
L	liquid phase
LG	two-phase
M	mixture
OL	superficial liquid velocity
OG	superficial gas velocity
p	particle
s	solid phase
wat	refers to water in Fig. 2

Superscripts

o	bulk condition
*	refers to interface condition

REFERENCES

Abbott, M. D., G. R. Moore, and J. L. Ralph, Paper S-5, 7th ABG Conf. (1967).

Abichandani, J. S., R. D. Mhaskar, and Y. T. Shah, Optimum performance of a non-adiabatic catalytic reactor, Indian J. Technol., *18*, 374 (1980).

Achwal, S. L. and J. B. Stepanek, Holdup profiles in packed beds, Chem. Eng. J., *12*, 69 (1976).

Acres, G. J. K. and B. J. Cooper, J. Appl. Chem. Biotechnol., *22*, 769 (1972).

Adlington, D. and E. Thompson, Desulfurization in fixed and fluidized bed catalyst systems, Proc. 3rd Eur. Symp. Chem. React. Eng., Pergamon Press, Oxford (1965), p. 203.

Akita, K. and F. Yoshida, Gas holdup and volumetric mass transfer coefficient in bubble columns, Process Des. Dev., *12*, 76 (1973).

Akita, K. and F. Yoshida, Bubble size, interfacial area and liquid phase mass transfer coefficients in bubble columns, Ind. Eng. Chem. Process Des. Dev., *13*, 84 (1974).

Arbiter, N., C. C. Harris, and R. F. Yap, Int. J. Min. Process., *3*, 257 (1976).

Armstrong, E. R., C. G. J. Baker, and M. A. Bergougnou, The effect of solid wettability on the characteristics of three phase fluidization, in *Fluidization Technology*, Vol. 1, D. L. Keairns, ed., Hemisphere, Washington, D.C., (1976a), p. 453.

Armstrong, E. R., C. G. J. Baker, and M. A. Bergougnou, Heat transfer and hydrodynamic studies on three phase fluidized beds, in *Fluidization Technology*, Vol. 1, D. L. Keairns, ed., Hemisphere, Washington, D.C., (1976a), p. 453.

Bach, H. F. and T. Pilhofer, Variation of gas holdup in bubble columns with physical properties of liquids and operating parameters of columns, Ger. Chem. Eng., *1*, 270 (1978).

Bakos, M. and J. C. Charpentier, Taux de rétention pour des écoulement gaz-liquide à co-courrant vers le bas dans des colonnes à garnissage arrosé et noyé, Chem. Eng. Sci., *25*, 1822 (1970).

Begovich, J. M. and J. S. Watson, Hydrodynamic characteristics of three phase fluidized beds, Fluidization Proc. 2nd Eng. Found. Conf., J. F. Davidson and D. L. Keairns, eds., (1978), p. 190.

Beimesch, W. E. and D. P. Desseler, Liquid-gas distribution measurements in the pulsing regime of two-phase concurrent flow in packed beds, AIChE J., *17*(5), 1160 (1971).

Beinhauer, R., Dynamische Messung des selativen Gasgehalts in blasensaulen Mittels Absorption von Rotgenstrahlen, Dr.-Ing. thesis, Tech. Univ. Berlin, (1971).

Bhatia, V. K. and N. Epstein, Paper 43, Proc. Int. Symp. Fluidization Appl., Toulouse, France (1974).

Bhatia, V. K., K. A. Evans, and N. Epstein, Effect of solids wettability on expansion of gas-liquid fluidized beds, Ind. Eng. Chem. Process Des. Dev., *11*, 151 (1972).

Botton, R., D. Cosserat, and J. C. Charpentier, Influence of column diameter and high gas throughputs on the operation of a bubble column, Chem. Eng. J., *16*, 107 (1978).

Brian, P. L. T. and H. B. Hales, Effects of transpiration and changing diameter on heat and mass transfer to spheres, AIChE J., *15*, 419 (1969).

Buchanan, J. E., Holdup in irrigated ring-packed towers below the loading point, Ind. Eng. Chem. Fundam., *6*, 400 (1967).

Buchholz, H., R. Buchholz, J. Lucke, and K. Schügerl, Bubble swarm behavior and gas absorption in non-Newtonian fluids on sparged columns, Chem. Eng. Sci., *33*, 1061 (1978).

Bunton, C. A., Reaction kinetics in aqueous surfactant solutions, Catal. Rev.-Sci. Eng., *20*, 1 (1979).

Burkel, W., Der Warmeubergang an Heiz- und Kuhlflachen in Begasten Flussigkeiten, Chem.-Ing.-Tech., *44*, 265 (1972).

Calderbank, P. H., Gas absorption from bubbles, Chem. Engr., *45*, CE209 (1967).

Chaudhari, R. V. and P. A. Ramchandran, Three phase slurry reactors, AIChE J., *26*, 177 (1980).

Chandrasekaran, D. and M. M. Sharma, Absorption of oxygen in aqueous solutions of sodium sulfide in the presence of activated carbon as catalyst, Chem. Eng. Sci., *32*, 669 (1977).

Charpentier, J. C., Recent progress in gas-liquid mass transfer in packed beds, Chem. Eng. J., *11*, 161 (1976).

Charpentier, J. C. and M. Favier, Some liquid holdup experimental data in trickle bed reactors for foaming and nonfoaming hydrocarbons, AIChE J., *21*, 1213 (1975).

Charpentier, J. C., C. Prost, and P. Legoff, Chute de pression pour des écoulements a co-courrant dans les colonnes a garnissage arrosé: Comparison avec le garnissage noyé, Chem. Eng. Sci., *24*, 1777 (1969).

Chou, T. S., F. L. Worley, Jr., and D. Luss, Transition to pulsed flow in mixed-phase cocurrent downflow through a fixed bed, Ind. Eng. Chem. Process Des. Dev., *16*, 424 (1977).

Clements, L. D., Dynamic liquid holdup in cocurrent gas-liquid downflow in packed beds, in *Two phase transport and Reactor Safety*, Vol. 1, T. N. Veziroglu and S. Kakac, eds., Hemisphere, Washington, D.C. (1978), p. 69.
Clements, L. D. and G. Halfacre, Liquid composition and flow regime effects in gas-liquid downflow in packed beds, in *Two Phase Transport and Reactor Safety*, Vol. 1, T. N. Veziroglu and S. Kakac, eds., Hemisphere, Washington, D.C. (1978), p. 69.
Clements, L. D. and G. Halfacre, Liquid composition and flow regime effects in gas-liquid downflow in packed beds, AIChE 72nd Annu. Meet., Houston, Tex. (1979).
Clements, L. D. and P. C. Schmidt, Two-phase pressure drop and dynamic liquid holdup in cocurrent downflow in packed beds, 69th Annu. AIChE Meet., Chicago (1976).
Clements, L. D. and P. C. Schmidt, Dynamic liquid holdup in two phase downflow in packed beds; air-silicone oil system, AIChE J., *26*, 317 (1980a).
Clements, L. D. and P. C. Schmidt, Two-phase pressure drop in cocurrent downflow in packed beds; air-silicone oil systems, AIChE J., *26*, 314 (1980b).
Cova, D. R., Catalyst suppression in gas-agitated tubular reactors, Ind. Eng. Chem. Process. Des. Dev., *5*(1), 20 (1966).
Dakshinamurthy, P., V. Subrahmanyam, and J. N. Rao, Bed porosities in gas-liquid fluidization, Ind. Eng. Chem. Process Des. Dev., *10*, 322 (1971).
Davidson, J. F., E. J. Cullen, D. Hanson, and D. Roberts, The holdup and liquid film coefficient of packed towers, Trans. Inst. Chem. Eng., *37*, 122 (1959).
Deckwer, W.-D., On mechanism of heat transfer in bubble column reactors, Chem. Eng. Sci., *35*(7), 1341 (1980).
Deckwer, W.-D, Y. Louisi, A. Zaidi, and M. Ralek, Hydrodynamic properties of the Fischer-Tropsch slurry process, Ind. Eng. Chem. Process Des. Dev., *19*, 699 (1980).
Dhanuka, V. R. and J. B. Stepanek, Gas and liquid holdup and pressure drop measurements in three phase fluidized bed, in *Fluidization*, J. F. Davidson, ed., Cambridge University Press, Cambridge (1978), p. 179.
Dhanuka, V. R. and J. B. Stepanek, Simultaneous measurement of interfacial area and mass transfer coefficient in three phase fluidized beds, AIChE J., *26*, 1029 (1980).
Dharwadkar, A. and N. D. Sylvester, Liquid solid mass transfer in trickle beds, AIChE J., *23*(3), 376 (1977).
Efremov, G. E. and I. A. Vakhurshev, A study of the hydrodynamics of three phase fluidized beds, Int. Chem. Eng., *10*, 37 (1970).
El-Temtamy, A. A., Y. O. Ed-Sharnoubi, and M. M. El-Halwagi, Liquid dispersion in gas-liquid fluidized beds; Part I. Axial dispersion, The Axially Dispersed Plug-Flow Model, Chem. Eng. J., *18*, 151 (1979).
Falk, J. C. in *Catalysis in Organic Synthesis*, P. N. Rylander and H. Greenfield, eds., Academic Press, New York (1976).
Fan, L. T., H. H. Hsu, and K. B. Wang, Mass transfer coefficient and pressure drop data of two-phase oxygen-water flow in bubble column packed with static mixers, J. Chem. Eng. Data, *20*, 26 (1976).
Farkas, E. J. and P. F. Leblond, Solid concentration profile in the bubble column slurry reactor, Can. J. Chem. Eng., *47*, 360 (1969).

Fendler, J. H. and E. J. Fendler, *Catalysis in miscellar and macromulecular systems*, Academic Press, New York (1975).

Field, R. W. and J. F. Davidson, Axial dispersion in bubble columns, Trans. Inst. Chem. Eng., *58*, 228 (1980).

Ford, L. H., Ph.D. thesis, University of London (1960).

Freedman, W. and J. F. Davidson, Holdup and liquid circulation in bubble columns, Trans. Inst. Chem. Eng., *47*, T251 (1969).

Friedlander, S. K., Mass and heat transfer to single spheres and cylinders at low reynolds number, AIChE J., *3*, 43 (1957).

Friedlander, S. K., A note on transport to spheres in stokes flow, AIChE J., *7*, 347 (1961).

Frossling, N., Beitr. Geophys., *52*, 170 (1938).

Fujie, K., M. Takaine, H. Kubota, and Y. Miyaji, Flow and oxygen transfer in cocurrent gas-liquid downflow, J. Chem. Eng. Jpn., *13*, 188 (1980).

Gelbe, H., A new correlation for liquid holdup in packed beds, Chem. Eng. Sci., *23*, 1901 (1968).

Gianetto, A., G. Baldi, and V. Specchia, Absorption in packed towers with cocurrent downflow high-velocity flows: I. Interfacial areas, Ing. Chem. (Milan), *6*, 125 (1970).

Goto, S. and J. M. Smith, Trickle-bed reactor performance: Part I. Holdup and mass transfer effects, AIChE J., *21*, 706 (1975).

Goto, S., J. Levec, and J. M. Smith, Ind. Eng. Chem. Process Des. Dev., *14*, 473 (1975).

Goto, S., S. Watabe, and M. Matusubara, The role of mass transfer in trickle bed reactors, Can. J. Chem. Eng., *54*, 551 (1976).

Govier, G. W. and K. Aziz, *The Flow of Complex Mixtures in Pipes*, Van Nostrand Reinhold, New York (1972).

Govindrao, V. M. H., On the dynamics of bubble column slurry reactors, Chem. Eng. J., *9*, 229 (1975).

Harriot, P., Mass transfer to particles: Part II. Suspended in a pipeline AIChE J., *8*, 101 (1962).

Hart, W. F., Heat transfer in bubble agitated systems—A general correlation Ind. Eng. Chem. Process Des. Dev., *6*, 218 (1976).

Hecht, V., J. Voigt, and K. Schügerl, Absorption of oxygen in countercurrent multistage bubble columns: II. Viscoelastic liquids comparison of systems with high viscosity, Chem. Eng. Sci., *35*, 1325 (1980).

Heilmann, Von W. and H. Hofmann, Proc. 4th Eur. Symp. Chem. React. Eng., Pergamon Press, Oxford (1971), p. 169.

Hikita, H., S. Asai, K. Tanigawa, K. Segawa, and M. Kitao, Gas holdup in bubble column, Chem. Eng. J., *20*, 59 (1980).

Hills, J. H., Radial nonuniformity of velocity and voidage in a bubble column, Trans. Chem. Eng., *52*, 1 (1974).

Hochman, J. M. and E. Effron, Two phase cocurrent downflow in packed beds, Ind. Eng. Chem. Fundam., *8*, 63 (1969).

Hofmann, H., Hydrodynamics, transport phenomena, and mathematical models in trickle-bed reactors, Int. Chem. Eng., *17*(1), 19 (1977).

Hofmann, H. P., Multiphase catalytic packed bed reactors, Cat. Rev. Sci. Eng. *17*(1), 71 (1978).

Hovmand, S. and J. F. Davidson, Chemical conversion in a slugging fluidized bed, Trans. Inst. Chem. Eng., *46*, T190 (1968).

Hsu, K. H., L. E. Erickson, and L. T. Fan, Oxygen transfer to mixed cultures in tower systems, Biotechnol. Bioeng., *17*, 499 (1975).

Hutton, B. E. T. and L. S. Leung, Cocurrent gas-liquid flow in packed columns Chem. Eng. Sci., *29*, 1681 (1974).

Imafuku, K., T. Y. Wang, K. Koide, and H. Kubota, The behavior of suspended solid particles in the bubble column, J. Chem. Eng. Jpn., *1*(2), 153 (1968).

Janakiraman, B. and M. M. Sharma, Chem. Eng. Sci., (1982, in press).

Jeffrey, D. J. and A. Acrivos, The rheological properties of suspensions of rigid particles, AIChE J., *22*(3), 417 (1976).

Jesser, B. W. and J. C. Elgin, Liquid holdup in packed towers, Trans. Inst. Chem. Eng., *39*, 277 (1943).

Joosten, G. E. H., J. G. M. Schilder, and J. J. Janssen, The influence of suspended solid material on the gas-liquid mass transfer in stirred gas-liquid contactors, Chem. Eng. Sci., *32*, 563 (1977).

Joshi, J. B., Axial dispersion in multiphase contactors—a unified correlation, Trans. Inst. Chem. Eng., *58*, 147 (1980).

Joshi, J. B., M. M. Sharma, Y. T. Shah, C. P. P. Singh, M. Ally, and G. E. Klinzing, Heat transfer in multiphase contactors, Chem. Eng. Commun., *6*, 1 (1980).

Joshi, J. B., J. Abichandani, Y. T. Shah, J. Reuther, and H. Ritz, Modeling of three phase reactors a case of oxydesulfurization of coal, AIChE J., *27*, 937 (1981).

Joshi, J. B., A. B. Pandit, and M. M. Sharma, Mechanically agitated gas liquid reactors, Chem. Eng. Sci., *37*(6), 813 (1982).

Juvekar, V. A. and M. M. Sharma, Absorption of CO_2 in a suspension of lime, Chem. Eng. Sci., *28*, 825 (1973).

Kamawura, K. and T. Sasano, Liquid-solid transfer in bubble columns, Kogaku Kagaku, *29*, 693 (1965).

Kara, S., Hydrodynamics and axial mixing in three phase bubble column, Ph.D. thesis, University of Pittsburgh (1981).

Kara, S., B. G. Kelkar, Y. T. Shah, and N. L. Carr, Hydrodynamics and axial mixing in three phase bubble column, Ind. Eng. Chem. Process Des. Dev., *21*(4), 584 (1981).

Kast, W., Untersuchugen zum Warmeubergang in Clasensaulen, Chem.-Ing.-Tech., *35*, 785 (1963).

Kato, Y., Kagaku Kogaku (Abr. English Ed.), *1*, 3 (1963).

Kato, Y., A. Nishiwaki, T. Fukuda, and S. Tanaka, The behavior of suspended solid particles and liquid in bubble column, J. Chem. Eng. Jpn., *5*, 112 (1972).

Kato, Y., A. Nishiwaki, T. Kago, T. Fukuda, and S. Tanaka, Gas holdup and overall volumetric absorption coefficient in bubble columns with suspended solid particles absorption rate of oxygen by an aqueous solution of sodium sulfite, Int. Chem. Eng., *13*, 528 (1973).

Kato, Y., K. Uchida, T. Kago, and J. Morooka, Liquid holdup and heat transfer coefficient between bed and wall in liquid-solid and gas-liquid-solid fluidized beds, Powder Technol., *28*, 173 (1981).

Kelkar B. G., Hydrodynamics and mixing in three phase bubble column, pH.D. thesis, University of Pittsburgh (1982).

Kim, S. D., C. G. Baker, and M. A. Bergougnou, Holdup and axial mixing characteristics of two phase and three phase fluidized beds, Can. J. Chem. Eng., *50*, 695 (1972).

Kim, S. D., C. G. Baker, and M. A. Bergougnoiu, Phase holdup characteristics of three phase fluidized beds, Can. J. Chem. Eng., *53*, 134 (1975).

Kirillov, V. A. and M. A. Nasamanyan, Int. Chem. Eng., *16*, 538 (1976).
Kobayashi, K., Y. Iida, and N. Kanegae, Distribution of local void fraction of air-water two phase flow in a vertical channel, Bull. ASME, *13*(62), 1005 (1970).
Kojima, H. and K. Asano, Hydrodynamic characteristics of suspension bubble column, Kagaku Kogaku Ronbunshu, *6*, 46 (1980).
Kölbel, H., E. Borchers, and K. Muller, Warmeubergang and Blasensaulen: II. Messungen an visosen Suspensionen, Chem.-Ing.-Tech., *30*, 729 (1958).
Kölbel, H., E. Borchers, and J. Martins, Warmeubergang an Blasensaulen: III. Messungen an gasdurchstromten Suspensionen, Chem.-Ing.-Tech., *32*, 84 (1960).
Kölbel, H., E. Borchers, and H. Langemann, Großenverteilung der Gasblasen in Blasensaulen, Chem.-Ing.-Tech., *33*, 668 (1961).
Krishnamurthy, S. and Y. T. Shah, Backmixing characteristics of a bubble column with side stream, Ind. Eng. Chem. Process Des. Dev., *18*(1), 187 (1979).
Kumar, S., S. N. Upadhyay, and V. K. Mathur, On the solubility of benzoic acid in aqueous carbonmethal cellulose solution, J. Chem. Eng. Data, *23*, 139 (1978).
Larkins, R. P., R. R. White, and D. W. Jeffrey, Two-phase concurrent flow in packed beds, AIChE J., *7*(2), 231 (1961).
Lee, J. C., Proc. 3rd Eur. Symp. Chem. React. Eng., Pergamon Press, Oxford (1965), p. 211.
Lemay, Y., G. Pineault, and J. A. Ruether, Ind. Eng. Chem. Process Des. Dev., *14*, 280 (1975).
Levich, V. G., *Physicochemical Hydrodynamics*, Prentice-Hall, Englewood, Cliffs, N.J. (1962).
Levins, R. E. and J. R. Glatonbury, Application of Kolmogoroffs theory to particle-liquid mass transfer in agitated vessels, Chem. Eng. Sci., *27*, 537 (1972).
Lin, G. H., B. S. Fang, and C. S. Wu, Oxygen transfer and mixing in a tower cycling fermentor, Biotechnol. Bioeng., *18*, 1557 (1976).
Mahajani, and M. M. Sharma, Effective interfacial area and liquid side mass transfer coefficient in trickle bed reactors, Chem. Eng. Sci., *34*, 1425 (1979).
Mahajani, V. V. and M. M. Sharma, Mass transfer in packed columns: cocurrent (downflow) operation: 1 in and 1.5 in. metal pall rings and ceramic intalox saddles: multifilament gauge packings in 20 cm. and 38 cm. i.d. columns, Chem. Eng. Sci., *35*, 941 (1980).
Mangartz, K.-H. and T. Pilhofer, Gas phase dispersion in bubble columns, Verfahrenstechnik (1980, in press).
Matrins, J., Warmeubergang an gasdurchstromten Suspensionen, Dipl. Arbeit, TU Berlin, (1959).
Meshelkar, R. A. and V. V. Chavan, J. Chem. Eng. Jpn., *6*, 160 (1973).
Massimilla, L., A. Sollmando, and E. Equillacc, Gas dispersion in solid liquid fluidized beds, Brit. Chem. Eng., *6*, 232 (1961).
Menger, F. M. and C. E. Portnoy, J. Am. Chem. Soc., *89*, 4698 (1967).
Mersmann, A., Zum Warmeubergang in Blasensaulen, Chem.-Ing.-Tech., *47*, 869 (1975).
Mhaskar, R. D., Y. T. Shah, and J. A. Paraskos, Optimum quench location for a hydrodesulfurization reactor with time varying catalyst activity: Part II. Cases of multiple gas quenches and a liquid quench, Ind. Eng. Chem. Process Des. Dev., *17*(1), 17 (1978).

Michell, R. W. and J. A. Furzer, Trickle flow in packed beds, Trans. Inst. Chem. Eng., *50*, 334 (1972).
Michelsen, M. L. and K. Østergaard, Holdup and mixing in coal-liquid fluidized beds, Chem. Eng. J., *1*, 37 (1970).
Midoux, N., M. Favier, and J. C. Charpentier, Flow pattern, pressure loss and liquid holdup data in gas-liquid downflow packed beds with foaming and nonfoaming hydrocarbons, J. Chem. Eng. Jpn., *2*, 350 (1976).
Mohunta, D. M. and G. S. Laddha, Prediction of liquid phase holdup in random packed beds, Chem. Eng. Sci., *20*, 1069 (1965).
Morsi, B. I. and J. C. Charpentier, Hydrodynamic and gas-liquid interface parameters with organic and aqueous liquids in catalytic and noncatalytic packings in trickle bed reactors, NATO School, Izmir, Turkey (1981).
Muller, K., Der Warmebergang an viscosen Suspensionen in Blasensaulen, Dipl. Arbeit, TU Berlin (1958).
Nakanoh, M. and F. Yoshida, Gas absorption by Newtonian and non-Newtonian liquids in a bubble columns, Ind. Eng. Chem. Process Des. Dev., *19*, 190 (1980).
Marayanan, S., V. K. Bhatia, and D. K. Guha, Suspension of solids by bubble agitation Can. J. Chem. Eng., *47*, 360 (1969).
Nguyen-Tien, K. and W.-D. Deckwer, Effect of liquid flow and particle concentration on gas-liquid mass transfer in a bubble column, Chem. Eng. Sci., (1981, in press).
Nishikawa, M., H. Kato, and K. Hashimoto, Heat transfer in aerated tower filled with non-Newtonian liquid, Ind. Eng. Chem. Process Des. Dev., *16*, 133 (1977).
Nunez, P., A. Calimli, J. Abichandnai, and Y. T. Shah, Multiple steady states in an adiabatic coal liquefaction reactor—role of preheater, Chem. Eng. Column, *13*(4–6), 231 (Jan. 1982).
Oels, U., J. Lucke, R. Bucholz, and K. Schürgerl, Influence of gas distributor type and composition of liquid on the behavior of a bubble column dioreactor, Ger. Chem. Eng., *1*, 115 (1978).
Orzaen, M. E. and L. E. Erickson, Oxygen transfer rates and efficiencies, in in K. Østergaard, *Gas-Liquid Particle Operations in Chemical Reaction Engineering*, Adv. Chem. Eng., *7*, 71 (1968).
Otake, T. and K. Okada, Liquid holdup in packed towers, Kagaku Kogaku, *17*, 176 (1963).
Østergaard, K., On bed porosity in gas liquid fluidization, Chem. Eng. Sci., *20*, 165 (1965).
Østergaard, K., On the growth at air bubbles formed at a single orifice in a water fluidized bed, Chem. Eng. Sci., *21*, 470 (1966).
Østergaard, K., Three phase fluidization, in *Fluidization*, J. F. Davidson and D. Harrison, eds., Academic Press, New York (1971).
Østergaard, K., AIChE Symp. Ser., *176*, 74, 82 (1978).
Østergaard, K. and M. L. Michelsen, Symp. Fundam. Appl. Fluidization, Tampa, Fla., preprint 34D (1968).
Østergaard, K. and W. Suchozebrski, Gas-liquid mass transfer in gas-liquid fluidized beds, Proc. 4th Eur. Symp. Chem. React. Eng., Pergamon Press, Oxford (1968), p. 21.
Østergaard, K. and P. I. Theisen, The effect of particle size and bed height on the expansion of mixed phase (gas-liquid) fluidized beds, Chem. Eng. Eng. Sci., *21*, 413 (1966).
Pal, S. K., M. M. Sharma, and V. A. Juvekar, Fast reactions in slurry reactors: catalyst particle size smaller than film thickness, oxidation of

aqueous sodium sulphide solutions with activated carbon particles as catalyst at elevated temperatures, Chem. Eng. Sci., *37*, 327 (1982).

Parulekar, S. J. and Y. T. Shah, Steady state behavior of gas-liquid solid fluidized bed reactors, Chem. Eng. J., *20*, 21 (1980).

Parulekar, S. J. and Y. T. Shah, Steady state thermal behavior of an adiabatic three phase fluidized bed reactor—coal liquid action under slow hydrogen consumption regime, Chem. Eng. J., *23*, 15 (1982).

Parulekar, S., S. Raghuram, and Y. T. Shah, Multiple steady states in an adiabatic gas-liquid-solid reactor, Chem. Eng. Sci., *35*, 745 (1980).

Patil, V. K. and M. M. Sharma, Packed tube columns hydrodynamics and effective interfacial area: poll rings and multifilament wire gauge packings, Can. J. Chem. Eng., *59*, 606 (1981).

Pilhofer, T., M. F. Bach, and K. H. Mangartz, Determination of third dynamic parameters in bubble columns design, ACS Symp. Ser., *65*, 372 (1978).

Pittsburgh Energy Research Center, Quarterly Report, 1975–1976; also D. Smith, A. Reznik, and A. Pontello, personal communication (1976).

Raghuram, S., and Y. T. Shah, Criteria for unique and multiple steady states for a gas-liquid reaction in an adiabatic CSTR, Chem. Eng. J., *13*, 81 (1977).

Ramachandran, P. A. and M. M. Sharma, Absorption with fast reaction in a slurry ocntaining sparingly soluble fine particles, Chem. Eng. Sci., *24*, 1681 (1969).

Ranz, W. E. and W. R. Marshall, Evaporation from drops, Chem. Eng. Prog. *48*, 141 (1952).

Regen, S. L., Triphase catalysis, J. Am. Chem. Soc., *97*, 5956 (1975).

Reiss, L. P., Cocurrent gas-liquid contacting in packed columns, Ind. Eng. Chem. Process Des. Dev., *6*, 486 (1967).

Reynier, J. P. and J. C. Charpentier, Holdup prediction in packed columns for cocurrent gas-liquid downflow, Chem. Eng. Sci., *26*, 1781 (1971).

Rigby, G. R. and C. E. Capes, Bed expansion and bubble wakes in three phase fluidization, Can. J. Chem. Eng., *48*, 343 (1970).

Riquarts, H. P., A physical model for axial mixing of the liquid phase for heterogeneous flow regime in bubble columns, Ger. Chem. Eng., *4*, 18 (1981).

Roy, N. K., D. K. Guha, and M. N. Rao, Suspension of solids in a bubbling liquid-critical gas flow rate for complete suspension, Chem. Eng. Sci., *19*, 215 (1964).

Saada, M. Y., Period. Polytech.-Chem. Eng., *19*, 317 (1975).

Sada, E., H. Kumazawa, and M. A. Butt, Removal of dilute sulfur dioxide by aqueous slurries of magnesium hydroxide particles, Chem. Eng. Sci., *32*, 972 (1977a).

Sada, E., H. Kumazawa, and M. A. Butt, Simultaneous absorption with reaction in a slurry containing fine particles, Chem. Eng. Sci., *32*, 1499 (1977b).

Sänger, P. and W.-D Deckwer, Liquid-solid mass transfer in aerated suspension, Chem. Eng. J., *22*, 179 (1981).

Sano, Y., N. Yamaguchi, and T. Adachi, Mass transfer coefficients for suspended particles in agitated vessels and bubble columns, J. Chem. Eng. Jpn., *7*, 255 (1974).

Sato, Y., T. Hirose, F. Takahashi, and M. Toda, Pressure loss and liquid holdup in packed bed reactor with cocurrent gas-liquid downflow in packed beds, J. Chem. Eng. Jpn., *6*(2), 147 (1973a).

Sato, Y., T. Hirose, F. Takahashi, M. Toda, and Y. Hashiguchi, Flow pattern and pulsation properties of co-current gas-liquid downflow in packed beds, J. Chem. Eng. Jpn., *6*(4), 315 (1973b).

Sato, Y., T. Hirose, and T. Ida, Kagaku Kogaku, *38*, 543 (1974).

Satterfield, C. N., Trickle bed reactors, AIChE J., *21*, 209 (1975).

Satterfield, C. N. and P. F. Way, The role of the liquid phase in the performance of a trickle bed reactor, AIChE J., *18*, 305 (1972).

Schmitt, K., New acetone chemistry in germany, Chem. Ind. Int., *3*, 2 (1960).

Schügerl, K., Proc. Int. Symp. Fluidization, Preprints, 9—4, Eindhoven (1967), p. 183.

Schürgerl, K. and J. Lucke, and U. Oels, Bubble column bioreactors, Adv. Biochem. Eng., T. K. Ghose, A. Fiechter, and N. Blakeborogh, eds., *7*, 1 (1977).

Schumpe, A., Die chemische Bestimmung von Phasengremg Hachen in Blasensaulen bi uneinheitlichen Blasengriben, Dr. thesis, university of Hannover (1981).

Schumpe, A., K. Nguyen-Tien, and W.-D. Deckwer, Beluften hochviskoser nichtNewtonscher Losungen in Blasensaulen, Chem.-Ing.-Tech., (1981, in press).

Shah, Y. T., *Gas-Liquid-Solid Reactor Design*, McGraw-Hill, New York, (1979).

Shah, Y. T., *Reaction Engineering in Direct Coal Liquefaction*, Addison-Wesley, Reading, Mass. (1981).

Shah, Y. T. and J. A. Paraskos, Criteria for axial disperison effects in adiabatic trickle bed hydroprocessing reactors, Chem. Eng. Sci., *30*, 1169 (1975).

Shah, Y. T. and M. M. Sharma, Desorption with and without chemical reaction, Br. Trans. Inst. Chem. Eng., *54*, 1 (1976).

Shah, Y. T. and P. C. Singh, Comportement termique d'un réacteur adiabatique pour la liquefaction de charbon, Entropie, No. 101, p. 48 (1981).

Shah, Y. T., R. D. Mhaskar, and J. A. Paraskos, Optimum quench location for a hydrosulfurization reactor with time varying catalyst activity, Ind. Eng. Chem. Process Des. Dev., *15*(3), 400 (1976).

Shah, Y. T., G. J. Stiegel, and M. M. Sharma, Backmixing in gas liquid reactors, AIChE J., *24*, 369 (1978).

Shah, Y. T., C. P. P. Singh, and J. B. Joseh, Gas-liquid-solid reactor design, Eng. Found. Conf. Computer Aided Process Design, Henniker, N.H. (July 1980); published in Chem. Process Des., 277 (1981).

Shah, Y. T., B. G. Kelkar, S. P. Godbole, and W.-D. Deckwer, Design parameter estimations for bubble column reactors, AIChe J., *28*(3), 353 (1982).

Sharma, M. M., Trans. Inst. Chem. Eng. (in press).

Sharma, M. M. and R. R. Mashelkar, Absorption with reaction in bubble columns, Inst. Chem. Eng. Symp. Ser., *28*, 10 (1968).

Shaykhutdinov, A. G., N. U. Bakinrow, and A. G. Usmanov, Determination and mathematical correlation of heat transfer coefficients under conditions of bubble flow, cellular and turbulent foam, Int. Chem. Eng., *11*, 641 (1971).

Sicardi, S., H. Gerhard, and H. Hoffmann, Flow regime transition in trickle bed reactors, Chem. Eng. J., *18*, 173 (1979).

Singh, C. P. P., Y. T. Shah, and N. L. Carr, Multiple steady states in coal liquefaction (SRC-II) reactors, C&M Division, Gulf R&D Co., Dept. No. 267RL025 (Feb. 1980).

Sittig, W., Umbersuchugen zur Vebesserung des Stoffaustauschs in Fermentern, Verfahrenstechnik (Mainz), *11*, 720 (1977).

Slesser, C. G., M. T. Allen, A. R. Cuming, V. Fawloski, and J. Schields, A dynamic oxygen transfer coefficient measurement method for column reactors, ESCREA, Brussels, 1968, Supplement to Chem. Eng. Sci., 41 (1968).

Smith, J. M., Physical Technology Laboratory, University of Technology, Delft, The Netherlands, personal communication (1980).

Smith, L. A. and M. N. Huddleston, New MTBE design now commercial, hydrocarbon Process., 121 (Mar. 1982).

Specchia, V. and G. Baldi, Pressure drop and liquid holdup for two phase concurrent flow in packed beds, Chem. Eng. Sci., *32*, 515 (1977).

Speechia, V. and G. Baldi, Heat transfer in trickle bed reactors, Chem. Eng. Commun., *3*, 483 (1979).

Specchia, V., S. Sicardi, and A. Gianetto, Abosrption in packed towers with concurrent upward flow, AIChE J., *20*, 646 (1974).

Sriram, K. and R. Mann, Dynamic gas disengagement: a new technique for assessing the behavior of bubble columns, Chem. Eng. Sci., *32*, 571 (1977).

Steiff, A. and P. M. Weinspach, Heat transfer in stirred and non-stirred gas-liquid reactors, Ger. Chem. Eng., *1*, 150 (1978).

Stewart, P. B. and F. Davidson, Three phase fluidization: water particle air, Chem. Eng. Sci., *19*, 319 (1964).

Stiegel, G. J. and Y. T. Shah, Can. J. Chem. Eng., *55*, 3 (1977a).

Stiegel, G. J. and Y. T. Shah, Backmixing and liquid holdup in a gas-liquid cocurrent upflow packed column, Ind. Eng. Chem. Process Des. Dev., *16*(1), 37 (1977b).

Subbarao, D. and V. K. Taneja, 3rd Eur. Conf. Mixing, Institute of Chemical Engineers (U.K.) (1979), p. 229.

Tallmadege, J. A., Packed bed pressure drop—an extension to higher Reynolds numbers, AIChE J., *15*, 1092 (1970).

Talmor, E., Two phase downflow through catalyst beds: Part I. Flow maps, AIChE J., *2396*), 868 (1977).

Turpin, J. L. and R. L. Huntington, Prediction of pressure drop for two phase, two-component concurrent flow in packed beds, AIChE J., *13* (6), 1196 (1967).

Uchida, S., K. Koide, and M. Shindo, Gas absorption with fast reaction into a slurry containing fine particles, Chem. Eng. Sci., *30*, 644 (1975).

Ueyama, K., S. Morooka, K. Koide, H. Kaji, and T. Miyauchi, Behavior of gas bubbles in bubble columns, Ind. Eng. Chem. Process Des. Dev. *14*, 492 (1980).

Vail, Y. K., N. K. Manakov, and V. V. Manshilin, Turbulent mixing in three phase fluidized bed, Int. Chem. Eng., *8*, 293 (1968).

Van Swaaij, W. P. M., J. C. Charpentier, and J. Villermaux, Residence time distribution in the liquid phase of trickle flow in packed columns, Chem. Eng. Sci., *24*, 1083 (1969).

Vasalus, I. A., E. M. Bild, D. N. Rundell, and D. F. Tatterson, Experimental techniques for studying third dynamics of H-COAL Reactor, in *Coal Processing Technology*, CEP Technical Manual 6, 226 (1980).

Vermeer, D. J. and R. Krishna, Hydrodynamics and mass transfer in bubble columns operating in the turbulent regime, Ind. Eng. Chem. Process Des. Dev., *20*, 475 (1981).

Viswanathan, S., A. S. Kakar, and P. S. Murti, Effect of dispersing bubbles into liquid fluidized beds on heat transfer and holdup at constant bed expansion, Chem. Eng. Sci., *20*, 903 (1965).

Voigt, J. and K. Schügerl, Absorption of oxygen in countercurrent multistage bubble columns: I. Aqueous solution with low viscosity, Chem. Eng. Sci., *34*, 1221 (1979).

Voigt, J., V. Hecht, and K. Schügerl, Absorption of oxygen in countercurrent multistage bubble columns: II. Aqueous solution with high viscosity, Chem. Eng. Sci., *35*, 1317 (1980).

Voyer, R. D. and A. I. Miller, Improved gas-liquid contacting in cocurrent flow, Can. J. Chem. Eng., *46*, 335 (1968).

Wallis, G. B., Interaction between fluids and particles, Inst. Chem. Eng. (London), 9 (1962).

Wallis, G. B., *One-Dimensional Two Phase Flow*, McGraw-Hill, New York (1969).

Wang, K. B. and L. T. Fan, Mass transfer in bubble columns packed with motionless mixers, Chem. Eng. Sci., *33*, 945 (1978).

Weekman, V. W., Jr., and J. E. Myers, Fluid flow characteristics of cocurrent gas-liquid flow in packed beds, AIChE J., *10*, 951 (1964).

Weekman, V. W., Jr., and J. E. Myers, Heat transfer characteristics of concurrent gas-liquid flow in packed beds, AIChE J., *11*, 13 (1965).

Weiland, P., Untersuchuunjg eins Airlift-Reaktors mit auberem Umtauf in Hinblik and seine Anwendumg als Bioreaktor, Dr. Inj. thesis, University of Dortmund (1978).

Woerde, H. M., L. J. Bostelaar, A. Hoek, and W. M. H. Sachtler, The effect of pore diffusion in enantioielective hydrogenation with supported nickel catalysts, J. Catal., *76*(2), 316 (1982).

Yamanaka, Y., T. Sekizawa, and H. Kubota, Age distribution of suspended solid particles in a bubble column, J. Chem. Eng. Jpn., *3*, 264 (1970).

Yan, T. Y., Dynamics of trickle bed hydrocracker with quench stream, Can. J. Chem. Process Des. Dev., *19*, 635 (1980).

Ying, D. H., E. N. Givens, and R. F. Weimer, Gas holdup in gas-liquid and gas-liquid solid flow reactors, Ind. Eng. Chem. Process Des. Dev., *19*, 635 (1980a).

Ying, D. H., R. Sivasubramanian, and E. N. Givens, Gas/slurry flow in coal liquefaction process, DOE Report FE-14801-3 (Jan. 1980b).

Yoshida, F., and K. Akita, Performance of gas bubble columns: volumetric liquid-phase mass transfer coefficient and gas hodlup, AIChE J., *11*, 9 (1965).

Zahradnik, J. and F. Kastanek, The effect of plates on some parameters of a heterogeneous bed in bubble-type flow reactors, Collect. Chech. Chem. Commun., *39*, 1419 (1974).

Zahradnik, J. and F. Kastanek, Gas holdup in uniformly aerated bubble column reactors, Chem. Eng. Commun., *3*, 413 (1979).

Zahradnik, J., F. Kastanek, and M. Rylek, Porposity of the heterogeneous bed and liquid circulation in multistage bubble-type column reactors, Collect Czechoslov. Chem. Commun., *39*, 1403 (1974).

Zaidi, A., Y. Louisi, M. Ralek, and W.-D. Deckwer, Ger. Chem. Eng., *2*, 94 (1979).

Zlokarnik, M., Sorption characteristics of slot injectors and their dependence of the coalescence behavior of the system, Chem. Eng. Sci., *34*, 1265 (1979).

Zwietering, T. N., Suspension of solid particles in liquid by agitators, Chem. Eng. Sci., *8*, 244 (1958).

11
Polymerization Reactors

MATTHEW TIRRELL and RAFAEL GALVAN *University of Minnesota, Minneapolis, Minnesota*

ROBERT L. LAURENCE *University of Massachusetts, Amherst, Massachusetts*

INTRODUCTION

A separate chapter on polymerization reactors in any treatise on chemical reaction engineering is required for several reasons. One is the steady and considerable glowth in production of polymeric products. In 1985, it is expected that production of the five leading thermoplastic polymers alone (polystyrene, polyvinyl chloride, polypropylene, and high- and low-density polyethylene) will exceed 100 billion pounds worldwide (Webber, 1985). This total doubles the production of 1975 and amounts to about $40 billion worth of product. At this scale, even relatively minor efficiencies created by good engineering are very significant.

Another challenge lies in the unique difficulties of performance optimization of polymerization reactors. To optimize a system, we need to know what optimal is—to define the objective. Polymerization engineering is complicated in this regard since most polymeric products end up as fabricated materials, so that criteria such as processibility and end-use (mechanical, etc.) properties are the real reactor performance measures. No truly analogous situation exists in the design of a reactor for a monomeric product.

We can contrast the polymer reactor with a reactor making ammonia, for example. We know the properties of ammonia. The only design problem is to make as much of it as fast and economically as possible. We want to maximize conversion and if there are parallel or sequential reactions in the synthetic scheme, maximize selectivity. The first point of contrast is that simply to specify the generic name of a polymer (e.g., polystyrene) is inadequate to specify its properties. The second point of contrast is that polymerization reactions present selectivity problems that are orders of magnitude more complex than those for the monomeric reactor. An elaborate set of sequential and parallel reactions occurs. The resulting distribution of polymeric product, which composes any commercial polymeric material, is of comparable importance to the chemical structure of the polymer in determining product performance in end-use properties. The product distribution, as well as the generic chemical formula, must therefore be described in any complete specification of a polymer product.

As in other reactor design situations for optimal yield and selectivity, the factors of reactor configuration—batch or continuous; mixing pattern; residence-time distribution; recycle; heat transfer—all play roles in determining the product distribution. The sensitivity of the product distribution to these factors is enormous. Polymers produced in a batch reactor can be very different from those produced in a continuous reactor under otherwise substantially identical chemical conditions.

A final point to be made regarding the distinguishing features of polymer reactors pertains to polymers at the other end of the production scale from the commodity polymers mentioned in the first paragraph. Those products not withstanding, many more polymeric products are low-volume specialty materials, frequently copolymers, designed to perform a specified function. Efficiency is not the principal design objective for these reactors (frequently, there is a high margin of profitability). Rather, the need is for reproducible production of a complex product distribution so that the polymer performs precisely as required.

Product distribution analysis and control is, therefore, the principal novel feature and objective of polymerization reactor engineering relative to other chemical reactors. This chapter is informative but not comprehensive. It is designed to provide an entry into the field by presenting the main ideas and techniques, by way of some examples and placing the emphasis where possible on the most recent ideas, techniques, problems, and challenges. This is made possible by referring the reader interested in more comprehensive information on this subject to several other excellent sources on polymerization reactor engineering (Ray, 1972; Ray and Laurence, 1977; Biesenberger and Sebastian, 1983; Platzer, 1970; Gerrens, 1982).

MATHEMATICAL TECHNIQUES

If the elementary reaction steps of a polymerization mechanism are known or can be postulated with an adequate degree of certainty, mathematical analysis is a (somewhat underutilized) tool available to the engineer for making predictions about product distributions.

There are two alternative generic starting points for the employment of mathematical analysis in polymerization. One can start from the mass balance equations on the reactor of interest. This leads to differential or algebraic difference equations, depending on reactor type, which must be solved to get the desired information on product distribution. Data are needed on the rate constants of the model. This is a systematic approach, readily adapted to the analysis of different reactor types. This method produces an infinite set of mass balance equations, one for each possible polymer chain length, 1 to ∞. The solution of any particular set of such equations may be difficult and involved, but there is a catalog of solution methods available (Ray and Laurence, 1977).

Alternatively, statistical methods can be useful (Flory, 1936; Schulz, 1935; Lowry, 1970). These rely on picturing the polymer chain as growing by a process of selecting monomers from the reaction mixture according to some statistical distribution. The success of this approach depends on a correct, and frequently intuitively based, correspondence of the polymerization process with a postulated stochastic process. It is therefore less easily adaptable to different reactor configurations since the appropriate correspondence may be difficult to find or, in fact, may not exist. On the

other hand, when statistical methods do work, they usually lead to results in a comparatively simple way and can sometimes reveal features of a polymerization system that a mass balance approach does not. The fully equipped polymerization engineer should be able to take advantage of both mass balance and statistical techniques.

Table 1 lays out several approaches to solving infinite sets of mass balance equations, illustrated with an example from anionic polymerization. All the techniques in that table have general utility beyond this particular mechanism. The basic example equation of Table 1 has been simplified through the use of several generally practical maneuvers.

To illustrate these generic tricks, we go back to the mechanism of anionic polymerization with no termination and instantaneous initiation (Szwarc, 1968; Nanda and Jain, 1964). In this case, only the propagation reaction is of interest.

$$P_j + M \xrightarrow{k_p} P_{j+1} \tag{1}$$

At the instant of introduction of initiator (concentration I_0) a certain number of growing chains (concentration I_0) are established. They add monomer with rate constant k_p until monomer is exhausted. The polymerization engineer's problem is to determine the distribution of polymer chain lengths at any specified instant of the reaction. Solution of the mass balance on P_j, for arbitrary j, gives the answer:

$$\frac{dP_j}{dt} = k_p MP_{j-1} - k_p MP_j \qquad P_1(0) = I_0, \text{ all other } P_j(0) = 0 \tag{2}$$

In equations like (2), it is frequently useful to scale time with the rate constant and monomer concentration in order to eliminate k_p from explicit consideration temporarily and to decouple Eq. (2) from the monomer balance:

$$\frac{dM}{dt} = -k_p M \sum_{j=1}^{\infty} P_j \qquad M(0) = M_0 \tag{3}$$

This scaling is done by defining a new time variable (Dostal and Mark, 1936):

$$\tau = \int_0^t k_p(t')M(t')\, dt' \tag{4}$$

or

$$d\tau = k_p M\, dt \tag{5}$$

Thus temperature variations and monomer depletion are eliminated from explicit consideration in calculation of the polymer chain-length distribution. Using Eq. (5), Eq. (2) becomes

TABLE 1 Techniques for Solving Polymerization Rate Equations, Illustrated with Anionic Polymerization

---

Basic equation: $\dot{P}_j = P_{j-1} - P_j$: $P_j\ (0) = I_0\delta(j - 1)$

Techniques

1. Analytical, successive substitution:

$$\dot{P}_1 = -P_1 \quad \Rightarrow \quad P_1(\tau) = I_0 e^{-\tau}$$

$$\dot{P}_2 = -P_2 + I_0 e^{-\tau} \quad \Rightarrow \quad P_2(\tau) = I_0 \tau e^{-\tau}$$

$$\dot{P}_3 = -P_3 + I_0 \tau e^{-\tau} \quad \Rightarrow \quad P_3(\tau) = \frac{1}{2} I_0 \tau^2 e^{-\tau}$$

$$\vdots \qquad\qquad \vdots$$

$$P_j(\tau) = \frac{1}{(j - 1)!} I_0 \tau^{j-1} e^{-\tau}$$

2. Numerical: conceptually identical to 1. Used when rate equations are either impossible to integrate analytically or when a general form for P_j cannot be found by successive substitution.

3. Laplace transform: converts the infinite set of differential-difference equations to a single difference equation

$$L[P_j(\tau)] = \bar{P}_j(\lambda) = \int_0^\infty P_j(\tau) e^{-\lambda\tau}\, d\tau$$

$$\bar{\dot{P}}_j(\lambda) = \int_0^\infty \dot{P}_j e^{-\lambda\tau}\, d\tau = \lambda P_j - \text{I.C.}$$

The rate equation becomes

$$\lambda\bar{P}_j = -\bar{P}_j + \bar{P}_{j-1} \qquad j \geqslant 2$$

$$\bar{P}_j = \frac{1}{\lambda - 1} \bar{P}_{j-1} \Rightarrow \bar{P}_j = I_0 \left(\frac{1}{\lambda - 1}\right)^j$$

From a compilation of Laplace transforms,

$$\left(\frac{1}{\lambda - 1}\right)^n = \int_0^\infty \frac{\tau^{n-1}}{(n - 1)!} e^{-\tau} e^{-\lambda\tau}\, d\tau$$

Thus

$$P_j(\tau) = -1(P_j) = I_0 \frac{\tau^{j-1}}{(j - 1)!} e^{-\tau}$$

TABLE 1 (Continued)

4. Generating function transformation: converts the infinite set to a single ordinary differential equation (ODE). In principle, can provide the total MWD. Moments are easily obtained otherwise. Definition:

$$G(s,\tau) = \sum_{j=1}^{\infty} s^j P_j(\tau)$$

Moments are obtained by noting (see technique 5 below)

$$\left. \frac{\partial^k G(s,\tau)}{(\partial \ln s)^k} \right|_{s=1} = \mu_k$$

Using $\Sigma_{j=1}^{\infty} s^j P_{j-1} = sG(s,\tau)$, the rate equation becomes

$$\dot{G}(s,\tau) = -(1 - s)G(s,\tau) \qquad G(s,0) = sI_0$$

$$G(s,\tau) = sI_0 e^{-(1-s)\tau}$$

Expanding the exponential in power series in s yields

$$G(s,\tau) = I_0 \sum_{j=1}^{\infty} s^j \frac{\tau^{j-1}}{(j - 1)!} e^{-\tau}$$

Comparing with the definition, we obtain

$$P_j(\tau) = I_0 \frac{\tau^{j-1}}{(j - 1)!} e^{-\tau}$$

5. Moment generation: uses the definition of the moments of the MWD, $\mu_k = \Sigma_{j=1}^{\infty} j^k P_j$, to convert the infinite set to a few ODEs *for the moments*. Used when no analytical solution is possible or when taking the logarithmic derivative of $G(s,\tau)$ is cumbersome, and the information given by moments suffices. The rate equation becomes

$$\dot{\mu}_k = \sum_{j=1}^{\infty} j^k P_{j-1} - \mu_k \qquad \mu_k(0) = I_0$$

Letting k = 0, 1, 2, . . . gives

$$\dot{\mu}_0 = 0 \qquad \Rightarrow \mu_0(\tau) = I_0$$

$$\dot{\mu}_1 = \mu_0 \qquad \Rightarrow \mu_1(\tau) = I_0(1 + \tau)$$

TABLE 1 (Continued)

$$\dot{\mu}_2 = 2\mu_1 + \mu_0 \Rightarrow \mu_2(\tau) = I_0(1 + 3\tau + \tau^2)$$

•
•
•

Frequently, only the three leading moments are needed.

6. Continuous variable approximation: assumes the chain sufficiently long such that the chain length j can be considered to be a continuous variable. Expanding P_{j-1} in a Taylor's series about P_j, the following is obtained:

$$P_{j-1} = P_j - \frac{\partial P_j}{\partial j} + \frac{1}{2}\frac{\partial^2 P_j}{\partial j^2} + \cdots$$

Truncating after the quadratic term, the rate equation (an infinite set) is converted to a partial differential equation for P_j:

$$\frac{\partial P_j}{\partial \tau} + \frac{\partial P_j}{\partial j} = \frac{1}{2}\frac{\partial^2 P_j}{\partial j^2}$$

with solution

$$P_j(\tau) = \frac{I_0}{(2\pi\tau)^{1/2}} \exp\left[\frac{(j - 1 - \tau)^2}{2\tau}\right]$$

This approximation is valid only for reasonably high molecular weight, 1/j small, for the Taylor's expansion to be valid. Equivalent to Poisson distribution for large τ.

7. Statistical techniques: historically, the first techniques used in polymerization problems. Can be classified as follows:

Direct approach	Flory (1953)
Markov chain theory approach	Lowry (1970)
Recursive approach	Lopez-Serrano et al. (1980)

$$\dot{P}_j = P_{j-1} - P_j \tag{6}$$

where the dot signifies a derivative with respect to τ. Techniques for solving this equation are enumerated in Table 1. Some additional results on the resulting product distribution are given by Peebles (1971).

Implicit in this reaction scheme, and necessary for this time-scaling procedure to work, is k_p independent of j. This is known as the *equal reactivity assumption*, meaning that the reactivity of a chain end is uninfluenced by the length of chain to which it is attached (Flory, 1953). This is an assumption which is very often, but not always, correct. It

has been examined for several different mechanisms and certain exceptions have been noted, the most important of which are the termination reaction in free radical polymerization (Coyle et al., 1985) and monomer addition at high conversion in polycondensation (Gupta and Kumar, 1985).

Certain solution techniques for polymerization equations work better in certain situations. Only experience teaches this, but some guidelines can be laid down. Analytical solutions (technique 1, Table 1) are the exception, not the rule, and are more the province of the academician than of the engineering practitioner. However, this does not mean that sequential computer numerical solution (technique 2, Table 1) should be called in immediately by those with pragmatic views. Moment generation via techniques 4 and 5 of Table 1 will lead to a set of moment equations that may be solvable analytically, even when the mass balance cannot, and may provide adequate, though incomplete predictive information about the MWD in the polymerization reactor. It is important for the engineer to decide in advance what level of analysis is appropriate, based on the level of information required from the analysis. It is frequently a good idea to set the simplest information first (i.e., moments) and then move on to attempt more involved analytical or numerical solutions.

The level and varieties of mathematics required for polymerization analysis are shown in Table 1. Analysis is not as arcane as it is sometimes made to seem. Solution of ordinary differential equations is fundamental. Partial differential equations arise in several ways: (a) when the continuous variable approximation (Bamford and Tompa, 1954; Bamford et al., 1958; Zeman and Amundson, 1963, 1965a,b) is made (technique 6, Table 1); (b) when diffusion limitations are considered (Nagel et al., 1980); (c) in some applications of generating functions. Difference equations, particularly of the form $P_j = \alpha P_{j-1}$, arise frequently. It is useful to realize that the solution to this first-order difference equation is $P_j = K\alpha^n$. In other words, this difference equation has a geometric distribution as its solution. Clearly, a *sine qua non* is facility with the ideas of probability and distribution functions. Series of various types, and their infinite (and, less often, partial) sums, arise frequently. Familiarity with the short list of mathematical areas above will provide the engineer with tools needed to employ modeling profitably. A final caveat on mathematical analysis is never to let it become more complex than necessary, nor let it become an end in itself. Two rules help to avoid this. One, mentioned above, is to have a good idea in advance of the kind of information you want out of analysis *before* wading into it too deeply. The second rule is to look always for a physical interpretation of any derived equation. Frequently, bearing in mind the two generic and complementary modeling approaches, mass balance and statistical, is helpful in this regard. For example, the difference equation mentioned above;

$$P_j = \alpha P_{j-1} \tag{7}$$

arises in many polymerization mass balances. In the anionic polymerization mechanism of Table 1, but occurring in a steady-state continuous stirred-tank reactor of average residence time θ, a balance on P_j with no P_j in the feed gives

$$\frac{P_j}{\theta} = k_p M P_{j-1} - k_p M P_j \tag{8}$$

[compare Eq. (2)], which is exactly the form of Eq. (7) with

$$\alpha = \frac{k_p M}{1/\theta + k_p M} \tag{9}$$

Alternatively, we could arrive at Eq. (7) by an equivalent, but at first sight perhaps different, statistical argument. We can regard the concentration P_j as the probability of forming a polymer of length j. P_j is then the product of two other probabilities: P_{j-1}, the probability of forming a polymer of length j − 1 times, and α, the reaction probability for adding a monomer to the growing polymer. Equation (9) then gives an explicit physical interpretation of this reaction probability. The growing chain has two choices: (a) add a monomer with rate k_pM, or (b) wash out of the reactor with rate 1/θ. Equation (9) gives the probability that it reacts. Very similar statistical arguments apply to several other polymerization situations (Lowry, 1970).

POLYMERIZATION REACTIONS

Classification of all polymerization reactions as either step growth or chain growth has been discussed by Lenz (1968). It is important to note that this is a classification of reactions, not of the polymers themselves, since there are instances of polymers that can be made via either step-growth or chain-growth routes. Generally speaking, however, most important polymers are made exclusively by one or the other.

The principal characteristic distinguishing step-growth and chain-growth reactions is that each polymer chain grows at a relatively slow rate over a much longer period of time for the former than for the latter. In step-growth polymerization there is usually a single type of chemical reaction that links polymer molecules of all sizes. The reactions are of the same mechanism that link monomers, oligomers, and polymers. In contrast, a chain-growth polymerization consists of distinctly different chemical steps: initiation, which creates a highly reactive polymerization locus; propagation in which monomer (*only*) adds to the active center, which grows in length; (*frequently*) termination, in which the active center is extinguished and this polymer chain no longer participates in the reaction; (*sometimes*) transfer, where the active locus is not extinguished permanently but rather transferred in location from one growing chain (which becomes dead) to another species in the medium (monomer, solvent, agent, impurity, etc.), which then propagates. The existence of the active center in chain-growth polymerization, in that not every chain has one, is another important distinction between step growth and chain growth.

In the remaining parts of this section we discuss the characteristic molecular weight distributions that arise in each of these types of polymerization in a batch reactor. This will further differentiate the two types of polymerization. Subsequent sections will deal with the modified distributions produced in other types of reactors.

Step-Growth Polymerization

The most important chemical reaction used for the preparation of polymer via step-growth polymerization is that of addition and elimination at the

carbonyl double bond of carboxylic acids and their derivatives (Lenz, 1968). These reactions are the routes to most nylons and polyesters. Other important step-growth reactions are the urethane-forming double-bond addition and carbonyl addition-substitution for polyacetals and phenol-formaldehyde. Many, but not all step-growth reactions produce a condensation product (e.g., water or glycol) that must be removed.

Schematically, for the purposes of molecular weight distribution analysis, the general step-growth reaction in its least encumbered form can be represented:

$$P_n + P_m \xrightarrow{k} P_{n+m} \qquad n, m = 1, \ldots, \infty \tag{10}$$

In step-growth polymerization attention is focused on the end groups of the polymer chain. Restricting our attention for the moment to bifunctional monomers, every polymer has exactly one end group of each chemical type involved in the linkage. Thus fractional conversion of end groups is directly related to the number of polymer molecules, and therefore also to the molecular weight and molecular weight distribution of the polymer. In the case of exact stoichiometric equality of the concentrations of the two functional groups, we expect that a second-order rate law will govern the increase in fractional conversion, p, of c_0 initial functional groups:

$$\frac{dp}{dt} = kc_0(1 - p)^2 \tag{11}$$

which integrates to

$$\frac{1}{1 - p} = c_0kt + 1 \tag{12}$$

Kinetic data from many step-growth polymerizations fit the rate equation (12) (Flory, 1953). One exception occurs in polyesterification with no added acid catalyst; in this case, acid functional groups participate also as catalysts and thus the rate law is third order. As we shall show below, the number-average degree of polymerization in this reaction is $1/(1 - p)$. So a result of this analysis in Eq. (12) is that the number-average molecular weight builds linearly with time over the course of the reaction.

The analysis of the molecular weight distribution in the simplest step-growth situation of Eq. (10) with equal stoichiometry of the two end groups is most easily done with statistical arguments. The fractional conversion p is alternatively interpreted as the probability that a functional group selected at random has reacted to form a linkage. If we inquire after the probability of finding one molecule containing n monomer units, we realize that this is a molecule having (n − 1) linkages and one free unreacted functional group of a particular type. The probability of having (n − 1) linkages is the product of (n − 1) separate probabilities that an ester linkage would have been formed, p^{n-1}. So the probability of finding the desired molecule with (n − 1) linkages and one unlinked end is

$$P_n = (1 - p)p^{n-1} \tag{13}$$

which is the geometric or "most probable" distribution. The average molecular weights and chain lengths resulting from this distribution are

given in Ray and Laurence (1977). These authors, following Abraham (1963) and others (Kilkson, 1964; Szabo and Leathrum, 1969), have shown how the mass balance approach for the mechanism of Eq. (10) leads to the same result as Eq. (13).

From an engineering viewpoint one of the most important characteristics of Eq. (13) is that fractional conversion completely specifies the entire molecular weight distribution and therefore all the average molecular weights. Conversion control is the only means of molecular weight control for this system available to the reactor engineer. The number- and weight-average degrees of polymerization resulting from the distribution equation (13) are $1/(1 - p)$ and $(1 + p)/(1 - p)$, respectively, which show that only as p nears 1 is a high-molecular-weight product produced. There are two significant exceptions to this absolute control of the molecular distribution by conversion. These are interfacial polymerization and nonlinear, that is, branching or cross-linking, step-growth polymerization both of which are discussed later.

While Eq. (13) gives an accurate global description of many linear step-growth polymerization processes, in practice there are important additional effects and complications that often arise. Stoichiometry is sometimes inadvertently or intentionally unbalanced. Monofunctional species (chain stoppers) may be added to gain an independent means of control over the MWD. Equal reactivity may not strictly apply; this is, monomer functional groups may react more rapidly than identical groups on longer chains, particularly as conversion gets high. Reversibility of reaction (10) can be an important consideration. Diffusion limitations on the rate of removal of condensation product or the rate of polymer coupling may dominate the kinetics at high conversion.

Two of these effects will be considered further here to give a more complete picture of step-growth polymerization and to illustrate the use of some of the mathematical modeling techniques of the preceding section. We shall examine, in turn, the effect of addition of monofunctional agents and the effect of reversibility of reaction (10), both for the case of the batch reactor. Reactor configuration effects are considered in a subsequent section. This section concludes with brief discussions of interfacial step-growth polymerization and nonlinear step-growth polymerization.

Effects of Chain Stoppers (Gupta and Kumar, 1985; Kilkson, 1964)

The practical situation is one in which a monofunctional compound such as acetic acid is added to a mixture of difunctional amines and acids, or amino acids, a common practice in nylon-6 manufacture. We now have two distinguishable kinds of polymers, those with two reactable ends,

$$P_n:\quad A\!-\!(BA)_{n-1}B \tag{14}$$

and those with one end blocked by the monofunctional agent,

$$P_{nx}:\quad A\!-\!(BA)_{n-1}X \tag{15}$$

We have, therefore, two kinds of polymer linkage reactions:

$$P_n + P_m \xrightarrow{k} P_{n+m} \tag{16}$$

$$P_n + P_{mx} \xrightarrow{k} P_{(n+m)x} \tag{17}$$

X-terminated molecules cannot react with one another. Mass balances on the two kinds of polymeric species read:

$$\frac{dP_n}{dt} = k \sum_{j=1}^{n-1} P_j P_{n-j} - kP_n(2P + P_x) \qquad P_n(0) = 0 \tag{18}$$

$$\frac{dP_{nx}}{dt} = k \sum_{j=1}^{n-1} P_{jx} P_{n-j} - kP_{nx}P \qquad P_{nx}(0) = 0 \tag{19}$$

where $P_{(x)} = \Sigma_{n=1}^{\infty} P_{n(x)}$. (The x in parentheses is used to denote situations where the same equation applies for P_n or P_{nx}.) The first terms on the right-hand sides of each equation give the rate of formation of $P_{n(x)}$ by linkages between smaller molecules. The remaining terms give the rates of dissappearance of $P_{n(x)}$ by reaction will all other species in the mixture. The factor of 2 in the second term of Eq. (18) is due to the fact that a P_n has two distinct ways to couple with any other P_m.

Equations (18) and (19) apply for $n \geqslant 2$. For $n = 1$, we have no formation reactions; thus

$$\frac{dP_1}{dt} = -kP_1(2P + P_x) \qquad P_1(0) = P_{10} \tag{20}$$

$$\frac{dP_{1x}}{dt} = -kP_{1x}P \qquad P_{1x}(0) = P_{1x0} \tag{21}$$

It is convenient to scale time in a manner similar to Eq. (4):

$$\tau = \int_0^t k\ dt \tag{22}$$

For the solution of the set of equations (18) to (22), we illustrate the generating function technique, as in technique 4, Table 1. We will need generating functions for the two distributions:

$$G(s,\tau) = \sum_{n=1}^{\infty} s^n P_n \tag{23}$$

$$G_x(s,\tau) = \sum_{n=1}^{\infty} s^n P_{nx} \tag{24}$$

Adding Eq. (18) to Eq. (20) and Eq. (19) to Eq. (21), multiplying the two results by s^n, and summing as indicated in Eqs. (23) and (24) gives two differential equations for the generating functions:

$$\frac{\partial G}{\partial \tau} = G^2 - G(2P + P_x) \qquad G(0) = sP_{10} \tag{25}$$

$$\frac{\partial G_x}{\partial \tau} = G_x G - G_x P \qquad G(0) = sP_{1x0} \tag{26}$$

Partial derivatives are indicated since $G_{(x)}$ is a function of both s and τ. Note that $G_{(x)}(1) = \Sigma_{n=1}^{\infty} P_{n(x)} = P_{(x)}$. Setting s = 1 in Eqs. (25) and (26) gives

$$\frac{dP}{d\tau} = -(P^2 + PP_x) \qquad P(0) = P_{10} \tag{27}$$

$$\frac{dP_x}{d\tau} = 0 \qquad P_x(0) = P_{x0} \tag{28}$$

illustrating that as molecular weight builds, the total number, P, of difunctional monomers decreases but, as expected, the total number, P_x, of monofunctional species remains constant, although these chains do grow in length. Kilkson (1964) showed that these equations are easily solved using normalized generating functions:

$$y = \frac{G}{P} \tag{29}$$

$$y_x = \frac{G_x}{P_x} \tag{30}$$

to give:

$$\frac{\partial y}{\partial \tau} = Py(y - 1) \tag{31}$$

$$\frac{\partial y_x}{\partial \tau} = Py_x(y - 1) \tag{32}$$

from which it is clear that the two generating functions, and therefore the two distributions, satisfy the proportionality

$$G = \frac{P}{P_x} G_x \tag{33}$$

Finally, to make contact and compare with the statistically derived result Eq. (13) for the simpler situation of no monofunctional agent, it is convient to solve Eqs. (31) and (32) in terms of conversion of end groups rather than time. Conversion, p, in this case is defined as

$$p = \frac{P_{10} - P}{P_{10} + P_{1x0}} \tag{34}$$

as the fractional conversion of all initial groups of the, say, acid type, including those on the monofunctional agent. The time rate of conversion can easily be obtained from Eq. (27):

$$\frac{dP}{d\tau} = (P_{10} + P_{1x0})\left(\frac{1}{r} - p\right)(1 - p) \tag{35}$$

where $r = 1 + (P_{1x0}/P_{10})$, always greater than 1, gives the stoichiometry of the initial addition of monofunctional agent. Dividing Eq. (31) by Eq. (35) gives

$$\frac{dy}{dp} = \frac{y(y-1)}{1-p} \qquad y = s \text{ at } p = 0 \tag{36}$$

leading to

$$y = \frac{s(1-p)}{1-ps} \tag{37}$$

and by Eq. (33),

$$y_x = \frac{s(1-p)}{1-ps} \tag{38}$$

so that the normalized distributions of monofunctional and difunctional chains are identical. Recognizing that $ax/(1 - bx) = a\,\Sigma_{n=1}^{\infty}\, b^{n-1}x^n$, we see by rewriting Eqs. (37) and (38) as power series in s that

$$\frac{P_n}{P} = \frac{P_{nx}}{P_x} = (1-p)p^{n-1} \tag{39}$$

In other words, even in the presence of monofunctional agents, we obtain a product with a "most probable" or geometric distribution like Eq. (13). The distribution is still uniquely determined by p, but in this case p is not exclusively a function of time as it is in Eq. (12). Here p is also dependent on the initial stoichiometry, r. For the simple step-growth case leading to Eq. (13),

$$p = 1 - \frac{P}{P_{10}} \tag{40}$$

With monofunctional agents,

$$p = \left(1 - \frac{P}{P_{10}}\right)\frac{1}{r} \tag{41}$$

The effect of molecular weight is important. In both cases

$$\overline{DP}_n = \frac{1}{1-p} \tag{42}$$

which for the case of added chain stoppers gives

$$\overline{DP}_n = \frac{1}{1 - (1 - P/P_{10})(1/r)} \tag{43}$$

In the high conversion limit $P \to 0$ and we find

$$\left[\overline{DP}_n\right]_{P=0} = \frac{r}{r - 1} \tag{44}$$

Thus final molecular weight can be influenced not only by reaction time and conversion but also by addition of monofunctional agent. However, this addition does not change the overall character or shape of the distribution very much, except to shift the mean to lower degree of polymerization.

The similarity and implications of these results for another case of practical importance, that of unbalanced stoichiometry of the two difunctional monomers, should also be clear.

Effects of Reversibility and Interchange Reactions

In practice, many step-growth polymerizations, polyesterification and polyamidation included, are run at elevated temperatures (250°C or higher) in order to force the reaction to completion and to drive off and extract low-molecular-weight condensation products (e.g., water or glycol). The same high temperatures also promote reversibility of the polymerization reaction as well as interchange reactions, such as transamidation and transesterification, between the polymer molecules formed (Lenz, 1968). It is of interest then to know what effect reversibility has on the molecular weight distribution. This example will also afford the opportunity to illustrate some additional useful techniques in polymerization analysis.

The reversible step-growth mechanism is written

$$P_n + P_m \underset{k'}{\overset{k}{\rightleftharpoons}} P_{n+m} \tag{45}$$

with the understanding that the larger species P_{n+m} may unlink any of its $n + m - 1$ links. The interchange reaction is a simultaneous unlinking and relinking by two polymers:

$$P_n + P_m \underset{k_i}{\overset{k_i}{\rightleftharpoons}} P_j + P_k \tag{46}$$

Both reactions should be expected to have a similar effect on the molecular weight distribution. The effect of both reversibility and interchange is to unlink a polymer at a randomly chosen link. Statistical arguments have been used to show that a most probable distribution of the form of Eqs. (13) and (39) is obtained when reversibility is an important factor. This is, in fact, an important general conclusion. Random scission processes, such as thermal or radiation degradation will also lead ultimately to a most probable distribution, even starting from a monodisperse product.

Reversibility presents some special difficulties to the mass balance approach to modeling (Gupta and Kumar, 1985). The mass balance on P_n resulting from the mechanism (42) is

$$\frac{dP_n}{dt} = \frac{k}{2} \sum_{j=1}^{n-1} P_j P_{n-j} - kP_n P - k'(n - 1)P_n + 2k' \sum_{j=n+1}^{\infty} P_j \tag{47}$$

The first two terms of this equation are identical (aside from the factor-of-2 difference in the definition of the rate constants here) to the first two terms in Eq. (18). The third term gives the rate of dissappearance of P_n by a P_n unlinking any of its (n − 1) bonds. The fourth term gives the rate of appearance of P_n by larger molecules unlinking, the factor of 2 accounting for the fact that any polymer has two sites at which it can unlink to give a polymer of some specified shorter length. Mathematically, the nP_n term and the partial sum term due to reversibility pose the difficulties. Applying the generating function operation of Eq. (23) yields

$$\frac{\partial G}{\partial t} = \frac{k}{2} G(G - 2P) - k' \left(s \frac{\partial G}{\partial s} - G - \frac{G - sP}{1 - s}\right) \tag{48}$$

Rather than solve this nonlinear first-order partial differential equation, we shall use it to derive moment equations (see Table 1) by successive differentiation of Eq. (48) with respect to ln s and taking the limit as $s \to 1$.

$$\frac{d\mu_0}{dt} = \frac{dP}{dt} = -P_{10} \frac{dp}{dt} = -k\mu_0^2 + k'(\mu_1 - \mu_0) \tag{49}$$

$$\frac{d\mu_1}{dt} = 0 \tag{50}$$

$$\frac{d\mu_2}{dt} = k\mu_1^2 + \frac{k'}{3}(\mu_1 - \mu_3) \tag{51}$$

with initial conditions $\mu_k(0) = P_{10}$. Equation (50) expresses the fact that the total number of monomer units, $\Sigma_{n=1}^{\infty} nP_n$, in the reaction mixture remains constant. Evidently, Eq. (49) and (50) can be solved together to give conversion p and degree of polymerization $\overline{DP}_n$ as functions of time. It is found that the most probable distribution relation between $\overline{DP}_n$ and p, Eq. (42), is obeyed.

Higher moments present the mathematical difficulties. The set of moment equations is not close in that μ_2 depends on μ_3, the equation for μ_3 would involve μ_4 and so on. This arises from the third term of Eq. (48). Terms like this also arise in analysis of other chain scission reactions. One route to analysis of this type of problem is the moment expansion procedure discussed by Bamford and Tompa (1954) and Hulbert and Katz (1964). This is an approximation procedure which for the purposes of obtaining relations among the moments, replaces the unknown true distribution by an expansion in Laguerre polynomials about a standard distribution

for which the relations among the moments can be calculated exactly. These known relations enable one to express, for example μ_3, in terms of the lower moments. Application of the Hulburt-Katz procedure gives

$$\mu_3 = \frac{\mu_2}{\mu_0\mu_1}(2\mu_0\mu_2 - \mu_1^2) \tag{52}$$

In fact, this moment closure approximation has been examined in several reversible polymerization problems and found to give very accurate numerical results, 1% error when three to five terms are used (Bamford and Tompa, 1954; Gupta and Kumar, 1985), for the moments when compared with the much more laborious complete numerical solution. It could therefore be recommended as a generally useful approximation provided that there is no a priori reason to believe that the distribution will have a very unusual shape.

In summary, the effect of reversibility can be treated by the means described here. Reversibility does not change the conclusion reached from analysis of irreversible linear step-growth polymerization that the product MWD is most probable with the mean dictated completely by fractional conversion of end groups. Reversibility does, however, alter the fractional conversion itself. Means of driving the equilibrium toward higher conversion and therefore higher molecular weight are frequently employed.

Interfacial Polycondensation

The preceeding discussion has illustrated the intimate connection between conversion, stoichiometry, and molecular weight distribution in linear step-growth polymerization. Reaction engineers should always seek additional control handles on MWD. In step-growth polymerization, interfacial reaction provides an alternative to the strict conversion and stoichiometry constraints. This has been exploited in polyamide production. The principles of the technique have been explained by Kilkson (1968). It relies on what he has termed the *concept of slow influx*.

Consider polymerization of two difunctional monomers of the AA and BB types (e.g., hexamethylene diamine and adipic acid or its derivatives), where we start with exact stoichiometric equivalent amounts of the two monomers. A "normal" polycondensation would be done homogeneously, adding all of both the monomers at the beginning. This is the situation of strict conversion and stoichiometry control, where most of the analysis of this section applies. Imagine now the following alternative. All of the BB monomer is placed in the reactor and slow addition of AA is begun. As each increment of AA enters the large excess of BB, it will be immediately reacted on both ends by BB. This process can be continued until the full stoichiometric equivalent amount of AA has been added, consuming the last B group. If the addition has been very slow, the polymer at that point, will be of essentially infinite molecular weight since all the AA and BB units have been linked into the polymer. Addition of further AA, in stoichiometric excess, now serves only to dilute this high-molecular-weight product, not decrease molecular weight as it would in the case of the homogeneous reaction with unbalanced stoichiometry [see Eq. (44)].

The point is that molecular weight is now influenced strongly by the monomer addition rate, not exclusively by conversion and stoichiometry. In interfacial polycondensation, the slow influx of monomer is accomplished

by imposing diffusion limitations. Each monomer is dissolved in one of a pair of mutually immiscible solvents. Reaction occurs at the interface, where the monomer supply is limited by diffusion.

In practice, interfacial polymerization is done under mild reaction conditions. Therefore, reactive monomers must be used. Diacid chlorides are used instead of diacids. Stirring and addition of emulsifier to create and stabilize large interfacial areas are also important considerations in this type of reactor.

Nonlinear Step-Growth Polymerization

Not a great deal of literature is available on reactor design for nonlinear step-growth polymerization. Two factors probably account for this. One is that nonlinear step-growth polymers have never been produced in very large volumes. The other is that many of these materials are thermosetting materials, neither soluble nor fusible once made, so that the "reactor" is frequently simply a mold of the desired shape. Examples of important nonlinear step-growth polymerizations include phenolic resin manufacture, urea-formaldehyde resin manufacture, epoxies, and many polyurethanes.

Nonlinear polymerization requires the presence of some monomers with functionality greater than 2. The amount of multifunctional monomer is another variable that influences MWD. For example, the number-average degree of polymerization is given by (Flory, 1953)

$$\overline{DP}_n = \frac{2}{2 - pf} \tag{53}$$

where f is the average functionality of the monomers containing the branching group. For example, consider a polymerization of AB, BB, and a trifunctional A monomer, A_3. If there is one A_3 in every 10 original monomers containing A functional groups, then $f = 2.1$. For linear step growth $f = 2$ and Eq. (53) reverts to the linear result: $1/(1 - p)$.

Notice that for $f > 2$ in Eq. (53) $\overline{DP}_n$ becomes infinite at conversions less than 1. For example, if $f = 2.1$, $\overline{DP}_n$ diverges at $p = 0.952$. Thus even a small amount of multifunctional monomer increases the molecular weight greatly at conversions less than 1.

Divergence of the molecular weight (i.e., gelation) is the most spectacular feature of nonlinear polymerization. The "gel" point where this occurs is most succinctly expressed in terms of the branching probability α, defined by Flory (1953) and discussed in detail by Billmeyer (1971). α is the probability in a reaction mixture of linking two multifunctional monomers. This will depend in general upon conversion and stoichiometry of branched and linear monomers. Consult Flory (1953) or Billmeyer (1971) for detailed calculations for specific cases. Gelation occurs at a critical value of α:

$$\alpha_c = \frac{1}{f - 1} \tag{54}$$

which corresponds to the divergence of the weight-average molecular weight.

Analysis of the MWD up to and after gelation is beyond the scope of this chapter. We simply conclude with a few important observations. Weight-average molecular weight goes to infinity at the gel point, not the

number-average molecular weight. This means that there is a "sol" fraction of finite-sized molecules present even after gelation. Historically, statistical methods of calculating MWD, such as those of Flory (1953) and Stockmayer (1943), have been employed. These methods neglect the possibility of cyclization, that is, different branched units being connected by two or more chains, which is redundant from the point of view of the connectivity required for gelation. There is much interest currently in determining how important this process is in the structure of nonlinear polymers (Family and Landau, 1984). Certainly as a first approximation the Flory-Stockmayer method is excellent. However, it remains to be decided if there are subtle but important influences of cyclization on the gel point, mechanical properties, and scattering.

Finally, the more chemical engineering-oriented mass balance approach has only recently been applied to nonlinear polymerization (Ver Strate et al., 1980; Hendriks et al., 1983). This has the advantage of more ready generalization to different reactor configurations, such as continuous reactors. In addition, the time evolution of the MWD is more explicit in the mass balance treatment. Incorporation of cyclization in these methods is also problematic.

Chain-Growth Polymerization

Chain-growth polymerization is based on the application of a free radical or ionic reaction to a polymerization chain or macromolecule which results from the propagation of one kinetic chain reaction (Lenz, 1968). Chain-growth polymerization is initiated by a reactive species R produced from an initiator or catalyst. A chain reaction can be initiated by a radical producing initiator such as azobisbutyronitrile or benzoyl peroxide. It can be initiated thermally, photochemically, with cationic and anionic initiators, or with true catalysts that are not consumed by the reaction

$$I \longrightarrow nR_1$$

(In chain-growth polymerization we must distinguish between growing chains, which we designate R, and dead chains with no active center attached, which we designate P.) The reaction chain propagates by reaction of a monomer with the active species or radical. The propagation reaction is simply represented by

$$R_j + M \longrightarrow R_{j+1}$$

This representation does not detail the various methods or mechanisms of propagation or growth of the active chain. Lenz (1968) treats this question in considerable detail. The kinetic chain can be transferred by several reactions

$$R_j + M \longrightarrow P_j + R_1 \qquad \text{transfer to monomer}$$

$$R_j + S \longrightarrow P_j + R_1 \qquad \text{solvent}$$

$$R_j + A \longrightarrow P_j + R_1 \qquad \text{agent}$$

$$R_j + P_k \longrightarrow P_j + R_k \qquad \text{polymer}$$

Transfer reactions terminate the active chain and initiate another one. We will find that transfer reactions do not affect the shape of the distribution, although they affect strongly the average degree of polymerization. The polymer transfer reaction in free radical polymerizations leads to branching (Bamford and Tompa, 1954; Graessley et al., 1969; Nagasubramanian and Graessley, 1970a,b).

Sooner or later the propagating polymer chain is terminated not by a transfer reaction, but by annihilation of active groups. Two active groups react with each other in radical polymerization by (a) combination, and (b) disproportionation. A shorthand representation of the termination reactions is

$$R_j + R_k \longrightarrow P_{j+k} \qquad \text{combination}$$

$$R_j + R_k \longrightarrow P_j + P_k \qquad \text{disproportionation}$$

The termination reactions occur between two large macromolecular active species. Such a reaction runs the risk of being strongly diffusion controlled. This leads to a phenomenon termed the *gel effect* or *Trommsdorff effect* (Tulig and Tirrell, 1981). An extensive discussion of the ramifications of this phenomenon is particularly important in the design of polymerization reactors involving chain-growth polymerization taken to high conversions. Ionic and heterogeneous catalyzed polymerizations suffer termination by small molecule poisons present accidentally or purposefully to control molecular weight.

Free Radical

Of the several types of polymerizations classed as chain growth, free radical initiation has found more general industrial application. This may be largely historical, but it remains that a significant fraction of polymer produced today are made by radical mechanisms.

For the sake of simplicity, consider a batch chain-growth polymerization involving only initiation, propagation, and combination termination. We must describe the evolution of monomer, M; active polymer species of chain length j, R_j; dead or inactive polymer of chain length j, P_j; and initiator, I. The reaction sequence is given as

$$I \xrightarrow{k_1} 2R_1 \qquad \text{initiation} \tag{55}$$

$$M + R_j \xrightarrow{k_2} R_{j+1} \qquad \text{propagation} \tag{56}$$

$$R_j + R_k \xrightarrow{k_3} P_{j+k} \qquad \text{combination termination} \tag{57}$$

The material balance on the species of interest are

$$\frac{dI}{dt} = -k_1 I \tag{58}$$

$$\frac{dM}{dt} = -k_2M \sum_{n=1}^{\infty} R_n \tag{59}$$

$$\frac{dR_j}{dt} = 2fk_1I\delta(j - 1) + k_2M(R_{j-1} - R_j) - k_3R_j \sum_{n=1}^{\infty} R_n \tag{60}$$

$$\frac{dP_j}{dt} = \frac{k_3}{2} \sum_{n=1}^{\infty} R_kR_{j-k} \qquad (j > 2) \tag{61}$$

Note that $P_j = 0$ for $j < 2$. We use the Dirac delta function $\delta(j - 1)$ to designate a term that occurs only when the argument is zero, in this case when $j = 1$.

The assumptions implicit in this formulation are that:

1. The reaction velocity constants are independent of molecular size.
2. A fraction f of the initiator form active radicals.
3. Isothermal reaction conditions prevail.

The generating function can be used to collapse the infinite set of differential equations to a finite set:

$$\frac{dI}{dt} = -k_1I \tag{62}$$

$$\frac{dM}{dt} = -k_2MG(1) \tag{63}$$

$$\frac{\partial G(s)}{\partial t} = 2fk_1Is - k_2M(1 - s)G(s) - k_3G(s)G(1) \tag{64}$$

$$\frac{\partial H(s)}{\partial t} = k_3G^2(s) \tag{65}$$

$$\frac{\partial G(1)}{\partial t} = 2fk_1I - k_3G^2(1) \tag{66}$$

where G and H are the generating functions for the R and P distributions, respectively. These equations can be solved formally subject to initial conditions at $t = 0$:

$$G(s,0) = H(s,0) = 0 \qquad M(0) = M_0, \quad I(0) = I_0 \tag{67}$$

The procedure is to solve first for I, then for G(1), M, and finally for G(s) and H(s). In doing so, most analyses of free radical polymerization use an assumption known as the quasi-steady-state approximation (QSSA).

Bamford et al. (1958) state this fundamental assumption of radical chain-growth polymerization as "the assumption that the concentration of radical intermediates remains constant during the polymerization." More specifically,

$$\frac{d}{dt}\sum_{j=1}^{\infty} R_j \cong 0 \tag{68}$$

or

$$\left|\frac{d}{dt}\sum_{j=1}^{\infty} R_j\right| \ll \left|\frac{dM}{dt}\right| \tag{69}$$

or that the reaction has proceeded so far that the net rate of change of concentration of radical intermediates is very much less than both their rate of production and their rate of destruction.

One can examine more rigorously what the QSSA really means by attempting a rigorous solution of some of the equations and comparing with the result of assuming Eq. (68).

The equations governing the evolution of total polymer radical [Eq. (64) with s = 1], the initiator [Eq. (62)], and the monomer [Eq. (63)] can be solved for an isothermal batch polymerization starting with only monomer and initiator. The equations are

$$\frac{dI}{dt} = -k_1 I \tag{70}$$

$$\frac{dG(1)}{dt} = 2fk_1 I - k_3 G^2(1) \tag{71}$$

$$\frac{dM}{dt} = -k_2 M G(1) \tag{72}$$

At t = 0, $M = M_0$, $I = I_0$, and G(1) = 0. The solution to Eq. (70) can be obtained readily:

$$I = I_0 \exp(-k_1 t) \tag{73}$$

The equation for total polymer radical can be transfermed into a classical Riccatti equation with the following definitions:

$$\chi = \exp\left(-\frac{k_1 t}{2}\right) \qquad y = \frac{G(1)}{I_0}$$
$$\alpha = \frac{2k_3 I_0}{k_1} \qquad \gamma^2 = 4\alpha f \tag{74}$$

Equation (71), then, becomes

$$\frac{dy}{d\chi} + \frac{\alpha}{\chi} y^2 = \frac{\gamma^2 \chi}{\alpha} \tag{75}$$

Solution methods for Riccatti equations are well presented in Davis (1962). The substitution

$$y = \frac{\chi}{\alpha v} \frac{dv}{d\chi} \tag{76}$$

into Eq. (75) leads to

$$\frac{d^2 v}{d\chi^2} + \frac{1}{\chi}\frac{dv}{d\chi} - \gamma^2 v = 0 \tag{77}$$

The solution is straightforward and leads to the following expression for y:

$$y = \frac{\chi\gamma}{\alpha} \frac{K_1(\gamma)I_1(\gamma\chi) - K_1(\gamma\chi)I_1(\gamma)}{K_1(\gamma)I_0(\gamma\chi) - K_0(\gamma\chi)I_1(\gamma)} \tag{78}$$

where K_i and I_i are the modified Bessel functions of order i. We can make a comparison of this exact solution to that solution resulting from the quasi-steady-state approximation (QSSA). If the assumption characterized by Eq. (68) is applied to Eqs. (70) and (71), the result is

$$y_s = \frac{\gamma\chi}{\alpha} \tag{79}$$

This suggests that QSS conditions prevail if $y \cong y_s$, which requires

$$\frac{K_1(\gamma)I_1(\gamma\chi) - K_1(\gamma\chi)I_1(\gamma)}{K_1(\gamma)I_0(\gamma\chi) - K_0(\gamma\chi)I_1(\gamma)} \cong 1 \tag{80}$$

Saidel and Katz (1967) and Bamford (1965) have shown this to be a reasonable assumption for values that can be assigned to γ for typical radical chain-growth polymerization. The QSSA leads to the following representation of the kinetics:

$$\frac{dI}{dt} = -k_1 I \tag{81}$$

$$\frac{dM}{dt} = -k_2 M G(1) = -k_2 \left(\frac{2fk_1}{k_3}\right)^{1/2} I^{1/2} M \tag{82}$$

$$\frac{dG(1)}{dt} \cong 0 = 2fk_1 I - k_3 G^2(1) \tag{83}$$

$$\frac{dG(s)}{dt} \cong 0 = 2fk_1Is - k_2M(1 - s)G(s) - k_3G(s)G(1) \tag{84}$$

$$\frac{dH(s)}{dt} = \frac{k_3}{2} G(s)^2 \tag{85}$$

The quasi-steady-state assumption yields

$$G(1) = \left(\frac{2fk_1}{k_3} I\right)^{1/2} \tag{86}$$

and

$$G(s) = \frac{as}{1 - bs} \tag{87}$$

where

$$\begin{aligned} a &= (2fk_1I)[k_2M + (2fk_1k_3I)^{1/2}]^{-1} \\ b &= k_2M[k_2M + (2fk_1k_3I)^{1/2}]^{-1} \end{aligned} \tag{88}$$

Notice that this is exactly the same form of the generating function as Eq. (37) found in step-growth polymerization. This is an important general conclusion concerning free radical polymerization: the distribution of growing chains is most probable. This is the same distribution as is found in step-growth polymerizations, but the parameters of the distribution are controlled by entirely different factors. Here the mean of the MWD changes with conversion since a and b depend on M and I. The distribution of the terminated chains is slightly more complex and is discussed below.

Monomer consumption is given by the solution of Eq. (82), which can be rewritten as

$$\frac{d}{dt}(\ln M) = -k_2\left(\frac{2fk_1I_0}{k_3}\right)^{1/2} \exp -\frac{k_1t}{2} \tag{89}$$

The solution of the monomer equation can be expressed as

$$\ln \frac{M}{M_0} = -\frac{2k_2}{k_1}\left(\frac{2fk_1I_0}{k_3}\right)^{1/2}\left[1 - \exp\left(-\frac{k_1t}{2}\right)\right] \tag{90}$$

Note that as $t \to \infty$,

$$\ln \frac{M_\infty}{M_0} = -\frac{2k_2}{k_1}\left(\frac{2fk_1I_0}{k_3}\right)^{1/2} \tag{91}$$

This is sometimes termed dead-end polymerization. Appropriate choice of initiator type and amount will avoid this.

The rate of polymerization is given by

$$\frac{dM}{dt} = -k_2\left(\frac{2fk_1}{k_3}\right)^{1/2} I^{1/2}M \tag{92}$$

Here we see the various influences on the kinetics. In chemically initiated radical chain polymerization the rate is proportional to the 1/2 power of the initiator concentration and the first power in monomer concentration. The termination rate constant, k_3, occurs in the expression as a $-1/2$ power term. Should for some reason the termination reaction velocity constant decrease, the polymerization rate will increase. This is the effect previously mentioned as the *Trommsdorff* or *gel effect*. The reaction behaves much in the fashion of an autocatalytic reaction.

Regarding the chain-length distribution of the dead chains, we get that from the distribution of growing chains via Eq. (85). We saw that

$$G(s) = \frac{2fk_1Is}{k_2M(1 - s) + (2fk_1k_3I)^{1/2}} \tag{93}$$

so that for the combination termination scheme

$$\frac{dH(s)}{dt} = \frac{k_3}{2}\left(\frac{as}{1 - bs}\right)^2 \tag{94}$$

with a and b defined by Eq. (88).

Equation (94) can be readily inverted by expanding the right-hand side as a power series in s in a fashion similar to what was done to Eqs. (37) and (38) with the result

$$\frac{dP_j}{dt} = \frac{k_3a^2}{2} b^{j-2}(j - 1) \tag{95}$$

Since a and b are functions of time, readily determined from the solutions to Eqs. (81) and (82), the entire distribution for P_j can be directly obtained by integration of Eq. (95).

Our discussions earlier suggested that many polymerization schemes are best understood in terms of the moments of the distribution. The moment equations can be readily obtained from the equation for the transform of the chain-length distribution. The moment-generating property

$$\mu_k = \lim_{s\to1} \frac{d^kG(s)}{d(\ln s)^k} \tag{96}$$

gives the following set of moment equations from the original rate equations:

$$\frac{d\mu_0}{dt} = \frac{k_3a^2}{2(1 - b)^2} \tag{97}$$

$$\frac{d\mu_1}{dt} = \frac{k_3 a^2}{(1 - b)^3} \tag{98}$$

$$\frac{d\mu_2}{dt} = \frac{k_3 a^2}{(1 - b)^4} (2 + b) \tag{99}$$

One can see that the instantaneous number- and weight-average degree of polymerization have the simple form (for b near 1, which is necessary for high-molecular-weight polymer)

$$\frac{d\mu_1}{d\mu_0} = \frac{2}{1 - b} \tag{100}$$

$$\frac{d\mu_2}{d\mu_1} = \frac{3}{1 - b} \tag{101}$$

The polydispersity of any polymer formed in an instant [the ratio of Eqs. (100) and (101)] is constant and always 3/2 for the mechanism of combination termination used in this discussion of radical chain-growth polymerization. Notice that b is very similar in its definition to the reaction probability p of step-growth polymerization [see Eq. (13)], or the α of Eq. (9).

We have discussed a very simple mechanism for radical chain polymerization. It is worth examining briefly what happens when other reactions occur, chain transfer and branching, for example.

Consider the following mechanisms:

$$I \xrightarrow{k_1} 2R_1 \qquad \text{initiation}$$

$$R_j + M \xrightarrow{k_2} R_{j+1} \qquad \text{propagation}$$

$$R_j + M \xrightarrow{k_3} P_j + R_1 \qquad \text{transfer to monomer}$$

$$R_j + A \xrightarrow{k_4} P_j + R_1 \qquad \text{transfer to solvent or agent}$$

$$R_j + R_k \xrightarrow{k_5} P_{j+k} \qquad \text{combination termination}$$

The material balances for such a polymerization in a batch reaction are given by

$$\frac{dI}{dt} = -k_1 I \tag{102}$$

$$\frac{dM}{dt} = -(k_2 + k_3)M \sum_{j=1}^{\infty} R_j \tag{103}$$

$$\frac{dR_j}{dt} = \left(2fk_1I + k_3M \sum_{j=1}^{\infty} R_j + k_4A \sum_{\ell=1}^{\infty} R_\ell\right)\delta(j-1) + k_2M(R_{j-1} - R_j) - \left(k_3M + k_4A + k_5 \sum_{n=1}^{\infty} R_k\right)R_j \tag{104}$$

$$\frac{dP_j}{dt} = \frac{k_5}{2}(j-1) \sum_{k=1}^{\infty} R_{j-k}R_k + (k_3M + k_4A)R_j \tag{105}$$

These same equations can be put in the generating function format:

$$\frac{dI}{dt} = -k_1I \tag{106}$$

$$\frac{dM}{dt} = -(k_2 + k_3)MG(1) \tag{107}$$

$$\frac{dG(s)}{dt} = [2fk_1I + k_3MG(1) + k_4AG(1)]s - k_2MG(s)(1-s) - (k_3M + k_4A)G(s) - k_5G(1)G(s) \tag{108}$$

$$\frac{dG(1)}{dt} = 2fk_1I - k_5G^2(1) \tag{109}$$

$$\frac{dH(s)}{dt} = \frac{k_5}{2} G^2(s) + (k_3M + k_4A)G(s) \tag{110}$$

The QSSA can lead to considerable simplification since

$$G(1) = \left(\frac{2fk_1I}{k_5}\right)^{1/2} \tag{111}$$

$$G(S) = \frac{[2fk_1I + (k_3M + k_4A)G(1)]s}{k_5G(1) + k_3M + k_4A + k_2M(1-s)} \tag{112}$$

The distribution produced by this mechanism is not significantly different from that prescribed by the simpler mechanism. Only the mean values are different. G(s) can again be written in the form

$$G(s) = \frac{as}{1 - bs} \tag{113}$$

where

$$a = \frac{2fk_1I + (k_3M + k_4A)G(1)}{k_5G(1) + k_3M + k_4A + k_2M} \tag{114}$$

$$b = \frac{k_2M}{k_2M + k_3M + k_4A + k_5G(1)}$$

Equation (110) can be greatly simplified to yield

$$\frac{dH(s)}{dt} = \frac{k_5}{2}\left(\frac{as}{1 - bs}\right)^2 + (k_3M + k_4A)\,\frac{as}{1 - bs} \tag{115}$$

The equation for H(s) can be inverted readily to give the equation for P_j:

$$\frac{dP_j}{dt} = k_5a^2\left(\frac{j - 1}{2}\, b^{j-2} + \beta b^{j-1}\right) \tag{116}$$

where

$$\beta = \frac{k_3M + k_4A}{k_5a} \tag{117}$$

Again we should note that the quantities a, b, and β are functions of time, readily evaluated from the conservation equations for M, S, and I.

Now it might help to examine the moment equations. Let the moments of the radical and polymer distributions be defined as

$$\mu_j \equiv \sum_{n=1}^{\infty} n^jP_n \qquad \text{and} \qquad \sigma_j \equiv \sum_{n=1}^{\infty} n^jR_n \tag{118}$$

We have seen that the distribution for polymer radicals has a most probable distribution [i.e., Eq. (113)], so that the moments of the polymer radical distributions are

$$\sigma_0 = \frac{a}{1 - b} \tag{119}$$

$$\sigma_1 = \frac{a}{(1 - b)^2} \tag{120}$$

$$\sigma_2 = \frac{a(1 + b)}{(1 - b)^3} \tag{121}$$

The equations for the moments of the distribution can be readily derived from Eq. (110) in terms of the moments of the radical polymer distribution

$$\frac{d\mu_0}{dt} = k_5\left(\frac{1}{2}\sigma_0^2 + \beta a \sigma_0\right) \tag{122}$$

$$\frac{d\mu_1}{dt} = k_5(\sigma_0\sigma_1 + \beta a \sigma_1) \tag{123}$$

$$\frac{d\mu_2}{dt} = k_5(\sigma_2\sigma_0 + \sigma_1^2 + \beta a \sigma_2) \tag{124}$$

The instantaneous weight- and number-average degrees of polymerization then assume the form

$$\frac{d\mu_1}{d\mu_0} = \frac{\sigma_1(\sigma_0 + \beta a)}{\sigma_0(\sigma_0/2 + \beta a)} \tag{125}$$

$$\frac{d\mu_2}{d\mu_1} = \frac{\sigma_2}{\sigma_1} + \frac{\sigma_1}{\sigma_0 + \beta a} \tag{126}$$

The relations for σ_0, σ_1, and σ_2 afford considerable simplification:

$$\frac{d\mu_1}{d\mu_0} = \frac{1}{1-b}\left[\frac{1 + \beta(1-b)}{1/2 + \beta(1-b)}\right] \tag{127}$$

$$\frac{d\mu_2}{d\mu_1} = \frac{1}{1-b}\left[1 + b + \frac{1}{1 + \beta(1-b)}\right] \tag{128}$$

A little exercise will show that as $\beta \to 0$ (no transfer reactions) the instantaneous polydispersity has the same value, 3/2, as the simple mechanism and that in the limit of dominant transfer $\beta \to \infty$, the polydispersity approaches 2 from below.

Volume Change in Radical Polymerization

Volume change is a common effect in undiluted or bulk polymerization. For the case of styrene, Hui and Hamielec (1972) used a linear correlation between volume and conversion,

$$V = V_0(1 + \varepsilon X)$$

where ε is the fractional change in volume of the system between 0 and 100% conversion. ε depends also on temperature and may be calculated reasonably accurately from the individual properties of monomer and polymer, by assuming volume additivity and disregarding molecular weight dependency. For the case of styrene, values for ε range approximately from −0.16 at 60°C to −0.22 at 180°C (volume contraction).

The effect of incorporating volume change in the kinetic model can be evaluated quantiatively. While there are no appreciable changes in the conversion time relation, it does affect the prediction of average molecular

weights. Errors due to neglect of contraction in the polydispersity may approach 50% at higher conversions.

Anionic Polymerization

Anionic polymerization is important industrially for various types of copolymers and also because the same basic kinetic scheme is found in many heterogeneously catalyzed polymerizations. In the course of illustrating many of the techniques of molecular weight distribution analysis, the simplest version of anionic polymerization (instantaneous initiation, no termination) has already been examined. This mechanism, which can be closely approximated in practice with very pure systems, produces the narrowest possible MWD for any as-polymerized, unfractionated material.

It is instructive to examine what can happen if some of the conditions of simple ideal anionic polymerization are not achieved. We look briefly here at the case of slow initiation.

The kinetic scheme for this mechanism might be expressed as

$$I + M \xrightarrow{k_1} R_1 \tag{129}$$

$$R_j + M \xrightarrow{k_2} R_{j+1} \tag{130}$$

This is still a simple mechanism readily described as before. The species balances are

$$\frac{dI}{dt} = -k_1IM \tag{131}$$

$$\frac{dM}{dt} = -k_1IM - k_2M\sum_{n=1}^{\infty} R_n \tag{132}$$

$$\frac{dR_j}{dt} = k_1IM\delta(j-1) - k_2M(R_j - R_{j-1}) \tag{133}$$

This set of equations can be reformulated using both a generating function and the transformation:

$$\tau = \int_0^t k_2M\ dt \tag{134}$$

and Eqs. (131) to (134) become

$$\frac{dI}{d\tau} = -\alpha I \tag{135}$$

$$\frac{dM}{d\tau} = -\alpha I - G(1) \tag{136}$$

where $\alpha = k_1/k_2$,

$$\frac{dG(1)}{d\tau} = \alpha I \tag{137}$$

$$\frac{dG(s)}{d\tau} = \alpha Is - (1 - s)G(s) \tag{138}$$

The initial conditions are transformed to $I(0) = I_0$, $M(0) = M_0$, and $G(s,0) = 0$. A solution to Eq. (138) can be readily written as

$$G(s,\tau) = \frac{\alpha I_0 s}{1 - \alpha - s}\left(e^{-\alpha\tau} - e^{-(1-s)\tau}\right)$$

The inversion of G(s) is straightforward, if lengthy, and has been presented by Nanda and Jain (1964):

$$R_j = \alpha I_0 e^{-\tau}(1 - \alpha)^{-j}\sum_{k=j}^{\infty} \frac{[(1 - \alpha)\tau]^k}{k!} \tag{139}$$

This distribution is often termed the *Gold distribution* and for values assigned to the parameter α the distribution can vary from very narrow (Poisson distributed) to very broad at large α for small α.

Cationic Polymerization (Plesch, 1963)

As in most chain-growth polymerizations, in cationic reactions the chain must be initiated

$$I + M \xrightarrow{k_1} R_1 \tag{140}$$

This schematic representation is typical of initiation by proton donor (protonic acids). Lewis acids form another class of initiators. For such systems one generally needs a cocatalyst. The initiation is very rapid and reversible:

$$I \rightleftharpoons R_1 \tag{141}$$

There are, of course, other means of ionic catalysis, but we will, in our discussions, restrict ourselves to these two.

The active chain propagates

$$R_j + M \xrightarrow{k_2} R_{j+1} \tag{142}$$

The propagation reaction can be complex as a result of many intramolecular rearrangements. In our discussion we avoid these but maintain an awareness of their existence.

Transfer reactions occur commonly in cationic polymerizations. They are often called termination reactions. We avoid that usage since the kinetic chain is not destroyed.

$$R_j + M \xrightarrow{k_3} R_1 + P_j \tag{143}$$

Spontaneous transfer termination can also occur and is described by

$$R_j \xrightarrow{k_4} P_j + R_1 \tag{144}$$

The kinetic chain can be terminated (by combination of the carbonium ion with the gegenion) and is best represented by

$$R_j \xrightarrow{k_5} P_j \tag{145}$$

Other transfer reactions are important and should be considered in any model. The analysis of cationic polymerization is much less complete than for anionic polymerization due to the more complex kinetic scheme. Equations (140) to (145) represent a typical minimum set of reaction equations that should be analyzed.

Heat and Mass Transfer Effects in Chain-Growth Polymerization

Polymerization reactions are typically strongly exothermic. Moreover, polymers are typically poor thermal conductors. Mass transport by diffusion is also a slow process in polymer systems. Heats of reaction of 15 to 20 kcal/mol are common, as are thermal diffusivities of 10^{-4} cm^2/s and mass diffusivities, for small molecules in polymers, of 10^{-8} cm^2/s. For polymers diffusing in polymer, 10^{-14} cm^2/s is not an unusually low value of the diffusion coefficient.

Two examples of transport effects that stand out for their practical importance are (a) the autoacceleration or Trommsdorff effect in high-conversion-bulk free radical polymerization, and (b) the heterogeneous catalyzed polymerization of ethylene, propylene, and α-olefins. These will be discussed briefly in turn.

Trommsdorff Effect (Cardenas and O'Driscoll, 1976, 1977a,b; Ross and Laurence, 1976; Dionisio and O'Driscoll, 1979; Marten and Hamielec, 1979; Tulig and Tirrell, 1981; Balke et al., 1981; Soh and Sundberg, 1981a–d; Chiu et al., 1983)

In a free radical polymerization conducted in bulk or solution, it is found that monomer consumption follows the exponential approach toward complete conversion suggested by Eq. (92). However, when the reaction has proceeded to the point that the medium contains 20 to 40% polymer (i.e., 20 to 40% fractional conversion of monomer in bulk polymerization, later in solution polymerization), there is, in many free radical polymerizations, an abrupt increase in the rate of polymerization. This occurs, as suggested under Eq. (92), because the termination rate constant takes a drastic drop.

The practical implications of this phenomenon are enormous. Due to the high reaction exothermicity and the relatively low thermal conductivity of polymeric media, dispensing with this added heat production is difficult and can lead to significant, even dangerous temperature excursions. The book by Biesenberger and Sebastian (1983) discusses thermal runaway phenomena extensively. Even without large temperature changes, the highly nonlinear kinetic behavior can lead to multiple steady states and other uncommon dynamic phenomena otherwise normally associated with nonisothermal operation. In addition to these operational control problems, this autoacceleration changes the nature of the polymeric product significantly. Reduction in the termination rate produces long-lived free radicals and therefore higher-molecular-weight polymer. What is found as the acceleration sets in is that the molecular weight distribution of the polymer being produced at that instant of time shifts abruptly to higher average molecular weight. As this higher-molecular-weight polymer accumulates, the aggregate molecular weight distribution becomes quite broad and bimodal.

The physical basis of this phenomenon is well understood qualitatively. It clearly results from an abrupt slowing of the polymer diffusion process. As polymer concentration builds in the reaction medium, diffusion of large molecules, like the growing radicals, is going to be retarded more than that of monomers. Therefore, the termination reaction, which involves two large molecules, is inhibited more strongly than the propagation reaction and the overall rate of polymerization goes up.

Converting this qualitative understanding to practical, accurate, quantitative, predictive modeling is difficult, but also very important, and is the subject of much current activity and recent progress. The key points of any model of this effect are (a) how the onset of the autoacceleration is handled, and (b) how the rate constants are supposed to vary with polymer concentration and molecular weight in the reaction medium. It has been clear for some time that the onset of autoacceleration is *not* the onset of diffusion control of the termination reaction. Even at zero polymer concentration (i.e., zero conversion of monomer to polymer), it has been demonstrated that the termination reaction rate is diffusion controlled, as it scales directly with monomer or solvent viscosity. What happens at the onset, then, is an abrupt *change in the character* of the diffusion control. Recent work has gathered convincing evidence that the onset of acceleration corresponds to the onset of entanglement in the polymerizating mixture.

A fundamental approach to the modeling of diffusion-controlled reaction in this onset regime is very difficult since the behavior of the polymer diffusion coefficient itself is not clearly understood here with respect to its dependence on concentration and molecular weight. Several authors have proposed similar, ad hoc means of dealing with the onset, which is to treat it as a transition, much like the transition from unentangled to entangled behavior seen in rheological properties with increasing concentration. With minor variations among them, all suggest that the acceleration onset occurs at a concentration, c_0, that varies with molecular weight of the polymer like M^{-a}, where the exponent a is between 1/2 and 1 (Tulig and Tirrell, 1982). Exponents in this regime are found for the appearance of entanglement in rheological properties with increasing concentration. Ultimately, a smooth transition function from dilute to entangled behavior is needed, but this will require advances in our fundamental understanding of entanglement.

There is a wider diversity of approach to how the variation of k_t with conversion and molecular weight is handled *after* the onset. This part of

the model is the important part for prediction of molecular weight distribution and the resultant polymer properties. The onset part of the model by contrast has more bearing on the control of the reactor. There are several effective empirical or semiempirical approaches to treating the k_t variation (Ross and Laurence, 1976; Cardenas and O'Driscoll, 1976, 1977a,b; Marten and Hamielec, 1979). Among these several have been developed from the notion that the termination rate constant should be inversely proportional to the viscosity of the reaction medium (Marten and Hamielec, 1979). This hypothesis leads to a particular dependence of k_t on polymer concentration (conversion) and molecular weight. Given that a number of adjustable parameters are unavoidable in the construction of models like this, these models can be made to fit experimental data well.

Recently, this kind of model has been criticized on theoretical grounds and an alternative hypothesis proposed, namely that k_t should be proportional to the diffusion coefficient (D) of the free radicals rather than the viscosity of the entire reaction medium (Tulig and Tirrell, 1982). This model implies that the concentration and molecular weight dependence of k_t should follow that of D. Experimentally, these dependences of D are currently in the process of examination. The reptation model (de Gennes, 1976a,b) of polymer diffusion suggests that D varies with the inverse square power of molecular weight and with concentration to some power between -3 and -1.75, depending on the quality of the solvent. Experimental evidence to date supports the predicted molecular weight dependence of D but suggests that the concentration dependence is much stronger (Tirrell, 1984). A model with k_t depending on molecular weight and conversion as suggested by the reptation model has been developed and shown to fit kinetic, molecular weight, and molecular weight distribution data well (Coyle et al., 1985). However, this model, like its predecessors, contains adjustable parameters, so that it is difficult to make a convincing argument for its superiority based on its predictive power. The bottom line is that any model where there is an abrupt switch in k_t from a mild to a strong dependence on molecular weight and concentration can be made to fit the data reasonably well. The behavior, especially conversion versus time, is simply not very sensitive to the exact forms chosen for these dependences.

On the other hand, there is some recent evidence to suggest that the model based on the reptation theory is more fundamentally sound. Experimental measurements of k_t have been made in solutions of polymethyl methacrylate in methyl methacrylate monomer where the molecular weight and concentration of polymer in the reaction medium can be varied independently (Tirrell and Tulig, 1983). In this way, one can examine k_t at fixed concentration and vary the molecular weight of the added polymer and therefore the viscosity of the reaction medium. On doing this, one sees that k_t is only very weakly dependent on viscosity and molecular weight of added polymer. This is in contrast with early models but exactly what is expected from the reptation theory, which focuses on diffusion of the radicals. This diffusion rate is expected to be independent of the molecular weight of surrounding chains as long as they are sufficiently long and concentrated to be entangled. Only the molecular weight of the radicals plays a role in the diffusivity.

This area is one of active research currently and should therefore be watched closely by the engineer who wishes to keep abreast. On the practical side, it should be mentioned that there are also several completely

empirical models of autoacceleration cited above that have been used successfully for certain modeling purposes. A fundamentally based theory for the reaction of two macromolecules is clearly more desirable in the long run. Such a theory would find application not only to the problem of autoacceleration in free radical polymerization but also possibly to high-conversion step-growth polymerization and certain branching and cross-linking reactions.

Heterogeneous Polymerization

In several kinds of chain-growth polymerizations the active center may be in a separate phase, frequently in the form of droplets or particles, from other components of the reaction medium. Emulsion (Penlidis et al., 1985), precipitation, and heterogeneously catalyzed polymerizations are all examples of this. (Suspension polymerization, while multiphasic, contains all reactive components in one phase and therefore has more in common with bulk and solution polymerizations than with heterogeneous polymerization.) These polymerizations share several features. One is that the active center becomes isolated somewhat by this phase separation and therefore very long-lived. Very high molecular weights can be produced and frequently chain transfer additives are used to control molecular weight. A second common feature is that reactants have to diffuse into droplet or particle form in order to participate in the reaction, opening up the possibility of diffusion limitations on the rate. Also, heat has to diffuse out of and be carried away from the reaction particles, presenting the possibility of local temperature excursions.

These problems have received the most attention in heterogeneously catalyzed polymerizations, an important class of which are the Ziegler-Natta (Z–N) catalyzed polymerizations (and copolymerizations) of ethylene, propylene, other α-olefins, and sometimes, dienes. The class of Z-N catalysts includes transition metal halides (Ti, V, Zr, etc.) complexed with aluminum alkyls or alkyl halides (Boor, 1979). These catalysts have been developed to the point that very high rates and yield of polymer (per weight of catalyst) are obtained (Goodall, 1981). This means that the initial porous catalyst particle is rapidly fragmented by and buried inside the evolving polymer.

Interesting morphologies (Boor, 1979) of polymer particles are formed in this process. How they are formed is incompletely understood, but it is known that some morphologies are more desirable than others for subsequent processing of the polymer. The morphology of the polymer particle is obviously very important in regulating diffusion of monomer to the active site. A very common kind of morphology is one where the polymer particle is composed of an agglomeration of smaller microparticles, each of which presumably surrounds a fragment of the original catalyst particle. Diffusion from outside therefore involves moving through the pores of the macroparticle (between the microparticles) and penetration through the polymer coating of the microparticle to reach the active site.

Several groups have developed diffusion-reaction models for Z-N polymerizations, some of which have attempted to account explicitly for the morphology described above. Stimulus for this modeling has been provided by the experimental fact that Z-N polymerizations, almost without exception, produce a broad distribution of molecular weights, even though the chemistry at the active site resembles in many ways the anionic polymerization mechanism and therefore should produce a relatively narrow molecular weight distribution. Debate has developed about the origin of the broad

molecular weight distribution. Possibilities are that (a) diffusion limitations produce a variable supply of monomer to different active sites (Schmeal and Street, 1971; Nagel et al., 1980; Laurence and Chiovetta, 1983); (b) multiplicity of types and activity of active sites (Caunt, 1981; Keii et al., 1984; Barbe et al., 1983); and (c) time variation of site activity (i.e., deactivation). Diffusion limitations certainly occur, but whether they are strong enough alone to produce the broad molecular weight distributions seen is doubtful. Active site heterogeneity and deactivation have both been documented and are very likely partly responsible for the broad molecular weight distributions. For example, a recent analysis (Galvan and Tirrell, 1985) has shown that the Thiele modulus,

$$\alpha = \frac{a^2 k}{D} \tag{146}$$

where a is a characteristic size of the polymer particle, k is the turnover rate at the active site, and D is the diffusion coefficient for monomer, must exceed 10 for there to be significant molecular weight distribution broadening by diffusion. Data are limited and the model does not correspond exactly to reality in that the chemical mechanism is oversimplified but, speaking approximately, this seems to be on the high end of what might be found in practice.

Clearly, the relative importance of diffusion limitations will depend on whether the reaction is occurring in a slurry in liquid diluent or in a gas phase. Furthermore, Z-N copolymerization, such as that employed to produce EPDM rubber or linear low-density polyethylene, may experience diffusion effects on the composition and sequence distributions, even when the effects on the molecular weight distribution are not particularly important.

OPTIMIZATION AND CONTROL

Several unique problems are encountered in the optimal design and control of polymerization reactors, as discussed briefly earlier. One set of problems arises from the fact that the performance criteria for such reactors are not readily defined. Further, such criteria are not universal for a class of polymerization reactors. Polymerization reactors produce materials whose quality is assessed in terms of strength, stiffness, processability, and so on. Even if these ultimate qualities can be reduced to more abstract and readily quantifiable measures (molecular weight distribution, copolymer composition distribution, sequence distribution), it is unavoidable that (a) these measures are distributions (or characteristic values thereof) themselves, and (b) different measures react differently to changes in the reaction environment. Thus it is not at all apparent at first sight how to formulate properly a meaningful optimization problem. This problem is especially severe in the case of copolymerization (Ray and Gall, 1969; Tirrell and Gromley, 1981; Bejger et al., 1981; Garcia-Rubio et al., 1982). Distributions of molecular weight, composition, and sequence are all important, and what optimizes one may degrade another.

Obviously, the informed judgment of a designer is required at some stage in the optimization process in order to set priorities among objectives that may be in conflict. The ability to do this relies on accurate information

about the process from data and modeling. The procedure that has been used virtually exclusively in polymerization reactor optimization problems is to formulate a single objective functional which combines all identifiable performance measures (some of which may be in conflict in the sense that they react in opposite directions to manipulations of the control variables) with weighting factors on each measure chosen a priori. This choosing of weighting factors is one means of exercising judgment.

Several more restricted (i.e., single objective) optimization problems in polymerization have been treated (Ray and Gall, 1969; Hicks et al., 1969; Wu et al., 1982). Examination of these cases helps build experience and an intuitive basis for polymerization optimization. Minimization of the breadth of the molecular weight distribution in free radical homopolymerization has been treated several times. Effective manipulated variables are addition rate of monomer or initiator and temperature. All of these studies reach the conclusion that the narrowest possible distribution of molecular weights is obtained when the reactor produces a constant average molecular weight at every instant in time. Examination of Eqs. (125) and (126) shows that if initiator concentration is approximately constant, average molecular weight will tend to decline with increasing conversion because of declining monomer concentration. Adding monomer can therefore be used to keep molecular weight constant. Increasing temperature will reduce the instantaneous average molecular weight, so a decreasing temperature policy could be used to control a downward drift in molecular weight. Even this simple example, however, demonstrates the types of conflicts inherent in polymerization reactor optimization: decreasing temperature produces a better molecular weight distribution, but at the expense of a slower reaction and therefore slower production of polymer. Initiator addition can be effective if conditions are such that initiator concentration is declining (dead-end polymerization), thereby drifting molecular weight upward.

Copolymerization problems encompass an even larger set of objectives than homopolymerization. In optimal design and control of copolymerization reactors, it is useful, indeed important, to understand the trade-offs among the objectives quantitatively before selecting an "optimum" solution. Tsoukas et al. (1982) emphasized this in their study of semibatch copolymerization. They examined the objectives of simultaneous minimization of the breadths of the composition and molecular weight distribution using temperature and monomer feed rate as manipulated variables. The principal conclusion of their study was that under no circumstances would it be wise to optimize with the weight very heavily on one or the other of these objectives. In every copolymerization situation that they examined there was a policy of adjustment of the manipulated variable (temperature and monomer flow rate) away from the optimum for either individual objective, which would produce significant improvement in the other objective with very little damage to the former. The practical lesson is that even when a particular one of all the possible objectives seems to be dominant, it pays to examine other relevant objectives to see if some improvement in these can be obtained in addition, at little or no extra cost. This is even more valid if there are more control variables available to manipulate.

The examples above have been of optimal control designed to combat drift dispersion, that is, product nonuniformity due to reaction conditions that change in time. This kind of dispersion occurs in batch reactors (or with axial position in tubular continuous reactors). There is another type of dispersion, termed statistical dispersion, about which little can be

done via optimal control. This kind of dispersion results from the polymerization mechanism itself and from the reactor in which the polymer is made. For example, Eqs. (127) and (128) show that 1.5 is the minimum possible value of polydispersity for a free radical polymerization. This could be produced in a CSTR operating at steady state or in a batch reactor under time-invariant conditions. Drift broadens the distribution.

Choice of reactor is, of course, another element of polymerization optimization. Identical reactions run in different reactor configurations will produce products with different molecular weight distributions. This is easily seen on comparing the results of Table 1 for batch anionic polymerization with those of Eqs. (7) to (9) for anionic polymerization in a CSTR. The former produces a Poisson distribution, the latter a most probable distribution. The relative broadening in the CSTR is due to the fact that flow out of the reactor introduces an effective termination in the reaction scheme that is absent in ideal batch anionic polymerization. This is one example of what might be called *Denbigh's rule* (Denbigh, 1947). This rule could be phrased in the following way: when the mean lifetime of a growing chain is long compared to the mean residence time in the CSTR, the CSTR will produce a broader MWD than the batch reactor. This rule applies and is borne out in practice in anionic chain-growth and in step-growth polymerizations. The converse of Denbigh's rule is not always true. Free radical chain-growth polymerization is an example where the growing chains are very short-lived compared to the mean residence time. Here the shift from batch to CSTR usually produces narrowing due to the elimination of drift dispersion. In the absence of significant drift in the batch situation, narrowing cannot be achieved in the CSTR since with such short lifetimes, the growing chains do not have the opportunity to sense whether they have been made in batch or CSTR. Much work on other ideal reactor configurations, such as segregated continuous stirred-tank reactors and recycle reactors, has shown how these modify molecular weight distribution as well (Mecklenburgh, 1970; Nauman, 1974). Real reactor behavior can sometimes be represented approximately by these ideal reactors.

On-line control and optimization is a topic of significant current interest, facilitated greatly by the rapid evolution of computing equipment. Particular interest is in adaptive, self-tuning controllers. The idea is to use a model of the process in the controller. A reduced, aphysical model is used with parameters that do not necessarily have any physical significance themselves. However, algorithms are used for continually revising the parameter values based on comparison of process data with the designed control trajectory. In this way the parameters of the controller model are tuned to give optimal performance. This has been shown to be especially effective for non-steady-state situations where tuning of PID controllers is difficult. Self-tuning regulators can, however, also be used to set the gains of PID controllers (Bejger, 1982; Summers, 1985).

On-line optimization requires more computing power and has not yet been accomplished to any significant extent. This requires a better model of the process in the on-line computer that can compute updated optimal control policies based on new data. The problem with this sort of scheme is not the available computing power, it is the availability of sensors to monitor the process in an informative way. Currently, on-line data on polymerization reactors is limited largely to temperatures, crude estimates of medium viscosity (e.g., agitator motor torque) and perhaps some measures

of polymer content or conversion (Del Pino et al., 1982). Quantities such as molecular weight and its distribution, copolymer composition, and sequence distribution and particle size distribution in emulsion polymerization must in turn be estimated from these. Work is progressing on all of these fronts: on-line light scattering and viscosity for molecular weight, surface tension and dielectric relaxation for particle size distribution, and gas chromatography for copolymer composition. On-line gel permeation chromatography for molecular weight distribution remains a desirable but difficult and distant goal.

REACTOR SELECTION

The polymer "property" that all practicing engineers really strive to optimize is the bottom-line cost per unit production, subject to performance constraints. Meeting these performance constraints has been the subject of most of the first part of this chapter; however, cost can have important influence on the choice of polymerization reactor. Simplicity and energy efficiency are major goals along this route.

Considerations beyond cost and polymer product performance enter in as well. Viscosity of the reaction mixture is a primary consideration. Efficient mixing depends in part on power input, which in turn depends on viscosity (Oldshue, et al., 1982). An important caveat here is to take proper account of possible non-Newtonian rheological properties of the reaction mixture (Middleman, 1977, Chap. 13). Flow properties largely determine agitator size, power, and design. Motionless mixers are also an important possibility to consider (Middleman, 1977, p. 327). Conveying and pumping equipment must be designed around the flow properties of the reaction mixture. Extruders can be used as pumps and mixers. In many cases, the flow properties in the reactor have determined historically the choice of reactor for particular purpose. The rationale for these choices is found in the patent literature, which has been reviewed nicely up to the late 1970s by Gerrens (1982).

Separation processes that may have to be accomplished *after* the polymerization reaction can influence the choice of reactor. Two principal separation operations follow polymerization processes: elimination of monomers or diluents and separation of solid polymer. Devolatilization, to reduce residual monomer content to negligible levels, is an important example of the former. Extruders, tower, and wiped film reactors are all used in processes where devolatilization is a chief concern. There are also many common design concerns between devolatilization and reactor designs for driving certain step-growth polymerizations to very high conversion by removal of condensation product (pressure reduction, surface area regulation). Effective separation of solid product has been one of the important innovations in certain reactors, such as the Union Carbide fluidized-bed reactor process for polyolefins. Loop reactors incorporating a settling portion are also useful for separating solid polymer from slurries.

Temperature and its control are always considerations. Heats of polymerization are typically high, as mentioned earlier, so that maintaining desired temperature is not always a simple matter. Temperature can become spatially nonuniform and globally out of control. The typical consistency

of the reaction medium is again a factor. Normal good heat transfer design for large heat transfer surface, with coils or corrugated surfaces, is not effective in polymerization reactors since these surfaces create dead zones and accumulate material. This can produce nonuniform molecular weight distribution; foul the heat transfer equipment, exacerbating the heat removal difficulties; and make the reactor very difficult to clean. For these reasons, smooth heat transfer surfaces are usually preferable, with the best possible agitation to sweep the fluid clearly over the surface.

An example of these considerations can be seen in high-pressure ethylene polymerization. This reaction is done commercially in both tubular and stirred autoclave reactors. Tubular reactors have a high surface-to-volume ratio, which is good for heat transfer; they have, however, no mixing, which may produce segregation and inhomogeneity. Stirred-tank reactors have a comparatively low surface-to-volume ratio, although temperature can be manipulated somewhat using the feed temperature. Generally, these reactors have better mixing.

The desired form (pellet, powder, bead, etc.) can influence the reactor design. For example, suspension polymerizers produce beads that may be directly useful in processing. On the other hand, round beads can be dangerous if spilled and may not have suitable bulk flow properties. Extruder reactors (Stuber and Tirrell, 1985) are able to produce pellets, sheets, or coatings quite easily and directly. Safety considerations always place inviolable constraints on polymerization process design. Venting to the atmosphere in a safe and environmentally sound manner must always be designed for. Clearly, the entire process, not exclusively the reaction, must be considered in any useful polymerization process optimization. The articles by Platzer (1970) and Gerrens (1982) give additional useful information and insight into the practical aspects of reactor selection.

CONCLUSION

Polymerization reactors are rich in challenges and opportunities for the chemical engineer. This richness derives from the complexity of product distributions in even the simplest polymer reactor, the difficulties of handling viscous, exothermic reaction masses where informative measurements are extremely limited, and finally, from the fact that a polymer reactor is designed to follow a molecular architectural blueprint for a *material*. Deviation from this blueprint is reflected directly not in the behavior of the reactor but in the performance of the material. This is currently a characteristic nearly unique to polymer reactors, but may become an incresingly important matter to deal with in other materials fabrication reactors, such as chemical vapor deposition reactors for microelectronics applications.

This chapter, while covering a fair amount of ground, is at best an introduction to the techniques of analysis and experience on the behavior of polymerization reactors. In particular, the very important topics of copolymerization, emulsion polymerization, and varieties of continuous polymerization have received only superficial attention here. Our hope is that in the techniques and examples we have presented, the reader will find the motivation and means to attack new areas of polymerization engineering.

NOTATION

C_0	initial concentration of functional groups in polycondensation
DP_n	number-average degree of polymerization
D	diffusion coefficient of monomer in polymer
f	functionality of reacting mixture (in polycondensation); initiation efficiency (in free radical polymerization)
$G(s, \tau)$	generating function for growing chains (see Table 1 for definition)
$H(s, \tau)$	generating function for "dead" chains
I	initiator concentration
k	reaction rate constant
M	monomer concentration
p	conversion of functional groups
P_j	concentration of growing chains of length j (in polycondensation and anionic polymerization); concentration of "dead" chains of length j (in free radical polymerization)
r	stoichiometric ratio
R_j	concentration of growing chains of length j (in free radical polymerization)
t	time
V	volume of polymerizing media
X	conversion in free radical polymerization
y	normalized generating function, defined in Eq. (29)
Y_x	normalized generating function for chains with a stopper, defined in Eq. (30)

Greek Letters

α	probability of propagation; parameter in Gold distribution; Thiele modulus, depending on the context
γ	defined in Eq. (74)
δ	Dirac delta function
ε	fractional change in volume due to reaction
θ	residence time
μ_k	kth moment of the MWD of "dead" chains (in free radical polymerization) or growing chains (in anionic polymerization) (see Table 1 for definition)
σ_k	kth moment of the MWD of growing radicals in free radical polymerization
τ	reduced time, defined in Eq. (4)
χ	defined in Eq. (74)

REFERENCES

Abraham, W. H., Path dependent distribution of molecular weight in linear polymers, Ind. Eng. Chem. Fundam., *2*, 221 (1963).

Balke, S., L. Garcia, and R. Patel, Conversion prediction in high conversion in free radical polymerizations, AIChE Meet., New Orleans (Nov. 1981).

Bamford, C. H., Free radical polymerization with rapidly decaying initiators, Polymer, *6*, 63 (1965).

Bamford, C. H. and H. Tompa, The calculations of molecular weight distribution from kinetic schemes, Trans. Faraday Soc., *50*, 1097 (1954).

Bamford, C. H., W. G. Barb, A. D. Jenkins, and P. F. Onyon, *The Kinetics of Vinyl Polymerization by Radical Mechanisms*, Butterworth, London (1958).

Barbe, P. C., L. Noristi, G. Baruzzi, and E. Marchetti, Microscopic analysis of polyolefin initial formation on $TiCl4/MgCl_2$ supported catalysts, Makromol. Chem. Rapid Commun., *4*, 249 (1983).

Bejger, T. P., Real-time control of polymerization reactors, Ph.D. thesis, University of Minnesota (1982).

Bejger, T. P., M. Tirrell, and G. Stephanopoulos, Batch copolymerization reactors: self-tuning optimization and control, AIChE Meet., New Orleans (Nov. 1981).

Biesenberger, J. A. and D. H. Sebastian, *Principles of Polymerization Engineering*, Wiley-Interscience, New York (1983).

Billmeyer, F. W., *Textbook of Polymer Science*, 2nd ed., Wiley-Interscience, New York (1971).

Boor, J., *Ziegler Natta Catalysts and Polymerizations*, Academic Press, New York (1979).

Cardenas, J. N. and K. F. O'Driscoll, High conversion polymerization: I. Theory and application to methyl methacrylate, J. Polym. Sci. Chem. Ed., *14*, 883 (1976).

Cardenas, J. N. and K. F. O'Driscoll, High conversion polymerization: II. Influence of chain transfer on the gel effect, J. Polym. Sci. Chem. Ed., *15*, 1883 (1977a).

Cardenas, J. N. and K. F. O'Driscoll, High conversion polymerization: III. Kinetic behavior of ethyl methacrylate, J. Polym. Sci. Chem. Ed., *15*, 2097 (1977b).

Caunt, A. D., The determination of active centres in olefin polymerisation, Br. Polym. J., *13*, 22 (1981).

Chiu, W. Y., G. M. Carratt, and D. S. Soong, A computer model for the gel effect in free-radical polymerization, Macromolecules, *16*, 348 (1983).

Coyle, D. J., T. J. Tulig, and M. Tirrell, Computer analysis of high conversion free radical polymerization on models using the finite element method, Ind. Eng. Chem. Fundam., (1985, in press).

Davis, H. T., *Introduction to Nonlinear Differential and Integral Equations*, Dover, New York (1962).

de Gennes, P. G., Dynamics of entangled polymer solutions: I. The Rouse model, Macromolecules, *9*, 587 (1976a).

de Gennes, P. G., Dynamics of entangled polymer solutions: II. Inclusion of hydrodynamic interactions, Macromolecules, *9*, 594 (1976b).

Del Pino, O., J. A. Romagnoli, and J. M. Castro, Predicting conversion using temperature as the measured variable in the reaction injection

molding process: Part II Implementation of the on-line corrective algorithm, Polym. Eng. Sci., *23*, 895 (1983)

Denbigh, K. G., Continuous reactions: Part II. The kinetics of steady state polymerization, Trans. Faraday, Soc., *43*, 643 (1947).

Dionisio, J. M. and K. F. O'Driscoll, High conversion copolymerization of styrene and methyl methacrylate, J. Polym. Sci. Polym. Lett. Ed., *17*, 701 (1979).

Family, F. and D. P. Landau, eds., *Kinetics of Aggregation and Gelation*, North-Holland, New York (1984).

Flory, P. J., Molecular size distribution in linear condensation polymers, J. Am. Chem. Soc., *58*, 1877 (1936).

Flory, P. J., *Principles of Polymer Chemistry*, Cornell University Press, Ithaca, N.Y. (1953).

Galvan, R. and M. Tirrell, Orthogonal collocation applied to analysis of heterogeneous Ziegler-Natta polymerization, Comp. Chem. Eng. (1985, in press).

Garcia-Rubio, L. H., J. F. MacGregor, and A. E. Hamielec, Modeling and control of copolymerization reactors, ACS Symp. Series, *197*, 87 (1982).

Gerrens, H., How to select polymerization reactors: Part II, Chemtech, 434 (July 1982).

Goodall, B. L., Super high activity supported catalysts for the stereospecific polymerization of α-olefins: history, development, mechanistic aspects and characterization, Ind. Symp. on Transition Metal Catalyzed Polymerization: Unsolved Problems, Mich. (Aug. 1981).

Graessley, W. W., W. C. Uy, and A. Gandhi, Molecular weight distribution at high conversions in free radical polymers, Ind. Eng. Chem. Fundam., *8*, 696 (1969).

Gupta, S. K. and A. Kumar, *Reaction Engineering of Step Growth Polymerization*, Pergamon Press, Elmsford, N.Y. (1985).

Hendricks, E. M., M. H. Ernst, and R. M. Ziff, Coagulation equations with gelation, J. Stat. Phys., *31*, 519 (1983).

Hicks, J., A. Mohan, and W. H. Ray, The optimal control of polymerization reactors, Can. J. Chem. Eng., *47*, 590 (1969).

Hui, A. W. and A. E. Hamielec, Thermal polymerization of styrene at high conversions and temperatures. An experimental study, J. Appl. Polym. Sci., *16*, 749 (1972).

Hulburt, H. M. and S. Katz, Some problems in particle technology. A statistical mechanical formulation, Chem. Eng. Sci., *19*, 555 (1964).

Keii, T., Y. Doi, E. Suzuki, M. Tamura, M. Murata, and K. Saga, Propene polymerization with a magnesium chloride-supported Ziegler Catalyst: 2. Molecular weight distribution, Makromol. Chem., *185*, 1537 (1984).

Kilkson, H., Effect of reaction path and initial distribution on molecular weight distribution of irreversible condensation polymers, Ind. Eng. Chem. Fundam., *3*, 281 (1964).

Kilkson, H., Generalization of various polycondensation problems. Concept of slow influx, Ind. Eng. Chem. Fundam., *7*, 354 (1968).

Laurence, R. L., and M. G. Chiovetta, Heat and mass transfer during olefin polymerization from the gas phase, in *Proceedings of Berlin Workshop on Polymer Reactor Engineering*, K. H. Reichert, ed., Carl Hanser Verlag, Munich (1983).

Lenz, R. W., *Organic Chemistry of Synthetic High Polymers*, Wiley-Interscience, New York (1968).

Lopez-Serrano, F., J. M. Castro, C. W. Macosko, and M. Tirrell, A recursive approach to copolymerization statistics, Polymer, *21*, 263 (1980).

Lowry, G. G., *Markov Chains and Monte Carlo Calculations in Polymer Science*, Marcel Dekker, New York (1970).

Marten, F. L. and A. E. Hamielec, High conversion diffusion-controlled polymerization, ACS Symp. Series, *104*, 43 (1979).

Mecklenburgh, J. C., The influence of mixing on the distribution of copolymerization compositions, Can. J. Chem. Eng., *48*, 279 (1970).

Middleman, S., *Fundamentals of Polymer Processing*, McGraw-Hill, New York (1977).

Nagasubramanian, K. and W. W. Graessley, Continuous reactors in free radical polymerization with branching: I. Theoretical aspects and preliminary considerations, Chem. Eng. Sci., *25*, 1549 (1970a).

Nagasubramanian, K. and W. W. Graessley, Continuous reactors in free radical polymerization with branching: II. Experimental results on vinyl acetate polymerization, Chem. Eng. Sci., *25*, 1559 (1970b).

Nagel, E. J., V. A. Kirillov, and W. H. Ray, Prediction of molecular weight distributions for high-density polyolefins, Ind. Eng. Chem. Prod. Res. Dev., *19*, 372 (1980).

Nanda, V. S. and R. K. Jain, The statistical character of anionic polymers, J. Polym. Sci., *A2*, 4583 (1964).

Nauman, E. B., Mixing in polymer reactors, J. Macrom. Sci. Rev., *C10*, 74 (1974).

Oldshue, J. Y., D. O. Mechler, and D. W. Grinnell, Fluid mixing variables in suspension and emulsion polymerization, Chem. Eng. Prog., 68 (May 1982).

Peebles, L. H., *Molecular Weight Distributions in Polymers*, Wiley, New York (1971).

Penlidis, A., J. F. MacGregor, and A. E. Hamielec, Dynamic modeling of emulsion polymerization reactors, AIChE J., *31*, 881 (1985).

Platzer, N., Design of continuous and batch polymerization process, Ind. Eng. Chem., *62*, 6 (1970).

Plesch, P. H., *The Chemistry of Cationic Polymerization*, Macmillan, New York (1963).

Ray, W. H., On the mathematical modeling of polymerization reactors, J. Macromol. Sci. Rev., *C8*, 1 (1972).

Ray, W. H. and C. E. Gall, The control of copolymer composition distributions in batch and tubular reactors, Macromolecules, *2*, 425 (1969).

Ray, W. H. and R. L. Laurence, Polymerization reaction engineering, in *Chemical Reactor Theory*, N. R. Amundson and L. Lapidus, eds., Prentice-Hall, Englewood Cliffs, N.J. (1977).

Ross, R. T. and R. L. Laurence, Gel effect and free volume in the bulk polymerization of methyl methacrylate, AIChE Symp. Ser., *72*(160), 74 (1976).

Saidel, G. M. and S. Katz, Moments of the size distribution in radical polymerization, AIChE J., *13*, 319 (1967).

Schmeal, W. R. and J. R. Street, Polymerization in expanding catalyst particles, AIChE J., *17*, 1188 (1971).

Schulz, G. V., Uber die Beziehung zwischen Reaktionsgeschwindigkeit und Zusammentetzung des Reaktionsprodktes bei Makropolymerisations vorgängen 122. Mitteilung über hochpolymere Verbindugen, Z. Phys. Chem. (Leipzig), *B30*, 379 (1935).

Soh, S. K. and D. C. Sundberg, Diffusion-controlled vinyl polymerization: I. The gel effect, J. Polym. Sci. Chem. Ed., *20*, 1299 (1982a).

Soh, S. K. and D. C. Sundberg, Diffusion-controlled vinyl polymerization: II. Limitations on the gel effect, J. Polym. Sci. Chem. Ed., *20*, 1315 (1982b).

Soh, S. K. and D. C. Sundberg, Diffusion-controlled vinyl polymerization: III. Free volume parameters and diffusion-controlled propagation, J. Polym. Sci. Chem. Ed., *20*, 1331 (1982c).

Soh, S. K. and D. C. Sundberg, Diffusion-controlled vinyl polymerization: IV. Comparison of theory and experiment, J. Polym. Sci. Chem. Ed., *20*, 1345 (1982d).

Stockmayer, W. H., Theory of molecular size distribution and gel formation in branched-chain polymers, J. Chem. Phys., *11*, 45 (1943).

Stuber, N. P. and M. Tirrell, Studies of continuous methylmethacrylate polymerization in a twin-screw extruder, Poly. Process Eng., *3*, 71 (1985).

Summers, L., Composition control of suspension copolymerization at high conversion, M.S. thesis, U iversity of Minnesota (1985).

Szabo, T. T., and J. F. Leathrum, Analysis of condensation polymerization reactors, J. Appl. Polym. Sci., *13*, 477 (1969).

Szwarc, M., *Carbanions, Living Polymers and Electron Transfer Processes*, Wiley-Interscience, New York (1968).

Tirrell, M., Polymer self-diffusion in entangled systems, Rubber Chem. Technol., *56*, 523 (1984).

Tirrell, M. and K. Gromley, Composition control of batch copolymerization reactors, Chem. Eng. Sci., *36*, 367 (1981).

Tirrell, M. and T. J. Tulig, Diffusion-controlled reactions in polymers: the influence of polymer concentration and molecular weight in the reaction environment, in *Proceedings of Berlin Workshop on Polymer Reaction Engineering*, K. H. Reichert, ed., Carl Hanser Verlag, Munich (1983).

Tsoukas, A., M. Tirrell, and G. Stephanopoulos, Multiobjective dynamic optimization of semibatch copolymerization reactors, Chem. Eng. Sci., *37*, 1785 (1982).

Tulig, T. J. and M. Tirrell, Toward a molecular theory of the Trommsdorff effect, Macromolecules, *14*, 1501 (1981).

Tulig, T. J. and M. Tirrell, On the onset of the Trommsdorff effect, Macromolecules, *15*, 459 (1982).

Ver Strate, G., C. Cozewith, and W. W. Graessley, Branching by copolymerization of monovinyl and divinyl monomers in continuous-flow stirred reactors, J. Appl. Polym. Sci., *25*, 59 (1980).

Webber, D., Polymer output reached a new high in 1984, Chem. Eng. News, *63*, 12 (1985).

Wu, G. Z. A., L. A. Denton, and R. L. Laurence, Batch polymerization of styrene-optimal temperature histories, Polym. Eng. Sci., *22*, 1 (1982).

Zeman, R. J. and N. R. Amundson, Continuous models for polymerization, AIChE J., *9*, 297 (1963).

Zeman, R. J. and N. R. Amundson, Continuous polymerization models, I, Chem. Eng. Sci., *20*, 331 (1965a).

Zeman, R. J. and N. R. Amundson, Continuous polymerization models, II, Chem. Eng. Sci., *20*, 637 (1965b).

12
Biological Reactors

LARRY E. ERICKSON *Kansas State University, Manhattan, Kansas*

GREGORY STEPHANOPOULOS* *California Institute of Technology, Pasadena, California*

INTRODUCTION

According to recorded history, some of the first reactors were biological reactors. Fermentors for the production of alcoholic beverages such as beer and wine were designed, constructed, and used more than 5000 years ago (Pederson, 1979, pp. 1–24; Rose, 1959; Amerine, 1964). According to Pederson (1979, p. 14), the Babylonians made beer as early as 5000–7000 B.C. The baking of leavened bread may have started in Egypt some 3500 years ago (Pederson, 1979, p. 4). Cheese manufacture also appears to have been carried out by the Egyptians about 3000 B.C. (Pederson, 1979, p. 6). The science of fermentation advanced significantly in 1857 when Pasteur showed that microorganisms were present and active in fermentations (Aiba et al., 1973, p. 4). Another major advance was the production of penicillin in deep aerated tanks during World War II. It was during this period that Elmer Gaden, the father of modern biochemical engineering, did his Ph.D. research on oxygen transfer to aerobic cultures (Hixon and Gaden, 1950). Most of the science and engineering advances in biological reactor design have appeared during the period from World War II to the present. During this period several books have appeared in which biological reactor design is an important consideration (Aiba et al., 1973; Bailey and Ollis, 1977; Pirt, 1975; Wang et al., 1979; Aiba and Nagai, 1975; and Bekers, 1980). Journals have started which are devoted to biochemical engineering, such as *Biotechnology* and *Bioengineering*, and many articles on biological reactors are contained in the volumes of *Advances in Biochemical Engineering*.

Biological reactors include a wide variety of types, both human-made and natural. Thus one can find aerobic and anaerobic reactors, well-mixed or stratified reactors, enzymatic reactors with soluble or immobilized enzymes, immobilized cell reactors of various configurations, and others that will be discussed in this chapter. Biological reactor design has been influenced by several important factors. Strain selections and mutations placed time-varying specifications on the biological reactors while promoting batch or

---

**Current affiliation*: Department of Chemical Engineering, Massachusetts Institute of Technology, Cambridge, Massachusetts

fed-batch operation. Rheological conditions may change significantly during the course of a batch fermentation (van Suijdam et al., 1982). Furthermore, market forces frequently dictate that the same fermentor be used for the production of different products by different microorganisms during its operational life. All these considerations make the problem of optimal design and operation of bioreactors very important indeed.

Anaerobic reactors are used in the production of ethanol, acetone, butanol, many food products, and anaerobic waste treatment. Temperature and pH control are usually important considerations. Reactor design has received limited attention except for the case of anaerobic digestion, where mixed culture interactions have been considered extensively (Bailey and Ollis, 1977; Gould, 1971; Sundstrom and Klei, 1979). A completely mixed reactor is frequently employed.

The design of aerobic reactors, frequently employed for antibiotic production, has primarily considered oxygen mass transfer and oxygen limitation on reactor performance. Gas, liquid, and solid phases are present. In hydrocarbon fermentations, hydrocarbon transport may also limit growth (Erickson et al., 1969, 1970, 1975; Shah et al., 1972a,b; Erickson and Humphrey, 1969a,b; Gutierrez and Erickson, 1978). In aerobic ferementations in which the organic substrate is insoluble, four phases are important for biological reactor design. The utilization of solid substrates, such as cellulose, by microorganisms also requires that transport limitations to growth be considered (Lee et al., 1980; Avgerinos et al., 1981). Transport and surface area limitations are also important in the bacterial leaching of ores (Gormely et al., 1975; Chang and Myerson, 1982). In the modeling of these multiphase growth processes, basic concepts of reaction engineering, such as residence time, segregation, surface adsorption, transport limitation, and reaction kinetics, have been applied.

Enzymatic reactors is another important category of bioreactors. Enzymes are presently used and hold a strong promise for future applications as catalysts of biochemical reactions. In soluble or immobilized form they are analogous to other homogeneous or heterogeneous catalysts and principles of reaction and diffusion theories can be applied to optimize bioreactor operation. Enzyme reaction engineering studies have resulted in the development of membrane reactors to retain the enzymes (Deeslie and Cheryan, 1982; Cheryan and Deeslie, 1980; and Vieth et al., 1976), fixed-bed reactors with immobilized enzymes or immobilized whole cells (Lee et al., 1976; Vieth et al., 1976; and Tyagi and Ghose, 1982), and fluidized- and semifluidized-bed reactors (Vieth et al., 1976; Barker et al., 1981; Fan and Hsu, 1981). Immobilized enzymes have also been used to develop enzyme electrodes and other enzyme-based sensors in which reactors are used for instrumentation (Chen et al., 1982; Barker and Somers, 1978). Several conferences and books have been devoted to enzyme reaction engineering and the supporting science of enzyme and whole cell immobilization (Wingard, 1972; Pye and Wingard, 1974; Pye and Weetall, 1978; Weetall and Royer, 1980; Wingard et al., 1976, 1980; Zaborsky, 1973).

Media sterilization, enzyme deactivation, and the thermal processing of foods in canning is another important application of biological reaction engineering. The activation energy in the Arrhenius expression for temperature dependence of the rate of sterilization is very high (Aiba et al., 1973, pp. 242–243). Thus high-temperature short-time processing of foods optimizes food quality. The design of continuous processes which closely approximate plug flow requires the application of residence-time distribution analysis (Aiba et al., 1973, pp. 258–264; Bailey and Ollis, 1977, pp. 551–554).

There are many different types of biological reactors and biochemical reactions; however, in many cases, the basic principles of reaction engineering described in earlier chapters may be applied in designing the reactor system. To avoid unnecessary duplication, attention will focus on those aspects that are characteristic of biological systems and processes. The stoichiometry, kinetics, and energetics of these systems will first be reviewed, for they appear to differ from those of simple chemical systems. Subsequently, the various types of bioreactor configurations will be presented along with the design equations for the simple ones. Mass transfer considerations, especially as they apply to oxygen transfer in aerobic processes and procedures for bioreactor design, will occupy the last two sections of this chapter. Much of the discussion will be associated with aerobic microbial growth and product formation, for most of the current biological reaction engineering research is focused on the optimization of aerobic fermentation processes.

MICROBIAL GROWTH

The stoichiometry, kinetics, and energetics of a biochemical process must be known and understood in order to design the reactor system. These basic concepts as applied to microbial growth processes with product formation are reviewed in this section. Regularities are introduced because of the simplification achieved through their application in process analysis and design.

Stoichiometry and Nutrition

In direct analogy to the case of a single chemical reaction occurring in a chemical reactor, the processes of growth and product formation are represented by a chemical reaction in which ammonia (or another nitrogen source) with a carbon-energy source is converted into biomass, product, carbon dioxide, and water:

$$\underset{\text{carbon-energy source}}{CH_mO_l} + \underset{\text{ammonia (nitrogen source)}}{aNH_3} + bO_2 \longrightarrow \underset{\text{cell biomass}}{y_bCH_pO_nN_q} + \underset{\text{product}}{y_pCH_rO_sN_t} + y_dCO_2 + cH_2O \tag{1}$$

Reaction (1) is true for the case of a single carbon-energy source (or substrate) and a single product. If complex mixtures of organic substrates and products are involved, approximate formulas of the type indicated above are written for the substrate and product. It is further assumed that cell biomass can be represented by the formula $CH_pO_nN_q$ and that the biomass composition remains constant during the course of the fermentation. Although deviations occur, the above is a rather reasonable assumption. In addition to carbon, hydrogen, oxygen, and nitrogen, on a dry basis, there is also 3 wt % phosphorus, 1% sulfur, 1% potassium, 1% sodium, less than 1% calcium, magnesium, chlorine, and iron, and trace amounts of other elements, such as copper and zinc. All the latter are not represented in formula (1).

It should be noted that reaction (1) is written on the basis of one carbon atom of substrate. Therefore, y_b, y_p, and y_d give the fraction of organic substrate converted to biomass, product, and carbon dioxide, respectively; that is,

$$y_b + y_p + y_d = 1.0 \tag{2}$$

is a carbon balance for the process. Similarly, a nitrogen balance gives

$$a = y_b q + y_p t \tag{3}$$

One can also write balances for hydrogen and oxygen involving all the stoichiometric coefficients of reaction (1). Since the water involved in the process is not measurable, it is, in general, desirable to eliminate the stoichiometric coefficient of water c between the hydrogen and oxygen balances. The result is Eq. (4), which may also be interpreted as an available electron balance (Minkevich and Eroshin, 1973; Erickson et al., 1978a):

$$\gamma_s - 4b = y_b \gamma_b + y_p \gamma_p \tag{4}$$

Equation (4) introduces the concept of the degree of reductance of biomass, product, and substrate, γ_b, γ_p, and γ_s, respectively. The degree of reductance of a substance is defined as the number of equivalents of available electrons per gram atom of carbon of the substance which would be transfered to oxygen upon oxidation of the substance to CO_2, H_2O, and NH_3. Thus

$$\gamma_b = 4 + p - 2n - 3q \tag{5}$$

$$\gamma_p = 4 + r - 2s - 3t \tag{6}$$

$$\gamma_s = 4 + m - 2\ell \tag{7}$$

where the valences C = 4, H = 1, O = − 2, and N = −3 have been used. The concept of the degree of reductance is a very useful tool in establishing a unifying treatment of the biological processes of microbial metabolism. Furthermore, some regularities concerning the degree of reductance of biomass have been observed (Minkevich and Eroshin, 1973; Mikevich et al., 1977; Erickson, 1980b). By examining various biomass compositions, these authors found that γ_b is relatively constant at an average value of 4.291 with a coefficient variation of 4%. Another regularity was also observed concerning the weight fraction of carbon in biomass, σ_b. The same authors observed that σ_b has an average value of 0.462 with a coefficient of variation of 5%.

In a typical bioreactor situation the chemical formulas of the substrate and product are known and that of biomass can be determined by an elemental analysis. The only remaining unknowns, then, are the stoichiometric coefficients of reaction (1), namely a and b and y_b, y_p, and y_d. Notice that these coefficients are directly related to the yields of the process. The available equations are the carbon balance [Eq. (2)], the

nitrogen balance [Eq. (3)], and the electron balance [Eq. (4)]. By measuring the respiratory quotient (i.e., the ratio y_d/b) an additional equation becomes available and the system can be determined in the absence of product formation. If product(s) is being formed an additional measurement(s) is needed to close the system. In summary, the balances above can be used (a) to determine the stoichiometry of a process, in conjunction with the necessary measurements; (b) to monitor on-line the state of a bioreactor to provide the appropriate control action (Wang et al., 1977; Stephanopoulos and San, 1982); and (c) to test the consistency of measurements (Ferrer and Erickson, 1979) and identify the measurement that most likely introduces large error (Wang and Stephanopoulos, 1982).

The stoichiometry of a biological process is an important element in writing the proper macroscopic material balances over a bioreactor, which is the subject of the following section. However, in addition to mass, heat is also exchanged in most of the aerobic fermentations. One can write the following enthalpy balance for the heat of reaction ΔH_r:

$$\Delta H_r^\circ = (\Delta H_c^\circ)_{substrate} - y_b(\Delta H_c^\circ)_{biomass} - y_p(\Delta H_c^\circ)_{product} \tag{8}$$

where the products of combustion are taken to be $CO_2(g)$, $H_2O(\ell)$, and NH_3 in a dilute aqueous solution, and each of the values of ΔH_c° are for 1 atom of carbon.

Equation (8) can be utilized in two different ways. If the stoichiometric coefficients y_b and y_p are known, the heat released during growth can be calculated and the corresponding cooling requirements determined. Conversely, if the reaction heat can be measured, Eq. (8) provides an additional balance for the determination of the unknown yields. The heat of growth can be measured on-line by using flow microcalorimeters (Eriksson and Holme, 1973). Alternatively, a correlation may be used that relates the heat evolved to the amount of oxygen consumed. Thus, approximately, as shown in Fig. 1,

$$\Delta H_r^\circ = -27(4b) \tag{9}$$

where ΔH_r° is in kcal/g atom of carbon in the substrate. This approximation is based on the early work of Thornton (1917), Kharasch (1929), and Kharasch and Sher (1925) and the more recent work of Minkevich and Eroshin (1973), Roels (1981), Cooney et al. (1969), Minkevich et al. (1977), Erickson et al. (1978b), and Patel and Erickson (1981).

For aerobic fermentations the chemical energy per equivalent of available electrons is about the same for the organic substrate, the biomass, and the extracellular product. Minkevich and Eroshin (1973) and Minkevich et al. (1977) found a value of 26.95 kcal/g equiv of available electrons (112.8 kJ/Q equiv) for the quantity of energy given up when dry biomass is oxidized to $CO_2(g)$, $H_2O(\ell)$ and NH_3 in a dilute aqueous solution. Per gram atom of carbon, this yields

$$(\Delta H_c^\circ)_{biomass} = -26.95\gamma_b \tag{10}$$

Now, Eq. (4) can be divided by γ_s to yield

$$\varepsilon + \eta + \xi_p = 1.0 \tag{11}$$

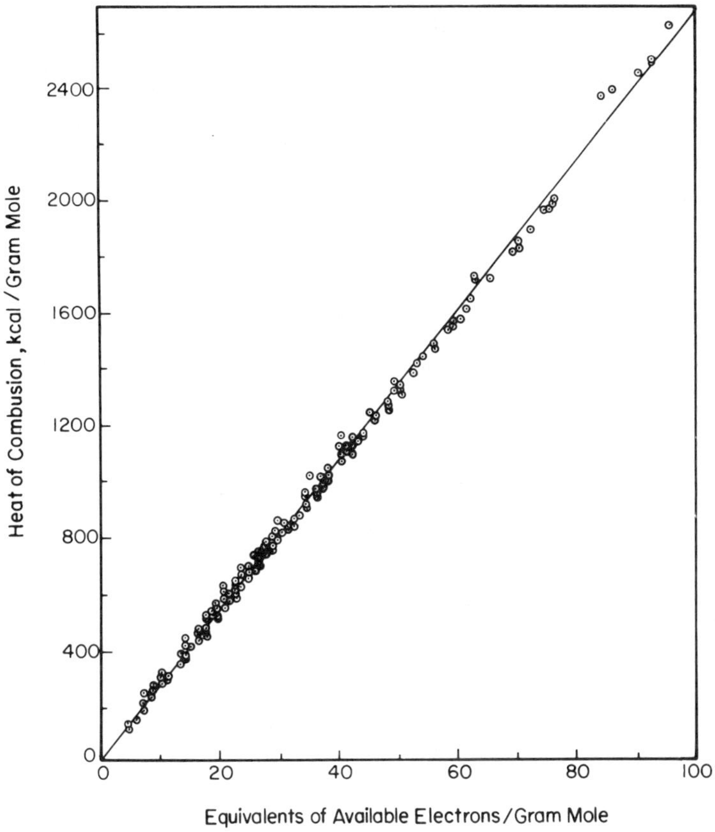

FIGURE 1 Linear variation of heat of combustion of organic compounds with the number of equivalents of available electrons. (From Patel and Erickson, 1981.)

where the three quantities $\varepsilon = 4b/\gamma_s$, $\eta = y_b\gamma_b/\gamma_s$, and $\xi_p = y_p\gamma_p/\gamma_s$ are fractions of the available electrons transferred to oxygen, incorporated into biomass, and incorporated into extracellular product, respectively. In view of the regularity mentioned earlier, according to which the chemical energy per equivalent of available electrons is the same, the quantities above may also be interpreted as the fractions of substrate energy evolved as heat, incorporated into biomass, and incorporated into extracellular products, respectively. In this regard, the electron balance Eq. (4) obtains the meaning of an energy balance.

It should be noted that the subject of stoichiometry and yields is essential in bioreactor design. In biological processes the stoichiometry is unknown and, unlike chemical systems, may vary during the course of the process. Consequently, the establishment of the proper stoichiometry, and the ability to follow any variations on-line, constitute the first objective in the analysis of a biological process. The second is the determination of the process kinetics, and this is the subject of the next section.

Kinetics

The kinetics of biological processes deals with the subject of the determination of the rates of substrate consumption, biomass growth, and product formation. Typically, in a chemical reaction, these rates are interrelated through the reaction stoichiometry. However, the variation of yields in biological processes complicates the situation.

The starting point is the specific rate of biomass growth, μ. A variety of models exists of varying degree of complexity. Thus one can find lumped-parameter models which treat the biomass as one entity, structured models that distinguish the various types of biomass, segregated models that recognize the individual nature of each microorganism, and the possible combinations among them. This discussion focuses on lumped-parameter models. Various structured and segregated models can be found in Fredrickson and Tsuchiya (1977) and Bailey and Ollis (1977). The typical trade-off with these models is that the number of parameters involved and the mathematical complexity increases dramatically, but more information becomes available and better dynamic response is achieved.

The lumped-parameter models express the specific growth rate as a function of key substances of the abiotic medium. To be sure, a large number of factors affect the growth of a culture. By keeping the concentrations of all but one above the saturation level, the specific growth rate becomes a function of the concentrations of this rate-limiting substance. Figures 2 and 3 depict the two basic functional relationships between μ and S.

The characteristics of the function shown in Fig. 2 can be represented by the rectangular hyperbola of Eq. (12):

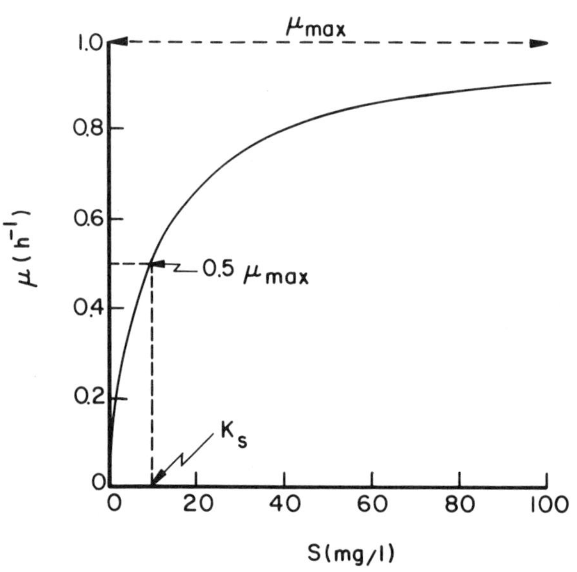

FIGURE 2 Specific growth rate μ plotted as a function of substrate concentration S according to the Monod model, where μ_{max} = 1.0 h^{-1} and K_s = 10 mg/L. Note that μ = 0.5 μ_{max} when S = K_s. (From Pirt, 1975, p. 9.)

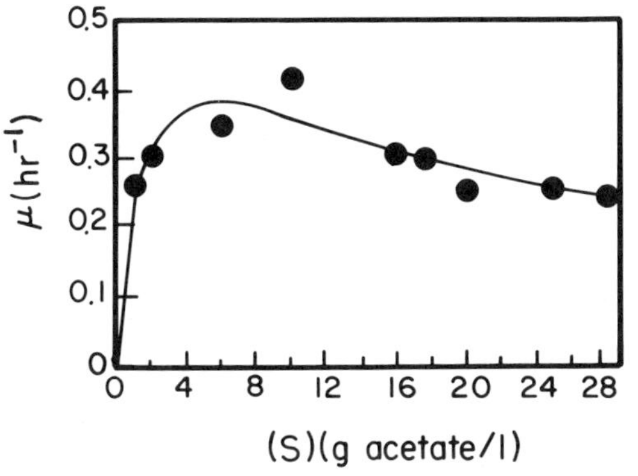

FIGURE 3 Substrate inhibition for growth of *Candida utilis* on sodium acetate. (From Aiba and Nagai, 1975, p. 188; Cama and Edwards, 1970.)

$$\mu(S) = \frac{\mu_{max} S}{K_s + S} \tag{12}$$

The model is the celebrated Monod model. It is an empirical model, patterned after the Michaelis-Menten kinetics of enzymatic reactions, which was found to describe adequately the growth kinetics of several microbial systems (Monod, 1949). This model has also been used repeatedly in a variety of applications ranging from activated sludge kinetics to the analysis of dynamics of mixed culture systems. It contains the basic features exhibited by most microbial systems, namely linear dependence at low S values and saturation at high S values. Furthermore, it is a simple, two-parameter model and in the absence of indications to the opposite, it is used to fit data or to study the general dynamic characteristics of a reactor system.

The meaning of the two parameters μ_{max} and K_s is shown in Fig. 2. These parameters are usually estimated by plotting available data in the form $1/\mu$ versus $1/S$. For many organic substrates the saturation constant K_s is between 0.1 and 20 mg/L (Bailey and Ollis, 1977, pp. 346, 347; Pirt, 1975, pp. 10–12). Maximum specific growth rates are usually larger for bacteria than yeast. Values between 0.3 and 1.0 h^{-1} are frequently reported; however, both lower and higher values are found (Pirt, 1975, p. 145).

The maximum specific growth rate, μ_{max}, depends on temperature and pH. Microorganisms are classified as thermophiles, mesophiles, and psychrophiles with respect to their optimum temperature for growth (Bailey and Ollis, 1977, p. 344). The optimum temperature for most mesophylic organisms is about 37°C; as shown in Fig. 4, at temperatures lower than about 37°C, the common Arrhenius temperature dependence is observed (Bailey and Ollis, 1977, pp. 132–135, 344; Prokop and Erickson, 1972). Values of the activation energy have been found to be significantly different for growth (8.2 kcal/mol), respiration (17.8 kcal/mol), and penicillin production (26.8 kcal/mol) for *Penicillin chrysogenum* (Aiba et al., 1973, pp. 108–110; Calam et al., 1971).

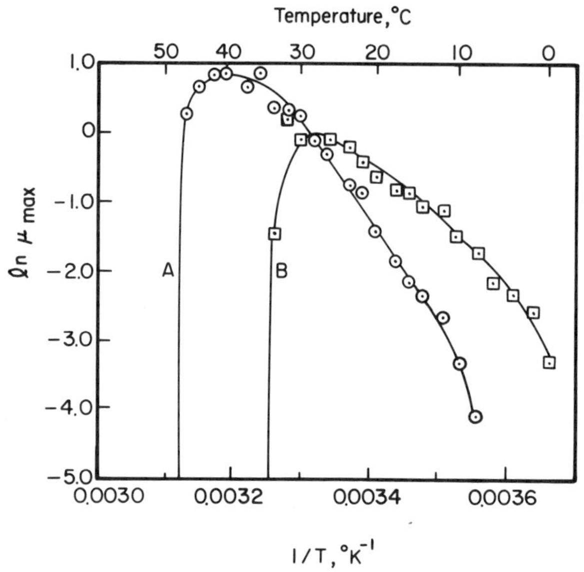

FIGURE 4 Arrhenius plot for the effect of temperature on the maximum specific growth rate of a mesophile (*E. coli*, curve A, E = 25.5 kcal/g mol) and a psychrophile (*Pseudomonas* strain 21-3C, curve B, E = 16.2 kcal/g mol). (Data of Ingraham, 1958.)

As shown in Fig. 5, the optimum pH for growth depends on the organism. Bacteria frequently grow in the pH range 4 to 8, yeasts grow in the pH range 3 to 6, molds from 3 to 7, and higher eucaryotic cells from 6.5 to 7.5 (Wang et al., 1979, pp. 89–90). A rather narrow window of optimal growth exists around the optimal pH with the growth rate dropping rapidly outside this window. However, there exist organisms, such as yeast, *Escherichia coli*, and others, which maintain optimal growth over a wider pH variation. If it can be assumed that pH affects growth by inactivating some rate-limiting enzymes through protonation-deprotonation reactions, then the primary effect of pH variations is the decrease of μ_{max}. Changes in K_S may also be present, but they are secondary importance. Temperature and pH control are commonly included in the design of fermentors.

Some systems exhibit rate inhibition at high substrate concentrations. This situation is depicted in Fig. 3 and described by the model

$$\mu(S) = \frac{\mu_{max} S}{K_S + S + S^2/K_i} \tag{13}$$

Equation (13) has a maximum and exhibits steady-state multiplicity and very interesting dynamic features when considered in a CSTR situation (San and Stephanopoulos, 1983).

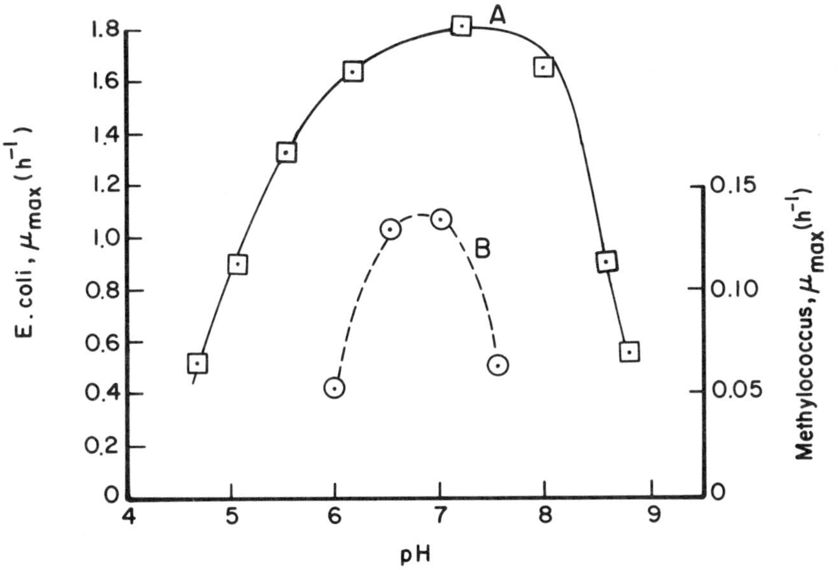

FIGURE 5 Effects of medium pH on maximum bacterial specific growth rates: A, *E. coli* in anaerobic casein hydrolysate medium (from data of Gale and Epps, 1942); B, *Methylococcus capsulatus* grown on methane (from data of Harwood and Pirt, 1972). (From Pirt, 1975, p. 145.)

Other models of microbial kinetics have also appeared in the literature. These models contain various features that occasionally were found to be important, such as multiple substrate limitation, product inhibition, dependence of μ on biomass concentration, and others. A detailed description of these models is given by Fredrickson and Tsuchiya (1977) and in tabular form by Ollis (1977). It should be also pointed out that even though S usually stands for an organic limiting substrate, it can represent any limiting compound. Johnson (1967), for example, found a value of $K_S = 0.0234$ mg/L for a Monod model with oxygen rate limiting. Also, expressions completely analogous to those presented above have been developed for the velocity of enzymatic reactions.

The rate of nutrient consumption is modeled by the product of the specific growth rate and the biomass concentration divided by an appropriate yield. Various possibilities regarding these yields and their variation with time will be presented when the bioreactor design equations are derived in a later section.

Product formation is often the objective of carrying out a biological process. The kinetics of product formation depend on the product or products that are produced. Some products are associated with the process of growth, whereas some others are not. The specific (≡ per unit amount of biomass) product formation rate in the first case is proportional to the specific growth rate; it is constant in the second case of non-growth-associated product. Intermediate situations also exist. Leudeking and Piret (1959) introduced a successful mathematical model which includes all the foregoing possibilities:

$$\frac{1}{X}\frac{dP}{dt} = \alpha + \beta\mu \qquad (14)$$

The model above has been applied to a number of fermentations where the parameters α and β, determined experimentally, indicated the extent of growth association. Some flexibility can be introduced in the model by providing different parameter values for the different phases of growth. This certainly extends the applicability of the model but does not alter the fact that it is an empirical one and results based on this model cannot be extrapolated in general.

In constructing product formation models with a mechanistic basis, the classification of the various products in rational categories can be useful. Gaden (1955, 1959) has proposed a classification scheme which includes type I products, which are directly associated with primary energy metabolism, such as ethanol; type II products, which are indirectly associated with primary energy metabolism, such as citric acid; and type III products, which are complex molecules that are not directly associated with primary energy metabolism, such as penicillin (Bailey and Ollis, 1977, pp. 371–375). Wang et al. (1979, pp. 1–45), divide microbial product formation into biosynthesis of primary metabolites, secondary metabolites, and enzymes with a fourth category of bioconversions. Primary metabolites are end products of low moelcular weight that are used as building blocks by cells. Secondary metabolites are products such as antibiotics which are not required for growth; they are usually synthesized late in the growth cycle. Secondary metabolites are usually type III fermentations according to Gaden's classification. Finally, Deindoerfer (1960), has organized fermentations patterns into four categories, according to the similarity of the formation of fermentation products to those produced by simple, simultaneous, consecutive, and stepwise reactions.

In type I fermentations, the last term in Eq. (14) is usually important and product formation is primarily growth associated. In type II fermentations, product formation kinetics are frequently complex, with maximum product formation rates often occurring late in the fermentation at low specific growth rates.

The rate of product formation, the product yield, and the concentration of the product in the final broth are important economic considerations.

Product inhibition is an important consideration in many fermentations because this limits the final product concentration. Pirt (1975, pp. 170–185) has considered the effect of product inhibition on growth and has presented the expressions

$$\mu = \frac{\mu_{max}S}{S + K_s(1 + I/K_i)} \qquad (15)$$

for competitive product inhibition and

$$\mu = \frac{\mu_{max}S}{(S + K_s)(1 + I/K_i)} \qquad (16)$$

for noncompetitive product inhibition, respectively. The concentration of the product which inhibits growth is I, and K_i is the kinetic saturation

constant. For the production of ethanol, Aiba et al. (1968) found that ethanol inhibits growth by noncompetitive inhibition.

Product formation kinetics have been developed and used in reactor design and operation for many industrial fermentations. Constantinides et al. (1970a,b) have used kinetic data for penicillin production to optimize temperature profiles in batch fermentations to achieve improved yields of about 15%. Kinetic models for the production of gluconic acid from glucose have been developed and used to optimize pH and temperature (Constantinides and Rai, 1974). The optimization resulted in improved yields and rates. Rai and Constantinides (1973) found a time saving of up to 60% by conducting experiments near the optimal temperature and pH. In these examples, the kinetic models are first developed and then employed to optimize batch operation.

One of the deficiencies of the Monod model is its inability to model the dynamics of growth processes. Structured models in which microbial composition is considered are needed to properly represent the dynamic variations associated with growth which occur because of sudden changes in environmental conditions. Harder and Roels (1982) have recently reviewed structured models; in much of the work they review, RNA is considered as a variable in the model. While structured models may be extensively used in future reactor design work, they have not yet become widely accepted in industrial practice (Harder and Roels, 1982).

Another developing area which may be much more important in the future is that of growth kinetic models for mixed cultures. Fredrickson (1977) has reviewed the types of interactions and associated dynamics in mixed cultures. Among others the interaction of competition is receiving increasing attention and related work is summarized in a recent review (Fredrickson and Stephanopoulos, 1981). One of the useful kinetic models of mixed cultures is that of Andrews and Graef (1971), which has been used in anaerobic digestor control. Bailey and Ollis (1977, pp. 635–743) present several kinetic models for mixed cultures and discuss mixed culture applications. Harrison (1978) point out that the number of mixed cultures in commercial use which are well studied and understood, and deliberately constituted, is extremely small. Fredrickson and Tsuchiya (1977) have reviewed mixed culture kinetic models including much of their own work.

Energetics

In the design of a fermentation process, the energetics must be considered and energy balances must be made for the process. Aerobic processes may be limited by the availability of carbon in the organic substrate or by the quantity of energy in the organic substrate (Erickson, 1981; Linton and Stephenson, 1978). Figures 6 and 7 show the variation of the biomass carbon yield, y_b, and the biomass energetic yield, η, as a function of the reductance degree of the organic substrate. For low values of γ_s, these results show that the growth process is energy limited; however, at a value of γ_s above 4, the results indicate that the availability of carbon becomes important. For example, Veselov et al. (1974) reported substantially larger yields with CO_2 in the gas fed to the fermentor where *n*-alkanes were used as the organic substrate ($\gamma_s = 6.1$). Figures 6 and 7 are for growth processes without product formation; the process of product formation may also be carbon limited or energy limited (Erickson, 1981).

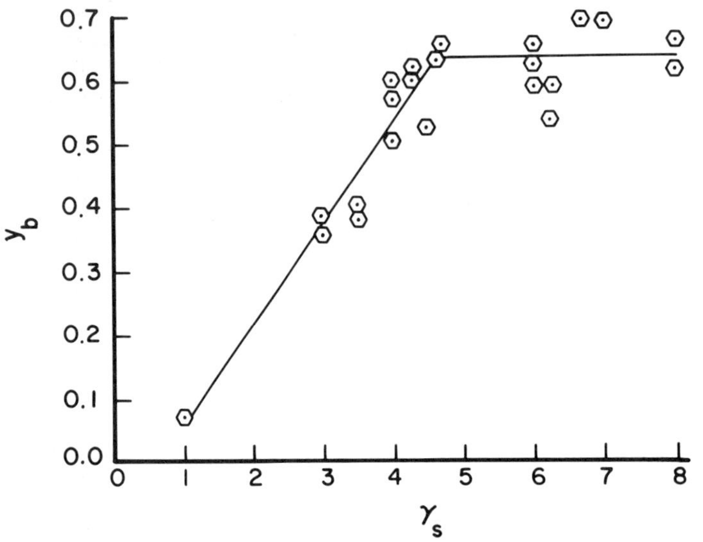

FIGURE 6 Variation of biomass carbon yield with reductance degree of organic substrate. (From Erickson, 1981.)

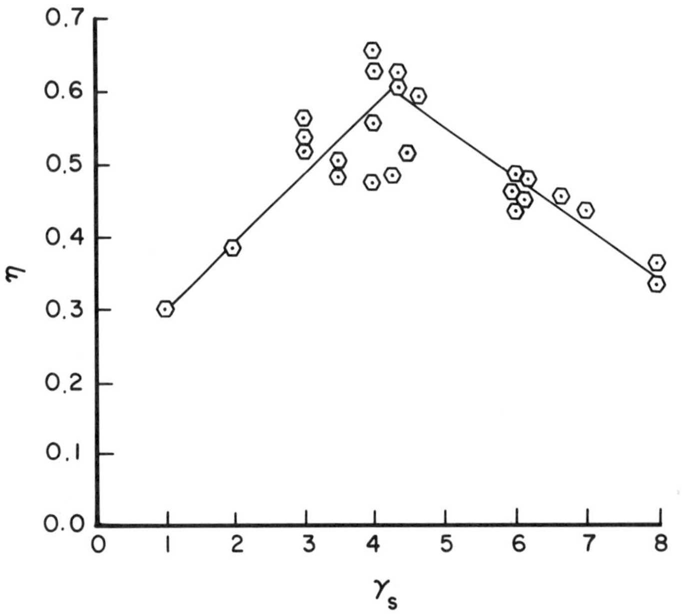

FIGURE 7 Variation of biomass energetic yield with reductance degree of organic substrate. (From Erickson, 1981.)

The energetics of growth and product formation may be viewed as a process in which available electrons in the organic substrate are rearranged into available electrons in biomass and products with similar energy per available electron. Energy in the form or ATP is used to achieve the rearrangement. The parameters Y_{ATP}^{max} and $Y_{P/ATP}^{max}$ are measures of the quantity of biomass and product, respectively, that can be formed from organic substrates per mole or ATP. Thus, under energy-limited growth conditions, the maximum biomass energetic yield (Erickson, 1980c), η_{max}, and the maximum product energetic yield, ξ_P^{max}, are related to Y_{ATP}^{max} and $Y_{P/ATP}^{max}$ and the efficiency by which ATP is formed (P/O) by

$$\eta_{max} = \frac{(\sigma_b/\gamma_b/12)Y_{ATP}^{max}}{(\sigma_b\gamma_b/12)Y_{ATP}^{max} + \{2/[(P/O)_s + (P/O)_o]\}} \tag{17}$$

$$\xi_p^{max} = \frac{(\sigma_p\gamma_p/12)Y_{P/ATP}^{max}}{(\sigma_p\gamma_p/12)Y_{P/ATP}^{max} + \{2/[(P/O)_s + (P/O)_o]\}} \tag{18}$$

(P/O) is the gram moles of ATP high-energy bonds formed per gram atom of oxygen consumed and the subscripts s and o refer to substrate-level phosphorylation and oxidative phosphorylation, respectively. For glucose as the organic substrate, the theoretically estimated value of $Y_{ATP}^{max} = 28.8$ g of cells/mol ATP may be used with the value of 3.167 mol of ATP/g atom of oxygen for $(P/O)_s + (P/O)_o$ to obtain $\eta_{max} = 0.88$ as the theoretical maximum biomass energetic yield for aerobic growth with glucose as the organic substrate (Erickson, 1980c; Stouthamer, 1979). For polysaccharide production from glucose, Erickson (1980c) has found a maximum product energetic yield of $\xi_p^{max} = 0.95$. Actual yields should be smaller than the theoretical maximum yield.

By using available electron concepts, Erickson (1979) obtained the following expressions for the energetic yields in terms of η_{max}, m_e, and ξ_p^{max}:

$$\frac{1}{\eta} = \frac{1}{\eta_{max}} + \frac{m_e}{\mu} + \frac{\xi_p}{\eta\xi_p^{max}} \tag{19}$$

$$\frac{\varepsilon}{\eta} = \frac{1-\eta_{max}}{\eta_{max}} + \frac{m_e}{\mu} + \frac{\xi_p}{\eta}\,\frac{1-\xi_p^{max}}{\xi_p^{max}} \tag{20}$$

$$\frac{d}{\eta} = \frac{1-(\gamma_s/\gamma_b)\eta_{max}}{\eta_{max}} + \frac{m_e}{\mu} + \frac{\xi_p}{\eta}\,\frac{1-(\gamma_s/\gamma_p)\xi_p^{max}}{\xi_p^{max}} \tag{21}$$

These equations may be used with parameter estimation methods to estimate η_{max}, m_e, and ξ_p^{max} from experimental data. The parameter, η_{max}, is the true growth yield based on that fraction of the available electrons in the organic substrate that are associated with growth, ξ_p^{max} is the true product yield based on that fraction of the available electrons that are associated with product formation, and m_e is the maintenance parameter.

Values of the parameters η_{max} and m_e have been estimated for microbial growth processes without product formation (Solomon et al., 1981, 1982; Ferrer and Erickson, 1979, 1980). Equivalent parameters have also been estimated by others (Pirt, 1975, p. 69; Stouthamer, 1979; Heijnen and Roels, 1981). Erickson (1980a,c), Erickson and Hess (1981), and Oner et al. (1983) have considered the true product yield ξ_p^{max} in estimating yield parameters. Solomon et al. (1982) and Oner et al. (1983) have developed parameter estimation methods in which all the data are used simultaneously in estimating parameters.

TYPES OF BIOLOGICAL REACTORS

Stirred Tanks

Well-stirred reactor systems have traditionally been employed for carrying out the great majority of fermentation processes. Providing a homogeneous reactor environment has not been always an easy task. Large residence times in continuous operation or batch operation are certainly positive factors in maintaining a liquid phase that is well mixed with respect to the nutrients. On the other hand, the large size of these reactors and the complex and time-varying rheological behavior of a multiphase reaction system make homogenization in these reactors an energy consuming and, in general, difficult proposition. Nevertheless, very large (200 m^3) fermentation vessels have been built and operated in the pharmaceutical, food, and chemical industries mainly because of the relative simplicity in their design and operation.

Because of the need to avoid mutations and maintain the superior qualities of the genetically developed strain, batch or fed-batch operations are the rule in most applications. The intrinsic property of continuous reactor systems to select for the best grower and not for the best product producer has hindered the expansion of continuous operation. Nevertheless, continuous culture operations are of great interest in research laboratories, for they provide a time-invariant environment at steady state which facilitates greatly the study of a biological process. Furthermore, some industrial operations employ continuous reactors, such as the single-cell protein facility of ICI in Billingham (total reactor volume of about 2300 m^3), all waste treatment processes, and others. It should be noted that it is relatively common to follow a batch process with a period of fed-batch or continuous operation. Also, batch cultivation is the optimal start-up procedure for continuous cultivation or fed-batch cultivation in most cases (Yamane et al., 1979).

Recently, many batch operations have been transformed into fed-batch (semicontinuous) by introducing the nutrient gradually into the reactor instead of charging everything at the beginning of the operation. The rationale is to optimally control the feed to maximize a composite performance index.

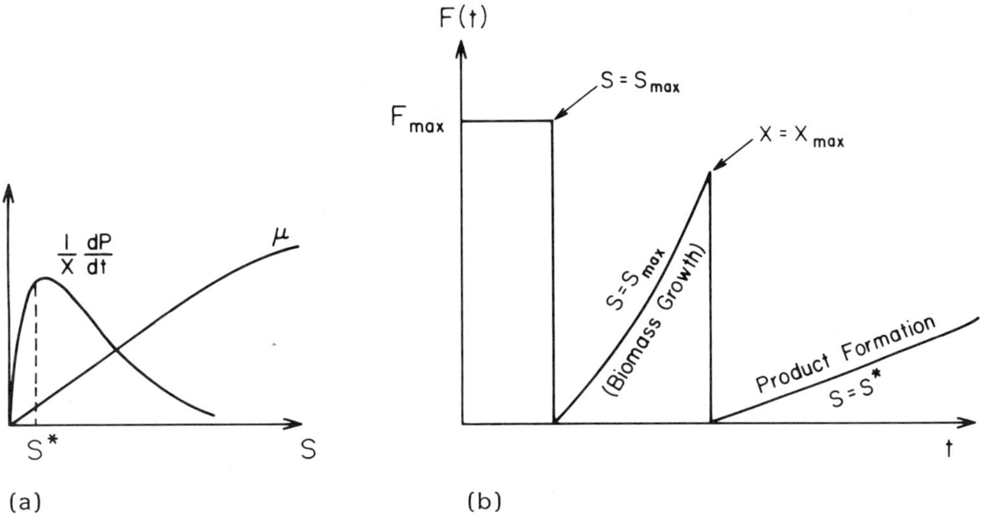

FIGURE 8 Schematic representation of an optimal fed-batch operation: (a) mutual disposition of specific growth and product formation rates; (b) feeding schedule. S_{max} is the maximum substrate concentration allowed to prevent undesirable side reactions. X_{max} is the maximum cell concentration allowed by mixing and oxygen transfer requirements.

For the case of penicillin fermentation, for example, for which the specific growth rate and the specific penicillin formation rate are mutually disposed as shown in Fig. 8a, the optimal feed policy is shown schematically in Fig. 8b. According to this strategy, the fermentation is carried out in two phases. During the first, cell biomass is built up to the maximum level allowed as quickly as possible. The second is a product formation phase during which the feed is controlled as an optimal level of high penicillin production and sufficient biomass growth to make up for the dilution effects. Other advantages of fed-batch operation include flexibility in the introduction of growth factors, deliberate variation of the feed concentration, and similar control actions.

The mass balance equations for all three modes of operation are as follows:

$$\frac{dX}{dt} = \mu X - DX \tag{22}$$

$$\frac{dS}{dt} = D(S_o - S) - \frac{\mu X}{Y_s^{max}} - m_s X - \frac{(\alpha + \beta\mu)X}{Y_{P/S}^{max}} \tag{23}$$

$$\frac{dP}{dt} = -DP + (\alpha + \beta\mu)X \tag{24}$$

where D is the dilution rate—equal to zero for batch operation and to F/V_ℓ for continuous operation and for fed-batch operation (notice that in the latter case, $F = dV_\ell/dt$).

In aerobic fermentations a continuous airstream flows through the reactor to provide the necessary oxygen for growth. Balances for the liquid- and gas-phase concentrations of the two main gaseous components, oxygen and carbon dioxide, are as follows:

$$\frac{dC_o}{dt} = (k_L a')(C_o^* - C_o) - Q_{O_2} X - DC_o + DC_o^f \tag{25}$$

$$\frac{dC_T}{dt} = -(k_L a')_c(C_c - C_c^*) + Q_{CO_2} X - DC_T + DC_T^f \tag{26}$$

$$\frac{V_g}{RT}\frac{dp_o}{dt} = \frac{Q_{in}p_o^{in}}{RT} - \frac{Q_{out}p_o}{RT} - (k_L a')_o(C_o^* - C_o)V_\ell \tag{27}$$

$$\frac{V_g}{RT}\frac{dp_c}{dt} = \frac{Q_{in}p_c^{in}}{RT} - \frac{Q_{out}p_c}{RT} + (k_L a')_c(C_c - C_c^*)V_\ell \tag{28}$$

where Q_{O_2} and Q_{CO_2} are the specified oxygen uptake and carbon dioxide evolution rates, respectively, defined by

$$Q_{O_2} = \frac{\mu}{Y_o^{max}} + m_o + \frac{\alpha + \beta\mu}{Y_{P/O}^{max}} \tag{29}$$

$$Q_{CO_2} = \frac{\mu}{Y_D^{max}} + m_D + \frac{\alpha + \beta\mu}{Y_{P/D}^{max}} \tag{30}$$

The dilution rate D in Eqs. (25) and (26) is defined as before. The values of the non-growth- and growth-associated product formation coefficients, α and β, respectively, depend on nutrient concentration, temperature, and pH. The true growth yield parameters, Y_s^{max}, Y_o^{max}, and Y_D^{max}, are related to the true biomass energetic yield parameter, η_{max}, as has been shown by Erickson (1979). The maintenance parameters, m_s, m_o, and m_D, are also related because of the stoichiometry of the maintenance process (Erickson, 1979; Erickson et al., 1979). Erickson (1979) has also shown how the true product yield parameters, $Y_{P/S}^{max}$, $Y_{P/O}^{max}$, and $Y_{P/D}^{max}$, are related to the true product energetic yield, ξ_p^{max}.

A stoichiometry difficulty associated with the use of the Luedeking and Piret model for product formation in the previous equations has been discussed by Fredrickson and Tsuchiya (1977). These authors point out that products come from some precursors. If the precursor is the growth-limiting substrate, then, according to Eq. (24), product is formed at a finite rate even after the precursor has been exhausted and is not present in the system. If the precursor is the biomass itself, this should be accounted for by adding a sink term in Eq. (22). In either case, some modification

is needed in the equations or the parameters if the models above are going to be valid during all phases of growth and product formation.

Steady-state operation can be achieved in a continuous culture system. The steady-state concentrations of biomass, substrate, product, O_2, and CO_2 in the liquid phase are obtained by setting the left-hand sides of Eqs. (16), (22), and (26) equal to zero. The possibility of multiple steady states exists, especially for the more complicated models of the specific growth rate. Fredrickson and Tsuchiya (1977) have analyzed the dynamics of various pure and mixed culture systems. Stephanopoulos (1980) presented a method for quickly analyzing the dynamics of two-population mixed culture systems. This method makes use of topological arguments and Poincaré index.

Various modifications and extensions of the previous model equations have appeared in the literature. The purpose of this work has been twofold: first, to model various departures from the ideal situation of a perfectly mixed vessel, such as wall growth, film formation, and incomplete mixing. Second, to describe some interesting process or hardware configurations, such as cell recycle, vessels in series, film reactors, and so on. Some of this work has been reviewed by Ollis (1977).

Other Design Considerations

The design equations of the preceding section were developed with the assumption of well-mixed liquid and gas phases. Even though macroscopic mixing is a good assumption, microscopic mixing is often hard to achieve. Dunlop and Ye (1982) have pointed out that in many fermentors the smallest turbulent eddy has a size of about 50 μm. This, contrasted with an average size of about 1 to 2 μm for a bacterium, indicates that various cells may be enclosed in an environment where nutrients are rapidly depleted. Increasing the power input has a minimal effect, for the size of the smallest eddy is inversely proportional to the one-fourth power of the energy input. The same authors conclude that the length scale of turbulence and the rates at which eddies are being created and destroyed must be quantified.

Supplying sufficient oxygen and removing the produced carbon dioxide in the case of aerobic fermentation is probably the single most important issue in fermentor design. Various air supply mechanisms, impeller designs and modifications of the bioreactor interior (baffles, etc.) have been proposed to facilitate the transport of these gases. Much of the development in this area is proprietary.

The equilibrium concentration C_o^* is usually provided by applying Henry's law as the phase equilibrium relationship:

$$P_o = H_o C_o^* \tag{31}$$

with H_o being the Henry's law constant for oxygen (Perry and Chilton, 1973, pp. 3–98; Schumpe et al., 1978). Similarly, for carbon dioxide,

$$p_c = H_c C_c^* \tag{32}$$

where H_c is the Henry's law constant for carbon dioxide (Stumm and Morgan, 1970, p. 148; Perry and Chilton, 1973, pp. 3–96).

Carbon dioxide in the liquid phase is present as dissolved CO_2, H_2CO_3, HCO_3^-, and CO_3^{2-}. The fraction of bicarbonate and carbonate ions which are present at equilibrium depends on pH (Stumm and Morgan, 1970, pp. 118–129). At low pH and up to pH 5, most of the CO_2 is present as dissolved CO_2; for the pH range 7 to 9, the bicarbonate ion is strongly favored, and at pH 11 and above the carbonate ion is strongly favored, as can be deduced by considering the equilibrium constants:

$$K_1 = \frac{[H^+][HCO_3^-]}{[CO_2] + [H_2CO_3]} = 10^{-6.3} \tag{33}$$

$$K_2 = \frac{[H^+][CO_3^{2-}]}{[HCO_3^-]} = 10^{-10.3} \tag{34}$$

at 30°C (Pirt, 1975, pp. 76–77; Stumm and Morgan, 1970, pp. 148–150). Under gas/liquid-phase equilibrium conditions, the total quantity of carbon in the liquid phase is given by

$$C_T = [CO_2, H_2CO_3, HCO_3^-, CO_3^{2-}] = C_c\left[1 + \frac{K_1}{[H^+]} + \frac{K_1K_2}{[H^+]^2}\right] \tag{35}$$

increases with pH above 5 for fixed concentration of CO_2 in the gas phase as shown in Fig. 9.

Under conditions where the bicarbonate ion in the broth is being converted to H_2CO_3 and the H_2CO_3 to $CO_2 + H_2O$, the rates of chemical reaction should be considered. The reversible chemical reaction

$$H_2CO_3 \rightleftharpoons HCO_3^- + H^+$$

is very fast and may be at equilibrium. Kern (1960) gives a value of

$$K_a = \frac{[H^+][HCO_3^-]}{[H_2CO_3]} = 2.5 \times 10^{-4} \tag{36}$$

for the equilibrium constant at 25°C. The chemical reaction

$$H_2CO_3 \underset{k_2}{\overset{k_1}{\rightleftharpoons}} CO_2 + H_2O$$

is much slower. Kern (1960) reports average values of $k_1 = 20\ s^{-1}$ and $k_2 = 0.03\ s^{-1}$ at 25°C, with activation energies of 16.5 and 15 kcal/mol, respectively (Stumm and Morgan, 1970, p. 153). If both dissolved CO_2 and bicarbonate ion are present within the cell and both diffuse through

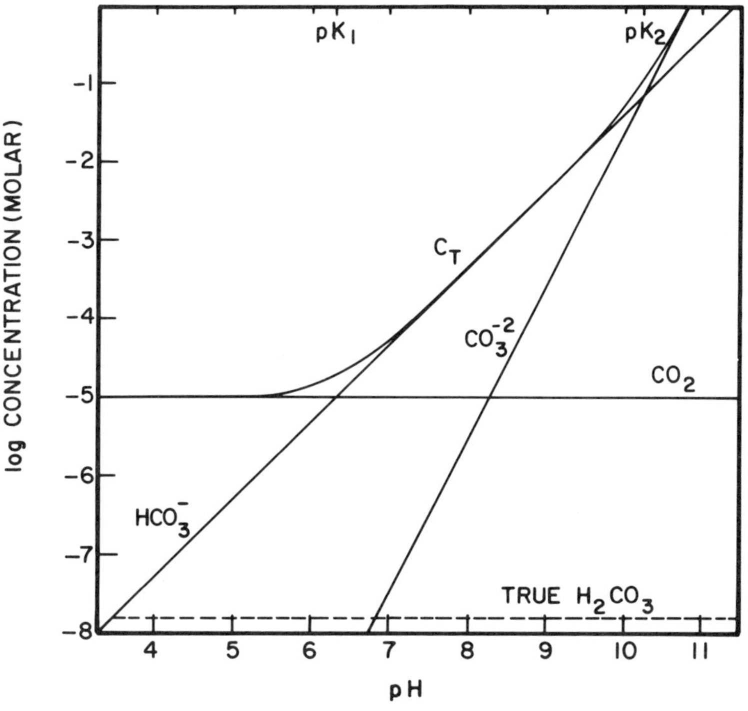

FIGURE 9 Aqueous carbonate equilibrium constant P_{CO_2}. Water is equilibrated with the atmosphere ($P_{CO_2} = 10^{-3.5}$ atm) and the pH is adjusted with strong base or strong acid. (From Stumm and Morgan, 1970, p. 127.)

the membrane rapidly, the bicarbonate ion will need to be converted to H_2CO_3 and the H_2CO_3 to CO_2 in the bulk liquid as a part of the CO_2 removal process.

It is clear from the published literature that rate processes associated with CO_2 removal from the broth may be of interest in some industrial fermentations (Ishizaki et al., 1971a–c, 1973; Barford and Hall, 1979; Yagi and Yoshida, 1977). Because of rate limitations, concentrations of CO_2, H_2CO_3, and HCO_3^- may substantially exceed the values that would be in equilibrium with the exhaust gas from the fermentor.

As microbial cultures are very sensitive systems, they pose high demands on the monitoring and control capabilities of a bioreactor. A schematic of the instrumentation available for on-line bioreactor monitoring is shown in Fig. 10. Off-line analysis by liquid chromatography or enzymatic assay methods can also be employed for the measurement of substances associated with the process of growth as substrates or products. Presently, on-line measurement is basically limited to off-gas analysis (O_2 and CO_2) by mass spectrometer or paramagnetic O_2 and infrared CO_2 analyzers, and to the concentrations of dissolved oxygen and carbon dioxide. These measurements can be utilized together with the elemental balances developed earlier to provide on-line estimates for the reactor state variables, which, in turn, can be used for control or other purposes.

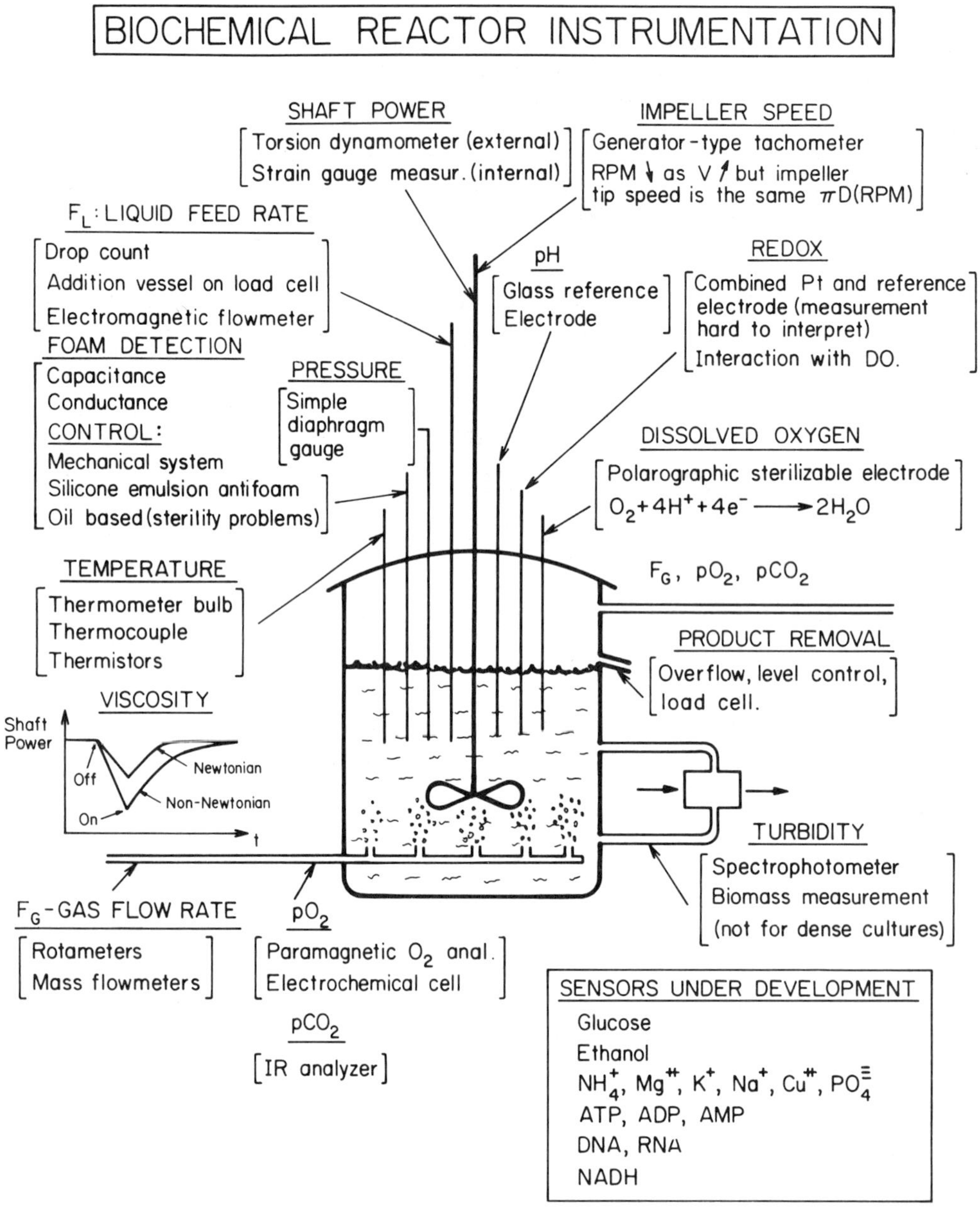

FIGURE 10 Schematic of biochemical reactor instrumentation. Temperature and pH controls are not shown. Viscosity measurements can be obtained from the slope of the shaft power versus time graph. Redox measurements are usually hard to interpret.

On-line, automated microanalytical methods, as well as various enzyme probes, are presently being developed for the purpose of providing the means for monitoring many other reactor variables on-line.

Bioreactor control is presently concerned with the control of the abiotic culture environment as the latter is described by the temperature, the pH, the dissolved O_2 concentration, and the presence of foam. All the foregoing parameters are maintained at set points which are optimal for the process. Manipulative variables are the coolant flow rate for the temperature, the rate of an acid or alkali addition for pH, agitation speed and gas flow rate for dissolved O_2, and mechanical foam breakers or agent addition for foam. In fed-batch operation the feed rate of nutrients is also controlled according to a predetermined schedule or on-line measurement or estimate. Certain interaction exists between the various control actions, but the multivariable control characteristics of these systems has not been studied. Most of the control is presently carried out by local analog controllers. However, as the reactor systems and processes grow more complex and the monitoring-estimation capabilities increase, more sophisticated control policies are expected to be introduced in bioreactor operation.

Other Bioreactor Designs

Although the stirred tank has been employed extensively in the past, it is considered inadequate for more demanding applications. The main factors against it are technical, but there are also economic and biological reasons for developing novel bioreactor schemes (Schügerl, 1982b). Schügerl (1982b) summarizes the main reasons for the introduction of new bioreactors as (a) the inability of stirred-tank scaleup to very large size because of design difficulties and heat removal problems, often accompanied by high power requirements and high energy and utility costs; (b) reduction of specific energy and capital costs; (c) avoidance of cell damage; and (d) increase of yields and reduction of substrate losses.

Information on these new systems is scarce. Most of the available knowledge comes from laboratory and pilot plant units and is summarized in a comprehensive review by Schügerl (1982b). The discussion below follows his paper.

The various types of bioreactors are classified into the following categories according to the method of energy introduction:

1. Reactors with mechanically moved internals
2. External liquid pumping
3. Reactors with gas compression

The tranditional submerged-culture stirred tanks belong to the first category (Fig. 11). They can have various types of agitation without (1.1) and with (1.2) loop. Other types are mechanically stirred loop reactors (1.3); self-aspirated aerated reactors without (1.4) and with (1.5) loop; mechanically stirred, self-aspirated, aerated, horizontal loop reactors (1.6); cascade reactors with rotating mixing elements (1.7), with axially moving mixing elements (1.8), or with pulsating liquid (1.9); and finally surface reactors such as thin-layer reactors (1.10), disk (1.11), and paddlewheel reactors (1.12).

Types of reactors in which the energy input is by means of an external pump are shown in Fig. 12. Such submerged-type reactors include plunging

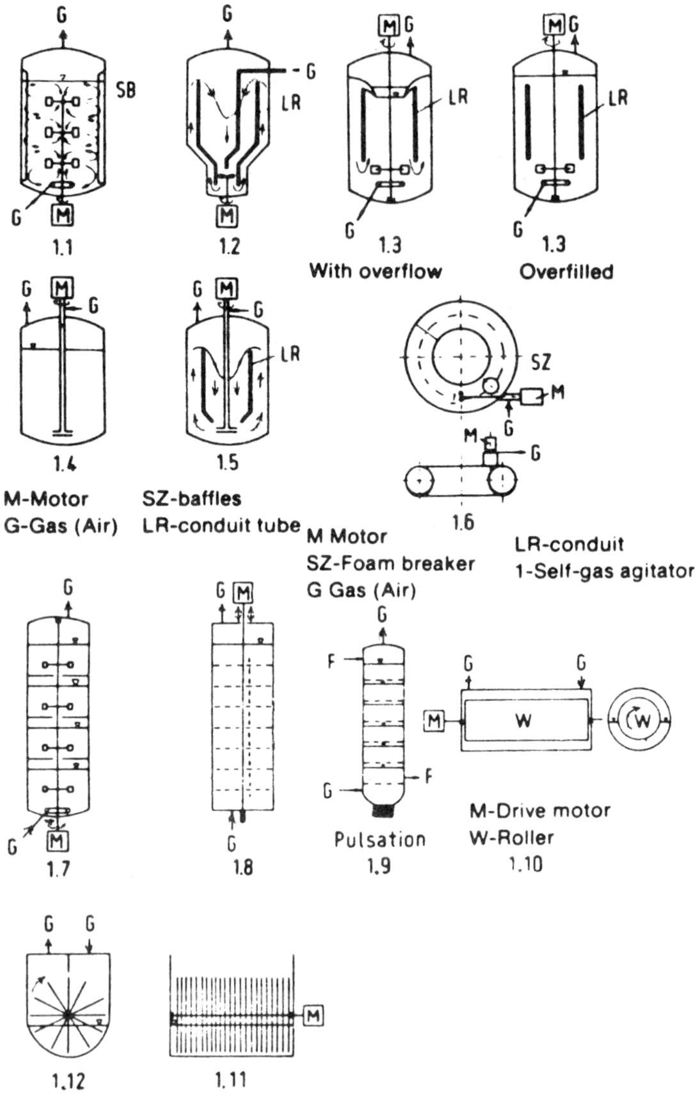

FIGURE 11 Biochemical reactors in which energy input is by mechanically moved internals. (From Schügerl, 1982b.)

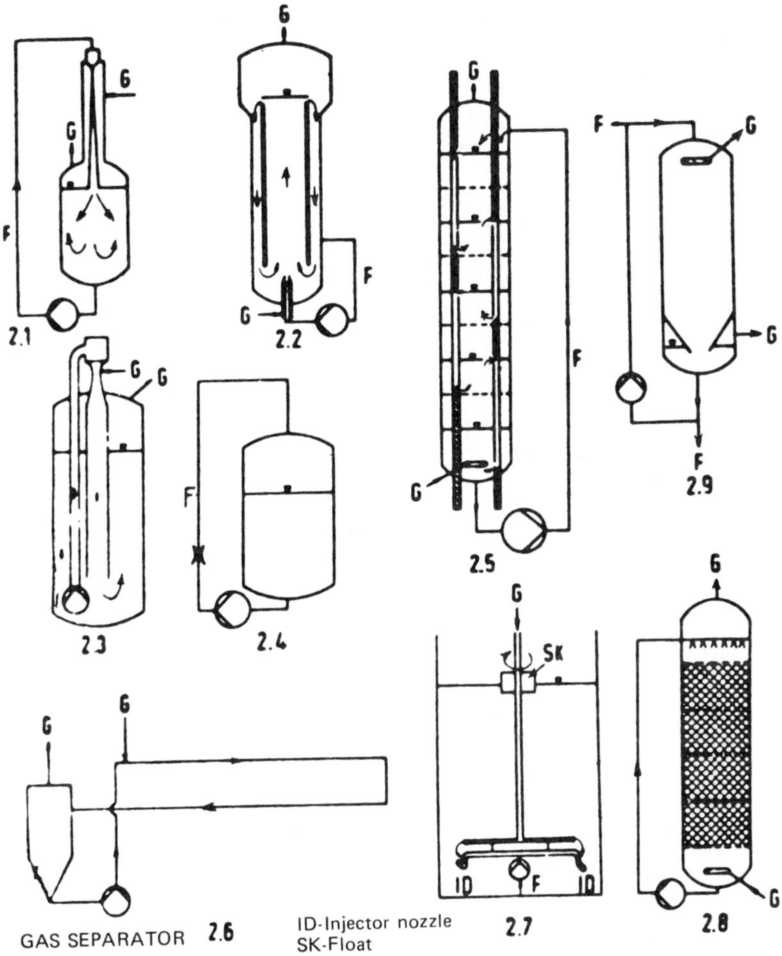

FIGURE 12 Biochemical reactors in which energy input is by circulation using an external pump. (From Schügerl, 1982b.)

jet reactors (2.1), jet-loop reactors (2.2), plunging-channel reactors (2.3), nozzle loop reactors (2.4), perforated-tray or sieve-tray cascade reactors (2.5), tubular-loop-reactors (2.6), reactors with rotating injector nozzles (2.7), packed-bed (percolating) reactors with counterflow of the phases (2.8), and bubble column, downflow reactors (2.9). Packed-bed reactors can also be classified as surface reactors.

The third category, shown in Fig. 13, includes submerged-type reactors in which the energy is introduced by means of compression. In this category, along with the simple single-stage bubble column (3.1), are the single-stage mammoth loops with centrally arranged draft tubes (3.2), externally attached tubular loops (3.3), vertical partition wall (3.4), and downflow loop (3.5). Between the loop reactors (3.2 and 3.4) and the downflow loop reactors (3.5) there is, of course, a certain overlapping. Bubble columns with static internals belong to this group, including concurrent bubble columns with stage-separating trays (3.6), concurrent

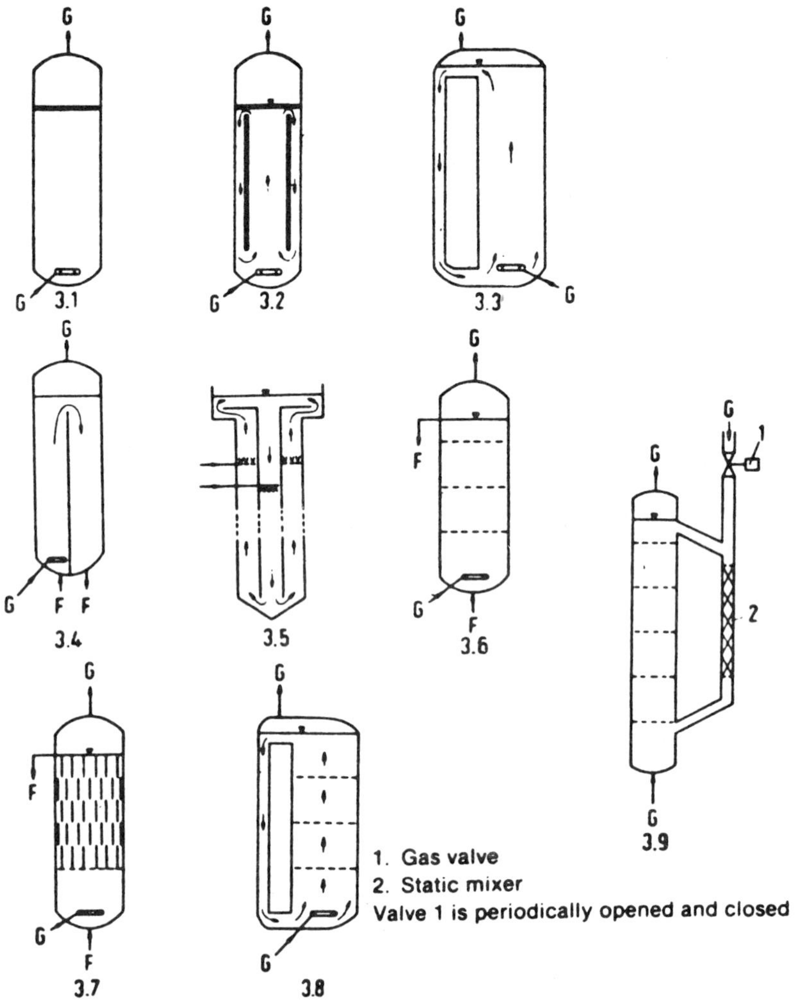

FIGURE 13 Biochemical reactors in which energy input is by compression. (From Schügerl, 1982b.)

bubble columns with static mixers (3.7) and mammoth loop reactors with stage-separating trays (3.8), and bubble columns with stage-separating trays with external tubular loops filled with static mixers and with pneumatically imposed liquid pulsation (3.9). Schügerl (1982b) also provides tables with the characteristics of these reactors and references for more details.

MASS TRANSFER IN AEROBIC FERMENTATIONS

In multiphase processes with chemical reactions, it is common for mass transfer to be an important consideration. The transfer of oxygen to aerobic cultures and the transfer of carbon dioxide from the culture to the exhaust gas are probably the two most important mass transfer considerations in

biochemical engineering. Oxygen transfer from the gas phase into the liquid phase has been the subject of much research (Moo Young and Blanch, 1981; Schügerl, 1981; Aiba et al., 1973, pp. 163–217; Bailey and Ollis, 1977, pp. 411–496). The high viscosity of many fermentation broths has been one of the important reasons for this research. With bacteria and yeasts, oxygen transfer from the gas phase to the liquid phase is often the most important mass transfer consideration; however, it is relatively easy to design industrial equipment in which the desired mass transfer rates are achieved. As the size of the microorganism increases, oxygen transfer from the liquid phase to the microorganism becomes more important. For mold fermentations, it is important to consider oxygen transfer both from the gas to the liquid and from the liquid to the microbial biomass. For mold pellets, transport within the pellet is also important.

The transport of CO_2 from the broth and the transport of O_2 to the broth should both be considered in the analysis of the mass transfer associated with an aerobic fermentation. The same gas flow rate is common to both of these processes. Furthermore, the pressure in the fermentor affects the dissolved O_2 and CO_2 concentrations, but in opposite directions. Increasing the pressure enhances oxygen transfer, but it negatively influences the rate of CO_2 release.

Mass Transfer Coefficient

Mass transfer in gas-liquid systems is treated in detail in Chapter 10. In addition, mass transfer in biochemical reactors is reviewed by Moo Young and Blanch (1981). In most fermentors, the mass transfer coefficient in the liquid phase depends on the Sherwood, Grashof, and Schmidt numbers because the density difference is the dominant force that causes fluid motion between the bubble and the liquid. For small rigid bubbles less than 2.5 mm diameter (Calderbank and Moo Young, 1961; Calderbank, 1958, 1959, 1967),

$$\mathrm{Sh} = 2.0 + 0.31\mathrm{Gr}^{1/3}\mathrm{Sc}^{1/3} \tag{37}$$

and for bubbles greater than 2.5 mm in diameter which are much less rigid,

$$\mathrm{Sh} = 0.42\mathrm{Gr}^{1/3}\mathrm{Sc}^{1/2} \tag{38}$$

where

$$\mathrm{Sh} = \frac{k_L d_b}{D} \tag{39}$$

$$\mathrm{Gr} = \frac{d_b^3 \rho_L g(\rho_L - \rho_g)}{\mu_c^2} \tag{40}$$

$$\mathrm{Sc} = \frac{\mu_c}{\rho D} \tag{41}$$

Equations (37) and (38) appear to be applicable to a wide variety of geometries and mixing configurations because the density difference causes fluid motion near the gas-liquid interface. Fukuda et al. (1978, 1979) have measured k_L in an airlift fermentor and a perforated plate column and found that Eq. (38) with a constant of 0.50 instead of 0.42 fits their data very well. This small difference may be due to some properties of the solutions which are not considered in the correlations and experimental error.

Equation (37) may also be used to estimate mass transfer coefficients to microbial biomass. For bacteria and yeast, the expression Sh = 2.0 provides a useful estimate of the mass transfer coefficient. This expression shows that mass transfer coefficients are relatively large for very small objects such as bacteria and yeasts. For mold pellets and other suspended particles with densities close to that of the continuous phase where agitation in a stirred mixing vessel creates the dominant force for relative fluid motion between the phases (Calderbank, 1967)

$$k_L = 0.13 \frac{[(P/V)\mu_c/\rho_c^2]^{1/4}}{Sc^{2/3}} \tag{42}$$

where P/V is the power per unit volume supplied by the agitator.

Surfactants are frequently present in fermentors. Surfactants provide a barrier for diffusion through the film at the interface; they also act to give the bubble more rigidity. The mass transfer coefficient may decrease significantly because of surfactants (Aiba et al., 1973, pp. 184–185).

Swarms of bubbles and non-Newtonian liquid phase flow behavior also may affect the value of the liquid-phase mass transfer coefficient. Corrections for these factors have been reviewed by Moo Young and Blanch (1981) and Schügerl (1981).

Interfacial Area

The gas-liquid interfacial area is an important variable in aerobic microbial growth processes. Because the interfacial area is affected by a large number of physical properties and other variables, very few correlations that accurately predict the interfacial area under fermentation conditions are available. However, knowledge of the effect of various physical properties, equipment designs, and operating conditions on the interfacial area is of considerable importance. The bubble size distribution and the gas-phase holdup or volume fraction are directly related to the interfacial area per unit volume of dispersion; that is,

$$a = \frac{6\varepsilon_g}{d_s} \tag{43}$$

where ε_g is the gas phase volume fraction and d_s the Sauter mean bubble diameter;

$$d_s = \frac{\Sigma f_i d_i^3}{\Sigma f_i d_i^2} \tag{44}$$

The bubble size frequency distribution, f_i, depends on the physical properties of the broth, equipment design, and operating conditions. Equation (43) shows that changes in gas holdup as well as changes in the bubble size distribution will result in changes in the interfacial area.

The bubble size distribution depends on the gas-liquid interfacial tension; liquid-phase density, viscosity, and ionic strength; gas flow rate and holdup; gas distributor; rate of turbulent energy dissipation; and equipment design. The chemical composition of the liquid phase also appears to be important. Calderbank (1967) measured bubble size distributions and interfacial areas in stirred tanks and sieve tray columns. Recently, Bhavaraju et al. (1978) investigated the effects of viscosity and turbulence on bubble size and interfacial area. Schügerl (1981) has reviewed the literature with respect to bubble size interfacial area and mass transfer in highly viscous broths. The sparger or gas distributor design and the gas flow rate through the sparger determine the initial bubble size distribution in the region of the sparger. For moderately high gas flow rates, Bhavaraju et al. (1978) found that

$$\frac{d_s}{d_o} = 3.23 Re_{OL}^{-0.1} Fr_o^{0.21} \tag{45}$$

where d_o is the nozzle diameter and

$$Re_{OL} = \frac{4\rho_L Q_o}{\pi d_o \mu_L} \tag{46}$$

is the orifice Reynolds number based on the volumetric flow rate through the orifice, Q_o, and orifice diameter, d_o. The orifice Froude number is

$$Fr_o = \frac{Q_o^2}{d_o^5 g} \tag{47}$$

The rate of coalescence influences the bubble size distribution. The ionic strength and chemical composition of the broth significantly influence the coalescence frequency and the bubble size distribution (Schügerl, 1981; Schügerl et al., 1978; Zlokarnik, 1978); however, equations that properly incorporate these variables into expressions for the interfacial area or bubble size distribution for fermentation broths are lacking. This is especially true for high-viscosity broths.

The level of turbulence greatly influences bubble breakup throughout the fermentation broth. As viscosity increases a point is reached where bubble breakup due to turbulence is unimportant. Bhavaraju (1978) have used the criterion that the scale of the energy containing eddies should be

200 times larger than the scale of the energy dissipating eddies for turbulent breakup to be important. They found that

$$d_s = 0.7 \frac{\sigma^{0.6}}{(P/V)^{0.4}\rho_L^{0.2}} \left(\frac{\mu_L}{\mu_G}\right)^{0.1} \quad (48)$$

for the Sauter mean diameter in a gas sparged vessel when turbulent breakup was important. For mixing vessels with agitators, Calderbank (1967) found that

$$d_s = 2.25 \frac{\sigma^{0.6}\rho_g^{0.4}}{(P/V)^{0.4}\rho_L^{0.2}} \left(\frac{\mu_L}{\mu_G}\right)^{0.25} \quad (49)$$

Schügerl (1981) found that in viscous fermentation broths small bubbles have too long a mean residence time in the broth. As equilibrium conditions are approached, the concentration driving force decreases and oxygen transfer from the small bubbles to the liquid slows considerably. Renewal of these small bubbles is very important in highly viscous broths. The vertical circulation pattern of the airlift fermentor and various multistage column designs may help achieve improved small bubble renewal rates; however, further investigation is needed. Because some of the interfacial area is not effective, Schügerl has chosen to show correlations in terms of the combined volumetric mass transfer coefficient.

A number of expressions have been presented for the gas holdup in aerated mixing vessels (Calderbank, 1967; Hughmark, 1980; Bailey and Ollis, 1977, p. 436; Aiba et al., 1973, p. 178). For bubble columns, Hikita et al. (1980) and Moo Young and Blanch (1981) have recently reviewed the literature on gas holdup. Schürgel (1981) has divided the holdup in viscous media into that associated with very small bubbles and that associated with intermediate and large bubbles. He has presented the results in a series of graphs.

The interfacial area may be obtained using expressions for the gas holdup, ε_g, and the Sauter mean bubble diameter, d_s, in Eq. (43) or by using correlations for interfacial area. Calderbank (1967), Hughmark (1980), Moo Young and Blanch (1981), and Joshi et al. (1982) have presented correlations for interfacial area, for example. However, because of the difficulty of obtaining correlations in which all the important variables are included, most fermentor design is carried out using correlations involving the volumetric mass transfer coefficient.

Volumetric Mass Transfer Coefficient

It is important to realize that two different volumetric mass transfer coefficients are commonly found in the biochemical engineering literature for gas-liquid mass transfer. The interfacial area per unit volume of gas-liquid dispersion, a, and the interfacial area per unit volume of liquid, a', are related; that is,

$$a'(1 - \varepsilon_G) = a \quad (50)$$

Equation (43) leads directly to the interfacial area per unit volume of dispersion, a; however, in studies where the liquid phase is completely mixed and the quantity of oxygen added to the liquid phase is measured, k_La' arises naturally, as shown in Eq. (25), for example. In reviewing the literature, one must be very careful to determine if the results presented are for k_La or k_La'.

The flow models selected for data analysis should also be examined in reviewing data in the literature. The correctness of estimates of the volumetric mass transfer coefficient depends on the selection of proper flow models which appropriately fit the experimental system when significant variations in concentration occur with position.

Bubble Columns

In bubble columns [Fig. 13, (3.1)] the energy to provide turbulence and bubble breakup is contained in the entering gas; that is, the power

$$P = Q\rho_G \frac{RT}{M} \ln \frac{p_1}{p_2} \tag{51}$$

is directly related to the volumetric gas flow rate Q. The volumetric mass transfer coefficient k_La has been correlated by Akita and Yoshida (1973, 1974), who found that

$$\frac{k_L d_t}{D} ad_t = 0.6 Sc^{0.5} Bo^{0.62} Ga^{0.31} \varepsilon_G^{1.1} \tag{52}$$

where the Bond number is

$$Bo = \frac{g d_t^2 \rho_L}{\sigma}$$

and the Galileo number is

$$Ga = \frac{g d_t^3}{\nu_L^2} = \frac{\rho_L^2 g d_t^3}{\mu_L^2}$$

These authors also correlated the gas holdup and found

$$\frac{\varepsilon_G}{(1 - \varepsilon_G)^4} = 0.20 Bo^{0.125} Ga^{0.083} Fr \tag{53}$$

where the Froude number is

$$Fr = \frac{\mu_G}{\sqrt{g d_t}}$$

with u_G the superficial gas velocity. These two correlations were tested in large-diameter bubble columns using CO_2 and water and it was found that the effect of column diameter disappeared above about 0.60 m. For diameters above 0.60 m, the authors and Kataoka et al. (1979) recommended using 0.6 m in Eqs. (52) and (53). The authors recommend using these correlations for superficial gas velocities up to 1500 m/h and gas holdup up to 30%. These equations are for dispersions in which the viscosity is sufficiently low for turbulence to play a dominant role in determining the bubble size distribution and interfacial area. Liquids up to 0.21 poise were considered by the authors.

Deckwer (1981) has tested Eq. (52) using some of his data and other literature data (see Fig. 14). He concludes that the correlation is good in some cases, but that with some spargers larger values of k_La are found. The effects of salts and organic substances on coalescence are not included in the correlation.

Schügerl et al. (1977) have investigated gas holdup, interfacial area, and mass transfer in bubble columns for a wide variety of model fermentation media using porous plate, perforated plate, injector nozzle, and ejector nozzle gas distributors. They point out that higher volumetric mass transfer coefficients than those predicted by Eq. (52) can be achieved by using distributors that produce a bubble size distribution in the entrance zone which is smaller than the equilibrium bubble size distribution associated with the turbulence due to the energy dissipation from Eq. (51). Additional energy is required at the distributor to produce these small bubbles. Under the conditions of their experiments, the holdup, interfacial area, and

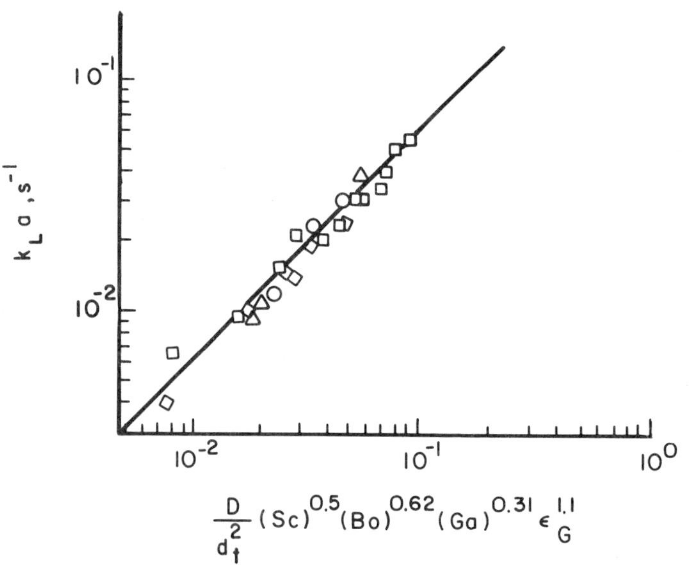

FIGURE 14 Volumetric mass transfer data of Towell et al. (1965) for d_t = 40.6 cm, △; Deckwer et al. (1974) for d_t = 20 cm, □; Kastanek (1977) for d_t = 100 cm, ◇; and Kataoka et al. (1979) for d_t = 550 cm, ○ plotted against the correlation of Akita and Yoshida (1973) using d_t = 60 cm for the larger-diameter columns.

volumetric mass transfer coefficients varied significantly with the various media selected. Deckwer (1981) points out that these results are not well correlated by Eq. (52).

Moo Young and Blanch (1981) have reviewed mass transfer correlations for bubble columns and found that volumetric mass transfer coefficients are directly proportional to the superficial gas velocity. These results are for low viscosity broths in which turbulence plays an important role in determining the bubble size distribution. As viscosity increases, a point is reached where the behavior at the gas distributor becomes more important (Bhavaraju et al., 1978). This occurs as the scale of the energy dissipating eddies becomes larger. At higher viscosities, the results for bubble columns are not in good agreement. Henzler (1981) correlated the data of Buchholz et al. (1978a,b) for a 3.9-m-high bubble column with 0.14-m diameter for apparent viscosities ranging from 46 to 217 mPa·s. The result is (see Fig. 15)

$$\frac{k_L a'}{u_G}\left(\frac{\bar{\nu}^2}{g}\right)^{0.333} Sc^{0.5} = 0.06\left[\frac{u_G}{(g\bar{\nu})^{0.333}}\right]^{-0.9} \qquad (54)$$

Deckwer et al. (1982) point out that values of $k_L a'$ predicted by this expression are at least five times larger than those predicted by Nakanoh and Yoshida (1980). The results of Deckwer et al. (1982) for aqueous carboxymethyl cellulose (CMC) solutions are correlated by the equation

$$k_L a = 2.08 \times 10^{-4} u_G^{0.59} \bar{\mu}^{-0.84} \qquad (55)$$

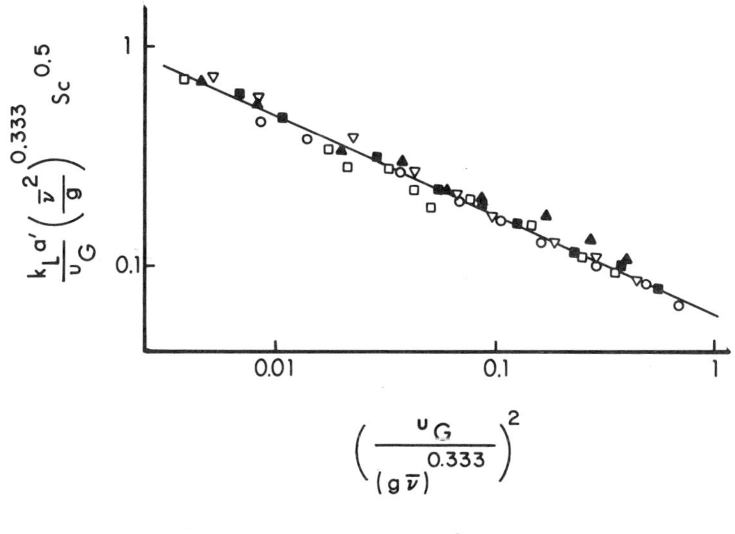

FIGURE 15 Bubble column data of Buchholz et al. (1978a,b) as correlated by Henzler (1981) to obtain Eq. (54). For aqueous solutions with the following apparent viscosities: ○, 46 to 65 mPa·s; ■, 64 to 95 mPa·s; ▽, 92 to 140 mPa·s; ▲, 109 to 172 mPa·s; and □, 130 to 217 mPa·s.

where u_G is mean superficial gas velocity in cm/s and $\overline{\mu}$ is the apparent viscosity in Pa·s. This correlation predicts results which are smaller than those of Nakanoh and Yoshida (1980). There are significant differences which may explain these results. Buchholz et al. (1978a) used a sintered plate sparger with a mean hole diameter of 0.0175 mm, while Deckwer et al. (1982) used a sintered plate with 0.20-mm holes. The results of Deckwer et al. (1982) and the results of Voigt et al. (1980) for CMC solutions are in fairly good agreement for gas velocities above 2 cm/s. Plates of 0.5-, 1.0-, and 3.0-mm hole diameter were used by Voigt et al. (1980) in a multistage column; however, only one stage was used for the experiments with CMC. Henzler (1980) correlated the results of Voigt et al. (1980) to obtain:

$$k_La' = Cu_G^{0.6}\nu^{-1.03} \tag{56}$$

or

$$\frac{k_La}{u_G}\left(\frac{\nu^2}{g}\right)^{0.333} Sc^{0.5} = 0.075\left[\frac{u_G}{(g\nu)^{0.333}}\right]^{-0.4} \tag{57}$$

The results of Deckwer et al. (1982) and those of Buchholz et al. (1978a) show that the design of the gas distributor may be important for viscous solutions.

The measure of Bhavaraju et al. (1978) for the existence of an internal subrange in the energy spectrum of turbulence is

$$\frac{1}{\eta_T} = \frac{d_t\rho_L^{0.5}}{\mu_L^{0.75}}\left(\frac{P}{V}\right)^{0.25} \tag{58}$$

where $1/\eta_T$ should be larger than 200 based on their data. Table 1 shows values of $1/\eta_T$ for some of the experiments of Deckwer et al. (1982). Some of the values of $1/\eta_T$ at high gas superficial velocity and low CMC concentration are greater than 200. This suggests that turbulence should be important. However, at 1.3 and 1.6 wt % CMC, $1/\eta_T$ is less than 200 at all superficial velocities investigated. The form of Eq. (55) suggests that turbulent forces are important in the experiments of Deckwer et al. (1982). The exponent of 0.59 for superficial velocity shows that the power input is important; however, the value of the exponent is smaller than the value of about 1.0 for low-viscosity broths. The dependence on viscosity differs from that found by Akita and Yoshida in Eq. (52). Equation (55) may be contrasted with Eq. (54) where the distributor appears to have significantly affected the bubble size distribution. In Eq. (54), k_La' is proportional to the 0.1 power of superficial velocity.

Viscoelastic fluids in bubble columns have been considered by Hecht et al. (1980) and Nokanoh and Yoshida (1980). The correlation of Nakanoh and Yoshida is

$$\frac{k_La'd_t^2}{D} = 0.09Sc^{0.5}Bo^{0.75}Ga^{0.39}Fr^{1.0}(1 + 0.13D_e^{0.55})^{-1} \tag{59}$$

TABLE 1 Ratio of the Scale of Energy-Containing Eddies to Energy-Dissipating Eddies for Bubble Column Experiments of Deckwer et al. (1982)

Superficial gas velocity (cm/s)	CMC (wt %)	Ratio $1/\eta_T$
0.4	1	139
1.0	1	185
2.0	1	229
3.0	1	260
4.0	1.3	167
0.4	1.6	55
2.0	1.6	103
4.0	1.6	135

where the Deborah number is De = $u_b\lambda/d_s$, where u_b is the bubble rise velocity, λ the characteristic time of the viscoelastic material, and d_s the Sauter mean bubble diameter. In Eq. (59), k_La is proportional to the superficial gas velocity to the 1.0 power. This may be due to the fact that Nakanoh and Yoshida used a single nozzle that was 4 mm in diameter in their 14.55-cm-diameter column. Under these conditions, almost all of the surface area must be generated by turbulent forces.

Airlift Columns

Airlift tower fermentors (several different designs are shown in Fig. 13) have received considerable attention during the last 15 years. The vertical liquid circulation that occurs because of the upflow and downflow sections of the airlift column helps to keep surface-active agents distributed throughout the broth. The cocurrent upflow section is similar to a bubble column in some respects; however, the cocurrent liquid flow influences the two-phase flow behavior. The transition from bubble flow occurs at a much larger superficial gas velocity in an airlift fermentor then in a bubble column (Hatch, 1973). Because of this, the power dissipated per unit volume in the upflow section of the airlift can be considerably greater under bubble regime operating conditions compared to an aeration tower. Blenke (1979) has reviewed his work and that of others in airlift reactors, jet loop reactors, and propeller loop reactors; he has shown that forced liquid circulation may be used to increase substantially gas holdup and interfacial area. Correlations for liquid circulation rate, gas holdup, and volumetric oxygen transfer coefficient have been presented by Bello et al. (1981); for oxygen transfer,

$$k_La = b\frac{(P_G/V)^{0.8}}{(1 + A_d/A_r)^{1.2}} \tag{60}$$

where b is 0.000507 for water and 0.000502 for 0.15 M NaCl solution, P_G is the power (watts), V is the total reactor dispersion volume (m^3), and A_d/A_r is the ratio of the area of the downcomer to the riser. Blenke (1979) found that the interfacial area per unit volume of liquid, a', increased with power per unit volume of liquid in the bubble flow regime; to correlate his data in the bubble flow regime, exponents ranging from 0.85 to 1.0 are needed on power per unit volume. At superficial gas velocities above about 30 cm/s, the dependence on power per unit volume decreases; this may be due to slug flow. Orazem and Erickson (1979) found that the oxygen transfer coefficient increased linearly with superficial gas velocity in a two-stage airlift tower. Hatch (1973) correlated the volumetric mass transfer coefficient for the upflow region, head region, and downflow region. At the same sparger gas flow rate, the highest volumetric mass transfer coefficients are in the head region; those in the draft tube are slightly higher than those in the downflow region. Moo Young and Blanch (1981) have reviewed some of their own work and that of Lin (1976) and found a slightly greater dependence on gas superficial velocity with values of the exponent ranging from 1.06 to 1.29.

Hatch (1973) has shown that volumetric mass transfer coefficients in the head region are approximately twice as large as those for an aeration tower under similar sparger gas flow rates. In the downflow section the gas holdup must be less than that in the upflow section because the liquid circulation rate depends on the difference in dispersion density between the two regions. The most common operating condition results in a liquid circulation rate sufficient to entrain small bubbles but not large ones in the liquid downflow. Under fermentation conditions where coalescence rates are relatively small, the bubbles in the downflow appear to be segregated with little coalescence occurring. Very little attention has been directed to research in the downflow section of the airlift fermentor (Orazem et al., 1979, 1980). Blenke (1979) and Fujui et al. (1980) have investigated some related cocurrent downflow phenomena.

The effect of viscosity on oxygen transfer in airlift biological reactors has received very little study. Several investigators have used airlift fermentors and obtained good results under viscous conditions (Konig et al., 1982; Malfait et al., 1981; Barker and Worgan, 1981); however, further research is needed to better appreciate the oxygen transfer limitations associated with high viscosity in an airlift fermentor. Because higher values of power per unit volume may be obtained in an airlift compared to a bubble column, the airlift column should be superior to a bubble column under high-viscosity conditions.

The effect of the ratio of the downflow area to upflow area on oxygen transfer has been investigated by Hatch (1973, 1975). Based on the performance ratio, the optimum area ratio is about 0.8.

Columns with Motionless Mixers

As shown in Fig. 13, multistage columns with sieve plates, columns with Koch motionless mixers, and columns with other stationary internals that

promote mixing have been investigated (Schügerl, 1982b). Kitai et al. (1969, 1972) showed that volumetric mass transfer coefficients in multistage perforated plate columns were larger than those in aeration towers at high superficial gas velocities. The column internals help to maintain bubble flow conditions by removing large bubbles and slugs. K. Hsu et al. (1975, 1977) investigated oxygen transfer in multistage columns with sieve trays and with Koch motionless mixers; they found that Koch mixers and perforated plates gave fairly similar results. For both, the performance ratio was higher when column internals were employed compared to an aeration tower. These results are very much in agreement with those given above for bubble columns, where it was pointed out that the gas distributor could positively influence the volumetric mass transfer coefficient.

Columns with Koch motionless mixers and liquid circulation have been considered by Fan et al. (1975), H. Hsu et al., (1975a), and Wang (1977). Wang (1977) has developed correlations for the volumetric mass transfer coefficient which includes the superficial liquid velocity u_L, as well as the superficial gas velocity u_G; he found for 7 AY whole Koch mixers that

$$k_L a = 0.120 u_L^{0.624} \frac{u_G}{1.99 u_G + 47.1} \tag{61}$$

where $k_L a$ is in units of s^{-1} and the superficial velocities are in cm/s. Vertical circulation within the fermentor is often needed to distribute surface-active agents which are carried to the liquid surface by air bubbles. Foaming problems were encountered in early designs which did not allow for vertical circulation. Considerable research and the commercial application of columns with motionless mixers has been reported in Latvia. This work has been reviewed by Viesturs et al. (1980, 1981). Better oxygen transfer efficiencies are reported for 100-m^3 columns with stationary internals.

Schügerl and coworkers have investigated oxygen transfer in multistage columns with perforated plates at low viscosity (Voigt and Schügerl, 1979, 1981) and high viscosity (Schügerl, 1981; Schügerl et al., 1982a; Voigt et al., 1980; Hecht et al., 1980). The optimum spacing between stages depends on the coalescence rate of the bubbles. For rapid rates of coalescence, the bubble size distribution generated by the distributor needs to be renewed more frequently, and a shorter distance between plates is suggested (Voigt and Schügerl, 1979). For slow rates of coalescence, bubble size distributions generated by the distributor are retained for a longer time; under these conditions the contribution of the distributor to oxygen transfer can be greater. Under actual fermentation conditions larger volumetric mass transfer coefficients are found for the multistage column (Voigt and Schügerl, 1981).

Appropriate correlations that account for the contributions of both the distributor and bulk turbulence to the interfacial area and volumetric mass transfer coefficient are lacking. The effect of coalescence must be considered in order to relate the bubble breakup at a gas distributor to the volumetric mass transfer coefficient throughout the fermentor. The power dissipated at the gas distributor should also be considered. Zlokarnik (1978, 1979) has correlated the volumetric mass transfer coefficient with the power of a liquid jet of an injector nozzle. He has found (Zlokarnik, 1978) that

$$k_L a' \left(\frac{\nu}{g^2}\right)^{0.33} = 0.00023 \left[\frac{P/V_L}{\rho(\nu g^4)^{0.333}}\right]^{0.8} \tag{62}$$

or that the volumetric mass transfer coefficient is proportional to the power per unit volume to the 0.8 power. Blenke (1979) has also shown that power supplied at the nozzle of his jet loop reactor should be considered in correlating interfacial area; that is,

$$a = 5.4 \times 10^3 u_G^{0.4} \left(\frac{P_L}{V}\right)^{0.66} \tag{63}$$

where a is in m^{-1}, u_G is in m/s, and P_L/V is kW/m^3 of dispersion volume. Zlokarnik (1979) has shown how the coalescence rate influences the volumetric mass transfer coefficient in large-scale systems with injector nozzles.

Mechanically Agitated Systems

Mechanically agitated fermentors (Fig. 11) with air sparged into the vessel below the agitator are widely used in industry. Volumetric mass transfer coefficients for such systems have been reviewed by Moo Young and Blanch (1981), Joshi et al. (1982), and Schügerl (1981). Moo Young and Blanch (1981) and Joshi et al. (1982) have presented tables of correlations for the volumetric mass transfer coefficient from the literature. The agitation power per unit volume, the superficial gas velocity, and the physical properties of the broth affect the volumetric mass transfer coefficient. Miller (1974), Zlokarnik (1978), Yagi and Yoshida (1975), and Henzler (1980), for example, have investigated volumetric mass transfer coefficients in agitated vessels. Yagi and Yoshida developed the correlation

$$\frac{k_L a' d_I^2}{D} = 0.060 \left(\frac{d_I^2 N \rho_L}{\mu_a}\right)^{1.5} \left(\frac{d_I N^2}{g}\right)^{0.19} \left(\frac{\mu a}{\rho D}\right)_L^{0.5} \times \left(\frac{\mu_a u_G}{\sigma}\right)^{0.6} \left(\frac{N d_I}{u_G}\right)^{0.32} [1 + 2(\lambda N)^{0.5}]^{-0.67} \tag{64}$$

using both Newtonian and non-Newtonian fluids. The dimensionless groups are the modified Sherwood number, impeller Reynolds number, Froude number, Schmidt number, gas flow number, aeration number, and Deborah number. Henzler (1980) developed the correlation

$$\frac{k_L a' V_L}{Q} Sc^{0.3} = 0.045 \left[\frac{P_G}{Q \rho_L (g\nu)^{0.667}}\right]^{0.5} \tag{65}$$

using data for water, glucose, and glycerin. For aqueous CMC solutions, he found that

$$\frac{k_L a' V_L}{Q} Sc^{0.3} = 0.082 \left[\frac{P_G}{Q \rho_L (g\nu)^{0.667}} \right]^{0.6} \tag{66}$$

Zlokarnik (1978) has examined the effect of coalescence on the volumetric mass transfer coefficient in agitated vessels.

DESIGN OF BIOLOGICAL REACTORS

There are many factors that must be considered in the design of biological reactors. Providing the desired environment for the microbial population at reasonable cost may be accomplished with an available fermentor or by design of a new fermentor. When a new fermentor is designed, many more decisions must be made. Decisions that may be made with an existing mechanically agitated fermentor include the agitation rate, the gas flow rate, the operating pressure in the fermentor, the liquid volume, the nutrient composition of the feed, pH, temperature, method of foam control, and mode of operation (batch, fed batch, or continuous). For a new fermentor, equipment design must also be considered. In both cases, the rate of oxygen transfer in aerobic fermentations is usually one of the important concerns.

Economic considerations must be carefully considered in the design process. Some of the important costs include the capital costs for equipment, the cost of organic substrates and other nutrients, energy cost for aeration, cooling water or refrigeration costs, product separation costs, waste treatment costs, and personnel costs. An economic optimum exists because of the following considerations: The rate of the fermentation process should be high to reduce fermentor capital costs; however, aeration and cooling costs increase as the rate increases. Microbial growth is usually most efficient at relatively high specific growth rates. Cell and product concentrations should be as large as possible to reduce separation costs; however, aeration, cooling, and nutrient costs (yield) must also be considered. An optimum gas flow rate and operating pressure need to be selected in the design process. For a given production rate of the product, the net present value of the capital and operating costs should be minimized.

Mechanically Agitated Fermentors

The design of mechanically agitated fermentors is treated by Aiba et al. (1973, p. 171), Bailey and Ollis (1977, p. 443), and Wang et al. (1979, p. 157). Relationships between the power number, $P_{no} = P/\rho_c N^3 d_I^5$, and the impeller Reynolds number for agitated liquids are presented in Fig. 16. Recently, Schügerl (1981) has presented similar graphs for non-Newtonian viscous broths. Hughmark (1980) correlated the gassed to ungassed power ratio and found that for flat-blade turbine impellers,

$$\frac{P_G}{P} = 0.10 \left(\frac{Q}{NV_L} \right)^{-0.25} \left(\frac{N^2 d_I^4}{g d_i V_L^{0.667}} \right)^{-0.20} \tag{67}$$

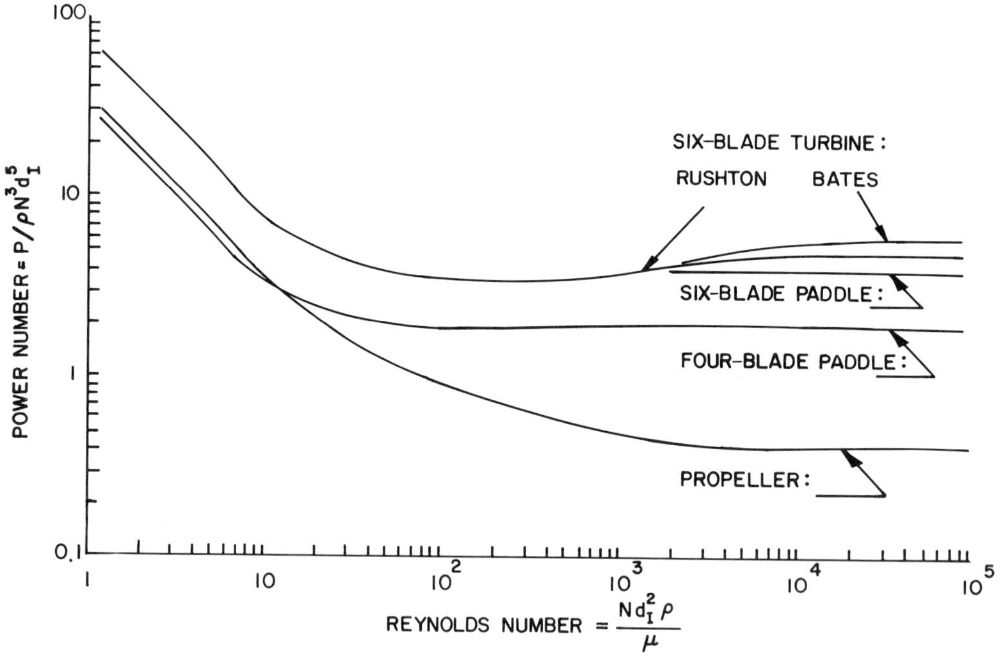

FIGURE 16 Correlation of impeller Reynolds number and power number for different impellers in ungassed liquids. (From Wang et al., 1979, p. 160.)

Other correlations have been presented by Luong and Volesky (1979), Hassan and Robinson (1977), and Oyama and Endoh (1955). Luong and Volesky (1979) and Schügerl (1981) have investigated non-Newtonian CMC solutions, while Haasan and Robinson investigated electrolytic solutions. Joshi et al. (1982) has recently reviewed the available correlations and has recommended the correlation of Hughmark (1980), which is presented above.

For a mechanically agitated fermentor, the optimal design for continuous production of microbial biomass is considered to illustrate the economic decisions that must be made. The specific growth rate μ, the biomass concentration in the fermentor X, the agitation speed N, and the gas flow rate Q, may be considered as independent decision variables which are to be selected to minimize the net present value of capital and operating costs per unit of biomass produced. The biomass energetic yield η is related to the specific growth rate,

$$\frac{1}{\eta} = \frac{1}{\eta_{max}} + \frac{m_e}{\mu} \tag{68}$$

and the mass yield is $Y_s = \sigma_s \gamma_s \eta / \sigma_b \gamma_b$ in units of grams of biomass per gram of organic substrate. The oxygen transfer rate may be found from the available electron balance to be

$$\frac{\mu X}{Y_o} = \frac{\sigma_b \gamma_b (1 - \eta)\mu X}{48\eta} \tag{69}$$

where Y_o is grams of biomass produced per gram mole of oxygen consumed, and $\mu X/Y_o$ is gram moles of O_2 transferred per unit volume of liquid per unit time. The heat that needs to be removed is directly related to the oxygen transfer rate; that is, Q_o = 112.8 kJ/equivalent of oxygen or 451.2 kJ/g mole of oxygen; thus

$$\frac{\mu X}{Y_Q} = 451.2\,\frac{\mu X}{Y_o} = \frac{Q_o \sigma_b \gamma_b (1 - \eta)\mu X}{12\eta} \tag{70}$$

For Monod kinetics, the exit organic substrate concentration is

$$S = \frac{\mu K_s}{\mu_{max} - \mu} \tag{71}$$

and from the yield equation $X = Y_s(S_o - S)$,

$$S_o = \frac{X}{Y_s} + S \tag{72}$$

The computational sequence is as follows for a desired productivity rate P_1, where

$$FX = P_1 \tag{73}$$

where F is the liquid volumetric flow rate. First, values of the independent decision variables μ and X are selected. Then η and Y_s are found using Eq. (68). Next, the oxygen transfer rate

$$k_L a'(C_o^* - C_o) = \frac{\mu X}{Y_o} \tag{74}$$

is calculated using Eq. (69). Equation (70) is used to estimate the heat to be removed per unit volume per unit time. The volumetric liquid flow rate may be found using Eq. (73). Since the specific growth rate equals the dilution rate, the required liquid volume is

$$V_L = \frac{F}{\mu} \tag{75}$$

Equations (71) and (72) may be used to find the organic substrate concentrations S and S_o.

From the required liquid volume, the total dispersion volume may be estimated using the relationship

$$V = \frac{V_L}{1 - \varepsilon_G} \tag{76}$$

where the gas holdup ε_G is to be estimated initially and calculated more precisely later from the expression (Hughmark, 1980)

$$\varepsilon_G = 0.74\left(\frac{Q}{NV_L}\right)^{0.5}\left(\frac{N^2 d_I^4}{g d_i V_L^{0.667}}\right)^{0.5}\left(\frac{d_b N^2 d_I}{\sigma V_L^{0.667}}\right)^{0.25} \tag{77}$$

where the bubble diameter d_b may be estimated from available correlations, such as Eq. (49).

The dimensions of the fermentor may now be selected; the height-to-diameter ratio, the number of impellers, and the type of impeller are independent variables. The gas flow rate and the agitation speed may also be viewed as independent variables which may be optimally selected based on economics. The gas flow rate must be more than sufficient to supply the oxygen needs of the fermentation; that is, for air,

$$Q_{in} \geqslant \frac{RT\mu XV_L}{0.21 p_{in} Y_o}$$

For 30% consumption of oxygen, for example,

$$Q_{in} = \frac{RT\mu XV_L}{(0.3)(0.21) p_{in} Y_o} \tag{78}$$

The optimization of oxygen transfer efficiency in the agitated vessel for the required oxygen transfer rate involves selecting the height-to-diameter ratio, number of impellers, type of impeller, gas flow rate, and agitation speed. For standard impeller designs, this fixes the impeller diameter and allows one to find the power number using Fig. 3 and the calculated impeller Reynolds number. The ungassed power P may then be found from the power number. The gassed power may be found from Eq. (67). Equation (64) or (65) may be used to calculate the volumetric mass transfer coefficient, $k_L a'$. Equation (74) may now be used to calculate the driving force $(C_o^* - C_o)$, where the dissolved oxygen concentration C_o is usually chosen to be small; that is, about 0.03 m mol/L or less. Henry's law relates C_o^* to the total pressure p and the gas-phase oxygen mole fraction y; that is,

$$yp = HC_o^* \tag{79}$$

The mean mole fraction of oxygen in the exit gas is found from the gas-phase oxygen balance,

$$y = \frac{n_{in} y_{in} - \mu XV_L / Y_o}{n} \tag{80}$$

For a completely mixed gas phase, the mole fraction oxygen in the fermentor equals the exit mole fraction, and the required total pressure may be

calculated using Eq. (79). This is the average pressure within the dispersion.

The capital costs, operating costs, and net present value of the costs per unit of productivity may be calculated. The separation costs depend primarily on the liquid flow rate F and biomass concentration X. The fermentor costs depend on the total dispersion volume V, the pressure p, and the number of agitators. The size of the agitator motor and the power costs associated with it may be determined. The size of the gas compressor and the power requirements may be determined. The cost of heat transfer may be determined from the cooling requirements. The cost of organic substrate may be determined using the quantity of substrate FS_0 per unit time.

The optimization problem involves selecting the values of the independent variables so as to minimize the net present value of the costs per unit of productivity. Thus values of specific growth rate, biomass concentration, height-to-diameter ratio, number of impellers, type of impeller, impeller speed, and gas flow rate may be varied in the search for an economic optimum. Of these variables, the height-to-diameter ratio, number of impellers, and type of impeller would probably be selected based on prior experience.

The example above is presented to illustrate how the many variables are related to each other in fermentor design. The simplest flow models have been used by assuming complete mixing in both phases. For the gas phase a flow model composed of two completely mixed tanks in series may be more appropriate (Joshi et al., 1982). The assumption of complete mixing of the liquid phase is appropriate. For simplicity, the possibility of CO_2 inhibition is not included.

Bubble Columns

When a bubble column is designed to be used as an aerobic fermentor, several important considerations should be taken into account. The superficial gas velocity should be selected such that the bubble two-phase flow regime is present. The pressure gradient due to the liquid head should be taken into account in the design. Axial dispersion should be considered. The effect of segregation may be important because small bubbles have longer residence times than larger bubbles and they also have greater interfacial area per unit volume than large bubbles. The gas-phase molar flow rate may change with position due to gas absorption; however, in carbohydrate fermentations the respiratory quotient is nearly unity and the molar flow rate is approximately constant.

The design of bubble columns has been considered by Deckwer (1976, 1977) and Deckwer et al. (1974, 1977). Shah et al. (1978) have reviewed axial dispersion or backmixing in bubble columns and other gas-liquid reactors. Deckwer (1976, 1977) has presented mathematical models which take into account axial dispersion and the pressure gradient. Deckwer recommends including the effect of the pressure gradient when the pressure at the bottom of the column is more than 30% larger than the pressure at the top above the liquid phase. Deckwer (1977) found that the effective value of the volumetric mass transfer coefficient was smaller in large industrial-scale bubble columns compared to smaller columns. Segregation could account for this. Despite the limitations of Deckwer's success, the approach used has considerable merit; however, computationally, numerical solution of the system of differential equations is required.

A simpler, less exact approach is to consider dispersion of oxygen in the axial direction in the liquid phase and either a complete mixing or plug-flow model for the gas phase. The complete-mixing model is a conservative one since the exit gas-phase mole fraction of oxygen is used for the entire vessel. If gas-phase segregation is important, some bubbles will have a lower mole fraction of oxygen than the value predicted by the complete-mixing model. Since the liquid-phase superficial velocity is zero (or close to zero), the oxygen balance is taken to be

$$D_L \frac{d^2C_o}{dx^2} + k_La'(C_o^* - C_o) - \frac{\mu X}{Y_o} = 0 \tag{81}$$

where X is the height above the bottom of the column:

$$C_o^* = \frac{p_T y}{H_o}[1 + \alpha(1 - z)] \tag{82}$$

and

$$\alpha = \frac{\rho_L g(1 - \varepsilon_G)L}{p_T} \tag{83}$$

is the ratio of the liquid head to the pressure at the top of the column, p_T. The boundary conditions are

$$\frac{dC_o}{dx} = 0 \qquad \text{at } x = 0 \text{ and } x = L$$

A dimensionless distance $\bar{x} = x/L$ and dimensionless concentration $\bar{c} = C_o/C_{oT}^*$ may be introduced, where $C_{oT}^* = p_T y/H_o$ is the liquid-phase oxygen concentration in equilibrium with the gas phase at the top of the column. The dimensionless groups

$$B_1 = \frac{D_L}{L^2 k_L a'} \tag{84}$$

and

$$B_2 = \frac{\mu X}{Y_o k_L a' C_T^*} \tag{85}$$

appear in the dimensionless expression

$$B_1 \frac{d^2\bar{c}}{d\bar{x}^2} - \bar{c} = B_2 - 1 - \alpha + \alpha\bar{x} \tag{86}$$

which has the solution

$$\bar{c} = A_1 e^{-M_1\bar{x}} + A_2 e^{M_1\bar{x}} + (1 + \alpha - B_2) - \alpha\bar{x} \tag{87}$$

where

$$A_1 = \frac{\alpha(e^{M_1} - 1)}{M_1(e^{-M_1} - e^{M_1})} \tag{88}$$

$$A_2 = \frac{\alpha(e^{-M_1} - 1)}{M_1(e^{-M_1} - e^{M_1})} \tag{89}$$

and

$$M_1 = \sqrt{\frac{L^2 k_L a'}{D_L}} = \frac{1}{\sqrt{B_1}} \tag{90}$$

Figure 17 shows the effect of dispersion in the axial direction on the dimensionless dissolved oxygen profile along the length of the column for $\alpha = 1$ and $B_2 = 1$. Figure 17 is also a plot of

$$\frac{\bar{c}}{\alpha} - \frac{1 - B_2}{\alpha} = \frac{A_1}{\alpha} e^{-M_1\bar{x}} + \frac{A_2}{\alpha} e^{M_1\bar{x}} + 1 - \bar{x} \tag{91}$$

for all values of α and B_2 in which $\bar{c} > 0$ for $\bar{x} = 1$. Note that zero-order kinetics are used for the dependence of μ on $\bar{c}$; that is, as long as $\bar{c} > 0$, the specific growth rate does not depend on the dissolved oxygen concentration. The column should be designed such that $\bar{c} > 0$ at the top of the column.

Values of the axial dispersion coefficient D_L have been correlated by Deckwer et al. (1974, 1982) and others (Shah et al., 1978). Deckwer et al. (1974) found that

$$D_L = 2.7 d_t^{1.4} u_G^{0.3} \tag{92}$$

where units of centimeters and seconds are used.

A design methodology analogous to that used for mechanically agitated vessels may now be visualized. Four independent variables may be optimally selected based on economic considerations. The specific growth rate, μ, and biomass concentration, X, may be selected as independent variables if continuous biomass production is again considered as the example of interest. Complete mixing is assumed for the organic substrate and biomass in the column; this should be appropriate unless it is necessary to consider the dispersion of the organic substrate fed to the column.

After values of μ and X are assumed, values of η and $\mu X/Y_o$ may be found from Eqs. (68) and (69). The rate of heat removal may be found

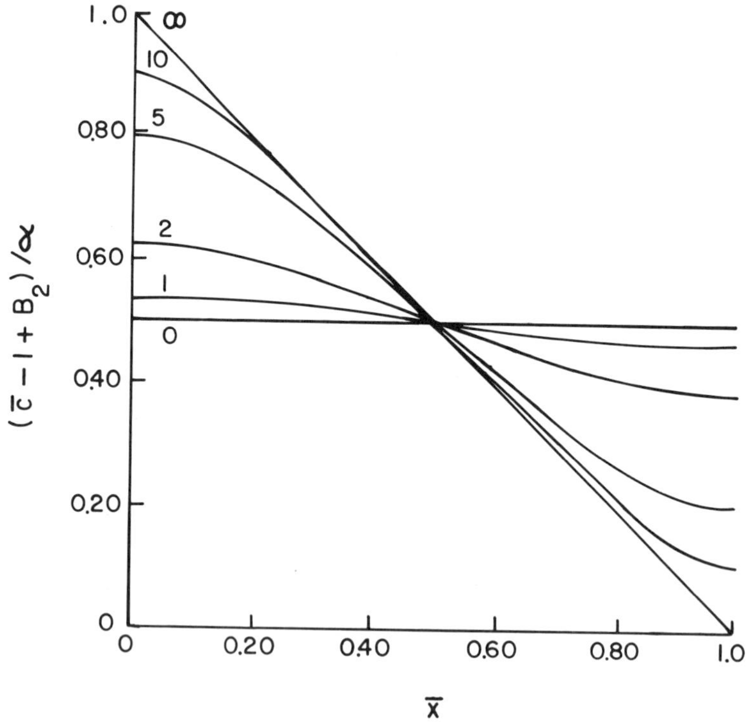

FIGURE 17 Effect of axial dispersion and pressure gradient on dimensionless dissolved oxygen concentration in a bubble column. The parameter is $\sqrt{L^2 k_L a'/D_L}$. Results are limited to conditions where $\bar{c} \geq 0$.

using Eq. (70) and the organic substrate concentrations from Eqs. (71) and (72). The liquid flow rate may be found from Eq. (73) and the liquid volume from Eq. (75). The fraction of oxygen in the gas phase which is consumed and the superficial gas velocity may be selected as the other two independent variables. The oxygen balance for a constant molar flow rate, n_T,

$$n_T(y_{in} - y) = \frac{\mu X V_L}{Y_o} \tag{93}$$

may be used to find n_T. A trail-and-error iteration is required at this point since the pressure, column diameter, and gas superficial velocity are all related to the molar gas flow rate. Assuming a pressure, the column diameter may be found from the relationship

$$p_T u_G \pi d_t^2 = 4 n_T R T$$

The gas holdup may now be estimated from an appropriate correlation such as Eq. (53). The total dispersion volume may be found from Eq. (76) and

the height L of the column may also be found. The volumetric mass transfer coefficient k_La' may be found from an appropriate correlation, such as Eq. (52). The axial dispersion coefficient D_L may be found using Eq. (92). The value of $\bar{c}$ at $\bar{x} = 1$ may be found from Eq. (87) and compared to the desired value. The trial-and-error iteration involves adjusting the pressure until a desirable value of $\bar{c}$ is obtained. The optimal values of μ, X, u_G, and y are to be selected based on economic considerations.

Airlift Fermentors

Airlift fermentors have a well-defined loop for liquid circulation which is driven by the gas flow. The liquid circulation rate depends on the design of the airlift column. The extent to which the gas is separated from the liquid at the top of the column prior to entering the downflow section can significantly affect the liquid circulation rate. In the work of Hatch (1973) and Gutierrez and Erickson (1977), for example, the dispersion enters the downflow section with more gas than can be carried downward by the liquid. Some of the large gas bubbles must rise against the downflow of the dispersion and escape. It is necessary for this to happen because the average gas velocity is much smaller than the liquid velocity in the downflow section, and only a fraction of the gas in the dispersion entering the downflow region can be carried downward. For this disengagement process to occur, the actual liquid velocity in the downflow section must be larger than the rise velocity of small bubbles and smaller than the rise velocity of large bubbles; that is, it should be about 10 to 30 cm/s. Hatch (1973) reported superficial liquid velocities ranging from 10 to 30 cm/s and Orazem et al. (1979) reported values ranging from 17 to 21 cm/s. Hatch (1973) points out that the liquid circulation velocity increases with gas superficial velocity in the upflow section until the liquid circulation velocity is limited by the gas bubble disengagement process at the top of the downflow section. Further increases in gas superficial velocity lead to liquid recirculation within the upflow section similar to that found in bubble columns.

Hsu and Dudukovic (1980) and Merchuk and Stein (1981b) have investigated the liquid velocity in airlift columns, but in both cases the design was such that the downflow section was almost free of gas. Merchuk and Stein found that the actual liquid velocity could be correlated with the superficial gas velocity to the 0.41 power. They observed liquid velocities above 100 cm/s at superficial gas velocities above 20 cm/s. They found that their data could be correlated by the expression of Zuber and Findley (1965); that is,

$$\frac{u_G}{\varepsilon_G} = C_A u_M + v_D \tag{94}$$

where $C_A = 1.03$ and $v_D = 0.33$ m/s. That is, $v_D = 0.33$ m/s is approximately the gas velocity relative to the superficial velocity of the mixture, u_M.

Hsu and Dudukovic (1980) presented an overall momentum balance for the airlift loop and presented a correlation for the two-phase friction factor. Their results do not correlate well with the popular Lockhart-Martinelli correlation (Perry and Chilton, 1973, pp. 5-41).

Hatch (1969) found that the optimum ratio of upflow to downflow area was about 0.83. Since the upflow liquid velocity increases with superficial gas velocity, one would expect the optimum area ratio to decrease as superficial gas velocity increases.

Hatch (1973) has measured the axial dispersion coefficient D_L in the airlift upflow section and found that D_L is constant at about 58 cm^2/s for low superficial gas velocities where liquid recirculation within the upflow section is small. Hatch concludes that the liquid phase may be modeled using a plug-flow model for both the upflow and downflow sections. Hatch recommends a complete-mixing model for the head region. Merchuk and Stein (1981a) have modeled the airlift fermentor using plug flow in the upflow and downflow regions and complete mixing in the head region. Ho et al. (1977) have modeled the airlift fermentor using a tanks-in-series model in the upflow and downflow regions and a completely mixed tank for the head region.

The design methodology for the airlift fermentor is analogous to that described above for bubble columns. The fraction of oxygen consumed, the feed gas superficial velocity in the draft tube, and the ratio of upflow to downflow area may be selected as independent variables together with μ and X. A trial-and-error calculation is required to find the operating pressure that results in an acceptable dissolved oxygen concentration. A simulation procedure analogous to that of Ho et al. (1977) or Merchuk and Stein (1981a) is needed to determine the dissolved oxygen level in the various parts of the airlift fermentor for a particular operating pressure.

The available correlations for use in the design of airlift fermentors are not as complete and extensive as those for agitated tanks and bubble columns. Further research is needed to develop such correlations. Airlift fermentors have been compared to bubble columns and agitated tanks and the mass transfer efficiency of airlift columns has been demonstrated (Orazem and Erickson, 1979; Hatch, 1975).

A variety of equipment arrangements involving vertical circulation due to airlift designs have been reported. The length of the vertical divider ranges from the short downcomers of Viesturs et al. (1980, 1981) to the very long downcomers of the deep shaft fermentors (Takamatsu et al., 1981a,b; Kubota et al., 1978; Hosono et al., 1979; Hines et al., 1975). Further research is needed to develop optimal designs for biological applications of these systems.

Tower Reactors with Motionless Mixers

The design methods of this chapter can be applied to various tower fermentors with internals such as Koch motionless mixers, sieve trays, and downcomers. Appropriate flow models should be selected for the proposed system and the effect of pressure should be considered. Correlations for the volumetric mass transfer coefficient and the gas holdup are needed. The work of Viesturs et al. (1980, 1981) has resulted in the industrial-scale application of such systems; they have compared the performance of several large-scale systems experimentally. Further research is needed to optimize mathematically such designs.

ACKNOWLEDGMENT

This work was partially supported by the National Science Foundation.

NOTATION

a	stoichiometric coefficient in Eq. (1); interfacial area
a'	interfacial area per unit volume of liquid
A	area
b	stoichiometric coefficient in Eq. (1)
B_1	dimensionless group defined by Eq. (84)
B_2	dimensionless group defined by Eq. (85)
Bo	Bond number ($gd_t^2 \rho_L/\sigma$)
$\bar{c}$	dimensionless dissolved oxygen, C_o/C^*_{OT}
C_c	molar carbon dioxide concentration in liquid phase
C_o	molar oxygen concentration in liquid phase
C^*_{OT}	$p_T y/H_O$; equilibrium oxygen concentration in liquid at the top of the column
d_b	bubble diameter
d_i	width of impeller blade
d_I	impeller diameter
d_o	orifice diameter
d_s	Sauter mean diameter [see Eq. (44)]
d_t	tower diameter
D	dilution rate F/V_ℓ (hr^{-1}); molecular diffusivity
D_L	dispersion coefficient for liquid phase
De	Deborah number ($u_b\lambda/ds$)
F	volumetric flow rate of liquid phase
Fr	$u_G/\sqrt{gd_t}$; Froude number
Fr_o	$Q_o^2/d_o^5 g$; orifice Froude number
g	gravitational acceleration
Ga	gd_t^3/u_L^2; Galileo number
Gr	$d_b^3 \rho_L g(\rho_L - \rho_g)/\mu_L^2$
H_c	Henry's law constant for CO_2
H_o	Henry's law constant for oxygen
H^+	hydrogen ion concentration
I	inhibitor concentration
k_L	liquid-phase mass transfer coefficient
K_i	saturation constant for product inhibition
K_s	saturation constant of Monod kinetic model
m_D	maintenance coefficient based on CO_2, g mol CO_2(g biomass)·h
m_e	maintenance coefficient, g equiv available electrons substrage/g

m_o	maintenance coefficient based on oxygen, g mol O_2(g biomass^{-1} · h^{-1}
m_s	maintenance coefficient, g substrate/(g biomass)·h
M	molecular weight
n_T	molar flow rate of gas
N	impeller speed
p_c	partial pressure of CO_2
p_o	partial pressure of oxygen
p_T	total pressure
P	product concentration, g/L
P	power
P_1	productivity in Eq. (73)
P_G	gassed power
P_{no}	power number, $P/\rho_c N^3 d_I^3$
Q	volumetric gas flow rate
Q_{CO_2}	specific rate of CO_2 evolution, g mol/(g biomass)·h
Q_{O_2}	specific rate of oxygen utilized, g mol O_2/(g biomass)·h
Q_o	heat evolved per equivalent of available electrons transferred to oxygen, kJ/equiv
R	gas constant
Re_{OL}	$4\rho_L Q_o/\pi d_o \mu_L$; orifice Reynolds number
Sc	$\mu_L/\rho_L D$; Schmidt number
Sh	$k_L d_b/D$; Sherwood number; use d_s for bubble size distribution
T	absolute temperature
u_b	bubble rise velocity
u_G	gas-phase superficial velocity
u_L	liquid-phase superficial velocity
u_M	superficial velocity of gas-liquid mixture
V	dispersion volume
V_g	gas volume
V_ℓ	liquid volume
x	height from bottom of the column
X	biomass concentration, g/L
y	mole fraction oxygen in gas phase
y_b	biomass carbon yield in Eq. (1)
y_d	carbon dioxide yield in Eq. (1)
y_p	product yield in Eq. (1)

Y_{ATP}^{max}	true biomass yield, g biomass/g mol ATP
Y_D	biomass yield based on CO_2, g biomass/g mol CO_2
Y_D^{max}	true biomass yield based on CO_2, g biomass/g mol CO_2
Y_o	biomass yield based on oxygen, g biomass/g mol oxygen
Y_o^{max}	true biomass yield based on oxygen, g biomass/g mol oxygen
$Y_{P/ATP}^{max}$	true product yield based on ATP, g product/g mol ATP
$Y_{P/D}^{max}$	true product yield based on CO_2, g product/g mol CO_2
$Y_{P/O}^{max}$	true product yield based on oxygen, g product/g mol oxygen
$Y_{P/S}^{max}$	true product yield based on substrate, g product/g substrate
Y_Q	biomass yield based on heat evolved, g biomass/kJ
Y_s	biomass yield based on organic substrate, g biomass/g substrate
Y_s^{max}	true biomass yield based on substrate, g biomass/g substrate

Greek Letters

α	pressure ratio defined by Eq. (83)
α	rate coefficient for product formation, g product/(g biomass)·h
β	rate coefficient for product formation, g product/g biomass
γ_b	reductance degree of biomass [see Eq. (4)]
γ_p	reductance degree of product [see Eq. (7)]
γ_s	reductance degree of substrate [see Eq. (6)]
ΔH_c°	standard heat of combustion per gram atom of carbon
ΔH_r°	heat of reaction per gram atom of substrate carbon
ε	fraction of available electrons transferred to oxygen
ε_G	volume fraction of dispersed gas phase
η	biomass energetic yield
η_{max}	true biomass energetic yield
η_T	Kolmogoroff scale [see Eq. (58)]
λ	characteristic time of viscoelastic material
μ	specific growth rate, h^{-1}
μ_a	apparent viscosity
μ_G	viscosity of gas
μ_L	viscosity of liquid

μ_{max}	maximum specific growth rate
ν	kinematic viscosity
$\overline{\nu}$	apparent kinematic viscosity
ξ_p	product energetic yield
ξ_p^{max}	true product energetic yield
ρ	density
σ	surface tension
σ_b	weight fraction carbon in biomass
σ_p	weight fraction carbon in product
σ_s	weight fraction carbon in substrate

Subscripts

b	biomass
c	continuous phase; CO_2
g	gas phase
G	gas phase
in	inlet condition
L	liquid phase
ℓ	liquid phase
o	oxygen; orifice
p	product
s	organic substrate

Superscript

*	equilibrium concentration

REFERENCES

Aiba, S., and S. Nagai, *Biochemical Engineering Reaction Rate Theory*, Scientific Technology, Kanazawa City, Japan (1975). (In Japanese.)

Aiba, S., M. Shoda, and M. Nagatani, Kinetics of product inhibition in alcohol fermentation, Biotechnol. Bioeng., *10*, 845 (1968).

Aiba, S., A. E. Humphrey, and N. F. Millis, *Biochemical Engineering*, 2nd ed., Academic Press, New York (1973).

Akita, K. and F. Yoshida, Gas holdup and volumetric mass transfer coefficients in bubble columns, Ind. Eng. Chem. Process Des. Dev., *12*, 76 (1973).

Akita, K. and F. Yoshida, Bubble size, interfacial area, and liquid phase mass transfer coefficient in bubble columns, Ind. Eng. Chem. Process Des. Dev., *13*, 84 (1974).

Amerine, M. A., Wine, Sci. Am., *211*(2), 46 (1964).

Andrews, J. F. and S. P. Graef, Dynamic modeling and simulation of the anerobic digestion process, Adv. Chem., *105*, 126 (1971).

Avgerinos, G. C., H. Y. Fang, I. Biocic, and D. I. C. Wang, A novel single step microbial conversion of cellulosic biomass to ethanol, Adv. Biotechnol., *2*, 119 (1981).

Bailey, J. E. and D. F. Ollis, *Biochemical Engineering Fundamentals*, McGraw-Hill, New York (1977).

Barford, J. P. and K. J. Hall, Investigation of the significance of a carbon and redox balance to the measurement of gaseous metabolism of *Saccharomyces Cerevisiae*, Biotechnol. Bioeng., *21*, 609 (1979).

Barker, S. A. and P. J. Somers, Enzyme electrodes and enzyme based sensors, Top. Enzyme Ferment. Biotechnol., *2*, 120 (1978).

Barker, T. W. and J. T. Worgan, The application of airlift fermentors to the cultivation of filamentous fungi, Eur. J. Appl. Microbiol. Biotechnol., *13*, 77 (1981).

Barker, G. C. J., A. Margaritis, and M. A. Bergougnou, Fluidization principles and applications in biotechnology, Adv. Biotechnol., *1*, 635 (1981).

Bekers, M. E., *Biotechnology of Microbial Synthesis*, Zinatne, Riga, Latvia (1980). (In Russian.)

Bello, R. A., C. W. Robinson, and M. Moo Young, Mass transfer and liquid mixing in external circulation loop contractors, Adv. Biotechnol., *1*, 597 (1981).

Bhavaraju, S. M., T. W. F. Russell, and H. W. Blanch, The design of gas sparged devices for viscous liquid systems, AIChE J., *24*, 454 (1978).

Blenke, H., Loop reactors, Adv. Biochem. Eng., *121* (1979).

Broun, G. B., G. Manecke, and L. B. Wingard, *Enzyme Engineering*, Vol. 4, Plenum Press, New York (1978).

Buchholz, H., R. Huchholz, J. Lucke, and K. Schürgel, Bubble swarm behavior and gas absorption in non-Newtonian fluids in sparged columns, Chem. Eng. Sci., *33*, 1061 (1978a).

Buchholz, H., R. Buchholz, J. Lucke, and K. Schügerl, Gassing of highly viscous culture media in bubble column reactors, Chem.-Ing.-Tech., *50*, 227 (1978b).

Calam, C. T., N. Driver, and R. H. Bowers, Studies in the production of penicillin, respiration and growth of *Penicillium chysogenum* in submerged culture, in relation to agitation and oxygen transfer, J. Appl. Chem., *1*, 209 (1951).

Calderbank, P. H., Physical rate processes in industrial fermentations, Trans. Inst. Chem. Eng., *36*, 443 (1958).

Calderbank, P. H., Physical rate processes in industrial fermentations, Part II, Trans. Inst. Chem. Eng., *37*, 173 (1959).

Calderbank, P. H., Mass transfer in frementation equipment, Biochem. Biol. Eng. Sci., *1*, 102 (1967).

Calderbank, P. H. and M. B. Moo Young, The continuous phase heat and mass transfer properties of dispersions, Chem. Eng. Sci., *16*, 39 (1961).

Cama, F. J. and V. H. Edwards, Single cell protein from chemical wastes: the growth of *Candida utilis* of sodium acetate, J. Ferment. Technol., *48*, 787 (1970).

Chang, Y. C. and A. S. Myerson, Growth models of the continuous bacterial leaching of iron pyrite by *Thiobacillus ferroxidans*, Biotechnol. Bioeng., *24*, 889 (1982).

Chen, A. K., J. A. Startzmann, and C. C. Liu, Potentiometric Quantitation of gylcerol using immobilized glycerol dehydrogenase, Biotechnol. Bioeng., *24*, 971 (1982).

Cheryan, M. and W. D. Deeslie, Production of protein hydrolyzates in ultrafilteration enzyme reactors, in *Ultrafiltration Membranes and Applications*, A. R. Cooper, ed., Plenum Press, New York (1980), p. 591.

Constantinides, A. and V. R. Rai, Application of the continuous maximum principle to fermentation processes, Biotechnol. Bioeng. Symp., *4*, 663 (1974).

Contantinides, A., J. L. Spenser, and E. L. Gaden, Optimization of batch fermentation processes: I. Development of mathematical models for batch penicillin fermentations, Biotechnol. Bioeng., *12*, 803 (1970a).

Constantinides, A., J. L. Spenser, and E. L. Gaden, Optimization of batch fermentation process: II. Optimum temperature profiles for batch penicillin fermentations, Biotechnol. Bioeng., *12*, 1081 (1970b).

Cooney, C. L., D. I. C. Wang, and R. I. Mateles, Measurement of heat evolution and correlation with oxygen consumption during microbial growth, Biotechnol. Bioeng., *11*, 269 (1969).

Deckwer, W. D., Non-isobaric bubble columns with variable gas velocity, Chem. Eng. Sci., *31*, 309 (1976).

Deckwer, W. D., Adsorption and reaction of isobutene in sulfuric acid: III. Chem. Eng. Sci., *32*, 51 (1977).

Deckwer, W. D., Physical transport phenomena in bubble column bioreactors: I. Gas-liquid mass transfer, Adv. Biotechnol., *1*, 465 (1981).

Deckwer, W. D., R. Buckhart, and G. Zoll, Mixing and mass transfer in tall bubble columns, Chem. Eng. Sci., *29*, 2177 (1974).

Deckwer, W. D., U. Allenbach, and H. Bretschneider, Adsorption and reaction of isobutene in sulfuric acid: II, Chem. Eng. Sci., *32*, 43 (1977).

Deckwer, W. D., K. N. Tien, A. Schumpe, and Y. Serpemen, Oxygen mass transfer into aerated CMC solutions in a bubble column, Biotechnol. Bioeng., *24*, 461 (1982).

Deeslie, W. D. and M. Cheryan, A CSTR-hollow fiber system for continuous hydrolysis of proteins, factors affecting the long term stability of the reactor, Biotechnol. Bioeng., *24*, 69 (1982).

Deindoerfer, F. H., Fermentation kinetics and model processes, Adv. Appl. Microb., *2*, 321 (1960).

Dunlop, E. H. and S. J. Ye, Micromixing in biological reactors, Annual AIChE Meet., Los Angeles (1982).

Erickson, L. E., Energetic efficiency of biomass and product formation, Biotech. Bioeng., *21*, 725 (1979).

Erickson, L. E., Analysis of micribial growth and product formation with nitrate as nitrogen source, Biotechnol. Bioeng., *22*, 1929 (1980a).

Erickson, L. E., Biomass elemental composition and energy content biotechnol. Bioeng., *22*, 451 (1980b).

Erickson, L. E., Growth and production energetic yields of *Rhodopseudomonas sphaeroides* in dark and aerobic chemostat cultures, J. Ferment. Technol., *58*, 53 (1980c).

Erickson, L. E., Energetic yields associated with hydrocarbon fermentations, Biotechnol. Bioeng., *23*, 793 (1981).

Erickson, L. E. and J. L. Hess, Analysis of growth and polysaccharide yields in chemostat cultures of *Rhizobium trifolii*, Ann. N.Y. Acad. Sci., *369*, 81 (1981).

Erickson, L. E. and A. E. Humphrey, Growth models of cultures with two liquid phases: II. Pure substrate in dispersed phase, Biotechnol. Bioeng., *11*, 467 (1969a).
Erickson, L. E. and A. E. Humphrey, Growth models with two liquid phases: III. Continuous cultures, Biotechnol. Bioeng., *11*, 489 (1969b).
Erickson, L. E., A. E. Humphrey, and A. Prokop, Growth models of cultures with two liquid phase: I. Substrate dissolved in dispersed phase, Biotechnol. Bioeng., *11*, 449 (1969).
Erickson, L. E., L. T. Fan, P. S. Shah, and M. S. Chen. Growth models of cultures with two liquid phase: IV. Cell adsorption, drop-size distribution and batch growth, Biotechnol. Bioeng., *12*, 713 (1970).
Erickson, L. E., T. Nakahara, and A. Prokop, Growth in cultures in two liquid phases: hydrocarbon uptake and transport, Process Biochem. *10*(5), 9 (1975).
Erickson, L. E., I. G. Minkevich, and V. K. Eroshin, Application of mass and energy balance regularities in fermentation, Biotechnol. Bioeng., *20*, 1595 (1978a).
Erickson, L. E., S. E. Selga, and U. E. Viesturs, Application of mass and energy balance regularities to product formation, Biotechnol. Bioeng., *20*, 1623 (1978b).
Eriksson, R. and T. Holme, The use of microcalorimetry as an analytical tool for microbial processes, Biotechnol. Bioeng. Symp., No. 4, 581 (1973).
Fan, L. T. and E. H. Hsu, Semifluidized bed bioreactor, Adv. Biotechnol., *1*, 663 (1981).
Fan, L. T., H. H. Hsu, and K. B. Wang, Mass transfer coefficient and pressure drop data of two phase oxygen water flow in bubble column packed with static mixers, J. Chem. Eng. Data, *20*, 26 (1975).
Ferrer, A. and L. E. Erickson, Evaluation of data consistency and estimation of yield parameters in hydrocarbon fermentations, Biotechnol. Bioeng., *21*, 2203 (1979).
Ferrer, A. and L. E. Erickson, Data consistency, yield, maintenance, and hysteresis in batch cultures of *Candida lipolytica* cultured on *n*-hexadecane, Biotechnol. Bioeng., *22, 421 (1980)*.
Fredrickson, A. G., Behavior of mixed cultures of microorganisms, Annu. Rev. Microbial., *31*, 63 (1977).
Fridrickson, A. G., and G. Stephanopoulos, Microbial competition, Science, *213*, 972 (1981).
Fredrickson, A. G. and H. M. Tsuchiya, Microbial kinetics and dynamics, *Chemical Reactor Theory. A Review*, L. Lapidus and N. R. Amundson, eds., Prentice-Hall, Englewood Cliffs, N.J. (1977), p. 405.
Fujui, K., M. Takaine, H. Kubota, and Y. Miyaji, Flow and oxygen transfer in gas-liquid downflow, J. Chem. Eng. Jpn., *13*, 188 (1980).
Fukuda, H., T. Shiotani, W. Ohada, and H. Morikawa, Oxygen transfer in a new tower bioreactor containing a draft tube and perforated plates, J. Ferment. Technol., *56*, 619 (1978).
Fukuda, H., T. Nakashima, and H. Morikawa, Bubble size and mass transfer coefficient in a tower bioreactor containing a draft tube and perforated plates, J. Ferment. Technol., *57*, 554 (1979).
Gaden, E. L., Fermentation kinetics and productivity, Chem. Ind., *154* (1955).
Gaden, E. L., Fermentation process kinetics, J. Biochem. Microb. Technol. Eng., *1*, 413 (1959).

Gale, E. F. and H. M. R. Epps, The effect of the pH of the medium during growth on the enzymatic activities of bacteria (*Escherichia coli* and *Micrococcus lysodeikticus*) and the biological significance of the changes produced, Biochem. J., *36*, 600 (1942).

Gormely, L. S., D. W. Duncan, R. M. R. Branion, and K. L. Pinder, Continuous culture of *Thiobacillus ferroxidance* on a zinc sulfide concentrate, Biotechnol. Bioeng., *17*, 31 (1975).

Gould, R. F., Anaerobic biological treatment processes, Adv. Chem., *105* (1971).

Gutierrez, J. R. and L. E. Erickson, Hydrocarbon uptake in hydrocarbon fermentations, Biotechnol. Bioeng., *19*, 1331 (1977).

Gutierrez, J. R. and L. E. Erickson, Continuous culture of *Candida lipolytica* on *n*-hexadecane, Biotechnol. Bioeng., *20*, 1833 (1978).

Harder, A. and J. A. Roels, Application of simple structured models in bioengineering, Adv. Biochem. Eng., *21*, 55 (1982).

Harrison, D. E. F., Mixed cultures in industrial fermentation processes, Adv. Appl. Microbiol., *24*, 129 (1978).

Hardwood, J. H. and S. J. Pirt, Quantitative aspects of growth of the methane oxidizing bacterium *Methylococcus capsulatus* on methane in shake flask and continuous chemostat culture, J. Appl. Bacteriol., *35*, 597 (1972).

Hassan, I. T. M. and C. W. Robinson, Stirred-tank mechanical power requirement and gas holdup in aerated aqueous phase, AIChE J., *23*, 48 (1977).

Hatch, R. T., Mass transfer characteristics of an airlift fermentator, M.S. thesis, Massachusetts Institute of Technology, (1969).

Hatch, R. T., Experimental and theoretical studies of oxygen transfer in the airlift fermentor, Ph.D. dissertation, Massachusetts Institute of Technology (1973).

Hatch, R. T., Fermentor design, in *Single Cell Protein*, Vol. 2, MIT Press, Cambridge, Mass. (1975), p. 46.

Hecht, V., J. Voigt, and K. Schügerl, Absorption of oxygen in countercurrent multistage bubble columns: III. Viscoelastic liquids, Chem. Eng. Sci., *35*, 1325 (1980).

Heijnen, J. J. and J. A. Roels, A macroscopic model describing yield and maintenance relationships in aerobic fermentation processes, Biotechnol. Bioeng., *23*, 739 (1981).

Henzler, H. J., Aeration of relatively viscous liquids, Chem.-Ing.-Tech., *52*, 643 (1980).

Henzler, H. J., Gas-liquid contacting in highly viscous liquids, Adv. Biotechnol., *1*, 497 (1981).

Hikita, H., S. Asai, K. Tanigawa, K. Segawa, and M. Kitao, Gas holdup in bubble columns, Chem. Eng. J., *20*, 59 (1980).

Hines, D. A., M. Bailey, J. C. Ousby, and F. C. Roesler, The ICI deep shaft aeration process for effluent treatment, Inst. Chem. Eng. Symp. Ser., *41*, D1 (1975).

Hixon, A. W. and E. L. Gaden, Oxygen transfer in submerged fermentation, Ind. Eng. Chem., *42*, 1792 (1950).

Ho, C. S., L. E. Erickson, and L. T. Fan, Modeling and simulation of oxygen transfer in airlift fermentors, Biotechnol. Bioeng., *19*, 1503 (1977).

Hosono, Y., K. Fujie, and H. Kubota, Operational characteristics evaluation of liquid pump type deep shaft aerator, J. Chem. Eng. Jpn., *12*, 136 (1979).

Hsu, Y. C. and M. P. Dudukovic, Gas holdup and liquid recirculation in gas lift reactors, Chem. Eng. Sci., *35*, 135 (1980).

Hsu, H. H., K. B. Wang, and L. T. Fan, Oxygen transfer and absorption efficiencies in bubble columns, Water Sewage Works, *122*(2), 34 (Feb. 1975a).

Hsu, K. H., L. E. Erickson, and L. T. Fan, Oxygen transfer to mixed cultures in tower systems, Biotechnol. Bioeng., *17*, 499 (1975b).

Hsu, K. H., L. E. Erickson, and L. T. Fan, Pressure drop, gas holdup, and oxygen transfer in tower systems, Biotechnol. Bioeng., *19*, 247 (1977).

Hughmark, G. A., Power requirements and interfacial area in gas-liquid turbine agitated systems, Ind. End. Chem. Process Des. Dev., *19*, 638 (1980).

Ingraham, J. L., Growth of psychrophilic bacteria, J. Bacteriol., *76*, 75 (1958).

Ishizaki, A., H. Shibai, Y. Hirose, and T. Shiro, Studies on the ventilation in submerged fermentations: Part I. Dissolution and dissociation of carbon dioxide in the model system, Agric. Biol. Chem., *35*, 1733 (1971a).

Ishizaki, A., Y. Hirose, and T. Shiro, Studies on the ventilation in submerged fermentations: Part II. Quantitative studies and dissociation of carbon dioxide in cultured system, Agric. Biol. Chem., *35*, 1852 (1971b).

Ishizaki, A., Y. Hirose, and T. Shiro, Studies on the ventilation in submerged fermentations: Part III. Behavior of carbon dioxide in the fermentation and dynamic characteristics of respiration in submerged culture, Agric. Biol. Chem., *35*, 1860 (1971c).

Ishizaki, A., H. Shibai, Y. Hirose, and T. Shito, Estimation of aeration and agitation conditions with respect to oxygen supply and ventilation, Agric. Biol. Chem., *37*, 107 (1973).

Johnson, M. J., Aerobic growth at low oxygen concentrations, J. Bacteriol., *94*, 101 (1967).

Joshi, J. B., A. B. Pandit, and M. M. Sharma, Mechanically agitated gas-liquid reactors, Chem. Eng. Sci., *37*, 813 (1982).

Kastanek, F., The volume mass transfer coefficient in a bubble bed column, Collect. Czech. Chem. Commun., *42*, 2491 (1977).

Kataoka, H., H. Takeuchi, K. Nakao, H. Yagi, T. Tadaki, T. Otake, T. Miyauchi, K. Washimi, K. Watanabe, and F. Yoshida, Mass transfer in large bubble columns, J. Chem. Eng. Jpn., *12*, 105 (1979).

Kern, D. M., The hydration of carbon dioxide, J. Chem. Educ., *37*, 14 (1960).

Kharasch, M. S., Heats of combustion of organic compounds, Bur. Stand. J. Res., *2*, 359 (1929).

Kharasch, M. S. and B. Sher, The electronic conception of valence and heats of combustion of organic compounds, J. Phys. Chem., *29*, 625 (1925).

Kitai, A., H. Tone, and A. Ozaki, Performance of a perforated plate column as a multistage continuous fermentor, Biotechnol. Bioeng., *11*, 911 (1969).

Kitai, A., R. Okamoto, and A. Ozaki, Continuous culture using a perforated plate column, Ferment. Technol. Today, Society of Fermentation Technology, Japan (1972), p. 147.

Konig, B., K. Schügerl, and C. Seewald, Strategies for penicillin fermentation in tower loop reactors, Biotechnol. Bioeng., *24*, 259 (1982).

Kubota, H., Y. Hosono, and K. Fujie, Characteristic evaluations of ICI airlift type deep shaft aerator, J. Chem. Eng. Jpn., *11*, 319 (1978).

Lee, D. D., Y. Y. Lee, P. J. Reilly, E. V. Collins, and G. T. Tsao, Pilot plant production of glucose with glucoamylase immobilized to porous silica, Biotechnol. Bioeng., *18*, 253 (1976).

Lee, Y. H., L. T. Fan, and L. S. Fan, Kinetics of hydrolysis of insoluble cellulose by cellulase, Adv. Biochem. Eng., *17*, 131 (1980).

Lin, C. H., B. S. Fang, C. S. Wu, H. Y. Fang, T. F. Kuo, and C. Y. Hu, Oxygen transfer and mixing in a tower cycling fermentor, Biotechnol. Bioeng., *18*, 1557 (1976).

Linton, J. D. and R. J. Stephenson, A preliminary study on growth yields in relation to the carbon and energy content of various organic growth substrates, FEMS Microbial. Lett., *3*, 95 (1978).

Luedeking, R. and E. L. Piret, A kinetic study of the lactic acid fermentation batch processes at controlled pH, J. Biochem. Microb. Technol. Eng., *1*, 393 (1959).

Luong, H. T. and B. Volesky, Mechanical power requirements of gas-liquid agitated systems, AIChE J., *25*, 893 (1979).

Malfait, J. L., D. J. Wilcox, D. G. Mercer, and L. D. Barker, Cultivation of a filamentous mold in a glass pilot scale airlift fermentor, Biotechnol. Bioeng., *23*, 863 (1981).

Merchuk, J. C. and Y. Stein, A distributed parameter model for an airlift fermentor. Effects of pressure, Biotechnol. Bioeng., *23*, 1309 (1981a).

Merchuk, J. C. and Y. Stein, Local holdup and liquid velocity in airlift reactors, AIChE J., *27*, 377 (1981b).

Miller, D. N., Scale-up of agitated vessels gas-liquid mass transfer, AIChE J., *20*, 445 (1974).

Minkevich, I. G. and V. K. Eroshin, Productivity and heat generation of fermentation under oxygen limitation, Folia Microbiol., *18*, 376 (1973).

Minkevich, I. G., V. K. Eroshin, T. A. Alekseeva, and A. T. Tereshchenko, Elemental composition and energy content of microbial biomass, Mikrob. Prom., *2*(144), 1 (1977). (In Russian.)

Monod, J., The growth of bacterial cultures, Annu. Rev. Microbiol., *3*, 371 (1949).

Moo Young, M. and H. W. Blanch, Design of biochemical reactors. Mass transfer criteria for simple and complex systems, Adv. Biochem. Eng., *19*, 1 (1981).

Nakanoh, M. and F. Yoshida, Gas absorption by Newtonian and non-Newtonian liquids in a bubble column, Ind. Eng. Chem. Process Des. Dev., *19*, 190 (1980).

Ollis, D. F., Biological reactor systems, in *Chemical Reactor Theory, A Review*, L. Lapidus and N. R. Amundson, eds., Prentice-Hall, Englewood Cliffs, N.J. (1977), Chap. 8.

Oner, M. D., L. E. Erickson, and S. S. Yang, Estimation of true growth and product yields in aerobic cultures, Biotechnol. Bioeng., *25*, 63 (1983).

Orazem, M. E. and L. E. Erickson, Oxygen transfer rates and efficiencies in one and two stage airlift towers, Biotechnol. Bioeng., *21*, 69 (1979).

Orazem, M. E., L. T. Fan, and L. E. Erickson, Bubble flow in the downflow section of an airlift tower, Biotechnol. Bioeng., *21*, 1579 (1979).

Orazem, M. E., L. T. Fan, and L. E. Erickson, Characterization of two directional dispersed phase flow through time series analysis, Chem. Eng. Commun., *6*, 161 (1980).

Oyama, Y. and K. Endoh, Power characteristics of gas-liquid contacting mixers, Chem. Eng. Jpn., *19*, 2 (1955).

Patel, S. A. and L. E. Erickson, Estimation of heats of combustion of biomass from elemental analysis using available electron concepts, Biotechnol. Bioeng., *23*, 2051 (1981).

Peterson, C. S., *Microbiology of Food Fermentations*, 2nd ed., AVI, Westport, Conn. (1979).

Perry, R. H. and C. H. Chilton, *Chemical Engineers' Handbook*, 5th ed., McGraw-Hill, New York (1973).

Pirt, S. J., *Principles of Microbe and Cell Cultivation*, Wiley, New York (1975).

Prokop, A. and L. E. Erickson, Growth models of cultures with two liquid phases: VII. Substrate dissolved in dispersed phase; Effect of dispersed phase volume and temperature, Biotechnol. Bioeng., *14*, 571 (1972).

Pye, E. K. and H. H. Weetal, *Enzyme Engineering*, Vol. 3, Plenum Press, New York, (1978).

Pye, E. K. and L. B. Wingard, *Enzyme Engineering*, Vol. 2, Plenum Press, New York (1974).

Rai, V. R. and A. Constantinides, Mathematical modeling and optimization of the gluconic acid fermentation, AIChE Symp. Ser., *69*(132), 114 (1973).

Roels, J. A., The application of macroscopic principles to microbial metabolism, Annu. N.Y. Acad. Sci., *369*, 113 (1981).

Rose, A. H., Beer, Sci. Am., *200*(6), 90 (1954).

San, K. Y. and G. Stephanopoulos, Optimal control policy for substrate inhibited kinetics with enzyme deactivation in an isothermal CSTR, AIChE J., *29*, 417 (1983).

Schügerl, K., Oxygen transfer into highly viscous media, Adv. Biochem. Eng., *19*, 71 (1981).

Schügerl, K., Characterization and performance of single and multistage tower reactors with outer loop for cell mass production, Adv. Biochem. Eng., *22*, 94 (1982a).

Schügerl, K., New bioreactors for aerobic processes, Int. Chem. Eng., *22*, 591 (1982b).

Schügerl, K., J. Lücke, and U. Oels, Bubble column bioreactors, Adv. Biochem. Eng., *7*, 1 (1977).

Schumpe, A., I. Adler, and W. D. Deckwer, Solubility of oxygen in electrolyte solutions, Biotechnol. Bioeng., *20*, 145 (1978).

Shah, P. S., L. E. Erickson, L. T. Fan, and A. Prokop, Growth models of cultures with two liquid phase: VI. Parameter estimation and statistical analysis, Biotechnol. Bioeng., *14*, 533 (1972a).

Shah, P. S., L. T. Fan, I. C. Kao, and L. E. Erickson, Modeling of growth processes with two liquid phases: a review of drop phenomena, mixing, and growth, Adv. Appl. Microbiol., *15*, 367 (1972b).

Shah, Y. T., G. J. Stiegel, and M. M. Sharma, Backmixing in gas-liquid reactors, AIChE J., *24*, 369 (1978).

Solomon, B. O., L. E. Erickson, and J. L. Hess, Application of data consistency tests and new parameter estimation methods to microbial growth on corn dust in batch culture, Biotechnol. Bioeng., *23*, 2333 (1981).

Solomon, B. O., L. E. Erickson, J. L. Hess, and S. S. Yang, Maximum likelihood estimation of growth yields, Biotechnol. Bioeng., *24*, 633 (1982).

Stephanopoulos, G., Dynamics of mixed cultures of microoorganisms: Some topological considerations, AIChE J., *26*, 802 (1980).

Stephanopoulos, G. and K. Y. San, On-line estimation of the state of biochemical reactors, ACS Symp. Ser., *196*, 155 (1982).

Stouthamer, A. H., The search for correlation between theoretical and experimental growth yields, Int. Rev. Biochem., *21*, 1 (1979).

Stumm, W. and J. J. Morgan, *Aquatic Chemistry*, Wiley, New York (1970).

Sundstrom, D. W. and H. E. Klei, *Wastewater Treatment*, Prentice-Hall, Englewood Cliffs, N.J. (1979).

Takamatsu, T., S. Shioya, and H. Nakatani, Analysis of a deep U-tube aeration system from the viewpoint of oxygen transfer, J. Ferment. Technol., *59*, 287 (1981a).

Takamatsu, T., S. Shioya, H. Nakatani, T. Fujimoto, and T. Kawasugi, Oxygen transfer and biological effects in a deep tank aeration system, Adv. Biotechnol., *1*, 581 (1981b).

Thornton, W. M., The relation of oxygen to the heat of combustion of organic compounds, Philos. Mag., *33*, 196 (1917).

Towell, G. D., C. P. Strand, and G. H. Ackerman, Mixing and mass transfer in larger diameter bubble columns. Prepr. AIChE-Inst. Chem. Eng. Joint Meet., London, Chem. Eng. Symp. Ser., *10*, 97 (1965).

Tyagi, R. D., and T. K. Ghose, Studies on immobilized *Saccharomyces cerevisiae*: I. Analysis of continuous rapid ethanol fermentation in immobilized cell reactor, Biotechnol. Bioeng., *24*, 781 (1982).

van Suijdam, J. C., H. Hols, and N. W. F. Kossen, Unstructured model for growth of mycelial pellets in submerged cultures, Biotechnol. Bioeng., *24*, 1977 (1982).

Veselov, I. Y., A. D. Gololobov, E. R. Davidov, L. G. Eliseeva, N. N. Latysheva, V. V. Rachinsky, and I. P. Uvarov, Effect of heterotrophic fixation of carbon dioxide on the enzyme activity of *Candida* yeast cultivation on *n*-alkanes and glucose, Prikl, Biokhim. Mikrobiol., *10*, 697 (1974). (In Russian.)

Viesturs, U. E., I. A. Sturmanis, V. V. Krikis, V. D. Prokopenko, and L. E. Erickson, Investigation of fermentors with various contacting devices, Biotechnol. Bioeng., *22*, 799 (1980).

Viesturs, U. E., I. A. Sturmanis, V. V. Krikis, V. D. Prokopenko, and L. E. Erickson, Fermentors with the power introduced by aerating gas, Biotechnol. Bioeng., *23*, 1171 (1981).

Vieth, W. R., K. Venkatasubramanian, A. Constantinides, and B. Davidson, Design and analysis of enzyme flow reactors, Appl. Biochem. Bioeng., *1*, 221 (1976).

Voigt, J. and K. Schügerl, Absorption of oxygen in countercurrent multistage bubble columns: I. Aqueous solutions with low viscosity, Chem. Eng. Sci., *34*, 1221 (1979).

Voigt, J. and K. Schügerl, Comparison of single and three stage tower loop reactors, Eur. J. Appl. Microbiol. Biotechnol., *11*, 97 (1981).

Voigt, J., V. Hecht, and K. Schügerl, Absorption of oxygen in countercurrent multistage bubble columns: II. Aqueous solutions with high viscosity, Chem. Eng. Sci., *35*, 1317 (1980).

Wang, K. B., Performance of bubble columns packed with Koch motionless mixers, Ph.D. dissertation, Kansas State University (1977).

Wang, N. S. and G. Stephanopoulos, Application of macroscopic balances to the identification of gross measurement error, Biotechnol Bioeng., *25*, 2177 (1983).

Wang, D. I. C., C. L. Cooney, A. L. Demain, P. Dunnill, A. E. Humphrey, and M. D. Lilly, *Fermentation and Enzyme Technology,* Wiley, New York (1979).

Wang, H. Y., C. L. Cooney, and D. I. C. Wang, Computer-aided baker yeast fermentations, Biotechnol. Bioeng., *19*, 69 (1977).

Weetall, H. H. and G. P. Royer, *Enzyme Engineering*, Vol. 5, Plenum Press, New York, (1980).

Wingard, L. B., *Enzyme Engineering, Biotechnology and Bioengineering Symposium,* Vol. 3, Wiley, New York (1972).

Wingard, L. B., E. Katchalski-Katzir, and L. Goldstein, *Applied Biochemistry and Bioengineering,* Vol. 1: *Immobilized Enzyme Principles,* Academic Press, New York (1976).

Wingard, L. B., I. V. Berezin, and A. A. Klyosov, *Enzyme Engineering,* Plenum Press, New York (1980).

Yagi, H. and F. Yoshida, Gas absorption by Newtonian and non-Newtonian fluids in sparged agitated vessels, Ind. Eng. Chem. Process Des. Dev., *14*, 488 (1975).

Yagi, H. and F. Yoshida, Desorption of carbon dioxide from fermentation broth, Biotechnol. Bioeng., *19*, 801 (1977).

Yamane, T., E. Sada, and T. Takamatsu, Start-up of chemostat: Application of fed batch culture, Biotechnol. Bioeng., *21*, 111 (1979).

Zaborsky, O., *Immobilized Enzymes,* CRC Press, Cleveland, Ohio (1973).

Zlokarnik, M., Sorption characteristics for gas-liquid contacting in mixing vessels, Adv. Biochem. Eng., *8*, 133 (1978).

Zlokarnik, M., Sorption characteristics of slot injectors and their dependency on the coalescence behavior of the system, Chem. Eng. Sci., *34*, 1265 (1979).

Zuber, N. and J. A. Findlay, Average concentration in two phase flow systems, J. Heat Transfer, *87*, 453 (1965).

13
Analysis and Design of Photoreactors

ELIANA R. DE BERNARDEZ, MARÍA A. CLARIÁ, and
ALBERTO E. CASSANO *Instituto de Desarrollo Tecnológico para la Industria Química–INTEC, Universidad Nacional del Litoral–Consejo Nacional de Investigaciones, Científicas y Técnicas, Santa Fe, Argentina*

SIGNIFICANCE AND SCOPE

The advantages of photochemical reactions over thermal or catalytic reactions have frequently been quoted: selectivity, negligible side reactions, low operating temperature, possibility of carrying out some reactions that would be thermodynamically infeasible by thermal means due to unfavorable equilibrium conditions, liquid-phase operation, and so on.

Despite these advantages, photoreactors have been employed industrially only when either thermal or conventional catalytic reaction paths have proved to be very difficult. The reasons are: lack of suitable reactor models and design methods, lack of quantitative information about the pertinent physical and chemical parameters, limited variety of equipment available, and the understandable reluctance of successful industrial companies to publish design and scaleup procedures. However, many reactions are suitable for photochemical production.

The most common industrial photoreactions are those that proceed via chain mechanisms. Among these, *photochlorinations* are undoubtedly the most important processes due to their high-energy yields. Thermal clorinations, carried out at high temperatures, produce a variety of different molecular species which have to be separated; this results from the broad energy distribution of the thermal field and its effects on reactions with different activation energies. In contrast to this, in many cases, the photochemical process may produce only a single product. The synthesis of γ-hexachlorocyclohexane has been one of the photochlorinations of greatest economic significance.

An industrial process for the preparation of volatile primary mercaptans, which are commonly used as additives to natural gas, is based on the anti-Markovnikov addition of hydrogen sulfide to terminal olefins, a free radical chain reaction initiated by light. *Photopolymerizations* are another example. Such chains, started by light, have high energy yields; in addition, no fragments of initiator molecules remain and production rates by means of the flow process are extremely fast.

The *photooximation* of cyclohexane, a reaction with low energy yield, is currently used to prepare caprolactam, which in its turn is utilized for the production of nylon-6; nylon-12 is prepared by a similar route. This reaction involves a free radical mechanism but is not a chain reaction. Several companies have experimented with this process; however, the first large-scale use has been reported by the Toyo Rayon Co. of Japan.

The *photooxidation* of olefins has been investigated extensively in relation to the "smog" problem that appears in the atmosphere by the photochemical reaction between several atmospheric contaminants. Also, a number of organic compounds, epoxides, aldehydes, and ketones have been produced in the laboratory by reaction of oxygen atoms (photochemically produced) with olefins. Generally, these reactions have high yields and side reactions are minor (at least for terminal olefins).

There are photochemical reactions with low energy yields that are generally limited in application to the synthesis of high-priced, low-volume chemicals such as *pharmaceuticals* or perfumes. The photosynthesis of vitamin D_2 is the most important example in the pharmaceutical industry. It is based on the ergosterin photolysis and is the only way to obtain this vitamin.

We should add that, very often, photochemical processes have been proposed with the following objectives:

1. Decomposition of water contaminants refractory to other ways of destruction
2. Sterilization processes
3. Bacteriological treatment of drinkable water
4. Destruction of very dangerous chemicals that may have been produced either voluntarily or by error, particularly when there are no other ways of making them harmless

Several more applications can be included, but perhaps the most important development in the field of photochemistry is the recent emphasis on *photo-assisted catalysis* (Childs and Ollis, 1980). Heterogeneous photo-assisted catalysts are substances that require photo-excited electrons or holes in the course of a heterogeneously catalyzed reaction. As an example we can mention several semiconducting oxides, such as TiO_2 and ZnO; the former, when irradiated, becomes active in the selective oxidation of alkanes to ketones and aldehydes as well as in the dechlorination of polychlorinated biphenyls; on the other hand, carbon monoxide oxidation occurs over illuminated ZnO. Several photo-assisted semiconductor electrodes, such as crystals of $SrTiO_3$ and the perovskite $KTaO_3$, can electrolyze water without an applied electromotive force. Many other applications could be listed. Photo-assisted catalysis may become practically useful if a relatively cheap source of light with appropriate wavelength is available, the sun being the obvious source. The prospects for solar energy conversion following this path are particularly appealing.

Despite all these potentialities, photoreactor design is still a novel field in reaction engineering, with many important problems still not properly solved. In fact, up to the present, no reliable design methods have been made available to the practicing engineer. We believe that a significant part of the responsibility for this lack of know-how lies mostly in considering its analysis as a kinetic variation of a conventional reactor,

when in fact it must necessarily include the radiation field as the key viewpoint. Consequently, if some progress is to be made, the design of a photoreactor cannot avoid all the complexities that radiation problems involve.

First, we present a definition of the problem and the requisite theory. Later, the design methods are shown through four applications, three of which have already been tested in bench-scale experiments with excellent agreement. Finally, in two short sections we propose some new reactors not yet theoretically modeled, and we end with what we believe are the most important problems still unsolved in the field.

TYPES OF REACTORS

When dealing with the design of the equipment for carrying out a photochemical reaction, several aspects must be considered. Some of them are common to the design of conventional thermal reactors, such as the kinetic characteristics of the reactions involved, the phases of the system, the necessity of temperature control, the requirements about the material of construction, and so on. Others arise specifically from the selection of the appropriate radiation source for the reaction under study (i.e., the spectral distribution of the emitted light and the geometrical configuration of the reactor-lamp arrangement). In addition, a certain number of important photochemical reactions share the common characteristic of presenting products and reactants with highly corrosive or dissolving properties; a typical example is the chlorination of hydrocarbons where the presence of chlorine, hydrochloric acid, and chlorinated solvents creates difficulties in the selection of materials for equipments and seals. In the case of heterogeneous systems, the mixing pattern of the reactor is extremely important. Another important aspect in the design is the requirement for temperature control in strongly exothermic or endothermic reactions. We may add that in some particular cases the design can be improved by adding a reflecting device; its use complicates the analysis of the radiation field.

Many of these difficulties can be overcome by choosing an appropriate configuration of the photoreactor system. Among a wide range of possible reactor-lamp-reflector configurations, those most widely studied in the literature are the annular reactor, the cylindrical reactor with elliptical reflector, and the cylindrical reactor with parabolic reflector; they are described briefly below. Other types of configurations can be used in commercial applications; nevertheless, they arise from different combinations of the three mentioned above.

Annular Reactor

Annular photoreactors are an excellent approach to what is perhaps the most practical type of photochemical reactor to be used for commercial purposes. The utilization of energy can be the maximum expected and, moreover, they represent the common case of a reaction vessel with a tubular lamp placed at its axis by means of an immersion well.

Figure 1 shows the main features of the system. The reacting stream is contained in the annular space that surrounds the lamp. If the reactor

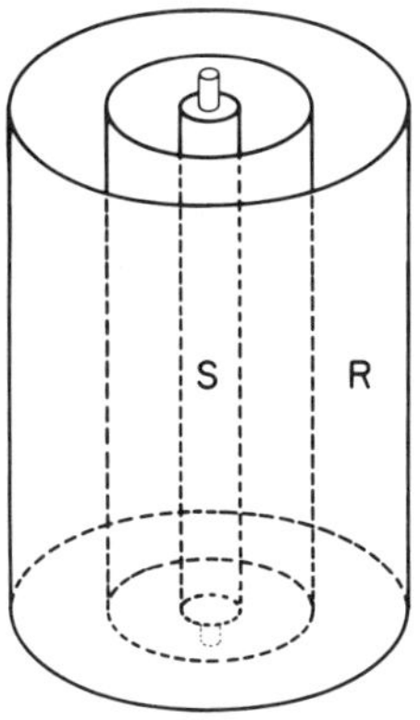

FIGURE 1 Annular reactor.

vessel is separated from the lamp, the annular space between them can be used to cool the lamp and/or control the operating temperature of the reactor. However, it must be noted that the requirement of light transmission introduces some limitations in the heat transfer possibilities because at least the inner reactor wall must be made of quartz or glass. Since light transparency is not required in the outer reactor wall, it provides much more freedom for design purposes, such as cooling, reflection of nonabsorbed light, and so on. Hence if one desires to improve the absorption efficiency of the reactor, a reflector surface covering the outer reactor wall can be used.

In dealing with laboratory research, the annular photoreactor cannot be recommended for continuous-flow operations due to its unavoidably large cross-sectional area, which requires the handling of large volumetric flow rates. On the contrary, this characteristic makes it very appropriate for commercial-scale operations.

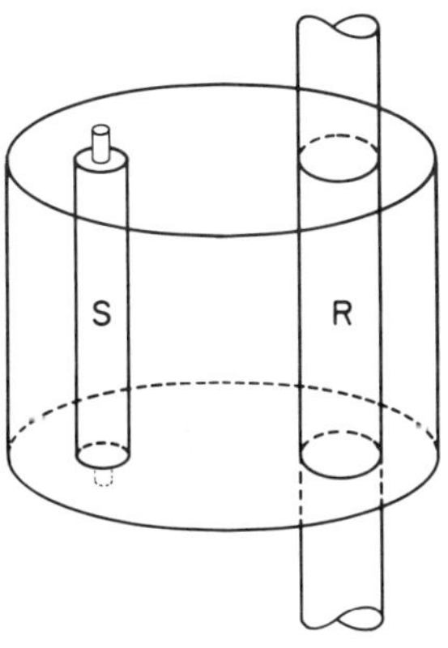

FIGURE 2 Cylindrical reactor with elliptical reflector.

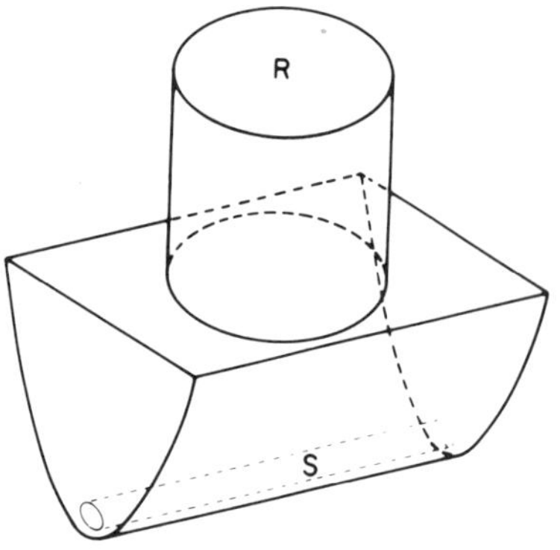

FIGURE 3 Cylindrical reactor irradiated from the bottom.

Cylindrical Reactor with Elliptical Reflector

This system consists of a reactor made with a cylindrical tube placed at one of the focuses of a cylindrical reflector of elliptical cross section. A tubular radiation source is placed at the other focus. This particular arrangement is often called an *elliptical photoreactor* (Fig. 2).

The possibility of using cylindrical tubular reactors with a small cross-sectional area together with the generally accepted (but only partially true) concept of the existence of a uniform irradiation from outside has resulted in a rather extensive application of this reactor for laboratory and bench-scale research work.

On the other hand, since the incidence efficiency of this type of reactor—defined as the capacity of concentrating the energy within the reactor boundaries—is not very high, this is not an advisable device for commercial-scale operation unless other energetic considerations compel its use.

Cylindrical Reactor with Parabolic Reflector

Another possibility of isolating the reaction system from the radiation source (which could also simplify the solution of the well-known problem of wall deposites, generally more severe at the radiation entrance wall) consists of the use of a cylindrical reactor irradiated from the bottom by a tubular source located at the focal axis of a cylindrical reflector of parabolic cross section (Fig. 3). Since the cylindrical reactor may be a perfectly stirred tank reactor, this device is especially appropriate to carry out liquid or gas-liquid reactions where vigorous stirring is required. This type of reaction is indicated for both laboratory and commercial scale work and can be used in batch, semibatch, or continuous operations. Problems of corrosion and sealing can be more easily handled in this system.

PHOTOREACTOR DESIGN: STATEMENT OF THE PROBLEM

To design a photochemical reactor, an engineer should have the following information (a) the reaction; (b) the radiation energy demand (generally

qualitatively defined), which can be satisfied with a radiation source (lamp) and perhaps a reflector; and (c) a choice of possible reactor configurations (sometimes with some restrictions). Then the following questions must be answered:

1. Which is the best source-reactor-reflector arrangement?
2. Which is the optimal value of the spectral distribution and overall radiation power output to convert reactants into the desired product?
3. Which is the optimal feed composition?
4. How much conversion can be achieved?
5. Which is the optimal size of the reactor?
6. How much selectivity can be obtained?

The answers to these questions must be treated in a logical sequence which starts with the knowledge of the rate of a photochemical reaction. The description of the rate makes it necessary to distinguish between *dark-* and *radiation-activated steps*. To treat dark reactions one uses the same methodology as for thermal reactions; the main hindrance appears when evaluating the rate of the initiation step. It is known that the rate of this step is related to the *local volumetric rate of energy absorption* (LVREA). The evaluation of the LVREA is performed stating first a balance of radiant energy at steady state for a control volume. For simplicity, a nonemitting medium is generally assumed. Afterward, it is necessary to incorporate a radiation source model and a constitutive equation in the absorption term.

The source models have been treated by several research teams, which have developed different proposals to predict radiation profiles inside photoreactors. Conceptual approaches may be classified into two well-defined groups: (a) those which assume a given radiation distribution in the vicinity of the reactor (*incidence models*) and (b) those which start from a proposal of a model for the source (*emission models*). There is no way of using incidence models without experimentally adjustable parameters. The reader interested in this approach can find a critical review in a previous work (Alfano et al., 1985). However, since we are looking for an a priori design of photoreactors, attention will only be paid to the emission models. Their formulation will be presented later.

Once we know how to formulate the rate of a photochemical reaction, we can progress further ahead with the design problem. Then the mass, energy and momentum balances must be established. The expression for the conservation of momentum is similar to that of conventional reactors. Due to the energetical features of photochemical reactions (very low and sometimes null "equivalent" activation energy for the global reaction), it is very frequent to find that the reaction behavior can be accurately described with the mass balance and the energy balance reduced exclusively to radiation. However, in some cases, when highly exothermic or endothermic reactions are present, the design of the reactor will unavoidable require the complete energy equation of change. In any event, even for the simplest case, the mass and radiation balances are normally coupled through concentrations of the radiation-absorbing species.

The nature of this coupling is well understood. The local volumetric rate of energy absorption (LVREA) is generally a function of the radiation field inside the space of reaction, the concentration of the absorbing species, and other physicochemical parameters. If species i which is being considered in the mass balance is an *absorbing species*, a coupling between the

LVREA and the corresponding mass balance equation will take place. Simultaneously, the concentration of the absorbing species (i) goes inside an integral expression, accounting for the radiation attenuation by absorption. Thus the general problem of modeling a photoreactor presents an integro-differential mathematical nature. This degree of complexity of the problem should not be surprising since the physical situation indicates that the progress of the reaction at a point depends on the LVREA (which is, among other variables, a function of the radiation attenuation suffered by each bundle along its path inside the reactor); in its turn, the attenuation suffered by each pencil of radiation (due to its concentration dependence) is a function of the reaction extent reached at each one of the points constituting the trajectory of the bundle. We discuss later how to model the absorption process in the reactor by one or more species, as well as evaluating the LVREA and presenting several applications using the three main reactor configurations described previously.

To solve the design problem, additional information about the lamp, the reactor, the reaction, and perhaps, if it is used, the reflector will be needed. We should state beforehand that we assume that the reaction is already known. This means that kinetic data are available as well as qualitative information regarding radiation requirement. Because of this, the system will impose some constraints on the selection of radiation sources and reactor configurations. At the same time the market availability of ultraviolet- and visible-light lamps and quartz (or Pyrex glass) tubes and plates, if they are needed, will provide definite guidelines about the type of reactor-lamp combination better suited for each process. In any case, a design will generally start with a tentatively chosen radiation source. More often than not, the choice of the lamp considerably reduces the freedom concerning most reactor dimensions.

Thus the problem must begin with a tentative list of needed information, generally oriented by one's qualitative knowledge of the reaction and possible reactor configurations. Even if the information is complete, very often the design will have to be done in an iterative way because of the limitations imposed on the construction of the reactor by the above-mentioned market restrictions. Let us list what is needed:

1. About the *tentative* radiation source:
   a. Its power consumption and power output.
   b. The spectral distribution of its power output.
   c. Its geometrical dimensions.
   d. Its operating conditions.
2. About the *tentative* reactor (assuming that some decision has been taken about its shape and type of operation):
   a. Its geometrical dimensions.
   b. In some designs, the transmission characteristic of the wall through which the radiation should enter the reactor.
   c. The spectral distribution of the wall transmission characteristics.
   d. Information about ranges of operating conditions (e.g., pressure, temperature, required production, etc.).
3. About the reaction:
   a. A kinetic sequence with the following characteristics:
      (1) An initiation step that is activated by absorption of radiation.

(2) One or more "dark" reactions. From the industrial point of view the most important reactions are those recognized by the name of chain reactions. Starting from an atomic species or a free radical formed in the initiation step, they involve the following constituents:
   (a) One or more propagation steps.
   (b) One or more termination steps.

   In some cases it is possible to work with a global kinetics that relates the absorption of radiation to the formation of products. Nevertheless, if it is available, it is always better to work with microkinetic information such as that mentioned in (1) and (2).

b. The radiation absorption characteristics of reactants, inerts, and products.
c. The *primary quantum yield* of the initiation step (i.e., how many atomic or free radical species are effectively produced for each absorbed photon by the appropriate reactant in the first reaction of the kinetic sequence). If only global kinetic expressions are available, an *overall quantum yield* may be necessary, which is defined according to the reaction scheme (moles of product formed or reactant consumed/energy absorbed).

4. About the *tentative* reflector (if it is used and assuming that its shape is given):
   a. Its geometrical dimensions.
   b. Its reflectivity
   c. The spectral distribution of its reflectivity.

Here we have pointed out mainly the particular design variables and the parameters of photochemical reactors. In addition, we must know all other physicochemical properties or characteristic parameters usually needed for thermal reactors (densities, diffusivities, thermal properties, etc.).

In the next sections we derive the theory and show the methods used to reach an a priori design of a photoreactor, when the problem is posed as indicated above.

EMISSION MODELS

We begin our analysis of the emission models by defining the fundamental properties of the radiation field. Then the balance of radiant energy for an absorbing medium will be derived with which we will finally be able to obtain an expression for the rate of the initiation step for a photochemical reaction. The concepts introduced here are based on well-accepted definitions in the areas of radiation engineering and radiation gas dynamics. However, we shall point out some differences between the conventional thermal applications of radiation (e.g., furnaces) and those that involve the use of visible, ultraviolet, and even shorter wavelength radiations.

Definitions

Specific Intensity

Radiation engineering generally studies the energy exchange between surfaces (Fig. 4). To characterize the amount of energy departing from a

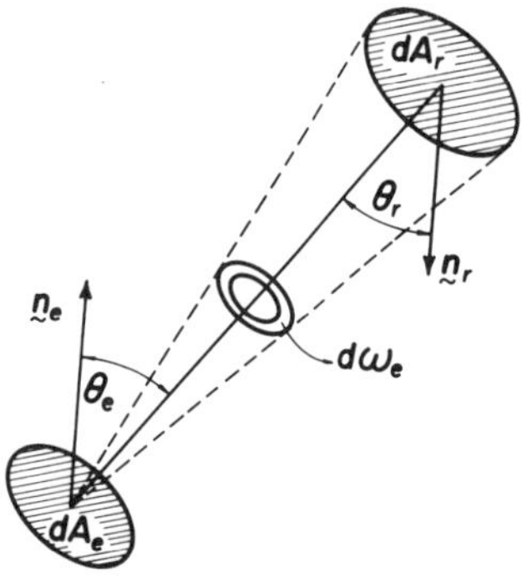

FIGURE 4 Geometric representation of the SEES model.

surface, the concept of specific intensity of radiation is used. The specific intensity of radiation is defined as the amount of energy in the frequency range between ν and $\nu + d\nu$, leaving a surface per unit area normal to the pencil or rays, per unit solid angle and unit time. From now on we will simplify our notation and call energy in the frequency interval between ν and $\nu + d\nu$ simply energy of frequency ν. To illustrate the rise of the specific intensity concept, let $d\psi_\nu$ be the radiant energy of frequency ν per unit time and unit area, leaving a given surface dA_e in the direction θ_e and contained within a solid angle $d\omega_e$ (Fig. 4); then

$$I_\nu = \frac{d\psi_\nu}{d\omega_e f(\theta, \phi)} \tag{1}$$

where $f(\theta, \phi)$ takes into account the directional characteristics of the emission (isotropic, diffuse, etc.). We must extend this concept to photoreactor engineering, where the energy exchange may involve not only surfaces but volumes as well.

When the radiation source is a fluorescent lamp, only the external surface of the lamp participates in the emission process. Hence the energy exchange phenomenon involves two surfaces, one for emission and the other for reception (Fig. 4). On the other hand, if the radiation source is a nonfluorescent arc lamp, the emission is volumetric (i.e., the entire volume of the lamp participates in the emission process). In this case (Fig. 5) the energy exchange phenomenon involves an emitting volume (dV_e) and a receiving surface (dA_r). The first proposal of an extense model was published by Irazoqui et al. (1973). It was applied to a nonfluorescent arc lamp and it is known as the volumetric emission extense source model (VEES). Stramigioli et al. (1975, 1977) presented the superficial emission extense source model (SEES) to describe the performance of fluorescent lamps. Their formulation was an extension of the VEES model.

In spite of the fact that radiation sources have finite spatial dimensions, many authors have used a third model, which considers the lamp as a line (Figure 6). This model, first proposed by Jacob and Dranoff (1966, 1968) is known as the spherical emission linear source model (SELS). As we shall see later, the SELS model can be used only in some special—but very important—reactor designs.

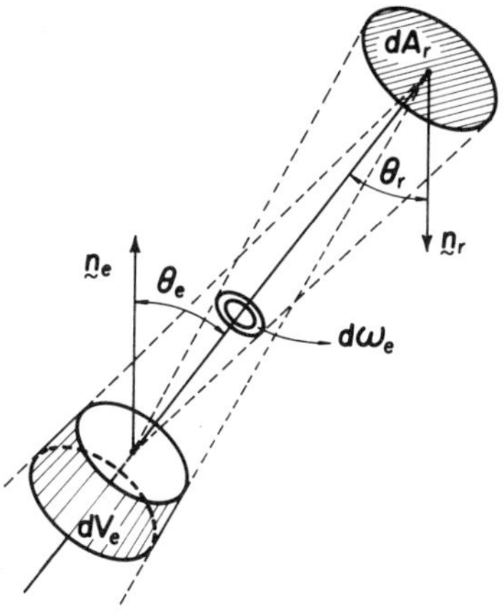

FIGURE 5 Geometric representation of the VEES model.

We can now derive a formulation of the specific intensity of radiation which will account for the type of energy exchange involved; then we have:

1. Emitting surface–receiving surface (SEES model)
2. Emitting volume–receiving surface (VEES model)
3. Emitting line–receiving surface (SELS model)

When formulating the equations we make the following assumptions:

1. The emitters of the radiation source are uniformly distributed inside its extension (be it a volume, a surface, or a line).
2. Any elementary extension of the source emits per unit time, at any frequency, an amount of energy proportional to its extension.
3. The emission characteristics are constant along the z direction (this assumption does not imply a constant radiation field in the z direction).

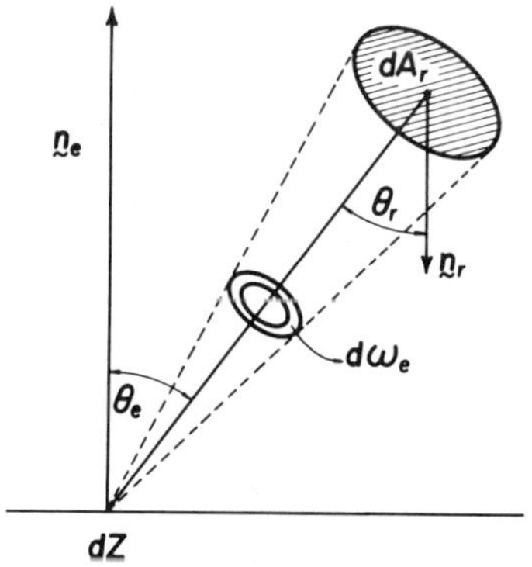

FIGURE 6 Geometric representation of the SELS model.

Maintaining the expression (1) for the definition of I_ν, we may introduce the differences in the energy exchange process in the expression for $d\psi_\nu$. Let us use a generic expression for an "energy flux" ($d\Psi_\nu$) which includes the possibility of emission by either lines, surfaces, or volumes. The generic extent of the lamp will be indicated by $d\Gamma_e$, which may have dimensions of length, area, or volume; depending on the case. Then

$$d\Psi_\nu = \frac{dE_\nu}{d\Gamma_e} \tag{2}$$

where $d\Psi_\nu$ is the amount of energy of frequency ν per unit time that leaves an infinitesimal extent of the lamp, characterized by $d\Gamma_e$. This distinction is the definition of the specific intensity of radiation allows us to achieve complete consistency in a unified theory. Considering Eq. (2), the equation for I_ν becomes

$$I_\nu = \frac{d^{(b)}E_\nu}{d^{(2)}\omega_e f(\theta,\phi)\, d^{(d)}\Gamma_e} \tag{3}$$

Here a superscript in parentheses indicates the order of the infintesimal quantity. Values of (d) and the expressions of $d^{(d)}\Gamma_e$ for each emission model are given in Table 1. It is obvious that $b = d + 2$.

At this point it should be noted that in a diactinic medium (a homogeneous nonabsorbing medium) the specific intensity is constant along the propagation path, and hence independent of distance. On the other hand, in absorbing and/or nonhomogeneous media the specific intensity will change (due to absorption, scattering, reflection, and refraction) according to the physical properties of the medium.

In the following analysis we consider three special cases:

1. When the emission is independent of the direction, it is called *isotropic* emission; then $f(\theta,\phi) = 1$.
2. When the emission is *diffuse*, it follows what is known as the "cosine law;" then $f(\theta,\phi) = \cos\theta_e$.
3. When the directional characteristics of the emission are different from cases 1 and 2, the model needs the specification of function $f(\theta,\phi)$.

From now on we consider only the first two cases, isotropic and diffuse emission. To simplify the treatment of the equations, we can reduce the two emission cases into a unique expression:

$$I'_\nu = \frac{d^{(b)}E_\nu}{d^{(2)}\omega_e d^{(d)}\Gamma_e} \tag{4}$$

For isotropic emission, $I'_\nu = I_\nu$. In the case of diffuse emission,

$$I'_\nu = I_\nu \cos\theta_e \tag{5}$$

TABLE 1 Final Equations for the Three Emission Models

Model	d	$d^{(d)}\Gamma_e$	Γ_e^*	I'_ν [Eq. (4)]	$\lvert d^{(d)}q_\nu \rvert$ [Eq. (9)]
SELS	1	dz	L_L	$\frac{dE_\nu}{d\omega_e\, dz} = \frac{\rho^2\, dE_\nu}{dA_r \cos\theta_r\, dz}$	$\frac{I'_\nu\, dz}{\rho^2}$
SEES	2	$dA_e = \rho^2\, d\omega_r$	$2\pi r_L L_L$	$\frac{dE_\nu}{d\omega_e\, dA_e} = \frac{\rho^2\, dE_\nu}{dA_r \cos\theta_r\, dA_e}$	$\frac{I'_\nu\, dA_e}{\rho^2} = I'_\nu\, d\omega_r$
VEES	3	$dV_e = \rho^2\, d\omega_r\, d\rho$	$\pi r_L^2 L_L$	$\frac{dE_\nu}{d\omega_e\, dV_e} = \frac{\rho^2\, dE_\nu}{dA_r \cos\theta_r\, dV_e}$	$\frac{I'_\nu\, dV_e}{\rho^2} = I'_\nu\, d\omega_r\, d\rho$

However, since the product $I_\nu \cos\theta_e$ is independent of direction, in both cases I'_ν is always isotropic. The distinction between the two types of emission will be introduced in the following derivations only when necessary. Table 1 summarizes the final expressions of I'_ν for each emission model.

Finally, for polychromatic sources, the radiation intensity is obtained integrating the specific intensity over the whole frequency range, as follows:

$$I = \int_{\nu=0}^{\nu=\infty} I_\nu d\nu \tag{6}$$

Specific Energy Flux Density Vector

The amount of energy of frequency ν passing through a surface of extent dA_r (Figs. 4, 5, and 6), per unit time may be expressed as

$$d^{(b)}E_\nu = dA_r \underset{\sim}{n} \cdot d^{(c)}\underset{\sim}{q}_\nu \tag{7}$$

Isotropic emission, e_ν^a [Eq. (29)]	Isotropic emission κ_ν	Diffuse emission, e_ν^a [Eq. (33)]	Diffuse emission (κ_ν)
$\kappa_\nu \mu_\nu \int_0^{L_L} \frac{dz}{\rho^2} \times \exp\left(-\int_{\rho_0^*}^{\rho^*} \mu_\nu \, d\rho'\right)$	$\frac{E_{L,\nu}}{4\pi L_L}$	$\kappa_\nu \mu_\nu \int_0^{L_L} \frac{\sin\theta \, dz}{\rho^2} \times \exp\left(-\int_{\rho_0^*}^{\rho^*} \mu_\nu\right) d\rho'$	$\frac{E_{L,\nu}}{\pi L_L}$
$\kappa_\nu \mu_\nu \int_\phi \int_\theta \sin\theta \times \exp\left(-\int_{\rho_0^*}^{\rho} \mu_\nu \, d\rho'\right) d\theta \, d\phi$	$\frac{E_{L,\nu}}{8\pi^2 r_L L_L}$	$\kappa_\nu \mu_\nu \int_\phi \int_\theta \sin^2\theta \cos\phi \times \left(-\int_{\rho_0^*}^{\rho^*} \mu_\nu \, d\rho' \; d\theta \, d\phi\right)$	$\frac{E_{L,\nu}}{2\pi^2 r_L^2 L_L}$
$\kappa_\nu \mu_\nu \int_\phi \int_\theta \int_\rho \sin\theta \times \left(-\int_{\rho_0^*}^{\rho^*} \mu_\nu \, d\rho'\right) d\rho \, d\theta \, d\phi$	$\frac{E_{L,\nu}}{4\pi^2 r_L^2 L_L}$	$\kappa_\nu \mu_\nu \int_\phi \int_\theta \int_\rho \sin^2\theta \times \cos\phi \times \exp\left(-\int_{\rho_0^*}^{\rho^*} \mu_\nu \, d\rho'\right) \times d\rho \, d\theta \, d\phi$	$\frac{E_{L,\nu}}{\pi^2 r_L^2 L_L}$

where $d^{(c)}\underset{\sim}{q}_\nu$ is the specific energy flux density vector. We do not need to make a distinction between the definition of $d^{(c)}\underset{\sim}{q}_\nu$ for each model because it is a properly related to the receiving surface. From Eq. (7) if follows that

$$d^{(b)}E_\nu = dA_r \cos\theta_r \; |d^{(c)} \underset{\sim}{q}_\nu| \tag{8}$$

where $|d^{(c)}\underset{\sim}{q}_\nu|$ is the modulus or the specific energy flux density vector. For both Eqs. (7) and (8) it is clear that $c = b - 2 = d$.

The relationship between the specific intensity and the modulus of the specific energy flux density vector can be derived relating Eqs. (4) and (8):

$$d\omega_e = \frac{dA_r \cos\theta_r}{\rho^2} \tag{9}$$

where we used the geometrical relationship

$$d\omega_e = \frac{dA_r \cos\theta_r}{\rho^2} \tag{10}$$

In spite of the fact that $|d^{(d)}\underset{\sim}{q}_\nu|$ and I'_ν have different physical meanings, they are often confused in the photoreactor engineering literature. In a diactinic medium while I'_ν remains constant, Eq. (9) shows that the modulus of the specific energy flux density vector changes with distance. Table 1 summarizes the expressions of Eq. (9) for each emission model.

Attenuation Coefficient

The attenuation coefficient is a property of the substance through which the radiation travels and is related to the process of radiation absorption. It is defined through a constitutive equation commonly referred to as Lambert's "law," which can be derived from molecular theory. The constitutive equation states that for an absorbing and homogeneous medium, the contribution of absorption to the gradient of the specific intensity is

$$\left.\frac{dI_\nu}{d\rho}\right|^a = -\mu_\nu I_\nu \tag{11}$$

Equation (11) is valid for any type of emission model with proper use of the definition of I_ν. The left-hand side of Eq. (11) is the directional derivative of I_ν (i.e., the component of the gradient of the scalar function I_ν along the direction of propagation of the ray). Hence Eq. (11) may be rewritten

$$\underset{\sim}{j} \cdot \underset{\sim}{\nabla} I_\nu = -\mu_\nu I_\nu \tag{12}$$

where $\underset{\sim}{j}$ is a unit vector in the direction of each ray. Since $f(\theta,\phi)$ is independent of distance ρ, Eq. (12) is also valid for I'_ν.

Radiation Balance

We shall state the steady-state radiation balance for a nonemitting, nondispersing homogeneous control volume (Fig. 7). The difference between the energy flux density of frequency ν impinging on dA_r at position ρ with direction (θ,ϕ), and the energy flux density of frequency ν leaving the fixed control volume with the same direction, at position $\rho + d\rho$, is due to the absorption of radiation:

$$\left. d^{(b)}E_\nu \right|_\rho - \left. d^{(b)}E_\nu \right|_{\rho+d\rho} = d_\rho^{(b+1)}E_\nu^a \tag{13}$$

where on the right-hand side the subscript ρ indicates differentiation with respect to ρ.

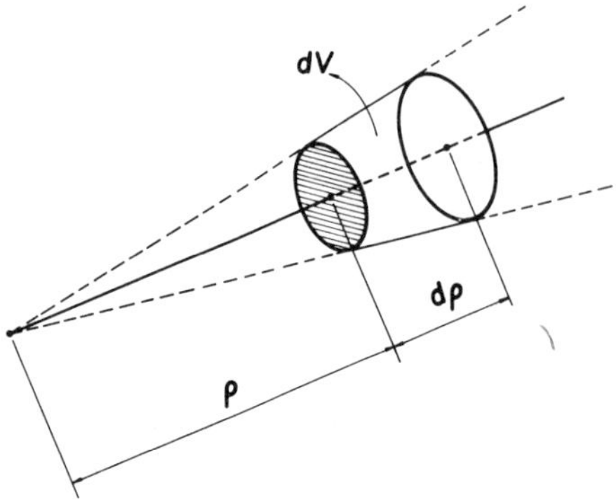

FIGURE 7 Radiation balance in an elementary absorbing volume.

Introducing Eqs. (8) and (10) into Eq. (13), we have

$$d_\rho^{(b+1)}E^a = -d\omega_e d_\rho \left[\rho^2 \left| d^{(d)}\underset{\sim}{q}_\nu \right| \right] \tag{14}$$

Combining Eqs. (14) and (9) yields

$$d_\rho^{(b+1)}E_\nu^a = -d\omega_e \; d_\rho I'_\nu d^{(d)}\Gamma_e \tag{15}$$

Substituting the constitutive equation (11) into Eq. (15) gives us

$$d_\rho^{(b+1)}E_\nu^a = d\omega_e \; I'_\nu \mu_\nu d\rho \; d^{(d)}\Gamma_e \tag{16}$$

Finally, considering Eqs. (9), (14), and (16), the steady-state radiation balance becomes

$$\frac{1}{\rho^2}\,\frac{2}{d\rho}\left(\rho^2 \left| d^{(d)}\underset{\sim}{q}_\nu \right| \right) = -\mu_\nu \left| d^{(d)}\underset{\sim}{q}_\nu \right| \tag{17}$$

Rate of the Activation Step

As mentioned earlier, the rate of the activation step is proportional to the local volumetric rate of energy absorption (LVREA). The LVREA is defined as follows:

$$d^{(f)}e_\nu^a = \frac{d^{(b+1)}E_\nu^a}{dV_r} \tag{18}$$

with f = b + 1 − 3 = b − 2 = d. Combining Eq. (14) with Eq. (18) and considering that $dV_r = \rho^2 \, d\omega_e \, d\rho$, we obtain

$$d^{(d)}e_{\nu}^{a} = \frac{1}{\rho^2}\frac{1}{d\rho}(\rho^2 | d^{(d)}q_{\nu} |) \tag{19}$$

Finally, introducing the radiation balance (17) into Eq. (19), we have

$$d^{(d)}e_{\nu}^{a} = \mu_{\nu} | d^{(d)}\underset{\sim}{q}_{\nu} | \tag{20}$$

To evaluate $d^{(d)}e_{\nu}^{a}$ we need to know $| d^{(d)}\underset{\sim}{q}_{\nu} |$. From the radiation balance and integrating inside the reacting space (where the absorption of the radiation process takes place), we obtain (Fig. 8)

$$\rho^2 | d^{(d)}\underset{\sim}{q}_{\nu} | = \rho_o^2 | d^{(d)}\underset{\sim}{q}_{\nu} |_o \exp\left(-\int_{\rho_o^*(\theta,\phi)}^{\rho^*(\theta,\phi)} \mu_{\nu}\, d\rho'\right) \tag{21}$$

In Eq. (21), $| d^{(d)}\underset{\sim}{q}_{\nu} |_o$ is the modulus of the specific energy flux density vector for a point $W(\rho_o^*, \theta, \phi)$ at the surface of entrance of radiation to the reactor volume. To evaluate it, assuming no absorption between the source surface and point W, we use the equations valid for diactinic media, that is,

$$\rho^2 | d^{(d)}\underset{\sim}{q}_{\nu} | = I'_{\nu,o}\, d^{(d)}\Gamma_e \tag{22}$$

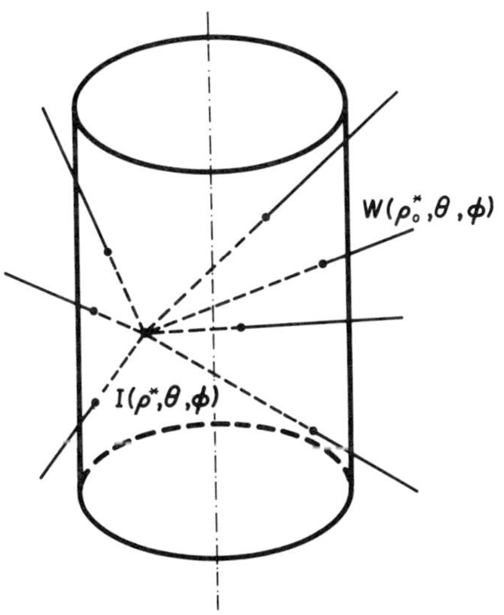

FIGURE 8 Schematic representation of the attenuation process.

Because I'_ν remains constant in a diactinic medium, Eq. (22) may be written

$$\rho_o^2 \left| d^{(d)} \underset{\sim}{q}_\nu \right|_o = I'_{\nu,o} \, d^{(d)}\Gamma_e \tag{23}$$

Combining Eqs. (20), (21), and (23), we obtain

$$d^{(d)}e_\nu^a = \mu_\nu \frac{I'_{\nu,o}}{\rho^2} \exp\left(- \int_{\rho_o^*}^{\rho^*} \mu_\nu \, d\rho' \right) d^{(d)}\Gamma_e \tag{24}$$

Finally, the local volumetric rate of energy absorption at frequency ν is obtained by integrating Eq. (24):

$$e_\nu^a = \mu_\nu \int_{\Gamma_e} \frac{I'_{\nu,o}}{\rho^2} \exp\left(- \int_{\rho_o^*}^{\rho^*} \mu_\nu \, d\rho' \right) d^{(d)}\Gamma_e \tag{25}$$

In Eq. (25), $I'_{\nu,o}$ must be related to the power output of the radiation source. From Eq. (4) we have

$$d^{(b)}E_\nu = I'_{\nu,o} \, d\omega_e \, d^{(d)}\Gamma_e \tag{26}$$

for a diactinic medium. Integrating Eq. (26) for the whole range of solid angles and for the whole extension of the emission source, we obtain the lamp power output at frequency ν:

$$E_{L,\nu} = \int_{\omega_e} \int_{\Gamma_e} I'_{\nu,o} \, d\omega_e \, d^{(d)}\Gamma_e \tag{27}$$

where we have made the following assumptions:

1. In the case of volumetric emission, each elementary volume is transparent to the emission of its surroundings.
2. In the case of three-dimensional emission sources, they are supposed to be bounded by mathematical surfaces of zero thickness. Thus any bundle of radiation coming out of the lamp does not change its intensity or direction when it crosses this boundary.

To perform the integration in Eq. (27), we may distinguish the two emission cases:

1. *Isotropic emission.* Here $I'_{\nu,o} = I_{\nu,o}$, which is constant (independent of distance). Then Eq. (27) becomes

$$E_{L,\nu} = I'_{\nu,o} 4\pi \Gamma_e^* \tag{28}$$

where the expression of Γ_e^* for each emission model is condensed in Table 1. Finally, we have

$$e_\nu^a = \frac{\mu_\nu E_{L,\nu}}{4\pi \Gamma_e^*} \int_{\Gamma_e} \frac{1}{\rho^2} \exp\left(-\int_{\rho_o^*}^{\rho^*} \mu_\nu \, d\rho'\right) d^{(d)}\Gamma_e \tag{29}$$

2. *Diffuse emission.* In Eq. (27) we must replace $I'_{\nu,o}$ by $I_{\nu,o} \cos\theta_e$ and then perform the integration:

$$E_{L,\nu} = I_{\nu,o} \int_{\omega_e} \cos\theta_e \, d\omega_e \int_{\Gamma_e} d^{(d)}\Gamma_e \tag{30}$$

where $I_{\nu,o}$ is constant (independent of distance). Here, since the intensity of radiation follows the cosine law, the emission solid angle is 2π. Then

$$E_{L,\nu} = I_{\nu,o} \pi \Gamma_e^* \tag{31}$$

Finally, we have

$$I'_{\nu,o} = I_{\nu,o} \cos\theta_e = \frac{E_{L,\nu}}{\pi \Gamma_e^*} \cos\theta_e \tag{32}$$

and

$$e_\nu^a = \frac{\mu_\nu E_{L,\nu}}{\pi \Gamma_e^*} \int_{\Gamma_e} \frac{\cos\theta_e}{\rho^2} \exp\left(-\int_{\rho_o^*}^{\rho^*} \mu_\nu \, d\rho'\right) d^{(d)}\Gamma_e \tag{33}$$

The final expressions of e_ν^a valid for diffuse emission for each model can be found in Table 1, where $\cos\theta_e$ has been replaced by the appropriate expression for each model. Equations (29) and (33) are valid for monochromatic light if we deal with polychromatic radiation, these expressions must be integrated over the entire frequency spectrum, that is,

$$e^a = \int_{\nu=o}^{\nu=\infty} e_\nu^a \, d\nu \tag{34}$$

MODELING THE ABSORPTION PROCESS

As mentioned in preceding sections, the main difference between a photochemical and a thermal reaction is due to the presence of a radiation-activated step. The rate of reaction of this step is proportional to the local volumetric rate of energy absorption (LVREA). Hence the evaluation of the LVREA constitutes the most important stage in the analysis and design of a photochemical reactor.

Table 1 illustrates that for each emission model, the LVREA is a function of the spatial variables, of the physical properties and geometrical characteristics of the lamp-reactor system, and some physicochemical properties of the reacting mixture. The LVREA dependence on the geometrical characteristics of the lamp-reactor system will be studied in detail in the next section, and it comes from the limits of integration of the LVREA expression for each emission model. The physical properties of the lamp are represented by the parameter κ_ν in the LVREA equations; both are described in Table 1 for each model. Finally, the attenuation coefficient μ_ν involves the dependence on the physicochemical properties of the reacting system. This section is devoted to the evaluation of the attenuation coefficient and the local volumetric rate of energy absorption.

Attenuation Coefficient

In the evaluation of this coefficient it is necessary to make a distinction between homogeneous and heterogeneous media. The latter are encountered when one attempts to model systems in which gas bubbles are dispersed in a liquid, or solid particles are incorporated into a gas or liquid phase as in photocatalytic applications. The distinction arises from the fact that the presence of heterogeneities introduces unavoidable distortions in the radiation field.

Homogeneous Media

When the medium is homogeneous, the attenuation of radiation changes continuously. As mentioned before, this variation in the specific intensity due to absorption inside an elementary volume of homogeneous media may be expressed by a constitutive equation which involves an attenuation coefficient μ_ν. It is a function of the frequency of radiation ν and the state variables such as temperature, composition, and in some cases pressure. The exact form of the attenuation coefficient function can be determined from the microscopic theory by studying the absorption of a photon by atoms or molecules exposed to a radiation field. The reader is referred to the specific literature [e.g., Pai (1966) or Vincenti and Kruger (1965)].

Macroscopically, the attenuation coefficient is considered to be a linear function of the concentration of the absorbing species:

$$\mu_\nu = \alpha_\nu C \tag{35}$$

where α_ν is the absorption coefficient or absorptivity. When more than one species participate in the absorption process, Eq. (35) may be written as

$$\mu_\nu = \sum_i \mu_{i,\nu} = \sum_i \alpha_{i,\nu} C_i \tag{36}$$

Both Eqs. (35) and (36) are strictly valid for dilute solutions. The range of validity of these equations can be experimentally tested by studying the variation of μ_ν with concentration by means of a spectrophotometer. This type of experiment also allows us to obtain values of the absorption coefficient necessary to model the attenuation process. On the other hand,

for many chemical compounds, values of the absorption coefficient and the range of validity of Eqs. (35) and (36) can generally be found in standard handbooks of physics and chemistry. The absorption characteristics of many other species and mixtures of different compounds may be found in specialized publications in the field of spectroscopy. Later we will see an example of a photochemical reaction with an inorganic complex in which the characteristics of the absorbing species change with composition and the data required by Eq. (35) are not available. Instead, the attenuation coefficient is evaluated as a function of the composition of the absorbing medium.

Heterogeneous Media

Since homogeneous systems have been the main objective of the research performed so far in photochemical reactor analysis, the information available in the literature about modeling the attenuation in heterogeneous media is particularly scarce. A critical review about this subject can be found in Alfano et al. (1985). The problem can be approached in two different ways: (a) by modeling the attenuation process taking into account the scattering of radiation produced by the heterogeneities, or (b) by using the existing formulation of homogeneous media with an effective attenuation coefficient. About the first strategy, it can be ascertained that the use of a rigorous radiation source model having no adjustable parameters, together with the complexity provided by the heterogeneities of the medium, have not yet been attempted. On the other hand, the simplified method using an effective attenuation coefficient is appropriate to carry out the calculation of the rate of radiant energy absorption in heterogeneous systems with a good degree of approximation. Based on these considerations, if one attempts to design and/or analyze an industrial photochemical reactor, the second strategy must be followed for the time being.

At present, several correlations have been proposed to evaluate the effective attenuation coefficient in gas-liquid dispersions. Among them, we can particularly mention two:

1. The first one, proposed by Otake et al. (1981), is a simple empirical expression which accounts for the absorption effects produced by the liquid phase and the reflection, refraction, and transmission effects provoked by the gaseous phase. Considering that the latter are proportional to the specific surface area (a'_ν), $\mu_{eff,\nu}$ may be represented by the following correlation:

$$\mu_{eff,\nu} = \mu_\nu(1 - \varepsilon_g) + ka'_v \quad (37)$$

where μ_ν is the attenuation coefficient of the liquid phase, ε_g the holdup of the dispersed phase, and k is an empirical coefficient depending on the optical properties of the system. Otake et al. (1981) plotted the experimental information obtained by them and other authors, finding that the values can be well correlated by using k = 0.125 for $\mu_{eff,\nu}$, μ_ν, and a'_ν in the same units (cm^{-1}).

2. The other correlation proposed by Yokota et al. (1981), in which the effective attenuation coefficient becomes a function of the attenuation coefficient in the liquid phase, the bubble diameter, and the gas holdup. The expression is

$$\mu_{eff,\nu} = \mu_{\nu}(1 + h\varepsilon_g) \tag{38}$$

and

$$h = \left(\frac{3.6}{d_b}\right)^{0.66} \tag{39}$$

where d_b is the bubble diameter in mm for μ_ν and $\mu_{eff,\nu}$ expressed in m^{-1}. Both correlations may be used in the case of μ_ν smaller than 40 m^{-1}. When μ_ν is greater than 40 m^{-1}, the dispersing effect of radiation due to the presence of bubbles may normally be neglected.

Later we will see an example of the design of a two-phase photochemical reactor where the absorption process is modeled using an effective attenuation coefficient. However, no reliable design procedures have yet been developed for photocatalytic reactors (solid-fluid photoreactors).

Local Volumetric Rate of Energy Absorption

As derived in the preceding section, the integrated expression for the local volumetric rate of energy absorption may be expressed as

$$e_\nu^a = \mu_\nu |\underset{\sim}{q}_\nu| \tag{40}$$

for monochromatic radiation for each point I (r_I, β_I, z_I) inside the reactor volume. In Eq. (40) the modulus of the energy flux density vector $|\underset{\sim}{q}_\nu|$ represents the energy flux arriving from all directions in space and from the whole volume of the lamp; μ_ν is the attenuation coefficient, which is a function of the concentration of the absorbing species at point I.

Figure 8 illustrates the attenuation process undergone by each radiation bundle before reaching point I inside the reactor volume. To evaluate $|\underset{\sim}{q}_\nu|$ at point I, every direction (θ, ϕ) must be considered. For each direction (θ, ϕ) in the most general case, the ray travels from the emission source without changing its specific intensity until it reaches the reactor wall, represented by point $W(\rho_o^*, \theta, \phi)$ in Fig. 8. In the trajectory between $W(\rho_o^*, \theta, \phi)$ and $I(\rho^*, \theta, \phi)$ of Fig. 8, the specific intensity changes due to absorption. This change is represented by the exponential factor

$$\exp\left(-\int_{\rho_o^*(\theta,\phi)}^{\rho^*(\theta,\phi)} \mu_\nu \, d\rho'\right) \tag{41}$$

in the expression of $|\underset{\sim}{q}_\nu|$ for each model. Due to the concentration dependence of the attenuation coefficient, the modulus of the radiation flux density vector, and hence the LVREA, becomes a functional of the concentration field inside the reactor. This functional characteristic of the LVREA makes the problem of modeling a photochemical reactor more complex than expected. The complexity is due to the coupling between the radiation field evaluation and the resolution of the mass balance equations. The

of the reaction at each point I depends on the local value of the LVREA, but the LVREA value is a function of the extent of the reaction reached at all points in space through which every ray arriving at point I has previously traveled.

There are some special situations where this complexity is reduced. Working with a perfectly mixed reactor, the absence of stable species concentration gradients inside the reactor volume makes the attenuation coefficient uniform in space. In this case the LVREA is no longer a function of the concentration field but is still spatially dependent through its dependence with $|\underset{\sim}{q}_\nu|$.

This condition may also be attained by working with a *photosensitized* reaction independently of the mixing conditions in the reactor. In this special kind of reaction, the absorbing species does not change its concentration (i.e., it is not consumed). In this case, due to the spatial and temporal uniformity of the concentration field of the absorbing species, the link "progress of the reaction attenuation of radiation" is destroyed. In the liquid phase, photosensitized reactions generally involve large conjugate molecules such as benzophenone. In gas-phase work, sulfur dioxide, benzene, and other organic molecules are used. In addition, metal vapors have been particularly useful in gas-phase photosensitized reactions, such as Na, Zn, Cd, and Hg, of which the latter has been applied most extensively.

Another situation where the degree of complexity is lessened is in the modeling of *black body reactors*. In this type of reactor, almost all the absorption of radiation occurs in a thin layer close to the entrance of light. This characteristic may be attained by working with a strongly absorbing medium. In this case the attenuation can be modeled as occurring on the reactor wall (i.e., at the "black" surface). Consequently, the evaluation of the radiant field is reduced to estimate the energy flux density on the reactor wall.

EVALUATION OF THE LVREA FOR DIFFERENT REACTORS

The rate of initiation of a photoreaction is proportional to the LVREA (i.e., $\Omega_{act} = \Phi e^a$); the porportionality constant is the *primary quantum yield.*

The definition of this yield will depend on the nature of the process concerned. The primary process may be defined, including both the initial act of absorption and those immediately following processes which are determined by the properties of the initially excited electronic state. In almost all cases the primary process leads to a dissociation of the absorbing molecule. For such cases the primary quantum yield may be defined as

$$\Phi = \frac{\text{number of molecules dissociated by the primary process}}{\text{number of quanta of radiation absorbed}}$$

Φ cannot be measured easily; to determine its value we must gain information about the relative rates of the other possibilities more likely involved in the primary process (recombination, physical quenching, etc.). Sometimes, these rates may be obtained by inspection of the absorption spectrum. Then we may be able to estimate the value of Φ. To illustrate this we can consider the photochemical dissociation of chlorine molecules: referring to the potential curves for chlorine (Herzberg, 1950), the inequivocally show

that upper states are unstable and the molecule dissociates very rapidly. Then this process will predominate over all others and we can surely say that, almost independently of the frequency considered, $\Phi = 1$. Following the statement of the second photochemical law (Turro, 1967), the maximum possible value of Φ is unity; therefore, for the great majority of cases its value may be taken as unity.

In addition, we may use the *overall quantum yield* as

$$\Phi_{ov} = \frac{\text{number of molecules of a given reactant or product finally decomposed or formed}}{\text{number of quanta absorbed}}$$

This property will generally give an indication of the nature of the reaction mechanism. If Φ_{ov} is large, a chain mechanism is indicated; if it is small, either deactivation or recombination is suggested, although other possibilities must not be overlooked. It can clearly be seen that while Φ can never be larger than unity, Φ_{ov} can attain very large values. (Depending on the chain length, it can be as large as 10^6.)

Expressions for the LVREA for different emission models were presented earlier. There we mentioned that the integration limits will depend on the geometrical configuration of the lamp-reactor system and, perhaps, the reflector. They constitute the major complexity in the evaluation of the radiation field. For the three types of reactors described earlier, they will be derived applying the VEES and the SELS models. The SEES model requires the same equations for the lamp limiting values as the VEES model, with the exception of the ρ coordinate, which need not be integrated because the emission is produced by the surface of the lamp only. The final equations for the LVREA and $|\underset{\sim}{q}_\nu|$ also differ in the value of the characteristic constant of the radiation source κ_ν (see Table 1).

To simplify the treatment, we will derive the values of $|\underset{\sim}{q}|$ for monochromatic radiation and isotropic emission. The extension to polychromatic radiation and diffuse emission is straightforward.

In addition to the assumption already described for the emission model and the lamp, we must add the following ones about the reactor (Irazoqui et al., 1973):

1. The inner wall of the reactor is bounded by a cylindrical mathematical surface without thickness (i.e., no reflection or refraction occurs).
2. The opaque zones at the top and bottom parts of the reactor do not reflect or emit radiation.

The modulus of the radiation flux density vector at any point in the reacting space, $I(r,\beta,z)$, is the energy flux arriving from all directions in space and from the whole extension of the lamp. It can be obtained after integration of

$$|\underset{\sim}{q}| = \kappa \int_{\Gamma_e} \frac{1}{\rho^2} \exp\left(-\int_{\rho_o^*}^{\rho^*} \mu \, d\rho'\right) d^{(d)}\Gamma_e \qquad (42)$$

The problem that will always be present along this section is centered on the complexity of the limits for the integrals of Eq. (42).

Annular Photoreactor

VEES Model

In this case Eq. (42) is written as follows (see Table 1):

$$|\underset{\sim}{q}_\nu| = \kappa \int_\theta \int_\phi \int_\rho d\theta\, d\phi\, d\rho \sin\theta \exp\left(-\int_{\rho_o^*(\theta,\phi)}^{\rho^*(\theta,\phi)} \mu\, d\rho'\right) \tag{43}$$

To obtain the integration limits, one starts from the point of reception $I(r,\beta,z)$ and investigates all the arriving rays with directions (θ,ϕ). One must find those that intersect the boundaries of the surface of the cylinder representing the lamp (Fig. 9).

The limits for variable ρ are the intersections of this coordinate with the front and rear parts of the lamp in any θ,ϕ direction:

$$\rho_{1,2}(\theta,\phi) = \frac{r\cos\phi \pm (r^2\cos^2\phi - r^2 + r_L^2)^{1/2}}{\sin\theta} \tag{44}$$

The limits for variable θ are obtained considering that the limiting rays coming from the lamp and reaching the generic point (r,β,z) must satisfy two restrictive conditions: (a) a ray limits the θ angle when its equation has a common solution with the equation of the circumference that defines the nonemitting zones at the upper and lower parts of the lamp; and (b) the ray will be a limit when the common solution above corresponds to the intersection of the ray with that portion of the circumference which, limited by two generatrix lines corresponding to the ϕ limiting angles, is closer to the generic point (r,β,z). The second condition is needed because for any plane at constant ϕ, there are two θ angles that satisfy the first restriction for both the upper and lower boundaries (Fig. 10).

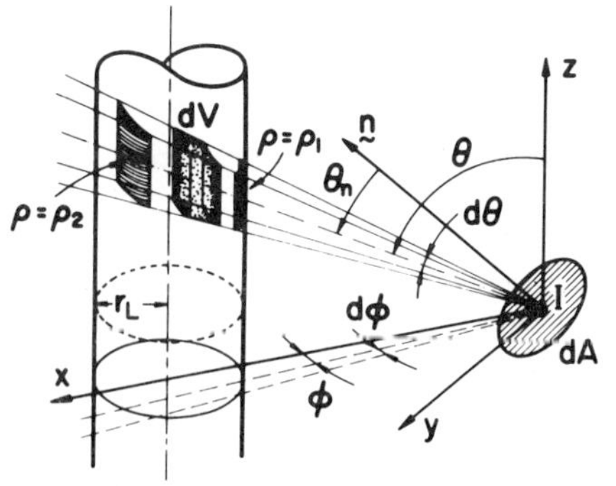

FIGURE 9 Geometry and coordinate systems for direct radiation

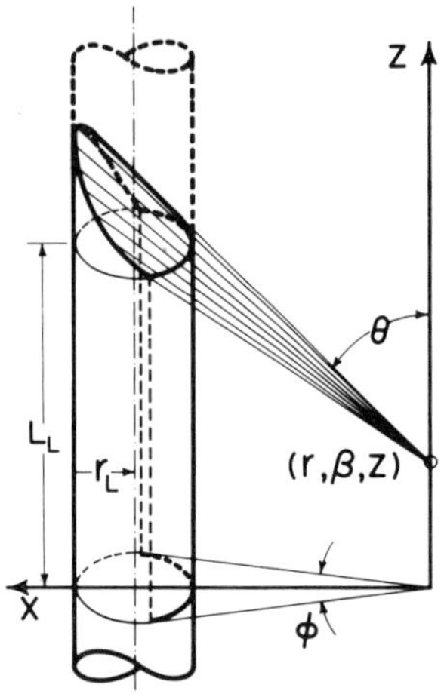

FIGURE 10 Limits of θ and ϕ for direct radiation.

In this way, for θ_1 and θ_2 we obtain the following expressions

$$\theta_1(\phi) = \tan^{-1} \frac{r \cos \phi - [r^2(\cos^2\phi - 1) + r_L^2]^{1/2}}{L_L - z} \tag{45a}$$

$$\theta_2(\phi) = \tan^{-1} \frac{r \cos \phi - [r^2(\cos^2\phi - 1) + r_L^2]^{1/2}}{-z} \tag{45b}$$

A more rigorous approach should require computation of the reactor and the lamp wedges. The three-dimensional characteristic of the emission process causes the reactor to be partially irradiated in the incoming and outgoing regions (Fig. 11a). This fact increases the effective reaction volume. This effect is more noticeable when the inner reactor wall is close to the lamp. The expressions for the θ-limiting angles should be modified in the partially irradiated regions; the reader is referred to Romero et al. (1983) for a detailed analysis of the reactor wedge computation. On the other hand, Fig. 11b shows that the calculations should be corrected to account for the effective lamp length. If no corrections were introduced, we would be computing as emission volumes portions of the cylinder that correspond to regions in the space where there is no emission at all. This effect is also significant when the inner reactor wall is very close to the lamp; for a more detailed analysis, the reader is referred to De Bernardez and Cassano (1985).

The limiting rays in the ϕ direction can be obtained by imposing the restriction that both intersections of the ρ coordinate with the lamp boundary must coincide, that is,

$$\rho_1 = \rho_2 \tag{46}$$

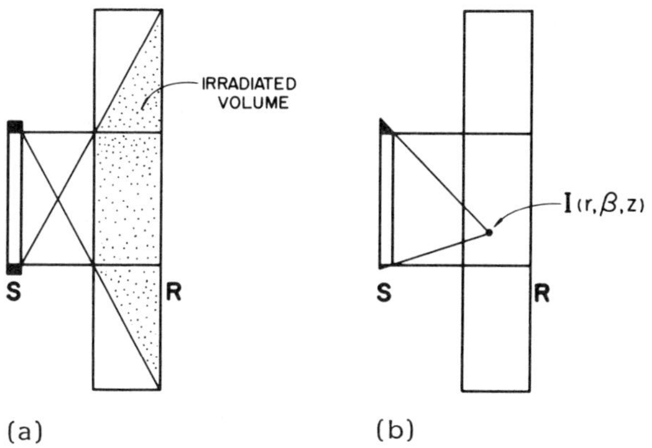

FIGURE 11 (a) Reactor wedges; (b) lamp wedges.

With Eq. (44) and since ϕ can take values only in the closed interval $[-\pi/2, \pi/2]$,

$$-\phi_1 = \phi_2 = \cos^{-1}\frac{(r^2 - r_L^2)^{1/2}}{r} \tag{47}$$

With these limits, Eq. (45) can be integrated to obtain

$$|\underset{\sim}{q}| = 2\kappa\int_{\phi_1}^{\phi_2} d\phi\ (r_L^2 - r^2\sin\phi)^{1/2}\int_{\theta_1(\phi)}^{\theta_2(\phi)} d\theta\ \exp\left(-\int_{\rho_o^*(\theta,\phi)}^{\rho^*(\theta,\phi)} \mu\ d\rho'\right) \tag{48}$$

SELS Model

In this case Eq. (42) becomes (see Table 1):

$$|\underset{\sim}{q}| = \kappa\int_o^{L_L}\frac{dz}{\rho^2}\ \exp\left(-\int_{\rho_o^*(\theta)}^{\rho^*(\theta)} \mu\ d\rho'\right) \tag{49}$$

Let us call

$$\rho(\theta)\cos\theta = r \tag{50}$$

$$\rho(\theta)\sin\theta = z \tag{51}$$

Then we have

$$\frac{dz}{\rho^2} = \frac{d\theta}{r} \tag{52}$$

and finally,

$$|\underset{\sim}{q}| = \frac{\kappa}{r} \int_{\theta_1}^{\theta_2} d\theta \exp\left(-\int_{\rho_o^*(\theta)}^{\rho^*(\theta)} \mu \, d\rho'\right) \tag{53}$$

where θ_1 and θ_2 can be obtained from Eqs. (44) and (45), making $r_L = 0$ and $\phi = 0$. Again, a more rigorous approach should consider the reactor wedges (see Fig. 11a), as it was mentioned when the VEES model was discussed.

Cylindrical Photoreactor with Elliptical Reflector

Four additional assumptions may be added to those given earlier (Cerdá et al., 1973):

1. The reflector is a perfect elliptical cylinder.
2. The lamp is located in such a way that its centerline passes through one of the focuses of the elliptical reflector (F_1).
3. Specular reflection occurs with an average reflection coefficient that is independent of wavelength and direction. (These restrictions could be easily relaxed and are used here for simplicity.)
4. The reflected radiation comes only from the elliptical reflector (i.e., the top and bottom parts of the cylinder do not reflect radiation).

The radiation impinging at any point inside the elliptical cylinder is made up of two parts: *direct radiation* and *reflected radiation*. Excluding direct radiation, a ray could arrive at a point of reception with a given direction by a single reflection process or as a result of successive contacts with the elliptical mirror. In the second case, for points located close to the focus F_2, the ray will describe a broken line touching the surface of the reflector at least three times. The question is whether it is always possible for a ray coming out of the lamp to undergo so many reflections and still arrive at the point of interest. It can be answered by following the ray path in the reverse way (i.e., starting from the point of reception). A multiply reflected ray will be effective at a given point in space if and only if, traveling the reverse trajectory previously described and in spite of its progressive vertical displacement, it still intersects the surface of the lamp boundary that is seen from the points of incidence. Through the same reasoning, a multiply reflected ray would have no real existence if this intersection falls outside the length of the lamp.

The problem consists in finding the mathematical relationship between the incident radiant energy flux and the reactor, lamp and reflector dimensions, the optical properties of the reflector and the reactor wall, and the physical properties of the radiation coming out of the source. The point of reception (I) on the reactor volume is referred to a cylindrical coordinate system located at the centerline of the cylindrical reactor with origin at the reflector bottom (r_I, β_I, ζ_I).

VEES Model

Direct Radiation: For this case, the limits of integration for variable ρ may be obtained in a similar way as for the annular reactor. The results are (see Fig. 12)

$$\rho_{1,2}(\theta,\phi) = \frac{\Delta \cos \phi \pm [\Delta^2 \cos^2\phi - (\Delta^2 - r_L^2)]^{1/2}}{\sin \theta} \tag{54}$$

where

$$\Delta = [(2c + r_I \cos \beta_I)^2 + r_I^2 \sin^2\beta_I)]^{1/2} \tag{55}$$

$$\theta_1(\phi) = \tan^{-1} \frac{\Delta \cos \phi - (r_L^2 - \Delta^2 \sin^2\phi)^{1/2}}{(1/2)(L_{Rf} + L_L) - \zeta_I} \tag{56}$$

$$\theta_2(\phi) = \tan^{-1} \frac{\Delta \cos \phi - (r_L^2 - \Delta^2 \sin^2\phi)^{1/2}}{(1/2)(L_{Rf} - L_L) - \zeta_I} \tag{57}$$

$$\phi_1 = \sin^{-1} \frac{r_L}{\Delta} \tag{58}$$

$$\phi_2 = 2\pi - \phi_1 \tag{59}$$

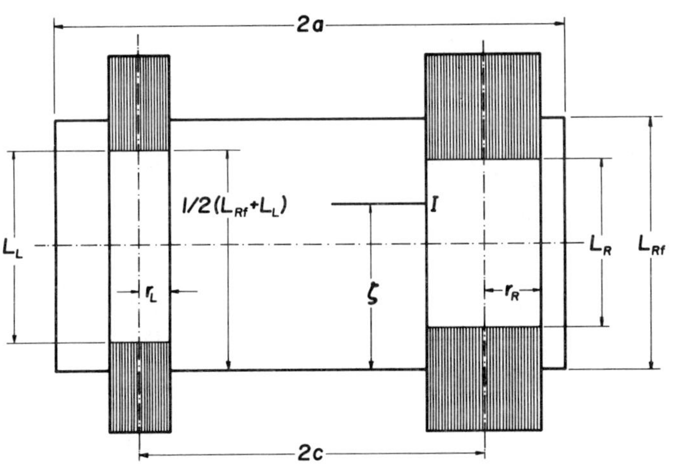

FIGURE 12 Variables used in the determination of the limits of integration.

For direct radiation the final expression after integration of Eq. (43) is:

$$|\underset{\sim}{q}| = 2\kappa \int_{\phi_1}^{\phi_2} d\phi \, (r_L^2 - \Delta^2 \sin^2\phi)^{1/2} \int_{\theta_1(\phi)}^{\theta_2(\phi)} d\theta \, \exp\left(-\int_{\rho_o^*(\theta,\phi)}^{\rho^*(\theta,\phi)} \mu \, d\rho'\right) \tag{60}$$

Indirect Radiation with Only One Reflection: The methodology of analysis for reflected radiation is to follow a ray starting from the point of reception in order to identify the point of reflection, and from this (applying the laws of optical physics) to search in the direction of emission. To understand the analysis it is necessary to locate several key points (Fig. 13): first, the point of incidence of the reflected ray (point of reception I); second, the point of reflection (P); and finally, the two points of intersection of the direction of emission with the lamp cylinder (E_1 and E_2). Any ray coming from the lamp will be indicated by an incidence position vector at the point of reflection (ρ_E) and a reflection position vector at the point of reception (ρ_I). The former has its origin at P and the latter at I. The intersections of ρ_E with the cylinder of the lamp tube define two vectors, $\rho_{E,1}$ and $\rho_{E,2}$. The details of the geometrical derivations and definitions of the variables are explained in the appendix to this chapter. The results are

$$\rho_{E_{2,1}} = \frac{D \cos \xi \pm (r_L^2 - D^2 \sin^2 \xi)^{1/2}}{\sin \theta} \tag{61}$$

$$\theta_1 = \tan^{-1} \frac{\rho_I' + D \cos \xi - (r_L^2 - D^2 \sin^2 \xi)^{1/2}}{(1/2)(L_{Rf} + L_L) - \zeta_I} \tag{62}$$

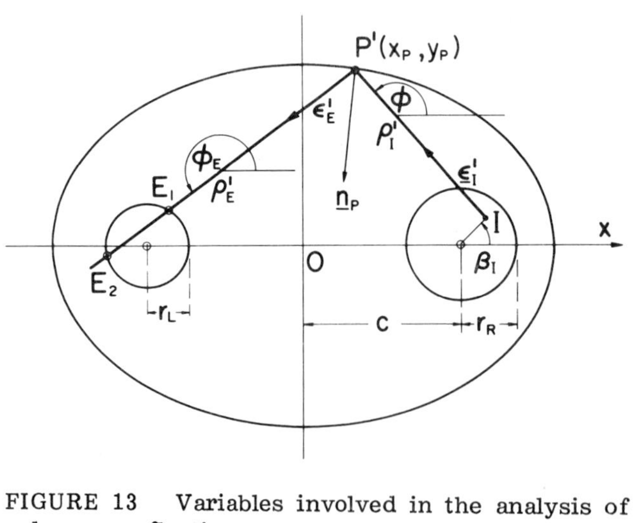

FIGURE 13 Variables involved in the analysis of indirect radiation with only one reflection.

$$\theta_2 = \tan^{-1} \frac{\rho_I' + D\cos\xi - (r_L^2 - D^2 \sin^2\xi)^{1/2}}{(1/2)(L_{Rf} - L_L) - \zeta_I} \tag{63}$$

$$D^2 \sin^2\xi(\phi_{1,2}) = r_L^2 \tag{64}$$

Then the integration of Eq. (43) in the ρ coordinate gives us the following expression for the modulus of the radiation flux density due to indirect radiation with only one reflection:

$$|\underset{\sim}{q}| = 2\kappa\Gamma_{Rf} \int_\phi d\phi (r_L^2 - D^2 \sin^2\xi)^{1/2} \int_{\theta_1(\phi)}^{\theta_2(\phi)} d\theta \exp\left(-\int_{\rho_o^*(\theta,\phi)}^{\rho^*(\theta,\phi)} \mu\, d\rho'\right) \tag{65}$$

where the attenuation of the radiant beams at the mirror surface was taken into account using the reflection coefficient of the elliptical mirror Γ_{Rf}.

Indirect Radiation with Two or More Reflections: As explained earlier, the mathematical analysis is accomplished by considering the broken trajectories of the rays arriving at the reaction surface after more than one reflection. The derivation will not be presented here. Details may be found in Cerdá et al. (1977).

Asymmetries: An additional problem with this system is originated by the azimuthal asymmetry. This is due to the overlapping of two effects: the arrival of direct radiation (only on the points situated in front of the lamp) and the existence of umbra and penumbra zones, giving rise to nonuniformities in the indirect radiation. When the characteristics of the reactor generate important asymmetries, the addition of angular terms to the mass balances transforms all problems from two- to three-dimensional.

A previous work (De Bernardez and Cassano, 1982) showed that for this system the uniform angular distribution will be obtained only if one works with a reactor diameter/lamp diameter ratio lower than 0.5 and with an eccentricity as low as possible (about 0.4). If one wishes to work with higher eccentricities, the reactor diameter/lamp diameter ratio should be lower than 0.2. It is recommended that values of the distance between focuses of the ellipse be as large as possible.

SELS Model

Direct Radiation: The expression of the modulus of the energy flux density vector for this model is similar to that obtained for the annular reactor, but r must be replaced by Δ, with Δ given by Eq. (55)

$$|\underset{\sim}{q}| = \frac{\kappa}{\Delta} \int_{\theta_1}^{\theta_2} d\theta \exp\left(-\int_{\rho_o^*(\theta)}^{\rho^*(\theta)} \mu\, d\rho'\right) \tag{66}$$

The limits in the θ coordinate can be obtained from Eqs. (56) and (57), making $\phi = 0$ and $r_L = 0$.

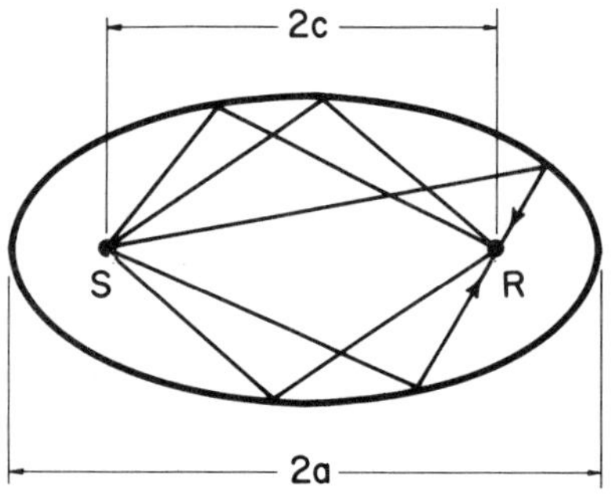

FIGURE 14 Projections of reflected ray trajectories.

Indirect Radiation with Only One Reflection: All reflected rays must lie on planes that also contain the reactor axis. A ray which after reflection reaches the reactor centerline may be projected on a plane that goes through the point of emission and is perpendicular to the lamp axis. The length of this projection is equal to 2a (Fig. 14).

Each one of these planes allows us to define a circumference constructed with the locus of all the end points of rectified rays (Fig. 15). This can be done in this way because the reflector is a cylinder and the lamp is assumed to be a line parallel to the generatrix line of the mirror. This circumference may be repeated for each point of emission of the lamp, giving rise to a cylinder of circular cross section of radius equal to 2a. This cylinder may be thought like a virtual surface of emission that acts as a substitute for the linear lamp. Taking a planar cross section going through the reactor focal axis of the elliptical cylinder, we finally obtain a situation like the one depicted in Fig. 16. Points such as those

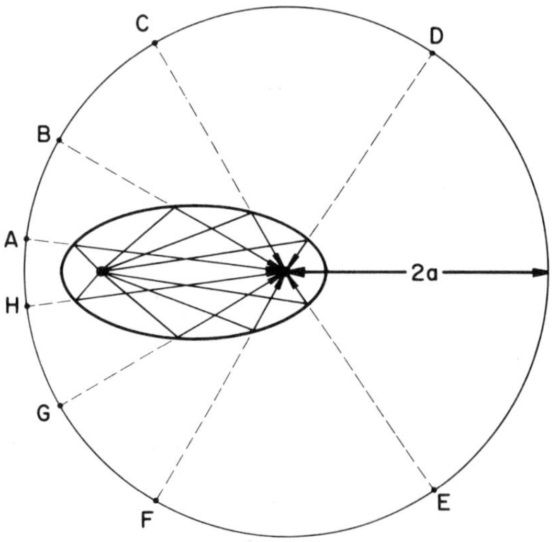

FIGURE 15 Locus of end points of rectified rays.

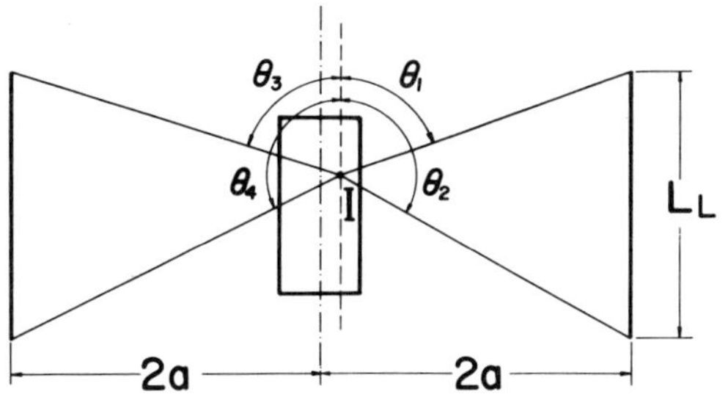

FIGURE 16 Limiting angles for the θ coordinate.

designated in Fig. 15 with letters A, B, C, D, and so on, when converted into straight lines as is done for one cross section in Fig. 16, represent the way in which the linear lamp would be seen by observers located at the reactor centerline if they looked in the indicated directions. The projection of the actual distance travelled by any ray with direction θ on the lamp centerline is

$$\frac{2a}{\sin\theta}\cos\theta = 2a\cot\theta \tag{67}$$

The key point, as in any emission model, is to know the limits of the θ angles (i.e., out of all those θ directions arriving at the reactor axis which are really coming out of the "line" source). If the ray effectively comes out of the lamp (i.e., if it is a direction that transports energy), its projection given by Eq. (67) must lie within the lamp longitudinal dimensions. This allows us to define the limiting angles for θ in the following manner:

$$\theta_1 = \tan^{-1}\frac{2a - r_I}{(1/2)(L_{Rf} + L_L) - \zeta_I} \tag{68}$$

$$\theta_2 = \tan^{-1}\frac{2a - r_I}{(1/2)(L_{Rf} - L_L) - \zeta_I} \tag{69}$$

$$\theta_3 = \tan^{-1}\frac{2a + r_I}{(1/2)(L_{Rf} + L_L) - \zeta_I} \tag{70}$$

$$\theta_4 = \tan^{-1}\frac{2a + r_I}{(1/2)(L_{Rf} - L_L) - \zeta_I} \tag{71}$$

To compute the reflected radiation at point I, we have split the integral into two parts to account for energy contributions from the whole closed reflecting surface of the elliptical mirror. The final expression for the modulus of the radiation flux density of the reflected radiation with the linear model is

$$|\underset{\sim}{q}| = \kappa\Gamma_{RF}\left[\frac{1}{2a - r_I}\int_{\theta_1}^{\theta_2} d\theta \exp\left(-\int_{\rho_o^*(\theta)}^{\rho^*(\theta)} \mu\ d\rho'\right)\right.$$

$$\left. + \frac{1}{2a + r_I}\int_{\theta_3}^{\theta_4} d\theta \exp\left(-\int_{\rho_o^*(\theta)}^{\rho^*(\theta)} \mu\ d\rho'\right)\right] \qquad (72)$$

It should be stressed that since it is a linear model, we have lost the integration in the ϕ coordinate, which must always be taken into account when an extense model is used (with finite dimensions for r_L).

Cylindrical Photoreactor with Parabolic Reflector

Here again we may have direct and reflected radiation. For the analysis we will require two coordinate systems. The first one ($x_F - y_F - z_F$) is defined with the origin at point F located at the center of the bottom of the empty cylindrical reactor, with y_F, consequently, being its centerline (Fig. 17). An auxiliary coordinate system parallel to the former is also needed. Its center is located at the point of incidence designated by I in Fig. 17 and it can be moved all over the space under analysis.

VEES Model

Direct Radiation: For this case, the limits of integration may be obtained in a similar way as for the annular and elliptical reactors. The results are

$$\rho_{1,2}(\theta,\phi) = -\frac{\Delta_1 \sin\phi + \Delta_2 \cos\phi}{\sin\theta} \pm \frac{[r_L^2 - (\Delta_2 \sin\phi - \Delta_1 \cos\phi)^2]^{1/2}}{\sin\theta} \qquad (73)$$

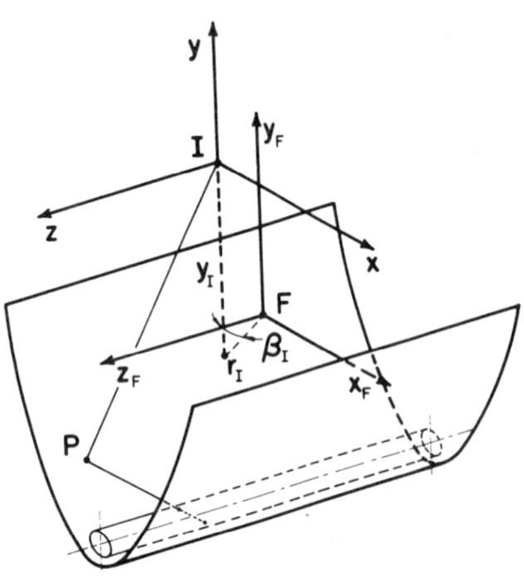

FIGURE 17 Geometry of the tubular source with parabolic reflector.

where

$$\Delta_1 = y_I + \ell - a \tag{73}$$

$$\Delta_2 = r_I \sin \beta_I \tag{74}$$

$$\theta_1(\phi) = \tan^{-1} \frac{\rho'_1}{L_L/2 + z_I} \tag{75}$$

$$\theta_2(\phi) = \tan^{-1} - \frac{\rho'_1}{L_L/2 + z_I} \tag{76}$$

$$\rho'_1 = -\Delta_1 \sin \phi - \Delta_2 \cos \phi - [r_L^2 - (\Delta_2 \sin \phi - \Delta_1 \cos \phi)^2]^{1/2} \tag{77}$$

The limits for ϕ are obtained from the solution of the implicit equation

$$r_L^2 - (\Delta_2 \sin \phi - \Delta_1 \cos \phi)^2 = 0 \tag{78}$$

and the final expression for $|\underset{\sim}{q}|$ is

$$|\underset{\sim}{q}| = 2\kappa \int_{\phi_1}^{\phi_2} d\phi \, [r_L^2 - (\Delta_2 \sin \phi - \Delta_1 \cos \phi)^2]^{1/2} \int_{\phi_1(\phi)}^{\phi_2(\phi)} d\theta \times \exp\left(- \int_{\rho_o^*(\theta,\phi)}^{\rho^*(\theta,\phi)} \mu \, d\rho' \right) \tag{79}$$

Indirect Radiation: The methodology of analysis for this type of radiation is similar to that employed with the elliptical photoreactor and will not be repeated here. The expression resulting for the modulus of the radiation flux density due to indirect radiation is identical to Eq. (65).

The limits for the ρ variable are the same as those which were obtained for the elliptical photoreactor, that is,

$$\rho_{2,1} = \frac{D \cos \xi \pm (r_L^2 - D^2 \sin^2 \xi)^{1/2}}{\sin \theta} \tag{80}$$

But the meaning of the variables in this case is

$$D = (x_L^2 + y_L^2)^{1/2} \tag{81}$$

$$\sin \xi = - \frac{1}{D} (x_L \sin \phi_E - y_L \cos \psi_E) \tag{82}$$

$$\cos \xi = -\frac{1}{D}(x_L \cos \phi_E + y_L \sin \phi_E) \tag{83}$$

where

$$x_L = \Delta_2 + x_P \tag{84}$$

$$y_L = \Delta_1 + y_P \tag{85}$$

$$\phi_E = \tan^{-1} \frac{(1 - G^2)\sin \phi - 2G \cos \phi}{(G^2 - 1)\cos \phi - 2G \sin \phi} \tag{86}$$

$$G = \frac{-2a}{x_P + \Delta_2} \tag{87}$$

The limits for the θ variable are represented by the following expressions:

$$\theta_1(\phi) = \tan^{-1} \frac{\rho_I' + D \cos \xi - (r_L^2 - D^2 \sin^2 \xi)^{1/2}}{L_L/2 - z_I} \tag{88}$$

$$\theta_2(\phi) = \tan^{-1} - \frac{\rho_I' + D \cos \xi - (r_L^2 - D^2 \sin^2 \xi)^{1/2}}{L_L/2 + z_I} \tag{89}$$

and

$$\rho_I' = \frac{2a \tan \phi - \Delta_2 - 2[a^2(1 + \tan^2\phi) + a(\Delta_1 - \Delta_2 \tan\phi)]^{1/2}}{\cos \phi} \tag{90}$$

To obtain the limits for the ϕ variable, we must solve (as in the case of the elliptical photoreactor) the following implicit equation:

$$r_L^2 = D^2 \sin^2 \xi \tag{91}$$

Finally, it should be mentioned that in certain cases it may be necessary to put a protecting cover over the emission system which can also be used to support the cylindrical reactor (see Figs. 18 and 19). If this is the case, there will be totally and partially irradiated zones within the reaction space. Consequently, for each incidence point I, we have to determine which portion of the parabolic reflector or source is capable of illuminating the point. From the mathematical point of view, the integration interval for each point I must be adjusted. To do this, the limiting angles determined by the circular hole of the lamp-reflector cover system must be calculated (see Figs. 18 and 19), and afterward compared with the limiting angles defined for the integration of the source volume [Eqs. (88) to (91)]. For more detailed information, the reader is referred to Alfano et al. (1984b).

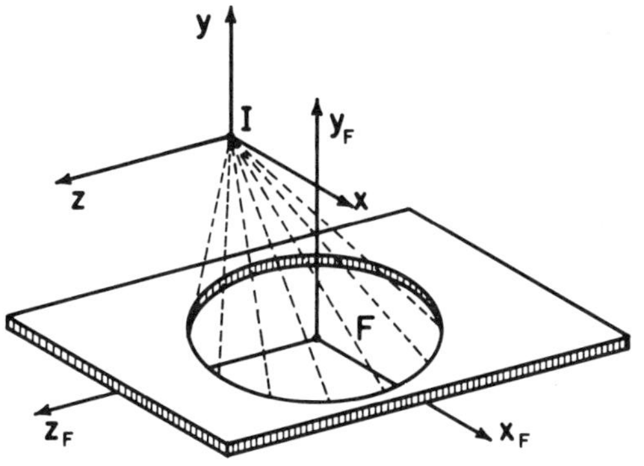

FIGURE 18 Illustrative representation of changes needed in the limits for θ due to the shape of the emitting system.

SELS Model

Direct Radiation: The expression for this case is the same as the one given for the annular reactor, but replacing r by Δ with Δ given by

$$\Delta = (\Delta_1^2 + \Delta_2^2)^{1/2} \tag{92}$$

The limits in the θ coordinate can be obtained from Eqs. (75) and (76) making $r_L = 0$ and

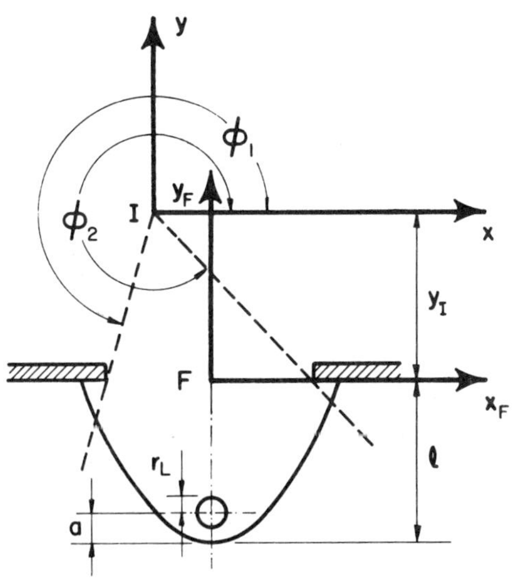

FIGURE 19 Illustrative representation of changes needed in the limits for ϕ due to the shape of the emitting system.

$$\tan \phi = \frac{\Delta_1}{\Delta_2} \tag{93}$$

Indirect Radiation: Following the same approach as the one derived to obtain $|\underset{\sim}{q}|$ for the elliptical reflector, we get

$$|\underset{\sim}{q}| = \kappa\Gamma_{Rf}\int_{\theta_1}^{\theta_2} d\theta \frac{1}{\rho_I' + \rho_E'} \exp\left(-\int_{\rho_o^*(\theta)}^{\rho^*(\theta)} \mu \, d\rho'\right) \tag{94}$$

where

$$\theta_1 = \tan^{-1}\frac{\rho_I' + \rho_E'}{L_L/2 - z_I} \tag{95}$$

$$\theta_2 = \tan^{-1}\frac{\rho_I' + \rho_E'}{L_L/2 + z_I} \tag{96}$$

$$\rho_I' = \frac{4a(\Delta_1 + a) - \Delta_2^2}{4a} \tag{97}$$

abd

$$\rho_E' = [(\Delta_1 - \rho_I')^2 + \Delta_2^2]^{1/2} \tag{98}$$

Inability of the Linear Models to Predict the Radiation Field When Curved Reflecting Surfaces Are Part of the Emitting System

From previous theoretical developments it is clear that the linear models are much simpler to use than the extense ones. The computing time is considerably reduced, and consequently they should always be the starting equations for photoreactor design. Cerdá et al. (1978) and Romero et al. (1983) showed that for certain cases the SELS model provides good results. For the annular type of reactor, where only direct radiation is involved, errors are never larger than 15%. A recent work (Clariá et al., 1986b) showed that under no conditions can linear models be used if curved reflecting surfaces combined with tubular lamps are employed. This work confirmed experimental results of Alfano et al. (1984), regarding the validity of linear models when a reflector is used. We will not reproduce here the details of these papers, but we shall present a brief summary of their main results as well as of their conclusions.

Figures 20 and 21 show computed and experimental results for two reactions (chlorination of ethane and photodecomposition of uranyl oxalate) carried out in an elliptical photoreactor. Plots include predictions with the SELS model and the VEES model. It is obvious that the error of the linear model is intolerable for any practical purpose. [The details are given in Clariá et al. (1985b; 1986a,b).]

Table 2 clearly shows that when only direct radiation is considered the differences between the linear model and the extense one are negligible.

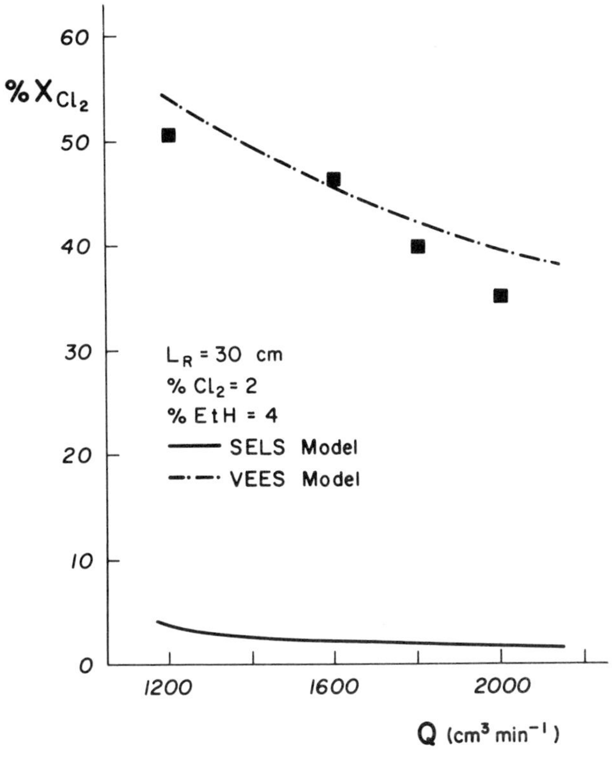

FIGURE 20 VEES model and SELS model theoretical predictions compared with experimental results for ethane photochlorination.

Even though there exist conditions where this approximation is not good, it can be safely used for most engineering purposes (see Clariá et al., 1986b).

Table 3 shows experimental verifications performed in a system which has a cylindrical reflector of parabolic cross section under the following experimental conditions: measurements of total and only direct radiation were performed with a precise thermopile; and they were done with the radiation probe receiving the sum of both direct and indirect radiation and with the same sensing device but the reflector covered with a copper sheet completely masked with carbon deposition. The data were obtained for different positions in space and the results compared with predictions made the SELS model and the VEES model. In all cases previsions were taken in the computation of the radiation field to account for the orientation and finite size of the sensor (thermopile). Clearly, in cases like this, the linear models are useless. [The details may be found in Alfano et al. (1984).]

Without going into the details of the theoretical proof of this inability of the linear models to predict reflected radiation, let us explain the reasons in an intuitive way. When the tubular lamp is placed parallel to the generatrix straight line of the cylindrical reflector and it is modeled as a straight line, the source does not "recognize" the curvature of the reflecting surface. Hence the concentrating effect of the curved mirror, be it elliptical or parabolic, cannot be predicted.

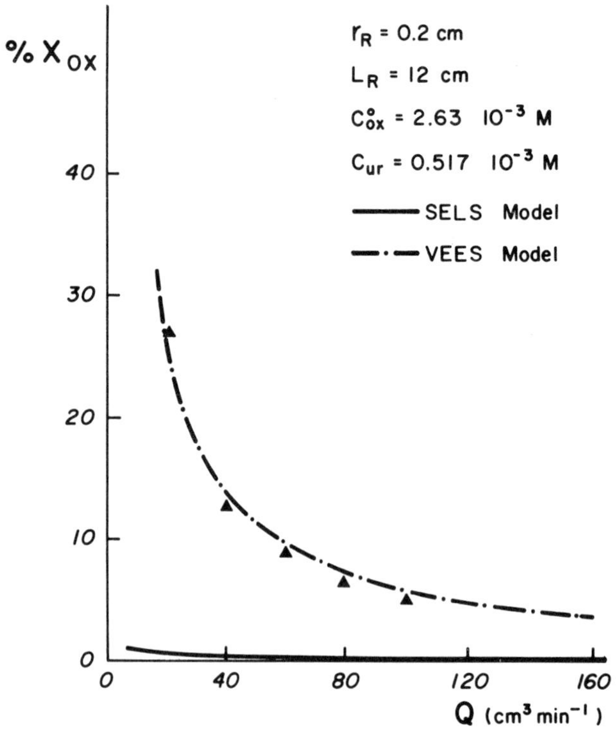

FIGURE 21 VEES model and SELS model theoretical predictions compared with experimental results for uranyl oxalate photodecomposition.

TABLE 2 Comparison of the SELS and VEES Models for Direct Radiation (Elliptical Reflector System)[a]

	$\lvert q \rvert_D \times 10^7$ (einstein/cm^2·s)	
r_L/c	SELS	VEES
0.018	0.031	0.031
0.022	0.047	0.046
0.025	0.058	0.058
0.030	0.083	0.082
0.033	0.102	0.100
0.037	0.127	0.123
0.040	0.143	0.139
0.050	0.215	0.205

[a] e, 0.66; r_R, 0.5 cm; L_L, 50 cm; L_R, 50 cm; r_I, $r_R/2$; β_I, 0; z_I, $L_R/2$; E_L, 300 W (3000 Å).

TABLE 3 Direct-to-Total Radiation Ratio: Comparison of The SELS and VEES Models (Cylindrical Reactor Irradiated from the Bottom)[a]

Cylindrical coordinate	Contribution of direct radiation to total radiation: Prediction of SELS model (%)	Prediction of VEES model (%)	Experimental values (%)
$y_I = 12.02$ cm $r_I = 2.80$ cm $\beta_I = \pi/4$	69.0	27.7	29.2
$y_I = 12.02$ cm $r_I = 1.69$ cm $\beta_I = 0$	69.3	28.1	30.8
$y_I = 12.02$ cm $r_I = 2.99$ cm $\beta_I = 0$	69.2	28.0	31.2
$y_I = 12.02$ cm $r_I = 4.35$ cm $\beta_I = 0$	69.1	27.7	31.6

[a] a, 2.1 cm; ℓ, 8.4 cm; r_L, 0.95 cm; L_L, 15.2 cm; lamp, GE UA3-360.

Depending on the significance of direct radiation on the total radiation flux density, the error may go from more than 100%, as in the case of the parabolic reflector, to up to more than two orders of magnitude for the case of the elliptical one.

In conclusion: for most practical purposes, when only direct radiation is involved, the linear models can be used. When reflecting curved surfaces are present, only the extense source type of model can be used. The finite size of the radius of the tubular lamp must be incorporated in the model, and in this way the extent of the emission source will be able to recognize the curvature of the mirror and its concentrating effects.

PHOTOREACTOR DESIGN: APPLICATIONS

For the photoreactor design problem stated earlier, all the theory has now been developed. In this section we show how it is possible to perform the design of a photoreactor. To do so, we have chosen four applications. The first deals with the design of an annular photoreactor with simple kinetics. The methodology is applied to a bench-scale reactor and the

theoretical results are compared with experimental data. The second is a gas-phase photochlorination, a typical chain reaction. Again the theory is presented and applied to a bench-scale photoreactor inside an elliptical reflector. Results are also compared with experimental data. A third problem deals with a heterogeneous reaction. Problems related to the gas absorption, the reaction rates, the radiation field in a two-phase system, and the mixing states must be taken into account to formulate a rigorous model. The method is applied to a chain reaction and one must search for simplifications that could reduce the design problem to a tractable one. Once more, predictions for a bench-scale system are compared with experimental results. Finally, the optimal design of an annular reactor for consecutive photochlorinations is illustrated in the fourth application. Here the whole process of designing a photoreactor is presented in detail, starting with the information that a practicing engineer would have at hand to begin work.

Simple Kinetics in a Continuous Annular Photoreactor with Almost Monochromatic Radiation

Consider the isothermal steady-state photodecomposition of a solution of uranyl oxalate flowing through an annular reactor irradiated by an ultraviolet arc lamp located at its axis. For the bench-scale lamp and reactor characteristics listed below, we wish to study the incidence of the volumetric flow rate on the oxalic acid exit conversion.

Reactor characteristics: r_{in} = 2.3 cm; r_{ou} = 2.7 cm; L_R = 30 cm. Lamp characteristics: GE G15T8 (15 nominal watts); L_L = 43 cm; r_L = 1.25 cm. Uranyl oxalate initial composition (UO_2SO_4): 1×10^{-6} mol/cm^3, $H_2C_2O_4$: 5×10^{-6} mol/cm^3.

The uranyl oxalate photochemical decomposition is a well-known chemical actinometer. Chemists have found that the rate of oxalic acid decomposition is proportional to the local volumetric rate of energy absorption, where the proportionality factor is the overall quantum yield. It represents a good example of a photochemical reaction that can be modeled using an overall kinetics. To find the relationship between the oxalic acid exit conversion and the volumetric flow rate, we must solve the mass balance equation for the oxalic acid.

We can introduce the following assumptions and/or simplifications in the general equation: (a) negligible thermal effects; (b) steady state ($\partial C_{ox}/\partial t = 0$); (c) physical properties such as velocities, diffusivities, and densities are constant through the reaction volume; (d) axial laminar flow; (e) Newtonian fluid; (f) azimuthal symmetry; and (g) axial diffusion negligible when compared with the convective flux. If the hypothesis above are fulfilled, the mass balance equation for the oxalic acid reduces to

$$v_z(r)\,\frac{\partial C_{ox}}{\partial z} = D_{ox}\,\frac{1}{r}\,\frac{\partial}{\partial r}\left(r\,\frac{\partial C_{ox}}{\partial r}\right) + \Omega_{ox} \qquad (99)$$

with the following boundary conditions:

$$r = r_{in}, \quad \forall_z, \quad \frac{\partial C_{ox}}{\partial r} = 0 \qquad (100)$$

$$r = r_{ou}, \quad \forall_z, \quad \frac{\partial C_{ox}}{\partial r} = 0 \tag{101}$$

and

$$\forall_r, \quad z = 0, \quad C_{ox} = C^\circ_{ox} \tag{102}$$

We assume that the velocity profile is fully developed; for example, allowing the existence of an entrance region, we can express

$$v_z(r) = 2\langle v\rangle \frac{1 - (r/r_{in})^2 + [(1 - \chi^2)/\ln(1/\chi)]\ \ln(r/r_{in})}{1 + \chi^2 - [(1 - \chi^2)/\ln(1/\chi)]} \tag{103}$$

where $\chi = r_{in}/r_{ou}$.

Finally, to solve the mass balance equation, we need an expression for the reaction rate Ω_{ox}. We know that Ω_{ox} can be expressed by

$$\Omega_{ox} = -\Phi_{ov} e^a \tag{104}$$

Now we need some data about the absorption characteristics of the uranyl oxalate solution. Light is absorbed by the uranyl oxalate complexes, which are decomposed into water, carbon dioxide, carbon monoxide, and formic acid, while the uranyl ion remains unchanged. The overall reactions are

$$UO_2^{2+} + H_2C_2O_4 + h\nu \longrightarrow UO_2^{2+} + CO_2 + CO + H_2O$$

$$UO_2^{2+} + H_2C_4O_4 + h\nu \longrightarrow UO_2^{2+} + CO_2 + HCOOH$$

The relative significance of each reaction path depends on the pH of the solution. Nevertheless, the overall rate of oxalic acid decomposition is proportional to the LVREA and independent of the reaction path.

Eighty-five percent of the radiation power output of the lamp falls into a single wavelength (λ = 2537 Å). The absorption characteristics of the uranyl-oxalate complex show the following significant variations

λ (Å)	μ (cm^{-1})	Φ_{ov} (mol/einstein)
2250	16.71	0.56
2950	1.96	0.57
3450	0.17	0.51

It is then clear that the reactor can be safely modeled as operating with monochromatic radiation.

The chemical nature of the absorbing species changes with the composition of the solution. However, a useful data for solving the modeling problem would be a plot of the attenuation coefficient versus the oxalic acid concentration, at constant uranyl ion concentration (Fig. 22). The

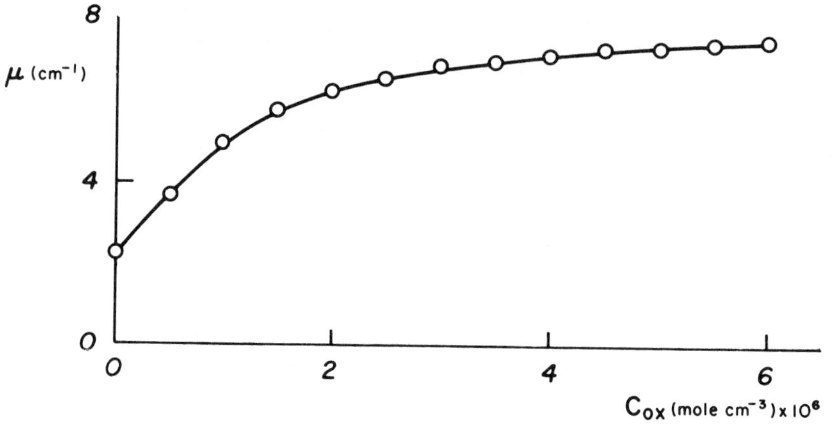

FIGURE 22 Attenuation coefficient as a function of the oxalic acid concentration.

plot shows that the attenuation coefficient is not a linear function of the oxalic acid concentration, which agrees with the fact that the oxalic acid is not the actual absorbing species. The experimental data were fitted to give an analytical expression for the concentration dependence of the attenuation coefficient:

$$\mu(cm^{-1}) = 2.362 + 3.372 \times 10^{6}C_{ox} - 9.582 \times 10^{11}C_{ox}^{2} + 1.286 \times 10^{17}C_{ox}^{3} - 6.527 \times 10^{21}C_{ox}^{4} \quad (105)$$

valid for $C_{ur} = 1 \times 10^{-6}$ mol/cm^3, $\lambda = 2537$ Å, and C_{ox} expressed in mol/cm^3.

Figure 22 shows that when the molar ratio of oxalic acid to uranyl ion is 4 or greater, the attenuation coefficient is almost constant (i.e., practically independent of the oxalic acid concentration). Hence, starting with a solution of composition UO_2SO_4 1×10^{-6} mol/cm^3 and $H_2C_2O_4$ 5×10^{-6} mol/cm^3, if the reaction proceeds without exceeding a maximum permissible conversion value of 20%, the attenuation coefficient can be considered to be constant during the reaction. This characteristic of the actinometer is generally pointed out in the literature without further analysis. Although a constant value of the attenuation coefficient allows us to simplify calculation of the local volumetric rate of energy absorption, we are not able to use this property in modeling a continuous, segregated flow reactor. In these reactors, as a result of the unavoidable concentration gradients, there are regions where the requirement of maximum 20% conversion may be violated although the average exit conversion does not exceed that value.

Figure 22 also shows that the value of the attenuation coefficient is greater than 0.4 cm^{-1} for the whole range of oxalic acid concentration. Hence, considering the criterion drawn earlier for the absorption process in heterogeneous media, we can neglect the distortion effect produced by the presence of CO_2 and CO bubbles.

The oxalic acid mass balance represented by Eq. (99) to (102) cannot be solved analytically. Any of the numerical methods recommended for parabolic differential equations can be used, combined with an appropriate integration technique (Gauss, Simpson, etc.) for calculating e^a.

Since the GE G15T8 is a nonfluorescent ultraviolet arc lamp, the appropriate model to describe the radiation field inside the reactor is the VEES model with isotropic emission. However, as pointed out earlier, the SELS model can also be used as a good approximation to predict the radiation field inside an annular reactor because only direct radiation is present. Using the SELS model with isotropic emission, the computing time required to evaluate the LVREA is lessened.

Either with the VEES model or with the SELS model the solution strategy must account for the functional characteristics of the LVREA (i.e., in order to evaluate it we need to know the oxalic acid concentration field inside the whole reactor volume). Generally, the numerical procedure follows the direction of the flow; in this case, when solving each point, we should compute bundles of radiation traveling through regions already calculated (upstream section) together with bundles coming from regions whose concentration field has not been calculated yet (downstream section). Then, to solve the problem, we must follow an iterative scheme, until reaching a "steady-state" oxalic acid concentration field. Finally, after finding the "steady-state" concentration field, we estimate the oxalic acid exit conversion as follows:

$$X_{ox} = 1 - \frac{\langle C_{ox} \rangle}{C^0_{ox}} \tag{106a}$$

where

$$\langle C_{ox} \rangle = \frac{2\pi}{Q} \int_{r_{in}}^{r_{ou}} r v_z(r) C_{ox}(r, L_R)\, dr \tag{106b}$$

Figure 23 shows a plot of the oxalic acid exit conversion as a function of the volumetric flow rate calculated using the SELS model together with Eq. (105) for the attenuation coefficient μ (solid line). In the same figure experimental results obtained in our laboratory are depicted.

This example shows that it is possible to perform an a priori design of a photochemical reactor if we use an appropriate emission model to estimate the radiation field. The agreement between the experimental results and the theoretical predictions shown in Fig. 23 corroborates what we have said. Later we will see that an a priori design of more complex reactions and reactors can be performed with a similar degree of accuracy. For more detailed information about the example, the reader may refer to De Bernardez and Cassano (1985).

Complex Kinetics in an Elliptical Photoreactor with Polychromatic Radiation

A gaseous mixture of N_2, Cl_2, and C_2H_6 is fed into a bench-scale elliptical photoreactor (described in Fig. 2) whose characteristics are listed below. For the same assumptions made in the preceding example (steady state,

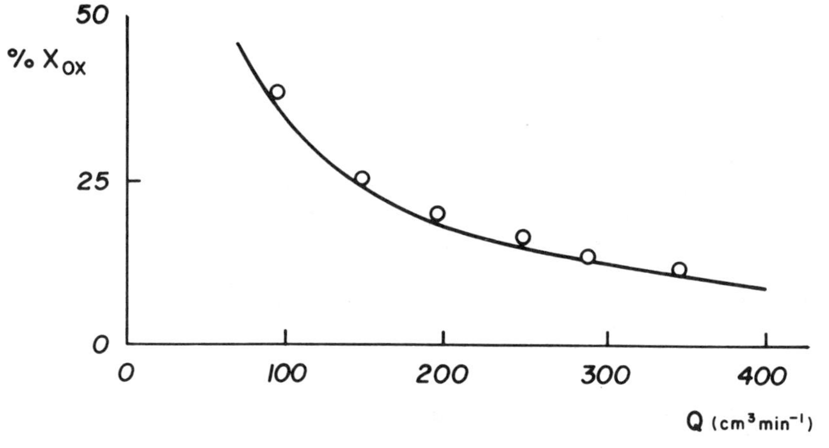

FIGURE 23 Comparison between theoretical predictions and experimental results for the photodecomposition of oxalic acid. Effect of the volumetric flow rate.

laminar flow, isothermal operation, etc.) we wish to study the incidence of the feed composition on the chlorine conversion for low-percental values of the chlorine fraction in the feed. (These conditions may be convenient to overcome undesirable thermal effects and avoid side reactions and consequently improve product selectivity.)

Reactor characteristics: $L_R = 30$ cm, $r_R = 0.2$ cm; reflector characteristics: $L_{Rf} = 59$ cm, a = 53.5 cm, e = 0.4; lamp characteristics: GE G30T8 (30 nominal watts), $L_L = 81.3$ cm, $r_L = 1.25$ cm.

Under unrestricted design conditions, the existence of direct radiation and umbra and penumbra zones for the reflected one originates angular asymmetries in the radiation field. They induce the same problem in the concentration field. However, it has been shown (De Bernardez and Cassano, 1982) that when the ellipse eccentricity is low (≤ 0.4) and the ratio of r_R/r_L is small (<0.5) azimuthal asymmetries inside the reactor can be safely neglected. This condition implies that a three-dimensional mass balance can be avoided, making computations considerably simpler. To take advantage of this convenience, the conditions above have been fulfilled in this example. Ré and Cassano (1986) showed that under normal practical conditions ($r_R > r_L$), the three-dimensional modeling of the reactor is almost indespensable. With this exception only, the design method here described can be applied to any reactor size and configuration.

The proposed low chlorine concentration in the feed, together with the presence of an inert gas, allow us to safely assume that:

1. The main reaction products will be hydrochloric acid and monochloroethane. This assumption can be easily confirmed with experimental results.
2. Under the stated operating conditions thermal effects will be small. Hence isothermal behavior is applicable.

The overall reaction is

$$C_2H_6 + Cl_2 \xrightarrow{h\nu} C_2H_5Cl + HCl$$

Several research groups have made detailed studies of the chlorination kinetics of saturated hydrocarbons. They have proposed similar reaction mechanisms. Gosselain et al. (1956) indicated a general mechanism for the gas-phase halogenation of saturated hydrocarbons for both thermal and photochemical reactions that include the possibility of using nonpolar solvents. The reaction mechansim has all the possible reaction steps that are feasible under low chlorine concentrations and temperatures below 600K.

$$Cl_2 \xrightarrow{\Phi e^a} 2Cl\cdot \qquad \text{Initiation}$$

$$\left.\begin{array}{l} Cl\cdot + EtH \underset{k_{p-2}}{\overset{k_{p2}}{\rightleftharpoons}} ClH + Et\cdot \\ Cl_2 + Et\cdot \underset{k_{p-3}}{\overset{k_{p3}}{\rightleftharpoons}} ClEt + Cl\cdot \end{array}\right\} \text{Propagation steps}$$

$$\left.\begin{array}{l} Cl\cdot + Cl\cdot + M \xrightarrow{k_{t4}} Cl_2 + M \\ Cl\cdot + Et\cdot \xrightarrow{k_{t5}} ClEt \\ Et\cdot + Et\cdot \xrightarrow{k_{t6}} C_4H_{10} \end{array}\right\} \text{Homogeneous termination steps}$$

The authors also reported that *heterogeneous termination reactions* (at the wall) may be safely neglected. This was also precisely shown by Clariá et al. (1985a).

Under these conditions, eight species are present, which will be identified, for simplicity, in the following manner:

Species	Cl_2	Cl	EtH	EtCl	Et	HCl	C_4H_{10}	N_2
Number	1	2	3	4	5	6	7	8

From the momentum balance under the described operating conditions (fully developed laminar, Newtonian and imcompressible flow), we have

$$v_z(r) = 2\langle v\rangle \left[1 - \left(\frac{r}{r_R}\right)^2\right] \tag{107}$$

Neglecting thermal effects, we only need the radiation balance for the LVREA and the mass balance equations. They will be derived under the

assumptions and simplifications described in the previous application.

The mass balance for species i is

$$v_z(r)\frac{\partial C_i}{\partial z} = \frac{D_{im}}{r}\frac{\partial}{\partial r}\left(r\frac{\partial C_i}{\partial r}\right) + \Omega_i \tag{108}$$

with

$$z = 0, \quad \forall_r, \quad C_i = C_i^o \tag{108a}$$

$$\forall_z, \quad r = 0, \quad \frac{\partial C_i}{\partial r} = 0 \tag{108b}$$

$$\forall_z, \quad r = r_R, \quad \frac{\partial C_i}{\partial r} = 0 \tag{108c}$$

If heterogeneous reactions were present, the boundary conditions for the species at $r = r_R$ should be changed to

$$\forall_z, \quad r = r_R, \quad D_{jm}\frac{\partial C_j}{\partial r} = \sum_j \nu_{jh}\Omega_h \tag{108c'}$$

where the summation must include all the heterogeneous steps where species j participates; Ω_h is the rate of the h heterogeneous termination step and ν_{jh} is the stoichiometric coefficient of species j in the h heterogeneous step.

In Eq. (108) for the eight homogeneous steps included in the reaction mechanism we have

$$\Omega_i = \sum_{r=1}^{r=8} \nu_{ir}\Omega_r \tag{109}$$

Hence the rates of formation of species i are

$$\Omega_1 = -\Phi e^a - k_{p3}C_1C_5 + k_{t4}C_8C_2^2 + k_{p-3}C_2C_4 \tag{110}$$

$$\Omega_2 = 2\Phi e^a - k_{p2}C_2C_3 + k_{p-2}C_5C_6 + k_{p3}C_1C_5 \\ -k_{p-3}C_2C_4 - 2k_{t4}C_2^2C_8 - k_{t5}C_2C_5 \tag{111}$$

$$\Omega_3 = -k_{p2}C_2C_3 + k_{p-2}C_5C_6 \tag{112}$$

$$\Omega_4 = k_{p3}C_1C_5 - k_{p-3}C_2C_4 + k_{t5}C_2C_5 \tag{113}$$

$$\Omega_5 = k_{p2}C_2C_3 - k_{p-2}C_5C_6 - k_{p3}C_1C_5 - k_{t5}C_2C_5 \\ - 2k_{t6}C_5^2 + k_{p-3}C_2C_4 \tag{114}$$

$$\Omega_6 = k_{p2}C_2C_3 - k_{p-2}C_5C_6 \tag{115}$$

$$\Omega_7 = k_{t6}C_5^2 \tag{116}$$

De Bernardez and Cassano (1986) and Clariá et al. (1985a) showed that in these systems the local or microscopic steady-state approximation for intermediate unstable species can be applied without error. Hence Eq. (111) and (114) can be set equal to zero:

$$\Omega_i = 0 \quad \text{for} \quad i = 2 \text{ and } 5 \tag{117}$$

These two algebraic equations are still coupled with the mass balance equations through the reaction rate term. In addition, Eq. (110) and (111) clearly show the coupling of the mass balance equations with the radiation balance through the rate of the activation step.

Since a curved reflecting surface is present, extense models (VEES or SEES) must necessarily be used. On the other hand, the GE G30T8 is a nonfluorescent ultraviolet (UV) arc lamp, so the appropriate model to describe the emission process is the VEES with isotropic emission.

The reacting system is homogeneous (gas phase) and chlorine is the only absorbing species in the wavelength of lamp the emission of the lamp. Hence

$$e_\nu^a = \mu_\nu |\underset{\sim}{q}_\nu| = \alpha_\nu C_1 |\underset{\sim}{q}_\nu| \tag{118}$$

where α_ν is the absorption coefficient of chlorine.

From the emission characteristics of the lamp (see Table 4) approximately 90% of the output energy falls into the UV region. Out of this amount, about 95% is emitted in a single wavelength (2537 Å). However, we will not consider monochromatic radiation because the absorption by chlorine in this line is low, having instead much higher values at longer wavelengths. Thus

$$e^a = \int_{\nu=0}^{\nu=\infty} d\nu\, e_\nu^a = \int_{\nu=0}^{\nu=\infty} d\nu\, \alpha_\nu C_1 |\underset{\sim}{q}_\nu| \tag{119}$$

Since the total energy flux density vector results from the contributions of direct and indirect radiation (the latter with one or more reflections), in Eq. (119) we have

$$|\underset{\sim}{q}_\nu| = |\underset{\sim}{q}_\nu|_D + |\underset{\sim}{q}_\nu|_{In,1} + \cdots + |\underset{\sim}{q}_\nu|_{In,n} \tag{120}$$

When the hypothesis of azimuthal simmetry is used, it can also be shown that if c is sufficiently large, only indirect radiation with one reflection is significant. Hence we can safely consider only this contribution. Its expression was derived earlier [Eqs. (40) and (65)]:

TABLE 4 Spectral Distribution of Power Output for the GE G30T8

Wavelength interval (Å)	E_λ (W)	$E_\lambda \times 10^7$ (einstein/s)
2200–2400	—	—
2400–2600	8.34	174.20
2600–2800	0.01	0.226
2800–2900	0.02	0.476
2900–3000	0.03	0.739
3000–3100	0.02	0.510
3100–3200	0.16	4.211
3200–3400	0.01	0.276
3400–3600	0.01	0.292
3600–3800	0.17	5.255
3800–4000	0.01	0.326
4000–5000	0.66	24.814
5000–6000	0.33	15.164
6000–7000	—	—

Source: General Electric Co. (1967).

$$e^a = \frac{2\Phi \langle \Gamma_{Rf} \rangle \langle \Upsilon_R \rangle}{4\pi V_L} C_1 \int_{\nu=0}^{\nu=\infty} d\nu \, E_{L,\nu} \alpha_\nu \int_\phi d\phi \, (r_L^2 - D^2 \sin^2 \xi)^{1/2} \times \int_{\theta_1(\phi)}^{\theta_2(\phi)} d\theta \quad \exp\left(- \int_{\rho_o^*(\theta,\phi)}^{\rho^*(\theta,\phi)} d\rho' \, \alpha_\nu C_1 \right) \tag{121}$$

For the wavelength range of this application, the quartz transmittance, the average reflection coefficient and the primary quantum yield are considered independent of ν. For the latter, in addition, its value is taken equal to 1.

In Eq. (121), only $E_{L,\nu}$ (lamp specific power output) and α_ν (chlorine absorptivity) are wavelength dependent. Hence we may write

$$e^a = \frac{2\Phi \langle \Gamma_{Rf} \rangle \langle \Upsilon_R \rangle}{4\pi V_L} C_1 \int_\phi d\phi \, (r_L^2 - D^2 \sin^2 \xi)^{1/2} \int_{\theta_1(\phi)}^{\theta_2(\phi)} d\theta \times \int_{\nu=0}^{\nu=\infty} d\nu \, E_{L,\nu} \alpha_\nu \quad \exp\left(- \int_{\rho_o^*(\theta,\phi)}^{\rho^*(\theta,\phi)} \alpha_\nu C_1 \, d\rho' \right) \tag{122}$$

Let us separately analyze the wavelength-dependent integral:

$$A = \int_{\nu=0}^{\nu=\infty} d_{\nu}\, E_{L,\nu}\, \alpha_{\nu} \exp\left(-\int_{\rho_o^*(\phi,\theta)}^{\rho^*(\theta,\phi)} \alpha_{\nu} C_1\, d\rho'\right) \tag{123}$$

$$= \int_{\lambda=0}^{\lambda=\infty} d\lambda\, E_{L,\lambda}\, \alpha_{\nu} \exp\left(-\int_{\rho_o^*(\theta,\phi)}^{\rho^*(\theta,\phi)} \alpha_{\lambda} C_1\, d\rho'\right) \tag{124}$$

$$= \int_{\lambda_1=2200Å}^{\lambda_2=5000Å} d\lambda\, E_{L,\lambda}\, \alpha_{\lambda} \exp\left(-\int_{\rho_o^*(\theta,\phi)}^{\rho^*(\theta,\phi)} \alpha_{\lambda} C_1\, d\rho'\right) \tag{125}$$

In Eq. (125) the wavelength interval has been reduced to the region in which we simultaneously have significant absorption by chlorine and emission from the lamp (see Tables 4 and 5). Furthermore, considering that the absorption by choroine is continuous (Herzberg, 1950) and the lamp emission is discrete in very well defined lines, Eq. (125) can be written

TABLE 5 Chlorine Absorption Coefficient

λ (Å)	$[\alpha_{Cl_2}]_{\ell n} \times 10^{-3}$ (cm^2/mol)	λ (Å)	$[\alpha_{Cl_2}]_{\ell n} \times 10^{-3}$ (cm^2/mol)
2450	1.08	3750	41.13
2550	3.33	3850	23.49
2650	4.96	3950	14.23
2750	10.62	4050	9.89
2850	28.78	4150	7.22
2950	55.98	4250	5.73
3050	92.33	4350	4.09
3150	127.33	4450	2.94
3250	145.46	4550	2.02
3350	146.59	4650	1.34
3450	123.11	4570	0.88
3550	93.47	4850	0.66
3650	64.51	4950	0.21

Source: Calvert and Pitts (1966).

$$A = \sum_{i=1}^{n} \int d\lambda\, E_{L,i} \delta(\lambda - \lambda_i) \alpha_i \exp\left(- \int_{\rho_o^*(\theta,\phi)}^{\rho^*(\theta,\phi)} d\rho'\, \alpha_i C_1 \right) \tag{126}$$

$$= \sum_{i=1}^{n} E_{L,i} \alpha_i \exp\left(- \int_{\rho_o^*(\theta,\phi)}^{\rho^*(\theta,\phi)} d\rho'\, \alpha_i C_1 \right) \tag{127}$$

where n are all the lamp spectral lines of emission and the summation must be taken over all the n values were $\alpha_i \neq 0$. δ is the Dirac function; $E_{L,i}$ and α_i are the values of E_L and α at $\lambda = \lambda_i$. In going from Eq. (126) to Eq. (127) we have made

$$E_{L,\lambda} = E_{L,i} \delta(\lambda - \lambda_i) \tag{128}$$

This is equivalent to substitute a polychromatic radiation source by n hypothetical monochromatic devices, which simultaneously occupy the same physical space. Each of them emits a single frequency which corresponds to one of the emission lines of the polychromatic source.

The mass balance equations are solved in a r–z plane. However to evaluate the LVREA we must consider bundles of radiation coming from every direction in space and from the whole volume of the lamp. Consequently, rays not belonging to the r–z plane should also be considered in the calculation. This makes it necessary to evaluate the real distance traveled by each energy bundle and the sequence of concentrations found along its path; considering the azimuthal symmetry, it is possible to find the required information by performing a cylindrical projection on the rays on the plane r–z (Fig. 24). [For further details see Romero et al. (1983).]

Substituting Eq. (127) into Eq. (122) and solving Eq. (108) with Eqs. (110) to (117) we can treat the problem with any conventional numerical procedure. We should only recall that to obtain the $|\underset{\sim}{q}_\nu|$ value at each point, we need to use the methodology indicated in Fig. 24 and the computation of the radiation field as described earlier. The solution provides the concentration field for each species i. Finally, the averaged exit conversion is defined as

$$X = \frac{C_1^\circ - \langle C_1 \rangle}{C_1^\circ} = 1 - \frac{\langle C_1 \rangle}{C_1^\circ} \tag{129}$$

and the value of $\langle C_1 \rangle$ may be obtained as it was illustrated in the preceding application [Eq. (106b)].

Clariá et al. (1985a,b) solved this system and compared the predicted values with experimental results obtained in a bench-scale reactor. Some of their results are shown in Figs. 25 and 26, where the incidence of the feed composition upon the chlorine conversion is analyzed.

Considering that no adjustable parameters have been used and the values of $\langle \Gamma_{Rf} \rangle$, $\langle \Upsilon_R \rangle$, and $E_{L,\nu}$ are taken from the specifications provided by the manufacturing companies, we think that the results are extremely good.

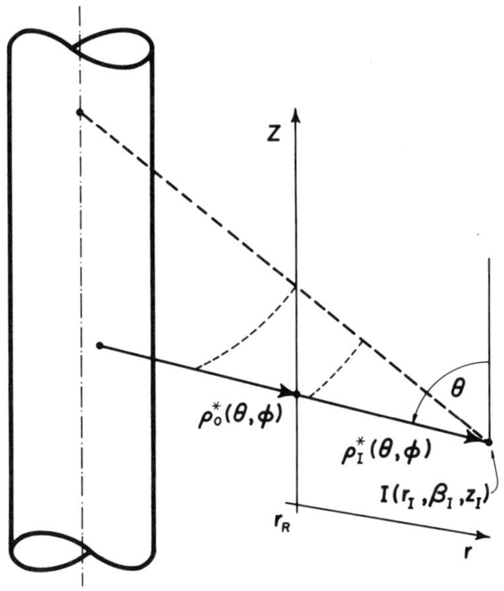

FIGURE 24 Cylindrical projection to compute the three-dimensional attenuation in a two-dimensional grid.

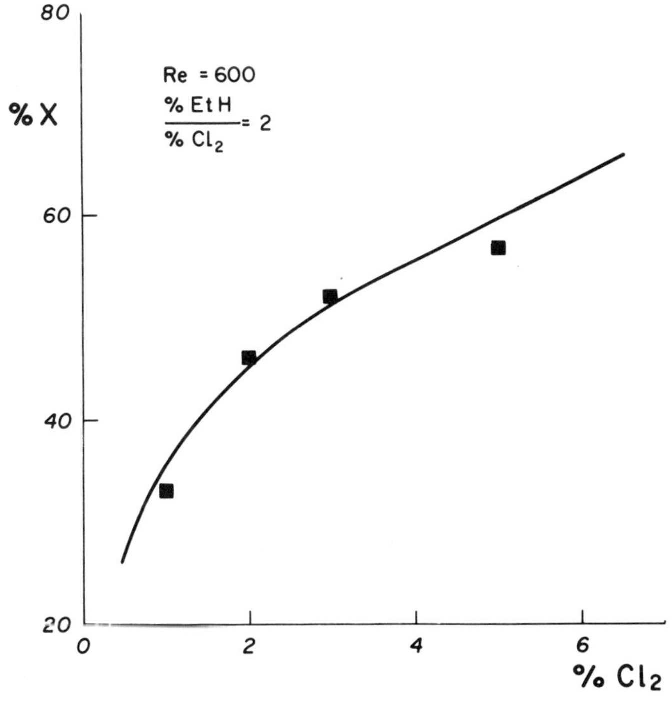

FIGURE 25 Comparison of theoretical predictions and experimental results for the chlorination of ethane. Effect of the chlorine inlet concentration.

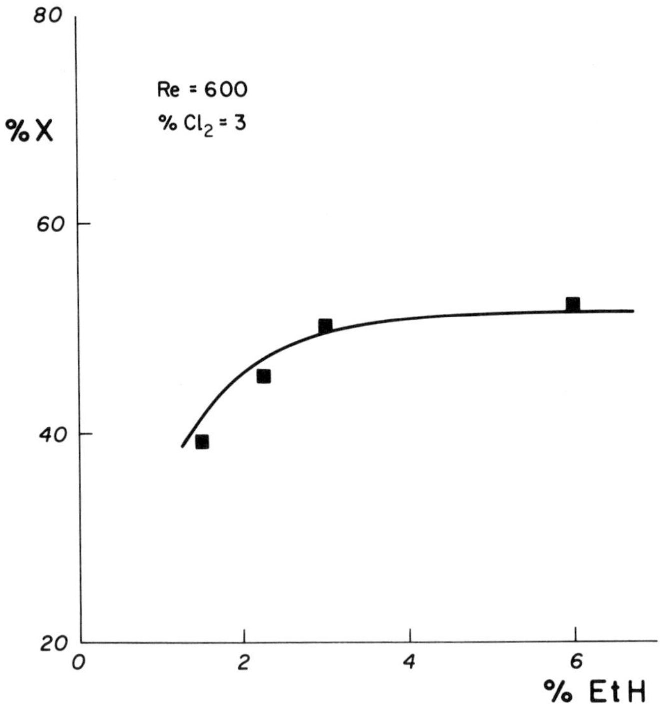

FIGURE 26 Comparison of theoretical predictions and experimental results for the chlorination of ethane. Effect of the ethane inlet concentration.

Complex Kinetics in a Heterogeneous Semibatch Reactor Irradiated from the Bottom

The photochlorination of trichloroethylene is carried out in a semibatch stirred tank reactor irradiated from the bottom by a tubular source located at the focal axis of a cylindrical reflector of parabolic cross section. The liquid phase is a homogeneous solution of trichloroethylene and carbon tetrachloride, while the gas phase is a mixture of chlorine and nitrogen. The overall reaction is

$$C_2HCl_2 \xrightarrow{+Cl_2} C_2HCl_5 \xrightarrow[-\,HCl]{+Cl_2} C_2Cl_6 \tag{130}$$

We wish to analyze (a) the possibility of transforming all the C_2HCl_3 initially present in the liquid phase into C_2HCl_5 only, and (b) the effect of the inlet nitrogen concentration upon selectivity, conversion, and reaction time.

The modeling problem must be consider the following aspects: the radiation energy distribution inside the reactor, the reaction rate, the mixing pattern, and the gas absorption rate. Since the reaction is considered to occur only in a perfectly stirred liquid phase, the overall phenomenon involves the following elementary steps:

1. Diffusion of chlorine from the bulk of the gas phase to the interface, where physical equilibrium is assumed
2. Diffusion of chlorine from the interface to the bulk of the liquid phase
3. Chemical reaction within the liquid phase

It is useful to point out here that the chemical reaction proceeds through through a chain mechanism; hence not only are stable species present but highly reactive intermediates as well. Since the liquid phase is considered to be perfectly stirred, we can visualize the following mixing characteristics:

1. The concentration field of the stable species is uniform inside the whole reactor volume.
2. The rate of initiation step (e^a) is no longer a functional of the concentration field as in the previous applications but is still spatially dependent
3. The spatial distribution of the radiation field together with the short lifetime of the highly reactive intermediate species suggest that concentration gradients of these species should be expected. Moreover, they can be considered to be born, live, and die in the same place. Alfano and Cassano (1986a,b) showed that this characteristic corresponds to reality, and perfect mixing for the short-lived intermediates cannot be assumed. Even though the liquid phase is perfectly stirred and concentration gradients of the stable species are not present, the reaction rate is a function of position due to the spatial distribution of the radiation field and the short-lived intermediate concentration field. Hence the mass transfer (with chemical reaction) coefficient may be spatially dependent.

For simplicity, the species ill be identified in the following way:

Cl_2	C_2HCl_3	C_2HCl_5	C_2Cl_6	HCl	$C_2HCl_4\cdot$	$C_2Cl_5\cdot$	$Cl\cdot$	N_2
1	2	3	4	5	6	7	8	9

An order-of-magnitude analysis f the characteristic reaction time and the characteristic absorption time in the liquid film shows that

$$t_{Dif} = \frac{Di}{(k_L^o)^2} \cong 10^{-2} s \tag{131}$$

and

$$t_{rc} = \frac{C_1^{if}}{\Omega_1(C_1^{if}, \underset{\sim}{x} = \underset{\sim}{x}_{max})} \cong 10s \tag{132}$$

where we have estimated t_{rc} at the interface conditions and in the spatial position where e^a is maximum.

This means that the reaction is slow compared with the diffusional rate. This physical situation was characterized in the literature as the slow reaction regime (Astarita, 1967) and in this case

$$k_L = k_L^\circ \tag{133}$$

That is, the mass transfer with chemical reaction coefficient is equal to the physical mass transfer absorption coefficient and hence independent of the spatial position.

The chlorine absorption rate is

$$N_1 = k_L^\circ a_v (C_1^{if} - C_1) = \frac{C_1^{if} - C_1}{t_{ed}} \tag{134}$$

where

$$t_{ed} = \frac{1}{k_L^\circ a_v} = \text{effective diffusion time} \tag{135}$$

C_1^{if} is the chlorine concentration at the gas-liquid phase, and C_1 is the chlorine concentration in the bulk of the liquid phase.

The mass balance in the gas phase is posed under the following assumptions: (a) pseudo steady state, (b) no reaction in the gas phase, (c) negligible gas-phase resistance to mass transfer, (d) well-mixed gas phase, and (e) no accumulation or significant reaction in the liquid film. Hence we can write

$$F_1^\circ - F_1 = N_1 V = k_L^\circ \, a_v V (C_1^{if} - C_1) = \frac{V}{t_{ed}} (C_1^{if} - C_1) \tag{136}$$

Assumption c must be made mainly because no reliable correlations can be used to predict the mass transfer rates. It is usually made in problems related to gas absorption with chemical reaction in spite of the fact that we known that even if the gas-side resistance to mass transfer may occasionally be negligible, it is never zero.

Since the inlet gas stream is a mixture of chlorine and nitrogen,

$$F_9 = F_9^\circ$$

$$F_1 = \frac{F_9 y_1}{1 - y_1} = \frac{F_9^\circ y_1}{1 - y_1} \tag{137a}$$

If we assume equilibrium conditions at the interface,

$$C_1^{if} = \frac{p_1^{if}}{H} = \frac{y_1 P}{H} \tag{137b}$$

Solving Eq. (136) together with Eq. (137) we obtain

$$C_1^{if} = \frac{1}{2}\left\{\left(\frac{F^\circ t_{ed}}{V} + \frac{P}{H} + C_1\right) - \left(\frac{F^\circ t_{ed}}{V} + \frac{P}{H} + C_1\right)^2 - 4\frac{P}{H}\left(C_1 + \frac{y_1^\circ F^\circ t_{ed}}{V}\right)^{1/2}\right\} \tag{138}$$

Finally, the mass balance equations in the bulk of the liquid phase are

$$\frac{dC_1(t)}{dt} = \langle \Omega_1(\underline{x},t)\rangle(t) + N_1(t) \tag{139a}$$

$$\frac{dC_i(t)}{dt} = \langle \Omega_i(\underline{x},t)\rangle(t) \qquad i = 2, \ldots, 5 \tag{139b}$$

$$\frac{\partial C_j(\underline{x},t)}{\partial t} = \Omega_j(\underline{x},t) \qquad j = 6, 7, 8 \tag{139c}$$

where we have considered perfect mixing for the stable species and non-mixing for the highly reactive intermediate species.

The system of represented by Eqs. (139) must be solved together with the following initial conditions:

$$C_i(0) = C_i^\circ \qquad i = 1, \ldots, 5 \tag{140a}$$

$$C_j(0) = 0 \qquad j = 6, 7, 8 \tag{140b}$$

The reaction scheme involves the following steps:
Initiation:

$$Cl_2 \xrightarrow{\Phi e^a} 2Cl\cdot \tag{141a}$$

Propagation:

$$C_2HCl_3 + Cl\cdot \underset{k_2'}{\overset{k_2}{\rightleftharpoons}} C_2HCl_4\cdot \tag{141b}$$

$$C_2HCl_4\cdot + Cl_2 \underset{k_3'}{\overset{k_3}{\rightleftharpoons}} C_2HCl_5 + Cl\cdot \tag{141c}$$

$$C_2HCl_5 + Cl\cdot \underset{k_4'}{\overset{k_4}{\rightleftharpoons}} C_2Cl_5\cdot + HCl \tag{141d}$$

$$C_2Cl_5\cdot + Cl_2 \underset{k_5'}{\overset{k_5}{\rightleftharpoons}} C_2Cl_6 + Cl\cdot \tag{141e}$$

Termination:

$$2Cl\cdot \xrightarrow{k_6} Cl_2 \tag{141f}$$

$$2C_2HCl_4\cdot \xrightarrow{k_7} \text{products} \tag{141g}$$

$$2C_2Cl_5\cdot \xrightarrow{k_8} \text{products} \tag{141h}$$

$$Cl\cdot + C_2HCl_4 \xrightarrow{k_9} \text{products} \tag{141i}$$

$$Cl\cdot + C_2Cl_5 \xrightarrow{k_{10}} \text{products} \tag{141j}$$

$$C_2Cl_5\cdot + C_2HCl_4\cdot \xrightarrow{k_{11}} \text{products} \tag{141k}$$

where the possibility of wall reactions has been disregarded.

Considering that the rate of each step of the mechanism has the form of the mass action law, and applying the local or microscopic steady-state approximation for the highly reactive intermediate species together with the long-chain approximation, we obtain

$$\Omega_1 = -\Phi e^a - k_3C_1C_6 + k_3'C_3C_8 - k_5C_7C_1 + k_5'C_4C_8 + k_6C_8^2 \tag{142}$$

$$\Omega_2 = -k_3C_1C_6 + k_3'C_3C_8 \tag{143}$$

$$\Omega_3 = k_3C_1C_6 - k_3'C_3C_8 - k_4C_3C_8 + k_4'C_7C_5 \tag{144}$$

$$\Omega_4 = k_4C_3C_8 - k_4'C_7C_5 \tag{145}$$

$$\Omega_5 = k_4C_3C_8 - k_4'C_7C_5 \tag{146}$$

$$\Omega_6 = \Omega_7 = \Omega_8 = 0 \tag{147}$$

and

$$C_6 = \frac{k_2C_2 + k_3'C_3}{k_2' + k_3C_1} \quad C_8 = K_1C_8 \tag{148}$$

$$C_7 = \frac{k_4C_3 + k_5'C_4}{k_4'C_5 + k_5C_1} \quad C_8 = K_2C_8 \tag{149}$$

$$C_8 = \left(\frac{\Phi e^a}{k_6 + K_1^2 k_7 + K_2^2 k_8 + K_1 k_9 + K_2 k_{10} + K_1 K_2 k_{11}} \right)^{1/2} \tag{150}$$

To complete the reaction rate expressions, the rate of the initiation step (Φe^a) must be evaluated in a heterogeneous medium. Earlier we anticipated that a suitable strategy for the modeling problem is to make use of the LVREA expressions valid for homogeneous media but including an effective attenuation coefficient to account for the distortions in the radiation field due to the presence of gas bubbles. Since a reflecting surface is present, we cannot use a linear model; hence either the SEES model for fluorescent lamps or the VEES model for arc lamps will be required.

Regarding the kinetic sequence, we observe that the overall behavior is the one corresponding to a series-parallel reaction system, that is

$$A \xrightarrow{+Cl_2} B \xrightarrow[-HCl]{+Cl_2} C$$

where the first path ($A \rightarrow B$) is an addition reaction and the second path ($B \rightarrow C$) is a substitution reaction. It is known that the characteristic time for both reactions should be different, that is, $t_{A \rightarrow B} \ll t_{B \rightarrow C}$. Under this condition it would be possible to obtain high selectivity values for the intermediate product B (C_2HCl_5), as it is observed in homogeneous consecutive reactions. However, with single-phase systems, only in exceptional situations could we expect to achieve 100% selectivity. In a heterogeneous medium it is possible to improve selectivity by operating on the process conditions so as to change the rate-controlling step as follows.

Let $t_{A \rightarrow B}$ be the characteristic reaction time for the kinetic path $A \rightarrow B$, and $t_{B \rightarrow C}$ be the one corresponding to the path $B \rightarrow C$. Let t_{ed} be the effective diffusion time calculated as $(k_L a_V)^{-1}$ and t_{Dif} the characteristic diffusion time estimated with Eq. (131). Under the slow reaction regime ($t_{Dif} \ll t_{A \rightarrow B} \ll t_{B \rightarrow C}$), the following possibilities are present (Astarita et al., 1983):

1. $t_{ed} \ll t_{A \rightarrow B} \ll t_{B \rightarrow C}$. Both reactions are under the kinetic subregime and the overall expected behavior would be similar to that observed in the homogeneous system $A \rightarrow B \rightarrow C$.
2. $t_{A \rightarrow B} \ll t_{ed} \ll t_{B \rightarrow C}$. The first path $A \rightarrow B$ is under the diffusional subregime while the second is under the kinetic subregime. In this case it would be possible to obtain 100% selectivity of C_2HCl_5 with almost 100% conversion of C_2HCl_3 because in practical terms reactions $A \rightarrow B$ and $B \rightarrow C$ do not take place simultaneously.
3. $t_{A \rightarrow B} \ll t_{B \rightarrow C} \ll t_{ed}$. Both paths are under the diffusional subregime. However, for the reacting system under analysis and the type of reactor being modeled (perfectly mixed stirred tank reactor), condition (3) cannot be practically achieved. Hence this situation will not be analyzed here.

Let us suppose that condition (2) is fulfilled in the semibatch stirred-tank reactor under analysis. Hence the reaction would proceed under the diffusional subregime until almost 100% selectivity of C_2HCl_5 (B) together with almost 100% conversion of C_2HCl_3 (A) are attained. After that, the overall behavior changes from the diffusional to the kinetic subregime and reaction B → C becomes more significant. If B is the desired product the reaction operation should be frozen (by turning off the lamp) before the transition from the diffusional to the kinetic subregime takes place. Under the diffusional subregime we may expect the following characteristics for the concentration distribution:

1. Dependent on the intensity of mixing of the liquid phase
2. Proportional to the interfacial area
3. Independent of the liquid volume
4. Independent of the emission system used to activate the reaction

The latter characteristic allows us to use a low-energy emission source, decreasing the operating costs. However, there is a minimum permissible value of the energy output below which the characteristic reaction time of the path A → B becomes greater than the effective diffusion time and the overall behavior changes from the diffusional subregime to the kinetic one [i.e., from case (2) to case (1)], reducing the possibility of achieving 100% selectivity of C_2HCl_5 with 100% conversion of C_2HCl_3.

To illustrate the previous qualitative analysis we will report some of the results obtained by Alfano and Cassano (1986a,b), who carried out bench-scale experiments in a photoreactor with the following characteristics. Reactor: radius, 0.145 m; height, 0.20 m; bottom material, quartz or Pyrex glass. Lamp: Ge UA3-360; radius, 0.0095 m; length, 0.152 m. Reflector: characteristic constant of the parabola, 0.021 m; distance from vertex of parabolic reflector to bottom of reactor, 0.084 m; cover radius, 0.013 m. Operating conditions: temperature, 298 K; pressure, 101.3 kPa; volume of liquid, 0.002 m^3; stirring speed, 800 rpm (nominal).

Theoretical predictions were performed with the VEES model with isotropic emission (the GE UA3-360 emission source is a nonfluorescent UV arc lamp) and using Yokota et al.'s (1981) correlation [Eqs. (38) and (39)] to evaluate the effective attenuation coefficient. It is interesting to point out that any emission model can be used when the reaction is under the diffusional subregime but only extense source models (VEES or SEES) can be safely used under the kinetic subregime.

The existence of both subregimes needs to be investigated to find the real possibilities of a very high selectivity. To do so, it is necessary to change the mass transfer characteristics of the system (to search for the diffusional subregime) and the kinetics of the reaction by changing the radiation field, for example (to search for the kinetic subregime). Both tests can be computationally and experimentally performed and were carried out by Alfano and Cassano (1986a,b). Their results showed that the first reaction was sensitive to the stirring speed, while the second one (starting from pentachloroethane) was almost insensitive to changes in the mass transfer characteristics, but indicated a noticeable response to changes in the light transmission properties of the reactor bottom. Computational and experimental results clearly support the idea of two different reacting conditions. The details can be found in the papers cited.

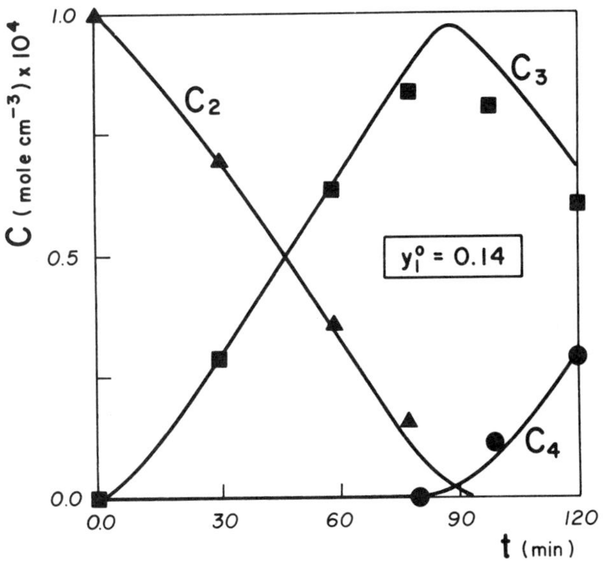

FIGURE 27 Comparison of theoretical predictions and experimental results for the chlorination of trichloroethylene (chlorine fraction in the feed, 14%).

Figure 27 shows an example obtained for the case in which the reaction is conducted using 86% of nitrogen in the feed. Since the first reaction is in the diffusional subregime the good correspondence between computed and experimental values only indicates that the mass flow rates of reactant gases have been properly modeled and the energy input to the reactor is sufficient. In the second reaction we known that it is kinetically controlled and the agreement must be ascribed to the quality of the radiation field prediction as well as the kinetics and reactor modeling. It can be observed that C_2HCl_5 selectivity is always 100% until almost all C_2HCl_3 has been consumed (t = 90 min); then C_2HCl_5 concentration goes down because the substitution reaction ($C_2HCl_5 + Cl_2 \rightarrow C_2Cl_6 + HCl$) becomes significant. For this particular reactor-source-reflector configuration, condition 2 is certainly fulfilled and, in practice, both reaction paths A → B and B → C do not take place simultaneously. Stopping the reaction at t = 90 min (by turning off the lamp), it is possible to transform all the C_2HCl_3 initially present in the liquid phase into C_2HCl_5 only. Figure 28 shows the results for the case in which 60% of nitrogen is used in the feed. Similar conclusions can be drawn from the selectivity and conversion viewpoints. However, the time necessary to achieve 100% C_2HCl_5 selectivity together with almost 100% C_2HCl_3 conversion has been reduced to 35 min.

These results clearly show that it is possible to achieve 100% selectivity for the intermediate species together with almost 100% conversion of C_2HCl_3 and that the presence of nitrogen does not affect the final product distribution, but the time needed to reach a given value of conversion. The time of reaction can be sharply reduced by using pure chlorine in the feed but, obviously, it is possible that under such conditions the reaction control will be much more difficult.

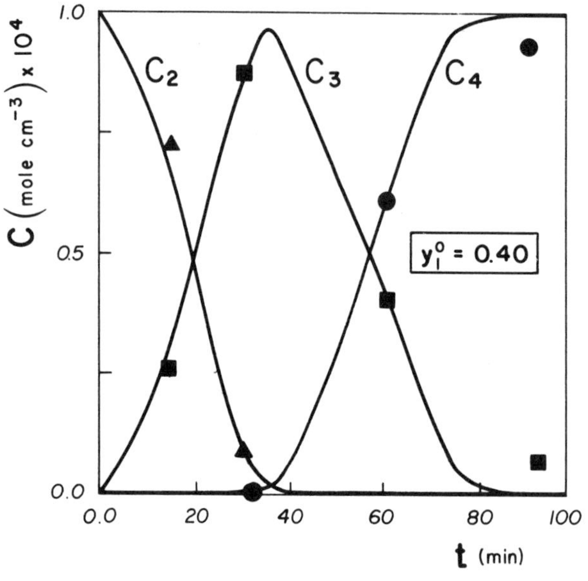

FIGURE 28 Comparison of theoretical predictions and experimental results for the chlorination of trichloroethylene (chlorine fraction in the feed, 40%).

Consecutive Reactions in a Continuous Annular Photoreactor

A gaseous mixture of Cl_2, CH_4, and N_2 is flowing through an annular photoreactor whose inner and outer walls are cooled to avoid large temperature rises, precluding undesirable overheating and by-product formation. Moreover, with the same purpose, it is recommended that the chlorine inlet molar fraction be maintained no higher than 50%. The aim is to perform a preliminary design of the reactor, including the selection of the emitting system for an optimal production of chloroform. For safety reasons, the maximum reaction temperature and pressure are T = 363 K and P = 101.3 kPa.

In dealing with the design of this photochemical reactor we must answer the following questions: (a) What is the size of the reactor? (For the annular reactor, this means the values of r_{in}, r_{ou}, and L_R.) (b) What are the operating conditions? (These include inlet composition, flow rate, etc.) (c) Which is the appropriate emission system to carry out the reaction? (This involves the spectral distribution of power output, overall power output, and values of r_L and L_L.) The first two questions are common in conventional thermal reactor design; the last arises only in photochemical reactor design and constitutes one of the most important stages in the design strategy.

To provide an answer to the foregoing questions, we have to solve the modeling equations several times because the number of variables involved in the design problem is large. Hence it would be recommended to find a simple, even though only approximate model to describe the performance of the photoreactor. This simple model will allow us to (a) choose an emission system, and (b) find a starting set of values for the main variables to carry

out the final design using a reliable (but more complex) model. The simple model should have the following characteristics:

A simplified kinetic scheme with reaction rate expressions valid for a wide range of operating conditions
A simplified model to describe the reactor operation
A simplified model for the emission process

Simplified Kinetic Scheme

We can use the entire reaction mechanism for which the rate expressions for the individual steps are known and obtain global kinetic expressions by introducing some typical simplifications, such as the local steady state approximation for highly reactive intermediate species.

The reaction mechanism for the chlorination of methane can be summarized as follows:

Initiation step:

$$Cl_2 + h\nu \xrightarrow{1} 2Cl\cdot \tag{151}$$

Propagation steps:

$$CH_4 + Cl\cdot \xrightleftharpoons{2} CH_3\cdot + HCl \tag{152}$$

$$CH_3\cdot + Cl_2 \xrightleftharpoons{3} CH_3Cl + Cl\cdot \tag{153}$$

$$CH_3Cl + Cl\cdot \xrightleftharpoons{4} CH_2Cl\cdot + HCl \tag{154}$$

$$CH_2Cl\cdot + Cl_2 \xrightleftharpoons{5} CH_2Cl_2 + Cl\cdot \tag{155}$$

$$CH_2Cl_2 + Cl\cdot \xrightleftharpoons{6} CHCl_2\cdot + HCl \tag{156}$$

$$CHCl_2\cdot + Cl_2 \xrightleftharpoons{7} CHCl_2 + Cl\cdot \tag{157}$$

$$CHCl_3 + Cl\cdot \xrightleftharpoons{8} CCl_3\cdot + HCl \tag{158}$$

$$CCl_3\cdot + Cl_2 \xrightleftharpoons{9} CCl_4 + Cl\cdot \tag{159}$$

Homogeneous termination steps:

$$CH_3\cdot + Cl\cdot \xrightarrow{10} CH_3Cl \tag{160}$$

$$CH_2Cl\cdot \xrightarrow{11} CH_2Cl_2 \tag{161}$$

$$CHCl_2\cdot + Cl\cdot \xrightarrow{12} CHCl_3 \tag{162}$$

$$CCl_3\cdot + Cl\cdot \xrightarrow{13} CCl_4 \tag{163}$$

$$Cl\cdot + Cl\cdot \xrightarrow{14} Cl_2 \tag{164}$$

$$CH_3\cdot + CH_3\cdot \xrightarrow{15} CH_3 - CH_3 \tag{165}$$

$$CH_3\cdot + CH_2Cl\cdot \xrightarrow{16} CH_3 - CH_2Cl \tag{166}$$

$$CH_3\cdot + CHCl_2\cdot \xrightarrow{17} CH_3 - CHCl_2 \tag{167}$$

$$CH_3\cdot + CCl_3\cdot \xrightarrow{18} CH_3 - CCl_3 \tag{168}$$

$$CH_2Cl\cdot + CH_2Cl\cdot \xrightarrow{19} CH_2Cl - CH_2Cl \tag{169}$$

$$CH_2Cl\cdot + CHCl_2\cdot \xrightarrow{20} CH_2Cl - CHCl_2 \tag{170}$$

$$CH_2Cl\cdot + CCl_3\cdot \xrightarrow{21} CH_2Cl - CCl_3 \tag{171}$$

$$CHCl_2\cdot + CHCl_2\cdot \xrightarrow{22} CHCl_2 - CHCl_2 \tag{172}$$

$$CHCl_2\cdot + CCl_3\cdot \xrightarrow{23} CHCl_2 - CCl_3 \tag{173}$$

$$CCl_3\cdot + CCl_3\cdot \xrightarrow{24} CCl_3 - CCl_3 \tag{174}$$

Heterogeneous termination steps:

$$CH_3\cdot + \text{wall} \xrightarrow{25} \text{products} \tag{175}$$

$$CH_2Cl\cdot + \text{wall} \xrightarrow{26} \text{products} \tag{176}$$

$$CHCl_2\cdot + \text{wall} \xrightarrow{27} \text{products} \tag{177}$$

$$CCl_3\cdot + \text{wall} \xrightarrow{28} \text{products} \tag{178}$$

$$Cl\cdot + \text{wall} \xrightarrow{29} \text{products} \tag{179}$$

We introduce the following simplifications:

1. Disregard the heterogeneous termination steps [Eqs. (175) to (179)] as shown by De Bernardez and Cassano (1986).
2. Disregard the homogeneous termination steps that lead to form chlorinated derivatives of ethane [Eqs. (165) to (174)], which are known to be negligible (Kurtz, 1972; De Bernardez and Cassano, 1986).
3. Apply the local steady-state approximation to the unstable intermediate species together with the long-chain approximation in the remaining reaction system [Eqs. (151) to (164)].

Thus the complex chain mechanism is reduced to the following kinetic scheme:

$$CH_4 + Cl_2 \underset{}{\overset{R-1}{\rightleftharpoons}} CH_3Cl + HCl \tag{180}$$

$$CH_3Cl + Cl_2 \underset{}{\overset{R-2}{\rightleftharpoons}} CH_2Cl_2 + HCl \tag{181}$$

$$CH_2Cl_2 + Cl_2 \underset{}{\overset{R-3}{\rightleftharpoons}} CHCl_3 + HCl \tag{182}$$

$$CHCl_3 + Cl_2 \underset{}{\overset{R-4}{\rightleftharpoons}} CCl_4 + HCl \tag{183}$$

The reaction rate expressions become a nonlinear function of the concentration of the stable species, the individual kinetic constants of each step of the mechanism, and the local volumetric rate of energy absorption. Considering that the kinetic expression for each elementary step of the mechanism has the form of the mass action law, the final expressions are

$$\Omega_1 = (k_2C_2 - k_2'C_7K_1)\left(\frac{\Phi e^a}{K}\right)^{1/2} \tag{184}$$

$$\Omega_2 = (k_4C_3 - k_4'C_7K_2)\left(\frac{\Phi e^a}{K}\right)^{1/2} \tag{185}$$

$$\Omega_3 = (k_6C_4 - k_6'C_7K_3)\left(\frac{\phi e^a}{K}\right)^{1/2} \tag{186}$$

$$\Omega_4 = (k_8C_5 - k_8'C_7K_4)\left(\frac{\Phi e^a}{K}\right)^{1/2} \tag{187}$$

where we have identified each species in the following way:

Cl_2	CH_4	CH_3Cl	CH_2Cl_2	$CHCl_3$	CCl_4	HCl	N_2
1	2	3	4	5	6	7	8

and in Eqs. (184) to (187) we have

$$K_1 = \frac{k_2C_2 + k_3'C_3}{k_2'C_7 + k_3C_1} \tag{188}$$

$$K_2 = \frac{k_4C_3 + k_5'C_4}{k_4'C_7 + k_5C_1} \tag{189}$$

$$K_3 = \frac{k_6C_4 + k_7'C_5}{k_6'C_7 + k_7C_1} \tag{190}$$

$$K_4 = \frac{k_8C_5 + k_9'C_6}{k_8'C_7 + k_9C_1} \tag{191}$$

and

$$K = k_{14} + k_{10}K_1 + k_{11}K_2 + k_{12}K_3 + k_{13}K_4 \tag{192}$$

It should be noticed that we have obtained kinetic expressions for the overall reactions [Eqs. (184) to (192)] which do not need any adjustable parameter if an appropriate emission model is used for evaluating the radiation field. Generally, the kinetic expressions found in the literature make use of phenomenological rate constants obtained by fitting experimental data. If appropriate microkinetics data are available, empirical rates do not seem appropriate for modern design purposes, since they are valid only for restricted ranges of operating conditions.

Simplified Model for the Reactor Operation

The simplest model to be used for the prediction of the performance of a continuous reactor is undoubtedly the plug-flow model. However, Romero et al. (1983) have shown that this model is a very poor approximation when strong radial gradients in the radiation field are present, which could be the case of the annular photoreactor if we have a medium-to-large optical density. Moreover, it may provide totally incorrect predictions of the global kinetic behavior. Nevertheless, we can use the maximum gradient model under the following assumptions: (a) isothermal conditions, (b) flat velocity profile, (c) negligible axial dispersion, and (d) negligible radial dispersion. For steady-state flow, the species i mass balance equation becomes

$$v \frac{\partial C_i(r,z)}{\partial z} = \Omega_i[\underline{C}(r,z), e^a(\underline{C},r,z)] \qquad i = 1, \ldots, 7 \tag{193}$$

and

$$\underline{C} = (C_1 \cdots C_7)^T \tag{194}$$

with

$$C_i(0,r) = C_i^\circ \qquad i = 1, \ldots, 7 \tag{195}$$

where

$$\Omega_i = \sum_{j=1}^{4} \nu_{ij}\Omega_j \tag{196}$$

and ν_{ij} is the stoichiometric coefficient of species i in reaction j.

Simplified Emission Model

The first historical step in the development of emission models considered a line source with emission in parallel planes perpendicular to the lamp axis (PELS model). This model is very attractive because of its simplicity but it has very little physical meaning since it requires that all points of the source emit radiation in planes perpendicular to the lamp axis, with zero probability of emission in other directions. As for any type of linear model, it is not recommended for evaluating the radiation field when reflecting surfaces are present. But, on the other hand, it is believed to give an approximate estimation of the radiation field when only direct radiation is present (as in the case of annular photoreactors). However, this statement is partially true. At point $I(r, \beta, z)$ inside the reactor volume (Fig. 8) the radiation arrives from all directions in space and from the whole volume of the lamp. The attenuation path length changes with the direction and may be much longer than the characteristic radiation path length $(r_{ou} - r_{in})$. Besides, the radiation field at point $I(r, \beta, z)$ in the annular reactor results from the contributions of all rays coming from the hemisphere containing the lamp volume. As can be observed in Fig. 29, this important physical situation is not accounted for when the PELS model is used. In fact, at point $I(r, \beta, z)$ inside the reactor, the radiation field results from the contribution of only one ray departing from the lamp; in addition, the attenuation path length is a representation of just the minimum length existing in the real process.

Another failure of the PELS model to predict the real radiation field arises from the fact that it does not account for the reactor wedges (see Fig. 11a) which may be very significant in an annular reactor whose inner wall is close to the lamp wall. With these disadvantages the PELS model is inappropriate to carry out the design of a photoreactor particularly when the characteristic radiation path length is one of the design parameters. Hence the simplest model to predict the radiation field is the SELS model, which at least considers the three-dimensional characteristic of the emission and attenuation processes.

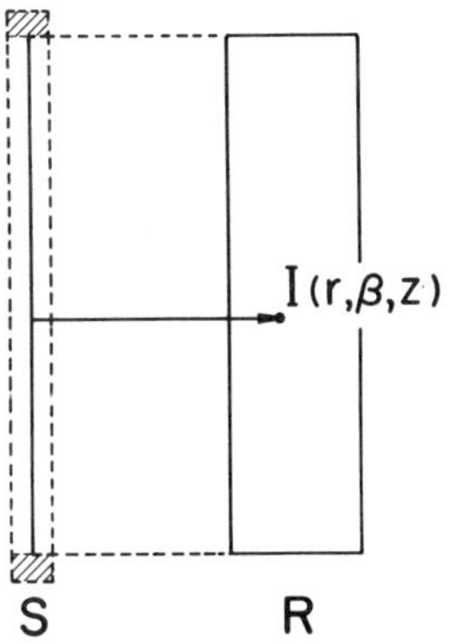

FIGURE 29 Schematic representation of the attenuation process for the PELS model.

Rigorous Model

For the final computation the model was perfected by incorporating the following improvements:

1. *Regarding the kinetic scheme*: The long-chain approximation was not used.
2. *Regarding the reactor model*: (a) The parabolic velocity profile was included, (b) radial diffusion was not neglected, and (c) the multicomponent diffusion coefficients were calculated with the Stefan-Maxwell relationships.
3. *Regarding the emission model*: Extense models (VEES and SEES models) were used instead of the SELS model.

Design Problem

Although production is generally the goal of an optimal design, when dealing with consecutive reactions, the recommended criterion is selectivity. However, in order to decrease the cost of the product separation it may be advisable to operate the reactor with maximum conversion for at least one of the reactants.

A suitable strategy for solving the design problem consists in stating an optimization programme in which the independent variables are the design parameters and the objective function results from a combination of both criteria: selectivity and conversion.

The design problem, which is described by Eqs. (193) to (196), initially has the following degrees of freedom: inlet composition, C_1°, C_2°, and C_8°; volumetric flow rate, Q; inner and outer reactor radiuses, r_{in} and r_{ou}; reactor length, L_R; and lamp characteristics: radius, length, power output, and spectral distribution of its power output.

However, only some of these variables are appropriate to perform an optimization problem. Although they are extremely important in the design of a photochemical reactor, the emission system characteristics cannot be varied continuously because only a limited number of lamp types are available in the market. Hence, in spite of the fact that the parameters related to the lamp enter into the mass balance equations, they cannot be used as optimization variables. Nevertheless, the problem may be solved for different lamps in order to select the most appropriate one. On the other hand, operating conditions and geometrical characteristics of the reactor can be continuously varied. Thus they can be included in the optimization.

Operating the reactor at constant pressure and temperature, the total inlet concentration is constant. Hence

$$C_1^\circ + C_2^\circ + C_8^\circ = C^\circ = \text{constant}$$

and the total number of independent variables is reduced. At the same time, one could set an upper limit to the chlorine inlet fraction: for example, not greater than 50% to avoid excessive heating. Thus if we adopt a constant chlorine inlet concentration,

$$C_1^\circ = 0.5C^\circ$$

the only independent variable representing the inlet composition will be C_2°.

In order to make use of the maximum amount of energy coming out of the lamp, the reactor length is generally taken equal to the lamp length. The inner reactor radius is adopted as close to the lamp radius as possible, leaving a small distance for cooling the lamp and controlling the operating reactor temperature if needed. Then, in our preliminary design problem we will adopt a constant value for the inner radius, and equal values for the lamp and reactor lengths. Finally, the independent variables are: methane inlet concentration, C_2°; volumetric flow rate, Q; and outer reactor radius, r_{ou}.

In the preceding paragraphs we mentioned the possibility of solving the design problem by means of an optimization technique. The following statement of this optimization problem results:

$$\text{Maximize } S_i = \frac{\langle C_i(L_R)\rangle}{\sum_{j=3}^{6} \langle C_j(L_R)\rangle} \qquad i = 3, \ldots, 6 \tag{197}$$

Subject to

$$a_k \leqslant v_k \leqslant b_k \qquad k = 1, 2, 3 \tag{198}$$

$$X_\ell \geqslant p_\ell \qquad \ell = 1, 2 \tag{199}$$

$$h_m(\underline{v}, \underline{C}, \underline{t}) = 0 \qquad m = 1, \ldots, 7 \tag{200}$$

where

S_i = selectivity of the desired chloromethane derivative

$\langle C_i(L_R)\rangle$ = average exit concentration of species i calculated with Eq. (106b)

$\underline{v}$ = independent variable array (Q, r_{ou}, etc.)

$\underline{a}$ = lower bound for the independent variable array

$\underline{b}$ = upper bound for the independent variable array

X_ℓ = ℓ-reactant conversion

p_ℓ = minimum permissible value for ℓ-reactant conversion

$\underline{C}$ = chloromethane concentration array

$\underline{t}$ = array that accounts for the fixed design and operating parameters (temperature, pressure, r_{in}, C°, lamp characteristics, etc.)

Constraints (198) define a feasible region for the optimization problem (i.e., a range of variation for the independent variables). On the other hand, constraints (199) allow us to establish a minimum permissible value for both reactant conversions so as to assure appreciable exit production. Finally, equality constraints (200) represent the solution of the species mass balance equations which give a relationship between the independent

variables $\underline{v}$ and the dependent variables $\underline{C}$ necessary to evaluate S_i. Equations (197) to (200) define a nonlinear multivariable constrained optimization problem which can be solved using any of the several algorithms available in the literature.

To illustrate the design methodology, we solved the optimization problem for chloroform production using a modified Powell algorithm. The statement of the optimization problem becomes:

$$\text{Maximize } S_5 = \frac{\langle C_5(L_R)\rangle}{\Sigma_{j=3}^{6}\langle C_j(L_R)\rangle} \tag{201}$$

subject to

$$0 \leqslant C_2^\circ \leqslant 0.5C^\circ \tag{202a}$$

$$0.00005 \leqslant Q \leqslant 0.00100 \text{ m}^3/\text{s} \tag{202b}$$

$$0.04 \leqslant r_{ou} \leqslant 0.12 \text{ m} \tag{202c}$$

$$X_1 \geqslant 0 \tag{202d}$$

$$X_2 \geqslant 0.995 \tag{202e}$$

and

$$h_k(\underline{v}, \underline{C}, \underline{t}) = 0 \qquad k = 1, \ldots, 7 \tag{203}$$

and $\underline{t}$ includes

$$P = 101.3 \text{ kPa} \tag{204a}$$
$$T = 363 \text{ K} \tag{204b}$$
$$C^\circ = 33.6 \text{ mol/m}^3 \tag{204c}$$
$$r_{in} = 0.03 \text{ m} \tag{204d}$$

and lamp data. The upper and lower bounds used in constraints (202b) and (202c) are typical values for annular photoreactor operation.

Finally, to complete the required information we must decide about the characteristics of the lamp. Since methane photochlorination is a chain-type reaction, it should be expected that low-energy radiation sources would be required. This statement was corroborated by De Bernardez and Cassano (1986), who studied the incidence of lamp parameters on selectivity and conversion in methane photochlorination. The authors found that the reaction can be carried out using sources of low-energy output such as germicidal or black-light lamps. Both types of lamps represent a very attractive solution from an economic point of view. In addition, they are available in the market in a variety of sizes and output powers. To illustrate the lamp selection procedure, we will solve the optimization problem for two different sources which are amenable to comparison:

Lamp	Emission model	Power input (W)	Power output (W)	Spectral range of emission (Å)	Length (m)
Germicidal	VEES isotropic	30	8.3	2537	0.91
Black light	SEES diffuse	30	5.4	3200–4200	0.91

Both lamps have identical values of energy input and length, but the output power and spectral distribution are different. The germicidal lamp emits at a wavelength range where chlorine absorption is poor (see Table 5), while the emission of the black lamp falls in a wavelength range where the absorption by chlorine is significantly stronger. Hence a higher reactant conversion should be expected when the black light lamp is used. However, the amount of energy emitted by the germicidal lamp is greater and consequently there exists the possibility of obtaining a larger product output for the same energy consumption. The final result will show which of the two characteristics causes a better improvement in conversion. In other words, we wish to obtain information about the relative significance of the quality and the quantity of the energy emitted. Similarly, it is impossible to anticipate the effect of both properties (power and spectral distribution) on chloroform selectivity.

Table 6 summarizes the results of the design problem for both lamps. Due to the characteristics of chlorine light absorption combined with the spectral distribution of the energy output of both lamps, we can state the following conclusions:

TABLE 6 Optimization of Chloroform Production: Design Problem Results

	Germicidal	Black light
S_{CH_3Cl}	11.3	3.9
$S_{CH_2Cl_2}$	25.0	18.9
S_{CHCl_3}	40.4	47.0
S_{CCl_4}	23.3	30.2
X_{Cl_2}	21.8	60.3
X_{CH_4}	89.8	95.8
P_{CHCl_3} (g/s) $\times 10^3$	7.4	12.8
C_2^o (mol/m^3)	1.32	2.83
r_{ou} (m)	0.050	0.044
Q(m^3/s) $\times 10^5$	13.0	8.5

1. The germicidal lamp allows us to use greater equivalent diameters, but to obtain high methane conversions, the system is forced to use low hydrocarbon concentration in the feed. This considerably reduces the output production. With germicidal lamps, chloroform selectivity would be approximately 40%.
2. Chloroform production and selectivity are greater in the case of black-light lamp, indicating that in spite of the differences in power output (more than 50%), what really counts is the quality of the supplied radiation. With this lamp we get higher selectivity (close to 50%), very high methane conversion (almost 96%), and can obtain 73% more output production of chloroform for the same energy input and reactor length. Perhaps what is even more surprising is that all these advantages are obtained with an equivalent diameter that is 30% smaller.

Table 6 also shows that the minimum permissible value of methane conversion could not be reached. However, by working with lamps of greater energy input we can increase the length of the lamp (and that of the reactor), the amount of energy per unit length of the lamp, and obviously, the mean residence time. This possibility could certainly allow one to fulfill the required exit conversion.

To increase the total amount of chloroform produced, several reactors (in parallel) should be used. Other possibilities are the multitube and the multitube-multilamp reactors.

These results clearly illustrate that a photochemical reactor provides new variables for optimization. Although reactant feed composition ratio and mean residence time are still very important parameters, the optical thickness (combination of absorptivity, characteristic radiation path, and absorbing species concentration) as well as the output characteristics of the radiation source have a significant effect not only in the exit conversion but also in the overall production and selectivity.

REACTOR CONFIGURATIONS FOR INDUSTRIAL APPLICATIONS

Single Lamp-Multitube Reactor

This arrangement (Fig. 30) is a suitable reactor for medium-pressure, highly exothermic or endothermic gas-phase reactions. It provides good heat transfer characteristics with the surrounding fluid. The individual reactors may have rather small diameters and consequently, well-tested Pyrex glass or quartz tubes can stand pressures up to 1500 kPa. It could be used for liquid-phase reactors as well. In any case, applications should be restricted to rather fast reactions (or equivalently low flow rates) because the reactor tubes will necessarily have length limitations (mainly motivated by market restrictions in the lamps and safe operating conditions for the reactor).

Multilamp Reactor

This reactor (Fig. 31) can be recommended for liquid-phase reactions with small-to-medium heat transfer requirements, which could be improved by adding cooling coils inside the vessel. The walls of the tank could have

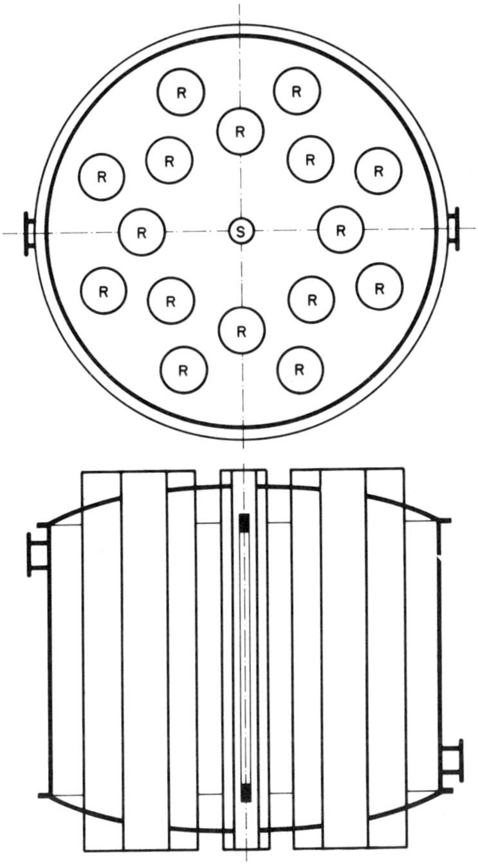

FIGURE 30 Single lamp-multitube reactor.

reflecting properties, increasing the efficiency of the radiation emitted by the lamps. Again, it can be used for low and medium reaction pressures, since restrictions are placed in the diameter of the tubes surrounding the radiation sources. This reactor is very appropriate for reactions where energy requirements are rather high or if, in spite of having low-energy requirements, the optical thickness of the medium is large. In the second case the same input energy could be distributed in several radiation sources, obtaining a more useful radiation field distribution inside the tank. It is also suitable for multiphase reactions since very good stirring is easily achieved. It is not appropriate for gas-phase reactions.

Pseudoannular Reactor

This reactor (Fig. 32) is a modified version of an annular photoreactor with the purpose of increasing its heat transfer possibilities and its capabilities for low and medium operating pressures. It is also very suitable for reactions that may need medium-to-long mean residence times. It could equally well be used for gas- or liquid-phase reactions.

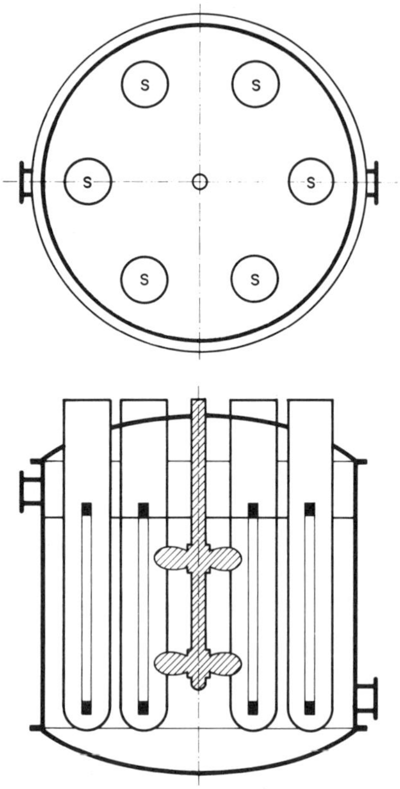

FIGURE 31 Multilamp reactor.

Multilamp-Multitube Reactor

This reactor (Fig. 33) is perhaps the best arrangement for large-scale production. The system has good pressure and heat transfer capabilities, large flexibility regarding energy requirements or optical properties of the reacting medium, and if there is adequate connection between tubes, good flexibility to overcome large residence-time requirements. It could not be used efficiently for multiphase reactions, this limitation being its main hindrance.

Flat-Plate Reactor

This reactor (Fig. 34) is suitable for reacting media of very high optical thickness. Its main application could be for decontamination of highly refractory materials to other technologies. In this case the "cleaning" of the reactor feed (to increase transparency from the optical point of view) could be infeasible for economic or technical reasons. Hence impurities will greatly increase the absorption characteristics of "reactants." A reduction in the radiation path length would be the only solution and the flat-plate reactor may be the most practical way to achieve it. Both plates (upper and bottom parts) should be made with practical provisions for

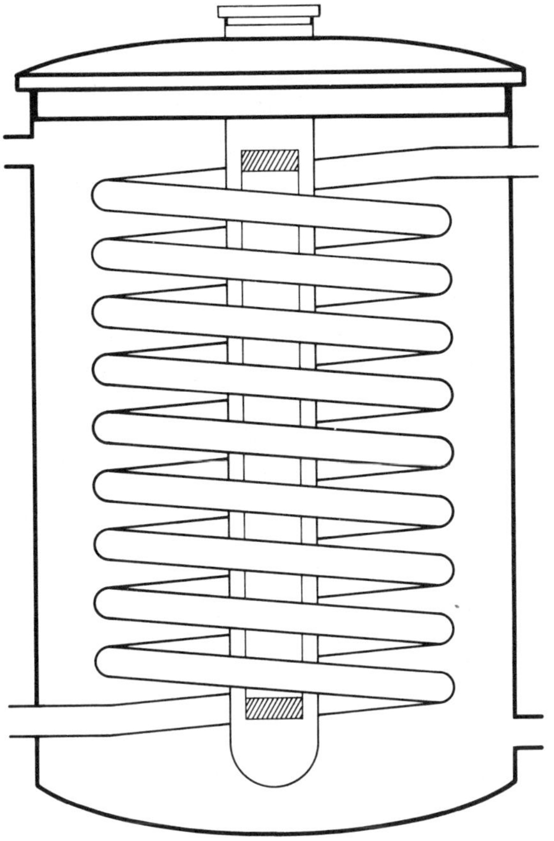

FIGURE 32   Pseudoannular reactor.

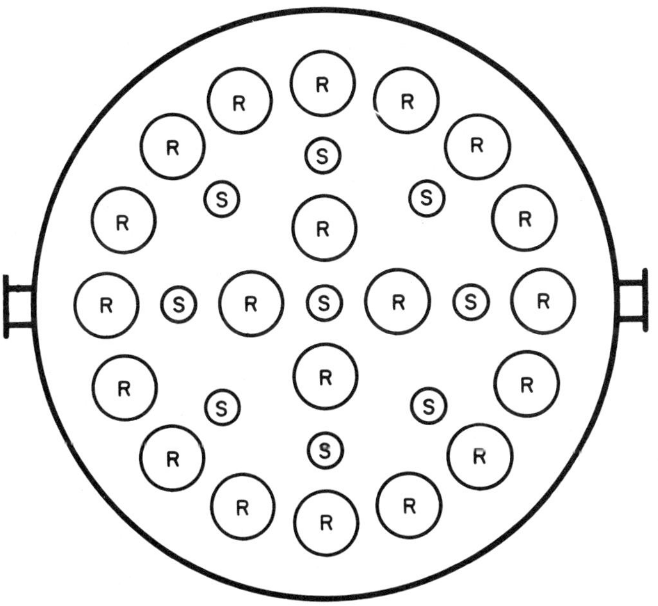

FIGURE 33   Multitube-multilamp reactor.

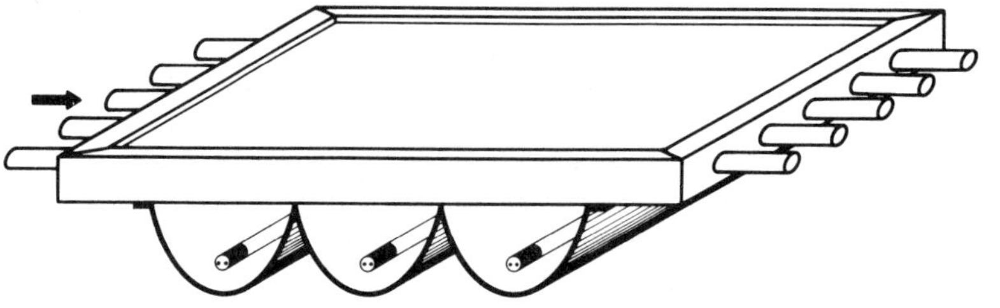

FIGURE 34 Flat-plate reactor.

fast assembling and reassembling purposes to facilitate unavoidable cleaning routines. A second important application may be photo-assisted catalysis. The modeling of this reactor is an easy extension of our reactor irradiated from the bottom (Fig. 34).

Finally, it should be noted regarding wall deposite difficulties that mechanical cleaning would be difficult only in the pseudoannular reactor. If chemical methods are used this reactor is one of the easiest to clean.

UNSOLVED PROBLEMS

With regard to photoreactor design, assuming that the reactions are well known, we recognize several problems where lack of information or reliable ways of handling them can be observed. The main ones are: modeling the absorption in heterogeneous systems, dark deposits at the reactor wall through which the radiation must enter into the reacting space (which reduces light transmission and affects yield), and problems related to pressurized gas-phase reactors.

Several research groups have made important contributions to the area of modeling the absorption in heterogeneous media (Otake et al., 1981; Yokota et al., 1981). However, we should point out that the only duly proved contributions related to the modeling of heterogeneous systems still require experimentally adjustable parameters. The use of rigorous models combined with the modeling of the distortions produced by the heterogeneities constitutes a particularly attractive problem where additional research work should be performed. These problems have great significance in the field of photocatalysis where an opaque solid may be present.

The reactions at the wall of the reactor are a serious disadvantage for the use of photochemical reactors in industrial processes and the main cause of frequently shutdowns for cleaning routines. It is known that some of these difficulties have already been overcome by the Toyo Rayon Co. of Japan in their cyclohexane photonitrosation commercial-scale process. Some additional work on the subject, proposing novel reactor designs not yet under the stage of providing results of economic significance, has been proposed by Lucas (1971). The author proposed a "two-zone segregated reactor" where all deposites are avoided since all secondary reactions after the initiation step occur in a place different from the one where the radiation is supplied to the system. Nevertheless, appropriate design of the reactor could provide continuous cleaning by a mechanical scraping of the wall where the radiation enters the reactor, thus supplying a practical

solution to the problem. Hopefully, the time will come when fundamental research on surface chemistry will be able to provide a definite answer.

Since photochemical gas-phase reactors must have at least one quartz or Pyrex glass wall, there exist some difficulties in working under pressure. It could be thought as a materials science problem but mainly a challenge for new reactor designs, which could range from multitubular reactors of small diameter (often used in industry for heat-release reasons) to the introduction of new technologies such as lasers combined with optical fibers (see Marinangeli and Ollis, 1977, 1980).

APPENDIX: CYLINDRICAL REACTOR WITH ELLIPTICAL REFLECTOR. LIMITING VALUES OF THE COORDINATES FOR INDIRECT RADIATION WITH ONLY ONE REFLECTION

The limiting values for the ρ coordinate can be obtained as follows (see Fig. 13). Let $\underset{\sim}{\varepsilon}_I$ be the unit vector representing the direction of a ray that impinges at any point I on the reaction space with direction θ,ϕ given by

$$\underset{\sim}{\varepsilon}_I = \sin\theta \cos\phi \underset{\sim}{i} + \sin\theta \sin\phi \underset{\sim}{j} + \cos\theta \underset{\sim}{k} \tag{A1}$$

This ray has previously rebounded on a point P of the elliptical mirror, whose coordinates referred to point I are:

$$x_P = \rho_I \sin\theta \cos\phi$$

$$y_P = \rho_I \sin\theta \sin\phi$$

$$z_P = \rho_I \cos\theta$$

where

$$\rho_I = \frac{-(hb^2 \cos\phi + ka^2 \sin\phi) + ab[\sigma^2 - (h \sin\phi - k \cos\phi)^2]^{1/2}}{\sigma^2 \sin\theta} \tag{A2}$$

and

$$\sigma^2 = b^2 \cos^2\phi + a^2 \sin^2\phi$$

$$h = c + r_I \cos\beta_I \tag{A3}$$

$$k = r_I \sin\beta_I$$

The expression for the unit normal vector to the reflector at point P can be written as follows:

$$\underset{\sim}{n}_P = \frac{\dfrac{x_P + h}{a^2}\underset{\sim}{i} + \dfrac{y_P + k}{b^2}\underset{\sim}{j}}{\left[\dfrac{(x_P + h)^2}{a^4} + \dfrac{(y_P + k)^2}{b^4}\right]^{1/2}} \tag{A4}$$

The unit vector $\underset{\sim}{\varepsilon}_E$ representing the direction of the ray before reflection (i.e., emerging from the lamp) can be obtained from the law of reflection:

$$\underset{\sim}{n}_P \times \underset{\sim}{\varepsilon}_E = \underset{\sim}{n}_P \times \underset{\sim}{\varepsilon}_I \tag{A5}$$

Solving this vectorial equation, the following expression for $\underset{\sim}{\varepsilon}_E$ is obtained:

$$\underset{\sim}{\varepsilon}_E = \sin\theta\cos\phi_E \underset{\sim}{i} + \sin\theta\sin\phi_E \underset{\sim}{j} + \cos\theta \underset{\sim}{k} \tag{A6}$$

where

$$\phi_E = \tan^{-1}\frac{(m^2-1)\sin\phi + 2m\cos\phi}{2m\sin\phi - (m^2-1)\cos\phi} \tag{A7}$$

and

$$m = \frac{a^2(y_P + k)}{b^2(x_P + h)} \tag{A8}$$

The two intersections of direction θ, ϕ_E with the bounding surface of the radiation source are precisely the limits of ρ. They are:

$$\rho_{E_{2,1}} = \frac{D\cos\xi \pm (r_L^2 - D^2\sin^2\xi)^{1/2}}{\sin\theta} \tag{A9}$$

where

$$D = [(x_P + h + c)^2 + (y_P + k)^2]^{1/2} \tag{A10}$$

$$\sin\xi = -\frac{1}{D}[(x_P + h + c)\sin\phi_E - (y_P + k)\cos\phi_E] \tag{A11}$$

$$\cos\xi = -\frac{1}{D}[(x_P + h + c)\cos\phi_E + (y_P + k)\cos\phi_E] \tag{A12}$$

The limiting values of the θ coordinate can be defined by means of (Fig. 12)

$$\begin{aligned}(\rho_I + \rho_{E,1})\cos\theta_1 &= \tfrac{1}{2}(L_{Rf} + L_L) - \zeta_I \\ (\rho_I + \rho_{E,1})\cos\theta_2 &= \tfrac{1}{2}(L_{Rf} - L_L) - \zeta_I\end{aligned} \tag{A13}$$

Using Eqs. (A2) and (A9), the integration limits as a function of ϕ result:

$$\theta_1 = \tan^{-1}\frac{\rho_I' + D\cos\xi - (r_L^2 - D^2\sin^2\xi)^{1/2}}{(1/2)(L_{Rf} + L_L) - \zeta_I} \tag{A14}$$

$$\theta_2 = \tan^{-1} \frac{\rho_I' + D \cos \xi - (r_L^2 - D^2 \sin^2 \xi)^{1/2}}{(1/2)(L_{Rf} - L_L) - \zeta_I} \tag{A15}$$

where

$$\rho_I'(\phi) = \frac{-(hb^2 \cos \phi + ka^2 \sin \phi) + ab[\sigma^2 - (h \sin \phi - k \cos \phi)^2]^{1/2}}{\sigma^2}$$

ρ_I' depends only on ϕ. The same can be said for D, ϕ_E, and ξ.

The limiting rays in the ϕ direction for any θ angle must be tangent to the lamp boundaries that is,

$$\rho_{E,1} = \rho_{E,2} \tag{A16}$$

With Eq. (A9) the following expression is obtained:

$$D^2 \sin^2 \xi = r_L^2 \tag{A17}$$

Solving this implicit equation numerically, the limiting values of ϕ can be determined.

ACKNOWLEDGMENTS

This work reports the photoreactor design methodology developed from research done with my former students, some of them now full professors, such as Horacio A. Irazoqui, Jaime Cerdá, and Jacinto L. Marchetti, and others that are still on their way to get their doctor's degrees, including Orlando M. Alfano, Roberto L. Romero, Roberto M. Ré, and my two coauthors, Eliana R. De Bernardez and María A. Clariá.

The theory has been verified with bench-scale experiments. Thanks must be given to Pedro G. Guarino, Mario J. J. Didier, Juan C. García, Antonio C. Negro, Omar Brizuela, José L. Giménez, and José L. Giombi for their contributions to this part of the work. Credit must also be given to Yolanda Pereyra for her excellent typing of this monography and to Elsa I. Grimaldi for the thoroughness of her review of the English used in the manuscript.

NOTATION

a	ellipse semimajor axis or characteristic constant of the parabola, cm
a_v	interfacial area per unit liquid volume, cm^{-1}
a_v'	interfacial area per unit liquid plus gas volume, cm^{-1}
A	area, cm^2
b	ellipse semiminor axis, cm
c	half distance between focuses of the ellipse, cm
C	concentration, mol/cm^3

d_b bubble diameter, cm

D diffusivity, cm^2/s

e ellipse eccentricity, dimensionless

e^a local volumetric rate of energy absorption, einstein/cm^3•s

E radiant energy flow rate, einstein/s

F molar flow rate, mol/s^1

G value defined by Eq. (87) dimensionless

h Planck's constant, J•s. or value defined by Eq. (A3), cm

H Henry's constant, kPa•cm^3/mol

I intensity of radiation (dimensions are source-model dependent)

I' isotropic intensity of radiation (dimensions are source-model dependent

$\underset{\sim}{j}$ unit vector

k value defined by Eq. (A3), cm

k_i reaction rate constant for step i, s^{-1} (first order) or cm^3/mol•s (second order)

k_L mass transfer with chemical reaction absorption coefficient, cm/s

k_L^o physical mass transfer absorption coefficient, cm/s

L length, cm

ℓ distance from vertex of parabolic reflector to the bottom of the reactor, cm

m value defined by Eq. (A8), dimensionless

$\underset{\sim}{n}$ unit normal vector

N mass absorption rate, mol/cm3•s

p partial pressure, kPa, or production, g/s

P total pressure, kPa

$\underset{\sim}{q}$ radiation flux density vector, einstein/cm^2•s

Q volumetric flow rate, cm^3/s

r radial coordinate, cm

R indicates reactor

Re Reynolds number, dimensionless

S selectivity, dimensionless; also indicates radiation source

t time, s

v velocity, cm/s

V volume, cm^3

x rectangular coordinate, cm

X fractional conversion, dimensionless

y	rectangular coordinate, cm or gas-phase molar fraction
z	rectangular coordinate, cm

Greek Letters

α	absorption coefficient, cm^2/mol
β	cylindrical coordinate, rad
Γ	emitting extension (dimensions are source-model dependent)
Γ_{Rf}	reflection coefficient, dimensionless
δ	Dirac function
Δ	value defined by Eqs. (55) and (92), cm
$\Delta_{1,2}$	values defined by Eq. (74), cm
ε_g	dispersed-phase holdup, dimensionless
ζ	axial coordinate, cm
θ	spherical coordinate, rad
κ	characteristic emission property of the lamp (dimensions are source-model dependent)
λ	wavelength, cm
μ	attenuation coefficient, cm^{-1}
ν	frequency, s^{-1}; also stoichiometric coefficient
ξ	angle defined by Eqs. (A11) and (A12) rad
ρ	spherical coordinate, cm
σ	value defined by Eq. (A3), cm
Υ	reactor wall transmittance, dimensionless
ϕ	spherical coordinate, rad
Φ	primary quantum yield, mol/einstein
Φ_{ov}	overall quantum yield, mol/einstein
χ	inner radius/outer radius, dimensionless
ψ	energy flow rate per unit area, $einstein/s \cdot cm^2$
Ψ	radiant energy flow per unit of source extent (dimensions are source model dependent)
ω	solid angle, Sr
Ω	reaction rate, $mol/cm^3 \cdot s$

Subscripts

act	relative to an activation step
D	denotes direct radiation
Dif	denotes a property related to diffusion

e	relative to emission
ed	denotes a property related to effective diffusion
eff	denotes an effective property
E	denotes a property of an emerging ray from the source
F	relative to a fixed coordinate system
i	denotes species i
in	relative to the inner wall of the annular reactor
I	denotes an incident point or an incident ray property
In	denotes indirect radiation
L	denotes a lamp property
m	relative to a multicomponent mixture
max	denotes maximum value
o	denotes a wall property
ou	relative to the outer wall of the annular reactor
ox	relative to oxalic acid
P	denotes point of reflection
r	relative to reception
rc	relative to chemical reaction
R	denotes a reactor property
Rf	denotes a reflector property
T	denotes total value
ur	relative to the uranyl ion
z	relative to the z axis
1	denotes a lower limit of integration
2	denotes an upper limit of integration

Greek Letters

λ	denotes a wavelength dependence
ν	denotes a frequency dependence
ρ	relative to the ρ coordinate

Superscripts

a	relative to absorption
if	denotes interface property
o	initial value
'	value projected on the x-y plane
*	relative to the attenuation path

REFERENCES

Alfano, O. M. and A. E. Cassano, Modeling of a two-phase photoreactor irradiated from the bottom: I. Theory (1986a, submitted for publication.)

Alfano, O. M. and A. E. Cassano, Modeling of a two-phase photoreactor irradiated from the bottom: II. Experiments (1986b, submitted for publication.)

Alfano, O. M., R. L. Romero, and A. E. Cassano, Radiation field inside a cylindrical photoreactor irradiated from the bottom: Theory and experiments, in *Frontiers in Chemical Reaction Engineering*, Vol. 1, Doraiswamy and R. A. Mashelkar, eds., Wiley Eastern, New Delhi (1984), p. 506.

Alfano, O. M., R. L. Romero, and A. E. Cassano, Modeling of radiation transport and energy absorption in photoreactors, Adv. Transport Process, *4*, 201 (1985).

Astarita, G., *Mass Transfer with Chemical Reaction*, Elsevier, Amsterdam (1967).

Astarita, G., D. W. Savage, and A. Bisio, *Gas Treating with Chemical Solvents*, Wiley, New York (1983).

Calvert, J. G. and J. N. Pitts, *Photochemistry*, Wiley, New York (1966).

Cerdá, J., H. A. Irazoqui, and A. E. Cassano, Radiation fields inside an elliptical photoreflector with a source of finite spatial dimensions, AIChE J., *19*, 963 (1973).

Cerdá, J., J. L. Marchetti, and A. E. Cassano, Radiation efficiencies in elliptical photoreactors, Lat. Am. J. Heat Mass Transfer, *1*, 33 (1977).

Cerdá, J., J. L. Marchetti, and A. E. Cassano, The use of simple radiation models for the case of direct irradiation of photochemical reactors, Lat. Am. J. Chem. Eng. Appl. Chem., *8*, 15 (1978).

Childs, L. P. and D. F. Ollis, Is photocatalysis catalytic? J. Catal., *68*, 383 (1980).

Clariá, M. A., H. A. Irazqui, and A. E. Cassano, A priori design of the photochemical monochlorination of ethane: I. Theory (1985, submitted for publication.)

Clariá, M. A., H. A. Irazoqui, and A. E. Cassano, A priori design of the photochemical monochlorination of ethane: II. Computed and experimental results (1985b, submitted for publication.)

Clariá, H. , A. Irazoqui, and A. E. Cassano, Modeling and experimental validation of the radiation field inside an elliptical photoreactor (1986a, submitted for publication.)

Clariá, M. A., H. A. Irazoqui, and A. E. Cassano, The use of linear and extense source models in photoreactor design (1986b, submitted for publication.)

De Bernardez, E. R. and A. E. Cassano, Azimuthal asymmetries in tubular photoreactors, Lat. Am. J. Heat Mass Transfer, *6*, 333 (1982).

De Bernardez, E. R. and A. E. Cassano, A priori design of a continuous annular photochemical reactor. Experimental validation, J. Photochem. *30*, 285 (1985).

De Bernardez, E. R. and A. E. Cassano, Optimal selection of the radiation source for a consecutive—chain type—reactions in a continuous photoreactor Ind. Eng. Chem. Process Des. Dev. (1986, in press.)

General Electric Co., Tech. Bull. T.P. 122 (1967).

Gosselain, P. A., J. Adam, and P. Goldfinter, La spécificité des halogénations atomiques: II. Les règles générales qui déterminent le mécanisme et a. spécificité des halogénations atomiques, Bull. Soc. Chem. Belg., *65*, 533 (1956).

Herzberg, G., *Molecular Spectra and Molecular Structure*, Vol. I: *Spectra of Diatomic Molecules*, 2nd ed., Van Nostrand Reinhold, New York (1950).

Irazoqui, H. a., J. Cerdá, and A. E. Cassano, Radiation profiles in an empty annular photoreactor with a source of finite spatial dimensions, AIChE J., *19*, 460 (1973).

Jacob, S. M. and J. S. Dranoff, Radial scale-up of perfectly mixed photochemical reactors, Chem. Eng. Prog. Symp. Ser., *62*, 47 (1966).

Jacob, S. M. and J. S. Dranoff, Design and analysis of perfectly mixed photochemical reactors, Chem. Eng. Prog. Symp. Ser., *64*, 54 (1968).

Kurtz, B. E., Homogeneous kinetics of methyl chloride chloration, Ind. Eng. Chem. Process Des. Dev., *11*, 332 (1972).

Lucas, G., A new concept in oximation by the nitrosyl chloride molecule, Inf. Chim., *16*, 33 (1971).

Marinangeli, R. E. and D. F. Ollis, Photoassisted heterogeneous catalysis with optical fibers: I. Isolated single fiber, AIChE J., *23*, 415 (1977).

Marinageli, R. E. and D. F. Ollis, Photoassisted heterogeneous catalysis with optical fibers: II. Nonisothermal single fiber and fiber bundle, AIChE J., *26*, 1000 (1980).

Otake, T., S. Tone, K. Higuchi, and K. Nakao, Light intensity profile in gas-liquid dispersion. Applicability of effective absorption coefficient, Kagaku Kogaku Ronbunshu, 7, 57 (1981).

Pai, S., *Radiation Gas Dynamics*, Springer-Verlag, Berlin (1966).

Ré, R. M. and A. E. Cassano, Three-dimensional modeling of a tubular photoreactor when the radiation field has no angular symmetry (1986, submitted for publication.)

Romero, R. L., O. M. Alfano, J. L. Marchetti, and A. E. Cassano, Modeling and parametric sensitivity for an annular photoreactor with complex kinetics, Chem. Eng. Sci., *38*, 1593 (1983).

Stramigioli, C., F. Santarelli, and F. P. Foraboschi, Photosensitized reactions in an annular photoreactor, Ing. Chim. Ital., *11*, 143 (1975).

Stramigioli, C., F. Santarelli, and F. P. Foraboschi, Photosensitized reactions in an annular continuous photoreactor, Appl. Sci. Res., *33*, 23 (1977).

Turro, N. J., *Molecular Photochemistry*, W. A. Benjamin, New York (1965).

Vincenti, W. G. and C. H. Kruger, Jr., *Introduction to Physical Gas Dynamics*, Wiley, New York (1965).

Yokota, T., T. Iwano, H. Deguchi, and T. Tadaki, Light absorption rate in a bubble column photochemical reactor, Kagaku Kogaku Ronbunshu, 7, 156 (1981).

14
Electrochemical Reaction Engineering

GARY G. TROST,* VICTORIA EDWARDS, and JOHN S. NEWMAN
University of Califronia and Lawrence Berkeley Laboratory, Berkeley, California

INTRODUCTION

Electrochemistry is involved to a significant extent in the present-day industrial economy. Examples are found in primary and secondary batteries and fuel cells; in the production of aluminum, chlorine, caustic soda, and other chemicals; in electroplating, electromachining, and electrorefining; and in corrosion. The electrochemical industries presently consume 6% of the total electric energy generated in the United States (Darlington and Woo, 1982; Mantell, 1960). The battery industry amounts to 0.2% of the total U.S. manufacturing (U.S. Department of Commerce, 1980, p. 14; National Research Council, 1982, p. 183). Over 3% of the total U.S. electric energy is consumed by the aluminum industry (Darlington and Woo, 1982; Mantell, 1960), and the chlorine industry consumes 2% of the total U.S. electric energy (Darlington and Woo, 1982; Mantell, 1960; Beck, 1976; Bockris et al., 1981). The economic importance of corrosion is evident from the amount of money spent to protect what has been built and to replace what has been destroyed. It has been estimated (Wenglowski, 1966) that corrosion prevention products amount to 3% of the U.S. electrochemical industry. The quantity of iron destroyed by corrosion is between 25 and 33% of the total production (Van Muylder, 1981).

In spite of the economic significance of electrochemistry, engineering design procedures for electrochemical systems have not been developed as thoroughly as for the mass transfer operations such as distillation. Nevertheless, the fundamental laws governing electrochemical systems are known. The purpose of this chapter is to review the design and analysis of certain electrochemical systems in relation to these fundamental laws.

One of the difficulties in designing an electrochemical process is that there are several different types of reactors that can be used. These reactors can be broadly classified as either batch or continuous reactors. For example, many batteries are batch reactors, while fuel cells and many electrolysis cells are continuous reactors. Naturally, there are several different reactor geometries within each category. For example, chlorine

*Current affiliation: Raychem Corporation, Menlo Park, California

and caustic soda are produced in any of three types of cells (Bockris et al., 1981). The diaphragm cell contains a vertical dimensionally stable anode, where chlorine is evolved, and steel screen cathodes, where hydrogen gas and hydroxyl ions are produced. The electrode compartments are separated by a porous diaphragm that allows passage of electric current. The second type of chlor-alkali cell is the membrane cell. This cell is similar to the diaphragm cell with the exception of the separator, which, in this case, is a cation-exchange membrane, which inhibits transfer of chloride and hydroxyl ions, and does not permit bulk flow of electrolyte. The third type of cell, the mercury cell, contains no diaphragm, but achieves separation of electrode products by the use of a mercury cathode, where sodium ion is reduced and forms a sodium-mercury amalgam. The amalgam is then passed to a second cell, where the amalgam is reacted with water to form sodium hydroxide, hydrogen, and pure mercury, which is recycled to the electrolyzer.

Another economically significant electrochemical reactor is the Hall-Heroult cell for the production of aluminum. This process is operated in the batch mode at high temperature (approximately 1000°C) in a bath of aluminum oxide or alumina (Al_2O_3) dissolved in fused cryolite (Na_3AlF_6), to which aluminum fluoride is added. The thermal insulation is adjusted to provide sufficient heat loss to freeze a protective coating of electrolyte on the inner walls, but not on the bottom, which must make electrical contact with the molten aluminum cathode. A crust of frozen electrolyte and alumina covers the molten electrolyte. Electric current enters the cell through carbon anodes and flows through the electrolyte, forming CO_2 at the anode and aluminum at the cathode. Steel collector bars joined to the carbon lining at the bottom conduct electric current from the cell. Some reviews of these and other industrial electrochemical processes can be found in Bockris et al. (1981), Kuhn (1971), and Houghton and Kuhn (1974).

INTRODUCTION TO DESIGN

Any chemical process requires the evaluation of various alternative routes, whether they be chemical or electrochemical. Two major tasks in designing an electrochemical reactor are first to narrow down the choices of reactor configurations and then to devise an economically optimum final design. As an example of eliminating some reactor geometries, consider metal-ion removal from dilute waste streams. In this case it would be impractical to use two plane parallel electrodes because the reactor volume would be very high, whereas a porous electrode, with its high surface area per unit volume, would be a good candidate for this application. Porous electrodes are also more suitable for gas-evolving reactions. On the other hand, the channel configuration might be very useful for removing valuable metals, such as silver, from concentrated solutions because the metal would be recovered directly as a sheet, rather than being incorporated in a porous structure.

After choosing some electrode configurations to evaluate, one would want to estimate the capital and operating costs and the product composition, or, in the case of a battery, the energy and power output, as a function of the feed conditions, the cell dimensions, and the operating conditions. This can sometimes be a very difficult task because there are so many phenomena that can interact, such as mass transfer, ohmic potential drop,

thermodynamics, and electrode kinetics. In many cases, however, the problem can be greatly simplified by realizing which phenomena dominate the system behavior. This approach is expedient in choosing design alternatives.

In either the crude design or the final design, it is desirable to maximize profit or minimize cost. Thus one should be aware of the various economic trade-offs that are important in electrochemical systems. Perhaps the most obvious trade-off is between capital and operating costs. Usually, the degree of conversion is specified, and one can meet that specification either by building a large cell, with a high capital cost, or by using more power with a smaller cell. There are also other trade-offs; for example, in a channel flow cell, one can reduce the ohmic drop and increase the mass transfer rate by making the interelectrode gap very thin, but this will increase the pressure drop and hence the pumping cost. In addition, it will lead to fabrication problems and the possibility of shorts.

GOVERNING PHENOMENA

If an electrochemical reactor is to be compared to an alternative, be it chemical or electrochemical, detailed knowledge for the design and scaleup is needed. To design an electrochemical system, one needs to consider many effects, such as migration and diffusion of charged species, fluid dynamics, thermodynamics, and reaction kinetics on electrode surfaces. An additional system variable, not present in analyses of most chemical reactors, is the electric potential. It is the difference in potential between the electrode and the solution that governs which electrochemical reactions will occur. In addition, gradients of electric potential and differences in potential between phases constitute major driving forces for mass transfer and for electrochemical reactions. Thus a knowledge of the current and potential distribution is useful in designing a reactor to carry out a desired electrochemical reaction.

To describe mass transfer in an electrochemical system, one must consider not only ordinary diffusion, but also migration of charged species in an electric field. Because it is the flux of charged species that produces an electric current, one must include the contribution of migration to the species flux to predict correctly the current flow.

In addition to mass transfer, one must consider thermodynamics in electrochemical systems. Thermodynamics can be used to describe the properties of electrolytic solutions and their dependence on composition, temperature, and pressure. Thermodynamics also provides a framework for describing reaction equilibria, which manifest themselves in equilibrium cell potentials. Furthermore, the driving forces for irreversible processes are conveniently expressed in thermodynamic terms.

Departures from equilibrium conditions are inherent in electrochemical applications. Electrode kinetics concerns the nonequilibrium driving force, called surface overpotential, necessary to make heterogeneous electrode reactions proceed at appreciable rates.

In addition to heterogeneous reactions, there are other interfacial phenomena at electrode surfaces that can affect the current-potential behavior. The most prominent of these is the formation of a thin double layer near electrode surfaces. This double layer may be on the order of 10 to 100 Å (1 to 10 nm) in thickness.

Because there are so many interacting phenomena, it can sometimes be very difficult to design an electrochemical system. In the present age of computers, however, many problems that were intractable 30 years ago can be handled today.

In spite of the power of computers, however, it is still useful to simplify the problem if certain phenomena dominate the system behavior. For example, there are some systems where it is possible to neglect concentration variations near the electrodes. The current distribution is then determined by the ohmic potential drop in the solution and by electrode overpotentials. (The ohmic drop is simply the voltage difference that arises when current flows through the resistive solution.) Mathematically, this means that the potential satisfies Laplace's equation, and many results of potential theory, developed in electrostatics, the flow of inviscid fluids, and steady heat conduction in solids, are directly applicable. These can be called "potential-theory problems" (Newman, 1968). The electrode kinetics provides boundary conditions which are usually different from those encountered in other applications of potential theory.

There are also systems where the ohmic potential drop can be neglected. The current distribution is then determined by the same principles which apply to heat transfer and nonelectrolytic mass transfer. These can be called "convective-transport problems" (Newman, 1968).

Some systems do not fall into either of the categories mentioned above. Fortunately, however, some of these complex systems can be treated realistically by using the computer to model the interacting phenomena.

FUNDAMENTAL EQUATIONS

To calculate the current and potential distribution, which is necessary to design an electrochemical system, one needs a set of fundamental equations applicable to electrochemical systems. These equations include a description of the movement of mobile ionic species, an expression for current flow, material balances, electroneutrality, and equations from fluid mechanics. In addition, the differential equations describing the electrolytic solution require boundary conditions describing the thermodynamics and kinetics of electrode reactions.

In dilute systems, the flux of each species is given by (Newman, 1973a)

$$\underset{\sim}{N}_i = -z_i u_i F c_i \nabla \Phi - D_i \nabla c_i + c_i \underset{\sim}{v} \tag{1}$$

where Φ is the electrostatic potential and z_i is the number of proton charges carried by ion i. This movement is due, first, to the motion of the fluid with the bulk velocity $\underset{\sim}{v}$. However, the movement of the species can deviate from this average velocity by diffusion if there is a concentration gradient, ∇c_i, or by migration if there is an electric field, $-\nabla \Phi$, and the species is charged. The migration term is peculiar to electrochemical systems or systems containing charged species.

The second and third terms on the right side of Eq. (1) are the usual terms required to describe diffusion and convection in nonelectrolytic systems. The species will diffuse from regions of high concentration to regions of lower concentration. The three terms on the right in Eq. (1) thus represent three mechanisms of mass transfer: migration of a charged species

in an electric field, molecular diffusion due to a concentration gradient, and convection due to the bulk motion of the medium.

Note that Eq. (1) only applies to dilute solutions. A more general form of Eq. (1) is (Newman, 1973a)

$$c_i \nabla \mu_i = RT \sum_j \frac{c_i c_j}{D_{ij} c_T} (\underset{\sim}{v}_j - \underset{\sim}{v}_i) \tag{2}$$

where μ_i is the electrochemical potential of species i. This equation must be applied to systems, such as molten salt systems, where the activity coefficients are not close to unity, or to systems where the diffusion flux of each species can be affected by more than one concentration gradient. The driving forces for diffusion and migration are both included in the gradient of the electrochemical potential in Eq. (2). Further details of concentrated solution theory can be found in Newman (1973a).

After writing the appropriate flux equation, one can write an expression for the current, which is, of course, due to the motion of charged particles in an electrolytic solution.

$$\underset{\sim}{i} = F \sum_i z_i \underset{\sim}{N}_i \tag{3}$$

Here $\underset{\sim}{i}$ is the current density expressed in amperes per square centimeter, and $z_i F$ is the charge per mole.

Next we need to state a material balance for any component i:

$$\frac{\partial c_i}{\partial t} = -\nabla \cdot \underset{\sim}{N}_i + R_i \tag{4}$$

Since reactions are frequently restricted to the surfaces of electrodes, the bulk reaction term R_i is often zero in electrochemical systems.

Finally, we can say that the solution is electrically neutral:

$$\sum_i z_i c_i = 0 \tag{5}$$

Such electroneutrality is observed in all solutions except in the thin double-charge layer near electrodes and other boundaries.

These equations provide a consistent description of transport processes in electrolytic solutions. Equations (1) and (2) state that species in the solution can move by migration, diffusion, and convection. Equation (3) merely says that the sum of fluxes of charged species constitutes an electric current. Equation (4) is a material balance for a species, and Eq. (5) is the condition of electroneutrality. Although the specific description may be refined, any theory of electrolytic solutions will need to consider these physical phenomena.

Note that in order to solve the equations describing the mass transfer, it is necessary to know the convective velocity $\underset{\sim}{v}$. This velocity can be

found from the equations of fluid mechanics, such as the Navier-Stokes equation

$$\rho(\partial \underset{\sim}{v}/\partial t + \underset{\sim}{v}\cdot\nabla \underset{\sim}{v}) = -\nabla p + \mu\nabla^2 \underset{\sim}{v} + \rho \underset{\sim}{g} \tag{6}$$

and the continuity equation

$$\nabla \cdot \underset{\sim}{v} = 0 \tag{7}$$

Next one needs boundary conditions describing the differences in potential between phases. These boundary conditions describe the processes occurring at the electrode interface.

One process that can occur is the adsorption of ions onto the surface. (Usually, anions are specifically adsorbed.) A double-charge layer is formed because the ions of one sign are attracted from the solution to the adsorbed ions of the opposite sign. This attractive force, however, is balanced by thermal agitation, which tends to make the ions wander. Therefore, the double-charge layer has a diffuse part in the solution. That is, the charge density decreases with distance from the interface. The thickness of this diffuse layer of excess charge density is characterized by the Debye length,

$$\lambda = \left(\frac{\varepsilon RT}{F^2 \sum_i z_i^2 c_{i\infty}} \right)^{1/2} \tag{8}$$

The Debye length is typically on the order of 10 Å. The major effect of the double layer on the overall behavior of the interface is that it superimposes a capacitive effect on top of the electrode kinetics of the electrode reaction itself. This means that when the potential of the electrode is varied, the current that flows is partly due to charging the double-layer capacity, and partly due to the charge-transfer reaction. One can measure the capacitance of the double layer in several ways (Newman, 1973a; Grahame, 1947). Typically, the double-layer capacity is 10 to 40 $\mu F/cm^2$ (Bard and Faulkner, 1980), but this capacitance is a function of potential. The double layer is discussed further in Newman (1973a), Grahame (1947), Bard and Faulkner (1980), Delahay (1965), and Parsons (1954).

Although the existence of the double layer should not be neglected, it is the charge-transfer reactions at the electrodes which are usually of primary importance in electrochemical systems of practical interest. The kinetics of these reactions relate the potential driving force to the rate of reaction. Thus a knowledge of the electrode kinetics is necessary to formulate boundary conditions for the differential equations describing the movement of ionic species.

For the purpose of assessing potential variations in a solution, it is convenient to conceive of placing reference electrodes into the solution at appropriate locations, usually just outside the diffuse double layer or in the bulk solution. Figure 1 illustrates the placement of these reference electrodes (White, 1977). The letter s designates reference electrodes of the same kind as the working electrode, and the letter g refers to reference

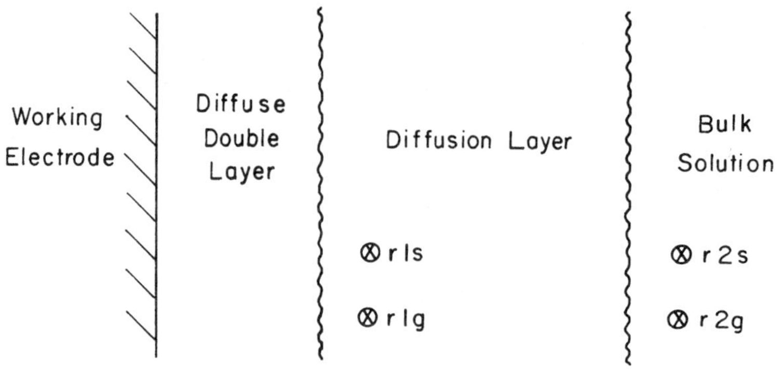

FIGURE 1 Reference electrodes, which may be imaginary, positioned in the bulk solution and within the diffusion layer. (From White, 1977.)

electrodes of a given kind. It is important to realize that a reference electrode of a "given kind" cannot be a reference electrode of *any* kind. Instead, the reference electrode must be carefully selected to be as reversible and reproducible as possible. The selection of reference electrodes is discussed in Newman (1973a). One must also realize that a measurement made by a reference electrode of a given kind must be corrected for any liquid-junction potentials that might exist between the solution in question and that within the reference-electrode compartment (see Newman, 1973a, Sec. 40).

In the absence of concentration variations, the two reference electrodes shown in Fig. 1 would measure the ohmic potential drop between the two points labeled 1 and 2. In the presence of concentration variations, the potential difference between the two reference electrodes will be composed of an ohmic portion and what is called a concentration overpotential. The concentration overpotential, η_c, reflects the thermodynamic effect of the difference in concentration between the interface and the bulk solution. An additional contribution to the overall cell potential is the driving force required to make the electrode reactions proceed at appreciable rates. This surface overpotential, η_s, is defined as the potential of the working electrode relative to a reference electrode of the same kind, placed in the solution adjacent to the surface of the working electrode, but just outside the diffuse double layer. In terms of the notation of Fig. 1, this surface overpotential for a reaction j would be written

$$\eta_{sj} = V - V_{r\ 1s} = V - V_{r\ 1g} - (V_{r\ 1s} - V_{r\ 1g}) \tag{9}$$

or

$$\eta_{sj} = V - \Phi_o - U_{j,o} \tag{10}$$

where $\Phi_o = V_{r\ 1g}$ is the potential just outside the double layer, as measured by a reference electrode of a given kind, and

$$U_{j,o} = V_{r\ 1s} - V_{r\ 1g} \tag{11}$$

$$U_{j,o} = \left(U_j^\theta - \frac{RT}{n_jF}\sum_i s_{ij} \ln \frac{c_{io}}{\rho_o}\right) - \left(U_{re}^\theta - \frac{RT}{n_{re}F}\sum_i s_{i,re} \ln \frac{c_{i,re}}{\rho_o}\right) \tag{11}$$

Here the subscript re denotes the reference electrode reaction. The stoichiometric coefficient, s_{ij}, refers to species i in a reaction j, written in the form

$$\sum_i s_{ij} M_i^{z_i} \rightarrow n_j e^- \tag{12}$$

where n_j denotes the number of electrons transferred, M_i refers to species i, and z_i is the charge number of species i.

For a large class of electrochemical reactions, the current density depends exponentially on the surface overpotential according to the Butler-Volmer equation

$$i_j = i_{oj}\left[\exp\left(\frac{\alpha_{aj}F}{RT}\eta_{sj}\right) - \exp\left(\frac{-\alpha_{cj}F}{RT}\eta_{sj}\right)\right] \tag{13}$$

where i_{oj} is the exchange current density for reaction j and depends on the concentrations c_i. Usually, this dependence is expressed as

$$i_{oj} = i_{oj,ref} \prod_i \left(\frac{c_{io}}{c_{i,ref}}\right)^{\gamma_{ij}} \tag{14}$$

In Eq. (13), the first exponential term corresponds to the forward reaction rate for an anodic process, and the second term represents the reverse reaction rate. Note that if the anodic surface overpotential is large, the reverse reaction term can be neglected. This approximation, known as the Tafel approximation, produces a linear plot of surface overpotential vs. the logarithm of current density. A more thorough discussion of overpotentials can be found in Newman (1973a) and White (1977).

We have now introduced enough of the elements to be in a position to discuss the composition of the overall cell potential. This is due partly to the ohmic potential drop in the solution. In addition, there is a potential loss associated with the concentration variations in the solution near electrodes, which we have termed the concentration overpotential. Finally, there is the surface overpotential due to the limited rates of the electrode reactions. The sum of these is the cell potential, which can be written

$$\begin{aligned} V &= \Phi(\text{anode}) - \Phi(\text{cathode}) \\ &= \Phi(\text{anode}) - \tilde{\Phi}_1 + (\tilde{\Phi}_1 - \tilde{\Phi}_2) - \Phi(\text{cathode}) + \tilde{\Phi}_2 \\ &= \eta_s(\text{anode}) + \eta_c(\text{anode}) + \Delta\tilde{\Phi}_{ohm} - \eta_c(\text{cathode}) - \eta_s(\text{cathode}) \end{aligned} \tag{15}$$

where $\tilde{\Phi}_1$ is the value that the potential adjacent to the anode would have if there were no concentration variations in the solution. The terms η_c and

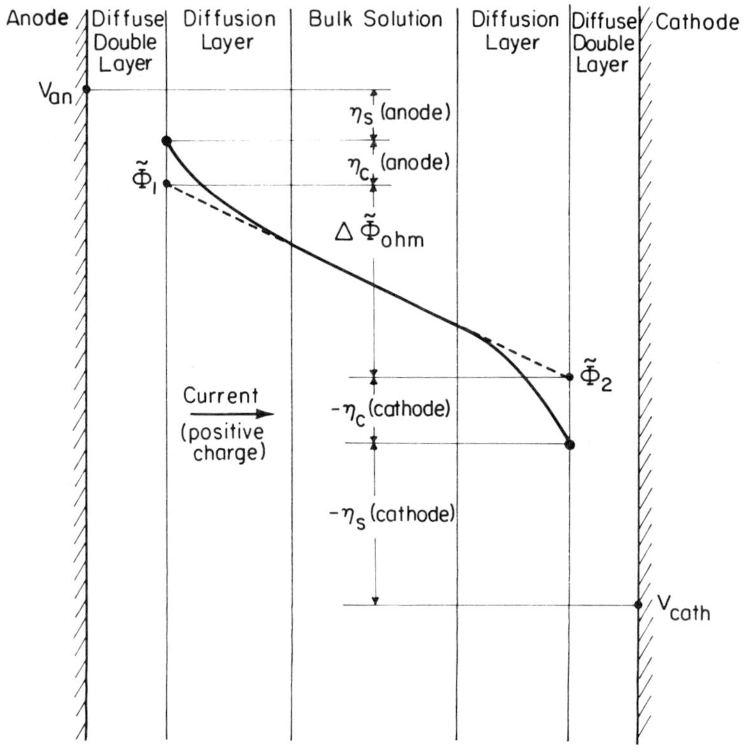

FIGURE 2 Overpotentials in an electrolytic cell. The solid line represents potential distribution in the absence of concentration variations.

η_s for the cathode enter with negative signs because of the conventions that have been adopted. Since they are generally negative, they make a positive contribution to the cell potential. Thus none of these terms, ohmic drop or overpotentials, represents a source of energy. An example of how the various overpotentials contribute to the overall cell potential is shown in Fig. 2. The dashed line represents what the potential in the solution would be if all concentrations were at their bulk values.

Note that this decomposition presumes a bulk solution and a diffusion layer adjacent to each electrode. In situations where this does not prevail, such as a cell with overlapping diffusion layers, or a porous electrode, it is necessary to start over from the basic equations, but the fundamentals remain the same.

CHANNEL FLOW CELLS

Channel flow between two parallel planar electrodes is used in many industrial electrochemical processes, such as metal refining, energy storage, and electro-organic synthesis. Specific examples include copper refining, some zinc-halogen energy storage cells, and the Monsanto process for conversion of acrylonitrile to adiponitrile.

Channel flow cells are very useful, since they provide continuous production, they are simple to operate and maintain, and they do not require a high capital investment (Fitzjohn, 1975). In addition, the analysis of channel flow cells has been relatively well developed (Sakellaropoulos, 1979; Sakellaropoulos and Francis, 1979, 1980; Parrish and Newman, 1969, 1970; Cabán and Chapman, 1976; Lee and Selman, 1982; Jorné, 1982).

A channel flow cell consists of two parallel plates, which serve as the anode and the cathode for electrochemical reactions. The electrolyte flows past the electrodes, and the current flow is perpendicular to the fluid flow.

In general, a thin-gap cell with multiple reactions occurring on each electrode is difficult to analyze. The electrodes cannot be treated separately, and mass transfer, thermodynamics, and the kinetics of more than one reaction must be considered simultaneously. This problem can be solved, but for design purposes, it is useful to see if any simplifications can be made. One assumption that is commonly used is that the diffusion layers are thin. In this case, the electrodes can be treated separately. This is known as the Lévêque approximation (Newman, 1973a,b), and it is valid as long as

$$\frac{L}{h} < 0.01 \mathrm{ReSc} \tag{16}$$

where

$\mathrm{Re} = 2\langle v\rangle h/\nu$

$\mathrm{Sc} = \nu/D$ (about 1000 for most aqueous systems)

L = length of electrodes

h = spacing between electrodes

Thus we see that the thin-diffusion-layer approximation breaks down only if the electrodes are very long, the gap is very thin, or the velocity is very low.

Parrish and Newman (1970) used the Lévêque approximation to obtain the current and concentration distributions in a channel flow cell. They examined the case of a metal deposition and dissolution reaction. Intuitively, one would expect that mass transfer would not be important near the front of each electrode, where the ions in solution are still unreacted. Farther down the channel, however, the solution becomes depleted of metal ions. Mass transfer then begins to affect the cathodic deposition reaction. The anodic dissolution reaction, however, is not affected by the concentration of ions in the solution. Therefore, we would not expect the anodic reaction to become mass-transfer limited.

In the absence of mass transfer, the current distribution is governed by the ohmic potential drop and the reaction kinetics. This current distribution is called the secondary current distribution. If there are no kinetic limitations, however, the current distribution is called the primary distribution. To calculate the primary current distribution, one would solve Laplace's equation for the potential, assuming that the potential adjacent to each electrode is held at a constant value. For two electrodes embedded in planar channel walls, the primary current density is infinite at the electrode edges. The secondary current distribution is similar to the primary distribution, but it is more uniform because there are kinetic limitations. Therefore, the current density is not infinite anywhere, but it is still higher at the electrode edges.

If mass transfer dominates the current distribution, on the other hand, the current density is high only at the front of the electrode, and it decreases with distance down the channel.

For an anodic dissolution reaction, one would expect that mass transfer would not be important. Therefore, the current distribution should resemble the secondary current distribution. Parrish and Newman found that this is indeed the case. The cathodic current distribution, however, should behave like a secondary current distribution near the front of the electrode, but like a mass-transfer-limited current distribution near the back. In fact, Parrish and Newman's numerical results show maxima in the cathodic current densities. These maxima arise from a compromise between the secondary current distribution and the mass-transfer-limited current distribution.

Throughout their analysis, Parrish and Newman assumed a single reaction at each electrode. Very often, this simplifying assumption is valid (Pickett, 1979). Occasionally, however, multiple reactions must be considered (Sakellaropoulos, 1979; Sakellaropoulos and Francis, 1979, 1980; White and Newman, 1977). The assumption of a single reaction may be checked by identifying a single parameter which characterizes the manner in which a side reaction tends to obscure the limiting-current plateau for a main, or desired, reaction (White and Newman, 1977). It is the magnitude of the side reaction relative to the main reaction at potentials in the neighborhood of the limiting-current plateau which is important.

Figure 3 illustrates this concept for the example of copper deposition as the main reaction and hydrogen evolution as the side reaction. The

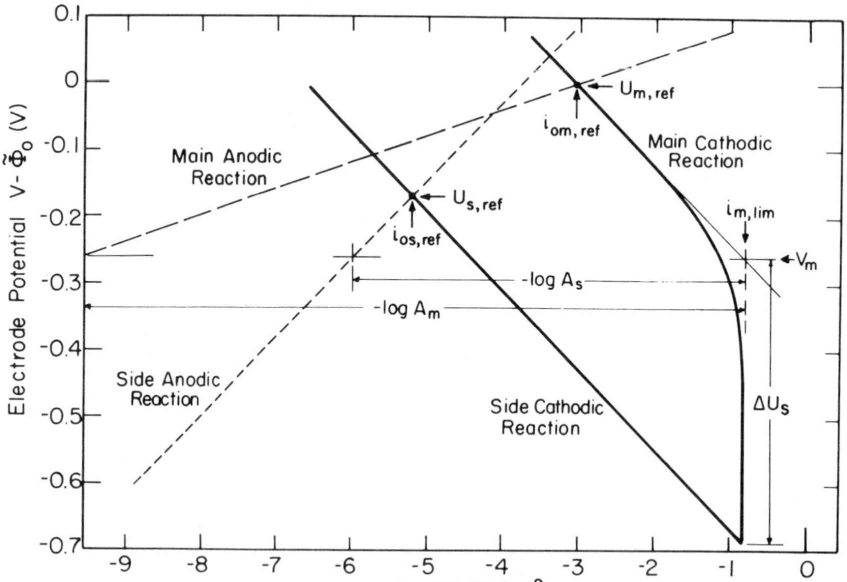

FIGURE 3 Qualitative sketch of the current-potential curves for a main and a side reaction showing some of the parameters defined in the text. (From White, 1977.)

solid line, labeled "main cathodic reaction," is the current-potential curve of interest. The curves in the figure are plots of the potential difference, $V - \tilde{\Phi}_o$, versus the logarithm of the current density, where V is the electrode potential and $\tilde{\Phi}_o$ is the potential in the solution just outside the double layer as measured by a reference electrode of the same kind as the main reaction. Note that the potential difference $V - \tilde{\Phi}_o$ is related to the surface overpotential by Eq. (10).

$$\eta_{sj} = V - \tilde{\Phi}_o - U_{j,o} \tag{10}$$

where $U_{j,o}$ is defined by Eq. (11).

Figure 3 shows the following four curves:

1. Tafel approximation of main anodic reaction (copper dissolution)
2. Tafel approximation of side anodic reaction (production of hydrogen ions)
3. Tafel approximation of side cathodic reaction (hydrogen evolution)
4. Main cathodic reaction (copper deposition)

The heavy line for the main cathodic reaction (curve 4) has two asymptotes: at high $|V - \tilde{\Phi}_o|$, the reaction is mass-transfer controlled; at low $|V - \tilde{\Phi}_o|$, the reaction is kinetically controlled.

The Tafel approximation is applied to the cathodic reactions, since it is unlikely that the backward terms in the Butler-Volmer equation are important in the neighborhood of the limiting-current plateau, where $-\eta_{s,m}$ is large. The anodic reactions are plotted as Tafel lines to illustrate that the limiting-current plateau is not affected by either anodic reaction. Although the Tafel approximation is not valid near $i = i_o$, we are interested in the neighborhood of the limiting-current plateau, which is far from i_o. Therefore, we only examine the portion of the Tafel curve near the limiting-current plateau, where the backward terms in the Butler-Volmer equation can be neglected.

As long as there are no mass-transfer limitations, the plot of $V - \tilde{\Phi}$ versus log $|i|$ will remain linear. Note that the solid line for the main cathodic reaction does start to bend over as mass transfer comes into play. The side cathodic reaction, hydrogen evolution, however, has no mass transfer limitations because hydrogen evolution occurs from a solution with excess hydrogen ion in the supporting electrolyte.

To plot a Tafel line, one needs not only the slope, but also the value of $V - \tilde{\Phi}_o$ at a single current density, i. If this current density is chosen as the exchange current density at the reference concentrations, then

$$V - \tilde{\Phi}_o = U_{m.ref} \tag{17}$$

where $U_{m,ref}$ is given by

$$U_{m,ref} = U_m^\theta - U_{re}^\theta - \frac{RT}{n_m F} \sum_i s_{im} \ln \frac{c_{i,ref}}{\rho_o} + \frac{RT}{n_{re} F} \sum_i s_{i,re} \ln \frac{c_{i,re}}{\rho_o} \tag{18}$$

The Tafel approximation can also be written for the main reaction in the anodic direction. Again, at $i = i_{om,ref}$, $|V - \tilde{\Phi}_o|$ is $U_{m,ref}$. In fact, the anodic and cathodic Tafel lines, for any reaction j, intersect at $i_{oj,ref}$ and $U_{j,ref}$, if the surface concentrations are at their reference values. Therefore, for any reaction j, if one knows the exchange current densities $i_{oj,ref}$, and the anodic and cathodic transfer coefficients, α_{cj} and α_{aj}, one can plot lines similar to those shown in Fig. 3.

The parameters A_m and A_s, also shown in Fig. 3, characterize the magnitude of the anodic part of a reaction, at the electrode potential for which the main reaction is beginning to reach the limiting current. This parameter is defined (White and Newman, 1977), for reaction j, as

$$A_j = \left|\frac{i_{oj,ref}}{i_{m,lim}}\right|^{1+\alpha_{aj}/\alpha_{cj}} \exp\left(\frac{\alpha_{aj}F}{RT}\,\Delta U_j\right) \tag{19}$$

It should be noted that if the surface concentrations are not at the reference values, the Tafel lines will be shifted. For example, if the concentration of dissolved hydrogen is less than $c_{H_2,ref}$, the Tafel line for the anodic side reaction will be shifted to the left. The exact placement of this line, in this case, however, does not affect the length of the limiting-current plateau for the main reaction. In fact, neither anodic reaction affects the length of the limiting-current plateau. Thus, since the limiting-current region is of greatest interest, the anodic reactions are not important.

The electrode potential, V_m, for which the main reaction is beginning to reach the limiting current is, in the absence of an ohmic potential drop, approximately equal to

$$V_m = U_{m,ref} - \frac{RT}{\alpha_{cm}F} \ln \frac{-i_{m,lim}}{i_{om,ref}} \tag{20}$$

At the potential defined by Eq. (20), the current density for the side reaction divided by the limiting current density for the main reaction is proportional to $\exp(-\alpha_{cs}F\,\Delta U_s/RT)$, where

$$\Delta U_s = U_{m,ref} - U_{s,ref} - \frac{RT}{\alpha_{cs}F} \ln \frac{i_{os,ref}}{-i_{m,lim}} + \frac{RT}{\alpha_{cm}F} \ln \frac{i_{om,ref}}{-i_{m,lim}} \tag{21}$$

corresponds to the length of the limiting-current plateau, as shown in Fig. 3. The parameter ΔU_s shows that both the exchange current density and the open-circuit potential determine the relative significance of a side reaction. If the exchange current density is very small or the open-circuit potential is quite negative, ΔU_s will be large, and the side reaction will not obscure the limiting-current plateau.

CELL OPTIMIZATION

The goal in designing any reactor is to maximize the profit or to minimize the total cost by making an appropriate choice of operating variables and

cell dimensions. Naturally, there are constraints, such as mass transfer and current-potential relationships, as well as feed and product specifications. Usually, however, there are few enough constraints that there is enough flexibility to perform a cost minimization.

An example of the trade-off between capital and operating costs in a fuel cell has been discussed by Newman (1979). Ibl and coworkers (Ibl, 1977; Ibl and Adam, 1965; Ibl and Robertson, 1973) have also treated the design and optimization of various electrochemical systems. Alkire et al. (1982) have discussed the optimization of electrochemical systems having many variables. We shall discuss here the example of optimization of potential and hydrogen utilization in an acid fuel cell.

A fuel cell for commercial power generation consists of a device to reform a hydrocarbon fuel to yield hydrogen and an electrochemical reactor, where this hydrogen is combined with oxygen from the air to produce water vapor and electric power. In this example, we shall perform a simplified cost analysis, where only the capital costs and the fuel costs will be included. The fuel cost, denoted C_f, will include all operating costs which can be associated with or set proportional to the consumption of fuel. Thus the reformer costs will be included in C_f. The trade-off between capital and operating costs results because there is an optimum fuel utilization beyond which the increase in electrode area costs more than can be justified by the increase in electric power produced.

To perform the optimization, one must first write an expression for the cost to be minimized. In this example, we shall write the cost in terms of the hydrogen utilization u and the cell potential V, where utilization is defined as the fraction of the hydrogen in the feed that is consumed in the electrochemical reactor. To write the cost in terms of only the variables u and V, one needs to know how the cell potential depends on current density and gas consumption. Once the cost is written in terms of u and V the optimization is performed by simply setting the derivatives of the cost with respect to u and V equal to zero and solving for u and V.

To calculate the relationship among cell potential, current density and gas composition, one can assume that the air cathode obeys Tafel kinetics. At the hydrogen electrode, on the other hand, the surface overpotential is negligible and the Nernst equation applies, where the anode potential differs from the potential of the adjacent hydrogen reference electrode mainly because the hydrogen partial pressure differs from the reference pressure of 1 atm. Finally, one can calculate the ohmic potential drop within the cell and combine the three equations to obtain the relationship among cell potential, current density, and gas composition,

$$V = U' + \frac{RT}{F} \ln \frac{p_{O_2} p_{H_2}^{1/2}}{I} - R'I \tag{22}$$

The partial pressures of oxygen and hydrogen appearing in Eq. (22) are at the catalyst layers and must be obtained from the values in the flowing streams by allowance for the gas-phase mass-transfer resistance through the inert gas in the electrode substrates. These partial pressures are then substituted into Eq. (22) to obtain the desired current-potential relationship.

Next, one needs an expression for the total cost. In this example, the total cost is the fuel cost plus the capital cost. Let C_f be the value of a

unit of hydrogen in the feed, and K_uC_f be the value of hydrogen in the exit fuel stream after a utilization u, where C_f is expressed in \$/C. Let the capital costs be expressed by C_a, having units of \$/s·cm^2, based on superficial electrode area. C_a is obtained by multiplying the total capital cost (\$/cm^2) by a factor such as 0.3 yr^{-1}, representing the effect of interest, depreciation, and taxes. Components with a 5-year life and a 10% interest rate might give rise to this factor. The total cost can now be expressed as

$$\frac{\text{cost}}{\text{time}} = C_f 2FF_a X^o_{H_2} - C_f K_u 2FF_a X^{exit}_{H_2} + C_a A_{tot} \tag{23}$$

where

F = Faraday's constant, 96,487 C/equiv

F_a = flow rate of inerts on anode side, mol/s

$X^{exit}_{H_2}$ = ratio of moles of H_2 to moles of inerts in the exit stream

$X^o_{H_2}$ = ratio of moles of H_2 to moles of inerts in the feed stream

A_{tot} = superficial electrode area, cm^2

The first term on the right is the cost of the hydrogen feed stream, the second term is the resale value of the hydrogen exit stream, and the last term is the capital cost. We are concerned with minimizing the cost per unit of electric energy, which is the ratio of the cost per unit time to the electric power produced, where

$$\text{Power} = P = \int_0^{A_{tot}} VIdA \tag{24}$$

Here I is the superficial current density, expressed in A/cm^2. As a step toward writing the total cost in terms of u and V, we can relate the flow rate and consumption of hydrogen to the electric current by Faraday's law,

$$2FF_a(X^o_{H_2} - X_{H_2}) = 2FF_a X^o_{H_2} u' = \int_0^A I\, dA \tag{25}$$

where u' denotes the value of the utilization u for a local mole ratio X_{H_2} of hydrogen. Now we can write the total area as

$$A_{tot} = 2FF_a X^o_{H_2} \int_0^u \frac{du'}{I} \tag{26}$$

and the electric power from Eq. (24) becomes

$$P = 2FF_a X^o_{H_2} \int_0^u V\, du' \tag{27}$$

By means of these relationships, the power cost takes the form

$$C_e = \frac{C_f - C_f K_u (1 - u) + C_a \int_0^u (du'/I)}{\int_0^u V\, du'} \tag{28}$$

This equation has been written so that the extensive quantities—power required, electrode area A_{tot}, and flow rate of the fuel stream—have been eliminated in favor of the utilization u, an intensive variable.

We restrict ourselves here to the case where the cell potential V is a constant for all values of the utilization. Now we can set the derivatives of the power cost in Eq. (28) with respect to u and V equal to zero. The differentiation with respect to u gives, after some manipulation,

$$\frac{C}{I} = \frac{I - K_u}{u} + \frac{C}{u} \int_0^u \frac{du'}{I} + (1 - u) \frac{dK_u}{du} \tag{29}$$

where $C = C_a/C_f$ is the ratio of the capital cost to the fuel cost and has the units of A/cm^2. The differentiation with respect to V gives

$$1 - K_u(1 - u) + C \int_0^u \frac{du'}{I} = -VC \int_0^u \left(\frac{\partial I}{\partial V}\right)_u \frac{du'}{I^2} \tag{30}$$

The results of this optimization show that when depleted hydrogen can be sold at cost, that is, $K_u = 1$, it is economical to operate with the richest fuel and airstreams available, as expected. Thus the optimum utilization obtained from Eq. (29) is u = 0. Substituting this into Eq. (30) gives

$$-\frac{\partial V}{\partial I}\left(1 + \frac{I}{C}\right) = \frac{V}{I} \tag{31}$$

This equation, when coupled with the current-potential relationship gives the optimum current density or cell potential. For C, the ratio of capital to fuel cost, approaching infinity, Eq. (31) says that the power maximum, where $dVI/dI = 0$, is the optimum. As fuel costs become more important and C decreases, optimum operation is at a current density below (and a potential above) the power maximum. The limit $K_u = 1$ will give the most favorable power cost.

$$C_e = \frac{C_f}{V}\left(1 + \frac{C}{I}\right) \tag{32}$$

The other extreme is $K_u = 0$, where waste hydrogen has no value. This will be the worst case and will yield the highest optimum utilization, for a given cost ratio C.

POROUS ELECTRODES

In the discussion above we have considered a flow channel as a possible reactor configuration. We wish now to focus attention on porous electrodes as another class of electrochemical reactors.

Porous electrodes can be used in flow-through configurations as in fuel cells, redox energy storage systems, and chemical reactors or in closed configurations as in many primary and secondary batteries. Thus porous electrodes find potential applications as electrochemical reactors in a variety of areas. Flow-through porous electrodes are reviewed by Newman and Tiedemann (1978). The mass transfer of reacting species within the fixed bed and the ohmic potential variation throughout the bed are treated in detail. A second review article treats porous electrodes with regard to battery applications (Newman and Tiedemann, 1975). A literature survey is given in each review.

Two distinguishing features of porous electrodes are the intimate contact of the electrode with the solution (and possibly a gaseous phase) and the high surface area–to-volume ratio that can be obtained. The high surface area–to-volume ratio is important to applications where the intrinsic rate of the heterogeneous, electrochemical reaction is slow. In processes using double-layer adsorption, the high surface area is again important. Dilute reactants in solution require the close proximity of solution and electrode to enhance mass transfer to the electrode surface. In battery or fuel-cell applications, porous electrodes offer a means for storing the soluble reactants in close proximity to the electrode surface. For non-conducting reactants of low solubility, another solid phase (as in batteries) or gas phase (as in fuel cells) may be incorporated into the system.

Flow-through porous electrodes could find applications in the area of metal processing. This includes the purification, electrowinning, and possibly electroplating of aluminum, copper, magnesium, sodium, manganese, nickel, gold, silver, and chromium. The electrorefining of aluminum from an aluminum-manganese alloy might involve a flow-through porous anode in order to prevent the dissolved manganese from reaching the cathode. Solutions too dilute to treat with electrowinning processes may be treated with a flow-through porous electrode. Dilute aqueous metal removal and recovery of copper (Bennion and Newman, 1972), silver (Van Zee and Newman, 1977), and mercury (Matlosz and Newman, 1982) have been studied in fixed-bed, flow-through porous electrodes. Copper was recovered from feed streams of 667 ppm with effluent concentrations less than 1 ppm, with simultaneous production of a concentrated stream of 47,660 ppm. Copper recovery has also been investigated in fluidized-bed reactors (Fleischmann et al., 1971; Germain and Goodridge, 1976). Gold can readily be recovered from plating baths which have deteriorated, and streams from processing photographic emulsions can be reduced to less than 1 ppm silver. Mercury contamination in brine solutions has been reduced to concentrations as low as 5 ppb from feed streams of 50 ppm. Reduction of contaminants usually requires processing of large volumes of dilute solutions. For metal removal, consideration must be made for the periodic removal of material from the matrix. Oxidation of organic contaminants, however, can proceed with no retention of solids.

The use of porous electrodes in electro-organic syntheses may provide an economical alternative to other chemical routes. The electrochemical synthesis may give higher yields under less severe operating constraints. The ability to control the electrode potential allows optimization for a particular reaction product, while minimizing side reactions or multiple products. A bibliography of electro-organic syntheses has been recently compiled by Swann and Alkire (1980).

Energy storage and conversion systems may use flow-through porous electrodes. Flow-redox systems, zinc-chloride hydrate storage systems, and fuel cells are some examples. It has also been shown that the performance of some primary or secondary batteries could be enhanced by a fresh supply of electrolyte (Wierschem and Tiedemann, 1980; Liebenow, 1897). A promising zinc-bromine secondary battery system uses a flowing electrolyte.

Flow-through porous electrodes also find applications in fundamental studies. Appel and Newman (1976) apply a limiting-current method for the measurement of mass-transfer coefficients at very low Reynolds numbers. Fedkiw and Newman (1982) summarize mass transfer results of several workers. The results are correlated by a dual-sized, straight-pore model for the bed's pore volume.

Governing Equations

The many potential applications of porous electrodes warrant the need for a mathematical description of the system. One can then scaleup a system or predict the result of a change in operating parameters. This modeling can lead to designs that optimize or maximize the desired process. Less detailed modeling can be a guide in screening various alternatives such as newly proposed battery systems. It is important, then, to develop guidelines as to the behavior of the porous electrodes.

Porous electrodes are inherently different from planar electrodes due to the intimate contact of the solution and matrix phases. Here the current flows within the matrix and the solution phases and exchanges between the matrix and solution nonuniformly throughout the bed. An electrical analog that can help picture the inherent complications is seen in Fig. 4. This figure shows two porous electrodes operating as an electrolytic cell. The subscripts 1 and 2 refer to the matrix and solution phases respectively. Subscripts a and c refer to the anode and cathode, and R_s represents the resistance due to the separator. When current flows through the porous electrode, the electrical double-layer capacity (represented by C_d) is charged. This is a net flow of current through the solution causing a change in solution composition near the interface. In parallel to this process, net current flows via a Faradaic reaction where R_F represents the charge-transfer resistance of the electrochemical reaction. These processes occur nonuniformly throughout the volume of the porous bed. It is cautioned that this figure is meant to serve as a guide in thinking about the distribution of reactions in porous electrodes, but is not a substitute for modeling the porous electrode with the appropriate governing equations. Mass transfer of reacting species, for example, is not considered in the electrical analog.

The mathematical description of porous electrodes assumes a macrohomogeneous system. The porous bed is represented as the superposition of two continua, a matrix and a solution. The actual geometric detail of the

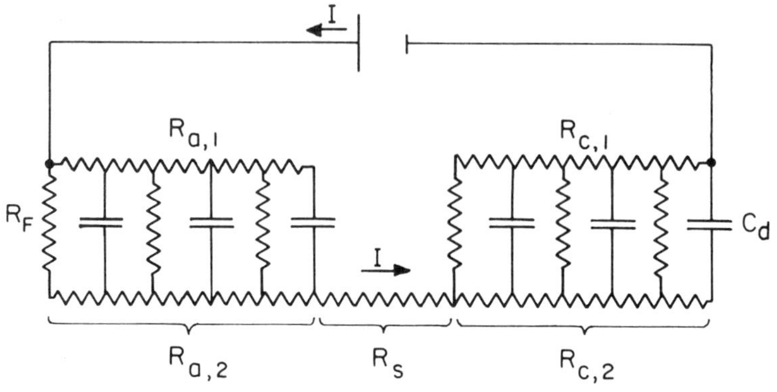

FIGURE 4 Simplified electrical analog of a porous anode and cathode (subscripts a and c, respectively) showing the resistance in the matrix phases (subscript 1) and solution phases (subscript 2). Also shown are the Faradaic charge-transfer resistance (R_F), electrical double-layer capacity (C_d), and separator resistance (R_S).

bed is ignored. Average physical parameters such as porosity and surface area are used. This type of analysis leads to a volume-averaged approach in the governing differential equations (Gray, 1975; Dunning, 1971).

A schematic of a section of a porous bed is shown in Fig. 5. The volume-averaged material balance (Dunning, 1971) of a species i within a flooded porous bed in the absence of homogeneous chemical reactions takes the form [compare Eq. (4)]

$$\frac{\partial(\varepsilon c_i)}{\partial t} = aj_{in} - \nabla \cdot \underset{\sim}{N}_i \tag{33}$$

Here three different averages are used. ε is the porosity or void volume fraction. a is the specific interfacial area (surface area of pore walls per unit volume of bed). c_i is the concentration averaged over the volume

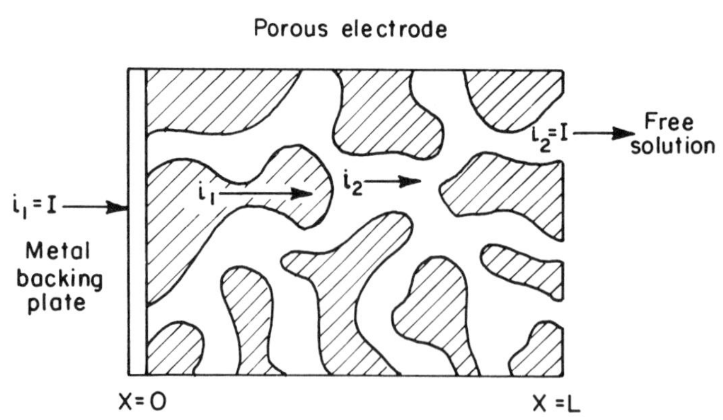

FIGURE 5 Schematic of a one-dimensional porous electrode.

of solution in the pores. εc_i is then the superficial concentration averaged over the bed volume (matrix plus pores). j_{in} is the normal component of the pore wall flux of species i into the solution relative to the velocity of the pore wall, averaged over the interfacial area. $\mathbf{N}_i$ is the average flux of species i in the pore solution averaged over the cross-sectional area of the pore plus matrix.

With the flux $\mathbf{N}_i$ referenced to the cross-sectional area of the pore plus matrix, the superficial current density $\mathbf{i}_2$ in the solution phase is given as

$$\mathbf{i}_2 = F \sum_i z_i \mathbf{N}_i \tag{34}$$

Similarly, $\mathbf{i}_1$, the current density in the matrix phase, is defined to refer to the superficial area and not to the area of an individual phase.

The matrix and solution phases are taken to be electrically neutral. For the solution, then,

$$\sum_i z_i c_i = 0 \tag{35}$$

We have assumed that the electrical double layer at the matrix pore interface is a small volume compared to our averaging volume. This assumption may break down for very dilute solutions and high-surface-area electrodes. Electroneutrality requires that the divergence of the total current density $(\mathbf{i}_1 + \mathbf{i}_2)$ is zero; charge leaving the matrix must enter the solution. Thus

$$\nabla \cdot \mathbf{i}_1 + \nabla \cdot \mathbf{i}_2 = 0 \tag{36}$$

A combination of Eqs. (33) to (35) gives

$$\nabla \cdot \mathbf{i}_2 = aF \sum_i z_i j_{in} = ai_n \tag{37}$$

where i_n is the average transfer current density from the matrix to the solution phase. $\nabla \cdot \mathbf{i}_2$ is the transfer current per unit volume of the electrode (A/cm^3) and is positive for anodic currents. For a single electrode reaction represented as

$$\sum_i s_i M_i^{z_i} \longrightarrow ne^- \tag{38}$$

Faraday's law becomes

$$aj_{in} = -\frac{as_i}{nF} i_n = -\frac{s_i}{nF} \nabla \cdot \mathbf{i}_2 \tag{39}$$

Substitution into Eq. (33) and neglect of double-layer charging yield

$$\frac{\partial(\varepsilon c_i)}{\partial t} = -\nabla\cdot\underset{\sim}{N}_i - \frac{s_i}{nF}\nabla\cdot\underset{\sim}{i}_2 \tag{40}$$

A kinetic polarization equation relating the local rate of reaction (transfer current density) to the surface concentrations and interfacial potential drop is needed. For the porous electrode, Eq. (13) becomes

$$\nabla\cdot\underset{\sim}{i}_2 = ai_o\left[\exp\left(\frac{\alpha_a F}{RT}\eta_s\right) - \exp\left(\frac{-\alpha_c F}{RT}\eta_s\right)\right] \tag{41}$$

Porosity changes can be taken into account by a solid-phase material balance for a single electrode reaction:

$$\frac{\partial\varepsilon}{\partial t} = -A_0\nabla\cdot\underset{\sim}{i}_2 \tag{42}$$

where

$$A_0 = -\sum_{\substack{\text{solid}\\\text{phases}}}\frac{s_i M_i}{\rho_i nF} \tag{43}$$

Transport processes are needed to complete our description. Ohm's law for the matrix phase is

$$\underset{\sim}{i}_1 = -\sigma\nabla\Phi_1 \tag{44}$$

where σ is an effective matrix conductivity dependent on the composition of solid phases, the manner in which the granules of the conducting phases are connected, and the volume fraction of conducting phase or phases.

For a dilute electrolytic solution in the pores, the flux of a mobile solute can be attributed to diffusion, dispersion, migration, and convection [compare Eq. (1)]:

$$\frac{\underset{\sim}{N}_i}{\varepsilon} = -(D_i + D_a)\nabla c_i - z_i u_i F c_i\nabla\Phi_2 + \frac{\underset{\sim}{v}c_i}{\varepsilon} \tag{45}$$

where u_i is a corrected ionic mobility and D_i is an ionic diffusion coefficient corrected for the tortuosity of the pores. D_a represents the effect of axial dispersion. A discussion of the effect of axial dispersion on the average mass transfer coefficient is found in the review article by Newman and Tiedemann (1978) (see also Fedkiw and Newman, 1978). The current density in the solution phase can now be represented by

$$\underset{\sim}{i}_2 = -\kappa\nabla\Phi_2 - \varepsilon F\sum_i z_i D_i\nabla c_i \tag{46}$$

where $\kappa = \varepsilon F^2 \Sigma_i z_i^2 u_i c_i$. The second term in Eq. (46) represents the diffusion potential. As a consequence of electroneutrality, convection and dispersion make no contribution to the current density.

We should also recognize here that other forms of the transport equations may be necessary to describe a system. The appropriate equations for a concentrated binary electrolyte are given in Newman and Tiedemann (1975). Two binary molten salts are treated by Pollard and Newman (1979). When necessary, the full multicomponent transport equations can be used. [Compare the discussion of Eq. (2).]

In summary, the equations presented above have been found to describe adequately porous electrodes in many cases. A certain level of complexity is necessary in order to treat the simultaneous interaction of the physical processes. Equations (33) to (35) and (45) govern the transport, conservation, and electrical neutrality in the solution phase. Equation (44) covers the transport in the matrix phase. Equations (36) and (41) couple the species in the bulk phase to the electrochemical processes occurring at the interface. These equations are normally considered boundary conditions in systems not involving porous electrodes, but here they are applied throughout the volume of the bed.

Let us now illustrate an important design principle for flow-through porous electrodes (Newman and Tiedemann, 1978; Bennion and Newman, 1972). Consider the case of reducing a species i at the limiting current in an excess of supporting electrolyte. Equation (33) becomes

$$\frac{dN_i}{dx} = aj_{in} \tag{47}$$

for a one-dimensional, steady-state material balance. In the absence of migration in an electric field for the reacting species, the superficial flux of species i in the direction of the superficial fluid velocity is [from Eq. (45)]

$$N_i = -\varepsilon(D_i + D_a)\frac{dc_i}{dx} + c_i v \tag{48}$$

The local flux to the wall is given by a local mass transfer coefficient k_m such that

$$j_{in} = k_m(c_{io} - c_i) = -k_m c_i \tag{49}$$

where the wall concentration c_{io} has been set equal to zero at limiting current. Substitution of Eqs. (48) and (49) into the continuity equation (47) gives

$$v\frac{dc_i}{dx} = \varepsilon(D_i + D_a)\frac{d^2 c_i}{dx^2} - ak_m c_i \tag{50}$$

Equation (50) governs the concentration distribution of species i throughout the reactor. It is solved subject to the Danckwerts (1953) and Wehner and Wilhelm (1956) boundary conditions for the concentration of species i at the

inlet (x = 0) and the outlet (x = L) of the reactor. This formulation assumes that an inert packing extends from the active portion of the bed in both the upstream and downstream directions. A result of the constraints that the concentration of species i cannot increase without bounds and that the concentration and flux should be continuous is that

$$\frac{dc_i}{dx} = 0 \qquad \text{at } x = L \tag{51}$$

For the upstream boundary condition, an inlet concentration of c_o and continuous concentration and flux give

$$c_o v = c_i v - \varepsilon(D_i + D_a)\frac{dc_i}{dx} \qquad \text{at } x = 0 \tag{52}$$

Under this condition, the concentration at the inlet of the bed will be less than c_o because some of the reactant will have diffused ahead to the active portion of the bed. The solution to Eq. (50) subject to boundary conditions (51) and (52) is

$$\theta = \frac{e^{-y/B} + (D'/B^2)e^{By/D'} \exp[-\alpha L(1/B + B/D')]}{B + (D'/B^2)(1 - B)\exp[-\alpha L(1/B + B/D')]} \tag{53}$$

where we have introduced the quantities

$$\theta = \frac{c_i}{c_o}, \qquad y = \frac{ak_m x}{v}, \quad D' = \frac{\varepsilon\, ak_m}{v^2}(D_i + D_a)$$
$$B = \frac{1 + \sqrt{1 + 4D'}}{2}, \qquad \alpha = \frac{ak_m}{v} \tag{54}$$

Note that Eq. (53) simply reduces to

$$c_i = c_o e^{-\alpha x} \qquad \text{or} \qquad \theta = e^{-y} \tag{55}$$

when the effects of axial dispersion and diffusion are ignored.

Substitution of Faraday's law for the local flux to the wall in Eq. (49) governs the behavior of the superficial current density in the solution. Equation (50) becomes

$$v\frac{dc_i}{dx} = \varepsilon(D_i + D_a)\frac{d^2c_i}{dx^2} - \frac{s_i}{nF}\frac{di_2}{dx} \tag{56}$$

For an upstream counterelectrode, the current density i_2 must go to zero at the back of the electrode as all the current has been transferred to the matrix and current collector. Using this boundary condition and integrating Eq. (56) give

$$v(c_i - c_L) = \varepsilon(D_i + D_a)\frac{dc_i}{dx} - \frac{s_i}{nF}i_2 \tag{57}$$

where the subscript L refers to the exit of the reactor. Ohm's law for the solution [Eq. (46)] is taken as

$$i_2 = -\kappa \frac{d\Phi_2}{dx} \tag{58}$$

The diffusion potential has been ignored here, a good assumption when an excess of supporting electrolyte is present. Integration of Eq. (58) subject to Eqs. (57) and (54) gives the local variation of potential through the bed. This result is expressed as

$$\Phi_2(L) - \Phi_2(0) = \frac{nFvc_o}{s_R\kappa}\frac{v}{ak_m}\left[B^2\theta_L e^{\alpha L/B} - \frac{D'}{B} - (\alpha L + 1 + D')\theta_L\right] \tag{59}$$

Figure 6 shows the nature of the potential variation through the electrode bed. The potential in the matrix is constant through the length of the bed if the matrix conductivity is very high. The ohmic potential drop in the solution causes the variation in solution potential. The potential driving force at the back of the electrode $\Phi_1 - \Phi_2(L)$ must be large enough to ensure limiting current, while the potential difference at the front of the electrode must not be large enough to have secondary reactions, such as hydrogen evolution. Thus we have a maximum allowable ohmic potential drop in our reactor. As the bracketed quantity in Eq. (59) is of order unity, the coefficient of this quantity represents the magnitude of the potential variation in the porous bed. Thus we see that, for given values of c_o and κ, side reactions limit the maximum flow rate through the bed.

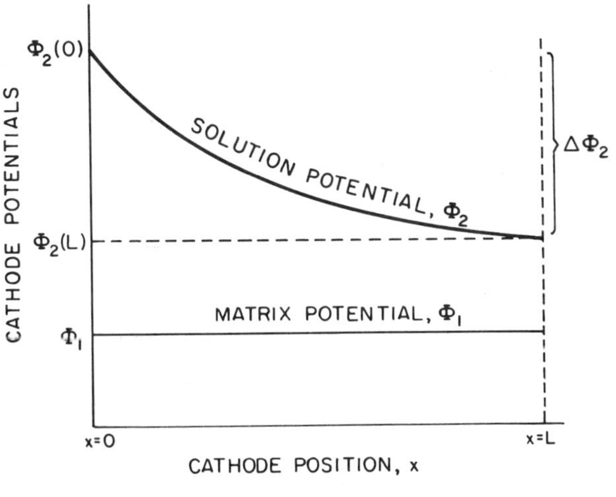

FIGURE 6 Variation of solution and matrix potentials as a function of position through a porous cathode. Matrix conductivity is infinite.

Since v/ak_m represents the order of magnitude of the distance through the reactor where the reaction occurs to an appreciable extent, a limit on the velocity therefore limits the useful thickness of the bed. $\Phi_2(L) - \Phi_2(0)$ can be assigned a maximum value based on data that might be taken on a rotating-disk electrode. Equation (59) can then be solved for the velocity v. What is left to be specified, then, is the length of the electrode. The length of the electrode is governed by the desired degree of conversion. Equation (53) can be solved for L if y is replaced by αL and θ is replaced by θ_L, the desired conversion. In the case of neglecting the effects of axial dispersion and diffusion,

$$L = \frac{v}{ak_m} \ln \frac{c_o}{c_L} \tag{60}$$

As mentioned above, other configurations of anode placement and current collector placement can be considered (Trainham and Newman, 1978). Qualitatively, an upstream counterelectrode will yield the highest reactant concentration and the highest potential driving force at the front of the electrode. A downstream counterelectrode will have the maximum reactant concentration where the potential driving force is smallest. Thus the latter configuration might appear to give a more uniform reaction distribution throughout the bed. It is important then to consider the best configuration to achieve a given objective. Calculations show that the best configuration for achieving low effluent concentrations is with an upstream counterelectrode. This rule applies for both high or low ratios of σ/κ. In fact, a limiting-current distribution cannot be achieved for a system utilizing a downstream counterelectrode (except for short reactors). For very high values of σ/κ, the matrix potential is constant, and the placement of the current collector is not important. For moderate matrix conductivities the optimum placement of the current collector depends on the particular chemical system being investigated and the actual value of σ/κ.

We have considered above that often a flow-through porous electrode may be ohmically limited. This suggests an alternative configuration seen at the bottom of Fig. 7. The flow-by porous electrode configuration has the advantage that the flow of current is perpendicular to the fluid flow. The electrodes can be made thin to minimize the ohmic potential drop in the electrodes, and they can be made long to achieve high conversions. The disadvantage of the flow-by configuration is the necessity of including a separator to avoid anolyte and catholyte mixing. The mathematical analysis is also inherently more complicated due to the two-dimensional nature of the problem. Two-dimensional modeling work has been done by Alkire and Ng (1974). They treat a cylindrical packed-bed electrode surrounded by a concentric counterelectrode. Trainham and Newman (1981) present an engineering model that compares the performance on an economic basis of the two porous electrode configurations for redox energy storage. The results of the computer optimization show that the flow-by configuration is superior in this case, where the dimensionless quantity $\varepsilon nFD_0c_0/s_R\kappa\,\Delta\Phi_2$ is large. For dilute solutions (small values of $\varepsilon nFD_0c_0/s_R\kappa\,\Delta\Phi_2$) the flow-through configuration may continue to have merit. Fedkiw (1981) presents a comparison of the performance of flow-through and flow-by electrodes operated at limiting current. For a given maximum ohmic potential drop and desired conversion, he concludes that a flow-by electrode with a length-to-width aspect

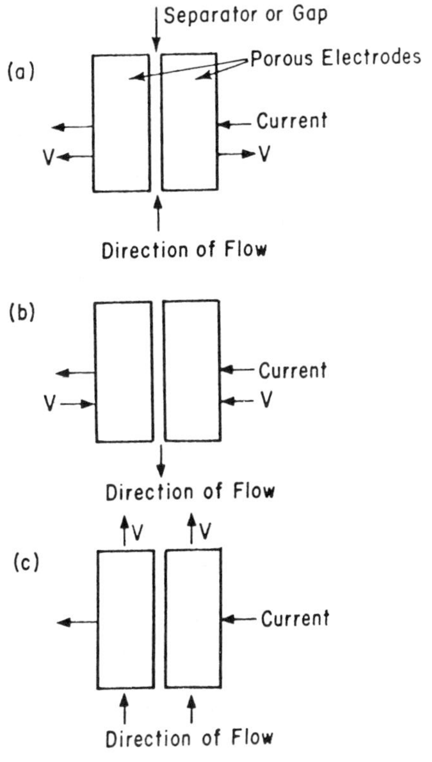

FIGURE 7 Various configurations of electrode placement relative to the direction of fluid flow. (a) and (b) are flow-through configurations (current and fluid flow are parallel), and (c) is the flow-by configuration (current flows perpendicular to fluid flow).

ratio greater than 5 will have a higher processing rate than a flow-through electrode. An ultimate design of a flow-by electrode must recognize that it can be run below the limiting current.

It is clear that the flow-by system holds sufficient promise that more detailed mathematical modeling and scaleup criteria are needed. The zinc-bromine secondary battery system as well as flow-redox energy storage systems have evolved to flow-by configurations.

Battery Applications

Let us now focus attention on the use of porous electrodes for battery applications. Several levels of mathematical sophistication can be used to examine battery systems. Initially we should like to develop guidelines for screening prospective systems. As development of a new or old system continues, we then resort to more sophisticated mathematical models for our design and scaleup. Before going into details, let us consider some general aspects of battery systems (Tiedemann, 1978).

We wish to examine desirable characteristics for batteries so that we may define inherent limitations or areas needing development for new and existing systems. One recognizes that listing a set of desirable

characteristics implies a particular application. In examining a new or old system we wish to look at advantages and disadvantages in terms of cost (perhaps amortized cost/cycle), energy and power density, shelf life, reversibility, energy and coulombic efficiencies, material utilization efficiency, cycle life, reliability, portability, safety, and availability of materials.

One of the first characteristics of a battery to be considered is its voltage and specific energy. Knowledge of the overall reaction allows calculation of the open-circuit potential from thermodynamic data. Dividing the potential by the equivalent weight (kg/C) yields the theoretical specific energy of the electrode pair. Thus more energetic electrode couples with lower equivalent weight will increase the theoretical specific energy. From a practical viewpoint, the advantage of more energetic couples must be weighed against the more severe materials compatibility requirements in the more corrosive environment. For new batteries the specific energy quoted is often a theoretical energy based only on the mass of active materials. The theoretical specific energy of the lead-acid battery is 218 W•h/kg based on the active materials PbO_2, Pb, and H_2SO_4. Including the weight of the solvent (for an initial concentration of 5 molal H_2SO_4) decreases the energy to approximately 100 W•h/kg. Addition of the weights of current collectors, excess active materials, separator, container, and connecting posts decreases this value further to approximately 40 W•h/kg. A rule of thumb that the final specific energy is 25 to 30% of the theoretical specific energy has been found to be approximately true for several systems. This rule should be used with caution, but it does illustrate the performance penalties associated with battery packaging and scaleup.

Knowledge of the specific energy of the system can be used to assess the promise of a given system. Data on the specific power of the system are also needed to determine the ability of the battery to deliver energy at different rates. The maximum specific power that a battery can deliver can be approximately calculated by $(U/2)^2/R_i$ where U is the open-circuit potential and R_i is the sum of the internal area specific resistances ($\Omega \cdot cm^2$) times the loading density of material (kg/cm^2). The area specific resistance can be estimated on the basis of electrolyte conductivity and separator thickness. Battery designs with large current densities should keep overall area specific resistances down to approximately 1 $\Omega \cdot cm^2$ and keep area specific resistances of the separator itself down to about 0.2 $\Omega \cdot cm^2$. These calculations become more refined as we obtain more information on a system. Higher specific energies and specific powers are important in applications where total weight or volume are important. In electric vehicle applications, for example, a battery may not have enough energy or power to carry itself and its support structure.

Three system efficiencies characterize the performance of a battery. Coulombic efficiency is a measure of the reversibility of the electrodes or the presence of side reactions. It is given by the ratio of the number of coulombs released during discharge to the number of coulombs required to charge the system back to its initial state. Energy efficiency is calculated by multiplying the coulombic efficiency by the ratio of average discharge voltage to average charging voltage. Differences in charge and discharge voltages are due to irreversibilities associated with the ohmic potential drop in the solution and matrix, and overpotentials associated with electrode kinetics and mass transfer resistances. Coulombic efficiencies can often approach 100%, while energy efficiencies are typically 50 to 80%. The method of charging and discharging the battery will affect both of these numbers.

The material utilization efficiency is the ratio of the actual coulombs passed during discharge to the theoretical amount of active material available. Utilizations of 70 to 80% are typical, and again are dependent on the method of discharge as well as any imposed cutoff voltage. Material utilizations often decrease with cycling. Among the factors that can lead to material utilizations less than 100% are the isolation of active materials, solubility problems with reactants and products, and, as mentioned above, any imposed voltage cutoff. Isolation of active material can occur when an insoluble, insulating reaction product covers the active material. Highly nonuniform reaction distributions can lead to pore blockage. Differences in the molar volume of reactants and products cause porosity changes which can lead to matrix fracture.

The solubility of reactants and products is important to material utilization as well as cycle life and shelf life. Dunning et al. (1971) identify a range of approximately 5×10^{-5} to 11.4×10^{-5} molar for the desirable solubility range of a sparingly soluble reactant. The lower limit of solubility is based on the need to diffuse the reactant to the active sites from sparingly soluble crystallites. The ability to store the reactant in close proximity to the active site was listed as one of the advantages of porous electrodes. The upper limit of solubility can determine the shelf life of the cell because of self-discharge processes. In general, a soluble reactant on discharge will react if it can diffuse to the other electrode, while a soluble reactant on charge will be substantially inert if it diffuses to the other electrode. A soluble discharge reactant can be permanently incorporated into the other electrode and not be recoverable on charge. The magnitude of solubility also influences the redistribution of active material. The relatively high solubility of ZnO in KOH leads to marked concentration changes on cycling, and these can couple with the fluid flow to produce zinc shape change (Choi et al., 1976a,b).

We have mentioned above that the method of charging can affect the efficiencies of a given system. Ideally, the charging process will restore the battery to its state at the onset of the discharge cycle. In general, the charging process plays an important role in the cycle life and performance of a battery. Two common charging methods that may be used are constant current and constant voltage. These can be combined with voltage or current limits to help minimize side reactions, or for thermal management. Side reactions, however, often do occur on charge. If these reactions occur preferentially on one electrode, an imbalance in the state of charge occurs. The ability of the system to accept overcharge then becomes important. More sophisticated charging procedures can be developed, but may be prohibitively expensive depending on specific applications. In some redox energy storage systems, for example, a third electrode is used to correct for imbalances in the state of charge.

Porous battery electrodes can be constructed in a variety of configurations. Often individual electrodes are constructed with a highly conducting, inert substrate (grid) that mechanically holds the porous matrix in place. It should be strong enough to withstand volume changes on cycling and be inert over the operating temperature and voltage ranges. For a battery plate with poorly conducting active materials, the grid is necessary to act as a current collector to conduct electrons to or from the reaction site. For moderately conducting active materials it can act as a secondary current collector. In electrode configurations using grids, the current is collected from each plate and is connected in parallel with an interconnecting bus.

The optimal design of the current collecting grid and intercell connectors is a scaleup problem that will be considered later. Another battery configuration that can be considered is a bipolar arrangement. Such an arrangement eliminates the need for a separate current collecting grid for each electrode and may possibly minimize the voltage and weight performance penalties associated with the grid. Here the positive and negative active materials are put on opposite sides of an inert, conducting substrate. Current flows straight through the cell stack and is collected at the ends. The materials constraints for the conducting substrate are severe; it must be substantially inert to both the oxidizing and reducing environment. Corrosion of the substrate will lead to short circuits.

Materials compatibility is often a major problem in new or old battery systems. Materials constraints may limit cycle life or prohibitively increase the cost of a system. For example, corrosion of the current collecting grid is cited as a major failure mechanism for the lead-acid battery. Active material in ambient temperature Li electrodes becomes electronically isolated from its substrate, presumably due to reaction with impurities and the electrolyte itself (Brummer et al., 1980). In general, we look for a battery system where the active materials are compatible with each other and the other support materials required in the battery. We require the electrolyte, separator, battery container, interconnecting bus and post, current collectors, and so on, to be stable over the operating voltage and temperature ranges of our system while not catalyzing side reactions or otherwise reacting adversely with the system.

A separator is a major component of the battery package. It is required to separate the positive and negative electrodes in the battery, so that they may be in close proximity but not short together. A common failure mechanism in early battery development is shorting of the positive and negative electrodes by dendritic growth of material through the separator (Choi et al., 1976a,b). Separators must be electronic insulators, but have relatively high ionic conductivities. A separator with a high area specific resistance ($\Omega \cdot cm^2$) leads to a high ohmic potential drop with subsequent poor performance. Separators may also be required to have other desirable properties. Often they should be specifically conducting to only certain ionic species (Choi et al., 1976a,b) or must contain additives that increase electrolyte wetting.

Costs, of course, must always be considered within a particular application. Besides the costs of raw materials, some other factors can be considered. Costs associated with safety, recycling (particularly if availability of materials is low), and associated environmental factors need to be considered. A common basis for comparison of various alternatives of secondary batteries is the amortized cost of the system in \$/W·h·cycle.

Scaleup of porous electrodes is not straightforward. Simply increasing the thickness of an electrode, for example, does not necessarily bring about proportional increases in performance. We wish to develop a micromodel of the porous-electrode system so that we can predict the effects of changing parameters such as electrode thicknesses on the behavior of the system. Having this information in hand then leads us to couple the micromodeling to further scaleup considerations of plate area and lengths of intercell connectors.

In examining the micromodeling of porous electrodes let us first consider the zero-time behavior (Newman and Tobias, 1962) where concentrations are assumed to be uniform throughout the pore volume. Further,

let us ignore double-layer charging effects. Four dimensionless ratios govern the current distribution. These can be stated as a dimensionless current density,

$$\delta = \frac{\alpha_a FIL}{RT}\left(\frac{1}{\kappa} + \frac{1}{\sigma}\right) \tag{61}$$

a dimensionless exchange current,

$$\nu^2 = (\alpha_a + \alpha_c)\frac{Fai_o L^2}{RT}\left(\frac{1}{\kappa} + \frac{1}{\sigma}\right) \tag{62}$$

the ratio of transfer coefficients in the polarization equation, α_a/α_c, and the ratio of effective solution and matrix conductivities κ/σ. δ and ν^2 are ratios describing the competing effects of ohmic potential drop and slow electrode kinetics. For large values of δ or ν^2, the ohmic effect dominates with a nonuniform reaction distribution. For small values of κ/σ, the reaction occurs preferentially near the electrode-separator boundary at the expense of the region near the backing plate.

The nonuniformity of the reaction distribution for Tafel kinetics can be seen in Fig. 8. For Tafel kinetics the current distribution depends only

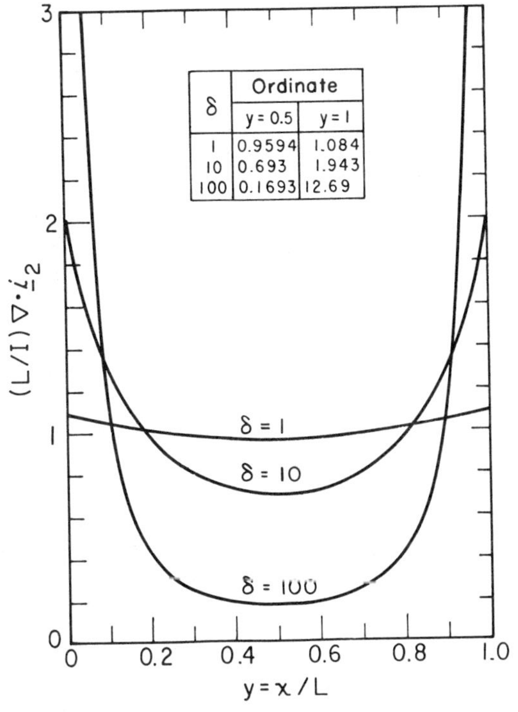

FIGURE 8 Reduced current distributions for Tafel kinetics with equal matrix and solution conductivities.

on the parameter δ and the ratio κ/σ. Curves for linear polarization would exhibit similar behavior as in Fig. 8. For linear behavior, the distribution becomes nonuniform for large ν and is independent of the total current. For both cases the ratio of κ/σ serves to shift the reaction distribution from one face to the other.

The distance to which the reaction can penetrate the electrode determines how thick an electrode can be utilized. This penetration depth is characterized by

$$\frac{L}{\nu} = \left[\frac{RT\kappa\sigma}{(\alpha_a + \alpha_c)ai_oF(\kappa + \sigma)}\right]^{1/2} \tag{63}$$

Electrodes much thinner than the penetration depth behave like plane electrodes with an enhanced surface area. Electrodes much thicker are not fully utilized. For high current levels, in the Tafel range, the ratio L/δ will be more characteristic of the penetration of the reaction.

To continue to follow the discharge behavior of a porous electrode through the transient behavior, we need to consider the time derivative in Eq. (33). Porous electrodes used in primary and secondary batteries invariably involve solid reactants and products, and the matrix is changed during discharge. Consequently, no steady state is strictly possible. We may nonetheless examine a steady-state operation of a porous electrode. Just above we have considered the irreversibilities associated with electrode kinetics and ohmic potential drop. As the reaction proceeds, reactant is depleted at the pore-solution interface. This then represents an additional irreversibility. Newman and Tobias (1962) also treat a redox reaction in a porous electrode. Convection is assumed to be absent, and migration is neglected due to an excess of supporting electrolyte. The stoichiometric coefficients of the reactant and product species are taken to be +1 and −1. For a redox reaction Eq. (41) is often written as

$$\nabla \cdot \underset{\sim}{i}_2 = ai_o\left\{\frac{c_1}{c_1^o} \exp\left[\frac{\alpha_a F}{RT}(\Phi_1 - \Phi_2)\right] - \frac{c_2}{c_2^o} \exp\left[\frac{-\alpha_c F}{RT}(\Phi_1 - \Phi_2)\right]\right\} \tag{64}$$

Now i_o is a constant representing the exchange current density at the composition c_1^o, c_2^o. The potential difference $\Phi_1 - \Phi_2$ is equal to η_s plus an additive term which depends on the local solution composition. [Compare this with Eqs. (10) and (11).] A new dimensionless group $\gamma_i = s_iIL/nF\varepsilon D_ic_i^o$ can be formed due to the introduction of the diffusion coefficient of each species and a characteristic concentration. Another special case that can be treated is deposition from a binary electrolyte. The binary electrolyte formulation can be applied to sulfuric acid in lead-acid batteries or to the polysulfide in the sodium-sulfur cell if the melt is taken to be composed of Na_2S and S. This formulation also applies to systems with concentrated KOH electrolyte, such as in Ni-Fe and Ni-Zn cells, although the solubility of ZnO must be ignored.

Often a system cannot be approximated by one of the limiting cases presented above. Full treatment of the complicated factors governing the behavior of the porous electrode requires the use of high-speed digital

computation. Newman and Tiedemann (1975) suggest a computational method for battery electrodes involving a binary electrolyte. In general, reactant species are depleted during the course of discharge, and time must be included as a variable. Thus the coupled equations are solved simultaneously at each time step. Pollard and Newman (1981) treat the transient behavior of the lithium-aluminum, iron sulfide high-temperature battery for a constant current discharge. Concentration distributions across the cell sandwich are presented at various times throughout the discharge.

In summary we can list a number of factors that can affect the performance of porous electrodes:

1. Charge and discharge methods affect battery efficiencies and cycle life.
2. The solubility of reactants and products can limit cycle life and shelf life.
3. Higher current densities yield higher overpotentials, and thus a given cuttoff potential is reached sooner.
4. The pores may become constricted or even plugged with solid reaction products. A nonuniform reaction distribution will accentuate this problem at the mouth of the pores.
5. Utilization of the solid fuel can be limited by covering of the reaction surface with reaction products.
6. Rates of mass transfer between crystallites and the reaction surface may become more limiting as the discharge exhausts the front part of the electrode. This could account for changes in the apparent limit of utilization with current density.

Until now we have considered the mathematical modeling of porous battery electrodes. Experimental data are needed, of course, to ensure that our understanding of the system is substantially correct. In constructing an experimental cell we want to eliminate any scaleup effects not included in the mathematical modeling so that we can directly compare experimental and theoretical results. The scaleup effects of current collectors and interconnecting bus and post will then be considered separately.

The experimental system can be arranged in a monocell configuration with one positive and one negative plate, or as a bicell with a single positive electrode and two negative electrodes. The bicell arrangement represents a "section" of a positive and two half negatives that would be found in the scaled-up battery. The construction and symmetry of discharge of this cell would be similar to the scaled-up version. The monocell's main advantage over the bicell may be in the ease and cost of construction.

A schematic of a bicell is shown in Fig. 9. This figure shows heavy, highly conducting current collecting sheets in the center of the positive electrode and at the back of the two half negative electrodes. These current collectors promote a uniform current distribution across the face of the electrode by minimizing the ohmic potential drop in the current collecting sheet. This is important for comparison to one-dimensional micromodeling results or for use as data in subsequent scaleup calculations. Separate voltage and current taps should be used to eliminate any error in voltage readings due to ohmic potential drop in the cell leads. Reference electrodes should be used in the experimental cell so that the total cell potential can be decomposed into contributions associated with the positive and negative electrodes. Although we will see that this decomposition is not necessary

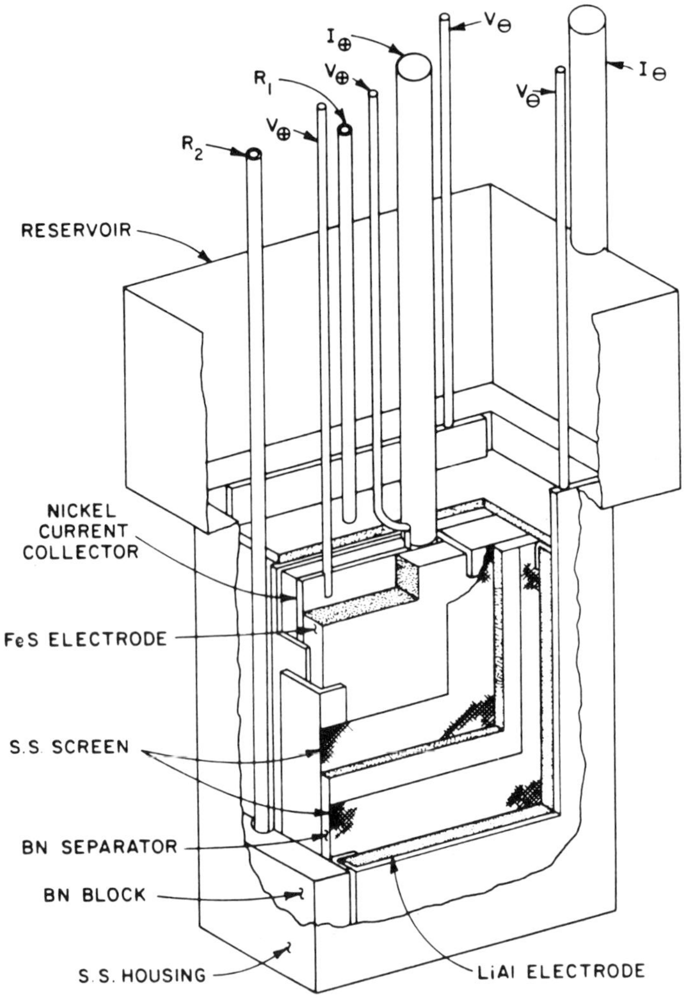

FIGURE 9 Bicell design of a positive and two half negative electrodes designed to promote a uniform current distribution across the electrode face.

for our scaleup calculations, research efforts at improving the battery need to be largely directed to the limiting electrode. Batteries are often designed so that the positive electrode limits the battery capacity. This should then be the case with the experimental cell as well. In general it is important to use the same electrode thickness, amount of active material, excess electrolyte, temperature, separator material, and so on, that is being considered for the scaled-up version. Micromodeling results can be used as a guide in the selection of some of these parameters. The cross-sectional area of the test cell is not important since our experiment is designed to be one dimensional.

The apparent open-circuit potential is measured during discharging or charging with a current-interruption technique. The cell potential during

current flow ($\Phi_p - \Phi_n$) and after 15 seconds* of interruption is interpreted according to the equation

$$i = Y(U - \Phi_p + \Phi_n) \tag{65}$$

where i is the current density from negative to positive plate (A/cm2), Y is the conductance of a cell element ($\Omega^{-1} \cdot cm^{-2}$), U is the apparent open-circuit potential (V), Φ_p is the potential of the positive plate (V), and Φ_n is the potential of the negative plate (V). This relationship assumes a linear polarization curve; however it can also be regarded as a step in the linearization of a nonlinear problem. Values of U and Y can be determined as a function of the state of charge for a given constant current density. Data for a lithium-aluminum, iron sulfide high-temperature cell taken at Argonne National Laboratory (Barney et al., 1981) are shown in Fig. 10. The use of reference electrodes allows the decomposition of the cell potential and specific conductance into values for the individual electrodes. These values are related to the cell values by

$$U = U_+ - U_- \tag{66}$$

and

$$\frac{1}{Y} = \frac{1}{Y_+} + \frac{1}{Y_-} \tag{67}$$

These results form a basis both for comparison to micromodeling results and now for scaling up the plate size.

Cost, weight, and volume considerations dictate that the current collectors will not be the heavy plates used in our test cell. The mass of the current collector that should be used for a given plate area is a scaleup consideration that is subject to optimization. Our goal is to develop a discharge curve for the plate as a whole (with current collectors) based on individual cell elements shown in Figure 10. Two common configurations for the current collector are the sheet current collector and a grid current collector. Tiedemann and Newman treat the nonuniform current and potential distributions in composite sheet electrodes (Tiedemann and Newman, 1979c) and in battery plates with grid configurations (Tiedemann and Newman, 1979a).

The lead-acid battery uses a current collecting grid with the active material pressed between the ribs. A honeycomb grid has been used in

---

*When the total external current is interrupted, we can identify three transients: relaxation of the double-layer capacity, a local equilization of charge and concentration from front to back of the electrode, and a reduction of concentration gradients in the whole cell by diffusion across the separator. In the current-interruption technique, we wish to wait long enough for double-layer charging to relax. A characteristic time for this is L^2aC/κ. The apparent open-circuit potential will continue to rise as the other transients continue. We choose 15 s here so that we may more closely approximate the results that would be obtained with a 15-s power test.

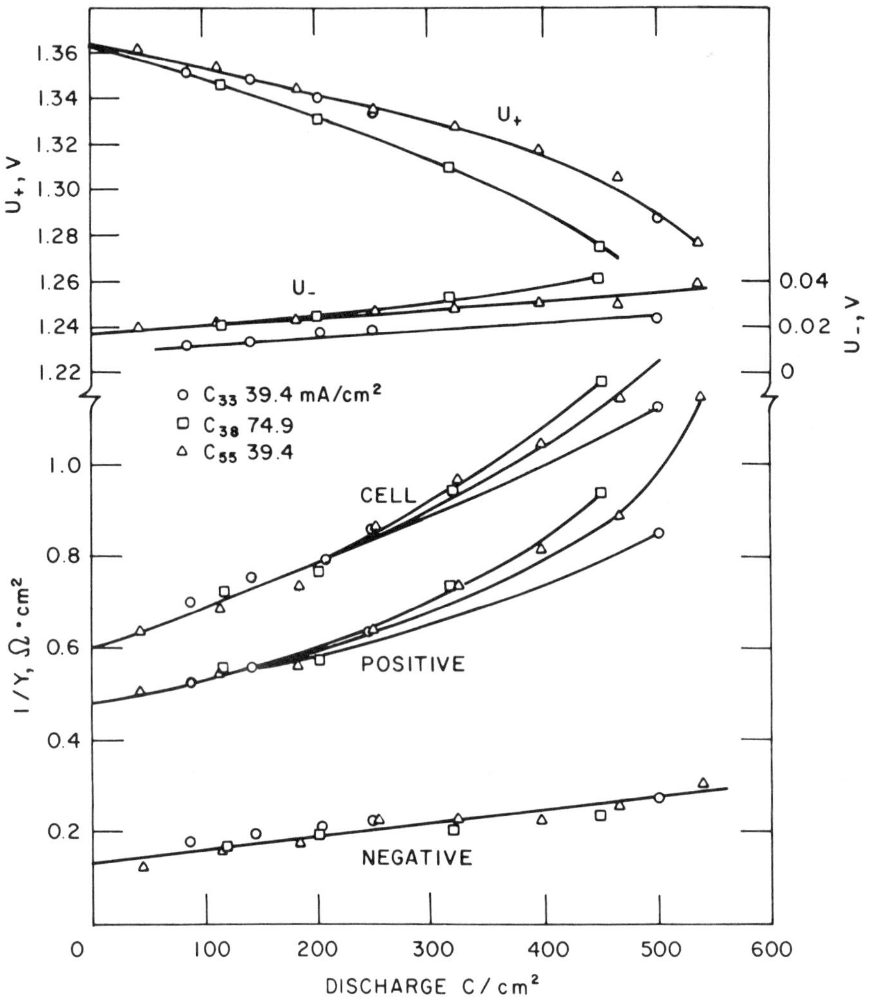

FIGURE 10 Electrochemical characteristics (U and Y) on discharge for L_p = 0.32 cm and 30% excess capacity in the negative electrode.

experiments with the lithium-aluminum, iron sulfide battery. We choose as an operational current-collector model a rectilinear grid with horizontal and vertical elements. A one-dimensional micromodel (Tiedemann and Newman, 1979b) or data as in Fig. 10 is coupled to a two-dimensional model of the grid. Equation (65) can still be used where the current, area, and potentials are now local values for node points on the grid. The polarization parameters, U and Y, are curve fit as functions of depth of discharge and local current density. Kirchhoff's law is used with Eq. (65) to solve for the local potential distribution across the face of the plate during discharge.

Results of this analysis give the overall plate behavior as a function of state of charge. The current distribution across the face of the electrode is nonuniform at the beginning of discharge and becomes more nearly uniform

as discharge proceeds because of the dependence of the electrochemical resistance and apparent open-circuit potential on the state of charge. The overall behavior can now be represented by

$$U - V = \Delta V = \left(R_g + \frac{1}{Y}\right) i \tag{68}$$

where ΔV is the voltage displacement from open circuit, and R_g is the resistance of the grid. Since $1/Y$ and R_g vary through discharge in ways that depend on the specific system, a general formulation of results cannot be made. However, we can consider the zero-time behavior of a system with constant polarization parameters and formulate the problem. For the primary variables ΔV the voltage displacement from open circuit, I the total current leaving the grid, A the area of the plate, Y the conductance of the cell element, M the mass of the grid, σ the grid conductivity, ρ the grid density, and L_p the positive electrode thickness, four dimensionless groups govern the system. One of these, L_p^2/A, does not have direct relevance to the problem. We are left with

$$\Pi_1 = \frac{I}{2AY\Delta V}, \qquad \Pi_2 = \frac{M\sigma}{\rho YA^2}, \qquad \Pi_3 = \frac{M}{\rho L_p A}$$

Π_1 represents a ratio of overall conductance $I/2\Delta V$ to electrochemical conductance AY. Stated another way, Π_1 is the ratio of the actual current leaving the tab to the current that would leave the tab if there were no ohmic resistance in the grid. The factor of 2 in Π_1 reflects the fact that the total current leaving the grid tab is being collected from both sides of the plate. Π_2 represents a ratio of grid conductance $M\sigma/\rho A$ to electrochemical conductance AY. Π_3 is the volume fraction of grid material.

Other minor dimensionless parameters must now be chosen before the results of the grid model can be plotted. These include the ratio of the tab width to plate width, relative tab position, aspect ratio of plate height to plate width, ratio of total grid material on horizontal elements to that in vertical elements, and the number of horizontal and vertical elements. Further, results can be presented for an infinitely conducting negative grid, a symmetric negative grid with equal conductivity, or a complete description of a positive and negative with different conductivities.

Figure 11 is a plot of dimensionless plate current versus dimensionless plate area for the beginning of discharge (Tiedemann and Newman, 1979a). The volume fraction of grid material has been taken to be zero so that L_p does not enter as a parameter in this figure. Here the horizontal and vertical grid elements have the same amount of grid material, and a symmetric negative grid with equal conductivity is used. Figure 12 is an example of an improved grid. Extra conducting material has been added to the two columns of vertical elements below the tab and the horizontal elements across the top of the grid. The two center vertical elements are heavier by a factor of 11, and the top horizontal elements have four times the mass of the base elements. Figure 13 is a dimensionless graph of plate current versus plate area for this grid design. Here the negative grid has been assumed to be infinitely conducting. A comparison of Figs. 11 and 13 shows the improvement in overall conductance for the improved grid design. This improvement is actually due to two major effects, the

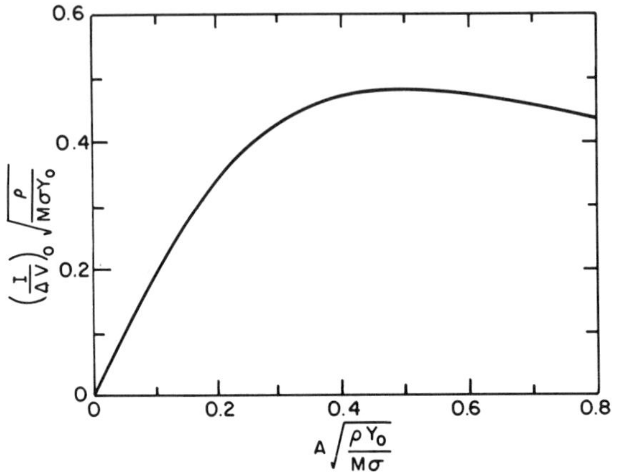

FIGURE 11 Dimensionless plate current as a function of the dimensionless plate area. Height-to-width ratio equals 0.8. Ten percent tab is located 30% from the edge of the plate. L_p is taken to be ∞. (From Tiedemann and Newman, 1979a.)

reduction of ohmic potential drop in the improved grid design and the neglect of ohmic potential drop in the negative grid.

Figure 14 is a dimensionless correlation of the same data as in Fig. 13. The dimensionless groups have been adjusted so that all the data closely follow the same curve. Here it is recognized that the grid necessarily

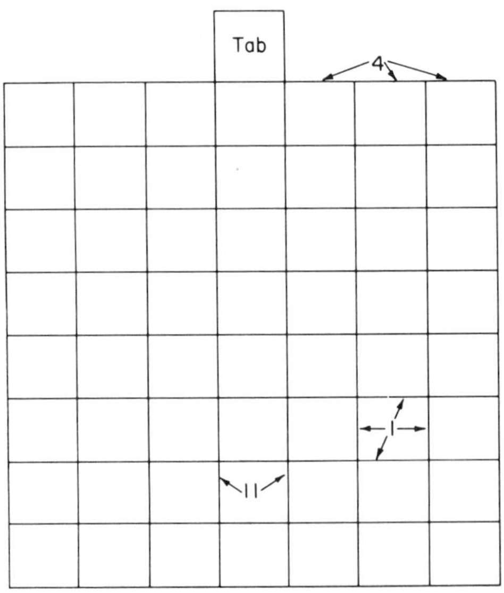

FIGURE 12 Improved grid design with extra conducting material in the two vertical elements and top horizontal elements.

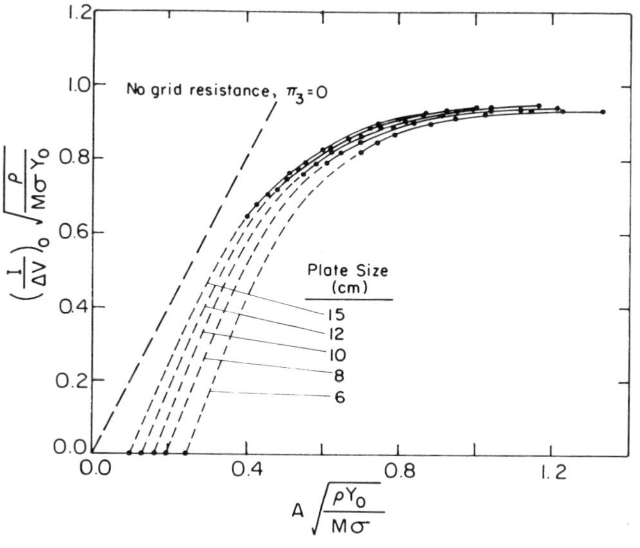

FIGURE 13 Dimensionless plate current as a function of the dimensionless plate area for the improved grid design.

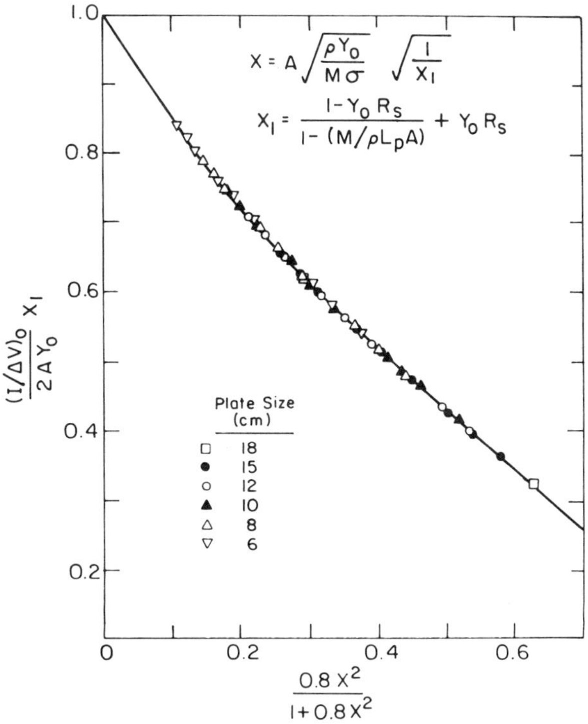

FIGURE 14 Dimensionless plate conductance versus dimensionless plate area based on a reduced area for current flow through the matrix bed.

displaces some active material, reducing the area for current within the bed. The active bed cross-sectional area is therefore $A - M/\rho L_p$. The area for current flow through the separator is A. Thus we correct the electrochemical conductance AY_o by an additive contribution of separator resistance based on area A, and the rest of the electrochemical resistance based on area $A - M/\rho L_p$. Thus $1/Y_o$ is replaced by

$$\frac{1}{Y_o}\left(\frac{1 - Y_oR_s}{1 - M/\rho L_pA} + Y_oR_s\right)$$

in plotting Fig. 14. The value of the separator resistance, R_s ($\Omega^{-1}\cdot cm^{-2}$), can be estimated on the basis of electrolyte conductivity. The data plotted in this manner fall very closely on the same curve and have unity intercepts on both the ordinate and abscissa.

As pointed out above, Fig. 14 is strictly valid for zero-time behavior. Our alternative to performing complex grid calculations to map out the overall conductance during discharge is to assume Fig. 14 is valid throughout discharge, replacing Y_o with Y and $(I/\Delta V)_o$ by $I/\Delta V$. Here we are recognizing the dependence of the polarization parameters on the average depth of discharge, but still assume that they are constant across the electrode face. Y and U actually depend on the local current density and local depth of discharge. This method should give a good approximation to the actual behavior of the plate and reduces to an exact solution to the complete grid formulation as the current densities become more uniform. A computer program can now be developed that uses data as in Figs. 10 and 14 to generate scaleup predictions for energy and power.

For a given plate size and grid weight, we can now calculate the specific energy and specific power through discharge. The calculations vary the average depth of discharge to a specified cutoff potential for a 4-h discharge rate. The results of these calculations, using the correlation in Fig. 14 and the data in Fig. 10, are given in Fig. 15. We immediately see a conflict in trying to optimize simultaneously the battery system for maximum energy and maximum power. Instead, we choose an intermediate grid weight that gives a ratio of power to energy equal to the ratio of maximum power to maximum energy. Alternative compromises could be selected based on the particular applications at hand.

Having selected a natural system ratio for determining the grid compromise, we now recognize that our small plates must still be connected in parallel with an interconnecting bus, and that the cell must be connected in series to another cell with an interconnecting post. Two important new parameters—the length of interconnecting bus and length of interconnecting post—must now be considered. Figure 16 shows a symmetric arrangement of interconnecting buses for three positive electrodes. L_{bus} is approximately equal to the sum of the widths of the positive electrode, negative electrode, and two separators. The post represents the cell terminal plus intercell connector that extends from one cell to the next cell. In our calculations we take L_{post} to represent only that part of the post associated with one cell (i.e., the length of the terminal plus half of the length of intercell connector). The lengths of interconnecting buses for the positive and negative plates are taken to be the same, as are the post lengths. Given the post and bus lengths needed to connect our cells, an optimization of the distribution of conducting material among the parallel

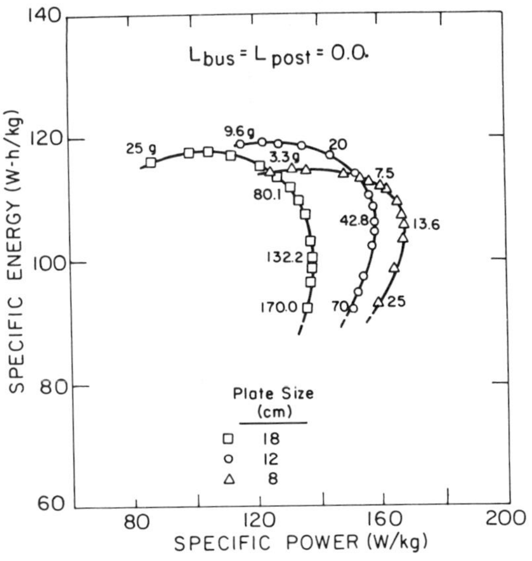

FIGURE 15 Maximum specific energy versus maximum specific power halfway through discharge.

bus connections and the post connections can be made independently of other scaleup considerations. The important quantity that results from this optimization is

$$L_{bus}\left(\frac{\rho_{bus}}{\sigma_{bus}}\right)^{1/2} + L_{post}\left(\frac{\rho_{post}}{\sigma_{post}}\right)^{1/2}$$

If the density of the bus and post are taken to be equal, we see only the sum of the bus and post length is important to the scaleup predictions. We also note that the number of positive electrodes per cell is subject to variation. The number used will affect the final results for the power and energy, but does not affect the optimum distribution of material over the bus and post per cell "section."

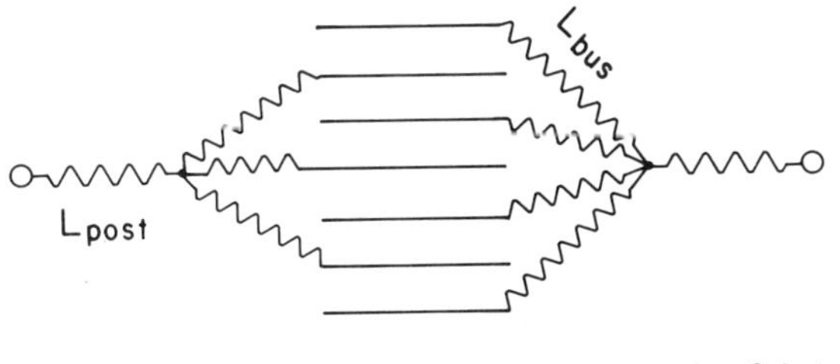

FIGURE 16 Symmetric arrangement of interconnecting buses and interconnecting posts for three positive electrodes.

For a given plate size and grid weight, the calculated specific energies versus specific powers will form a loop in the same manner as in Fig. 15 as the mass of the bus and post is varied from very small values to larger values. Thus, for each plate area and grid weight we can find a mass of bus and post that maximizes either the specific energy or specific power. As we investigate different grid weights (and thus different optimum post and bus weights) we can find values of the mass of grid, bus, and post that yield the maximum power. Similarly, we search for values of grid weights and bus and post weights that yield the maximum energy. Now our compromise ratio is the value of the maximum power to this maximum energy. Thus for each plate size our compromise design is the values of the mass of grid, bus, and post that give a power-to-energy ratio equal to the compromise ratio. More than one combination of grid and bus and post weight will yield a power-to-energy ratio equal to the compromise value. The largest values of energy and power are selected from this collection of results to represent the best compromise design for a given plate area.

Figure 17 represents the composite prediction of specific energy versus specific power as they relate to the size of plates and lengths of bus and post. Recall that each point given represents the best compromise of specific energy and specific power for each plate area as just discussed above. The performance penalty (weight and voltage) of including the interconnecting bus and post is readily seen. This graph can now be used to select the plate area which will give adequate performance at a suitable cost. The grid weights and plate areas used to generate Fig. 17 are plotted in Fig. 18. Also shown are the grid weights that correspond to maximum power and maximum energy. This figure, then, yields the grid weight that gives the maximum power, maximum energy, or compromise for

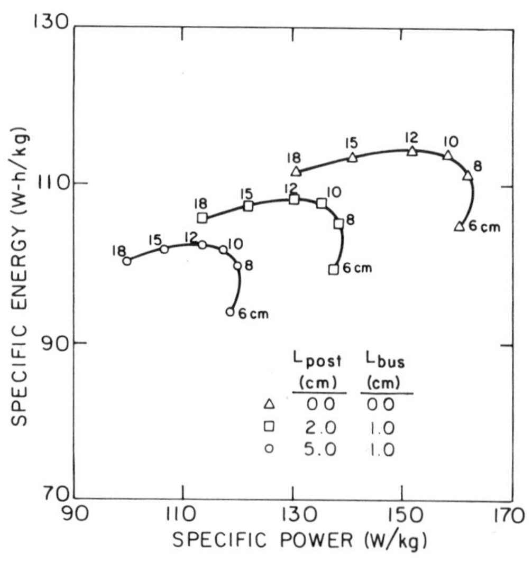

FIGURE 17 Composite, compromise predictions of the specific energy and specific power for various post and bus lengths and electrode areas.

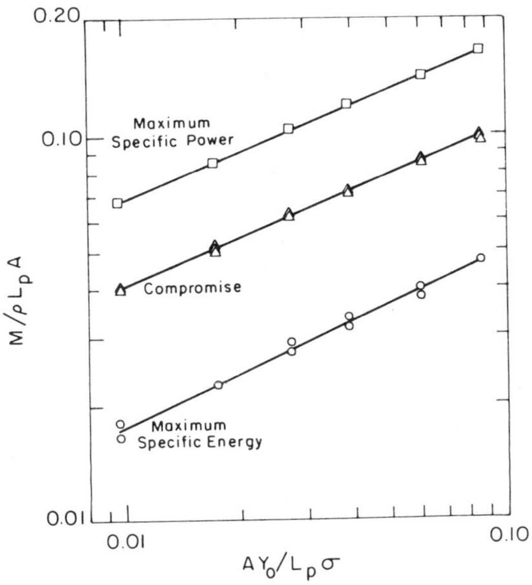

FIGURE 18 Dimensionless grid weight versus dimensionless plate area along the maximum specific energy, maximum specific power, or the compromise design.

each plate area and length of bus and post. It is seen that the optimum grid weight is not greatly dependent on the length of the bus and post, and that this dependence is actually undetectable for maximum specific power. These results suggest that we can improve our grid design without concern over interaction with the optimum post and bus lengths, and that improvements may be independent of depth of discharge as well. We also see that the optimum grid weight should be roughly increased in proportion to $A^{3/2}$.

We should also note here that no consideration has been given to any minimum current-collector weight necessary to support the active materials through cycling. In general, for smaller cells it may be stated that a current collector that is large enough to contain the active material will have a weight large enough to approach or exceed the value corresponding to the design for maximum power. More detailed scaleup considerations are needed for bigger plate sizes. Finally, we should note that the value of the grid weight corresponding to maximum specific energy will depend on the discharge time. As we decrease the discharge time from the 4-h value used here, the grid weight will move upward toward the value for maximum specific power.

In the results presented above, the delivered ampere-hour capacity of the battery has been a dependent variable. An alternative design sequence is to specify a value for the total delivered capacity. In this case the area of the plate is varied until the optimum values of grid weight and bus and post weight are found so that the specified capacity is reached at the cutoff voltage.

In summary, we have shown how to account for the performance penalties associated with the grid and interconnecting bus and post.

NOTATION

a	specific interfacial area per unit volume of porous electrode, cm^{-1}
A	cross-sectional area of the plate, cm^2
A_j	parameter characterizing magnitude of anodic reaction at the potential where the main reaction is beginning to reach limiting current [see Eq. (19)]
A_0	constant defined by Eq. (43)
A_{tot}	total superficial electrode area, cm^2
B	quantity defined in Eq. (54)
c_i	concentration of species i per unit volume of solution, mol/cm^3
c_{io}	local surface concentration of species i, mol/cm^3
$c_{i,re}$	concentration of species i in the reference electrode compartment, mol/cm^3
$c_{i,ref}$	reference concentration of species i, mol/cm^3
$c_{i\infty}$	concentration of species i in the bulk, mol/cm^3
c_L	exit concentration of reactant, mol/cm^3
c_o	feed concentration to flow-through electrode, mol/cm^3
C	ratio of capital cost to fuel cost, A/cm^2
C_a	capital cost coefficient, $\$/cm^2 \cdot s$
C_d	electrical double-layer capacity, F/cm^2
C_e	electric energy cost, \$/J
C_f	fuel cost coefficient, \$/C
D'	dimensionless dispersion coefficient defined in Eq. (54)
D_a	dispersion coefficient, cm^2/s
D_i	diffusion coefficient of species i, cm^2/s
D_{ij}	diffusion coefficient for interaction of species i with species j, cm^2/s
F	Faraday's constant, 96487 C/equiv
F_a	flow rate of inerts on anode side, mol/s
g	acceleration of gravity, cm/s^2
h	spacing between electrodes, cm
$\underset{\sim}{i}$	current density, A/cm^2
$\underset{\sim}{i}_1$	superficial current density in the matrix, A/cm^2
$\underset{\sim}{i}_2$	superficial current density in the pore phase, A/cm^2
i_j	current density for reaction j, A/cm^2
$i_{m,lim}$	limiting current density for the main reaction, A/cm^2

i_n	transfer current per unit of interfacial area, A/cm^2
i_o	exchange current density, A/cm^2
$i_{oj,ref}$	exchange current density for reaction j with reference concentrations, A/cm^2
I	current density, A/cm^2, or total current leaving tab, A
j_{in}	pore-wall flux of species i, mol/cm^2
k_m	coefficient of mass transfer between flowing solution and electrode surface, cm/s
K_u	resale value of unused but depleted hydrogen, as a fraction of value in feed
L	length of planar electrodes, or length of porous electrode, cm
L_{bus}	length of interconnecting bus, cm
L_p	thickness of positive electrode, cm
L_{post}	length of interconnecting post, cm
M	current collector weight, g
M_i	molecular weight of species i, g/mol
n	number of electrons transferred in electrode reaction, equiv/mol
n_j	number of electrons transferred in reaction j, equiv/mol
$\tilde{N}_i$	flux of species i, or superficial flux of species i, $mol/cm^2{\cdot}s$
p	pressure, $dyne/cm^2$
p_{H_2}	partial pressure of hydrogen at the anode, bar
p_{O_2}	partial pressure of oxygen at the cathode, bar
P	electric power produced, W
R	universal gas constant, 8.3143 J/mol·K
$R_{a,1}$	anode matrix resistance, $\Omega{\cdot}cm^2$
$R_{a,2}$	anolyte solution resistance, $\Omega{\cdot}cm^2$
$R_{c,1}$	cathode matrix resistance, $\Omega{\cdot}cm^2$
$R_{c,2}$	catholyte solution resistance, $\Omega{\cdot}cm^2$
R_F	Faradaic charge-transfer resistance, $\Omega{\cdot}cm^2$
R_g	resistance of grid, $\Omega{\cdot}cm^2$
R_i	rate of homogeneous production of species i, $mol/cm^3{\cdot}s$, or the sum of internal area specific resistances, $\Omega{\cdot}cm^2$
R_s	separator resistance, $\Omega{\cdot}cm^2$
R'	effective ohmic resistance, $\Omega{\cdot}cm^2$
Re	Reynolds number
s_i	stoichiometric coefficient of species i in electrode reaction
s_{ij}	stoichiometric coefficient of species i in reaction j

Sc	Schmidt number
t	time, s
T	absolute temperature, K
u	fractional utilization of hydrogen
u_i	mobility of species i, $cm^2 \cdot mol/J \cdot s$
U	apparent open-circuit potential, V
U'	constant in Eq. (22), V
$U_{j,o}$	theoretical open-circuit potential for reaction j at the composition prevailing locally at the electrode surface relative to a reference electrode of a given kind, V
$U_{j,ref}$	theoretical open-circuit potential evaluated with reference concentrations, V
U_j^θ	standard electrode potential of reaction j, V
U_{re}^θ	standard potential for the reference electrode reaction, V
$\underset{\sim}{v}$	solution velocity or superficial fluid velocity, cm/s
$\langle v \rangle$	average solution velocity, cm/s
V	electrode potential or cell potential, V
V_m	electrode potential where main reaction begins to reach limiting current, V
x	distance through porous electrode, cm
x_{H_2}	ratio of moles of H_2 to moles of inerts
y	$= ak_m x/v$, dimensionless distance
Y	electrochemical conductance, $\Omega^{-1} \cdot cm^{-2}$
Y_o	electrochemical conductance at start of discharge, $\Omega^{-1} \cdot cm^{-2}$
z_i	valence or charge number of species i

Greek Letters

α	$= ak_m/v$, reciprocal reaction distance, cm^{-1}
α_a	transfer coefficient in anodic direction
α_{aj}	anodic transfer coefficient for reaction j
α_c	transfer coefficient in cathodic direction
α_{cj}	cathodic transfer coefficient for reaction j
γ_{ij}	exponent in composition dependence of exchange current density
δ	dimensionless current density defined in Eq. (61)
ΔU_s	approximate length of the limiting current plateau for the main reaction before the onset of the side reaction, V
ΔV	voltage displacement from open circuit, V
$\Delta \tilde{\Phi}_{ohm}$	ohmic potential drop in the absence of concentration variations, V

ε	porosity or void volume fraction, dimensionless, or permittivity, F/cm
η_{cj}	concentration overpotential for reaction j, V
η_s	surface overpotential, V
η_{sj}	surface overpotential for reaction j, V
θ	$= c_i/c_o$, dimensionless concentration
θ_L	$= c_L/c_o$, dimensionless exit concentration
κ	effective conductivity of solution, $\Omega^{-1}\cdot cm^{-1}$
λ	Debye length, cm
μ	viscosity, g/cm·s
μ_i	electrochemical potential of species i, J/mol
ν	kinematic viscosity of the solution, cm^2/s
ν^2	dimensionless exchange current given by Eq. (62)
Π_1	$= I/2AY\Delta V$, dimensionless plate current
Π_2	$= M\sigma/\rho YA^2$, ratio of grid to electrochemical conductance, dimensionless
Π_3	$= M/\rho L_p A$, volume fraction of grid, dimensionless
ρ	density, g/cm^3
ρ_o	pure solvent density, kg/cm^3
σ	effective conductivity of the matrix phase, or the conductivity of the current collecting grid, $\Omega^{-1}\cdot cm^{-1}$
Φ	electrostatic potential, V
Φ_1	electric potential in the matrix, V
Φ_2	electric potential in the solution, V
Φ_n	potential of the negative plate, V
Φ_o	potential in the bulk solution extrapolated to the electrode surface, V
Φ_p	potential of the positive plate, V

Subscripts

g	reference electrode of a given kind
i	species i
j	reaction j
m	main reaction
o	electrode surface, solvent, reactor inlet, or beginning of discharge
ohm	ohmic

r1	reference electrode located just outside the double layer
r2	reference electrode in the bulk solution
re	reference electrode
ref	reference concentrations
s	side reaction, or reference electrode of the same kind as the working electrode
∞	bulk solution
1	porous electrode matrix phase
2	porous electrode solution phase

Superscripts

exit	exit stream
$\circ$	feed stream

REFERENCES

Alkire, R. and P. K. Ng, Two-dimensional current distribution within a packed-bed electrochemical flow reactor, J. Electrochem. Soc., *121*, 95 (1974).

Alkire, R. C., G. D. Cera, and M. A. Stadtherr, Optimization of electrolytic cells having large numbers of variables, J. Electrochem. Soc., *129*, 1225 (1982).

Appel, P. W. and J. Newman, Application of the limiting current method to mass transfer in packed beds at very low Reynolds numbers, AIChE J., *22*, 979 (1976).

Bard, A. J. and L. R. Faulkner, *Electrochemical Methods, Fundamentals and Applications*, Wiley, New York (1980).

Barney, D. L. and others, *Lithium/Iron Sulfide Batteries for Electric-Vehicle Propulsion and Other Applications*, Progress Report for October 1979—September 1980, Argonne National Laboratory, Argonne, Ill., ANL-80-128 (Feb. 1981), p. 71.

Beck, T. R., *Proceedings of the Workshop on Energy Conservation in Industrial Electrochemical Processes*, Argonne National Laboratory Report ANL/OEPM-77-1 (Aug. 1976), p. 37.

Bennion, D. N. and J. Newman, Electrochemical removal of copper ions from very dilute solutions, J. Appl. Electrochem., *2*, 113 (1972).

Bockris, J. O'M., B. E. Conway, E. Yeager, and R. E. White, eds., *Comprehensive Treatise of Electrochemistry*, Vol. 2: *Electrochemical Processing*, Plenum Press, New York (1981).

Brummer, S. B., K. M. Abraham, V. R. Koch, and G. L. Holleck, *Review of the Status of Ambient-Temperature Secondary Lithium Batteries for Applicability to Load Levelling and Electric Vehicles*, EIC Corporation, Newton, Mass. (July, 1980).

Cabán, R. and T. W. Chapman, Rapid computation of current distribution by orthogonal collocation, J. Electrochem. Soc., *123*, 1036 (1976).

Choi, K. W., D. N. Bennion, and J. Newman, Engineering analysis of shape change in zinc secondary electrodes: I. Theoretical, J. Electrochem. Soc., *123*, 1616 (1976a).

Choi, K. W., D. Hamby, D. N. Bennion, and J. Newman, Engineering analysis of shape change in zinc secondary electrodes: II. Experimental, J. Electrochem. Soc., *123*, 1628 (1976b).

Danckwerts, P. V., Continuous flow systems. Distribution of residence times, Chem. Eng. Sci., *2*, 1 (1953).

Darlington, W. B. and M. Y. C. Woo, *Report of the Electrolytic Industries 1981*, Report sponsored by the Industrial Electrolytic Division of the Electrochemical Society (May 12, 1982).

Delahay, P., *Double Layer and Electrode Kinetics*, Interscience, New York (1965).

Dunning, J. S., Analysis of Porous Electrodes with Sparingly Soluble Reactants, Dissertation, University of California, Los Angeles (1971).

Dunning, J. S., D. N. Bennion, and J. Newman, Analysis of porous electrodes with sparingly soluble reactants, J. Electrochem. Soc., *118*, 1251 (1971).

Fedkiw, P. S., Ohmic potential drop in flow-through and flow-by porous electrodes, J. Electrochem. Soc., *128*, 831 (1981).

Fedkiw, P. and J. Newman, Low Péclet number behavior of the transfer rate in packed beds, Chem. Eng. Sci., *33*, 1043 (1978).

Fedkiw, P. S. and J. Newman, Mass-transfer coefficients in packed beds at very low Reynolds numbers, Int. J. Heat Mass Transfer, *25*, 935 (1982).

Fitzjohn, J. L., Electro-organic synthesis, Chem. Eng. Prog., *71*, 85 (1975).

Fleischmann, M., J. W. Oldfield, and L. Tennakoon, Fluidized bed electrodes: Part IV. Electrodeposition of copper in a fluidized bed of copper-coated spheres, J. Appl. Electrochem., *1*, 103 (1971).

Germain, S. and F. Goodridge, Copper deposition in a fluidised bed cell, Electrochim. Acta, *21*, 545 (1976).

Grahame, D. C., The electrical double layer and the theory of electrocapillarity, Chem. Rev., *41*, 441 (1947).

Gray, W. G., A derivation of the equations for multi-phase transport, Chem. Eng. Sci., *30*, 229 (1975).

Houghton, R. W. and A. T. Kuhn, Mass-transport problems and some design concepts of electrochemical reactors, J. Appl. Electrochem., *4*, 173 (1974).

Ibl, N., Optimization of copper refining, Electrochim. Acta, *22*, 465 (1977).

Ibl, N. and E. Adam, Optimierung in der elektrochemischen Verfahrenstechnik, Chem.-Ing.-Tech., *37*, 573 (1965).

Ibl, N. and P. M. Robertson, Optimisation of use of surplus night electricity in electrochemical processes, Electrochim. Acta, *18*, 897 (1973).

Jorné, J., Mass transfer in laminar flow channel with porous wall, J. Electrochem. Soc., *129*, 1727 (1982).

Kuhn, A. T., ed., *Industrial Electrochemical Processes*, Elsevier, New York (1971).

Lee, J. and J. R. Selman, Effects of separator and terminal on the current distribution in parallel-plate electrochemical flow reactors, J. Electrochem. Soc., *129*, 1670 (1982).

Liebenow, C., Über die Berechnung der Kapazität eines Bleiakkumulators bei variabeler Stromstärke, Z. Elektrochem., *4*, 58 (particularly p. 63) (1897).

Mantell, C. L., *Electrochemical Engineering,* 4th ed., McGraw-Hill, New York (1960), p. 391.
Matlosz, M. and J. Newman, Use of a flow-through porous electrode for removal of mercury from contaminated brine solutions, Proc. Symp. Transport Processes Electrochem. Syst., The Electrochemical Society, Vol. 82-10 (1982), p. 53.
National Research Council, *Assessment of Research Needs for Advanced Battery Systems,* NMAB-390, National Materials Advisory Board, Commission on Engineering and Technical Systems (1982), p. 183.
Newman, J., Engineering design of electrochemical systems, Ind. Eng. Chem., *60*(4), 12 (1968).
Newman, J., *Electrochemical Systems,* Prentice-Hall, Englewood Cliffs, N.J. (1973a).
Newman, J., The fundamental principles of current distribution and mass transport in electrochemical cells, in *Electroanalytical Chemistry*, Vol. 6, A. J. Bard, ed., Marcel Dekker, New York (1973b).
Newman, J., Optimization of potential and hydrogen utilization in an acid fuel cell, Electrochim. Acta, *24*, 223 (1979).
Newman, J. and W. Tiedemann, Porous-electrode theory with battery applications, AIChE J., *21*, 25 (1975).
Newman, J. and W. Tiedemann, Flow-through porous electrodes, in *Advanced Electrochemistry and Electrochemical Engineering,* Vol. 11 (1978), p. 353.
Newman, J. S. and C. W. Tobias, Theoretical analysis of current distribution in porous electrodes, J. Electrochem. Soc., *109*, 1183 (1962).
Parrish, W. R. and J. Newman, Current distribution on a plane electrode below the limiting current, J. Electrochem. Soc., *116*, 169 (1969).
Parrish, W. R. and J. Newman, Current distributions on plane, parallel electrodes in channel flow, J. Electrochem. Soc., *117*, 43 (1970).
Parsons, R., Equilibrium properties of electrified interfaces, in *Modern Aspects of Electrochemistry,* Vol. 1 (1954), p. 103.
Pickett, D. J., *Electrochemical Reactor Design,* 2nd ed., Elsevier, New York (1979).
Pollard, R. and J. Newman, Transport equations for a mixture of two binary molten salts in a porous electrode, J. Electrochem. Soc., *126*, 1713 (1979).
Pollard, R. and J. Newman, Mathematical modeling of the lithium-aluminum, iron sulfide battery: I. Galvanostatic discharge behavior, J. Electrochem. Soc., *128*, 491 (1981).
Sakellaropoulos, G. P., Criteria for selective path promotion in electrochemical reaction sequences, AIChE J., *25*, 781 (1979).
Sakellaropoulos, G. P. and G. A. Francis, Electrochemical reactor analysis: selectivity of multiple competing reactions, J. Electrochem. Soc., *126*, 1928 (1979).
Sakellaropoulos, G. P. and G. A. Francis, Selectivity and reactor analysis for parallel electrochemical reactions, J. Chem. Technol. Biotechnol., *30*, 102 (1980).
Swann, S., Jr. and R. Alkire, *Bibliography of Electro-Organic Syntheses 1801—1975,* The Electrochemical Society, Princeton, N.J. (1980).
Tiedemann, W., Electrochemical Engineering of Batteries, Short Course Lecture Notes, University of California, Los Angeles (May 1978).
Tiedemann, W. H. and J. Newman, Current and potential distribution in lead-acid battery plates, Proc. Symp. Battery Des. Optimization, The Electrochemical Society, Vol. 79-1 (1979a), p. 39.

Tiedemann, W. H. and J. Newman, Mathematical modeling of the lead-acid cell, Proc. Symp. Battery Des. Optimization, The Electrochemical Society, Vol. 79-1 (1979b), p. 23.

Tiedemann, W. H. and J. Newman, Nonuniform current and potential distribution in composite sheet type battery electrodes, Abstr. 169, Los Angeles, Calif., meeting of the Electrochemical Society (Oct. 1979c).

Trainham, J. A. and J. Newman, The effect of electrode placement and finite matrix conductivity on the performance of flow-through porous electrodes, J. Electrochem. Soc., *125*, 58 (1978).

Trainham, J. A. and J. Newman, A comparison between flow-through and flow-by porous electrodes for redox energy storage, Electrochim. Acta, *26*, 455 (1981).

U.S. Department of Commerce, *1980 Annual Survey of Manufactures*, M80(AS)-1, p. 14.

Van Muylder, J. Thermodynamics of corrosion, in *Comprehensive Treatise of Electrochemistry*, Vol. 4: *Electrochemical Materials Science*, J. O'M. Bockris, B. E. Conway, E. Yeager, and R. E. White, eds., Plenum Press, New York (1981).

Van Zee, J. and J. Newman, Electrochemical removal of silver ions from photographic fixing solutions using a porous flow-through electrode, J. Electrochem. Soc., *124*, 706 (1977).

Wehner, J. F. and R. H. Wilhelm, Boundary conditions of flow reactor, Chem. Eng. Sci., *6*, 89 (1956).

Wenglowski, G. M., An economic study of the electrochemical industry in the United States, in *Modern Aspects of Electrochemistry*, Vol. 4 (1966), p. 251.

White, R. E., Simultaneous reactions on a rotating disk electrode, Dissertation, University of California, Berkeley (Mar. 1977).

White, R. and J. Newman, Simultaneous reactions on a rotating-disk electrode, J. Electroanal. Chem., *82*, 173 (1977).

Wierschem, G. L. and W. H. Tiedemann, Cycling behavior of lead-acid batteries for electric vehicles, Abstr. 102, Hollywood, Fl., meeting of the Electrochemical Society (Oct. 1980).

15
Reactor Steady-State Multiplicity and Stability

MASSIMO MORBIDELLI *Politecnico di Milano, Milan, Italy*

ARVIND VARMA *University of Notre Dame, Notre Dame, Indiana*

RUTHERFORD ARIS *University of Minnesota, Minneapolis, Minnesota*

INTRODUCTION

Although it is titled broadly, the scope of this chapter is more limited. We attempt to provide a review of concepts and results related to steady-state multiplicity and dynamic behavior in two types of reactors: the continuous-flow stirred-tank reactor (CSTR) and the empty tubular reactor (note that in qualifying it as "empty," we mean that there is no special reference made to packing found in a packed-bed reactor; see Chapter 6 for the design of such reactors). In thus limiting the scope, for reasons of brevity, we therefore make no direct mention of similar problems that arise and that have been attacked for other types of reactors and for single-catalyst pellets. These results are discussed in other chapters.

Despite the limitation in scope, the two reactor types that are covered are the most widely treated in the literature and offer a wide spectrum of results. Many of these results also carry over to other reactors. The CSTR and the empty tubular reactor thus serve as paradigms for understanding steady-state multiplicity and dynamic behavior of reactors in general.

In its simplest form, the CSTR is perfectly well-mixed; thus there is no spatial distribution of reagent concentrations or of temperature. In the simplest model for the tubular reactor, a plug of fluid flows through the tube and reaction occurs along the length of the tube; there is no fluid mixing in the axial direction while there is perfect mixing in the radial direction—the so-called plug-flow model. Since the reaction does occur along the length of the tube, there is a distribution of reagent concentrations—and also of temperature for nonisothermal cases—along the axial direction. Fluid mixing can be taken into account by including axial dispersion superimposed on the plug flow, and as noted earlier (Varma and Aris, 1977), the axial dispersion model in the limits of small and large dispersion does indeed yield plug-flow and CSTR behavior, respectively.

This chapter may be considered an update of our previous review (Varma and Aris, 1977). In the interim period, numerous additional results have appeared in the literature, and this chapter provides us with an opportunity to collect them in one place.

STEADY-STATE MULTIPLICITY

Continuous-Stirred-Tank Reactor

The mass and energy balances for a CSTR under steady-state conditions can be written as follows (see Varma and Aris, 1977):

$$q(C_0 - C) - Vf(C,T) = 0 \tag{1}$$

$$q\rho C_p(T_0 - T) + (-\Delta H)Vf(C,T) - Ua(T - T_c) = 0 \tag{2}$$

where all the symbols are explained in the notation section. In deriving Eqs. (1) and (2), it has been assumed for simplicity that:

1. The flow rate q and the reactor volume V are constant.
2. The fluid thermal and physical parameters, ρ, C_p, and ΔH are independent of temperature and composition.
3. The temperature of the cooling medium, T_c, is constant.

Equations (1) and (2) can be made dimensionless by introducing the following quantities:

$$u = \frac{C}{C_0} \qquad v = \frac{T}{T_m} \qquad \gamma = \frac{E}{RT_m}$$

$$Da = \frac{Vf(C_0,T_m)}{qC_0} \qquad \delta = \frac{Ua}{q\rho C_p} \tag{3}$$

$$\bar{\beta} = \frac{(-\Delta H)C_0}{\rho C_p T_m(1 + \delta)} \qquad r = \frac{f(C,T)}{f(C_0,T_m)}$$

The reference temperature T_m is given by the weighted-average value between the inlet and the cooling temperatures:

$$T_m = \frac{T_0 + \delta T_c}{1 + \delta} \tag{4}$$

As shown by Kauschus et al. (1978), this choice allows the simultaneous treatment of the adiabatic and the nonadiabatic case and reduces by one the number of dimensionless parameters.

Substituting Eqs. (3) and (4), Eqs. (1) and (2) reduce to

$$1 - u - Dar(u,v) = 0 \tag{5}$$

$$1 - v + \bar{\beta}Dar(u,v) = 0 \tag{6}$$

These can be suitably combined to give the dimensionless concentration

$$u = \frac{1 + \bar{\beta} - v}{\bar{\beta}} \tag{6}$$

which substituted in Eq. (6) leads to

$$1 - v + \bar{\beta}\,\mathrm{Dar}\left(\frac{1 + \bar{\beta} - v}{v}, v\right) = 0 \tag{8}$$

The problem has then been reduced to the solution of the nonlinear algebraic equation (8) in the only unknown v.

A priori bounds of v, in the case of an exothermic reaction ($\bar{\beta} > 0$), can readily be derived from Eq. (7), noting that from physical arguments that $0 \leqslant u \leqslant 1$ and

$$1 \leqslant v \leqslant (1 + \bar{\beta}) \tag{9}$$

while for an endothermic reaction ($\bar{\beta} < 0$) the same inequality holds with the signs reversed.

In the following sections, Eq. (8) is examined for various expressions of the dimensionless reaction rate r, to define, a priori, the number of all possible steady states for any given set of parameter values.

Positive nth-Order Kinetics

Using the reaction rate expression $f = k(T)C^n$ (in dimensionless form, $r = u^n \exp[\gamma(1 - 1/v)]$), Eq. (8) can be rewritten as follows:

$$\frac{\bar{\beta}^{n-1}}{\mathrm{Da}} = \frac{(1 + \bar{\beta} - v)^n}{v - 1} \exp\left[\gamma\left(1 - \frac{1}{v}\right)\right] \equiv F(v) \tag{10}$$

Since in the case of endothermic reactions Eq. (10) admits a unique solution for all Da, in the following we shall consider only exothermic reactions (i.e., $\bar{\beta} > 0$).

From inspection of Eq. (10), $F(1) = +\infty$, $F(1 + \bar{\beta}) = 0$, and $F(v) > 0$ for $v \in [1, 1 + \bar{\beta}]$, so that uniqueness is guaranteed for all Da if and only if (Luss, 1971)

$$\frac{dF}{dv} \leqslant 0 \qquad \text{for all } v \in [1, 1 + \bar{\beta}] \tag{11}$$

If condition (11) is violated, multiple steady states will occur for some value of Da. By differentiating F(v) it can be seen that Eq. (11) leads to

$$(n - 1)v^3 + (\bar{\beta} + \gamma + 1 - n)v^2 - \gamma(2 + \bar{\beta})v + \gamma(1 + \bar{\beta}) \geqslant 0 \tag{12}$$

Based on this relation, Van den Bosch and Luss (1977) derived strong sufficient criteria for both uniqueness and multiplicity, and subsequently, Tsotsis and Schmitz (1979) defined the corresponding exact necessary and sufficient conditions. These can be obtained in a more compact form by rewriting Eq. (12), according to Leib and Luss (1981), as follows:

$$\gamma \leqslant \frac{(1 + \bar{\beta}Z)^2[1 + Z(n - 1)]}{\bar{\beta}Z(1 - Z)} \equiv G(Z; n, \bar{\beta}) \qquad \text{for all } Z \in (0,1) \tag{13}$$

where the quantity $Z = (v - 1)/\bar{\beta}$ has been introduced. From inspection of Eq. (13) it appears that $G(0) = G(1) = +\infty$, $G(Z) > 0$ for all $Z \in (0,1)$, and $G' = 0$ for values of Z satisfying

$$-\bar{\beta}(n - 1)Z^3 + Z^2(n - 1)(2\bar{\beta} + 1) + (2 + \bar{\beta})Z - 1 = 0 \tag{14}$$

Since Eq. (14) exhibits one and only one root $\bar{Z}$ in the interval $Z \in (0,1)$ for any value of $\bar{\beta}$ and $n > 0$, it can be concluded that when

$$\gamma \leqslant \bar{G} \equiv G(\bar{Z}) \tag{15}$$

there is uniqueness for *all* values of Da. On the other hand, if condition (15) is violated, the system admits multiple solutions in some range of Da. The boundaries of such range are called lower and upper *bifurcation points*: Da_* and Da^*. They can be calculated by substituting in Eq. (10) the two roots Z_- and Z_+ (i.e., v_- and v_+) of Eq. (13) [i.e., $\gamma = G(Z)$], thus obtaining the following relationships:

$$Da_* = \frac{\bar{\beta}^{\,n-1}}{F(v_+)} \qquad Da^* = \frac{\bar{\beta}^{\,n-1}}{F(v_-)} \tag{16}$$

Thus for $\gamma > \bar{G}$ the system admits multiplicity for $Da_* < Da < Da^*$, and uniqueness for $Da < Da^*$ or $Da > Da^*$. The functions F(v) and G(Z) are sketched in Fig. 1, where all the above-mentioned features are indicated. In Fig. 2 the bifurcation curves in the $\bar{\beta} - \gamma$ plane are shown for kinetics of various positive orders.

First-Order Kinetics: In the classical case of first-order reaction, Eq. (14) can be solved analytically, leading to $\bar{Z} = 1/(2 + \bar{\beta})$. Then, using Eqs. (13) and (15), the well-known necessary and sufficient condition for uniqueness can be derived:

$$\gamma\bar{\beta} \leqslant 4(1 + \bar{\beta}) \tag{17}$$

If condition (17) is violated, then multiplicity occurs for $Da \in [Da_*, Da^*]$, where the bifurcation points are given by Eq. (16) with

$$v_{\pm} = \frac{\gamma(2 + \bar{\beta}) \pm \{\gamma\bar{\beta}[\gamma\bar{\beta} - 4(1 + \bar{\beta})]\}^{1/2}}{2(\gamma + \bar{\beta})} \tag{18}$$

Kauschus et al. (1978) have obtained equivalent conditions in terms of only two parameters: $a = (1 + \bar{\beta})/\gamma\bar{\beta}$ and $b = \gamma\bar{\beta}\,Da$. This allows us to characterize the entire steady-state behavior of the system with a single plot in the a-b plane, as shown in Fig. 3. Parameter values inside the region enclosed by the two curves representing the bifurcation points lead to three steady states, while values outside this region can lead to uniqueness. The point where the two curves merge and vanish is called the *cusp point* ($a = 1/4$; $b = 4e^{-2}$).

It is worth stressing that the shape of the multiplicity pattern, such as that depicted in Fig. 1a, is strongly dependent on the definition of the

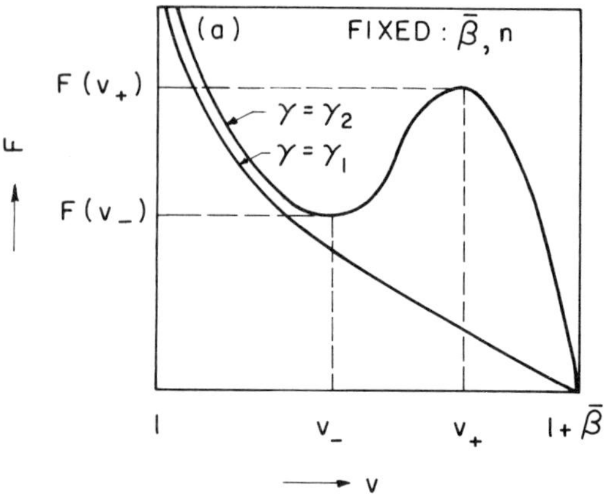

(a)

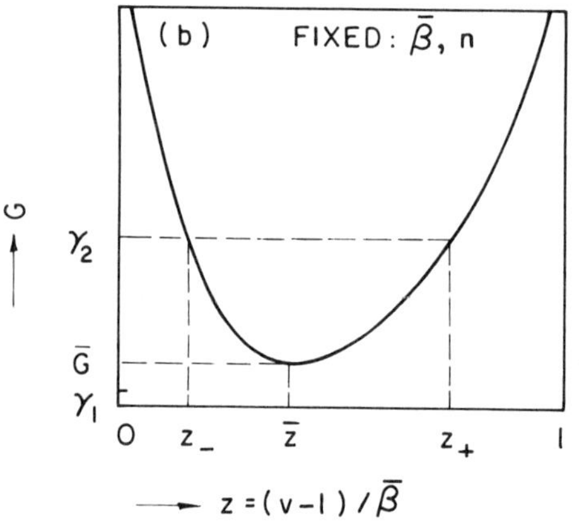

(b)

FIGURE 1 Schematic representation of functions (a) F(v) and (b) G(z). $\gamma = \gamma_1$, uniqueness; $\gamma = \gamma_2$, three or one steady state.

dimensionless parameters used. For practical applications it is convenient to study the multiplicity pattern with respect to the reactor residence time, which can easily be varied in a continuous fashion by varying the reactor inlet flow rate. After the first works of Furusawa and Nishimura (1968) and Hlaváček et al. (1970), Uppal et al. (1976) reported a detailed study of the multiplicity pattern in terms of the reactor residence time.

Golubitsky and Keyfitz (1980) showed through singularity theory that the multiplicity pattern, in terms of conversion/residence-time plots, may assume one of the six different shapes shown in Fig. 4, in addition to the obvious uniqueness case. Knowledge of these multiplicity patterns is very important when discussing the problem of attainability of a given steady state. For example, stable steady states on an isola (see Fig. 4c) that lies entirely along with a stable continuous branch cannot be reached by changing only the reactor inlet flow rate (see Huang and Varma, 1980).

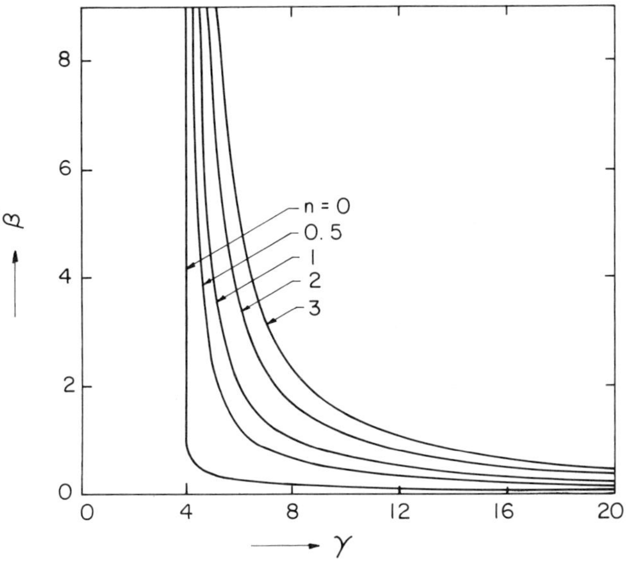

FIGURE 2 Bifurcation curves for reactions of various positive order: Below the curve, uniqueness; above the curve, three or one steady state.

Balakotaiah and Luss (1981) developed an a priori procedure for the definition of the multiplicity pattern. Although it requires a different definition of parameters δ and $\bar{\beta}$ such that the reactor flow rate appears only in Da, the spirit of this technique can be illustrated briefly using Eqs. (10) and (11) with n = 1. As mentioned above, these equations can

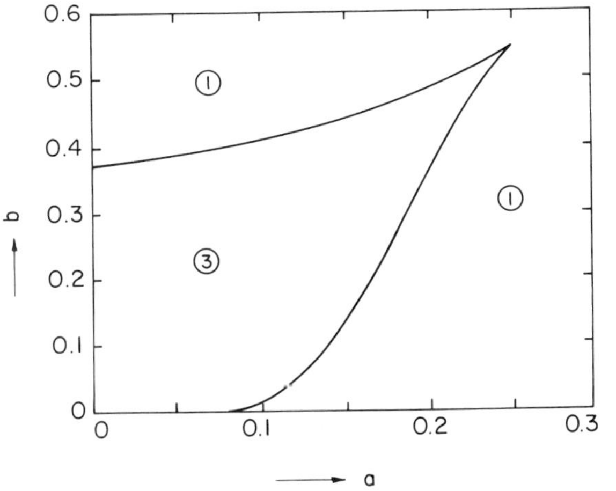

FIGURE 3 Multiplicity behavior for a first-order reaction: $a = (1 + \bar{\beta})/\gamma\bar{\beta}$, $b = \gamma\bar{\beta}\,Da$. Inside the curves, three steady states; outside, one steady state.

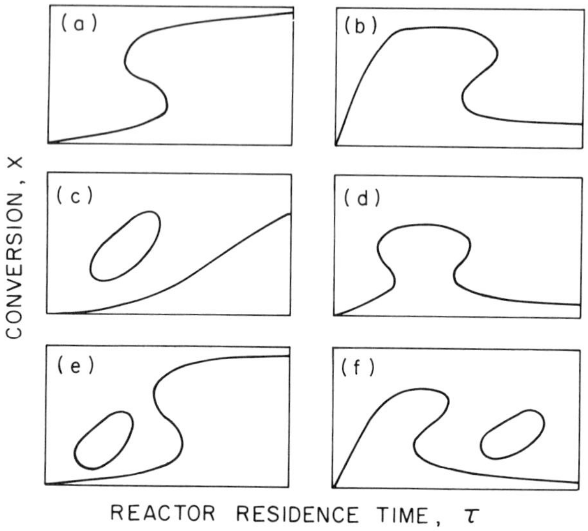

FIGURE 4 Schematic representation of the multiplicity patterns in terms of outlet conversion versus residence time for a first-order reaction.

be solved simultaneously to give the bifurcation values Da_* and Da^*. Regarding the function $F = F(v, Da, \underline{p})$, where $\underline{p}$ represents the vector of all parameters but Da, it can be stated on the basis of the implicit function theorem that the transition from zero to two or four bifurcation points can occur only when

$$\begin{vmatrix} \dfrac{\partial F}{\partial v} & \dfrac{\partial F}{\partial \, Da} \\ \dfrac{\partial^2 F}{\partial v^2} & \dfrac{\partial^2 F}{\partial v \, \partial Da} \end{vmatrix} = 0 \tag{19}$$

which, using Eqs. (10) and (11), can be replaced by either one of the two conditions

$$\frac{\partial F^2}{\partial v^2} = 0 \qquad \text{or} \qquad \frac{\partial F}{\partial \, Da} = 0 \tag{20}$$

Each of these conditions identifies a surface in the parameter space, named "hysteresis variety" and "isola variety," respectively. According to Golubitsky and Keyfitz (1980), crossing the first surface through a continuous change of one of the parameters causes the appearance or disappearance of an S (Fig. 4a) or inverse S (Fig. 4b) pattern; similar behavior, but relative to the isola pattern (Fig. 4c), is observed for the second surface. These two surfaces are represented in the α–B plane shown in Fig. 5 for the particular case $\gamma = \infty$ and $T_c = T_0 = T_m$. They divide the plane in five regions, each corresponding to a specific multiplicity pattern.

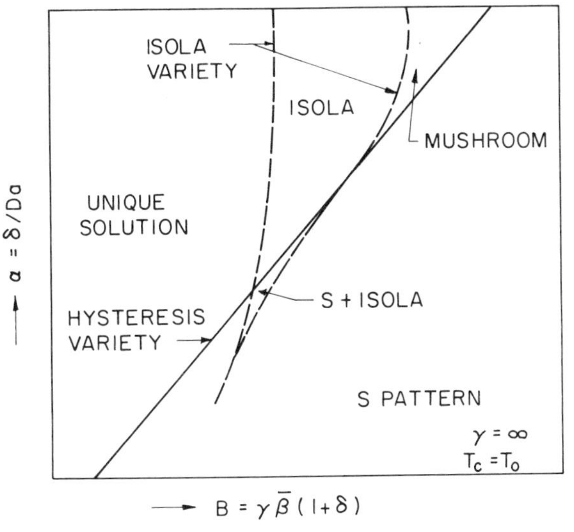

FIGURE 5 Schematic of the regions in the parameter plane $\alpha = \delta/\mathrm{Da}$ versus B which lead to the different multiplicity patterns of Fig. 4.

For example, the location of the mushroom pattern region is quite obvious since by crossing the isola variety it must degenerate into an isola, while crossing the hysteresis variety it must degenerate into a S-shape pattern. Note that the inverse S-shape pattern never occurs in the case illustrated in Fig. 5, since it requires very strong cooling (i.e., $T_c < T_0$).

Zeroth-Order Kinetics: This particular case differs from the general one in that G(1) is positive but finite, which reflects the well-known peculiarity of these reactions that they reach complete conversion with a positive reaction rate. Since Eq. (14) admits the solution $\bar{Z} = 1/\bar{\beta}$, it can be seen from Eq. (13) that the uniqueness condition for all Da is given by

$$\gamma < 4 \qquad \text{for } \bar{\beta} \geq 1 \tag{21a}$$

$$\gamma < \frac{(1 + \bar{\beta})^2}{\bar{\beta}} \qquad \text{for } \bar{\beta} < 1 \tag{21b}$$

Nonmonotone Kinetics

An unusual feature of some reactions in catalysis, as well as in other fields, such as biocatalysis, is to exhibit kinetics which are not monotonically increasing with the reactant concentration. This situation occurs when the reactant can somehow inhibit the rate of reaction. For example, in the case of oxidation of CO or various hydrocarbons (see Voltz et al., 1973) or hydrogenation of olefins (see Sinfelt, 1964; Rogers et al., 1966) over noble metals, the reactant can be strongly adsorbed on the catalyst active sites, thus inhibiting the development of the reaction. Similar behavior is encountered in biocatalysis for certain substrate-inhibited enzymatic reactions (see Laidler and Bunting, 1973).

Of particular interest is the case of oxidation (hydrogenation) reactions, where oxygen (hydrogen) is present in large excess. This corresponds to the well-known Langmuir-Hinshelwood kinetics $f = kC/(1 + KC)^2$, where k is the reaction rate constant and K the reactant adsorption constant. In dimensionless form

$$r = \exp\left[\gamma\left(1 - \frac{1}{v}\right)\right] \frac{(1 + \sigma)^2 u}{(1 + \sigma u)^2} \tag{22}$$

Tsotsis et al. (1982) have examined in detail the following general form of nonmonotone kinetics:

$$r = \exp\left[\gamma\left(1 - \frac{1}{v}\right)\right] \frac{(1 + \sigma)^q u^p}{(1 + \sigma u)^q} \tag{23}$$

which reduces to Eq. (22) for p = 1 and q = 2.

Under similar conditions, the intrinsic high nonlinearity of nonmonotone kinetics leads in general to more complex multiplicity patterns than those characteristic of monotone kinetics. In fact, in the latter case, the only source of nonlinearity is in the Arrhenius temperature dependence of the reaction rate constant, while in the former multiplicity can occur even under isothermal conditions. Thus for this type of reaction we shall examine separately the isothermal and the nonisothermal cases.

Isothermal Reactor: In this case v = 1, and only the mass balance Eq. (5) needs to be considered. Introducing the kinetic expression (23), Eq. (5) can be rewritten in the form

$$\frac{1}{Da} = \frac{(1 + \sigma)^q}{(1 - u)} \quad \frac{u^p}{(1 + \sigma u)^q} \equiv F(u) \tag{24}$$

Since $F(0) = 0$ and $F(1) = \infty$, it is seen that the necessary and sufficient condition for uniqueness is (Luss, 1971)

$$\frac{dF}{du} \geqslant 0 \qquad \text{for all } u \in [0,1] \tag{25}$$

Tsotsis et al. (1982) found that condition (25) is satisfied if and only if

$$\sigma \leqslant \frac{p(q - p + 1) + q + 2[pq(q - p + 1)]^{1/2}}{(p - q)^2} \tag{26}$$

which in the case of a bimolecular Langmuir-Hinshelwood reaction (p = 1, q = 2) reduces to

$$\sigma \leqslant 8 \tag{27}$$

As expected, it appears that multiplicity under isothermal conditions occurs only for sufficiently strong inhibition. If Eq. (27) is violated, then in the interval

$$Da_* \equiv \frac{1}{F(u_-)} < Da < Da^* \equiv \frac{1}{F(u_+)} \tag{28}$$

where u_+ and u_- are the roots of $dF/du = 0$, the reactor exhibits three steady states. Outside this interval the steady state is unique. Thus the most complex pattern of multiplicity is again of the type 1-3-1.

Nonisothermal Reactor: Similar to the nth-order kinetics case, the necessary and sufficient condition for uniqueness is given by Eq. (11), with F(v) defined as follows:

$$\frac{\bar{\beta}^{p-1}}{Da} = \frac{(1 + \sigma)^q}{v - 1} \frac{(1 + \bar{\beta} - v)^p}{[1 + \sigma(1 + \bar{\beta} - v)]^q} \exp\left[\gamma\left(1 - \frac{1}{v}\right)\right] \equiv F(v) \tag{29}$$

The critical condition $dF/dv = 0$ leads in this case to a quartic equation in v, which implies that F(v) can have none, one, or two pairs of extreme points (one maximum and one minimum). These three possible situations are illustrated in Fig. 6. It is apparent that by continuously varying the Damkhöler number Da, the following multiplicity patterns can be obtained: 1-3-1 (b), 1-3-1-3-1 (c1), or 1-3-5-3-1 (c2). As expected, this case exhibits the most complex multiplicity pattern among those examined up to now. This is due to superimposition of the temperature nonlinearity with the nonmonotonic concentration dependence of the reaction rate. In fact, when the latter tends to vanish, as for $q \leqslant p - 1$, the system again exhibits at most the multiplicity pattern 1-3-1.

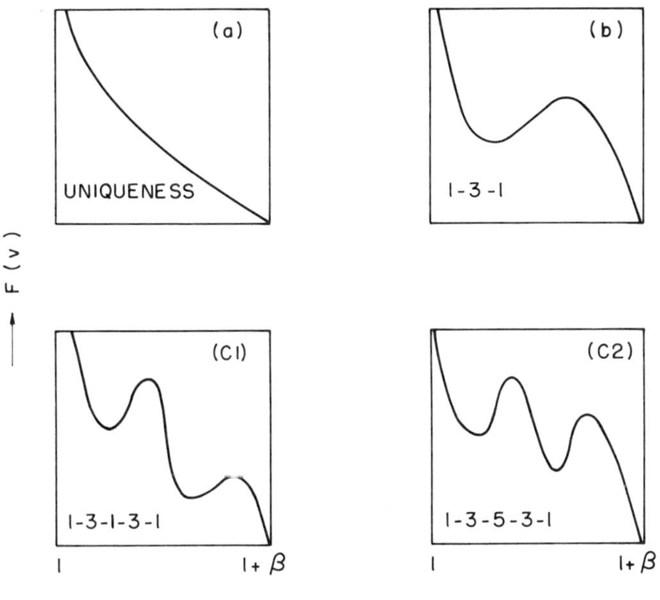

FIGURE 6 Sketches of function F(v) corresponding to any possible multiplicity pattern for a nonisothermal CSTR with nonmonotone kinetics [Eq. (23)].

Tsotsis et al. (1982) have examined in detail this quartic equation using Sturm's theorem to obtain a priori criteria for the number of roots in the interval $v \in [1, 1 + \overline{\beta}]$. Their procedure gives the exact number of steady states of the reactor, but it is too complex to be reported here in detail. Some simple sufficient conditions for uniqueness by these authors are:

For an overall negative order (i.e., $q > p$) and endothermic reaction, the condition is given by Eq. (26).

For an overall positive order ($p > q$), the condition is given by $\gamma \leqslant 4$ or $\overline{\beta} < 0$ (endothermic reaction).

For the specific case of $p = 1$ and $q = 2$, Brown et al. (1984) used singularity theory to locate two butterfly singularities, one in the exothermic case and the other in the endothermic case. Around these singular points, a maximum of five steady states exist. In Fig. 7a, uniqueness and multiplicity regions in the $\overline{\beta}-\gamma$ plane are shown for various values of the inhibition parameter σ. There is multiplicity, for some Da, to the right of the curves lying on the $\overline{\beta}$-positive half plane and to the left of those lying on the $\overline{\beta}$-negative half plane. For the particular case $\sigma = 7$, all the possible multiplicity patterns are indicated in Fig. 7b.

More Complex Configurations

For all systems examined above, a complete characterization of the multiplicity pattern has been obtained. Following an increasing complexity order, the next systems to be examined would include either several reactors in series with a single reaction (so as to approach the behavior of tubular

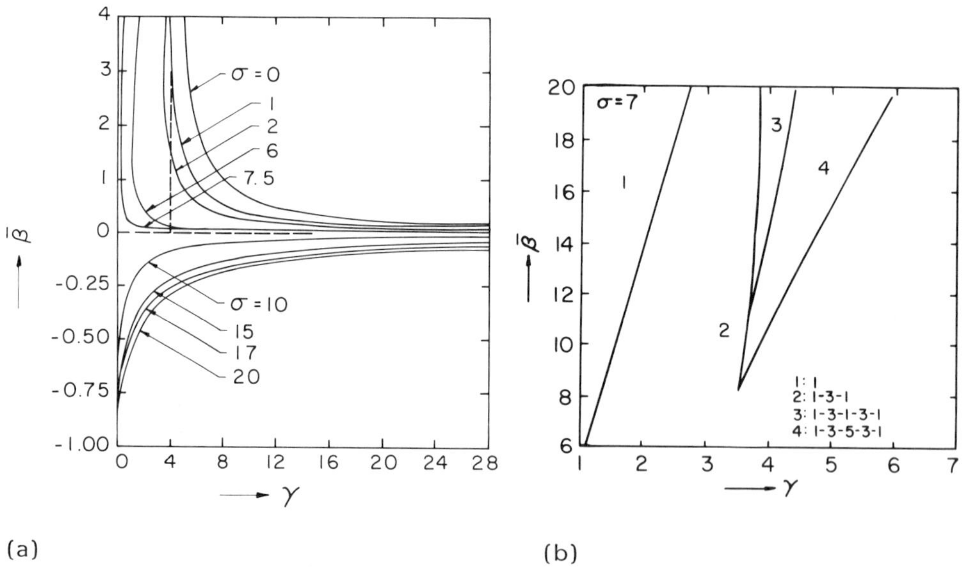

(a) (b)

FIGURE 7 (a) Uniqueness and multiplicity regions in the β versus γ plane for various values of the inhibition parameter σ. (b) Multiplicity patterns in the case $\sigma = 7$.

reactors) or a single reactor with a more complex kinetic scheme—say, a series and/or parallel sequence of first-order reactions. In both cases the number of equations of the model and the number of characteristic dimensionless parameters of the system increase rapidly. This leads to a rapid increase in the complexity of the multiplicity study. In particular, we now have to deal with a multidimensional parameter space, which prevents an exhaustive representation of the system behavior in terms of phase plane. The work in this area is then oriented toward the analysis of some sections of the parameter space, which are usually selected either on the basis of some specific practical interest or to identify all the possible qualitative behavior of the system.

In the following, two cases of this type that have received particular attention in the literature are examined. The aim is to give the reader a feeling for the rapidly increasing complexity and to indicate some of the techniques used to attack these problems.

Series of Continuous Stirred-Tank Reactors: The interest in examining a series of CSTR relies on its well-known use for modeling tubular reactors. The steady-state mass and energy balance for the generic ith reactor of the series can be written as follows:

$$u_{i-1} - u_i - Da_i r_i = 0 \qquad 1 = 1, 2, \ldots, N \qquad (30)$$

$$v_{i-1} - v_i + \beta Da_i r_i - \delta_i (v_i - v_{ci}) = 0 \qquad i = 1, 2, \ldots, N \qquad (31)$$

where the same parameters defined by Eq. (3) have been used, but taking as reference temperature T_0 instead of T_m, so that $u_0 = 1$, $v_0 = 1$. Moreover,

$$\beta = \bar{\beta}_i (1 + \delta_i) \quad \text{and} \quad v_{ci} = \frac{T_{ci}}{T_0} \qquad (32)$$

In the case of N nonisothermal CSTRs in series with a first-order reaction, Varma (1980) has shown that at most $(2^{N+1} - 1)$ steady states are possible, of which however at most $(N + 1)$ are stable. It is worth noting in passing that this same conclusion applies also to this same system configuration in the case of a nonmonotone isothermal reaction of the type in Eq. (22). This result can be intuitively understood, noting that each reactor in the cascade can exhibit three steady states in correspondence to each steady state exhibited by the previous one. To the number thus obtained, we should substract all those steady states that imply physically unrealistic extinction phenomena along the sequence of reactors (i.e., the transition from a high-conversion steady state to a low-conversion steady state).

A detailed study of the behavior of this system for the particular case of only two reactors in series has been performed independently by Kubicek et al. (1980) and Svoronos et al. (1982). Even for this simple case the total number of parameters is quite elevated (Da_i, δ_i, β, γ, v_{ci} = eight independent parameters), so that most of the analysis has been dedicated to the case of two identical CSTRs (i.e., $Da = Da_i$ and $\delta_i = \delta$) in the limit of tho Frank-Kamenetskii approximation of the Arrehnius temperature dependence of the rate constant, which allows us to eliminate the dimensionless activation energy parameter γ. To simplify the treatment, in the

following we treat only the case $v_{ci} = 1$ (i.e., cooling temperature equal to T_0 for both reactors). After all these approximations, Eqs. (30) and (31) can be solved, leading to

$$Da = \Delta[u_1;\phi] \equiv \frac{1 - u_1}{u_1} \exp\left[-\frac{4}{\phi}(1 - u_1)\right] \tag{33a}$$

$$Da = \Gamma[u_1,u_2;\phi,\delta] \equiv \frac{u_1 - u_2}{u_2} \exp\left\{\frac{-4}{\phi/u_1}\left[\frac{u_1 - u_2}{u_1} + \frac{1 - u_1}{u_1(1 + \delta)}\right]\right\} \tag{33b}$$

where

$$\phi = \frac{4(1 + \delta)}{\gamma\beta} \tag{34}$$

A particularly important feature of this system is that the behavior of the second reactor is dictated by the same equation as the first one, but with proper scaling of the unknown and the involved parameters. This is apparent, noting from Eqs. (33a) and (33b) that

$$\Gamma[u_1,u_2;\phi,\delta] = \Delta\left(\frac{u_2}{u_1}, \frac{\phi}{u_1}\right) \exp\left(-\frac{4}{\delta}\,\frac{1 - u_1}{1 + \delta}\right) \tag{35}$$

Since the reactor equations can be solved in cascade, this allows us immediately to characterize the behavior of the second reactor once that of the first is known (i.e., the function Δ has been investigated fully). Namely Γ exhibits the same behavior as Δ, but scaled in the interval $0 \leqslant u_2 \leqslant u_1$ and with increased value of the parameter ϕ/u_1.

The multiplicity pattern of the first reactor is obtained as in the preceding section:

For $\phi > 1$: uniqueness for all Da

$$\text{For } \phi < 1: \begin{cases} Da_* < Da < Da^* & \text{multiplicity} \\ Da < Da_* \text{ or } Da > Da^* & \text{uniqueness} \end{cases} \tag{36}$$

where $Da_* = \Delta(m_+)$, $Da^* = \Delta(m_-)$, and

$$m_{\pm} = \frac{1 \pm (1 - \phi)^{1/2}}{2} \tag{36a}$$

Note that Eqs. (36) and (36a) correspond to Eqs. (17) and (18) in the limit of very large activation energy.

Similarly, for the second reactor it can be shown that

For $\phi/u_1 > 1$: uniqueness for all Da

$$\text{For } \phi/u_1 < 1: \begin{cases} Da_{**} < Da < Da^{**} & \text{multiplicity} \\ Da < Da_{**} \text{ or } Da > Da^{**} & \text{uniqueness} \end{cases} \tag{37}$$

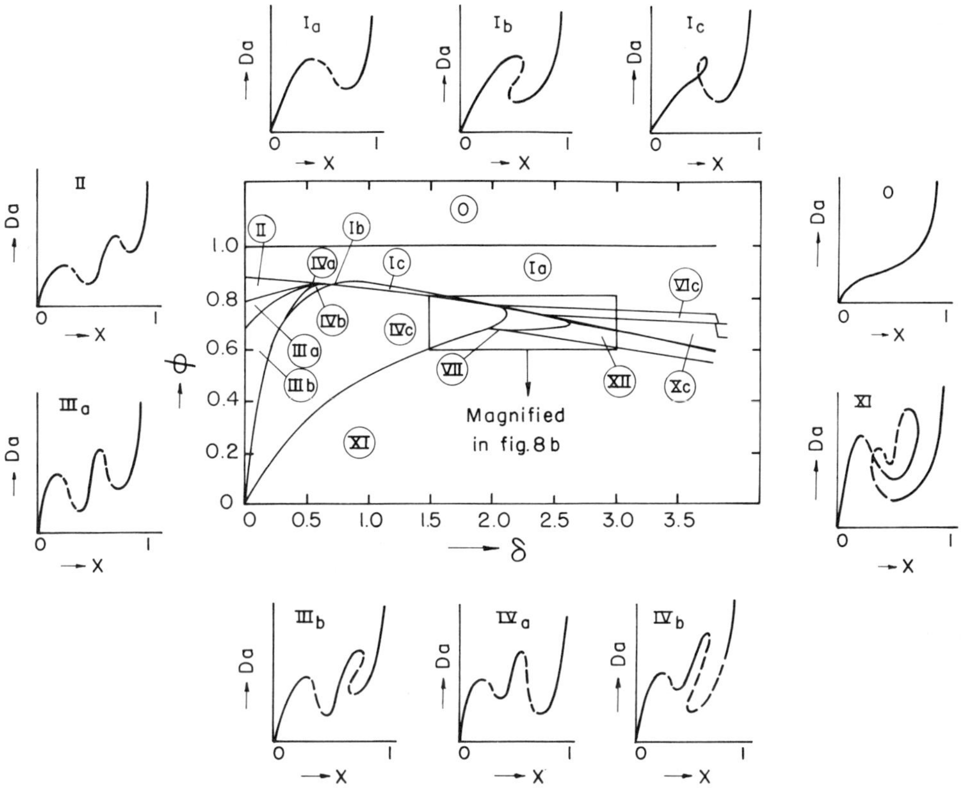

(a)

FIGURE 8 Regions in the $\phi = 4(1 + \delta)/\gamma\beta$ versus β parameter plane, surrounded by the sketches of the corresponding multiplicity pattern in the Da versus outlet conversion plane. Solid line: stable steady state; dashed line: unstable steady state (parameter values: $v_{c1} = v_{c2} = 1$, $Da_1 = Da_2 = Da$, $\delta_1 = \delta_2 = \delta$).

where $Da_{**} = \Gamma(u_1, n_+)$, $Da^{**} = \Gamma(u_1, n_-)$, and

$$n_{\pm} = \frac{2 - u_1 \pm u_1(1 - \phi/u_1)^{1/2}}{2} \tag{37a}$$

The impact of the above-mentioned scaling property of the equations examined is evident by noting that Eqs. (37) and (37a) reduce to Eqs. (36) and (36a), respectively, as $u_1 = u_0 = 1$.

Since space prevents us from analyzing this problem in further detail, we shall limit our discussion to the final answers for the specific case under examination. In Fig. 8 the multiplicity pattern is reported through a proper subdivision of the ϕ–δ plane into 23 regions. For each of these, a sketch of the corresponding Da versus $(1 - u_2)$ curve is shown on the side of the figure. It appears that the maximum number of steady states is seven

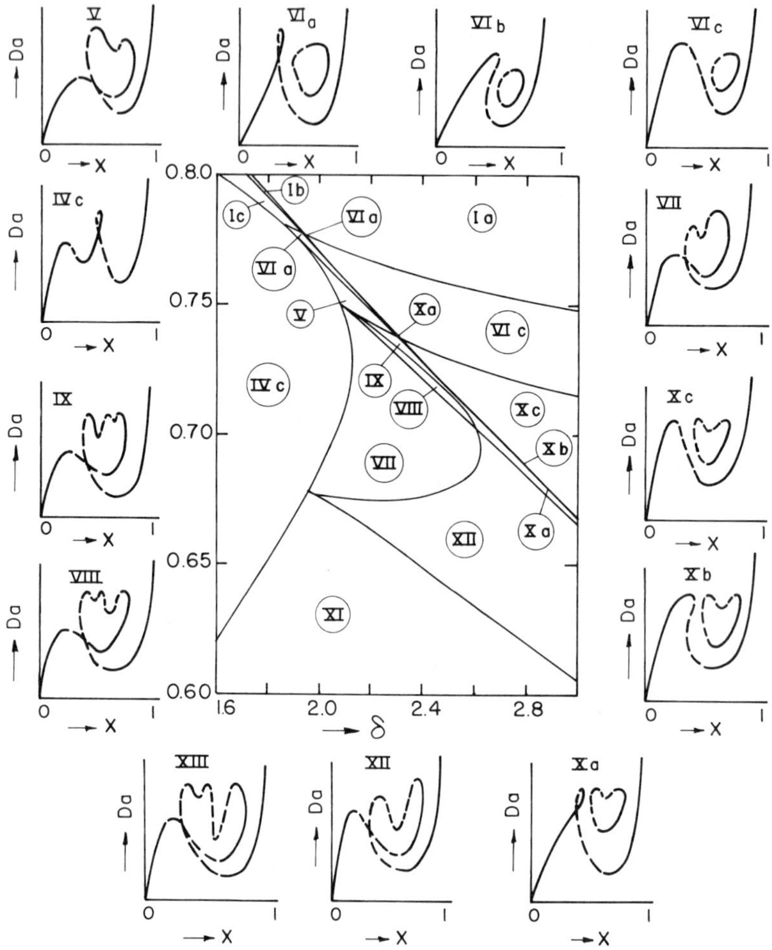

(b)

FIGURE 8 (Continued)

(of which no more than three can be stable), and the total number of possible multiplicity patterns is much larger than that in the case of a single reactor.

A more detailed treatment of this problem, including the case $v_{c1} \neq v_{c2} \neq 1$, has been reported by Svoronos et al. (1982). Other interesting features, such as the effect of two different Da values for the reactors or the influence of the activation energy, have been reported by Kubicek et al. (1980). Moreover, in the latter work the effect of recirculation from the second to the first reactor is also examined. It is found that for increasing recirculation ratios, the multiplicity region tends to disappear.

Complex Kinetic Schemes: When more than one reaction is considered, the variety of possible configurations is enormously—in fact, infinitely—enlarged. The only way in which to present some of the modes of attacking

the problem of multiplicity in these cases is to take a simple example, and the obvious extensions of the single reaction are the parallel reactions, $A \to B$, $A \to C$, and the sequential $A \to B \to C$; we shall concentrate on the latter. Since there are three reaction parameters (the Damköhler number, the heat of reaction or Prater number, and the Arrhenius number) for a single reaction, we should expect six parameters for $A \to B \to C$. In addition, there is in both cases the cooling rate parameter and the dimensionless coolant temperature. We shall see that the latter may often be ignored, and that in steady-state problems the cooling rate parameter can be combined with the heat of reaction, but this leaves six for the steady state and seven for stability questions. In the face of so many parameters we can hardly hope to explore every nook and cranny of parameter space, but some progress may be made through the use of singularity theory. The work of Balakotaiah and Luss (1981–1984) has been outstanding in this area, and one cannot get a better account than is given in their original papers.

If C_A, C_B, and T denote the two concentrations and the temperature respectively, the steady-state mass and heat balances give the three equations:

$$q(C_{A0} - C_A) - Vk_1(T)C_A = 0 \tag{38a}$$

$$q(C_{B0} - C_B) + Vk_1(T)C_A - Vk_2(T)C_B = 0 \tag{38b}$$

$$q\rho C_p(T_0 - T) - UA(T - T_c) + (-\Delta H_1)Vk_1(T)C_A + (-\Delta H_2)Vk_2(T)C_B = 0 \tag{38c}$$

where all the symbols are as defined in the notation section. It is customary to take $C_{B0} = 0$ and use the dimensionless quantities

$$u_1 = \frac{C_A}{C_{A0}} \qquad u_2 = \frac{C_B}{C_{A0}} \qquad \Theta = \frac{\gamma(T - T_0)}{T_0}$$

$$\gamma = \frac{E_1}{RT_0} \qquad Da = \frac{Vk_1(T_0)}{q} \qquad B = \frac{(-\Delta H_1)C_{A0}\gamma}{\rho C_p T_0} \qquad \lambda = \frac{E_2}{E_1} \tag{39}$$

$$\alpha = \frac{\Delta H_2}{\Delta H_1} \qquad \nu = \frac{k_2(T_0)}{k_1(T_0)} \qquad \delta = \frac{UA}{q\rho C_p}$$

Then

$$1 - u_1 - Dau_1 \exp\left(\frac{\gamma\Theta}{\Theta + \gamma}\right) = 0 \tag{40a}$$

$$-u_2 + Dau_1 \exp\left(\frac{\gamma\Theta}{\Theta + \gamma}\right) - \nu Dau_2 \exp\left(\frac{\lambda\gamma\Theta}{\Theta + \gamma}\right) = 0 \tag{40b}$$

$$-\Theta + BDau_1 \exp\left(\frac{\gamma\Theta}{\Theta + \gamma}\right) + B\alpha\nu Dau_2 \exp\left(\frac{\lambda\gamma\Theta}{\Theta + \gamma}\right) - \gamma(\Theta - \Theta_c) = 0 \tag{40c}$$

Replacing T_0 by $(T_0 + \delta T_c)/(1 + \delta)$, we have the same equations but with $\Theta_c = 0$. This is satisfactory unless either q or U is taken to be the distinguished parameter. If $T_c = T_0$, this difficulty disappears, and this assumption is commonly made.

Since the equations are linear in u_1 and u_2 the steady state is usually found by solving Eq. (40a) for u_1 in terms of Θ, substituting in Eq. (40b) and solving for u_2 to give, on substitution in Eq. (40c), a single equation in Θ. This may be written

$$(1 + \delta)\Theta = BH_1(\Theta)[1 + \alpha H_2(\Theta)] \tag{41}$$

where

$$H_1(\Theta) = \mathrm{Da}\exp\left(\frac{\gamma\Theta}{\gamma + \Theta}\right) \Big/ \left[\left(1 + \mathrm{Da}\exp\frac{\gamma\Theta}{\gamma + \Theta}\right)\right]$$
$$H_2(\Theta) = \nu\,\mathrm{Da}\exp\left(\frac{\lambda\gamma\Theta}{\Theta + \gamma}\right) \Big/ \left[\left(1 + \nu\,\mathrm{Da}\exp\frac{\lambda\gamma\Theta}{\Theta + \gamma}\right)\right] \tag{42}$$

This is the form which Bilous and Amundson (1955) used and it has the simple physical interpretation of a heat balance, $(1 + \delta)\Theta$ being proportional to the rate of cooling, and H_1 and H_1H_2 to the two rates of heat generation. Of course, α and B or both can be negative if one step is endothermic and the other exothermic. Arguing that H_1 and H_2 are curves with a steplike transition between values near zero and values near 1, most workers in this field have assumed that at most five steady states can be obtained. Actually, the subtleties of shape of the right side allow seven, but only in a very restricted region of parameter space (Farr, 1985). This phenomenon is lost when γ is allowed to become large and the exponential approximation is used (Jorgensen et al., 1984). Formally, we let $\gamma \to \infty$ in the two exponentials, simplifying them to e^{Θ} and $e^{\lambda\Theta}$, but this ignores the presence of γ in Θ and B. The true nature of this so-called "positive exponential approximation" is obscure, but it is probably best to think of it as implying that the deviations of temperature from T_0 are not large.

Though Eq. (41) is a physically appealing form of the equation that has to be solved for the steady state, it is in better shape for computation when regarded as a quadratic for Da given the steady-state temperature Θ. Thus the multiplicity question is to be investigated by exploring the nature of the equation

$$F \equiv \mathrm{Da}^2\nu B^*(1 + \alpha)e^{(1+\lambda)\Theta} - \nu\mathrm{Da}\Theta e^{\lambda\Theta}(1 + \mathrm{Da}e^{\Theta})$$
$$+ B^*\mathrm{Da}e^{\Theta} - \Theta(1 + \mathrm{Da}e^{\Theta}) = 0 \tag{43}$$

where $B^* = B/(1 + \delta)$. Certain combinations emerge from tis form and we can take

$$\varepsilon = \mathrm{Da}e^{\Theta} \qquad A = (1 + \alpha)\nu B^*\mathrm{Da}^{1-\lambda} \qquad \Gamma = \nu\mathrm{Da}^{1-\lambda}$$

giving

$$A\,\varepsilon^{1+\lambda} + B^*\varepsilon - \Gamma\Theta\varepsilon^{\lambda}(1 + \varepsilon) - \Theta(1 + \varepsilon) = 0 \tag{44}$$

If $\lambda = 1$ and Θ is specified, this is a quadratic for ε given A, B*, and Γ, and its positive root will give the corresponding Da. The usual form of multiplicity diagram is Θ versus Da, and this may take one of a number of forms shown in the bottom row of diagrams of Fig. 9. In the upper part of that figure all parameters except B* and Da are fixed, and the curve PQRST is the locus of turning points of the Θ-Da curve [i.e., the points where $d(Da)/d\Theta = 0$]. The branches of this curve are numbered for the order in which they are encountered in traversing the Da-Θ curves shown below in the direction of increasing Da. The form of the Da-Θ curve is shown in the lower part of the figure for the appropriate range of B*. For example, over the square C, any vertical with increasing Da encounters the branches of the turning-point curve in the order 2143. There are thus two regions of three possible steady states. For D these overlap and there is an interval of five states with three on either side. The Da-Θ diagrams for the transitions are shown as vixillae attached to the vertical. Thus the B/C transition comes about by a so-called "hysteresis point," where the tangent at a point of inflection is vertical. On the other hand, the

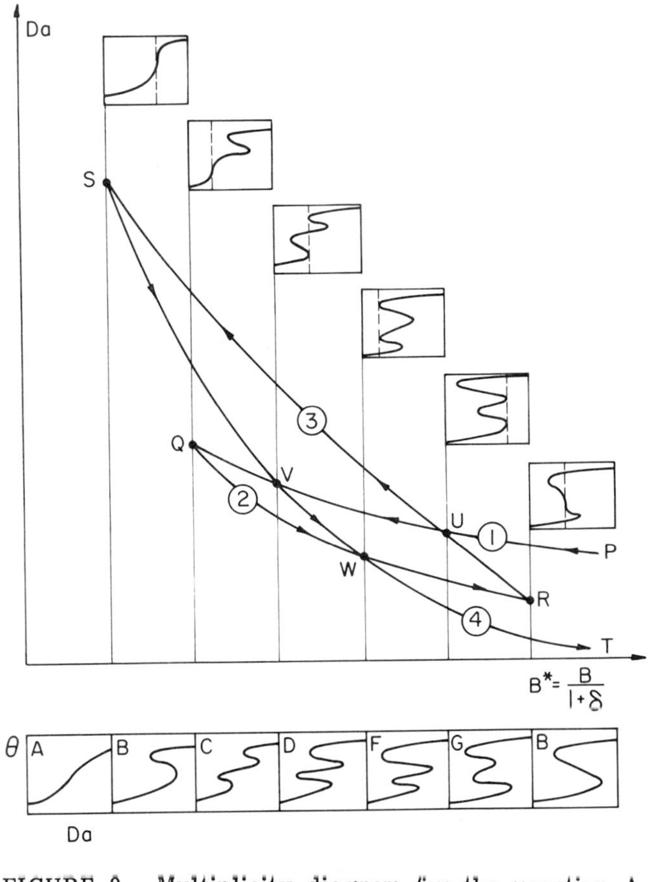

FIGURE 9 Multiplicity diagram for the reaction A → B → C occurring in a nonisothermal CSTR.

C/D transition is made when two turning points are vertically above one another.

A point on the Da-Θ diagram is given by a solution of $F = 0$ according to Eq. (43). A turning point is given by $F = F_\Theta = 0$, and when B* and the other parameters are held constant, this pair of equations has from zero to four solutions for Da and Θ. For example, Fig. 9 shows that for sufficiently small B* there are no roots, while for sufficiently large B* there are just two. The transitions occur at particular values of B*. Those that involve vertical tangents at inflation points Q,R,S are obtained by solving $F = F_\Theta = F_{\Theta\Theta} = 0$ for Θ, Da, and B*. The double limit points U, V, and W are solutions for Θ_1, Θ_2, Da, and B* of the four equations $F(\Theta_1) = F_\Theta(\Theta_1) = F(\Theta_2) = F_\Theta(\Theta_2) = 0$, where $F(\Theta_1) = F(\Theta, Da, B^*; \alpha, \gamma, \lambda, \nu)$ with $\Theta = \Theta_1$. Θ_1 and Θ_2 are the ordinates of the turning points that share a common Da. The surface in parameter space given by eliminating Θ and Da between $F = F_\Theta = F_{\Theta\Theta} = 0$ is called the "hysteresis variety." If Θ_1, Θ_2, and Da are eliminated between $F(\Theta_1) = F_\Theta(\Theta_1) = F(\Theta_2) = F_\Theta(\Theta_2) = 0$ the "double limit variety" is obtained. Isolas arise when either the curve loops back on itself or an entirely new branch is born from an isolated point. In both cases the direction of the curve $F = 0$ is indeterminate, so both F_Θ and F_{Da} vanish simultaneously. This does not happen in this particular case, but if Θ and Da can be eliminated between $F = F_\Theta = F_{Da} = 0$, we have the "isola variety."

Following the precepts of singularity theory we look for the points where as many as possible of the derivatives vanish. Balakotaiah and Luss (1982a,b) showed that there are points at which F and its first four derivatives with respect to Θ vanish. For $\lambda = 1$, $\gamma = \infty$, these are $(\Theta, Da, B^*, \alpha, \nu) = (3, 0.1858, 3.2154, 0.8660, 0.0718)$ and $(3, 0.0133, 44.7846, -0.8660, 13.9282)$. A point where F and its first four derivatives vanish is known as a "butterfly point" and F is there "contact equivalent" (see the Appendix to this chapter) to $x^5 - \Lambda = 0$. This means that x is some transformation of Θ and Λ of the Damköhler number Da, and that a full unfolding is given by three other parameters α_1, α_2, and α_3 in the form

$$x^5 - \alpha_3 x^3 - \alpha_2 x^2 - \alpha_1 x - \lambda = 0 \qquad (45)$$

Singularity theory (subject to appropriate tests) shows that α_1, α_2, and α_3 will be smooth functions of B*, α, and ν, although it would be a work of some magnitude to find them. However, the qualitative picture in the neighborhood of the "organizing center" must be that of the butterfly. Thus, for $\alpha_3 < 0$ there are only two possible Θ-Da diagrams, namely A and B. The former is obtained when the parameters α_1 and α_2 lie beneath the parabolic curve shown in the left half of the top part of Fig. 10. Otherwise, the common phenomenon of three steady states in the intermediate interval of the Damköhler number is obtained. The situation with $\alpha_3 > 0$ is more complicated and is shown on the right at the top of Fig. 10. The various Da-Θ diagrams shown below by the letters A, . . . , G are obtained for $\alpha_3 > 0$ when α_1 and α_2 lie in the correspondingly marked region. Referring back to Fig. 9, where the sequence A, B, C, D, F, G, B is obtained for increasing B*, we see that it would be quite consistent with this if α_1 were a monotone-increasing function of B*, and that α and ν were such that α_2 and α_3 were positive, for then the same sequence would be obtained passing upward along the dashed curve, PQ.

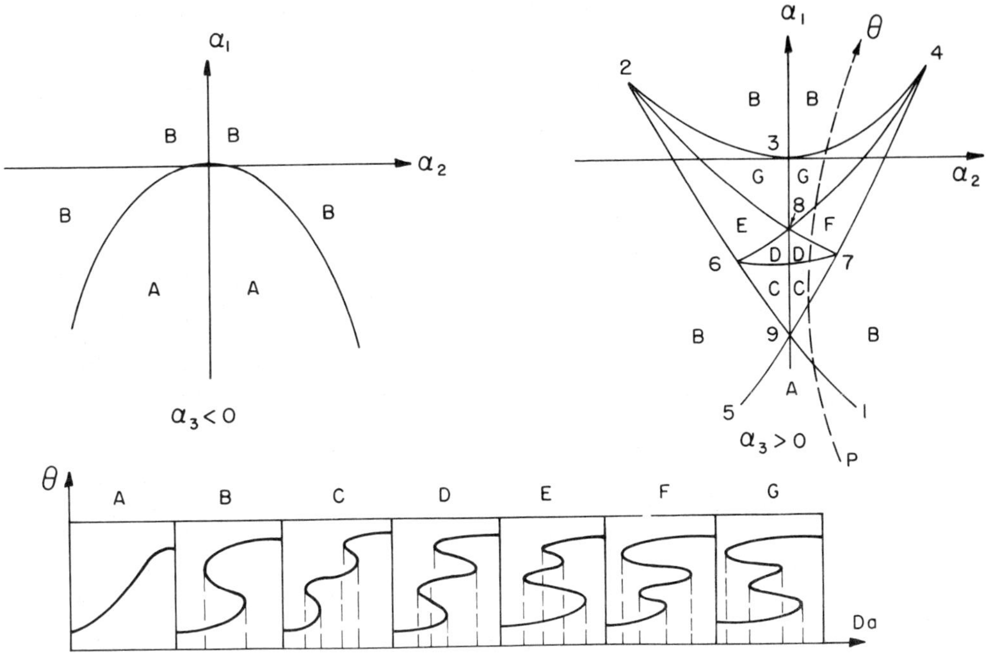

FIGURE 10 Qualitative bifurcation diagrams for the reaction A → B → C occurring in a nonisothermal CSTR, in the neighborhood of the organizing center.

By working away from the singularity the various regions of multiplicity can be determined computationally. Thus the curve 1,9,2,3,4,9,5 for $\alpha_3 > 0$ and the parabolic for $\alpha_3 < 0$ are obtained by generating the hysteresis variety, while 2,8,7,6,8,4 is the double-limit variety. The loci on which the points 2, 4, and 9 lie arise when two hysteresis points coincide (i.e., $F = F_\Theta = F_{\Theta\Theta} = F_{\Theta\Theta\Theta} = 0$), while 6, 7, and 8 are where a turning point lies vertically above or below a hysteresis point [i.e., $F(\Theta_1) = F_\Theta(\Theta_1) = F_{\Theta\Theta}(\Theta_1) = F(\Theta_2) = F_\Theta(\Theta_2) = 0$]. In this way the multiplicity regions can be built up with some confidence that we have a comprehensive picture.

Further examples are given in the papers of Balakotaiah and Luss already referred to, and no doubt we shall see many more studies using singularity theory (see Polizopoulos and Takoudis, in press). The sequence of two stirred tanks, for example, has been considered by Dangelmayr and Stewart (1984).

Tubular Reactor

The steady-state mass and energy balances in a tubular reactor can be written as follows (Varma and Aris, 1977):

$$D_e \frac{d^2C}{dz^2} - v\frac{dC}{dz} - f(C,T) = 0 \tag{46}$$

$$k_e \frac{d^2C}{dz^2} - \bar{v}\rho C_p \frac{dT}{dz} + (-\Delta H)f(C,T) - \frac{UP}{A}(T - T_c) = 0 \tag{47}$$

together with Danckwerts' boundary conditions (BCs)

$$D_e \frac{dC}{dz} = \bar{v}(C - C_0) \qquad \text{at } z = 0 \tag{48a}$$

$$k_e \frac{dT}{dz} = \bar{v}\rho C_p(C - C_0) \qquad \text{at } z = 0 \tag{48b}$$

In deriving these equations the following assumptions have been implicitly made, in addition to assumptions 2 and 3 reported earlier for the CSTR:

1. Radial gradients of temperature and concentration are negligible.
2. The fluid spatial velocity is constant in the axial direction.
3. The dispersive fluxes of mass and heat can be described through Fick and Fourier law, respectively, with effective mass and heat diffusion coefficients.

We shall not pursue any further the classical question of relaxing the abovementioned restrictions in order to improve the model capability of simulating experimental reactor behavior (see Chapter 6). Our aim is instead to indicate the multiplicity pattern of this model, usually referred to as the "axial dispersion model," in connection with various reacting systems. Note in passing that the occurrence of multiple steady states in tubular reactors has been experimentally verified in several works, which have been extensively reviewed by Jensen and Ray (1982). Introducing the dimensionless quantities

$$\begin{gathered} u = \frac{C}{C_0} \qquad v = \frac{T}{T_0} \qquad v_c = \frac{T_c}{T_0} \qquad s = \frac{z}{L} \\ Da = \frac{Lf(C_0,T_0)}{\bar{v}C_0} \qquad \delta = \frac{UPL}{A\bar{v}\rho C_p} \\ \beta = \frac{(-\Delta H)C_0}{\rho C_p T_0} \qquad r = \frac{f(C,T)}{f(C_0,T_0)} \\ Pe_m = \frac{\bar{v}L}{D_e} \qquad Pe_h = \frac{\bar{v}L\rho C_p}{k_e} \qquad \gamma = \frac{E}{RT_0} \end{gathered} \tag{49}$$

the model equations reduce to

$$\frac{1}{Pe_m}\frac{d^2u}{ds^2} - \frac{du}{ds} - Da\,r(u,v) = 0 \tag{50}$$

TABLE 1 A Priori Bounds on the Steady-State Temperature of the Axial Disperison Model with Irreversible Reaction

	Adiabatic ($\delta = 0$)	Nonadiabatic ($\delta > 0$)	
Exothermic reaction ($\beta > 0$)	$1 \leqslant v \leqslant 1 + \beta$	$v_c \geqslant 1$	$1 \leqslant v \leqslant 1 + \beta + \delta(v_c - 1)$
		$v_c < 1$	$v_c \leqslant v \leqslant 1 + \beta$
Endothermic reaction ($\beta < 0$)	$\max[0, 1 - \lvert\beta\rvert] \leqslant v \leqslant 1$	$v_c \geqslant 1$	$\mathrm{Max}[0, 1 - \lvert\beta\rvert] \leqslant v \leqslant v_c$
		$v_c < 1$	$\max[0, 1 - \lvert\beta\rvert - \delta(1 - v_c) \leqslant v \leqslant 1$

Source: Varma (1977).

$$\frac{1}{Pe_h}\frac{d^2v}{ds^2} - \frac{dv}{ds} + \beta Dar(u,v) - \delta(v - v_c) = 0 \tag{51}$$

with BCs

$$\frac{du}{ds} = Pe_m(u - 1) \qquad \frac{dv}{ds} = Pe_h(v - 1) \qquad \text{at } s = 0 \tag{52a}$$

$$\frac{du}{ds} = 0 \qquad \frac{dv}{ds} = 0 \qquad \text{at } s = 1 \tag{52b}$$

The a priori bound for the steady-state reactant concentration is obvious from physical arguments: $0 \leqslant u(s) \leqslant 1$ for all $s \in [0,1]$, while for the temperature a more detailed analysis is required. This has been developed, based on the strong maximum principle for elliptic equations, by Varma (1977) and the obtained bounds are summarized in Table 1.

Uniqueness and Multiplicity Criteria

As a direct consequence of the Danckwerts BCs, the axial dispersion model reduces to the CSTR and the plug-flow models for very large and very small axial dispersion (i.e., $Pe \to 0$ and $Pe \to \infty$), respectively. The multiplicity pattern of the first limit has been fully characterized in the preceding section, at least for simple reacting systems. On the other hand, the plug-flow model does not admit any multiple steady state, since it is constituted by initial value ordinary differential equations.

However, while the multiplicity behavior of the axial dispersion model is well characterized for limiting Pe values, it is still not so for intermediate situations (i.e., finite Péclet number values). In particular, there is a lack of a priori criteria, based on the values of the relevant parameters, for the definition of the reactor uniqueness or multiplicity. Although several attempts have been made in the literature, based on comparison theorems, fixed-point arguments, perturbation methods, lumping procedures, and other techniques, a necessary and sufficient criterion for steady-state uniqueness has not been found.

Next, we summarize briefly the most widely used sufficient or necessary, or neither but simply approximate, bounds of the multiplicity region which have been reported in the literature. Since no proofs will be reported, the interested reader should refer to the original papers or to a previous review (Varma and Aris, 1977).

Special Case of Adiabatic Reactor: The case of adiabatic reactor ($\delta = 0$) with equal mass and heat Péclet numbers ($Pe_m = Pe_h = Pe$) is of particular mathematical interest, since it allows to combine Eqs. (50) and (51) [in the same way we did for the CSTR case and leading to the same u-v relation given by Eq. (7)], reducing the problem to the single ordinary differential equation (ODE)

$$\frac{d^2v}{ds^2} - Pe\frac{dv}{ds} + DaPe\beta r\left(\frac{1 + \beta - v}{\beta}, v\right) = 0 \tag{53}$$

with BCs given by the second of Eqs. (52a) and (52b).

Luss (1971) showed that the same condition [Eq. (11)] which is necessary and sufficient for uniqueness in lumped systems gives a sufficient condition in distributed systems of the single-catalyst-particle type. Since Eq. (53) can be reduced to the latter form (Varma and Amundson, 1973c), the following sufficient condition for stability is obtained:

$$\frac{dF}{dv} \leqslant 0 \qquad \text{for any } v \in [1\ ,\ 1+\beta] \tag{54}$$

where $F(v) = r(v)/(v - 1)$, and the temperature bound for an exothermic reaction from Table 1 has been used.

An earlier sufficient criterion developed by Luss and Amundson (1967b) using comparison methods which ensures uniqueness is

$$Da \leqslant \frac{\mu}{\beta r'(v)} \qquad \text{for any } v \in [1,\ 1+\beta] \tag{55}$$

where $\mu = \lambda_1/Pe + Pe/4$, and λ_1 is the smallest root of the equation

$$\tan\sqrt{\lambda} = \frac{Pe\sqrt{\lambda}}{\lambda - (Pe^2/4)} \qquad 0 < \lambda_1 < \pi^2 \tag{55a}$$

Similarly, two additional sufficient uniqueness criteria developed by Luss (1971) for a slab catalyst particle can readily be applied to the case under examination:

$$Da \leqslant \frac{1}{F(v_+)} \tag{56}$$

$$Da \geqslant \frac{\mu}{F(v_-)} \tag{57}$$

where v_+ and v_- represent the local maximum and minimum of F(v). In the case of an irreversible first-order reaction, F(v) and $v_\pm$ are given by Eqs. (10) (with n = 1) and (18) respectively. Depending on the value of the Péclet number, either condition (55) or condition (56) is more sharp.

General Nonadiabatic Case: Using comparison methods, Varma and Amundson (1972b) have developed the following two sufficient criteria for uniqueness, which are restricted to reaction rates of positive order. The first applies only to the case $Pe_m = Pe_h = Pe$; it requires that

$$Da \leqslant \frac{\mu + \delta}{\beta r'(v) + [\delta(a_1 - a_2)/2\mu]} \qquad \text{for any } v \in [1,\ 1+\beta] \tag{58}$$

where μ is the same parameter as in Eq. (55), and

$$a_1 = \operatorname{Sup}\frac{\partial r}{\partial u} > 0 \qquad a_2 = \operatorname{Inf}\frac{\partial r}{\partial u} \geqslant 0 \tag{58a}$$

Note that Eq. (58) reduces to Eq. (55) for $\delta = 0$. The second criterion is more general since it applies to $Pe_h \neq Pe_m$ also:

$$\mathrm{Da} \leqslant \frac{[-\beta\mu_m a_3 + (\mu_h + \delta)a_2] + \Delta^{1/2}}{a_3\beta[2a_2 + a_1 \exp(|\mathrm{Pe}_h - \mathrm{Pe}_m|)]} \tag{59}$$

where

$$\Delta = [\beta\mu_m a_3 + (\mu_h + \delta)^2 + 2\beta\mu_m(\mu_m + \delta)a_1 a_3 \exp(|\mathrm{Pe}_h - \mathrm{Pe}_m|)] \tag{59a}$$

$$a_3 = \mathrm{Sup}\ \frac{\partial r}{\partial v}$$

and μ_m and μ_h are evaluated as μ in the preceding criterion, but using Pe_m and Pe_h in Eq. (55), respectively. Note that for the nonadiabatic case, no upper bound to the region of uniqueness is available.

From the results reported above, at least some qualitative conclusions can be drawn: for sufficiently large Péclet number Pe and heat transfer coefficient δ, and for sufficiently small Damkhöler number Da and reactor length L, the axial dispersion model exhibits a unique steady state for any value of all the other parameters.

Approximate Analysis of Multiplicity Pattern

As shown above, no exact necessary and sufficient criteria for the definition of the multiplicity pattern of the axial dispersion model are available. Hlaváček and Hofmann (1970a,b) devleoped a lumping procedure that allows us to approximate the original set of nonlinear ODEs, with a set of nonlinear algebraic equations, whose multiplicity pattern can be determined using the methods described in the CSTR section.

The lumping procedure consists of approximating the linear space differential operator, say in Eq. (53), with the first term of the series solution of the corresponding transient problem where the nonlinear reaction term has been neglected, that is,

$$\frac{d^2v}{ds^2} - \mathrm{Pe}\frac{dv}{ds} \cong -\left(\lambda_1 + \frac{\mathrm{Pe}^2}{4}\right)v \tag{60}$$

where λ_1 is given as the smallest root of Eq. (55a). Thus for the particular case of Eq. (53), the problem reduces to

$$\left(\lambda_1 + \frac{\mathrm{Pe}^2}{4}\ v\right) - \mathrm{DaPe}\beta r\left(\frac{1 + \beta - v}{\beta}\ ,\ v\right) = 0 \tag{61}$$

whose multiplicity pattern can be easily determined. Note that this procedure can also be applied readily to the more general case [i.e., Eqs. (50) and (51)], and it is independent of the form of the reaction rate expression r(u,v).

It may be observed that as expected from the intrinsic nature of the lumping approximation, Eq. (61) reduces to the correct CSTR model as $\mathrm{Pe} \to 0$. However, this is not true for the other limit condition (i.e., $\mathrm{Pe} \to \infty$), where the characteristic uniqueness of the plug-flow model is not correctly predicted. Therefore, this procedure should be used with care, since as we will see in more detail in the case of a first-order

reaction, it can lead to erroneous conclusions about the reactor steady-state multiplicity.

A different lumped model can be obtained by approximating the space differential operator through a one-point collocation method. Using Legendre polynomials, so that the collocation point is located at the reactor center, we obtain

$$\frac{d^2v}{ds^2} - Pe\frac{dv}{ds} = -\frac{2 + Pe}{2 + 3Pe/4} Pev \tag{62}$$

which then leads to algebraic nonlinear equations of the same form as Eq. (61). Jensen and Ray (1982) have noted that this lumping procedure exhibits similar limiting behavior in terms of Pe as the previous one, and thus it basically suffers from the same shortcomings.

Irreversible First-Order Reaction in a Nonadiabatic Reactor

This particular case has been examined numerically in detail by Varma and Amundson (1973c), Hlaváček et al. (1973), and Kubicek et al. (1979), who found multiplicity patterns of the type 1-3-1, 1-3-1-3-1, and 1-3-5-3-1.

More recently, Kapila and Poore (1982) have developed an analytical asymptotic solution of the model in the limit of very large activation energy (i.e., $\gamma \to \infty$). This solution is achieved by dividing the reactor into zones of various reaction intensity, where a simple analytic representation of the solution can be given, then completing the solution through the asymptotic matching technique. It has been concluded that whereas the adiabatic model can exhibit at most three steady states, the nonadiabatic one can yield up to seven. These findings have been confirmed by the numerical results reported by Heinemann and Poore (1982). Some of the obtained results are shown in Fig. 11, in terms of the maximum temperature along the reactor as a function of the Damköhler number. It appears that with all other parameters fixed, for increasing values of the dimensionless activation energy γ, the model yields unique solution ($\gamma = 25$), 1-3-1 multiplicity ($\gamma = 40$), 1-3-5-3-1 multiplicity ($\gamma = 50, 60, 75$), and finally 1-3-5-7-5-3-1 multiplicity ($\gamma = 100, 125$). Note that in all cases the intermediate steady states are all unstable; only the lower and upper ones are asymptotically stable.

The inaccuracy of the lumping approximation [Eq. (60)] becomes evident in the case of a first-order reaction, since the lumped model can, at most, predict a 1-3-1 multiplicity pattern. Moreover, it has been shown (Varma and Amundson, 1973c) to predict uniqueness when the distributed nonadiabatic model exhibits multiplicity, and vice versa. However, in the adiabatic case with equal Péclet numbers, numerical comparisons seem to indicate that the lumped model yields a truly sufficient condition for uniqueness, although no proof is available. Jensen and Ray (1982) have compared the lumping approximation (60) and (62) with the exact numerical calculations for adiabatic and nonadiabatic reactors. As shown in Fig. 12, when the multiplicity regions are compared, both approximation may lead to erroneous conclusions. In general, Eq. (62) seems to be more accurate.

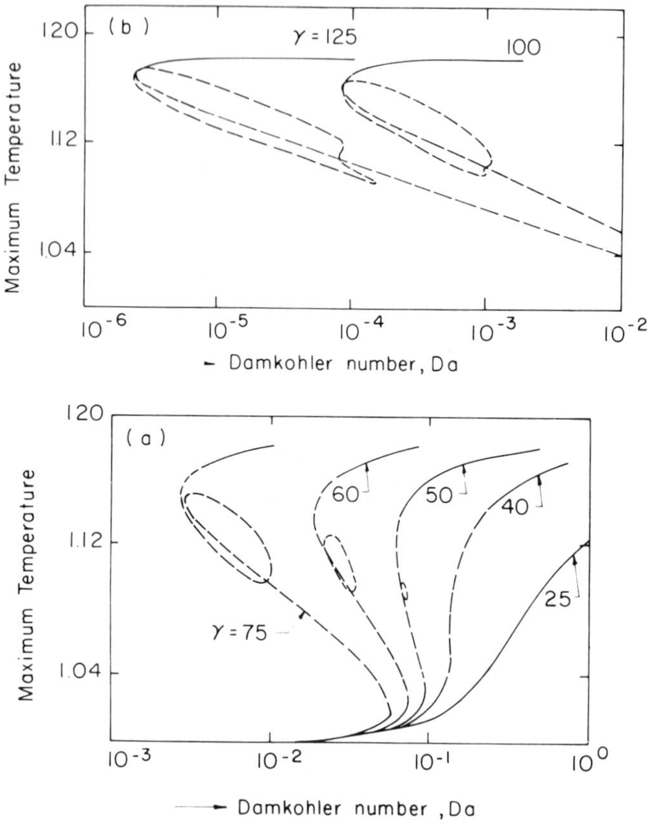

FIGURE 11 Maximum dimensionless temperature as a function of Da for various activation energy values. (a) γ = 25, 40, 50, 60, 75; (b) γ = 100, 125 (parameter values: $Pe_m = Pe_h = 1.0$, $\delta = 4.0$, $\beta = 0.5$, $v_c = 1.0$).

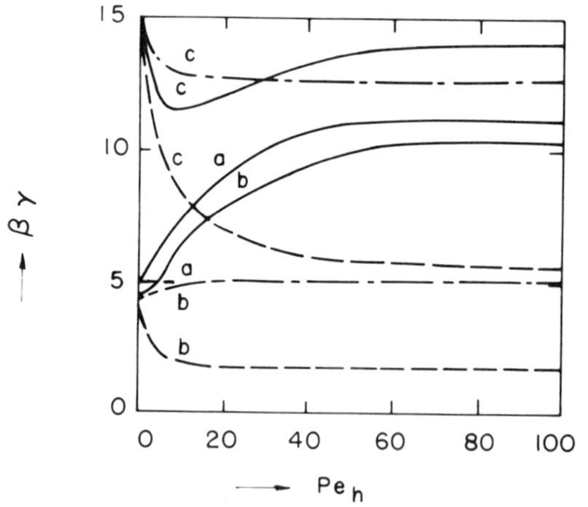

FIGURE 12 Comparison of exact and approximate regions of multiple steady states in the $\beta\gamma$ versus Pe_h plane; exact (——), Eq. (62) (—•—•—), Eq. (60) (— — —); $\delta = 0$, $Pe_m = Pe_h$ (curve a); $\delta = 0$, $Pe_m = 3Pe_h$ (curve b); $\delta = 2$, $Pe_m = Pe_h$ (curve c).

Nonmonotone Isothermal Reaction

Nonmonotone kinetics were examined in an earlier section. We refer here to the case of a substrate-inhibited enzyme reaction, of some importance in biocatalysis, examined in detail by DeVera and Varma (1979). The isothermal axial dispersion model in this case reduces to Eq. (50), with boundary conditions given by the first of Eqs. (52a) and (52b), with

$$r(u) = \frac{(1 + \bar{K} + 1/\bar{K}_s)u}{\bar{K} + u + u^2/\bar{K}_s} \tag{63}$$

where $\bar{K}$ and $\bar{K}_s$ represent the adsorption and substrate inhibition dimensionless constants, respectively. Note that Eq. (63) is quite similar to the biomolecular Langmuir-Hinshelwood reaction rate (22) examined previously in the context of a CSTR.

In Fig. 13 the reactor outlet concentration u(1) is shown as a function of the Damköhler number Da for various Pe. It appears that the reactor exhibits at most a 1-3-1 multiplicity pattern. In Fig. 14 the region of multiple steady states in the Pe-Da plane is shown for fixed values of the kinetic constants $\bar{K}$ and $\bar{K}_s$.

The calculated bifurcation values of Da have been compared with the uniqueness criteria reported above. It is found that Eqs. (56) and (57) (rewritten for the isothermal model under examination) give conservative bounds by a factor ranging between 1.6 to 3.3 and 1 to 2.1, respectively. On the other hand, unlike the case of the adiabatic axial disperison model with first-order reaction, the lumping approximation fails in predicting the region of steady-state uniqueness close to the cusp point.

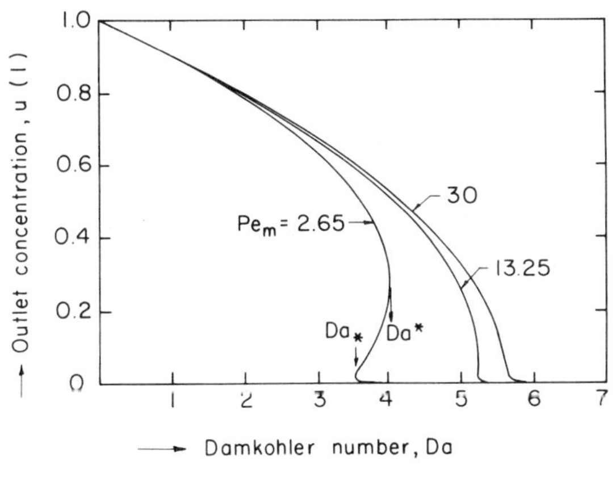

FIGURE 13 Outlet concentration as a function of Damköhler number for various Pe_m values in the case of isothermal substrate-inhibited enzyme reaction (63) (parameter values: $\bar{K} = 0.01$, $\bar{K}_s = 0.1$).

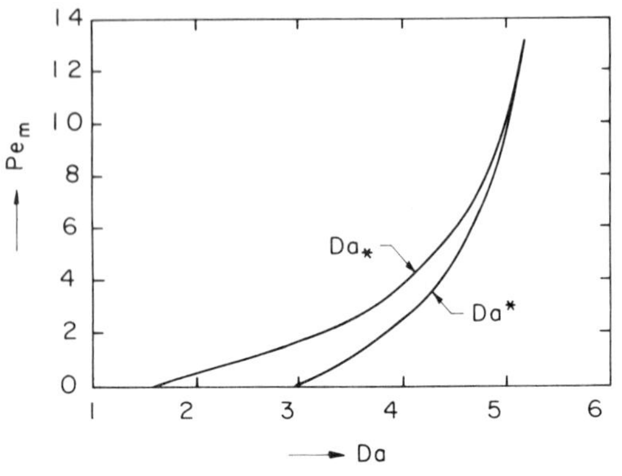

FIGURE 14 Region of multiple steady states in the Pe_m versus Da plane. Parameter values as in Fig. 13.

DYNAMIC BEHAVIOR

In the preceding section we examined several reacting systems and have shown that under specific conditions these may exhibit one or more steady states. It is now important to characterize the stability character of these steady states, which determines whether or not they can be reached in practice by the reactor, through a suitable starting procedure. However, in general, the analysis of all the reactor steady states and their stability character does not exhaustively describe the reactor dynamic behavior. In particular, the occurrence of periodic or aperiodic solutions has to be investigated separately. In this section we focus on steady-state stability and periodic solutions.

Significant contributions have been made to the solution of these problems in recent years, so that at least from a methodological point of view, definitive answers are available in some instances. However, a complete picture of reactor dynamics has been obtained in only a few very simple cases. Maintaining the same structure as the preceding section, we examine first the CSTR and then the axial dispersion model tubular reactor (i.e., a lumped-parameter model and a distributed-parameter model). Subsequently, examples of *aperiodic* solutions will be given through the analysis of somewhat more complex reacting systems.

Continuous Stirred-Tank Reactor

Stability

The transient mass and energy balances can be readily derived from Eqs. (1) and (2) by adding on the right-hand side the mass and energy accumulation terms, respectively. However, in making the resulting equations dimensionless, it is convenient to use the quantities most widely used in the literature, which are somewhat different from those given by Eq. (3). In particular, using the inlet temperature as reference temperature $T_m = T_o$, and

$$x = \frac{(C_0 - C)}{C_0} \qquad \Theta = \frac{(T - T_0)\gamma}{T_0} \qquad \tau = \frac{V}{q}$$

$$t = \frac{t'}{\tau} \qquad B = \frac{(-\Delta H)C_0\gamma}{\rho C_p T_0} \tag{64}$$

we obtain

$$\frac{dx}{dt} = -x + \text{Dar} \tag{65}$$

$$\text{Le}\,\frac{d\Theta}{dt} = -\Theta + \text{BDar} - \delta(\Theta - \Theta_c) \tag{66}$$

with initial conditions

$$x = x^{in} \qquad \Theta = \Theta^{in} \qquad \text{at } t = 0 \tag{67}$$

where Da, δ, and r are as defined in Eq. (3), while the Lewis number Le represents the ratio between the time constants of the thermal and material transport, respectively. As discussed in detail by Ray and Hastings (1980), Le is defined as

$$\text{Le} = \frac{(V\rho C_p)_T}{V_m \rho C_p} \tag{68}$$

where V_m and $(V\rho C_p)_T$ represent the total mass and thermal capacitances of the reaction unit, respectively.

The first criterion for the stability of the steady state of a CSTR was developed in a pioneering paper by van Heerden in 1953. It can be obtained from the steady-state version of Eqs. (65) and (66) (i.e., setting the left-hand side equal to zero) by solving explicitly for x,

$$x = \frac{\Theta + \delta(\Theta - \Theta_c)}{B} \tag{69}$$

and substituting back in Eq. (66) to obtain an equation in the only unknown Θ:

$$R(\Theta) \equiv \Theta + \delta(\Theta - \Theta_c) = \text{BDar}\left(\frac{\Theta + \delta(\Theta - \Theta_c)}{B}, \Theta\right) \equiv G(\Theta) \tag{70}$$

The left- and right-hand sides of Eq. (70) represent the heat removed from the reactor through flow and the cooling medium, and the heat produced inside the reactor by the exothermic reaction, respectively. These two functions are shown in Fig. 15, for those cases where the reactor exhibits one (curve 1) and three (curve 2) steady states. On physical grounds, it can be argued that the stability of each steady state depends on the relative slopes of these two curves. If at steady state the slope

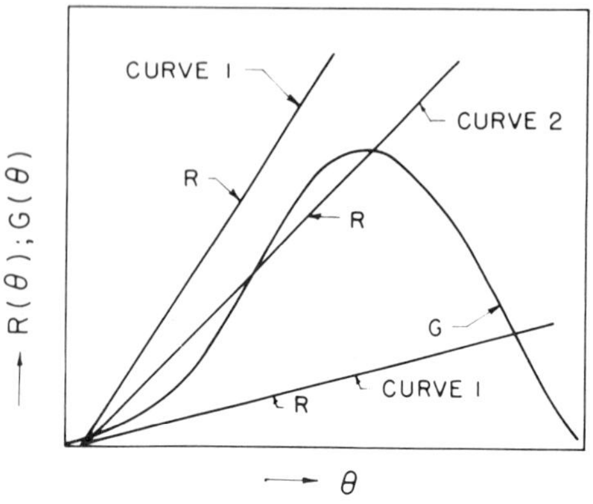

FIGURE 15 Schematic representation of the heat removal $R(\Theta)$ and the heat generation functions $G(\Theta)$, defined by Eq. (70) in a CSTR.

of the heat generation curve is larger than that of the heat removal curve, the steady state is unstable, because for small perturbation of the temperature value the system evolves away from the steady state itself. Based on this observation, van Herdeen concluded that the middle steady state in the case of multiplicity is unstable, while all the others, including the case of a unique steady state, are stable.

The first rigorous analysis of this problem was presented by Bilous and Amundson (1955), who applied the first method of Liapunov. It consists of linearizing Eqs. (65) and (66) around the steady state, introducing the two deviation variables

$$y_1 = x - x_s \qquad y_2 = \Theta - \Theta_s \tag{71}$$

and analyzing the stability character of the linearized system. In matrix form this can be written as follows:

$$\underline{\underline{C}} \frac{d\underline{y}}{dt} = \underline{\underline{J}}\,\underline{y} \tag{72}$$

where $\underline{y} = (y_1 y_2)^T$, $\underline{\underline{C}}$ is the capacitance matrix given by

$$\underline{\underline{C}} = \begin{bmatrix} 1 & 0 \\ 0 & Le \end{bmatrix} \tag{73a}$$

and $\underline{\underline{J}}$ is the Jacobian evaluated at steady state defined as

$$\underline{\underline{J}} = \begin{bmatrix} Dar_x - 1 & Dar_\Theta \\ BDar_x & -1 - \delta + BDar_\Theta \end{bmatrix} \tag{73b}$$

where r_x and r_Θ indicate the partial derivatives of the reaction rate, r at steady state with respect to conversion x and temperature Θ, respectively.

In the case of an irreversible first-order reaction $r = (1 - x)\exp[\Theta/(1 + \Theta/\gamma)]$, the Jacobian reduces to

$$\underline{\underline{J}} = \begin{bmatrix} \dfrac{1}{x - 1} & \dfrac{x}{(1 + \theta/\gamma)^2} \\ \dfrac{Bx}{x - 1} & \dfrac{-1 - \delta + Bx}{(1 + \Theta/\gamma)^2} \end{bmatrix} \tag{74}$$

The stability of the linear system (72) is dictated by the eigenvalues of matrix $\underline{\underline{C}}^{-1}\underline{\underline{J}}$. Assuming that Le = 1 (i.e., $\underline{\underline{C}} = \underline{\underline{I}}$), these coincide with the eigenvalues of $\underline{\underline{J}}$, which are given by

$$\lambda_{\pm} = \frac{\text{tr } \underline{\underline{J}} \pm \Delta^{1/2}}{2} \qquad \Delta = (\text{tr } \underline{J})^2 - 4 \det \underline{\underline{J}} \tag{75}$$

The effect of the Lewis number Le will be examined in detail later.

The necessary and sufficient condition for asymptotic stability (i.e., stability to "small" perturbations from the steady state) is that both eignevalues have negative real part, that is,

$$\det \underline{\underline{J}} > 0 \tag{76a}$$

and

$$\text{tr } \underline{\underline{J}} < 0 \tag{76b}$$

From standard linear differential equation theory the following classification of the steady state of the linear system (72) is obtained:

1. $\det \underline{\underline{J}} < 0$: unstable saddle point (77a)
2. $\det \underline{\underline{J}} > 0$, $\text{tr } \underline{\underline{J}} > 0$: unstable node ($\Delta > 0$) or focus ($\Delta < 0$) (77b)
3. $\det \underline{\underline{J}} > 0$, $\text{tr } \underline{\underline{J}} < 0$: stable node ($\Delta > 0$) or focus ($\Delta < 0$) (77c)
4. $\det \underline{\underline{J}} > 0$, $\text{tr } \underline{\underline{J}} = 0$: center (77d)

Liapunov's theorem assures us that the stability character of the nonlinear system (65) and (66) is the same as that of the linear one for the first three cases. For the fourth, further analysis is required.

A few comments on the comparison between the rigorous stability condition (76) and the slope condition previously reported are now in order. It should first be noted that condition (76a), using Eqs. (70) and (73), is actually identical to the slope condition, since it requires the slope of the heat removal function to be larger than that of the heat generation function. Thus van Heerden criterion does actually give only a necessary condition for stability. This implies that when three steady states occur, the middle one is certainly unstable, while the lower and upper ones *can* be stable. These two, as well as the single steady state in the case of uniqueness, are in fact asymptotically stable if condition (76b), usually referred as the dynamic condition, is also satisfied.

Finally, it can be shown (see Varma and Aris, 1977) that in the case of an adiabatic reactor (i.e., $\delta = 0$) condition (76a) implies condition (76b) automatically. In this case the slope condition becomes necessary and sufficient for asymptotic stability.

Periodic Solutions

To complete the picture of the dynamic behavior of the reactor, it is necessary to characterize fully the occurrence of periodic solutions. A very large number of papers have appeared in the literature using several mathematical methods, such as Liapunov's direct method, averaging techniques, Fourier analysis, and perturbation methods [see Hlaváček et al. (1970) and Uppal et al. (1974) for detailed references]. The most convenient approach, which allows us to answer the problem exhaustively, has been presented by Poore (1973).

This approach is based on the Hopf-Friedrichs bifurcation theorem (Friedrichs, 1965), which states that bifurcation to a periodic solution occurs *only* from the center of the linearized system, and under the restriction that the trace of the matrix $\underline{\underline{D}} = d\underline{\underline{J}}/d\varepsilon$ (where ε is the bifurcation parameter) does not vanish. Moreover, this theorem allows the construction of a function $\delta(x)$ whose sign at the bifurcation point allows to estimate *both* the direction of the periodic solution *and* its stability character. It should be mentioned that in this type of bifurcation, often referred to as *Hopf bifurcation*, a pair of complex conjugate eigenvalues change their real part sign by crossing the imaginary axis, thus implying a change in the system stability character, since all the other eigenvalues are required to have negative real part. This should not be confused with the so-called *static bifurcation*, where one real eigenvalue changes its sign by passing through the origin. This is the bifurcation we have been referring to in the section on steady-state multiplicity.

Poore (1973) applied this theory to the case of a CSTR with first-order irreversible reaction and Le = 1, in the limit of infinite activation energy. To investigate all possible qualitative phase-plane behavior for the system, he considered the parameter plane B versus δ (with fixed $\gamma = \infty$, $\Theta_c = 0$, and Le = 1) taking Da as the bifurcation parameter (i.e., eventually the only parameter that changes, while all the others are kept constant, to explore the bifurcations and/or jumps of the system). The phase plane is then divided into six regions by three curves, which are defined conceptually as follows:

1. *Curve M*. Below this curve m_1 and m_2 [defined as conversion values corresponding to the Da bifurcation values shown in Fig. 16 (i.e., where det $\underline{\underline{J}} = 0$)] are complex, thus det $\underline{\underline{J}} > 0$; above the curve, det $\underline{\underline{J}} < 0$ for $m_1 < x < m_2$ and det $\underline{\underline{J}} >$ otherwise. This curve thus represents the uniqueness condition for any Da, and is analytically given, for a first-order reaction, by Eq. (17).
2. *Curve S*. It is based on the same concepts as curve M, but with respect to stability insteady of multiplicity. Indicating with s_1 and s_2 the conversion values where the stability character changes (i.e., tr $\underline{\underline{J}} = 0$) (see Fig. 16), it follows that below the curve s_1 and s_2 are complex, thus tr $\underline{\underline{J}} < 0$; above the curve for $s_1 < x < s_2$, tr $\underline{\underline{J}} > 0$; while for $x < s_1$ and $x > s_2$, tr $\underline{\underline{J}} < 0$.

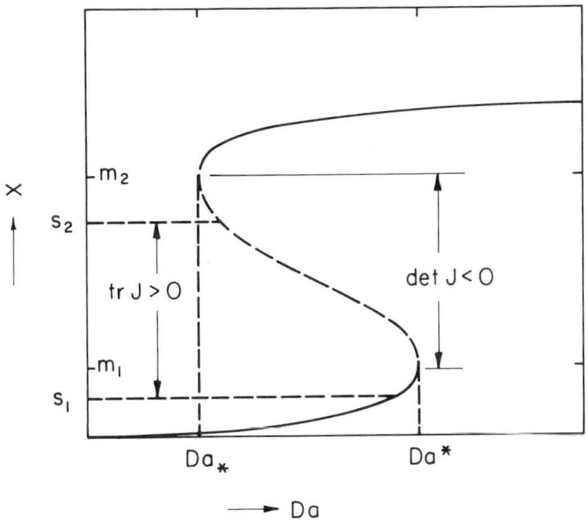

FIGURE 16 Sketch of outlet conversion as a function of Da at steady state.

3. *Curve SM*. This curve establishes the relative location of the solutions m_1, m_2, and s_1, s_2. In particular, it represents the locus of points where the m's and s's have the same values (i.e., det $\underline{\underline{J}} = 0$ and tr $\underline{\underline{J}} = 0$).

For the case examined by Poore (1973) these curves are described by particularly simple expressions:

$$\text{Curve M:} \quad B = 4(1 + \delta) \tag{78a}$$

$$\text{Curve S:} \quad B = 3 + \delta + 2\sqrt{2 + \delta} \tag{78b}$$

$$\text{Curve SM:} \quad B = \frac{(1 + \delta)^3}{\delta} \tag{78c}$$

Also, for the above-mentioned function $\delta(x)$ a simple expression is derived for $x = s_1$ and s_2:

$$\delta_i = (-1)^{i+1}[\omega_i^2(b_i - 1) + 2b_i - Bs_i - (2b_i - Bs_i)^2] \qquad i = 1, 2 \tag{79}$$

where $\omega_i^2 = b_i(Bs_i^2 - b_i)$ and $b_i = Bs_i - 1 - \beta$. It is found that for bifurcation at $s_1(s_2)$, $\delta_1 > 0(\delta_2 > 0)$ indicates bifurcation to a stable (unstable) periodic solution for Da values larger than the bifurcation value, and vice versa for $\delta_1 < 0(\delta_2 < 0)$.

The six regions in the β-δ plane obtained using the three curves given by Eq. (78) are shown in Fig. 17. The bifurcation behavior corresponding to each region is illustrated as a function of Da in the side figures, where the capital letters refer to the various shapes of phase portraits shown in Fig. 18 (here the symbols filled and open circles indicate stable and unstable limit cycles, respectively, and their distance to the steady-state

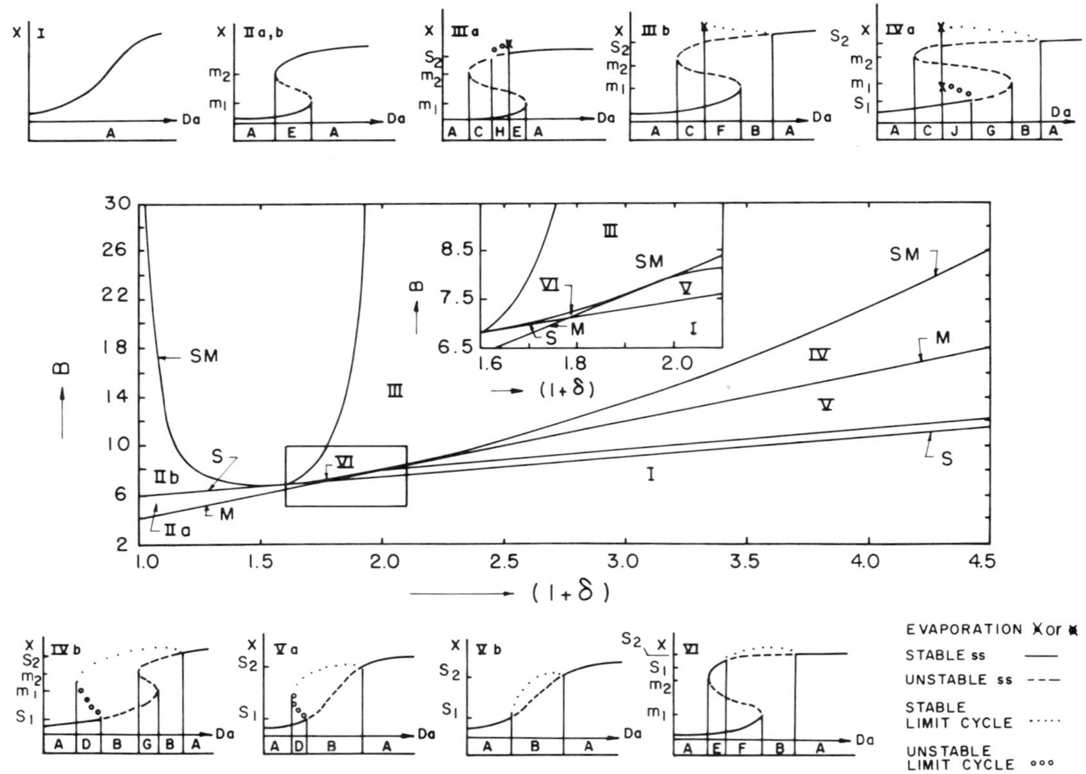

FIGURE 17 Regions in the parameter plane B versus (1 + δ), surrounded by the sketches of the corresponding bifurcation diagrams in the X versus Da plane (parameter values: $\gamma = \infty$, $\Theta_c = 0$, Le = 1). Irreversible first-order reaction in a CSTR.

TYPICAL PHASE PLOTS

CASE		A	B	C	D	E	F	G	H	J
NODE or FOCUS	STABLE	1	0	1	1	2	1	0	2	1
	UNSTABLE	0	1	1	0	0	1	2	0	1
SADDLE POINT		0	0	1	0	1	1	1	1	1
LIMIT CYCLES	STABLE	0	1	0	1	0	1	1	0	1
	UNSTABLE	0	0	0	1	0	0	0	1	1
TOTAL INVARIANTS		1	2	3	3	3	4	4	4	5

ABSCISSA–CONCENTRATION ORDINATE–TEMPERATURE

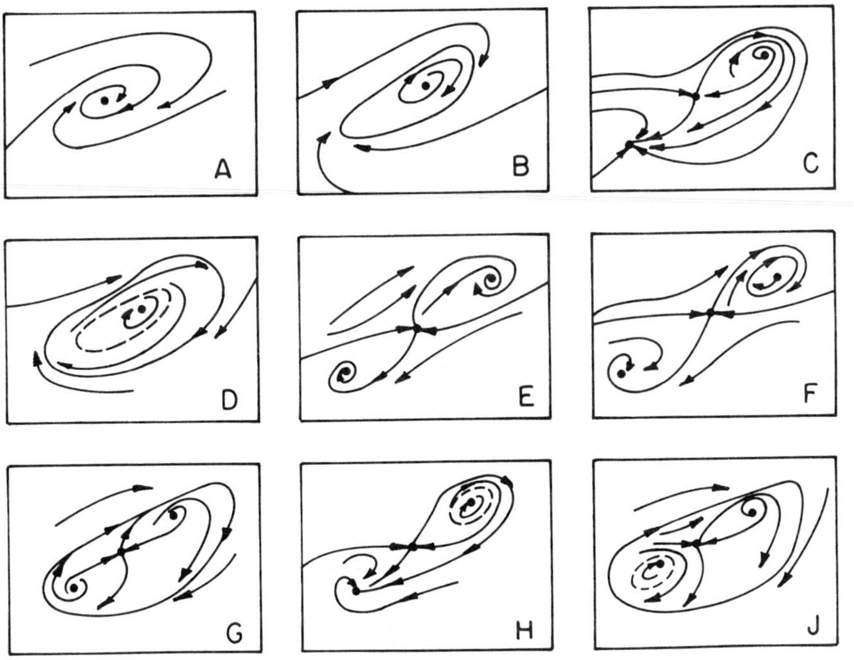

FIGURE 18 Phase portraits characteristic of various regions in the parameter plane shown in Fig. 17.

curve represents the maximum amplitude of the oscillation). These figures have been taken from Uppal et al. (1974), who extended Poore (1973) results to the case of finite activation energy, and completely defined the CSTR dynamics by following the branches of periodic solutions up to completion through accurate numerical integration.

The correspondence between the regions in the parameter plane and the corresponding bifurcation diagram follows directly from the relative position of each region with respect to the above-mentioned curves. For example, since bifurcation to a periodic solution occurs only for tr $\underline{\underline{J}} = 0$ and det $\underline{\underline{J}} > 0$, it follows that bifurcation at $x = s_1$ occurs only in regions IV ($0 < s_1 < m_1 < m_2 < s_2 < 1$), V ($0 < s_1 < s_2 < 1$, m, and m_2 are complex), and VI ($0 < m_1 < m_2 < s_1 < s_2 < 1$), while bifurcation at $x = s_2$ can occur also in region III ($0 < m_1 < s_1 < m_2 < s_2 < 1$).

Several interesting dynamic features of the CSTR have been evidenced by this analysis. For example, for a set of parameter values inside region IIIb in Fig. 17, it appears that starting with large Da and decreasing it, the system goes from a high-conversion steady state to a limit cycle (transition A-B), whose amplitude increases up to a Da value, where the limit cycle evaporates and the system precipitates to a low-conversion steady state (transition F-C). This behavior has been likened by Poore (1973) to the flickering of the flame when turning off a bunsen burner.

Another interesting observation concerns the mechanisms for the limit cycle disappearance pointed out by Uppal et al. (1974). Three of these have been observed: shrinking into a steady state, increasing amplitude up to merging with a separatrix line, and coalescence of a stable and unstable limit cycle.

Also, about the relative position of limit cycles and steady states some general conclusions can be drawn. Poore (1973) has shown that any steady state in the uniqueness region has index +1, while in the multiplicity region the middle steady state has index −1, and the lower and upper ones have again index +1 (see Gavalas, 1968). Since any periodic solutions must encircle a number of steady states such that the sum of their indices gives +1, it can be concluded that a limit cycle can either include the steady state in the case of uniqueness, or include only the upper steady state or only the lower or all the three steady states in the case of multiplicity. This conclusion is confirmed by the phase-plane portraits shown in Fig. 18. Along this same line, Poore (1973) also proved that if a unique unstable steady state exists, then a stable periodic orbit must also exist surrounding the steady state.

As we have already discussed in the multiplicity section, the most convenient way to change the Da value in practice is to change the feed stream flow rate. However, this also affects the cooling parameter δ, defined by Eq. (3). So Uppal et al. (1976) redefined the dimensionless quantities, isolating the feed stream flow rate in the residence-time parameter, and explicitly accounting for it as follows:

$$\text{Da} = \text{Da}^\circ \tau \qquad \delta = \delta^\circ \tau \tag{80}$$

The entire bifurcation analysis has then to be repeated, now taking τ instead of Da as a bifurcation parameter. As mentioned earlier, the multiplicity pattern with respect to the reactor residence time is quite complex, giving rise to the isola and mushroom shapes, which do not appear when the Damkhöler number is taken as a bifurcation parameter. Therefore, it

is not surprising that more complex features and up to 17 different types of dynamic behavior have been found by Uppal et al. (1976).

It should be noted that the number of possible phase portraits is unchanged, since these depend only on the parameter steady-state values. Thus in this case also, the nine portraits shown in Fig. 18 cover all possible phase-plane shapes.

Effect of the Lewis Number

The Lewis number has been defined in Eq. (68) as the ratio between the thermal and the mass capacitances of the reactor. Up to this point it has been assumed that Le = 1; however, while for homogeneous tubular reactors the Lewis number is very close to 1, for packed-bed reactors this is not true, and in general it tends to be much larger than unity.

Clearly, the Lewis number does not affect the value of the reactor steady state, only its stability character. In fact, Le appears only as a multiplier of the thermal accumulation term in the transient model (65) and (66). The linearized system (72) can be rewritten as

$$\frac{d\underline{y}}{dt} = \underline{\underline{A}}\underline{y} \tag{81}$$

where $\underline{\underline{A}} = \underline{\underline{C}}^{-1}\underline{\underline{J}}$ involves the lewis number.

In general, decreasing values of Le enlarge the instability region of the reactor. For example, in the case of a first-order reaction with $\gamma = \infty$, the conversion values $s_{1,2}$, where tr $\underline{\underline{A}} = 0$, can be readily calculated from Eqs. (74) and (81) as follows:

$$s_{1,2} = \frac{B + 1 + \delta}{2B} \pm \frac{1}{2B}\left[(B + 1 + \delta)^2 - 4BLe - 4B(1 + \delta)\right]^{1/2} \tag{82}$$

where it appears that as Le $\rightarrow$ 0, s_1 decreases while s_2 increases. Since tr $\underline{\underline{A}} > 0$ for $s_1 < x < s_2$, it follows that decreasing values of Le expand the region where the dynamic stability condition (76b) is violated. The numerical findings of various authors also go in this direction, showing that models of this type tend to exhibit periodic solutions for sufficiently low Le values [see Aris (1975) for the case of a catalyst particle].

Perelson (1976) extending a previous work by Luss (1974), has examined the effect of the capacitance matrix on the stability of the linear system (81) in a more general way. It was observed that $\underline{\underline{C}}$ being a positive definite symmetric matrix, if $\underline{\underline{J}}$ is nonsingular and symmetric, then $\underline{\underline{A}}$ has all real eigenvalues. If $\underline{\underline{J}}$ is nonsingular and nonsymmetric, as in the case under examination, the eigenvalues of $\underline{\underline{A}}$ can be either real or complex conjugate, depending on the value of the capacitance parameter. Thus taking one of the elements of $\underline{\underline{C}}$ as bifurcation parameter (in our case, Le), and applying the above-mentioned Hopf-Friedrichs bifurcation theorem, for any given steady state of the reactor the value of Le at which periodic orbits bifurcate can be identified.

In particular, for a system equivalent to the adiabatic CSTR model, Perelson (1976) found that such a critical value of Le is given by

$$Le_c = (x - 1)\left[1 - \frac{Bx}{(1 + \Theta/\gamma)^2}\right] \tag{83}$$

where x and Θ are evaluated at steady-state conditions, and periodic solutions can occur only for $Le < Le_c$. Since the occurrence of Hopf bifurcation requires det $\underline{\underline{A}} > 0$, it can be shown that $Le_c < 1$. This implies that for adiabatic systems of the type examined, oscillations can occur only for $Le < 1$.

This result has been previously reported by Ray et al. (1974), in the context of multiplicity and stability of catalytic wires. Since wires are characterized by Lewis number values much larger than 1, it appears that they cannot oscillate on their own. These findings agree with the analysis of Edwards et al. (1974) on flickering of catalytic wires. This phenomenon, characterized by the occurrence of a localized region of very high luminosity whose intensity changes with time, has been attributed to fluctuations of the concentrations in the gaseous stream, which can be significantly enhanced by the catalytic wire.

Ray and Hastings (1980) have applied bifurcation analysis to the system of Eqs. (65) and (66) taking as bifurcation parameter the Lewis number Le. The procedure is identical to that applied by Uppal et al. (1974, 1976) taking as bifurcation parameters the Damköhler number and the reactor residence time, respectively. By inspection of Eqs. (65) and (66) it is readily seen that by replacing the parameters, B, δ, and Θ_c with the new quantities

$$\hat{\alpha} = \frac{1 + \delta}{Le} \qquad \hat{B} = \frac{B}{Le} \qquad \hat{\Theta}_c = \frac{\Theta_c}{Le} \tag{84}$$

Eqs. (65) and (66) reduce to the same form used by Uppal et al. (1974) in the case of $Le = 1$. Thus, with these parameter changes, the results obtained previously still hold. In particular, the $\hat{B}-\hat{\alpha}$ parameter plane [previously, $B - (1 + \delta)$] is again divided into the same six regions, each leading to a different shape of the conversion versus Damköhler number bifurcation diagram. This diagram is shown in Fig. 19 for the case $\gamma = \infty$ and $\Theta_c = 0$, in terms of the new dimensionless parameters accounting for the effect of Lewis number.

A quite interesting feature of the dynamics of this system can be observed in Fig. 19. Since the reactor steady state for $\Theta_c = 0$ is affected only by the ratio $B/(1 + \delta)$, it follows that the same x-Da bifurcation diagram corresponds to all the points on any straight line going through the origin. For a given straight line, by decreasing Le we move toward the upper-right region of the diagram. So, depending on the given value of the ratio $B/(1 + \delta)$, the curve may enter region III (curve 1), V (curve 2), or even the small region VI. In any event, in either one of these regions, the system can exhibit stable periodic orbits. Thus it appears that for decreasing values of Le, lower than the critical value Le_c where the straight line enters one of the above-mentioned regions, we have persistent stable periodic solutions.

Using the same procedure adopted for deriving Eq. (78b), the following expression for Le_c, valid for $\gamma = \infty$ and $\Theta_c = 0$, is obtained:

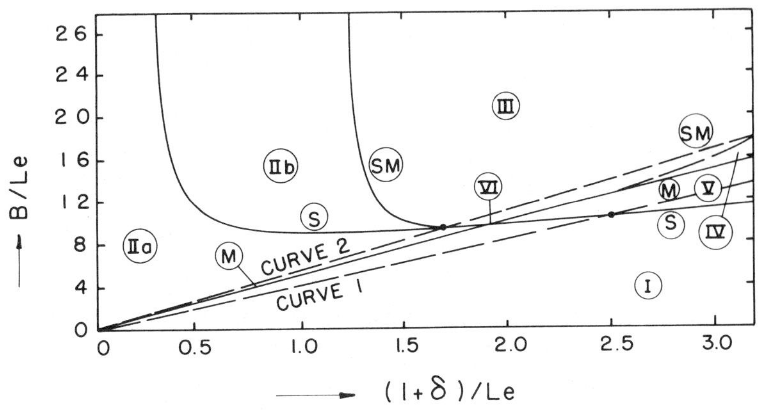

FIGURE 19 Effect of Lewis number on reactor dynamics (parameter values as in Fig. 17).

$$Le_c = \left[\frac{(1 + \psi)^2}{4\psi} - 1\right](1 + \delta) \qquad (85)$$

where $\psi = B/(1 + \delta)$.

It should be noted that the specific form of the bifurcation diagram x versus Le depends on the particular choice of the values for $B/(1 + \delta)$ and Da. A detailed analysis of all possible situations has been reported by Ray and Hastings (1980).

These authors have also pointed out that in some cases the periodic orbits asymptotically approach relaxation oscillations as $Le \to 0$. In this case the characteristic thermal time is so much smaller than that for mass that the energy balance (66) can be assumed in pseudo-steady-state conditions with respect to the mass balance (65). Under these conditions, relaxation oscillations can occur only if the function $\Theta = \Theta(x)$, given by the pseudo-steady-state approximation of Eq. (66) is multivalued (see Gilles et al., 1978). For example, this is not the case for $\gamma = \infty$, which therefore does not exhibit this peculiar type of dynamic behavior.

Some Comparison with Experimental Data

As a conclusion of the analysis of the dynamic behavior of a CSTR, it is convenient to report some experimental verification of the presented results. It should be understood that what we are really verifying with such a comparison is the CSTR model and its ability to simulate real systems.

Several experiments have been reported in the literature of multiplicity, oscillations, and sensitivity in CSTRs, for both homogeneous and heterogeneous reactions. Exhaustive reviews have been presented by Schmitz (1975) and Sheintuch and Schmitz (1977). We shall here content with few references, specifically concerning homogeneous reactions in a CSTR, which explicitly prove some of the features previously described. In particular, we refer to the experiments reported by Schmitz and coworkers using a laboratory well-mixed continuous reactor where the second-order reaction between sodium thiosulfate and hydrogen peroxide occurs.

Several of the phase portraits shown in Fig. 16 have been experimentally found using this system. Vejtasa and Schmitz (1970) used an adiabatic reactor, which we have shown (at least for a first-order reaction) to belong either to region I or II in Fig. 17. Accordingly, no oscillations were found and only phase portraits of type A and E were reported. Chang and Schmitz (1975a,b) used a nonadiabatic unit, which allowed them to observe oscillatory behavior. Besides type A and E, a phase portrait of type B was also reported.

Subsequently, the authors of the most significant latest theoretical and experimental contributions in this area joined together with the aim of experimentally verifying additional dynamic features of the above-mentioned reaction system (Schmitz et al., 1979). Applying the Hopf-Friedrichs bifurcation theorem, the dynamic behavior of a CSTR with a second-order reaction was explored. Again, the same six regions characteristic of the first-order reaction were found. The qualitative agreement between the above-mentioned previous experimental results and the model predictions was complete. Moreover, a fourth type of phase portrait, C was experimentally found, where three steady states are present, but only the low conversion one is stable. This result was achieved thanks to the dual nature, theoretical and experimental, of this work. Unfortunately, due to limitations intrinsic of any experimental apparatus, not all the dynamic features found theoretically were possible to be reproduced experimentally.

An interesting result regarding the effect of the Lewis number has also been reported by these authors. The experimental unit was, in fact, equipped with a jacket, surrounding the reactor wall, which could be filled with water in order to increase the thermal capacity of the reactor and thus also the corresponding value of the Lewis number. It was shown that repeating the experiment relative to the phase portrait of type C, but now filling the jacket of the reactor with water, a stable high conversion steady state was obtained. This confirms the previously discussed stabilizing effect of increasing values of the Lewis number.

Tubular Reactor

Stability

The transient behavior of a tubular reactor is described by the axial dispersion model through the following set of parabolic partial differential equations (PDEs):

$$\frac{\partial u}{\partial t} = \frac{1}{Pe_m}\frac{\partial^2 u}{\partial s^2} - \frac{\partial u}{\partial s} - Dar(u,v) \qquad (86)$$

$$Le\frac{\partial v}{\partial t} = \frac{1}{Pe_h}\frac{\partial^2 v}{\partial s^2} - \frac{\partial v}{\partial s} + \beta Dar(u,v) - \delta(v - v_c) \qquad (87)$$

where the dimensionless parameters

$$t = \frac{t'q}{V} \qquad Le = \frac{\varepsilon\rho C_p + (1-\varepsilon)\rho_s C_{ps}}{\varepsilon\rho C_p} \qquad (88)$$

have been added to those previously defined in Eq. (49). The Lewis number is defined as the ratio between the total heat capacitance of the reacting system (fluid + solid) and that of the fluid. In the case of homogeneous reactors Le = 1, while it rises up to a few hundred in the case of catalytic reactors.

Boundary conditions of Eqs. (86) and (87) are the same reported for the steady-state model [Eq. (52)], while the initial conditions are

$$u = u^{in}(s) \qquad v = v^{in}(s) \qquad \text{at } t = 0 \tag{89}$$

It is worth point out that the BCs (52) proposed by Danckwerts in 1953 are valid also under transient conditions only when no mass or heat dispersion occurs in the inlet and outlet tubes to and from the reactor (Wehner and Wilhelm, 1956; van Cauwenberghe, 1966). In the following we shall always refer to the Danckwerts boundary conditions, and eventually discuss the effect on reactor stability of the degree of mixing in the regions before and after the reacting region.

In the preceding section it was shown that Eqs. (86) and (87) can admit up to seven steady-state solutions. It is then necessary to discuss numerical procedures and a priori criteria to define the stability character of a steady state. Unfortunately, the mathematical apparatus for dealing with stability of parabolic PDEs is not as well assessed as for ODEs. Therefore, except for very special situations, strong statements about steady-state stability, such as those reported for the CSTR, cannot be made.

Even though no rigorous mathematical proof is available (of the type of Liapunov's first theorem for ODEs), it seems quite reasonable to relate the local stability character of a given steady state of nonlinear system of parabolic PDEs to that of the corresponding system obtained through linearization about the steady state. Introducing the perturbation variables

$$\tilde{u}(s,t) = u(s,t) - u_s(s) \tag{90a}$$

$$\tilde{v}(s,t) = v(s,t) - v_s(s) \tag{90b}$$

where the index s refers to steady-state condition, and performing the usual change of variables (Amundson, 1965; Varma and Aris, 1977),

$$y_1(s,t) = \tilde{u}(s,t) \exp\left(-\frac{Pe_m s}{2}\right) \tag{91a}$$

$$y_2(s,t) = \tilde{v}(s,t) \exp\left(-\frac{Pe_h s}{2}\right) \tag{91b}$$

the following linearized version of Eqs. (86), (87), (52), and (89) can be obtained:

$$\frac{\partial y_1}{\partial t} = \frac{1}{Pe_m}\frac{\partial^2 y_1}{\partial s^2} - \left(\frac{Pe_m}{4} + Dar_u\right)y_1 - Dar_v y_2 \exp\left[\frac{(Pe_h - Pe_m)s}{2}\right] \tag{92}$$

$$\text{Le}\,\frac{\partial y_2}{\partial t} = \frac{1}{\text{Pe}_h}\,\frac{\partial^2 y_2}{\partial s^2} - \left(\frac{\text{Pe}_h}{4} + \delta - \beta\text{Dar}_v\right)y_2 + \beta\text{Dar}_u y_1 \exp\left[\frac{(\text{Pe}_m - \text{Pe}_h)s}{2}\right] \tag{93}$$

with BCs

$$\frac{\partial y_1}{\partial s} - \frac{\text{Pe}_m}{2}\,y_1 = 0 \qquad \frac{\partial y_2}{\partial s} - \frac{\text{Pe}_h}{2}\,y_2 = 0 \qquad \text{at } s = 0 \tag{94a}$$

$$\frac{\partial y_1}{\partial s} + \frac{\text{Pe}_m}{2}\,y_1 = 0 \qquad \frac{\partial y_2}{\partial s} + \frac{\text{Pe}_h}{2}\,y_2 = 0 \qquad \text{at } s = 1 \tag{94b}$$

and ICs

$$y_1 = y_1^{in} \qquad y_2 = y_2^{in} \qquad \text{at } t = 0 \tag{95}$$

Equations (92) to (94) can be rewritten in the following matrix form:

$$\underline{\underline{C}}\,\frac{\partial \underline{y}}{\partial t} = \underline{\underline{A}}\,\frac{\partial^2 \underline{y}}{\partial s^2} + \underline{\underline{B}}\,\underline{y} \tag{96}$$

with BCs

$$-\frac{\partial \underline{y}}{\partial s} + \underline{\underline{D}}\,\underline{y} = 0 \qquad \text{at } s = 0 \tag{97a}$$

$$\frac{\partial \underline{y}}{\partial s} + \underline{\underline{D}}\,\underline{y} = 0 \qquad \text{at } s = 1 \tag{97b}$$

and IC

$$\underline{y} = \underline{y}^{in} \tag{98}$$

Its solution is given by an infinite series expansion of the form

$$y_i = \sum_{j=1}^{\infty} C_{ij} Y_{ij}(s) \exp(\lambda_j t) \tag{99}$$

where λ_j are eigenvalues of the associated eigenvalue problem

$$\lambda \underline{\psi} = \underline{\underline{A}}\,\frac{d^2 \underline{\psi}}{ds^2} + \underline{\underline{B}}\,\underline{\psi} \tag{100}$$

that is, the values of λ for which Eq. (100) admits a nontrivial solution.

From Eq. (99) it is seen that if all eigenvalues have negative real parts, the steady state is stable; if one of the eigenvalues has positive real part, the steady state is unstable. This is a quite straightforward

conclusion, but its application can be quite difficult, since in general the number of eigenvalues is infinite, and they can of course be either real or complex. Therefore, in general it is necessary to resort to suitable numerical techniques. The standard procedure is to discretize Eq. (100) along the space variable s and then compute with a suitable algorithm the eigenvalues associated with the obtained system of algebraic equations. Since the eigenvalues thus obtained approximate the corresponding largest eigenvalues of the original infinite set, they allow us to identify the stability character of the steady state under examination. Particularly fortunate is the case of an adiabatic reactor with $Pe_m = Pe_h$, for it leads to a linearized system (96) whose linear spatial operator is self-adjoint.

In the following we first analyze this last case, which has received a great deal of attention in the literature. Then we consider the general case of a nonadiabatic reactor and discuss some of the stability conditions which have been obtained using various methods, mainly based on comparison theorems and Liapunov direct method. Finally, some numerical techniques for solving the eigenvalue problem (100) are indicated.

Special Case of Adiabatic Reactor: As mentioned above, in the case of adiabatic axial dispersion model with $Pe_m = Pe_h$ and Le = 1, the stability problem can be greatly simplified. A standard procedure used in the earlier studies is to introduce the so-called "residual enthalpy," defined as

$$w(s,t) = y_2(s,t) + \beta y_1(s,t) \tag{101}$$

By suitably combining Eqs. (92) and (93), under the assumptions noted above, it is found that

$$\frac{\partial w}{\partial t} = \frac{1}{Pe}\frac{\partial^2 w}{\partial s^2} - \frac{Pe}{4} w \tag{102}$$

with BCs

$$\frac{\partial w}{\partial s} - \frac{Pe}{2} w = 0 \quad \text{at } s = 0 \tag{102a}$$

$$\frac{\partial w}{\partial s} + \frac{Pe}{2} w = 0 \quad \text{at } s = 1 \tag{102b}$$

which, using the IC

$$w^{in} = 0 \tag{103}$$

admits only the trivial solution w = 0. This implies that if we limit the analysis to so-called adiabatic perturbations (i.e., $w^{in} = 0$ leads to $\tilde{v} + \beta\tilde{u} = 0$), then during the transient the following relation holds:

$$v + \beta u = 1 + \beta \tag{104}$$

This allows us to reduce the original system of equations to the single PDE

$$\frac{\partial v}{\partial t} = \frac{1}{Pe}\frac{\partial^2 v}{\partial s^2} - \frac{\partial v}{\partial s} + \beta Dar\left(\frac{1 + \beta - v}{\beta}, v\right) \tag{105}$$

By linearizing Eq. (105) and introducing the change of variable (91b), a single linear PDE is obtained whose space operator is self-adjoint. The stability of the system is then related to the eigenvalues of the problem:

$$\lambda\psi = \frac{1}{Pe}\frac{d^2\psi}{ds^2} - \left(\frac{Pe}{4} + Dar_u - Da\beta r_v\right)\psi \tag{106}$$

with the BCs (102) with w replaced by ψ. In this case it is known that the eigenvalues λ_n form an infinite set of real, discrete values.

Using these properties Amundson (1965) derived a necessary and sufficient condition for stability which requires the solution of an initial value ODE whose coefficients depend only on the steady-state profiles.

The final answer to the stability question was given by Luss and Amundson (1967a), who exploited the possibility of reducing the problem to a single PDE. Using topological concepts they proved that in the case of multiple steady states the low- and high-conversion steady states are asymptotically stable, while the middle one is unstable. Moreover, using the maximum principle, it was also proved that if a unique steady state exists, it is always globally stable.

Again using the maximum principle, Luss and Lee (1968) determined finite regions of stability for each steady state. These allow us a priori to determine the response of the reactor to large perturbations and even the final steady state reached from a given initial condition, without performing any transient calculation. In particular, it was found that the unstable-steady-state profile separates the regions of stability of the two stable steady states. For cases where the initial profile does not entirely lie in one or the other of the two stability regions, it is necessary to integrate the transient model. If at any time during the transient the profile is found to entirely lie inside one stability region, the above-mentioned criterion can be applied again.

It is worth pointing out that in the particular case under examination the axial dispersion model is very similar to the catalyst particle model in the case of Le = 1. Thus the results obtained for one model can be directly extended to the other. In fact, some of the stability conditions reported below, as well as for the above-mentioned work by Luss and Lee (1968), were originally derived for the catalyst particle.

Even though the stability picture for this special case of adiabatic reactor is quite clear at this point, there are two reasons for reporting results obtained with other techniques. The first is that the above-mentioned stability conditions were obtained considering only adiabatic perturbations, which gives a character of conditional stability to the final answer. Luss and Lee (1968) have shown using Eq. (102) that the residual enthalpy monotonically decreases in time, thus indicating a more general validity of the stability criteria developed. Nevertheless, other techniques are available which do not require the use of the residual enthalpy concept at all (Padmanabhan et al., 1971). The second reason is that since the above-mentioned methods apply only to a single PDE, they can be used only in this particular case.

Wei (1965) was the first to apply the direct method of Liapunov to the catalyst particle. Berger and Lapidus (1968) and Padmanabhan et al. (1971) extensively used this method, including more general situations such as the catalyst particle with Le ≠ 1. Varma and Amundson (1972a) applied the method of comparison functions. Both these two last approaches do not require any restriction on the residual enthalpy.

A sufficient condition for stability was first developed by Luss and Amundson (1967a); it is identical to the uniqueness condition (55). This relation has also been derived by several of the above-mentioned authors using various methods. For example, Aris (1969) showed that it could be derived using either the maximum principle or the first or second Liapunov methods. Finally, this expression was also obtained by Gupalo and Ryazantsev (1969) using the Sturm-Liouville eigenvalue problem properties, again limited to adiabatic perturbations.

The issue of the appropriateness of the Danckwerts boundary conditions (52) in the transient case has been addressed in detail by Parulekar and Ramkrishna (1984) for the cases of both zero and nonzero residual enthalpy. Even though these boundary conditions are valid under transient conditions only in the case of no mass or heat dispersion outside the reactor, the stability results were found to be qualitatively quite similar. Besides the multiplicity characteristics, which are obviously unchanged, Parulekar and Ramkrishna (1984) have proved again that a unique steady state is globally asymptotically stable. A sufficient stability condition very similar to Eq. (55) was also obtained in this case, the only difference arising from the effect of the initial state outside the reactor on the reactor transient, which is obviously nonzero only in the presence of dispersion. A detailed analysis of the effects of various mixing intensities before and after the reaction region, as well as of the stability regions, is also presented.

General Nonadiabatic Case

A priori conditions: In the most general case of nonadiabatic tubular reactor, or adiabatic with either Le ≠ 1 or $Pe_m \neq Pe_h$, the heat and energy balances cannot be uncoupled. Thus we have to deal with a system of PDEs which, when linearized, produces a non-self-adjoint space operator. Under these conditions most of the above-mentioned methods either fail or become exceedingly complicated, with only the partial exception of the direct Liapunov method.

One of the very few direct extensions of the results in the preceding section is in the case of an adiabatic reactor with $Pe_m = Pe_h$ and Le ≠ 1. Gavalas (1968) has shown that if a steady state is unstable for Le = 1, it is so also for any other Le ≠ 1.

Nishimura and Matsubara (1969) developed a sufficient condition which requires the integration of a set of initial value linear ODEs whose coefficients depend on the steady-state profiles under examination. Varma and Amundson (1972b), using the method for optimizing the Liapunov functional proposed by Padmanabhan et al. (1971), have shown that the same conditions which guarantee steady-state uniqueness also guarantee asymptotic stability. Thus Eqs. (58) and (59) are sufficient conditions for asymptotic stability for the nonadiabatic reactor with $Pe_m = Pe_h$, Le = 1 and $Pe_m \neq Pe_h$, Le ≠ 1, respectively (note that for a homogeneous reactor Le = 1 implies that $Pe_m = Pe_h$).

More conservative sufficient conditions for stability, obtained using different choices of the Liapunov functional, have been reported by various

authors (Clough and Ramirez, 1972; Liou et al. 1972a,b; Varma and Amundson, 1973a; McGreavy and Soliman, 1973). The effect of recycle on stability has been studied by Matsuyama (1970) for the adiabatic tubular reactor, and by McGowin and Perlmutter (1971b) and Liou et al. (1974) for the nonadiabatic case.

Numerical methods: Due to the lack of sharp sufficient conditions for stability, it is often necessary to resort to numerical techniques. The first approach is the direct integration of the complete transient model. It was used in the earliest work of Raymond and Amundson (1964) and Hlaváček et al. (1971). Since then, many sophisticated numerical methods for solving system of parabolic PDEs have been developed. Some of them have been used in general computer codes, such as PDEPACK (Madsen and Sincovec, 1975) and PDEL (Cardenas and Karplus, 1970). Most of these techniques are based on the method of lines, where the space variable is discretized using a suitable technique such as finite difference or weighted residuals. The obtained system of ODEs is then integrated in time using a suitable marching technique. Since the system of ODEs can be stiff for certain range of parameter values, implicit integration methods (at least partly) are usually recommended. More detailed information about these numerical techniques is reported by Finlayson (1972) and by Michelsen and Villadsen (1978), while a detailed analysis of the application of the orthogonal collocation method to the particular model under examination has been reported by Georgakis et al. (1977a).

Another approach consists of evaluating the eigenvalues for the linearized system of PDEs (92) to (95) given by the non-self-adjoint eigenvalue problem (100). It should be noted that although Eq. (100) is linear, it can only be solved numerically, because some of the elements of matrix $\underline{\underline{B}}$ involve the partial derivatives of the reaction rate at steady state along the reactor, which in general do not admit analytical representation. Usually, a discretization method is used to transform the parabolic PDEs into a system of ODEs. The most common techniques used for this are: finite differences (Kirkby and Schmitz, 1966; Heinemann et al., 1979, 1980; Heinemann and Poore, 1981), Galerkin method (Kuo and Amundson, 1967; McGowin and Perlmutter, 1971a,b; Varma and Amundson, 1973b; Jensen and Ray, 1982), and orthogonal collocation (McGowin and Perlmutter, 1971a; Georgakis et al., 1977a; Jensen and Ray, 1982). The size N of the final system of ODEs depends on the number of terms in the trial solution in the Galerkin method, or of collocation points in the orthogonal collocation method. The final step of the procedure involves evaluation of all the eigenvalues of the discretized system, which can be accomplished through standard routines (Michelsen and Villadsen, 1978).

Based on regularity considerations of the solution, it is expected that by increasing the number of discretization elements, N such that $N \to \infty$, the system of ODEs gives a perfect description of the original system of PDEs. In particular, the N eigenvalues obtained approximate the first N largest eigenvalue of the non-self-adjoint problem. This process is illustrated by the results reported in Tables 2 and 3. In the first one are shown the estimates of the largest eigenvalue as a function of the number of trial functions used in the Galerkin's method (McGowin and Perlmutter, 1971a). The orthogonal set of trial functions used is based on the solution of the same model without chemical reaction. This explains why convergence becomes slower for increasing values of the outlet conversion. In Table 3

TABLE 2 Estimate of the Largest Eigenvalue as a Function of the Number of Trial Functions in Galerkin's Method, n(Pe_m = Pe_h = 10; δ = 0.1)

x	0.10	0.35	0.90	0.995
n = 1	−2.237			
n = 2	−2.227	−0.02909	−0.2498	
n = 3	−2.227			
n = 4	−2.227	0.4212	0.3404	−1.413
n = 6		0.4537	0.5382	−0.6879
n = 8		0.4553	0.9819	−0.5102
n = 10		0.4554	1.275	−0.1721
n = 12		0.4554	1.386	−0.1277
n = 14		0.4554	1.412	−0.1144
n = 16			1.415	−0.1002
n = 18			1.415	−0.0991
n = 20				−0.0986
n = 22				−0.0982
n = 24				−0.0982

are shown the first 15 eigenvalues estimated using the orthogonal collocation methods with 7 (i.e., N = 15) and 15 (i.e., N = 31) collocation points, respectively (Georgakis, et al., 1977b). It appears that while convergence already seems to have been achieved for the largest eigenvalues (which are those that determine the reactor dynamics) this is not true yet for the smaller ones.

About the discretization procedure, McGowin and Perlmutter (1971a) have compared the Galerkin and the orthogonal collocation methods. They found that the first is more rapidly convergent, and also converges monotonically, while the second converges in a damped oscillatory manner. In general, they recommended Galerkin's method, unless the model contains mixed boundary conditions, as in the presence of recycle. These observations have been confirmed by Jensen and Ray (1982).

Finally, it is worth commenting on the correctness of this entire procedure, since as mentioned above, a rigorous mathematical relation between the stability of the original nonlinear system and that of the linearized system is not available for distributed parameter models. Varma and Amundson (1973b) successfully checked the stability of the system predicted through the linearization method in several cases by direct integration of the original model. Since these results have been analyzed in detail by Varma and Aris (1977), this argument is not pursued further here. Some of the results obtained regarding the stability of multiple steady states are discussed in

TABLE 3 Estimates of the First 15 Eigenvalues Using 7 and 15 Collocation Points

	7 collocation points	15 collocation points
j	λ_j	λ_j
1	+3.001	+3.007
2	−0.972	−0.977
3	−4.273 + 3.322i	−4.292 + 3.321i
4	−4.273 − 3.322i	−4.292 − 3.321i
5	−6.793	−6.766
6	−11.02 + 1.937i	−11.03 + 1.981i
7	−11.02 − 1.937i	−11.03 − 1.981i
8	−19.51 + 2.097i	−19.28 + 1.857i
9	−19.51 − 2.097i	−19.28 − 1.857i
10	−29.14 + 1.210i	−33.20 + 1.969i
11	−29.14 − 1.210i	−33.20 − 1.969i
12	−71.84 + 3.602i	−50.95 + 1.933i
13	−71.84 − 3.602i	−50.95 − 1.933i
14	−85.42 + 1.352i	−72.66 + 1.931i
15	−85.42 − 1.352i	−72.66 − 1.931i

the next section, where the occurrence of periodic solutions is also taken into account.

Bifurcation and Dynamic Behavior

As we have noted in the case of a CSTR, the most efficient and sharpest technique available for investigating multiplicity and dynamic behavior is bifurcation analysis. A rigorous mathematical treatment of the extension of this method to distributed systems can be found in Crandall (1977) and Iooss and Joseph (1980). From the application point of view, the main conclusion is that Hopf bifurcation theorem can be extended to the case of parabolic PDEs, by approximating these with an infinite set of ODEs using one of the discretization techniques discussed above. In other words, the bifurcation behavior of the system of Eqs. (86), (87), (52), and (89) is determined by the eigenvalues of the discretized system of linearized ODEs given by Eqs. (96) to (98) [i.e., the solution of the non-self-adjoint eigenvalue problem (100)]. As in the case of lumped systems, two types of bifurcation are possible: a static bifurcation, which occurs when one real eigenvalue is zero, and a Hopf bifurcation, which occurs when a pair of complex eigenvalues has zero real part (i.e., they are purely imaginary). Again, in order to have bifurcation to periodic solution it is necessary that

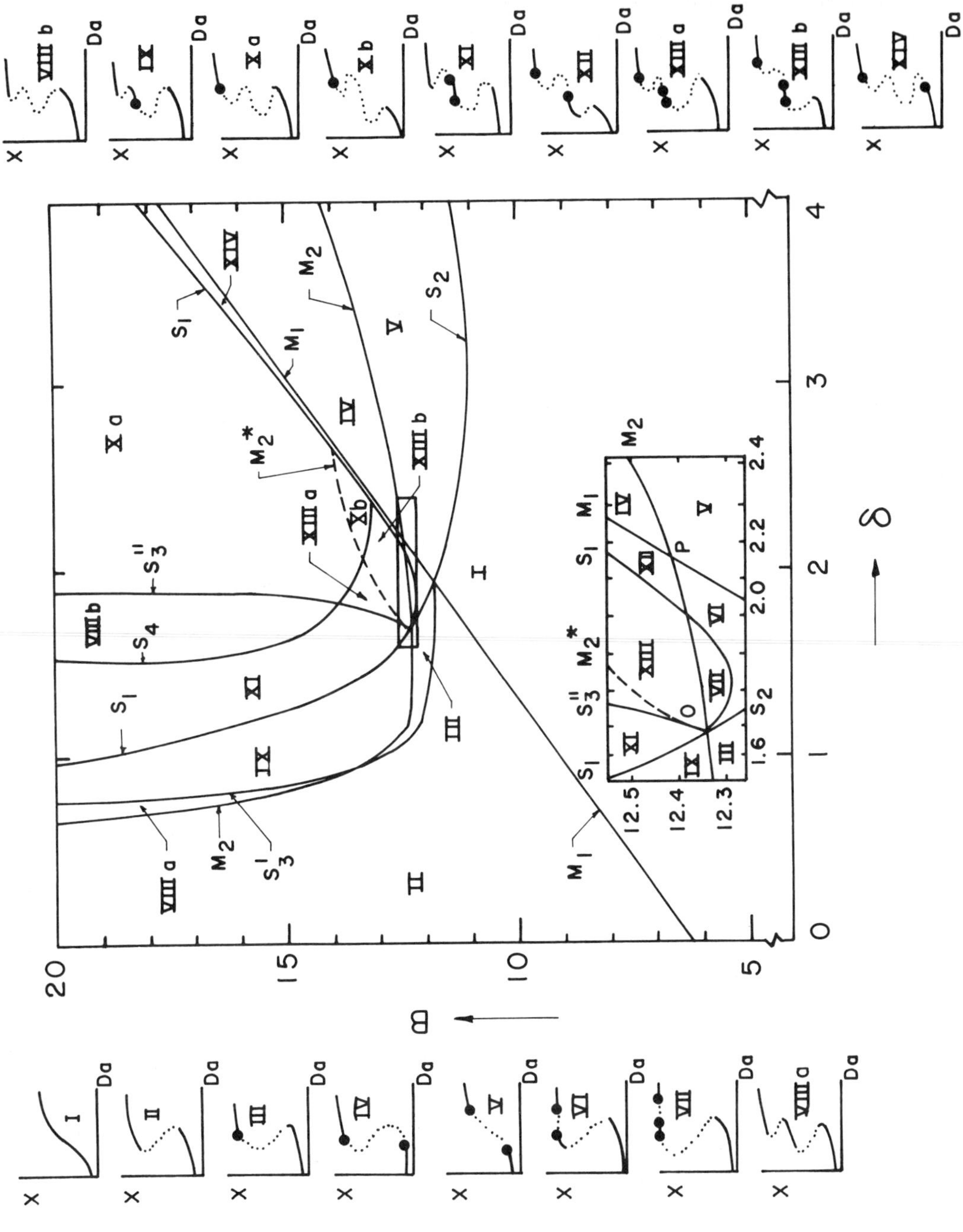

Da
X
B
δ
20
15
10
5
0
1
2
3
4
M1
M2
S1
S2
M2*
S3''
S3'
S4
I
II
III

all the other eigenvalues, besides the pair of purely imaginary ones, have a negative real part, so that the system actually undergoes a change in its stability character.

Two different types of numerical techniques are usually employed for locating such bifurcation points. For static bifurcation points one needs to solve the steady-state problem and calculate the value of the bifurcation parameter at which one of the eigenvalues is zero. One way is to solve the steady state with the usual techniques for any given value of the bifurcation parameter, and use the Euler-Newton continuation method to satisfy the bifurcation condition. To avoid numerical instabilities near the bifurcation points where the Jacobian is singular, it is necessary to adopt the solution normalization method proposed by Keller (1977). A detailed description of this method, with specific reference to premixed laminar flame models, is reported by Heinemann et al. (1979).

A different method has been used successfully by Jensen and Ray (1982) in the bifurcation analysis of tubular reactor models. This method is based substantially on the same idea proposed previously by Weisz and Hicks (1962) for one-dimensional problems, and then applied to multidimensional problems by Sorensen et al. (1973). It consists of fixing one value of the unknown solution (e.g., at one collocation point) and taking in its place the Damkhöler number (or Thiele modulus for catalyst particles) as unknown. The problem so formulated usually admits a unique solution, and it is then possible to solve it for any given Da, even in the multiplicity region.

The Hopf bifurcation points can be obtained by evaluating the eigenvalues corresponding to a given value of the bifurcation parameter through the method indicated in the preceding section, and then searching for the bifurcation parameter value that satisfies the bifuraction condition through standard techniques, such as section, Newton-Raphson, and so on. Kubicek (1979) has suggested including the Hopf bifurcation condition in the nonlinear system of algebraic equations which solve the steady-state problem (after discretization) and then solving it all together with a Newton-Raphson method. Jensen and Ray (1982) have suggested using necessary conditions for the characteristic polynomial to have purely immaginary roots to improve the efficiency of the bifurcation point search.

The bifurcation theory has recently been applied to the axial dispersion model in the case of a first-order reaction by Heinemann and Poore (1981) and Jensen and Ray (1982). Some of the above-mentioned numerical methods have been applied in these two works, and the reported agreement between the obtained Hopf bifurcation points constitutes a check on their correctness.

In the first of these two papers the problem of stability of the periodic solution was also considered. Formulas of the same type as those illustrated for the CSTR [see Eq. (79)] were reported. They allow us, on the basis of only steady-state information, to derive both the direction and the stability character of the bifurcating periodic solution.

FIGURE 20 Regions in the parameter plane B versus δ, surrounded by the sketches of the corresponding bifurcation diagrams in the x versus Da plane (parameter values: $Pe_m = Pe_h = 5$, $Le = 1$, $\Theta_c = 0$, $\gamma = 20$). Irreversible first-order reaction in a tubular reactor with axial dispersion.

Due to the larger number of parameters involved, a picture of dynamic behavior as complete as that reported for the CSTR is now unattainable. We shall therefore be content with examining some of the novel features exhibited by this system. For the special case of adiabatic reactors the situation is again quite clear. Namely, since the eigenvalue problem in this case is self-adjoint, the eigenvalues are all real and no bifurcation to periodic solutions can occur. To summarize: this reactor can exhibit at most three steady states, with alternate stability character (i.e., stable-unstable-stable), and can never exhibit self-sustained oscillations.

On the other hand, when the adiabaticity condition is removed, the dynamic behavior becomes much more complex. Up to seven steady states with various combinations of stability character can be found, including the case where *all* steady states are unstable, which obviously implies the occurrence of periodic solutions.

In Fig. 20 the same type of representation of the parameter space as in Fig. 15 for the CSTR has been adopted to illustrate the dynamic behavior of the axial dispersion model. In particular, the case where $Pe_m = Pe_h = 5$ is examined. This situation is thus intermediate between the two well-known ideal cases of the CSTR and the plug-flow reactor (Cohen and Poore, 1974): the latter always admits a unique globally stable steady state, since it is described by a system of initial value ODEs. In Fig. 20, it appears that besides curves M_1, S_1, and S_2, which correspond to curves M, S, and SM, respectively, for the CSTR, one extra multiplicity curve M_2 appears together with three more Hopf bifurcation curves: S_3', S_3'', and S_4. Curve M_2 bounds the region where two additional steady states occur (the fourth and fifth) and curves S_3' and S_3'' envelope the region where two more Hopf bifurcation points appear. Curve S_4 represents the locus where two Hopf bifurcation points coalesce. On the whole 14 regions characterized by qualitatively different bifurcation diagrams of outlet conversion versus Da are found. The first six of these are the same as those of the CSTR, while the others are quite different, as is evident from the sketches shown in the side of Fig. 20. It is noticeable that, besides the occurrence of five steady states in regions VII and XIII, the bifurcation diagram exhibits three Hopf bifurcation points. As a consequence, quite different phase portraits from any of those found for the CSTR (see Fig. 18) are generated.

A similar analysis has been performed taking the Lewis number as the bifurcation parameter, in the same manner as Ray and Hastings (1980) did for the CSTR. This leads to the discovery of more exotic bifurcation diagrams, the most peculiar being perhaps those shown in Fig. 21, which include for a very small Da range (on the right of the lower Hopf bifurcation point) the possibility of having five unstable steady states surrounded by a stable limit cycle. More interesting on practical grounds are the reported critical values of the Lewis number (i.e., the Le value above which no oscillations are possible). Jensen and Ray (1982) have shown that this value is less than 1 for any $Pe_m = Pe_h > 14$, which indicates that even in homogeneous reactors (Le $\cong$ 1) oscillations can occur only in very short reactors. Oscillations in a packed bed, where the Lewis number is of the order of few hundreds, are then quite unlikely, at least as far as the phenomena described by the pseudohomogeneous axial dispersion model with a simple reaction rate expression are concerned.

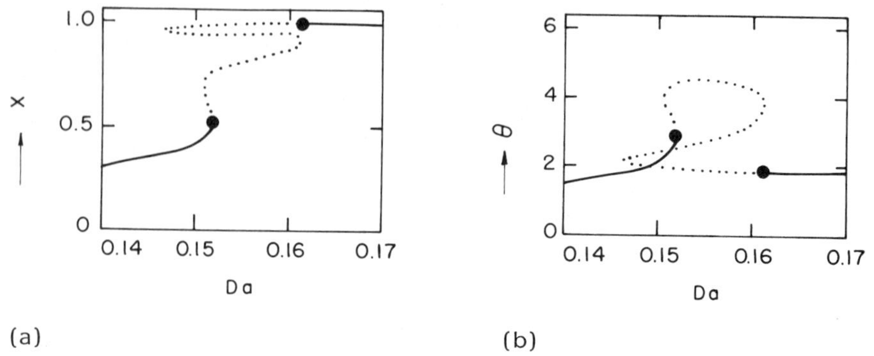

FIGURE 21 Bifurcation diagrams (a) x versus Da, and (b) Θ versus Da characteristic of region XIV (parameter values: Le = 0.8, δ = 3, B = 15, other parameters as in Fig. 20).

COMPLEX TRANSIENTS IN CSTRs

As we have already observed from the steady-state multiplicity point of view, the complexity of behavior of a CSTR increases significantly for even a slight increase in the reacting system complexity. This was observed earlier in the cases of an irreversible first-order reaction in two CSTRs in series, and of two such reactions occurring in series in a CSTR. The aim of this section is to indicate some of the exotic dynamic behavior which can occur in these systems. For completeness, a few comments on the stability character of the multiple steady states discussed earlier will be made first.

First-Order Reaction in a Series of Two CSTRs

We have already mentioned that a series of N CSTRs, with a first-order irreversible reaction, can exhibit at most $(2^{N+1} - 1)$ steady states, of which no more than (N + 1) can be stable. The stability of two reactors in series has been examined by Kubicek et al. (1980) and Svoronos et al. (1982), by applying Liapunov's first method to the transient mass and energy balances, which can be readily derived from Eqs. (30) and (31). Since the Jacobian of this system is equal to the product of the Jacobian of the two single reactors, it follows that a necessary and sufficient condition for stability is that each of the reactors is itself stable. This conclusion does not hold if a recycle stream is present. Through a detailed analysis of the eigenvalues of these two Jacobians, Svoronos et al. (1982) determined the stability character of each possible steady state, as shown in Fig. 8.

Other novel features have been shown by Kubicek et al. (1980), who explored different regions of the parameter space and took the reactor residence time as a bifurcation parameter. As an example, two bifurcation diagrams (outlet temperature versus residence time) obtained for different set of heat transfer parameter values δ_1° and δ_2° are shown in Fig. 22. The first one (Fig. 22a) exhibits three ignition and three extinction points—according to the occurrence of five steady states. In the second one

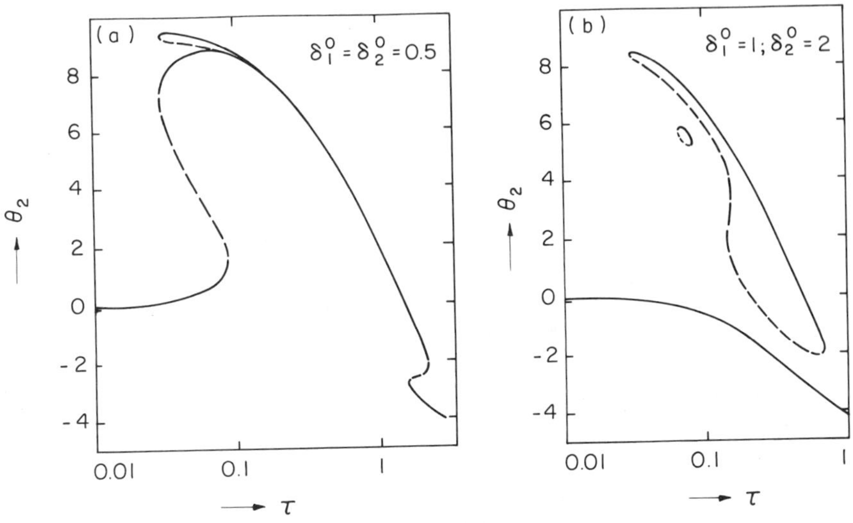

FIGURE 22 Characteristic bifurcation diagram (outlet temperature versus residence time) for two CSTRs in series with a first-order reaction (parameter values: $Da_1^o = Da_2^o = 0.5$, $\gamma = 20$, $B = 10$, $v_{c1} = v_{c2} = 0.75$).

(Fig. 22b) two isolas are present, leading to a total of three separate curves on the same bifurcation diagram. Since the smaller isola lies entirely in the shadow of the larger one, it can never be reached from any of the other two curves simply by changing the reactor residence time.

Svoronos et al. (1982) have applied Hopf-Friedrichs bifurcation theory to this model and reported an analytical relationship for the direction and stability of the bifurcating periodic orbits. In the case of a single reactor, this reduces to the same equation [Eq. (79)] as that developed by Poore (1973). It has been found that while in this case also, periodic solutions cannot occur in the adiabatic system, the dynamic behavior is in general more complex than in a single CSTR. For example, it is possible that self-sustained oscillations in the first reactor may force oscillations in the second, which, per se, would have been at a stable steady state.

Two Consecutive Irreversible First-Order Reactions in a CSTR

The multiplicity pattern of this reacting system was discussed earlier. Since stability analysis of these steady states can be performed with the usual methods, we shall concentrate immediately on the dynamic behavior. First note that the transient model is given by three nonlinear ODEs which can readily be obtained from the steady-state mass and energy balances (40) by adding on the right-hand sides the accumulation terms for the mass of A, mass of B, and energy, respectively. We have already discussed the extension of bifurcation theory from two to N dimensions, in the context of its application to a system of N ODEs arising from the discretization of one or more parabolic PDEs. From this point of view, the three-dimensional system now under examination can be regarded as the simplest case (i.e., N = 3).

Hlaváček et al. (1972) first reported oscillations in a CSTR with two exothermic first-order reactions, by direct integration of the transient model. Halbe and Poore (1981) applied Hopf-Friedrichs bifurcation theory and performed an extensive search in the parameter space to identify all possible shapes of the first reactant conversion versus Damköhler number bifurcation diagram. Ten qualitatively different bifurcation diagrams were found, which included situations where up to five steady states and four Hopf bifurcation points were present simultaneously. Relationships were also reported for the a priori definition of the direction and stability of the bifurcating periodic orbits.

The determination of the Hopf bifurcation points is only the first step toward defining the dynamic behavior of this system, whose detailed understanding requires a detailed analysis of each periodic branch. This involves an identification of the structure of the system dynamics as the bifurcation parameter changes away from the Hopf bifurcation point. Analysis of this type usually requires tedious integration of the transient model equations. Doedel and Heinemann (1983) have applied a continuation method for the calculation of the entire branch of periodic solutions, both stable and unstable. This method is based on a transformation of the original initial value problem into a boundary value problem, where the period is taken as unknown and correspondingly the periodicity condition is added. The final equations are then discretized using orthogonal collocation, and the resulting system of nonlinear algebraic equations is solved through Newton's method. This procedure has been implemented by Doedel (1981) on a computer code which also provides the steady-state solution branches using Keller's (1977) arc-length method for avoiding singularity problems in the Jacobian at the static bifurcation point, as well as the Hopf bifurcation points and the bifurcating periodic solution, using the usual perturbation techniques. This appears to be quite a useful tool for the complete study of the dynamics of a given system, since it gives the entire bifurcation diagram (i.e., both steady state and periodic solution branches, both stable and unstable). The stability character of periodic solutions is determined through the corresponding Floquet multipliers.

Using this program, Doedel and Heinemann (1983) have completed the 10 bifurcation diagrams previously reported by Halbe and Poore (1981) by computing the entire periodic solution branches. One of the most peculiar bifurcation diagrams so obtained is sketched in Fig. 23a, where four Hopf bifurcation points occur: in all of them bifurcation occurs to the left; the first and third are unstable, while the second and fourth are stable. Now, starting at large Da and decreasing it, the system undergoes a soft bifurcation into a periodic stable orbit and follows this branch until it becomes unstable. Here a hard bifurcation occurs to another periodic solution, with smaller oscillations, and eventually the system reaches the lowest stable steady state through one more hard bifurcation. It may be noted that in this diagram the classic S-shaped multiplicity pattern is reproduced with periodic instead of steady-state solutions.

Another interesting and peculiar bifurcation diagram is shown in Fig. 23b. Here the periodic solution branches emanating from the third and fourth Hopf bifurcation points are not connected, but terminate at an infinite-period bifurcation point (indicated by the symbol +). Note that due to the stability character of these two branches and the relative position of the respective termination points, there is always a stable steady state or periodic solution available for the system for any Damköhler number value.

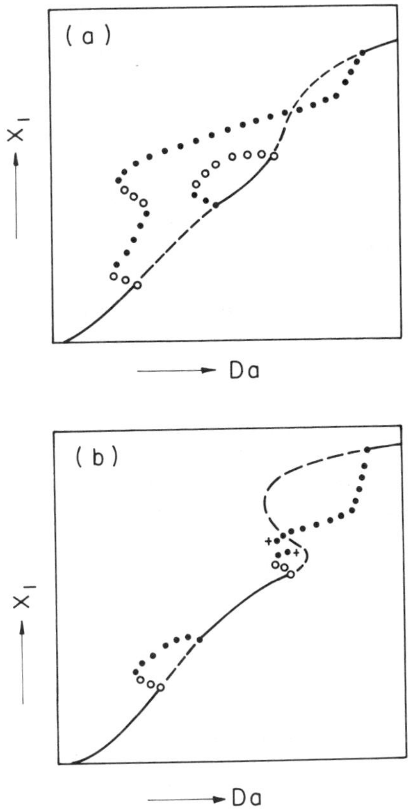

FIGURE 23 Qualitative sketches of two bifurcation diagrams (outlet conversion of the first reactant versus Damköhler number) characteristic of a CSTR with two consecutive exothermic reactions.

The detailed analysis of periodic solutions does not exhaust all possible features of the dynamics of the system under examination, because it can also exhibit nonperiodic solutions (i.e., chaotic behavior). This peculiar behavior, which according to Poincaré-Bendixson theorem cannot occur in a two-dimensional system (cf. a CSTR with one reaction), is instead quite common in models of larger dimensionality, just as periodic oscillations are common in two-dimensional systems. Kahlert et al. (1981) found chaos in a CSTR with two consecutive reactions: the first exothermic and the second endothermic. The direct integration of the transient model equation leads, for a certain set of parameters, to the trajectory in the phase space shown in Fig. 24. The presence of a well-defined attraction region, shaped as a folded-over rubber strip, is an indication of the chaotic behavior of the system.

The way the system approaches this situation is through a typical Feigenbaum (1979) cascade of periodic solution bifurcations. In particular, with reference to Fig. 25 and for the parameter values reported in the caption, it is seen that the system admits two Hopf bifurcation points, both bifurcating to the right: the first stable and the second unstable. The periodic solution branch can be constructed through subsequent

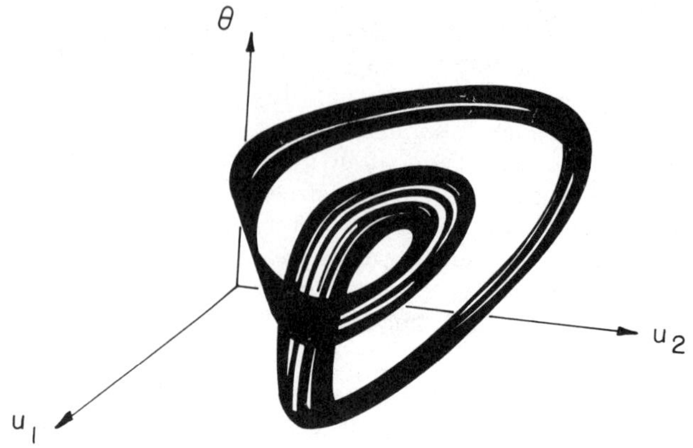

FIGURE 24 Shape of the attraction region in the chaotic regime in the case of one exo and one endothermic reaction (parameter values: $\gamma = \infty$, $\Theta_c = 0$, $\nu = 0.5$, $\alpha = -0.426$, $\lambda = 1$, Da = 0.26, B = 57.77, $\delta = 7.995$).

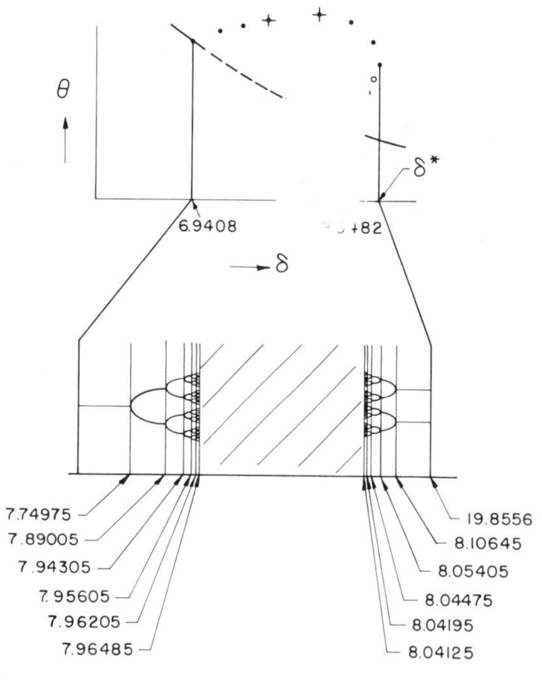

FIGURE 25 Bifurcation diagram: outlet conversion of the first reactant versus heat transfer parameter, other parameters as in Fig. 24.

numerical integrations starting from the left Hopf bifurcation point and increasing the value of δ. It is found that by increasing δ the period of the stable oscillations changes continuously up to a critical δ value where suddenly a doubling in the oscillation period occurs. In other words, the periodic orbit becomes unstable and simultaneously a new one, stable and with doubled period, appears. This new type of bifurcation, which is often referred to as *Poincaré bifurcation*, occurs again and again at successively closer δ intervals, up to a new critical δ value where chaotic behavior begins. For a particularly simple one-dimensional case, Feigenbaum has shown that the δ values at which period doubling occurs tend to approach a convergent geometric series, such that the ratio

$$\rho = \frac{\delta_i - \delta_{i-1}}{\delta_{i+1} - \delta_i} \tag{107}$$

is constant and equal to the universal value 4.6642016.

Chemburkar et al. (in press) have further investigated this system by analyzing in detail the periodic orbits that branch off from the hard Hopf bifurcation point δ^* (all the numerical values reported here are taken from this last source). The same type of bifurcation pattern has also been found in this region, again ending in an infinite period termination point, as qualitatively sketched in Fig. 25. By plotting the number of peaks per period (which is proportional to the oscillation period) as a function of the bifurcation parameter, as shown in Fig. 26, it clearly appears that there exists a window of δ values where the system exhibits chaotic behavior.

It should be mentioned that the way this system approaches the chaotic region is qualitatively similar to that described by Feigenbaum (1979), but the geometric series of the bifurcation δ values is not perfectly reproduced (at least after the few doublings investigated), and the computed values of the ratio ρ [Eq. (107)] are somewhat smaller than the natural Feigenbaum's constant (ranging from 3 to 3.5).

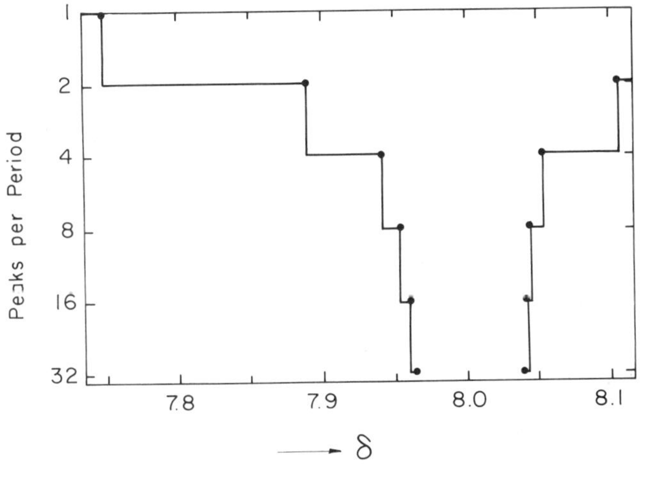

FIGURE 26 Window of chaotic behavior.

Similar behavior has been found by Jorgensen and Aris (1983), who studied the same reacting system, but considering two exothermic reactions. In this case some more complex dynamic behavior has been identified. A useful tool for enlightening the bifurcation pattern of the system is offered by Floquet multipliers. Let us define $\underline{\xi}$ as the three-dimensional vector of periodic solutions of the transient model characterized by a period value $\bar{\tau}$. Its stability can be determined by considering the solution of the perturbation problem

$$\frac{d\underline{\underline{\sigma}}}{dt} = \underline{\underline{J}}\,\underline{\underline{\sigma}} \tag{108}$$

with the initial condition $\underline{\underline{\sigma}} = \underline{\underline{I}}$, the identity matrix, where $\underline{\underline{\sigma}}$ is a 3 × 3 matrix and $\underline{\underline{J}}$ is the Jacobian of the original system evaluated along the periodic solution $\underline{\xi}$. Introducing now the eigenvalues μ of the matrix $\underline{\underline{\sigma}}$, it can be shown that one of these is equal to unity (this gives a useful check for the accuracy of the adopted numerical integration technique) and the system is asymptotically stable if and only if all the other eigenvalues are such that

$$|\mu| < 1 \tag{109}$$

In the bifurcation study of periodic solutions it is then quite convenient also to compute such eigenvalues—the so-called Floquet multipliers. In the system under examination the doubling of a periodic solution corresponds to the situation where one of the multipliers crosses the unit circle in the complex plane at −1, and then immediately reappears inside the circle (i.e., the orbit becomes unstable and a new one stable and with double period emerges).

Some of the results obtained by Jorgensen and Aris (1983) are shown in Fig. 27, where the heat transfer parameter δ is used as a bifurcation parameter. The various types of dynamic regimes have been indicated by a capital letter. Regime A corresponds to the case where one unstable steady state is present together with a stable limit cycle. Regimes B and C correspond to the case where doubling of the oscillation period occurs (B) up to the emergence of a chaotic region (C). All these features are quite similar to those described above for the case of one endothermic and one exothermic reaction. Regimes D, E, and F correspond to interlude situations separating the occurrence of a chaotic behavior. Even though some bifurcations of periodic orbits, involving doubling of the period, have been detected in these regimes, the connection with the Feigenbaum "route to chaos" appears to be quite weak. Finally, regime G is characterized by quite complex behavior, where periodic solutions and chaos seem to coexist. It should be observed that at this point the numerical accuracy of the integration technique becomes critical, and only some significant advances in the theoretical insight can lead to a better understanding of system behavior. For illustrative purposes, some of the periodic solutions characteristic of regime B and the attraction region characterizing regime C are shown in Figs. 28 and 29, respectively.

In conclusion, the dynamic behavior of a CSTR approaching chaos is not yet well defined, and more research in this area is needed. For example, considering a different kinetic scheme, constituted by two independent first-order exothermic reactions:

$$A \rightarrow B \qquad C \rightarrow D$$

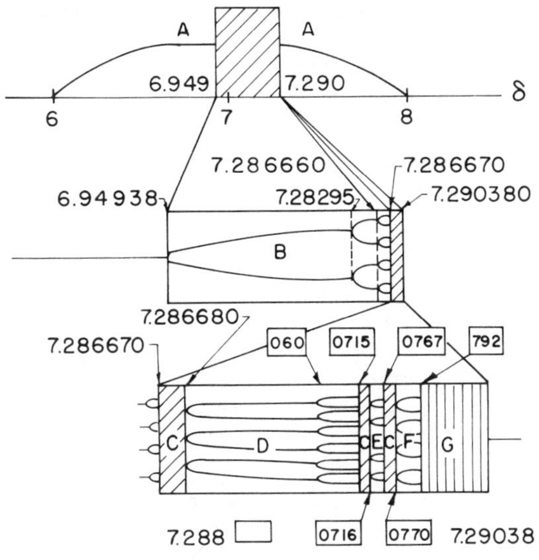

FIGURE 27 Bifurcation pattern, with δ as a bifurcation parameter, in the case of two exothermic reactions (parameter values: $\gamma = \infty$, $\Theta_c = 0$, $\nu = 0.01$, $\alpha = 1$, $\lambda = 1$, Da = 0.55, B = 17.5).

Lynch et al. (1982) have defined the occurrence of chaos, which is reached by the system through a way different from that discussed so far. Moreover, a useful numerical technique for the identification of the occurrence of chaos, based on the autocorrelation analysis, has been applied in this work.

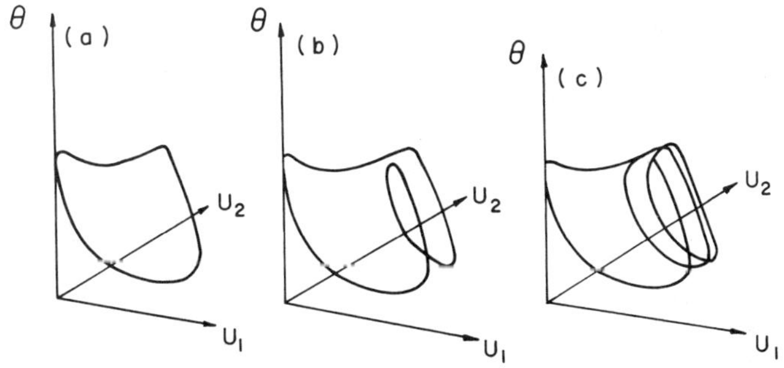

FIGURE 28 Sequence of trajectories in the phase space showing the first two doubling of the periodic orbit in regime B. (a) δ = 6.94938; (b) δ = 7.28294; (c) δ = 7.28665 (other parameters as in Fig. 27).

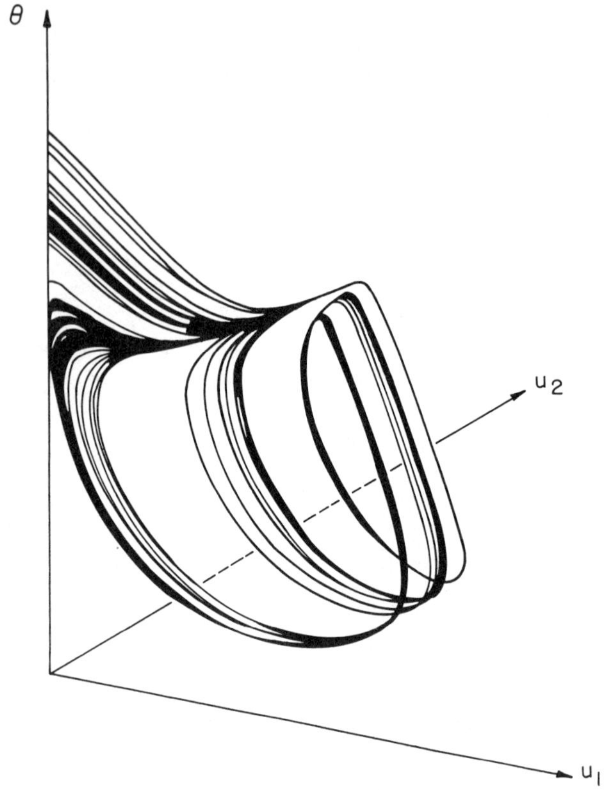

FIGURE 29 Chaotic trajector in regime C; δ = 7.286670 (other parameters as in Fig. 27).

PARAMETRIC SENSITIVITY AND RUNAWAY IN TUBULAR REACTORS

A quite common problem in mathematical modeling, as well as in the experimental analysis of any real-life device, is to establish the effect of each input parameter on any of the model outputs. Quite often there exists regions in the parameter space where small variations of the input parameters cause large variations of the outputs; these are usually called regions of *parametric sensitivity*. This concept was first introduced in the study of tubular reactors by Bilous and Amundson (1956). When a sufficiently exothermic reaction occurs within nonadiabatic tubular reactors, the temperature profile at steady state exhibits a maximum, usually called a hot spot. In the region of parametric sensitivity the magnitude of the hot spot can undergo very large increases even for very small variations of one or more of the input parameters; this phenomenon is usually referred to as temperature *runaway*. This is illustrated qualitatively in Fig. 30, where for a particular example (see van Welsenaere and Froment, 1970) the temperature profile variation as a function of equal changes of the inlet temperature is shown before and after the runaway boundary.

The prevention of excessive temperature peaks along the reactor is very important in practical applications not only for safety reasons but also

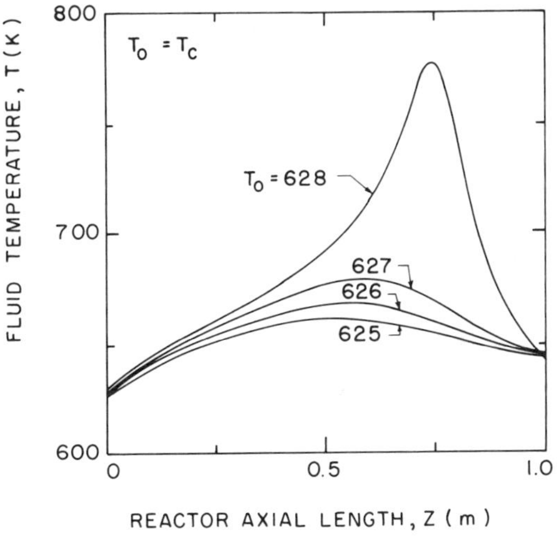

FIGURE 30 Examples of temperature profiles along the reactor axis for various inlet temperature values inside and outside the runaway region.

because these may adversely affect the conversion of exothermic equilibrium reactions, selectivity of the process, and in the case of catalytic reactors, the catalyst activity and durability. Therefore, most of the research in this area has been devoted to providing graphical or analytical representations of the region of runaway, so that such behavior could be avoided immediately at the earlier stage of reaction design. To attain such a representation, two problems need to be solved:

1. Formulation of an a priori criterion of parametric sensitivity, which has to be based on intrinsic reactor behavior with no reference to any specific situation
2. Implementation of the criterion to produce a simple representation, either graphical or analytical, of the region of runaway

Most contributions in the literature have examined the case of a plug-flow reactor with an irreversible nth-order reaction, whose mathematical model is given by

$$\frac{du}{ds} = -\text{Da}u^n \exp\left(\frac{\Theta}{1 + \Theta/\gamma}\right) \tag{110}$$

$$\frac{d\Theta}{ds} = \text{BDa}u^n \exp\left(\frac{\Theta}{1 + \Theta/\gamma}\right) - \delta\Theta \tag{111}$$

with IC

$$u = u_0 \qquad \Theta = \Theta_0 \qquad \text{at } s = 0 \tag{112}$$

where in the definition of the quantities Θ, Da, γ, and B the cooling temperature T_c has been taken as reference temperature T_r (i.e., $\Theta = (T - T_c)E/RT_c^2$, and so on].

The first a priori criterion for runaway was proposed by Barkelew (1959) as a result of an empirical analysis of a large number of temperature profiles along the reactor axis. Dente and Collina (1964) defined as runaway the occurrence of a region along the reactor where the second derivative of temperature with respect to distance is positive. This is physically sould since it identifies runaway with an "acceleration" in the temperature-reactor axis plane. According to this criterion, regions in the parameter space where runaway occurs can be identified, and the boundaries of this are given by the locus of parameter values where the third derivative of temperature versus reactor axis is zero. Van Welsenaere and Froment (1970) have adopted this same definition of runaway. They produced simple explicit analytical expressions for the boundaries of the runaway region by introducing a second criterion, based on the occurrence of a maximum in the locus of the maxima in the temperature-conversion plane, which according to them is substantially equivalent to the previous criterion.

Some additional insight into this problem is given by studies reported earlier in the context of thermal explosion theory. Hlaváček et al. (1969) pointed out that the lumped mass and heat balances in a symmetrically heated reactive material are, in fact, identical to Eqs. (110) and (111), where the reactor length coordinate is replaced by time. The earlier criteria for ignition or explosion were based on the occurrence of a region with positive second-order derivative in the temperature versus time profile (see Semenov, 1928), which coincides with the criterion proposed by Dente and Collina (1964) in the reactor context. Subsequently, Adler and Enig (1964) proposed a somewhat more intrinsic definition of ignition or explosion based again on the occurrence of a positive second-order derivative but in the temperature-conversion (instead of temperature-time) plane. It was also shown that runaway in the temperature-conversion plane implies runaway in the temperature-time plane, but not vice versa. The relevant equation, which can easily be derived from Eqs. (110) and (112) by introducing the conversion $x = 1 - u$, is simply given by

$$\frac{d\Theta}{dx} = B - \frac{\delta}{Da} \frac{\Theta}{(1 - x)^n} \exp\left(\frac{-\Theta}{1 + \Theta/\gamma}\right) \tag{113}$$

with IC

$$\Theta = \Theta_0 \quad \text{at } x = 0 \tag{114}$$

This criterion has been used in the context of tubular reactors by Morbidelli and Varma (1982), who considered reactions of any positive order with no limitation on the activation energy or inlet temperature values. The runaway region was explored using the isocline method, originally introduced by Chambré (1956), which provides quite an efficient procedure for calculating the region boundaries, avoiding tedious trial and error.

The runaway region in the parameter plane B versus δ/Da (i.e., heat of reaction versus heat transfer parameter) is shown in Figs. 31 and 32 for various values of the reaction order, activation energy, and inlet temperature. For parameter values above the line (i.e., large values of the heat of reaction parameter B and low values of the heat transfer parameter δ/Da), the reactor undergoes runaway, while it does not runaway for parameter values below the line. As expected on physical grounds, it is

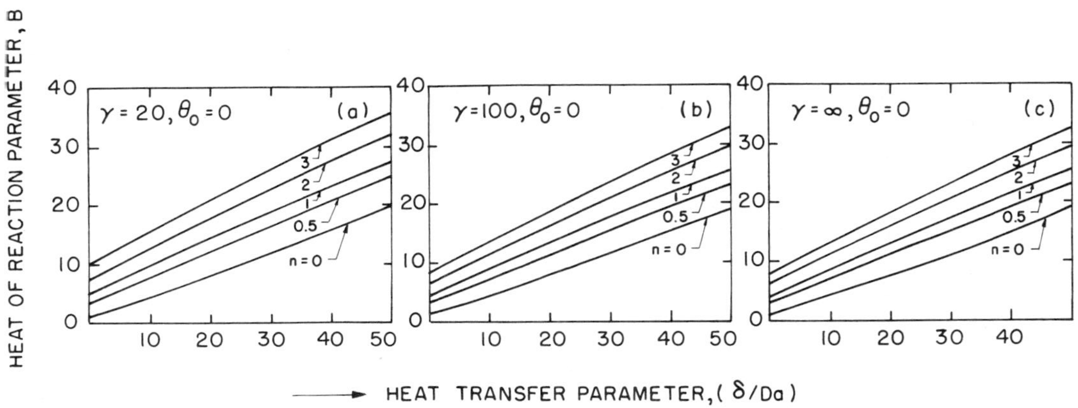

FIGURE 31 Effect of the reaction order on the runaway region for three values of the activation energy; the runaway region is above the curve.

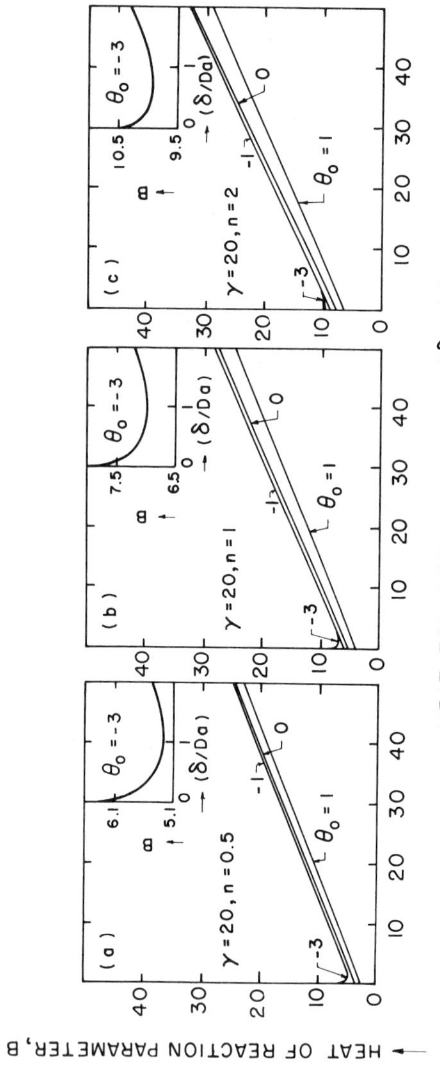

FIGURE 32 Effect of the inlet temperature on the runaway region for three reaction orders; the runaway region is above the curve.

also apparent that runaway is more likely for larger values of the activation energy γ and the inlet temperature Θ^i and for smaller reaction order n.

A detailed comparison between the runaway region predicted by the latter criterion and those predicted by the earlier criteria proposed by Barkelew (1959), Dente and Collina (1964), and van Welsenaere and Froment (1970) has been reported by Morbidelli and Varma (1985). It was found that in general the criterion based on the temperature-conversion plane leads to the less conservative results, and that the discrepancy becomes more significant for low values of the activation energy γ and the heat transfer parameter δ/Da. This is apparent from Fig. 33, where the runaway regions calculated by van Welsenaere and Froment (1970) (dashed line) and Morbidelli and Varma (1982) (solid line) are compared.

On the other hand, this criterion provides estimates of the nonrunaway region which in fact correspond to safe operating condition for the reactor. For example, from inspection of the temperature versus conversion profiles shown in Fig. 34 for increasing values of the heat of reaction parameter, it appears that the critical value B_c = 14.31 indeed represents a reasonable boundary between safe and runaway operating conditions. Note that the critical value of B_c is defined as the boundary between runaway and no-runaway, and thus corresponds to the situation where the third as well as the second derivative of temperature with respect to conversion vanish. In conclusion, this criterion can be recommended for designing tubular reactors operating outside the runaway region, without requiring serious limitations on the operating conditions.

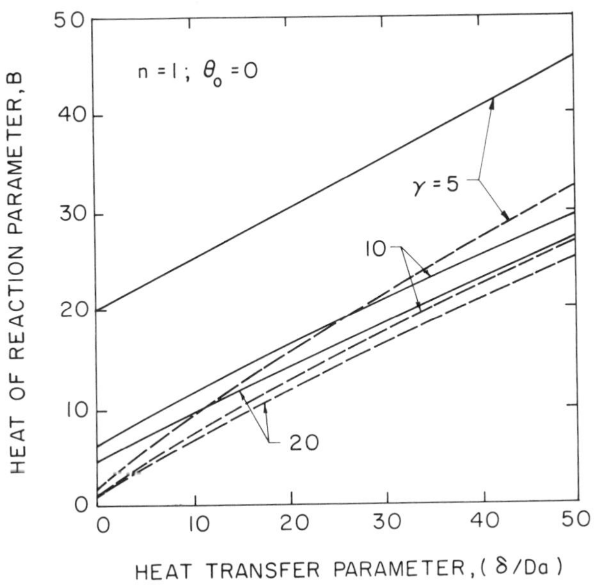

FIGURE 33 Comparison between the runaway region calculated according to the criterion of Morbidelli and Varma (1982) (solid line), and the one by van Welsenaere and Froment (1970) (dashed line).

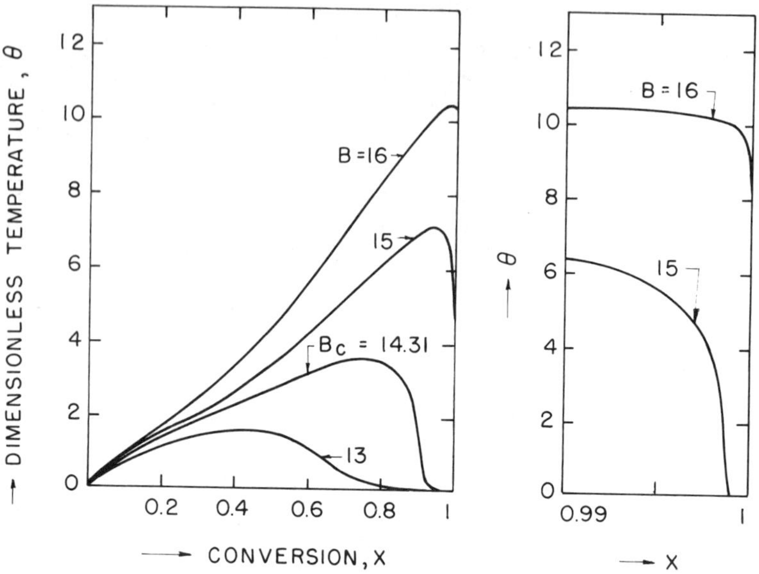

FIGURE 34 Temperature versus conversion profiles inside and outside the runaway region (parameter values: $n = 1$, $\gamma = 20$, $\delta/Da = 20$, $\Theta_0 = 0$).

Following a procedure adopted by Gray and Lee (1965) and Hlaváček et al. (1969), Morbidelli and Varma (1985) have proposed the following analytic representation of the runaway region boundaries:

$$\frac{\delta}{Da} = \lambda eB\left[1 - \left(\frac{B_c^\circ}{B}\right)^{2/3}\right] \tag{115}$$

where

$$\lambda = \frac{8.7}{7.66 + n^{0.6}} \tag{116}$$

is an empirical factor introduced to improve accuracy. B_c° represents the critical B value in the case of an adiabatic reactor (i.e., $\delta = 0$). Using the isocline method, the following explicit expression can be derived:

$$B_c^\circ = \bar{\Theta} - \Theta_0 - \frac{n\bar{\Theta}(1 + \bar{\Theta}/\gamma)^2}{(1 - \bar{\Theta}/\gamma)^2 - \bar{\Theta}} \tag{117}$$

where $\bar{\Theta}$ is the unique solution of the equation

$$(n - 1)\bar{\Theta}^4 + 2\gamma(n - 1)(2 - \gamma)\bar{\Theta}^3 + [2(n - 1)(3 - \gamma) - \gamma(\gamma - 2)]\gamma^2\bar{\Theta}^2$$
$$+ 2[2(n - 1) + \gamma]\gamma^3\bar{\Theta} + (n - 1)\gamma^4 = 0 \tag{118}$$

in the internal $\overline{\Theta} \in [\Theta_{-},\Theta_{+}]$, where $\Theta_{\pm} = [(\gamma - 2) \pm \sqrt{\gamma(\gamma - 4)}]\gamma/2$. To facilitate the use of Eq. (115), the values of B_c°, calculated from Eqs. (117) and (118) as a function of γ, are shown in Fig. 35 for various reaction orders.

In Fig. 36 a comparison between the runaway region predicted by Eq. (115) and that numerically calculated by Morbidelli and Varma (1982) is shown. The agreement seems to be quite satisfactorly for any value of the reaction order and the activation energy. It should be mentioned that less satisfactory results are obtained for inlet temperature values Θ_0 significantly different from zero. However, since this situation is quite unlikely in industrial applications, this point has not been pursued further.

Several attempts have been made in the literature to extend this sensitivity analysis to more complex models to account for other phenomena which may also be of importance in establishing runaway operating conditions of industrial reactors. In particular, Agnew and Potter (1966) have extended Barkelew's (1959) criterion to account for heat and mass axial as well as radial dispersion; Soria Lopez et al. (1981) extended van Welsenaere and Froment (1970) criterion to the case where the cooling fluid temperature changes along the reactor; finally, the effect of side reactions on the runaway region has been investigated by Dente et al. (1966a) in the case of consecutive reactions and by Dente et al. (1966b) and Westerterp and Ptasinski (1984) in the case of parallel reactions. It should be mentioned that runaway behavior is most likely to occur in catalytic reactors, where the intrinsic heterogeneous nature of the reacting system can play a very important role. However, the effect of inter and intraparticle mass and heat transfer resistances on runaway has not received much attention so far in the literature. McGreavy and Adderley (1973) have analyzed the effect of transport limitations on a single catalyst particle;

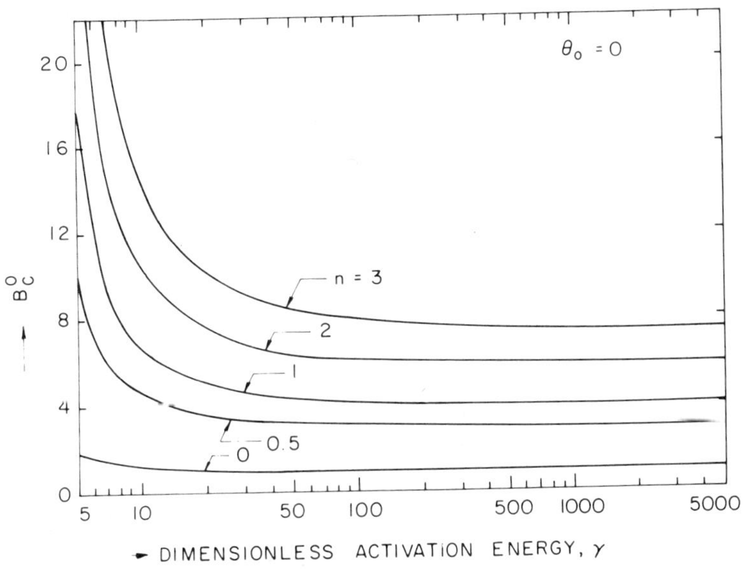

FIGURE 35 Values of the adiabatic critical parameter B_c° from Eq. (117) as a function of γ for various orders of reaction and $\Theta_0 = 0$.

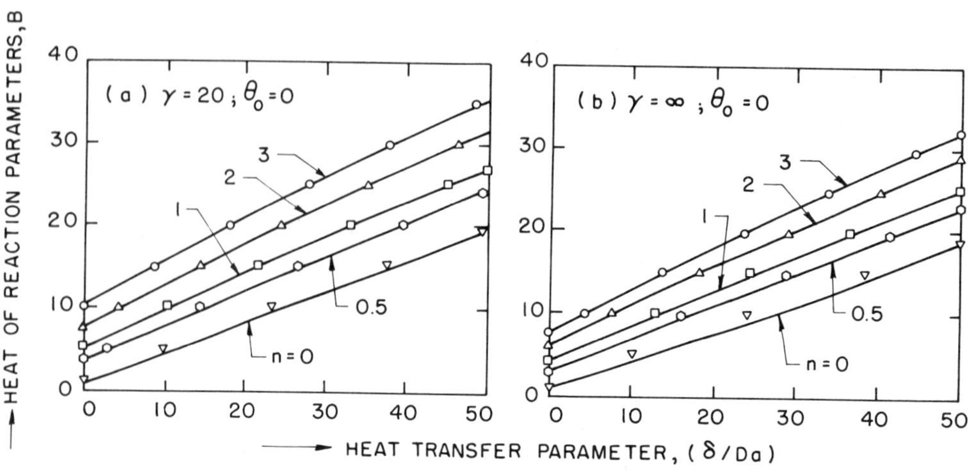

FIGURE 36 Comparison of the runaway region predicted by Eq. (115) (dotted line) and the exact one calculated by Morbidelli and Varma (1982) (solid line).

Rajadhyaksha et al. (1975) have extended van Welsenaere and Froment (1970) criterion to account for transport limitations in four particular regimes of reactor operation; finally, the case of multitubular reactors has been investigated by McGreavy and Dunobbin (1984).

APPENDIX: SINGULARITY THEORY

Singularity theory provides a systematic way of tackling the multiplicity question. It is concerned with three basic ideas about maps or functions which bear directly on the search for the pattern of solutions of a nonlinear equation. The first of these is stability and we say, that a map is stable if small perturbations do not really change its character. Thus the graph of $u = x^2$ is merely shifted up to down by a perturbation $u = x^2 + \alpha$; it is distorted by the perturbation $u = x^2 + 2\beta x$, but by writing this as $u = (x + \beta)^2 - \beta^2$ we see that its qualitative shape is in no way changed. On the other hand, $u = x^3$ is a monotonic curve which a perturbation $u = x^3 - \alpha x$ will change into a curve with a maximum and minimum (i.e., qualitatively different) if $\alpha > 0$, however small it may be. Singularity theory provides a way of determining when a map is stable. For example, the two dimensional mapping $u = x^3 - xy$, $v = y$ is stable, for its cusp-like character is unchanged by small perturbations such as $u = x^3 + 3\alpha x^2 - xy$, $v = y + \beta$, and so on. Now if $y = \alpha$ is fixed, $u = x^3 - x\alpha$ and for $\alpha = 0$, $u = x^3$. So $f(x) = x^3$ is *embedded* in $F(x,\alpha) = x^3 - \alpha x$ or F is an *unfolding* of f. The second great topic of singularity theory is the question of how to unfold an unstable map such as f(x) into a stable map such as $F(x,\alpha)$ using the least number of parameters. For example, $x^3 - \alpha$ will not do because it is still unstable and $x^3 + 3\beta x^2 + \alpha x$ is uneconomical, for it may be written as $(x + \beta)^3 + (\alpha - 3\beta)(x + \beta) - \beta^3 + 3\beta^2 - \alpha\beta$ and β is otiose. The third question is one of *genericity*, that is, of whether there are enough stable maps to approximate an arbitrary map.

For multiplicity problems we need a systematic way of finding the boundaries of parameter space across which the number of roots of a nonlinear equation changes. These are sometimes called *catastrophe surfaces* or *sets* after René Thom, a pioneer of singularity theory, classified the elementary catastrophes and gave their canonical unfoldings. In the literature there are several terms that it will be helpful to explain first. Suppose that A is a subset of R^n, the Euclidean space of n dimensions, U an open set in R^n, and V is any set; then if f is a function f: $A \cap U \rightarrow V$ we denote the pair (U,f) by $\underset{\sim}{F}$. f may be specified further as continuous, as differentiable, or in whatever way we wish, but two elements of $\underset{\sim}{F}$ [e.g., (U_1,f_1) and (U_2,f_2)] will be considered equivalent if in some subsome subset of $U_1 \cap U_2$ they define the same map (i.e., $f_1|_U = f_2|_U$) for some $U \in U_1 \cap U_2$. The equivalence class defined by this relation is called a germ and this is clearly a way of making precise the fact that we are going to concentrate on local properties and so do not need to consider the function outside the immediate neighborhood of a particular point. Any smooth function can be expanded in a formal power series about a particular point (say, the origin) and this is called its *Taylor series*:

$$\sum_{r=0} \frac{1}{r!}(D^r f)_0 x^r$$

(By *formal* we mean that we are not saying anything about the convergence of the series.) If this series is truncated at the kth term we have the *k jet* of the function

$$j^k f(x) = \sum_{r=0}^{k} \frac{1}{r!}(D^r f)_0 x^r$$

which is a polynomial that behaves like f "to order k" in the neighborhood of the origin. Clearly, this may be a local approximation at best and may not "fit" very well. For example, $j^1 \sin x = x$ quickly diverges from sin x and it takes j^9 to give a decent approximation over the interval $(0,\pi)$, or more generally, j^{8n+1} to approximate over $(0,n\pi)$ after which the k jet goes off to $\pm\infty$. [It should be remembered that the formal Taylor series may not always converge to the function, as, for example, with $f(x) = 0, x \leqslant 0$, $f(x) = \exp(-1/x)$, $x > 0$, whose formal Taylor series is identically zero.] If the k jet of a function vanishes, it is said to be of *order* (k + 1) at the origin and the function $f - j^k f$, which has been deliberately made of order (k + 1), is sometimes called its *Tayl*.

The term contact equivalent also occurs frequently and is a precise way of saying that two functions are locally of essentially the "same shape," so that we can learn a lot about the one by studying the other. In particular, we can rely on a canonical form to give us a correct picture. Formally, $F(x,\lambda)$ and $G(y,\mu)$ are *contact equivalent* if there is a smooth and invertible change of coordinates $x \rightarrow y$ given by $y = Y(x,\lambda)$, $x = X(y,\mu)$, a reparameterization $\mu = M(\lambda)$, $\lambda = L(\mu)$, and a positive scale change $S(x,\lambda)$ or $T(y,\mu)$ such that

$$G(y,\mu) = T(y,\mu)F(X(y,\mu),L(\mu))$$

or

$$F(x,\lambda) = S(x,\lambda)G(Y(x,\lambda),M(\lambda))$$

For example, $x^2 \pm \lambda x$ is contact equivalent to $y^2 - \mu^2$, for we have only to put $X = y \pm \mu$, $L = -2\mu$, $T = 1$ to make the transformation.

Golubitsky [in two striking papers, one with Keyfitz (1980) and one with Schaeffer (1979)] has given a very useful test for the contact equivalence of an equation $F(x,\lambda)$ to one of a number of canonical forms, and his method has been fruitfully applied by Balakotaiah and Luss (1981, 1982a,b, 1983, 1984) and others. It will be best explained by taking a concrete example. We will use the autocatalytic scheme that Gray and Scott (1983, 1984) have shown to have many of the features of the single exothermic reaction without the algebraic complexity that the Arrhenius temperature dependence induces.

We consider the reactions $A + 2B \rightarrow 3B$, $B \rightarrow C$ taking place in a stirred tank and so, with obvious notation, have

$$\theta\dot{a} = a_f - a - \theta k_1 ab^2$$

$$\theta\dot{b} = b_f - b + \theta k_1 ab^2 - \theta k_2 b$$

With a view to using the residence time θ as a parameter, we shall want to vary we set

$$x = \frac{a}{a_f} \qquad y = \frac{b}{a_f} \qquad \tau = k_2 t$$

$$\alpha = k_2\theta \qquad \beta = \frac{k_1 a_f^2}{k_2} \qquad \gamma = \frac{b_f}{a_f}$$

Then

$$\alpha\dot{x} = 1 - x - \alpha\beta xy^2$$

$$\alpha\dot{y} = \gamma - y + \alpha\beta xy^2 - \alpha y$$

The steady state is obtained when $\dot{x} = \dot{y} = 0$. From $\dot{x} = 0$ we have $x = (1 + \alpha\beta y^2)^{-1}$ and substituting this in $\dot{y} = 0$ gives the cubic

$$F(y;\ \alpha,\beta,\gamma) \equiv \alpha(1 + \alpha)\beta y^3 - \alpha\beta(1 + \gamma)y^2 + (1 + \alpha)y - \gamma = 0$$

This appears to be the neatest form, although cubics can also be found for x and $z = 1 - x = (1 + \alpha)y$.

Thus the steady states will be the roots of this cubic equation $F = 0$. Since $F(0) \leqslant 0$ and $F \rightarrow +\infty$ as $y \rightarrow \infty$, there is always one real root but there may also be more than one. We shall have two steady states when there is a double root (i.e., $F = F_y = 0$), and this represents the

transition between one and three. There will be a unique triplet root if $F = F_y = F_{yy} = 0$. Let us record the derivatives for future use:

$$F \equiv \alpha(1+\alpha)\beta y^3 - \alpha\beta(1+\gamma)y^2 + (1+\alpha)y - \gamma$$

$$F_y \equiv 3\alpha(1+\alpha)\beta y^2 - 2\alpha\beta(1+\gamma)y + (1+\alpha)$$

$$F_{yy} \equiv 6\alpha(1+\alpha)\beta y - 2\alpha\beta(1+\gamma)$$

$$F_{yyy} \equiv 6\alpha(1+\alpha)\beta$$

$$F_\alpha = (1+2\alpha)\beta y^3 - \beta(1+\gamma)y^2 + y$$

$$F_{\alpha y} = 3(1+2\alpha)\beta y^2 - 2\beta(1+\gamma)y + 1$$

$$F_{\alpha\alpha} = 2\beta y^3$$

$$F_\beta = \alpha(1+\alpha)y^3 - \alpha(1+\gamma)y^2$$

$$F_\gamma = -\alpha\beta y^2 - 1$$

Were we to eliminate y between $F = 0$ and $F_y = 0$ we would have a complicated equation between α, β, and γ representing the surfaces in α-β-γ space that separate regions of one steady state from regions of three. Steady-state diagrams giving y as a function of, say, α could be constructed by passing through this space on a line of constant β and γ.

However, Golubitsky's method calls for going to the highest-order singularity we can find. Clearly, we could have $F = F_y = F_{yy} = 0$ simultaneously and this would be a line in α-β-γ space. We cannot make $F_{yyy} = 0$ except by putting α or $\beta = 0$, and this is pointless. But it is possible to find a point where $F = F_y = F_{yy} = F_\alpha = 0$, namely $\alpha = 1$, $\beta = 256/27$, $\gamma = 1/8$, $y = 3/16$, and we find that although we had no right to expect it and for reasons which will become clear later, $F_{\alpha y}$ also vanishes here but $F_{\alpha\alpha}$ does not. These are precisely the conditions for the "winged cusp" singularity, which Golubitsky and Keyfitz found to be the organizing center for the stirred tank with a single exothermic reaction and this accounts for the similarity of behavior noted by Gray and Scott (1984). The canonical form of the winged cusp is $x^3 \pm \lambda^2$ and Golbuitsky's theorems imply that there is a contact transformation that will put it in this form. He also gives the universal unfolding as $x^3 + (\alpha_2\lambda + \alpha 3)x + \alpha_1 + \lambda_2$. This canonical form allows us to describe every possible qualitative shape of the x-λ curve. This is shown in Fig. 37, where in the center a surface in α_1-α_2-α_3 space has been drawn. It really consists of two parts: the so-called hysteresis variety, H, which is like a trough lying on its side and pinched into a line, OA, the α_1 axis in the plane $\alpha_3 = 0$; and the bifurcation surface, B which is like a breaking wave with a cusp line COD. It is clear that O, the organizing center, is a very singular point, for the hysteresis trough is pinched down to a cusp, while the cusp of the

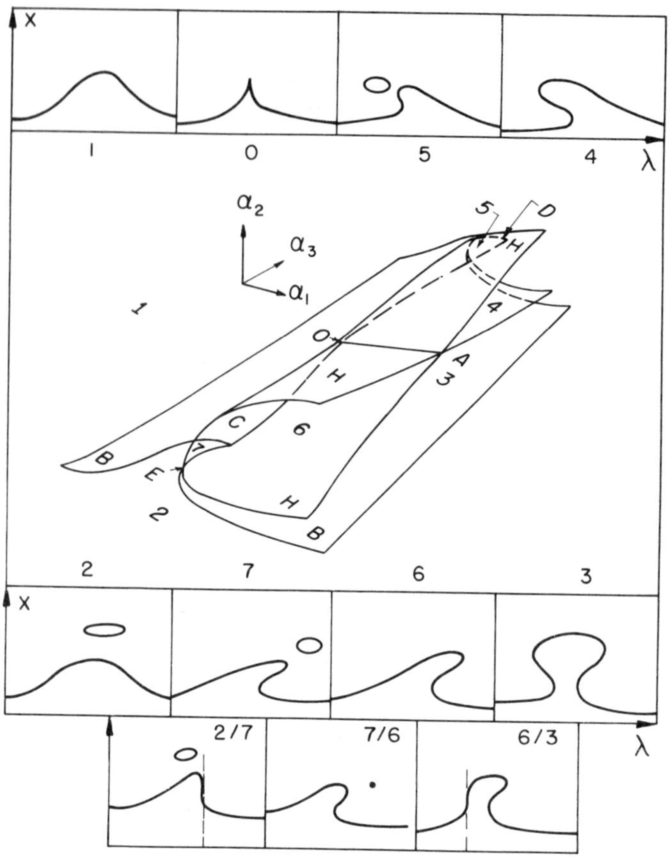

FIGURE 37 Qualitative bifurcation diagrams for a winged-cusp singularity.

bifurcation wave passes through it. The seven regions into which these surfaces divide the α_1-α_2-α_3) space correspond to the seven types of bifurcation diagrams in the x-λ plane. The hysteresis variety is the locus of triples (α_1,α_2,α_3) for which the x-λ diagram has a point of inflection with vertical tangent, as in the 2/7 and 6/3 diagrams at the bottom. The bifurcation variety is the locus of (α_1,α_2,α_3) points for which a third solution is just being born by the vanishing of the imaginary part of a pair of complex roots. This is represented by the dot in the 7/6 diagram. The point E represents a point at which both the 2/7 and the 7/6 transitions take place at once (i.e., the dot is vertically above the point of vertical inflection); the locus of points like E might be denoted by 2/7/6. Notice that diagrams with one number, such as 2, 7, or 6, belong to a three-dimensional region of α space; those like 2/7, 7/6 are two-dimensional surfaces, while those like 2/7/6 are one-dimensional lines. The organizing center, O, is a point.

The strength of the theory is that it exhaustively catalogs the various bifurcation diagrams that can arise. It is systematic and Golubitsky and Keyfitz (1980, pp. 326–327) give a useful table of conditions for finding

the various organizing centers. Its weakness is that it demands considerable algebraic labor in most cases, and an excessive amount if the contact transformation that leads to the canonical form is sought.

In the case of Gray's autocatalytic reaction there are only two additional parameters, for α plays the role of λ and this leaves only β and γ to be related to α_1, α_2, α_3. This implies that the system moves in a two-dimensional subspace of α_1-α_2-α_3 space and it could be that one or more of the seven possible diagrams is absent. Gray and Scott (1984) give mushroomlike curves such as 3 in the figure, and these seem to be the most complicated form that their system can attain, for there is an inherent degeneracy of the hysteresis variety that does not allow one side of the mushroom to straighten out without the other following suit. Thus there is never the breaking wave of 4, 5, 6, or 7. An attempt has been made to use three parameters by taking the order of the autocatalytic reaction as a further parameter, but aside from the difficulty of regarding it as continuous, this does not seem to avoid the degeneracy.

In contrast to the autocatalytic scheme, which has two parameters for its partial unfolding, the stirred tank with the irreversible, first-order reaction has a universal unfolding through the dimensionless parameters representing the adiabatic temperature rise, coolant temperature, and heat transfer coefficient. This is shown by Balakotaiah and Luss (1984), who give a diagram of the isola and hysteresis varieties in the neighborhood of the organizing center.

NOTATION

a	external heat transfer area
A	cross-sectional area of the tubular reactor
$\underline{\underline{A}}$	matrix, $\underline{\underline{C}}^{-1}\underline{\underline{J}}$
B	dimensionless heat transfer coefficient, $(-\Delta H)\gamma C_0/\rho C_p T_c$
C	concentration
$\underline{\underline{C}}$	capacitance matrix defined by Eq. (73a)
C_p	specific heat
D_e	axial dispersion coefficient of mass
Da	Damköhler number, $V f(C_0,T_r)/qC_0$(CSTR) or $Lf(C_0,T_r)/\bar{v}C_0$ (tubular reactor)
Da°	Da/τ
E	reaction activation energy
f	reaction rate
$\underline{\underline{J}}$	Jacobian matrix defined by Eq. (73b)
k	Arrhenius reaction rate constant
k_e	axial dispersion coefficient of heat
K	reactant adsorption constant
L	reactor length

Le	Lewis number, defined by Eq. (68) (CSTR) or (88) (tubular reactor)
n	reaction order
P	perimeter of tubular reactor
Pe	Péclet number; $Pe_h = \bar{v}L\rho C_p/k_e$, $Pe_m = \bar{v}L/D_e$
q	volumetric flow rate
r	dimensionless reaction rate, $f(C,T)/f(C_0,T_r)$
R	universal gas constant
s	dimensionless axial coordinate, z/L
t	dimensionless time, qt'/V
t'	time
T	temperature
T_m	reference temperature, defined by Eq. (4)
T_r	reference temperature: T_m, T_0, or T_c
u	dimensionless concentration, C/C_0
U	overall heat transfer coefficient
v	dimensionless temperature, T/T_r
$\bar{v}$	average fluid velocity
v_s	superficial velocity
V	reactor volume
x	conversion, 1 − u
z	axial coordinate
Z	$(v - 1)/\bar{\beta}$

Greek Letters

α	$\Delta H_2/\Delta H_1$
β	dimensionless heat of reaction, $(-\Delta H)C_0/\rho C_p T_0$
$\bar{\beta}$	dimensionless heat of reaction, $(-\Delta H)C_0/\rho C_p T_m(1 + \delta)$
γ	dimensionless activation energy, E/RT_r
δ	dimensionless heat transfer coefficient, $Ua/q\rho C_p$ (CSTR) or $UPL/A\bar{v}\rho C_p$ (tubular reactor)
$\delta°$	δ/τ
ΔH	enthalpy change of reaction
Θ	dimensionless temperature, $(T - T_r)\gamma/T_r$
λ	E_2/E_1
ν	$k_2(T_0)/k_1(T_0)$
ρ	density

σ dimensionless inhibition parameter, KC_0

τ reactor residence time, V/q(CSTR) or $L/\overline{v}$ (tubular reactor)

ϕ dimensionless parameter defined by Eq. (34)

Subscripts

A reactant

c cooling

0 feed

ACKNOWLEDGEMENT

Massimo Morbidelli was the recipient of a NATO-CNR Fellowship (Bando number 215.15/3) during the writing of this chapter.

REFERENCES

Adler, J., and J. W. Enig, The critical conditions in thermal explosion theory with reactant consumption, Comb. Flame, *8*, 97 (1964).

Agnew, J. B. and O. E. Potter, The design of stable packed tubular reactors for highly exothermic reactions, Trans. Inst. Chem. Eng., *44*, T216 (1966).

Amundson, N. R., Some further observations on tubular reactor stability, Can. J. Chem. Eng., *43*, 49 (1965).

Aris, R., On stability criteria of chemical reaction engineering, Chem. Eng. Sci., *24*, 149 (1969).

Aris, R., *The Mathematical Theory of Diffusion and Reaction in Permeable Catalysts*, Vol. 2, Oxford University Press, Oxford (1975).

Balakotaiah, V., and D. Luss, Analysis of the multiplicity patterns of a CSTR, Chem. Eng. Commun., *13*, 111 (1981); *19*, 185 (1982).

Balakotaiah, V. and D. Luss, Exact steady-state multiplicity criteria for two consecutive or parallel reactions in lumped-parameter systems, Chem. Eng. Sci., *37*, 433 (1982a).

Balakotaiah, V. and D. Luss, Structure of the steady-state solutions of lumped-parameter chemically reacting systems, Chem. Eng. Sci., *37*, 1611 (1982b).

Balakotaiah, V. and D. Luss, Multiplicity features of reacting systems. Dependence of the steady-states of a CSTR on the residence time, Chem. Eng. Sci., *38*, 1709 (1983).

Balakotaiah, V. and D. Luss, Global analysis of the multiplicity features of multi-reaction lumped-parameter systems, Chem. Eng. Sci., *39*, 865 (1984).

Barkelew, C. H., Stability of chemical reactors, Chem. Eng. Prog. Symp. Ser., *25*(55), 37 (1959).

Berger, A. J. and L. Lapidus, An introduction to the stability of distributed systems via a Liapunov functional, AIChE J., *14*, 558 (1968).

Bilous, O. and N. R. Amundson, Chemical reactor stability and sensitivity. AIChE J., *1*, 513 (1955).

Bilous, O. and N. R. Amundson, Chemical reactor stability and sensitivity, 2, AIChE J., *2*, 117 (1956).

Cardenas, A. F. and W. J. Karplus, PDEL—A language for partial differential equations, Commun. ACM, *13*(3), 184 (1970).

Chambré, P. L., On the characteristics of a nonisothermal chemical reactor, Chem. Eng. Sci., *5*, 209 (1956).

Chang, M. and R. A. Schmitz, An experimental study of oscillatory states in a stirred reactor, Chem. Eng. Sci., *30*, 21 (1975a).

Chang, M. and R. A. Schmitz, Feedback control of unstable states in a laboratory reactor, Chem. Eng. Sci., *30*, 837 (1975b).

Chemburkar, R., O. E. Rossler, and A. Varma, Chem. Eng. Sci. (in press)

Clough, D. E. and W. F. Ramirez, Local stability of tubular reactors, AIChE J., *18*, 393 (1972).

Cohen, D. S. and A. B. Poore, Tubular chemical reactors: the lumping approximation and bifurcation of oscillatory states, SIAM J. Appl. Math., *27*, 416 (1974).

Crandall, M. G., in *Applications of Bifurcation Theory*, P. H. Rabinowitz, ed., Academic Press, New York (1977).

Danckwerts, P. V., Continuous flow systems. Distribution of residence times, Chem. Eng. Sci., *2*, 1 (1953).

Dangelmayr, G. and I. Stewart, Classification and unfolding of sequential bifurcations, SIAM J. Math. Anal., *15*, 423 (1984).

Dente, M. and A. Collina, Il comportamento dei reattori chimici a flusso longitudinale nei riguardi della sensitività, Chim. Ind., *46*, 752 (1964).

Dente, M., G. Buzzi Ferraris, A. Collina, and A. Cappelli, La sensitivita del regime dei reattori chimici a flusso longitudinale, 2, Chim. Ind., *2*, 47 (1966a).

Dente, M., A. Collina, A. Cappelli, and G. Buzzi Ferraris, La sensitività del regime dei reattori chimici a flusso longitudinale, 3, Chim. Ind., *2*, 55 (1966b).

DeVera, A. and A. Varma, Substrate-inhibited enzyme reaction in a tubular reactor with axial dispersion, Chem. Eng. Sci., *34*, 275 (1979).

Doedel, E. J., AUTO: A program for the automatic bifurcation analysis of autonomous systems, Congr. Number, *30*, 265 (1981) (Proc. 10th Manitoba Conf. on Numer. Math. and Computing).

Doedel, E. J. and R. F. Heinemann, Numerical computation of periodic solution branches and oscillatory dynamics of the stirred tank reactor with A → B → C reactions, Chem. Eng. Sci., *38*, 1493 (1983).

Edwards, W. M., J. E. Zuniga-Chaves, F. L. Worley, and D. Luss, Causes, magnitude, and effects of temperature fluctuations (flickering) of catalytic wires and gauzes, AIChE J., *20*, 572 (1974).

Farr, W. W., Ph.D. thesis, University of Minnesota (1985).

Feigenbaum, M., The universal metric properties of nonlinear transformations, J. Stat. Phys., *21*, 669 (1979).

Finlayson, B. A., *The Method of Weighted Residuals and Variational Principles*, Academic Press, New York (1972).

Friedrichs, K. O., *Advanced Ordinary Differential Equations*, Gordon and Breach, New York (1965).

Furusawa, T. and H. Nishimura, Steady-state properties of an autothermal and non-adiabatic continuous stirred-tank reactor in changing residence time, J. Chem. Eng. Jpn., *1*, 180 (1968).

Gavalas, G., *Nonlinear Differential Equations of Chemically Reacting Systems*, Springer-Verlag, Berlin (1968).

Georgakis, C., R. Aris, and N. R. Amundson, Studies in the control of tubular reactors: 1, Chem. Eng. Sci., *32*, 1359 (1977a).

Georgakis, C., R. Aris, and N. R. Amundson, Studies in the control of tubular reactors: 2, Chem. Eng. Sci., *32*, 1371 (1977b).

Gilles, E. D., G. Eigenburger, and W. Ruppel, Relaxation oscillations in chemical reactors, AIChE J., *24*, 912 (1978).

Golubitsky, M. and B. L. Keyfitz, A qualitative study of the steady state solutions for a continuous flow stirred tank chemical reactor, SIAM J. Math. Anal., *11*, 316 (1980).

Golubitsky, M. and D. Schaeffer, A theory of imperfect bifurcation via singularity theory, Commun. Pure Appl. Math., *32*, 31 (1979).

Gray, P. and P. R. Lee, Thermal explosions and the effect of reactant consumption on critical conditions, Combust. Flame, *9*, 201 (1965).

Gray, P. and S. K. Scott, Autocatalytic reactions in the isothermal continuous stirred tank reactor. Isolas and other forms of multistability, Chem. Eng. Sci., *38*, 29 (1983).

Gray, P., and S. K. Scott, Autocatalytic reactions in the isothermal continuous stirred tank reactor. Oscillations and instabilities in the system A + 2B → 3B; B → C, Chem. Eng. Sci., *39*, 1087 (1984).

Gupalo, Y. P. and Y. S. Ryazantsev, On the stability of the steady-state conditions in a tubular reactor with axial heat and mass transfer, Chem. Eng. Sci., *24*, 1441 (1969).

Halbe, D. C., and A. B. Poore, Dynamics of the continuous stirred tank, reactor with reactions A → B → C, Chem. Eng. J., *21*, 241 (1981).

Heinemann, R. F. and A. B. Poore, Multiplicity, stability, and oscillatory dynamics of the tubular reactor, Chem. Eng. Sci., *36*, 1411 (1981).

Heinemann, R. F. and A. B. Poore, The effect of activation energy on tubular reactor multiplicity, Chem. Eng. Sci., *37*, 128 (1982).

Heinemann, R. F., K. A. Overholser, and G. W. Reddien, Multiplicity and stability of premixed laminar flames: an application of bifurcation theory, Chem. Eng. Sci., *34*, 833 (1979).

Heinemann, R. F., K. A. Overholser, and G. W. Reddien, Multiplicity and stability of the hydrogen-oxygen-nitrogen flame: the influence of chemical pathways and kinetics on the transitions between steady states, AIChE J., *26*, 725 (1980).

Hlaváček, V. and H. Hofmann, Modeling of chemical reactors: XVI. Steady state axial heat and mass transfer in tubular reactors. An analysis of the uniqueness of solutions, Chem. Eng. Sci., *25*, 173 (1970a).

Hlaváček, V. and H. Hofmann, Modeling of chemical reactors: XVII. Steady state axial heat and mass transfer in tubular reactors. Numerical investigation of multiplicity, Chem. Eng. Sci., *25*, 187 (1970b).

Hlaváček, V., M. Marek, and T. M. John, Modelling of chemical reactors: XII. Tubular non-isothermal non-adiabatic packed bed reactor analysis of a one-dimensional approximation, Collect. Czech. Chem. Commun., *34*, 3868 (1969).

Hlaváček, V., M. Kubicek, and J. Jelinek, Modeling of chemical reactors: XVIII. Stability and oscillatory behavior of the CSTR, Chem. Eng. Sci., *25*, 1441 (1970)

Hlaváček, V., H. Hofmann, and M. Kubicek, Modeling of chemical reactors: XXIV. Transient heat and mass transfer in tubular reactors. Stability considerations II. Chem. Eng. Sci., *26*, 1629 (1971).

Hlaváček, V., M. Kubicek and K. Visnak, Modeling of chemical reactors: XXVI. Multiplicity and stability analysis of a continuous stirred tank reactor with exothermic consecutive reactions A → B → C, Chem. Eng. Sci., *27*, 719 (1972).

Hlavácek, V., H. Hofmann, J. Votruba, and M. Kubicek, Modeling of chemical reactors: XXVII. Steady state axial heat and mass transfer in tubular reactors. Effect of different values of Péclet numbers on the region of multiplicity, Chem. Eng. Sci., *28*, 1897 (1973).

Huang, D. T. J. and A. Varma, On the reference time in the multiplicity analysis of CSTRs, Chem. Eng. Sci., *35*, 1806 (1980).

Ioos, G. and D. D. Joseph, *Elementary Stability and Bifurcation Theory*, Springer-Verlag, New York (1980).

Jensen, K. F and W. H. Ray, The bifurcation behavior of tubular reactors, Chem. Eng. Sci., *37*, 199 (1982).

Jorgensen, D. V. and R. Aris, On the dynamics of a stirred tank with consecutive reactions, Chem. Eng. Sci., *38*, 45 (1983).

Jorgensen, D. V., W. W. Farr, and R. Aris, More on the dynamics of a stirred tank with consecutive reactions, Chem. Eng. Sci., *39*, 1741 (1984).

Kahlert, C., O. E. Rossler, and A. Varma, Chaos in a continuous stirred-tank reactor with two consecutive first-order reactions, one exo-, one endothermic, in *Modeling of Chemical Reaction Systems*, (K. H. Ebert, P. Deuflard, and W. Jager, eds.), Springer Series Chem. Phys., *18*, 355 (1981).

Kapila, A. K. and A. B. Poore, The steady state response of a nonadiabatic tubular reactor, Chem. Eng. Sci., *37*, 57 (1982).

Kauschus, W., J. Demont, and K. Hartman, On the steady states of continuous stirred tank reactors, Chem. Eng. Sci., *33*, 1283 (1978).

Keller, H. B., In *Applications of Bifurcation Theory*, P. H. Rabinowitz, ed., Academic Press, New York (1977).

Kirkby, L. L. and R. A. Schmitz, An analytical study of diffusion flame stability, Combust. Flame, *10*, 205 (1966).

Kubicek, M., Occurrence of oscillatory regimes in lumped parameter systems. Determination of Hopf's bifurcation points, Chem. Eng. Sci., *34*, 1078 (1979).

Kubicek, M., H. Hofmann, and V. Hlaváček, Modeling of chemical reactors: XXXII. Nonisothermal nonadiabatic tubular reactor one-dimensional model-detailed analysis, Chem. Eng. Sci., *34*, 593 (1979).

Kubicek, M., H. Hofmann, V. Hlaváček, and J. Sinkule, Multiplicity and stability in a sequence of two nonadiabatic nonisothermal CSTRs, Chem. Eng. Sci., *35*, 987 (1980).

Kuo, J. C. W. and N. R. Amundson, Catalytic particle stability studies. III. Complex distributed resistances model, Chem. Eng. Sci., *22*, 1185 (1967).

Laidler, K. J. and P. S. Bunting, *The Chemical Kinetics of the Enzyme Action*, Clarendon Press, Oxford (1973).

Leib, T. M. and D. Luss, Exact uniqueness and multiplicity criteria for an n-th order reaction in a CSTR, Chem. Eng. Sci., *36*, 210 (1981).

Liou, C. T., H. C. Lim, and W. A. Weigand, On local stability of tubular reactors, AIChE J., *18*, 875 (1972a).

Liou, C. T., H. C. Lim, and W. A. Weigand, Sufficient conditions for uniqueness and the local asymptotic stability of adiabatic tubular reactors, AIChE J., *18*, 1079 (1972b).

Liou, C. T., H. C. Lim, and W. A. Weigand, Liapunov functional method for analysis of nonadiabatic tubular reactors with axial mixing and recycle, Chem. Eng. Sci., *29*, 705 (1974).

Luss, D., Uniqueness criteria for lumped and distributed parameter chemically reacting systems, Chem. Eng. Sci., *26*, 1713 (1971).

Luss, D., The influence of capacitance terms on the stability of lumped and distributed parameter systems, Chem. Eng. Sci., *29*, 1832 (1974).

Luss, D. and N. R. Amundson, Some general observations on tubular reactor stability, Can. J. Chem. Eng., *45*, 341 (1967a).

Luss, D. and N. R. Amundson, Uniqueness of the steady state solutions for chemical reaction occurring in a catalyst particle or in a tubular reactor with axial diffusion, Chem. Eng. Sci., *22*, 253 (1967b).

Luss, D. and J. C. M. Lee, On global stability in distributed parameter systems, Chem. Eng. Sci., *23*, 1237 (1968).

Lynch, D. T., T. D. Rogers, and S. E. Wanke, Chaos in a continuous stirred tank reactor, Math. Modelling, *3*, 103 (1982).

Madsen, N. K. and R. F. Sincovec, PDEPACK: a new tool for simulation, Proc. 1975 Summer Comput. Simul. Conf., San Francisco (1975).

Matsuyama, H., Sufficient conditions for the existence of a unique and stable equilibrium state of an adiabatic tubular reactor with recycle, J. Chem. Eng. Japan, *3*, 248 (1970).

McGowin, C. R. and D. D. Perlmutter, A comparison of techniques for local stability analysis of tubular reactor systems, Chem. Eng. J., *2*, 125 (1971a).

McGowin, C. R. and D. D. Perlmutter, Tubular reactor steady state and stability characteristics: I. Effect of axial mixing, AIChE J., *17*, 831 (1971b).

McGowin, C. R. and D. D. Perlmutter, Tubular reactor steady state and stability characteristics: III. Effect of recycle, AIChE J., *17*, 842 (1971c).

McGreavy, C. and C. I. Adderley, Generalized criteria for parametric sensitivity and temperature runaway in catalytic reactors, Chem. Eng. Sci., *28*, 577 (1973).

McGreavy, C. and B. R. Dunbobbin, Operational stability in multitubular catalytic reactors, in *Frontiers in Chemical Reaction Engineering*, L. K. Doraiswamy and R. A. Mashelkar, eds., Vol. 2, Wiley Eastern, New Delhi, (1984), p. 398.

McGreavy, C. and M. A. Soliman, Stability of tubular reactors, AIChE J., *19*, 174 1973).

Michelsen, M. L. and J. Villadsen, *Solution of Differential Equation Models by Polynomial Approximation*, Prentice-Hall Englewood Cliffs, N.J. (1978).

Morbidelli, M. and A. Varma, Parametric sensitivity and runaway in tubular reactors, AIChE J., *28*, 705 (1982).

Morbidelli, M. and A. Varma, On parametric sensitivity and runaway criteria of pseudohomogeneous tubular reactors, Chem. Eng. Sci., *40*, 2165 (1985).

Nishimura, Y. and M. Matsubara, Stability conditions for a class of distributed-parameter systems and their applications to chemical reaction systems, Chem. Eng. Sci., *24*, 1427 (1969).

Padmanabhan, L., R. Y. K. Yang, and L. Lapidus, On the analysis of the stability of distributed reaction systems by Liapunov's direct method, Chem. Eng. Sci., *26*, 1857 (1971).

Parulekar, S. J. and D. Ramkrishna, Tubular reactor stability revisited without the Danckwerts boundary conditions, Chem. Eng. Sci., *39*, 455 (1984).

Perelson, A. S., A note on the qualitative theory of lumped parameter systems, Chem. Eng. Sci., *31*, 170 (1976).

Polizopoulos, I. and C. G. Takoudis, On the steady-state solutions of chemically reacting systems: I. Theory and examples, Chem. Eng. Sci. (1985, in press).

Poore, A. B., A model equation arising from chemical reactor theory, Arch. Rat. Mech. Anal., *52*, 358 (1973).

Rajadhyaksha, R. A., K. Vasudeva, and L. K. Doraiswamy, Parametric sensitivity in fixed bed reactors, Chem. Eng. Sci., *30*, 1399 (1975).

Ray, W. H. and S. T. Hastings, The influence of the Lewis number on the dynamics of chemically reacting systems, Chem. Eng. Sci., *35*, 589 (1980).

Ray, W. H., A. Uppal, and A. B. Poore, On the dynamic behavior of catalytic wires, Chem. Eng. Sci., *29*, 1330 (1974).

Raymond, L. R. and N. R. Amundson, Some observations on tubular reactor stability, Can. J. Chem. Eng., *42*, 173 (1964).

G. B. Rogers, M. M. Lih, and O. A. Hougen, Catalytic hydrogenation of propylene and isobutylene over platinum. Effect of noncompetitive adsorption, AIChE J., *12*, 369 (1966).

Schmitz, R. A., Multiplicity, stability and sensitivity of states in chemically reacting systems—a review, Adv. Chem. Ser., *148*, 156 (1975).

Schmitz, R. A., R. R. Bautz, W. H. Ray, and A. Uppal, The dynamic behavior of a CSTR: some comparisons of theory and experiment, AIChE J., *25*, 289 (1979).

Semenov, N. N., Zur Theorie des Verbrennungsprozesses, Z. Phys., *48*, 571 (1928).

Sheintuch, M., and R. A. Schmitz, Oscillations in catalytic reactions, Catal. Rev.-Sci. Eng., *15*, 107 (1977).

Sinfelt, J. H., Hydrogenolysis of ethane over supported platinum, J. Phys. Chem., *68*, 344 (1964).

Sorensen, J. B., E. W. Guertin, and W. E. Stewart, Computational models for cylindrical catalyst particles, AIChE J., *19*, 969, 1286 (1973).

Soria Lopez, A., H. I. DeLasa, and J. A. Porras, Parametric sensitivity of a fixed bed catalytic reactor, Chem. Eng. Sci., *36*, 385 (1981).

Svoronos, S., R. Aris, and G. Stephanopoulos, On the behavior of two stirred tanks in series, Chem. Eng. Sci., *37*, 357 (1982).

Tsotsis, T. T. and R. A. Schmitz, Exact uniqueness and multiplicity criteria for a positive-order Arrhenius reaction in a lumped system, Chem. Eng. Sci., *34*, 135 (1979).

Tsotsis, T. T., A. E. Haderi, and R. A. Schmitz, Exact uniqueness and multiplicity criteria for a class of lumped reaction systems, Chem. Eng. Sci., *37*, 1235 (1982).

Uppal, A., W. H. Ray, and A. B. Poore, On the dynamic behavior of continuous stirred tank reactors, Chem. Eng. Sci., *29*, 967 (1974).

Uppal, A., W. H. Ray, and A. B. Poore, The classification of the dynamic behavior of continuous stirred tank reactors-influence of the reactor residence time, Chem. Eng. Sci., *31*, 205 (1976).

van Cauwenberghe, A. R., Further note on Danckwerts' boundary conditions for flow reactors, Chem. Eng. Sci., *21*, 203 (1966).

van den Bosch, B. and D. Luss, Uniqueness and multiplicity criteria for an n-th order chemical reaction, Chem. Eng. Sci., *32*, 203 (1977).

van Heerden, C., Autothermic processes. Properties and reactor design, Ind. Eng. Chem., *45*, 1242 (1953).

van Welsenaere, R. J. and G. Froment, Parametric sensitivity and runaway in fixed bed catalytic reactors, Chem. Eng. Sci., *25*, 1503 (1970).

Varma, A., Bounds on the steady state concentration and temperature in a tubular reactor, Can. J. Chem. Eng., *55*, 629 (1977).

Varma, A., On the number and stability of steady states of a sequence of continuous-flow stirred tank reactors, Ind. Eng. Chem. Fundam., *19*, 316 (1980).

Varma, A. and N. R. Amundson, Global asymptotic stability in distributed parameter systems: comparison function approach, Chem. Eng. Sci., *27*, 907 (1972a).

Varma, A. and N. R. Amundson, Some problems concerning the non-adiabatic tubular reactor, Can. J. Chem. Eng., *50*, 470 (1972b).

Varma, A. and N. R. Amundson, Local stability of tubular reactors, AIChE J., *19*, 395 (1973a).

Varma, A. and N. R. Amundson, The nonadiabatic tubular reactor: stability considerations, Can. J. Chem. Eng., *51*, 459 (1973b).

Varma, A. and N. R. Amundson, Some observations of uniqueness and multiplicity of steady states in non-adiabatic chemically reacting systems, Can. J. Chem. Eng., *51*, 207 (1973c).

Varma, A. and R. Aris, Stirred pots and empty tubes, in *Chemical Reactor Theory. A Review*, L. Lapidus and N. R. Amundson, eds., Prentice-Hall, Englewood Cliffs, N.J. (1977).

Vejtasa, S. A. and R. A. Schmitz, An experimental study of steady-state multiplicity and stability in an adiabatic stirred tank, AIChE J., *16*, 410 (1970).

Voltz, S. E., C. R. Morgan, D. Liederman, and S. M. Jacob, Kinetic study of carbon monoxide and propylene oxidation on platinum catalysts, Ind. Eng. Chem. Prod. Res. Dev., *12*, 294 (1973).

Wehner, J. F. and R. H. Wilehlm, Boundary conditions for flow reactor, Chem. Eng. Sci., *6*, 89 (1956).

Wei, J., The stability of a reaction with intra-particle diffusion of mass and heat: the Liapunov methods in a metric function space, Chem. Eng. Sci., *20*, 729 (1965).

Weisz, P. B. and J. S. Hicks, The behavior of porous catalyst particles in view of internal mass and heat diffusion effects, Chem. Eng. Sci., *17*, 265 (1962).

Westerterp, K. R. and K. J. Ptasinski, Safe design of cooled tubular reactors for exothermic, multiple reactions: parallel reactions: I. Development of criteria, Chem. Eng. Sci., *39*, 235 (1984).

Index

and therefore,

$$U_{mf} = \frac{0.0008g(\rho_s - \rho_f)d_p^2}{\mu} \tag{24}$$

which is similar to Eq. (9) but with a slightly different constant.

Newton's regime applies at $R_e > 10^3$ when

$$C_{D_I} = 0.44 \tag{25}$$

From Eq. (1) it follows that

$$U_T = \left[\frac{3g(\rho_s - \rho_f)d_p}{\rho_f}\right]^{1/2} \tag{26}$$

and again from Eqs. (18) to (20) that

$$U_T = \sqrt{70}\, U_{mf} \tag{27}$$

$$= 8.4U_{mf} \tag{28}$$

Thus

$$U_{mf} = \frac{0.206g^{1/2}(\rho_s - \rho_f)^{1/2}d_p^{1/2}}{\rho_f^{1/2}} \tag{29}$$

It is now seen that, excepting those expressions that treat skin friction and from drag as additive, equations for predicting minimum fluidization velocity are of the form

$$U_{mf} = kg(\rho_s - \rho_f)^a d_p^b \rho_f^c \mu^d \tag{30}$$

and Table 1 compares the various equations in this way. In particular it shows that Leva's empirical equation is a compromise at a Reynolds number that is evidently near the average of practical systems examined

TABLE 1

	K	a	b	c	d	
Stokes	0.0008	1	2	0	−1	Increasing
Carman-Kozeny	0.0006	1	2	0	−1	Re_{mf}
Leva	(0.805)	0.94	1.82	0	−0.88	↓
Ergun	—	$\frac{1}{2}$	$\frac{1}{2}$	$-\frac{1}{2}$	0	↓
Newton	0.066	$\frac{1}{2}$	$\frac{1}{2}$	$-\frac{1}{2}$	0	↓

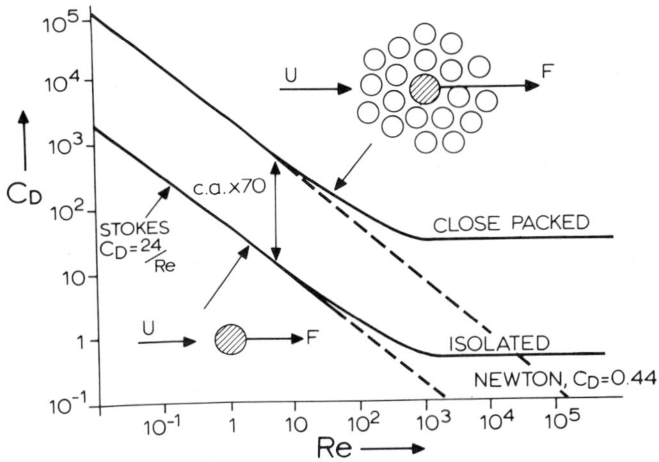

FIGURE 3 Drag coefficient as a function of Reynolds number.

where C_{D_I} and $C_{D_{mf}}$ refer to the drag coefficients under isolated and minimum fluidization (closed packed) conditions, respectively. The relationship between these two is revealed by Fig. 3.

As is further seen from Fig. 3,

$$C_{D_I} = f(Re_T) \tag{19}$$

and

$$C_{D_{mf}} = 70 \times f(Re_{mf}) \tag{20}$$

The form of this functional relationship is not easy to express mathematically because of the varying contributions of skin friction and form drag as Reynolds number changes. There are various empirical equations to describe it, but they are not helpful for the present purpose. The two end conditions can be treated analytically.

Stokes' regime lies at $Re < 0.2$ when form drag can be neglected and in this case

$$C_{D_I} = \frac{24}{Re} \tag{21}$$

It follows from Eq. (1) that

$$U_T = \frac{g(\rho_s - \rho_f)d_p^2}{18\mu} \tag{22}$$

which is a well-known expression. From Eqs. (18) to (20),

$$U_T = 70 \times U_{mf} \tag{23}$$

TABLE 2

Re_T	m	$U_T/U_{mf} = \varepsilon_{mf}^{-m}$	
0.1	4.65	71	70 from Eq. (23)
1	4.35	53	
10	3.53	26	
100	2.81	13	
1000	2.39	8.9	8.4 from Eq. (28)

Apart from the insight given by the approach above, it shows the relationship between U_{mf} and U_T, which is of considerable practical importance. It is fundamental to any consideration of a fluidized-bed chemical reactor that it should be operated at a fluid velocity greater than U_{mf} but not in excess of U_T if particles are to be retained in the bed. Practical powders are generally made up of a range of particle sizes, so there is a corresponding range of U_T (but not of U_{mf}). The above is therefore only a guide to an estimate of elutriation, but it is a useful one.

Intermediate values of the ratios given by Eqs. (23) and (28) can be derived from the work of Richardson and Zaki (1954). As is mentioned in the next section, liquid fluidized beds expand uniformly as liquid velocity increases, and the relationship between voidage and velocity can be described by

$$U = U_T \varepsilon^m \tag{31}$$

where m is constant for a given system but varies with Re_T. A particular case of Eq. (31) is

$$U_{mf} = U_T \varepsilon_{mf}^m \tag{32}$$

and this has been evaluated in Table 2 for the case of uniform spheres, where $\varepsilon_{mf} = 0.4$ and using values of m found by Richardson and Zaki (1954). The end conditions agree well with the previously described and quite independent experimental measurement.

Particulate Fluidization

The subject of this chapter is gas fluidized beds, but it is necessary briefly to refer to the special case of liquid fluidization because under certain conditions dry beds can behave as if liquid fluidized.

Once they have been fluidized, particles generally behave differently depending on whether the fluid stream is a liquid or a gas. Liquid fluidized beds expand uniformly as velocity is increased. The particles separate to accommodate more liquid flow between them while retaining the same drag force. As is implied by Fig. 3, the drag coefficient diminishes

as particle spacing increases, but $C_D U^2$ remains constant. The same relationship is expressed differently by Eq. (31). Between Eqs. (6) and (3) if follows that

$$\log \frac{U}{U_T} = m \log \left[1 - \left(\frac{M}{HA\rho_s}\right)\right] \tag{33}$$

from which it is seen how bed height varies with velocity when plotted against logarithmic coordinates.

Uniform expansion of this kind is generally referred to as *particulate fluidization* and for most purposes it can be regarded as of uniform particle concentration throughout. The particles are not stationary but mix fairly rapidly by a process akin to eddy diffusion. The (modified) eddies in the liquid carry particles with them causing both lateral and longitudinal dispersion.

Solids Classification

The observation that certain powders undergo considerable expansion beyond the point of minimum fluidization and only begin to bubble when another critical gas velocity, U_{mb}, is exceeded has led to the acceptance of the principle that such powders should be grouped together and their fluidization properties considered separately from those materials that show more "normal" behavior and begin to bubble at U_{mf}. Moveover, experience with a wide range of particulate solids in gas fluidized beds has shown there to be other modes of behavior that are different from those of the foregoing materials, and as a result it has been found convenient to divide all powders into four groups according to criteria of particle size and density. Such a scheme of classification was suggested originally by Geldart (1973) and it has now become firmly established. It classifies powders as belonging to Groups A, B, C, or D as given in Table 3. Although in practice the divisions between the groups are somewhat diffuse, making it difficult in some cases to assign a powder firmly to one group or another, the classification does nevertheless provide a broad guide to the type of behavior that may be expected from a given powder. It is useful to represent the four groups of materials on a "phase diagram" showing the density difference ($\rho_s - \rho_g$) plotted against particle diameter d_p (Fig. 4). Here the boundaries between groups C and A and groups B and D are entirely empirical. The A/B boundary, however, may be defined on the basis of whether or not a powder begins to form bubbles at its minimum fluidzation velocity. For group A materials, $U_{mf}/U_{mb} \leq 1$, while for group B, $U_{mf}/U_{mb} > 1$; a convenient dividing line would thus be one for which $U_{mf}/U_{mb} = 1$. Baeyens and Geldart (1973) correlated U_{mf} and U_{mb} with particle and gas properties for a wide range of commonly used powders and showed that for air at ambient conditions the A/B boundary is given approximately by the line

$$\bar{d}_p(\rho_s - \rho_g) = 2 \times 10^5 \tag{34}$$

where $\bar{d}_p$ (μm) is the surface mean diameter of the particles. Equation (34) is plotted as line XY on Fig. 4.

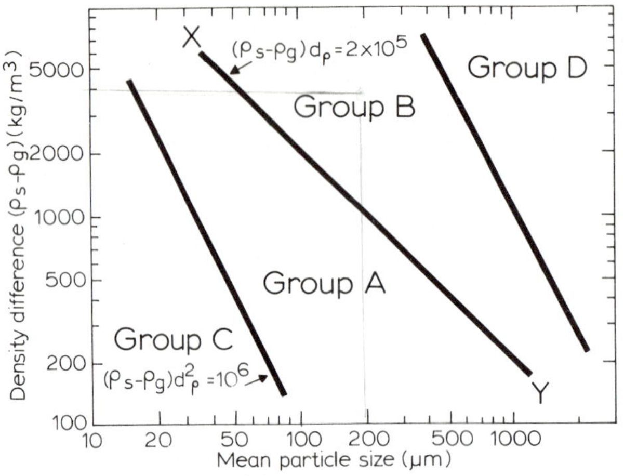

FIGURE 4 Powder classification diagram for fluidization with air at ambient conditions. (Based on Baeyens and Geldart, 1973.)

Gas Fluidization and the Two-Phase Theory

When the fluid velocity is increased above U_{mf}, gas fluidized beds begin to bubble and look remarkably like boiling opaque liquids. They are sometimes called *boiling beds*, which is certainly descriptive. In contrast to "particulate," this has been referred to as *aggregative* fluidization, but that is not a very descriptive name and *boiling* or *bubbling* is to be preferred.

The two-phase theory states essentially that the gas flow divides into that which flows interstitially among the closely packed particles and that which takes part in bubble flow. That is,

$$Q = Q_i + Q_B \tag{35}$$

As generally expressed, the two-phase theory further assumes that $Q_i = Q_{mf}$, which here will be referred to as the simple two-phase theory. This can conveniently be written

$$\frac{Q_B}{A} = U - U_{mf} \tag{36}$$

which gives an immediate estimate of the bubble flow for any operating velocity.

The two-phase theory is fundamental to most models of fluidized-bed chemical reactors. Gas in each of the two phases has a different residence time and pattern of contacting with the particles. In realistic situations there is generally some exchange between the phases.

Two obvious corrections are necessary to Eq. (36), although they are often small enough to be neglected. First, as the bed fills with bubbles the overall average height increases but the cross-sectional area available for interstitial flow is reduced. Second, there is necessarily a pressure

TABLE 3 Solids Classification for Fluidized Beds

Group	Range of particle size (μm)	Range of particle density (kg/m^3)	Behavior when fluidized	Examples
A	30–200	5000–800	"Abnormal" bed expansion between U_{mf} and U_{mb} followed by bed collapse $U_{mb} > U_{mf}$, $U_b > U_i$, $\varepsilon_i > \varepsilon_{mf}$	Cracking catalyst
B	50–1000	4000–1000	Normal behavior $U_{mb} = U_{mf}$, $\varepsilon_i \cong \varepsilon_{mf}$	Sand, glass ballotini
C	<30	Any value	Cohesive powders, difficult to fluidize, prone to channeling	Talc, flour, cement
D	>400	>1000	Prone to slugging, bubbles ill-defined, $U_b < U_i$	Steel shot, dried peas

Source: Baeyens and Geldart (1973).

drop across the bed and therefore gas must expand as it rises. Volumetric flow and therefore gas velocity consequently increases with height. As Table 2 shows, U_{mf} is largely independent of gas density (or varies inversely as its square root for large particles), so that the excess flow available for bubbles ($U - U_{mf}$) increases with height. It can be shown (Rowe, 1978b) that these two corrections lead to

$$\left(\frac{Q_B}{A}\right)_H = U_D\left[1 + \frac{H(1 - \varepsilon)\rho_s g}{p_H}\right] - \frac{U_{mf}(3 - 4f_B)}{3} \tag{37}$$

where the suffices D and H refer, respectively, to the distributor plane and a plane height H above it. The expansion correction is usually necessary only for deep beds of dense material or when operating at subatmospheric pressure. Even in a vigorously fluidized bed, f_B is rarely greater than about 30% and, especially when $U_D \gg U_{mf}$, this correction also is commonly very small.

The simple two-phase theory is essentially correct for many powders, particularly for those with mean particle size greater than about 100 μm. From consideration of the minimum fluidization conditions as implied, for example, in Eq. (11), it follows that the interstitial gas velocity must remain at U_{mf} if the voidage of the dense phase remains at ε_{mf}. It is difficult to imagine that particles may become closer spaced after fluidization, but they may separate more widely, thus causing the dense phase to be more permeable. It is thus to be expected and there are indeed cases when interstitial flow is greater than U_{mf} and the two-phase theory needs modification.

Three situations have been observed experimentally where $U_i > U_{mf}$.

1. Beds of fine particles of group A (Geldart, 1973) (typically the particle size is less than 100 μm).
2. When operating at high gas pressure
3. In the bottom several centimeters of the bed, where it appears to take measurable time for the flows to come to equilibrium

These special cases will be discussed in the following sections.

Properties of Bubbles

In many ways the bubbles in gas fluidized beds have similar properties to bubbles in a true liquid. They have a roughly similar shape, they rise at about the same velocity, disturb the surrounding medium in much the same way, and coalesce and break up. However, they differ in several points of detail and the analogy must be drawn with care. Bubbles are fundamental to the behavior of gas fluidized-bed reactors, so it is important properly to understand their properties as much is now known about them from research experiments.

In any realistic reactor there will be many bubbles, but it is helpful first to examine the properties of single bubbles in isolation. Because the dense phase of closely spaced powder particles is opaque, bubbles cannot normally be seen within the bed and special experimental techniques are required to reveal them. One such technique is the so-called *two-dimensional bed*, where the powder is confined between glass plates perhaps 1 cm apart. This effectively exposes a thin slice of the bed and the bubbles

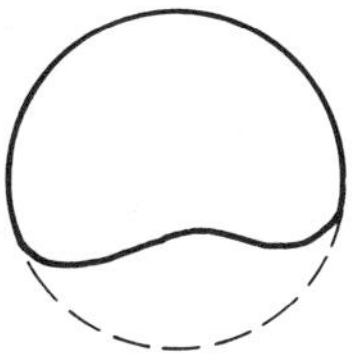

FIGURE 5 Undisturbed gas bubble in a fluidized bed.

can be seen by suitable lighting and recorded photographically (Rowe, 1971). Bubble shape, rise velocity, and some other properties are changed a little by this constraint within two dimensions; nevertheless, the technique reveals many essential features and gives much insight. Another technique with different limitations is x-ray ciné photography.

Undisturbed bubbles are spherical with an indented base, as shown in Fig. 5. The boundary is quite definite and the interior is empty of particles. They rise by particles flowing around them in streamlines rather as a bluff body moves through a true fluid at low Reynolds number.

Bubble rise velocity can be described by the Davies-Taylor expression (Davies and Taylor, 1950), which can be written

$$U_B = \sqrt{\frac{gd_B}{2}} \tag{38}$$

The standard deviation of a single observation is very high and even under ideal experimental conditions prediction of velocity for a particular bubble is subject to much error. Even with large numbers of observations it is not possible to confirm that Eq. (38) is the correct form and it is chosen simply on the basis of analogy with bubbles in liquids. From many data it is clear that the constant of proportionality in Eq. (38) is unity within the limits of experimental error. For the purpose of reaction engineering it is behavior of the average bubble that is of interest, and for this, Eq. (38) is adequate and as good as any other. From this equation it is seen that bubbles rise at a few tens of cm/s and the practical range of velocities is less than one order of magnitude.

The indentation at the bottom of a bubble is a wake of particles traveling with it and roughly circulating as indicated in Fig. 6. It is very like the attached vortices behind a bluff body moving through a liquid at $10 < Re < 100$. The wake grows as further particles are captured until it becomes unstably large, when a large portion is shed—again closely analogous to behavior in a true liquid. Pursuing this analogy suggests that the fluidized particles have a viscosity of 10^{-2} N·s/m^2, but it can be misleading to attribute a true Newtonian viscosity to the fluidized system. At high shear rates particles will come in contact with each other, and apparent viscosity will change abruptly to elasticity. The system has been treated as a rheological fluid (Rietema, 1967), but such an approach has not advanced our understanding very far.

The continuous growth and discontinuous shedding of wakes accounts in part for the variability of velocity. It also accounts for the fact that instantaneous pictures of bubbles show a variety of wake sizes. The